GUIDEBOOK
OF
ELECTRONIC
CIRCUITS

BOOKS by JOHN MARKUS

ELECTRONIC CIRCUITS MANUAL

ELECTRONICS DICTIONARY

ELECTRONICS FOR COMMUNICATION ENGINEERS

ELECTRONICS FOR ENGINEERS

ELECTRONICS MANUAL FOR RADIO ENGINEERS

ELECTRONICS AND NUCLEONICS DICTIONARY

GUIDEBOOK OF ELECTRONIC CIRCUITS

HANDBOOK OF ELECTRONIC CONTROL CIRCUITS

HANDBOOK OF INDUSTRIAL ELECTRONIC CIRCUITS

HANDBOOK OF INDUSTRIAL ELECTRONIC CONTROL CIRCUITS

HOW TO GET AHEAD IN THE TELEVISION AND
RADIO SERVICING BUSINESS

SOURCEBOOK OF ELECTRONIC CIRCUITS

TELEVISION AND RADIO REPAIRING

WHAT ELECTRONICS DOES

GUIDEBOOK OF ELECTRONIC CIRCUITS

Over 3,600 modern electronic circuits, each complete
with values of all parts and performance details,
organized in 131 logical chapters for quick
reference and convenient browsing

JOHN MARKUS

Consultant, McGraw-Hill Book Company
Senior Member, Institute of Electrical and Electronics Engineers

McGRAW-HILL BOOK COMPANY

New York   St. Louis   San Francisco   Dusseldorf   Johannesburg
Kuala Lumpur   London   Mexico   Montreal   New Delhi
Panama   Paris   Sao Paulo   Singapore
Sydney   Tokyo   Toronto

Library of Congress Cataloging in Publication Data

Markus, John.
    Guidebook of electronic circuits.

    1.   Electronic circuits—Handbooks, manuals, etc.
I. Title.
TK7867.M34   621.3815'3'0202   74-9616
ISBN   0-07-040445-3

1234567890 KPKP 7987654

The editors for this book were Tyler G. Hicks and Patricia A. Allen,
and the production was supervised by Stephen J. Boldish.
It was set in Spartan Heavy by the Fuller Organization.

It was printed and bound by Kingsport Press.

Contents

Preface vii
Abbreviations used ix
Semiconductor symbols used . . . xi
Addresses of sources used xiii
1. Alarm circuits 1
2. Amplifier circuits 5
3. Analog-digital converter circuits 21
4. Audio amplifier circuits 26
5. Audio compressor circuits 41
6. Audio control circuits 47
7. Automatic gain control circuits . 56
8. Automotive control circuits . . . 61
9. Automotive ignition circuits . . . 75
10. Auto theft alarm circuits 82
11. Battery-charging circuits 86
12. Burglar alarm circuits 92
13. Capacitance control circuits . . . 102
14. Cathode-ray circuits 108
15. Chopper circuits 115
16. Citizens band circuits 119
17. Clock circuits 124
18. Code circuits 129
19. Comparator circuits 140
20. Computer circuits 145
21. Contact bounce suppression cir-
    cuits 152
22. Converter circuits—d-c to d-c 155
23. Converter circuits—general . . . 163
24. Converter circuits—radio 169
25. Counter circuits 179
26. Current control circuits 194
27. Data transmission circuits 199
28. Digital-analog converter circuits 203
29. Digital clock circuits 208
30. Display circuits 220
31. Door lock circuits 234
32. Electronic music circuits 240
33. Filter circuits—active 252
34. Filter circuits—passive 265
35. Fire alarm circuits 270
36. Flasher circuits 275
37. Fluorescent lamp circuits 281
38. Frequency counter circuits . . . 286
39. Frequency divider circuits 294
40. Frequency modulation circuits . 300
41. Frequency multiplier circuits . . 309
42. Frequency synthesizer circuits . 317

43. Function generator circuits . . . 323
44. Gate circuits 331
45. High-voltage circuits 337
46. Hobby circuits 344
47. I-f amplifier circuits 350
48. Infrared circuits 356
49. Instrumentation circuits 363
50. Integrator circuits 370
51. Intercom circuits 375
52. Inverter circuits 380
53. Lamp control circuits 386
54. Laser circuits 391
55. Limited circuits 397
56. Liquid level control circuits . . . 402
57. Logarithmic circuits 405
58. Logic circuits 412
59. Measuring circuits—capacitance 418
60. Measuring circuits—current . . . 423
61. Measuring circuits—frequency . 430
62. Measuring circuits—general . . . 442
63. Measuring circuits—power . . . 453
64. Measuring circuits—resistance . 457
65. Measuring circuits—temperature 462
66. Measuring circuits—voltage . . . 469
67. Medical circuits 481
68. Memory circuits 490
69. Metal detector circuits 496
70. Modem circuits 500
71. Modular circuits 506
72. Motor control circuits 514
73. Multiplexer circuits 527
74. Multiplier circuits 536
75. Multivibrator circuits 542
76. Music-controlled light circuits . 554
77. Navigation circuits 560
78. Noise circuits 566
79. Operational amplifier circuits . . 573
80. Optoelectronic circuits 584
81. Oscillator circuits—a-f 591
82. Oscillator circuits—r-f 601
83. Phase control circuits 613
84. Phonograph circuits 620
85. Photoelectric circuits 628
86. Photography circuits 635
87. Power supply circuits 645
88. Protection circuits 650
89. Public address circuits 655

90. Pulse generator circuits 661
91. Pulse shaping circuits 674
92. Quadraphonic circuits 679
93. Radar circuits 686
94. Receiver circuits 691
95. Regulated power supply circuits 709
96. Regulator circuits 735
97. Remote control circuits 761
98. Repeater circuits 767
99. Sampling circuits 777
100. Science fair circuits 784
101. Servo circuits 789
102. Signal generator circuits 795
103. Single-sideband circuits 804
104. Siren circuits 813
105. Square-law circuits 817
106. Squelch circuits 821
107. Staircase generator circuits . . . 824
108. Stereo circuits 828
109. Surveillance circuits 846
110. Sweep circuits 851
111. Switching circuits 858
112. Tape recorder circuits 864
113. Telemetry circuits 874
114. Telephone circuits 879
115. Teleprinter circuits 884
116. Television circuits—color 896
117. Television circuits—general . . . 906
118. Television circuits—remote
     control 917
119. Temperature control circuits . . 921
120. Test circuits—general 930
121. Test circuits—solid state 935
122. Three-phase control circuits . . . 947
123. Timer circuits 952
124. Transceiver circuits 959
125. Transmitter circuits 968
126. Trigger circuits 988
127. Ultrasonic circuits 994
128. Voltage-controlled oscillator
     circuits 997
129. Voltage-level detector circuits 1004
130. Voltage reference circuits . . . 1011
131. Zero-voltage switching circuits 1015
     Name index 1019
     Subject Index 1027

Preface

More than 3,600 electronic circuits, published largely within the past 5 years in the United States and abroad, are presented here in 131 chapters, logically organized for convenient reference and browsing by electronic engineers, technicians, and experimenters. Each circuit has the values of all significant components, an identifying title, a concise description, performance data, and suggestions for applications. At the end of each description is a citation giving the title of the original article or book, its author, and the exact location of the circuit in the original source.

The circuits in this book are completely different from those in the highly successful predecessor volumes, "Sourcebook of Electronic Circuits" published by McGraw-Hill in 1968, and "Electronic Circuits Manual" published by McGraw-Hill in 1971. Entirely new chapters in this book provide easy access for the first time to circuits for Auto Theft Alarms, Burglar Alarms, Contact Bounce Suppression, Data Transmission, Intercoms, Liquid Level Control, Modems, Navigation, Quadraphonic Stereo, Science Fairs, Sirens, Surveillance, Three-Phase Control, and Zero-Voltage Switching. Significant new circuits appear in chapters found also in previous books, particularly for Battery Charging, Choppers, Flashers, Infrared, Lamp Control, Lasers, Medical Electronics, Metal Detectors, Optoelectronics, Regulated Power Supplies, Servos, Staircase Generators, and Ultrasonics.

This electronic circuit compendium serves as a highly effective desk-top information retrieval system for tested practical electronic circuits developed throughout the world. Engineering libraries, particularly in foreign countries, will find it a welcome substitute for the original sources when facing limitations on budgets, shelving, or search manpower. The circuits for this new book were located by cover-to-cover searching of back issues of U.S. and foreign electronic periodicals, the published literature of electronic manufacturers, and recent electronic books, together filling well over 100 feet of shelving. This same search would take weeks or even months at a large engineering library, plus the time required to write for manufacturer literature and locate elusive sources.

To find a desired circuit quickly, start with the alphabetically arranged table of contents at the front of the book. Note the chapters most likely to contain the desired type of circuit, and look in these first. Remember that most applications use combinations of basic circuits, so a desired circuit could be placed in any of several different chapters.

If a quick scan of these chapters does not give the exact circuit desired, use the index at the back of the book. Here the circuits are indexed in depth under the different names by which they are known. In addition, the index contains hundreds of cross-references that speed up your search. The author index will often help find related circuits after one potentially useful circuit is found, because authors tend to specialize in certain circuits.

In most cases, you can locate a desired circuit in a few minutes. It will be free of drafting errors, because the corrections pointed out in subsequently published errata notices have been made; this alone can save many frustrating hours of troubleshooting.

Values of important components are given for every circuit, because these help in reading the circuit and redesigning it for other requirements. The development of a circuit for a new application is speeded when design work can be started with a working circuit, instead of starting from scratch. Research and experimentation are thereby cut to a minimum, so even a single use of this circuit-retrieval book could pay for its initial cost many times over.

This book is organized to provide a maximum of circuit information per page, with minimum repetition. The chapter title at the top of each right-hand page, the text, and the original title in the citation should therefore be considered along with the description when evaluating a circuit.

Abbreviations are used extensively, to conserve space. Their meanings are given after the table of contents. Abbreviations on diagrams and in original article titles were unchanged and may differ slightly, but their meanings can be deduced by context.

Mailing addresses of all cited original sources are given at the front of the book, for convenience in writing for back issues or copies of articles when the source is not available at a local library. These sources will often prove useful for construction details, performance graphs, and calibration procedures.

To Mary Kay McCoy, student at Foothill College, goes credit for perseverance in learning the language of electronics and typing it exactly as dictated, hyphens and MHz's and all. And to Jack Quint, more active than ever in Florida retirement from Electronics magazine, goes credit for arranging the circuits in the greater part of this book, each unmistakably associated with its text.

To the original publications cited and their engineering authors and editors should go the major credit, however, for making possible this third encyclopedic contribution to electronic circuit design. The diagrams have been reproduced directly from the original source articles, by permission of the publisher in each case.

John Markus

Abbreviations Used

A	ampere	kc	kilocycle	preamp	preamplifier
a-c	alternating current	keV	kiloelectron-volt	prf	pulse repetition frequency
a-d	analog-digital	kHz	kilohertz	prr	pulse repetition rate
a-f	audio frequency	kV	kilovolt	prv	peak reverse voltage
afc	automatic frequency control	kW	kilowatt	ps	picosecond
aft	automatic fine tuning	L	inductance; inductor	psc	permanent split capacitor
agc	automatic gain control	las	light-activated switch	psk	phase-shift keying
alc	automatic level control	lascr	light-activated scr	put	programmable ujt
a-m	amplitude modulation	lascs	light-activated scs	pW	picowatt
bc	broadcast	ldr	light-dependent resistor	pwm	pulse-width modulation
bcd	binary coded decimal	led	light-emitting diode	R	resistance; resistor
bfo	beat-frequency oscillator	m	milli- (10^{-3})	RCTL	resistor-capacitor-transistor logic
b-w	black-and-white	M	mega- (10^6)	RDTL	resistor-diode-transistor logic
C	degrees Centigrade; capacitance; capacitor	mA	milliampere	r-f	radio frequency
CATV	community-antenna television	Mc	megacycle	rfi	radio-frequency interference
CB	citizens band	meg	megohm	rms	root-mean-square
CCTV	closed-circuit television	mH	millihenry	rom	read-only memory
cm	centimeter	MHz	megahertz	rpm	revolutions per minute
c-mos	complementary mos	mil	0.001 inch (10^{-3} inch)	rps	revolutions per second
cos-mos	complementary-symmetry mos	min	minute	rtty	radioteletype
cp	candlepower	ms	millisecond	RTL	resistor-transistor logic
cr	cathode-ray	mm	millimeter	sbs	silicon bilateral switch
cro	cathode-ray oscilloscope	modem	modulator-demodulator	SCA	Subsidiary Communications Authorization
crt	cathode-ray tube	mono	monostable multivibrator	scope	oscilloscope
C-T	center-tap	mos	metal-oxide-semiconductor	scr	silicon controlled rectifier
c-w	continuous-wave	mosfet	metal-oxide semiconductor fet	scs	silicon controlled switch
d-a	digital-analog	most	metal-oxide-semiconductor transistor	s	second
dB	decibel	μ	micro (10^{-6})	s/n	signal-to-noise
dBm	decibels above 1 mW	μA	microampere	spdt	single-pole double-throw
d-c	direct current	μF	microfarad	sq cm	square centimeter
deg	degree	μH	microhenry	ssb	single-sideband
diac	diode a-c switch	μm	micrometer	sus	silicon unilateral switch
dpdt	double-pole double-throw	μs	microsecond	s-w	short-wave
DTL	diode-transistor logic	mV	millivolt	sync	synchronizing
dvm	digital voltmeter	mvbr	multivibrator	T	tera- (10^{12})
dvom	digital vom	mW	milliwatt	td	tunnel diode
ecg	electrocardiograph	n	nano- (10^{-9})	t-r	transmit-receive
ECL	emitter-coupled logic	nA	nanoampere	triac	triode a-c switch
eeg	electroencephalograph	nF	nanofarad	TTL	transistor-transistor logic
emf	electromotive force	npn	negative-positive-negative	tv	television
emi	electromagnetic interference	ns	nanosecond	tvm	transistor voltmeter
fet	field-effect transistor	nW	nanowatt	twt	traveling-wave tube
FHP	fractional-horsepower	opamp	operational amplifier	uhf	ultrahigh frequency
f-m	frequency modulation	p	pico- (10^{-12})	ujt	unijunction transistor
fsk	frequency-shift keying	p-a	public address	V	volt
ft	feet	pA	picoampere	VA	voltampere
G	giga- (10^9)	pam	pulse-amplitude modulation	vco	voltage-controlled oscillator
GHz	gigahertz	pcm	pulse-code modulation	vdr	voltage-dependent resistor
G-M	Geiger-Muller	pep	peak envelope power	vfo	variable-frequency oscillator
hp	horsepower	pF	picofarad	vhf	very high frequency
hr	hour	pin	positive-intrinsic-negative	vlf	very low frequency
Hz	hertz	piv	peak inverse voltage	vom	volt-ohm-milliammeter
IC	integrated circuit	pll	phase-locked loop	vox	voice-operated transmission
i-f	intermediate frequency	p-m	phase modulation; permanent magnet	vswr	voltage standing wave ratio
igfet	insulated-gate fet	pnp	positive-negative-positive	vtvm	vacuum-tube voltmeter
ips	inches per second	pot	potentiometer	vu	volume unit
IR	infrared	p-p	peak-to-peak	W	watt
jfet	junction fet	ppi	plan position indicator	YIG	yttrium-iron-garnet
k	kilo- (10^3)	ppm	pulse per minute	Z	impedance
K	kilohm (,000 ohms); degrees Kelvin	pps	pulse per second		

Semiconductor Symbols Used

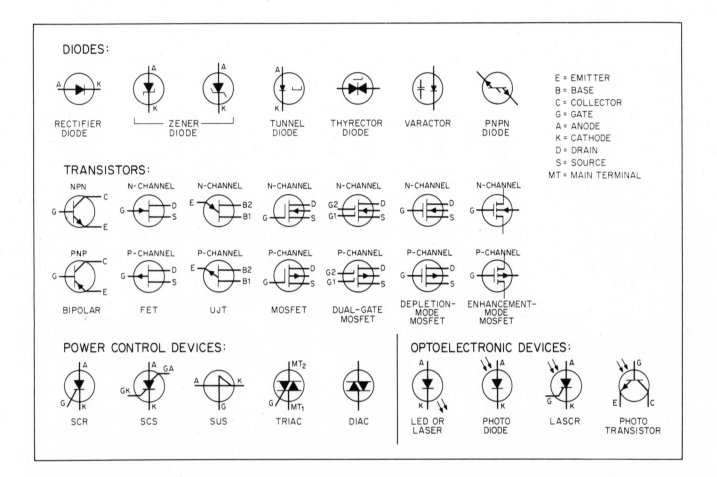

The commonest forms of the basic semiconductor symbols are shown here. Leads are identified where appropriate, for convenient reference. Minor variations in symbols, particularly those from foreign sources, can be recognized by comparing with these symbols while noting positions and directions of solid arrows with respect to other symbol elements.

Omission of the circle around a symbol has no significance. Arrows are sometimes drawn open instead of solid. Thicker lines and open rectangles in some symbols on diagrams have no significance. Orientation of symbols is unimportant; artists choose the position that is most convenient for making connections to other parts of the circuit. Arrow lines outside optoelectronic symbols indicate the direction of light rays.

On some European diagrams, the position of the letter k gives the location of the decimal point for a resistor value in kilohms. Thus, 2k2 is 2.2K or 2,200 ohms.

Substitutions can often be made for semiconductor and IC types specified on diagrams. Newer components, not available when the original source articles was published,

may actually improve the performance of a particular circuit. Electrical characteristics, terminal connections, and such critical ratings as voltage, current, frequency, and duty cycle, must of course be taken into account if experimenting without referring to substitution guides.

Semiconductor, integrated-circuit, and tube substitution guides can usually be purchased at electronic parts supply stores or ordered directly from publishers. For availability and prices of these guides, write to manufacturers such as Fairchild, General Electric, Motorola, National, RCA, Signetics, Sprague, Sylvania, and Texas Instruments. For books, write for the catalog of Howard W. Sams & Co., Inc. (their 1974 catalog lists "Transistor Substitution Handbook" at $2.95 and "Tube Substitution Handbook" at $1.95). The 1974 mail-order catalog of Allied Electronics Corp., 2400 Washington Blvd., Chicago, IL 60612, priced at $5, contains many pages of cross-reference data and connection diagrams for transistors, diodes, and integrated circuits, along with descriptions and prices of electronic components.

Addresses of Sources Used

In the citation at the end of each abstract, the title of a magazine is set in italics. The title of a book or report is placed in quotes. Each source title is followed by the name of the publisher of the original material, plus city and state. Complete mailing addresses of all sources are given below, for the convenience of readers who want to write to the original publisher of a particular circuit.

Most reports can be obtained without charge by writing to manufacturers. Give the complete citation, exactly as in the abstract.

Books can be ordered from their publishers, after first writing for prices of the books desired. Some electronics manufacturers also publish books and large reports for which charges are made. Some of the books cited as sources in this volume are also sold by electronic supply firms and bookstores. Locations of these firms can be found in the YELLOW PAGES of telephone directories under headings such as "Electronic Equipment and Supplies" or "Television and Radio Supplies and Parts."

Only a few magazines have back issues on hand for sale, but most magazines will make copies of a specific article at a fixed charge per page or per article. When you write to a magazine publisher for prices of back issues or copies, give the *complete* citation, *exactly* as in the abstract. Include a stamped self-addressed envelope to make a reply more convenient.

If certain magazines consistently publish the types of circuits in which you are interested, use the addresses below to write for subscription rates.

Admiral Corp., National Service Division, P.O. Box 845, Bloomington, IL 61701

Admiral Service News Letter, Admiral Corp., National Service Division, P.O. Box 845, Bloomington, IL 61701

American Radio Relay League, 225 Main St., Newington, CT 06111

Amperex Electronic Corp., Providence Pike, Slatersville, RI 02876

Analog Devices, Inc., Rte. 1 Industrial Park, Box 280, Norwood, MA 02062

Analog Dialogue, Analog Devices, Inc., Rte. 1 Industrial Park, Box 280, Norwood, MA 02062

Audio, 134 N. 13th Street, Philadelphia, PA 19107

Burr-Brown Research Corp., 6730 S. Tucson Blvd., Tucson, AZ 85734

Calectro, GC Electronics, 400 South Wyman St., Rockford, IL 61101

CB Magazine, 250 Park Ave., New York, NY 10017

Clairex Electronics, 560 South Third Ave., Mt. Vernon, NY 10550

Delco Electronics Division, General Motors Corp., Kokomo, IN 46901

Design Electronics—See *Electron*

Dialight Corp., 60 Stewart Ave., Brooklyn, NY 11237

Digital Design, 167 Corey Road, Brookline, MA 02146

Electrical Communication, ITT, 320 Park Ave., New York, NY 10022

Electromechanical Design, 167 Corey Road, Brookline, MA 02146

Electron, Dorset House, Stamford St., London SE1 9LU, England

The Electronic Engineer, One Decker Square, Bala-Cynwyd, Pa 19004

Electronic Engineering, 28 Essex St., Strand, London WC2, England

Electronics, 1221 Avenue of the Americas, New York, NY 10020

Electronic Servicing, 1014 Wyandotte St., Kansas City, MO 64105

Electronics World—See *Popular Electronics*

Elementary Electronics, 229 Park Ave. South, New York, NY 10003

Exar Integrated Systems, Inc., 750 Palomar Ave., Sunnyvale, CA 94086

Fairchild Semiconductor, 464 Ellis St., Mountain View, CA 94040

GE (General Electric Co.):
   Electronic Components Division, General Electric Co., 316 East Ninth St., Owensboro, KY 42301
   Semiconductor Products Dept., General Electric Co., Bldg. 7, Electronics Park, Syracuse, NY 13201
   Semiconductor Products Dept., Auburn, NY—Write to General Electric Co., Semiconductor Products Dept., Bldg. 7, Electronics Park, Syracuse, NY 13201

General Radio Experimenter, General Radio Co., 300 Baker Ave., Concord, MA 01742

Gould Ionics Inc., P.O. Box 1377, Canoga Park, CA 91304

Ham Radio, Communication Technology Inc., Greenvale, NH 03048

Harris Semiconductor, Linear Applications Engineering, P.O. Box 883, Melbourne, FL 32901

Hewlett Packard, 1501 Page Mill Rd., Palo Alto, CA 94304

Howard W. Sams & Co. Inc., 4300 West 62nd St., Indianapolis, IN 46206

Hybrid Systems, 87 Second Ave., Burlington, MA 01803

IEEE Publications, 345 East 47th St., New York, NY 10017

Intel Corp., 3065 Bowers Ave., Santa Clara, CA 95051

International Rectifier, Semiconductor Division, 233 Kansas St., El Segundo, CA 90245

Lloyd's Trading Co., 59 North Fifth St., Saddlebrook, NJ

Magnavox Co., Service Publications, Dept. 765, 1700 Magnavox Way, Fort Wayne, IN 46804

Matsushita Electric Corp. of America, 200 Park Ave., New York, NY 10017

McGraw-Hill Book Co., 1221 Avenue of the Americas, New York, NY 10020

Microwave Journal, Horizon House, 610 Washington St., Dedham, MA 02026

Monsanto Co., 10131 Bubb Rd., Cupertino, CA 95014

Motorola Semiconductors, Box 20912, Phoenix, AZ 85036

Mullard Limited, Mullard House, Torrington Place, London WC1E 7HD, England

Mullard Technical Communications, Mullard Limited, Mullard House, Torrington Place, London WC1E 7HD, England

National Semiconductor Corp., 2900 Semiconductor Drive, Santa Clara, CA 95051

Optical Electronics Inc., Box 11140, Tucson, AZ 85734

Panasonic—See Matsushita Electric Corp. of America

J. C. Penney, 1301 Sixth Ave., New York, NY 10020

Philips, Pub. Dept., Elcoma Division, Eindhoven, The Netherlands

Popular Electronics, One Park Ave., New York, NY 10016

Popular Science, 355 Lexington Ave., New York, NY 10017

Precision Monolithics, 1500 Space Park Drive, Santa Clara, CA 95050

QST, American Radio Relay League, 225 Main St., Newington, CT 06111

Radio-Electronics, 200 Park Ave. South, New York, NY 10003

RCA, Solid State Division, Route 202, Somerville, NJ 08876

RCA Review, David Sarnoff Research Center, Princeton, NJ 08540

Howard W. Sams & Co. Inc., 4300 West 62nd St., Indianapolis, IN 46206

Science and Electronics—See *Elementary Electronics*

73 Magazine, Peterborough, NH 03458

Short Wave Magazine, 55 Victoria St., London SW1H-0HF, England

Signetics Corp., 811 East Arques Ave., Sunnyvale, CA 94086

Siliconix, Inc., 2201 Laurelwood Rd., Santa Clara, CA 95054

Teledyne Philbrick, Allied Drive at Route 128, Dedham, MA 02026

Teledyne Semiconductor, 1300 Terra Bella Ave., Mountain View, CA 94040

Texas Instruments, Components Group, Box 5012, Dallas, TX 75222

Unitrode Corp., 580 Pleasant St., Watertown, MA 02172

Westinghouse, Semiconductor Division, Youngwood, PA 15697

Wireless World, Dorset House, Stamford St., London SE1, England

CHAPTER 1
Alarm Circuits

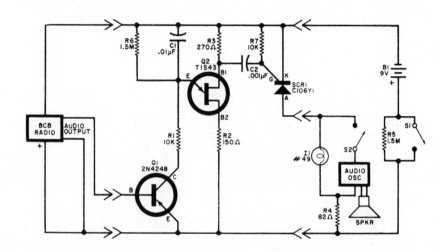

STORM WARNING—Senses approaching thunderstorm at distances up to 50 miles and gives visible or audible warning to head for shelter from rain. Uses inexpensive portable transistor radio tuned to low-frequency end of broadcast band (not to station) to pick up static noise associated with storm. Audio output of radio turns on Q1, allowing C1 to charge through R1 at rate determined by length of static burst and setting of radio volume control. When C1 charges enough, Q2 breaks down and discharges it, creating pulse that fires scr and turns on alarm. S1 must be opened to stop alarm.—J. E. Shepler, The Picnicker's Friend, *Popular Electronics*, July 1969, p 47–50.

MOISTURE-RAIN ALARM—Uses 4,000-ohm relay to close alarm circuit when raindrops fall on sensor S1, with setting of R4 determining amount of moisture required to trigger alarm. If humidity sensor is used instead, same circuit can be installed in basement to turn on dehumidifier automatically when basement gets too damp.—R. M. Brown and T. Kneitel, "49 Easy Transistor Projects," Howard W. Sams, Indianapolis, IN, 1972, p 34–35.

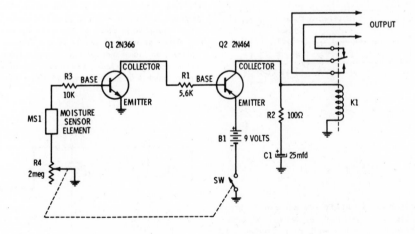

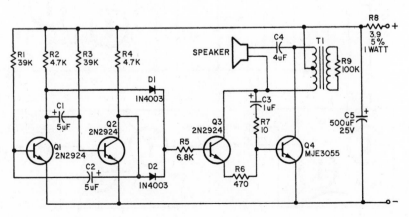

ELECTRONIC HORN—Produces ear-splitting burglar-scaring racket for up to 1 hour with single 9-V transistor battery, and will operate on almost any battery from 6 to 18 V. Mvbr Q1-Q2 drives amplifier Q4 through switching transistor Q3. Omitting one of diodes gives more interrupting sound. R9 can be as low as 39K to change tone. Changing R1 and R3 in range of 22K to 220K changes rate of interruption. Place alarm-actuating switch in one battery lead.—C. D. Rakes, Magnum Alarm, *Elementary Electronics*, March-April 1972, p 63–65 and 96.

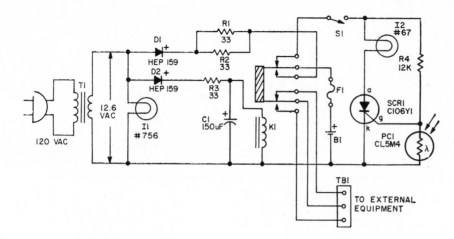

EMERGENCY LIGHT—Operates from 12-V lead-acid storage battery that is kept on trickle charge as long as a-c power is available. If power line fails, relay K1 drops out to turn on alarm, energize standby generator, or perform other switching. Independently, photocell PC1 senses darkness and triggers scr to turn on No. 67 12-V auto lamp as emergency light, if S1 is on. K1 is Guardian 900-905 12-V dpdt relay. B1 can be two 6-V motorcycle batteries in series.—R. F. Graf and G. J. Whalen, Sentry Shiner, *Elementary Electronics*, Sept.-Oct. 1970, p 53–56.

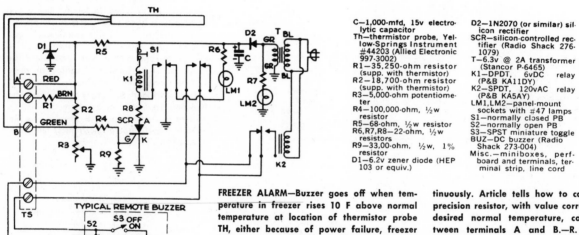

C—1,000-mfd, 15v electrolytic capacitor
Th—thermistor probe, Yellow-Springs Instrument #44203 (Allied Electronic 997-3002)
R1—35,250-ohm resistor (supp. with thermistor)
R2—18,700-ohm resistor (supp. with thermistor)
R3—5,000-ohm potentiometer
R4—100,000-ohm, ½ w resistor
R5—68-ohm, ½ w resistor
R6,R7,R8—22-ohm, ½ w resistors
R9—33,00-ohm, ½ w, 1% resistor
D1—6.2v zener diode (HEP 103 or equiv.)

D2—1N2070 (or similar) silicon rectifier
SCR—silicon-controlled rectifier (Radio Shack 276-1079)
T—6.3v @ 2A transformer (Stancor P-6465)
K1—DPDT, 6vDC relay (P&B KA11DY)
K2—SPDT, 120vAC relay (P&B KA5AY)
LM1,LM2—panel-mount sockets with #47 lamps
S1—normally closed PB
S2—normally open PB
S3—SPST miniature toggle
BUZ—DC buzzer (Radio Shack 273-004)
Misc.—miniboxes, perfboard and terminals, terminal strip, line cord

FREEZER ALARM—Buzzer goes off when temperature in freezer rises 10 F above normal temperature at location of thermistor probe TH, either because of power failure, freezer failure, or someone leaving door open. Buzzer operates on four pen-cell alkaline batteries, which will drive it over 12 hours continuously. Article tells how to calibrate with precision resistor, with value corresponding to desired normal temperature, connected between terminals A and B.—R. M. Benrey, Keep Your Cool—With This Homemade Freezer Alarm, *Popular Science*, March 1972, p 112–113.

ALL-PURPOSE ALARM—Sensor can be breakable wire, photocell, or water-level probes. Produces siren-like wail. K1 is 1,000-ohm 50-mW sensitive relay.—T. Brown, Build a General-Purpose Alarm, *Popular Electronics*, Sept. 1972, p 64–65.

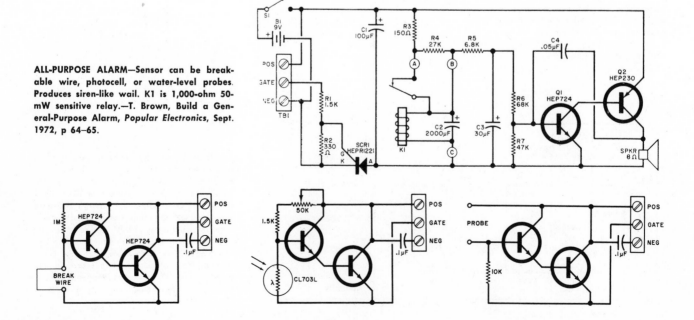

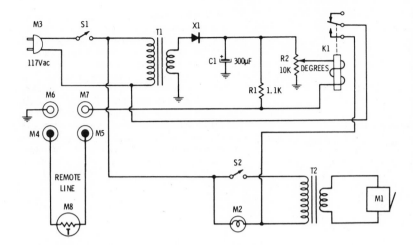

FROST ALARM—Uses 31D7 thermistor to sense when temperature drops close enough to freezing to endanger plants. Will work with thermistor up to 500' away from alarm circuit. X1 is 130-V 70-mA selenium rectifier, T1 has 25-V secondary and T2 has 6.3-V secondary driving 6-V a-c buzzer. K1 has 8K coil. To calibrate, take ice cube out of freezer, let it stand 5 minutes at room temperature of 70 F, then put it on thermistor and adjust R2 until buzzer sounds.—R. M. Brown and T. Kneitel, "49 Easy Entertainment and Science Projects," Vol. 2, Howard W. Sams, Indianapolis, IN, 1969, p 51–53.

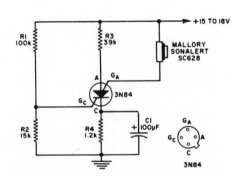

BEEPER—Single scs in self-triggering astable mvbr circuit has repetition rate of 2 pps, determined by value used for C1. Used to interrupt continuous piercing wail of d-c Sonalert at same rate, to improve attention-getting effectiveness. Two or more beepers set for different rates can be used for monitoring different situations and identifying each.—F. H. Tooker, "Beeping" a Sonalert, Electronics World, Dec. 1969, p 79.

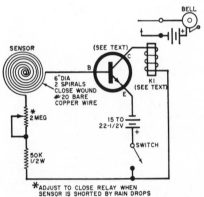

RAIN ALARM—Raindrops between closely separated wires or foil strips of sensor make transistor pull in relay and sound alarm bell or buzzer. Use any general-purpose transistor such as GE-5 or SK-3011. Relay has 5,000-ohm coil.—C. J. Schauers, Automatic Rain Switch, Popular Electronics, Sept. 1967, p 77.

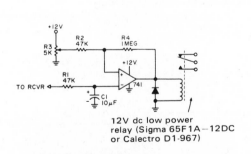

12V dc low power relay (Sigma 65F1A—12DC or Calectro D1-967)

CARRIER-OPERATED RELAY—Simple opamp circuit energizes relay whenever carrier is present in 2-meter f-m receiver, to connect speaker for emergency message. Receiver signal is taken at output of audio amplifier, which swings from normal +9 V down to about +8 V when station is received. R3 adjusts point at which relay pulls in. R4 prevents relay chatter.—P. A. Stark, 741 Op-Amp COR and Tone Decoder Circuits, 73, July 1972, p 83–88.

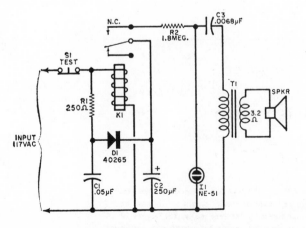

POWER FAILURE—Uses energy stored in electrolytic C2 to operate neon relaxation oscillator when a-c line voltage fails and relay drops out. Resulting tone from speaker lasts over 5 minutes without significant decrease in volume though with gradually decreasing frequency, and is loud enough to be heard in adjacent room if door is open. T1 is Stancor A-3856 or similar output transformer. Use at least 5-inch speaker with heavy magnet. No batteries are required.—F. H. Tooker, Batteryless Power Failure Alarm, Popular Electronics, Feb. 1968, p 43–44.

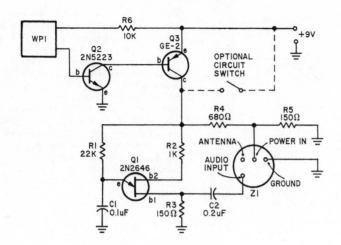

MOISTURE ALARM—Current passing though moisture-sensing plate WP1 is amplified by Q2 and Q3 to turn on a-f relaxation oscillator Q1. Resulting audio howl is broadcast on 95 MHz to any nearby f-m receiver by Z1, which is Lafayette Radio No. 19-55277 f-m wireless microphone transmitter module. Can be used as wet-diaper alarm if WP1 is placed in diaper. WP1 can be made from two conductive plates close together but not touching, or by scratching comb pattern on unused printed circuit board.—J. E. Lockridge, Diaper Snooper, Elementary Electronics, March-April 1972, p 44–46 and 102.

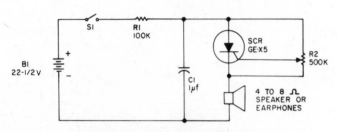

CLICKER—Simple relaxation oscillator generates clicking sound in speaker at rate determined by setting of R2. Can serve as alarm if S1 is switch that is closed by action of intruder.—"Electronics Experimenters Circuit Manual," General Electric, Owensboro, KY, 1971, 3rd Ed., p 161–166.

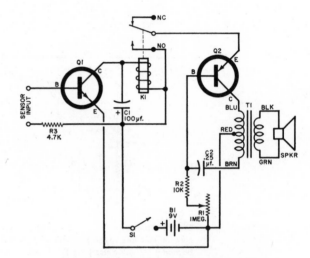

RAIN ALARM—Sounds audible alarm whenever rain provides conductivity between pattern of conductors separated by small insulated space. Q1 is any general-purpose npn transistor such as 2N229, and Q2 is general-purpose pnp such as 2N107. K1 is sensitive relay. Output transformer T1 has 500-ohm CT primary and 3.2-ohm secondary driving 4-ohm speaker.—E. Richardson, Reader's Circuit, Popular Electronics, June 1964, p 72.

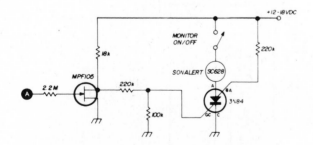

CARRIER ALARM—With monitor switch closed, appearance of carrier signal on channel being monitored by receiver makes fet trigger 3N84 and operates Sonalert alarm. Alarm continues until switch is opened, even though carrier disappears. Monitor works best when input terminal A is connected to grid of d-c amplifier in squelch circuit of receiver; this prevents triggering of 3N84 by noise.—M. Ronald, Solid-State Carrier-Operated Relay and Call Monitor, Ham Radio, June 1971, p 22–23.

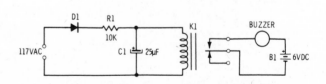

POWER-FAILURE ALARM—Buzzer operating from independent 6-V battery continues sounding when power line fails and relay K1 (3,000 to 5,000-ohm 3-mA sensitive type) drops out, as reminder that alarm clocks will not wake up household in time for work. D1 is 500-mA 200-piv silicon.—M. H. Friedman, "99 Electronic Projects," Howard W. Sams, Indianapolis, IN, 1971, p 32–33.

CHAPTER 2
Amplifier Circuits

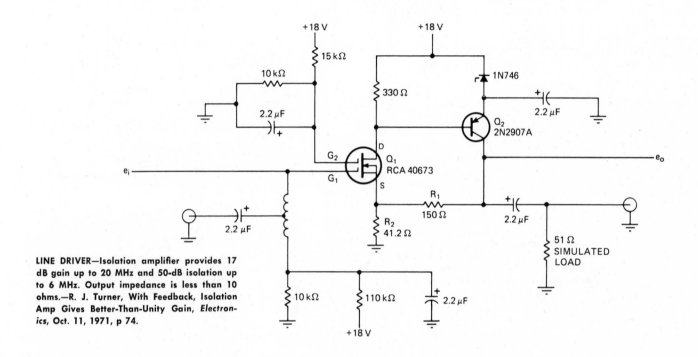

LINE DRIVER—Isolation amplifier provides 17 dB gain up to 20 MHz and 50-dB isolation up to 6 MHz. Output impedance is less than 10 ohms.—R. J. Turner, With Feedback, Isolation Amp Gives Better-Than-Unity Gain, *Electronics*, Oct. 11, 1971, p 74.

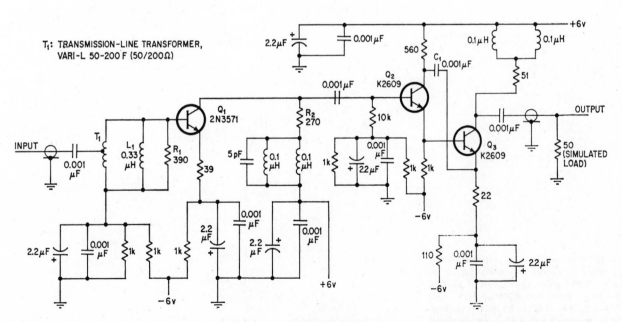

T_1: TRANSMISSION-LINE TRANSFORMER, VARI-L 50-200 F (50/200Ω)

LOW-NOISE PREAMP—Bootstrap connection between emitter of Q3 and collector of Q2 increases cutoff frequency from 150 MHz to 275 MHz in low-noise high-gain vhf preamp capable of detecting 5-μV signals without hanging up on strong signals.—R. J. Turner, Bootstrap Boosts Gain of Low-Noise RF Preamp, *Electronics*, May 24, 1971, p 57.

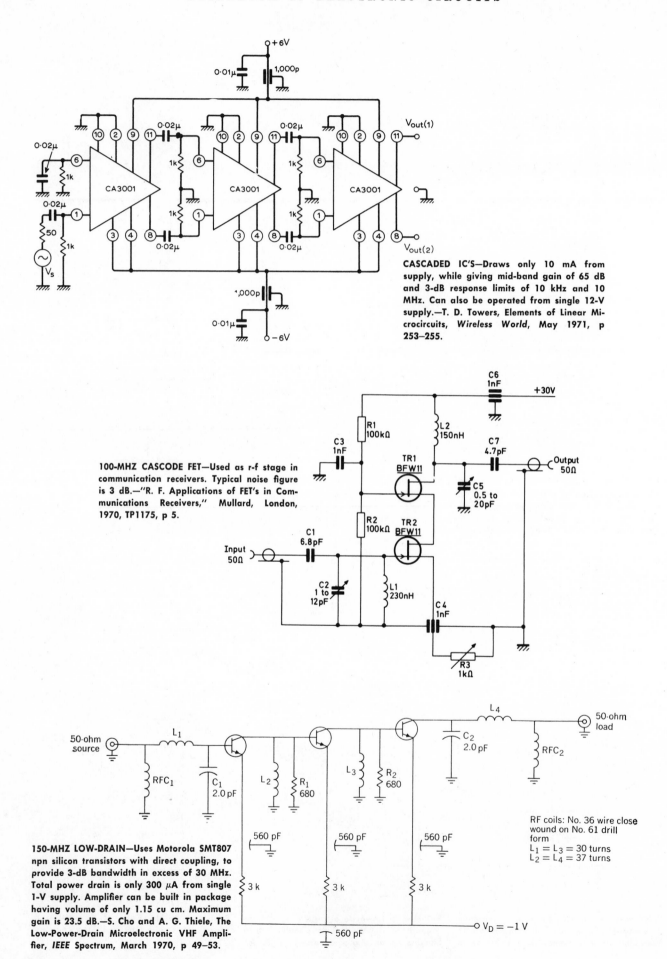

CASCADED IC'S—Draws only 10 mA from supply, while giving mid-band gain of 65 dB and 3-dB response limits of 10 kHz and 10 MHz. Can also be operated from single 12-V supply.—T. D. Towers, Elements of Linear Microcircuits, *Wireless World*, May 1971, p 253–255.

100-MHZ CASCODE FET—Used as r-f stage in communication receivers. Typical noise figure is 3 dB.—"R. F. Applications of FET's in Communications Receivers," Mullard, London, 1970, TP1175, p 5.

150-MHZ LOW-DRAIN—Uses Motorola SMT807 npn silicon transistors with direct coupling, to provide 3-dB bandwidth in excess of 30 MHz. Total power drain is only 300 μA from single 1-V supply. Amplifier can be built in package having volume of only 1.15 cu cm. Maximum gain is 23.5 dB.—S. Cho and A. G. Thiele, The Low-Power-Drain Microelectronic VHF Amplifier, *IEEE* Spectrum, March 1970, p 49–53.

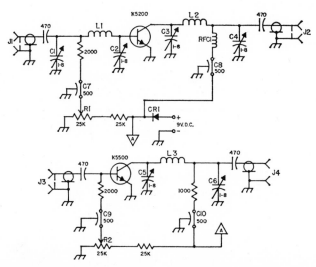

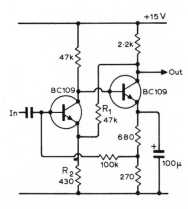

1,296-MHZ PREAMP—Used to extend coverage of crystal-mixer converter. Upper circuit may be used alone if combined gain of 19 dB with both stages (points A connected together and J2 connected to J3) is not needed. Article covers construction and adjustment.—D. Vilardi, A Two-Stage Transistor Preamplifier for 1,296 Mc, *QST*, Dec. 1968, p 40–42.

DARLINGTON—Uses separate npn transistors with feedback for stabilization. Gain depends on R1 and R2, and is about 100. Input impedance is about 68K, and open-loop gain about 2,000. Is noninverting.—J. L. Linsley Hood, The Liniac, *Wireless World*, Sept. 1971, p 437–441.

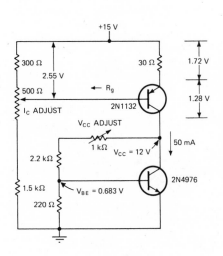

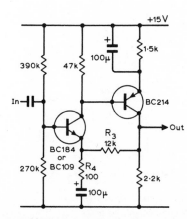

TEMPERATURE COMPENSATION—Uses collector resistance of lower transistor to provide temperature compensation over range of 0 to 70 C for 2N1132 high-frequency grounded-emitter power transistor, without excessive power dissipation in bias circuit of stage. Suitable for 300 to 3,000 MHz if power levels are at least 200 mW. Article gives design equation.—B. K. Erickson, "Temperature Compensation for High-Frequency Transistors, *Electronics*, May 24, 1973, p 102–103.

DARLINGTON—Uses individual npn and pnp transistors. Gain depends on R3 and R4, and is about 100. Input impedance is about 50K and open-loop gain about 2,000. Is noninverting.—J. L. Linsley Hood, The Liniac, *Wireless World*, Sept. 1971, p 437–441.

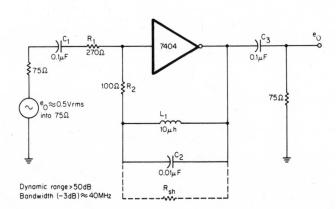

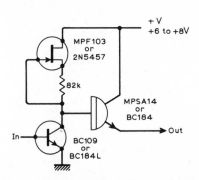

40-MHZ BANDWIDTH—Combining L-C tuned circuit with TTL inverter gives low-cost bandpass amplifier for use as frequency multiplier, tone filter, or i-f amplifier. Q of feedback path can be modified by adding Rsh, to change bandwidth.—C. A. Herbst, Linearized TTL Inverter Makes Bandpass Amplifier, *The Electronic Engineer*, Oct. 1971, p 68.

LINIAC—Linear inverting amplifying circuit consists of bipolar transistor connected as grounded-emitter amplifier, fet used as constant-current load, and monolithic Darlington as output amplifier. Configuration is for positive supply voltage. Gain is up to 4,000, with very low noise.—J. L. Linsley Hood, The Liniac, *Wireless World*, Sept. 1971, p 437–441.

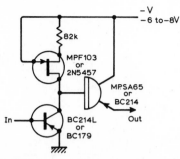

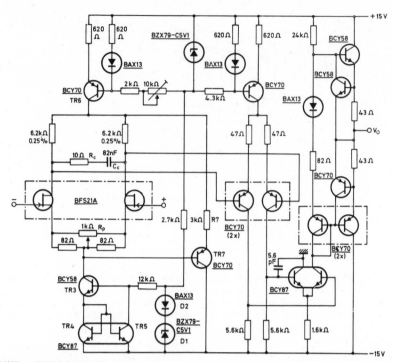

LINIAC—Linear inverting amplifying circuit for negative supply consists of bipolar transistor connected as grounded-emitter amplifier, fet used as constant-current load, and monolithic Darlington as output amplifier. Gain is up to 4,000, with very low noise.—J. L. Linsley Hood, The Liniac, *Wireless World*, Sept. 1971, p 437–441.

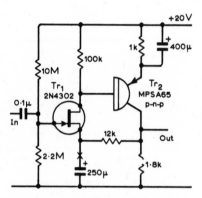

Feedback resistor inserted at X to provide feedback control of gain

FET WITH DARLINGTON—Use of monolithic Darlington as TR2 gives loop gain above 4,000 in noninverting mode, with high input impedance.—J. L. Linsley Hood, The Liniac, *Wirless World*, Sept. 1971, p 437–441.

3-MHZ UNITY-GAIN—Discrete components are combined with IC's to give high bandwidth and high slew rate. Voltage gain is over 90 dB and slew rate over 8 V/μs. Uses fet pair as differential input stage having over 100,000-meg differential input imped-ance in parallel with 4 pF and same common-mode input impedance. Output is short-circuit proof. Has unconditional stability for closed-loop gains down to unity.—"Field Effect Transistors," Mullard, London, 1972, TP1318, p 98–99.

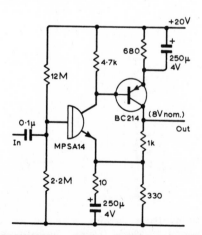

DARLINGTON WITH BIPOLAR—Monolithic Darlington pnp device connected as shown gives open-loop gain above 6,000 and input impedance of about 1.5 meg.—J. L. Linsley Hood, The Liniac, *Wireless World*, Sept. 1971, p 437–441.

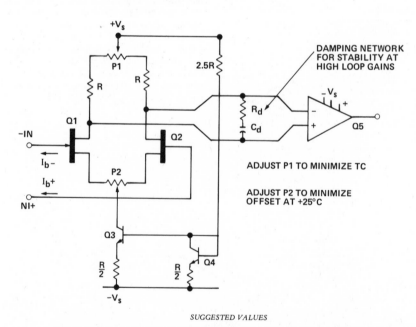

ADJUST P1 TO MINIMIZE TC

ADJUST P2 TO MINIMIZE OFFSET AT +25°C

SUGGESTED VALUES

I_b	Q1, Q2	R	P1	P2	Q3, Q4	Q5	R_D	C_D
50pA	AD3954A to AD3958	25kΩ	2kΩ AD79PR2k	1kΩ AD79PR1k	2N4880 Dual Transistor	AD741C	2.2kΩ	0.47μF
1pA	AD5906 to AD5909	150kΩ	10kΩ AD79PR10k	5kΩ AD79PR5k	2N4880 Dual Transistor	AD502J	15kΩ	0.05μF

FET PREAMP—Outputs of Q1 and Q2 are taken from drains, with Q3 and Q4 forming constant-current source unaffected by common-mode swing for inputs from —10 V to +5 V. Provides at least 20 dB more gain than circuit without current control. All solid-state units and opamp are made by Analog Devices.—Choosing and Using N-Channel Dual J-Fets, *Analog Dialogue*, Dec. 1970, p 4–9.

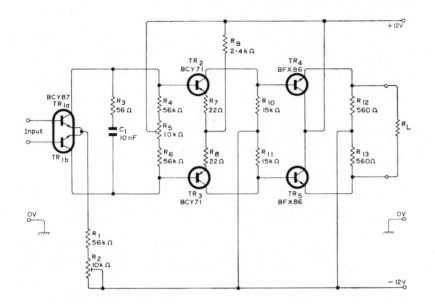

VOLTAGE GAIN OF 6,500—All stages operate in push-pull. First and second stages are long-tailed pairs. Operating voltages are chosen so output is zero when input signal is zero. Frequency response is shaped to drop approximately 6 dB per octave above about 1 kHz. Output resistance is 80 ohms and input resistance 180K.—"Circuits Using Low-Drift Transistor Pairs," Mullard, London, 1968, TP994, p 16–17.

VOLTAGE GAIN OF 70,000—Differential transistor-pair input stage uses constant-current source TR2 to stabilize combined emitter currents. R5 is used to adjust for zero output when input voltage is zero. Power supply lines are decoupled from 0-V line with 10-nF capacitors. Frequency response is shaped to fall about 6 dB per octave above about 500 Hz. Output resistance is 3 ohms and input is 200K. D1 is BZX61-C27.—"Circuits Using Low-Drift Transistor Pairs," Mullard, London, 1968, TP994, p 20–21.

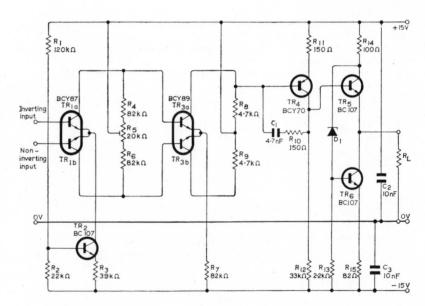

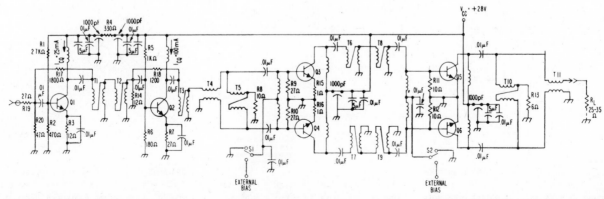

2–30 MHZ UNTUNED—Overall efficiency on c-w is above 47% and above 31% for two-tone 60-W pep output. Circuit uses broadband transmission-line transformers wound on ferrite toroids for impedance matching. Q1 and Q2 are 2N3866 biased class A. Q3 and Q4 are 2N3375 in class AB. Q5 and Q6 are 2N5071 operating class AB into 12.5-ohm load-line.—O. Pitzalis, R. E. Horn, and Ronald J. Baranello, Broadband 60-W HF Linear Amplifier, *IEEE Journal of Solid-State Circuits*, June 1971, p 93–103.

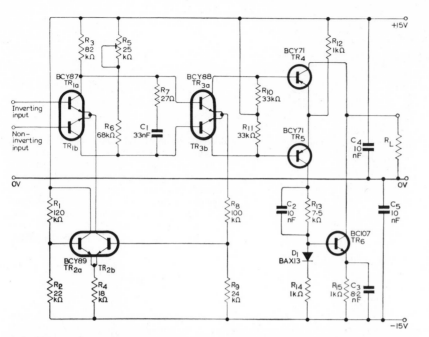

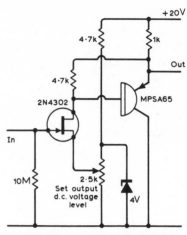

AMPLIFIER WITH VOLTAGE GAIN OF 100,000—Differential transistor-pair input stage uses transistor pair TR2 as constant-current source to keep combined emitter currents of TR1 constant when common-mode input signal is applied. Second stage is also differential, feeding long-tailed pair output stage TR4-TR5 served by variable-current source TR6

and silicon diode D1. Nominal output is 10 V at 2.5 mA for either polarity. Output resistance is 20K and input is 100K. Frequency response drops about 6 dB per octave above 100 Hz.—"Circuits Using Low-Drift Transistor Pairs," Mullard, London, 1968, TP994, p 22—23.

D-C BOOTSTRAP—Combination of fet and monolithic Darlington in phase-inverting configuration gives gain of about 250. Connection of load resistor between base and emitter of Darlington multiplies effective dynamic impedance of resistor at all frequencies down to d-c by figure approaching current gain of Darlington.—J. L. Linsley Hood, The Liniac, *Wireless World*, Sept. 1971, p 437—441.

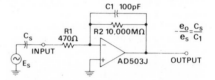

CHARGE AMPLIFIER—Use with crystal mikes and pickups, capacitance alarms, piezoelectric accelerometers, and other capacitive sources. Noise level is only 5.5 μV rms for band from 5 to 10,000 Hz. Frequency response is flat within 3 dB from 0.16 Hz to 800 kHz.—"AD503, AD506 I.C. Fet Input Operational Amplifiers," Analog Devices, Inc., Norwood, MA, Technical Bulletin, Aug. 1971.

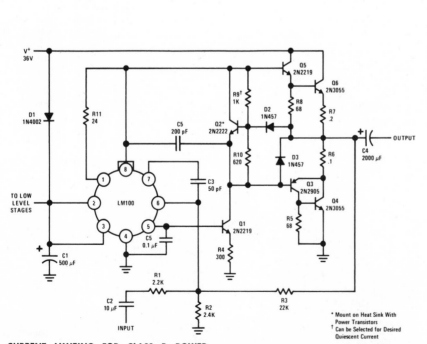

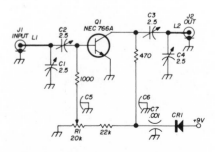

CURRENT LIMITING FOR CLASS B POWER AMPLIFIER—Uses LM100 voltage regulator as high-gain amplifier driving output transistors Q5-Q6 for positive-going output signals, while current source Q1 drives Q3-Q4 for negative-going signals. Q2 eliminates dead zone of class-B output stage. D2 and D3 pro-

vide output current limiting. Power-supply ripple is peak-detected by D1 and C1 to get increased positive output voltage swing during troughs of ripple.—R. J. Widlar, "New Uses for the LM100 Regulator," National Semiconductor, Santa Clara, CA, 1968, AN-8.

2,304-MHZ PREAMP—Provides 8 dB gain and 5.7 dB noise figure, using Nippon Electronics transistor. Chokes are 0.001-inch brass strip 3/32 inch wide; L1 is 5/16 inch long and L2 ⅜ inch long. Article covers construction.—D. Vilardi, Solid-State 2,304-MHZ Preamplifier, *Ham Radio*, Aug. 1972, p 20—23.

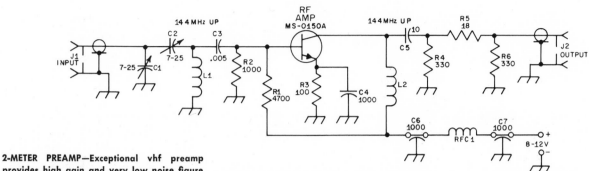

2-METER PREAMP—Exceptional vhf preamp provides high gain and very low noise figure on 144-, 220-, and 432-MHz bands without readjustment or change in circuitry. Transistor used is also known as MS175 by Texas Instruments. Operates into 50-ohm load. L1 is 8 turns No. 22 on 1K ½-W resistor, L2 5½ turns on 10K, and RFC1 10 turns No. 33 wound as toroid on ferrite bead. Gain is 23 dB and noise figure 1.45 without neutraliza-tion.—J. R. Hattaway and D. K. Belcher, A State-of-the-Art 2-Meter Preamplifier, *QST*, April 1971, p 92–93.

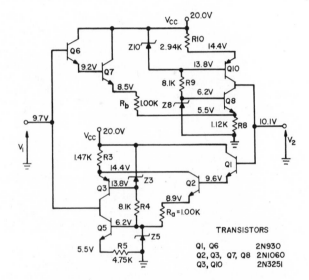

GYRATOR—Practical grounded gyrator uses discrete components, with d-c bias levels chosen to eliminate undesirable latch-up problems.—J. T. May and J. F. Pierce, Bias-Level Latch-Up Problems in Gyrators, *IEEE Journal of Solid-State Circuits*, June 1971, p 130–131.

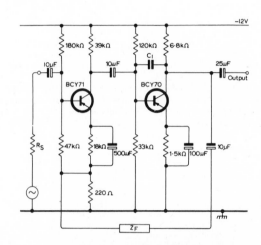

A-C PREAMP WITH FEEDBACK—Uses low-noise input transistor providing input resistance above 100K. With resistive feedback ZF, response is down 3 dB at 1 Hz and 250 kHz. Capacitor may be used in feedback network to reduce high-frequency interference. Gain is 1,000 without feedback and 100 with 22K feedback.—"P-N-P Practical Planar Transistors—BCY70 Family and BFX29 Family," Mullard, London, 1967, TP887, p 6.

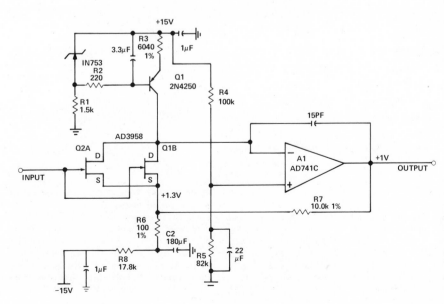

9 HZ–250 KHZ PREAMP WITH OPAMP—Provides voltage gain of 100 over entire band. Equally suited for low-impedance magnetic phono pickups and high-impedance piezoelectric transducers. Noise current and voltage are low.—Choosing and Using N-Channel Dual J-Fets, *Analog Dialogue*, Dec. 1970, p 4–9.

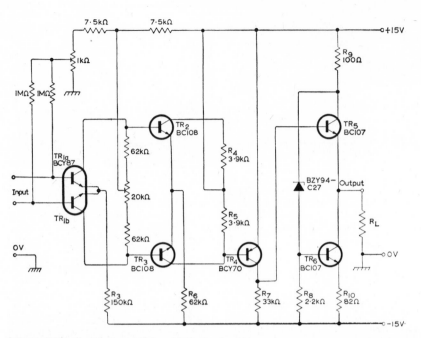

DIRECT-COUPLED WITH 150,000 VOLTAGE GAIN—Differential input uses matched pair of npn silicon planar transistors mounted to have practically identical junction temperatures. Report gives design calculations for all stages. Output resistance is 10 ohms and input resistance 240K.—"Circuits Using Low-Drift Transistor Pairs," Mullard, London, 1968, TP994, p 10–11.

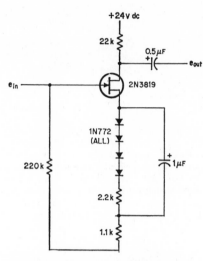

FET USING DIODES WITH FEEDBACK—Use of negative feedback around stabilizing diodes in gate-source loop of fet improves stability while reducing gain. Gain is nearly constant at 11 from —10 to 50 C and degrades only 3% at low of —30 C or high of 100 C.—D. F. DeKold, Diodes Stabilize FET Gain to 1% over 100°C Range, *Electronics*, June 7, 1971, p 82–83.

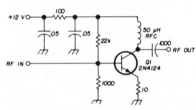

BUFFER FOR VFO—Provides constant high-impedance load for vfo, to minimize oscillator loading and give chirpless signal. Provides ample power for driver stage of transistor transmitter up to 5 W.—C. E. Galbreath, VFO Buffer Amplifier, *Ham Radio*, July 1971, p 66–67.

C_1 6 μF –5 volt electrolytic
C_2, C_3 1.0 μF 50 VDC paper
C_4 .02 μF 1000 V ceramic
R_1 1 K ohm ½ watt resistor
R_2 33 ohm ½ watt resistor
R_3 6.2 K ohm ½ watt resistor
R_4, R_5 15 ohm 2 watt resistor
R_6, R_7 6 K ohm 5 watt resistor
R_8, R_9 .27 ohm 2 watt resistor

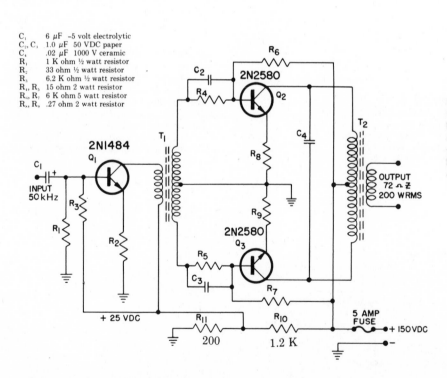

200 W AT 50 KHZ—Single class-A stage drives class-B output stage. Full output is obtained with 0.5 mW input. Power supply drain is 3 A at full power output, and idling current 150 mA. T1 primary is 40 turns No. 26 and secondary 12 turns No. 22 C-T bifilar wound on Ferroxcube 3C 206 F 440 core. T2 primary is 50 turns No. 20 C-T bifilar wound and secondary is 40 turns No. 20 on Allen Bradley U2375C127A ferrite core. Response is 3 dB down at 25 and 68 kHz, and efficiency is 54%. Overall gain is 65.5 dB.—"2N2580 VLF Silicon Power Amplifier," Delco, Kokomo, IN, July 1971, Application Note 23.

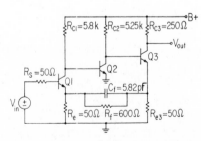

87-MHZ COMPUTER-DESIGNED—Values shown were obtained in 6-minute run on CDC-6400 computer, using program written to implement automated frequency-domain network design principle. Transistor parameters were fed into program as constants.—S. W. Director and R. A. Rohrer, Automated Network Design—The Frequency-Domain Case, *IEEE Trans. on Circuit Theory*, Aug. 1969, p 330–337.

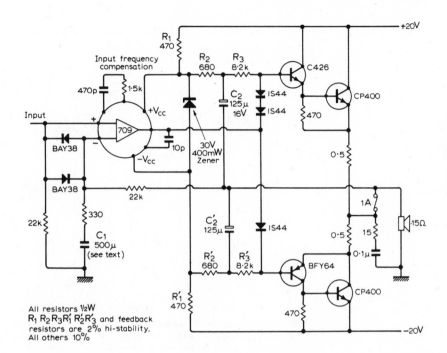

1.5 HZ–150 KHZ IC DRIVER—Bootstrapped opamp provides sufficient voltage swing to drive complementary class-B output stage. Distortion is very low because of 55 dB of negative feedback at low frequencies. For normal audio applications, 50 μF is sufficient for C1.—J. M. A. Wade, I.C. Driver for Power Amplifier, *Wireless World*, Nov. 1969, p 530.

100-MHZ PREAMP—Broadband general-purpose amplifier has gain of more than 40 dB below 5 MHz, dropping to 30 dB at 10 meters and 23 dB at 6 meters. Will improve noise figure and image rejection of any receiver. Operates with either 50- or 75-ohm receiver input impedance.—D. K. Bercher and A. Victor, A General Purpose Solid-State Preamplifier, *QST*, Sept. 1971, p 32.

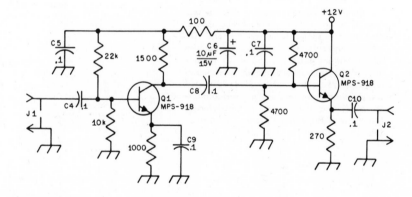

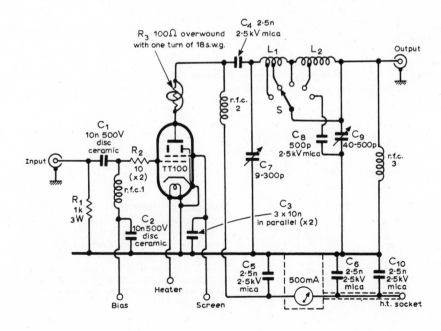

200-W LINEAR—Single-tube stage operates between 3.5 and 28 MHz. With push-pull pair connection, will provide 400 W output into 2,460-ohm anode-to-anode load impedance. Anode supply is 850 V for grid current of 2 mA, or 1 kV with no grid current. Tube can be cooled either with fan or with close-fitting heat shield attached to heat sink. Screen voltage is 216 V stabilized, and grid bias is —50 V. Article gives all coil data.—G. R. Jessop, 200-W Linear Amplifier, *Wireless World*, June 1971, p 273–275.

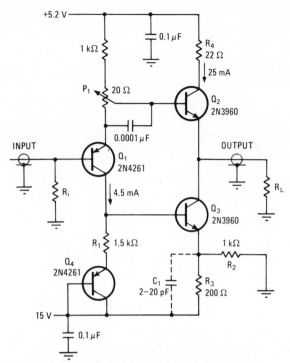

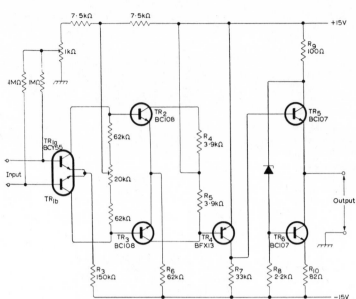

VOLTAGE FOLLOWER—Gives better perform-ance than IC opamp when isolation is needed with or without impedance transformation in fast analog circuits. Propagation delay is less than 1 ns in complementary circuits shown. Back-to-back emitter-followers Q1 and Q2 are complements, so their opposite-polarity offset voltages practically cancel. P1 permits fine offset zero adjustment. Q3 is variable-load current source, while diode-connected Q4 provides temperature compensation.—O. A. Horna, High-Speed Voltage-Follower Has Only 1-Nanosecond Delay, *Electronics*, July 19, 1973, p 115–116.

150,000 VOLTAGE GAIN—Output stage, hav-ing output resistance under 20 ohms, is shunt-compensated emitter-follower driven by BFX13 single-ended stage. Input uses two dif-ferential amplifiers, first of which is selected for low drift. Article gives design calcula-tions. Input resistance is greater than 200K. Zener is BZX61-C27.—"Directly Coupled Am-plifiers," Mullard, London, 1967, TP7000/1, p 31.

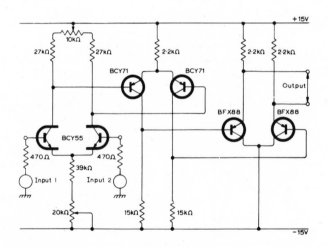

DIFFERENTIAL INPUT AND OUTPUT—Provides voltage gain of 8,000 for d-c and flat re-sponse within 3 dB up to 100 kHz. Provides zero output voltage for zero input. Nominal output is 10 V with either polarity. Used where extremely low drift is not required.—"P-N-P Practical Planar Transistors—BCY70 Family and BFX29 Family," Mullard, London, 1967, TP887, p 3.

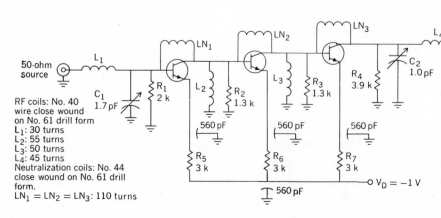

RF coils: No. 40 wire close wound on No. 61 drill form
L_1: 30 turns
L_2: 55 turns
L_3: 50 turns
L_4: 45 turns
Neutralization coils: No. 44 close wound on No. 61 drill form.
$LN_1 = LN_2 = LN_3$: 110 turns

150-MHZ NEUTRALIZED LOW-DRAIN—Maxi-mum transducer gain is 32 dB at center fre-quency, and 3-dB bandwidth is 11.4 MHz. All interstage networks are direct-coupled. Drain is only 300 μA from 1-V supply. Package vol-ume is only about 1.5 cu cm.—S. Cho and A. G. Thiele, The Low-Power-Drain Microelec-tronic VHF Amplifier, *IEEE Spectrum*, March 1970, p 49–53.

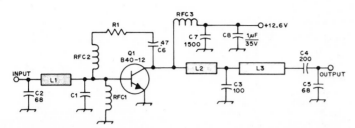

2-METER 40-W—Single CTC transistor on Thermalloy 6151B heatsink will operate into open or short-circuit without damage. Requires 10-W drive. Article covers design procedure based on use of Smith chart, and gives commercial sources and values for un-marked components. Harmonic attenuation is 37 dB for 2nd, 40 dB for 3rd, and 45 dB for 4th. Operating range is 144 to 148 MHz.—J. H. Johnson and R. Artigo, Fundamentals of Solid-State Power-Amplifier Design, *QST*, Nov. 1972, p 16–20.

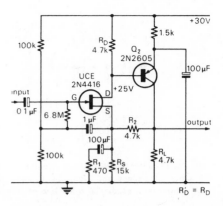

FET INPUT—Simple amplifier uses common-source fet input stage and common-emitter output stage. Cutoff is at about 1 MHz.—Some Applications of Field-Effect Transistors, *Electronic Engineering*, Sept. 1969, p 18–23.

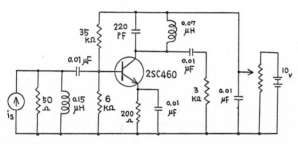

40-MHZ BANDPASS—Article gives mathematical design procedure for amplifier having given stability factor and other basic parameters for active two-port network. In circuit used as example, measured values of gain between 38 and 42 MHz agree closely with computed values.—Y. Miwa, K. Okuno, and T. Namekawa, High-Frequency Amplifier Design Using Nichols Chart, *IEEE Journal of Solid-State Circuits*, April 1972, p 195–201.

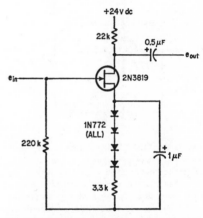

FET STABILIZATION—Diodes in gate-source loop of fet minimize gain variations. As temperature increases, forward voltage drop of each diode decreases, to give virtually constant gain of 27 from 0 to 40 C. Drop is only 6% at 100 C.—D. F. DeKold, Diodes Stabilize FET gain to 1% over 100°C Range, *Electronics*, June 7, 1971, p 82–83.

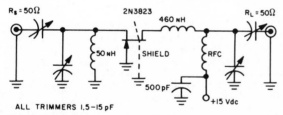

ALL TRIMMERS 1.5–15 pF

200-MHZ COMMON-GATE FET—Report covers theoretical and practical design considerations for high-frequency fet amplifiers. Gain is 14 dB and noise figure 3.2 dB.—R. C. Hejhall, "Field Effect Transistor RF Amplifier Design Techniques," Motorola Semiconductors, Phoenix, AZ, AN-423, 1971.

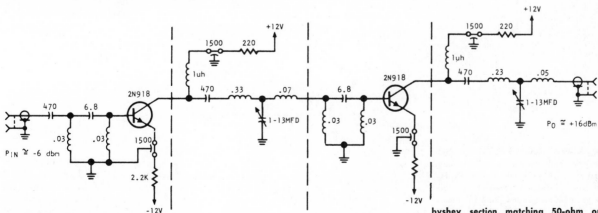

CONSTANT-GAIN BROADBAND UHF—Uses insertion-loss parameter impedance-matching filters to provide gain of about 25 dB between 150 and 325 Hz. First-stage output network is N=4, 40% bandwidth Butterworth filter which transforms output impedance of first transistor to input of second. Second-stage output network is N=4 Chebyshev section matching 50-ohm output.—R. V. Snyder, Broadband Impedance-Matching Techniques Applied to Design of UHF Transistor Amplifiers, *Proc. IEEE*, Jan. 1967, p 124–215.

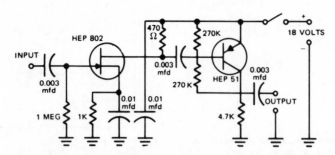

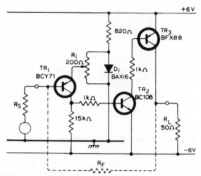

1—40 MHZ BROADBAND—Circuit features high input impedance and high signal-noise ratio. Gain drops gradually from 100 at 1 MHz to about 20 at 40 MHz.—"Tips on Using FET's," Motorola Semiconductors, Phoenix, AZ, HMA-33, 1971.

100-KHZ DIRECTLY COUPLED—Voltage gain without feedback is 1,000, with 5.5 V maximum output for either polarity. Response is flat within 3 dB up to 100 kHz. Coupling resistors protect transistors from surge currents when amplifier is switched on. R1 is adjusted to give zero output voltage for zero input. Can be used as summing amplifier if RF is added for resistive feedback; use at least 100K for good stability.—"P-N-P Practical Planar Transistors—BCY70 Family and BFX29 Family," Mullard, London, 1967, TP887, p 4.

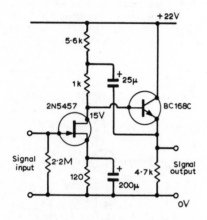

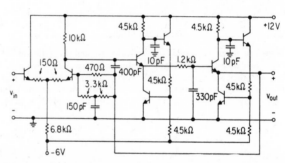

STANDARD BOOTSTRAP—Widely used basic circuit has many advantages, but distortion is sometimes higher than expected. Article reveals that cause is variation in output admittance with drain voltage of fet, known as Early effect in England. Article gives modified circuit that may also be used with class B amplifier.—J. H. Wilkinson, Fractional Bootstrap Feedback, *Wireless World*, Dec. 1971, p 600.

0.65-MHZ SELECTIVE AMPLIFIER—Provides constant Q of 50 within 5% over temperature range of —10 to +110 C. Consists of two unity-gain blocks, transconductance block, and bridged-T network. Q values up to 150 are possible, but with greater variation over temperature range. Circuit design is well suited for single-chip IC production. Design is applicable to center frequencies up to at least 10 MHz.—G. A. Rigby and D. G. Lampard, Integrated Frequency-Selective Amplifiers for Radio Frequencies, *IEEE Journal of Solid-State Circuits*, Dec. 1968, p 417—422.

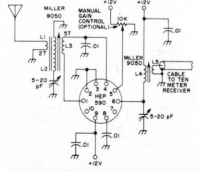

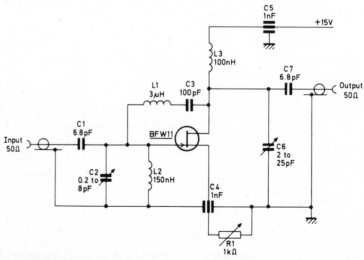

10-METER PREAMP—Motorola IC gives adjustable high gain, high selectivity, and minimum image response, either when used with 10-meter receiver or 10-meter converter. If gain control is not needed, ground pin 5 for maximum gain. Preamp has practically no oscillation-producing feedback.—B. Hoisington, IC Ten Meter Tuner for Use with Solid State VHF-UHF Converters, 73, Jan. 1973, p 99—103.

100-MHZ COMMON-SOURCE FET—Used in r-f stages of h-f, vhf, and uhf communication receivers. Noise figure is 2 dB for values shown.

—"R. F. Applications of FET's in Communications Receivers," Mullard, London, 1970, TP1175, p 3—4.

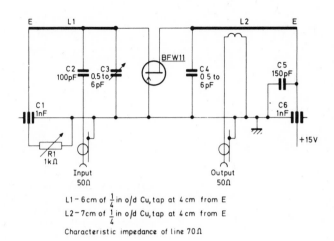

470-MHZ COMMON-GATE FET—Gives noise figure of 5.5 dB as r-f stage in communication receivers. Power gain is 11 dB.—"R. F. Applications of FET's in Communications Receivers," Mullard, London, 1970, TP1175, p 3–4.

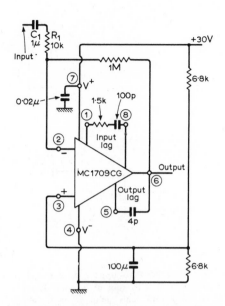

40-DB GAIN—Simple and practical a-c amplifier operates from single power supply, with resistive divider providing bias for inverting amplifier of opamp. Upper 3-dB frequency limit is 150 kHz.—G. B. Clayton, Operational Amplifiers, *Wireless World*, Oct. 1969, p 482–483.

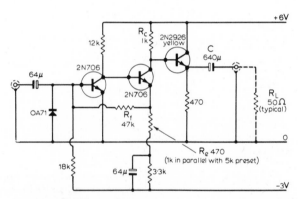

30 HZ TO 3.5 MHZ—Direct-coupled amplifier provides power gain of 20 dB and requires about 20K source impedance. Stability is excellent.—H. N. Griffiths, Simple Wideband Amplifier, *Wireless World*, Oct. 1969, p 478.

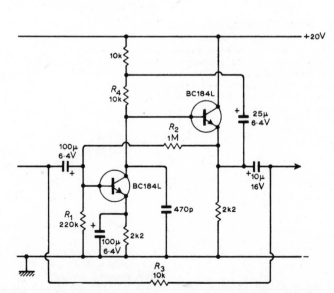

BOOTSTRAPPING GIVES 100 DB S/N—R1 and R2 provide d-c feedback and d-c stability in low-noise tone-control circuit, while R3 provides a-c feedback. Suitable for virtual ground applications.—C. R. Cathles, Bootstrapping, *Wireless World*, May 1972, p 225.

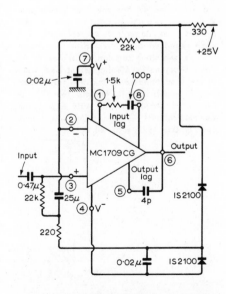

4-MEG INPUT IMPEDANCE—Provides 40 dB gain using single supply voltage and split-zener biasing system. Bootstrapping gives high input impedance.—G. B. Clayton, Operational Amplifiers, *Wireless World*, Oct. 1969, p 482–483.

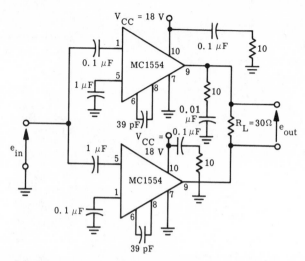

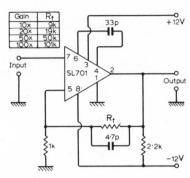

ADJUSTABLE GAIN—Changing value of Rf changes gain over range of 10 to 100, as shown in table, with opamp connected as shown.—General-Purpose Amplifier, *Wireless World*, Jan. 1970, p 16.

3-W DIFFERENTIAL—Upper IC amplifier is noninverting with gain of 9, while lower is inverting with same gain. Effective voltage gain of both is therefore 18. Connection permits output swing of 27 V p-p with supply voltage of only 18 V.—"The MC1554 One-Watt Monolithic Integrated Circuit Power Amplifier," Motorola Semiconductors, Phoenix, AZ, AN-401, 1972.

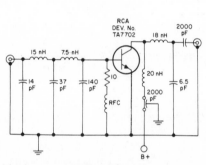

17 W AT 225–400 MHZ—With 4-W drive, uhf power module has only 1-dB power-gain variation over band. IC construction uses multi-section L-C ladder network at input, with Chebyshev low-pass characteristics that transform base-emitter junction resistance to 50 ohms. Designed for 50-ohm output load. Used in uhf broadband communication transmitter.—C. Kamnitsis and R. Minton, Thin-Film RF Integrated Amplifiers for UHF Broadband Communications Systems, *The Microwave Journal*, Dec. 1970, p 37–40.

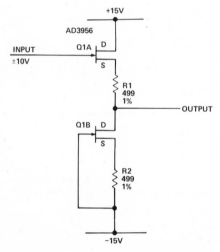

FET INPUT BUFFER FOR OPAMP—Low-drift follower is used primarily to unload very high impedance signal for low-cost opamp such as AD741C. Requires no selection of parts when driving 100K and up. No-load gain is typically 0.998.—Choosing and Using N-Channel Dual J-Fets, *Analog Dialogue*, Dec. 1970, p 4–9.

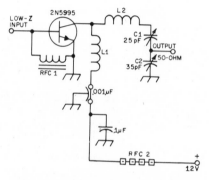

2-METER 10-W AMPLIFIER—Uses npn overlay r-f power transistor. Same circuit will deliver 35 W with 2N5996. Driving power is 0.75 W and 5 W, respectively. L1 is 9 turns No. 20 on 3/16 inch diameter and L2 3 turns No. 14 on 3/8 inch. RFC1 is 4 turns No. 30 wound as toroid on ferrite bead. RFC2 is four ferrite beads.—2-Meter Solid-State Amplifier, *QST*, March 1971, p 17.

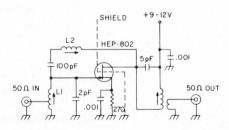

2-METER PREAMP—Use of fet in common-gate amplifier minimizes Miller effect, providing ideal compromise between selectivity and gain. Provides over 14 dB signal improvement in fixed-frequency 2-meter operation. Article gives coil data.—K. W. Sessions, Jr., Low Noise Economy Preamp For 2M, *73*, Nov. 1972, p 120–121 and 124.

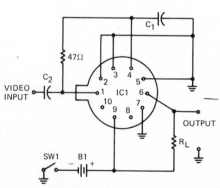

20–160 MHZ AMPLIFIER—Broadband IC circuit provides 25 dB gain at lower frequency limit, dropping gradually to 10 dB at higher limit. Uses HEP 590 IC with 6-V supply, 0.1-μF capacitors, and 1.8K load RL.—"Radio Amateur's IC Projects," Motorola Semiconductors, Phoenix, AZ, HMA-36, 1971.

1,296-MHZ PREAMP—Uses single Nippon Electric V766B low-noise transistor having gain above 13 dB and noise figure below 3 dB. Performance is comparable to that of parametric amplifier. CR1 is protective diode rated at least 10 mA. L1 and L2 are brass strips ⅜ inch wide. Beads are ferrite. Article gives construction details.—D. Vilardi, Low-Noise Transistor 1,296 MHz Preamplifier, *Ham Radio,* June 1971, p 50–54.

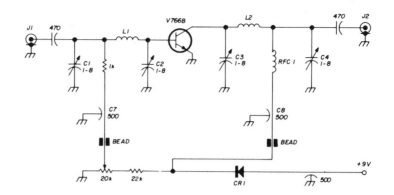

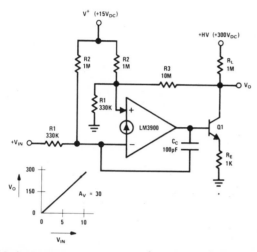

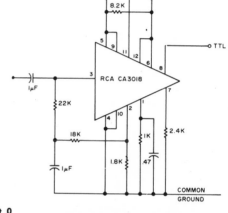

HIGH-VOLTAGE NONINVERTING—Uses Norton current-differencing opamp with common-mode bias resistors R2 to allow input voltages to go to 0 V d-c. Output voltage will then drop to about 0.3 V d-c. Gain is 30. Input voltage range of 0 to +10 V d-c makes output voltage increase from about 0 to +300 V d-c.—T. M. Frederiksen, W. M. Howard, and R. S. Sleeth, "The LM3900—A New Current-Differencing Quad of ± Input Amplifiers," National Semiconductor, Santa Clara, CA, 1972, AN-72, p 37–38.

LOW-LEVEL BROADBAND—Designed for inputs from 20 μV to 0.4 mV. Response is flat within 3 dB from 800 Hz to 32 MHz, as required for use in frequency counter going up to 32 MHz.—T. Johnson, Another Integrated Circuit Frequency Counter, 73, Jan. 1973, p 39–42 and 44–45.

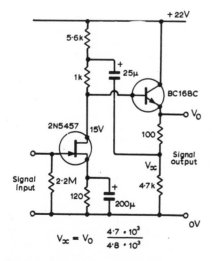

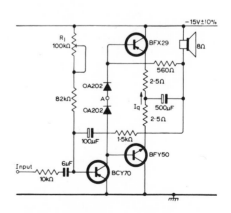

MODIFIED BOOTSTRAP—Reduces distortion of standard bootstrap by feeding only fraction (about 0.98) of input signal back. This means that output load seen by constant-current generator at drain of fet is kept low, to reduce nonlinear effects. May also be used in class B amplifier to reduce load seen by driver-stage transistor.—J. H. Wilkinson, Fractional Bootstrap Feedback, *Wireless World,* Dec. 1971, p 600.

550-MW PUSH-PULL—Transformerless output stage with driver is down 3 dB at 6 Hz and 500 kHz. Requires 2.1 V rms input for full output. Total harmonic distortion at 1 kHz is 3.2%.—"P-N-P Practical Planar Transistors—BCY70 Family and BFX29 Family," Mullard, London, 1967, TP887, p 6.

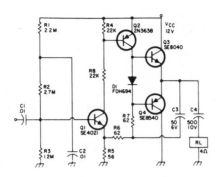

CLASS-A LOW-POWER—Article covers fundamental design problems of complementary amplifier, including temperature stability. Large amount of feedback is used to make nonlinearity of gain unimportant.—D. Campbell and R. Westlake, Design Concepts For Low-Power Amplifiers, 73, May 1971, p 36–38.

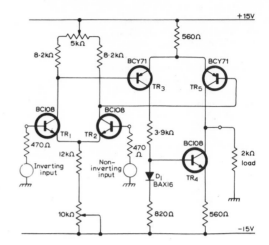

DIFFERENTIAL INPUT—Provides single-ended output, with voltage gain of 4,500 for d-c and flat response within 3 dB up to 100 kHz. Output voltage is zero for zero input. Nominal output is 10.5 V at 5 mA both positive and negative.—"P-N-P Practical Planar Transistors—BCY70 Family and BFX29 Family," Mullard, London, 1967, TP887, p 3.

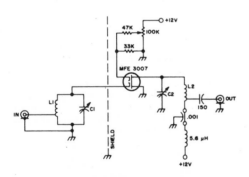

146-MHZ FET PREAMP—Provides high gain, along with excellent noise figure, when used ahead of older 2-meter receivers. Voltage gain is about 20 dB. L1 and L2 are both 3½ turns No. 18 on ½-inch forms, stretched out to ¾ inch. C1 and C2 are 1—28 pF trimmers.—C. Klinert, A Dual-Gate Fet Preamp For 2 Meters, 73, May 1971, p 20 and 22.

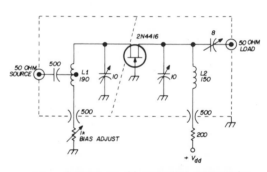

L1   6 turns, no. 16 enamelled copper wire, 3/8-inch inner diameter, input tapped 1¾ turns from ground

L2   5 turns, no. 16 enamelled copper wire, 3/8-inch inner diameter

COOLED PREAMP—Noise figure of 144-MHz common-gate fet amplifier is reduced to 0.8 dB by operating circuit in Dewar flask containing liquid nitrogen at 77 K. Holding dry ice against fet with tweezers gives 200K and noise figure of 1.2, as contrasted to 1.5 dB at room temperature of 300 K. Preamp cooling is most useful for weak-signal uhf reception, such as moonbounce signals.—J. Dietrich, Cooled Preamplifier for VHF-UHF Reception, Ham Radio, July 1972, p 36–38.

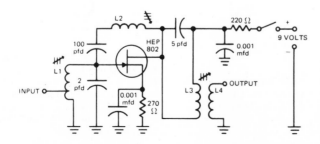

150-MHZ PREAMP—Provides 14 dB gain at low noise for use ahead of 2-meter receiver. L1 is 5¼ turns No. 26 tapped at 1¼ turn. L2 is 9½ turns No. 34. L3 is 5 turns No. 26, with 1¼ turn No. 26 over low end for L4. All coils are wound on brass-slug ceramic form.—"Tips on Using FET's," Motorola Semiconductors, Phoenix, AZ, HMA-33, 1971.

CHAPTER 3
Analog-Digital Converter Circuits

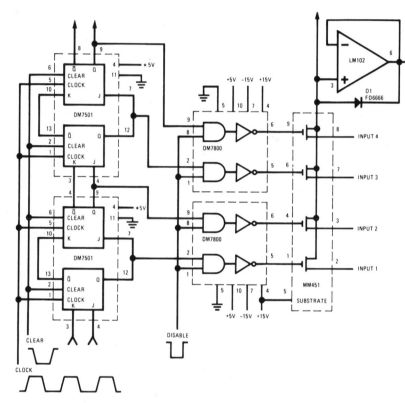

ANALOG COMMUTATOR—Low input current and fast slewing of LM102 voltage follower make it ideal as buffer amplifier in high-speed analog commutator for a-d converter. Arrangement shown is expandable four-channel commutator using two DM7501 dual flip-flops to form four-bit static shift register whose outputs drive DM7800 level translators for converting TTL logic levels to MOS drive voltages. Extra gate on input of translator can be used to shut off all analog switches. Bit enters register and cycles through at clock frequency, turning on each analog switch in sequence.—R. J. Widlar, "Integrated Voltage Follower," National Semiconductor, Santa Clara, CA, 1968, AN-5.

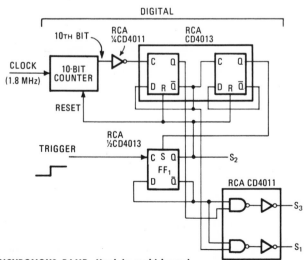

SYNCHRONOUS RAMP—Used in multichannel a-d converter in which digital words are developed to represent transducer outputs. When triggered, circuit generates linear ramp having time and voltage parameters that are independent of clock rate, supply voltage, and component tolerances. Uses low-cost complementary-mos RCA CD4016 quad analog switches instead of expensive ladder networks. Ramp output rises as 10-bit counter runs from its zero state to full-scale count.—D. M. Brockman, Synchronous Ramp Generator Maintains Output Linearity, *Electronics*, Jan. 18, 1973, p 170.

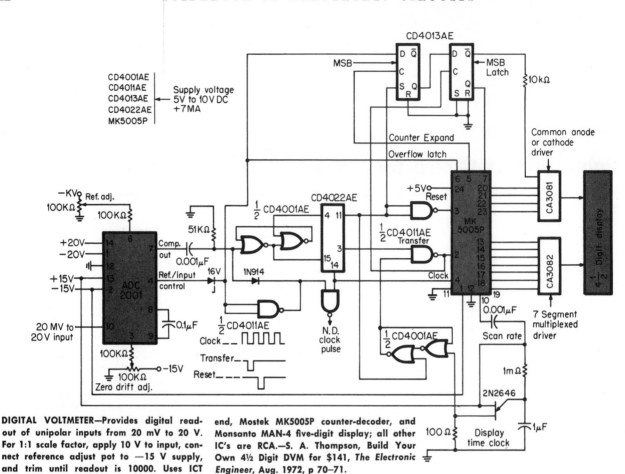

DIGITAL VOLTMETER—Provides digital read-out of unipolar inputs from 20 mV to 20 V. For 1:1 scale factor, apply 10 V to input, connect reference adjust pot to −15 V supply, and trim until readout is 10000. Uses ICT ADC2001 analog-digital converter as front end, Mostek MK5005P counter-decoder, and Monsanto MAN-4 five-digit display; all other IC's are RCA.—S. A. Thompson, Build Your Own 4½ Digit DVM for $141, *The Electronic Engineer*, Aug. 1972, p 70–71.

GRAY-CODE OUTPUT—Provides absolute value of weighted sum of several inputs, using only two opamps. Stages are cascaded, with analog output of stage j forming input to stage j + 1 to give *n*-bit gray-code analog-digital converter. Gray-code information is available at output terminal of opamp in each stage. Output of first opamp is suitable for driving npn transistor for interfacing with TTL logic systems. With values shown, bit 1 settles in less than 1 μs, and following bits settle at about 500 ns per bit.—M. A. Smither, Summing Proportioning Absolute Value Circuit, *IEEE Journal of Solid-State Circuits*, Dec. 1971, p 417–418.

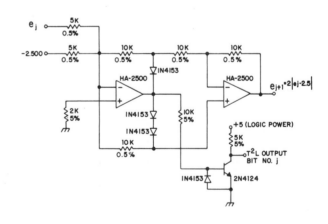

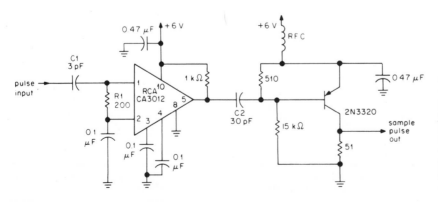

A-D CONVERTER TRIGGER—High-gain IC limiting amplifier detects zero crossings of input pulses having various shapes and rates, to give accurately timed pulse for triggering high-speed analog-digital converter. Use of passive differentiator R1-C1 eliminates noise and overload problems of active differentiators. Will operate up to 2 MHz.—R. P. Rufer, Simplified A-To-D Converter Triggering, *The Electronic Engineer*, Nov. 1968, p 74.

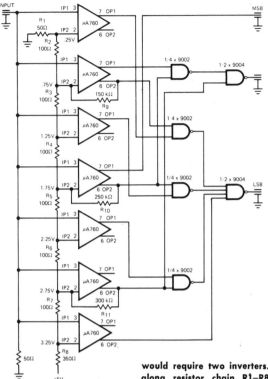

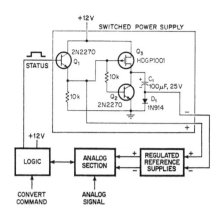

MINIMIZING BATTERY DRAIN—Analog section of a-d converter is switched off when no data is converted, holding maximum current drain at 20 mA from 12-V supply when using 5-kHz conversion rate developed for remote environmental data-gathering systems as on ocean buoys and space probes. Power is turned on after status pulse from logic section indicates conversion is to be made.— R. D. Moore and J. J. Pastoriza, Low-Power A-D Converter Is a Battery Life-Saver, *Electronics*, March 29, 1971, p 71–73.

HIGH-SPEED CONVERSION—Each of seven IC comparators has one input biased to voltage level at which digital equivalent of input signal should change. Use of both outputs of second and fourth comparators gives simpler and faster circuit, whereas single outputs would require two inverters. Voltages shown along resistor chain R1–R8 are theoretical values at which each comparator should switch. Maximum error is less than 0.1 V.— P. Holtham, "The µA760—A High Speed Monolithic Voltage Comparator," Fairchild Semiconductor, Mountain View, CA, No. 311, 1972, p 6.

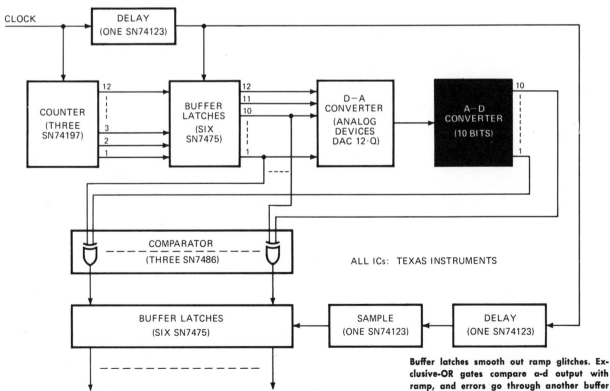

A-D CONVERTER TESTER—Provides go-no-go check for 10 most significant bits of analog-digital converter (black box) at cost of about $36 for IC's plus cost of one digital-analog converter. Test system generates 12-bit digital ramp with ripple-through binary counter. Buffer latches smooth out ramp glitches. Exclusive-OR gates compare a-d output with ramp, and errors go through another buffer for comparison with appropriately delayed clock.—C. J. Huber, Logic System Checks Out Analog-to-Digital Converter, *Electronics*, June 19, 1972, p 91.

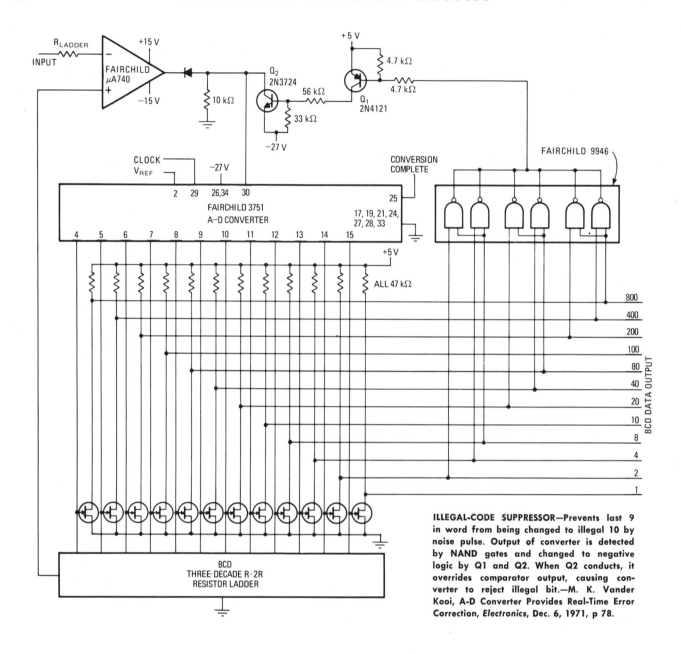

ILLEGAL-CODE SUPPRESSOR—Prevents last 9 in word from being changed to illegal 10 by noise pulse. Output of converter is detected by NAND gates and changed to negative logic by Q1 and Q2. When Q2 conducts, it overrides comparator output, causing converter to reject illegal bit.—M. K. Vander Kooi, A-D Converter Provides Real-Time Error Correction, *Electronics*, Dec. 6, 1971, p 78.

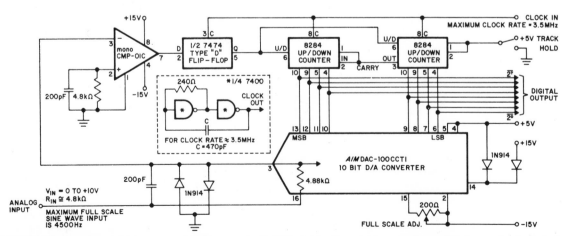

8-BIT TRACKING—Uses up-down counter, current-output a-d converter, and voltage comparator costing no more than simple ramptype tracking a-d converters while making digital data continuously available at·output. Requires no sample-hold circuit. Comparator examines output voltage for polarity, and always drives counter's code in direction which causes output voltage to approach zero. When balance is achieved, loop is locked and tracks analog input signal. Output of converter is then bcd equivalent of analog input. —"CMP-01 Fast Precision Comparator Series," Precision Monolithics, Santa Clara, CA, 1971.

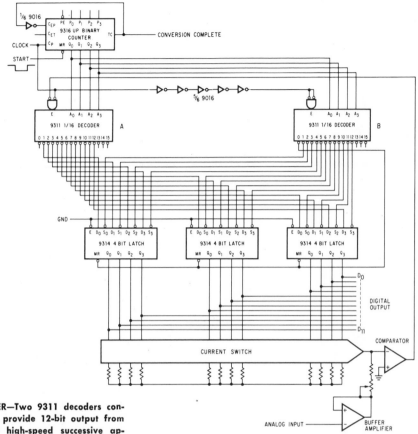

12-BIT CONVERTER—Two 9311 decoders connected as shown provide 12-bit output from analog input by high-speed successive approximation. Conversion is achieved in 13 clock periods. Report describes operation in detail.—"Applications of the 9311 1-out-of-16 Decoder," Fairchild Semiconductor, Mountain View, CA, No. 170, 1971, p 6–7.

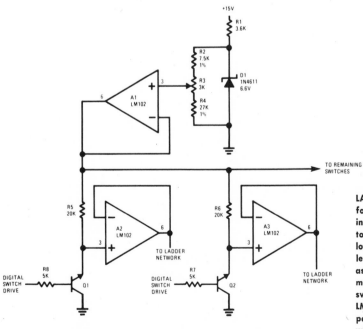

LADDER DRIVER—National LM102 IC voltage followers are used in switch circuit for driving ladder network in a-d converter. Transistor switches, connected in reverse mode for low saturation voltage, generate 0- and 5-V levels for ladder network. A2 and A3 serve as buffers for switch output. Circuit gives much lower output resistance than push-pull switches, along with center drive. Upper LM102 serves as buffer for temperature-compensated voltage reference, adjusted in value with R3.—R. J. Widlar, "Integrated Voltage Follower," National Semiconductor, Santa Clara, CA, 1968, AN-5.

CHAPTER 4
Audio Amplifier Circuits

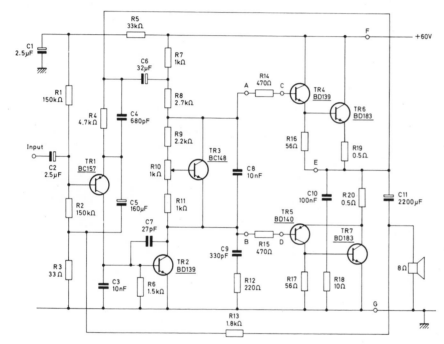

35-W HI-FI—Response is flat within 1 dB from 10 Hz to 100 kHz. Total harmonic distortion is 0.1% at 1 kHz. All transistors are silicon. Report gives optional circuit that can be connected between lettered terminals A—G to provide short-circuit protection.—"Technical Notes on 15, 25 and 35 W Audio Amplifiers," Mullard, London, 1970, TP1206, p 1–6.

15 OR 20-W HI-FI—Designed to operate class A and deliver 15 W to 8-ohm load, or class AB for delivering 20 W to 4-ohm load. Does not require protection agaist short-circuit. All transistors are silicon. Input sensitivity is 360 mV for 8-ohm load and 295 mV for 4-ohm load. Response is down 3 dB at 20 and 30,-000 Hz.—"Technical Notes on a 15 W Audio Amplifier," Mullard, London, 1970, TP1217, p 1–4.

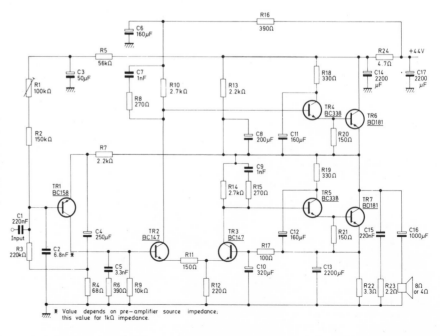

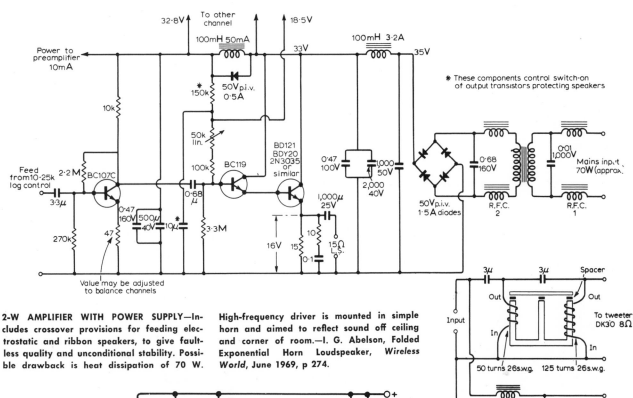

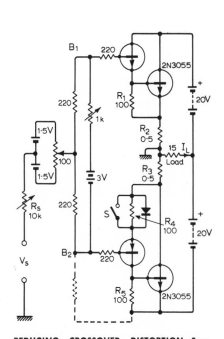

2-W AMPLIFIER WITH POWER SUPPLY—Includes crossover provisions for feeding electrostatic and ribbon speakers, to give faultless quality and unconditional stability. Possible drawback is heat dissipation of 70 W. High-frequency driver is mounted in simple horn and aimed to reflect sound off ceiling and corner of room.—I. G. Abelson, Folded Exponential Horn Loudspeaker, *Wireless World*, June 1969, p 274.

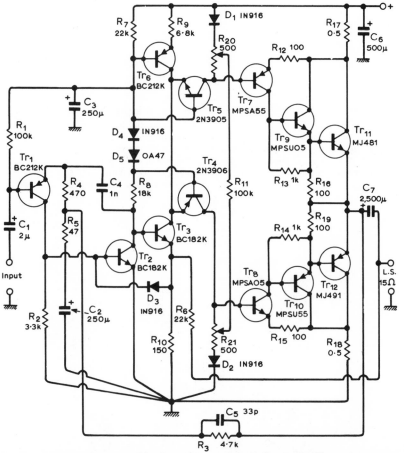

30 W to 100 KHZ—Class-B amplifier has 3-dB limits of 30 Hz and 100 kHz, with less than 0.01% total harmonic distortion throughout audio band for all power levels. Voltage gain is 100, and noise level is −120 dB below full power. Requires 60-V regulated d-c supply. Uses signal splitter between input low-level class-A amplifier and output stages. Developed for possible use as portable standard oscillator or low-distortion transmitter amplifier; performance is much better than is needed for hi-fi system.—P. Blomley, New Approach to Class B Amplifier Design, *Wireless World*, March 1971, p 127–131.

REDUCING CROSSOVER DISTORTION—Symmetrical class B circuit uses transistors chosen for maximum possible output impedance, to maximize feedback around complete amplifier and reduce distortion. Parasitics are suppressed by 220-ohm base resistors. Dotted resistor represents effect of finite output resistance of transistor connected to B2. Article analyzes circuit operation and performance.—P. J. Baxandall, Symmetry in a Class B, *Wireless World*, Sept. 1969, p 416–417.

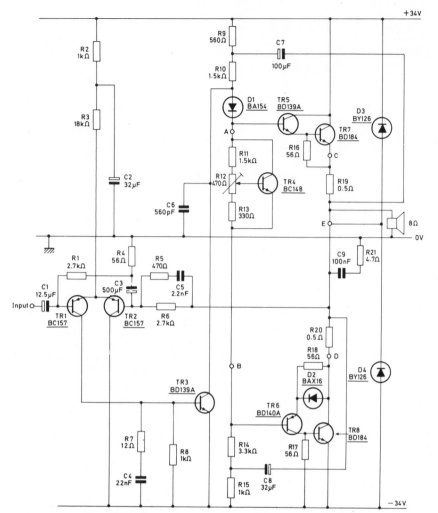

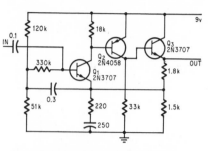

BOOTSTRAPPING GIVES LOW NOISE—Input-stage bootstrapping raises input impedance to several megohms, while emitter-follower Q3 lowers output impedance without affecting d-c stability. Noise figure is less than 1.5 dB. Article covers optimization of noise figure.—J. A. Roberts and N. A. Jolly, Audio Noise: Why Settle for More?, *Electronics*, Sept. 28, 1970, p 82–83.

50-W HI-FI—Total harmonic distortion is less than 0.025% at 1 kHz. Response is down 3 dB at 7 and 35,000 Hz. Split power supply improves low-frequency response. Report includes two different methods of protecting amplifier from short-circuit across output, using protection circuits connected to lettered terminals A—E.—"Technical Notes on a 50 W Audio Amplifier," Mullard, London, 1970, TP1218, p 1–6.

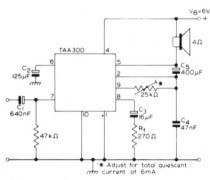

300-MW SINGLE-IC—Requires only 6-V supply to give full output power for input of 14 mV, using single Mullard IC. Distortion is under 0.7% up to 200 mW. Intended for use after detector in portable radios.—"TAA300 Integrated One-Watt Class B A.F. Amplifier," Mullard, London, 1968, TP1050, p 16–17.

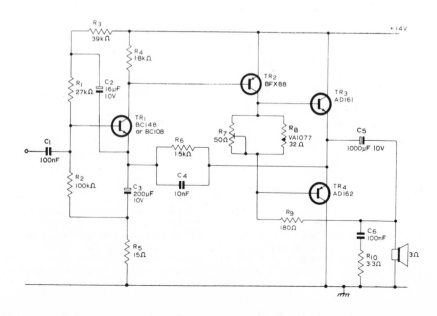

6 W FOR CAR RADIO—Transformerless class B audio amplifier has 52-mV sensitivity for full output, using complementary output pair. Response is down 3 dB at 74 Hz and 10.6 kHz. Distortion at full output is 1.2%.—"Transistor Audio and Radio Circuits," Mullard, London, 1972, TP1319, 2nd Ed., p 115–116.

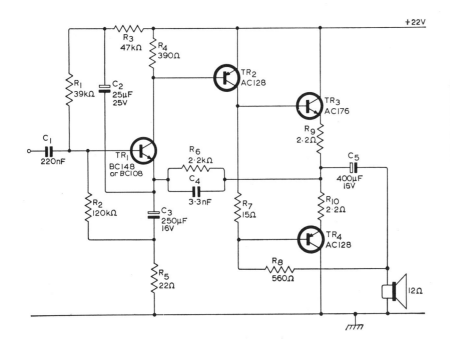

3-W DIRECTLY COUPLED—Draws 200 mA from power supply at rated output. Sensitivity is 70 mV for full power. Response is flat within 3 dB from 40 Hz to 20 kHz. Distortion at full power is 1.7%.—"Transistor Audio and Radio Circuits," Mullard, London, 1972, TP1319, 2nd Ed., p 39—40.

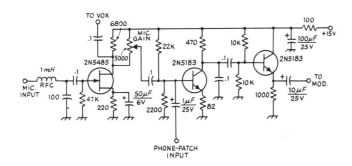

RESISTANCE-COUPLED SPEECH AMPLIFIER—designed for use with ssb, f-m, and a-m phone transmitters, for which it gives good frequency response over desired range of 200 to 3,500 Hz.—"The Radio Amateur's Handbook," American Radio Relay League, Newington, CT, 49th Ed., 1972, p 397—399.

10-W ZOBEL-NETWORK—Thermistor R8 compensates for temperature changes and maintains constant quiescent current for complementary matched-pair push-pull output stage. Zobel network CZ-RZ is used in parallel with speaker to present essentially pure resistive load to output transistors. RZ should be made equal to equivalent resistance RS of speaker, and CZ should be equal to equivalent inductance of speaker divided by square of equivalent resistance. Sensitivity is 80 mV for rated output, and 3-dB limits are 40 Hz and 20 kHz.—"Transistor Audio and Radio Circuits,"—Mullard, London, 1972, TP1319, 2nd Ed., p 41—43.

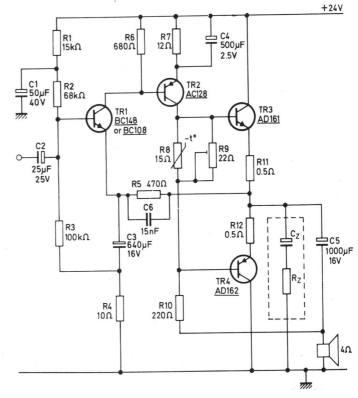

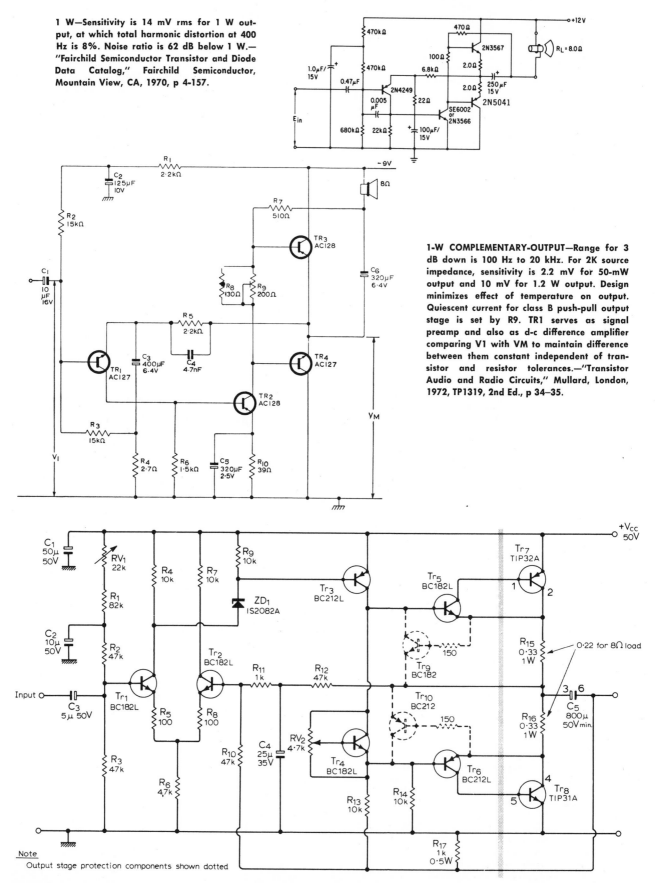

1 W—Sensitivity is 14 mV rms for 1 W output, at which total harmonic distortion at 400 Hz is 8%. Noise ratio is 62 dB below 1 W.—"Fairchild Semiconductor Transistor and Diode Data Catalog," Fairchild Semiconductor, Mountain View, CA, 1970, p 4-157.

1-W COMPLEMENTARY-OUTPUT—Range for 3 dB down is 100 Hz to 20 kHz. For 2K source impedance, sensitivity is 2.2 mV for 50-mW output and 10 mV for 1.2 W output. Design minimizes effect of temperature on output. Quiescent current for class B push-pull output stage is set by R9. TR1 serves as signal preamp and also as d-c difference amplifier comparing V1 with VM to maintain difference between them constant independent of transistor and resistor tolerances.—"Transistor Audio and Radio Circuits," Mullard, London, 1972, TP1319, 2nd Ed., p 34–35.

15 W UP TO 100 KHZ—Directly coupled design uses differential input amplifier and symmetrical output stage. Requires input of 312 mV, for which signal-noise ratio is 73 dB at 1 kHz. Intermodulation distortion is between 0.02% and 0.07%. Values shown are for 15-ohm load. Dotted components, optional, provide better protection than fuse for output stage.—I. Hardcastle and B. Lane, Low-Cost 15-W Amplifier, *Wireless World*, Oct. 1969, p 456–457.

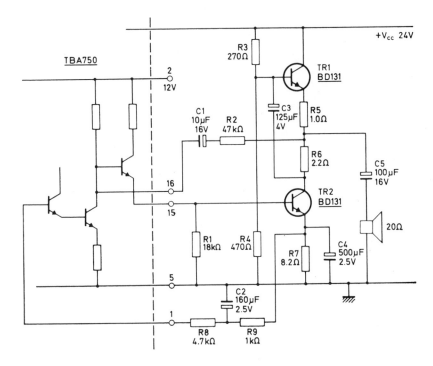

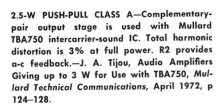

2.5-W PUSH-PULL CLASS A—Complementary-pair output stage is used with Mullard TBA750 intercarrier-sound IC. Total harmonic distortion is 3% at full power. R2 provides a-c feedback.—J. A. Tijou, Audio Amplifiers Giving up to 3 W for Use with TBA750, *Mullard Technical Communications*, April 1972, p 124–128.

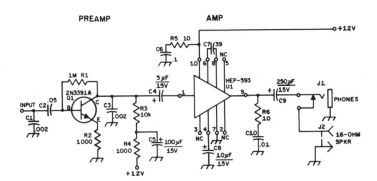

1-W IC—Transformerless circuit uses transistor amplifier ahead of IC for developing full power from detector output of practically any receiver circuit. Current drain from 12-V supply for peak signals is less than 200 mA for room-volume listening.—A 1-Watt Solid-State Audio Module, *QST*, May 1970, p 42–43.

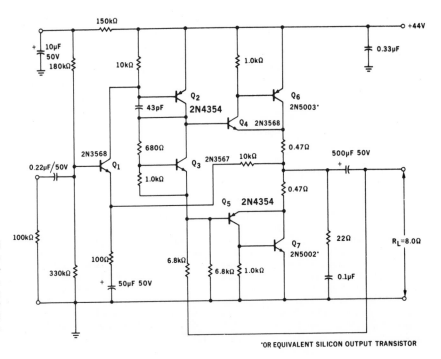

25-W OUTPUT—Uses 2N4254 pnp low-level low-noise diffused silicon epitaxial transistors for drivers Q2 and Q5. Maximum power dissipation of these transistors is 0.8 W each, but this can be boosted to 4 W by changing to FT4354 drivers.—"Fairchild Semiconductor Transistor and Diode Data Catalog," Fairchild Semiconductor, Mountain View, CA, 1970, p 4-129.

*OR EQUIVALENT SILICON OUTPUT TRANSISTOR

5 W FOR CAR RADIO—Designed for nominal 14-V supply, to operate from 12-V car battery under charge. Response is down 3 dB at 50 Hz and 10 kHz. Speaker is fed through center-tapped 30-mH 1-ohm choke having small winding to provide feedback to first stage; voltage ratio to secondary is 13:1.—"Transistor Audio and Radio Circuits," Mullard, London, 1972, TP1319, 2nd Ed., p 112–114.

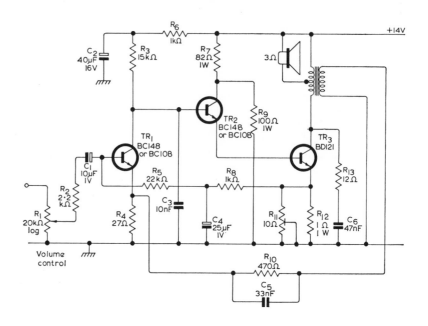

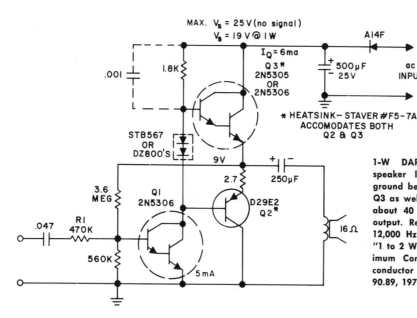

1-W DARLINGTON—Designed for 16-ohm speaker load, which can be returned to ground because circuit uses Darlington IC for Q3 as well as Q1. Open-loop voltage gain is about 40 dB. Requires 0.8 V input for full output. Response is 3 dB down at 35 and 12,000 Hz.—D. V. Jones and L. T. Anderson, "1 to 2 Watt Amplifier Circuits Requiring Minimum Components," General Electric, Semiconductor Products Dept., Auburn, NY, No. 90.89, 1970, p 8–9.

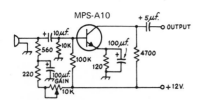

CARBON-MIKE PREAMP—Output of carbon mike is extremely high, so output of single-transistor amplifier stage provides more than enough output (several volts) for modulation of ssb and f-m phone transmitters and even for a-m transmitters using high-level plate modulation. Chief drawbacks of carbon mike are instability and nonlinear distortion.—"The Radio Amateur's Handbook," American Radio Relay League, Newington, CT, 49th Ed., 1972, p 397–398.

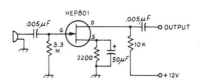

CRYSTAL-MIKE PREAMP—Output is adequate for p-a amplifier and for modulator of ssb, f-m, and a-m phone transmitters. Will also work with ceramic mikes and with high-impedance dynamic mikes. With ceramic mikes, particularly for ssb, mike load should be reduced to at least 0.25 meg to eliminate unwanted low-frequency response.—"The Radio Amateur's Handbook," American Radio Relay League, Newington, CT, 49th Ed., 1972, p 397–398.

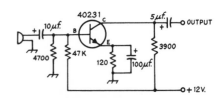

DYNAMIC-MIKE PREAMP—Designed for use with low-impedance dynamic mikes (high-impedance dynamic units take same preamp as crystal or ceramic mike). Output is adequate for modulator of ssb, f-m, and a-m phone transmitters, where frequency range desired is only 200 to 3,500 Hz.—"The Radio Amateur's Handbook," American Radio Relay League, Newington, CT, 49th Ed., 1972, p 397–398.

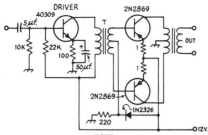

TRANSFORMER-COUPLED PHASE INVERTER—Used between stages of inverter only when power is to be transferred, for which resistance coupling is very inefficient. Driver and transformer T together provide equal voltages 180 deg out of phase for transistors of push-pull stage.—"The Radio Amateur's Handbook," American Radio Relay League, Newington, CT, 49th Ed., 1972, p 398–399.

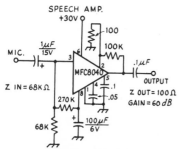

LOW-NOISE SPEECH AMPLIFIER—Uses Motorola IC for which noise level is typically 1 μV, for high-impedance mikes.—"The Radio Amateur's Handbook," American Radio Relay League, Newington, CT, 49th Ed., 1972, p 399.

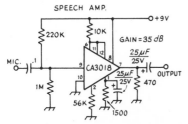

SPEECH AMPLIFIER FOR LOW-IMPEDANCE LOAD—Uses RCA IC providing 500 mW output power. Use with medium-impedance mike.—"The Radio Amateur's Handbook," American Radio Relay League, Newington, CT, 49th Ed., 1972, p 399.

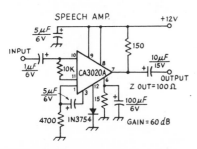

SPEECH AMPLIFIER—Use with high-impedance mike. RCA IC transistor array has high-gain Darlington feeding emitter-follower that gives low-impedance output.—"The Radio Amateur's Handbook," American Radio Relay League, Newington, CT, 49th Ed., 1972, p 399.

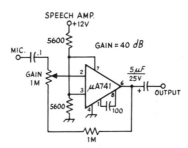

OPAMP SPEECH AMPLIFIER—Fairchild opamp has internal frequency compensation, permitting use as high-gain speech amplifier for high-impedance mike with minimum of external parts. With 709 series of opamps, also suitable for circuit, external frequency compensation is needed to prevent self-oscillation.—"The Radio Amateur's Handbook," American Radio Relay League, Newington, CT, 49th Ed., 1972, p 399.

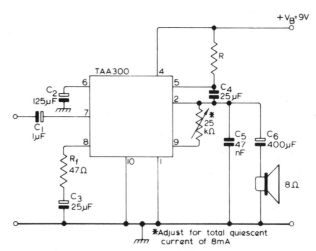

1-W SINGLE-IC—Uses Mullard IC as complete audio amplifier delivering full output power for input of 8.5 mV. Suitable for use after detector in portable radios or with ceramic phono pickup. Speaker is connected to common line. Value of R must be high with respect to load impedance because R is in parallel with load; typical value is 330 ohms when operating from a-c supply, but lower value can be used with battery supply.—"TAA300 Integrated One-Watt Class B A.F. Amplifier," Mullard, London, 1968, TP1050, p 16–17.

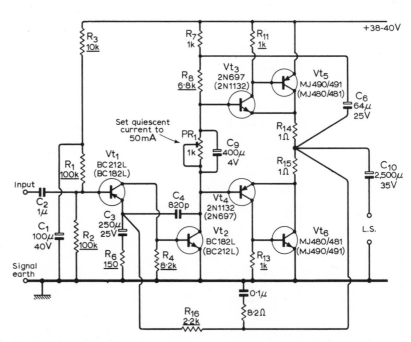

10-W DINSDALE—Modification of earlier circuit design uses silicon planar transistors. Square-wave response is free from overshoot or ringing, even with capacitive loads, beyond 20 kHz. To use with negative supply voltage, reverse electrolytic capacitor polarities and change to transistor types shown in parentheses.—J. L. Linsley Hood, Dinsdale Amplifier Mods, Wireless World, Feb. 1970, p 74.

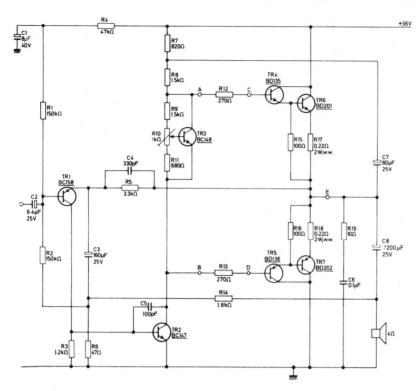

25-W HI-FI—Matched output transistors are driven by matched complementary pair. Diodes D1, D2, and D3 ensure that distortion in crossover region at low output powers is very low. Distortion at full output is only 0.1%, and sensitivity is 400 mV. Response is 1 dB down at 15 Hz and 100 kHz. Preset pot R1 is set to give symmetrical clipping. TR3 provides bias for output stage along with thermal and voltage compensation. Will operate from simple bridge-rectifier supply using single 2,500-μF 64-V electrolytic.—"Transistor Audio and Radio Circuits," Mullard, London, 1972, TP1319, 2nd Ed., p 129–132.

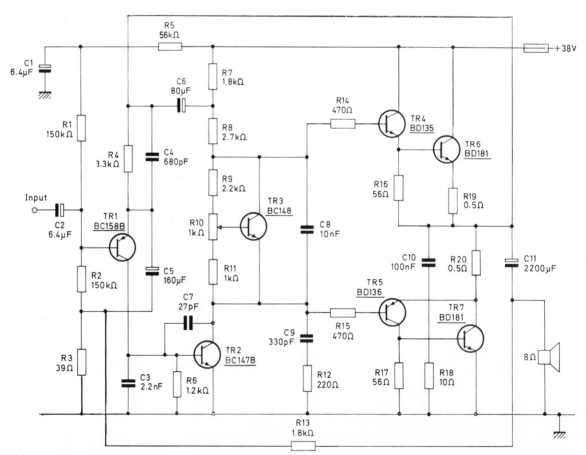

15-W HI-FI—Response is flat within 3 dB from 10 Hz to 100 kHz. Uses 800-mA fast fuse in supply lead for short-circuit protection. Total harmonic distortion is 0.1% at 1 kHz. All transistors are silicon.—"Technical Notes on 15, 25 and 35 W Audio Amplifiers," Mullard, London, 1970, TP1206, p 1–6.

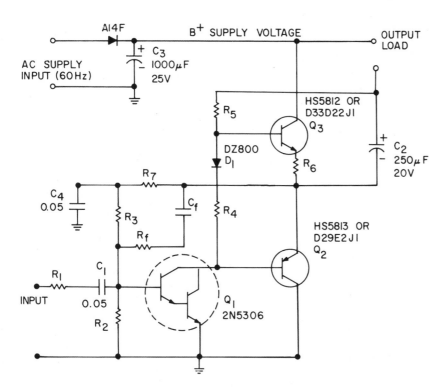

1 OR 2 W—Darlington Q1 combined with complementary output amplifier having separate a-c feedback for bass boost provides high input impedance for crystal or ceramic mike or pickup, along with good low-frequency response. Component values vary with output power and speaker impedance, as given in table. Typical distortion at full power is 2%. Response is flat within 3 dB from 90 to 10,000 Hz for 16-ohm load and from 130 to 15,000 Hz for 8-ohm load.—D. V. Jones and L. T. Anderson, "1 to 2 Watt Amplifier Circuits Requiring Minimum Components," General Electric, Semiconductor Products Dept., Auburn, NY, No. 90.89, 1970, p 6–7.

1-W NONINVERTING—Operates with single power supply. Signal source and load both have capacitive coupling. Voltage gain is 9 for flat response within 3 dB from 200 to 22,000 Hz. Total harmonic distortion is less than 0.75%. Report also gives connections for split-supply, inverting, and pulsed operation of same IC.—"The MC1554 One-Watt Monolithic Integrated Circuit Power Amplifier," Motorola Semiconductors, Phoenix, AZ, AN-401, 1972.

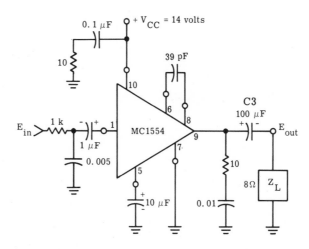

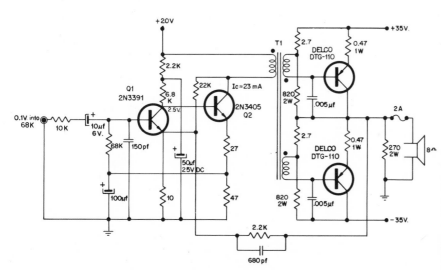

50-W HI-FI—Designed as slave power amplifier to follow tone control or preamp and feed 8-ohm load. T1 is Triad TY-160X. Response is flat within 1 dB from 20 to 20,000 Hz at 5 W output, and total harmonic distortion at 1 kHz is less than 2% up to full output.—"50 Watt Audio Power Amplifier Design," Delco, Kokomo, IN, Jan. 1967, Application Note 36.

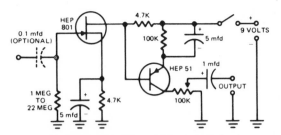

BASIC AMPLIFIER—Will amplify any low-level audio signal. Has high input impedance, current drain of only 200–400 μA, low output impedance, gain of 200 to 400, and frequency range of 10 Hz to 30 kHz. Use audio taper for 100K pot.—"Tips on Using FET's," Motorola Semiconductors, Phoenix, AZ, HMA-33, 1971.

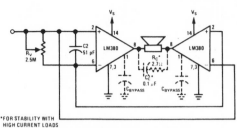

*FOR STABILITY WITH HIGH CURRENT LOADS

BRIDGE AMPLIFIER—Two National IC power audio amplifiers are used in bridge configuration shown when more power is desired than maximum of about 4 W from single IC. Circuit provides twice the voltage swing across load for given supply, thereby increasing power capability four times. Package dissipation will usually limit output power to twice that of single load, however.—J. E. Byerly and M. K. Vander Kooi, "LM380 Power Audio Amplifier," National Semiconductor, Santa Clara, CA, 1972, AN-69, p 5.

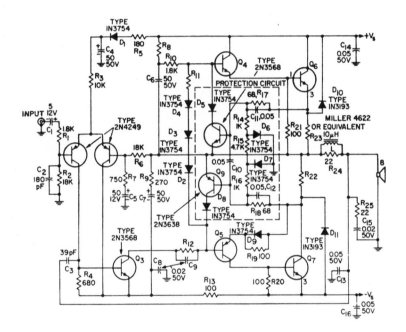

12–70 W UNIVERSAL AMPLIFIER DESIGN—Identical circuit serves for four different quasi-complementary-symmetry audio amplifiers. Each gives frequency response beyond 20 kHz with total harmonic distortion of 1%. Circuit incorporates short-circuit protection. Preamp is direct-coupled, with balanced-bridge circuit using pair of 2N4249 transistors. Book gives values of components for four different power output ratings. Supply voltages range from 19 V for 12 W to 42 V for 70 W.—"Silicon Power Circuits Manual," RCA, Harrison, NJ, SP-51, p 416.

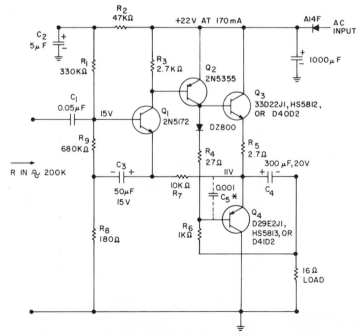

2 W WITH FOUR TRANSISTORS—Provides closed-loop gain of 35 dB. High degree of feedback through R8 keeps distortion under 0.5% from 40 to 20,000 Hz at full output power. Sensitivity is 100 mV rms for full power. C5 may be required to prevent high-frequency oscillation.—D. V. Jones and L. T. Anderson, "1 to 2 Watt Amplifier Circuits Requiring Minimum Components," General Electric, Semiconductor Products Dept., Auburn, NY, No. 90.89, 1970, p 8–10.

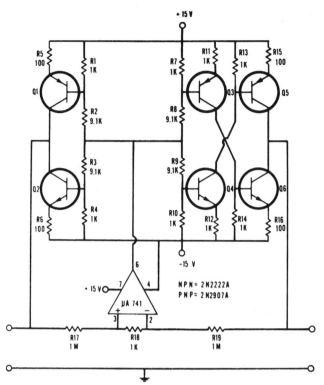

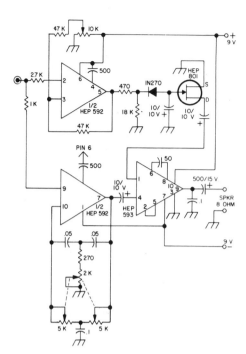

SEMIFLOATING SYNTHETIC INDUCTOR—Uses frequency-dependent characteristic of opamp to simulate inductor. Circuit gave Q of 78 at 237 Hz with 1-μF capacitive load at output port, when used in series resonant circuit. Article includes mathematical analysis.—P. E. Allen and J. A. Means, Inductor Simulation Derived from an Amplifier Rolloff Characteristic, *IEEE Transactions on Circuit Theory*, July 1972, p 395–397.

PHONE-JACK AMPLIFIER—Provides speaker volume for receiver or transceiver designed only for phones, along with agc and tunable a-f selectivity. Power output is 1 W. Designed for medium to high-impedance phone-jack input. Triple-section pot connected to one half of HEP592 tunes audio filter that passes all frequencies except that to which it is set. Since controls are in feedback circuit, effect is peaking of amplifier at frequency to which filter is set.—J. J. Schultz, IC Receiver Accessory, 73, Jan. 1971, p 76–80.

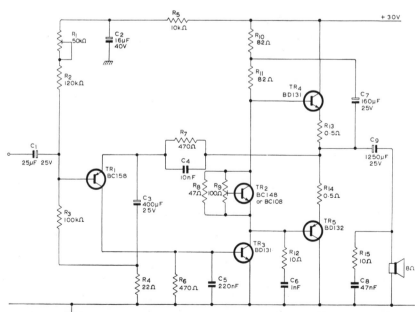

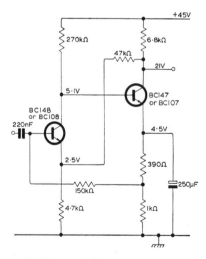

10-W HI-FI—Complementary pair of output transistors gives simpler circuit than quasi-complementary arrangement usually used in class B hi-fi amplifiers. Distortion at full output is less than 0.1%. Frequency response for 3 dB down at 2 W output is 20 Hz to 35 kHz. Preset pot R1 sets midpoint voltage for symmetrical clipping. Operates from bridge-rectifier supply using only single 2,500-μF electrolytic filter.—"Transistor Audio and Radio Circuits," Mullard, London, 1972, TP1319, 2nd Ed., p 122–124.

HIGH OUTPUT VOLTAGE—Provides voltage gain of 20 dB and maximum output voltage of 10 V. Total distortion at 1 kHz is 0.11% for maximum output, and 3-dB response is 20 Hz to 20 kHz. Input impedance is 140K and output is 200 ohms.—"Transistor Audio and Radio Circuits," Mullard, London, 1972, TP1319, 2nd Ed., p 173–174.

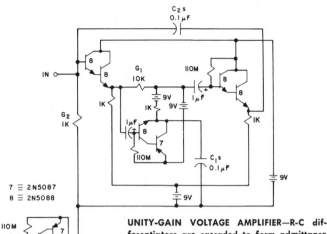

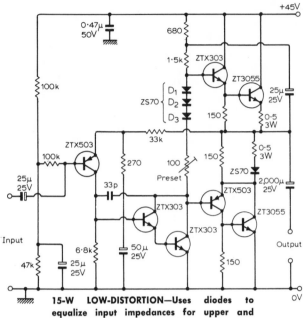

UNITY-GAIN VOLTAGE AMPLIFIER—R-C differentiators are cascaded to form admittance polynomial whose coefficients are determined by RC products. Article gives detailed synthesis procedure that results in selecting component values by inspection and permits tuning circuits with resistors. Provides peak attenuation of about 35 dB at 500 Hz.—D. Hilberman, Synthesis of Rational Transfer and Admittance Matrices with Active RC Common-Ground Networks Containing Unity-Gain Voltage Amplifiers, *IEEE Trans. on Circuit Theory*, Dec. 1968, p 431–440.

15-W LOW-DISTORTION—Uses diodes to equalize input impedances for upper and lower halves of quasi-complementary output stage, to overcome distortion at low listening levels. Operates from simple unstabilized supply. Will drive electrostatic speaker without instability.—I. M. Shaw, Quasi-complementary Output Stage Modification, *Wireless World*, June 1969, p 265–266.

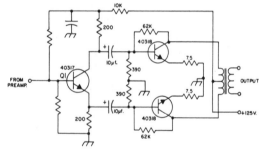

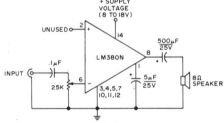

PHASE INVERTER—Voltage developed across 200-ohm emitter resistor of Q1 is equal to but 180 deg out of phase with voltage swing across collector resistor, so following two stages receive equal a-f voltages. Gain of Q1 is very low.—"The Radio Amateur's Handbook," American Radio Relay League, Newington, CT, 49th Ed., 1972, p 398–399.

2 W WITH SIX PARTS—Uses National IC operating over wide range of supply voltages, with only 7 mA quiescent drain at 18 V. Bandwidth extends to 65 kHz. Ideal for portable and mobile operation.—H. Balyoz, The Simplest Audio IC Yet! *73*, December 1972, p 49.

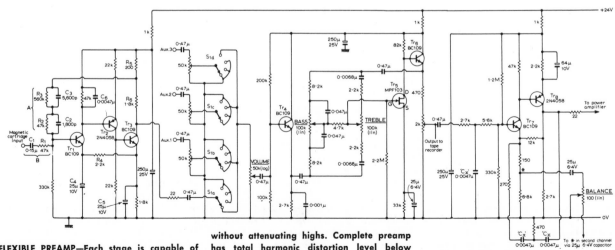

FLEXIBLE PREAMP—Each stage is capable of operating on its own, and has sufficiently low output impedance for use with shielded cable without attenuating highs. Complete preamp has total harmonic distortion level below 0.1% for 20–20,000 Hz, at any tone control setting and for up to 2 V rms output.—J. L. Linsley Hood, Modular Pre-amplifier Design, *Wireless World*, July 1969, p 306–310.

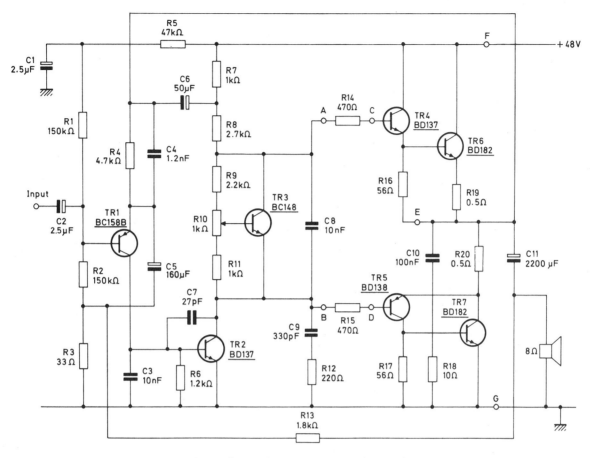

25-W HI-FI—Response is flat within 1 dB from 10 Hz to 100 kHz. Total harmonic distortion is 0.1% at 1 kHz. All transistors are silicon. Report gives optional circuit that can be connected between lettered terminals A–G to provide full short-circuit protection.— "Technical Notes on 15, 25 and 35 W Audio Amplifiers," Mullard, London, 1970, TP1206, p 1–6.

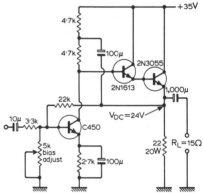

REDUCING DISTORTION—Direct coupling between driver and output stages of 3-W amplifier serves to reduce distortion at high output levels. Measured output voltage just before symmetrical clipping is 6.65 V rms into 15-ohm resistive load. Bias point for highest undistorted output power is 24 V across 22-ohm resistor.—J. Vanderkooy, Amplifier Efficiency, *Wireless World*, Aug. 1969, p 381.

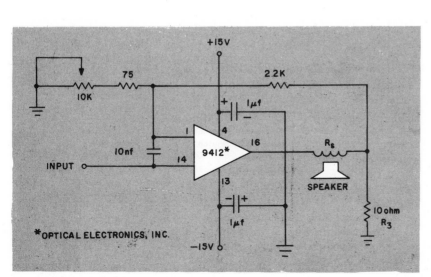

OPAMP DRIVES SPEAKER—Uses Optical Electronics Inc. IC opamp as voltage-current transducer for current-driving 8-ohm speaker to power output of 1 W peak. Response is essentially flat from d-c to almost 100 kHz. Power supplies should be bypassed at opamp as shown. Nonlinearity is low, measuring less than 2.8% at 1 kHz with voltage gain of 9.6.—J. D. Ricks, Op Amp Drives Loudspeaker, *Electro-Technology*, Dec. 1969, p 18.

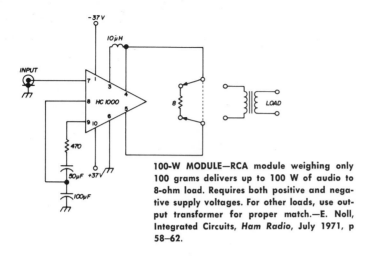

100-W MODULE—RCA module weighing only 100 grams delivers up to 100 W of audio to 8-ohm load. Requires both positive and negative supply voltages. For other loads, use output transformer for proper match.—E. Noll, Integrated Circuits, *Ham Radio*, July 1971, p 58—62.

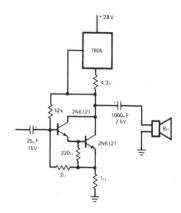

5-W CLASS A—Combination of Fairchild μA7805 voltage regulator and Darlington transistor pair provides 5 W rms across 8-ohm load. Distortion is only 0.6% at 1 W, with 20-dB voltage gain. Regulator keeps ripple low.—J. W. Chu and R. D. Ricks, "The μA7800 Series, Three-Terminal Positive Voltage Regulators," Fairchild Semi-conductor, Mountain View, CA, No. 312, 1971, p 6.

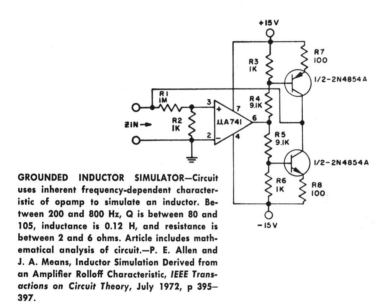

GROUNDED INDUCTOR SIMULATOR—Circuit uses inherent frequency-dependent characteristic of opamp to simulate an inductor. Between 200 and 800 Hz, Q is between 80 and 105, inductance is 0.12 H, and resistance is between 2 and 6 ohms. Article includes mathematical analysis of circuit.—P. E. Allen and J. A. Means, Inductor Simulation Derived from an Amplifier Rolloff Characteristic, *IEEE Transactions on Circuit Theory*, July 1972, p 395—397.

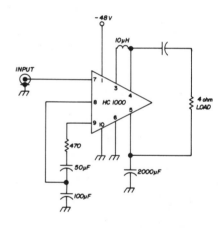

100 W—Single RCA module requires only negative supply voltage, weighs only 100 grams, and delivers up to 100 W to 4-ohm load.— E. Noll, Integrated Circuits, *Ham Radio*, July 1971, p 58—62.

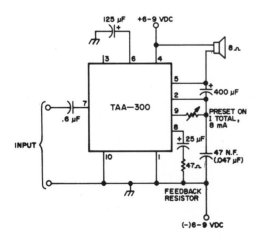

30—25,000 HZ SINGLE-IC—Amperex IC provides excellent bandwidth at 3-dB points while drawing only 180 mA for power output of 1 W. Can also be used as modulator for transmitter. Input impedance is about 10K. Replacing 47-ohm feedback resistor with 500K rheostat permits varying feedback and gain.—B. Hoisington, A Hi-Fi IC for Amateur Modulators and Receiver Audio, *73*, May 1972, p 32—34.

CHAPTER 5
Audio Compressor Circuits

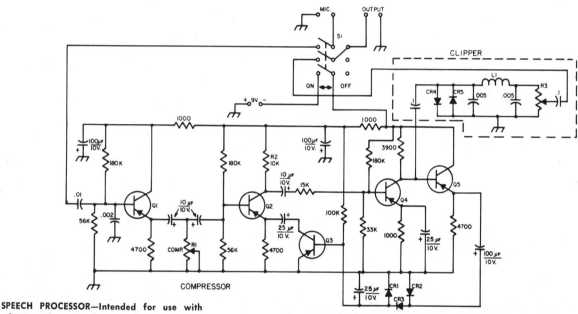

SPEECH PROCESSOR—Intended for use with ssb transmitter. Combines audio compression and clipping to produce up to 3 dB improvement in average transmitted output power. All transistors are 2N1375 or equivalent, and all diodes 1N270 or equivalent. R1 is 10K audio taper, R3 is 50K linear, and L1 is 3 to 3.5 H. To adjust, feed into hi-fi system having headphone output, and set R3 where voice quality deterioration is barely noticeable.— "The Radio Amateur's Handbook," American Radio Relay League, Newington, CT, 49th Ed., 1972, p 408–409.

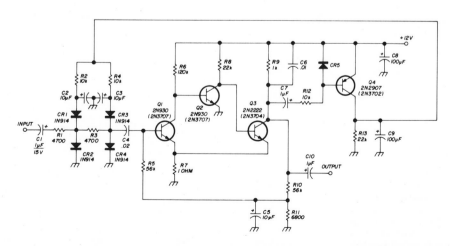

AUDIO AGC—Circuit is used between microphone and transmitter to prevent overmodulation and to minimize decreases in modulation when voice level drops or microphone is moved away. Ratio of maximum to minimum gain is 1,400. Can also be used in direct-conversion receivers in which most or all of gain is provided in audio amplifier. Q1-Q2-Q3 form amplifier, CR5 is detector, and Q4 is d-c amplifier that provides control voltage.— C. Hall, Audio AGC Principles and Practice, *Ham Radio*, June 1971, p 28–33.

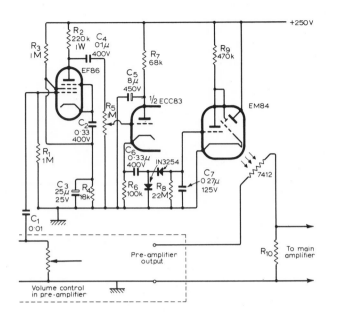

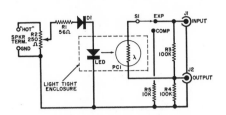

PHOTOELECTRIC LOUDNESS CONTROL—Type EM84 tuning indicator tube serves as audio-controlled light source mounted in dark box with photoresistor such as RCA 7412. Photoresistor is mounted close to tuning-indicator screen, midway between ends of fluorescent strips, to give nonlinear response required for volume expansion. Used with phonograph records to compensate for volume range restrictions imposed by recording process.—M. B. Catford, A Flexible Expander/Compressor, *Wireless World*, Dec. 1968, p 472–473.

COMPRESSOR-EXPANDER—Samples output of audio system at speaker, and feeds rectified a-f signal to led for optoelectronic coupling to photocell. With S1 set to EXPAND, photocell is connected across high end of voltage divider R3-R4; when light from led lowers resistance of photocell-R3 combination, audio output across R4 increases to give expansion. With S1 on COMPRESS, light lowers resistance of photocell-R5 combination and thereby lowers audio level at J2 to give signal compression. Circuit is used between preamp and power amplifier of audio system or between tape deck and preamp. Two units in stereo system can put new life into overly compressed recordings. D1 is 50-piv 1-A silicon diode. LED is Radio Shack 276-026, and PC1 is Radio Shack 276-116 cadmium sulfide cell.—C. Anderton, Simple Compressor-Expander, *Popular Electronics*, April 1973, p 50.

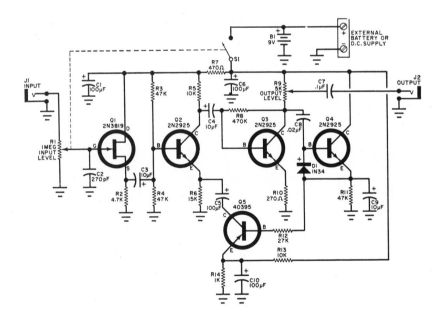

DISTORTIONLESS SPEECH COMPRESSOR—Eliminates need for watching recording level indicator of tape recorder. Use between mike and recorder input. Has 32-dB compression range. Can also be used with ham or CB transmitter, or with public-address system. For inputs under 1 mV, circuit operates as straight amplifier with voltage gain of 200. Greater inputs up to 40 mV are compressed by Q5 acting as current-sensitive variable resistor. Frequency response is flat from 10 Hz to over 20 kHz.—C. Caringella, Add "Comply" to Your Tape Recorder, *Popular Electronics*, Feb. 1968, p 47–52.

PREAMP WITH COMPRESSION—Designed for use with low-impedance transceiver mike to provide match with high-impedance input of transceiver. Uses clipper-type compressor which easily gives 20 dB compression without too much distortion, and also drops out undesirable voice components below 150 Hz. Frequency response is flat from 250 to 3,500 Hz and drops to —20 dB at 60 Hz and 20 kHz. Article covers construction and adjustment.—H. P. Fischer, $5 Preamp Compressor, *73*, Aug. 1972, p 75–76.

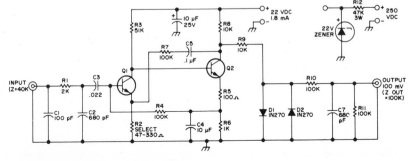

1—Q1 & Q2 ARE Si NPN TYPES, hfe 30–120, VCBo 45V OR EQUIV.
2—SELECT R2 FOR DESIRED GAIN.
3—IF NECESSARY, CHANGE R4 TO YIELD 10–12 VDC AT COLLECTOR OF Q2.
4—IF LOWER OUTPUT IS DESIRED, CHANGE R10–R11 RATIO.

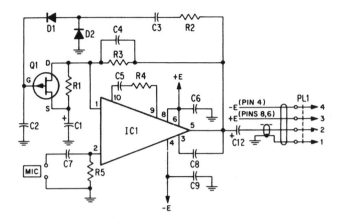

ANTIHOWLER—Audio compressor using IC opamp and fet keeps output of dynamic mike constant within 3 dB when built into or mounted at rear of mike. Reduces tendency of p-a system to break into microphonic howling by about 30 dB. Not suitable for recording music because it would give effect of player piano, with no expression. Goes into maximum compression even at signal level representing very low whisper, without introducing distortion. R1 is 100K, R2 4.7K, R3 1 meg, R4 100 ohms, and R5 47K. Q1 is 2N3820, IC1 is MC1433G, and diodes are 1N60. Batteries are 12.6-V mercury, C1 is 10 μF, C2 and C6 0.22 μF, C3 and C7 0.1 μF, C4 and C8 50 μμF, C5 0.002 μF, C9 0.05 μF, C10 and C11 100 μF, and C12 2 μF.—J. Ritchie, Compress-O-Phone, *Electronics Illustrated*, Nov. 1969, p 29–33 and 112.

FEEDBACK-LOOP GAIN—Compressor action of amplifier is considerably tightened by using second opamp to drive rectifier that provides fet control voltage proportional to signal level, while output is taken from first opamp. Output level changes only 2 dB for 30-dB change in input.—N. Doyle, "Some Useful Signal Processing Circuits Using FET's and Operational Amplifiers," Fairchild Semiconductor, Mountain View, CA, No. 243, 1971, p 3.

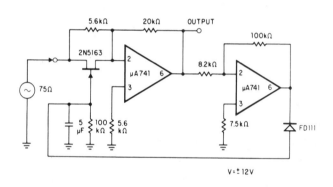

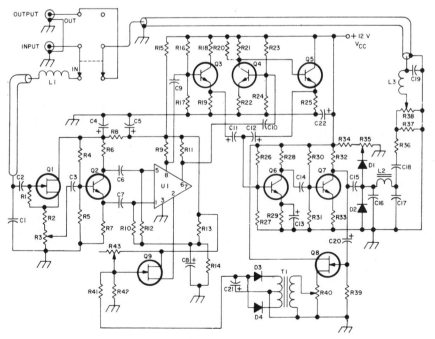

L1-L3 2.5 mH rf choke
L2 3.5 H miniature audio choke, UTC DOT-8
T1 GC Co. D1-728 transformer. For primary, use half of 500Ω CT secondary. For secondary, use 1000Ω CT primary.
Capacitors
C1, C19 0.001 μF
C2 0.005 μF
C3 0.05 μF
C4, C5, C8, C22, 100μF 15V
C6, C7, C9, C10, C15, C18 0.1μF
C11, C12, 10 μF 15V
C13 30 μF 15V
C14, 0.02μF
C16, C17 0.0015μF
C20, C21 6 μF 1.5V
Resistors (all except potentiometers 1/2 watt)
R1 2.2M
R2 91-R3 10K audio-taper pot
R4, R16, R23, R26 180K
R5, R17, R24 56K
R6, R7, R18, R19, R21, R22, R25, R34 4.7K
R8, R14, R15, R29, R33, R41 1.0K
R9, R11 10K
R10, R12, R37 47K
R13 2K
R28, R32, 3.9K
R30 120K
R35 5K pot
R36 24K
R38 50K audio taper pot   R42 30K
R39 100K   R20 See Text
R40 2K trimmer pot   R27, R31 33K

SPEECH PROCESSOR—Increases effectiveness of ssb voice transmitter by compressing, clipping, and filtering audio ahead of modulator. Attack time is less than 10 ms to insure reduction of initial peaks, and release time is about 300 ms to accommodate changes in voice level without having background noise come in too fast when voice is interrupted. Q2 provides equal and opposite a-f voltages for driving RCA CA3028A push-pull IC amplifier whose gain is controlled by Q9 which receives ALC voltage from Q8 and full-wave rectifier. Q1, Q8, and Q9 are 2N3819 or HEP 802. Q2-Q7 are 2N697 or HEP54. D1–D4 are 1N270.—R. Silberstein, An Improved Audio Speech Processor, 73, Jan. 1973, p 47–51.

SPEECH COMPRESSOR—Uses conventional low-noise mike amplifier Q1 to drive a-f amplifier Q2 having a-f gain control R1 which feeds compressed audio to succeeding a-f amplifier stages. Some of output from Q2 is amplified further by Q3 to feed 1,000-ohm primary of T, which has 10,000-ohm C-T feeding agc diodes that deliver negative agc voltage to base of Q2 to reduce its gain and provide compression depending on setting of R1. Action holds amplifier output at constant level. C1 and R2 have sufficiently long time constant to hold agc voltage steady between syllables and words.—"The Radio Amateur's Handbook," American Radio Relay League, Newington, CT, 49th Ed., 1972, p 402–403.

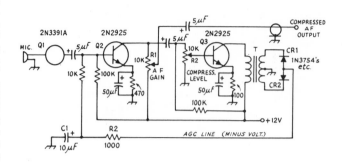

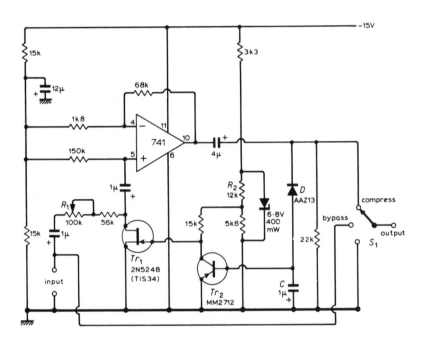

TAPE-RECORDER COMPRESSOR—Reduces gain during transients and unexpected crescendos, to allow higher average recording level without distortion. For stereo, use compressor in each channel and equalize operation of compressors by adjusting collector voltage with R1 and R2. Gain is unity at low levels, reducing as input approaches 350 mV, and holding output at fairly constant 400 mV for inputs from 450 mV to 4 V.—P. Hanson, Audio Dynamic Range Compressor, *Wireless World*, March 1973, p 143.

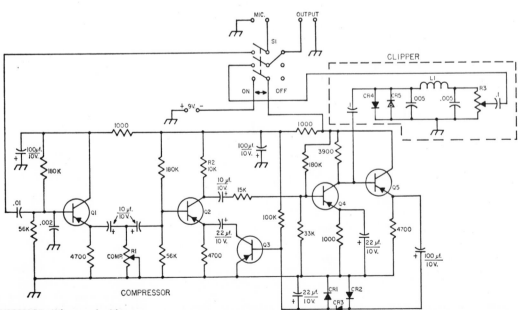

CLIPPER-COMPRESSOR—When used with transmitter, increases average r-f power without appreciably affecting audio quality, for greater output from ssb transmitter and higher modulation power for a-m transmitter. Clipper is added to output of speech compressor. Diodes are 1N270 or equivalent. L1 is about 3 H. All transistors are 2N1375 or equivalent. R1 is 10K audio taper, R2 is 10K, and R3 is 50K linear taper. Article covers construction and adjustment.—J. J. Spadaro, A Solid-State Speech Processor, QST, Nov. 1969, p 21–23.

IC SPEECH COMPRESSOR—Pin 3 of National Semiconductor IC audio compressor is used for input gain adjustment, while pin 4 receives agc voltage whenever output level exceeds preset norm determined by 10K compressor level control.—"The Radio Amateur's Handbook," American Radio Relay League, Newington, CT, 49th Ed., 1972, p 402–403.

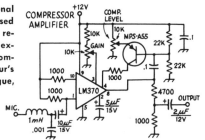

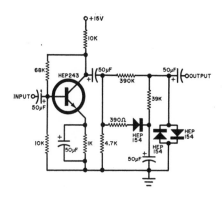

ORGAN VOLUME LEVELER—Handles volume changes occurring too fast for organist to correct with volume pedal, such as when changing from solo to rhythm or changing stops quickly. Responds to entire range of organ frequencies without adding coloring to voices and without clipping.—J. E. Rohen, Add a Compression Amplifier to Your Electronic Organ, Popular Electronics, Dec. 1972, p 105.

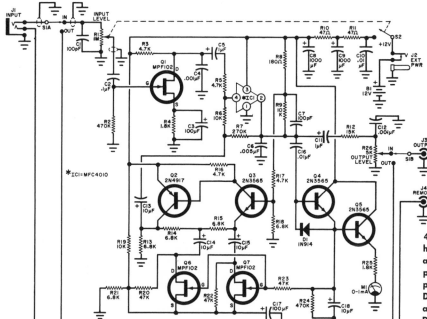

45-DB COMPRESSION—Consists essentially of high-quality audio amplifier having built-in agc loop to maintain constant-level nonclipping output. Provides about 46 dB gain, permitting use with any type of microphone. Developed for use with tape recorder or amateur radio transmitter.—C. Caringella, Distortionless Audio Compressor, Popular Electronics, Dec. 1971, p 25–30.

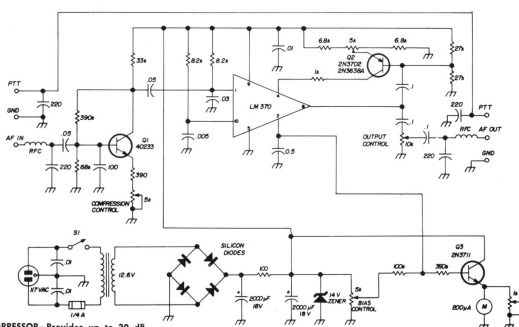

SPEECH COMPRESSOR—Provides up to 30 dB of peak limiting before distortion becomes objectionable, corresponding to over 4 dB intelligibility threshold improvement, when used with ssb transmitter or tape recorder. Uses National IC amplifier with built-in agc. PTT terminal goes to push-to-talk switch. Used with low-output 200-ohm dynamic mike that feeds low-noise low-gain stage Q1 which drives IC into compression range.—G. A. Nurkka, Integrated-Circuit Single-Sideband Speech Processor, Ham Radio, Dec. 1971, p 31–35.

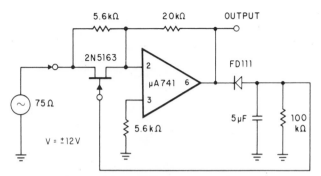

GAIN-CONTROLLED OPAMP—Output of opamp is rectified to get d-c proportional to signal level, for use as fet control voltage. Output level is thus stabilized for variations in input. Distortion at higher input signal levels is overcome by shunting fet with 5.6K resistor, so amplifier performs normally when control range of fet is exceeded. Circuit compresses input range of 20—50 dB to output range of 65—80 dB.—N. Doyle, "Some Useful Signal Processing Circuits Using FET's and Operational Amplifiers," Fairchild Semiconductor, Mountain View, CA, No. 243, 1971, p 3.

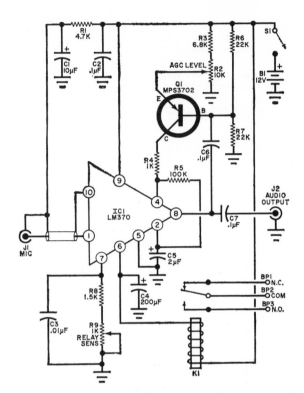

VOX MIKE—Operates without push-to-talk switch, and provides speech compression to compensate automatically for changes in voice level. May be connected to transmitter or tape recorder. R9 determines sensitivity of voice-operated relay K1, which is 1,640-ohm Sigma 65FIA-12DC or similar. Use 5-mV to 50-mV dynamic mike. Relay contacts may also be used to start tape recorder automatically.—R. A. Hirschfeld, Build The Voxor, *Popular Electronics*, Feb. 1970, p 81—84.

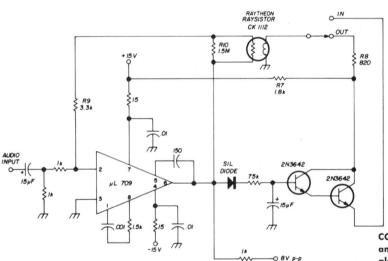

COMPRESSOR—Cadmium sulfide photocell and low-current pilot lamp can be used in place of Raysistor for optoelectronic coupling. Values shown give audio compression range of about 55 dB.—R. Factor, Multimode I-f System, *Ham Radio*, Sept. 1971, p 39—43.

AMPLITUDE LEVELER—Current-controlled bridge in feedback loop of opamp A1 develops differential voltage needed to keep output level steady. Incandescent lamps T1 and T2 act as current-variable resistors that balance bridge when input is 0.4 V. Suitable for amplitude leveling in test oscillators, communication equipment, and telemetry systems. Circuit applies zero feedback at optimum input level and produces positive or negative feedback above and below this level. Design equations are given in article. Operates over entire audio spectrum, for input voltages ranging from 0.02 to 2 V.—E. E. Pearson, Audio Amplitude Leveler Minimizes Signal Distortion, *Electronics*, May 24, 1973, p 104.

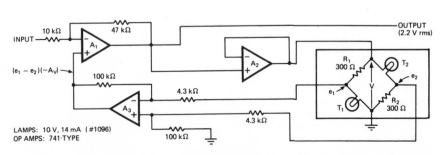

CHAPTER 6
Audio Control Circuits

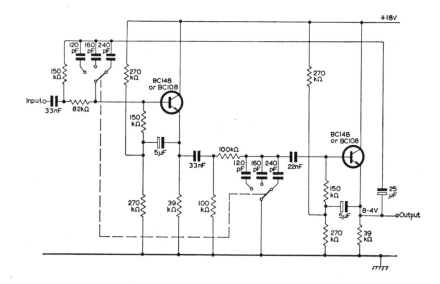

NOISE AND RUMBLE FILTER—Ganged switches provide freqency limits of 16, 12, and 7 kHz for noise filter, while rumble filter has fixed frequency limit of 45 Hz. Voltage gain is 0.95. Input and output impedances are 1.7 meg and 450 ohms. Provides both bass and treble cut by using one R-C network between transistors and another in feedback loop. Gives high slope of about 13 dB per octave.—"Transistor Audio and Radio Circuits," Mullard, London, 1972, TP1319, 2nd Ed., p 184–185.

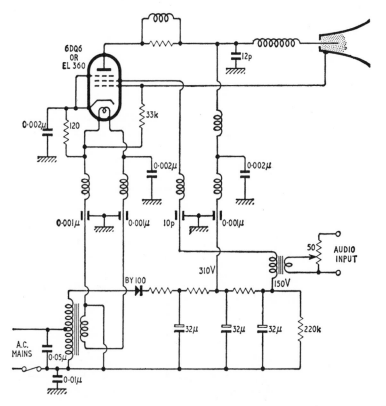

IONOVAC SPEAKER DRIVE—Based on principle of singing arc, in which r-f discharge is modulated with a-f to produce sound waves. Frequency is 27 MHz, generated by pentode having rated anode dissipation of 15 W, with screen modulation at about 33 V rms from 15-ohm input of 1 V rms. Frequency response is about 3 to 30 kHz.—International Audio Festival and Fair, *Wireless World*, June 1965, p 291–292.

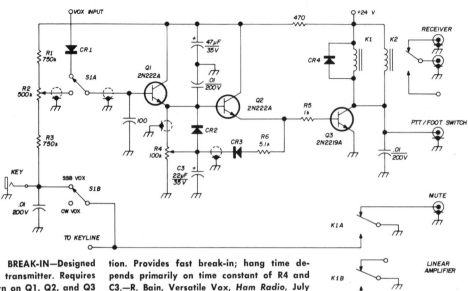

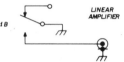

VOX WITH FAST C-W BREAK-IN—Designed for use with 9-MHz ssb transmitter. Requires positive vox input to turn on Q1, Q2, and Q3 and pull in both relays. Diode types are not critical. Article covers construction and operation. Provides fast break-in; hang time depends primarily on time constant of R4 and C3.—R. Bain, Versatile Vox, *Ham Radio*, July 1971, p 50–53.

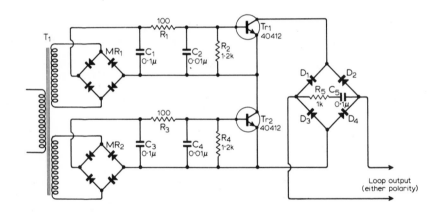

AUDIO SWITCH—Closes d-c circuit when sine or square-wave audio signal is applied to 10K primary of audio transformer T1 having two 600-ohm secondaries each providing about 1.7 V rms to bridge rectifiers, which can use OA81 or similar diodes. D1–D4 are silicon diodes rated 600 piv at 1 A. Input is about 2 kHz.— J. Du P. De Beer, Audio Switch, *Wireless World*, Feb. 1970, p 73.

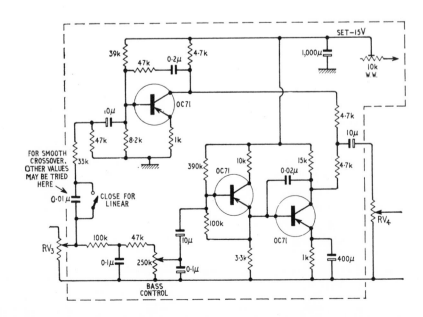

DEEP BASS BOOST—Provides boost only below about 120 Hz and therefore has no effect on speech. Can be inserted in audio amplifier. Requires large speaker.—R. G. Young, "1812" Bass Booster, *Wireless World*, March 1966, p 141.

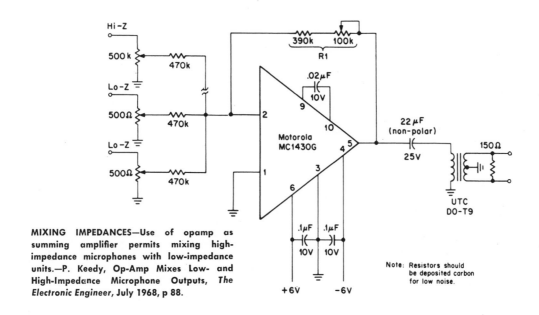

MIXING IMPEDANCES—Use of opamp as summing amplifier permits mixing high-impedance microphones with low-impedance units.—P. Keedy, Op-Amp Mixes Low- and High-Impedance Microphone Outputs, *The Electronic Engineer*, July 1968, p 88.

Note: Resistors should be deposited carbon for low noise.

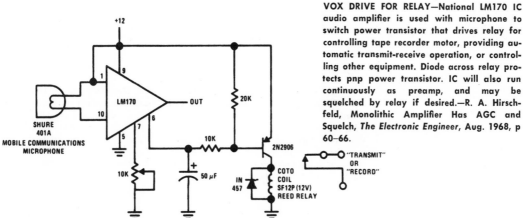

VOX DRIVE FOR RELAY—National LM170 IC audio amplifier is used with microphone to switch power transistor that drives relay for controlling tape recorder motor, providing automatic transmit-receive operation, or controlling other equipment. Diode across relay protects pnp power transistor. IC will also run continuously as preamp, and may be squelched by relay if desired.—R. A. Hirschfeld, Monolithic Amplifier Has AGC and Squelch, *The Electronic Engineer*, Aug. 1968, p 60–66.

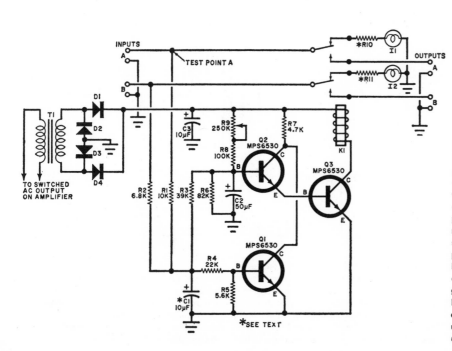

PROTECTING WOOFERS—Used with direct-coupled solid-state hi-fi power amplifiers to prevent voltage transients from damaging cones of expensive woofers and tweeters. Circuit is placed between power amplifier and speaker system. Timing circuit R9-R8-C2 prevents turn-on and turn-off transients from reaching speakers. Voltage-sensing circuit samples amplifier output voltage and completes circuit to speakers by pulling in relay K1 only when voltage is below safe d-c level of about 3 V with either polarity. If speakers get disconnected on high-level notes below 20 Hz, increase C1 to 20 μF. Article tells how to compute values of R10 and R11 for amplifier used.—P. Arthur, Build a Woofer Guard, *Popular Electronics*, July 1970, p 49–52.

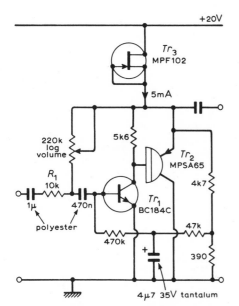

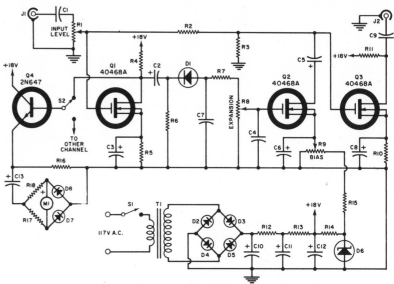

VARIABLE-GAIN VOLUME CONTROL—Used at preamp input to give large overload capability. Signal-noise ratio is 73 dB for 10-mV input. TR2 is Motorola Darlington transistor. Does not require stabilized supply. Maximum voltage gain of 22 drops almost to zero at minimum setting of pot.—A. Jenkins, Variable-Gain Volume Control, *Wireless World*, April 1972, p 186.

R1—250,000 ohm audio-taper pot
R2, R3—330,000 ohm, ½ W res.
R4, R11—3300 ohm, ½ W res.
R5, R10—270 ohm, ½ W res.
R6—10,000 ohm, ½ W res.
R7—10 megohm, ½ W res.
R8—3 megohm linear-taper pot
R9—2500 ohm linear-taper pot
R12—470 ohm, ½ W res.
R13—220 ohm, ½ W res.
R14, R15, R17, R18—1000 ohm, ½ W res.
R16—2200 ohm, ½ W res.
C1—0.05 μF paper capacitor
C2—5 μF, 25 V elec. capacitor
C3, C8—100 μF, 3 V elec. capacitor
C4—0.01 μF ceramic capacitor

C5, C6—5 μF, 10 V elec. capacitor
C7, C9—0.22 μF paper capacitor
C10, C11, C12—250 μF, 50 V elec. capacitor
C13—10 μF, 25 V elec. capacitor
S1—S.p.s.t. toggle switch
S2—S.p.d.t. slide switch
J1, J2—Pin or phono jack
M1—1 mA miniature d.c. meter
D1—Germanium diode (see text)
D2, D3, D4, D5—Silicon diode 200 p.i.v., 1 A (HEP 156)
D6—5 V zener diode (HEP 603)
D7, D8—1N34A germanium diode (HEP 134)
T1—24 V, 1 A filament trans.
Q1, Q2, Q3—RCA 40468A
Q4—2N647 (or HEP 641)

EXPANDER—Fet circuit provides up to 30 dB expansion of dynamic range for recorded or broadcast music. Gain increases linearly with input signal, to amplify strong signals more than weak signals and thereby restore original dynamic range that had been compressed to counteract system noise. Q1 amplifies input signal, and D1 converts result to d-c voltage for controlling Q2 in attenuator circuit. Q3 amplifies output of attenuator enough to drive most power amplifiers. Q4 drives meter circuit that simplifies correct setting of input levels. For stereo, duplicate circuits of Q1, Q2, and Q3.—R. Wilt, Low-Distortion Hi-Fi Volume Expander, *Electronics World*, June 1971, p 54–55.

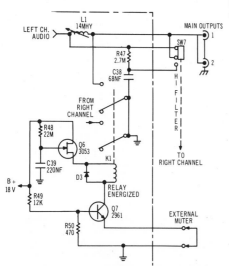

MUTING—Used in commercial stereo console to ground output of set for about 5 s while R-C circuit charges and turns on fet Q6. This delay is long enough for all turn-on transients to die out before speakers are energized. Circuit also provides external mute terminals to which spst switch can be connected for muting system from any remote location.—L. Feldman, Crown IC-150 Integrated Circuit Stereo Console, *Audio*, Jan. 1972, p 48, 50, 54–57.

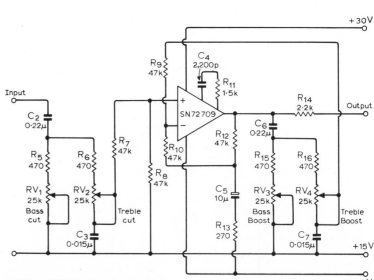

4-POT TONE CONTROL—Permits increasing crispness of speech in middle range without making higher-frequency sound seem unnaturally boosted. As increasing treble boost is applied, with other controls set flat, 6-dB-per-octave treble boost curve is brought progressively down frequency spectrum. Bass controls act in similar manner at other end of spectrum. Can be used to compensate for loudspeaker deficiencies at low frequencies. Linear pots give smoothest control.—P. B. Hutchinson, Tone Control Circuit, *Wireless World*, Nov. 1970, p 538–540.

AUDIO MIXER—Audio input voltages are converted to currents and combined to modulate emitter current of grounded-base transistor. Low impedance at emitter means low crosstalk between inputs. Frequency response is excellent.—W. G. Jung, Grounded-Base Transistor Acts as Current Summer, *Electro-Technology*, Nov. 1969, p 46.

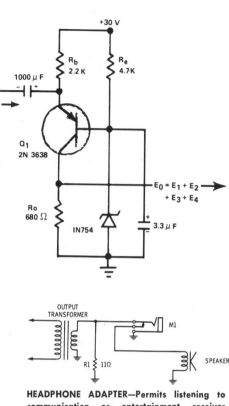

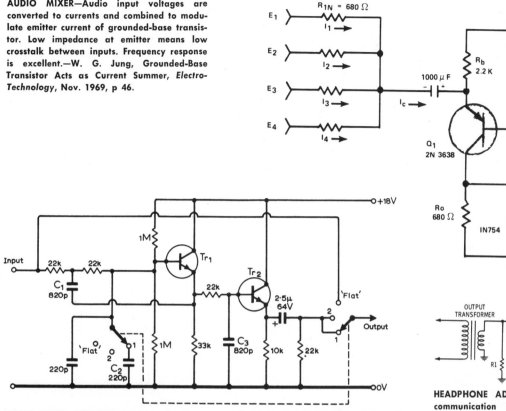

TREBLE TONE CONTROL—Position 2 attenuates highs at rate of 12 dB per octave to minimize transient distortion of active filter and resulting unpleasant nasal quality. Position 1 gives 18 dB per octave rolloff without seeming to cause noticeable transient distortion. Transistors are BC109 or similar.—H. P. Walker, Stereo Mixer, *Wireless World*, June 1971, p 295–300.

HEADPHONE ADAPTER—Permits listening to communication or entertainment receiver without disturbing rest of family. Insertion of magnetic phone in jack M1 disconnects speaker. R1 is added to load secondary of output transformer in receiver, to prevent damage to set when sepaker is disconnected.—R. M. Brown, "101 Easy CB Projects," Howard W. Sams, Indianapolis, IN, 1968, p 103–104.

VOICE-CONTROL PARAMETER EXTRACTOR—Circuit is essentially a-d converter using IC logic to recognize limited number of speech patterns based on amplitude, zero-crossing, and positive derivative detectors. Circuit is independent of speed with which word is pronounced, but definitely has many limitations. Presented as basis for further research. Table at lower left gives supply voltages for the five types of IC's used.—V. Biancomano, Limited Speech Recognition, *QST*, Oct. 1972, p 36–39.

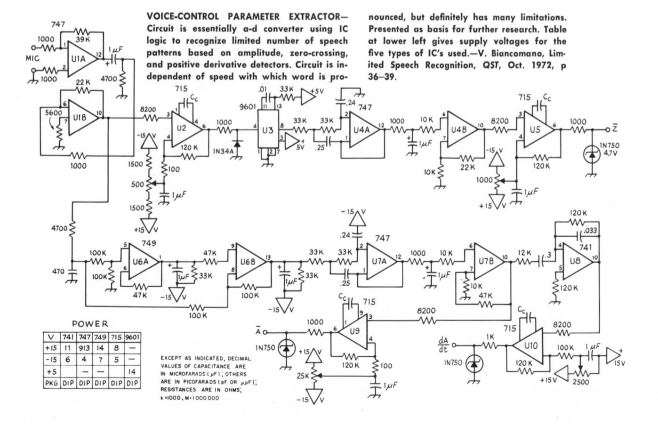

POWER

V	741	747	749	715	9601
+15	11	9 13	14	8	
-15	6	4	7	5	
+5	—	—			14
PKG	DIP	DIP	DIP	DIP	DIP

EXCEPT AS INDICATED, DECIMAL VALUES OF CAPACITANCE ARE IN MICROFARADS (µF); OTHERS ARE IN PICOFARADS (pF OR µµF); RESISTANCES ARE IN OHMS; k =1000, M=1000000

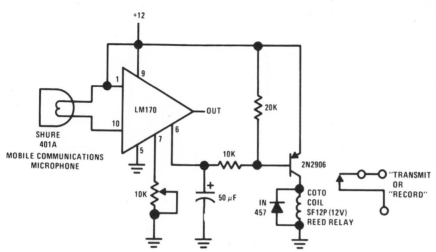

VOX PREAMP—Opamp is connected as voice-operated relay control for switching transmitter, tape recorder, or other electronic or electromechanical devices. Suitable for automatic transmit-receive operation. Will switch on motor of tape recorder at first syllable of dictation by slow-thinking author, to conserve tape. Any fast-acting relay may be used in place of reed relay shown. If circuit oscillates because power supply impedance is too high, place large capacitor across supply.—R. A. Hirschfeld, "AGC/Squelch Amplifier," National Semiconductor, Santa Clara, CA, 1969, AN-11.

TONE CONTROL WITH 20 DB GAIN—Provides 1 V nominal output for driving low-impedance power amplifier from 100-mV source. TR2 and TR3 reduce harmonic distortion of constant-current source TR1. Tone controls have impedance of about 10K in flat position.—J. N. Ellis, High-Level Low-Distortion Tone Control, *Wireless World*, Feb. 1972, p 78.

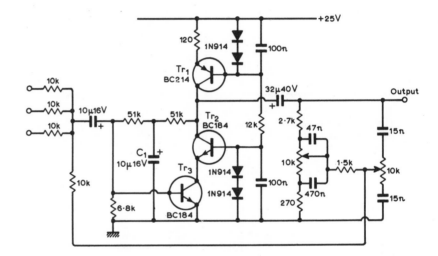

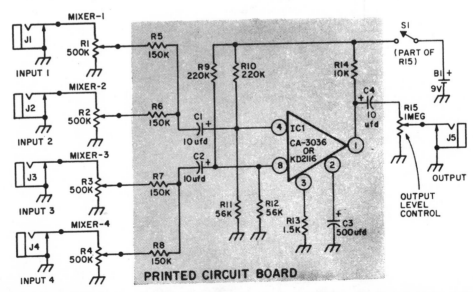

FOUR-INPUT MIXER—Provides individual controls for four guitar pickups along with preamplification, and feeds resulting combined signal to single input of power amplifier through master output level control R15. Uses RCA IC having Darlington-connected transistor pairs as differential amplifier, with two isolated inputs and one output. Operation from transistor battery gives hum-free active mixer.—J. J. Carr, Quad-Mix, *Elementary Electronics*, March-April 1971, p 53–58.

FOUR-CHANNEL MIXER—Preamp delivers constant 4.5-V p-p output for inputs not exceeding 0.25 V p-p, with flat response within 1 dB from 20 Hz to 1 MHz. All resistors are 330K and capacitors 10 μF. Add dashed circuit if required to keep input voltage under 0.25 V a-c.—"Tips on Using IC's," Motorola Semiconductors, Phoenix, AZ, HMA-32, 1972.

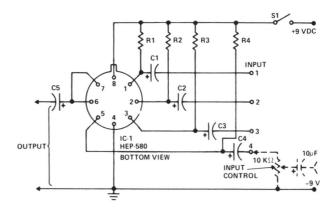

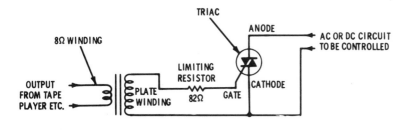

TONE-CONTROLLED SLIDE PROJECTOR—Audio tone of any frequency, even 60 Hz, from tape player or other source for about 1 s will trigger triac to start change cycle of automatic slide projector. Triac reading should be higher than power drawn by slide changer. Record narration and sound effects on one channel of stereo tape recorder, and record audio tone on other channel at each point where slide change is desired.—SCR's and Triacs—Testing and Theory of Operation, *Electronic Servicing*, Dec. 1971, p 42–46.

LOW-PASS AND HIGH-PASS—Consists of two R-C networks connected in series with buffer amplifier. Switches permit choice of 40, 80, 160, and 270 Hz for lower cutoff frequencies and 11, 9, 4.5, and 3.2 kHz for upper cutoff frequencies.—"Transistor Audio and Radio Circuits," Mullard, London, 1972, TP1319, 2nd Ed., p 182–183.

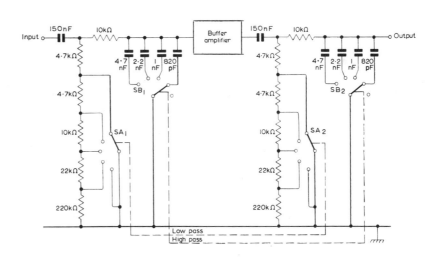

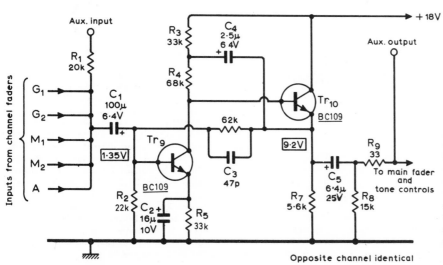

MIXER—Bootstrap capacitor C4 increases amplifier gain to over 4,000 and keeps harmonic distortion below 1% for 3-V rms output. Nominal signal level at output is about 350 mV, at which residual noise is 84 dB down for bandwidth of about 20 kHz. Use of 60 dB of negative feedback drops distortion below 0.01% and ensures proper mixing of signals from channel faders with no interaction. Uses virtual ground.—H. P. Walker, Stereo Mixer, *Wireless World*, May 1971, p 221–225.

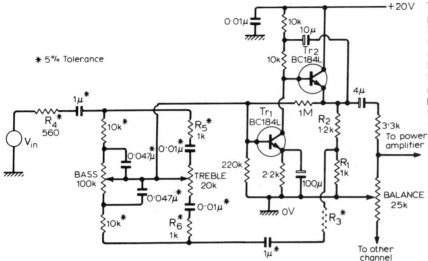

∗ 5% Tolerance

TONE CONTROL—Designed for use between preamp and power amplifier requiring at least 1 V rms. Uses bipolar npn silicon transistors in Baxandall configuration to minimize distortion. With controls at flat position, total harmonic distortion from 40 Hz to 20 kHz is less than 0.01% at 5 V rms output, and signal-noise ratio is 104 dB.—P. M. Quilter, Low Distortion Tone-Control Circuit, *Wireless World*, April 1971, p 199–200.

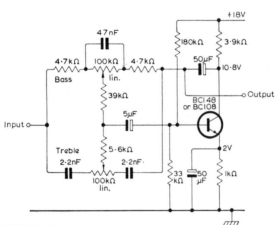

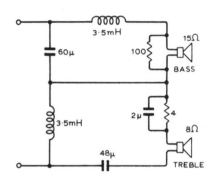

CROSSOVER NETWORK—Uses half-section series network to divide output of a-f amplifier between bass and treble horns of speaker system.—J. Greenbank, Horn Loudspeaker, *Wireless World*, Jan. 1972, p 14–15.

ACTIVE TONE CONTROL—Uses frequency-dependent feedback network between collector and base of transistor. Control range is —22 dB to 19 dB at 30 Hz and —19 dB to 19.5 dB at 20 kHz. Response is flat when both pots are centered. Distortion for weak signals is below 0.1% and rises to 0.85% at 12.5 kHz for 2-V output. Input and output impedances are 40K and 180 ohms.—"Transistor Audio and Radio Circuits," Mullard, London, 1972, TP1319, 2nd Ed., p 177–179.

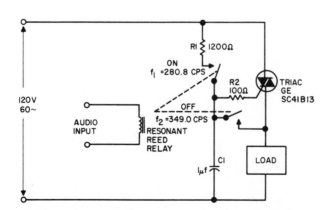

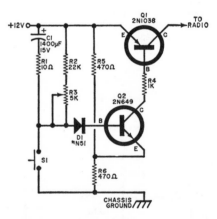

COMMERCIAL KILLER—When S1 is pushed momentarily, auto radio is turned off for period of up to 1 min, determined by setting of R3. Developed to silence long commercials that can be dangerously distracting in heavy traffic. Closing S1 allows C1 to charge, making Q2 nonconductive and removing base bias of Q1 so it also is nonconductive. Power supply to radio is thus blocked by Q1. Circuit shown is for negative ground.—E. M. McCormick, Reader's Circuit, *Popular Electronics*, Jan. 1970, p 86 and 98–99.

TONE SWITCH—Uses resonant-reed relay to trigger triac on at one audio frequency and turn it off with higher frequency. Triac shown can handle up to 1.2-kW load on 240-V line or half that on 120 V.—J. H. Galloway, "Using the Triac for Control of AC Power," General Electric, Semiconductor Products Dept., Auburn, NY, No. 200.35, 1970, p 5–6.

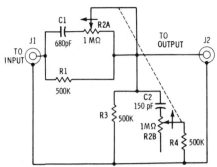

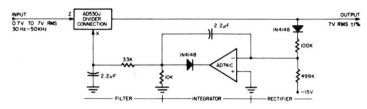

PRESENCE CONTROL—Ganged pots provide continuously variable increase in accentuation of voice frequencies around 1,000 Hz up to limit of 6 dB. Used to bring out voice of vocalist while keeping orchestral accompaniment in background. At maximum setting, vocalist sounds half as far away as with no accentuation.—L. Feldman, "Hi-Fi Projects for the Hobbyist," Howard W. Sams, Indianapolis, IN, 2nd Ed., 1972, p 30–32.

AUTOMATIC LEVEL CONTROL—Combination of single-IC analog multiplier connected as divider and AD741C opamp can be easily incorporated in tape recorders, telephone answering devices, and other audio equipment to compensate for variations in loudness. Also useful for stabilizing outputs of generators, alternators, and oscillators, as well as for controlling industrial processes. Circuit rectifies output signals, compares rectified signal with 15-V reference, filters difference with integrator, and applies resulting d-c output to divider as control signal.—"AD530 Complete Monolithic MDSSR," Analog Devices, Inc., Norwood, MA, Technical Bulletin, July 1971.

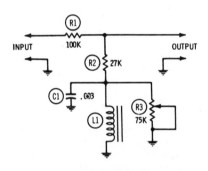

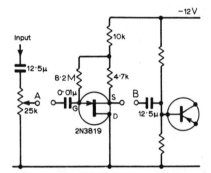

PRESENCE CONTROL—Simple passive filter circuit boosts voice frequencies without affecting bass and treble of musical accompaniment, to make singer stand out even though recorded at same level as accompanying instruments. R3 provides slight adjustment of presence effect. L1 is 10-mA 1.5-H filter choke.—R. Brown and T. Kneitel, "101 Easy Audio Projects," Howard W. Sams, Indianapolis, IN, 1971, p 157–158.

NOISY VOLUME CONTROL—Insertion of 2N3819 source follower between points A and B of conventional volume control circuit reduces noise caused by contact irregularities of control. If circuit uses positive supply line, entire source follower circuit should be inverted.—C. H. Banthorpe, Reducing Noise in Volume Controls, Wireless World, April 1971, p 204.

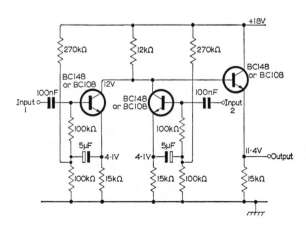

TWO-INPUT MIXER—Two inputs feed two resistors having common collector load resistor. Emitter-follower stage gives low output impedance of 70 ohms, while input impedance is 2.5 meg. Voltage gain for both inputs is unity. With one input shorted and 2 V output from other, distortion is 0.5%; voltages on each input must not exceed 1 V for same distortion.—"Transistor Audio and Radio Circuits," Mullard, London, 1969, TP1069/1, p 126–127.

CHAPTER 7
Automatic Gain Control Circuits

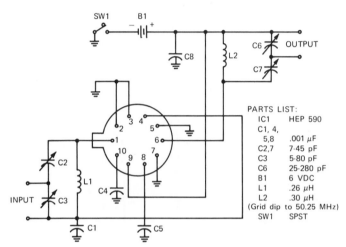

6-METER PREAMP—Provides 30 dB of signal gain with 0.6-MHz bandwidth. Agc may be added to pin 5. Ideal for use as receiver preamp.—"Radio Amateur's IC Projects," Motorola Semiconductors, Phoenix, AZ, HMA-36, 1971.

PARTS LIST:
IC1   HEP 590
C1, 4,
5,8   .001 μF
C2,7   7-45 pF
C3   5-80 pF
C6   25-280 pF
B1   6 VDC
L1   .26 μH
L2   .30 μH
(Grid dip to 50.25 MHz)
SW1   SPST

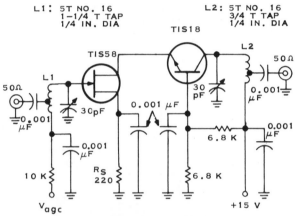

CASCODE WITH 1-MHZ BANDWIDTH—Provides high initial gain of common-source configuration and good gain-reducing agc action of common-gate amplifier. Loading is desired for desired bandwidth at maximum gain. Minimum gain with reverse agc is —13 dB and range with forward agc is greater than 40 dB.—C. L. Farell, "AGC Characteristics of FET Amplifiers," Texas Instruments, Dallas, TX, Bulletin CA-104, 1968, p 6–7.

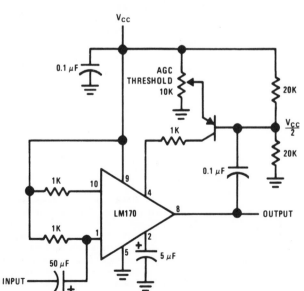

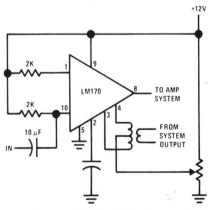

IC WITH TRANSISTOR—Negative-going peaks turn on pnp transistor (type is not critical) momentarily, reducing gain of IC amplifier. Large capacitor at pin 2 charges and holds enough d-c voltage to maintain gain at level set by agc threshold pot. Circuit thus provides agc with peak protection while preserving audio linearity.—R. A. Hirschfeld, Monolithic Amplifier Has AGC and Squelch, The Electronic Egineer, Aug. 1968, p 60–66.

FULL-WAVE AGC DETECTION—Both available agc inputs of opamp are used to provide full-wave output detection, for response to both positive and negative output peak voltages. D-c threshold for detector is set by 10K pot.—R. A. Hirschfeld, "AGC/Squelch Amplifier," National Semiconductor, Santa Clara, CA, 1969, AN-11.

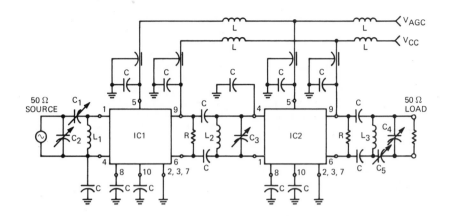

PARTS LIST:
IC1,2   HEP 590
C   .002 µF
C1,2,4,5 .9-35 pF
C3   2-8 pF
L1   0.42 µH
L2   0.68 µH
L3   0.55 µH
L   1 µH
R   510 Ω

45-MHZ WIDEBAND IC—Power gain is 30 dB at center frequency and 3 dB down at 38 and 52 MHz. Both IC amplifiers are agc-controlled by 0 to 4 V d-c.—"Radio Amateur's IC Projects," Motorola Semiconductors, Phoenix, AZ, HMA-36, 1971.

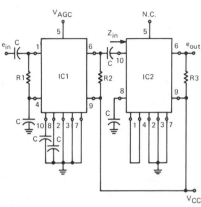

1–22 MHZ AMPLIFIER WITH AGC—With 0 V agc, voltage gain is 25.5 dB at 1 MHz and 3 dB down at 22 MHz. With 4 V agc, gain is about 12 dB. Uses HEP590 IC's. R1 is 47, R2 is 1.8K, R3 3.3K, and C 0.1 µF. Requires no tuned circuits.—"Radio Amateur's IC Projects," Motorola Semiconductors, Phoenix, AZ, HMA-36, 1971.

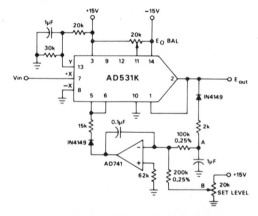

30-DB DYNAMIC RANGE—Maintains 3-V p-p output for inputs from 0.1 to over 12 V p-p, with better than 2% regulation over center half of range and less than 1% distortion. Response is 3 dB down at 30 Hz and 400 kHz. Input can be either single-ended or differential. Uses Analog Devices AD531K multiplier/divider IC with one additional opamp. —30 dB Automatic Gain Control with the AD531, *Analog Dialogue*, Vol. 7, No. 1, p 13.

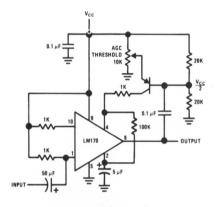

OPAMP CONTROL—External pnp transistor acts as negative peak detector for opamp, with agc threshold set by pot. Transistor is normally off. At level set by pot, negative peaks marginally turn on transistor. As input signal level increases, circuit automatically lowers gain to maintain constant p-p output voltage.—R. A. Hirschfeld, "AGC/Squelch Amplifier," National Semiconductor, Santa Clara, CA, 1969, AN-11.

HIGH-GAIN AMPLIFIER—Four IC video amplifiers provide voltage gain of 78 dB with 10-MHz bandwidth. Response is flat over full agc voltage range of 0–5 V on first stage. Report gives design equations.—"An I/C Wideband Video Amplifier with AGC," Motorola Semiconductors, Phoenix, AZ, AN-299, 1972.

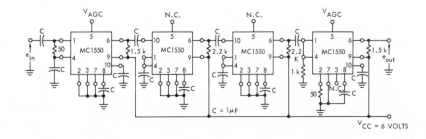

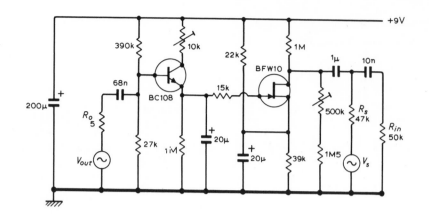

SHUNT AGC FOR A-F—Uses junction fet as voltage-controlled resistance in audio amplifier having 54 dB voltage gain and 3-dB limits of 250 and 3,000 Hz for voice operation. Threshold for agc is 3.3 V peak. Attack time is 20 to 50 ms and recovery time 15 s. Attenuation for 3-dB increase in amplifier output is 40 dB.—R. M. Lea, The Junction F.E.T. as a Voltage-Controlled Resistance, *Wireless World*, Aug. 1972, p 394–396.

SERIES AGC FOR A-F—Uses junction fet as voltage-controlled resistance in voice amplifier having 3-dB limits of 250 and 3,000 Hz. Provides 40 dB attenuation for 3-dB increase in amplifier output. Threshold is 3.3 V peak.— R. M. Lea, The Junction F.E.T. as a Voltage-Controlled Resistance, *Wireless World*, Aug. 1972, p 394–396.

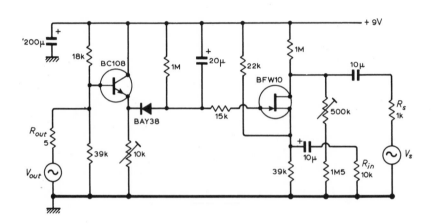

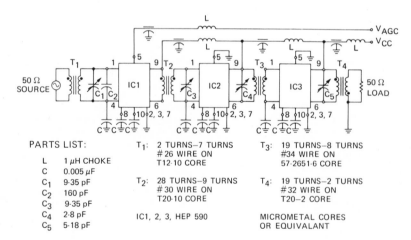

PARTS LIST:

L	1 μH CHOKE	
C	0.005 μF	
C_1	9-35 pF	
C_2	160 pF	
C_3	9-35 pF	
C_4	2-8 pF	
C_5	5-18 pF	

T_1: 2 TURNS–7 TURNS #26 WIRE ON T12-10 CORE

T_2: 28 TURNS–9 TURNS #30 WIRE ON T20-10 CORE

IC1, 2, 3, HEP 590

T_3: 19 TURNS–8 TURNS #34 WIRE ON 57-2651-6 CORE

T_4: 19 TURNS–2 TURNS #32 WIRE ON T20-2 CORE

MICROMETAL CORES OR EQUIVALANT

45-MHZ STAGGER-TUNED—Provides 70 dB of power gain, with agc control. Tuned by sweeping amplifier with sweep generator while observing output on cro, to give bandwidth of 6 MHz, with response 3 dB down at 42 and 48 MHz. Full agc occurs over range of 0–2.5 V.—"Radio Amateur's IC Projects," Motorola Semiconductors, Phoenix, AZ, HMA-36, 1971.

HANG AGC—Allows agc line in ssb or c-w communication receiver to remain at steady negative voltage following voice syllable or code character for predetermined time, after which it triggers discharge circuit which rapidly restores receiver gain to maximum in absence of signal. Requires high-impedance agc line, which makes it ideal for use with fet r-f amplifier. Values of R1 and C1 determine hang time.—J. L. Hartge, Solid-State Hang AGC Circuit for SSB and CW, *Ham Radio*, Sept. 1972, p 50–55.

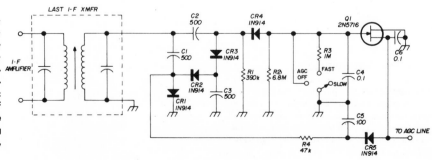

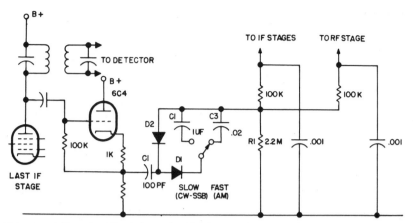

AGC FOR C-W AND SSB—Code and side-band reception are greatly improved by inserting 6C4 cathode follower between final i-f stage and conventional agc rectifier. Capacitor switching gives choice of slow or fast decay times. Addition of series diode gate doubles agc voltage available from cathode follower.—J. W. Herbert, Design for an Improved AGC System for CW and SSB Reception, 73, Jan. 1973, p 115 and 118–120.

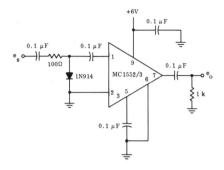

AGC WITH EXTERNAL DIODE—Diode connection shown controls gain even below normal unmodified gain of IC video amplifier. Diode is used as variable impedance in voltage divider network.—"A Wide Band Monolithic Video Amplifier," Motorola Semiconductors, Phoenix, AZ, AN-404, 1971.

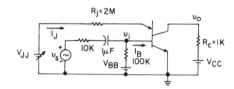

CONTROLLABLE-GAIN STAGE—Uses 3N81 pnpn tetrode. Article gives design equations. With millivolt input signals, distortion is negligible over circuit bandwidth of about 200 kHz. Used in applications where small-signal voltage gain must vary linearly with d-c signal.—N. C. Voulgaris and E. S. Yang, Linear Applications of a p-n-p-n Tetrode, *IEEE Journal of Solid-State Circuits*, Aug. 1970, p 146–150.

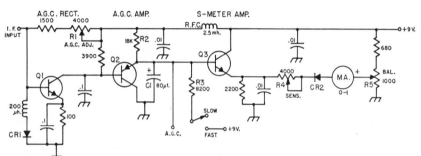

AGC WITH S METER—I-f input to 2N3693 agc rectifier Q1 is taken from capacitive divider across collector winding of last i-f transformer. R1 adjusts amount of delay introduced by voltage drop across CR1. Agc output is d-c voltage output of 2N3638 d-c amplifier Q2. Fast agc release is obtained by switching R3 in parallel with R2. Q3 is 2N3567, and diodes are 1N541 silicon.—Hammarlund HQ-215 Receiver, QST, Dec. 1968, p 50–54.

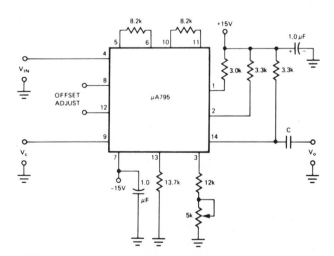

VOLTAGE-CONTROLLED GAIN—Positive control voltage of 0.05 to 5 V at Vc changes gain of IC analog multiplier over range of 100 or more. Negative control voltage inverts output and repeats gain control range. Signal input can be up to 5 V peak of either polarity. With a-c voltage control, gain can be varied at rate up to several hundred kHz.—"The µA795, A Low-Cost Monolithic Multiplier," Fairchild Semiconductor, Mountain View, CA, No. 211, 1971, p 2.

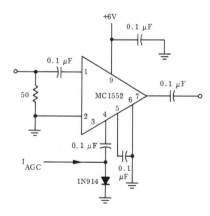

AGC WITH EXTERNAL DIODE—Diode connection shown for IC video amplifier provides wide range of gain control without using variable resistors. Diode type is not critical. Voltage gain increases from 60 to 350 V/V for agc control current increase from 10 to 4,000 µA.—"A Wide Band Monolithic Video Amplifier," Motorola Semiconductors, Phoenix, AZ, AN-404, 1971.

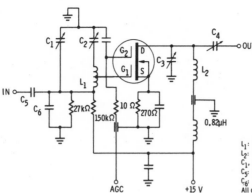

VHF AMPLIFIER—Uses Fairchild FT0601 dual-gate mosfet, which has separate gates to permit agc at front end of receiver for improving reception over wide range of input signal levels. Other advantages are reduced noise, stable center frequency, and elimination of cross-modulation distortion.—S. Sir, RF Applications of the Dual-Gate Mosfet, 73, Nov. 1970, p 22–25.

L_1: 15/64" ID coil, 0.8" long, 5t of gauge 18 wire, silver plated, tapped at 2.5t
L_2: 15/64" ID coil, 0.8" long, 4t of gauge 22 wire, silver plated, tapped at 2.5t
C_1, C_2, C_3, C_4: 1–10pF, Erie VAM010
C_5: 100pF, dip mica
C_6: 27pF, dip mica
All other capacitors: 1000pF, FT

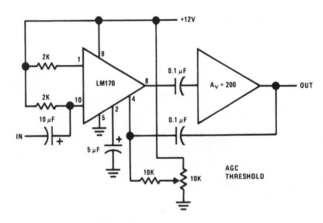

PEAK DETECTION WITH OPAMP—Tight regulation of output level of IC audio amplifier is achieved by taking control signals from output of external opamp having gain of 200. Positive peak excursions at output provide agc voltage for regulating gain.—R. A. Hirschfeld, Monolithic Amplifier Has AGC and Squelch, *The Electronic Engineer*, Aug. 1968, p 60–66.

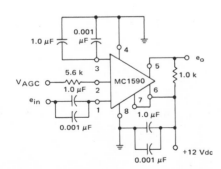

50-MHZ BANDWIDTH—Uses IC providing about 25 dB single-ended voltage gain with 100-ohm load and about 45 dB gain for 1K load. Has excellent agc capability. Cascading of two IC circuits will increase gain.—B. Trout, "A High Gain Integrated Circuit RF-IF Amplifier with Wide Range AGC," Motorola Semiconductors, Phoenix, AZ, AN-513, 1969.

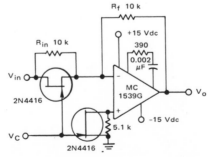

SIMPLE FET CONTROL—Nominal voltage gain of 30 dB can be made variable by changing ratio of Rf to Rin. Use of opamp with fet pair reduces distortion. Input voltage Vin is about 10 mV rms and control voltage VC 0–6 V d-c.—B. Botos, "Low Frequency Applications of Field-Effect Transistors," Motorola Semiconductors, Phoenix, AZ, AN-511, 1971.

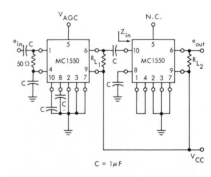

TWO-IC AMPLIFIER—Provides overall voltage gain of about 57 dB with bandwidth of about 10 MHz with only two IC video amplifiers. Agc voltage range for first stage is 0–6 V d-c. Bandpass is flat over full agc range. Vcc is 6 V d-c.—"An I/C Wideband Video Amplifier with AGC," Motorola Semiconductors, Phoenix, AZ, AN-299, 1972.

CHAPTER 8
Automotive Control Circuits

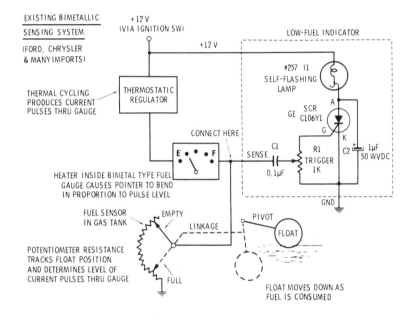

LOW-GAS ALARM—Designed for cars having bimetallic gage system. R1 can be set to make scr turn on at any desired reading of fuel gage. When fuel level falls to trigger point, lamp flashes until fuel is added. Mount lamp on dash.—R. F. Graf and G. J. Whalen, "Automotive Electronics," Howard W. Sams, Indianapolis, IN, 1971, p 150–151.

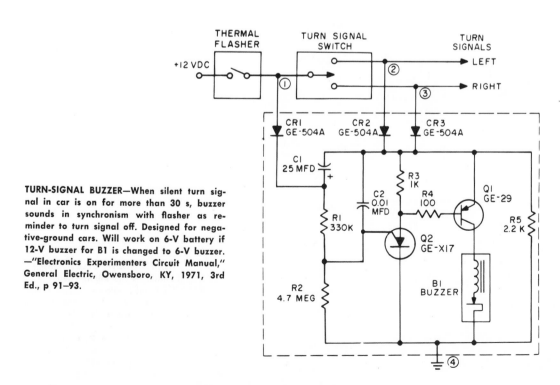

TURN-SIGNAL BUZZER—When silent turn signal in car is on for more than 30 s, buzzer sounds in synchronism with flasher as reminder to turn signal off. Designed for negative-ground cars. Will work on 6-V battery if 12-V buzzer for B1 is changed to 6-V buzzer. —"Electronics Experimenters Circuit Manual," General Electric, Owensboro, KY, 1971, 3rd Ed., p 91–93.

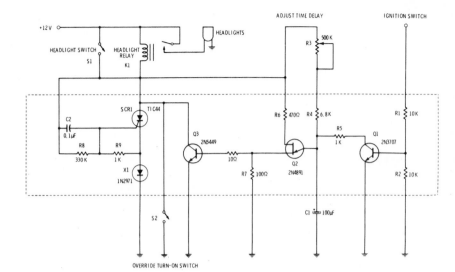

15-MIN HEADLIGHT-OFF DELAY—R3 gives range of 1 s to 15 min for delay after ignition switch in car is turned off, before SCR1 turns off and relay K1 drops out. Developed by Texas Instruments. Timing circuit uses ujt Q2. For manual turnoff, open S1 and momentarily close S2.—R. F. Graf and G. J. Whalen, "Automotive Electronics," Howard W. Sams, Indianapolis, IN, 1971, p 120–121.

POWER-DRAIN REMINDER—Horn or buzzer sounds if radio or any other current-drawing equipment is left on when ignition is turned off. Circuit draws current only during alarm conditions. If ignition is initially off, lights can be used without tripping alarm. Both 12-V relays operate when ignition is turned on, with K1C closing alarm circuit and K2B opening it. If lights are not needed when ignition is off, use spst relay for K1 and replace K1B with jumper.—V. A. Damora, Power-Drain Reminder, QST, Aug. 1969, p 48.

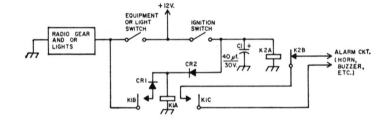

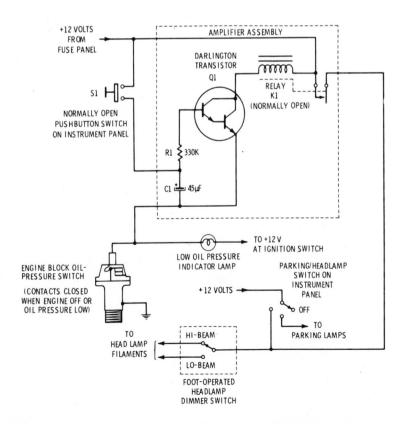

HEADLIGHT-OFF DELAY—Closing S1 delays shutoff of headlights about 1 minute after light and ignition switches are turned off. Circuit is armed by contacts on oil-pressure switch, which close when engine is off. Offered as option in 1970 General Motors cars.—R. F. Graf and G. J. Whalen, "Automotive Electronics," Howard W. Sams, Indianapolis, IN, 1971, p 115–117.

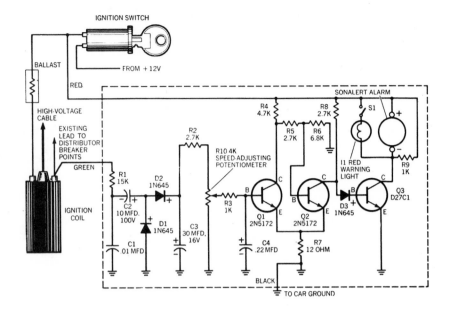

SPEED ALARM—Activates Sonalert alarm and red warning light whenever speed exceeds predetermined value determined by setting of R10. Circuit senses pulses from breaker points. Requires only three connections—to ignition coil, ignition switch, and car ground. Use 12-V 100-mA lamp and Mallory SC628P alarm.—G. J. Whalen and R. F. Graf, Save Yourself a Ticket with This Auto Speed Alarm, *Popular Mechanics*, June 1970, p 132–133.

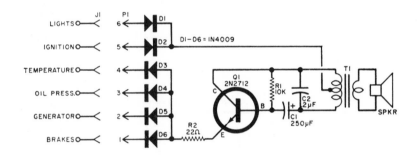

IDIOT-LIGHT ALARM—Beep tone warns driver that oil pressure, temperature, or other warning light has come on, and also warns if headlights or parking lights are left on when ignition is turned off. Circuit is for 12-V negative-ground system. Operates as modified Hartley oscillator that turns itself on and off to feed tone through transistor output transformer T1 to small 8-ohm speaker.—J. S. Simonton, Jr., Build the Automobile Omni-Alarm, *Popular Electronics*, Aug. 1968, p 51–53 and 105.

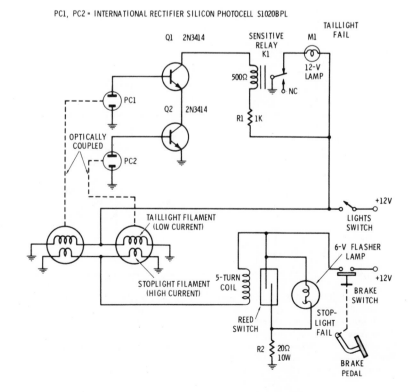

REAR-LAMP FAILURE ALARM—Turns on lamp M1, mounted on or near dash, if one of taillights fails. More serious failure of stoplight, which then gives erroneous turn-signal indication, makes M1 flash a warning. Uses silicon photocells mounted inside each rear-end light, with active face of each turned toward lamp so it is not affected by outside light sources. Failure of light on either photocell makes relay K1 drop out. Installation details are given.—R. F. Graf and G. J. Whalen, "Automotive Electronics," Howard W. Sams, Indianapolis, IN, 1971, p 297–299.

LOW-FUEL ALARM—Makes red self-flashing lamp I comes on when fuel level in tank reaches predetermined near-empty level depending on setting of R1. Applies only to GM cars, which use electromagnetic fuel gage.—R. F. Graf and G. J. Whalen, Fill 'Er Up! The Blinker Says So, *Popular Science*, Nov. 1970, p 95 and 121.

R1—5,000-ohm, ½w variable resistor
R2, R6—2,700-ohm, ½w carbon resistor
R3, R7—10,000-ohm, ½w carbon resistor
R4—4,700-ohm, ½w carbon resistor
R5—120-ohm, ½w carbon resistor
R8—1-ohm, 1w carbon resistor
Q1, Q2—Sprague 2N2923 or equiv.
Q3—Motorola HEP 242 or equiv.
LM—#257 self-flashing bulb with red tint (Inventive Electronics, Wykagyl Station, N.Y. 10804; $1 pp. for 2)

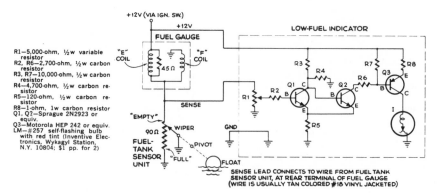

0-12,000 RPM TACHOMETER—Uses electromagnetic sensor requiring no mechanical coupling. Ferrous pointer on engine shaft sweeps past 3,000-turn search coil wound on permanent-magnet core 4 inches long and ⅛th inch in diameter, inducing pulse once per revolution. Pulse operates Schmitt trigger Tr2-Tr3 through preamp Tr1, and trigger in turn operates mvbr Tr4-Tr5 in which Tr5 is normally off. Each input pulse turns Tr5 on, to make reading of meter directly proportional to input frequency.—S. L. V. Chari and M. R. K. Rao, Electronic Tachometer, *Wireless World*, Feb. 1967, p 81–83.

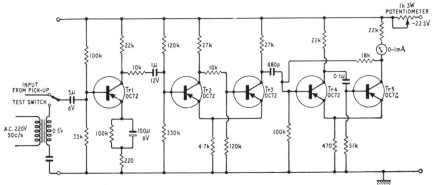

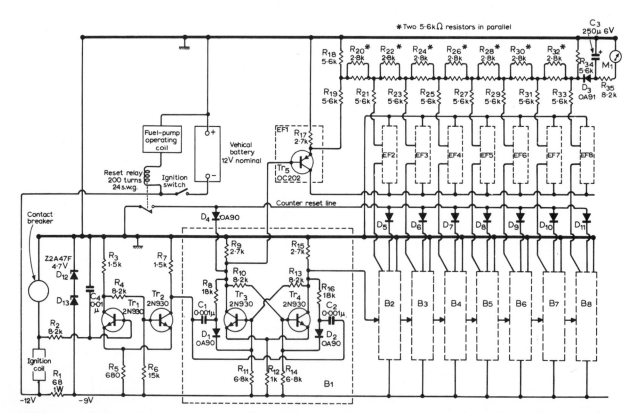

MILES-PER-GALLON METER—Provides continuous indication with accuracy within 10%. Although article gives elaborate calibration procedure required for indicating miles per gallon, can also be used without calibration to obtain maximum fuel economy by driving to get maximum meter deflection. Suitable only for cars having electric fuel pump and positive ground. Input to circuit is voltage across contact breaker (distributor contacts) of car, with low-pass filter R2-C4 inserted between breaker and Schmitt trigger TR1-TR2 to suppress damped oscillations due to inductive load of ignition coil. Eight-stage binary counter gives maximum count of 255. Counter is reset to zero each time fuel-pump solenoid is energized, with M1 (0.2 mA and 380 ohms) holding reading until next reset.—S. C. Hambly, Miles-Per-Gallon Meter, *Wireless World*, May 1971, p 218–220.

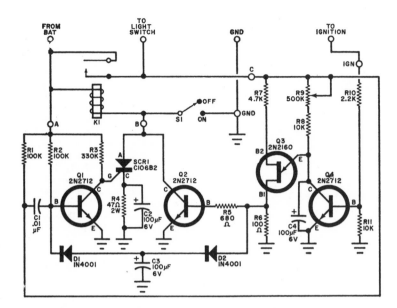

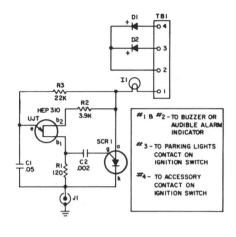

HEADLIGHT TURNOFF DELAY—Setting of R9 gives delay range of several seconds to several minutes between turning off of ignition and automatic turnoff of headlights after leaving car in dark driveway. Turning on headlights grounds junction of R1 and C1 through lights, making C1 discharge negative pulse that turns off Q1 momentarily, turning on SCR1 and pulling in 6-V relay K1 to activate timer. Closing ignition switch disables timer Q3-Q4. Opening ignition switch activates timer after delay set by charging time of C4, to reverse-bias SCR1 through Q2 and drop out relay, turning off headlights.—J. Stayton, Build the Time Out, *Popular Electronics*, Jan. 1970, p 52–53 and 58–59.

HEADLIGHT-ON ALARM—Designed for negative-ground cars. Ujt is relaxation oscillator with pulse frequency of about 1 kHz, determined by R3 and C1. Each pulse turns on scr, but alarm buzzer sounds only if lights or accessories are left on when driver opens door and thus turns on dome or courtesy light. Door switch is connected to jack J1. Scr is Motorola MCR1906-2. Article covers installation in car.—J. J. Carr, Teeny Genii, *Elementary Electronics*, July-Aug. 1970, p 79–81 and 98.

DWELL-TIME INCREASER—When points open to produce spark, time delay of scr starts. At end of delay, about 100 μs later, scr is turned on to simulate closing of breaker points and thus restore current flow through ignition coil. Circuit permits increasing dwell angle up to 70% at highway speeds and up to 40% at maximum speed. Used to improve gas mileage, boost power, and give smoother performance at high speed. R1 is 1.2K, R2 1K, R3 1.8K, capacitors 0.1 μF, and scr is 2N4103. Experiment with value of R1, down to 680 ohms, for smoothest engine performance at all speeds—R. M. Benrey, Dwell Extender for Your Car, *Popular Science*, Oct. 1971, p 104–105.

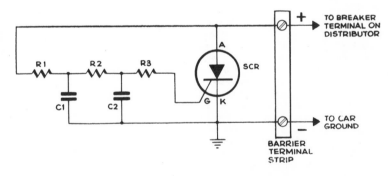

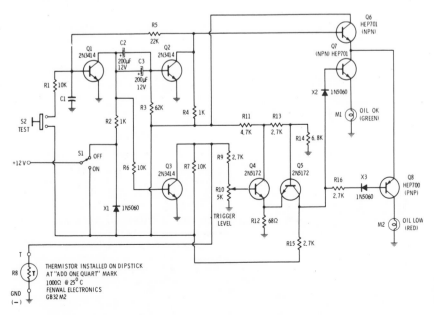

OIL-LEVEL CHECKER—Uses glass-bead thermistor mounted on oil-level dipstick at point indicating need for oil. Pressing test button starts 25-s test program. If oil is above thermistor, green lamp comes on, stays on half a minute, and goes out. If green lamp winks out and red lamp comes on, oil level is low and one quart should be added. Circuit passes current through thermistor, generating heat that dissipates in oil but not in air, giving resistance change that triggers circuit.—R. F. Graf and G. J. Whalen, "Automotive Electronics," Howard W. Sams, Indianapolis, IN, 1971, p 310-313.

LOW-FUEL INDICATOR
FOR ELECTROMAGNETIC GAUGE SYSTEM

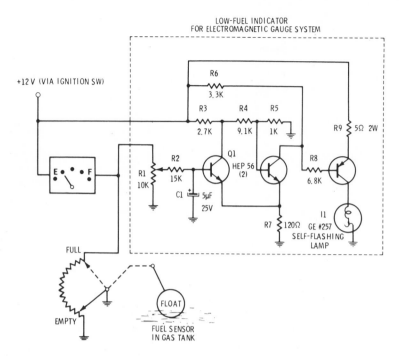

LOW-GAS ALARM—Designed for cars having electromagnetic gage system. Schmitt trigger senses d-c voltage level across R1, proportional to fuel level in tank. When fuel level reaches desired alarm point, as determined by float and setting of R1, Q1 is turned off and other two transistors turned on for energizing of self-flashing lamp mounted on or near dash.—R.F. Graf and G. J. Whalen, "Automotive Electronics," Howard W. Sams, Indianapolis, IN, 1971, p 151.

ICY-ROAD WARNING—Thermistor is mounted on car to sense ambient air temperature, to actuate alarm circuit that alerts driver when air temperature is in range of 32 and 36 F in wet weather at which ice will start forming on road. Uses quad opamp to generate varying-duty-cycle output for controlling flash duration of light-emitting diode. A2 operates as free-running mvbr. Flashes occur about once a second, starting at 36 F. Flash duration increases as temperature drops, with led on continuously at 32 F. To calibrate, place thermistor probe in ice slurry and adjust R2 to keep led on.—S. E. Summer, Ice Warning Indicator Monitors Road Conditions, *Electronics*, June 7, 1973, p 106.

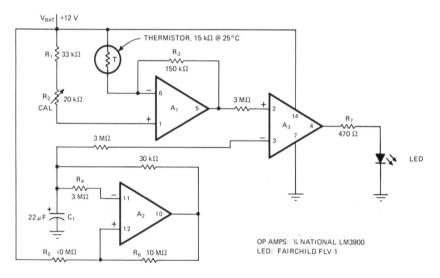

OP AMPS: ¾ NATIONAL LM3900
LED: FAIRCHILD FLV-1

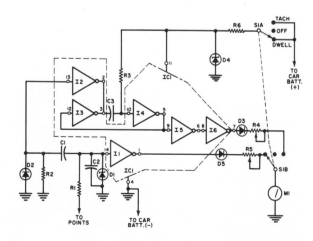

R1—1000 ohm, ½ W res.
R2—10,000 ohm, ½ W res.
R3—12,000 ohm, ½ W res.
R4, R5—5000 ohm trimmer (Mallory MTC-4)
R6—180 ohm, 1 W res. (12-V operation), or 56 ohm, ½ W res. (6-V operation).
C1—0.01 µF capacitor
C2—0.047 µF capacitor
C3—0.22 µF capacitor

D1, D4—3.9 V zener diode (1N4730)
D2—Germanium diode (1N34, 1N54, or HEP134)
D3, D5—Silicon diode (1N4001 or HEP154)
IC1—Hex inverter (Motorola MC789P)
M1—0-1 mA meter (Lafayette 99-5040)
S1—D.p.d.t. switch, center "off"

TACHOMETER WITH DWELL—Meter has 0–60 scale, for reading degrees dwell directly and rpm × 100. S1 gives choice of readings. Article covers construction and calibration for engines with 4, 6, and 8 cylinders.—J. Colt, Dynamic Dwell/Tachometer, *Electronics World*, May 1971, p 78–80.

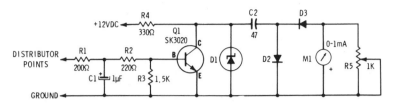

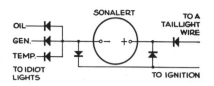

NEGATIVE-GROUND TACHOMETER—Accuracy is good enough for engine tuneup. Connect R1 to lead running between distributor and ignition coil. Calibrate against commercial tachometer at garage, adjusting R5 so meter gives correct reading. Full scale will be about 10,000 rpm. D1 is 9.1-V zener and other diodes are 100-mA 50-piv silicon.—H. Friedman, "99 Electronic Projects," Howard W. Sams, Indianapolis, IN, 1971, p 24—25.

IDIOT-LIGHT REMINDER—Simple circuit energizes Mallory Sonalert alarm whenever an idiot light is supposed to come on, even if light is burned out. Alarm also comes on as reminder to turn off headlights after ignition switch is turned off. Diodes can be any types rated over 25 V piv and over 20 mA forward current.—R. P. Tackett, Reminder to Turn Off Headlights, *Popular Science*, July 1971, p 8.

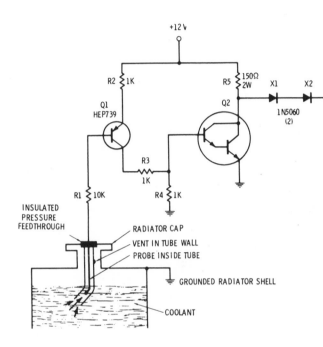

RADIATOR WATER-LEVEL ALARM—Uses feedthrough insulator installed in radiator cap, capable of withstanding pressure range of up to 20 lb per square inch in pressurized radiators. Metal probe goes through insulator down to liquid. Probe is surrounded by high-temperature tubing vented at cap end so liquid can rise inside and reach probe. When liquid level drops to low probe, collector voltage of Darlington stage Q2 rises and makes diodes conduct, so warning lamp M1 comes on.—R. F. Graf and G. J. Whalen, "Automotive Electronics," Howard W. Sams, Indianapolis, IN, 1971, p 302—304.

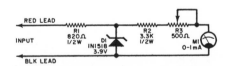

ZENER DWELLMETER—Provides accurate measurements of dwell angle. Black lead goes to chassis and red to points side of distributor coil for negative-ground cars; reverse for positive ground. Adjust R3 for full-scale reading initially with leads across car battery. Article covers conversion of meter reading to degrees for cars with 8, 6, and 4 cylinders.—S. Wald, Build a Batteryless Dwell Meter, *Popular Electronics*, Nov. 1967, p 40.

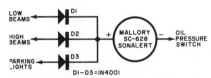

BATTERY PROTECTION—Based on fact that oil pressure switch is normally closed when engine is not running, and thus provides convenient current path to ground for energizing Sonalert alarm if any lights are left on when ignition key is removed.—S. J. Erst, Light Minder, *Popular Electronics*, Sept. 1972, p 91.

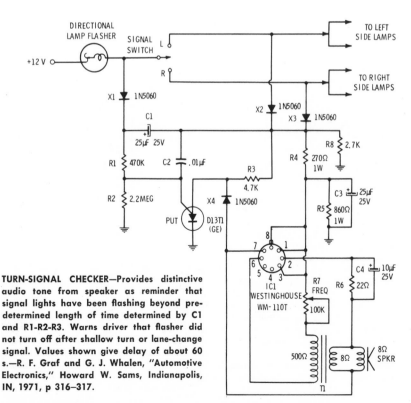

TURN-SIGNAL CHECKER—Provides distinctive audio tone from speaker as reminder that signal lights have been flashing beyond predetermined length of time determined by C1 and R1-R2-R3. Warns driver that flasher did not turn off after shallow turn or lane-change signal. Values shown give delay of about 60 s.—R. F. Graf and G. J. Whalen, "Automotive Electronics," Howard W. Sams, Indianapolis, IN, 1971, p 316—317.

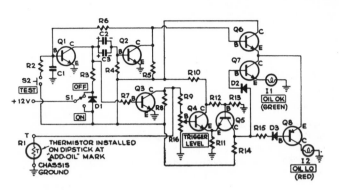

Q1, Q2, Q3—GE 2N3414 NPN
Q4, Q5—GE 2N5172 NPN
Q6, Q7—GE D27C1 NPN
Q8—GE D27D1
D1, D2, D3—GE 1N5060
R1—Glass-probe thermistor, 1,000 ohms @ 25° C, Fenwal Electronics type GB32M2
R2 through R15 below ½ watt, ±10%, carbon resistors
R2, R7—10,000 ohm
R3, R5, R8—1,000 ohm
R4—62,000 ohm
R6—22,000 ohm
R9, R12, R14, R15—27,000 ohm
R10—47,000 ohm
R11—68 ohm
R13—68,000 ohm
R16—5,000-ohm potentiometer Mallory, MTC-53L1
C1—1 mfd, 200-volt capacitor
C2, C3—200 mfd, 12 WVDC electrolytic

OIL-LEVEL TESTER—Thermistor mounted on dipstick at add-oil level will not rise as high in temperature when immersed in oil as when in air. Timing circuit Q1-Q2 applies correct heating current to thermistor for 25 s when test button S2 is pressed. Remainder of circuit then measures resistance of thermistor and turns on green or red 12-V lamp to indicate whether or not oil is needed. Entire unit is mounted under dash, with wires running to dipstick.—G. J. Whalen and R. F. Graf, Dipstick Oil-Level Indicator You Can Make, *Popular Science*, June 1969, p 116–119.

0–8,000 RPM TACH—Uses mono mvbr triggered by opening of breaker points. Filter R1-R3–C1-C2 differentiates breaker-point waveform and strips away ringing and overshoots. Since breaker-point pulse width varies with rpm but output of mvbr does not, average voltage at collector of Q2 is truly proportional to rpm, as also is reading of M1. Calibrate against accurate tachometer after checking both against 60-Hz line frequency. —R. F. Graf and G. J. Whalen, "Automotive Electronics," Howard W. Sams, Indianapolis, IN, 1971, p 141–143.

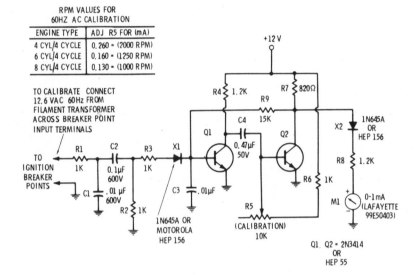

RPM VALUES FOR 60HZ AC CALIBRATION

ENGINE TYPE	ADJ R5 FOR (mA)
4 CYL/4 CYCLE	0.260 = (2000 RPM)
6 CYL/4 CYCLE	0.160 = (1250 RPM)
8 CYL/4 CYCLE	0.130 = (1000 RPM)

TO CALIBRATE CONNECT 12.6 VAC 60Hz FROM FILAMENT TRANSFORMER ACROSS BREAKER POINT INPUT TERMINALS

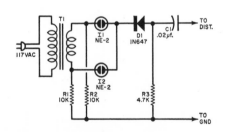

TACH TESTER—Pulses from ignition coil are keyed to 60-Hz power line to make neon lamps blink out calibration signal for automobile tachometers. Rate of flash decreases as engine speed approaches either 450 or 900 rpm for eight-cylinder car, or 600 or 1,200 rpm for 6-cylinder car. When each lamp flashes about once per second, engine speed is off by only 1.7%. When exactly at one of test speeds, one lamp stays on and the other off. Will not normally work with electronic ignition.—J. S. Shreve, Tachometer and Engine Idle Speed Calibrator, *Popular Electronics*, Feb. 1965, p 54–55 and 97.

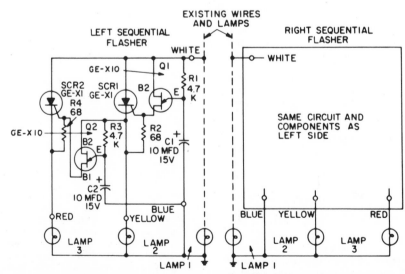

SEQUENTIAL FLASHER—Flashes banks of three lamps at right and left rear of car sequentially in direction of turn to enhance attention-getting value of turn signal. All lamps have same rating, not exceeding 4 A each.

Construction and installation details are given.—"Electronics Experimenters Circuit Manual," General Electric, Owensboro, KY, 1971, 3rd Ed., p 95–100.

Q1—GE 2N4989 SUS
Q2—GE 2N5172 transis-
tor
Q3—GE 2N3415 transis-
tor
Q4—GE D27D1 (D43C1)
transistor
C1—25-mfd, 20-volt elec-
trolytic capacitor, Mal-
lory MTA 25D20, for
operation below 0° F.
R1, R2—220-ohm, ½-watt
carbon resistor
R3, R4—10,000-ohm, ½-
watt carbon resistor
R5—470-ohm, ½-watt
carbon resistor
R6—22-ohm, 2-watt car-
bon resistor
L—#1156 lamp

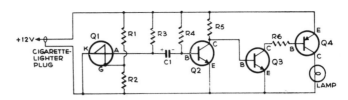

BREAKDOWN BLINKER—Three transistors and silicon unilateral switch Q1 flash 32-cp lamp behind red truck signal lens reliably as emergency warning when auto breaks down or tire needs changing. Will operate even when auto battery drops below 10 V. Use with 20-ft extension cord going to plug fitting into cigar lighter on dash. To reduce blinking rate and increase duration of flash, increase values of R3 and R4.—J. Stogel, Build a Better Blinker for Highway Safety, *Popular Science*, Aug. 1969, p 166–167 and 190.

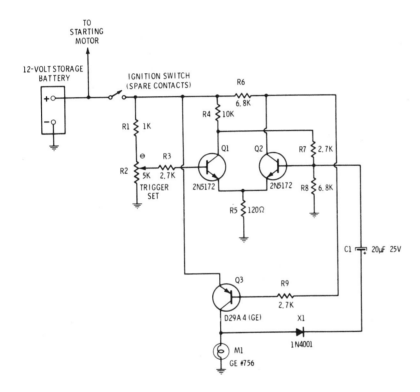

BATTERY MONITOR—Provides check of battery condition each time ignition is turned on, by means of spare contacts on ignition switch. Screwdriver-adjust control R2 is set so Q1 conducts if battery voltage across R1 and R2 is at least 9.2 V while starter is being used. Q2, Q3, and warning lamp M1 are then off. If voltage drops below 9.2 V, M1 comes on and is held on long enough to be seen because of feedback through X1 and C1 to base of Q2.—R. F. Graf and G. J. Whalen, "Automotive Electronics," Howard W. Sams, Indianapolis, IN, 1971, p 308–310.

BRAKE-FLUID MONITOR—Uses two probes sealed into master-cylinder reservoir cap and insulated at all points except at tips. Brake fluid is good conductor, so probes are normally grounded. Under these conditions, pushing S1 makes low-current test lamp M2 glow brightly while flasher M1 stays dark. When one of the master-cylinder chambers loses fluid, circuit triggers SCR1 on and shorts out M2, energizing flasher lamp when S1 is closed.—R. F. Graf and G. J. Whalen, "Automotive Electronics," Howard W. Sams, Indianapolis, IN, 1971, p 306–308.

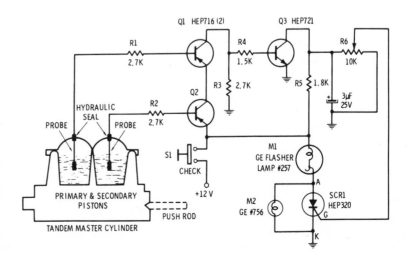

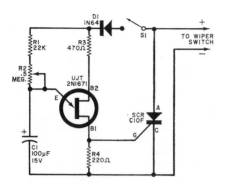

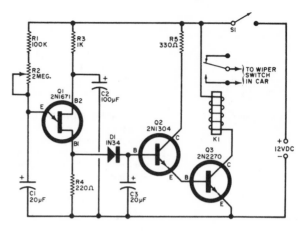

SLOW-KICK WIPER—Uses scr instead of relay to give adjustable slow sweep of wiper for so-so weather, to eliminate need for turning wiper on and off. Ujt operates as one-shot pulse generator whose rate is controlled by R2.—J. J. Albers, Reader's Circuit, *Popular Electronics*, Dec. 1968, p 70–71.

SLOW-KICK WIPER—With relay contacts connected to electric wiper switch, relay is energized momentarily at rate determined by setting of R2, to allow wipers to make one complete sweep in range of once every 2 s to once every 50 s. Prevents squeaking of wipers in light drizzle. K1 is 150 to 250-ohm 12-V relay.—D. K. Belcher, Slow Kick Your Windshield Wipers, *Popular Electronics*, March 1969, p 59–61.

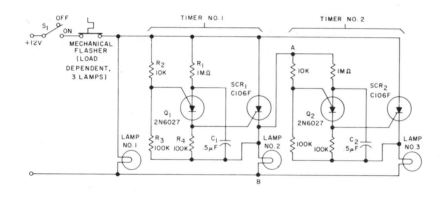

SEQUENTIAL AUTO TURN SIGNAL—Combination of mechanical flasher and two solid-state timers makes three lamps flash in sequence moving from left to right or right to left, as called for by control lever on steering wheel. Lamp 1 comes on first, followed by lamps 2 and 3. Flasher for other direction is identical, with positions of lamps 1 and 3 interchanged at rear of car.—"SCR Manual," General Electric, Syracuse, NY, 1972, 5th Ed., p 207–208.

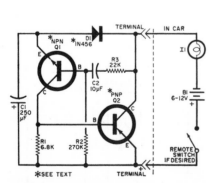

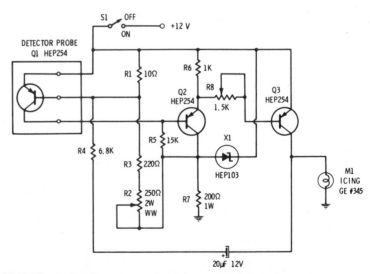

BLINKING IDIOT—Flasher circuit, connected in series with any warning lamp in car, converts its easily overlooked glow to attention-getting blinks. Transistor may be almost any type, except that maximum collector current of Q2 must be at least three times current required by lamp I1 being blinked. D1 is not critical. Flash rate can be slowed up by increasing value of C1 or C2.—M. Chan, Automatic Light Blinker, *Popular Electronics*, Aug. 1967, p 56.

ROAD-ICING ALARM—Germanium transistor Q1, mounted in waterproof case at front of car for sensing air temperature about 1 ft above road, changes emitter-base junction resistance linearly with temperature change from −25 to +25 C. R2 is carefully adjusted so circuit turns on warning lamp M1 when Q1 is 36 F (ice-forming region, corresponding to about 3 C).—R. F. Graf and G. J. Whalen, "Automotive Electronics," Howard W. Sams, Indianapolis, IN, 1971, p 299–301.

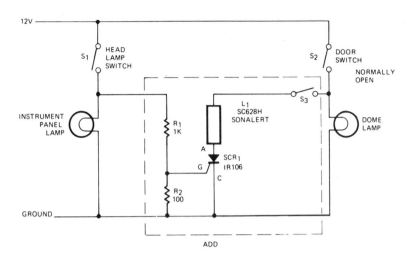

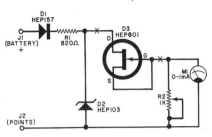

HEADLIGHT REMINDER—Activates Sonalert buzzer if headlight switch is on when door-switch closes as driver leaves car. Prevents running down battery by leaving headlights on, assuming driver hears buzzer and goes back to turn them off. Kill-switch S3 need be included only if there may be occasions when it is desired to leave headlights on while door is open. Closing door turns off scr and buzzer.—"Hobby Projects," International Rectifier, El Segundo, CA, Vol. II, p 18–20.

FET DWELLMETER—Will check performance of 4, 6, or 8-cylinder engines with negative-ground batteries and conventional, transistor, or capacitor-discharge ignition. Uses fet connected as constant-current diode to serve as resistance whose value increases automatically with voltage. Meter scale is calibrated as per instructions in article to read dwell time directly in degrees.—J. Saddler, Simplicity Dwell Meter, *Popluar Electronics*, Aug. 1969, p 33–35.

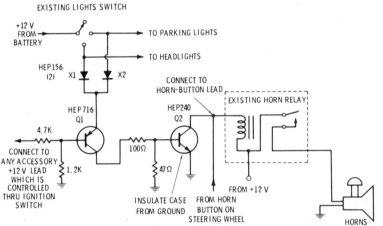

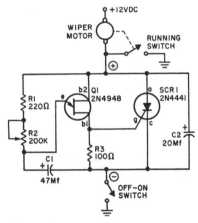

HEADLIGHT REMINDER—Sounds horn if ignition is turned off while headlights are on. Can be mounted in aluminum box on firewall under hood.—R. F. Graf and G. J. Whalen, "Automotive Electronics," Howard W. Sams, Indianpolis, IN, 1971, p 128–130.

WIPER-TIMER—Provides single-knob dashboard control of time wiper rests after a sweep, to match cleaning action to light drizzle. R2 controls amount of delay between sweeps. Circuit is connected in series with manual switch lead going to wiper motor.—J. E. Bjornholt, Wiper-Trol, *Elementary Electronics*, Jan.-Feb. 1971, p 35–36 and 39–40.

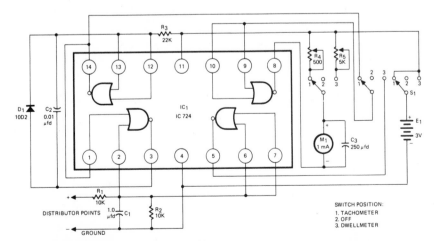

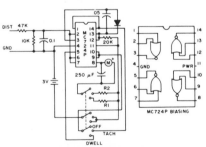

TACH-DWELLMETER—Single IC available from International Rectifier minimizes number of components required and simplifies construction. Permits fast, efficient tuneup of 4, 6, or 8-cylinder engines. Construction details are given, along with calibration procedure and tachometer test circuit. Two Eveready 1.4-V mercury cells in series can be used for E1. Can be used only on negative-ground cars.—"Hobby Projects," International Rectifier, El Segundo, CA, Vol. II, p 47–52.

TACHOMETER-DWELLMETER—Calibration is adjusted with R1 and R2. Adjust R1 to give 900-rpm reading on meter when 60-Hz voltage is fed in while switch is in TACH position. In DWELL position, adjust R2 to give full-scale reading with input terminals shorted, corresponding to 45 deg for eight cylinders, 60 deg for six, and 90 deg for four.—J. Kyle, The Logical Approach to Surplus Buying, 73, March 1970, p 80–92.

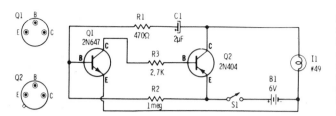

ANTENNA-TIP BLINKER—Helps locate your car in large parking lot at night after theater or ballgame is over. No. 49 pilot lamp mounted on top of radio antenna is connected to control unit inside car with two wires. Value of C1 determines blink rate produced by mvbr Q1-Q2.—H. Friedman, "99 Electronic Projects," Howard W. Sams, Indianapolis, IN, 1971, p 25–26.

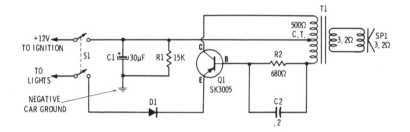

HEADLIGHT REMINDER—Produces own warning sound from speaker if lights are on after ignition is turned off. When lights and ignition are on, collector and emitter of Q1 are both positive so tone generator stays off. D1 is 500-mA 50-piv silicon rectifier.—H. Friedman, "99 Electronic Projects," Howard W. Sams, Indianapolis, IN, 1971, p 27–28.

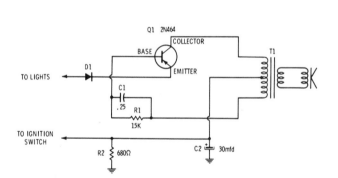

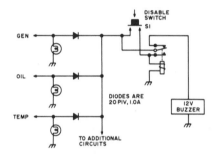

HEADLIGHT REMINDER—Speaker howls if headlights are accidentally left on after ignition is turned off. Will operate in cars having either 6-V or 12-V battery. D1 is 1N38B and T1 is output transformer with 400-ohm C-T primary and 8 or 11-ohm secondary for 8-ohm p-m speaker.—R. M. Brown and T. Kneitel, "49 Easy Transistor Projects," Howard W. Sams, Indianapolis, IN, 1972, p 33–34.

AUDIBLE IDIOT LIGHTS—Buzzer sounds when one of warning lights on dash comes on, to attract attention of driver. Disable switch shorts normally open contacts and energizes relay, to eliminate distracting sound of buzzer while driving to garage for repair. Once fault is corrected, alarm is reset automatically. Alarm will sound momentarily each time car is started, to indicate that it is working.—D. J. Holford, Educated Idiot Lights, 73, May 1970, p 78–79.

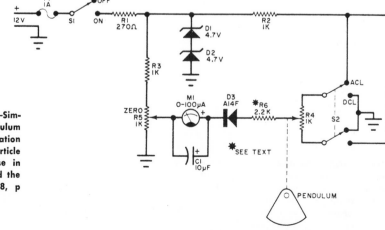

ACCELERATION AND BRAKING TESTER—Simple voltage-stabilized bridge with pendulum attached to shaft of R4 indicates acceleration or deceleration up to 1 G on meter. Article covers construction, calibration, and use in road testing of cars.—G. J. Whalen, Build the G-WHIZ, *Popular Electronics*, Sept. 1968, p 29–35 and 40.

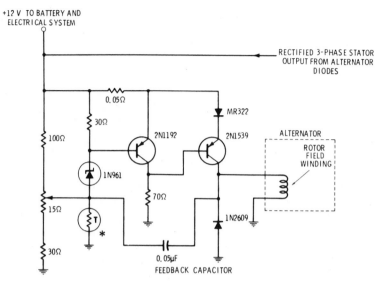

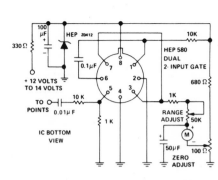

PRECISION TACH—Can be used on motorcycle or boat as well as auto. Single 50K pot provides range adjustment for any number of cylinders. Make zero adjustment for meter with ignition switch on but engine off. Calibrate against tach at service station. Uses HEP580 dual 2-input NOR gate and HEP Z0412 9.1-V zener.—"Tips on Using IC's," Motorola Semiconductors, Phoenix, AZ, HMA-32, 1972.

ALTERNATOR VOLTAGE REGULATOR—Solid-state regulator eliminates arcing contacts that often cause failure. Thermistor provides temperature compensation. Developed by Motorola.—R. F. Graf and G. J. Whalen, "Automotive Electronics," Howard W. Sams, Indianapolis, IN, 1971, p 47–48.

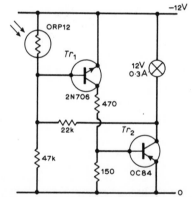

PARKING-LIGHT TURN-ON—Daylight keeps both transistors off. Prolonged darkness turns them on, energizing parking lamp of car. Circuit hysteresis is sufficient to prevent spurious triggering by passing vehicles.—P. Lacey, Automatic Car Parking Lamp, *Wireless World*, June 1972, p 274.

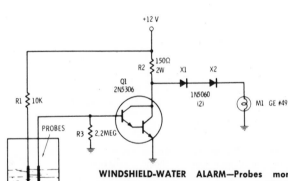

WINDSHIELD-WATER ALARM—Probes monitor water level in reservoir for windshield-wiper fluid, and turn on warning lamp M1 when absence of liquids between probes removes base current of transistor. Will work with detergent/antifreeze solutions and with most ordinary waters. Probes should extend within 1 to 2 inches from bottom.—R. F. Graf and G. J. Whalen, "Automotive Electronics," Howard W. Sams, Indianapolis, IN, 1971, p 301–304.

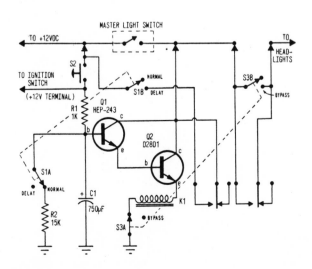

HEADLIGHT SWITCH-OFF—If headlights are not switched off within about 20 s after ignition in car is turned off, Darlington amplifier Q1-Q2 releases relay K1 to turn them off automatically. K1 is Guardian 200-12D. With S1 in delay position, lights stay on about 2.5 min after leaving car, for lighting driveway or dark path to doorway.—R. Michaels, Electronic Lightmaster, *Elementary Electronics*, Jan.-Feb. 1969, p 46–48.

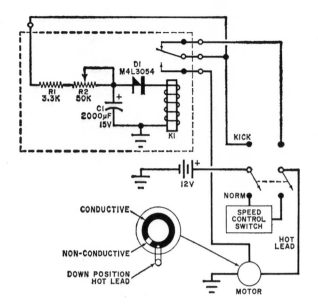

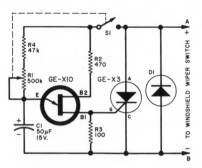

WIPER SPEED CONTROL—R1 adjusts wiper rate from once every 3 s to once every 30 s, to match requirements for mist or light rain. With multiple-speed wipers, connect adapter across switch terminals corresponding to highest wiper speed. Uses ujt as relaxation oscillator controlling scr.—S. Mentler, Variable-Rate Windshield Wiper, *Electronics World*, Oct. 1969, p 76.

VW SLOW-KICK WIPER—Circuit in dashed lines, for slowing up windshield wiper on misty days, will work on Volkswagen only if dpdt kick-normal switch is added and normally closed contact of relay K1 is connected to furnish power to wiper motor's down-position hot lead.—R. C. Gabbey, Slow That Kick, *Popular Electronics*, March 1971, p 88–89.

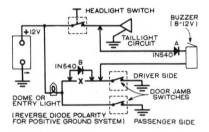

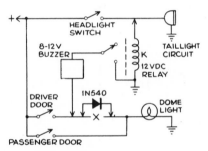

HEADLIGHT REMINDER—Buzzer sounds whenever driver's door is opened and headlight switch is on. Diodes prevent interaction between sensed lighting circuits via buzzer coil. Intended for cars having only one wire going to door-jamb switch (Ford uses two wires).—S. Claire, Two Automatic Reminders to Keep Yourself on the Ball, *Popular Science*, Sept. 1970, p 142.

HEADLIGHT REMINDER—Circuit is designed for use on Ford cars, which have two wires going to door-jamb switch rather than one as in most other cars. If driver's door is open while headlight switch is on, relay is energized and buzzer sounds.—S. Claire and G. F. Freesek, Two Automatic Reminders to Keep Yourself on the Ball, *Popular Science*, Sept. 1970, p 142.

CHAPTER 9
Automotive Ignition Circuits

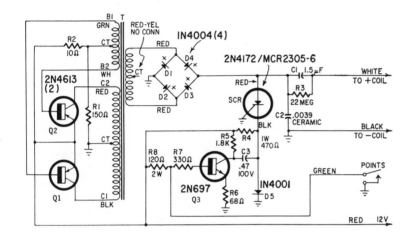

SCR C-D IGNITION—Uses special 12.6-to-250 V inverter transformer (Triad TY68-S) in d-c inverter to produce efficient drive for scr trigger. Delivers 40,000 V to spark plugs. Inverter operates at 2 kHz.— B. Ward, Solid-State C-D Ignition Under $25, *Radio-Electronics*, Feb. 1967, p 32–34.

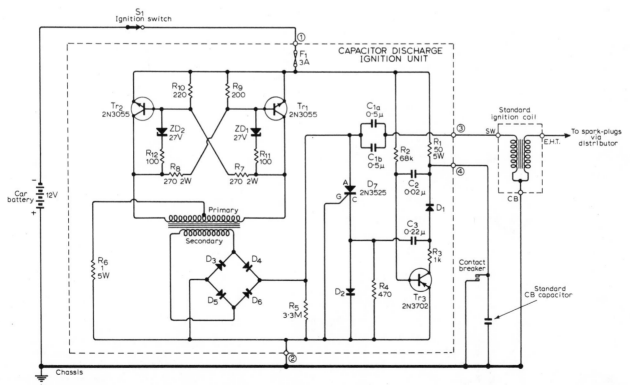

12-V POSITIVE-GROUND C-D—Article gives construction details and transformer winding data for electronic ignition system suitable for any car having 12-V coil ignition system regardless of number of cylinders. Circuit shown is for car with positive terminal of battery grounded to chassis; article also gives circuit for negative ground. T1 has 15:1 ratio of turns.—R. M. Marston, Capacitor-Discharge Ignition System, *Wireless World*, Jan. 1970, p 2–5.

75

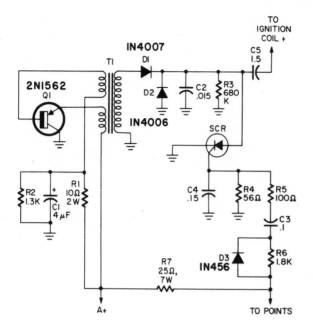

SIMPLIFIED C-D—COMPAC circuit for negative-ground cars uses d-c converter consisting of Q1, T1, D1, and D2. C5 is storage capacitor, with switching by scr when triggered by circuit going to points.—F. W. Holder, Add-On Electronics for Your Car, *Radio-Electronics*, April 1972, p 33–36, 74–75, and 83–84.

CAPACITOR-DISCHARGE SCR—Provides high efficiency at all engine speeds. Energy is stored at high voltage in capacitor, then dumped into ignition coil primary as short current pulse by triggering scr. Draws less than 1 A from negative-ground 12-V car battery at maximum rpm and less than 0.5 A at idling speeds. Diodes are GE-504A. L1 is 250 mH. Transistors are GE-X18. Complete winding data for T1 is given.—"Electronics Experimenters Circuit Manual," General Electric, Owensboro, KY, 1971, 3rd Ed., p 67–77.

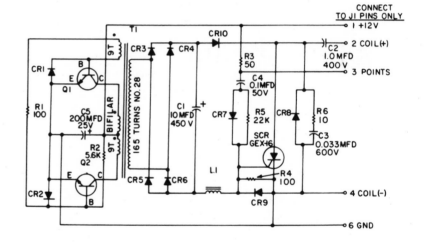

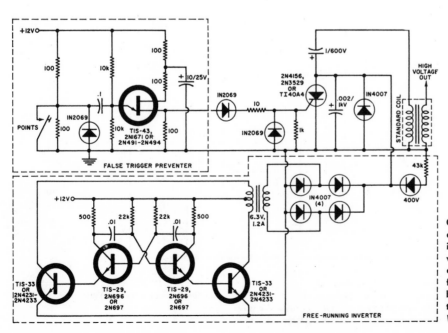

COUNTERACTING POINT BOUNCE—Uses ujt as one-shot scr trigger that cannot retrigger in less than 1 ms, to prevent false triggering when distributor points bounce. Circuit also eliminates need for special inverter transformer. Free-running inverter is 10-kHz astable mvbr.—R. L. Carroll, Improved Capacitive-Discharge Ignition System, *Electronics World*, Feb. 1971, p 78.

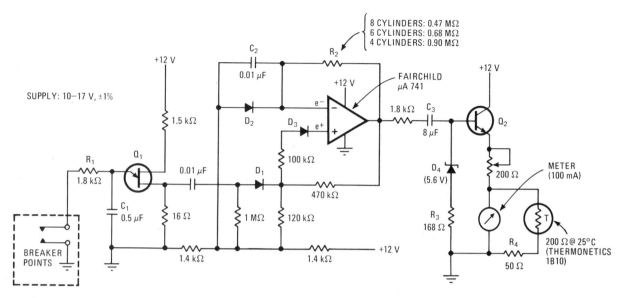

TACH for C-D IGNITION—Designed for use with capacitive-discharge ignition systems, where voltage waveform across breaker points consists of series of 14-V pulses rather than 200-V spikes of conventional ignition. Consists of input relaxation oscillator Q1 for suppressing point bounce, opamp mono mvbr for pulse generation, and buffer Q2 for driving meter which indicates motor rpm to 1% accuracy.—J. B. Young, Precision Auto Tachometer Squelches Point Bounce, *Electronics*, Dec. 18, 1972, p 106.

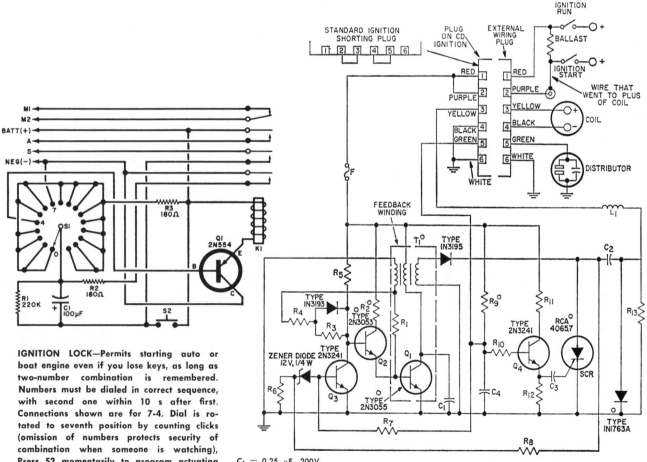

IGNITION LOCK—Permits starting auto or boat engine even if you lose keys, as long as two-number combination is remembered. Numbers must be dialed in correct sequence, with second one within 10 s after first. Connections shown are for 7-4. Dial is rotated to seventh position by counting clicks (omission of numbers protects security of combination when someone is watching), Press S2 momentarily to program actuating circuit Q1-K1, then rotate to position 4 and press S2 again to make relay pull in and complete ignition circuit. S goes to start solenoid. Top contacts of relay open grounding circuit of motor magnets. A goes to accessories.—J. Beiswenger, Electronic Combination Ignition Lock, *Popular Electronics*, July 1970, p 40–42.

C1 = 0.25 μF, 200V
C2 = 1 μF, 400V
C3 = 1 μF, 25V
C4 = 0.25 μF, 25V
F = 5A
L1 = 10 μH, 100T of No. 28
on a 2W resistor
R1 = 1000 ohms
R2 = 50 ohms, 5W
R3 = 22000 ohms
R4 = 1000 ohms
R5 = 10000 ohms
R6 = 15000 ohms
R7 = 8200 ohms
R8 = 0.39 megohm
R9 = 220 ohms, 1W
R10 = 1000 ohms
R11 = 68 ohms
R12 = 4700 ohms
R13 = 27000 ohms

SCR C-D NONFOULING—Single-ended self-oscillating swinging-choke inverter generates high-voltage output pulse having required rapid rate of rise for preventing fouling of spark plugs. Book analyzes circuit in detail and gives transformer winding data.—"Solid State Power Circuits," RCA, Somerville, NJ, SP-52, 1971, p 684–693.

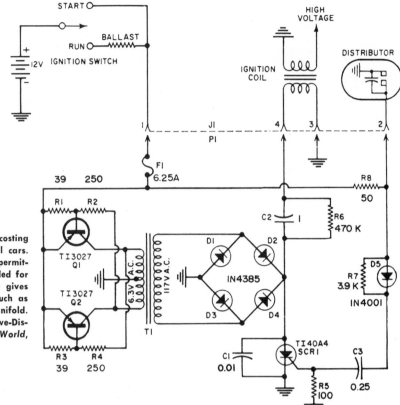

ELECTRONIC IGNITION—Uses parts costing around $20 yet should outlast several cars. Uses conventional coil and distributor, permitting quick changeover if plug is provided for restoring standard connections. Article gives construction details and precautions, such as keeping parts away from exhaust manifold. —B. F. Cawlfield, Low-Cost Capacitive-Discharge Ignition System, *Electronics World*, Nov. 1967, p 30–32.

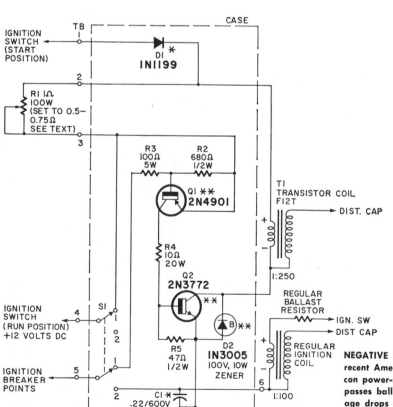

NOTES:
★★ HEAT SINK
★ DIST CAP

S1 MODE
POSITION 1 — TRANSISTOR SYSTEM
POSITION 2 — REGULAR SYSTEM

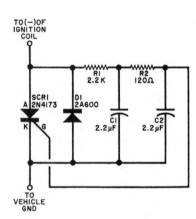

DWELL EXTENDER—Increases spark energy at high engine speed by lengthening time that breaker points remain closed. Requires only two connections to ignition system. Will work with any auto or boat coil-and-breaker-point system having negative ground. Use insulated heat sink on scr.—G. Meyerle, Build a Dwell Extender, *Popular Electronics*, Oct. 1969, p 51–54 and 112.

NEGATIVE GROUND—Suitable for almost all recent American autos. Uses high-voltage silicon power-switching transistors. Diode D1 bypasses ballast resistor R1 when battery voltage drops as low as 8 V during starting, to maintain full spark. T1 is Mallory F12T ignition coil.—B. C. Goldberg and R. G. Wilkins, Add Electronic Ignition to Your Car, *Radio-Electronics*, April 1969, p 32–34.

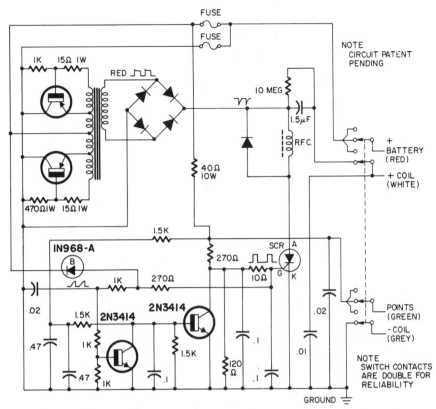

C-D FOR EMISSION-CONTROL DEVICES— Delta Mark Ten B capacitor-discharge ignition for negative-ground cars has improved input trigger circuit that controls spark duration as well as spark intensity over all ranges of operation for modern cars having emission control devices required by law.—F. W. Holder, Add-On Electronics for Your Car, *Radio-Electronics*, April 172, p 33–36, 74–75, and 83–84.

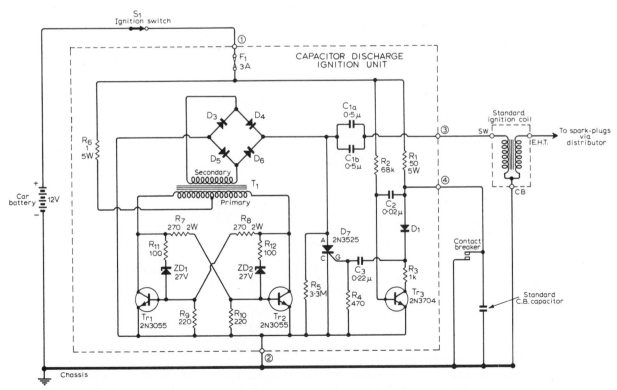

12-V NEGATIVE-GROUND C-D—Tr1, Tr2, T1, and D3–D6 form self-regulating voltage converter. Network including Tr3 is bounce-suppressing pulse shaper that fires thyristor D7 through C3. Although converter has natural oscillation frequency of about 50 Hz, it will give good spark generation above 660 Hz (above 20,000 rpm for 4-cylinder engine and above 10,000 rpm for 8-cylinder engine).— R. M. Marston, Capacitor-Discharge Ignition System, *Wireless World*, Jan. 1970, p 2–5.

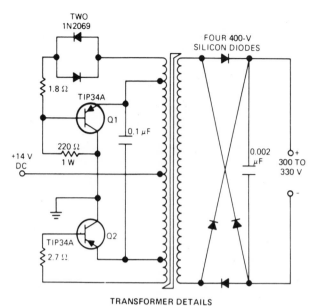

TRANSFORMER DETAILS
PRIMARY: 80 TURNS AWG 18 CENTER-TAPPED
DRIVE WINDINGS: 9 TURNS AWG 23 (EACH WINDING)
SECONDARY: 755 TURNS AWG 30
CORE: E175 BUTT-STACKED (E-I LAMINATIONS OF 14-MIL SILICON
　　　STEEL)
BOBBIN: E33A

14 V TO 300 V FOR AUTO IGNITION—Silicon transistors and single transformer provide economical voltage source for capacitor-discharge ignition system. Inverter frequency is 3 MHz. Efficiency is above 85%.—G. Ehni, "Economical High-Voltage Converters," Texas Instruments, Dallas, TX, CA-126.

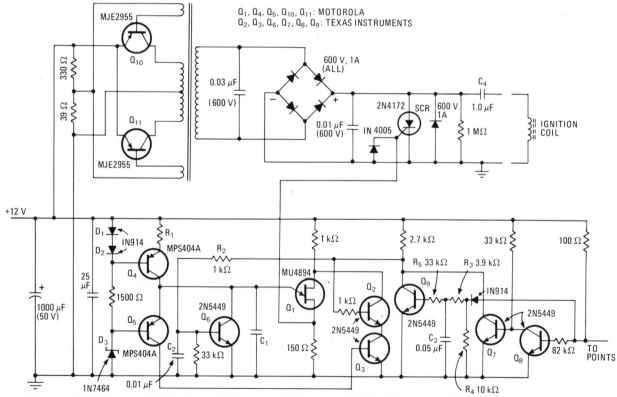

OVERSPEED PROTECTION—Capacitive-discharge ignition system improves engine performance by providing reliable scr triggering over broad range of temperatures and supply voltage values. Q3–Q5 and D1–D3 provide engine overspeed protection, with speed limit set by values of R1 and C1. Q10-Q11 with transformer and bridge rectifier form d-c/d-c inverter that charges C4 to about 375 V; this voltage level provides much greater spark energy than is available from standard ignition system. Conventional ignition coil serves as pulse transformer for boosting discharge voltage to about 40 kV for spark plugs.—J. P. Thomas, Automobile Ignition System Is Rugged and Reliable, *Electronics*, Jan. 18, 1973, p 171–172.

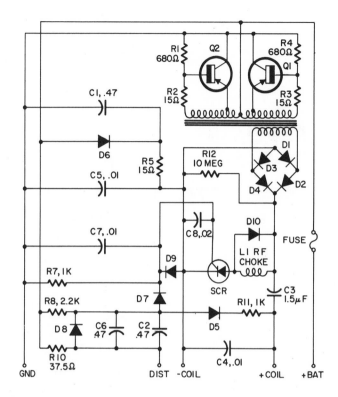

40,000-V C-D—Capacitor-discharge ignition for negative-ground autos improves starting and high-speed performance, and increases spark-plug life. Circuit shown is Delta Mark Ten, developed for cars that do not have emission control devices. Transistors (not identified in article) can be silicon power-switching types.—F. W. Holder, Add-On Electronics for Your Car, *Radio-Electronics*, April 1972, p 33–36, 74–75, and 83–84.

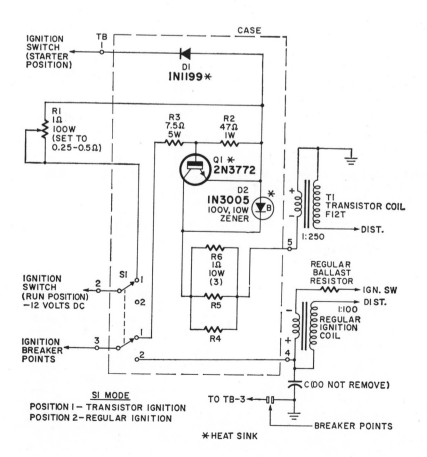

SI MODE
POSITION 1 – TRANSISTOR IGNITION
POSITION 2 – REGULAR IGNITION

∗ HEAT SINK

POSITIVE GROUND—Required for many European cars and some older US cars. Requires only one silicon power-switching transistor. Article covers construction and installation. T1 is Mallory coil.—B. C. Goldberg and R. G. Wilkins, Add Electronic Ignition to Your Car, *Radio-Electronics*, April 1969, p 32–34.

CHAPTER 10
Auto Theft Alarm Circuits

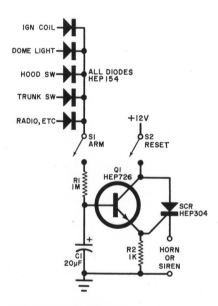

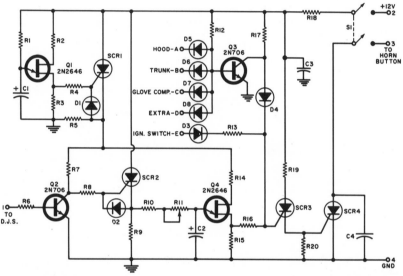

R1—400,000 ohm, ¼ W res.
R2, R14—1000 ohm, ¼ W res.
R3—47 ohm, ¼ W res.
R4, R16—5000 ohm, ¼ W res.
R5, R9—2600 ohm, ¼ W res.
R6, R7, R13, R17—22,000 ohm, ¼ W res.
R8—10,000 ohm, ¼ W res.
R10—100,000 ohm, ¼ W res.
R11—200,000 ohm pot
R12—47,000 ohm, ¼ W res.
R15—100 ohm, ¼ W res.
R18—47 ohm, ½ W res.
R19—470 ohm, ½ W res. R20—2200 ohm, ½ W res.

C1, C2—50 μF, 25 V elec. capacitor
C3, C4—0.25 μF, 100 V Mylar capacitor
SCR1, SCR2, SCR3—Silicon controlled
  rectifier (HEP 320, 2N5060)
SCR4—Silicon controlled rectifier
  (HEP 300)
D1, D2, D3, D5, D6, D7, D8—
  Germanium diode (1N34A or HEP 134)
D4—Silicon diode (1N4001 or HEP 154)
S1—D.p.d.t. switch
Q1, Q4—Unijunction transistor
  (2N2646 or HEP 310)
Q2, Q3—2N706, 2N718

BATTERY-DRAIN ALARM—One silicon diode is connected to each point at which 12-V battery voltage is applied when activated, as by radio, tape deck, door courtesy light, or ignition coil. Can be used with additional switches on trunk and hood. Anode of diode goes to hot terminal when car has negative ground as shown. Once tripped, scr keeps horn on until concealed reset switch is opened. R1 and C1 provide time delay long enough for driver to enter or leave car when reset switch is concealed inside car.—A. C. Caggiano, Build a Pair of Simple Alarms, *Popular Electronics*, Jan. 1973, p 32.

DELAYED ALARM—Provides up to 10 s delay for driver to turn alarm system off after opening car door, to permit concealing on-off switch inside car where it is not vulnerable to defeat by professional thief. Terminal 1 goes to door-jam switch that closes to turn on courtesy light when door is open. Q1 is pulse generator providing delay of between 12 and 22 s after closing of S1, to allow ample time for slow or elderly person to leave car and close door before alarm is activated.— S. Mentler, An Improved Vehicular Intrusion Alarm, *Electronics World*, Aug. 1971, p 38–39.

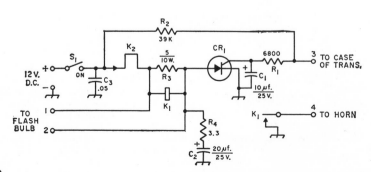

MOBILE EQUIPMENT ALARM—Pulses auto horn on and off and fires flashbulb when wire from terminal 3 to metal housing of grounded transceiver in car is broken by thief. K1 is energized intermittently when scr is turned on because K2, in series with supply, is auto turn-indicator flasher. Pulsing horn makes it distinguishable from stuck horn. Thief cannot stop horn by reconnecting wire, because it can be shut off only by flicking concealed on-off switch S1.—H. Lukoff, A Mobile Equipment Protective Alarm, *QST*, March 1967, p 16–19.

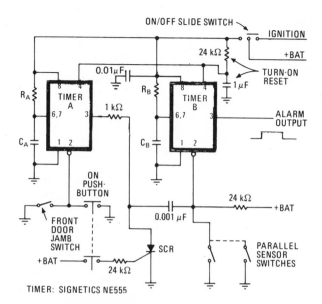

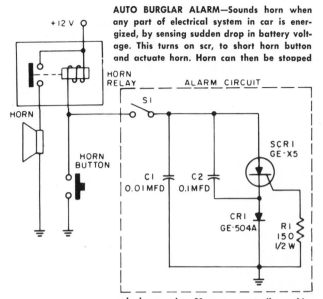

AUTO BURGLAR ALARM—Sounds horn when any part of electrical system in car is energized, by sensing sudden drop in battery voltage. This turns on scr, to short horn button and actuate horn. Horn can then be stopped only by opening S1 or momentarily pushing horn button. Designed for 12-V negative-ground cars.—"Electronics Experimenters Circuit Manual," General Electric, Owensboro, KY, 1971, 3rd Ed., p 87–89.

ARMING-DISARMING TIMERS—Uses two low-cost Signetics NE555 IC timers in circuit operating off car battery. Timer A provides delay of about 1.1 RACA for arming system and allowing driver to get out of car, and provides same delay for driver to enter locked car and turn off alarm switch concealed under dashboard. Scr prevents timer A from triggering alarm-activating timer B unless one of sensor switches is closed by burglar. Eliminates need for vulnerable external turn-off switch on car.—M. L. Harvey, Pair of IC Timers Sounds Auto Burglar Alarm, *Electronics*, June 21, 1973, p 131.

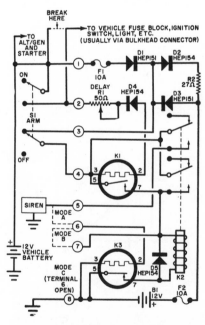

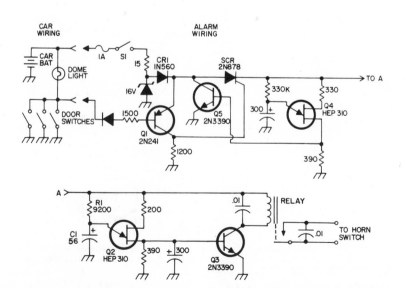

DOOR-SWITCH ALARM—Activated by closing of any of door switches used to turn on dome light. Built-in timer shuts off car horn in about 1½ min on assumption that thief has been scared off, and resets alarm. It is assumed that thief will have closed car door in attempt to shut off alarm himself, so it does not come on again at end of timed interval.—P. Lingane, Solid State Automobile Burglar Alarm, *73*, May 1973, p 97–98 and 100.

THEFT ALARM—Installation requires modification of only one wire in electrical system of car. Alarm is actuated by door switch that turns on dome light or by switches on hood or trunk if present. Operates from selfcontained battery (eight rechargeable alkaline D cells) that is continuously trickle-charged. Arming switch S1 disables all electrical systems in car but does not draw any current until triggered by intruder. Adjustable delay to 8 s gives driver time to arm and disarm. In mode A, siren sounds for 60 s after delay, then goes on and off until door is closed or arming switch is turned off. Mode B gives 60-s siren, and mode C gives continuous siren even if door is closed.—F. J. Dielsi, Vehicle Alarm System, *Popular Electronics*, June 1972, p 51–53.

THEFT ALARM—Requires no tell-tale lock switch outside car. Mercury-switch-operated lamp is installed under hood and on trunk lid (if not already on), so opening of these or of any door having light switch causes current drain on battery. Circuit senses resulting small negative-going pulse across battery and (assuming arming switch S1 inside car is closed) triggers scr on to energize horn relay K1 and sound horns. Simultaneously, ignition breaker points are shunted by X3 and close contacts of A2 so thief cannot start car even if he ignores horn. Action is intermittent, so battery is not run down too much even if horn keeps blasting for several hours after thief is scared away. Cutting horn wire does not remove short across points.—R. F. Graf and G. J. Whalen, "Automotive Electronics," Howard W. Sams, Indianapolis, IN, 1971, p 313–316.

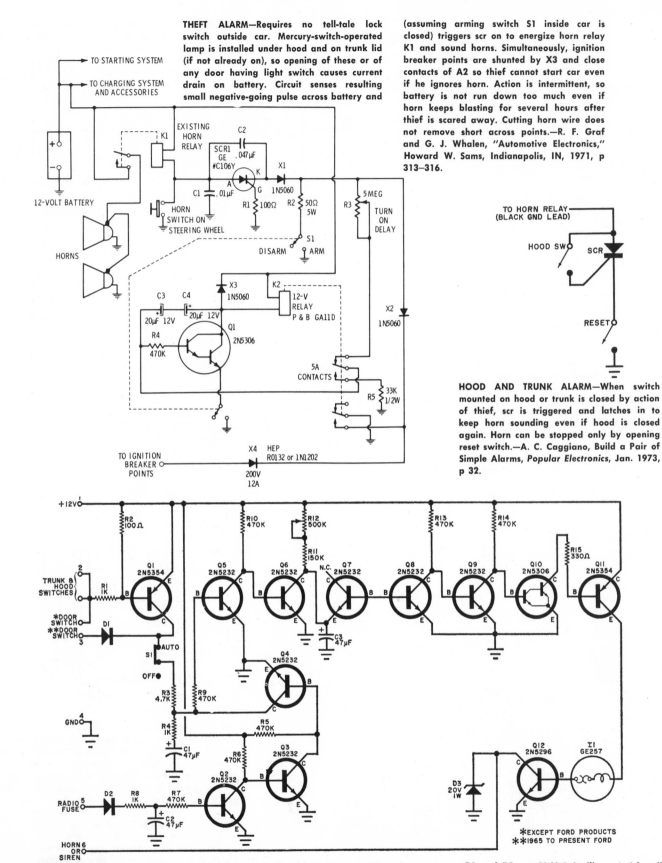

HOOD AND TRUNK ALARM—When switch mounted on hood or trunk is closed by action of thief, scr is triggered and latches in to keep horn sounding even if hood is closed again. Horn can be stopped only by opening reset switch.—A. C. Caggiano, Build a Pair of Simple Alarms, *Popular Electronics*, Jan. 1973, p 32.

TIMED AUTO ALARM—Hidden in glove compartment, alarm arms itself 2 minutes after ignition is turned off, and gives driver 10 to 45 s to turn on ignition after coming back and opening car doors. If he is too slow, horn or optional siren will shut off as soon as ignition key is inserted. When alarm is on, horn beeps once per second for about 2 min and then stops. If driver still can't get key in lock, alarm starts again. Use of ignition jumper wire turns on alarm after slight delay, embarrassing thief as he drives away. D1 and D2 are 50-V 1-A silicon. Article tells how to install in various cars.—G. Meyerle, Automatic Vehicle Burglar Alarm, *Popular Electronics*, April 1970, p 59–61, 66–68, and 73.

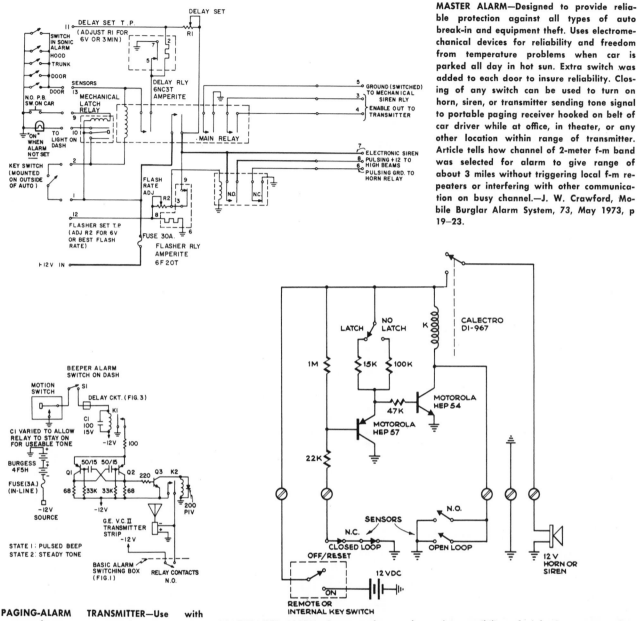

MASTER ALARM—Designed to provide reliable protection against all types of auto break-in and equipment theft. Uses electromechanical devices for reliability and freedom from temperature problems when car is parked all day in hot sun. Extra switch was added to each door to insure reliability. Closing of any switch can be used to turn on horn, siren, or transmitter sending tone signal to portable paging receiver hooked on belt of car driver while at office, in theater, or any other location within range of transmitter. Article tells how channel of 2-meter f-m band was selected for alarm to give range of about 3 miles without triggering local f-m repeaters or interfering with other communication on busy channel.—J. W. Crawford, Mobile Burglar Alarm System, 73, May 1973, p 19–23.

PAGING-ALARM TRANSMITTER—Use with master alarm system to send out tone signal on locally unused 2-meter f-m when anyone opens car door or attempts to steal equipment in car. Legal operation requires transmission of code identification of station each time alarm transmitter is energized; this can be done with automatic keyer, for use in place of 1,000-Hz tone. Intended for use primarily in high-risk crime areas.—J. W. Crawford, Mobile Burglar Alarm System, 73, May 1973, p 19–23.

BOAT-TRAILER ALARM—Covers various methods of applying auto alarm techniques to all types of boats, whether in water or on trailer in driveway. Circuit shown uses combination of open-loop and closed-loop sensors to minimize possibility of defeating sensor when spotted by thief. Latch switch locks alarm on until it is reset by owner.—E. Zadig, Simple Watchdogs Foil Boat Thieves, Popular Science, Aug. 1973, p 96–98.

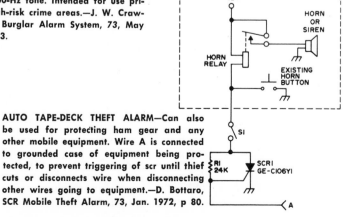

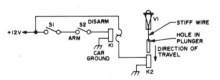

AUTO TAPE-DECK THEFT ALARM—Can also be used for protecting ham gear and any other mobile equipment. Wire A is connected to grounded case of equipment being protected, to prevent triggering of scr until thief cuts or disconnects wire when disconnecting other wires going to equipment.—D. Bottaro, SCR Mobile Theft Alarm, 73, Jan. 1972, p 80.

CHEMICAL GAS RELEASE—Closing of car-door courtesy-light switch as thief opens door energizes heavy-duty 12-V solenoid rigged to release On-Guard or other harmless chemical that provides temporary stinging and burning of eyes, nose, and throat as deterrent to theft. (Use with tear gas or Mace is illegal in many communities.)—R. L. Guard, Jr., The Theft Stopper, 73, Sept. 1971, p 52.

CHAPTER 11
Battery Charging Circuits

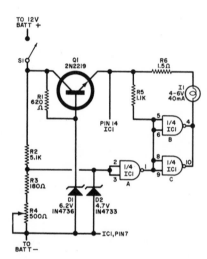

DRY-CELL CHARGER AND TESTER—Recharges accurately on current-time or ampere-hour basis, to avoid overcharging of expensive cells, and checks voltage of each cell under varying load and charging conditions. Switches S3–S7 have rheostat-controlled charging position, fixed trickle-charge position, and center-off position. Trickle-charge serves for long-term storage on charger. Values of R1—R10 are function of battery voltage, charge rate, and source voltage. Article tells how to use Ohm's law to calculate values.—H. H. Stover, Build a Combination Battery Charger and Tester, *Popular Electronics,* Aug. 1967, p 33–37.

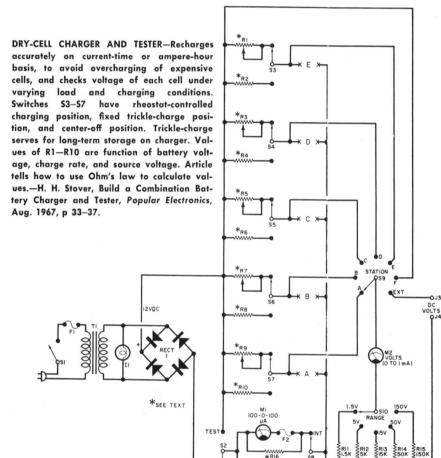

UNDERVOLTAGE LAMP—R4 sets level at which junction of R2 and R3 drops below 1.7 V and makes circuit turn on warning lamp I1 to indicate that 12-V storage battery needs recharging. Insures that car engine is restarted to charge battery while still having enough power to turn over starter. Ideal for ham field day operators using mobile equipment when car is parked and engine turned off. IC can be MC7401P or HEPC3001P.—C. Morris and J. Morris, Auto Battery Voltage Monitor, *Popular Electronics,* Aug. 1972, p 58–59.

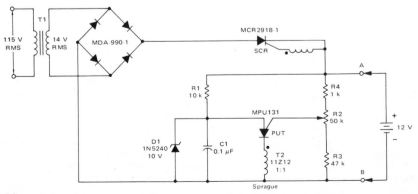

12-V PUT-SCR—Short-circuit-proof charger provides average charging current of 8 A to lead-acid auto storage battery. Circuit is not damaged by reversal of battery polarity. Charging voltage can be set between 10 and 14 V, with lower limit determined by D1 and upper by T1. Put is connected as relaxation oscillator; charging ceases when battery voltage reaches value determined by setting of R2 and oscillator stops.—R. J. Haver and B. C. Shiner, "Theory, Characteristics and Applications of the Programmable Unijunction Transistor," Motorola Semiconductors, Phoenix, AZ, AN-527, 1971.

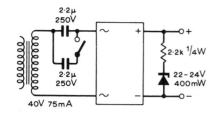

22mA into 18V
27mA " 6V } (HP7, VT4 etc.)

or

44mA into 18V
54mA " 6V } (SP11, VT9 etc.)

DRY-CELL CHARGER—Choose bridge rectifier (rectangle) for maximum charging voltage and current required. Capacitors act as current limiters. Will charge dry cells up to 18 V. Weekly recharging is recommended; it is pointless and dangerous to try to charge cells that have been discharged over several months.—K. W. Mawson, Dry Cell Charger, *Wireless World*, June 1971, p 269.

DRY-BATTERY CHARGER—Use high and common leads for charging batteries from about 20 to 90 V, and low with common for batteries under about 20 V. Charging current should be kept under 5 mA, as indicated by M1, for highest-voltage batteries, and proportionately less for lower-voltage batteries. K1 (equivalent of Allied Radio 41B5292) opens at charging current of 6 mA, lighting M2 and killing charging circuit. Use 6.3-V pilot lamps. M1 is 0–10 mA d-c meter and rectifiers are GE 504A. Common goes to minus terminal of battery.—R. M. Brown and T. Kneitel, "101 Easy Test Instrument Projects," Howard W. Sams, Indianapolis, IN, 1968, p 13–15.

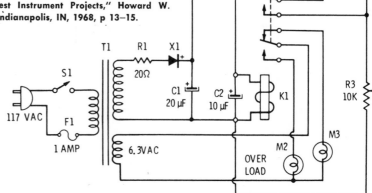

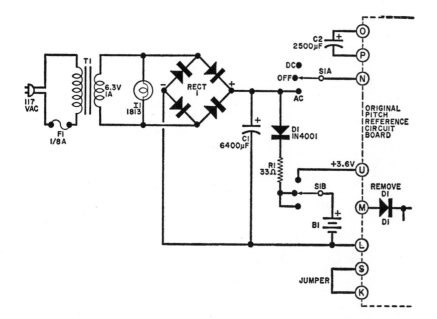

3.6-V CHARGER—Will keep three 1¼-V nickel-cadmium D cells charged when pitch reference for piano tuning can be operated from a-c line, while permitting normal battery operation otherwise. Rectifier can be four 1N4001, 1N2069, or similar.—E. M. Hoyt, Pitch Reference, *Popular Electronics*, June 1969, p 70–71.

NI-CD END-OF-CHARGE SENSOR—Permits charge times as short as 15 min at high rate for rechargeable nickel-cadmium batteries of home appliances and electric shavers, with automatic end-of-charge sensing of voltage rise. Circuit is experimental. Report covers advantages and drawbacks, and gives other methods of speeding charge time safely.—D. A. Zinder, "Fast Charging Systems for Ni-Cd Batteries," Motorola Semiconductors, Phoenix, AZ, AN-447, 1972.

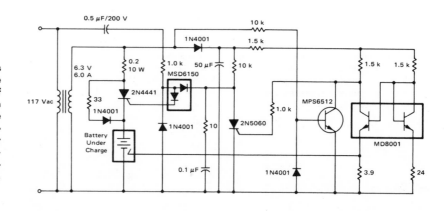

12 V WITH SHUTOFF—Charges auto or boat lead-acid storage battery at maximum design current until battery is fully charged, then automatically shuts off. Charger switches itself on automatically again whenever battery neds charging, as for emergency standby battery. Adjust R1 so charging just ceases when fully charged battery is connected.—"Electronics Experimenters Circuit Manual," General Electric, Owensboro, KY, 1971, 3rd Ed., p 83–85.

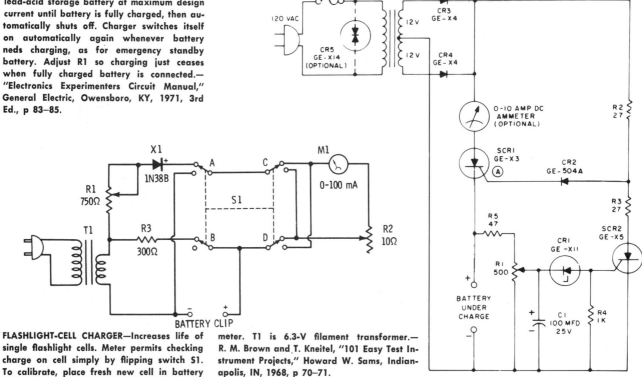

FLASHLIGHT-CELL CHARGER—Increases life of single flashlight cells. Meter permits checking charge on cell simply by flipping switch S1. To calibrate, place fresh new cell in battery clip and adjust R2 for full-scale deflection of meter. T1 is 6.3-V filament transformer.—R. M. Brown and T. Kneitel, "101 Easy Test Instrument Projects," Howard W. Sams, Indianapolis, IN, 1968, p 70–71.

HIGH-SPEED CHARGER—Charges at high rate and shuts off automatically to minimize overcharging. Charging time is accelerated by applying high-current reverse-polarity pulses to dissipate gas formed on plates during charging. Schmitt trigger Q1-Q2 and relay driver Q3 form burp control circuit for gas. Normal burp rate, set by R17, is once every 20 s. If charging battery in car, remove both battery connections to car first, to protect alternator and other components.—R. F. Graf and G. J. Whalen, Battery Charger, *Popular Science*, July 1970, p 101–103 and 118.

C1—100 MF, 25 WVDC electrolytic capacitor (Sprague TE1211)
C2, C3—12,000 Mfd, 40 WVDC electrolytic capacitor (Sprague Powerlytic 36D123G0 40BC2A)
C4—30 MF, 16 WVDC electrolytic capacitor (Sprague TE1158)
C5—350 MF, 16 WVDC

electrolytic capacitor (Sprague TE1166)
C6—0.22 MF, 200 WVDC Mylar capacitor (Sprague 192P22402)
D1, D2—Silicon rectifier diode, 20-amp rating, 200 PRV (GE-X4)
D3, D4, D5, D7—Silicon rectifier diode, 1-amp rating (GE A14F or

1N5059)
D6—8.2v zener diode (GE X-11 or 1N756)
I1, I2—12v indicator lamp (Industrial Devices B2990D5—green and B2990D6 —blue)
K—Hvy-duty 12 VDC relay, DPDT contacts, 8-amp min. rating (Potter &

Brumfield MR11D)
M—0-10-amp DC ammeter (EMICO RF2¼C-2303)
Q1, Q2—NPN transistor (GE 2N5172)
Q3—NPN power transistor (GE D27C1)
R1, R2—30-ohm, 5w wirewound resistor
R3—1K, ½w carbon resistor

R4—47-ohm, 1w carbon resistor
R5—500-ohm, 5w wirewound potentiometer (Mallory VW-500)
R6—15-ohm, 2w
R7—120-ohm, 1w
R8—4.7K, ½w
R9, R13, R14—2.7K, ½w
R10—6.8K, ½w
R11—68-ohm,

½ w   carb.
R12—10K, ½w resistors
R15—68K, ½w
R16—15K, ½w
R17—3 megohm, ½w potentiometer (Mallory U59)
Fuse—2-amp w/holder (cartridge type)
Case—3½x6x10" (Premier Metals Corp. PMC 1010)

SCR1—13-amp, 50 PIV (GE-X3)
SCR2—1.6-amp, 50 PIV GE-C106YI)
S—SPST toggle, 5-amp rating
T—Power transformer (Knight 6-K-65, Allied 54D2333)

NICKEL-CADMIUM CELL CHARGER—Will charge 6 to 7.5 V at constant current of 550 mA, and switch off automatically at preset voltage determined by setting of VR3.— F. Ballerini, Nickel-Cadmium Battery Charger, *Wireless World*, April 1971, p 204.

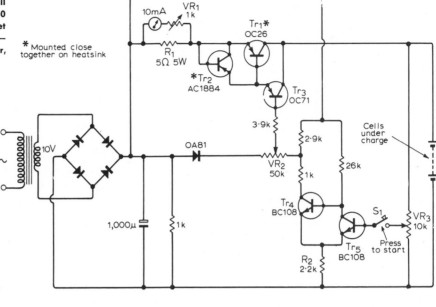

VOLTAGE MONITOR—Lamp comes on when battery needs replacing or recharging. Pot sets threshold voltage for triggering programable ujt which in turn fires scr. Zener keeps current drain under 300 μA during monitoring.—W. G. S. Brown and V. K. L. Huang, Voltage Monitor Is Easy on Both Battery and Budget, *Electronics*, Oct. 26, 1970, p 87.

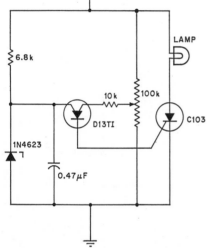

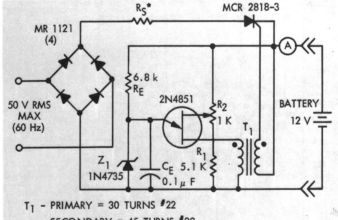

12 V AT 20 A WITH UJT CONTROL—Simple circuit uses ujt relaxation oscillator to turn off charger when battery is fully charged. Oscillator is activated only when battery voltage is low, and stops when voltage derived from battery under charge exceeds breakdown voltage of zener.—"Unijunction Transistor Timers and Oscillators," Motorola Semiconductors, Phoenix, AZ, AN-294, 1972.

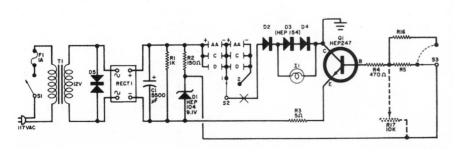

DRY-CELL CHARGER—Provides a choice of fixed charge rates for nearly all types of nickel-cadmium and zinc-carbon batteries, from 45 to over 500 mA. Optional 500-mA meter can be inserted at X in lead to S2. Charge rate control pot R17 is optional. Thyrector diode D5 (GE 6RS20SP1B1) is optional. RECT1 is 1-A 200-piv bridge rectifier such as HEP176. I1 is No. 49 lamp. Article gives charging times, currents, and other data on various types of cells.—A. A. Mangieri, AA-C-D Battery Charger, *Popular Electronics*, Oct. 1969, p 33–35 and 40–41.

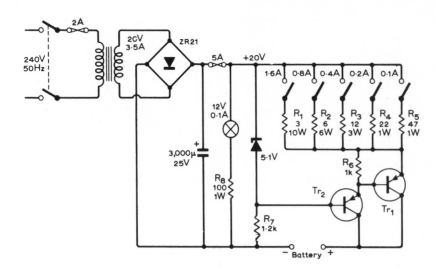

CONSTANT-CURRENT CHARGER — Charging currents up to 3.1 A are derived from five switched resistors held at constant voltage by zener and Darlington pair of transistors. Use 2N1021 germanium or similar power transistors. Output terminals may be short-circuited without damage. Will handle batteries up to 12V.—D. Allen, Constant-Current Battery Charger, *Wireless World*, April 1972, p 185–186.

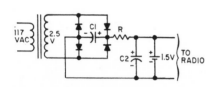

1.5-V CHARGER—Keeps filament battery in radio charged, so life becomes shelf life. Diodes can be International Rectifier type 8D4, and capacitors 6,000-μF 6-V. Try different values for R until current measured through battery is zero with receiver on. Serves in place of costly hum-free 1.5-V rectifier power supply.—H. Scott, Power Supply for Battery Radio, *Elementary Electronics*, July-Aug. 1971, p 74.

DRY-CELL CHARGER—Can be used with battery holders for C, D, and penlight cells. R1 provides large range of charging voltages, while I1 limits charging current and serves as indicator. Resistors R2–R12 prevent interaction between cells. T1 is 6.3-V 1-A filament transformer.—W. Temcor, New Life for Old Dry Cells, *Popular Electronics*, Feb. 1965, p 69–70.

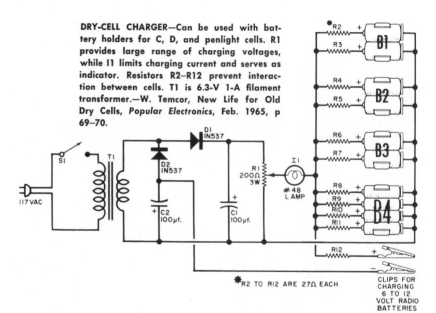

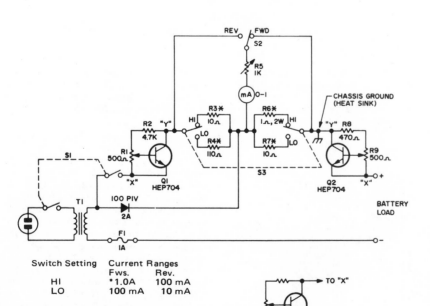

Switch Setting	Current Ranges	
	Fws.	Rev.
HI	*1.0A	100 mA
LO	100 mA	10 mA

*Maximum fwd. current available is about 0.7A with battery output terminals shorted.

REVERSE-CURRENT CHARGER—Control pots handle only base currents of current-limiting transistors used for both reverse and forward current control. Metering circuit provides monitoring over range of 2 to 500 mA. Use of dpst switch for S1 prevents batteries from discharging accidentally through reverse-current transistor and transformer secondary if charger is turned off without disconnecting batteries. Adjust reverse-current pot R1 to give about 10% of forward-current value.—F. J. Bauer, Jr., A Simple Reverse Current Battery Charger, *73*, Oct. 1971, p 55 and 57–58.

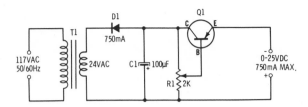

NICKEL-CADMIUM CHARGER—R1 adjusts output voltage over full range from 0 to 35 V d-c for charging practically any nickel-cadmium battery in commercial use. Maximum charging current is 750 mA. Q1 should be mounted on but insulated from metal cabinet serving as heatsink. Place ammeter in series with battery and adjust R1 to specified charging current. Q1 is any 40-W pnp power transistor.—H. Friedman, "99 Electronic Projects," Howard W. Sams, Indianapolis, IN, 1971, p 98–99.

STANDBY BATTERY CHARGER—Zener sets reference voltages for comparators of IC timer connected to maintain full charge on standby battery supply for instrument that is always connected to a-c power line. Settings of 25K pots determine turn-on and turn-off battery voltages. D2 prevents battery from discharging through timer when timer is off.—E. J. McGowan, Jr., IC Timer Automatically Monitors Battery Voltage, *Electronics*, June 21, 1973, p 130–131.

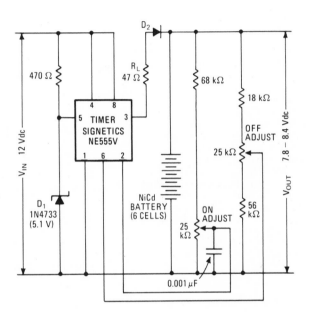

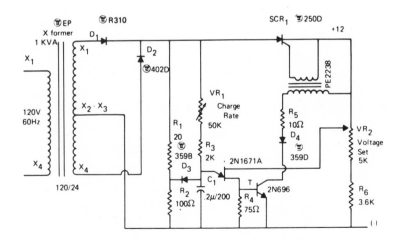

VOLTAGE AND RATE CONTROL—Phase-controlled scr converter automatically decreases output current as battery voltage increases, to protect battery from overcharge. Battery voltage determines phase delay angle at which scr is triggered. Will handle either 6-V or 12-V storage batteries.—"Silicon Controlled Rectifier Designers Handbook," Westinghouse, Youngwood, PA, 1970, 2nd Ed., p 7-62–7-64.

CHAPTER 12
Burglar Alarm Circuits

BEAMLESS BURGLAR ALARM—High-sensitivity photoelectric alarm detects movement of intruder in large area between ordinary lamp and photocell that can be over 25 feet away. Tubing is sealed to photocell to minimize effect of ambient light; length is determined by experimentation in range of 3 to 18 inches depending on distance and strength of light source. Article gives construction details. Circuit shows jacks for plugging in four remotely located photocells in tubes to increase area of protection. Mallory Sonalert can be connected directly to T1-T2. B1 is 9 V.—C. D. Rakes, Light Sensor, *Elementary Electronics*, March-April 1972, p 59–62.

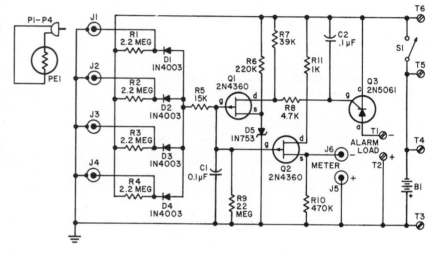

INVISIBLE-BEAM PHOTO-ALARM—Frequency-sensitive photoelectric circuit responds only to infrared beam chopped at frequency of twin-T feedback networks for pair of series-connected IC bandpass amplifiers. Amplifier output is fed to peak detector which holds Darlington Q4 in saturation, depriving scr of trigger voltage. Momentary interruption of invisible chopped beam by intruder triggers scr and alarm on, with alarm continuing until circuit is reset by closing scr-shunting switch momentarily. Motor-driven wheel with hole drilled near edge can be used for chopping beam from appropriately filtered light source. Chopping frequency should not be harmonically rated to 60 Hz.—J. Bliss, "Applications of Phototransistors in Electro-Optic Systems," Motorola Semiconductors, Phoenix, AZ, AN-508, 1971.

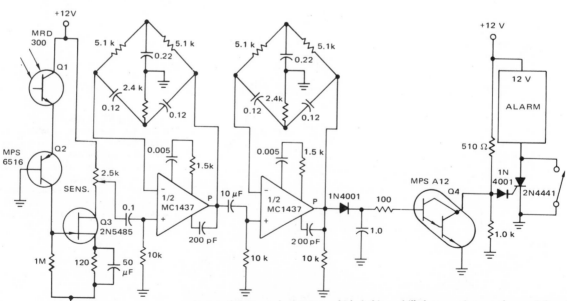

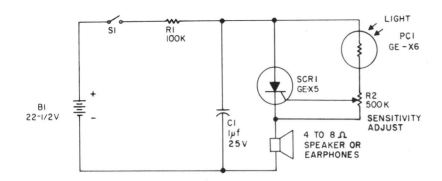

LIGHT-SENSITIVE CLICKER—Photocell (cadmium sulfide) in trigger circuit of scr responds to sweep of flashlight in darkroom and turns on relaxation oscillator to produce clicking sound in speaker at rate that increases with amount of light. Set R2 just below threshold of clicking in darkness. For fire alarm, replace PC1 with cadmium selenide cell responding to near infrared. For heat alarm, use lead sulfide cell whose coverage extends further into infrared region. S1 can be on R2.—"Electronics Experimenters Circuit Manual," General Electric, Owensboro, KY, 1971, 3rd Ed., p 161–166.

500-W NITE LITE—Will turn on up to five different 100-W lights in home automatically at night to discourage prowlers, and turn them off at dawn. If only single lamp up to 15 W is to be controlled, omit triac and connect A to A' and B to B'. PC1 is GE-X6 photoconductive cell and L1 is reed switch coil having 10,000 turns No. 39 enamel on ¼th inch form 2 inches long (GE coil C-2).—"Electronics Experimenters Circuit Manual," General Electric, Owensboro, KY, 1971, 3rd Ed., p 147–150.

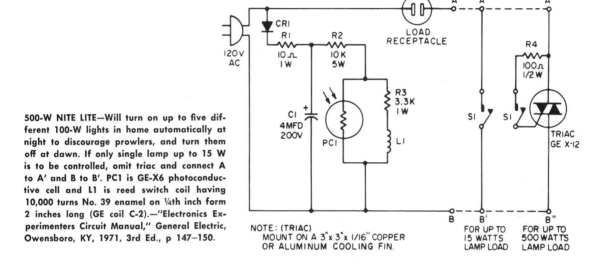

NOTE: (TRIAC)
MOUNT ON A 3"x 3"x 1/16" COPPER OR ALUMINUM COOLING FIN.

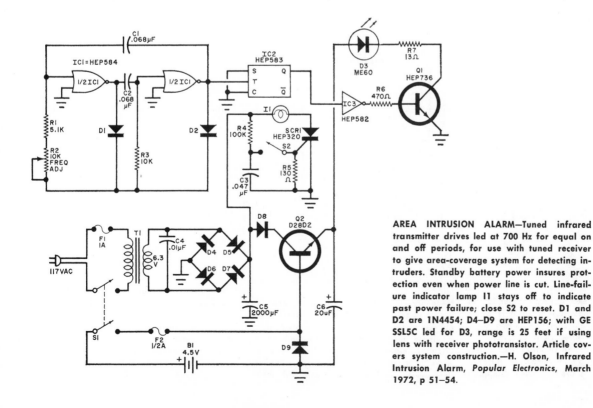

AREA INTRUSION ALARM—Tuned infrared transmitter drives led at 700 Hz for equal on and off periods, for use with tuned receiver to give area-coverage system for detecting intruders. Standby battery power insures protection even when power line is cut. Line-failure indicator lamp I1 stays off to indicate past power failure; close S2 to reset. D1 and D2 are 1N4454; D4–D9 are HEP156; with GE SSL5C led for D3, range is 25 feet if using lens with receiver phototransistor. Article covers system construction.—H. Olson, Infrared Intrusion Alarm, *Popular Electronics*, March 1972, p 51–54.

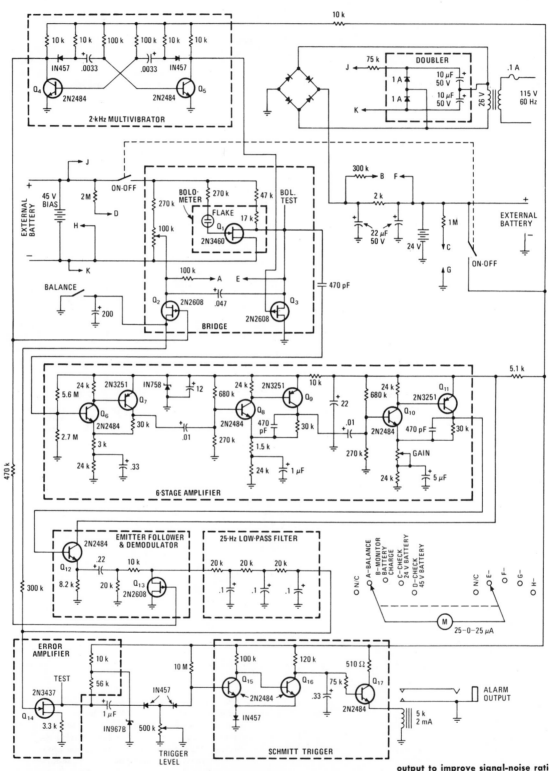

PASSIVE INTRUDER ALARM—Responds to movement of intruder up to 1,000 ft away from detector without signaling its own presence. Reticle and filter chop body-heat or other infrared radiation to insure that alarm is triggered only by movement of infrared source. False alarms are minimized. Chopping frequency is between 0 and 25 Hz, depending on angular speed of intruder across field, number of lines in reticle, and distances from reticle to target and to sensor bolometer. Chopped signal modulates 2-kHz square-wave mvbr signal feeding six-stage amplifier. Demodulator Q13 is synchronized with mvbr output to improve signal-noise ratio. Any output under 25 Hz having amplitude above trigger level set by 500K pot makes Schmitt trigger energize relay that can turn on lights, camera, gong, or r-f transmitter.—W. E. Osborne, *Long-Range Infrared Intruder Alarm Resists Fault Triggering, Electronics,* Nov. 20, 1972, p 111–113.

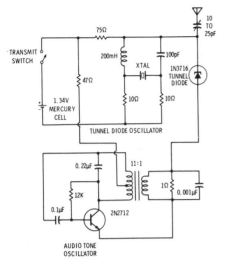

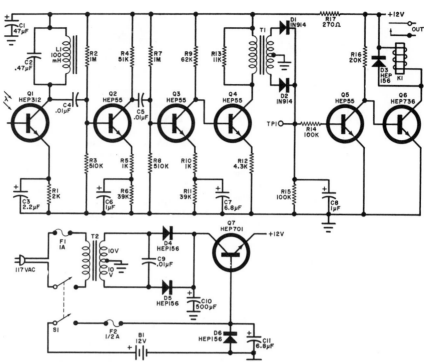

27-MHZ SECURITY TRANSMITTER—Can be used with any switch-closing burglar-alarm system for transmitting silent signal to appropriate headquarters. Will fit in housing no larger than pack of cigarettes, for use by cashiers, bank tellers, and guards having need to transmit silent alarm during holdup. Audio-tone oscillator frequency is chosen to identify source of emergency signal. Transmitter range is up to 250 ft, requiring use of communications-type receiver tuned to transmitter frequency, for passing on alarm signal over wire line or other means. Transmitter must meet FCC regulations, including antenna limit of 5 ft, frequency between 26.97 and 29.27 MHz, and power input to final stage not over 100 mW.—J. E. Cunningham, "Security Electronics," Howard W. Sams, Indianapolis, IN, 1972, p 100–101.

INFRARED RECEIVER—When using lens with phototransistor Q1, will detect intruders at range of 9 to 25 feet away from led transmitter without lens, depending on lens used. Lens on transmitting led increases range to over 80 feet but narrows beam. Consists of phototransistor driving audio amplifier, followed by rectifier that pulls in relay when intruder in covered area reflects 700-Hz chopped infrared to receiver. Operates from battery power if power line is cut, because regulator Q7 is referenced to standby battery.—H. Olson, Infrared Intrusion Alarm, *Popular Electronics*, March 1972, p 51–54.

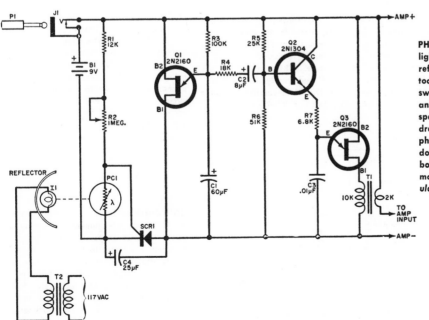

PHOTOELECTRIC ALARM—Interruption of light beam between pilot lamp in flashlight reflector and Clairex CL707HL or similar photocell triggers alarm, generating distinctive sweeping audio tone that can be fed into any low-cost transistor amplifier with speaker. Has no mechanical parts. Power drain is low before triggering. Lamp and photocell can be on opposite sides of single door or window, or mirrors can be used to bounce beam around entire room.—J. S. Simonton, Jr., "Cyclops" Intruder Detector, *Popular Electronics*, May 1968, p 41–44.

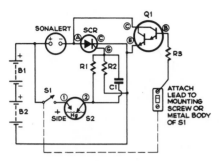

B1, B2—9v alkaline bat-
tery (Mallory MN-1400)
C1—0.1 MFD capacitor
(Sprague 192P)
Q1—Monolithic Darling-
ton transistor (GE
2N5306)
R1—1-meg., ½-watt
±10% carbon resistor
R2—1-k., ½-watt ±10%
carbon resistor
R3—39-k., ½-watt ±10%
carbon resistor

KNOCK-OVER ALARM—When placed near door of home, motel, or hotel, alarm is tipped over by anyone entering. Mercury switch S2, taken from any GE wall-type toggle switch, then closes and triggers GE C106Y2 scr which in turn energizes Mallory SC628P Sonalert. Resulting load beep can then be silenced only by touching screw of on-off switch and projecting case of mercury switch with moistened fingers while S1 is in on position. Can be mounted in frozen-juice can. Place on top of suitcase when making phone call at airport, to prevent theft of luggage.—R. F. Graf and G. J. Whalen, Electronic Watchdog in a Can, *Popular Science*, Dec. 1969, p 182, 184, and 186.

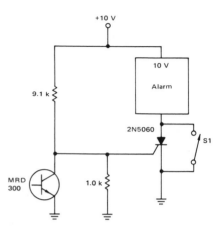

LIGHT-INTERRUPTION SCR ALARM—Use of sensitive-gate scr shown eliminates need for triggering relay. Alarm rings continuously when 125-footcandle minimum illumination on MRD300 photodiode is momentarily interrupted. Momentary-contact switch S1 is closed to reset system.—J. Bliss, "Applications of Phototransistors in Electro-Optic Systems," Motorola Semiconductors, Phoenix, AZ, AN-508, 1971.

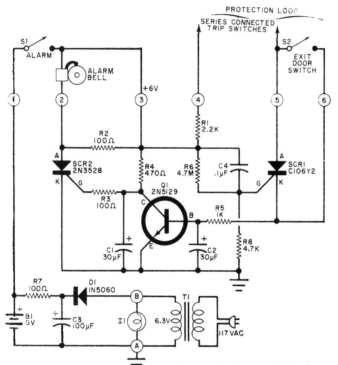

CONDUCTIVE-FOIL ALARM—Combines door and window switches with lengths of foil cemented on breakable glass windows in one series protection loop. If loop is opened or grounded, Q1 loses base bias and cuts off, to make SCR2 turn on and sound alarm. S1 must be opened to stop alarm. Rechargeable 6-V battery is trickle-charged as long as a-c line is plugged in. Alarm is fail-safe. S2 acts with SCR1 to permit leaving house without tripping alarm after S1 is closed. When S2 is closed after departure, SCR1 turns off but Q1 remains saturated and ready for action.—D. Meyer, Build the Homesteader, *Popular Electronics*, Oct. 1969, p 71–73 and 114.

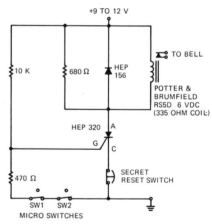

CIRCUIT-BREAKING ALARM—Opening of one of microswitches or breaking of similarly connected foil or thin wire triggers HEP320 scr to energize relay that turns on alarm or silent warning device. Battery drain is less than 1 mA on standby. Concealed switch must be pushed momentarily to stop alarm and reset circuit.—"Home Handyman's Construction Projects," Motorola Semiconductors, Phoenix, AZ, HMA-37, 1972.

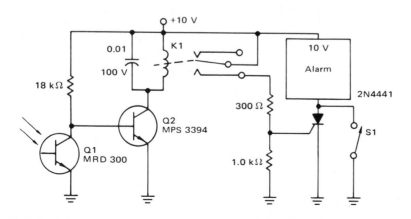

LIGHT-INTERRUPTION ALARM—Interruption of 125-footcandle minimum illumination on photodiode Q1 drops Q2 current below 5 mA and releases relay, to trigger scr on and energize alarm. Alarm rings continuously even if light is restored, until momentary-contact switch S1 is closed to reset system.—J. Bliss, "Applications of Phototransistors in Electro-Optic Systems," Motorola Semiconductors, Phoenix, AZ, AN-508, 1971.

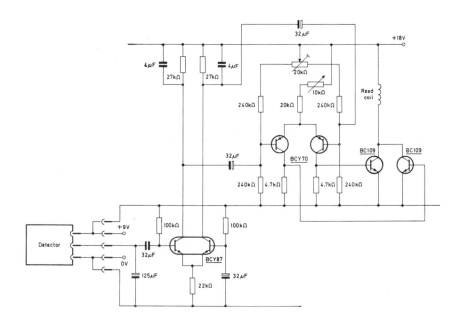

PASSIVE INTRUDER ALARM—Uses triglycine sulfate pyroelectric detector (Mullard 802CPY) which responds to radiation from objects slightly above and below room temperature. Arsenic trisulfide lens is used on detector to pass desired radiation while rejecting other wavelengths. Dual-transistor differential amplifier is incorporated in detector. Main amplifier drives reed relay which operates warning lamp, bell, or other signal device. Optical system provides 2 × 6 ft field of view at 200 ft, so intruder almost fills field at this range.—"Applications of Infrared Detectors," Mullard, London, 1971, TP1201, p 59–62.

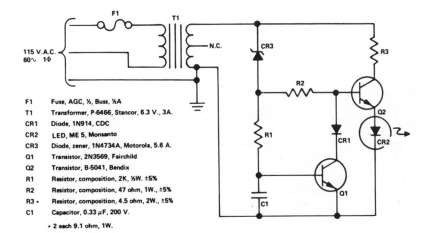

PULSED TRANSMITTER—Monsanto ME 5 infrared emitting diode CR2 is pulsed every 16.7 ms to 1 A for 1.5 ms, to give about 10% duty cycle. Used with suitable infrared receiver for which circuit is also given, to detect interruption of beam of modulated infrared light for intruder detection, counting moving objects, and other photoelectric-type applications. Lens is not necessary if range required is under 30'.—R. C. Bach, "Modulated IR Beam Control System," *Monsanto GaAsLite Tips*, Vol. 1, 1970.

F1	Fuse, AGC, ½, Buss, ½A
T1	Transformer, P-6466, Stancor, 6.3 V., 3A.
CR1	Diode, 1N914, CDC
CR2	LED, ME 5, Monsanto
CR3	Diode, zener, 1N4734A, Motorola, 5.6 A.
Q1	Transistor, 2N3569, Fairchild
Q2	Transistor, B-5041, Bendix
R1	Resistor, composition, 2K, ½W. ±5%
R2	Resistor, composition, 47 ohm, 1W., ±5%
R3 •	Resistor, composition, 4.5 ohm, 2W., ±5%
C1	Capacitor, 0.33 μF, 200 V.

• 2 each 9.1 ohm, 1W.

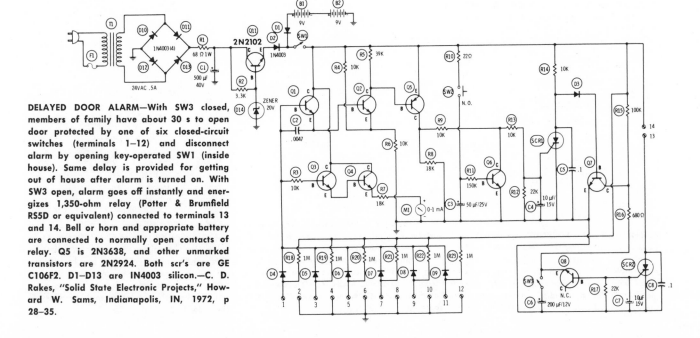

DELAYED DOOR ALARM—With SW3 closed, members of family have about 30 s to open door protected by one of six closed-circuit switches (terminals 1–12) and disconnect alarm by opening key-operated SW1 (inside house). Same delay is provided for getting out of house after alarm is turned on. With SW3 open, alarm goes off instantly and energizes 1,350-ohm relay (Potter & Brumfield RS5D or equivalent) connected to terminals 13 and 14. Bell or horn and appropriate battery are connected to normally open contacts of relay. Q5 is 2N3638, and other unmarked transistors are 2N2924. Both scr's are GE C106F2. D1–D13 are IN4003 silicon.—C. D. Rakes, "Solid State Electronic Projects," Howard W. Sams, Indianapolis, IN, 1972, p 28–35.

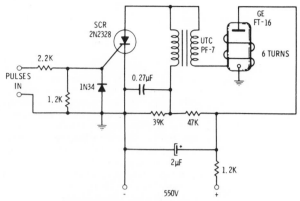

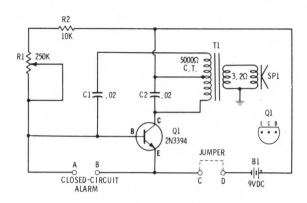

PHOTOFLASH TRIP—Positive-going pulse at input fires scr and flashtube, for scaring burglar when he touches safe or other protected metal object or for obtaining actual photograph of burglar. Requires sensing circuit of desired type, connected to trip pulse generator. Alternatively, circuit can be driven by low-frequency mvbr pulse generator to produce blinding and disconcerting repetitive flashes at rate of perhaps two per second. High-voltage supply need not be regulated.—J. E. Cunningham, "Security Electronics," Howard W. Sams, Indianapolis, IN, 1972, p 87–88.

MULTIPURPOSE TONE GENERATOR—Setting of R1 determines audio frequency generated by Hartley oscillator Q1. Serves as alarm for intruders when normally open door or window switches are connected to A and B, with jumper between C and D. For normally closed intruder switches, connect them to C and D and place jumper between A and B. For code practice, connect T between C and D, with jumper between A and B. To use as a-f signal generator, connect shielded test leads between speaker terminals.—H. Friedman, "99 Electronic Projects," Howard W. Sams, Indianapolis, IN, 1971, p 56–57.

DOPPLER ALARM—Uses Mullard CL8630 Gunn oscillator as 10.69-GHz self-oscillating mixer in intruder detector. Transmitted energy at known frequency is compared with frequency of reflected energy, and detected frequency difference due to moving object within predetermined range is used to operate alarm. Detected Doppler signal is fed into two-stage a-c amplifier which in turn drives trigger consisting of BC109 bipolar transistor and BFW11 fet. Detected movement turns on trigger and keeps it on for 10 s after receipt of last signal, for actuating any type of alarm used as load for trigger.—Gunn Oscillator Is Designed for Use in 10.69-GHz Miniature Doppler Radar Equipment, *IEEE Spectrum*, July 1970, p 88.

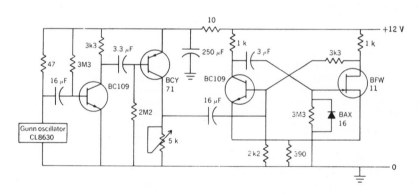

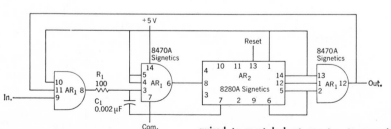

PULSED-INFRARED LOCKOUT COUNTER—Counter AR2 is driven by simple transformer circuit (also shown in article) for counting 60-Hz pulses. At count of nine, it triggers external relay circuit which is normally energized and not self-holding. Interruption of pulsed infrared beam removes logic signal required to reset lockout counter, to energize light-emitting diodes in relay circuit for indicating beam interruption. Used in infrared intrusion alarm that is highly immune to jamming, defeating, and disturbances.—New Infrared Photodiodes Make Possible Switching, Counting, and Detecting Applications, *IEEE Spectrum*, Dec. 1969, p 87.

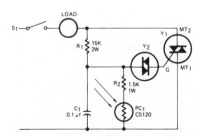

DARK-ON SWITCH—Cadmium sulfide cell and IR-IRD54 diac Y2 turn on IR-IRT82 triac Y1 at night, for energizing load which may be porch lights, night lights, or burglar alarms, for better home security. Maximum load is 1,000 W. Construction details are given.—"Hobby Projects," International Rectifier, El Segundo, CA, Vol. II, p 27–30.

D—Silicon rectifier diode (GE type IN 5059)
C1—20-mfd., 150 WVDC electrolytic capacitor (Sprague TE-1509)
C2—.22-mfd., 200 WVDC mylar capacitor (Sprague 192P22492)
C3—25-mfd., 50 WVDC electrolytic capacitor (Sprague TE13055)
SCR—Silicon-controlled rectifier (GE type C106B)
R1,R2—100-ohm, ½-watt, ±10% carbon resistors
R3—4.7-ohm, ½-watt, ±10% carbon resistor
R4—5-ohm linear-taper potentiometer (Mallory U12)
R5—5-ohm, 2-watt, ±10% carbon resistor
R6—Photoresistor (Clairex CL5M4)
R7—15-ohm, ½-watt, ±10% carbon resistor
Sonalert—Mallory SC-628P
S1—US-26 SPST switch
S2—SPST slide switch
RLY—Relay (LaFayette 99T6091)

SHADOW DETECTOR—R4 is set so scr does not quite receive enough positive pulses to cause regenerative feedback. Slightest reduction in light on photoresistor, as might be caused by burglar's shadow, makes scr draw enough current to energize relay and activate Sonalert alarm. Unit can also be positioned in home to warn when children are getting into mischief or entering dangerous area such as workshop. With S2 closed, alarm latches on; when open, alarm is momentary for duration of shadow. With normal daylight setting of R4, alarm will set itself off at darkness.—R. F. Graf and G. J. Whalen, *Alarm Keeps Its Electric Eye Out for Burglars*, Popular Science, June 1970, p 94–95.

PULSED-DIODE TRANSMITTER—Monsanto ME5 infrared-emitting diode is pulsed at 1 A for about 1.5 ms at intervals of 16.7 ms. Pulse width is proportional to capacitance of C1 between limits of 0.5 and 5 ms. Pulse magnitude is inversely proportional to value of R3 between limits of 0.3 and 3 A. Maximum receiver distance is about 9 meters without lenses. Pulsed-infrared system is difficult to jam or defeat and is virtually immune to disturbances.—*New Infrared Photodiodes Make Possible Switching, Counting, and Detecting Applications*, IEEE Spectrum, Dec. 1969, p 87.

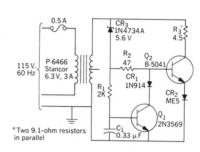

* Two 9.1-ohm resistors in parallel

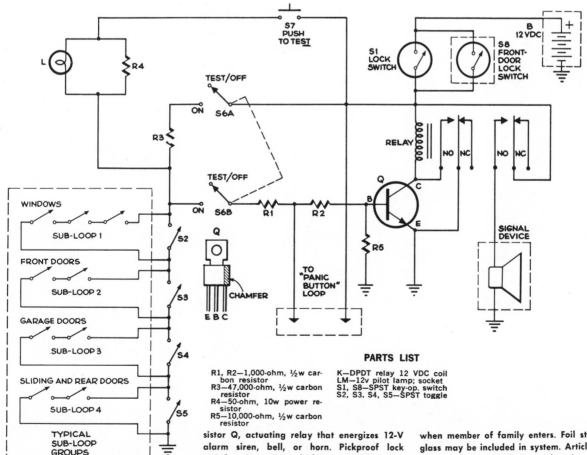

PARTS LIST

R1, R2—1,000-ohm, ½w carbon resistor
R3—47,000-ohm, ½w carbon resistor
R4—50-ohm, 10w power resistor
R5—10,000-ohm, ½w carbon resistor

K—DPDT relay 12 VDC coil
LM—12v pilot lamp; socket
S1, S8—SPST key-op. switch
S2, S3, S4, S5—SPST toggle

COMPLETE HOME SECURITY—Any open switch-type sensor or break in sensor wiring to door and window switches turns on transistor Q, actuating relay that energizes 12-V alarm siren, bell, or horn. Pickproof lock switches S1 or S8, whichever was closed to arm system, must be opened with key to turn off siren and reset circuit. Same switch must be opened to prevent alarm from going off when member of family enters. Foil strips on glass may be included in system. Article gives instructions for installation and testing of system. Q is GE D40C1.—R. M. Benrey, *How Electronics Can Guard Your Home*, Popular Science, Jan. 1971, p 92–95.

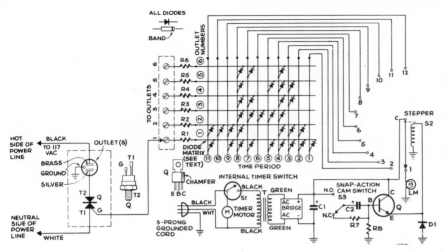

PROGRAMMED HOME-LIGHT CONTROL— Turns lights on automatically in all rooms of unoccupied home in patterns and times corresponding to normal family life. Conventional 24-hour timer from hardware store is modified by disconnecting interior wiring and bringing out individual leads from clock motor and internal switch mechanism for connecting to matrix of 1N914 diodes. Snap-action switch is closed by metal spacers cemented on rotating dial at times chosen for changes in lighting pattern, for advancing 12-position stepper switch S2. Diodes are wired between one of 11 time period lines and one of lines going to the six controlled electric outlets in home. Only single No. 20 stranded insulated hookup wire goes from control to GE SC40B triac in each output, for handling up to 250-W lamp. Q is GE D40D1, bridge is International Rectifier 50FB05L, T is Stancor P-6466, LM is No. 47 pilot, C1 is 1,-000 μF, C2 150 μF, R1–R7 100 ohms, and R8 560 ohms.—R. M. Benrey, People-at-Home Light Pattern, *Popular Science*, June 1971, p 98—100 and 118.

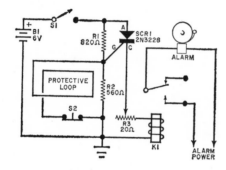

NO FALSE ALARMS—Designed to prevent false alarms caused by lightning or other electrical noise. Protective loop of foil or fine wire must be broken by intruder to fire scr and pull in Guardian 200 relay that sets off alarm. Once fired, scr remains conducting even if intruder restores loop, sounding alarm until circuit is reset by opening key-operated power switch S1. S2 is used to test system.— C. H. Goulden, Reader's Circuit, *Popular Electronics*, March 1970, p 97—98.

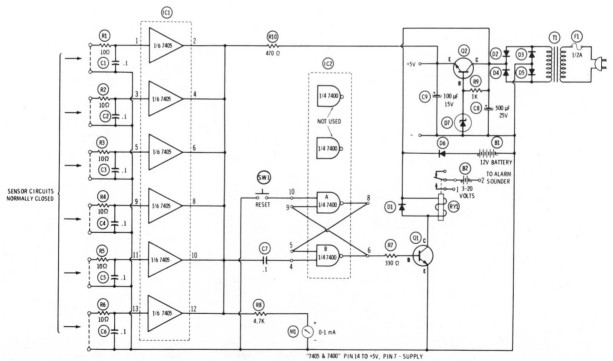

IC SENSOR CIRCUITS—Six separate input sensor circuits share 7405 hex inverter. Latch consisting of cross-coupled 2-input NAND gates of 7400 IC maintains alarm condition after sensor circuit is broken. Relay RY1 (Potter & Brumfield RS5D or similar) closes alarm circuit. Q1 is 2N3414 and Q2 is 2N2102. Diodes are 1N4003 silicon and zener D7 is 1N5232 (5.6 V). T1 has 12-V secondary at 700 mA.—C. D. Rakes, "Solid State Electronic Projects," Howard W. Sams, Indianapolis, IN, 1972, p 41—44.

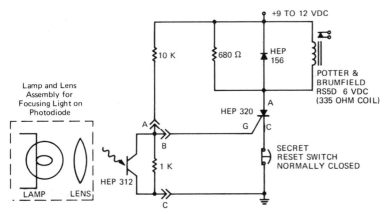

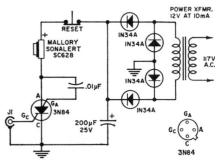

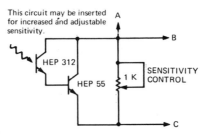

LIGHT-BEAM ALARM—Intruder walking through light beam interrupts illumination on HEP312 photodiode, to trigger HEP320 scr and energize relay controlling alarm bell. Concealed switch must be pushed momentarily to stop alarm and reset circuit. Transistor amplifier stage may be connected as shown to points A, B, and C for increased and adjustable sensitivity.—"Home Handyman's Construction Projects," Motorola Semiconductors, Phoenix, AZ, HMA-37, 1972.

This circuit may be inserted for increased and adjustable sensitivity.

LATCHING ALARM—Touching metal doorknob or other metal object connected to cathode-gate of scs by up to 25 feet of wire will trigger alarm and energize 2,800-Hz Sonalert. Reset button must be pressed to stop wail of alarm. Will not operate from batteries, nor will it be set off by person wearing gloves. —F. H. Tooker, Sensitive Intrusion Alarm, *Electronics World*, Jan. 1970, p 87.

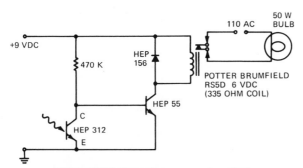

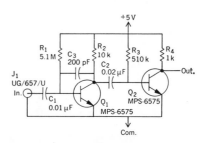

NIGHT LIGHT—Turns lamp on automatically at dusk, and off at sunrise, as deterrent to intruders when no one is home.—"Home Handyman's Construction Projects," Motorola Semiconductors, Phoenix, AZ, HMA-37, 1972.

PULSED-INFRARED RECEIVER—Signal from Monsanto MD2 silicon photodiode connected to J1 is amplified by Q1 and converted to saturated logic signal required for resetting lockout counter of intrusion alarm system. Will indicate interruption of pulsed infrared beam up to 9 meters from transmitter without using lenses.—New Infrared Photodiodes Make Possible Switching, Counting, and Detecting Applications, *IEEE Spectrum*, Dec. 1969, p 87.

CHAPTER 13
Capacitance Control Circuits

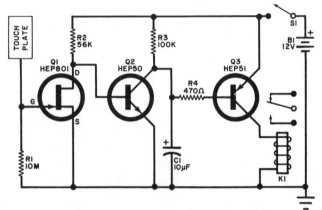

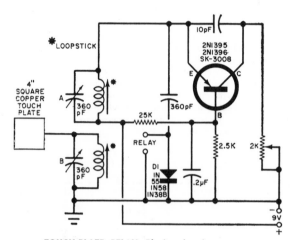

TOUCH SWITCH—When small metal plate is touched, signals picked up by body are amplified and detected to make 12-V d-c relay pull in and operate lamp, alarm, or other small electrical device connected to contacts.—Useful Circuits, *Popular Electronics*, Oct. 1971, p 88–89.

TOUCH-PLATE RELAY—Placing hand on copper plate detunes circuit, thereby energizing 2-mA 100-ohm relay connected to terminals shown. Mount loopsticks close to each other. Set B near minimum, set A about ¾ maximum, adjust pot so relay closes, and tune B until relay opens. Relay will now close when plate is touched.—C. J. Schauers, Capacitance Relay, *Popular Electronics*, March 1968, p 77.

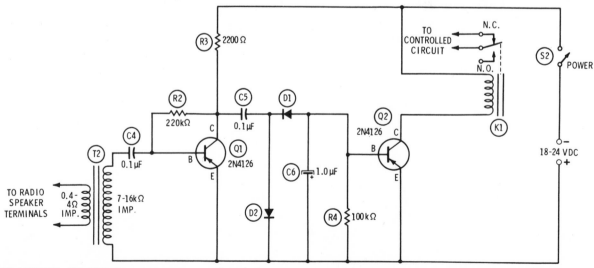

RELAY CONTROL FOR WIRELESS PROXIMITY DETECTOR—Used with capacitance control that radiates r-f signal in broadcast band when no one is near sensing plate. Power-supply hum modulation at speaker terminals is sufficient, after amplification by Q1 and rectification by 1N34A diodes, to energize 1,000-ohm d-c relay K1 through relay control transistor Q2. When person approaches or touches sensing plate of transmitter, carrier drops out and relay is released, closing alarm circuit. Any number of wireless proximity detectors may be used with single radio and relay control unit.—J. P. Shields, "How to Build Proximity Detectors and Metal Locators," Howard W. Sams, Indianapolis, IN, 1972, p 103–108.

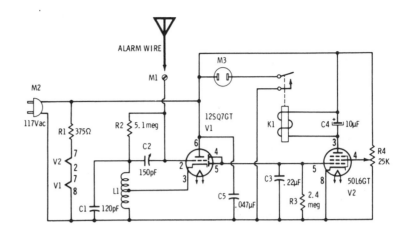

INTRUDER-DETECTOR—Energizes relay for turning on light or other device when person comes close to concealed or open wire serving as antenna. R4 and C2 adjust sensitivity for varying distance at which 5K relay is energized. L1 is 150 turns No. 30 enamel close wound on ½-inch diameter form.—R. M. Brown and T. Kneitel, "49 Easy Entertainment and Science Projects," Vol. 2, Howard W. Sams, Indianapolis, IN, 1969, p 21–23.

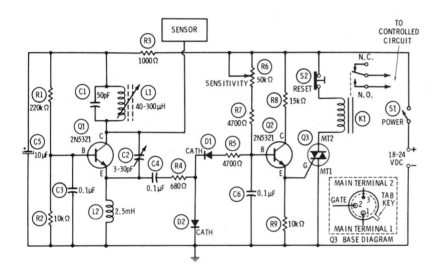

PROXIMITY DETECTOR WITH TRIAC—Uses RCA 40534 triac Q3 to control 1,350-ohm 12-V d-c relay having contacts rated 2 A, for turning on loud alarm or other device when person or metal object approaches sensing plate. Q1 is r-f oscillator whose output is rectified by 1N34A diodes for turning on Q2 and triggering triac. Relay remains energized until reset switch is pushed. Construction and adjustment details are given.—J. P. Shields, "How to Build Proximity Detectors and Metal Locators," Howard W. Sams, Indianapolis, IN, 1972, p 96–103.

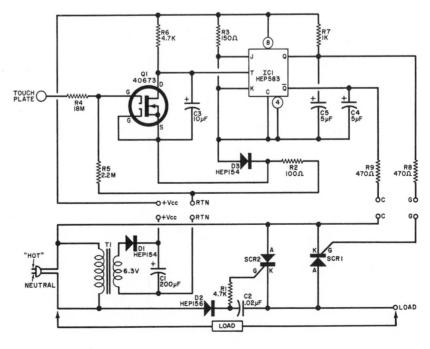

TOUCH SWITCH—Finger or any other part of body touching small metal plate feeds a-c signal pickup of body to insulated-gate mosfet Q1, making it conduct at 60 Hz and trigger IC flip-flop. This in turn produces trigger signals at points C and G for making scr's supply load power on alternate half-cycles. Will handle up to 450 W with IR106B1 or similar 0.5-mA 200-prv scr, and more with higher-power scr's. Load stays on until touch plate is touched again.—J. Nunley, Touch-Plate Power Switch, *Popular Electronics*, Aug. 1972, p 50–53.

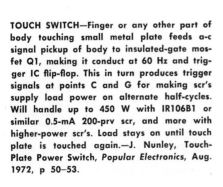

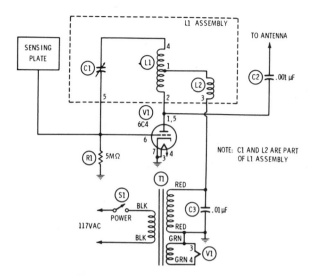

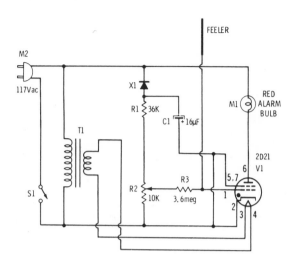

WIRELESS PROXIMITY DETECTOR—Simple transistor, consisting of loaded r-f oscillator, stops oscillating when object approaches sensing wire or metal plate. Used with relay control unit connected to speaker terminals of a-m radio tuned to transmitter frequency. Relay is energized as long as signal of transmitter is present. When person approaches and signal drops out, control relay releases and normally closed contacts complete alarm circuit. System is fail-safe, because alarm sounds for power failure or tampering with equipment. L1 assembly is J. W. Miller type 695. Power transformer T1 is Stancor PA-8421 with 6.3-V and 125-V secondaries.—J. P. Shields, "How to Build Proximity Detectors and Metal Locators," Howard W. Sams, Indianapolis, IN, 1972, p 103–108.

PROXIMITY DETECTOR—Flexible feeler wire concealed around doorway or window frame senses passage of intruder and trips thyratron which turns on silent alarm lamp that can be placed in bedroom or any other location. Adjust R2 until lamp almost lights by itself, then turn R2 back just enough so lamp comes on reliably whenever person passes through door. X1 is 70-mA 117-V selenium rectifier. T1 is 6-V filament transformer. For bell or buzzer alarm, use 5K relay in place of lamp.—R. M. Brown and T. Kneitel, "46 Easy Entertainment and Science Projects," Vol. 2, Howard W. Sams, Indianapolis, IN, 1969, p 61–62.

TOUCH TO TALK WITH LATCH—Touch of finger on contact plate energizes K1, which in turn pulls in latching relay K2 to keep transmitter turned on until contact plate is touched again. Use optional plate if capacitance of operator to ground is not enough; finger should bridge gap between plates.—C. Felstead, Touch To Talk, QST, Oct. 1968, p 20–21.

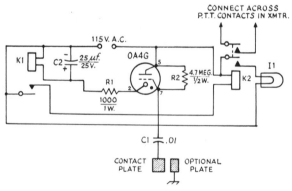

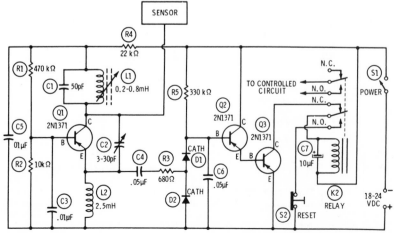

PROXIMITY DETECTOR—R-f output of loaded oscillator Q1 is rectified by 1N34A diodes for relay control transistors Q2 and Q3. K2 is 2,500-ohm d-c relay sufficiently sensitive to be energized when person or object approaches metal sensing plate without actually touching it. Once energized, relay is held closed by its latching contacts until reset button S2 is pressed. L1 is 0.2–0.8 mH adjustable ferrite-core coil, and L2 is 2.5-mH r-f choke. Construction and adjustment details are given.—J. P. Shields, "How to Build Proximity Detectors and Metal Locators," Howard W. Sams, Indianapolis, IN, 1972, p 91–93.

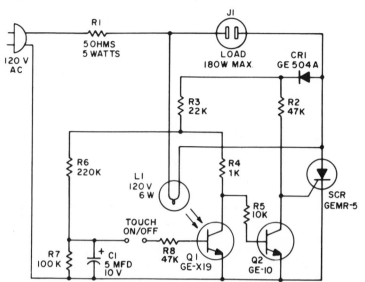

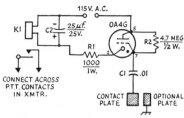

TOUCH SWITCH—One touch of spaced parallel bare copper wires on housing of switch box turns on lamp load of up to 180 W, and next touch turns off lamp. Eliminates fumbling for light switch in dark room, because control box can be placed on lamp table or any other convenient location. If lamp comes on when circuit is first plugged in, reverse wall plug. Construction details are given. Space wires about 1/16 inch apart in spiral or grid, anchored with epoxy cement and filed slightly, to form switch.—"Electronics Experimenters Circuit Manual," General Electric, Owensboro, KY, 1971, 3rd Ed., p 141–145.

TOUCH TO TALK—Finger on metal contact plate makes 0A4G gas triode conduct, energizing 2,500-ohm 60-mW sensitive d-c plate-circuit relay which turns on transmitter. Relay opens for switching back to receive when finger is lifted. Use optional plate if capacitance of operator to ground is not enough to pull in relay.—C. Frelstead, Touch To Talk, QST, Oct. 1968, p 20–21.

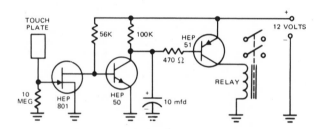

TOUCH SWITCH—Can be used to turn variety of devices on or off whenever person touches small piece of metal serving as touch plate. Relay is 12 V d-c. Plate should be close to gate terminal of HEP 801.—"Tips on Using FET's," Motorola Semiconductors, Phoenix, AZ, HMA-33, 1971.

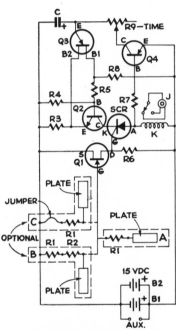

R1—10-megohm, ½w carbon resistor
R2—1.5-megohm, ½w carbon resistor
R3—47-ohm, ½w carbon resistor
R4—33,000-ohm, ½w carbon resistor
R5—10,000-ohm, ½w carbon resistor
R6—18,000-ohm, ½w carbon resistor
R7—1,000-ohm, ½w carbon resistor
R8—15,000-ohm, ½w carbon resistor
R9—1-megohm; ½w carbon potentiometer
C—50mfd, 25v electrolytic capacitor
Q1—N-channel FET (Int'l Rect. FE 100-C)
Q2—transistor G-E 2N657 or equiv.
Q3—2N1671-B Unijunction transistor
Q4—2N3638 transistor
SCR—Motorola HEP 320
K—SPDT, 12vDC relay (Calectro D1-963)
B1, B2—15v battery (Burgess K19)

TOUCH SWITCH—When metal plate is touched with finger, fet Q1 opens and allows scr to fire. This energizes relay K and initiates charging of capacitor C through time-delay control R9. When C is charged sufficiently to make Q3 conduct, current is removed from scr and relay drops out, resetting circuit. Values shown give delay up to 15 s; increasing value of C as high as 100 μF increases delay to several minutes, as might be required if relay contacts are connected to radio or tv speaker for killing commercials. Can also be used as doorbell button, to give good loud ring. (Three possible plate connections are shown; use only the one which works best.)—W. J. Hawkins, Action at Your Fingertips, Popular Science, March 1971, p 96–97 and 116.

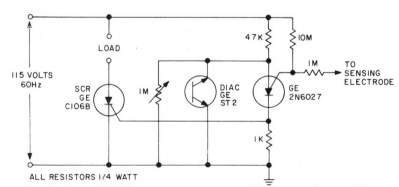

PROXIMITY SWITCH—When part of human body is within few inches of insulated metal plate serving as sensing electrode, resulting increase in capacitance to ground turns on 2N6027 put, to trigger scr on. Sensitivity is adjusted with 1-meg pot. Can also be used as electrically isolated touch switch. Will handle loads up to about 4-A rating of scr.—"SCR Manual," General Electric, Syracuse, NY, 1972, 5th Ed., p 224.

DOORKNOB ALARM—When hung on inside knob of door, alarm triggers loud bicycle-horn buzzer when anyone touches outside doorknob. Intended to frighten off intruder or annoying salesman, as well as to alert occupant. Sensitivity of alarm increases as battery runs down, so continuous operation when no one touches doorknob is warning that battery needs replacing. Will not work on metal doors unless knob mechanism is completely insulated from door. D1 and D2 are 1N60, D3 is 1N907, SCR1 is C6U, Q1 is 2N3394, Q2 is 2N3391, and L1 is 6.05–15.5 μH adjustable r-f coil. Connect bare wire loop to BP1 for hanging alarm on knob.—V. Kell, The Tipoff Intruder Alarm, *Electronics Illustrated*, Nov. 1969, p 69–72.

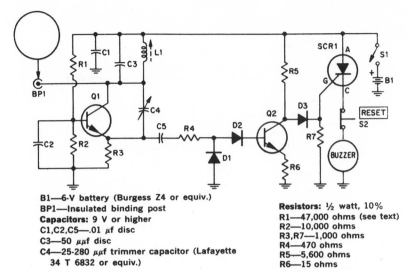

B1—6-V battery (Burgess Z4 or equiv.)
BP1—Insulated binding post
Capacitors: 9 V or higher
C1,C2,C5—.01 μf disc
C3—50 μμf disc
C4—25-280 μμf trimmer capacitor (Lafayette 34 T 6832 or equiv.)

Resistors: ½ watt, 10%
R1—47,000 ohms (see text)
R2—10,000 ohms
R3,R7—1,000 ohms
R4—470 ohms
R5—5,600 ohms
R6—15 ohms

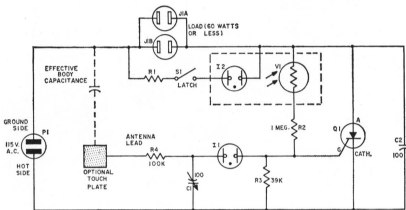

TOUCH TO TALK—Depends on body capacitance near antenna lead to trigger scr and supply current to lamps or other load up to 60 W plugged into outlets. With S1 closed, light from NE-2 neon I2 drops resistance of Clairex CL903 cadmium sulfide photocell enough to keep scr conducting for latch action. I1 is NE-83 neon and Q1 is GE C106B1 scr. Sensitivity depends on length of antenna lead or area of optional metallic plate.—J. J. Glauber, Touch Control, *QST*, June 1969, p 22–23.

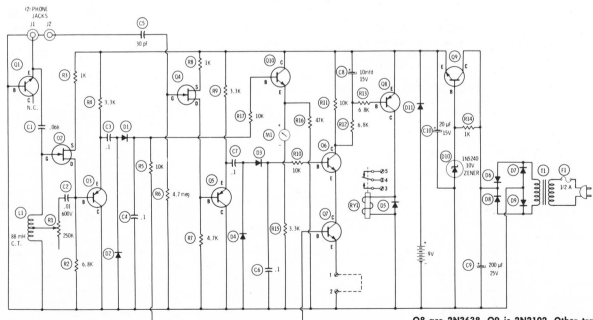

PROXIMITY ALARM—Two wires, both connected to metal safe or other metal object being protected, are run to jacks J1 and J2 of alarm circuit located at least 1 ft away. Cutting either or both wires makes 500-ohm 14.5-mA alarm relay RY1 drop out, energizing alarm circuit connected to terminals 4 and 5. Failure of a-c or battery power similarly triggers alarm. Range of sensitivity adjustment R1 can be varied from actual touch to detection of intruder up to 1 ft away from safe. Q2 and Q4 are 2N5485. Q3, Q5, and Q8 are 2N3638. Q9 is 2N2102. Other transistors are 2N2924. Diodes are 1N4003 silicon. M1 is 0–1 mA. Two wires run from terminals 1 and 2 to common ground provide further fail-safe action.—C. D. Rakes, "Solid State Electronic Projects," Howard W. Sams, Indianapolis, IN, 1972, p 35–41.

TOUCH-SWITCH PLATE—Can be used in place of standard wall switch, to eliminate fumbling for switch in dark. Plate can be same size as for standard switch, or larger to increase sensitivity. First touch closes circuit, and next touch opens it.—S. Hoberman, Touch Module—This New Version Does Many Jobs, *Popular Science*, Feb. 1973, p 124–125.

C1—220-mmf mica capacitor
C2—1-10 mmf var. capacitor
C3—4.7-mfd @ 25v electrolytic capacitor
R1—15,000-ohm, 1w resistor
R2—3300-ohm, ½w resistor
ZD—10v zener diode 1N302OB or equiv.
Q—NPN transistor 2N1304 or equiv.
SCR—silicon controlled rectifier (Texas Instr. T140A1 or equiv.)
NE—neon lamp NE-2
K—24vDC latching relay (Cornell Dubilier 662-24)
Misc.—vector board, terminals strip, spacers

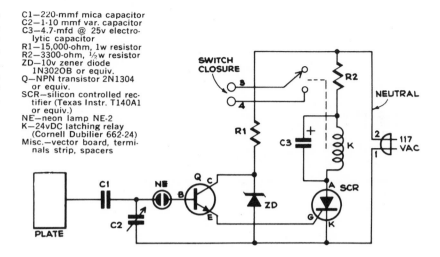

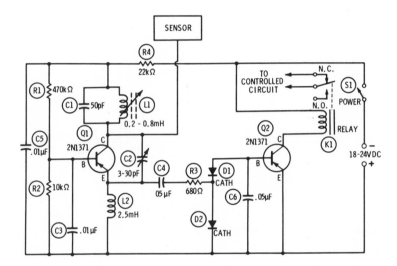

TOUCHSWITCH—Two-transistor loaded oscillator-amplifier energizes relay by capacitance effect when person or large object approaches or touches metal sensing plate. Portion of r-f voltage developed by Q1 is rectified by 1N34A or equivalent germanium diodes and applied to relay control transistor Q2. K1 is 1,350-ohm 12-V d-c relay. L1 is 0.2–0.8 mH adjustable ferrite-core coil, and L2 is 2.5-mH r-f choke. Construction and adjustment details are given.—J. P. Shields, "How to Build Proximity Detectors and Metal Locators," Howard W. Sams, Indianapolis, IN, 1972, p 87–91.

TOUCH CONTROL—Momentary touch on both contact plates triggers thyratron V1, energizing 2.5K relay K1 which in turn energizes and latches 115-V a-c locking relay K2 for applying power to stereo system, lamp, or any other equipment plugged into receptacle. Touch plates again to turn everything off. D1 is 65-mA selenium rectifier.—R. Brown and T. Kneitel, "101 Easy Audio Projects," Howard W. Sams, Indianapolis, IN, 1971, p 155–156.

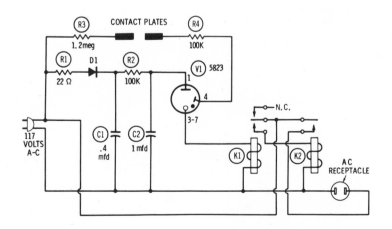

CHAPTER 14
Cathode-Ray Circuits

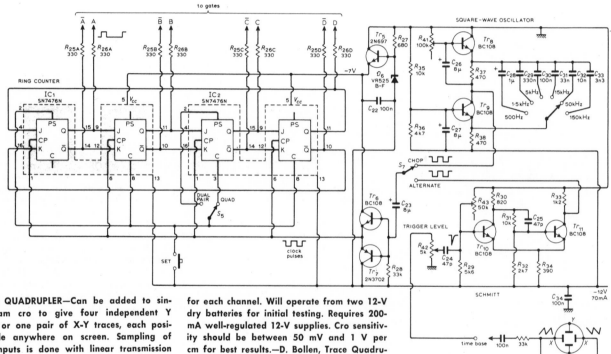

TRACE QUADRUPLER—Can be added to single-beam cro to give four independent Y traces or one pair of X-Y traces, each positionable anywhere on screen. Sampling of four inputs is done with linear transmission gates switched by J-K flip-flop ring counter. Switching frequency can be up to 2 MHz. Article includes circuit of input preamp needed for each channel. Will operate from two 12-V dry batteries for initial testing. Requires 200-mA well-regulated 12-V supplies. Cro sensitivity should be between 50 mV and 1 V per cm for best results.—D. Bollen, Trace Quadrupler for D.C. Scopes, *Wireless World*, May 1972, p 204–209.

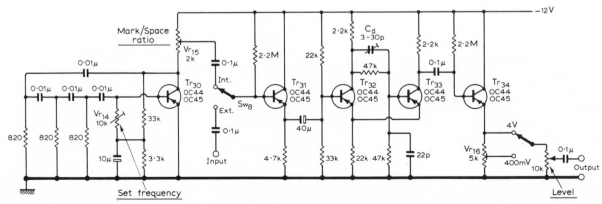

SQUARE-WAVE CALIBRATOR — Generates square wave with output amplitude variable between 50 mV and 4 V p-p from 10-kHz internal source of cro or at any frequency from 15 Hz to 20 kHz from external sine-wave source. Tr30 is 10-kHz sine-wave oscillator driving Schmitt trigger Tr32-Tr33 through buffer Tr31. Used in solid-state cro for work on color tv sets.—M. Phillips, Solid-State Oscilloscope, *Wireless World*, March 1969, p 110–115.

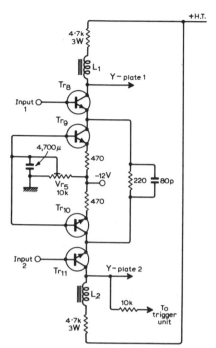

VERTICAL AMPLIFIER—Used between preamp and cathode-ray tube in solid-state cro designed for work on color tv sets. Should be located as close as possible to c-r tube base. Use heat sink for Tr8 and Tr11; article suggests 2.5 x 5 cm sheet of copper wrapped once around transistor, with remainder sticking out. Transistors can be low-power npn silicon like BC107.—M. Phillips, Solid-State Oscilloscope, *Wireless World*, March 1969, p 110–115.

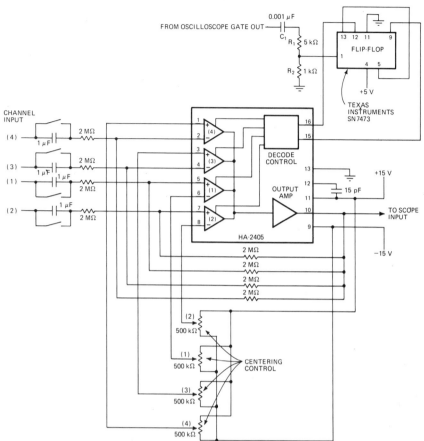

FOUR-CHANNEL DISPLAY—Converts single-channel or dual-channel scopes to provide four-channel display. Key element is Harris HA-2405 four-channel programmable opamp. Provides unity gain per channel, bandwidth of d-c to 5 MHz, maximum input voltage of ±10 V, and slew rate of 15 V per μs.—G. M. Wood, Converter for Oscilloscope Provides Four-Channel Displays, *Electronics*, July 31, 1972, p 76–77.

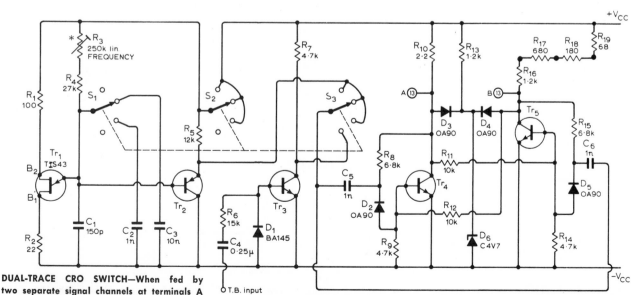

DUAL-TRACE CRO SWITCH—When fed by two separate signal channels at terminals A and B, switching waveform generator makes it possible to show two different signals of same frequency simultaneously on screen. TB input terminal connects to time base output of cro. All npn transistors are BC107, and pnp are BC157. Supply is 12 V. Article includes circuit for amplifier channel.—W. T. Cocking, Dual-Trace Oscilloscope Unit, *Wireless World*, Jan. 1972, p 19–25.

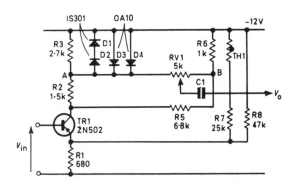

GAMMA CORRECTION—Used to modify applied signal in compliance with a gamma law, to improve crt display of signals having widely differing intensities. Provides constant output amplitude at point B, with RV1 serving to vary gamma.—S. L. Cachia, Control of Gamma in C.R.T. Displays Using Amplifiers with Exponential Negative Feedback, *The Radio and Electronic Engineer*, Nov. 1969, p 281–291.

END-OF-SWEEP ERASURE—Used with storage cro to erase trace of slow event automatically at end of each sweep. Circuit shown is designed for addition to Tektronix 564B scope, in which manual erasure is accomplished by grounding single charged capacitor. Circuit samples horizontal sawtooth at one of crt horizontal deflection plates, for triggering one-shot that closes two transistor switches, each in parallel with existing erase switch.—T. Richardson and A. R. Freeman, Adding Automatic Erasure to Storage Oscilloscopes, *Electronics*, Aug. 2, 1973, p 105–106.

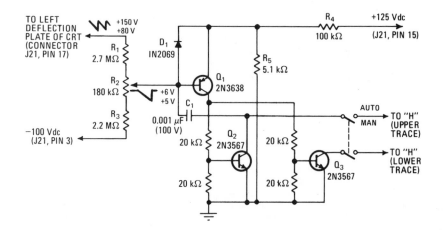

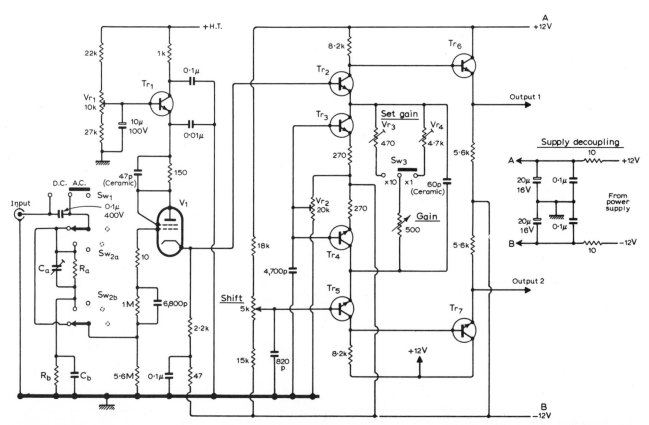

VERTICAL PREAMP—Response is flat from d-c to 6 MHz and is 3 dB down at 11 MHz. Gain can be increased 10 times with Sw3 but 3-dB bandwidth is then reduced to 1.5 MHz. Designed for troubleshooting of color tv sets. V1 is low-power triode or strapped pentode such as 6AK5. Transistors are all low-power silicon like BC107, with hfe of 125–500 and fT of about 85 MHz.—M. Phillips, Solid-State Oscilloscope, *Wireless World*, March 1969, p 110–115.

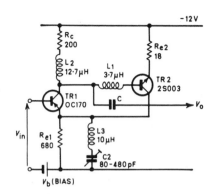

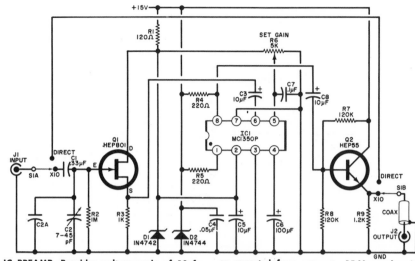

WIDE-BAND GAMMA CONTROL—Circuit maintains phase shift essentially at zero within feedback loop. Used to provide improved crt display of signals having widely differing intensities. Response is flat within 3 dB up to 8.7 MHz for signal input of 0.5 V. —S. L. Cachia, Control of Gamma in C.R.T. Displays Using Amplifiers with Exponential Negative Feedback, *The Radio and Electronic Engineer*, Nov. 1969, p 281–291.

IC PREAMP—Provides voltage gain of 10 for viewing low-level signals on scope. Output goes to vertical input of scope. Bandwidth is about 5 MHz. IC is 45-MHz tv i-f unit; alternative is HEPC6059P. Can be built into scope or operated from separate 15-V supply, circuit for which is given in article. If C2A is needed, try 33 pF.—I. Gorgenyi, A Preamplifier for Your Scope, *Popular Electronics*, Aug. 1972, p 47–49.

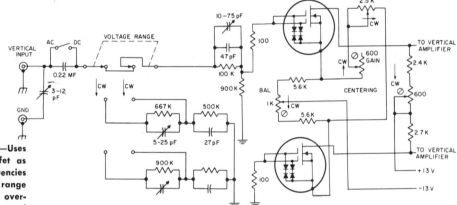

VERTICAL DIFFERENTIAL AMPLIFIER—Uses RCA 4841 silicon dual-insulated gate fet as differential amplifier covering all frequencies from d-c to 500 MHz. Wide dynamic range permits large-signal handling before overloading.—"Linear Integrated Circuits and MOS Devices," RCA, Somerville, NJ, 1973 Edition, SSD-201A, p 677.

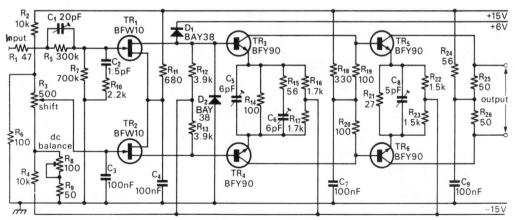

VERTICAL PREAMP—Effective noise signal is less than 0.2 mV (within width of electron beam of finest lab cro) from d-c to 300 MHz, with constant gain over entire range. Consists of three-stage differential amplifier with push-pull output. One input is used for signal, and other for adjusting d-c bias that controls vertical shift. For protection against accidental application of high voltage signal input, catching diodes D1 and D2 limit source voltage excursions of TR1.—Some Applications of Field-Effect Transistors, *Electronic Engineering*, Sept. 1969, p 18–23.

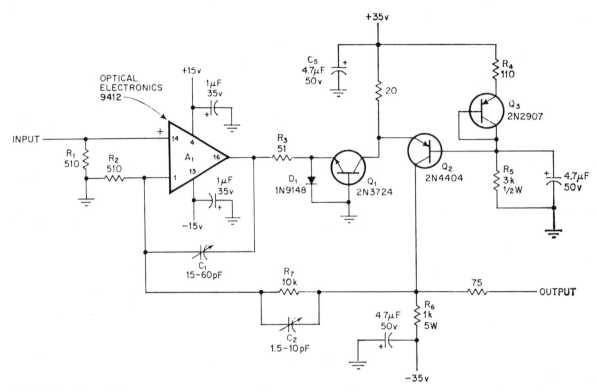

CRT CATHODE DRIVER—Delivers 50 V p-p, has response of 10 MHz, and is stable beyond 100 MHz. C1 is adjusted to roll off gain of A1 by 20 dB per decade from d-c to 10 MHz; gain is then unity until rolloff resumes again at about 350 MHz.—Walter A. Cooke, High-Voltage Amplifier Offers High Frequency, Too, *Electronics*, April 26, 1971, p 57.

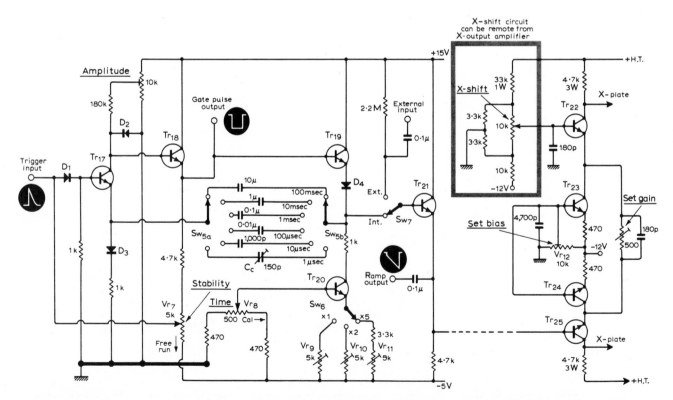

RAMP GENERATOR—Used in solid-state cro designed for work on color tv sets. Diodes are general-purpose germanium like OA81, and all transistors are low-power npn silicon like BC107. Horizontal output amplifier is included. Small step at start of ramp shows up as bright spot at start of trace. Sw6 switches resistors to provide multipliers for decade capacitors switched by Sw5.—M. Phillips, Solid-State Oscilloscope, *Wireless World*, March 1969, p 110–115.

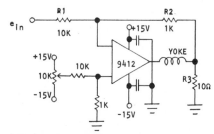

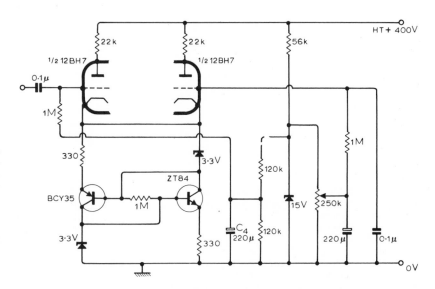

WIDEBAND DEFLECTION—Fast-deflection amplifier for computer terminal and other displays having no repetitive waveform meets requirements of high output current, short settling time, fast slewing rate, and 300-MHz gain-bandwidth product. Yoke current is 10 mA/V, so 20-V p-p input gives 200-mA p-p current. Current linearity is good. Bypass all power supply pins of module to common ground with 0.1 μF.—"A Wideband Deflection Amplifier," Optical Electronics, Tucson, AZ, Application Tip 10166, 1971.

ELECTROSTATIC DEFLECTION AMPLIFIER—Consists of double-triode connected as long-tailed pair, with two-transistor constant-current circuit in common cathode lead. Provides 460-V p-p output swing for 28-V p-p input swing.—D. E. Vaughan, Hybrid Push-Pull Deflection Amplifier, *Wireless World*, Sept. 1970, p 453.

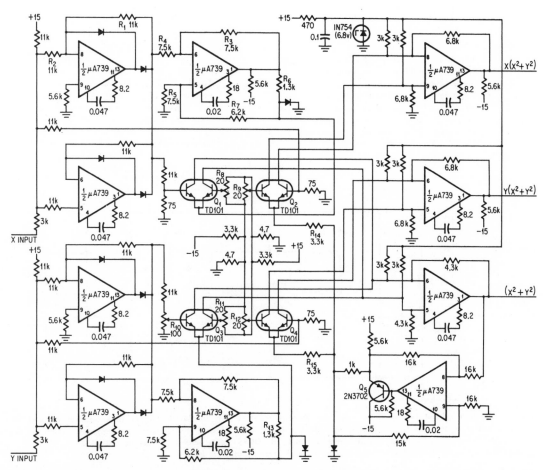

PINCUSHION AND BLURRING—Inexpensive IC opamps in correction circuit provide high-precision solution to problem of using spherical geometry of scanning beam on flat face of high-resolution cathode-ray display. Accuracy depends on matching of various resistors and transistors, and can be as good as 0.5% of full-scale accuracy.—J. L. Divilbiss and S. Franco, IC Op Amps Straighten Out CRT Graphic Displays, *Electronics*, Jan. 4, 1971, p 70–72.

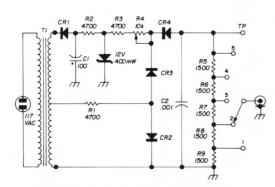

VOLTAGE CALIBRATOR—Provides outputs of 1, 2, 3, 4, and 5 V p-p for calibrating cro inexpensively to measure voltages of waveshapes displayed on screen. CR is 100 piv at 1 A; other diodes are 1-mA silicon, such as 1N4005. Adjust R4 to give −2.65 V d-c at TP; circuit will then be calibrated for 5-V p-p square wave.—F. E. Emerson, Oscilloscope Voltage Calibrator, *Ham Radio*, Aug. 1972, p 54–55.

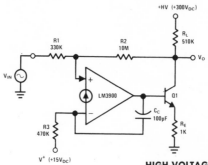

HIGH-VOLTAGE INVERTING—Provides output voltage swing from essentially 0 V to +300 V for input voltage range of 0 to 10 V d-c. Gain is 30. Q1 must have high breakdown voltage rating to withstand full supply voltage. Can be used in electrostatic crt deflection system.—T. M. Frederiksen, W. M. Howard, and R. S. Sleeth, "The LM3900—A New Current-Differencing Quad of ± Input Amplifiers," National Semiconductor, Santa Clara, CA, 1972, AN-72, p 37.

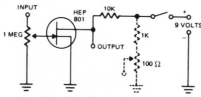

VTVM OR CRO AMPLIFIER—Can also be used as relay amplifier. Input of 0 to −1.5 V gives output of 0.4 to 9 V. To bring output down to 0, add dashed circuit and use terminal of pot as ground for output; adjust pot to give 0 output with input grounded.—"Tips on Using FET's," Motorola Semiconductors, Phoenix, AZ, HMA-33, 1971.

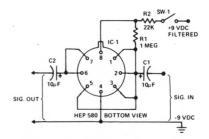

10 HZ–1 MHZ—Provides gain of 100 over bandwidth of entire frequency range, for 0.01 V p-p input. Can be used as preamp for cro, mike, phono, hearing aid, or stereo system.—"Tips on Using IC's," Motorola Semiconductors, Phoenix, AZ, HMA-32, 1972.

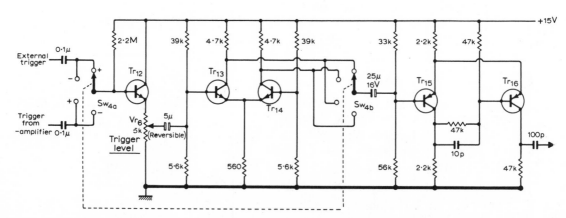

TIME-BASE TRIGGER—Used in solid-state cro designed for work on color tv sets. Emitter-follower Tr12 amplifies trigger and drives phase-splitting amplifier Tr13-Tr14 for polarity selection; these transistors can be low-power npn silicon like BC107. Tr15 and Tr16, which can be OC44 or OC45, form Schmitt trigger whose differentiated output is used to trigger time base of cro.—M. Phillips, Solid-State Oscilloscope, *Wireless World*, March 1969, p 110–115.

CHAPTER 15
Chopper Circuits

FLOATING AMPLIFIER—Noninverting IC chopper can float above ground if impedance between inverting terminal and ground is less than 5K. Used to measure low-level voltages produced by current signals passing through remotely located current shunt Rs.—P. Zicko, "Designing with Chopper Stabilized Operational Amplifiers," Analog Devices, Inc., Norwood, MA, Sept. 1970.

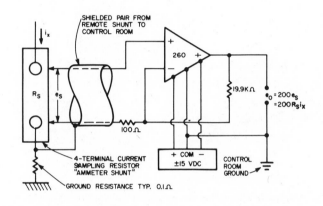

TWO-FET CHOPPER—Provides good coupling for d-c input signals, along with fast rise time and slow fall time, to minimize capacitive coupling into output.—B. R. Smith, Choppers: A Survey of Low-Level Switching, *The Electronic Engineer*, March 1968, p 59–70.

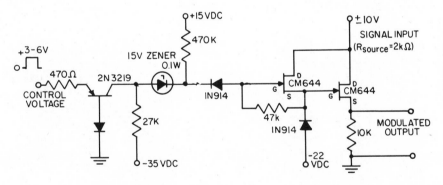

SERIES-SHUNT MOST—Circuit alternately switches input signal on and off at rate determined by clock pulse, while cancelling switching spikes. Can be used to generate amplitude samples of input signal if on/off ratio is made small enough. Provides +3 V to turn most's hard off and —20 V to turn them hard on. Can handle signals up to 5 V peak at input.—V. J. Phillips and J. Dunlop, A Series Shunt MOST Chopper with Complementary Drive Circuit, *Electronic Engineering*, Sept. 1969, p 29–31.

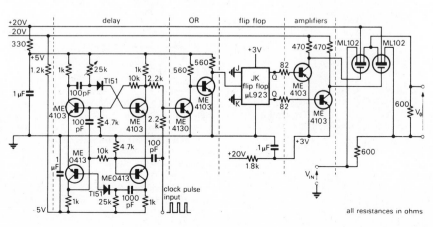

all resistances in ohms

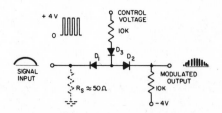

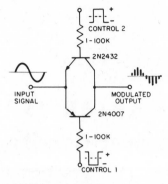

DIODE CHOPPER—In simplest chopper, D1 blocks when control voltage is 0, and is forward-biased when control is 4 V, to give chopping of input signal as long as its amplitude is less than drop across Rs. Diodes can be 1N999.—B. R. Smith, Choppers: A Survey of Low-Level Switching, *The Electronic Engineer*, March 1968, p 59–70.

COMPLEMENTARY-TRANSISTOR CHOPPER— Provides accurate series-shunt chopping. Not suitable for multiplexers. Provides return path for base current, to increase input impedance to control circuit.—B. R. Smith, Choppers: A Survey of Low-Level Switching, *The Electronic Engineer*, March 1968, p 59–70.

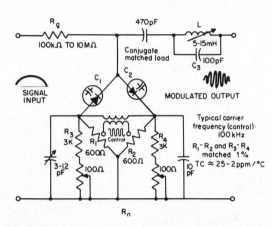

VARIABLE-C DIODE CHOPPER—Voltage-variable capacitor bridge modulates input signal and provides amplification. Circuit is essentially parametric amplifier.—B. R. Smith, Choppers: A Survey of Low-Level Switching, *The Electronic Engineer*, March 1968, p 59–70.

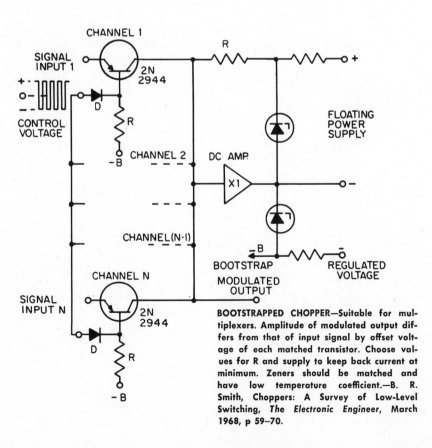

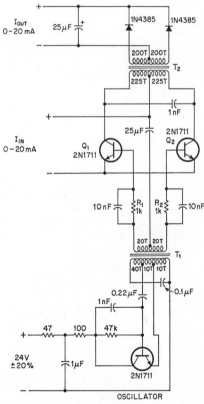

BOOTSTRAPPED CHOPPER—Suitable for multiplexers. Amplitude of modulated output differs from that of input signal by offset voltage of each matched transistor. Choose values for R and supply to keep back current at minimum. Zeners should be matched and have low temperature coefficient.—B. R. Smith, Choppers: A Survey of Low-Level Switching, *The Electronic Engineer*, March 1968, p 59–70.

D-C TRANSFER—Chopper Q1-Q2, acting on d-c input, is isolated by transformers from both oscillator and load. Used for transferring d-c to instrument outputs without degrading signal accuracy. Class-C Hartley oscillator operates at about 20 kHz. Can also be used to drive analog-digital converter.—S. R. Hjorth, AC Transfers DC Between Two Isolated Instruments, *Electronics*, April 12, 1971, p 88.

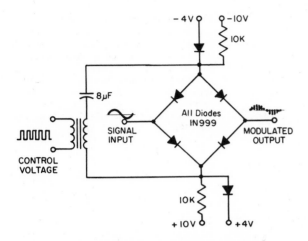

DIODE MODULATOR—Used in small multiplex systems that require good response time. Will chop signals up to 6 V p-p in either polarity, from 300 to 3,500 Hz.—B. R. Smith, Choppers: A Survey of Low-Level Switching, *The Electronic Engineer*, March 1968, p 59–70.

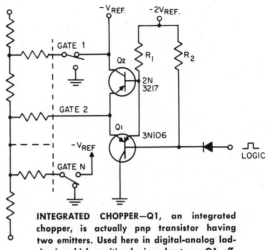

INTEGRATED CHOPPER—Q1, an integrated chopper, is actually pnp transistor having two emitters. Used here in digital-analog ladder in which positive logic pulse turns Q1 off, so its output lifts emitter of Q2 and turns it on to achieve switching on ladder. Choose values of R1 and R2 to keep Q1 on when logic input is 0.—B. R. Smith, Choppers: A Survey of Low-Level Switching, *The Electronic Engineer*, March 1968, p 59–70.

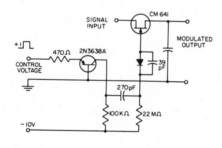

FET CHOPPER—Combines high impedance of fet for input signal with low control-voltage requirement of bipolar transistor. Low power dissipation makes circuit ideal for IC.—B. R. Smith, Choppers: A Survey of Low-Level Switching, *The Electronic Engineer*, March 1968, p 59–70.

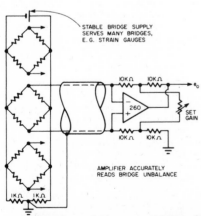

INVERTING CHOPPER FOR BRIDGES—Differential operation of Analog Devices Model 260 noninverting opamp permits handling push-pull signals of strain-gage bridge. Noninverting feature permits energizing of many bridges in parallel by single regulated d-c supply.—P. Zicko, "Designing with Chopper Stabilized Operational Amplifiers," Analog Devices, Inc., Norwood, MA, Sept. 1970.

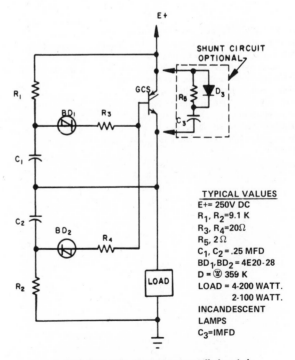

TYPICAL VALUES
E+= 250V DC
R_1, R_2=9.1 K
R_3, R_4=20Ω
R_5, 2 Ω
C_1, C_2 = .25 MFD
BD_1, BD_2 = 4E20-28
D = Ⓦ 359 K
LOAD = 4-200 WATT.
2-100 WATT.
INCANDESCENT
LAMPS
C_3=IMFD

1 KHZ AT 1 KW—Uses gate-controlled switch GCS as spst switch operating at 1 kHz with component values shown. Mvbr using four-layer diodes BD1 and BD2 determines operating frequency.—"Silicon Controlled Rectifier Designers Handbook," Westinghouse, Youngwood, PA, 1970, 2nd Ed., p 8-83–8-85.

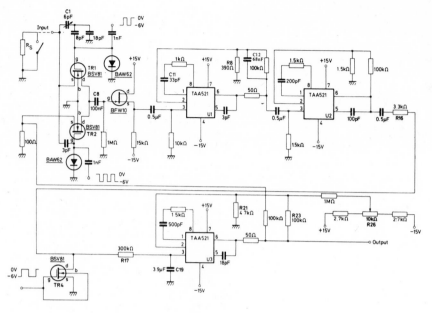

MOSFET CHOPPER IN D-C AMPLIFIER—TR1 and TR2 are driven by square-wave voltages in such a way that when one is on, other is off. Circuit is analyzed in detail. Chopping rate is 1 kHz. With chopper stabilization, offset voltage is less than 10 μV and offset drift 50 nV per deg C, while offset current is less than 100 pA and its drift 1.2 pA per deg C.—"Field Effect Transistors," Mullard, London, 1972, TP1318, p 110–115.

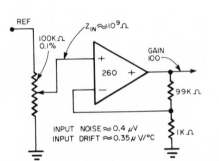

INVERTING CHOPPER OPAMP—Chopper-stabilized connection for Analog Devices Model 232 opamp provides buffering for output of precision pot. Accuracy is 0.1% or better, but noise and drift are several orders of magnitude greater than with alternative noninverting approach. Used in process control and instrumentation.—P. Zicko, "Designing with Chopper Stabilized Operational Amplifiers," Analog Devices, Inc., Norwood, MA, Sept. 1970.

NONINVERTING CHOPPER—Used as buffer for precision pot in process control and instrumentation systems. Inherently high resistance of circuit contributes to low noise and drift. Noise is only 0.1 μV for opamp and about 0.4 μV p-p for equivalent input. Common-mode error is negligible 0.003%.—P. Zicko, "Designing with Chopper Stabilized Operational Amplifiers," Analog Devices, Inc., Norwood, MA, Sept. 1970.

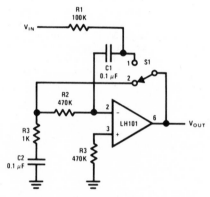

RESET STABILIZED AMPLIFIER—Variation of chopper-stabilized amplifier is operated with opamp having closed-loop gain of one. Circuit eliminates errors due to offset voltage and bias current. Output is pulse with amplitude equal to input voltage. Position 2 of S1 gives unity gain, with output voltage equal to sum of input offset voltage and drop across R2 due to input bias current. In position 1, voltage across C1 is inserted between output and inverting input of IC, so amplifier output changes by V-IN. Used at very low switching speeds—W. S. Routh, "Applications Guide for OP AMPS," National Semiconductor, Santa Clara, CA, 1969, AN-20.

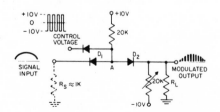

THREE-DIODE CHOPPER—Useful for multiplexers where several choppers must be operated from same supply. Adjust 20K pot to balance chopper, so no current flows through D1 when input signal is 0.—B. R. Smith, Choppers: A Survey of Low-Level Switching, *The Electronic Engineer*, March 1968, p 59–70.

CHAPTER 16
Citizens Band Circuits

EMERGENCY MONITOR—Speaker at right and components in dashed rectangle are in inexpensive CB transceiver tuned to citizens-band emergency channel 9. Combination relay amplifier and lamp is connected to transceiver output. With S1 set to LAMP, emergency message will make lamp come on; with S1 set to SPKR, relay is energized and message is heard from speaker. Closing S2 gives continuous monitoring of channel 9. T2 is Calectro D1-750, BR1 is HEP175, D1–D4 are 1N60, Q1–Q3 are any silicon npn transistor, Q4 is 2N3393, L1 is 12 V at 30 mA, and RY1 is P & B RS5D-6. C1 is 5 μF, C2 50 μF, C3 250 μF, and C4 100 μF. R1 is 47, R2 and R4 27K, R3 270K, R5 1 meg, R6 33, R7 10K, and R8 15K.—H. Friedman, Channel 9 Override Monitor, *Electronics Illustrated*, July 1972, p 31–34 and 99–100.

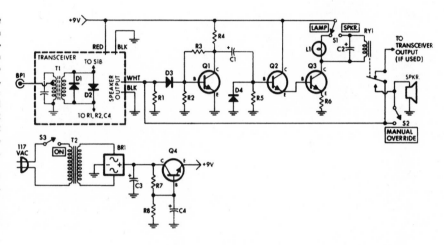

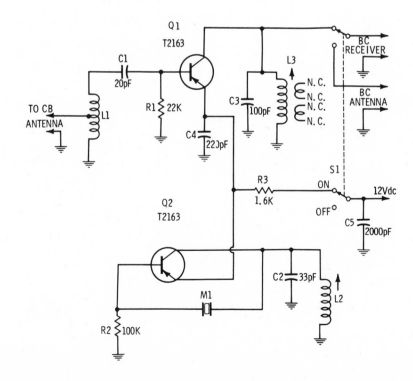

CONVERTER FOR CAR RADIO—When inserted between antenna and receiver of car radio, with switch S1 on, permits receiving CB calls. Flip S1 off to restore normal a-m reception. Keep all connections as short as possible. Initially, adjust L2 slowly to point giving reliable oscillation each time converter is turned on. Be sure to shield converter and use coax for its antenna and receiver connections. M1 is third-overtone 25.625-kHz crystal. L1 is 14 turns No. 20 enamel tapped 2 turns from ground, L2 is 9 turns No. 20 on same CTC LS3/B form as L1, and L3 is secondary of adjustable miniature bc oscillator coil.—R. M. Brown, "101 Easy CB Projects," Howard W. Sams, Indianapolis, IN, 1968, p 94–96.

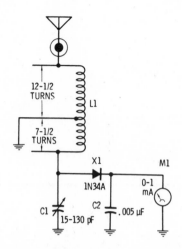

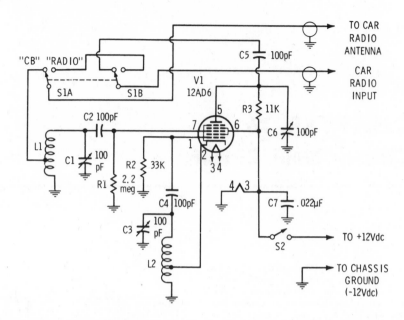

CB FIELD STRENGTH—Used for monitoring transmitter signal strength and improving transmitter tuning. L1 is No. 24 enamel on ⅝-inch form. Values shown for L1 and C1 are given for CB frequencies, but can be changed for other ham or communication frequencies.—R. M. Brown and T. Kneitel, "101 Easy Test Instrument Projects," Howard W. Sams, Indianapolis, IN, 1968, p 86–87.

AUTO RADIO CONVERTER—Can be used with single pushbutton setting of 12-V auto radio, because all CB tuning is done in converter. Adjust C6 once to optimize performance and give best signal-to-noise ratio. C3 serves as preset calibration control for main tuning control C1. L1 and L2 are each made with 5 turns No. 16 enamel spaced to ¾-inch on ⅝-inch diameter form, tapped 2 turns from grounded end.—R. M. Brown, "101 Easy CB Projects," Howard W. Sams, Indianapolis, IN, 1968, p 117–118.

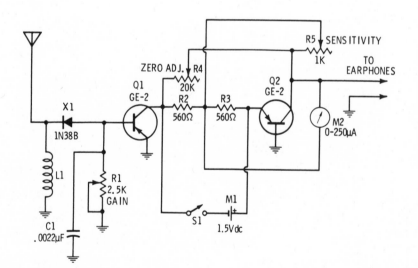

FIELD-STRENGTH METER—Short antenna wire, held close to antenna of CB rig, picks up actual r-f output. Adjust R4 to zero meter with no r-f present, then turn on CB set, adjust R5 for meter indication, adjust R1 for best volume, then make CB antenna matching adjustments for maximum volume and maximum meter reading. L1 is 10 mH.—R. M. Brown, "101 Easy CB Projects," Howard W. Sams, Indianapolis, IN, 1968, p 9–10.

TUNABLE CB RECEIVER—For earphone reception, requires only single 9-V transistor radio battery. Tune C1 to desired signal, then adjust preset control R3 for minimum distortion and maximum volume. L1 is one turn No. 26 interwound at collector end of L2 having 12 turns No. 14 on ½-inch form, spacewound. L3 is 50-H 4,400-ohm choke, and L4 is 1 mH.—R. M. Brown, "101 Easy CB Projects," Howard W. Sams, Indianapolis, IN, 1968, p 81–82.

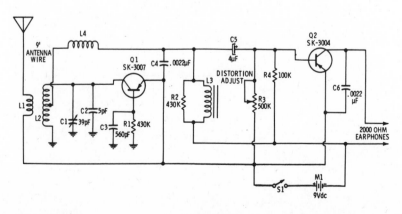

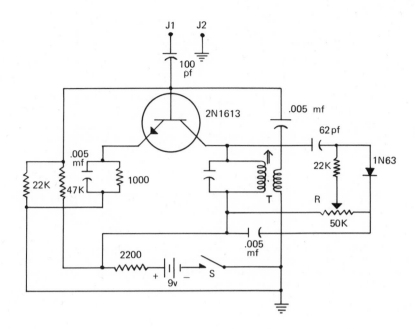

BFO FOR SSB WITH A-M CB—Single-transistor circuit can be built into citizens-band transceiver or mounted outboard. T is 455-kHz i-f transformer such as Miller 8901-B. Connect short piece of insulated wire to J1 and drape other end near CB detector diode or tube. T should have same i-f value as CB set if difference from 455 kHz.—BFO for Single Sideband, *CB Magazine*, Feb. 1972, p 44.

CB TUNER—Can be used for tuning power amplifier tank circuit and antenna trimmer of citizens-band transmitter, monitoring output power, and listening to modulation with headphones. Diodes can be 1N34A or other general-purpose types. Meter is 0-50 μA. With S in position shown, and transmitter on, adjust R3 for 30 μA on meter; power is then 3 W. Article also gives tuning instructions.—Transmitter "Tuner-Upper," *CB Magazine*, Aug. 1970, p 41.

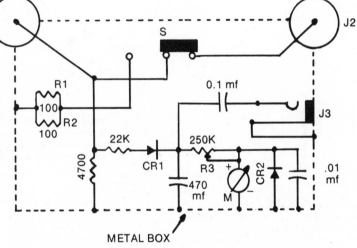
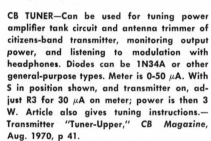

METAL BOX

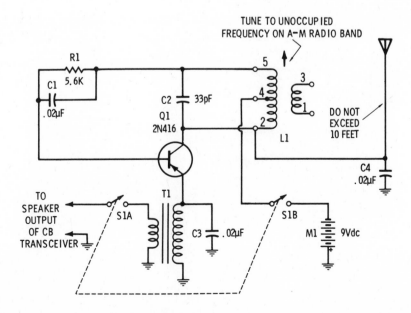

CB MONITOR—Feeds speaker output of CB transceiver by radio to transistor radio tuned to otherwise unused part of broadcast band, for monitoring CB transceiver 24 hours a day from any part of home. Keep antenna length under 10 ft to meet FCC regulations. L1 is transistor oscillator coil, such as J. W. Miller 2022, tuned to unoccupied bc frequency. Use volume control of CB receiver to get best modulation of new a-m carrier.—R. M. Brown, "101 Easy CB Projects," Howard W. Sams, Indianapolis, IN, 1968, p 119–120.

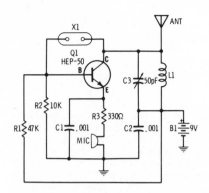

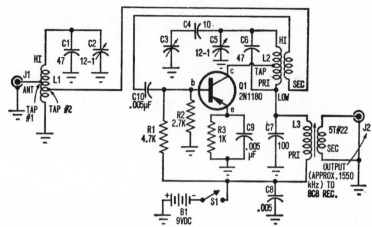

CB PAGER—Produces sufficiently strong citizens-band signal for paging person equipped with standard receiver within building. Frequency of crystal X1 should match that of transceivers. Antenna is three-section Walkie-Talkie type. Use carbon mike, which can be telephone transmitter. Adjust C3 for maximum output as indicated by S meter of transceiver within range.—H. Friedman, "99 Electronic Projects," Howard W. Sams, Indianapolis, IN, 1971, p 114–115.

CB CONVERTER—Simple oscillator converts 27-MHz signals at J1 into about 1,550-kHz signals at J2 for feeding into broadcast receiver. Article gives coil-winding data. C3 is 1–15 pF (E. F. Johnson 149-1) with one rotor blade removed.—C. Green, The CB Grabber, *Science and Electronics*, Jan. 1971, p 49–52.

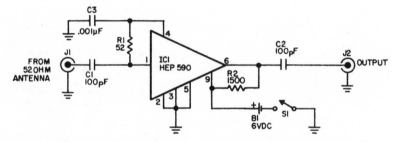

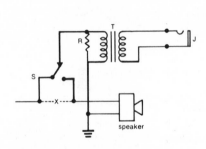

CB BOOSTER—Provides 3 dB additional r-f gain when used ahead of communications-type citizens-band receiver having antenna trimmer. Keep connecting leads short. Cannot be used with low-input-impedance CB transceiver.—CB Booster, *Elementary Electronics*, Nov.-Dec. 1971, p 80.

HIGH-IMPEDANCE PHONES FOR CB—Connect primary of 4-ohm output transformer to open-circuit headphone jack, for listening to citizens-band traffic in privacy with 2,000-ohm headphones. R is 100 ohms.—S. Meuron, Adding Headphones, *CB Magazine*, Feb. 1971, p 42.

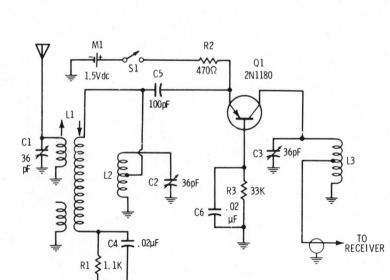

CB PREAMP—Used between antenna and CB receiver to provide pretuning and amplification of signals. L1 is Lafayette HP-66 or equivalent antenna transformer. To make L2 and L3, solder one end of length of No. 22 enamel very close to body of 2-meg 1-W resistor, wind 5 turns, twist ½-inch high, wind 18 more turns, then solder end of wire close to other end of R4. Ground 5-turn end of each.—R. M. Brown, "101 Easy CB Projects," Howard W. Sams, Indianapolis, IN, 1968, p 114–116.

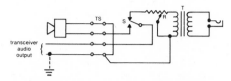

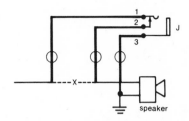

REMOTE PHONE JACK FOR CB—Make connections shown to citizens-band speaker when there isn't room inside transceiver for T, which is 4-ohm output transformer with primary connected to open-circuit jack. R is 4-ohm T-pad which protects transceiver and controls phone volume without affecting speaker. Use high-impedance (2,000-ohm) phones. Circuit can be used even if one side of speaker is not grounded.—S. Meuron, Adding Headphones, CB *Magazine*, Feb. 1971, p 42.

PHONES FOR CB—Low-impedance (8-ohm) headphones can be connected in place of speaker by connecting jack as shown, for listening to citizens-band traffic in privacy.—S. Meuron, Adding Headphones, CB *Magazine*, Feb. 1971, p 42.

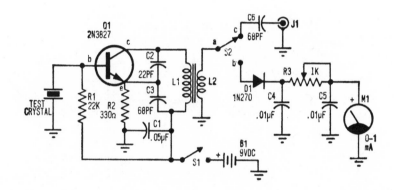

CRYSTAL ACTIVITY TESTER—Designed specifically for measuring relative activity of crystals used in citizens-band transmitters and transceivers. Will also show up other crystal defects, such as frequency jumping. Consists of Colpitts crystal oscillator feeding meter. With S2 at position c as shown, can be used as channel spotter or alignment generator for receiver whose antenna is plugged into J1. Crystal used must then be desired frequency.—E. Morris, CB Rock Rater, *Science and Electronics*, Jan. 1971, p 32–34.

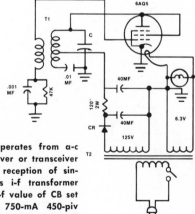

ONE-TUBE BFO FOR CB—Operates from a-c line, for use with a-m receiver or transceiver at home station to permit reception of single-sideband signals. T1 is i-f transformer that should be tunable to i-f value of CB set (usually 455 kHz). CR is 750-mA 450-piv diode. Place close to CB set and tune C until speech is understandable.—Make Your Own Beat Frequency Oscillator, CB *Magazine*, March 1971, p 45.

CHAPTER 17
Clock Circuits

CLOCK SYNCHRONIZER—Forces real-time counter clock to be in step with computer processor, so data is read only when counter is not changing. Requires only two TTL IC packages. Designed for PDP-8/E minicomputer with external I/O bus option, and can be applied directly to other DEC minicomputers. Article describes operation in detail.—J. Crapuchettes, TTL Interface Circuit Synchronizes Computer Clock, Electronics, July 5, 1973, p 97–98.

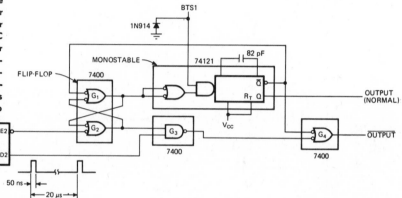

C-MOS CLOCK—Simple quartz-clock generator using RCA 4007AE c-mos inverters gives over-all frequency accuracy of 0.1%, sufficient for most digital systems, while providing adequate immunity from changes in temperature and supply voltage. Highest stability is obtained with 10-V supply for frequencies between 600 kHz and 5 MHz. Increase supply to 13.5 V for output of 5 to 10 MHz.—S. Das, C-MOS Holds Down Parts Count for Digital Clocks, Electronics, June 7, 1973, p 118–119.

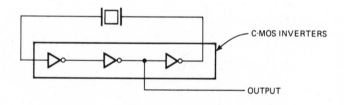

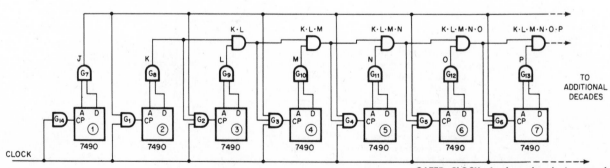

GATED CLOCK—A given decade in cascaded counter receives clock pulse only if less-significant decades before it have each counted to 9. 60-ns propagation delay remains same for any number of stages, for counting rates well above 10 MHz.—E. E. Pearson, Gated Clock Cuts Counter Delay to 60 NS, Electronics, June 21, 1971, p 64.

INPUT STATES REQUIRED TO ADVANCE DECADE COUNTS FOR (A)

DECADE	INPUT TO DECADE	DECIMAL COUNT ⑦⑥⑤④③②①
1 VIA G₁₄	CLOCK ONLY	0 0 0 0 0 0 0
2 VIA G₁	CLOCK · J	0 0 0 0 0 0 9
3 VIA G₂	CLOCK · J · K	0 0 0 0 0 9 9
4 VIA G₃	CLOCK · J · K · L	0 0 0 0 9 9 9
5 VIA G₄	CLOCK · J · K · L · M	0 0 0 9 9 9 9
6 VIA G₅	CLOCK · J · K · L · M · N	0 0 9 9 9 9 9
7 VIA G₆	CLOCK · J · K · L · M · N · O	0 9 9 9 9 9 9

124

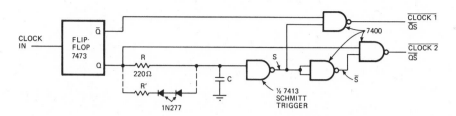

NONOVERLAPPING CLOCK TRAINS—Schmitt trigger using only one R-C time constant provides appropriate delay for staggering clock phase. Separation between output clocks remains constant despite changes in temperature, for clock inputs up to 10 MHz. Values of C from 0 to 1,000 μF produce interclock intervals of 30 ns to 1 s. Addition of dashed components gives equal interclock intervals. R' can range from 0 to 5K.—R. R. Osborn, *Schmitt Trigger Prevents Clock Train Overlap, Electronics,* July 3, 1972, p 86.

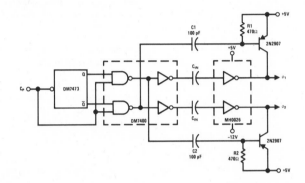

MINIMIZING CROSSTALK—Transistors are turned on just before transition of clocks to logic 1, to clamp voltage spikes that occur during transition and might otherwise cause crosstalk between clock lines. Report has broad discussion of clock line crosstalk.—B. Siegel and M. Scott, "Applying Modern Clock Drivers to MOS Memories," National Semiconductor, Santa Clara, CA, AN-76, 1973, p 5–6.

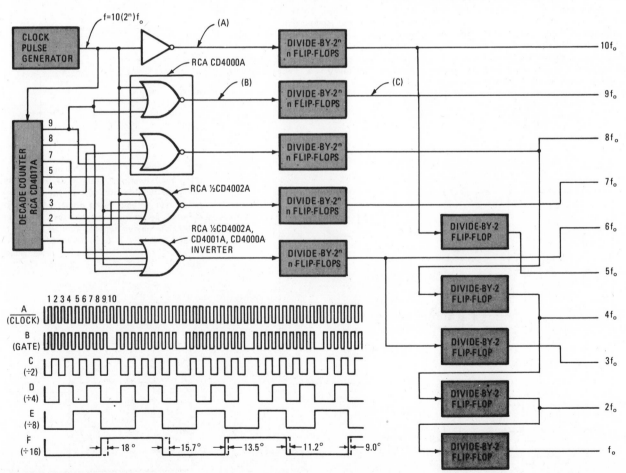

NINE-OVERTONE GENERATOR—Generates fundamental and first nine overtones (harmonics) by binary division of blanked pulse train, to give square-wave outputs from which phase distortion error has been minimized. Clock frequency must be 2^n times faster than highest harmonic. Except for fifth harmonic, clock signal must be properly gated before it can be divided, as described in article. Dashed lines on lowest waveform show correct locations of transitions for true square wave, with 18 deg as largest error. Maximum clock frequency is 5 MHz for RCA decade counter shown.—D. DeKold, *Binary Division Produces Harmonic Frequencies, Electronics,* Dec. 4, 1972, p 91–92.

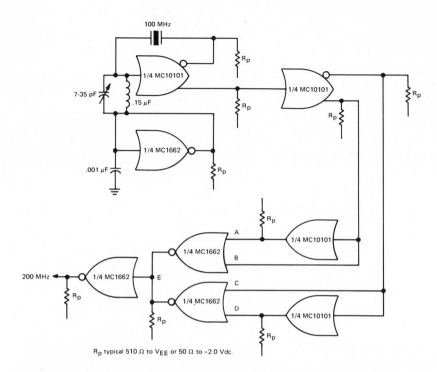

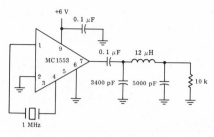

1-MHZ CRYSTAL—Wide bandwidth and output swing capability of IC video amplifier makes it ideal for use with 1-MHz series-mode crystal as master clock or local oscillator for many types of systems. Positive feedback is injected to input pin 1 through crystal. Uses brute-force pi filter at output to extract fundamental frequency, to eliminate need for controlling loop gain or amplitude of oscillator to reduce harmonic content.—"A Wide Band Monolithic Video Amplifier," Motorola Semiconductors, Phoenix, AZ, AN-404, 1971.

200-MHZ CRYSTAL USING DOUBLER—Used when stable oscillator is required in 200–300 MHz range as clock driver. Crystal is in series with feedback loop going to LC tank that tunes to 100-MHz harmonic of crystal. Tank also serves for calibrating circuit to exact frequency. Second section of IC serves as buffer and gives complementary 100-MHz signals for frequency doubler consisting of two MC10101 gates serving as phase shifters and two MC1662 NOR gates. Gates provide 90-deg delay to true and complement 100-MHz signals, for 50% duty cycle. Can also be used in frequency synthesizer.—C. Byers and B. Blood, "IC Crystal Controlled Oscillators," Motorola Semiconductors, Phoenix, AZ, AN-417A, 1972.

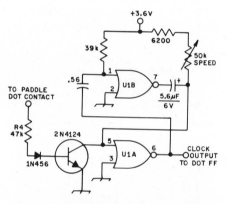

CLOCK FOR KEYER—Arrangement shown, using dual two-input gate such as HEP584, generates reliable clock signals for solid-state code keyer without being affected by energy radiated by relay contacts when switching 100-V lab power supply. Connect pin 4 to ground and pin 8 to +3.6V.—F. Getz, Jr., Integrated-Circuit Clock Oscillator for Solid-State Keyers, QST, March 1972, p 57.

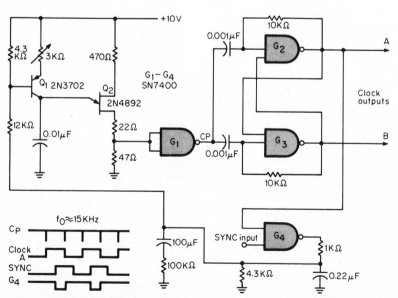

PLL SYNC—Low-cost phase-locked loop synchronizes two digital systems. Uses vco consisting of controlled constant-current generator Q1, ujt relaxation oscillator Q2, inverter G1, and pulsed binary G2-G3, connected as reference input to half-wave phase detector G4. Loop is locked when vco and sync input frequencies are identical and 90 deg out of phase.—C. A. Herbst, Simple Digital PLL Circuit Synchronizes Clock Signals, The Electronic Engineer, April 1971, p 67.

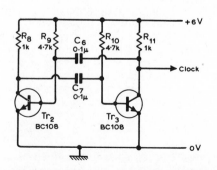

1.4-KHZ CLOCK—Simple astable mvbr serves as clock generator for Karnaugh map display generator.—B. Crank, Karnaugh Map Display, Wireless World, April 1971, p 185–189.

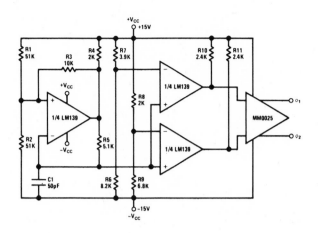

TWO-PHASE MOS CLOCK DRIVER—LM139 comparator square-wave generator at left is oscillator, with other comparators establishing desired phasing between two outputs of clock driver.—R. T. Smathers, T. M. Frederiksen, and W. M. Howard, "LM139/LM239/LM339 A Quad of Independently Functioning Comparators," National Semiconductor, Santa Clara, CA, AN-74, 1973, p 7–8.

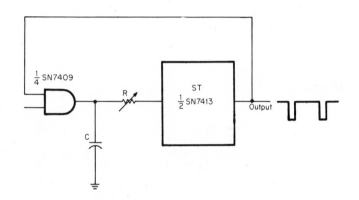

SCHMITT WITH ONE GATE—Inexpensive IC circuit gives stable clock generator having 80:1 ratio. Each 1 μF of C adds 1 ms to period between pulses. For flicker-free pulse trains, R should be under 400 ohms. Uses open-collector AND gate.—B. Carpenter, Clock Pulse Generator, The Electronic Engineer, May 1971, p 76.

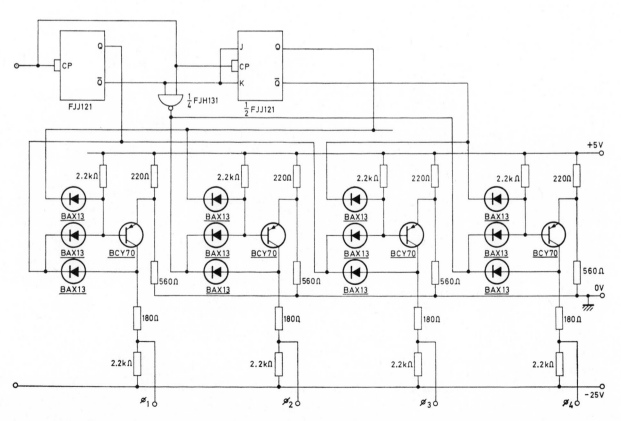

4-PHASE CLOCK GENERATOR—Consists of two-stage binary counter and four gates for decoding the four possible logic states. Will operate on inputs up to 4 MHz, to give 4-phase clock rate of 1 MHz. Pulse overlap is prevented by subtracting width of clock pulse from waveform ahead of decoding gates.— "MOS Integrated Circuits and Their Applications," Mullard, London, 1970, TP11081, p 89–93.

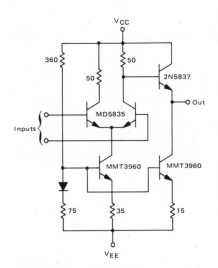

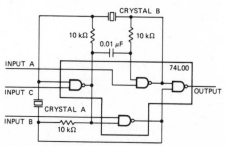

TWO-FREQUENCY CLOCK—Dual-frequency crystal-controlled clock oscillator uses all four gates in quad package to give remotely selected operation at frequency of crystal A or crystal B, either of which can be in excess of several MHz. Grounding input A activates crystal A, grounding input B activates crystal B, and grounding input C inhibits entire circuit. Output is TTL-compatible 40% square wave for either crystal.—H. L. Nurse, Quad NAND GATE Package Yields Two-Frequency Clock, *Electronics*, July 5, 1973, p 100.

HIGH-SPEED DRIVER—Required when operating flip-flops and other IC logic units above 300 MHz. Rise and fall times of output pulses are less than 1 ms when driving 50-pF load. Dual transistors shown can be replaced by 2N5835 single transistors.—B. Broeker, "Micro-T Packaged Transistors for High Speed Logic Systems," Motorola Semiconductors, Phoenix, AZ, AN-536, 1970.

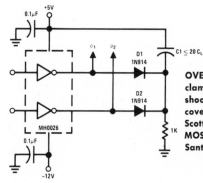

OVERSHOOT LIMITER—High-speed diode clamp network limits output waveform overshoot of clock driver to diode drop. Report covers design problems.—B. Siegel and M. Scott, "Applying Modern Clock Drivers to MOS Memories," National Semiconductor, Santa Clara, CA, AN-76, 1973, p 5.

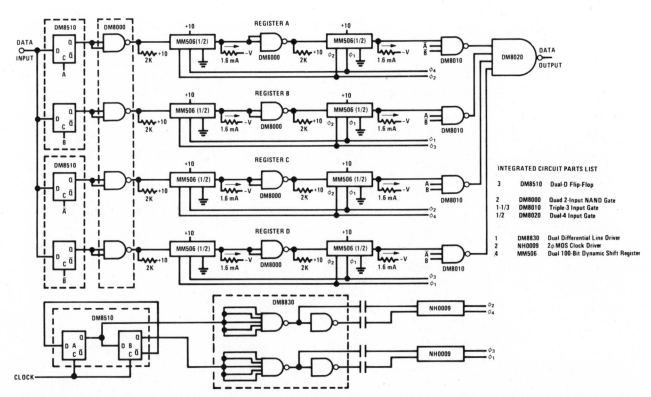

INTEGRATED CIRCUIT PARTS LIST

3	DM8510	Dual-D Flip-Flop
2	DM8000	Quad 2-Input NAND Gate
1-1/3	DM8010	Triple-3 Input Gate
1/2	DM8020	Dual-4 Input Gate
1	DM8830	Dual Differential Line Driver
2	NH0009	2φ MOS Clock Driver
4	MM506	Dual 100-Bit Dynamic Shift Register

16-MHZ DATA RATE—Combination of paralleling and mos/TTL interfacing permits operation of shift register as delay line for up to 16 MHz. Based on division of high-speed master clock into two-phase clock by flip-flops A and B, connected as sequence counter. Clock is divided again into four clock phases at DM8830 line driver, for distribution to system components. Used for storing data temporarily in digital system.—D. Mrazek "MOS Delay Lines," National Semiconductor, Santa Clara, CA, 1969, AN-25.

CHAPTER 18
Code Circuits

SEILER KEYED VFO—Diode keying circuit permits keeping IC warm at all times, to stop oscillator drift during warmup. Tuning is achieved by attaching shaft to slug of L1, and grounding slug to eliminate effect of hand capacitance. T1 is toroid with bifilar primary.—E. Noll, Keyed VFO, *Ham Radio*, July 1972, p 64–65.

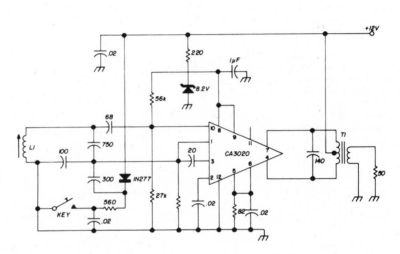

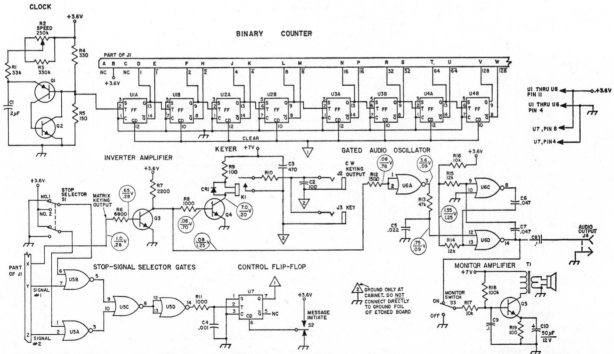

MESSAGE GENERATOR—At touch of button, matrix-controlled binary counter generates Morse Code message up to 255 bits long, with a dot or space equal to 1 bit and a dash to 3 bits. Ideal for identifying repeater or rtty transmissions. Oscillator Q1-Q2 generates clock pulses spaced apart by duration of 1 bit, with R2 adjusting speed between 10 and 50 words per minute. Article tells how to set up diode-resistor matrix which is plugged into jack J1 for generating desired message. Q1 is HEP52 or 2N4126. Q2 is HEP50 or 2N4123. Q3–Q5 are MPS3394 or equivalent. U1–U4 are MC790P dual J-K flip-flop. U5 and U6 are MC724P quad 2-input gate. U7 is MC882G J-K flip-flop. Message-stop signal from diode matrix toggles U7 to clear counter and prevent repetition of message. Article covers construction and use.—J. Hall, A Digital Morse-Code Message Generator, *QST*, June 1970, p 11–19 and 38.

129

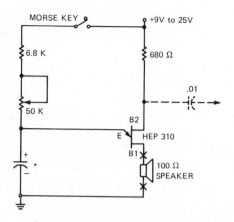

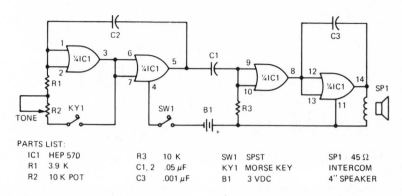

PARTS LIST:

IC1	HEP 570	R3	10 K	SW1	SPST	SP1	45 Ω
R1	3.9 K	C1, 2	.05 μF	KY1	MORSE KEY		INTERCOM
R2	10 K POT	C3	.001 μF	B1	3 VDC		4" SPEAKER

PRACTICE OSCILLATOR—For code practice, use 0.01-μF capacitor, not electrolytic. Can then be used also as audio signal generator, with frequency determined by 50K pot. If 10-μF electrolytic is used for capacitor, with switch in place of key, circuit will serve as metronome with tick intervals adjusted by 50K pot. Headphones shunted by 47-ohm resistor may be used in place of speaker. Dashed circuit shows output for signal generator.—"Home Handyman's Construction Projects," Motorola Semiconductors, Phoenix, AZ, HMA-37, 1972.

IC FOR CODE PRACTICE—Output of a-f oscillator-amplifier is 50 ohms, which calls for standard intercom-type speaker. Two units can be used up to 10 feet apart for sending and receiving practice by connecting keys in parallel. Can also serve as intercom for those seeking practical experience along with practice.—K. W. Session, Jr., Build An 8 Transistor Code Oscillator with Just 1 IC, 73, Feb. 1971, p 53–54.

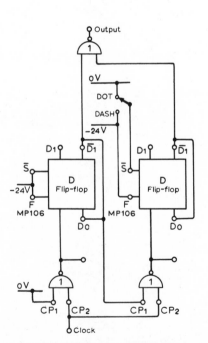

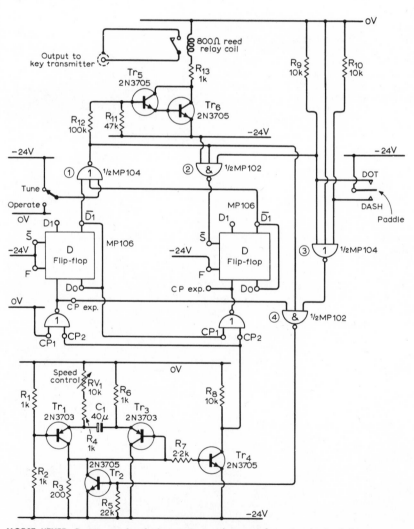

DOT-DASH GENERATOR—Paddle need only be touched momentarily on dot or dash side of key to trigger mvbr for completing waveform, for sending perfect Morse characters at high speeds. Uses Plessey MP106 counter-register-bistable IC's or equivalent. Article includes complete logic for generating spaces as well. Does not require regulated supply. Gates are part of IC.—C. I. B. Trusson and M. R. Gleason, Electronic Morse Keyer, Wireless World, Aug. 1970, p 379–381.

MORSE KEYER—Paddle need only be touched momentarily on dot or dash side of key to initiate generation of perfect Morse code waveforms separated by precisely correct spaces. All IC logic is Plessey or equivalent. Does not require regulated supply. Includes tune switch for holding transmitter on continuously while making tuning adjustments. A dot waveform is obtained from output of counter, so mark-space ratio is precisely 1:1 at all speeds. Mvbr is stopped between characters, so dot or dash generation commences instantly with no uncertainty.—C. I. B. Trusson and M. R. Gleason, Electronic Morse Keyer, Wireless World, Aug. 1970, p 379–381.

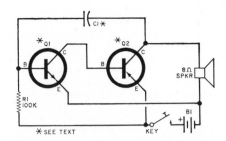

KEYED PRACTICE OSCILLATOR—Loudspeaker output permits teaching code to large group. Transistors are not critical; Q1 can be GE-8 and Q2 2N176. C1 is 25–50-V disc ceramic somewhere between 0.005 and 0.02 μF; larger value gives lower pitch. Use any battery; the higher the voltage, the greater the volume.—J. Kolodziej, Reader's Circuit, *Popular Electronics*, July 1969, p 85.

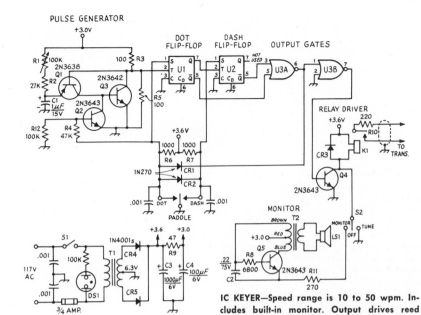

IC KEYER—Speed range is 10 to 50 wpm. Includes built-in monitor. Output drives reed relay, such as Magnecraft W102MX1, for keying transmitter. CR3 is any small-signal silicon diode. J-K flip-flops U1 and U2 are Fairchild μL923 or Motorola HEP583, and U3 is dual 2-input gate such as Fairchild μL914 or Motorola HEP584.—"The Radio Amateur's Handbook," American Radio Relay League, Newington, CT, 49th Ed., 1972, p 366–368.

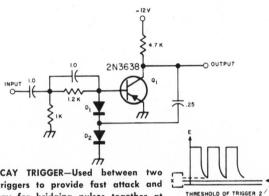

SLOW-DECAY TRIGGER—Used between two Schmitt triggers to provide fast attack and slow decay for bridging pulses together at their bases. Output trigger has threshold low enough to be in bridged-together portion of output wave of detector Q1. This makes second trigger produce pulse that is almost exactly as long as original dit or dah in c-w receiver, for control of audio oscillator to give tone that is not unpleasantly chopped.—D. Jensen, Some Ideas on Noise-Free CW Reception, 73, Feb. 1966, p 84 and 86–88.

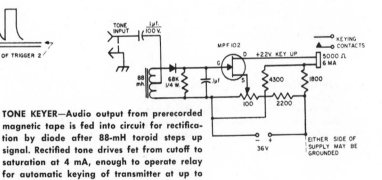

TONE KEYER—Audio output from prerecorded magnetic tape is fed into circuit for rectification by diode after 88-mH toroid steps up signal. Rectified tone drives fet from cutoff to saturation at 4 mA, enough to operate relay for automatic keying of transmitter at up to 30 wpm.—N. Foot, A FET Tone-Keyer, QST, Oct. 1969, p 103–104.

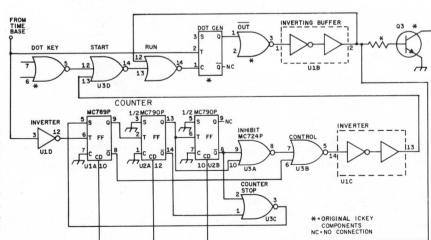

KEYER SPACING CONTROL—Addition of IC counter and control gates to dot-dash keyer permits automatic generation of correct letter spacing of three dot lengths and word spacing of seven dot lengths. Counter stop gate is used to stop counter after word space. Original ICKEY article appeared in QST, Nov. 1968. U1 is MC787P and U3 is MC724P.—K. Stone, Adding Letter and Word Spacing to ICKEY, QST, May 1972, p 48–49.

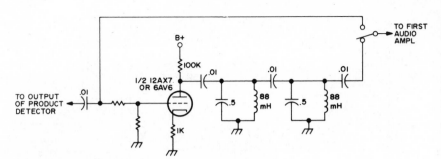

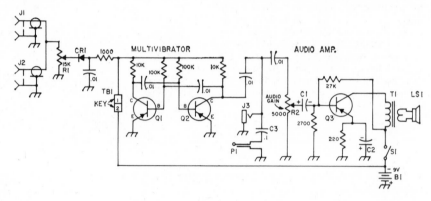

SIMPLE C-W FILTER—Values shown give center frequency of about 750 Hz and bandwidth of only 75 Hz at 6 dB down, for improving c-w reception in crowded band. High selectivity means receiver should be tuned slowly. Receiver should have good i-f noise blanker, to minimize ringing in presence of pulse noise. Tube can be changed to audio fet if desired. Audio leads should be shielded.—R. L. Grenell, An Ultra-Simple Selective Audio Filter, 73, Nov. 1971, p 49–50.

C-W MONITOR—Can be used with any transmitter, for monitoring of actual transmitted signal. Only connection required is to coaxial output lead of transmitter, for pickup of very small amount of r-f voltage. This voltage, after being rectified by 1N277 or 1N34A diode CR1, is used to power mvbr tone oscillator. Resulting audio drives phones directly or speaker through amplifier stage. All transistors are 2N406, SK3003, or equivalent. T1 is output transformer with 2,000–5,000 ohm primary and 4–10 ohm secondary driving small 4-ohm speaker. For code practice, connect key as shown.—L. G. McCoy, An R.F.-Actuated C.W. Monitor, QST, Nov. 1968, p 39–41.

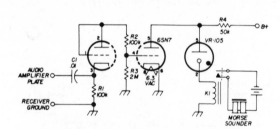

RADIO-DRIVEN SOUNDER—Connects to audio amplifier of any receiver picking up c-w signals, and drives Morse telegraph sounder. Has sentimental value for old-timers.—J. Proefrock, Radio-Controlled Morse Sounder, Ham Radio, Oct. 1971, p 66.

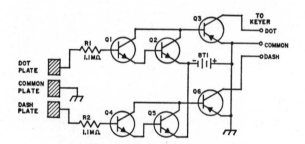

SOLID-STATE KEYING PADDLE—Uses 2N525 or HEP 253 switching transistors for Q3 and Q6 in place of conventional keying relays. Other transistors, in d-c amplifier stages, can be 2N1051 or HEP53. Uses 9-V battery.—W. D. Fredericks, Solid-State Switching for the Electronic Paddle, QST, April 1969, p 44.

KEYER—Requires only one IC dual opamp (Motorola MC1437 or equivalent), operating from dual power supply and providing negative-going keyed output. All diodes are 50-V small-signal silicon. Transistor may be required for coupling keyer to blocked-grid transmitter.—L. H. Vale, OAKEY—An Op-Amp Electronic Keyer, QST, Oct. 1972, p 40–44.

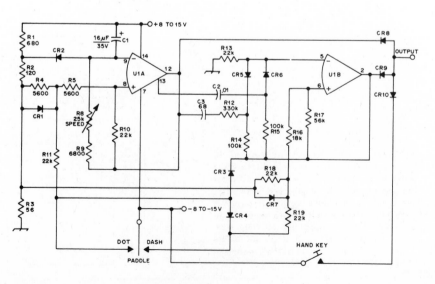

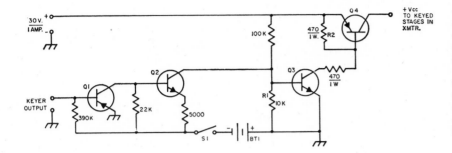

1-A KEYER SWITCH—Used between keyer and negative-ground transmitter drawing about 1 A at 30 V. With key closed, Q1 (SK3004) and Q2 (SK3020) are cut off so Q3 (SK3024) is biased into conduction and Q4 (SK3009) is turned on for keying of transmitter. Supply can be 9-V transistor radio battery because drain is only about 4 mA. Many other transistor types will work equally well.—R. R. Lucas, High-Current Switch for a Solid-State Keyer, QST, July 1969, p 47.

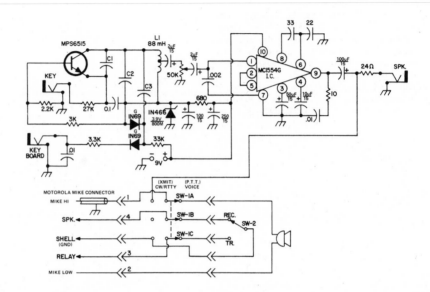

TWO TONES FOR PRACTICE—With speaker and key plugged in, circuit serves as 2,125-Hz code practice oscillator. With keyboard jack opened with unconnected plug, tone is 2,975 Hz. With switch in transmit position, can be used for keying transmitter. Can also be used for AFSK (automatic frequency-shift keying) and for modulated c-w at mike input, with speaker cut off. Any other switching arrangements may also be used.—F. D. Thomas, Customized AFSK-MCW and Code Practice Oscillator, 73, April 1972, p 36.

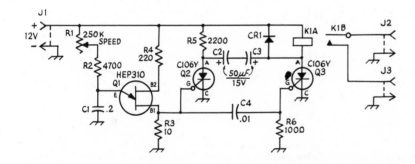

TEST KEYER—Sends dots automatically, for checking performance of amateur c-w transmitter by viewing waveforms of dot pulses on cro. K1 is miniature spst reed relay with 12-V coil, such as Magnecraft W101MX-2. Uses ujt oscillator which can be set for keying speeds between 25 and 50 wpm, driving mvbr using scr pair. For cathode-keyed transmitters, article gives various R-C combinations to be connected in parallel across relay contacts for spark suppression.—D. A. Blakeslee, The Dit Ditter, QST, July 1971, p 17 and 51.

RELAYLESS BREAK-IN KEYING—Simultaneously provides grid-block keying of transmitter and muting of receiver so incoming signals can be heard between dots and dashes. Designed for operation with electronic t-r switch or separate receiving antenna, to protect receiver front end. 2N2270 or equivalent a-f transistors Q3–Q4 form astable mvbr running continuously and producing about 5,000-Hz square wave. Other transistors are RCA 40264 or similar having 150-V breakdown. Diodes are 1N537. T1 is transistor output having 100-ohm primary and 10-ohm secondary.—M. L. Steine, Break-In Keying Without Relays, QST, Dec. 1967, p 26–27.

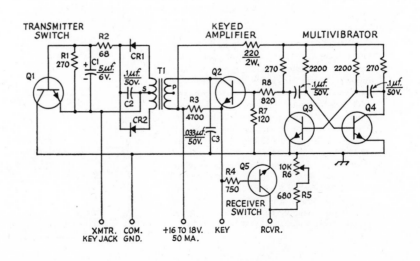

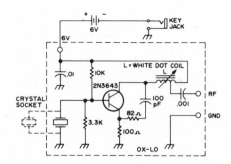

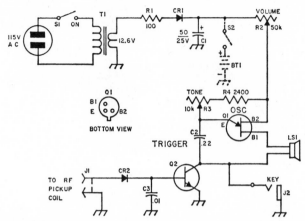

TRANSMITTER FOR CODE PRACTICE—Use without antenna, for transmitting signal across average room to receiver having about 4 feet of wire as antenna. Crystal can be either 80-meter or Novice-band. Short antenna on receiver gives experience in reading code with background noise present, when friend is operating key.—V. Fitzpatrick, A Pre-Novice Transmitter, 73, Jan. 1972, p 67–68.

C-W MONITOR—Can also serve as code practice oscillator. For monitoring output of transmitter while keying it, pickup coil having about two turns with same diameter as final amplifier tank coil is placed about ½ inch from tank coil. Monitor will easily follow signals up to 40 wpm. Will also operate from 9-V transistor radio battery. Diodes are 1N91, Q1 is 2N2160, and Q2 is 2N388 or equivalent. Use small 8-ohm speaker.—D. Butler, A Code-Practice Oscillator and CW Monitor, QST, Nov. 1969, p 24–25.

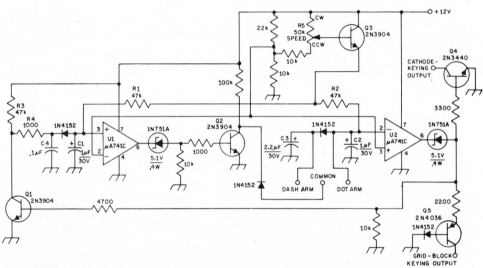

PORTABLE KEYER—Draws only 10 mA from 12-V battery, for minimizing battery weight on mountaineering and hiking trips. Provides variable speed and self-completing action. Output at Q4 is for keying positive voltage to ground at about 20 mA, and output at Q5 is for grid-block keying. Will operate on supplies from 9 to 30 V.—W. Hayward, An Integrated-Circuit QRP Keyer, QST, Nov. 1971, p 38–39.

C-W FILTER—Designed to eliminate high and low frequencies classed as CRUD (Continuous Random Unwanted Disturbances) without ringing. Based on 3-pole large-percentage bandpass filter having Butterworth response, with 600-Hz center frequency, for 600-ohm input and output impedances. Rectifiers are 1N4001 silicon. L1 is made up with six 88-MH toroids in series, while L2 and L3 are each an 88-MH toroid with 40 turns No. 30 enamel added in same direction as original winding. Switch out filter for phone. Bandwidth of filter is 362 Hz for 3 dB down and 1,160 Hz for 30 dB down. Attenuation is 50 dB for 120-Hz power-supply hum.—J. Hall and R. M. Myers, The Crud-O-Ject, QST, Feb. 1972, p 11–13.

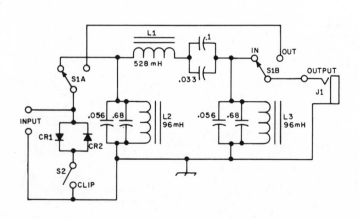

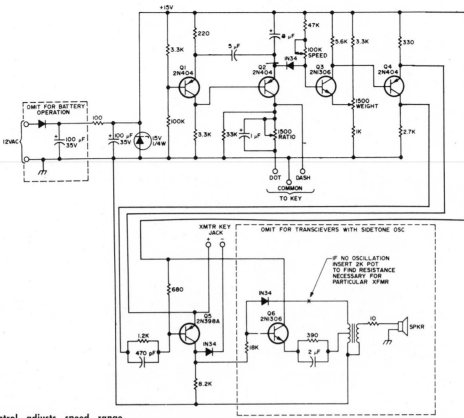

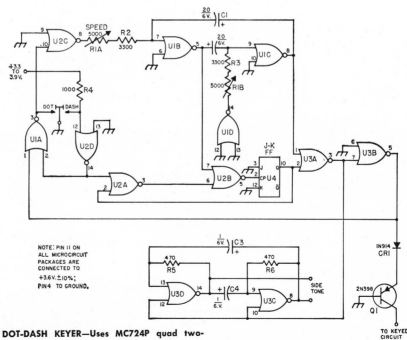

KEYER—Speed control adjusts speed range from 10 wpm upward, for producing self-completing dots and dashes. Optional amplifier for speaker can be used for teaching code to groups. Designed for use with transmitters that are grid-block keyed. For cathode keying, use reed relay in place of 8.2K collector resistor for Q5.—W. Pinner, Low Cost Automatic Keyer: A "First Project," 73, Nov. 1970, p 42–45.

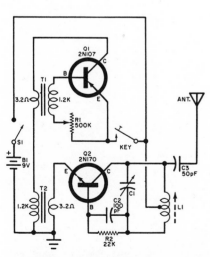

WIRELESS PRACTICE OSCILLATOR—Circuit is very low power transmitter that is tuned to dead spot on broadcast band of a-m radio, to permit code practice by two persons in different rooms of home. R-f oscillator Q2 operates continuously, and only tone oscillator Q1 is keyed; this minimizes frequency shift with keying. T1 and T2 are miniature output transformers, L1 is tapped ferrite loopstick, and C1 is 360-pF tuning capacitor.—A. Brookstone, Reader's Circuit, *Popular Electronics*, July 1968, p 81.

DOT-DASH KEYER—Uses MC724P quad two-input gates for U1–U3 and Motorola 723P single J-K flip-flop for U4. With switching resistor shown for Q1, keyed voltage should be under 100 V and current under 2 mA. Produces dots with required 50% duty cycle and dashes with 75% duty cycle. Output of gate U3A serves also to gate astable mvbr generating 550-Hz side tone. Speed range is 8 to 40 wpm. M. Jahn, Microcircuit Electronic Key, *QST*, Sept. 1969, p 32–35.

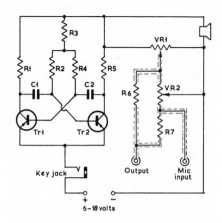

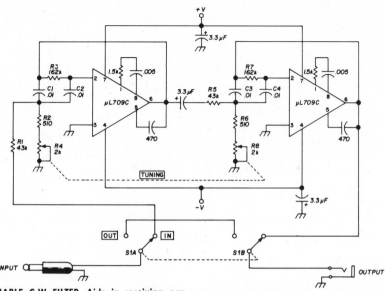

TONE-MODULATED PRACTICE SET—Provides tone that can be fed directly to speaker or used to feed either transmitter modulator or tape recorder through mixer that permits quick changeover from modulated c-w to a-m voice. C1 and C2 are 0.05 μF. R1 is 220 ohms; R2 and R4 10K; R3 5.6K; R5 1K; R6 and R7 220K; VR1 and VR2 250K; TR1 and TR2 OC71, NKT214, or similar.—J. Morris, Tone Modulated Oscillator, *The Short Wave Magazine*, April 1972, p 103–104.

TUNABLE C-W FILTER—Aids in receiving c-w signals through heavy interference. Gain is 1 at center frequency (0 dB), to prevent audio blasting when filter is switched in, and 3-dB bandwidth is 140 Hz. Supply voltages can be obtained from 9-V transistor battery if two

1K resistors are connected across it and their junction grounded. Tuning range of filter is about 750 to 1,600 Hz.—N. J. Nicosia, A Tunable Audio Filter for CW, *Ham Radio*, Aug. 1970, p 34–35.

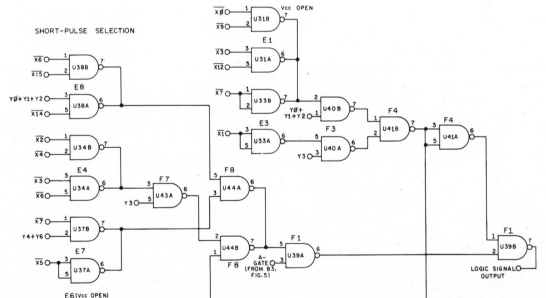

AUTOMATIC CQ—Uses only logic gates connected to preprogrammed matrix to generate Morse-code characters for CQ call or any other frequently repeated message. Circuit portion shown uses only RTL dual 2-input gates, such as Fairchild 914 or 9914, but same performance can be obtained with DTL or TTL logic. Article gives all connections required for programming "CQ CQ CQ DE VE2HN VE2HN".—H. H. Rugg, The VE2HN Digital CQer Using NAND Logic, Part 1: *QST*, Feb. 1972, p 33–40; Part 2: *QST*, March 1972, p 24–29 and 35.

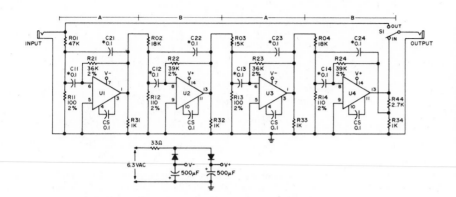

C-W FILTER—Uses relatively high-Q circuits tuned to slightly different frequencies to make 3-dB bandwidth greater than 50 Hz without giving up too much off-frequency attenuation, so signal comes in clearly even though receiver drifts a little. Circuit gives about 100 dB bandwidth. Opamps are μA739, but 709 can be used if amplifier is stabilized for unity gain. Since total gain of filter is unity, switching it in and out does not affect volume. Article covers design, construction and operation.—P. C. Cope, The CW EXcavator, 73, Sept. 1972, p 30–32 and 34–35.

TOUCH-CONTROLLED TRANSMITTER—Provides instantaneous c-w break-in whenever finger touches either dot or dash paddle of keyer. Each side of paddle is covered with conducting foil or paint connected to R1 of d-c amplifier string. Common or ground plate may not be needed. Battery can be 9 to 27 V, to match voltage rating of 10-mA reed relay used. All transistors are 2N967 or similar npn silicon. Amplifier should be shielded against r-f.—J. A. Foster, Touch-Controlled Break-In for C.W., QST, July 1969, p 45.

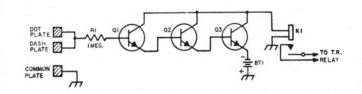

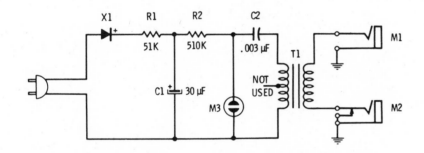

NEON TONE GENERATOR—With key plugged into one jack and phones into other, can be used as code practice oscillator. Audio tone across secondary of T1 can also be used for checking hi-fi system or for checking modulation of transmitter. X1 is GE 504A rectifier and M3 is NE-2E neon lamp. For variable audio pitch, use 1-meg pot for R2.—R. M. Brown and T. Kneitel, "101 Easy Test Instrument Projects," Howard W. Sams, Indianapolis, IN, 1968, p 23–24.

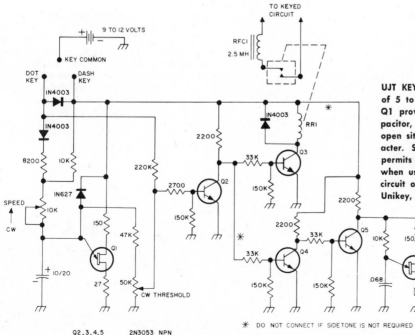

UJT KEYER—Provides adjustable speed range of 5 to 30 wpm. Diode going to emitter of Q1 provides discharge path for emitter capacitor, so capacitor is discharged if key-open situation occurs in middle of any character. Sidetone circuit Q4-Q5-Q6, optional, permits hearing actual code being transmitted when using ssb rig for c-w. Article explains circuit operation in detail.—J. Eschmann, The Unikey, 73, Nov. 1969, p 66–69.

* DO NOT CONNECT IF SIDETONE IS NOT REQUIRED

Q2,3,4,5	2N3053 NPN
Q1,6	2N2646
RR1	MAGNEREED W102X 12VOLT

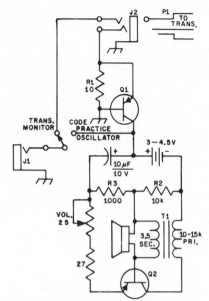

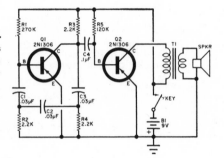

LOW-POWER C-W—Operates from single Burgess 2U6 or equivalent battery, for transmitting c-w to portable radio in same house for code practice. Use up to 6 feet of wire as antenna. L1 is 40–300 μH miniature bc antenna coil and T1 is audio transformer with 10,000-ohm primary and 2,000-ohm CT secondary. Add Cx (about 0.01 μF) if audio tone is too low. Adjust slug of L1 to tune transmitter to quiet part of bc band.—S. Daniels, No-Ticket Rig, *Science and Electronics*, Jan. 1971, p 53–54.

CODE OSCILLATOR AND MONITOR—With switch in position shown, keying of transmitter can be monitored on speaker. Choose R1 to give proper bias for keying circuit used in transmitter; 10 ohms is correct for 150-mV keying circuit. With switch in other position and key plugged into J1, circuit serves as code practice oscillator. Transistors are GE-3 or Sylvania ECG-104. Correction: Emitter of Q1 should be connected to negative side of 10-μF capacitor and positive side connected to chassis ground (QST, Jan. 1972, p 57).—A. Bremner, Jr., Code-Practice Oscillator and Monitor, QST, Dec. 1971, p 41.

750-HZ OSCILLATOR—Can be used as self-contained audio signal source after code is mastered. Makes good substitute for human voice when experimenting with microphone placement. Uses R-C phase-shift oscillator. Output transformer T1 has 2,000-ohm primary and secondary to match speaker.—C. A. Rankin, Reader's Circuit, *Popular Electronics*, March 1968, p 85.

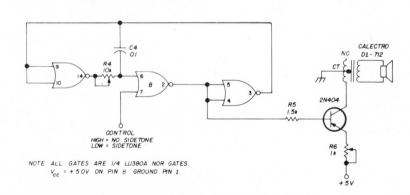

NOTE ALL GATES ARE 1/4 LU380A NOR GATES.
Vcc = +5.0V ON PIN 8 GROUND PIN 1

800-HZ SIDETONE OSCILLATOR—Uses three gates of Signetics IC. Output level, controlled by R6, is adequate for roomful of people in code practice classes. R4 adjusts tone frequency. Transformer has 500-ohm CT primary and 8-ohm secondary. Use with clock oscillator that forms dots and dashes, to give electronic keyer.—J. G. Curtis, Electronic Keyer Oscillators, *Ham Radio*, June 1970, p 44–45.

RELAY DRIVER FOR KEYER—Used in place of switching transistor when keying transmitter having voltages above 100 V and currents above 40 mA. Keyed current is 8 mA and open-circuit voltage across key line under 40 V, safe enough for practically all transistor-output keyers. Driver may be keyed with either npn or pnp transistor if polarity of key line is observed. Uses hermetically sealed 4,500-ohm 8-mA relay with mercury-wetted spdt contacts, such as Western Electric 275B. To improve operation, connect 0.001-μF capacitor across key line and 0.01 μF between positive key line and chassis. Use 1N4001 silicon diode across relay coil to protect keying transistor.—A Relay Driver for Use with Solid-State Keyers, QST, Oct. 1971, p 22–23.

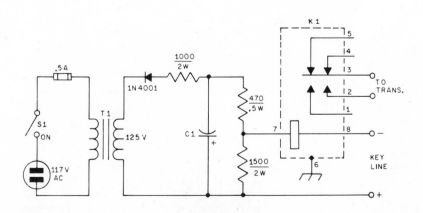

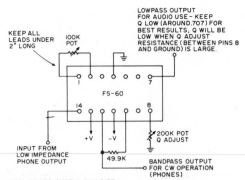

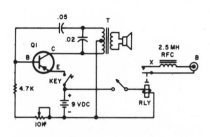

C-W FILTER—Simple circuit using Kinetic Technology IC active audio filter has unity gain regardless of bandwidth. Once signal is peaked, interference can be eliminated with Q control. Improves selectivity and intelligibility of c-w reception on any receiver.—A. F. Stahler, Improved Circuitry for KT1 CW Filter, 73, Dec. 1972, p 107 and 109.

KEYING MONITOR—Combination code practice oscillator and keying relay allows amateur operator to listen to his own keying as it is being transmitted. Any pnp audio transistor can be used. Use very small relay capable of operating at keying speeds without chattering, for controlling transmitter directly or through heavy-duty relay having required contact current rating.—E. Spencer, The Combo, 73, Jan. 1970, p 108.

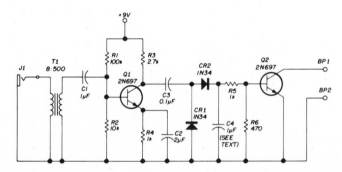

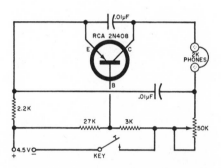

TONE-CONTROLLED SWITCH—Converts tape-recorded output of code oscillator to on-off switching impulses for driving c-w transmitter or for punching paper tapes for radiotele-type. Simplifies changing of call signals and other prerecorded frequently repeated messages.—J. R. Huffman, Memo-Key, *Ham Radio*, June 1972, p 58–60.

PRACTICE OSCILLATOR—Circuit is inexpensive and works fine with phones, from three flashlight cells.—C. J. Schauers, Code-Practice Oscillator, *Popular Electronics*, July 1968, p 68.

CHAPTER 19
Comparator Circuits

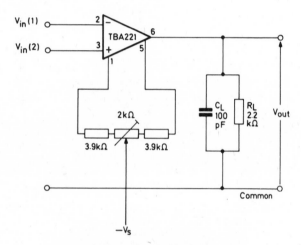

IC VOLTAGE COMPARATOR—Provides either positive or negative output voltage depending on relative levels of input voltages. Input signals should not exceed limiting values of differential and common-mode input voltage of Mullard IC. Output is limited on all but very small values of input voltage difference, because amplifier is operated open-loop.— "TBA221/222 Monolithic Internally Compensated Operational Amplifiers," Mullard, London, 1970, TP1238, p 10–11.

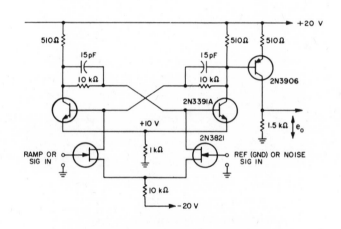

LOW-NOISE AMPLITUDE COMPARATOR—Balanced input circuit uses TI 2N3821 n-channel fet's. Hysteresis is 0.3 V at either gate, maintained constant to over 500 kHz. When used with precision ramp generator, p-p noise was no worse than 10 μV, corresponding to 3 μV rms.—R. W. Frank, Input Noise—Its Influence on Counter and Pulse-Generator Performance and its Measurement, *General Radio Experimenter*, Feb. 1966, p 3–7.

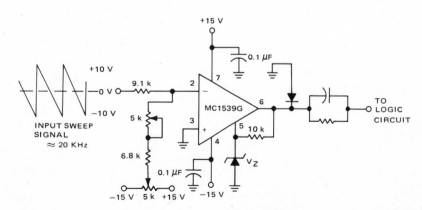

VOLTAGE COMPARATOR—Very high slewing rate and other characteristics of second-generation version of 709 opamp make MC1539 highly suitable for comparator applications. Choose values for parallel RC output network to match input of logic being fed.—E. Renschler, "The MC1539 Operational Amplifier and Its Applications," Motorola Semiconductors, Phoenix, AZ, AN-439, 1972.

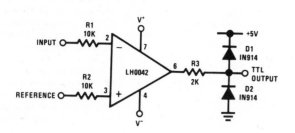

PRECISION VOLTAGE COMPARATOR—Uses fet opamp to make input current low and constant, essentially independent of input voltage as long as input and reference signals are not more than 4 V apart. Will drive three standard TTL loads or 30 National low-power TTL loads. Increase R3 to save power if full fanout is not required.—R. K. Underwood, "New Design Techniques for FET Op Amps," National Semiconductor, Santa Clara, CA, 1972, AN-63, p 9.

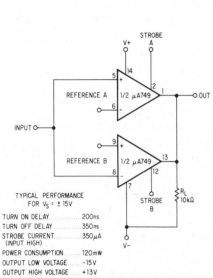

TYPICAL PERFORMANCE
FOR $V_S = \pm 15V$

TURN ON DELAY	200ns
TURN OFF DELAY	350ns
STROBE CURRENT	350µA
(INPUT HIGH)	
POWER CONSUMPTION	120mW
OUTPUT LOW VOLTAGE	−15V
OUTPUT HIGH VOLTAGE	+13V

DUAL-LEVEL COMPARATOR—Will operate on supply voltages from 4 to 15 V. For strobing, connect output lag pin to voltage within 0.5 V of positive supply. Outputs can then be independently strobed. Connection between output pins 1 and 13 serves to OR outputs.—D. K. Long, "Applications of the µA749 Dual Operational Amplifier," Fairchild Semiconductor, Mountain View, CA, No. 268, 1971.

$V_R(t) = E_R \cos \omega_o t$ C_c = COUPLING CAPACITOR
$V_S(t) = E_S \cos(\omega_o t + \phi)$ C_B = BYPASS CAPACITOR

PHASE COMPARATOR—Uses Exar XR-S200 multiplier and opamp sections, with high-level reference signal as one input and low-level information signal as other input. Since response of IC is symmetrical, inputs are interchangeable. For low input levels, conversion gain is proportional to amplitude of input signal. For high levels, above 40 mV rms, conversion gain is constant at about 2 V per radian. Frequency range is 0.1 Hz to 30 MHz.—"XR-S200 Multi-Function Integrated Circuit," Exar Integrated Systems Inc., Sunnyvale, CA, June 1972, p 3.

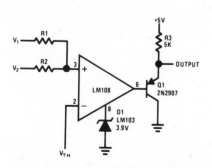

OUTPUT BUFFER—Designed for voltages of opposite polarity, but can also be used as differential comparator that goes through transition when input voltages are equal. Buffer transistor Q1 increases fanout to about 20 with standard DTL or TTL logic.—R. J. Widlar, "IC Op Amp Beats FETs on Input Current," National Semiconductor, Santa Clara, CA, 1969, AN-29, p 15.

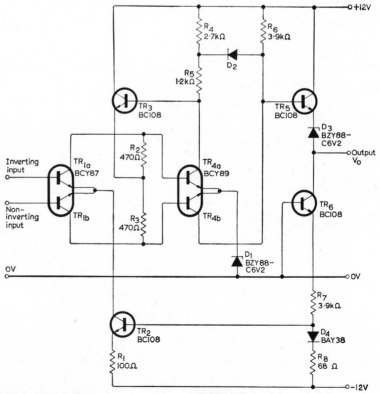

VOLTAGE COMPARATOR—Uses transistor pairs to minimize temperature effects. Output voltage range is −0.7 V to +3 V, for driving logic circuits. Voltage gain at zero d-c input is 2,000, with flat frequency response to 200 kHz. Offset voltage is less than 1 mV. D2 is BAW62.—"Circuit Using Low-Drift Transistor Pairs," Mullard, London, 1968, TP994, p 26–28.

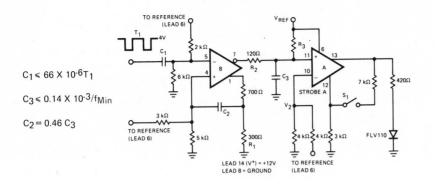

$C_1 \leqslant 66 \times 10^{-6} T_1$

$C_3 \leqslant 0.14 \times 10^{-3} / f_{Min}$

$C_2 = 0.46 \, C_3$

MIMIMUM FREQUENCY DETECTOR—Functionally independent comparator sections A and B of μA750 IC are connected as shown to make FLV110 led turn on when frequency of square-wave input drops to minimum value determined by R3 and C3. Circuit uses input pulse conditioning to achieve predictable operation with varying input pulse levels and widths. Minimum frequency is equal to 1/0.69R3C3. Comparator B is used as one-shot, triggering on negative-going edges. If R3 is 10K, equations give capacitor values required for stable operation.—H. Ebenhoech and R. Ricks, "The μA750 Dual Comparator Subsystem," Fairchild Semiconductor, Mountain View, CA, No. 315, 1972, p 7.

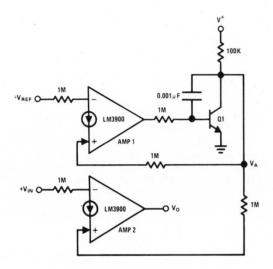

PRECISION COMPARATOR—Use of two Norton current-differencing opamps permits comparison of negative input voltages, which can be more closely matched initially and which track well with temperature changes. Current established by reference voltage at inverting input of opamp 1 makes Q1 adjust VA to supply this current. Equal current then flows into noninverting input of opamp 2. Precision is further improved by adding differential input stage to one of opamps, as covered in report.—T. M. Frederiksen, W. M. Howard, and R. S. Sleeth, "The LM3900—A New Current-Differencing Quad of ± Input Amplifiers," National Semiconductor, Santa Clara, CA, 1972, AN-72, p 29–30.

ONE-SHOT WITH COMPARATOR—Used in applications where pulse is required if d-c input signal exceeds predetermined value determined by ratio of R6 to R5 and value of supply voltage, or about 80% of supply voltage for values shown. Input voltage must fall to less than trip voltage prior to termination of output pulse. Can be used to generate reset pulse for recycling free-running oscillator.—T. M. Frederiksen, W. M. Howard, and R. S. Sleeth, "The LM3900—A New Current-Differencing Quad of ± Input Amplifiers," National Semiconductor, Santa Clara, CA, 1972, AN-72, p 28.

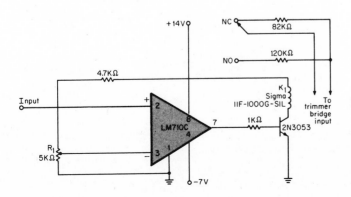

ACTIVE TRIMMER—Used to trim circuit to specific voltage or current, instead of resistance value. To operate, set resistance bridge on trimmer for 100K. Relay K1 is normally closed, placing 82K across bridge. Since this is below preset bridge value, trimmer starts to trim. When comparator input voltage reaches desired level preset by R1, comparator pulls in K1, to place 120K across bridge and shut off trimmer.—J. Agnew, Convert to Active Circuit Trimming, The Electronic Engineer, April 1971, p 66.

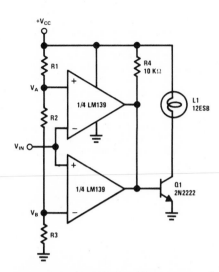

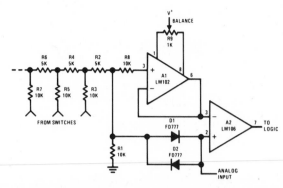

LIMIT COMPARATOR—Provides range of input voltages between which output devices of both LM139 comparators will be off. If input voltage becomes greater than VA or less than VB, one of comparators will switch on and short base of Q1 to ground, making lamp go out. If Q1 is changed to pnp transistor, lamp is normally off and comes on when input voltage gets out of range set by values of R1, R2, and R3.—R. T. Smathers, T. M. Frederiksen, and W. M. Howard, "LM139/LM239/LM339 A Quad of Independently Functioning Comparators," National Semiconductor, Santa Clara, CA, AN-74, 1973, p 5.

LADDER-NETWORK COMPARATOR—Uses LM102 voltage follower to buffer output of ladder network and drive one input of LM106 comparator. Analog signal is fed to other input of comparator. Clamp diode makes circuit faster by clamping output of ladder so it never differs more than 0.7 V from analog input. R9 balances offset of both buffer and comparator.—R. J. Widlar, "'Fast Voltage Comparators with Low Input Current," National Semiconductor, Santa Clara, CA, 1969, LB-6.

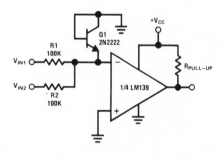

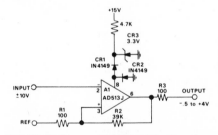

MEDIUM-SPEED COMPARATOR—Clamping diodes CR1 and CR2 limit output swing to range of −0.5 to +4 V. When negative input is driven from ±10 V to 10 mV beyond trigger level, output switches state within 10 μs and rise time is 0.5 μs.—"AD513, AD516 I.C. Fet Input Operational Amplifiers," Analog Devices, Inc., Norwood, MA, Technical Bulletin, Aug. 1971.

COMPARING OPPOSITE POLARITIES—LM139 comparator is connected to compare magnitudes of two input voltages which have opposite polarities. If input 1 is positive and greater than negative input 2, output will go low. If input 1 is positive and less than that of input 2, output will go high. Supply is 5 V.—R. T. Smathers, T. M. Frederiksen, and W. M. Howard, "LM139/LM239/LM339 A Quad of Independently Functioning Comparators," National Semiconductor, Santa Clara, CA, AN-74, 1973, p 5–6.

LAMP DRIVER—Comparator opamp drives Q1 for switching lamp L1, with R2 limiting cold-lamp current surge. Opamp has large differential input voltage range and easily clamped output for driver applications.—R. J. Widlar, "Operational Amplifiers," National Semiconductor, Santa Clara, CA, 1968, AN-4.

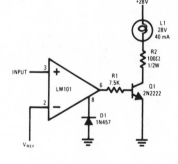

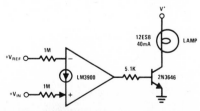

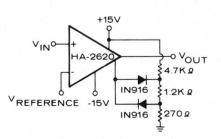

POWER COMPARATOR—Noninverting connection of Norton current-differencing opamp drives transistor capable of handling 40-mA current of 12-V panel lamp.—T. M. Frederiksen, W. M. Howard, and R. S. Sleeth, "The LM3900—A New Current-Differencing Quad of ± Input Amplifiers," National Semiconductor, Santa Clara, CA, 1972, AN-72, p 29.

HIGH-IMPEDANCE COMPARATOR—Simple circuit capable of driving up to 10 logic gates uses opamp having input impedance of about 500 meg and input current of 1 mA. Minimum output current of 15 mA is obtainable with output swing of up to ±10 V. Common-mode range is ±11 V and differential input range ±12 V.—G. G. Miler, "A Simple Comparator Using the HA-2620," Harris Semiconductor, Melbourne, FL, No. 509/A, 1970.

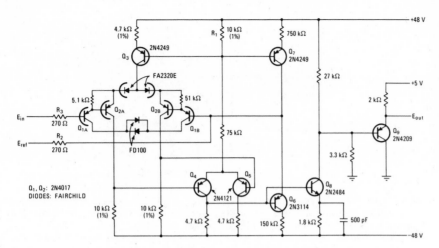

LSI TESTER—Developed for testing large-scale integrated-logic arrays. Analog comparator has fast response, high input impedance, wide common-mode voltage range, and accuracy of 0.1%. Input stage uses matched Darlington pairs Q1 and Q2. Output transistors Q8 and Q9 provide level shift for interfacing with DTL or TTL. Output signal is high (logic 1) when input voltage is lower than reference, and 0 when input is higher than reference.—G. Niu, Precision Comparator Circuit Satisfies LSI Testing Needs, *Electronics*, April 26, 1973, p 122–123.

LONG TIME—2N4393 fet is operated as Miller integrator having gain of about 60. Time constants up to 1 min can be achieved because equivalent capacitance looking into gate of fet is C times gain and gate source resistance can be as high as 10 meg.—"FET Circuit Applications," National Semiconductor, Santa Clara, CA, 1970, AN-32, p 1.

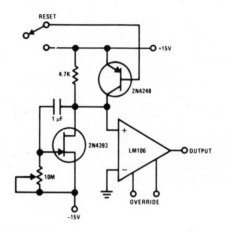

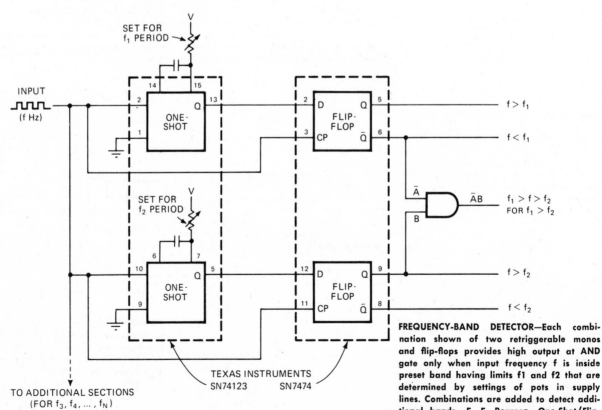

FREQUENCY-BAND DETECTOR—Each combination shown of two retriggerable monos and flip-flops provides high output at AND gate only when input frequency f is inside preset band having limits f1 and f2 that are determined by settings of pots in supply lines. Combinations are added to detect additional bands.—E. E. Pearson, One-Shot/Flip-Flop Pairs Detect Frequency Bands, *Electronics*, April 24, 1972, p 104.

CHAPTER 20
Computer Circuits

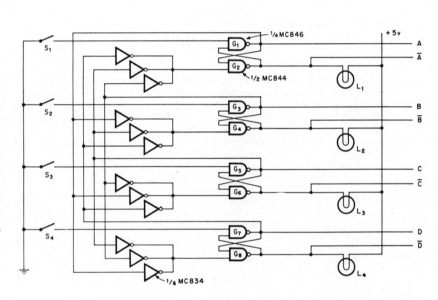

IDIOT-PROOF LOGIC—Control switches are backed up with simple mutually exclusive logic that honors only one switch closure at a time, to insure that relatively unskilled operators cannot set controls to conflicting modes of operation that would damage sensitive equipment. If two or more switches are closed simultaneously, all uncomplemented outputs go low and others go high. If several are closed sequentially, only last switch will enable its function and light its lamps.—I. P. Breikss, Low-Cost Digital ICs Prevent Operator Errors, *Electronics*, Aug. 17, 1970, p 89.

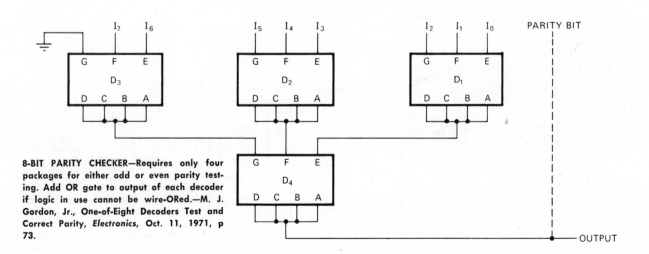

8-BIT PARITY CHECKER—Requires only four packages for either odd or even parity testing. Add OR gate to output of each decoder if logic in use cannot be wire-ORed.—M. J. Gordon, Jr., One-of-Eight Decoders Test and Correct Parity, *Electronics*, Oct. 11, 1971, p 73.

DECODER	LOGIC	PARITY	PACKAGE PIN NUMBERS						
			A	B	C	D	E	F	G
MOTOROLA MC 4038 P	TTL	ODD	13	2	4	11	15	6	1
		EVEN	3	5	12	14	15	6	1
MOTOROLA MC 1150 L	MOS	ODD	2	5	12	13	15	9	7
		EVEN	4	3	11	14	15	9	7

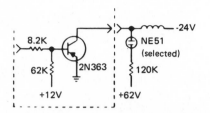

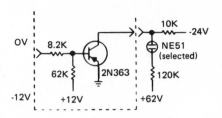

LAMP DRIVER—Used as inverter for flip-flop having outputs of 0 and —12 V. With 0 V, transistor is cut off and full 86 V is across neon to turn it on. When transistor conducts, neon end of 10K resistor is less than 1 V above ground, voltage across lamp is below minimum needed to maintain ionization, and lamp goes out.—C. N. Thompson, Jr., Neons Diagnose a Computer, *Electromechanical Design*, Jan.-Feb. 1971, p 24–25.

RELAY DRIVER FOR FLIP-FLOP—Transistor serves as inverter for energizing relay when input from flip-flop is —12 V, and releasing relay when flip-flop output is 0 V. Neon lamp glows only when transistor is cut off and relay is de-energized. Used to speed diagnosis of computer malfunctions and give status information.—C. N. Thompson, Jr., Neons Diagnose a Computer, *Electromechanical Design*, Jan.-Feb. 1971, p 24–25.

DIGIT DETECTOR—Detector circuit can be constructed as 12-detector array on 100-mil centers, using beam-lead monolithic chips on tantalum-film glass substrate, for use with plated-wire or other high-speed memory systems. Each detector will dynamically sense minimum bipolar 1 and 0 signals of 1 mV in either polarity within longitudinal noise of 150 mV, and generate two-rail logic output of 3 V in 15 ms. Gated flip-flop Q5-Q6 is connected to output of emitter-follower by D1 and D2. Bandwidth of differential amplifier Q1-Q2 is greater than 45 MHz, with differential gain of 15 and common-mode gain of 0.1.—D. J. Lynes, High-Speed DC Coupled Digit Detector, *IEEE Trans. on Computers*, Jan. 1969, p 43–47.

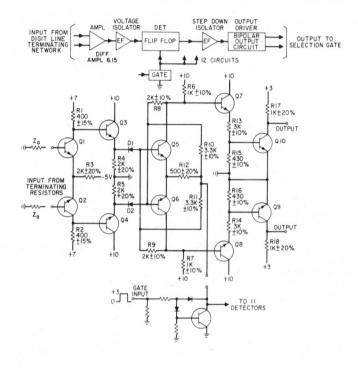

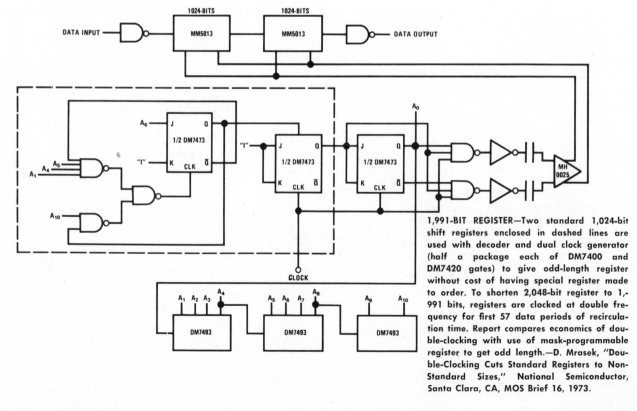

1,991-BIT REGISTER—Two standard 1,024-bit shift registers enclosed in dashed lines are used with decoder and dual clock generator (half a package each of DM7400 and DM7420 gates) to give odd-length register without cost of having special register made to order. To shorten 2,048-bit register to 1,991 bits, registers are clocked at double frequency for first 57 data periods of recirculation time. Report compares economics of double-clocking with use of mask-programmable register to get odd length.—D. Mrasek, "Double-Clocking Cuts Standard Registers to Non-Standard Sizes," National Semiconductor, Santa Clara, CA, MOS Brief 16, 1973.

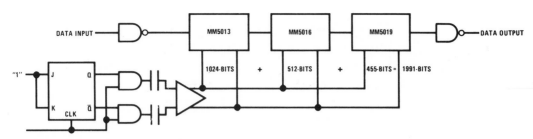

ODD-LENGTH REGISTER—Gives 1,991-bit register by using standard 1,024-bit and 512-bit registers with MM5019 which is mask-programmed to order for 455 bits out of available maximum of 512.—D. Mrazek, "Double-Clocking Cuts Standard Registers to Non-Standard Sizes," National Semiconductor, Santa Clara, CA, MOS Brief 16, 1973.

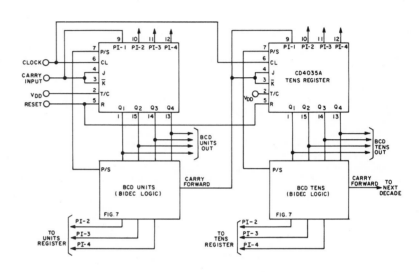

BINARY-BCD CONVERTER—Based on use of cos-mos four-stage parallel-in/parallel-out shift registers providing four outputs to binary-decimal logic units, identically connected as shown in book, for feeding units and tens registers.—"COS/MOS Digital Integrated Circuits," RCA, Somerville, NJ, SSD-203A, 1973 Ed., p 176.

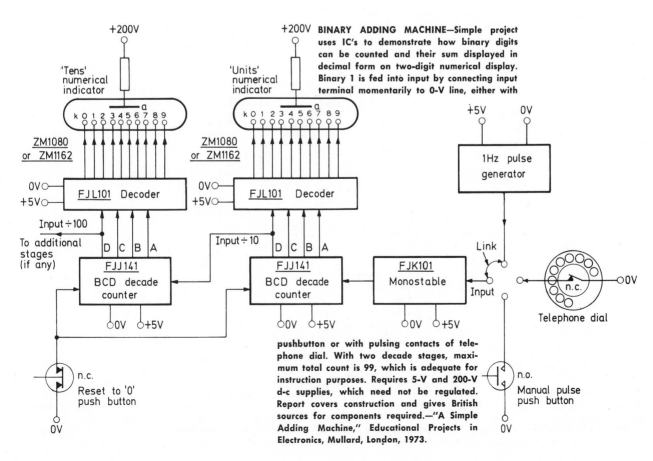

BINARY ADDING MACHINE—Simple project uses IC's to demonstrate how binary digits can be counted and their sum displayed in decimal form on two-digit numerical display. Binary 1 is fed into input by connecting input terminal momentarily to 0-V line, either with pushbutton or with pulsing contacts of telephone dial. With two decade stages, maximum total count is 99, which is adequate for instruction purposes. Requires 5-V and 200-V d-c supplies, which need not be regulated. Report covers construction and gives British sources for components required.—"A Simple Adding Machine," Educational Projects in Electronics, Mullard, London, 1973.

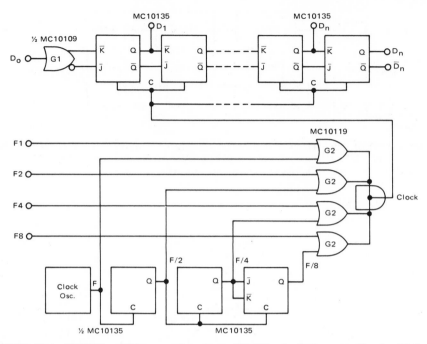

VARIABLE DIGITAL DELAY—Uses MC10135 dual master-slave d-c coupled J-K flip-flops with separate asynchronous set and reset inputs and common clock inputs, capable of shifting at rates above 100 MHz, to form shift register that can be used as variable digital delay line having delay range of 7 to 27 ns. Uses 50-MHz clock. Report describes operation in detail and gives output waveforms.—J. M. DeLaune, "Using Shift Registers As Pulse Delay Networks," Motorola Semiconductors, Phoenix, AZ, AN-565, 1972.

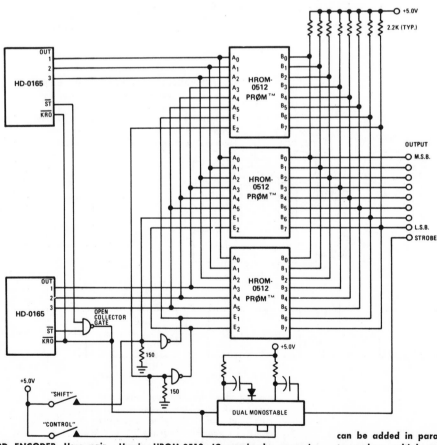

UNIVERSAL KEYBOARD ENCODER—Uses pair of Harris HD-0165 IC keyboard encoders to generate universal code which can be translated to any of several entirely different output codes at flip of switch by means of three Harris HROM-0512 IC read-only memories wired in parallel, with each storing 64 8-bit words. One of memories contains words for unshifted mode, second for shifted mode, and third for control mode. Additional memories can be added in parallel with enable inputs to produce multiple codes, such as ASCII and EBCDIC, either simultaneously or selectively.—D. Jones, "Designing with the HD-0165 Keyboard Encoder," Harris Semiconductor, Melbourne, FL, No. 204, 1971.

COMPARING MOST-SIGNIFICANT BITS—Eight-gate arrangement acts as three-state latch for comparing input words A and B when serially transferred and most-significant bit arrives first. First difference establishes relative magnitude, and all other differences in that word can be ignored. Useful in preliminary data sorting and number ranging. Gate at left uses only ¼ of SN7486N.—E. J. Murray, Simple Logic Circuits Compare Binary Numbers, *Electronics*, March 27, 1972, p 94.

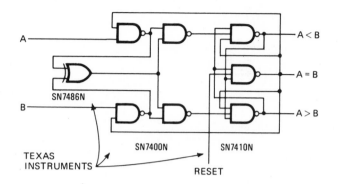

COMPARING LEAST-SIGNIFICANT BITS—Only six gates are required for comparison of two binary numbers being transferred serially when least-significant bit is transmitted first. Last difference between coincident word bits then determines which is the larger word. Arrangement is three-state latch providing three comparisons for input words A and B. Useful in preliminary data sorting and number ranging.—E. J. Murray, Simple Logic Circuits Compare Binary Numbers, *Electronics*, March 27, 1972, p 94.

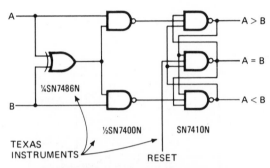

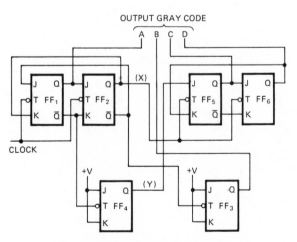

FLIP-FLOPS: TEXAS INSTRUMENTS SN7473

GRAY CODE WITHOUT GLITCHES—Six J-K flip-flops generate Gray code in such a way that only one signal transition occurs at any instant, so output signals can be ANDed without glitches. Maximum operating frequency is limited by delay of flip-flops. All flip-flops must be cleared initially, for correct code generation.—C. Moser, Gray-Code Generator Avoids Output Glitches, *Electronics*, Sept. 11, 1972, p 109.

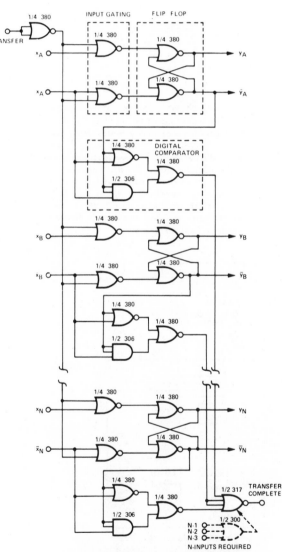

ASYNCHRONOUS TRANSFER—Transfers information stored at input locations XA, XB. . .XN of input register to output register Y. On arrival of TRANSFER command, input NOR gates (serving as AND gates) will connect X to Y, to allow Y to assume state of X. Digital comparators compare state of each Y position to state of each X position. When all positions agree, output for control circuit appears at TRANSFER COMPLETE terminal.—"Signetics Digital Utilogic 2/600 TTL-DTL Data Book," Signetics Corp., Sunnyvale, CA, 1972, p 68–69.

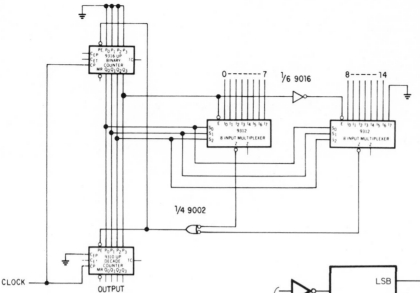

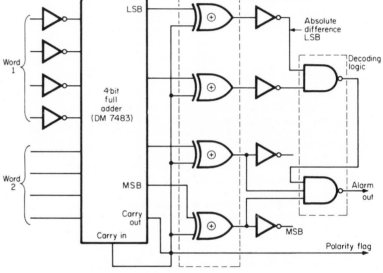

KEYBOARD TO BINARY—Produces four-bit binary coded output corresponding to each key of 15-key keyboard. Two 9312 eight-input multiplexers, two-input gate G1, and inverter form 16-input multiplexer which transmits low signal from depressed key to both 9310 and 9316. 9316 counter continuously counts modulo 16 while stepping multiplexer through each keyboard input in search of signal from depressed key. Report covers operation in detail.—"The 9309 and 9312 Multiplexers," Fairchild Semiconductor, Mountain View, CA, No. 181, 1971, p 11–13.

NOISE-IGNORING PARITY CHECKER—Uses absolute-value subtractor to produce difference between two binary words being compared. Decoding logic shown gives 1 only when difference exceeds two least significant digits, thereby ignoring least significant bit dither and system noise while detecting larger errors. Polarity flag can be used to indicate which word is larger.—R. W. Lewallen, Check Parity of Noisy Signals, *The Electronic Engineer*, Sept. 1971, p 92.

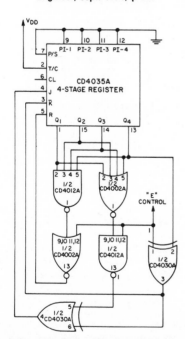

DOUBLE SEQUENCE GENERATOR—Based on use of CD4035A cos-mos four-stage parallel-in/parallel-out shift register and associated logic shown. Control line E permits generation of two different state sequences at outputs Q1—Q4, depending on whether line is 0 or 1. Book gives example of set of state sequences and example of binary-bcd conversion.—"COS/MOS Digital Integrated Circuits," RCA, Somerville, NJ, SSD-203A, 1973 Ed., p 178.

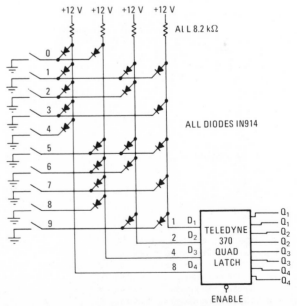

ENCODER WITH HOLDING REGISTER—Used when excess-three output of keyboard must be true rather than complementary and must be entered and held only when enable line is low. Holding latch consists of four type-D flip-flops.—D. Guzeman, Diode Switching Matrices Make a Comeback, *Electronics*, Jan. 17, 1972, p 76–77.

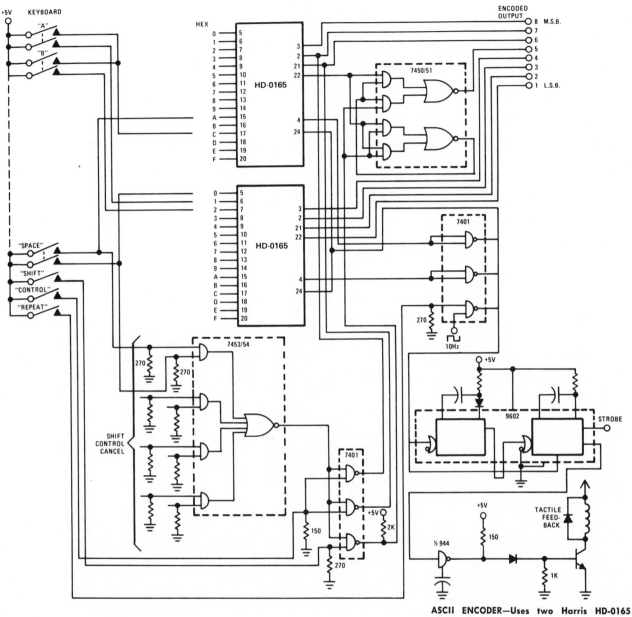

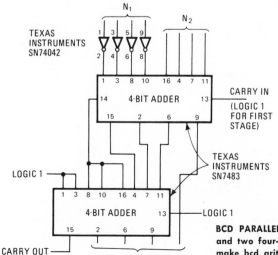

ASCII ENCODER—Uses two Harris HD-0165 IC's each having 20 inputs connected to switches of standard teletypewriter or crt terminal, with associated logic elements to provide standard ASCII code output for capital letters, symbols, and control functions. Can be adapted for lower-case letters by increasing complexity of logic at output to provide for shift function. Optional REPEAT key pulses strobe output at about 10 pps as long as it and another key are held down.—D. Jones, "Designing with the HD-0165 Keyboard Encoder," Harris Semiconductor, Melbourne, FL No. 204, 1971.

BCD PARALLEL SUBTRACTION—Four inverters and two four-bit adders connected as shown make bcd arithmetic almost as simple as binary arithmetic. First adder subtracts one digit from the other, while second adder converts difference to bcd form and, if necessary, generates borrow for subsequent stage. Diagram shows stage for one digit of parallel subtractor. Stages can be cascaded or one can feed shift register.—P. K. Bice, Two Adders Form BCD Subtractor, *Electronics*, Feb. 14, 1972, p 78.

CHAPTER 21
Contact Bounce Suppression Circuits

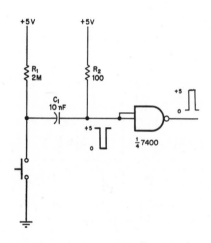

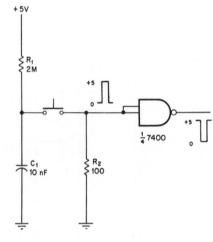

CONTACT BOUNCE SUPPRESSOR—Gate circuit has intrinsic dead time of 20 ms, to prevent switch from generating more than one logic pulse because of contact bounce occurring 5 μs to 5 ms after first closure in most mechanical switches. Gives positive-going output pulse.—G. Fontaine, Gate Suppresses Pulses from Switch Contact Bounces, *Electronics*, March 15, 1971, p 76.

BOUNCE ELIMINATOR—IC is connected as bistable flip-flop to deliver single clean output pulse for each operation of mechanical switch S1, as required for reliable operation of RTL digital circuits.—D. Lancaster, Build the No-Bounce Pushbutton, *Popular Electronics*, March 1970, p 51–53.

CONTACT BOUNCE SUPPRESSOR—Gate circuit has intrinsic dead time of 20 ms, to prevent switch from generating more than one logic pulse because of contact bounce occurring 5 μs to 5 ms after first closure in most mechanical switches. Gives negative-going output pulse.—G. Fontaine, Gate Suppresses Pulses from Switch Contact Bounces, *Electronics*, March 15, 1971, p 76.

KEY BOUNCE SUPPRESSION—Designed for use with Harris HD-0165 keyboard encoder to suppress multiple strobe pulses resulting from key bounce or arcing on either make or break. Eliminates need for mercury-wetted or other more expensive switches on keyboard. Circuit is dual retriggerable mono mvbr, one of which serves entire keyboard. Only requirement is that key be held down longer than few ms for data entry; it is unlikely that deliberate key depression would ever be shorter interval. Optional tactile feedback circuit in report increases keying accuracy at high speeds by actuating solenoid hammer or speaker which produces tick sound each time data entry actually occurs.—D. Jones, "Designing with the HD-0165 Keyboard Encoder," Harris Semiconductor, Melbourne, FL, No. 204, 1971.

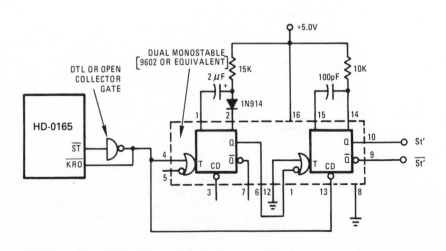

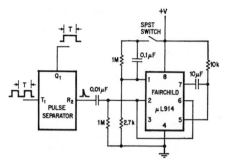

NOISELESS PUSHBUTTON—Used to provide noise-free reset pulses for pulse separator by shaping and cleaning up switch inputs for digital circuits. Will operate from d-c to 75 kHz with values shown, and up to 1 MHz by changing RC timing network of μL914 dual RTL gate.—L. E. Baker, Flip-Flop Pair Synchronizes Pulses and Floats Clocks, *Electronics*, May 24, 1971, p 56.

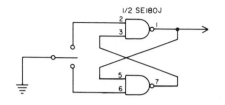

CROSS-COUPLED GATES—Gates form elementary latch that changes state only when switch position is changed, with absolutely no response to contact bounce after switch is changed. Eliminates common cause of errors in digital systems.—"SE100J-Series Design Data & Applications," Signetics Corp., Sunnyvale, CA, AN106, 1966, p 20.

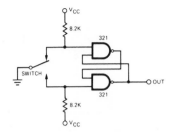

SWITCH CONTACT LATCH—Eliminates false pulses caused by contact bounce of switch. Uses quad 2-input NAND gates operating from 5-V supply.—"HiNIL High Noise Immunity Logic," Teledyne Semiconductor, Mountain View, CA, 1972, p 21.

RTL SINGLE-PULSE—Logic tester generates only one pulse each time S1 is operated, regardless of contact bounce. All IC's are two-input NOR gates such as 9914. Q2 is 2N3904. With 2.8-V d-c applied, pulse is 60 μs wide.—W. Simciak, Logic Pulser, *QST*, March 1970, p 39.

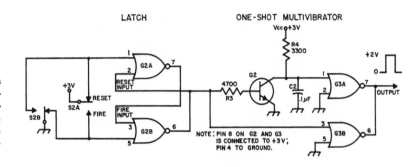

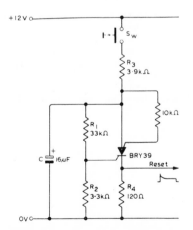

COUNTER RESET—When Sw is closed, scs circuit generates output pulse of fixed shape as required for resetting all decades of ring counter simultaneously. Once switch is closed even momentarily, spurious pulses due to contact bounce have no effect. Pulse quickly decays to very low level determined by ratio of R3 and R4, and ceases completely when switch is released.—D. J. G. Janssen, Circuit Logic with Silicon Controlled Switches, *Mullard Technical Communications*, March 1968, p 57–64.

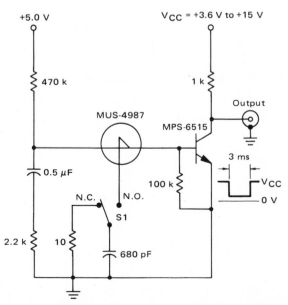

SINGLE-PULSE DIGITAL TESTER—Produces only single pulse, free of mechanical contact bounce, for each closing of switch. Useful for testing digital circuits. Inverter stage may be added if necessary.—L. J. Newmire, "Theory, Characteristics and Applications of Silicon Unilateral and Bilateral Switches," Motorola Semiconductors, Phoenix, AZ, AN-526, 1970.

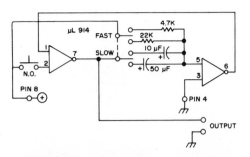

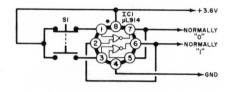

BOUNCELESS PUSHBUTTON—Dual two-input gate gives only one output pulse each time pushbutton is pressed. Toggle switch gives choice between long (1 s) and short pulse lengths. Supply uses two C cells, from which drain is so low that no switch is needed.—A. S. Joffe, IC Logic Doesn't Have to be Obscure, 73, Nov. 1972, p 217–219.

BOUNCELESS SWITCH—Eliminates erratic count that occurs when trigger is generated by conventional noisy mechanical switch directly. Uses spdt pushbutton to operate set-reset IC flip-flop. Each operation of pushbutton generates single clean pulse.—Logic Switch, *Popular Electronics*, Sept. 1969, p 67.

BOUNCELESS PUSHBUTTON—Circuit changes state at first bounce of pushbutton, and stays that way until first bounce in opposite direction. Used in automatic call-letter generator for f-m repeater. Uses two sections of 7404 IC hex inverter.—C. Klinert, A Simple and Inexpensive ID, 73, Oct. 1972, p 64, 66–67, and 69–70.

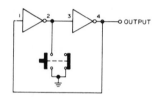

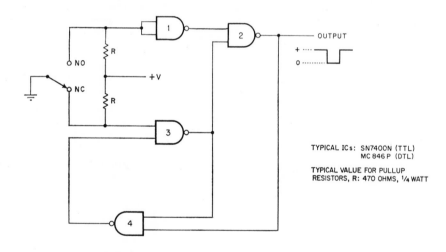

TYPICAL ICs: SN7400N (TTL)
MC 846 P (DTL)

TYPICAL VALUE FOR PULLUP
RESISTORS, R: 470 OHMS, ¼ WATT

BOUNCELESS—Quadruple dual-input positive NAND gate, utilizing gates 3 and 4, forms set-reset flip-flop that eliminates errors caused by contact bounce when mechanical switches are interfaced with high-speed logic. With TTL IC, place 220-pF capacitor between outputs of gates 3 and 4; with DTL, connect capacitor between gate 4 output and ground.—A. J. Laurino, Single IC Pulser Eliminates Contact Bounce, *Electronics*, Nov. 9, 1970, p 79.

CHAPTER 22
Converter Circuits—D-C to D-C

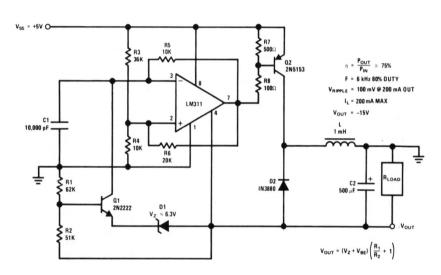

+5 V TO −15 V—LM311 is connected as free-running mvbr with high duty cycle at about 6 kHz, driving switching transistor Q2. Q1 and zener are in extra loop that modifies oscillator duty cycle to give desired output level of −15 V at maximum of 200 mA, with ripple of about 100 mV. Efficiency is about 75%, line regulation better than 3% for 5—10 V input, and load regulation better than 3% for 0 to 100 mA.—H. H. Mortensen, "+5 to −15 Volts DC Converter," National Semiconductor, Santa Clara, CA, 1972, LB-18.

$$\eta = \frac{P_{OUT}}{P_{IN}} \geq 75\%$$
F = 6 kHz 80% DUTY
V_{RIPPLE} = 100 mV @ 200 mA OUT
I_L = 200 mA MAX
V_{OUT} = −15V

$$V_{OUT} = (V_Z + V_{BE})\left(\frac{R_1}{R_2} + 1\right)$$

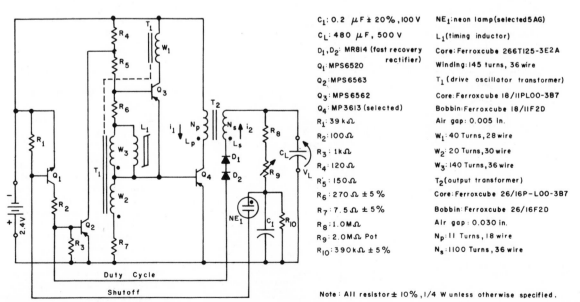

C_1: 0.2 μF ± 20%, 100 V	NE$_1$: neon lamp (selected 5 AG)
C_L: 480 μF, 500 V	L_1 (timing inductor)
D_1, D_2: MR814 (fast recovery rectifier)	Core: Ferroxcube 266T125-3E2A
	Winding: 145 turns, 36 wire
Q_1: MPS6520	
Q_2: MPS6563	T_1 (drive oscillator transformer)
Q_3: MPS6562	Core: Ferroxcube 18/11PLOO-3B7
Q_4: MP3613 (selected)	Bobbin: Ferroxcube 18/11F2D
R_1: 39 kΩ	Air gap: 0.005 in.
R_2: 100 Ω	W_1: 40 Turns, 28 wire
R_3: 1kΩ	W_2: 20 Turns, 30 wire
R_4: 120 Ω	W_3: 140 Turns, 36 wire
R_5: 150 Ω	T_2 (output transformer)
R_6: 270 Ω ± 5%	Core: Ferroxcube 26/16P-LOO-3B7
R_7: 7.5 Ω ± 5%	Bobbin: Ferroxcube 26/16F2D
R_8: 1.0 MΩ	Air gap: 0.030 in.
R_9: 2.0 MΩ Pot	N_p: 11 Turns, 18 wire
R_{10}: 390 kΩ ± 5%	N_s: 1100 Turns, 36 wire

Note: All resistor ± 10%, 1/4 W unless otherwise specified.

2.4 V TO 500 V FOR PHOTOFLASH—Charges 480-μF energy-storage capacitor to 500 V from two rechargeable nickel-cadmium cells (60 joules). Charging time is 20 s. Saturable magnetic core L1 sets on period of blocking oscillator having variable duty cycle. Load-voltage sensor is neon oscillator, which pulses inverter off each time neon conducts. During charge, inverter is on most of the time. When CL is charged enough, neon stays on and oscilla- tor turns off. Article describes circuit design in detail.—L. T. Rees, Charging Energy-Storage Capacitors from Low-Voltage Sources, The Electronic Engineer, Jan. 1969, p 50—56.

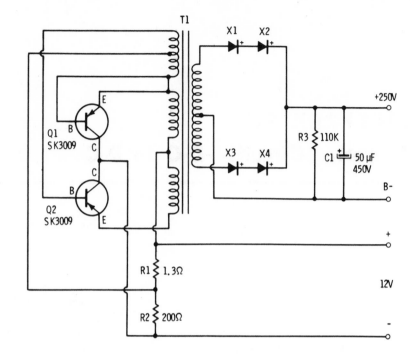

12 V TO 250 V—Provides up to 125 mA output, sufficient for low or medium-power mobile ham transmitter. Transistors should be mounted on but insulated from chassis serving as heatsink. Rectifiers are GE 504A diodes. Power transformer T1 is Chicago Transformer DCT-1 having 117-V primary and 280-V 125-mA secondary.—R. M. Brown and T. Kneitel, "101 Easy Test Instrument Projects," Howard W. Sams, Indianapolis, IN, 1968, p 118–119.

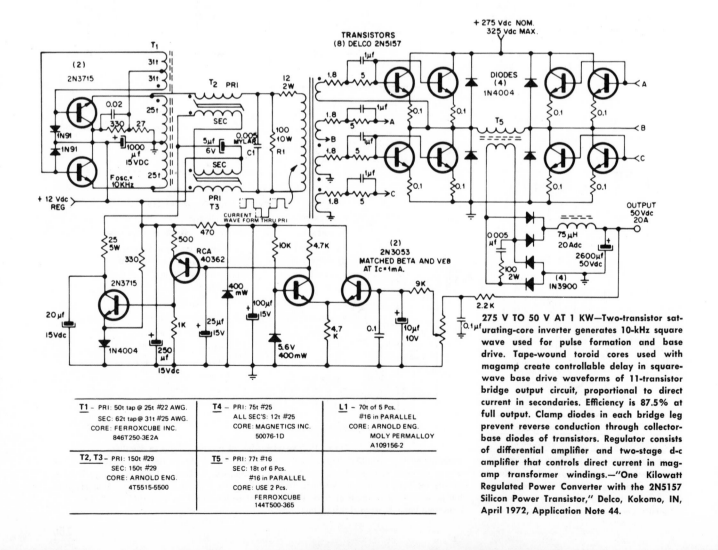

275 V TO 50 V AT 1 KW—Two-transistor saturating-core inverter generates 10-kHz square wave used for pulse formation and base drive. Tape-wound toroid cores used with magamp create controllable delay in square-wave base drive waveforms of 11-transistor bridge output circuit, proportional to direct current in secondaries. Efficiency is 87.5% at full output. Clamp diodes in each bridge leg prevent reverse conduction through collector-base diodes of transistors. Regulator consists of differential amplifier and two-stage d-c amplifier that controls direct current in magamp transformer windings.—"One Kilowatt Regulated Power Converter with the 2N5157 Silicon Power Transistor," Delco, Kokomo, IN, April 1972, Application Note 44.

T1 – PRI: 50t tap @ 25t #22 AWG. SEC: 62t tap @ 31t #25 AWG. CORE: FERROXCUBE INC. 846T250-3E 2A	**T4** – PRI: 75t #25 ALL SEC'S: 12t #25 CORE: MAGNETICS INC. 50076-1D	**L1** – 70t of 5 Pcs. #16 in PARALLEL CORE: ARNOLD ENG. MOLY PERMALLOY A109156-2
T2, T3 – PRI: 150t #29 SEC: 150t #29 CORE: ARNOLD ENG. 4T5515-5500	**T5** – PRI: 77t #16 SEC: 18t of 6 Pcs. #16 in PARALLEL CORE: USE 2 Pcs. FERROXCUBE 144T500-365	

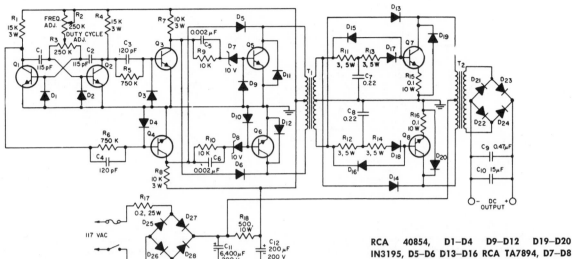

1 KW OFF LINE—Uses power transistors operating directly from rectified a-c line without need for bulky power transformer. Provides 100-V d-c output with overall efficiency above 85%, at operating frequency of 20 kHz. Q1–Q4 are 2N3440, Q5–Q6 2N6176, Q7–Q8 RCA 40854, D1–D4 D9–D12 D19–D20 IN3195, D5–D6 D13–D16 RCA TA7894, D7–D8 TRW 1N4740, D17–D18 RCA TA7900, D21–D24 RCA TA7985, and D25–D28 1N1195A. Report describes operation in detail.—"Power Transistors and Power Hybrid Circuits," RCA, Somerville, NJ, 1973 Ed., p 664–668.

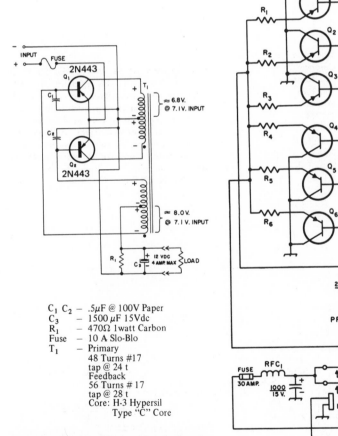

C₁ C₂ — .5μF @ 100V Paper
C₃ — 1500 μF 15Vdc
R₁ — 470Ω 1watt Carbon
Fuse — 10 A Slo-Blo
T₁ — Primary
48 Turns #17
tap @ 24 t
Feedback
56 Turns #17
tap @ 28 t
Core: H-3 Hypersil
Type "C" Core

6 V TO 12 V AT 4 A—Inverter circuit using 300-Hz oscillator doubles voltage of 6-V storage battery without using rectifier. Efficiency is high for all loads up to maximum of 4 A. Voltage multiplication is achieved by adding feedback voltage to supply voltage.—"6 to 12 Volt DC to DC Converter without Rectifiers," Delco, Kokomo, IN, Feb. 1972, Application Note 17.

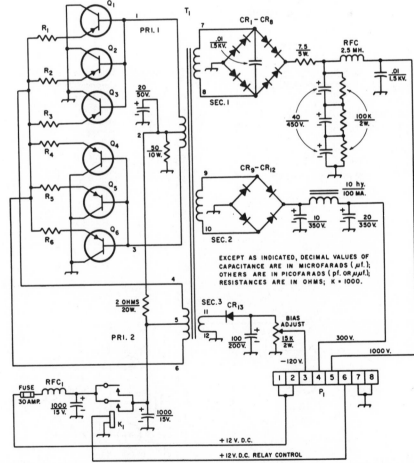

EXCEPT AS INDICATED, DECIMAL VALUES OF CAPACITANCE ARE IN MICROFARADS (μf.); OTHERS ARE IN PICOFARADS (pf. OR μμf.); RESISTANCES ARE IN OHMS; K = 1000.

12 V TO 1,000 V AT 500 W—Permits operating mobile tube-type transceiver from auto storage battery. CR13 is 400-piv 200-mA silicon and other diodes are 600-piv 750-mA silicon. Transistors are 2N1109 or similar. R1–R6 are each 12 inches of No. 20 wire wound on ¼-inch diameter forms. RFC1 is 14 turns No. 12 close-wound on ¾-inch form. Instructions are given for hand-winding T1 on 3-inch OD toroidal core.—"The Mobile Manual For Radio Amateurs," The American Radio Relay League, Newington, CT, 4th Ed., 1968, p 152–154.

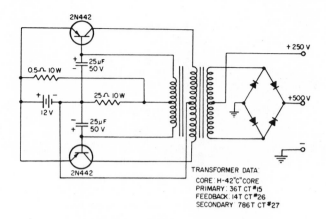

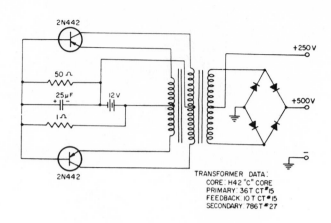

12 V TO 500 V—Delivers up to 100 W at 800 Hz, using common-emitter connection for inverter. Provides additional 250-V d-c output. —"Square Wave Oscillator Power Supplies," Delco, Kokomo, IN, Feb. 1972, Application Note 8-B.

12 V TO 500 V—Delivers up to 100 W at 800 Hz, using common-base connection for inverter. Provides additional 250-V d-c output. —"Square Wave Oscillator Power Supplies," Delco, Kokomo, IN, Feb. 1972, Application Note 8-B.

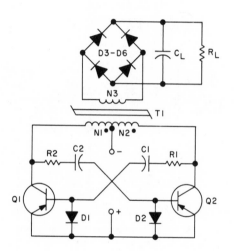

Transformer:	Stackpole #55–420
	Ceramag 24 A
N_1–N_2:	20 Turns
N_3:	50 Turns
Q_1–Q_2:	D43C5
D_1–D_2:	1N914
R_1–R_2:	560 Ω
C_1–C_2:	1 µf
R_L:	200 Ω
CL:	0.5 µf/100 V
D_3–D_6	A114F

28 V TO 68 V AT 23 W—Provides 23 W at 68 V d-c for load with efficiency of 76% when operating from 28-V d-c supply at 10 kHz. C1 and C2 serve to prevent both transistors from getting turned on in d-c mode.—L. E. Donovan, "An Assortment of High Frequency, Transistor Inverters/Converters Utilizing Saturating Core Transformers," General Electric, Semiconductor Products Dept., Auburn, NY, No. 200.57, 1970, p 9.

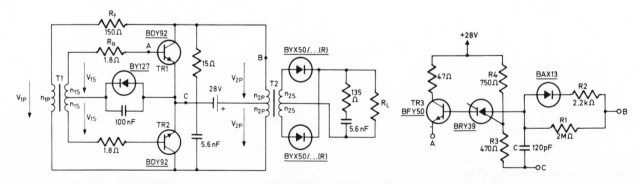

28 V to 250 V—Use of BDY92 silicon power switching transistors permits power outputs above 100 W at operating frequency of 25 kHz. High frequency permits use of smaller transformer and simpler filter, for reduction in weight and volume. Auxiliary starter cir- cuit, shown separately, connects to points A, B, and C of oscillator and applies full 28-V supply voltage to B if oscillator does not start immediately upon switch-on. This discharges C through R1 to initiate action that makes TR3 conduct and apply positive start- ing pulse to base of TR1. Design procedure is covered in report.—"BDY90 Series High Power, High Frequency DC/DC Converters," Amperex Electronic Corp., Slatersville, RI, Application Report S-154, Dec. 1971.

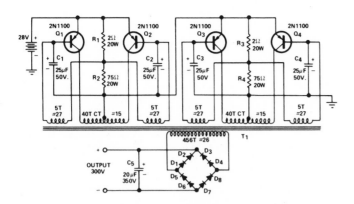

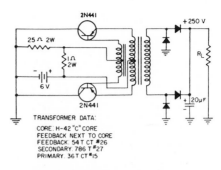

28 V TO 300 V AT 250 W—Series arrangement of converter circuits permits operation at higher supply voltages. All transistors operate as switches controlled by bias voltages induced in feedback windings. Operating frequency of 650 Hz is determined by number of turns in primary windings, supply voltage, and core. Use Westinghouse RH-79 toroid or equivalent core. All primary and feedback windings should be bifilar. All diodes are Sarkes Tarzian M500 or equivalent silicon.—"Series Connected DC to DC Converter," Delco, Kokomo, IN, Nov. 1971, Application Note 12-B.

6 V TO 250 V—Delivers up to 50 W at operating frequency of 400 Hz. Inverter uses common-collector connection.—"Square Wave Oscillator Power Supplies," Delco, Kokomo, IN, Feb. 1972, Application Note 8–B.

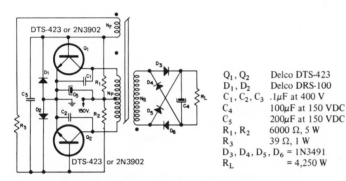

Q_1, Q_2	Delco DTS-423
D_1, D_2	Delco DRS-100
C_1, C_2, C_3	.1μF at 400 V
C_4	100μF at 150 VDC
C_5	200μF at 150 VDC
R_1, R_2	6000 Ω, 5 W
R_3	39 Ω, 1 W
D_3, D_4, D_5, D_6	= 1N3491
R_L	= 4,250 W

180 W AT 94% EFFICIENCY—Output power to load is 180 W maximum, at voltage approximately equal to number of turns on secondary winding NS. Primary has 157 turns No. 20 on each half, bifilar wound on Magnetics 51001-2A core, with 7-turn feedback winding. Input is 150 V d-c, and inverter frequency is about 1,200 Hz.—"High Efficiency DC-DC Converter with Silicon Transistors," Delco, Kokomo, IN, July 1971, Application Note 32.

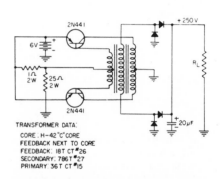

6 V TO 250 V—Delivers up to 50 W at 400 Hz. Inverter uses common-emitter connection.—"Square Wave Oscillator Power Supplies," Delco, Kokomo, IN, Feb. 1972, Application Note 8–B.

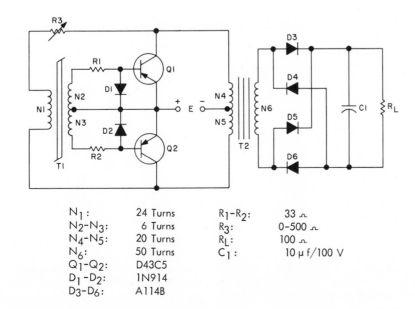

N_1:	24 Turns	R_1-R_2:	33 Ω	
N_2-N_3:	6 Turns	R_3:	0-500 Ω	
N_4-N_5:	20 Turns	R_L:	100 Ω	
N_6:	50 Turns	C_1:	10 μf/100 V	
Q_1-Q_2:	D43C5			
D_1-D_2:	1N914			
D_3-D_6:	A114B			

28 V TO 70 V AT 50 W—Efficiency is 85%, using two-transformer inverter operating at 18 kHz. Transformers use Stackpole Ceramag 24A No. 55-420 core. T2 is nonsaturating, permitting use of saturating base-drive transformer T1 having much smaller size and reducing losses in saturating core. Adjust R3 until desired inverter frequency is reached—L. E. Donovan, "An Assortment of High Frequency, Transistor Inverters/Converters Utilizing Saturating Core Transformers," General Electric, Semiconductor Products Dept., Auburn, NY, No. 200.57, 1970, p 12–13.

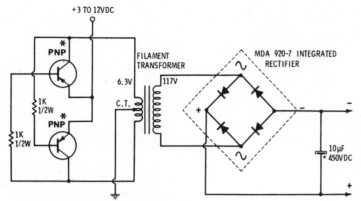

3 V TO 300 V—Developed for use with portable laser pulse generator. Transistors of blocking oscillator should be mounted on heat sinks for high-power applications.—R. W. Campbell and F. M. Mims III, "Semiconductor Diode Lasers," Howard W. Sams, Indianapolis, IN, 1972, p 107–108.

✱ FOR LOW CURRENT OPERATION, USE ANY COMMON PNP TRANSISTORS. FOR HIGH CURRENT OPERATION, USE PNP POWER TRANSISTORS SUCH AS HEP232 (MOTOROLA) OR 2N3614.

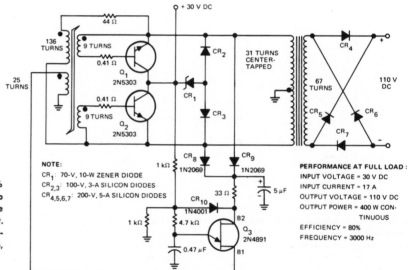

30 V TO 110 V AT 400 W—Efficiency is 80% at inverting frequency of 3 kHz. Use of two transformers gives good stability for large load changes and reduces transformer cost. Report gives design data.—G. Ehni, "Low-Cost 400-Watt Converter," Texas Instruments, Dallas, TX, CA-123.

NOTE:
CR_1: 70-V, 10-W ZENER DIODE
$CR_{2,3}$: 100-V, 3-A SILICON DIODES
$CR_{4,5,6,7}$: 200-V, 5-A SILICON DIODES

PERFORMANCE AT FULL LOAD:
INPUT VOLTAGE = 30 V DC
INPUT CURRENT = 17 A
OUTPUT VOLTAGE = 110 V DC
OUTPUT POWER = 400 W CONTINUOUS
EFFICIENCY = 80%
FREQUENCY = 3000 Hz

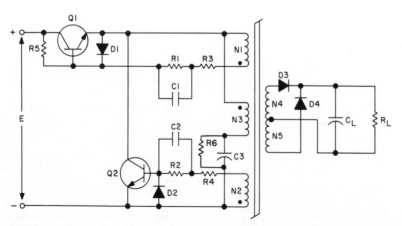

R_1-R_2:	1000 Ω
R_3-R_4:	10 Ω
R_5:	250 K Ω
C_1-C_2:	.01 μf
C_3:	20 μf
C_4:	.5 μf
N_1-N_2:	5 t
N_3:	50 t
N_4-N_5:	25 t
R_L:	200 Ω
R_6:	39 K Ω

180 V TO 78 V AT 30 W—Efficiency is 82% at inverter frequency of 12.4 kHz. Q1-Q2: D44R1; D1-D2: A14F; D3-D4: A114B. Transformer uses Stackpole Ceramag 24A No. 55- 420 core. Transistor breakdown voltage rating need be only 1.2 times input voltage. Will operate on supplies up to 500 V with transistors specified.—L. E. Donovan, "An Assortment of High Frequency, Transistor Inverters/Converters Utilizing Saturating Core Transformers," General Electric, Semiconductor Products Dept., Auburn, NY, No. 200.57, 1970, p 9.

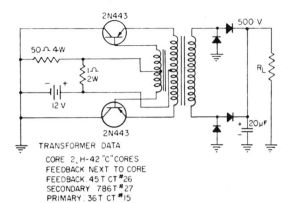

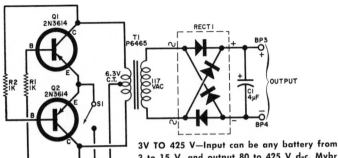

12 V TO 500 V—Delivers up to 100 W at 400 Hz, using common-collector connection for inverter.—"Square Wave Oscillator Power Supplies," Delco, Kokomo, IN, Feb. 1972, Application Note 8-B.

3V TO 425 V—Input can be any battery from 3 to 15 V, and output 80 to 425 V d-c. Mvbr uses filament transformer as saturable device for stepping up a-c voltage. Load should be at least 10,000 ohms. Will deliver more than 40 mA, corresponding to power output of 16 W.—J. Colt, Build a D.C. Transformer, *Popular Electronics*, March 1970, p 35 and 40–43.

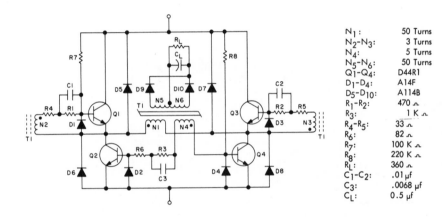

N_1:	50 Turns
N_2-N_3:	3 Turns
N_4:	5 Turns
N_5-N_6:	50 Turns
Q_1-Q_4:	D44R1
D_1-D_4:	A14F
D_5-D_{10}:	A114B
R_1-R_2:	470 Ω
R_3:	1 K Ω
R_4-R_5:	33 Ω
R_6:	82 Ω
R_7:	100 K Ω
R_8:	220 K Ω
R_L:	360 Ω
C_1-C_2:	.01 μf
C_3:	.0068 μf
C_L:	0.5 μf

180 V TO 167 V AT 77 W—Efficiency is 84%, using 23-kHz inverter frequency with bridge circuit. Designed for high-power applications with high input voltages, because transistors require breakdown voltage rating only slightly higher than input voltage. Requires no large capacitors. Transformer uses Stackpole Ceramag 24A No. 55-420 core.—L. E. Donovan, "An Assortment of High Frequency, Transistor Inverters/Converters Utilizing Saturating Core Transformers," General Electric, Semiconductor Products Dept., Auburn, NY No. 200.57, 1970, p 11.

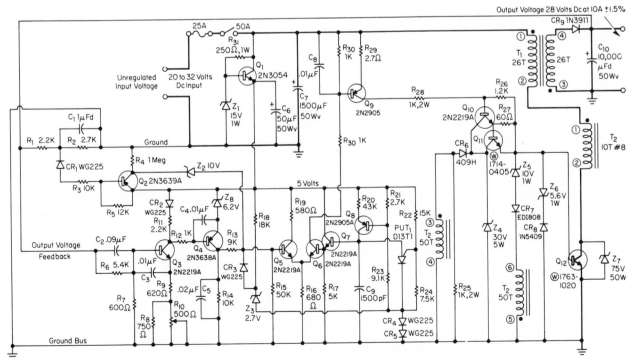

20–32 V D-C TO 28 V D-C—Uses 12-kHz fixed-frequency chopper to provide pwm square-wave voltage drive to primary of output transformer T1. Duty cycle of Q12 is controlled to stabilize output at 28 V d-c for loads up to 10 A even though input drops down to 20 V. Circuit efficiency is above 80% for outputs up to 6A, but drops as load increases further. Must be started without load, to prevent circuit from entering current-limiting mode in which rated output voltage is never reached. Circuit includes protection against output short-circuit.—"Type 1763 Transistor Chopper," Westinghouse, Youngwood, PA, Application Data 54-680, 1969.

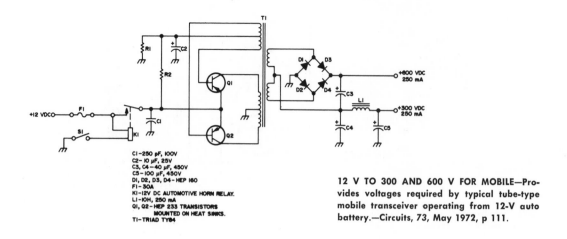

C1—250 pF, 100V
C2— 10 μF, 25V
C3, C4—40 μF, 450V
C5—100 μF, 450V
D1, D2, D3, D4— HEP 160
F1— 30A
K1—12V DC AUTOMOTIVE HORN RELAY.
L1—10H, 250 mA
Q1, Q2— HEP 233 TRANSISTORS
MOUNTED ON HEAT SINKS.
T1— TRIAD TY84

12 V TO 300 AND 600 V FOR MOBILE—Provides voltages required by typical tube-type mobile transceiver operating from 12-V auto battery.—Circuits, 73, May 1972, p 111.

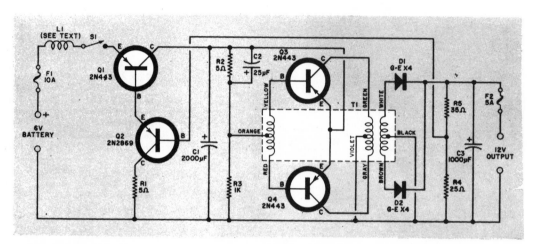

AUTO 6 V TO 12 V—Permits using 12-V cartridge players and other electronic equipment in cars having 6-V electrical system. Can handle up to 40 W. Q3 and Q4, in feedback-type power oscillator circuit with saturable transformer EC-0401-IC (Milwaukee Electromagnetics), delivers a-c output to rectifier diodes. Portion of output voltage is fed back to conventional series regulator Q1-Q2 to hold output voltage essentially constant—B. Richards, Make Your Own 6 to 12 Volt "Up-Verter", Popular Electronics, Oct. 1967, p 67–70.

CHAPTER 23
Converter Circuits—General

CHANGING R TO L—Used to transform non-linear resistor into nonlinear inductor. Resistor is connected across port two of mutator circuit using opamps. Circuit requires voltage-controlled current source, also given in article.—L. O. Chua, Synthesis of New Nonlinear Network Elements, *Proc. IEEE*, Aug. 1968, p 1325–1340.

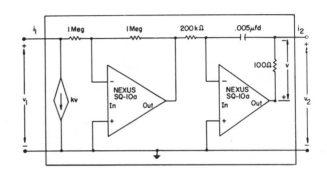

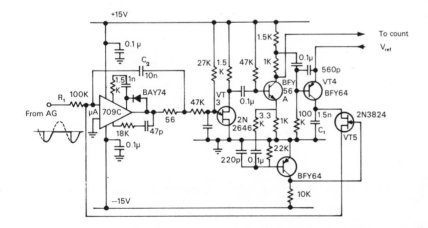

COUNTER DRIVE—Variation of voltage-frequency converter is used in power-factor meter to convert output of analog gate (AG) to proportional frequency up to 10 kHz for driving counter and display showing power factor as four-digit value on cold-cathode tubes, to accuracy of 0.1%. Conversion linearity of a-d converter shown is better than 0.08% for input voltage range of 0 to 10 V. —F. Bombi and D. Ciscato, Digital Power-Factor Meter Has High Order of Accuracy, *Electronic Engineering*, Feb. 1971, p 55–58.

ANALOG VOLTAGE TO PULSE WIDTH—Provides high precision at high speed, as required for pwm, delta code generation, a-d conversion, and motor speed control. Uses OEI 9694 opamps as fast comparators and 9710 fet-input opamp as fast integrator. Will handle pulse widths in 1-μs region. Linearity of pulse width to input voltage is better than 0.1% for 1-μs full-scale output. Minimum pulse width is 100 ns and maximum dynamic range 40 dB. Fifth 9694 sums outputs of two comparators and produces positive output.— "Voltage To Pulse Width Converter," Optical Electronics, Tucson, AZ, Application Tip 10230 1971.

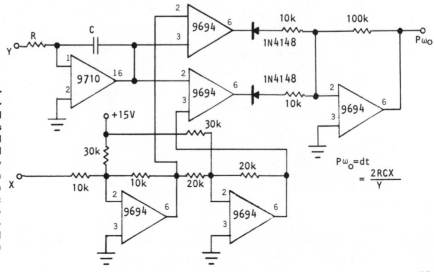

VOLTAGE TO PULSE WIDTH—Uses Burr-Brown Model 4013/25 unity-gain inverting sample-and-hold module as switched integrator serving as very linear and stable synchronized pulse-width modulator. Pulse train provides clocking signal. Output is pulse train synchronized to clock pulse input, with pulse width a linear function of input voltage. Choose values of ER, C1, and R1 for desired prr and dynamic range. If clock frequency is 1 kHz, input is 0.1–10 V, ER is 10 V, C1 is 0.01 μF, and R1 is 90K, output pulse width Tp is 0.09 E1 ms.—"Sample/Hold Modules," Burr-Brown, Tucson, AZ, 1969.

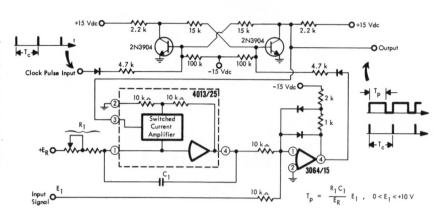

$$T_P = \frac{R_1 C_1}{E_R} E_1, \quad 0 < E_1 < +10\ V$$

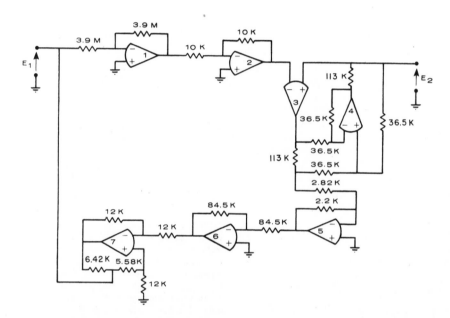

IMPEDANCE CONVERTER—Uses Burr-Brown 3057/01 opamps. Measured performance agrees substantially with calculated values from d-c up to about 10 kHz for resistive load between 20 and 200,000 ohms at E1 and between 1 and 10,000 ohms at E2. Versatility is chief advantage of circuit.—N. W. Cox, K. L. Su, and R. P. Woodward, A Floating Three-Terminal Nullor and the Universal Impedance Converter, *IEEE Trans. on Circuit Theory*, May 1971, p 399–400.

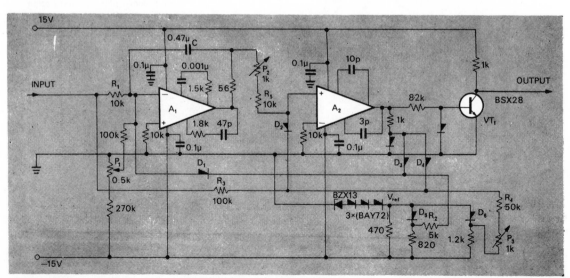

VOLTAGE-TO-FREQUENCY CONVERTER—Converts d-c input voltages between 1 mV and 10 V to constant-duration pulses whose repetition rate is proportional to input voltage. Nonlinearity error is less than 2 parts in 10,000. Both opamps are μA709C. First differential amplifier A1 operates as integrator, and A2 as level detector whose output changes sign when its input voltage crosses zero. Article includes error analysis.—F. Bombi, High-Performance Voltage-to-Frequency Converter Has Improved Linearity, *Electronic Engineering*, Dec. 1970, p 61–64.

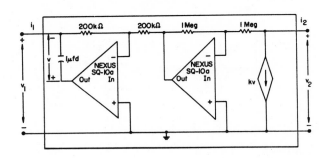

CHANGING R TO C—Nonlinear resistor connected across port two of mutator circuit is transformed into nonlinear capacitor. Circuit requires voltage-controlled current source, also given in article.—L. O. Chua, Synthesis of New Nonlinear Network Elements, *Proc. IEEE*, Aug. 1968, p 1325–1340.

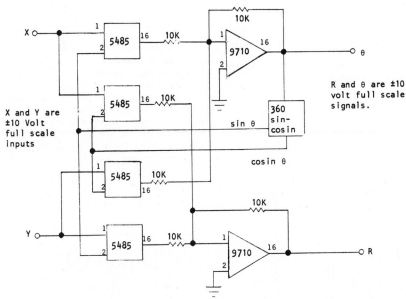

R and θ are ±10 volt full scale signals.

X and Y are ±10 Volt full scale inputs

RECTANGULAR TO POLAR—Uses OEI 5485 four-quadrant multipliers in circuit with opamps for generating common polar-type coordinates from standard X-Y input coordinates, over input bandwidth of d-c to 10 kHz. Overall error is 3% at d-c.—"Rectangular Coordinates to Polar Coordinates Scan Converter," Optical Electronics, Tucson, AZ, Application Tip 10156, 1971.

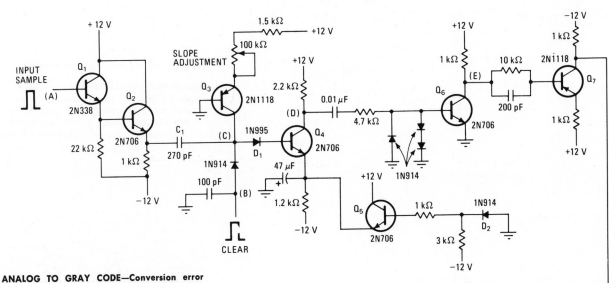

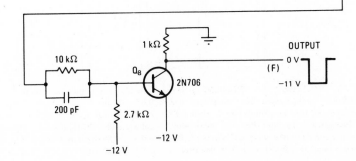

ANALOG TO GRAY CODE—Conversion error is less than 0.15 μs for full-scale output pulse width of 3.25 μs, corresponding to peak-to-peak video input level of 6.5 V. During first half of input, charge is removed from C1. During second half, charge proportional to sampled analog signal is placed on C1 by Q1 and Q2. When trailing edge of sample cuts off Q4, C1 is discharged linearly through current source Q3 and slope adjustment pot. Converter circuit thus generates output pulse with width proportional to analog input amplitude.—R. J. Turner, Height-to-Width Converter Digitizes Analog Samples, *Electronics*, April 10, 1972, p 99.

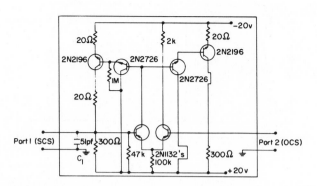

SIX-TRANSISTOR ROTATOR—Provides greater immunity to parameter variations than five-transistor rotator and better response at low frequencies than version using ordinary opamps. Nonlinear component connected to one port makes circuit act at other port as same component but with characteristic curve rotated about origin.—L. O. Chua, The Rotator—A New Network Component, *Proc. IEEE*, Sept. 1967, p 1566–1576.

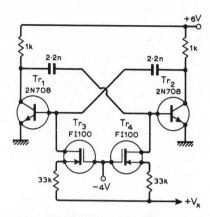

VOLTAGE-FREQUENCY CONVERTER—Combination of conventional and mos transistors, with operating conditions chosen to place mosfet's in their region of positive temperature coefficient, greatly improves temperature stability of free-running mvbr used as voltage-frequency converter. Mosfet's serve as constant-current source for transistors of mvbr.—S. Tesic, Stable Voltage-to-Frequency Converter, *Wireless World*, Feb. 1972, p 78.

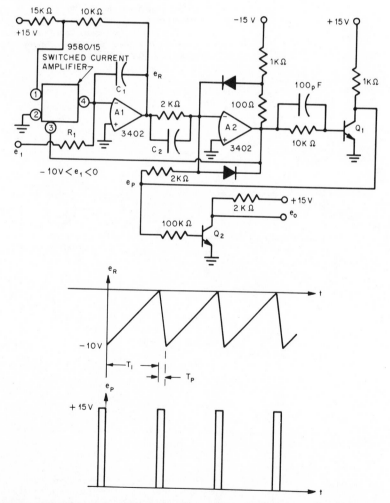

LINEAR VOLTAGE TO FREQUENCY—Uses Burr-Brown switched-current amplifier and two opamps for converting input voltage E1 to frequency with high linearity over wide range, up to 100 kHz when C1 is 100 pF. Dynamic range of 1,000:1 with 1% linearity is feasible. Reset time is about 3 μs if C1 is 0.01 μF and R1 10K, and frequency in Hz is then about 1,000 E1.—J. G. Graeme, G. E. Tobey, and L. P. Huelsman, "Operational Amplifiers," McGraw-Hill Book Co., New York, 1971, p 409–411.

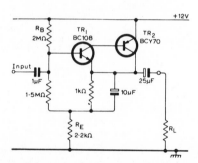

IMPEDANCE CONVERTER—Input resistance is about 1.5 meg and output resistance 40 ohms. Response is flat within 3 dB between 1.5 and 15,000 Hz for 1-meg source impedance.—"P-N-P Practical Planar Transistors—BCY70 Family and BFX29 Family," Mullard, London, 1967, TP887, p 5.

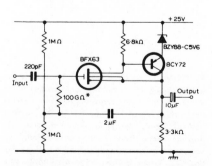

MOST IMPEDANCE CONVERTER—Uses metal-oxide-semiconductor transistor to give input resistance of 1 teraohm and output resistance of 50 ohms. Gain is about 0.98.—"P-N-P Practical Planar Transistors—BCY70 Family and BFX29 Family," Mullard, London, 1967, TP887, p 5.

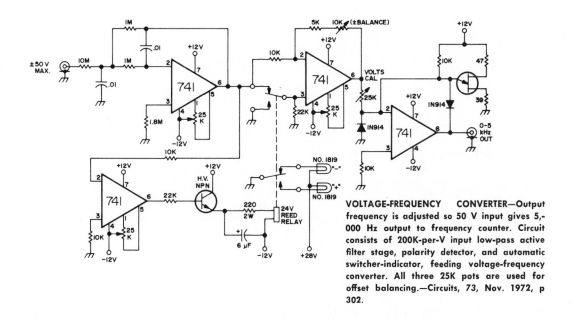

VOLTAGE-FREQUENCY CONVERTER—Output frequency is adjusted so 50 V input gives 5,-000 Hz output to frequency counter. Circuit consists of 200K-per-V input low-pass active filter stage, polarity detector, and automatic switcher-indicator, feeding voltage-frequency converter. All three 25K pots are used for offset balancing.—Circuits, 73, Nov. 1972, p 302.

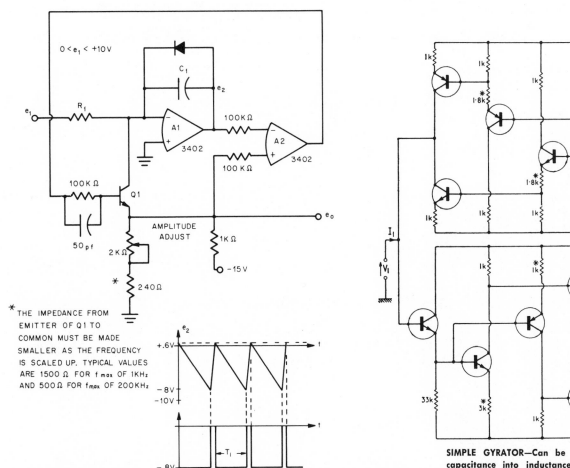

VOLTAGE TO FREQUENCY—Provides pulse-train output, using only two inexpensive wide-band fet-input opamps such as Burr-Brown 3402. Opamp should have good saturation characteristics, high input impedance, and good slew rate capability. Practical frequency range is limited by delay time of A2 as it switches, and is 20 to 100 μs for opamps shown. Output frequency is equal to input voltage E1 divided by 8.6R1C1. Add buffer to output if driving load less than 100K.—J. G. Graeme, G. E. Tobey, and L. P. Huelsman, "Operational Amplifiers," McGraw-Hill Book Co., New York, 1971, p 405–409.

SIMPLE GYRATOR—Can be used to convert capacitance into inductance or short-circuit into open circuit. Use silicon planar expitaxial transistors having gain-bandwidth products above 300 MHz, and use rheostats initially for starred resistors. Adjust for maximum un-distorted output from each amplifier treated separately. Requires well-filtered regulated supply.—F. Butler, Gyrators—Using Direct-Coupled Transistor Circuits, *Wireless World*, Feb. 1967, p 89–93.

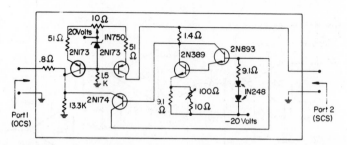

NEGATIVE-IMPEDANCE CONVERTER—Five-transistor rotator has two-terminal ports. When nonlinear resistor, inductor, or capacitor is connected to one port, circuit at other port behaves as new component whose characteristic curve is identical to that of original component but rotated by prescribed angle about origin. Article gives examples of rotated curves. One drawback of circuit is its high sensitivity to parameter variations. Circuit can be built with opamps having required good response at low frequencies.—L. O. Chua, The Rotator—A New Network Component, *Proc. IEEE*, Sept. 1967, p 1566–1576.

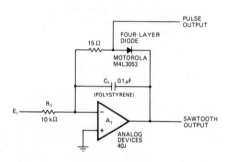

VOLTAGE TO FREQUENCY—Delivers 100 Hz per V with linearity of 0.2% for 0 to 10 V. Will operate up to 5 kHz. Uses diode to reset conventional d-c integrator to fixed potential. Can also be usd as vco. Gives choice of pulse and sawtooth outputs.—T. C. O'Haver, Voltage-Frequency Converter Uses Four-Layer Diode, *Electronics*, Dec. 6, 1971, p 76–77.

CHAPTER 24
Converter Circuits—Radio

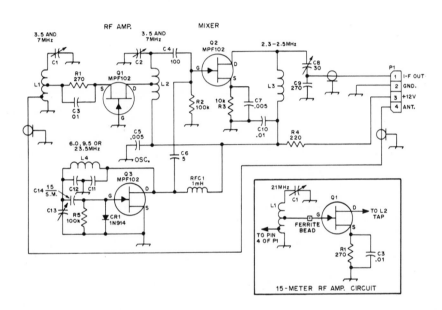

THREE-BAND PLUG-IN—Solid-state units are plugged into solid-state tunable i-f and a-f section to provide coverage of 80, 40, and 15 meters. Same circuit is used for all three bands, except for r-f amplifier wiring changes shown in inset for 15 meters. Article gives coil and capacitor values for all three bands.—J. Kaufmann and D. Demaw, A High-Performance Solid-State Receiver for the Novice or Beginner, QST, Oct. 1972, p 11–16 and 23.

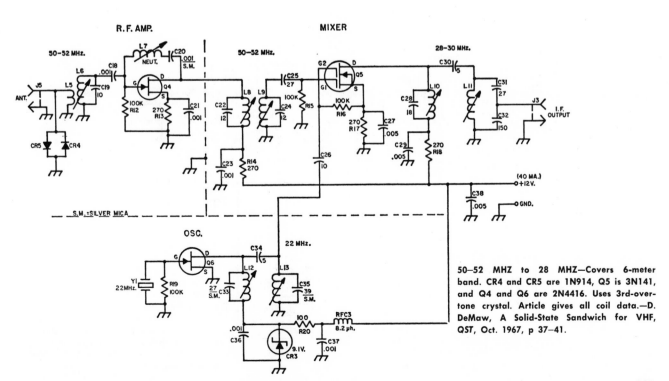

50–52 MHZ to 28 MHZ—Covers 6-meter band. CR4 and CR5 are 1N914, Q5 is 3N141, and Q4 and Q6 are 2N4416. Uses 3rd-overtone crystal. Article gives all coil data.—D. DeMaw, A Solid-State Sandwich for VHF, QST, Oct. 1967, p 37–41.

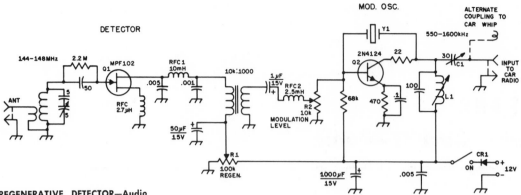

VHF SUPERREGENERATIVE DETECTOR—Audio output of detector is coupled to base of crystal-controlled oscillator Q2 operating in range of 550 to 1,600 kHz. Q2 operates as small transmitter, providing a-m bc signal that can be fed into antenna jack of any car radio. Frequency of crystal is chosen to fall in quiet part of broadcast band.—D. DeMaw, Quick-and-Easy Portable/Mobile Reception, QST, Feb. 1971, p 14–15.

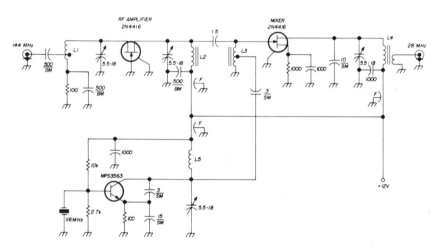

2-METER RADIO—Uses inexpensive jfet's in r-f amplifier and mixer. Crystal oscillator delivers 15 mW. BM units are button mica, and SM silver mica. Article gives coil data and construction tips.—B. Sutherland, Modular Two-Meter Converter, *Ham Radio*, Oct. 1970, p 64–68.

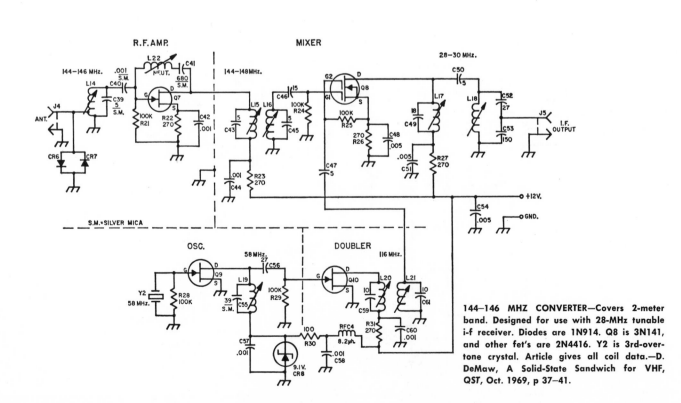

144–146 MHZ CONVERTER—Covers 2-meter band. Designed for use with 28-MHz tunable i-f receiver. Diodes are 1N914. Q8 is 3N141, and other fet's are 2N4416. Y2 is 3rd-overtone crystal. Article gives all coil data.—D. DeMaw, A Solid-State Sandwich for VHF, QST, Oct. 1969, p 37–41.

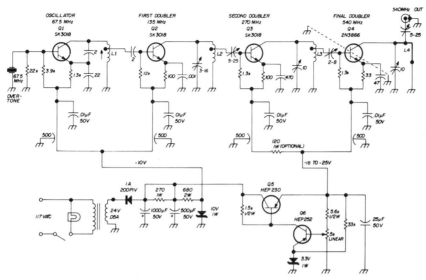

L1   11 turns no. 22 on ¼" slug-tuned form
L2   4¾ turns no. 18, 3/8" long, ¼" ID
L3   2½ turns no. 18, ¼" long, ¼" ID

L4   shim brass or copper, 3/8" wide, 23/32"
      long; length is in addition to tabs for
      soldering to capacitor and ground plane

2,304-MHZ HAM CONVERTER—Local-oscillator chain of converter multiplies 67.5-MHz output of crystal oscillator Q1 to 540 MHz with doubler stages. Article also covers construction of quadrupler using MV1622 varactor in trough line to give 2,160 MHz for driving Hewlett Packard HP2835 hot-carrier diode mixer to 1 mA. Mixing this with signal from 2,300-MHz amateur band gives required 144-MHz i-f value—W. Stanton, D. Moser, and D. Vilardi, Solid-State 2,304-MHz Converter, *Ham Radio*, March 1972, p 16–21.

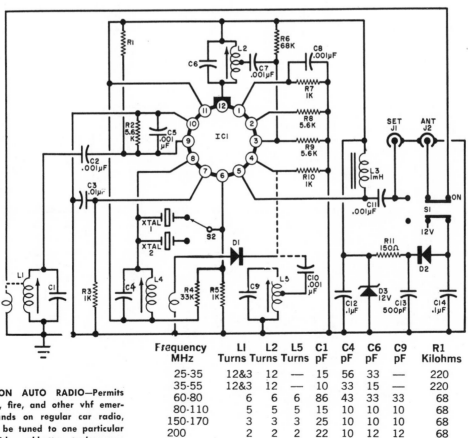

Frequency MHz	L1 Turns	L2 Turns	L5 Turns	C1 pF	C4 pF	C6 pF	C9 pF	R1 Kilohms
25-35	12&3	12	—	15	56	33	—	220
35-55	12&3	12	—	10	33	15	—	220
60-80	6	6	6	86	43	33	33	68
80-110	5	5	5	15	10	10	10	68
150-170	3	3	3	25	10	10	10	68
200	2	2	2	22	10	12	12	68

POLICE CALLS ON AUTO RADIO—Permits monitoring police, fire, and other vhf emergency service bands on regular car radio, which need only be tuned to one particular spot on dial. With pushbutton tuning, one button can be set up for police radio, because converter uses fixed-crystal tuning selected by S2. Article tells how to choose crystals and tune receiver. IC can be RCA-CA3018 or M6970. D1 is signal diode, and D2 is 1-A 400-V rectifier. Use No. 24 enamel for 3 and 12 turns of L1, either separate windings or tapped, and No. 18 for other coils.—L. E. Greenlee, VHF Frequency Converter, *Popular Electronics*, Dec. 1971, p 52–53 and 58.

POLICE AND FIRE BANDS—Used with car radio for monitoring emergency frequencies as well as amateur 2, 6, and 10-meter bands. To tune 28–54 MHz, wind L1 with 36 turns No. 30 on ¼ inch diameter form ½ inch long, tapped 11 turns from ground. To cover 100–200 MHz, wind L1 with 4 turns No. 16 on ¼ inch form, tapped 1 turn from ground. For both ranges, L2 is 36 turns No. 30 on ¼ inch form ½ inch long. Article tells how to determine crystal frequency for station to be monitored.—J. Crawford and G. Webber, Build the OTC, 73, Jan. 1966, p 74–75.

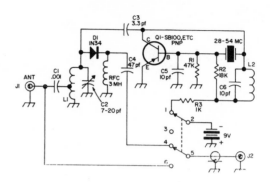

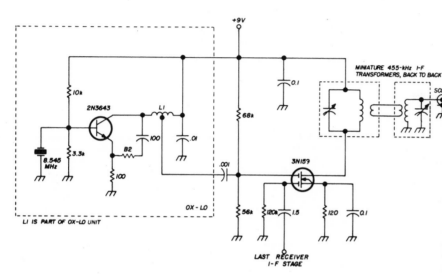

MOSFET CONVERTER—Designed to permit observation of 9-MHz i-f amplifier of transceiver on monitor scope capable of handling up to 6 MHz. Uses 8.545-MHz crystal oscillator to convert signals to 455 kHz for cro. Article covers construction.—R. A. Schiers, Jr., Mosfet Converter for Receiver Instrumentation, *Ham Radio*, Jan. 1971, p 62–64.

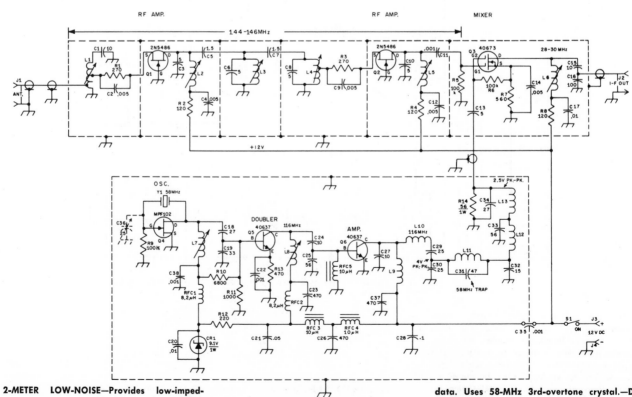

2-METER LOW-NOISE—Provides low-imped-ance output to tunable i-f receiver. Conver-sion gain is about 12 dB for mixer. Article covers construction, including coil-winding data. Uses 58-MHz 3rd-overtone crystal.—D. DeMaw, High Performance 2-Meter Converter, QST, June 1971, p 11–16 and 31.

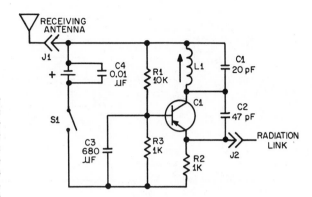

SW ON A-M—Inexpensive single-transistor converter covers range from 7 to 12 MHz with slug-tuned coil and converts to about 1,600 kHz for tuning with standard broadcast receiver. Transistor type is not critical. To pick up European and South American short-wave stations, use 25-ft length of hookup wire as antenna.—S. Daniels, Anti-Inflation Special, Elementary Electronics, Nov.-Dec. 1970, p 61-64.

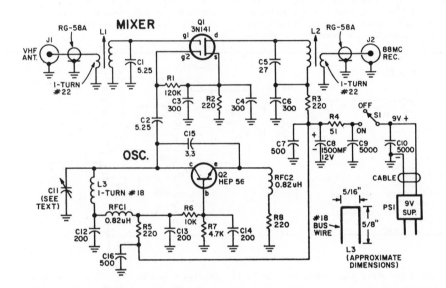

135–175 MHZ POLICE-CALL CONVERTER—Use with transistorized f-m receiver having good whip antenna, for listening to police calls on 155 MHz, fire on 154, tow trucks on 151, weather reports on 162, and other emergency and commercial radio services, including 2-meter ham band on 144–148 MHz. Article gives construction, alignment, and operating details. C11 is Lafayette 40F28411 or Hammarlund HFA-15-B modified to leave one rotor plate and one stator plate.—C. Green, VHF High-Band Converter, Elementary Electronics, Nov.-Dec. 1971, p 69–72.

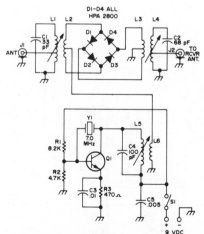

15-METER WITH HOT-CARRIER DIODES—Hewlett-Packard diodes D1–D4 make excellent low-noise mixers, with very little loss of signal. R-f amplifier is unnecessary. Input is tuned to 15 meters and output to 20 meters. Local oscillator Q1 uses 2N706 with 7-MHz crystal. Article gives coil data and adjustment procedure.—J. White, Hot Carrier Diode Converter, 73, Oct. 1972, p 96–97.

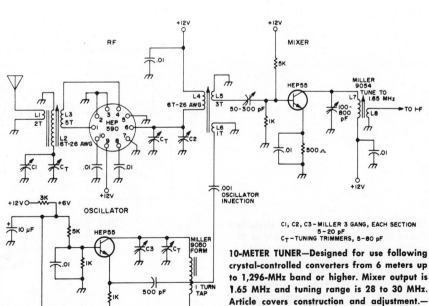

10-METER TUNER—Designed for use following crystal-controlled converters from 6 meters up to 1,296-MHz band or higher. Mixer output is 1.65 MHz and tuning range is 28 to 30 MHz. Article covers construction and adjustment.—B. Hoisington, IC Ten Meter Tuner for Use with Solid-State VHF-UHF Converters, 73, Jan. 1973, p 99–103.

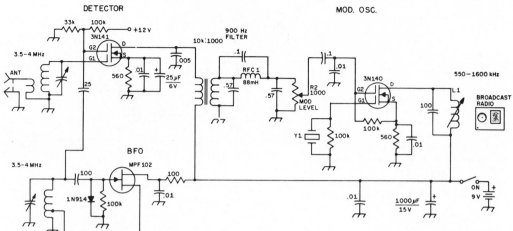

DIRECT-CONVERSION VHF—Dual-gate mosfet's are used both for 3.5–4 MHz c-w detector and modulated crystal oscillator feeding loopstick L1 which is loosely coupled to antenna of transistorized bc radio. Choose frequency for Y1 to place signal in quiet part of bc band. Output of Q1 goes through passive filter providing c-w selectivity at 900 Hz. If converter is used for ssb reception, change 0.57 μF capacitors to 0.22 μF to provide 2-kHz selectivity.—D. DeMaw, Quick-and-Easy Portable/Mobile Reception, QST, Feb. 1971, p 14–15.

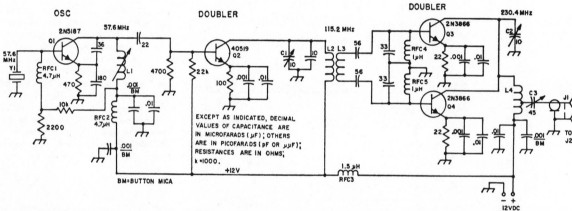

1,296-MHZ CONVERTER—Solid-state low-noise design consists of 57.6-MHz crystal oscillator and two doublers for delivering 230.4 MHz to output jacks J1-J2. Article gives all coil data, along with construction details of diode quintupler in trough line, filter, and mixer for combining 1,296-MHz input signal with quadrupled 1,152-MHz signal to obtain 144-MHz output for use with 2-meter converter.—D. W. Nelson, Modernizing a Classic 1,296-MHz Converter, QST, Dec. 1969, p 21–23.

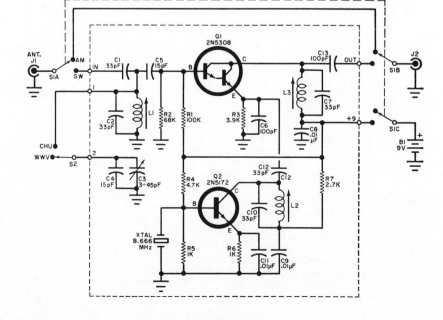

WWV AND CHU—Fixed-tuned crystal-controlled short-wave converter for auto or transistor a-m radio permits receiving highly accurate time signals and other information from NBS station WWV on 10 MHz or Canadian Dominion Observatory station CHU on 7.335 MHz at flick of switch. L1 and L2 are J. W. Miller 23A336RPC.—G. J. Whalen, Build the AccuraTime, Popular Electronics, July 1968, p 41–46.

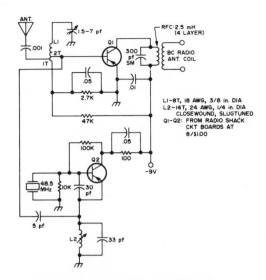

2-METER CONVERTER—Will work with pocket a-m radio. Tune receiver to frequency equal to that of desired 2-meter station minus 3 times crystal frequency of 48.5 MHz. If strong broadcast station happens to heterodyne with desired 2-meter signal, substitute slightly different crystal.—Circuits, 73, Feb. 1973, p 143.

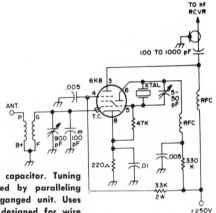

500-KHZ SHIPPING BANDS—Uses 3,100-kHz crystal for covering 400 to 1,000 kHz with 80-meter communication receiver. Antenna coil is bc type designed for use with wire antenna.—J. H. Smith, A Medium Frequency Converter for High Frequency Receivers, 73, Jan. 1972, p 73–74.

500-KHZ SHIPPING BANDS—Permits listening to passenger and cargo ships, weather bulletins, and commercial traffic on 500 kHz and below with communication receiver. For receiver tuning from 3.5 to 4.1 MHz on lowest band, converter permits tuning from 400 to 1,000 kHz with receiver on 80-meter band. Use 3,100-kHz crystal, adjusted to exactly this value with shunt trimmer capacitor. Tuning capacitor value is obtained by paralleling two or three sections of ganged unit. Uses standard bc antenna coil designed for wire antenna rather than loop or ferrite rod.—J. H. Smith, A Medium Frequency Converter for High Frequency Receivers, 73, Jan. 1972, p 73–74.

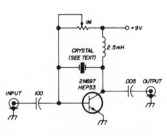

ham	WWV frequency (MHz)			
band	10	15	20	25
80	6-6.5	—	—	—
40	2.7-3.0	8-7.70	—	—
20	4-4.35	—	6-6.65	—
15	—	6-6.45	—	3.55-4.00

WWV—Simple circuit converts WWV signals above 5 MHz to amateur bands between 80 and 15 meters. Table gives range of crystal frequencies that can be used for particular ham band and WWV frequency.—D. Pongrance, Inexpensive WWV Converter, Ham Radio, Dec. 1970, p 55.

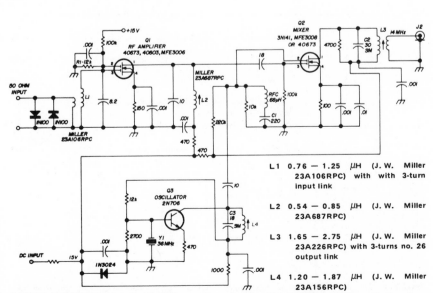

L1 0.76 — 1.25 μH (J. W. Miller 23A106RPC) with with 3-turn input link

L2 0.54 — 0.85 μH (J. W. Miller 23A687RPC)

L3 1.65 — 2.75 μH (J. W. Miller 23A226RPC) with 3-turns no. 26 output link

L4 1.20 — 1.87 μH (J. W. Miller 23A156RPC)

6-METER MOSFET—Output is 28 MHz at J2 for values shown. R-f stage Q1 improves reception of weak signals in presence of strong signals.—D. W. Nelson, Deluxe Mosfet Converters for Six and Two Meters, Ham Radio, Feb. 1971, p 41–47.

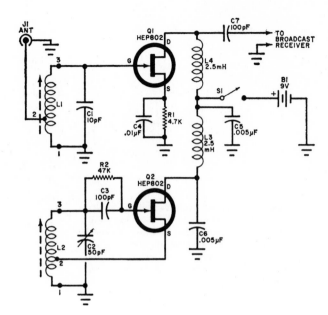

14–31 MHZ—Permits receiving radio amateurs, citizens band, and foreign stations in this range on any a-m broadcast radio. Signal of oscillator Q2 reaches mixer Q1 through mutual inductance between L1 and L2. Uses vernier dial on C2. Connect portable receiver with short length of wire running from C7 to vicinity of receiver antenna. L1 is 17 turns No. 28 enamel tapped 3½ turns from bottom, and L2 is 15 turns No. 28 enamel tapped 5 turns from bottom, each wound on its own Miller 20A000 coil form. Has brought in Paris, London, Moscow, and South Africa.—J. White, Shortwave Converter Makes Ideal Beginner's Construction Effort, *Popular Electronics*, Aug. 1969, p 75–79.

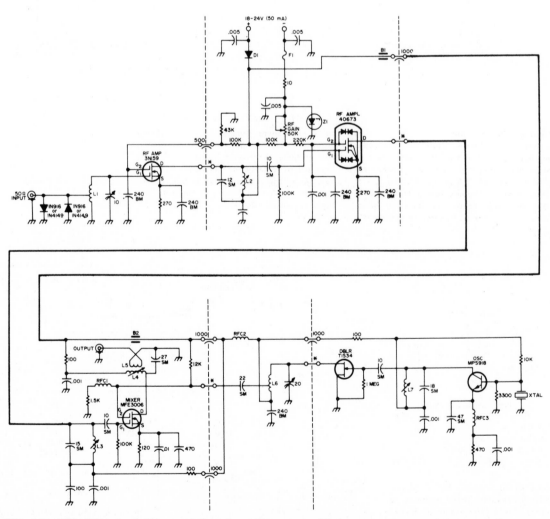

MOSFET FOR 144 MHZ—Wide bandwidth permits operation over entire 144–148 MHz range without repeaking. Has low noise figure, high gain, excellent stability, overload control, and excellent cross-modulation performance without requiring neutralization or stagger tuning. Article covers construction and adjustment, including all coil data. Mixer output coil is tuned to 16 MHz.—R. D. Morrison, A Versatile and Stable Mosfet Converter for 144 MHz, *73*, Sept. 1970, p 90–97.

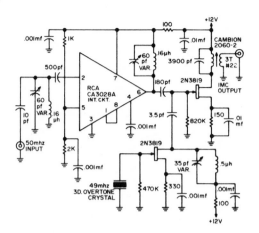

6-METER WITH IC FRONT-END—IC gives better gain characteristics than equivalent nuvistor or fet circuit and eliminates need for neutralization. Converter produces difference frequency of 1 MHz, which is in middle of broadcast band and ideal for use with auto radio. For 7-MHz difference frequency, use 43.5-MHz crystal and change 3,900-pF mixer drain tank capacitor to 100 pF. Overall gain is about 24 dB and bandwidth 100 kHz. Article covers construction and alignment.—E. Levy, Integrated Circuit 6 Meter Converter, 73, March 1971, p 16–17.

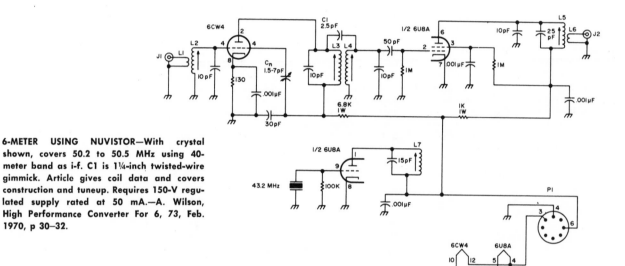

6-METER USING NUVISTOR—With crystal shown, covers 50.2 to 50.5 MHz using 40-meter band as i-f. C1 is 1¼-inch twisted-wire gimmick. Article gives coil data and covers construction and tuneup. Requires 150-V regulated supply rated at 50 mA.—A. Wilson, High Performance Converter For 6, 73, Feb. 1970, p 30–32.

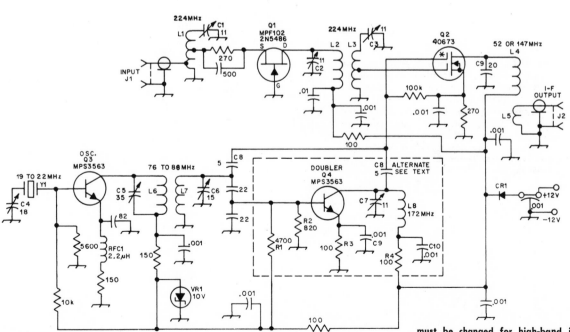

220-MHZ F-M CONVERTER—Use complete circuit when working into 144-MHz f-m receiver. Omit components in dashed rectangle if using with 50-MHz (2-meter) receiver. Q3 is oscillator and quadrupler, providing injection voltage to mixer Q2 for i-f of 147 MHz. Article covers construction, including coil-winding data, and tuneup. Connection point for C8 must be changed for high-band i-f. CR1 is 50-prv 200-mA silicon diode.—T. McMullen, 450 Cubic Centimeters of New Front End for Your FM Receiver, QST, June 1972, p 11–13 and 35.

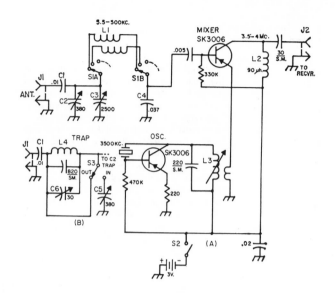

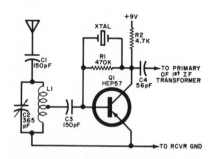

S-W ON TRANSISTOR RADIO—Use any crystal whose frequency is 455 kHz above or below desired short-wave band, and tune over band with C2. Requires outdoor antenna about 20 feet long. Feed into i-f of transistor portable radio. With a-c radio, choose one having power transformer, to avoid shock danger.—L. Lisle, A Simple SW Converter, *Popular Electronics*, March 1972, p 96.

5.5-KHZ VLF TO 80 METERS—Permits receiving ship-to-shore communications, along with long-wave c-w and rtty broadcasts to submerged nuclear submarines and a variety of other military and commercial broadcasts extending from the broadcast band down to the 18.5-kHz frequency of U.S. Navy Station NLK in Jim Creek, Wash. and even lower. Article gives inductance values required for coils to cover 5.5 to 1,100 kHz in 9 bands. Crystal frequency of 3,500 kHz makes converter work into 80-meter band of communications receiver, but other crystals may be used for other receivers. Trap circuit is needed to block WCC on 436 kHz when listening on 500 kHz. —W. H. Fishback, 600 to 20,000 METERS, *QST*, Sept. 1968, p 18–22.

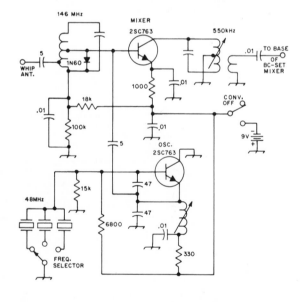

2-METER F-M—Uses three crystals for vhf reception, for which broadcast receiver serves as second section of double-conversion receiver. Output of converter is 500 kHz going to mixer base of bc set, for which oscillator is tuned to slope-detect f-m signal. Ferrite-bar antenna in hand-held set is used during a-m reception.—Sonar Sentry FM/A-M Monitor Receiver, Model FR-103SA, *QST*, Oct. 1972, p 53–55.

CHAPTER 25
Counter Circuits

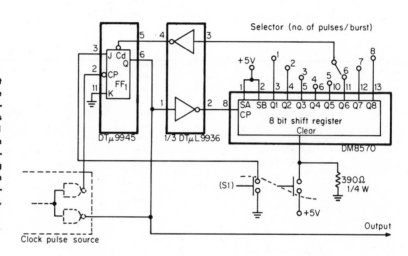

COUNTER TESTER—Selector switch can be set to provide finite number of pulses in response to electronic or mechanical trigger, as required for exercising or testing shift registers and counters. All Q outputs go low when S1 is opened. Clock input is then applied. When selected Q output goes high, it disables output and pulse train stops. To prevent slicing of first output pulse, clock is wire-ORed with output of flip-flop.—E. Sheffer, A Finite Number Pulse Source, *The Electronic Engineer*, May 1971, p 76.

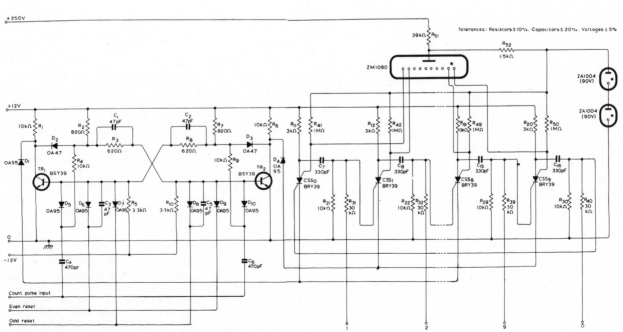

50-KHZ SCS RING—Circuit shows drive for one 0–9 numerical indicator tube. Cathode voltage of scs that is to be triggered next is made zero at same time as trigger voltage is initiated by turning off previous scs. Circuit incorporates facility for setting numerical indicator tube to any desired number. Requires negative-going 5-V count pulse with 10-μs duration.—"Electronic Counting," Mullard, London, 1967, TP874, p 81–83.

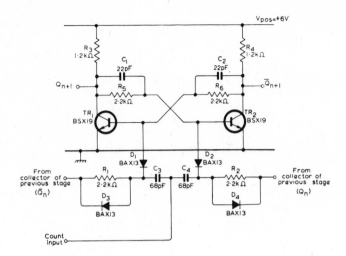

TWISTED-RING COUNTER—Also known as Johnson counter. Uses R-S bistable elements, one unit of which is shown. Requires fewer elements than ring counter, but output requires decoding. Five elements give cycle length of ten, for use as decade counter. Truth table is given. Maximum operating frequency is 20 MHz. Requires nominal 4-V drive pulse.—"Electronic Counting," Mullard, London, 1967, TP874, p 133–135.

TRIGGERED BISTABLE—Uses cross-coupling capacitors C3 and C4 to provide fast transfer of charge, thereby increasing switching speed. Negative edges of square-wave trigger input cause bistable to trigger, while positive edges are blocked by D1 and D2 and have no effect. Discharge time of C1 is limited by R1; diodes D3 and D4 prevent irregular triggering at highest counting rate by providing low-resistance discharge path for C1 and C2.—"Electronic Counting," Mullard, London, 1967, TP874, p 38.

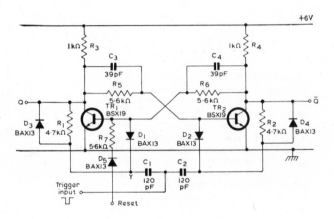

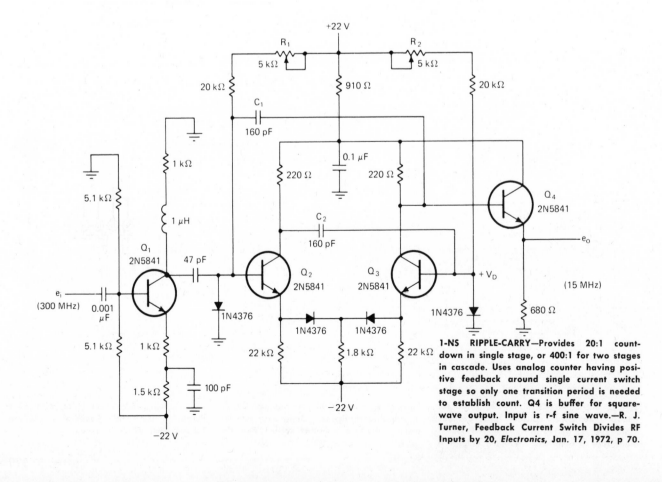

1-NS RIPPLE-CARRY—Provides 20:1 countdown in single stage, or 400:1 for two stages in cascade. Uses analog counter having positive feedback around single current switch stage so only one transition period is needed to establish count. Q4 is buffer for square-wave output. Input is r-f sine wave.—R. J. Turner, Feedback Current Switch Divides RF Inputs by 20, *Electronics*, Jan. 17, 1972, p 70.

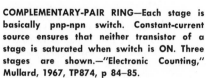

COMPLEMENTARY-PAIR RING—Each stage is basically pnp-npn switch. Constant-current source ensures that neither transistor of a stage is saturated when switch is ON. Three stages are shown.—"Electronic Counting," Mullard, 1967, TP874, p 84–85.

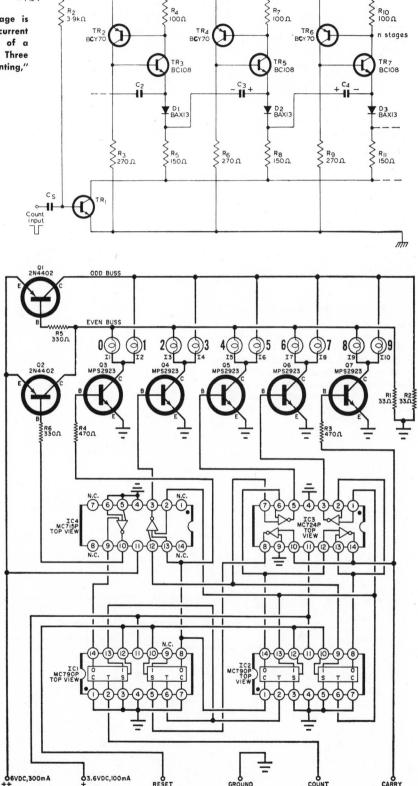

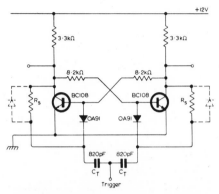

TRIGGERED BISTABLE—Counting frequency limit depends on value of Rs; maximum is 300 kHz for 4K, dropping to about 20 kHz for 60K, with 4-V trigger and 1-μs pulse width. Steering resistors Rs should completely discharge trigger capacitors in two pulse periods, but too low a resistor value distorts output waveform. Addition of recovery diodes (shown dotted) increases maximum counting frequency to 1.3 MHz.—"Electronic Counting," Mullard, London, 1967, TP874, p 208–209.

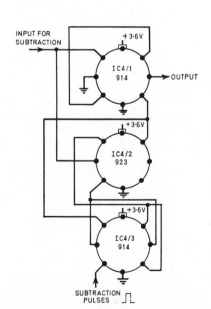

PULSE SUBTRACTOR—Used in magnetic-field stabilizer to subtract precise number of pulses from incident pulse train without any overlap or slicing of pulses.—R. A. Morris and R. G. Brown, A Proton Resonance Magnetic Field Stabilizer, *The Radio and Electronic Engineer*, April 1969, p 241–246.

DECIMAL COUNTER—Cost is under $12 per decade, using incandescent-lamp display and IC logic. Will handle counts up to 10 MHz, is resettable, and decades can be cascaded. In IC1 and IC2, rectangles represent flip-flops; other IC's contain multiple-input gates. Use 6.3-V 50-mA pilot lamps with lenses. De-signed to divide first by 2, then by 5. On each tenth input pulse, circuit provides output pulse for starting count on another decade and cycling to 0 indication. Article includes simple power supply circuit.—D. Lancaster, Build A Low-Cost Counting Unit, *Popular Electronics*, Feb. 1968, p 27–32.

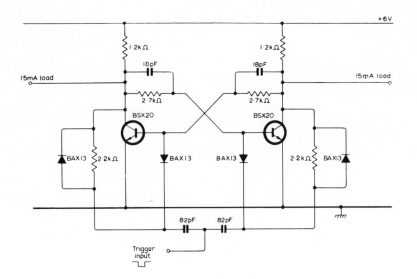

TRIGGERED BISTABLE—Maximum counting frequency with 15-mA resistive load is 30 MHz, but limit drops to 15 MHz when used as binary counter and to 13 MHz as one-gate decade. Minimum trigger input is 1.7 V.—"Electronic Counting," Mullard, London, 1967, TP874, p 174–175.

COUNTER DRIVE—Pulse-shaping circuit is triggered mono mvbr that produces 100-V negative-going pulse having 100-μs duration, used directly for guide-A cathodes of numerical indicator tube and fed through integrator circuit to generate required delayed pulse for guide-B cathodes. Input count pulse must be 5 V negative-going.—"Electronic Counting," Mullard, London, 1967, TP874, p 92–93.

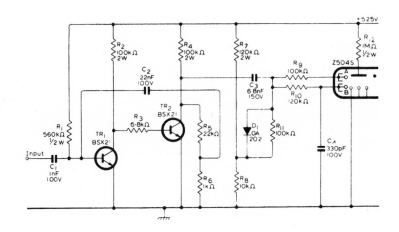

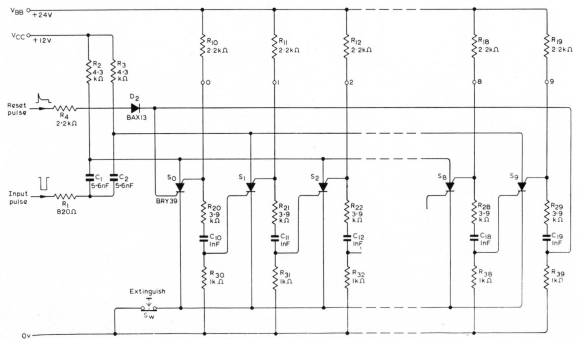

DECADE RING WITHOUT INDICATOR—Used when there is no need to display pulse count on numerical indicator tube, such as when driving control circuit for numerical printer. Requires 24-V supply for resistors that replace indicator tube, to prevent incorrect switching of an scs. Negative-going pulses can be taken from terminals 0 through 9.—D. J. G. Janssen, Circuit-Logic with Silicon Controlled Switches, *Mullard Technical Communications,* March 1968, p 57–64.

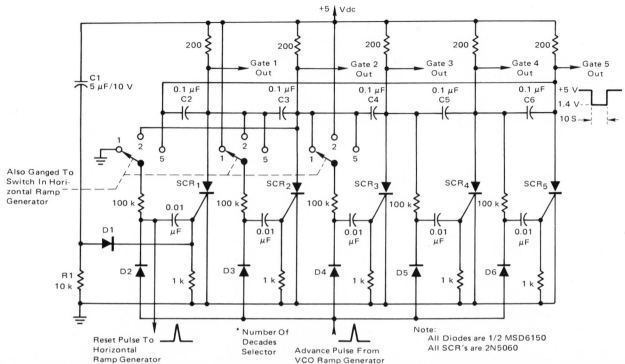

1-2-5 SCR RING—Selector switch allows user to select number of decades counter rings through. Circuit is extension of basic flip-flop, with scr's used to reduce cost. Used in 10 Hz–10 MHz lab sweep generator. Advance pulse for counter is derived from reset transi-

tion of vco sweep ramp generator of system. When power is applied, SCR1 always comes on first, through action of differentiation network C1-R1-D1. Additional section of decade selector switch, located in horizontal ramp generator, provides horizontal drive signal

having same amplitude in all positions, but with required 10, 20, or 50 s time duration required for switch setting.—R. A. Botos and L. J. Newmire, "A Synchronously Gated N-Decade Sweep Oscillator," Motorola Semiconductors, Phoenix, AZ, AN-540, 1971.

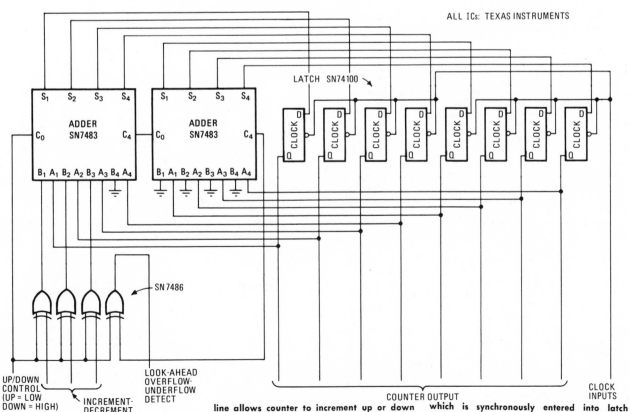

8-BIT UP-DOWN COUNTER—Used in differential analyzers and in X-Y deflection circuits for random-plot crt displays. Up-down control

line allows counter to increment up or down on each clock input pulse by any number from 1 to 7. Counter contains 8-bit latch driven by two 4-bit adders. Input to latch at any given time represents next counter state,

which is synchronously entered into latch upon receipt of clock input pulse.—R. J. Bouchard, Up/Down Synchronous Counter Takes Just Four MSI Packages, *Electronics*, March 29, 1973, p 84.

5-MHZ BINARY—Normal trigger pulse is 2 V with 20-ms rise time, and maximum trigger is 6 V. Stages may be cascaded. In stages operating below 500 kHz, D3 and D4 can be omitted for economy, and C1 and C2 can be made 560 pF for improved trigger sensitivity. —"P-N-P Practical Planar Transistors—BCY70 Family and BFX29 Family," Mullard, London, 1967, TP887, p 7.

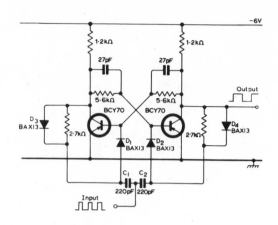

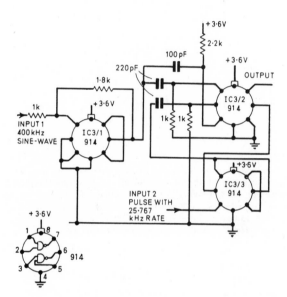

PULSE ADDER—Used in magnetic-field stabilizer to add precise number of pulses to incident pulse train without any overlap or slicing of pulses. First IC is Schmitt trigger that converts 400-kHz sinusoidal input to square wave. This is differentiated and resulting short positive pulses are applied to pin 1 of dual 2-input gate of second IC, to produce two sets of pulses with opposite polarity, for applying to third flip-flop IC. Connections make pulse cancellation or overlap impossible.—R. A. Morris and R. G. Brown, A Proton Resonance Magnetic Field Stabilizer, *The Radio and Electronic Engineer*, April 1969, p 241–246.

10-LAMP DISPLAY—Simple circuit demonstrates operation of digital counter by driving nine pilot lamps. Can also serve as fascinating sleep-inducing night-light for child's bedroom. All lamps are off for 0, PL1 is on for 1, PL2 and PL3 are on for 2, and so on until all lamps are on for 9. Pleasing change of patterns is obtained with 3 × 3 matrix, with lamps 4, 8, and 5 in first row, 2, 1, and 3 in second, and 6, 9, and 7 in third row. All diodes are HEP134 or HEP135. Although input can be pulses up to 100 kHz, best input frequency for display or night-light is about one pulse every 2.5 s, which can be generated with two HEP584 IC's connected as shown. After turn-on, circuit will begin counting accurately in less than nine counts. To reset initially or any other time, apply +3.6 V to RESET terminal.—F. H. Tooker, Novel Counter, Decoder, Readout, *Electronics World*, Dec. 1971, p 40–42.

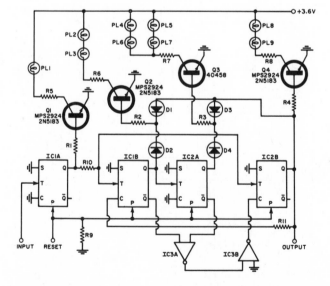

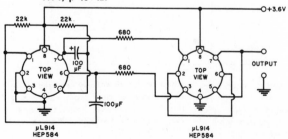

R1, R4—680 ohm, ½ W res.
R2, R3—390 ohm, ½ W res.
R5—56 ohm, ½ W res.
R6, R8—33 ohm, ½ W res.
R7—18 ohm, ½ W res.
R9—1000 ohm, ½ W res.
R10—120 ohm, ½ W res.
R11—330 ohm, ½ W res.

IC1, IC2—MC791P integrated circuit
IC3—μL914 or HEP584 integrated circuit
PL1-PL9—2 V at 0.06 A miniature lamp (No. 49)
Q1, Q2, Q4—MPS2924 or 2N5183 transistor
Q3—40458 transistor

RIPPLE COUNTER—Requires only ¾ of IC package per stage. High level on direction control line gives binary *up* count, while low level gives *down* count. Inhibit line should be low for counting, but high while performing *initial clear* and when reversing counting direction.—R. G. Burlingame, Bidirectional Ripple Counter Counts Up or Down, *The Electronic Engineer*, Nov. 1968, p 75.

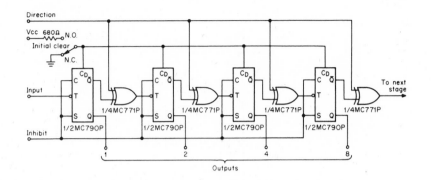

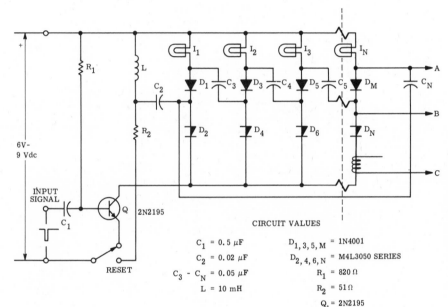

RING COUNTER WITH 4-LAYER DIODES—Transfers power sequentially to series of any number of lamp or other loads from 15 to 200 mA in response to pulse-train input signal. Input requirement is at least 6-mA negative current pulse lasting about 500 μs. After power is applied, press and release reset switch to initiate sequential switching operation.—J. Bliss and D. Zinder, "4-Layer and Current-Limiter Diodes Reduce Circuit Cost and Complexity," Motorola Semiconductors, Phoenix, AZ, AN-221, 1972.

CIRCUIT VALUES

$C_1 = 0.5 \ \mu F$

$C_2 = 0.02 \ \mu F$

$C_3 - C_N = 0.05 \ \mu F$

$L = 10 \ mH$

$D_{1,3,5,M} = 1N4001$

$D_{2,4,6,N} = M4L3050 \ SERIES$

$R_1 = 820 \ \Omega$

$R_2 = 51 \ \Omega$

$Q. = 2N2195$

PULSE-GOBBLING COUNTER—Gobbling technique is based on use of external flip-flop to hold pulse and thereby act as auxiliary counter. With MC10136 hexadecimal (0–15 binary) IC counters for program input, divide modulus M is 255 maximum. Maximum input frequency F-in is about 80 MHz, and F-out is input frequency divided by program input N, which is equal to divide modulus M. Report covers operation of counter in detail and gives waveforms.—T. Balph, "Programmable Counters Using the MC10136 and MC10137 MECL 10,000 Universal Counters," Motorola Semiconductors, Phoenix, AZ, AN-584, 1972.

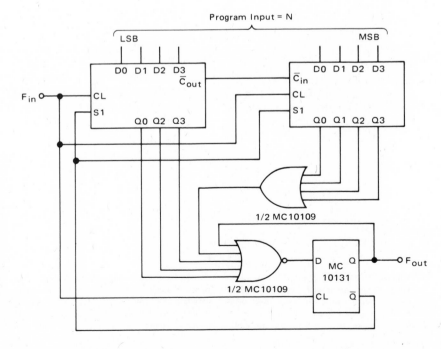

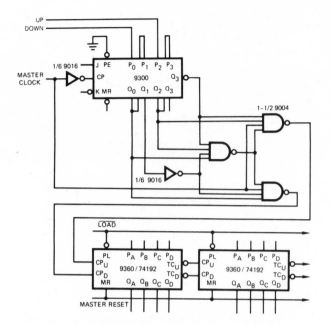

AVOIDING COINCIDENT PULSES—Used to synchronize asynchronous up/down input pulses of bidirectional counting, to avoid coincident pulses. Counters will decrement or increment when either up or down asynchronous input makes low-to-high transition. If both inputs make transitions simultaneously, counters are unchanged. Master clock should be at least twice frequency of inputs to avoid loss of input pulses. Outputs of 9360 counter are synchronized with master clock.—R. Montevaldo, "9360/74192, 9366/74193 Up/Down Counters," Fairchild Semiconductor, Mountain View, CA, No. 304, p 3.

4-STAGE TWISTED-RING—Used to generate phase-shifted waveforms. Truth table shows outputs of flip-flops for successive states. Uses standard J-K flip-flops.—P. J. Kindlmann, Counter Shifts Signal Phase Only One Way, *Electronics*, Feb. 15, 1971, p 86.

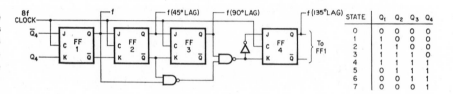

STATE	Q_1	Q_2	Q_3	Q_4
0	0	0	0	0
1	1	0	0	0
2	1	1	0	0
3	1	1	1	0
4	1	1	1	1
5	0	1	1	1
6	0	0	1	1
7	0	0	0	1

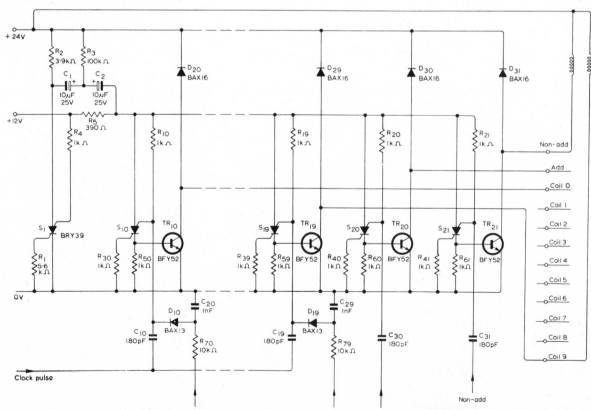

DRIVE FOR PRINT-OUT—Used with ring counter when numeric printout is required in place of display. Only stages for energizing print solenoids for 0 and 9 are shown. Operation resembles mono mvbr, with S1 (normally switched on) governing period. Delay in circuit prevents required scs from being switched on until next clock pulse occurs. Diodes D20–D31 protect transistors against voltage surges produced when coils are switched off. Circuit delivers energizing pulses of about 300 mA lasting 35 ms, suitable for slower printers.—D. J. G. Janssen, Circuit Logic with Silicon Controlled Switches, *Mullard Technical Communications*, March 1968, p 57–64.

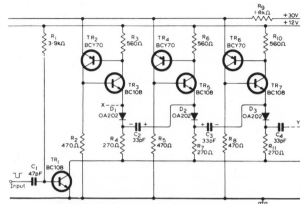

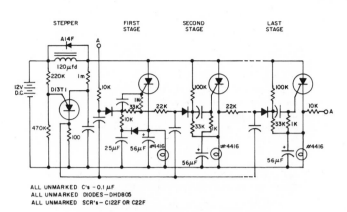

ALL UNMARKED C's – 0.1 μF
ALL UNMARKED DIODES – DHD805
ALL UNMARKED SCR's – C122F OR C22F

COMPLEMENTARY-PAIR RING—Any number of additional stages can be used beyond minimum of three shown. Connect Y of last stage to X of first stage. With 8-V minimum input pulse, will operate up to 1.5 MHz. Must be supplied from constant-current source to keep conducting transistor pair out of saturation, for faster switching.—"P-N-P Practical Planar Transistors—BCY70 Family and BFX29 Family," Mullard, London, 1967, TP887, p 8.

L-C COMMUTATED RING—Ring counter, also used as flasher, requires only L-C combination to provide commutation, eliminating need for costly polar capacitors. Supply can be 12V.—R. W. Fox, "Flashers, Ring Counters and Chasers," General Electric, Semiconductor Products Dept., Auburn, NY, No. 200.48, 1969, p 7–8.

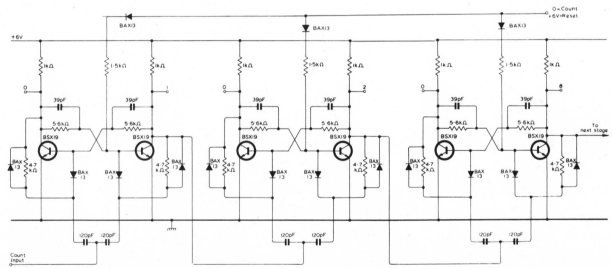

BINARY UP COUNTER—Consists of number of bistable elements cascaded so change of one element from 1 state to 0 state makes next element in chain change state. Capacitively steered bistable is triggered only by negative-going edge of input pulse. Reset input of each bistable element is connected to common reset line through isolating diode and current-limiting resistor. With four bistables for count of 16, this line becomes positive after 16th pulse and all bistables are reset to zero state for start of new count. Will count at speeds up to 5 MHz. Any number of stages may be cascaded. Third stage, omitted for simplicity, is identical to stages shown.—"Electronic Counting," Mullard, London, 1967, TP874, p 70–71.

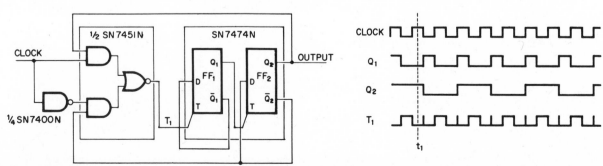

DIVIDE-BY-THREE—Odd-order countdown chain uses logic propagation delay and feedback to produce half-counts. Can be used whenever square wave is needed for time-sharing equipment or for split-phase operations that must be synchronized to variable master clock. Output is symmetric square wave because final stage divides evenly by two after first stage divides by 1.5.—E. J. Murray, Counting by Halves Simplifies Odd-Order Symmetric Counter, *Electronics*, July 19, 1971, p 74.

NEON STORE FOR COUNT—Bistable stage drives two neons that show status of stage at end of count for photoelectric readout. After count is stopped and appropriate neon is ignited, zero or negative voltage on store control line will hold neon on for display or readout while bistable is used for next count. Will operate up to 10 kHz.—"Electronic Counting," Mullard, London, 1967, TP874, p 147–149.

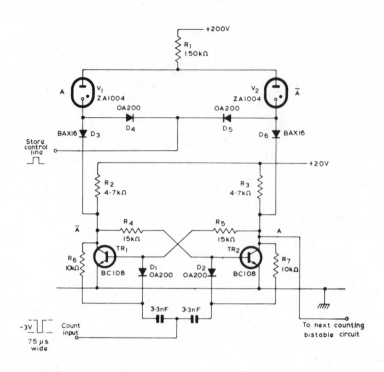

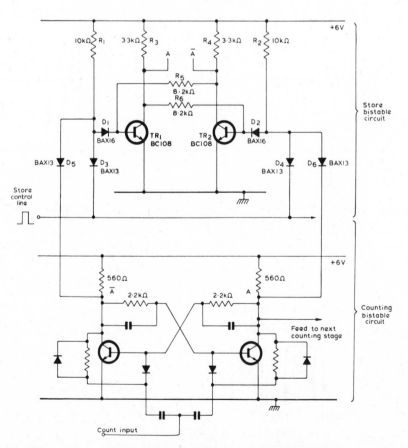

TRANSISTOR BISTABLE STORE—When voltage on store control line is low, both input gates of store bistable are closed and store is effectively disconnected from bistable counter. Count can therefore proceed without altering contents of store. When store control line is positive, gates are open and store bistable stage follows counter bistable stage outputs. Although following rate of 10 kHz is usually adequate, limit can be several MHz if high-speed switching transistors are used. Status of store is read with appropriate circuitry connected to terminals A.—"Electronic Counting," Mullard, London, 1967, TP874, p 149–150.

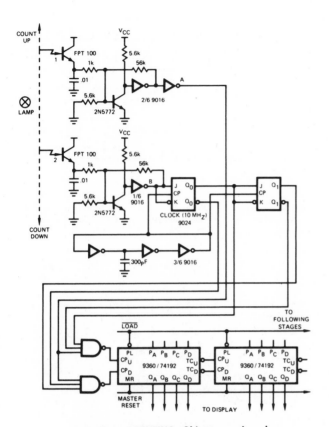

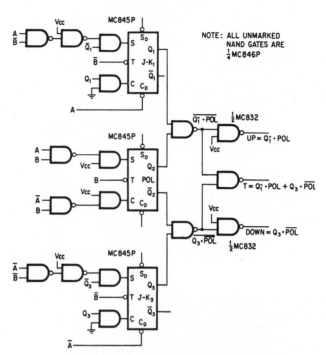

UP-DOWN COUNTING—Objects moving in either direction between light source and phototransistor pair are counted if they are large enough to cover both transistors simultaneously. Object passing from bottom to top will increment counter, while object passing from top to bottom will decrement counter. Hex inverters serve as clock generator and phototransistor amplifiers. Dual flip-flop and 3-input NAND gates route phototransistor signals to counters.—R. Montevaldo, "9360/74192, 9366/74193 Up/Down Counters," Fairchild Semiconductor, Mountain View, CA, No. 304, p 3.

NOISE IMMUNITY—Square-wave signals from incremental shaft encoders or other pulse sources are counted accurately with bidirectional logic arrangement that is essentially immune to noise and pulse jitter. Flip-flop J-K1 generates countup pulses and J-K3 countdown pulses. POL binary cancels pulses generated by systematic errors that can occur during reversal, by enabling or disabling appropriate pulse output line.— J. van Duijn, Digital Bidirectional Detector Keeps the Count Honest, Electronics, Dec. 21, 1970, p 55.

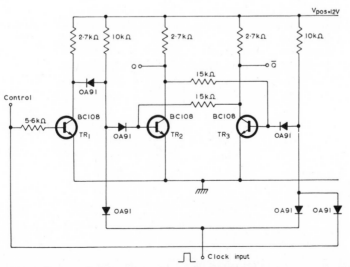

STATICISER—Uses one control line and one clock line to perform store operation similar to that of store bistable circuit. Control line information must be present until command pulse is removed. When control is 0, output Q is 0 and other output is 1. When control is 1, output Q is 1 and other output is 0. Used in synchronous counters.—"Electronic Counting," Mullard, London, 1967, TP874, p 44.

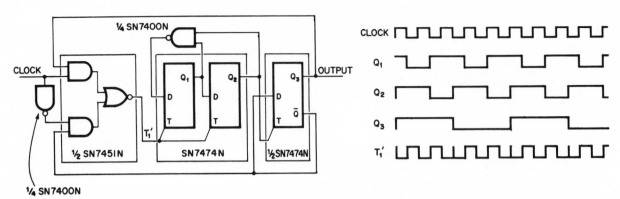

DIVIDE-BY-FIVE—Odd-order countdown chain uses logic propagation delay and feedback to produce half-counts. Derived from divide-by-

six configuration by using appropriate toggle controls, with divide-by-two final stage to give symmetric square-wave output.—E. J.

Murray, Counting by Halves Simplifies Odd-Order Symmetric Counter, *Electronics*, July 19, 1971, p 74.

GATE-DRIVEN LOAD—All stages use complementary scr's, with load in gate and triggering at anode. When 12-V supply is applied, first C13Y stage is triggered on by transistor-zener combination when anode line reaches 7 V, and power is delivered to first load. Each input pulse then turns off conducting scr and turns on next one. Load current should be limited to 50 mA.—R. W. Fox, "Flashers, Ring Counters and Chasers," General Electric, Semiconductor Products Dept., Auburn, NY, No. 200.48, 1969, p 7.

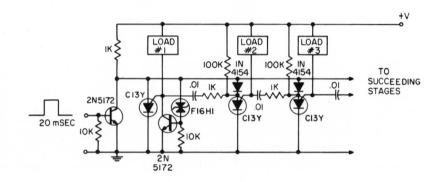

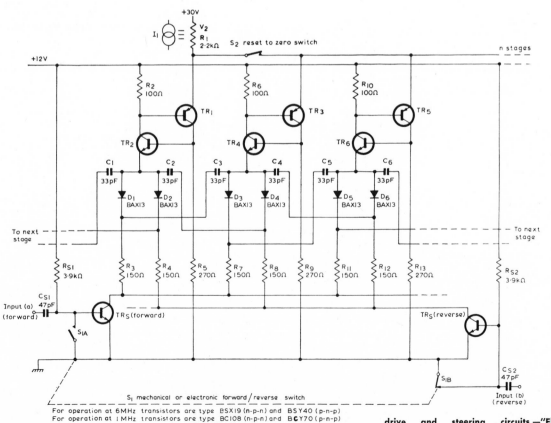

REVERSIBLE PNP-NPN—Maximum operating frequency is 1 or 6 MHz, depending on tran-

sistors used. Requires 2-V drive pulse. Counter is made reversible by providing additional

drive and steering circuits.—"Electronic Counting," Mullard, London, 1967, TP874, p 85–86.

2-STAGE TWISTED-RING—Does better job of generating phase-shifted waveforms than conventional flip-flop connection. Requires fewer logic units and has no phase-shift ambiguity. Can be used for motor drives, phase-sensitive detectors, and ssb modulators. Uses standard J-K flip-flops.—P. J. Kindlmann, *Counter Shifts Signal Phase Only One Way, Electronics*, Feb. 15, 1971, p 86.

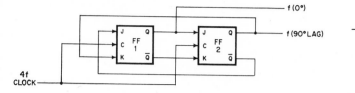

STATE	Q_1	Q_2
0	0	0
1	1	0
2	1	1
3	0	1

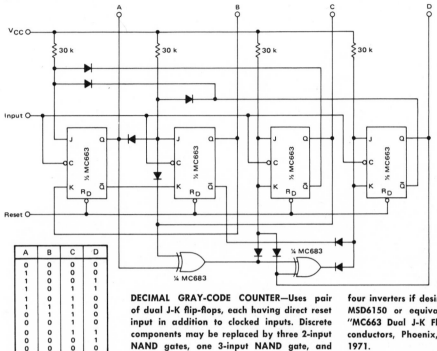

A	B	C	D
0	0	0	0
1	0	0	0
1	0	0	1
1	0	1	1
1	0	1	0
1	1	1	0
0	1	1	0
0	0	1	0
0	0	0	1
0	0	1	1
0	0	0	0

DECIMAL GRAY-CODE COUNTER—Uses pair of dual J-K flip-flops, each having direct reset input in addition to clocked inputs. Discrete components may be replaced by three 2-input NAND gates, one 3-input NAND gate, and four inverters if desired. Each diode is half of MSD6150 or equivalent. Supply is 15 V.—"MC663 Dual J-K Flip-Flop," Motorola Semiconductors, Phoenix, AZ, Data Sheet 9091, 1971.

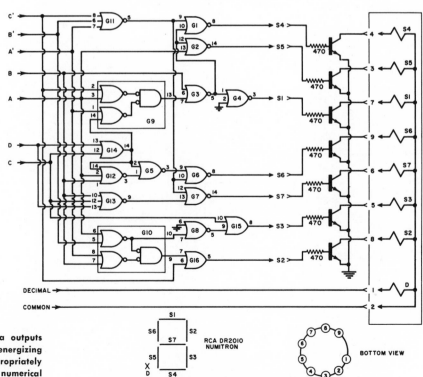

BCD DECODER—Converts seven data outputs of counter stage to voltages for energizing segments of single-digit display appropriately for showing digit corresponding to numerical input in bcd form. Decimal point is energized when appropriate.—D. L. Steinbach, *Digital Instruments, Electronics World*, Jan. 1971, p 35–38 and 56.

G1, G2, G3, G4, G5, G6, G7, G8—¼ quad 2-input gate (Motorola MC717P)
G9, G10—½ dual half adder (Motorola MC775P)
G11, G12, G13—⅓ triple 3-input gate (Motorola MC793P)
G14, G15, G16—⅓ quad 2-input "or" gate (Motorola MC9715P)
Transistors—Motorola MPS 5172

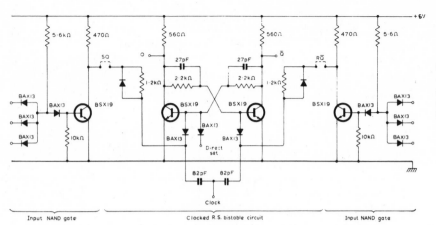

CLOCKED R-S BISTABLE WITH NAND GATES
—Two sets of gates are used for setting SQ and RQ terminals. Book includes truth table. All diodes are BAX13.—"Electronic Counting," Mullard, London, 1967, TP874, p 106–108.

TIME-DELAYED NEGATIVE FEEDBACK—Circuit is used as complementary driver for high-speed scaling circuit. Passive delay line in feedback path provides optimum positioning of bias voltage at base of nondriven transistor in current-mode pair. Q1 and Q2 are 2N3960. Emitter-base junction of 2N3638A transistor is used for each zener. Delay line is 3-inch length of 100-ohm coax, which provides optimum rise and fall times of about 1.5 ns for minimum pulse lengths expected. Input is Hewlett-Packard 215A pulse generator.—E. K. Reedy and J. F. Pierce, A Complementary Driver with a Subnanosecond Rise Time, *IEEE Journal of Solid-State Circuits*, Oct. 1969, p 293–295.

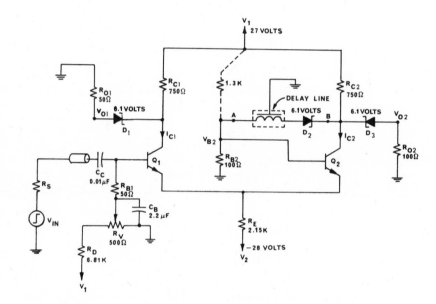

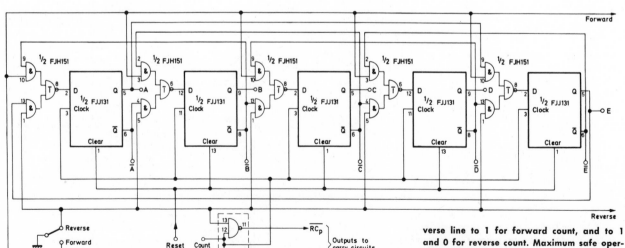

REVERSIBLE JOHNSON DECADE—Requires only 6⅓ TTL packs, including circuits for carry signal generation. Uses Johnson code, which is easier to decode to ten-line decimal than bcd code. Bidirectional performance is obtained by setting forward line to 0 and re-

verse line to 1 for forward count, and to 1 and 0 for reverse count. Maximum safe operating speed is 15 MHz. Report includes carry circuits which allow counter to be reversed at any time except during count input pulse.— "Reversible Decade Counter with the Minimum Number of TTL Packs," Mullard, London, 1969, TP1138, p 1–5.

SCR RING COUNTER—Uses complementary scr's. When 12-V supply voltage is applied, pnp transistor turns on and applies power to all C13Y scr's. Since none of these stages are on, collector voltage of 2N3415 rises and turns on C103Y scr. First input pulse turns off PN5354 momentarily, turning off C103Y. At end of pulse, first C13Y is turned on. Each additional pulse turns on next C13Y and turns off preceding scr, until last C13Y is turned off and C103Y is retriggered.—R. W. Fox, "Flashers, Ring Counters and Chasers," General Electric, Semiconductor Products Dept., Auburn, NY, No. 200.48, 1969, p 6–7.

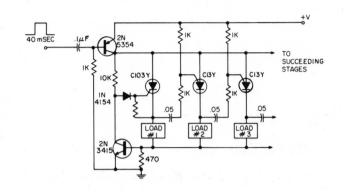

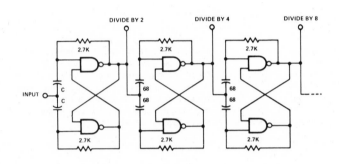

ALL GATES 1/4 387 EACH

BINARY RIPPLE COUNTER—Uses Signetics quad 2-input NAND gates as dividers, with each stage counting up as it changes state on each 1-to-0 transition of previous stage. Counter is implemented by connecting clock input of each stage to Q output of previous stage.—"Signetics Digital Utilogic 2/600 TTL-DTL Data Book," Signetics Corp., Sunnyvale, CA, 1972, p 60.

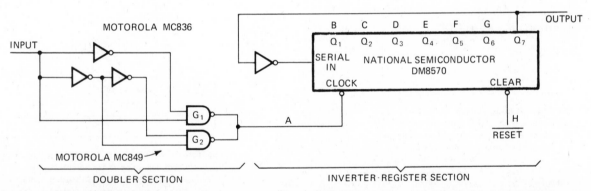

ODD-ORDER COUNTER—Achieves division of pulses by odd number, by first doubling input frequency and then dividing by twice desired divisor. In divide-by-seven counter shown, three inverters and two NAND gates form frequency doubler, with G1 giving pulse for each positive input transition and G2 giving pulse for each negative transition. Inverter and register then divide doubled pulse train by 14.—C. Gordon and T. Chau, Doubling the Divisor Yields Odd-Order Counter, Electronics, Sept. 13, 1971, p 88.

CHAPTER 26
Current Control Circuits

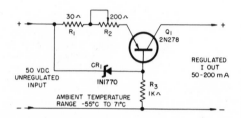

REGULATED 50–200 MA—R2 changes emitter current in grounded-base current-regulator circuit. R3 must draw enough current through reference diode so voltage drop across diode remains at 8 V as regulator is loaded.—"Current Regulator," Delco, Kokomo, IN, Aug. 1971, Application Note 4-B.

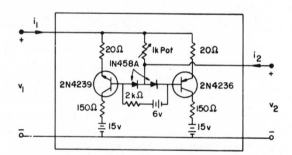

CURRENT SCALOR—Circuit is two-port network element in which independent variable is current rather than time. Cascading with voltage scalor results in power scalor.—L. O. Chua, Synthesis of New Nonlinear Network Elements, *Proc. IEEE*, Aug. 1968, p 1325–1340.

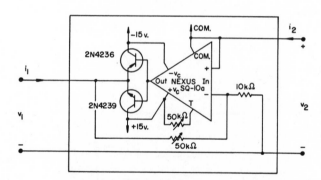

CURRENT SCALOR—Circuit is two-port network element in which independent variable is voltage rather than time. Cascading with current scalor results in power scalor.—L. O. Chua, Synthesis of New Nonlinear Network Elements, *Proc. IEEE*, Aug. 1968, p 1325–1340.

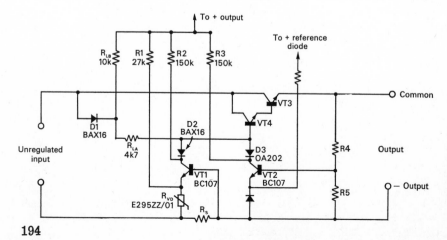

CURRENT LIMIT PROTECTION—Prevents accidental damage to supply. Used in dual 15-V 100-mA supply, only half of which is shown. Developed at Mullard Ltd. Comparator amplifier VT1 senses load current. VT2 is voltage comparator.—A. Gowthorpe, Economical Dual-Polarity Regulated Power Supplies, *Electronic Engineering*, March 1970, p 33–35.

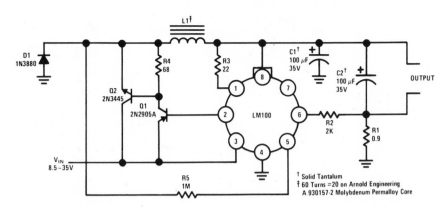

CURRENT SOURCE FOR FLOATING LOAD—Switching regulator principle is applied to IC voltage regulator for current control, to increase efficiency and reduce power dissipation in control transistors. For fixed load current, input power is roughly proportional to voltage across load. Current sense resistor R1 (or pot connected across it) sets output current.—R. J. Widlar, "New Uses for the LM100 Regulator," National Semiconductor, Santa Clara, CA, 1968, AN-8.

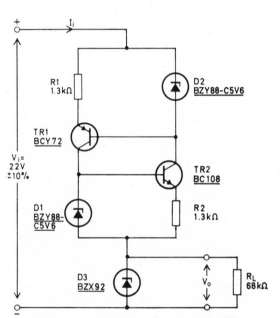

COMPLEMENTARY CONSTANT-CURRENT SOURCE—Provides constant 7.6 mA to voltage reference diode D3 in parallel with load. TR1 and voltage regulator diode D2 supply constant current of 3.8 mA to D1, while TR2 and D1 supply same constant current to D2. Circuit is highly stable and operates over wide input voltage range. Report gives performance curves.—"1N821 and BZX90 Series of High Stability Reference Diodes," Mullard, London, 1973, TP1339, p 8–9.

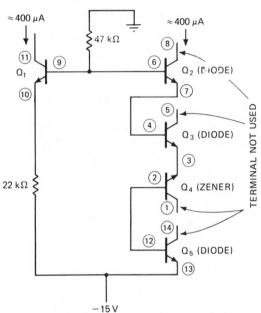

CONSTANT CURRENT FOR FET—Inexpensive RCA CA3046 five-transistor IC is connected to serve as temperature-compensated constant-current source for fet-input differential amplifier. Q1 operates as conventional transistor, Q4 as zener, and others as forward-biased diodes. Temperature coefficient of Q2 matches Q1, while positive coefficient of zener compensates for identical negative coefficients of Q3 and Q5.—A. Chace, IC Transistor Array Compensates For Temperature, *Electronics*, Dec. 6, 1971, p 77.

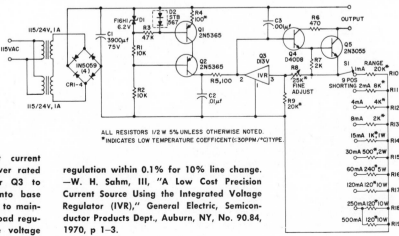

1–500 MA AT 0–50 V—Low-cost current source provides 0.1% regulation over rated range, using IC voltage regulator Q3 to transform sensed output current into base drive required for Darlington Q4-Q5 to maintain desired constant load current. Load regulation is typically 0.02% and line voltage regulation within 0.1% for 10% line change. —W. H. Sahm, III, "A Low Cost Precision Current Source Using the Integrated Voltage Regulator (IVR)," General Electric, Semiconductor Products Dept., Auburn, NY, No. 90.84, 1970, p 1–3.

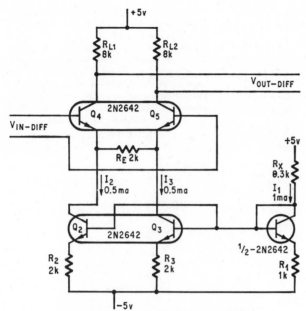

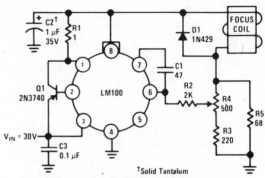

FOCUS CURRENT REGULATOR—Uses LM100 voltage regulator to maintain current through focus coil in tv set or cro. R1 protects regulator against shorts from any part of focus coil to ground, while D1 prevents inductive kickback of focus coil from reversing voltage on IC or pass element when input voltage is switched off.—R. J. Widlar, "New Uses for the LM100 Regulator," National Semiconductor, Santa Clara, CA, 1968, AN-8.

MATCHED 0.5-MA SOURCES—Provides high common-mode rejection with only one emitter resistor RE and matched collector resistors. Low-frequency differential voltage gain is 2RL/RE.—M. J. Callahan, Jr., Charts Speed the Designing of Constant Current Sources, *Electronics*, Aug. 17, 1970, p 92–95.

QUASI-CONSTANT CURRENT LOAD—Maximum current is 60 mA, and conduction will be maintained for voltages down to about 50 mV for 5-mA holding current. Used as load for scr in voltage sweep generator of transistor curve tracer.—K. E. Forward, A Voltage Sweep Generator, *Electronic Engineering*, March 1970, p 44–46.

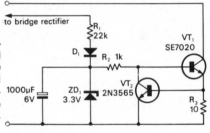

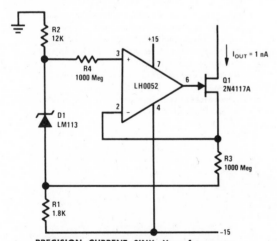

PRECISION CURRENT SINK—Uses fet opamp driving fet to generate precisely regulated current of 1 nA. Opamp output cuts off gate of Q1 as required to keep voltage across R3 equal to 1.220-V breakdown voltage of zener D1. Resulting current sink has worst-case initial accuracy of less than 2%. Technique is applicable to wide range of currents by properly scaling values of R3 and its balancing resistor R4.—R. K. Underwood, "New Design Techniques for FET Op Amps," National Semiconductor, Santa Clara, CA, 1972, AN-63, p 9.

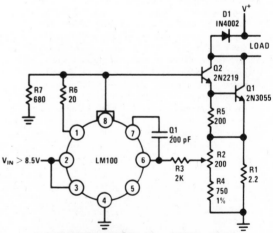

1-A CURRENT SOURCE—Uses LM100 voltage regulator to regulate emitter current of Darlington-connected transistors Q1-Q2. Output of IC is short-circuit protected with R6 to limit current when Q2 saturates. R7 provides minimum load current required by IC. D1 absorbs kickback of inductive loads when power is shut off. R2 adjusts load current. Maximum supply voltage is limited only by breakdown voltage of control transistors.—R. J. Widlar, "New Uses for the LM100 Regulator," National Semiconductor, Santa Clara, CA, 1968, AN-8.

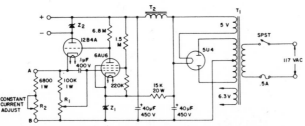

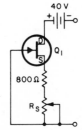

DIODE-TESTING SUPPLY—Provides adjustable constant current as required for measuring diode voltages. Zener Z2 prevents damage to component under test. Initially, positive and negative terminals are shorted and R1 adjusted until voltage between terminals A and B is 150 V d-c. Test circuit is then connected in place of short and R2 adjusted for desired constant current through test circuit.—"Germanium Alloy Power Transistor Electrical Test Procedures," Delco, Kokomo, IN, April 1972, Application Note 6-A.

T_1 — Secondary 720V CT, 125 mA, Stancor PM8410
T_2 — Choke 8H, 85 mA, 250 ohms, Stancor C 1709
Z_1 — Zener diode 10 watt, 51 volt 5% — 1N1368A
Z_2 — Zener diode 10 watt, 150 volt — 1N1812
R_1 — 500 K, 2 watt log taper — Ohmite CA5041
R_2 — 10 K, 2 watt log taper — Ohmite CA1041

ADJUSTABLE SOURCE—Single 2N4869 fet having long gate and internal impedance above 2 meg is used with 1-meg pot RS as current source that can be adjusted over range from 5 μA to 1 mA.—J. S. Sherwin, "The Field-Effect Transistor Constant Current Source," Siliconix, Santa Clara, CA, 1971.

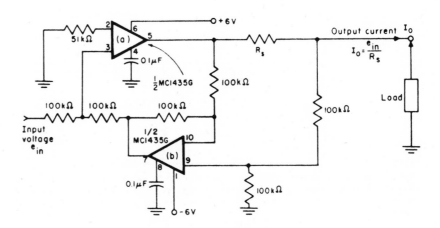

BIPOLAR SOURCE—Single-ended circuit is current source of either polarity for use with grounded load. Opamp (b) converts differential sense voltage across Rs into single-ended voltage applied as negative feedback to opamp (a), to monitor output current and regulate output voltage to within 1%. Will operate from d-c to 100 Hz.—R. C. Gerdes, Bipolar Current Source, *The Electronic Engineer*, April 1969, p 78.

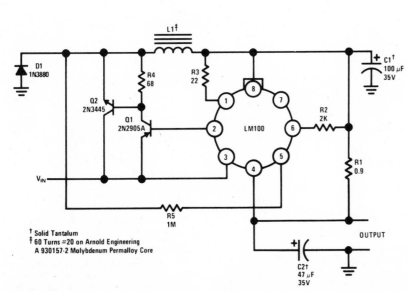

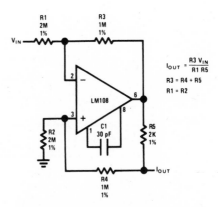

CURRENT SOURCE FOR GROUNDED LOAD—Switching regulator principle is applied to IC voltage regulator for current control, to increase efficiency and reduce power dissipation in control transistors. Load is inserted in ground line, with R1 setting output current.

Difference between input and load voltages cannot drop below 8.5 V with this circuit because lower voltage is insufficient to bias reference circuit of IC.—R. J. Widlar, "New Uses for the LM100 Regulator," National Semiconductor, Santa Clara, CA, 1968, AN-8.

BILATERAL CURRENT SOURCE—Supplies current proportional to input voltage and drives load referred to ground or to any voltage within output swing capability of amplifier. Gives negative output current for positive input voltage, but this can be reversed by grounding input and driving ground end of R2. Scale factor is unchanged as long as R4 is very much greater than R5.—R. J. Widlar, "IC Op Amp Beats FETs on Input Current," National Semiconductor, Santa Clara, CA, 1969, AN-29, p 14–15.

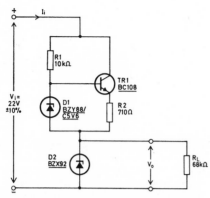

CONSTANT 7.6 MA—Circuit provides constant current to voltage reference diode D2 and load. D1 is voltage regulator diode acting with TR1, R1, and R2 as constant-current generator.—"1N821 and BZX90 Series of High Stability Reference Diodes," Mullard, London, 1973, TP1339, p 8–9.

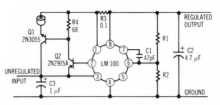

2 A WITH IC—Two external transistors provide additional current gain for National IC voltage regulator. Report gives optimum values of output voltage divider R1-R2 for outputs from 2 to 30 V; for 28 V, R1 is 33K and R2 is 2.4K. If required, use Ferroxcube K5-001-00/3B ferrite bead on emitter lead of Q1 to suppress parasitic oscillations. C2 and C3 are solid tantalum electrolytics.—R. J. Widlar, "Monolithic Voltage Regulator," National Semiconductor, Santa Clara, CA, 1967, AN-1, p 8.

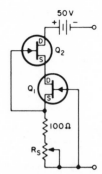

CASCADE FET SOURCE—Provides output current that can be adjusted from 2 μA to 1 mA with 1-meg pot RS. Internal resistance of cascade connection is greater than 10 meg. Q1 is 2N4340 and Q2 is 2N4341.—J. S. Sherwin, "The Field-Effect Transistor Constant Current Source," Siliconix, Santa Clara, CA, 1971.

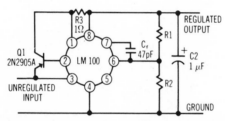

200-MA IC—Use of external transistor with National IC voltage regulator gives 1% regulation of load current. Report gives optimum values of output voltage divider R1-R2 for output voltages from 2 to 30 V; for 5 V, R1 is about 5K and R2 12K. C2 is solid tantalum electrolytic.—R. J. Widlar, "Monolithic Voltage Regulator," National Semiconductor, Santa Clara, CA, 1967, AN-1, p 7–8.

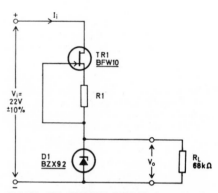

7.6-MA FET CURRENT SOURCE—Source resistor R1 is adjusted to give correct total current for voltage reference diode D1; typical value is 158 ohms, with range being 13.2 to 395 ohms depending on fet characteristics. Performance curves are given.—"1N821 and BZX90 Series of High Stability Reference Diodes," Mullard, London, 1973, TP1339, p 8–9.

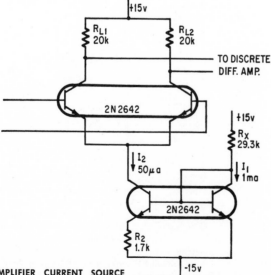

DIFFERENTIAL AMPLIFIER CURRENT SOURCE—Used as input stage to discrete amplifier, to give high common-mode input impedance and rejection ratio.—M. J. Callahan, Jr., Charts Speed the Designing of Constant Current Sources, *Electronics*, Aug. 17, 1970, p 92–95.

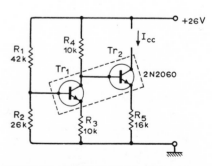

CONSTANT OUTPUT CURRENT—Choice of resistor values compensates for transistor drift, making output current constant through load connected to terminations marked Icc. Resistors should be wire-wound, with low temperature coefficients. Temperature coefficient of load current is typically 0.0015% per deg C over temperature range of 0 to 100 C.—M. Cadwallader, Constant Current Generator, *Wireless World*, Jan. 1970, p 12.

CHAPTER 27
Data Transmission Circuits

HIGH-SPEED LINE RECEIVER—Uses Monsanto MCT9 2-kV isolator having GaAs infrared-emitting diode and very high speed photo-transistor in axial package to provide high coupling efficiency for high-speed data transmission line.—"Phototransistor Opto-Isolator," Monsanto Co., Cupertino, CA, No. MCT9, 1972.

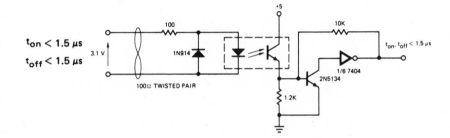

$t_{on} < 1.5 \ \mu s$

$t_{off} < 1.5 \ \mu s$

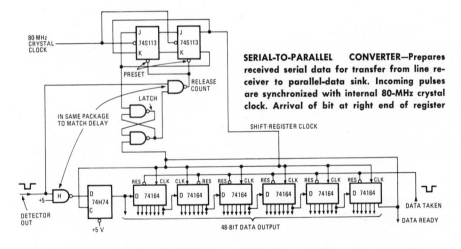

SERIAL-TO-PARALLEL CONVERTER—Prepares received serial data for transfer from line receiver to parallel-data sink. Incoming pulses are synchronized with internal 80-MHz crystal clock. Arrival of bit at right end of register signals that register is loaded. Handles 48-bit serial word, arriving at rates up to 20 MHz.—T. R. Blakeslee, Transformer-Coupled Transceiver Speeds Two-Way Data Transmission, *Electronics*, March 1, 1973, p 94–96.

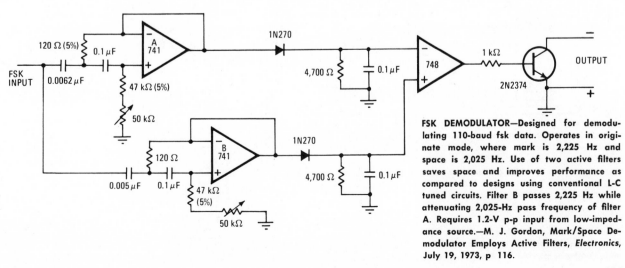

FSK DEMODULATOR—Designed for demodulating 110-baud fsk data. Operates in originate mode, where mark is 2,225 Hz and space is 2,025 Hz. Use of two active filters saves space and improves performance as compared to designs using conventional L-C tuned circuits. Filter B passes 2,225 Hz while attenuating 2,025-Hz pass frequency of filter A. Requires 1.2-V p-p input from low-impedance source.—M. J. Gordon, Mark/Space Demodulator Employs Active Filters, *Electronics*, July 19, 1973, p 116.

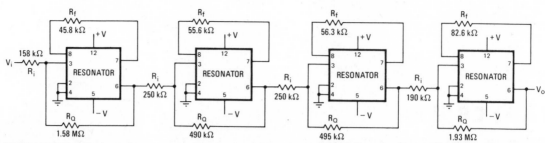

PACKAGED RESONATORS SIMPLIFY PHONE-LINE FILTER—Addition of only three resistors to each Integrated Electronics μAR 1700 regu-lator gives active fourth-order bandpass filter for low-speed data transmission over phone lines. Values shown give 3-dB bandwidth of 400 Hz at 2,125 Hz. Article gives design pro-cedure for any frequency.—R. Brandt, Active Resonators Save Steps in Designing Active Fil-ters, *Electronics*, April 24, 1972, p 106–110.

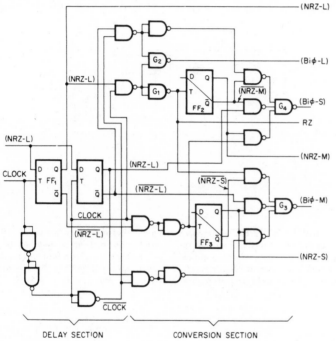

DELAY SECTION   CONVERSION SECTION

PCM CODE CONVERTER—Uses propagation delay of its logic elements to eliminate spikes during signal level transitions, allowing oper-ation at data bit rate equal to 10-MHz clock frequency. All logic circuits are series 54/74 TTL IC's. Uncomplicated arrangement of NAND gates and flip-flops gives seven differ-ent codes simultaneously for each non-re-turn-to-zero-level input code.—J. Jansen, Sev-en-Code PCM Converter Clocks at Data Out-put Rate, *Electronics*, July 5, 1971, p 57–58.

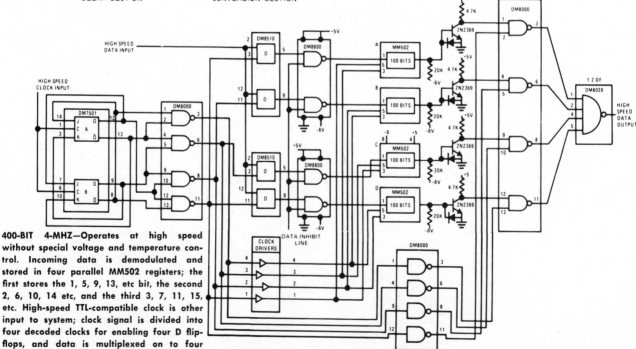

400-BIT 4-MHZ—Operates at high speed without special voltage and temperature con-trol. Incoming data is demodulated and stored in four parallel MM502 registers; the first stores the 1, 5, 9, 13, etc bit, the second 2, 6, 10, 14 etc, and the third 3, 7, 11, 15, etc. High-speed TTL-compatible clock is other input to system; clock signal is divided into four decoded clocks for enabling four D flip-flops, and data is multiplexed on to four lines driving DN8800 voltage translator units with 1-MHz data.—E. F. Kvamme, "MOS Memory Applications," National Semiconduc-tor, Santa Clara, CA, 1968, AN-7.

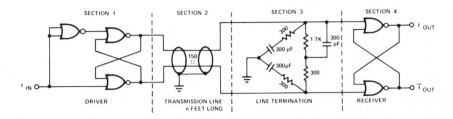

LINE DRIVER AND RECEIVER—Used for transmitting digital data over 150-ohm transmission line. Driver in first section is D-type latch producing complemented outputs with Signetics 380 NOR gates. Delta 150-ohm line termination feeds receiver consisting of Signetics 380 NOR gate latch. Will transmit and receive data on up to 1,500 feet of multiple-conductor telephone cable.—"Signetics Digital Utilogic 2/600 TTL-DTL Data Book," Signetics Corp., Sunnyvale, CA, 1972, p 72.

RESONATORS IN BANDPASS FILTER—Article gives design procedure for fourth-order bandpass filter that can be represented by four cascaded resonators. Each resonator, with its simulated inductance, requires three opamps connected as in lower diagram. Values shown give center frequency of 2,125 Hz, 3-dB bandwidth of 400 Hz, and over-all passband gain of 200 (46 dB), for low-speed telephone data communications.—R. Brandt, Active Resonators Save Steps in Designing Active Filters, *Electronics*, April 24, 1972, p 106–110.

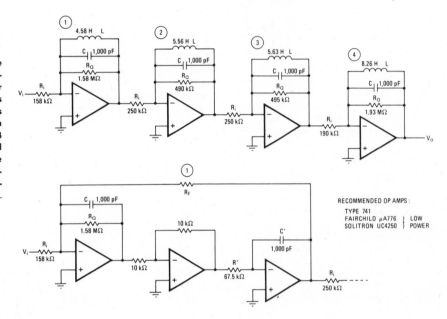

PARALLEL-BUFFER LINE DRIVER—Article shows how to calculate number of DTL buffer gates that must be paralleled to give desired output signal across termination of single-ended line. Components shown give 3.3-V output, with data rate up to 5 MHz.—R. M. Walker and R. A. Aldrich, Standard ICs for Digital Data Communication, *The Electronic Engineer*, March 1968, p 33–38.

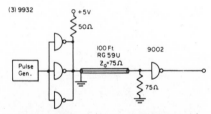

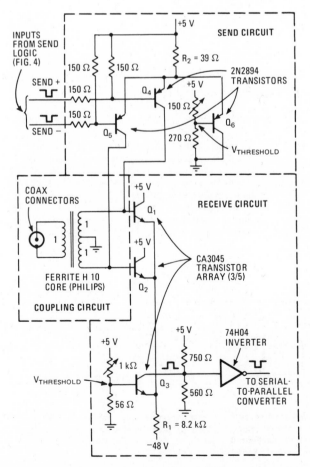

SERIAL DATA AT 20 MHZ—Combination line driver and receiver transmits serial data over ordinary two-wire parallel line up to several hundred meters in length at rates up to 20 MHz. Reduces cost of cumbersome parallel lines otherwise required for computer-controlled telephone exchange. Center tap on coupling transformer implements bipolar format. Both send and receive circuits use emitter-coupled transistors, which can be almost any general-purpose type since transistor saturation never occurs.—T. R. Blakeslee, Transformer-Coupled Transceiver Speeds Two-Way Data Transmission, *Electronics*, March 1, 1973, p 94–96.

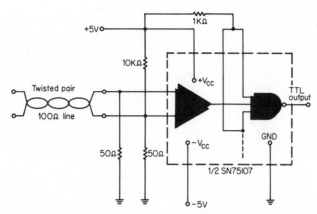

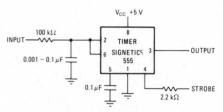

LINE RECEIVER—Low-cost type 555 IC timer is used as level-sensing device preceded by R-C integrator, to serve as noise-immune data line receiver in which system speed is limited by electromechanical devices. Output is TTL-compatible. Requires only one signal conductor, unshielded. Special driver is not required at sending end.—J. G. Pate, IC Timer Can Function as Low-Cost Line Receiver, *Electronics*, June 21, 1973, p 132.

DATA-LINE RECEIVER—Used when differential input voltage may be zero, input data line may be opened, or all drivers may be strobed off. To prevent oscillation at zero input despite high gain (typically 1,000), positive voltage is applied through 10K resistor, to give +25 mV level at minus input and output of logical 0. In this determinant condition of receiver it will not oscillate.—D. Pippenger, Line Receivers with Zero Differential Input, *The Electronic Engineer*, Dec. 1971, p 58.

50-KHZ DIGITAL RECEIVER—Adding diode as shown to any IC opamp and reducing supplies from 15 V to 5 V gives differential digital data receiver having TTL-compatible output. Frequency response limit of 50 kHz gives good high-frequency noise rejection.—R. Bogart and J. Buchanan, A Low-Cost Differential Line Receiver, *The Electronic Engineer*, May 1971, p 78.

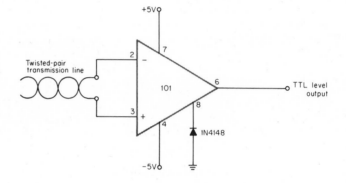

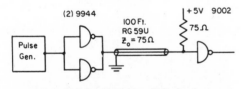

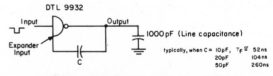

GATES DRIVE COAX—Paralleling of gates provides increased current for driving single-ended-terminal transmission line while maintaining high data transmission rate. Gives 5-V output across 75-ohm load, as compared to 2.5 V with single gate. Maximum data rate is over 5 MHz.—R. M. Walker and R. A. Aldrich, Standard ICs for Digital Data Communication, *The Electronic Engineer*, March 1968, p 33–38.

DRIVING UNTERMINATED LINE—Use of 50-pF feedback capacitor on 9932 DTL buffer makes rise and fall times of data pulses more nearly symmetrical and reduces generation of noise. Article gives design equations.—R. M. Walker and R. A. Aldrich, Standard ICs for Digital Data Communication, *The Electronic Engineer*, March 1968, p 33–38.

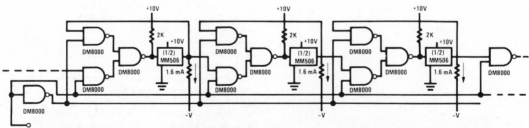

MOS DELAY—TTL gates allow mos shift registers to operate at maximum of 4 MHz rather than 1 or 2 MHz, for use as delay line for storing data temporarily in digital system. Negative voltage levels are not used because mos registers do not have negative inputs. More positive inputs are treated as logic 0 and more negative inputs as logic 1. Report covers operation of circuits in detail.—D. Mrazek, "MOS Delay Lines," National Semiconductor, Santa Clara, CA, 1969, AN-25.

CHAPTER 28
Digital-Analog Converter Circuits

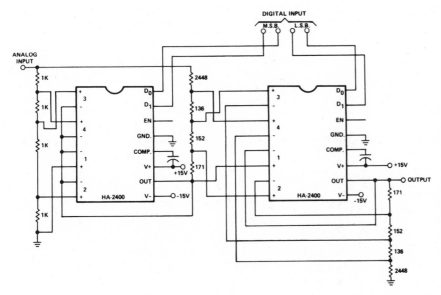

MULTIPLYING CONVERTER—Multiplies input voltage by N/16, where N is binary number from 0 to 15 formed by digital input. If analog input is fixed d-c reference, circuit becomes conventional 4-bit d-a converter. If analog input is variable or a-c signal, output is product of analog and digital signals. Circuit on left is programmable attenuator with weights of 0, ¼, ½, or ¾. Circuit on right is noninverting adder which adds weights of 0, 1/16, ⅛, or 3 1/6 to first output.—D. Jones, "The HA-2400 Pram Four Channel Operational Amplifier," Harris Semiconductor, Melbourne, FL, No. 514, 1972.

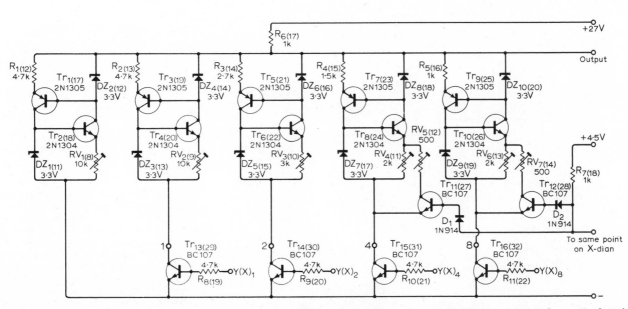

SIMULATOR—Generates X or Y staircase waveforms for analog output when natural 1-2-4-8 binary code is applied in form of logic levels 0 and 1 to inputs X1, X2, X4, and X8 of buffer amplifiers controlling bistables. First bistable (Tr1-Tr2) is connected to negative line and continuously draws 2 mA through R6 to give initial output of 25 V (2 V below supply). For input of 0110, X2 and X4 would cause additional 6 mA to flow through R6 and drop output 6 V lower. Numbers in parentheses are for identical X staircase generator.—B. S. Crank, Wireless World Logic Display Aid, *Wireless World*, June 1969, p 258–264.

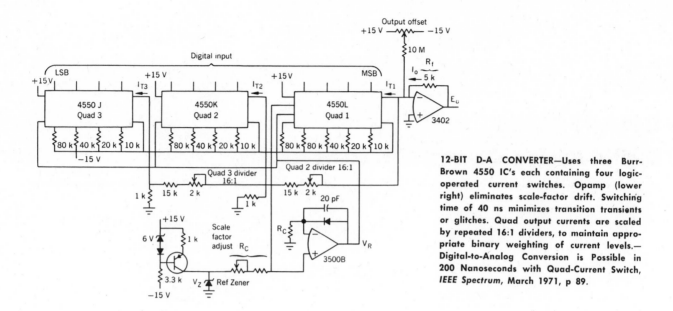

12-BIT D-A CONVERTER—Uses three Burr-Brown 4550 IC's each containing four logic-operated current switches. Opamp (lower right) eliminates scale-factor drift. Switching time of 40 ns minimizes transition transients or glitches. Quad output currents are scaled by repeated 16:1 dividers, to maintain appropriate binary weighting of current levels.—*Digital-to-Analog Conversion is Possible in 200 Nanoseconds with Quad-Current Switch,* IEEE Spectrum, March 1971, p 89.

10-BIT DAC—Uses two AD552 high-speed quint current switches, each consisting of five logic-operated current steering switches and reference transistor on single chip. Output current of second quint is attenuated to achieve 10-bit binary performance. Resistor ladder network can be derived from AD850 or AD852, using external 160K resistors for fifth and tenth bits, and 30.9K and 1K resistors for interquint attenuator.—"Monolithic Quint Current Switch," Analog Devices, Norwood, MA, Nov. 1972.

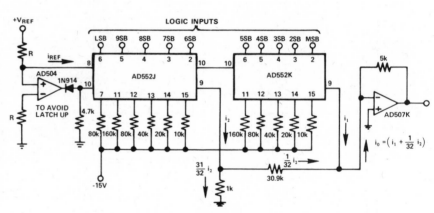

$$i_0 = \left(i_1 + \frac{1}{32} i_2 \right)$$

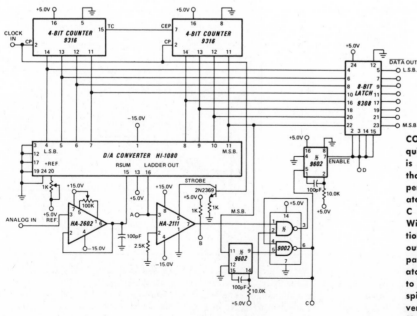

COUNTER-TYPE CONVERTER—Does not require sample-and-hold circuit because output is parallel digital number taken at instant that d-a and input signals are equal. Can perform 1,000 conversions per second. Accurate within one least significant bit from —55 C to 125 C, at clock rates up to 330 kHz. Will work up to 1 MHz with about one additional bit inaccuracy. Normally has negative output level, so positive input signal is compared by resistive summation at one comparator input. Comparator is strobed with clock to prevent premature triggering by switching spikes.—D. Jones, "Counter-Type D to A Converter," Harris Semiconductor, Melbourne, FL, No. 512, 1971.

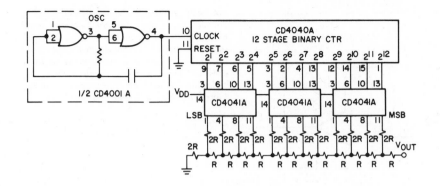

ULTRA-LOW POWER CONVERTER—Average dissipation per output is only 100 mW. For supply voltage of 5 V for VDD, value of R ranges from 3.5K to 896K for resolution range of 4 to 12 bits and accuracy within half of least significant bit.—"COS/MOS Quad True/Complement Buffer," RCA, Somerville, NJ, File No. 572, 1972.

SETTLING-TIME EXPANDER—Permits expansion of one portion of d-c level of interest in d-a converter when measuring time elapsed from change of digital input until transients have diminished enough so output is within half a least significant bit of final value. Scope pattern then serves as linearity test.—"Digital-to-Analog Converter Handbook," Hybrid Systems Corp., Burlington, MA, 1970, 2nd Ed., p 35–37.

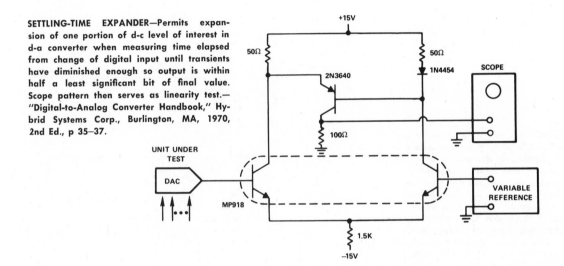

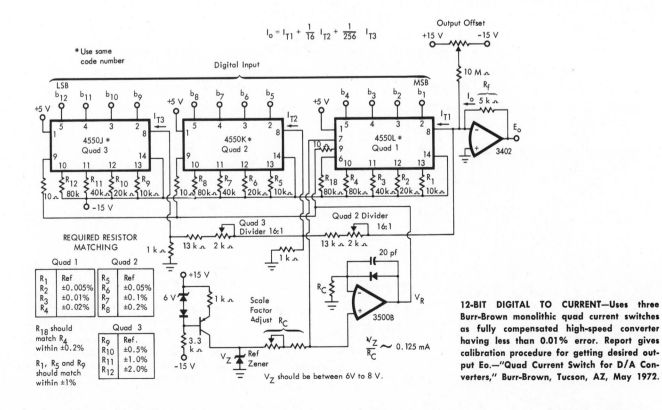

$$I_o = I_{T1} + \frac{1}{16} I_{T2} + \frac{1}{256} I_{T3}$$

REQUIRED RESISTOR MATCHING

Quad 1		Quad 2	
R_1	Ref	R_5	Ref
R_2	±0.005%	R_6	±0.05%
R_3	±0.01%	R_7	±0.1%
R_4	±0.02%	R_8	±0.2%

R_{18} should match R_4 within ±0.2%

R_1, R_5 and R_9 should match within ±1%

Quad 3	
R_9	Ref.
R_{10}	±0.5%
R_{11}	±1.0%
R_{12}	±2.0%

$$\frac{V_Z}{R_C} \sim 0.125 \text{ mA}$$

V_Z should be between 6V to 8 V.

12-BIT DIGITAL TO CURRENT—Uses three Burr-Brown monolithic quad current switches as fully compensated high-speed converter having less than 0.01% error. Report gives calibration procedure for getting desired output Eo.—"Quad Current Switch for D/A Converters," Burr-Brown, Tucson, AZ, May 1972.

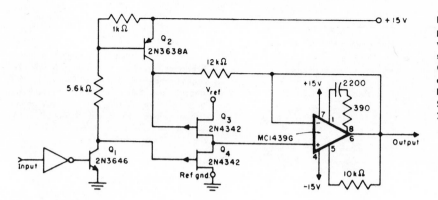

D-A LADDER SWITCH—Fet switch and opamp provide simple drive for digital-analog converter ladder networks, with offset voltage similar to that of bipolar chopper transistors. Circuit has negligible output impedance and does not load reference voltage source.—R. Peri, Low Cost, High Performance Analog Switch, *The Electronic Engineer*, April 1969, p 78.

9-BIT DAC—Uses cos-mos IC packages with 10-V logic level to perform switch function. Single 15-V supply provides positive bus for follower amplifier and feeds CA3085 voltage follower. Book gives complete circuit for voltage follower assembled from two CD4007AD packages and one CA3083 package. Resistor ladder arms 2, 3, 4, and 5, requiring ratio matches of 0.1% to 0.8% with resistor for bit 1, are built with series and parallel combinations of 806K 1% metal-oxide film resistors from same manufacturing lot; resistors for other arms then require only 1% accuracy to give 9-bit accuracy for entire system.—"COS/MOS Digital Integrated Circuits," RCA, Somerville, NJ, SSD-203A, 1973 Edition, p 301–302.

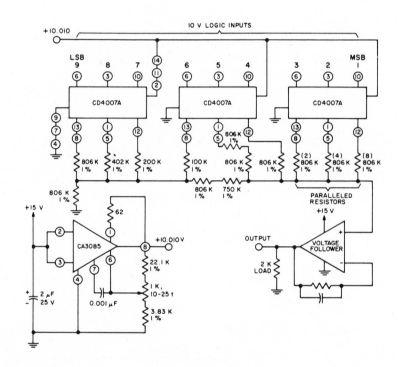

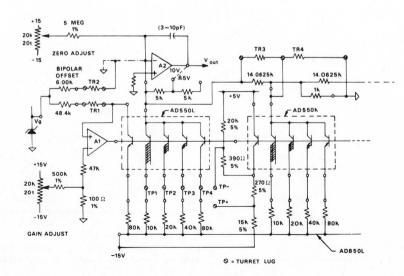

TRIMMING FOR 12-BIT ACCURACY—Circuit illustrates highly accurate and stable trimming of first 8 bits of 12-bit dac using Analog Devices converters. Opamps are not critical. Book gives adjustment procedures in detail, with emphasis on final fine trim of highest-order bits to achieve best possible resolution.—D. H. Sheingold, "Analog-Digital Conversion Handbook," Analog Devices, Norwood, MA, 1972, II-71–II-74.

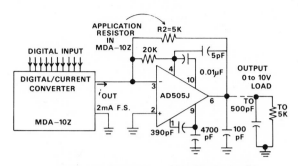

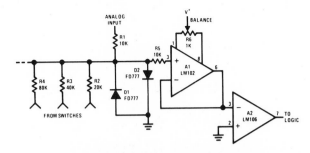

DIGITAL-TO-VOLTAGE CONVERTER—Combination of Analog Devices converter and opamp changes digital input to d-c output of 0 to 10 V for loads up to 5K in parallel with 500 pF. Typical settling time to 0.1% is 3 μs. May be used with variety of other digital-current converters as well.—H. Krabbe and R. S. Burwen, Stable Monolithic Op Amp Slews at 130V/μs, *Analog Dialogue*, Vol. 5, No. 4, p 3–5.

BINARY-WEIGHTED COMPARATOR—Analog input (from low source impedance) is fed into scaling resistor R1 selected so input voltage to LM102 voltage follower is zero when output of d-a network corresponds to analog input voltage. Output then changes to logical 1 when d-a output exceeds analog signal. To increase operating speed, change to silicon backward diodes. A2 is comparator having fanout of 10 with DTL or TTL, and gain of about 45,000.—R. J. Widlar, "Fast Voltage Comparators With Low Input Current," National Semiconductor, Santa Clara, CA, 1969, LB-6.

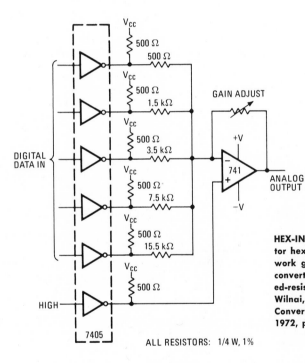

HEX-INVERTER SWITCH—Use of open-collector hex inverter as driving and switching network gives up to 5-bit word lengths for d-a converter. Circuit simplifies design of weighted-resistor high-resolution d-a converter.—A. Wilnai, Logic Driving Gates Double as D-A Converter Switches, *Electronics*, Aug. 14, 1972, p 110.

12-BIT DAC—Uses three Analog Devices AD551 high-speed quad current switches, each having four logic-operated current steering switches and reference transistor on single chip. Output current of each quad is scaled to achieve 12-bit binary performance using AD850 ladder. For bcd operation, required weighting is 10:1, available in AD851.—"Monolithic Quad Current Switch," Analog Devices, Norwood, MA, Sept. 1972.

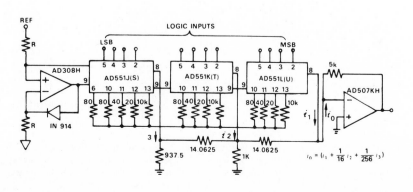

CHAPTER 29
Digital Clock Circuits

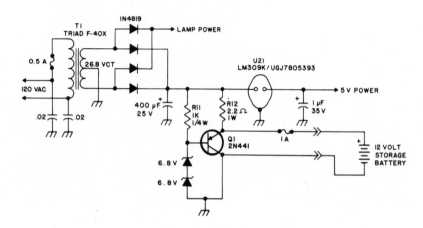

5 V FOR DIGITAL CLOCK—Includes 0.5-A charging circuit for standby storage battery which takes over when a-c line power fails. —M. McDonald, *Time-Frequency Measuring System, 73, Feb. 1973, p 49, 51, and 54–58.*

ALL GATES ¼ DTL 946, EXCEPT 3-INPUT NAND WHICH IS ⅓ DTL 962

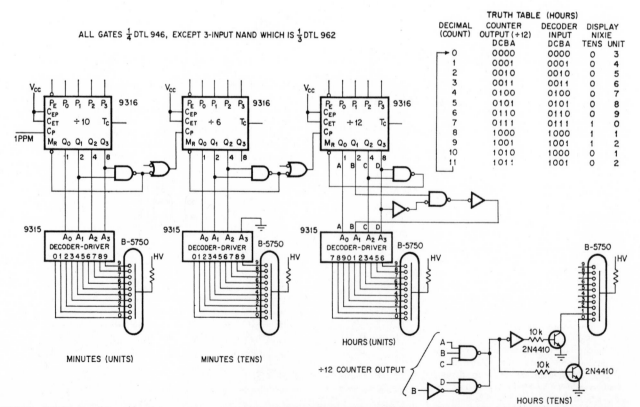

DECIMAL (COUNT)	COUNTER OUTPUT (÷12) DCBA	DECODER INPUT DCBA	DISPLAY NIXIE TENS UNIT	
0	0000	0000	0	3
1	0001	0001	0	4
2	0010	0010	0	5
3	0011	0011	0	6
4	0100	0100	0	7
5	0101	0101	0	8
6	0110	0110	0	9
7	0111	0111	1	0
8	1000	1000	1	1
9	1001	1001	1	2
10	1010	1000	0	1
11	1011	1001	0	2

TRUTH TABLE (HOURS)

MINUTES (UNITS)

MINUTES (TENS)

HOURS (UNITS)

÷12 COUNTER OUTPUT

HOURS (TENS)

DIGITAL CLOCK—Requires only three gates for decoding hours-counter outputs, when connections to hour-displaying Nixie tube are rearranged as shown. Designed for computer application. Seconds indicator can be added, along with AM and PM indicator if desired. —V. R. Clark, *Hitch in Time Saves Gates in 12-Hour Digital Clock, Electronics, April 26, 1971, p 55.*

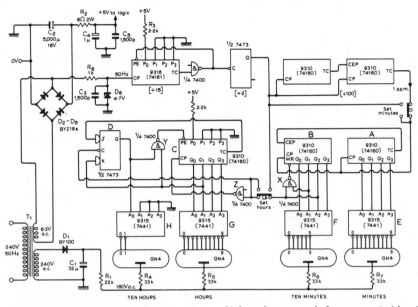

3-MODE DIGITAL CLOCK—Two-digit display gives hour first, then minutes, then seconds, every 10 s, going black between 10-s scans when set for automatic mode. On manual mode, pushing button makes clock show hours-minutes-seconds only once. Third mode shows only seconds, as often required in photolab or for timing microwave oven cooking. Circuit uses IC dividers, gates, flip-flops, and counters for converting 60-Hz power-line frequency to control signals required for digital display. When P2 is plugged normally into J2, clock displays hours, minutes, and seconds for 1 s each. When plugged into J3, each display lasts 2 s. Article tells how to set time with S2 and S4, along with construction details.—W. J. Hawkins, Time Cube—A Unique Digital Clock You Can Make, *Popular Science*, June 1972, p 99–102 and 133.

IC 1,2—7-segment decoder/driver, SN7446
IC 3,6—dual JK flip-flop, SN7476
IC 4—quad 2-input NOr gate, N7402
IC 5,20,21,22—dual 4-input NAND gate, SN7420
IC 7,9,11,23—decade counter, SN7490
IC 8,10,24—divide-by-12 counter, SN7492
IC 12-19, 26—quad 2-input NAND gate, SN7400
IC 25—monostable multivibrator, SN74121
S1—DP3P miniature toggle switch (center off)
S2—DPDT miniature toggle switch
S3—SPDT pushbutton (see text)
S4—SPDT pushbutton
D1,D2—7-segment display (MURA DA 133 D)
B—1.8A rectifier bridge (International Rectifier)
T—6.3v @ 1.2A filament transformer (Stancor) P-6134)
R1, R2—680-ohm, ¼ w resistor
R3—5-ohm, 5w power resistor
R4—33,000-ohm, ¼ w resistor
C1,C2—1,000 ufd @ 15v electrolytic capacitor
C3—0.68 ufd capacitor
Zd—5.1v zener diode, 1N751

DIGITAL 12-HOUR CLOCK—Display on indicator tubes recycles to 01.00 at 12.59 plus 1 minute. Uses 50-Hz power-line frequency as timing source. Two pushbuttons are provided for presetting clock by feeding in pulses at higher than normal frequency. Article describes operation in detail.—R. Buckley, A Digital Clock, *Wireless World*, Feb. 1971, p 54–56.

DISPLAY BOARD

CONTROL BOARD

HOURS BOARD

POWER BOARD

CHASSIS

POSITIVE LOGIC
1 = +
0 = -

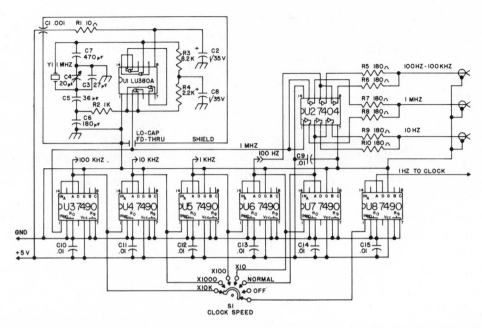

TIME BASE—Used in combination 24-hour digital clock and frequency counter. Article includes circuits for clock and required 5-V regulated supply, along with construction details. Clock recycles every 24 hours.—M. McDonald, Time-Frequency Measuring System, 73, Feb. 1973, p 49, 51, and 54–58.

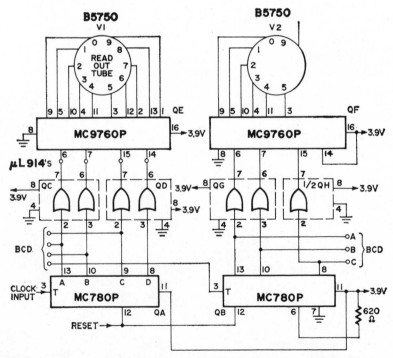

12 OR 24 HOURS WITH ALARM AND TIMER—Completely silent clock provides many optional features, including operation on either 50-Hz or 60-Hz power line, instant zero-reset for use as stop watch or elapsed-time meter, and as time switch for controlling such 24-hour functions as setting burglar alarm system, turning on home lights for several hours a day while on vacation, and turning on lawn-watering system before sunrise each day. Article gives all circuits and construction details. Stage shown divides by 60 for minutes and seconds sections.—L. Walker, Build Multipurpose IC Digital Clock, Radio-Electronics, Aug. 1970, p 46–51.

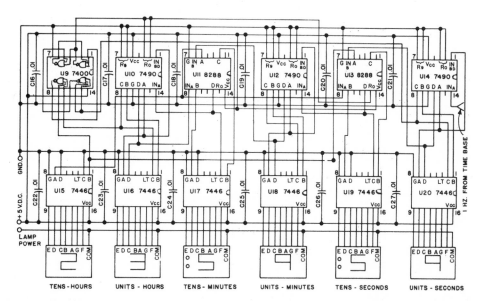

24-HOUR DIGITAL CLOCK—Includes 24-hour reset logic. Decoder turns on lamps required for each of the six display digits by grounding appropriate lamp lines under command of binary code fed in by counter. Article includes circuits for required time base and 5-V regulated supply, along with construction details. Can also be used as frequency counter. —M. McDonald, Time-Frequency Measuring System, 73, Feb. 1973, p 49, 51, and 54–58.

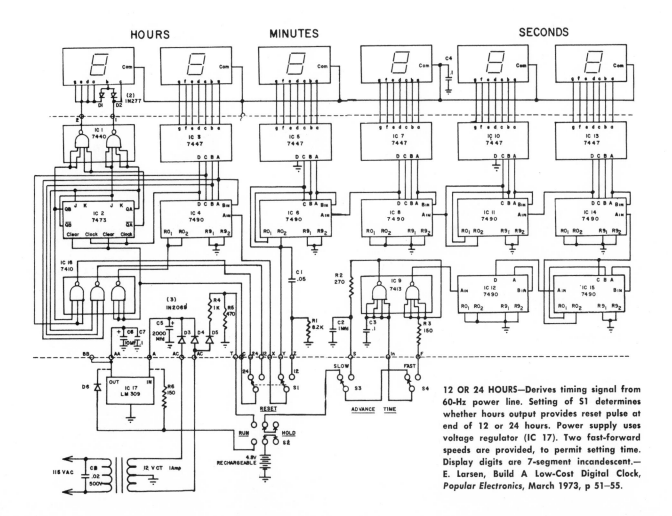

12 OR 24 HOURS—Derives timing signal from 60-Hz power line. Setting of S1 determines whether hours output provides reset pulse at end of 12 or 24 hours. Power supply uses voltage regulator (IC 17). Two fast-forward speeds are provided, to permit setting time. Display digits are 7-segment incandescent.— E. Larsen, Build A Low-Cost Digital Clock, *Popular Electronics*, March 1973, p 51–55.

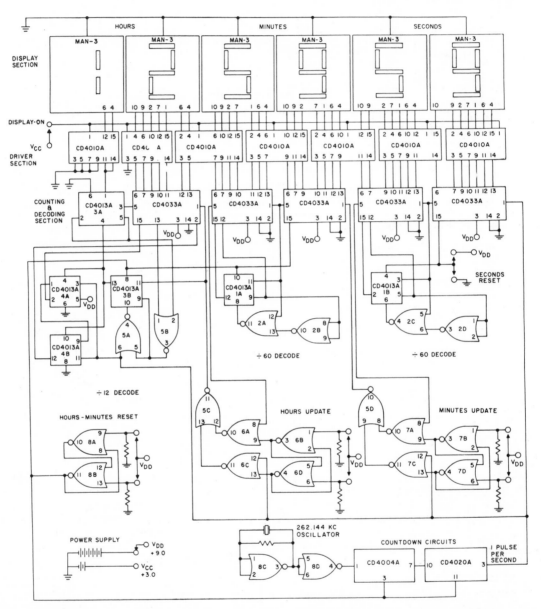

BATTERY-POWERED LED DISPLAY—Single 9-V battery drives all logic, with drain of about 7 mA. Separate pair of 15-V cells in series meets higher-current demands of MAN-3 digi-tal display units, which draw total of 120 mA when switch is pushed to activate them. Fila-ment power is off until time-reading display is required.—"COS/MOS Digital Integrated Circuits," RCA, Somerville, NJ, SSD-203A, 1973 Edition, p 446—447.

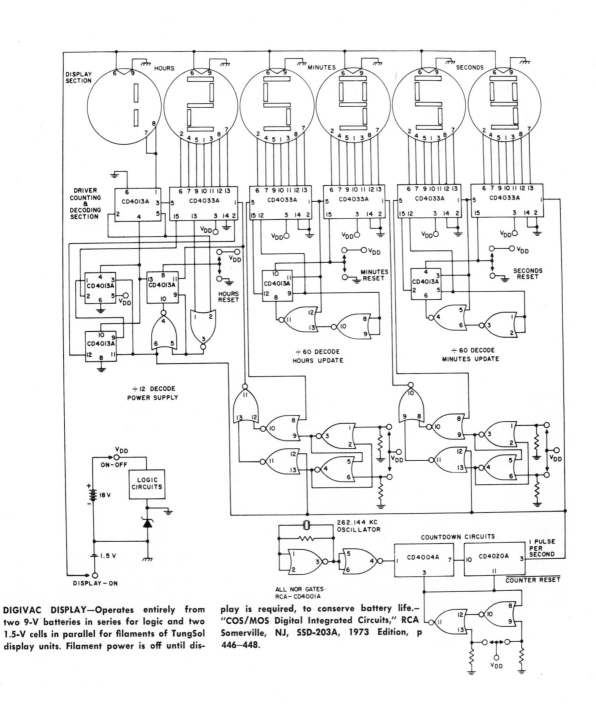

DIGIVAC DISPLAY—Operates entirely from two 9-V batteries in series for logic and two 1.5-V cells in parallel for filaments of TungSol display units. Filament power is off until display is required, to conserve battery life.—"COS/MOS Digital Integrated Circuits," RCA Somerville, NJ, SSD-203A, 1973 Edition, p 446–448.

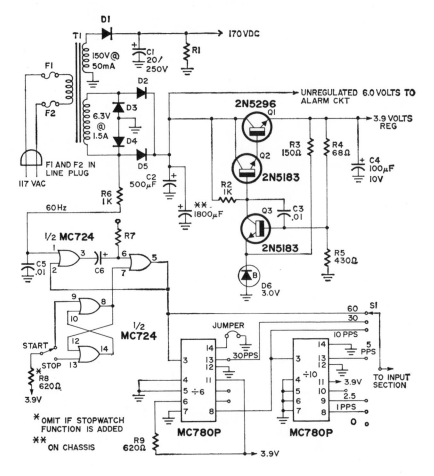

12 OR 24 HOURS WITH ALARM AND TIMER
—Circuit shown serves as power supply and clock pulse generator for universal digital clock operating from 60-Hz power line. Article gives all circuits and construction details. —L. Walker, Build Multipurpose IC Digital Clock, *Radio-Electronics*, Aug. 1970, p 46–51.

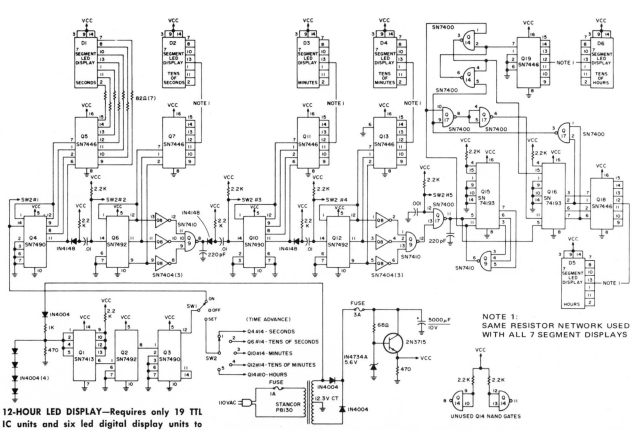

12-HOUR LED DISPLAY—Requires only 19 TTL IC units and six led digital display units to show hours, minutes, and seconds. Number of components is minimized by using up-down counters for hours. Article gives construction details.—W. F. Splichal, Jr., Build a Digital Clock with 19 Inexpensive ICS, 73, May 1973, p 55–60.

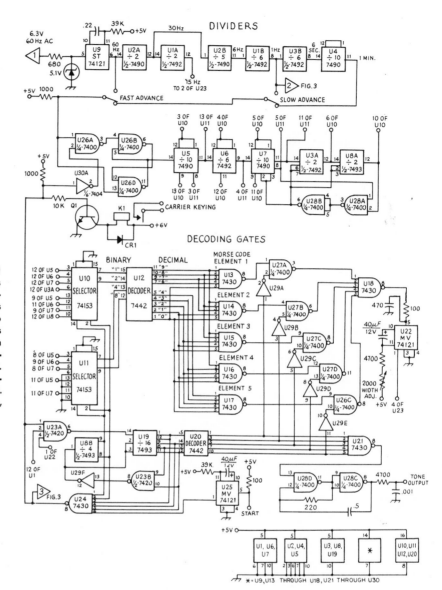

C-W CLOCK—Gives time (hours and minutes) in Morse code automatically on request, as required at end of each series of transmissions when using repeater. Basic frequency is derived from 60-Hz power line and fed into dividers to produce hour and minute units and tens. Dots and dashes are formed in gate U21 for keying 3-kHz square-wave output tone generated by gates U28D and U28C. Article covers construction and operation, including resetting after power failure.—E. C. Pienkowski, A Morse-Code Time Identifier, QST, Nov. 1972, p 34–37.

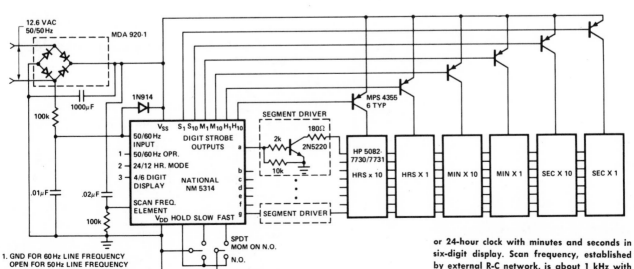

1. GND FOR 60Hz LINE FREQUENCY
   OPEN FOR 50Hz LINE FREQUENCY

2. GND FOR 12 HR. CLOCK
   OPEN FOR 24 HR. CLOCK

3. GND FOR 6 DIGIT DISPLAY (Seconds)
   OPEN FOR 4 DIGIT DISPLAY (Minutes and hours only)

HOURS-MINUTES-SECONDS—Uses National MM5314 clock chip to provide either 12-hour or 24-hour clock with minutes and seconds in six-digit display. Scan frequency, established by external R-C network, is about 1 kHz with values shown.—"5082-7730 Series Seven Segment Display," Hewlett-Packard, Palo Alto, CA, Application Brief, 1973.

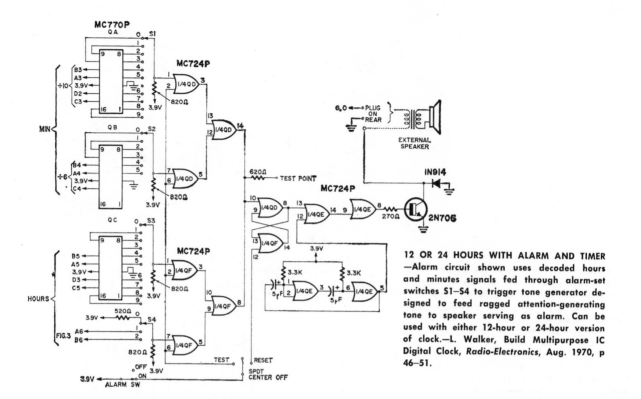

12 OR 24 HOURS WITH ALARM AND TIMER —Alarm circuit shown uses decoded hours and minutes signals fed through alarm-set switches S1–S4 to trigger tone generator designed to feed ragged attention-generating tone to speaker serving as alarm. Can be used with either 12-hour or 24-hour version of clock.—L. Walker, Build Multipurpose IC Digital Clock, *Radio-Electronics*, Aug. 1970, p 46–51.

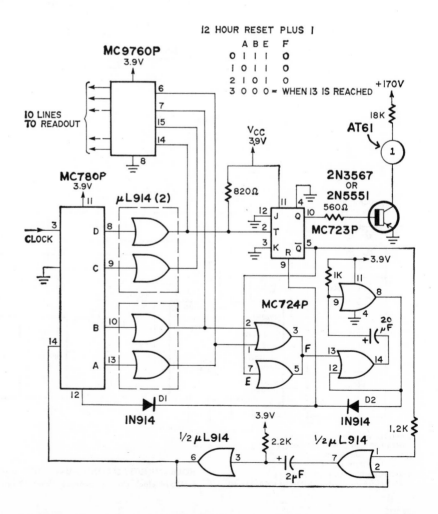

12 HOUR RESET PLUS 1

	A	B	E	F
0	1	1	1	0
1	0	1	1	0
2	1	0	1	0
3	0	0	0	0 = WHEN 13 IS REACHED

12 OR 24 HOURS WITH ALARM AND TIMER —Circuit shown provides for division by 12, to set clock to 1 when count of 13 is reached in 12-hour version. Article gives all circuits and construction details.—L. Walker, Build Multipurpose IC Digital Clock, *Radio-Electronics*, Aug. 1970, p 46–51.

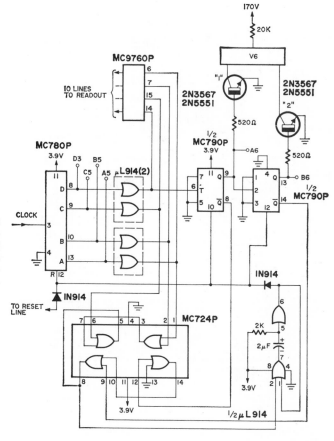

12 OR 24 HOURS WITH ALARM AND TIMER
—Circuit shown resets 24 to 00 for hours display when using 24-hour version of universal digital clock. Article gives all circuits and construction details.—L. Walker, Build Multipurpose IC Digital Clock, *Radio-Electronics*, Aug. 1970, p 46–51.

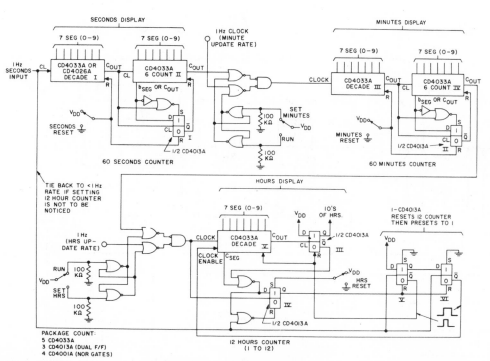

DIVIDERS FOR DIGITAL DISPLAY—Logic diagram gives divide-by-60, divide-by-60, and divide-by-12 counting circuits for 1-Hz seconds input. Switches are provided for updating minutes and hours at seconds rate and for resetting all three displays. Method of resetting to zero at 60th clock pulse for seconds display is described, with similar techniques serving for other displays.—"COS-/MOS Digital Integrated Circuits," RCA, Somerville, NJ, SSD-203A, 1973 Edition, p 445–446.

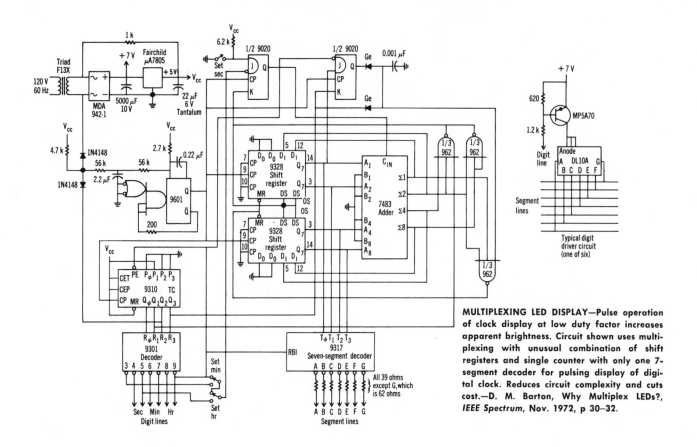

MULTIPLEXING LED DISPLAY—Pulse operation of clock display at low duty factor increases apparent brightness. Circuit shown uses multiplexing with unusual combination of shift registers and single counter with only one 7-segment decoder for pulsing display of digital clock. Reduces circuit complexity and cuts cost.—D. M. Barton, Why Multiplex LEDs?, *IEEE Spectrum*, Nov. 1972, p 30–32.

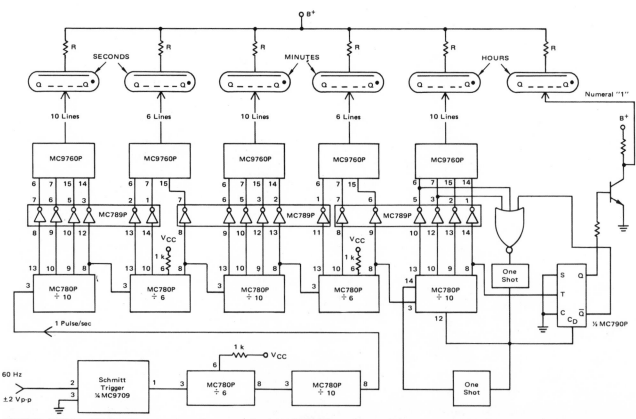

12-HOUR DIGITAL CLOCK—Illustrates use of Motorola MC9760P decoder IC for driving gas-filled decimal indicator tubes showing hours, minutes, and seconds. 60-Hz line frequency is shaped by Schmitt trigger before driving required dividers. One-shot mvbr's are used to clear last two hour counters after count of 12 and reset hours and units counters to one. Report covers clock design in detail.—J. M. Fallon, "MRTL Counting Elements," Motorola Semiconductors, Phoenix, AZ, AN-514, 1970.

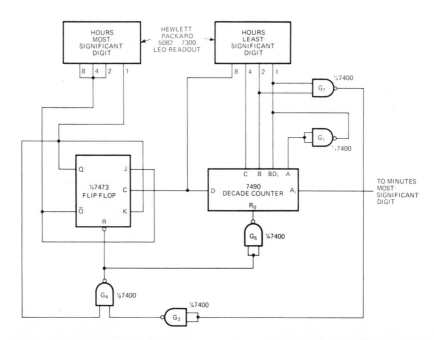

12:59 TO 1:00 TRANSITION—Simple logic arrangement makes transition from 13:00 to 1:00 occur so quickly that digital clock display indicates only transition from 12:59 to 1:00. Reset is accomplished with five dual-input NAND gates and one J-K flip-flop, operating as described in article.—J. Blackburn, Built-In LED Display Decoder Simplifies Digital-Clock Logic, *Electronics*, Feb. 1, 1973, p 116.

CHAPTER 30
Display Circuits

BCD TO LARGE 7-SEGMENT—Uses common-anode configuration of led segments on GaAsP chip, drawing about 20 mA per segment. Display is driven by decoder having active low output, produced from four bcd inputs.—"LED Readout Displays," Dialight Corp., Brooklyn, NY, Application Note AN-7210, 1972, p 3.

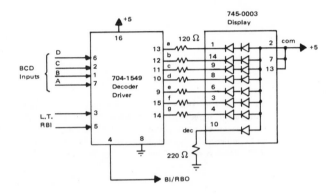

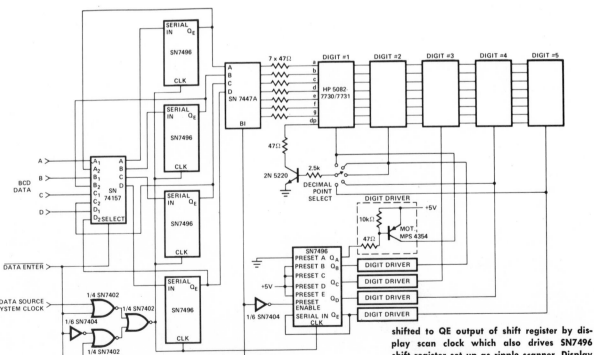

RECIRCULATING—Provides 5-digit strobed display by using one recirculating shift register memory for each bit of 4-bit bcd code. Data for each digit of display is sequentially shifted to QE output of shift register by display scan clock which also drives SN7496 shift register set up as ripple scanner. Display is blanked during data entry, after which DATA ENTER input is returned to high state and scanning begins under control of scan clock.—"5082-7730 Series Seven Segment Display," Hewlett-Packard, Palo Alto, CA, Application Brief, 1973.

FAIL-SAFE DIGITAL DISPLAY—Monitors currents through filaments of five segments of seven-segment incandescent digital display device and blanks out entire digit if any one of these filaments opens. Bottom and lower right segments are not monitored because failure of either is readily indicated by incomplete or distorted digit display. VBB is +5.25 V. All diodes are 1N270, R2 is 1K, and R3 is 2.4K.—"Linear Integrated Circuits and MOS Devices," RCA, Somerville, NJ, SSD-202A, 1973 Ed., p 252–254.

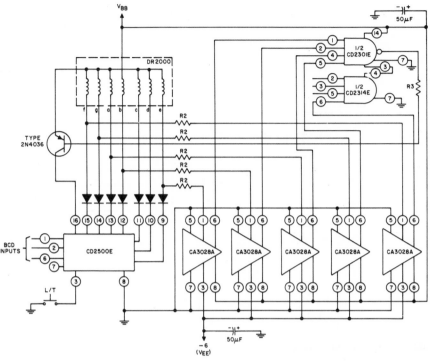

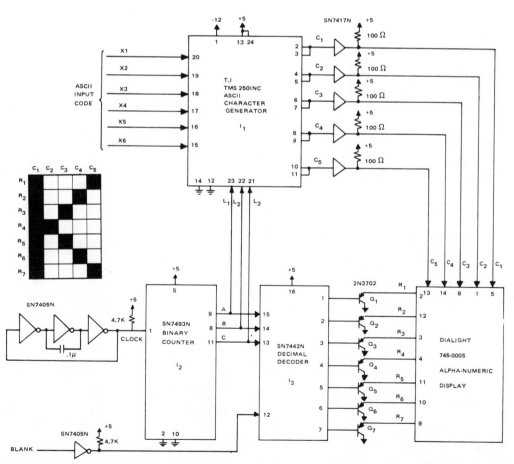

ASCII ALPHANUMERIC—Uses 5 × 7 led array with X-Y select and decimal point to give choice of 64 characters. Action is inherently dynamic, with display created by sequentially driving rows or columns one at a time fast enough so no flicker is detectable, generally greater than 100 Hz. Timing is controlled by clock driving binary counter whose outputs ABC determine row to be selected.—"LED Readout Displays," Dialight Corp., Brooklyn, NY, Application Note AN-7210, 1972, p 8–9.

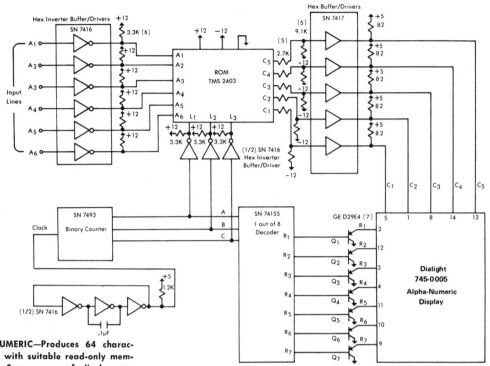

ASCII ALPHANUMERIC—Produces 64 characters when used with suitable read-only memory as shown. Seven rows of display are scanned sequentially one at a time, for activating gallium arsenide phosphide led's in accordance with signals on input lines A1—A6. Timing is controlled by clock driving binary counter whose outputs A, B, and C control row to be selected.—"Solid State LED 5 × 7 Dot Matrix Alpha-Numeric Display," Dialight Corp., Brooklyn, NY, Catalog RO5071, Dec. 1971.

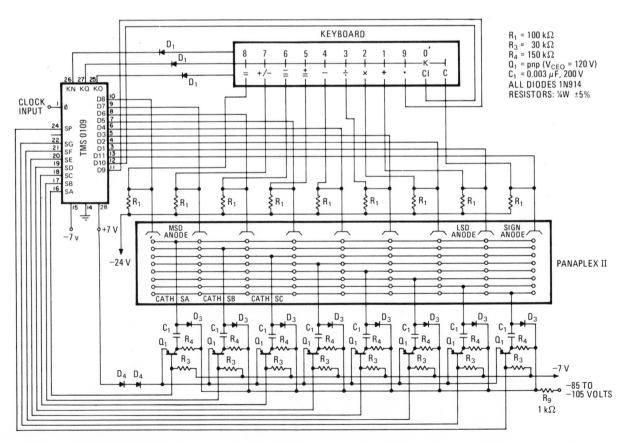

$R_1 = 100$ kΩ
$R_3 = 30$ kΩ
$R_4 = 150$ kΩ
$Q_1 = $ pnp ($V_{CEO} = 120$ V)
$C_1 = 0.003$ μF, 200 V
ALL DIODES 1N914
RESISTORS: ¼W ±5%

PANAPLEX II CALCULATOR DISPLAY—Serves as interface for Texas Instruments TMS 0109 one-chip calculator. Cathode blanking is provided by T1 chip. Anode is direct drive, with R1 serving as pulldown resistors from —24 V d-c bus. Cathode drivers use voltage doubler to develop required —180 V d-c ionization potential.—J. Y. Lee and E. Lord, MOS Chip Plus Level-Shifting Circuit Drives Gas-Discharge Display, *Electronics*, March 1, 1973, p 97—101.

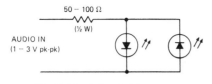

LEDs: FAIRCHILD FLV-100 RED,
OR MONSANTO MV-5094 RED/RED,
OR MONSANTO MV-5491 RED/GREEN

LED ZERO-BEAT DISPLAY—Light-emitting diodes connected with reverse polarity provide inexpensive visual indication of zero-beat frequency corresponding to exact tuning of receiver to transmitted signal. Above 1 kHz away from zero-beat, both led's appear to be on all the time. Within about 20 Hz of zero beat, led's flicker, then go out when within width of zero-beat-frequency notch which is about ±5 Hz. Intensity of led depends on low-frequency response of audio amplifier used.—C. R. Graf, Visual Zero-Beat Indicator Uses Reverse-Polarity LED's, *Electronics*, March 15, 1973, p 119.

GAS-DISCHARGE DISPLAY DRIVE—Provides sufficiently high voltage for triggering Panaplex II flat-cell gas-discharge display while limiting voltage on mos elements driving display. Self-regulating current amplifier Q1 maintains C2 at cathode bias plus voltage. C2 in turn recharges cathode level-shifting capacitors C1. Voltage on mos anode drivers and mos cathode segment drivers is limited to 25 V d-c.—J. Y. Lee and E. Lord, MOS Chip Plus Level-Shifting Circuit Drives Gas-Discharge Display, *Electronics*, March 1, 1973, p 97–101.

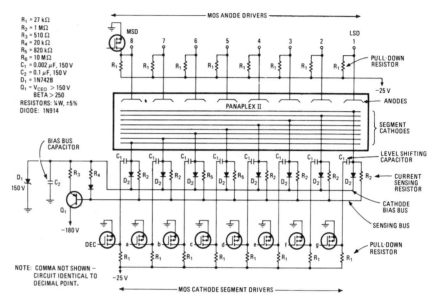

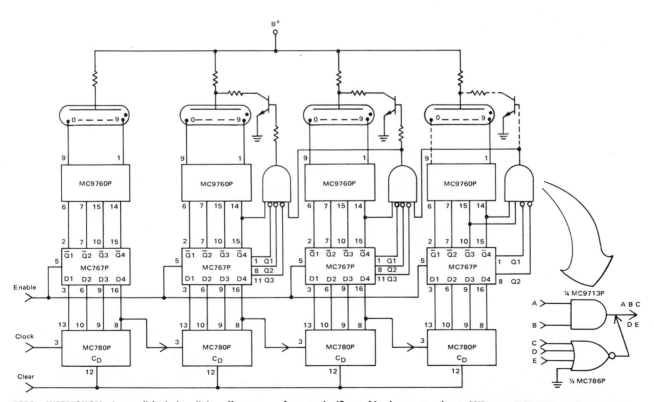

ZERO SUPPRESSION—Accomplished in digital-counter display by detecting zero on most significant bit and ANDing this with zero state at next most significant bit. Scheme is repeated for all but least significant place. Since zero of most significant bit does not contain information, it can be left out entirely by omitting dashed portion of circuit. Four-input and five-input AND gates are made by wiring output of expander to output of two-input AND gate MC9713P as shown at right. Expander inputs are driven by Q outputs of MC767P quad latches.—J. M. Fallon, "MRTL Counting Elements," Motorola Semiconductors, Phoenix, AZ, AN-514, 1970.

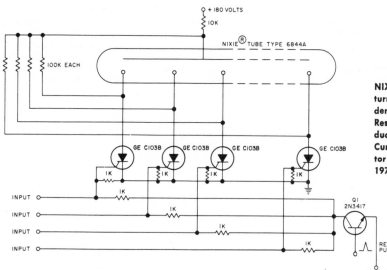

NIXIE DRIVER—Uses GE C103B scr's as gate turn-off switches, each turned on independently by positive-going pulse at its gate. Reset pulse turns on Q1 to turn off all conducting scr's.—D. R. Grafham, "Using Low Current SCR's," General Electric, Semiconductor Products Dept., Auburn, NY, No. 200.19, 1970, p 39.

MOS-INDICATOR INTERFACE—Used to feed resistive circuits from cathodes of numerical indicator tube with mos array. R1 ensures that nonconductive cathodes are held high, while R1 and D1 together prevent input of mos array from being damaged by positive inputs.—"MOS Integrated Circuits and Their Applications," Mullard, London, 1970, TP11081, p 101.

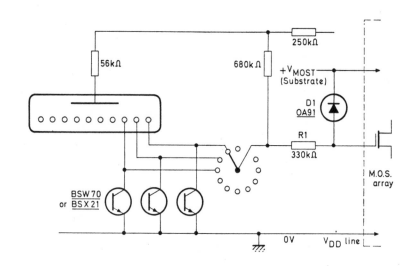

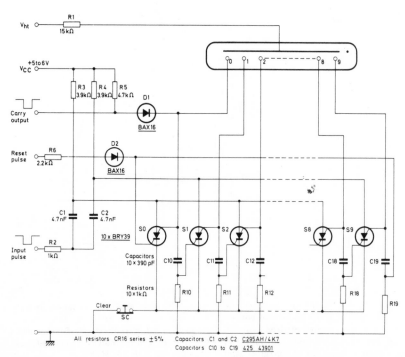

SCS RING COUNTER—Circuit uses ten BRY39 silicon controlled switches to drive numerical indicator tube. Several decades of circuit can be connected in series to form display system. Train of input pulses will switch scs's on and off sequentially, with cathode 9 of indicator tube producing carry pulse for switching next decade when going back to cathode 0. Latching property of scs keeps activated cathodes on when pulse train ends, for static display. —"Applications of the BRY39 Transistor," Mullard, London, 1973, TP1338, p 2–3.

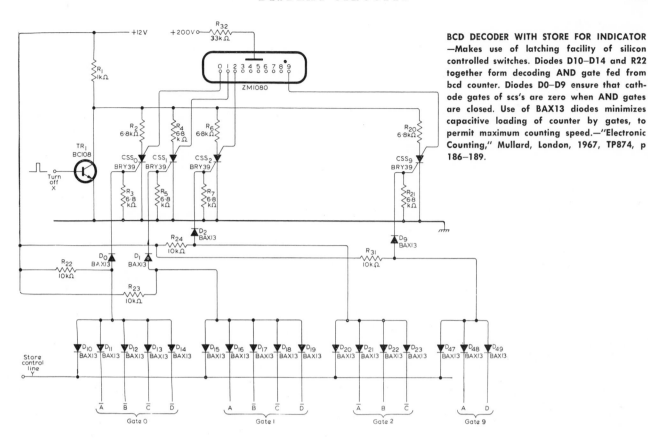

BCD DECODER WITH STORE FOR INDICATOR—Makes use of latching facility of silicon controlled switches. Diodes D10—D14 and R22 together form decoding AND gate fed from bcd counter. Diodes D0—D9 ensure that cathode gates of scs's are zero when AND gates are closed. Use of BAX13 diodes minimizes capacitive loading of counter by gates, to permit maximum counting speed.—"Electronic Counting," Mullard, London, 1967, TP874, p 186—189.

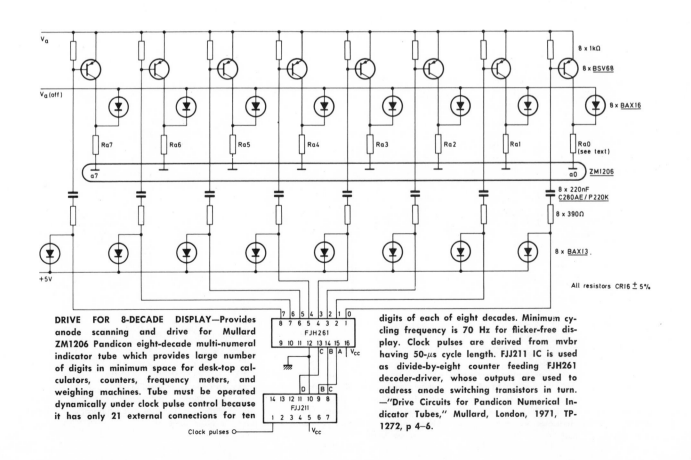

DRIVE FOR 8-DECADE DISPLAY—Provides anode scanning and drive for Mullard ZM1206 Pandicon eight-decade multi-numeral indicator tube which provides large number of digits in minimum space for desk-top calculators, counters, frequency meters, and weighing machines. Tube must be operated dynamically under clock pulse control because it has only 21 external connections for ten digits of each of eight decades. Minimum cycling frequency is 70 Hz for flicker-free display. Clock pulses are derived from mvbr having 50-μs cycle length. FJJ211 IC is used as divide-by-eight counter feeding FJH261 decoder-driver, whose outputs are used to address anode switching transistors in turn.—"Drive Circuits for Pandicon Numerical Indicator Tubes," Mullard, London, 1971, TP-1272, p 4—6.

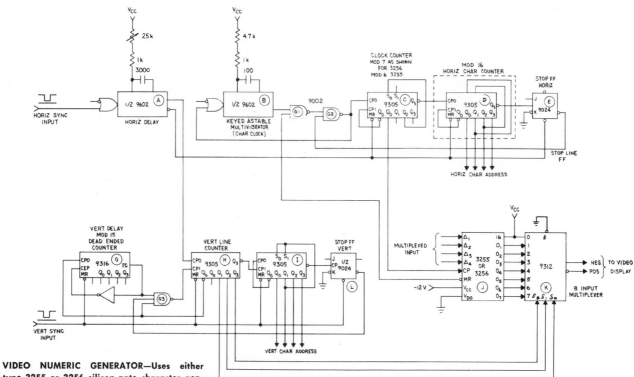

VIDEO NUMERIC GENERATOR—Uses either type 3255 or 3256 silicon-gate character generator having counter, column decode switch, and 560-bit read-only memory, to produce 12 rows each having 16 horizontal characters, for a display of up to 192 characters each formed on 5 × 7 dot matrix. Operates with standard tv raster scan. Numeric display can be superimposed on video picture. Report covers complete system, including video interface circuit.—E. G. Breeze, "AB-182 Horizontal Raster Scan Video Numeric Generator," Fairchild Semiconductor, Mountain View, CA, No. 182, 1971.

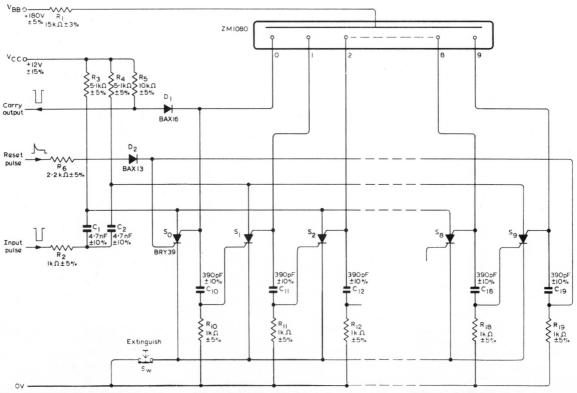

SCS RING COUNTER WITH DISPLAY—Design using silicon controlled switches is simpler and less expensive than bistable mvbr's for driving numerical indicator. At tenth pulse, S9 is switched off and S0 switched on again, while feeding carry signal for advancing next decade counter one step. Sw clears memory and extinguishes indicator tube. Circuit and tube are started again by feeding single positive-going pulse to reset terminal to light cathode 0 of display.—D. J. G. Janssen, Circuit Logic with Silicon Controlled Switches, *Mullard Technical Communications*, March 1968, p 57–64.

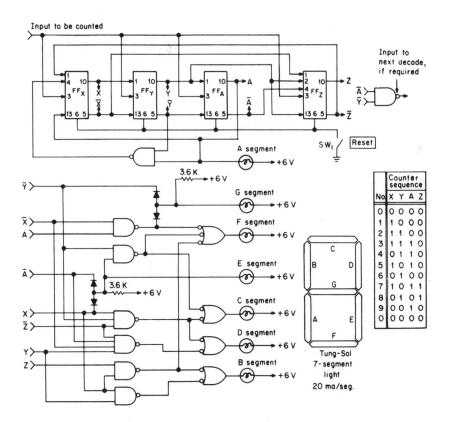

7-SEGMENT DRIVE—Four Westinghouse type 225 J-K flip-flops connected as shown minimize number of NAND and noninverting diode gates needed to drive numerical display. Momentary-contact switch SW1 (or logical 0 from peripheral logic) resets synchronous counter to 0000.—G. P. Scher, Counter Connection Minimizes Logic Count for Numerical Display, *The Electronic Engineer*, June 1968, p 72.

No	Counter sequence			
	X	Y	A	Z
0	0	0	0	0
1	1	0	0	0
2	1	1	0	0
3	1	1	1	0
4	0	1	1	0
5	1	0	1	0
6	0	1	0	0
7	1	0	1	1
8	0	1	0	1
9	0	0	1	0
0	0	0	0	0

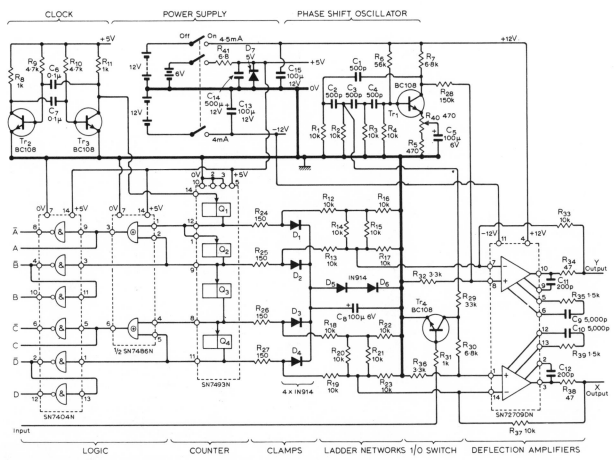

KARNAUGH MAP DISPLAY—Feeds cro to produce on screen diagram showing all possible combinations of number of two-state variables, as aid in teaching logic and Boolean algebra. Article covers construction and use.—B. Crank, Karnaugh Map Display, *Wireless World*, April 1971, p 185–189.

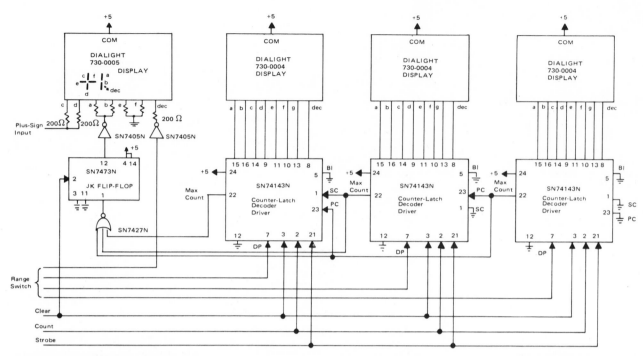

3 DIGITS WITH OVERFLOW—External range switch determines position of decimal point by applying high logic level to any one of four decimal inputs. Circuit is capable of detecting overflow. Clear input will blank overflow digit and set three digits of display to zero. After clearing, pulses at count input will be counted, decoded, and displayed. At every tenth count, maximum count output of least significant digit becomes a low, which enables parallel carry (PC) inputs. Second digit thus counts every ten pulses and most significant digit every 100 pulses. Strobe permits counter to acquire next count while current count is viewed.—"LED Readout Displays," Dialight Corp., Brooklyn, NY, Application Note AN-7210, 1972, p 7.

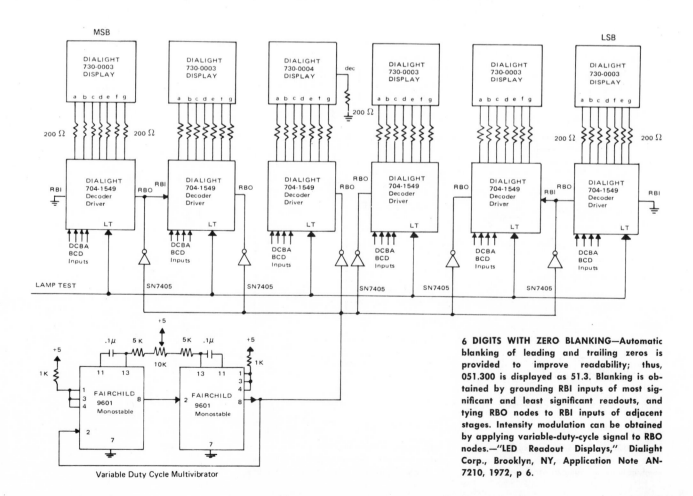

Variable Duty Cycle Multivibrator

6 DIGITS WITH ZERO BLANKING—Automatic blanking of leading and trailing zeros is provided to improve readability; thus, 051.300 is displayed as 51.3. Blanking is obtained by grounding RBI inputs of most significant and least significant readouts, and tying RBO nodes to RBI inputs of adjacent stages. Intensity modulation can be obtained by applying variable-duty-cycle signal to RBO nodes.—"LED Readout Displays," Dialight Corp., Brooklyn, NY, Application Note AN-7210, 1972, p 6.

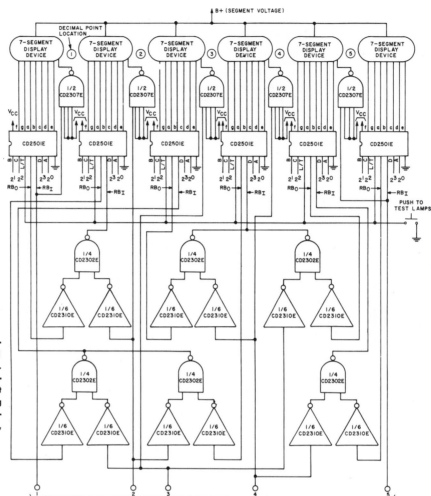

FLOATING DECIMAL—Line for desired position of decimal point is held at logical 1 while other four select inputs are logical 0. RCA CD2307E gate is used to energize decimal-point filament in display device. Circuit includes ripple-blanking circuits for blanking out no-value zeroes.—"Linear Integrated Circuits and MOS Devices," RCA, Somerville, NJ, SSD-202A, 1973 Ed., p 250–252.

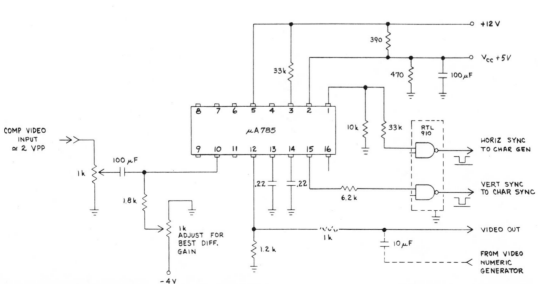

COMPOSITE VIDEO INTERFACE—Generates negative-going horizontal and vertical sync pulses from composite video signal for low-cost type 3256 16-character video generator which uses standard tv raster scan system. Each character is formed on 5 × 7 matrix occupying 14 horizontal scanning lines. Report covers complete testing.—E. G. Breeze, "AB-182 Horizontal Raster Scan Video Numeric Generator," Fairchild Semiconductor, Mountain View, CA, No. 182, 1971.

TRAPEZOID GENERATOR FOR CRT—High slewing rate of μA715 opamp is used advantageously to convert X and Y deflection output pulses of 3250 numeric character generator to trapezoidal waveforms required for electrostatic deflection of crt for 7-segment display.—"A Trapezoidal Deflection Circuit for Use with the 3250 Numeric Character Generator Using the μA715," Fairchild Semiconductor, Mountain View, CA, No. 127, 1969.

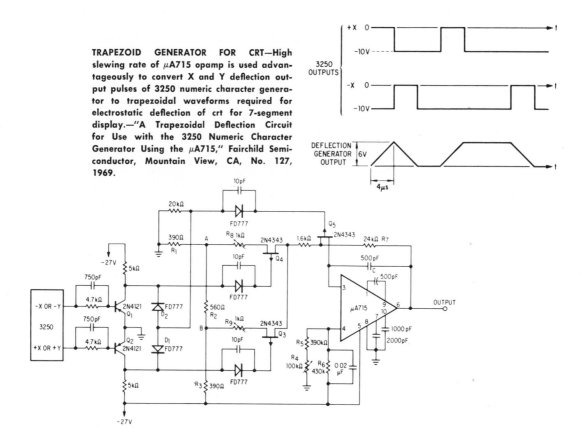

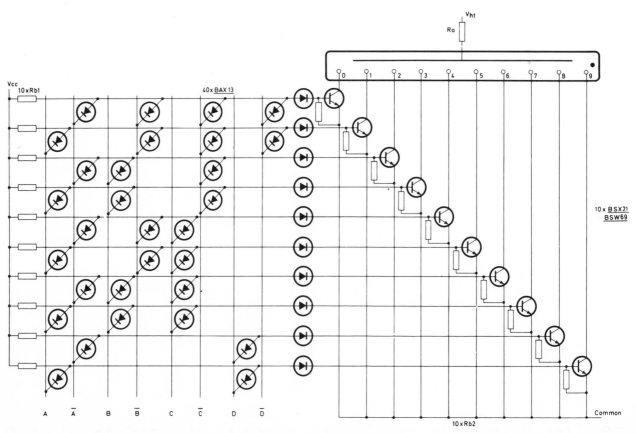

BCD-DECIMAL DECODER—Diode decoder serves for converting bcd inputs (bottom left) to corresponding decimal values for driving Mullard ZM1170 or similar numerical indicator tube. Report shows how to calculate all resistor values; if using BSX21 transistors, Rb1 is 8.2K and Rb2 1.8K.—"Numerical Indicator Tube Static Display Systems," Mullard, London, 1970, TP1203, p 11–12.

NIXIE DIMMER—Turns display tube on and off at variable rate, to give effect of dimming because human eye sees average brightness. Developed for locations requiring low ambient light levels, as in darkened airplane cockpit or radar room. Use logarithmic taper for dimming control R9.—J. J. Klinikowski, Controlling Numeric Display Tube Brightness and Blanking, *The Electronic Engineer*, Aug. 1967, p 44–47.

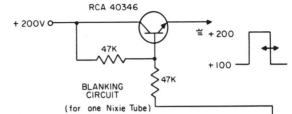

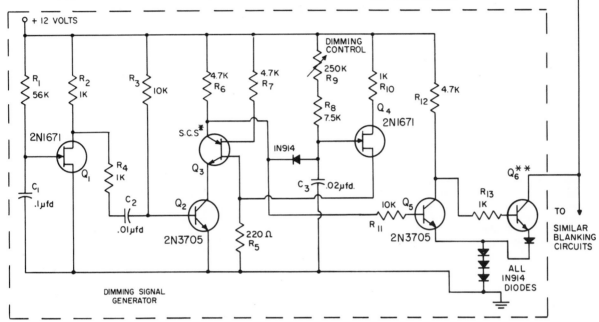

* Burroughs B7401   ** R.C.A. 40346 (selected for $V_{CBO} > 200V$)

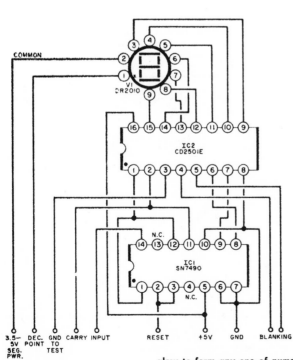

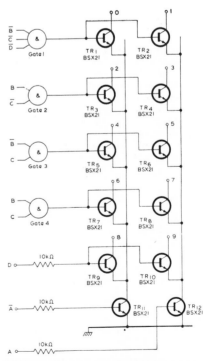

NUMITRON DRIVE—Decade counter IC1 and bcd-to-7-segment decode-driver IC2 make appropriate straight segments of wire in V1 glow to form any one of numerals from 0 to 9. V1 and IC2 are RCA products.—V. Wood, Numitron Readout, *Popular Electronics*, March 1970, p 73–75.

DECODER FOR BCD—Circuit minimizes duplication of elements by generating only five functions with gates and switching them with voltages on A terminals. Used for converting 1-2-4-8 input code to drive numerical indicator tube directly.—"Electronic Counting," Mullard, London, 1967, TP874, p 160–161.

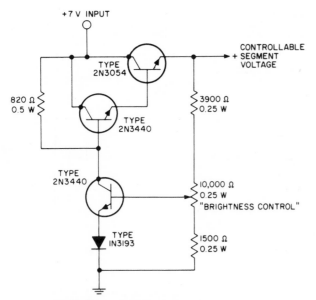

BRIGHTNESS CONTROL—Simple series voltage regulator operating from 7-V supply provides d-c output of 2.5 to 5 V for varying brightness of incandescent readout devices.—"RCA Numitron Display Devices," RCA, Somerville, NJ, NUM-421, 1972.

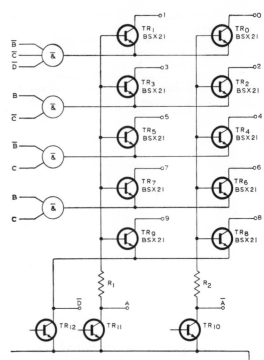

DECODER FOR BCD—Uses NAND gates for decoding 1-2-4-8 inputs for driving ten terminals of numerical indicator tube. TR10, TR11, and TR12 are parts of counting bistable circuits, while R1 and R2 limit base current of BSX21 drivers for indicator.—"Electronic Counting," Mullard, London, 1967, TP874, p 161.

ODD-EVEN DECODER—Requires only 14 diodes for converting bcd inputs (bottom left) to decimal values for driving Mullard ZM1170 or similar numerical indicator tube. Circuit decodes one out of five pairs of outputs using B, C, and D inputs, and selects correct one of pair by state of A input. Report shows how to compute resistor values; if using BSX21 transistors with supply voltage of 5 V, Rb should be 56 K. Value can be lower, such as 18K, to allow for low-gain transistors and other tolerances.—"Numerical Indicator Tube Static Display Systems," Mullard, London, 1970, TP1203, p 11–13.

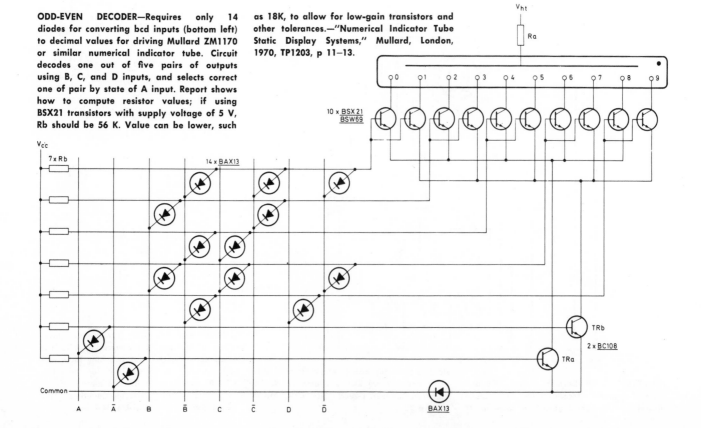

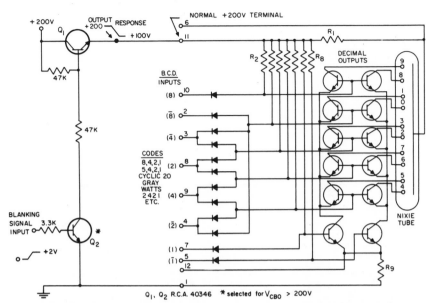

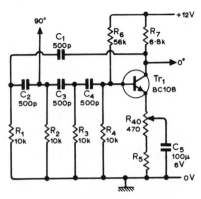

NIXIE BLANKER—Used to turn on or darken multi-digit displays in individual groups for emphasis. Circuit lowers cathode voltage of Nixie from normal 200 V to 100 V with Q1 and Q2.—J. J. Klinikowski, Controlling Numeric Display Tube Brightness and Blanking, *The Electronic Engineer*, Aug. 1967, p 44–47.

22-KHZ PHASE-SHIFT—Used to produce characters forming Karnaugh map display. Each RC section provides 45-deg shift, making it easy to pick off desired 90-deg signal as shown. R40 is used to adjust a-c gain of transistor to overcome losses in phase-shift network; adjust for good sine-wave output.—B. Crank, Karnaugh Map Display, *Wireless World*, April 1971, p 185–189.

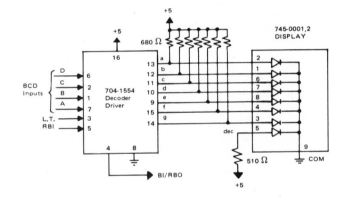

BCD TO 7-SEGMENT—Common-cathode display using single-digit gallium arsenide phosphide chip is typically driven by 5 mA per segment from TTL bcd inputs. Decoder driver has active high outputs. Combinations of high and low on bcd inputs create numerals 0–9 on display.—"LED Readout Displays," Dialight Corp., Brooklyn, NY, Application Note AN-7210, 1972, p 2.

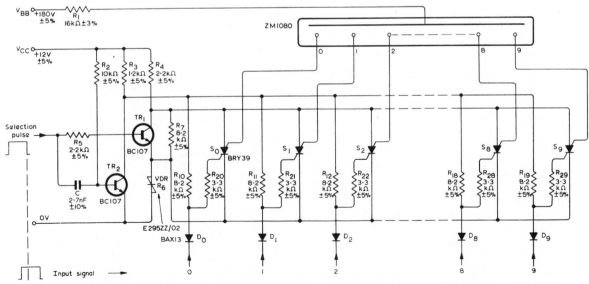

RING COUNTER DISPLAY WITH MEMORY—Used when numeric display must remain after input signal that selected it has ceased. Scs associated with each digit of numerical indicator tube acts both as switch and memory. Selection pulse clears memory and extinguishes previously displayed digit. Positive-going pulse at one of input signal terminals will now light corresponding cathode of display tube and hold display.—D. J. G. Jans-sen, Circuit Logic with Silicon Controlled Switches, *Mullard Technical Communications*, March 1968, p 57–64.

CHAPTER 31
Door Lock Circuits

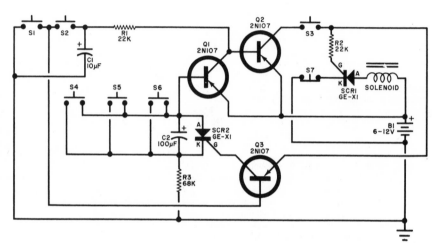

60-COMBINATION LOCK—Permutations of three switches, combined with three penalty switches S4–S6 and reset S7, make it unlikely that potential thief could try all possible combinations in time normally available. Each penalty switch deactivates lock circuit for about 20 s. Additional switches may be added in series to lengthen break-in time further.—W. Isengard, Useful Circuits, *Popular Electronics*, Dec. 1971, p 89 and 94.

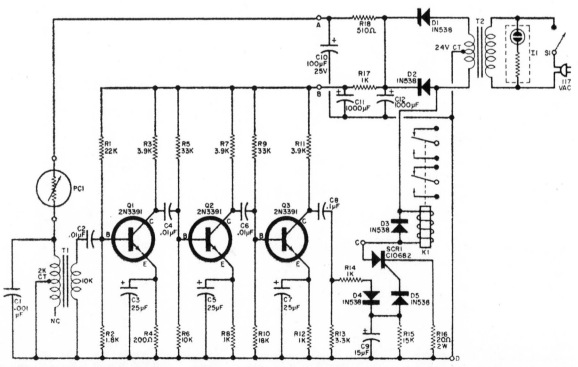

MODULATED-LIGHT LOCK—Receiver for 2-kHz flashing key is Clairex CL703L or similar photocell driving high-gain amplifier that pulls in 6-V relay K1 to energize solenoid of electric door lock. Article includes construction of 2-kHz flashing penlight key and more powerful light modulator for opening garage doors or serving as intrusion alarm.—J. S. Simonton, Jr., Freq-Out Electronic Lock, *Popular Electronics*, Oct. 1968, p 43–48 and 116–117.

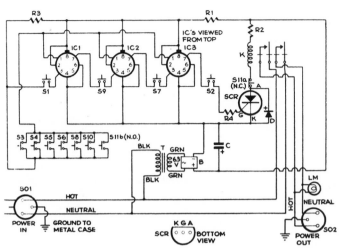

C—1,000-mfd, 15-volt, electrolytic capacitor
R1—180-ohm, 2-watt, carbon resistor
R2—5.6-ohm, 2-watt, carbon or wirewound resistor
R3—1,000-ohm, ½-watt, carbon resistor
R4—1,800-ohm, ½-watt, carbon resistor
B—Motorola HEP 175 full-wave bridge rectifier
IC1, IC2, IC3—Motorola HEP 583 flip-flop
SCR—Motorola HEP 320 Thyristor (SCR)
T—Stancor P-6465 (or equal) 6.3-volt AC @ .6 A filament transformer
K—Potter & Brumfield PR7DY relay; 6-volt DC coil, heavy duty, DPST normally-open contacts

A-C COMBINATION LOCK—Punching correct combination of four numbered buttons (S1—S10) triggers scr and pulls in relay K for applying power to outlet SO2 into which power tools of home workshop are plugged. Connections shown are for combination 1972. Code selected must have four different digits. At end of day, push S11 to open relay and kill power, after which code must be re-punched. Random pushing of buttons will not work because any incorrect button turns off all three IC flip-flops and rescrambles lock. D is 1N70 and LM is 120-V pilot lamp. Can also be used to control electric door lock.—R. M. Benrey, How Electronics Can Lock Power Tools, *Popular Science*, Dec. 1971, p 98–99 and 116.

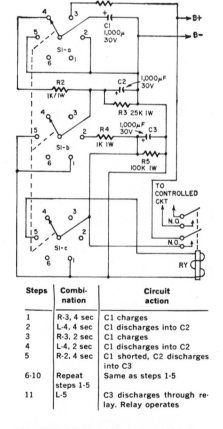

Steps	Combination	Circuit action
1	R-3, 4 sec	C1 charges
2	L-4, 4 sec	C1 discharges into C2
3	R-3, 2 sec	C1 charges
4	L-4, 2 sec	C1 discharges into C2
5	R-2, 4 sec	C1 shorted, C2 discharges into C3
6-10	Repeat steps 1-5	Same as steps 1-5
11	L-5	C3 discharges through relay. Relay operates

TIME-CONTROLLED LOCK—Relay which energizes electric door lock is closed only when C2 has been charged and discharged into C3 enough times so voltage across C3 is high enough to operate relay. With switch S1 wired as shown, clockwise R and counter-clockwise L rotations to numbered switch positions must be made in sequence shown in table, with each position held for approximate time indicated, to open lock. Supply voltage should be about 1.5 times minimum d-c operating voltage of relay used.—Combination Time Lock, *Radio-Electronics*, Sept. 1970, p 93.

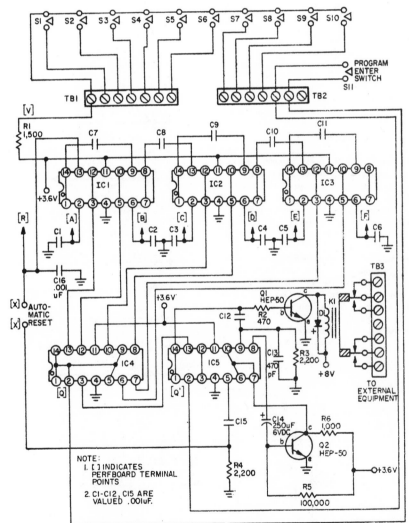

NOTE:
1. [ ] INDICATES PERFBOARD TERMINAL POINTS
2. C1-C12, C15 ARE VALUED .001uF.

NUMERIC 6-DIGIT LOCK—Odds against operating lock by chance are better than 1,-500,000 to one. Combination for wiring connection shown is 45-67-89. To operate, press S11 to make relay K1 pull in, press S1 to S10 in correct sequence, then press S11 again to trigger mono mvbr. After about 25 s delay, gates and flip-flops in circuit logic de-energize K1 to open lock. Article explains operation in detail and gives power supply, which includes provision for standby battery. Supply provides 8 V d-c for relay and 3.6 V regulated for logic. Switches can be remotely located. D1 is HEP154. IC 1-3 and 5 are MC724P and IC 4 is MC792P.—E. A. Morris, DIGITALOCK, *Science and Electronics*, Oct.-Nov. 1970, p 61–68 and 99–101.

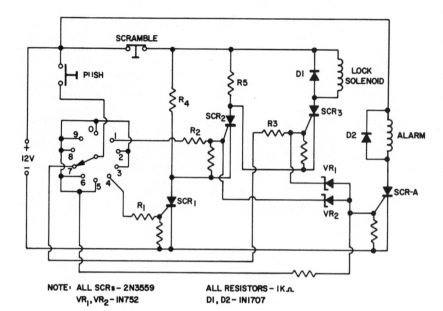

3-DIGIT LOCK—Series of scr's must be switched in correct sequence to energize solenoid of lock. If any part of sequence is incorrect, SCR-A is turned on, setting off alarm. Zeners VR1 and VR2 inhibit firing of SCR-A during correct operation of SCR2 and SCR3. Pushing scramble button switches off all scr's, making it necessary to start switching sequence over again. Connections shown give code of 417.—SCR-Operated Coded Lock, *Electro-Technology*, Oct. 1969, p 16–17.

NOTE: ALL SCRs – 2N3559
       VR₁, VR₂ – IN752

ALL RESISTORS – 1KΩ
D1, D2 – IN1707

FOUR-DIGIT LOCK—Gives 10,000 possible combinations. Code can be changed in a few seconds if change is desired. If incorrect combination is entered, by setting any of 10 pushbutton switches improperly, user must wait about 4 s before trying again. Setup of code involves connecting sequential input terminals of diodes D11–D14 to any four of terminals 0–9 as switch-diode matrix. Spring-loaded wire clips can be used for these connections. D1–D14 are 12-V piv, D15 is 6–8 V 1-W zener, D16 is 12-V 10-W zener, D17 is 35-piv 0.5-A silicon rectifier, and D18 is 20-piv 0.5-A silicon diode.—J. A. Nunley, Electronic Combination Lock, *Popular Electronics*, Aug. 1971, p 51–53 and 58.

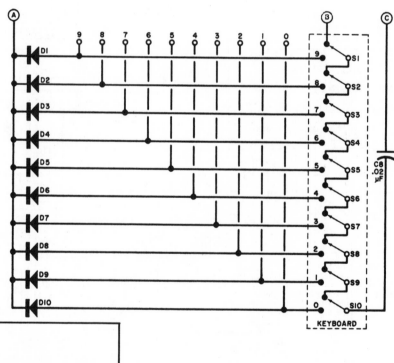

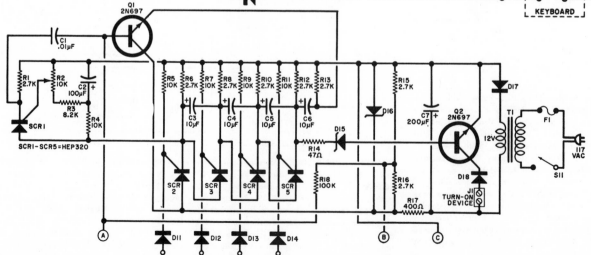

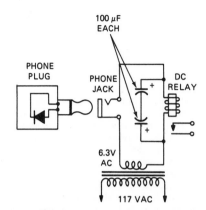

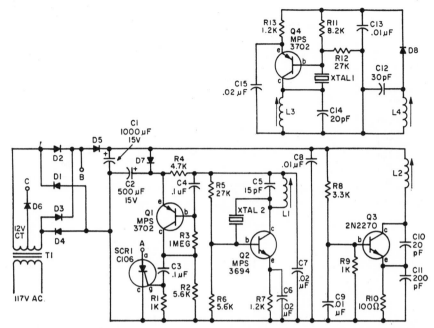

ELECTRONIC KEY—Diode in phone plug serves as key for opening electric door lock by rectifying stepped-down a-c voltage to provide d-c voltage required for relay. Lock cannot be defeated by inserting solid metal rod in jack, because relay will not operate on a-c. Diode should match relay current rating. Electrolytics should be rated 35 V or preferably 50 V if bulk is no factor.—M. Mandl, Switching Tricks with Silicon Diodes, *Radio-Electronics*, May 1972, p 54–56.

ELECTRONIC LOCK—Key is crystal oscillator in tiny box, held against plastic panel mounted in wall near door, with latch-releasing power circuit behind panel. When L2 is lined up with L4 and L1 with L3, slight difference in frequency of the two crystal oscillators appears as audio signal at base of Q1 for amplification and firing of scr. Door latch solenoid is connected between terminal A of scr and B if 12 V (between A and C if 6 V). Use HEP154 for D1–D7 and HEP134 or 1N34A for D8. Both crystals are 26.670 MHz. Coils are Lafayette 99E62028.—H. Cohen, The Elocktron, *Elementary Electronics*, Jan.-Feb. 1972, p 39–42 and 106.

MODULATED-LIGHT LOCK—Key is penlight containing 2-kHz IC oscillator driving switching transistor Q1 to make lamp flash at same rate. Use No. 222 flashlight lamp. Requires three Mallory M-76 hearing-aid 1.5-V cells. Article includes circuit and construction of matching receiver having photocell input, with relay to actuate solenoid of electric lock. —J. S. Simonton, Jr., Freq-Out Electronic Lock, *Popular Electronics*, Oct. 1968, p 43–48 and 116–117.

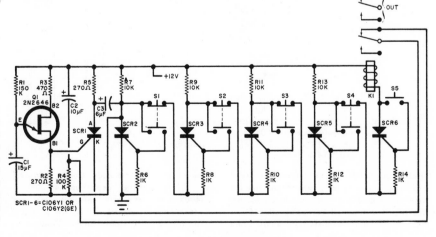

120-COMBINATION LOCK—Operates from 12-V storage battery of auto or boat. Five switches (S1–S5) must be operated in correct sequence within 3 s (adjustable) or process must be started over again. Combination is easily changed. Uses scr's in series, with last one pulling in relay that can operate electric door latch. Combination of circuit shown is 1-4-3-2-5.—D. E. Fahnstock, SCR Combination Lock, *Popular Electronics*, Nov. 1971, p 53–55.

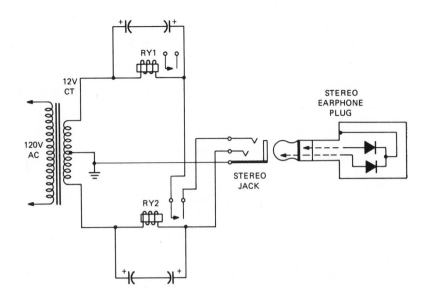

STEREO-PLUG KEY—Two diodes mounted in stereo earphone plug serve as key for opening electric door lock connected to contacts of normally open relay RY1. Lock opens with slight delay because contacts of RY2 must close to energize RY1. System cannot be cheated by pushing solid metal rod into stereo jack because this applies only a-c to RY2 and it doesn't close. Diode current ratings should be at least equal to relay current ratings. Capacitors are 100-μF 35-V.—M. Mandl, Switching Tricks with Silicon Diodes, *Radio-Electronics*, May 1972, p 54–56.

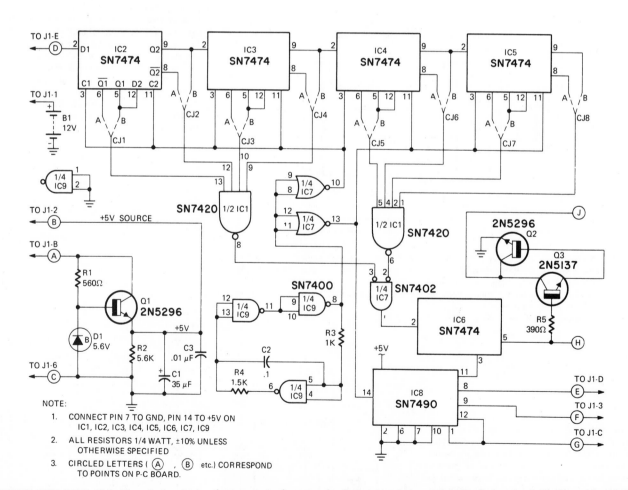

BINARY-CODE LOCK—Eight-bit code is wired into lock and key at time of construction from choice of 384 possible combinations. Article includes circuit of key using SN74151 IC mounted on small printed-circuit board using jumpers to implement code. Output transistor Q2 drives electric door opener having 1N4003 diode across its solenoid. Only one of each pair of dashed connections A and B is completed with eight jumpers required to set up desired code. Table in article shows how jumpers are used for lock and for key. —J. B. Wicklund, IC Key Opens Electronic Door Lock, *Radio-Electronics*, June 1972, p 41–43.

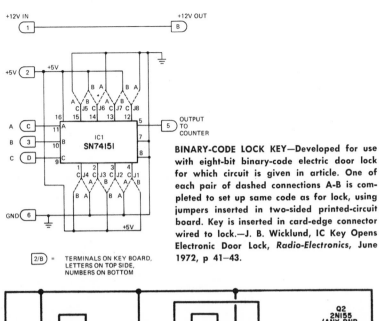

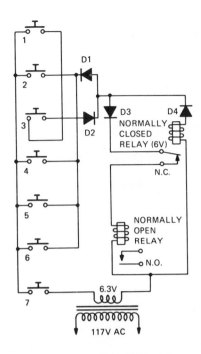

BINARY-CODE LOCK KEY—Developed for use with eight-bit binary-code electric door lock for which circuit is given in article. One of each pair of dashed connections A-B is completed to set up same code as for lock, using jumpers inserted in two-sided printed-circuit board. Key is inserted in card-edge connector wired to lock.—J. B. Wicklund, IC Key Opens Electronic Door Lock, *Radio-Electronics*, June 1972, p 41—43.

2/B = TERMINALS ON KEY BOARD, LETTERS ON TOP SIDE, NUMBERS ON BOTTOM

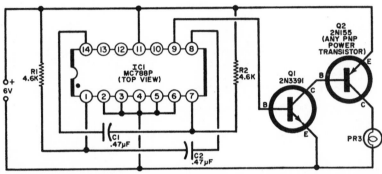

MODULATED-LIGHT LOCK—Higher-power 2-kHz modulated-light key for photoelectric lock gives sufficient range for use as intrusion alarm or garage-door opener. Supply can be 6-V lantern battery or suitable power supply. Mount PR-3 lamp in 3-inch or larger flashlight reflector to give maximum range. Article in-cludes circuit and construction of matching re-ceiver having photocell input, with relay to actuate solenoid of electric lock.—J. S. Si-monton, Jr., Freq-Out Electronic Lock, *Popular Electronics*, Oct. 1968, p 43—48 and 116—117.

COMBINATION LOCK—With connections shown, switches 1, 3, and 7 must be pushed simultaneously to energize normally open relay controlling electric door lock, garage-door opener, power-tool protection switch, or other device. If wrong combination is pressed or all buttons pressed at once, normally closed relay opens and prevents other relay from operating. Will work with a-c relay since diodes provide pulsating d-c. Use 100-μF capacitors across d-c relays for smoother operation, observing correct polarity for elec-trolytics. Diodes are 100-V 0.25-A silicon.—M. Mandl, Switching Tricks with Silicon Diodes, *Radio-Electronics*, May 1972, p 54—56.

CHAPTER 32
Electronic Music Circuits

GUITAR FUZZ AMPLIFIER—Insert between guitar pickup and its amplifier. Setting of R1 determines amount of fuzz generated, and R2 sets output level. Since R1 cannot completely eliminate fuzz effect, bypass path and S2 are required for fuzz-free sound. S2 can be foot-switch operated by guitar player.—H. Friedman, "99 Electronic Projects," Howard W. Sams, Indianapolis, IN, 1971, p 69—70.

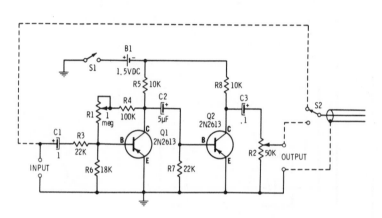

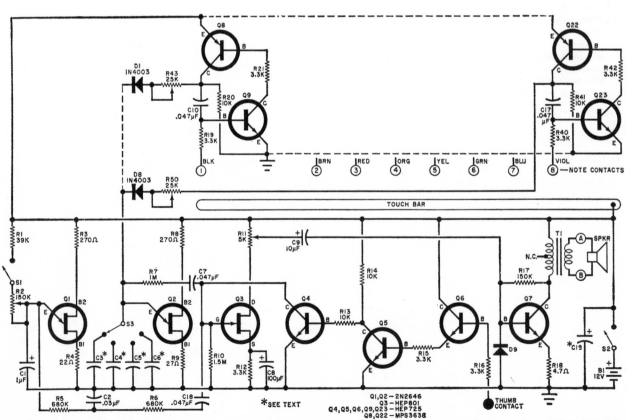

TOUCH-A-TONE—Produces range of four octaves, one below and three above middle C in organ-like tones. Has eight note contacts and octave selector switch, plus variable-rate tremolo control covering 6 to 36 beats per second. Only first and last note touch-switch circuits are shown; others are identical. Article tells how to combine capacitors to give values of 0.377 µF for C3, 0.19 µF for C4, 0.1 µF for C5, and 0.057 µF for C6, and gives octave tuning instructions. C19 is 250—500 µF 15-V electrolytic. T1 is 400:4-ohm output transformer.—C. D. Rakes, The Touch-A-Tone, *Popular Electronics*, March 1970, p 66—72 and 110.

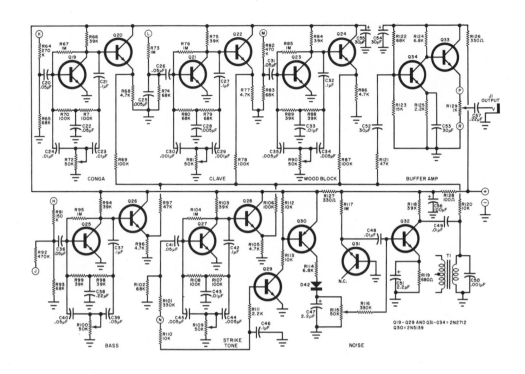

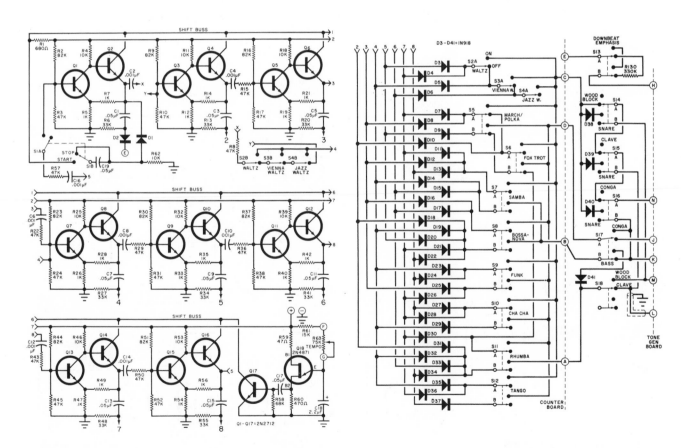

ELECTRONIC DRUMS—Uses variable-speed digital counter to give choice of 11 different beats from slow waltz to fast cha-cha, played by bass drum, wood block, clave, conga, snare drum, or a combination. Has variable tempo as well as volume control. Can be fed into any instrument power amplifier. Article covers construction in detail. All diodes are 1N918 or similar. Basic timing is provided by ujt relaxation oscillator Q18, which passes shift pulses through Q17 to eight-stage ring counter Q1-Q16. Outputs 2 through 8 of counter go to diode matrix arranged to pick up correct beats for selected rhythm. Switch-selected pulse groups go to tone generator through leads H-L. Snare drum simulation requires white noise generator Q31 along with Q27-Q30 and Q32 to get sound of striking of drumhead by stick and snares striking bottom drumhead. Article also gives power supply circuit.—J. S. Simonton, Jr., The Drummer Boy, *Popular Electronics*, July 1971, p 25–35, 40, and 97.

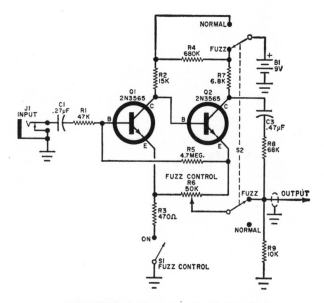

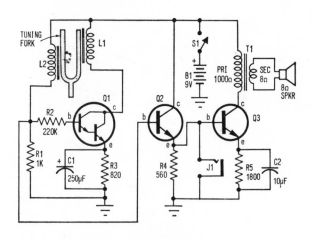

FUZZ TONE—Produces intentional distortion in audio range up to 3 kHz for electric guitar, to add stridency to single tones or simple chords. Use between guitar output and audio amplifier. S1, a part of R6, should be turned off when fuzz is not desired, to extend battery life.—A. Leo, Build Fuzz-Box for Electric Guitar, *Popular Electronics*, Feb. 1968, p 41—42 and 100—101.

TUNING-FORK AMPLIFIER—Eliminates passing of tuning fork among musicians in orchestra. Q1 and fork coils L1 and L2 form oscillator having frequency of fork, with Q2 and Q3 providing sufficient amplification of audio signal to drive speaker. Fork is usually 440 Hz (standard A). Q1 is 2N5306 Darlington. Q2 and Q3 are 2N5172. L1 and L2 are 1,000-ohm coils from 2,000-ohm headphones, removed with their magnets and mounted 1/16-inch from fork tines.—R. Michaels, Perpetual Motion Freq Standard, *Science and Electronics*, Oct.-Nov. 1969, p 79—82.

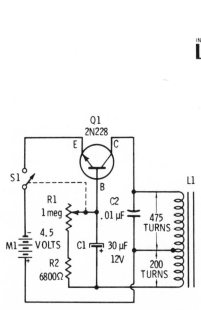

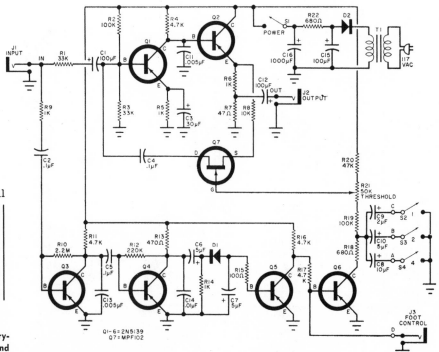

TRANSISTOR-RADIO METRONOME—Battery-operated circuit transmits metronome sound to radio near which it is placed. Tune radio to point having no station, and use R1 to adjust for desired timing. L1 has 675 turns No. 25 enamel random-wound on ¼-inch bolt, with tap at 475 turns.—R. M. Brown and T. Kneitel, "101 Easy Test Instrument Projects," Howard W. Sams, Indianapolis, IN, 1968, p 69—70.

VARIABLE ATTACK FOR GUITAR—Delay switches S2, S3, and S4 can be used singly or in combinations to give up to seven different delays for varying speed with which guitar sound is built up after string is plucked. Foot control switch permits grounding out delay selectively so some notes have zero delay. Fet Q7 controls negative feedback of high-gain audio amplifier Q1-Q2, and remaining transistors generate feedback control signal that acts with switches to determine amount of delay.—J. S. Simonton, Jr., Modify Your Electronic Guitar Sound, *Popular Electronics*, June 1970, p 53 and 58—61.

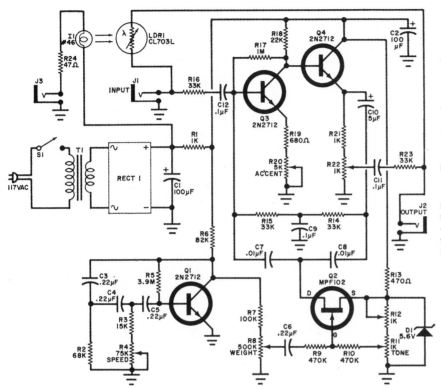

ORGAN UNDULATOR—Provides all-electronic simulator of effect produced in electronic organs by rotating massive diffuser in front of extra speaker to diffuse sound at gentle undulation of intensity at 8 to 12 Hz. Circuit shown is inserted between keyed tone generators and power amplifier of electronic organ. Uses wobbulated bandpass filter. Can also be adjusted to give vibrato and tremolo. Rectifier is Motorola MDA942A-1 or similar 1-A 50-piv full-wave rectifier bridge. T1 is 12.6-V 300-mA filament transformer. Light-dependent resistor is Clairex or similar. Footswitch is plugged into J3 to illuminate LDR1 and send input signal from J1 directly to J2 to bypass simulator.—J. S. Simonton, Jr., Leslie Effect Simulator, *Popular Electronics*, March 1971, p 51–53, 58–60.

7-TRANSISTOR THEREMIN—Played by moving hands back and forth near two metal plates to produce slowly changing musical notes at any desired volume, along with tremolo and vibrato. For serious music as well as amateur sound effects. Q2 is fixed-pitch oscillator, Q1 is variable-pitch, and Q4 is volume. Article covers construction and tuning. Fet output stage Q7 prevents loading of mixer Q3, while Q6 acts as voltage-variable resistor to control level of output. All coils are 50–300 μH adjustable.—L. E. Garner, Jr., Music a la Theremin, *Popular Electronics*, Nov. 1967, p 29–33 and 102–103.

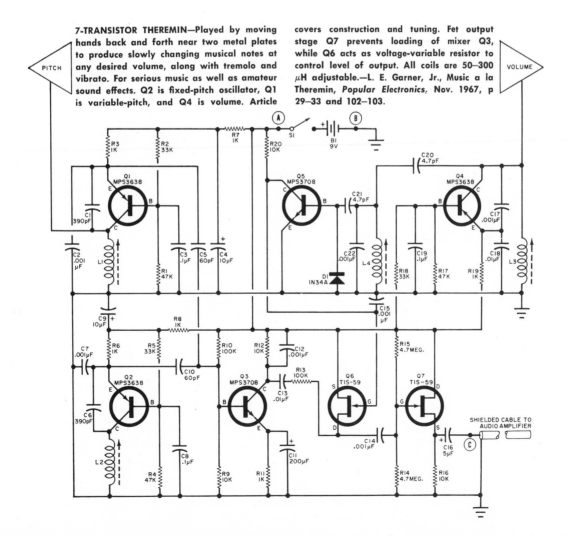

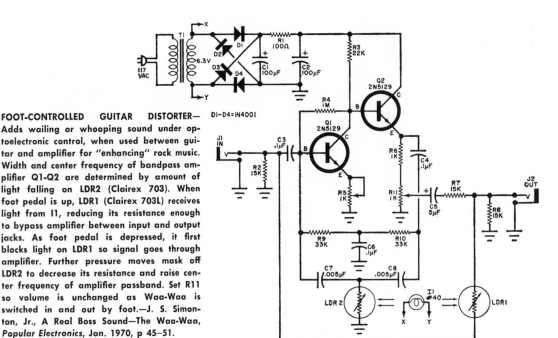

FOOT-CONTROLLED GUITAR DISTORTER— Adds wailing or whooping sound under optoelectronic control, when used between guitar and amplifier for "enhancing" rock music. Width and center frequency of bandpass amplifier Q1-Q2 are determined by amount of light falling on LDR2 (Clairex 703). When foot pedal is up, LDR1 (Clairex 703L) receives light from I1, reducing its resistance enough to bypass amplifier between input and output jacks. As foot pedal is depressed, it first blocks light on LDR1 so signal goes through amplifier. Further pressure moves mask off LDR2 to decrease its resistance and raise center frequency of amplifier passband. Set R11 so volume is unchanged as Waa-Waa is switched in and out by foot.—J. S. Simonton, Jr., A Real Boss Sound—The Waa-Waa, *Popular Electronics,* Jan. 1970, p 45–51.

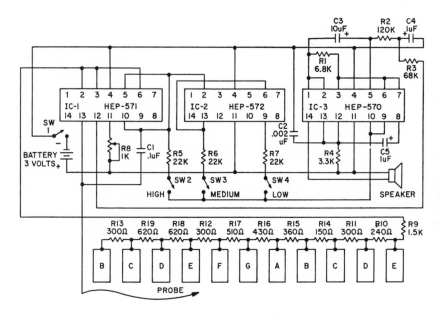

11 NOTES AND 3 VOICES WITH VIBRATO— Touching probe to one of eleven metal tabs places resistance in IC-1 tone oscillator that with C1 determines pitch or musical tone. IC-2 is dual J-K flip-flop, each used to divide input frequency exactly by 2. Closing SW3 gives half of input frequency, making all notes one octave lower. Closing SW4 makes notes two octaves lower. IC-3 is on continuously as vibrato oscillator and two-stage audio amplifier driving 8-ohm speaker. Article gives construction details.—D. Thorpe, Vibra-Tone, *Elementary Electronics,* July-Aug. 1972, p 59–61 and 95.

ACCENTED-BEAT METRONOME—Gives click rate of 50 to 160 clicks per minute according to setting of R2 in relaxation oscillator Q1, with certain clicks accented in each sequence to simulate down-beat at beginning of each musical bar. Can also be used in darkroom to tell photographer when to move on to next processing step. Dial of R13 is calibrated for accentuation every second beat, third beat, and so on up to every tenth beat. Q2 is power switching transistor driving speaker, and remaining four transistors determine accent rate.—Build Electronic Metronome With Accented Beat, *Popular Electronics,* Nov. 1968, p 59–61 and 66.

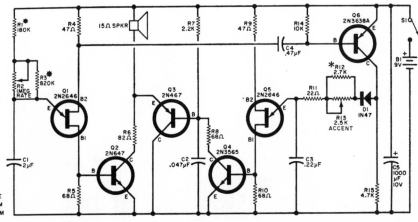

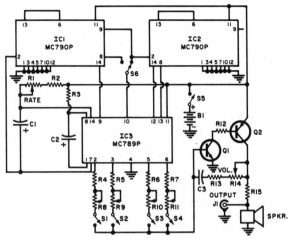

R1—10,000 ohm
linear-taper pot
R2—6800 ohm, 1/2 W res.
R3—3900 ohm, 1/2 W res.
R4, R5, R6, R7—100,000 ohm,
1/2 W res.
R8, R9, R10, R11—250,000 ohm
linear-taper pot
R12—390 ohm, 1/2 W res.
R13—22 ohm, 1/2 W res.
R14—500 ohm linear-taper pot
R15—4.7 ohm, 1/2 W res.
C1, C2—10 μF, 6 V elec.
capacitor

C3—0.02 μF capacitor
J1—RCA-type phono jack
S1, S2, S3, S4—S.p.s.t. sw.
(optional for R8, R9, R10, R11)
S5, S6—S.p.d.t. slide switch
B1—Two 1½-V "C" cells
in series
IC1, IC2—MC790P integrated
circuit (HEP 572)
IC3—MC789P integrated circuit
Q1—"n-p-n" transistor
(2N706, 2N4124, see text)
Q2—"p-n-p" transistor
(2N404, 2N4126, see text)

DIGITAL MUSIC—Uses two dual J-K flip-flops and one hex inverter to produce musical tones under control of six knobs. Any low-cost npn and pnp transistors may be substituted for those specified. Two transistors in IC3 are connected as mvbr whose frequency is controlled by R1. Square-wave output is fed through four flip-flops in IC1 and IC2, each dividing frequency by two, to get four different frequencies for routing through inverters in IC3 to pots R8, R9, R10, and R11 whose settings determine amount of each frequency added to others for feeding to vco Q1-Q2. R14 controls volume. Controls can be set to produce repetitive 16-note melodies.—B. Carlquist, Digital Music Maker, *Electronics World*, April 1971, p 60.

R-C ORGAN MASTER TONE—Complementary ujt oscillator Q1 and buffer Q2 deliver desired fixed tone above 2.5 kHz at parts cost of only about $1 with frequency stability within 3-cent band over 30 F temperature range. Q2 is 2N5172 and Q1 is GE D5K2 selected in three groups for interbase resistance value RBB which determines value of R4. Choose C1 and R3 for desired frequency; below 2.5 kHz, capacitor size will boost cost of parts somewhat. Report covers design of power supply and remainder of six-octave electronic organ using 3.52-kHz master oscillator and frequency dividers.—W. H. Sahm, III, "A Practical R-C Tone Generator System for Electronic Organs," General Electric, Semiconductor Products Dept., Auburn, NY, No. 90.90, 1970, p 2–4.

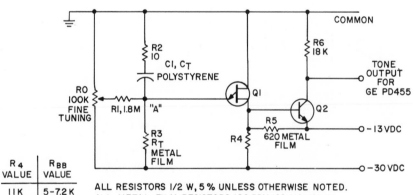

R_4 VALUE	R_{BB} VALUE
11K	5–7.2 K
18K	7–10.5K
36K	10–15 K

ALL RESISTORS 1/2 W, 5% UNLESS OTHERWISE NOTED.
ALL METAL FILM RESISTORS 100 PPM/°C MAXIMUM.

TOUCH-CONTACT VIBRATO—Five touch contacts and frequency-shifting lever in combination tone and vibrator generator permit creation of weird discordant musical sound effects or simple tunes. R4 varies vibrato oscillator over range of 2 to 20 Hz, to modulate output of tone generator oscillator. With thumb on T1, finger contact with T2, T3, T4, or T5 turns on associated tone generator. Exact frequency of selected tone is continuously variable, in three stages, from 130 to 2,000 Hz by R14 mechanically coupled to shift lever.—C. D. Rakes, Vibra Vox, *Elementary Electronics*, Jan.-Feb. 1970, p 31–35.

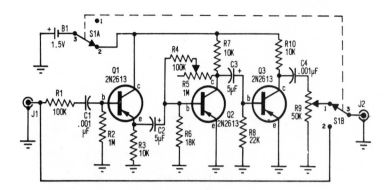

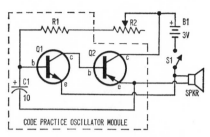

FUZZ GENERATOR—When connected between guitar and power amplifier, will amplify weak input signals with little distortion. On stronger signals, amount of distortion or fuzz effect increases with signal strength to give what is considered by some to be deep,

throaty saxophone sound for rock music. R5 is fuzz control and R9 volume control. Q1 operates as class C amplifier to give way-out sound effect economically.—O. K. Ryan, Fuzz-A-Tort, *Elementary Electronics*, Jan.-Feb. 1968, p 41–44.

METRONOME—Circuit shows how code practice oscillator module such as Lafayette 19T1513 can be modified to produce clicks in speaker at slower rate required for metronome. C1 in module is replaced with 10-μF 6-V electrolytic. Resistor in module is removed and new value determined experimentally with 200K pot; this gave 10K for R1 and 100K for R2 in module used.—S. Daniels, Cheapy Bippy, *Elementary Electronics*, July-Aug. 1969, p 61–62.

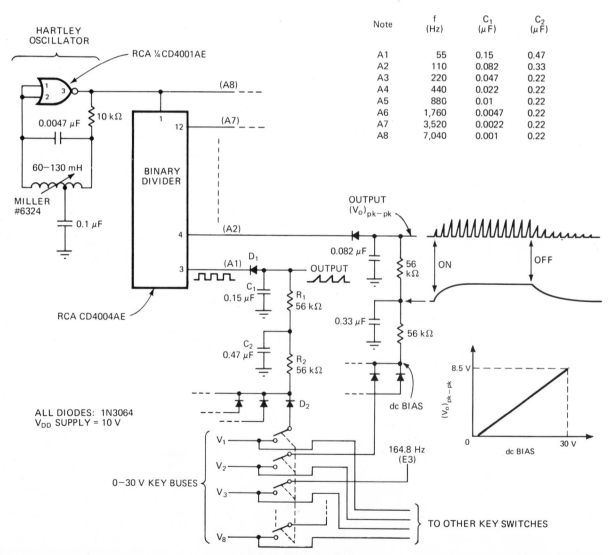

Note	f (Hz)	C_1 (μF)	C_2 (μF)
A1	55	0.15	0.47
A2	110	0.082	0.33
A3	220	0.047	0.22
A4	440	0.022	0.22
A5	880	0.01	0.22
A6	1,760	0.0047	0.22
A7	3,520	0.0022	0.22
A8	7,040	0.001	0.22

C-MOS LOGIC FOR ORGAN—Circuit shown generates eight frequencies of note A by changing values of C1 and C2 as given in table. Each of other six notes and five sharps requires its own Hartley oscillator and binary

divider network to generate all 96 organ frequencies. Diode gate at each divider output converts square wave to sawtooth to get necessary even harmonics. To create desired organ voices by summing several harmoni-

cally related tones, each gate output must be first filtered down to simplest voice of organ, which is flute.—R. Woody, C-MOS Sums Up Tones for Electronic Organ, *Electronics*, Nov. 20, 1972, p 114–116.

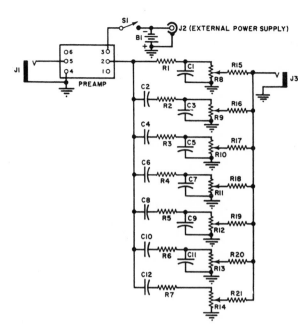

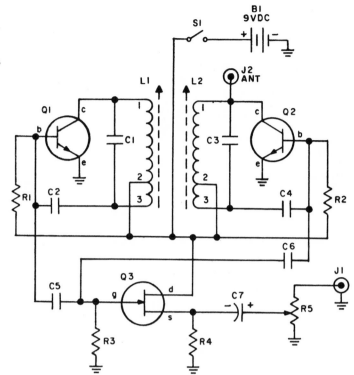

R1, R7—820,000 ohm res.
R2, R3, R4, R5, R6, R15, R16, R17,
 R18, R19, R20, R21—470,000 ohm
 res.
R8, R9, R10, R11, R12, R13, R14—
 500,000 ohm pot
C1—0.0015 μF capacitor
C2, C3—0.001 μF capacitor
C4, C5—500 pF capacitor

C6, C7—250 pF capacitor
C8, C9—120 pF capacitor
C10, C11—68 pF capacitor
C12—25 pF capacitor
S1—S.p.s.t. switch
J1, J3—Phone jack
J2—Phono jack
B1—9-volt battery

GUITAR TONE CONTROL—Uses seven individ-ually controlled R-C filters having center frequencies one octave apart, at 150, 300, 600, 1,200, 2,400, 4,800, and 9,600 Hz. Gives much more control over guitar sound than can be obtained with conventional bass and treble controls. Requires preamp with 30 dB gain to boost signal back to original level. Lowest control is low-pass for handling 80-Hz lowest note of guitar, and 9,600-Hz filter is high-pass for harmonics going beyond audibility.—I. E. Ashdown, "Multitone" Guitarist's Tone Control, *Electronics World*, Jan. 1971, p 68–69.

SQUEAL MUSIC—Simple solid-state modern version of Theremin generates musical sounds that vary in pitch as hand is moved toward or away from 12-inch length of stiff wire projecting out of small box. Can be fed into any audio amplifier. R5 can be mounted on foot pedal to serve as swell-to-great control. Article gives construction and adjustment details. L is Lafayette 32-41080 loopstick. B1 is 9 V, C1 and C3 270 pF, C2 and C4 300 pF, C5 and C6 100 pF, C7 10 μF, Q1 and Q2 HEP50 or 2N706, Q3 HEP801, R1 and R2 100K, R3 3.9 meg, R4 5.6K, and R5 10 meg. —S. Daniels, Ultra Simple Theremin, *Elementary Electronics*, May-June 1972, p 67–69.

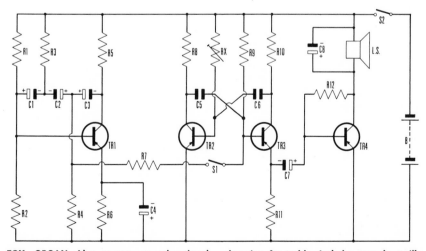

R1 82kΩ	R5 4·7kΩ	R9 12kΩ
R2 10kΩ	R6 1kΩ	R10 2·2kΩ
R3 4·7kΩ	R7 100kΩ	R11 100Ω
R4 4·7kΩ	R8 2·2kΩ	R12 39kΩ

Rx see text
All resistors ¼W, 10%

C1 2μF electrolytic	C5A 0·25μF (base circuit)
C2 2μF electrolytic	C6 0·1μF (treble circuit)
C3 2μF electrolytic	C6A 0·25μF (base circuit)
C4 250μF electrolytic	C7 2μF electrolytic
C5 0·1μF (treble circuit)	C8 2μF electrolytic

TOY ORGAN—Also serves as educational project. Designed to be built as virtually two complete organs, to permit playing bass accompaniment on one and treble on another. Circuit shown is for half of organ, either treble or bass, with only capacitor values changing for treble. Includes tremolo oscillator TR1, giving 6.8 Hz. Note generator is mvbr TR2-TR3, feeding output stage TR4, which delivers about 50 mW to 80-ohm speaker when using 9-V battery supply. Rx, shown for simplicity as variable resistor, must be constructed as number of preset pots which are shorted out by contacts on organ keys. TR1 is ACY17; other transistors are ACY20 or ACY21.—"An Electronic Organ," *Educational Electronic Experiments*, No. 12, Mullard, London.

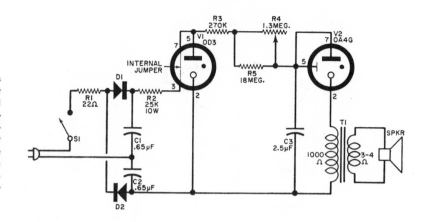

OLD RADIO BECOMES METRONOME—Chassis and speaker of junked tube-type broadcast radio readily accommodate gas tubes V1 and V2 along with transformerless voltage-doubler power supply, connected to give metronome continuously variable between at least 40 and 208 beats per minute. D1 and D2 are selenium rectifiers having five or more plates each, or 2-A 250-V silicon diodes.—K. Greif, Build a Cheapie Metronome, *Popular Electronics*, Jan. 1969, p 67–69.

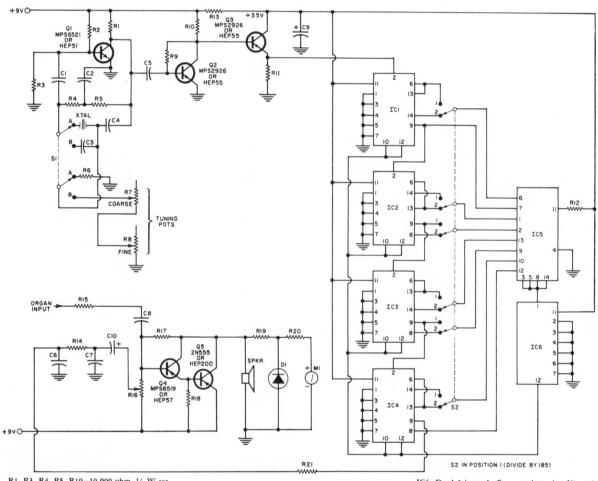

R1, R3, R4, R5, R10—10,000 ohm, ½ W res.
R2—2.7 megohm, ½ W res.
R6—1200 ohm, ½ W res.
R7—5000 ohm pot
R8—100 ohm pot
R9—510,000 ohm, ½ W res.
R11—2400 ohm, ½ W res.
R12—390 ohm, ½ W res.
R13—33 ohm, 2 W res.
R14, R21—15,000 ohm, ½ W res.
R15—100,000 ohm, ½ W res.
R16—50,000 ohm pot
R17—1 megohm, ½ W res.
R18—300 ohm, ½ W res.
R19—620 ohm, ½ W res.

R20—100 ohm, ½ W res.
C1, C5—0.1 μF capacitor
C2—820 pF capacitor
C3, C4—470 pF capacitor
C6, C7—0.02 μF capacitor
C8—1.0 μF capacitor
C9—100 μF, 6 V elec. capacitor
C10—5 μF, 6 V elec. capacitor
D1—Silicon rectifier (1N4001 or HEP154)
IC1, IC2, IC3, IC4—Dual J-K flip-flop IC package (MC790P or HEP572)
IC5—Quad 2-input gate "nand/nor" IC package (MC724P or HEP570)

IC6—Dual 2-input buffer, non-inverting IC package (MC788P or HEP571)
M1—0-1 mA meter
S1—D.p.d.t. switch
S2—6-pole, d.t. switch
Xtal.—81.4-kHz crystal
Spkr.—40-45 ohm speaker
Q1—Silicon "n-p-n" transistor (MPS6521 or HEP51)
Q2, Q3—Silicon "n-p-n" epitaxial, plastic-encapsulated transistor (MPS2926 or HEP55)
Q4—Silicon "p-n-p" transistor (MPS6519 or HEP57)
Q5—Germanium "p-n-p" power transistor (2N555 or HEP200)

DIGITAL TUNER FOR ORGAN—Accurate digital musical instrument tuner uses inexpensive IC's and transistors. With S1 in position A, oscillator Q1 is crystal-controlled at 81.4 kHz. This frequency is divided by 185 in IC divider when S2 is in position 2, to give accurate standard A output of 440 Hz. Changing S2 to position 2 gives division by 196; when given frequency is divided by 185 and then by 196, resulting frequencies are almost exactly 1 semitone apart. With S1 in position B, for variable tuning, octave from 220 to 440 Hz can be covered when divided by 185 or 196. Amplifier Q4-Q5 drives speaker to permit tuning for zero beat with organ input.—F. Maynard, The Electronic Strobotuner, *Electronics World*, March 1970, p 74–77.

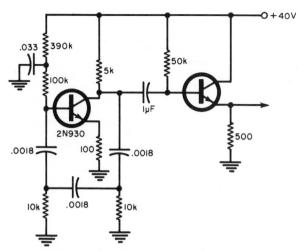

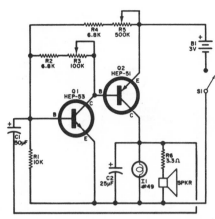

5 KHZ FOR DRUMS—Article gives design procedure, and tells how to adapt circuit for such electronic music as sounds of bongo drum, tom-tom, and bass drum. Both transistors can be 2N930.—J. L. Turino, Designing a Phase-Shift Oscillator, *Electronics World*, July 1971, p 34–35.

VISIBLE METRONOME—Light flashes in synchronism with audible tick from 3.2-ohm speaker. R5 controls minimum beat rate, and R3 increases this. Calibrate with stopwatch. —R. J. Fajardo, Reader's Circuit, *Popular Electronics*, May 1968, p 76–77.

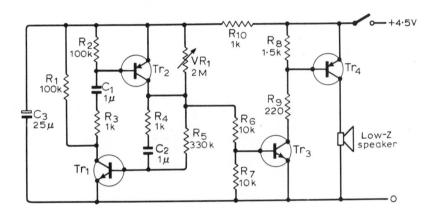

30–240 BEATS PER MIN—Complementary-pair mvbr Tr1-Tr2 generates timing pulses, with both transistors either conducting or cut off together. Conduction time of about 10 ms generates tick and off-time (0.25–2 s) interval between ticks. Tr3-Tr4 form simple low-quality amplifier driving speaker, since distortion is immaterial. Tr4 should be medium or high-power germanium pnp transistor capable of switching about 1 A, such as AC128 or OC83. Other transistors can be any medium or high-gain silicon planar types.—D. T. Smith, Electronic Metronome, *Wireless World*, Jan. 1970, p 35.

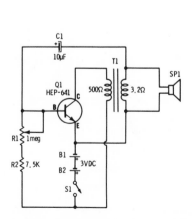

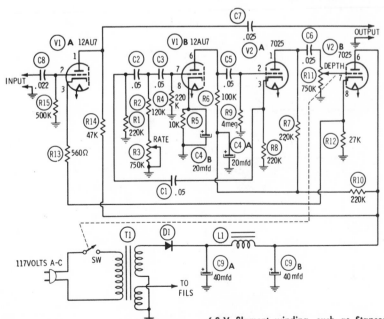

WIDE-RANGE METRONOME—R1 adjusts click rate of blocking oscillator Q1 over range of 3 to 300 per minute. Dial on shaft of R1 can be calibrated in beats per minute by comparison with standard metronome. Can also be used as audible interval timer in photography darkroom.—H. Friedman, "99 Electronic Projects," Howard W. Sams, Indianapolis, IN, 1971, p 70–71.

VIBRATO—Inserted between preamp and amplifier or recorder to add considerable vibrato depth to either speech or music. R3 controls rate of vibrato. Operates from a-c line, using 125-v isolation transformer having 6.3-V filament winding, such as Stancor PS-8415. D2 can be either 7025 or 6067. D1 is 130-V 20-mA selenium rectifier. L1 is 50-mA 3.5-H filter choke.—R. Brown and T. Kneitel, "101 Easy Audio Projects," Howard W. Sams, Indianapolis, IN, 1971, p 78–79.

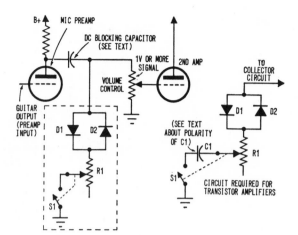

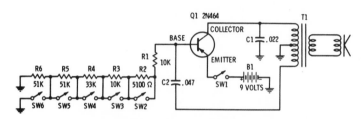

GUITAR FUZZ—For vacuum-tube amplifier, diodes connected as shown will clip that part of signal waveform exceeding about 0.7 V, depending on setting of R1, to provide fuzz effect. For transistor amplifier, C1 should be 100 μF, connected with same polarity as point to which D1 and D2 are connected. Diodes can be any low-voltage silicon, such as Lafayette 19T6001. R1 is 10K pot with on-off switch.—H. Friedman, 97-Cent Hard-Rock Fuzz Box, *Science and Electronics*, Oct.-Nov. 1969, p 67–70.

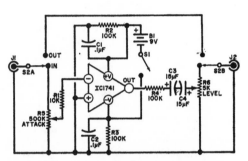

GUITAR FUZZ—Triggered square-wave output of opamp provides good fuzzy sound for electronic guitar. IC is connected as modified comparator that produces output only when signal applied to its inverting input is above certain level set by attack pot R5. S2 bypasses fuzz when desired. May also be used with organs and other instruments. Decay is much faster than conventional fuzz, due to triggering action.—C. Anderton, Build the Optimum Fuzz Adapter, *Popular Electronics*, July 1972, p 34–35.

TOY MUSIC BOX—All switches are unlabeled, including on-off switch SW1. Each switch gives different musical tone, and same switch can give variety of different tones depending on which other switches are also closed. Use spst toggle switches. T1 is Argonne AR-170 or equivalent output transformer with matching speaker. Child genius can even figure out how to play simple melodies.—R. M. Brown and T. Kneitel, "49 Easy Transistor Projects," Howard W. Sams, Indianapolis, IN, 1972, p 46–47.

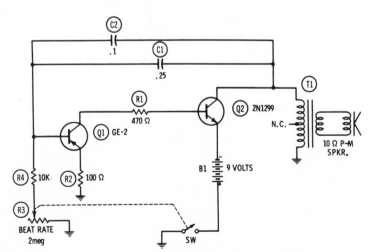

TRANSISTOR METRONOME—R3 controls rate of beats heard from speaker. Output transformer T1 has 2K primary and 10-ohm secondary.—R. Brown and T. Kneitel, "101 Easy Audio Projects," Howard W. Sams, Indianapolis, IN, 1971, p 156–157.

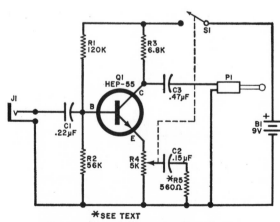

GUITAR TREBLE BOOST—Provides 20 dB boost at 3,000 Hz, for special effects or to highlight melody section. For 6 dB more boost, short R5. Use between guitar and preamp. Experiment to find most pleasing positions of pot R4.—A. Leo, Treble Boost for Your Guitar, *Popular Electronics*, Dec. 1969, p 59–60 and 92.

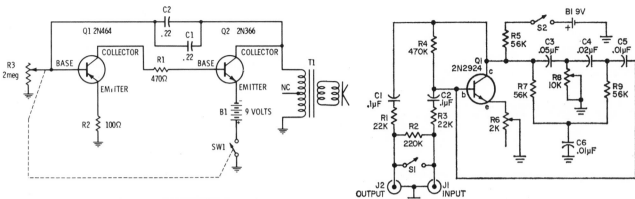

SLEEP-INDUCING METRONOME—R3 determines rate at which clicks or clacks are generated, for learning to play piano or for simulating patter of raindrops as sleep aid. T1 is Argonne AR-119 or equivalent output transformer driving matching p-m speaker.—R. M. Brown and T. Kneitel, "49 Easy Transistor Projects," Howard W. Sams, Indianapolis, IN, 1972, p 38–39.

GUITAR BASS BOOST—Connects between guitar and its power amplifier, to give choice of bass, treble, or midrange boost with single knob of R8. Circuit is phase-shift oscillator with gain lowered by trimmer pot R6 to point where transistor is amplifying rather than oscillating.—S. Daniels, U-Pick-It, *Elementary Electronics*, May-June 1971, p 59–60 and 102.

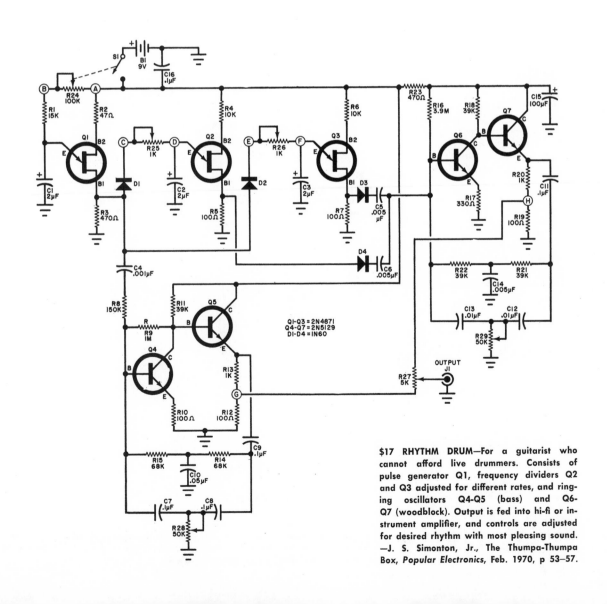

$17 RHYTHM DRUM—For a guitarist who cannot afford live drummers. Consists of pulse generator Q1, frequency dividers Q2 and Q3 adjusted for different rates, and ringing oscillators Q4-Q5 (bass) and Q6-Q7 (woodblock). Output is fed into hi-fi or instrument amplifier, and controls are adjusted for desired rhythm with most pleasing sound.—J. S. Simonton, Jr., The Thumpa-Thumpa Box, *Popular Electronics*, Feb. 1970, p 53–57.

CHAPTER 33
Filter Circuits—Active

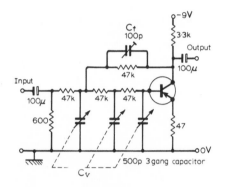

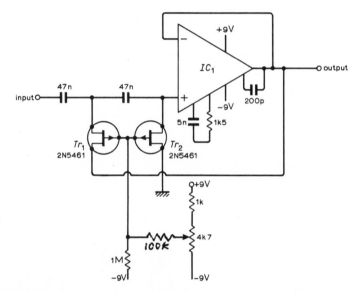

10–200 KHZ LOW-PASS—Ganged capacitors change cutoff frequency; each may be shunted by fixed capacitor to change range, at sacrifice of tuning range. Transistor (pnp) is not critical.—R. Cartwright, Variable Frequency Low-Pass Filter, *Wireless World*, Sept. 1969, p 409.

HIGH-PASS—Cutoff (3 dB down) occurs at 12 dB per octave from about 100 Hz to 10 kHz. Uses standard opamp mvbr such as μA709 or SN72709, adapted for voltage control by using matched fet's to give log frequency-voltage characteristic.—D. G. Malham, Voltage-Controlled High-Pass Filter, *Wireless World*, July 1972, p 349.

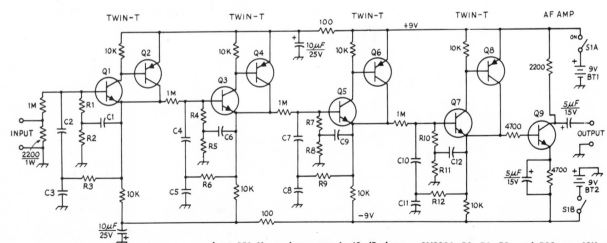

CASCADED TWIN-T C-W—Used to improve receiver selectivity for c-w. For unity-gain twin-T R-C filter sections, 6-dB bandwidth is about 150 Hz, and response is 40 dB down at 420 and 1,120 Hz. Audio amplifier after filter offsets small insertion loss of filter and permits low-level operation of receiver audio circuits that drive filter. Npn transistors (Q1, etc) are 2N3904, and pnp transistors are 2N3906. R3, R6, R9, and R12 are 68K, and other unmarked resistors are 100K. C1, C6, C9, and C12 are 0.005 μF, and other capacitors are 0.002 μF.—W. Hayward, An RC-Active Audio Filter for CW, *QST*, May 1970, p 51–54.

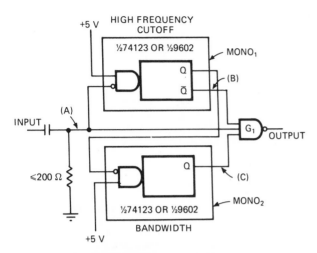

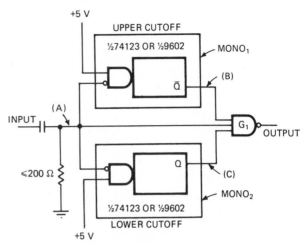

ADJUSTABLE-BANDWIDTH DIGITAL—Dual mono and three-input NAND gate of any type provide flexibility for digital frequency selection. Upper mono sets high-frequency cutoff and lower mono sets filter bandwidth. Input pulse reaches output only when upper mono has timed out and lower mono is high. —A. M. Volk, Two-IC Digital Filter Varies Passband Easily, *Electronics*, Feb. 15, 1973, p 106.

DIGITAL BANDPASS—Dual mono and three-input NAND gate of any type provide completely adjustable cutoff frequencies and excellent frequency stability, independent of input duty cycle. Output pulse length at first mono is set to period of highest frequency of interest while that for second mono is set for lowest frequency wanted.—A. M. Volk, Two-IC Digital Filter Varies Passband Easily, *Electronics*, Feb. 15, 1973, p 106.

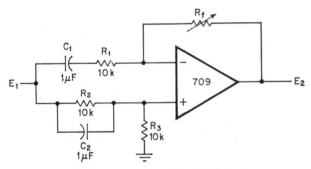

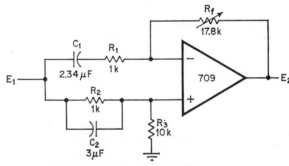

ADJUSTABLE ACTIVE R-C NETWORK—Rf moves zero-pair locations anywhere in complex plane. Circuit is easily modified for specific filter applications, and can also be used as compensating network in feedback systems. Article gives equation for transfer function.—R. D. Guyton, Active RC Network Has Two Movable Zeros, Fixed Poles, *Electronics*, April 26, 1971, p 58.

60-HZ NOTCH FILTER—Provides 70-dB rejection. Adjusting Rf moves zero-pair locations. Article gives equation for transfer function. —R. D. Guyton, Active RC Network Has Two Movable Zeros, Fixed Poles, *Electronics*, April 26, 1971, p 58.

500-HZ DIGITAL FILTER—Four J-K flip-flops with two NAND gates form self-starting, self-correcting ring counter that drives four-position sequential switch to make up main digital filter occupying only fraction of space needed by equivalent passive filter. Low-pass filter R1-C1 determines response curve; R2, C2, and Q2 form active low-pass filter that removes sampling ripple. With R1 51K, R2 5.1K, C1 2.2 μF, and C2 0.68 μF, measured 3-dB bandwidth at center (clock) frequency of 500 Hz is 2 Hz and equivalent skirt rolloff is 6 dB per octave.—M. H. Acuna, Digital Filter Replaces Bulky Components, *The Electronic Engineer*, Jan. 1969, p 95.

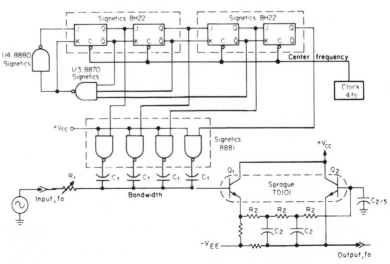

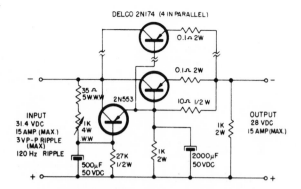

250:1 RIPPLE FILTER—Electronic filter supplements conventional L-C filter. If ripple is initially 250 mV p-p, circuit will bring it down to about 1 mV p-p for 28-V output.—"Electronic Ripple Filter," Delco, Kokomo, IN, July 1971, Application Note 28.

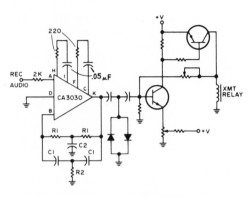

TONE DECODER—Twin-T active filter simplifies circuit for tone signaling and decoding. Filter values depend on filter frequency; R1 = 2R2, C1 = 0.5C2, and F = 1/6.28R1C1. —F. D. Williams, Active Filter Design and Use—Part IV, 73, Oct. 1972, p 43–44 and 46–47.

1,370-HZ SINGLE-AMPLIFIER—Provides near-optimum performance beyond 40 dB attenuation, with adjustment required on only one capacitor for alignment. Designed to give normalized three-pole low-pass Chebyshev transfer function with 1 dB ripple—R. E. Cooper and C. O. Harbourt, Active Network Synthesis for Reduced Sensitivity to Parameter Variations, *Frequency*, Jan.-Feb. 1967, p 34–35.

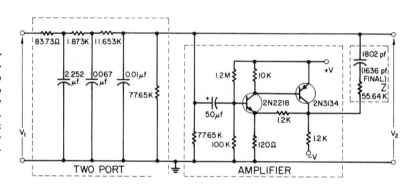

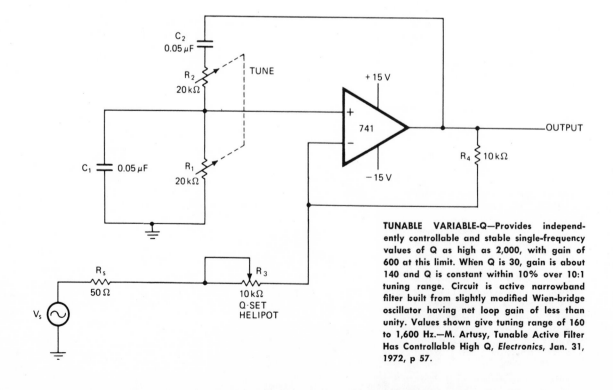

TUNABLE VARIABLE-Q—Provides independently controllable and stable single-frequency values of Q as high as 2,000, with gain of 600 at this limit. When Q is 30, gain is about 140 and Q is constant within 10% over 10:1 tuning range. Circuit is active narrowband filter built from slightly modified Wien-bridge oscillator having net loop gain of less than unity. Values shown give tuning range of 160 to 1,600 Hz.—M. Artusy, Tunable Active Filter Has Controllable High Q, *Electronics*, Jan. 31, 1972, p 57.

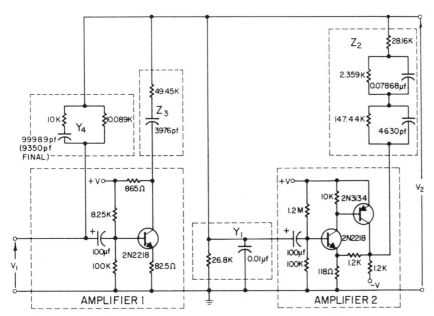

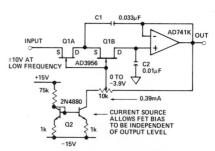

1,370-HZ TWO-AMPLIFIER—Provides near-optimum performance beyond 40 dB attenuation, with adjustment required on only one capacitor for alignment. Designed to give normalized three-pole low-pass Chebyshev transfer function with 1 dB ripple.—R. E. Cooper and C. O. Harbourt, *Active Network Synthesis for Reduced Sensitivity to Parameter Variations, Frequency,* Jan.-Feb. 1967, p 34–35.

ADJUSTABLE LOW-PASS—Cutoff frequency is adjustable between 1 and 10 kHz, with attenuation of 10 dB per octave, for filtering out small a-c noise components riding on slowly varying signal. Harmonic distortion of about 3% near cutoff is proportionately less at lower frequencies. Requires low-impedance source.—Choosing and Using N-Channel Dual J-Fets, *Analog Dialogue,* Dec. 1970, p 4–9.

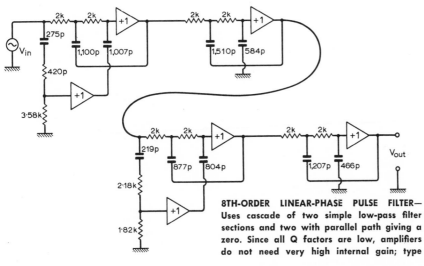

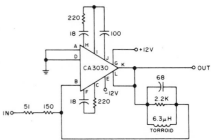

10-MHZ BANDPASS—Amplifier circuit using active filter provides gain of 20 dB at center frequency of 10 MHz, with 3-dB response about 1 MHz above and below. Effective Q is about 10.—F. D. Williams, *Active Filter Design and Use—Part IV, 73,* Oct. 1972, p 43–44 and 46–47.

8TH-ORDER LINEAR-PHASE PULSE FILTER—Uses cascade of two simple low-pass filter sections and two with parallel path giving a zero. Since all Q factors are low, amplifiers do not need very high internal gain; type 741 can be used. Intended for pulse transmission applications.—F. E. J. Girling and E. F. Good, *Active Filters, Wireless World,* April 1970, p 183–188.

3RD-ORDER PARALLEL-T LADDER—Provides bandpass response while using minimum number of opamps. Middle stage is sign-inverting and outer stages are not, for closing feedback loops in correct sense without need for separate inverting or adding stages. Q factors of tuned circuits are highly dependent on accurate balance in twin-T networks. All opamps can be type 741. Article is 14th of series dealing with design of all types of active filters.—F. E. J. Girling and E. F. Good, *Active Filters, Wireless World,* Oct. 1970, p 505–510.

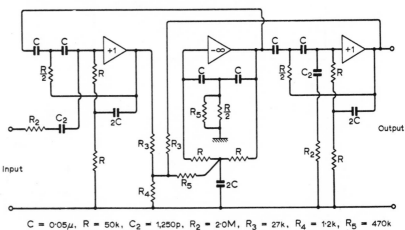

C = 0·05μ, R = 50k, C₂ = 1,250p, R₂ = 2·0M, R₃ = 27k, R₄ = 1·2k, R₅ = 470k

VARIABLE-VOLTAGE LOW-PASS—Level-shift circuit inserted in integrator feedback loop of IC multiplier permits d-c control of break point for filter. Rolloff frequency varies linearly with control voltage V_c applied to multiplier. For 0.5 V control, rolloff f_2 is 150 Hz, increasing to 2 kHz for 5 V.—"The μA795, A Low-Cost Monolithic Multiplier," Fairchild Semiconductor, Mountain View, CA, No. 211, 1971, p 3–4.

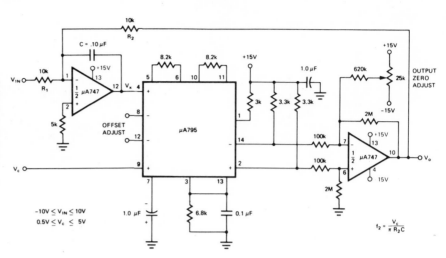

$$-10V \leq V_{IN} \leq 10V$$
$$0.5V \leq V_c \leq 5V$$

$$f_2 = \frac{V_c}{\pi R_2 C}$$

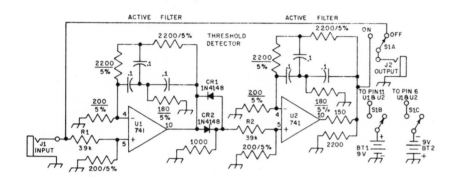

C-W FILTER—Uses active filters ahead of and following threshold detector to give extremely sharp filter skirts, without harmonics and discontinuities of threshold switching. For operating frequency of 1.1 kHz, 6-dB bandwidth is only 62 Hz.—C. B. Andes, Threshold Detectors in a CW Audio Filter, QST, Dec. 1971, p 20–21.

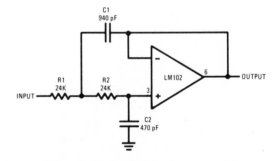

10-KHZ CUTOFF LOW-PASS—Uses IC voltage follower to provide filter characteristics of two isolated R-C filter sections, with buffered low-impedance output. Attenuation is 12 dB at twice cutoff frequency. Use silvered mica capacitors.—R. J. Widlar, "Integrated Voltage Follower," National Semiconductor, Santa Clara, CA, 1968, AN-5.

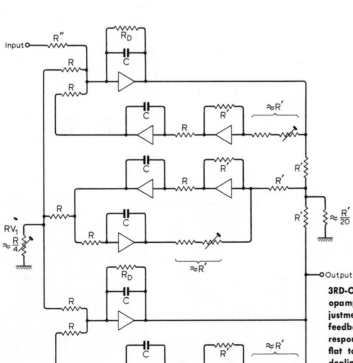

$$C = 0.01\mu, \quad R = R' = R'' = 100k, \quad R_D = 16R$$

3RD-ORDER BANDPASS—Uses type 741 opamps, connected to allow simultaneous adjustment of loop gains of two intermeshing feedback loops. Increasing coupling changes response from single-peak through maximally flat to overcoupled. Article is 14th of series dealing with design of all types of active filters.—F. E. J. Girling and E. F. Good, Active Filters, *Wireless World*, Oct. 1970, p 505–510.

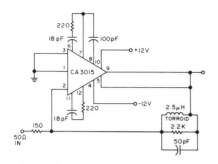

14 MHZ WITH 20 DB GAIN—Bandwidth (3 dB down) is less than 1 MHz. Useful as preamp or in other applications requiring active filter with wide response.—F. D. Williams, Active Filter Design and Use—Part IV, 73, Oct. 1972, p 43–44 and 46–47.

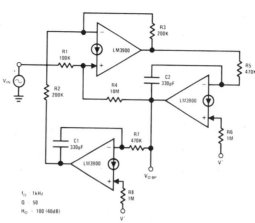

1-KHZ THREE-OPAMP BANDPASS—Known as bi-quad circuit because it can produce quadratic transfer function in both numerator and denominator. Gives Q of 50 and gain of 100. Report gives complete design procedure.—T. M. Frederiksen, W. M. Howard, and R. S. Sleeth, "The LM3900—A New Current-Differencing Quad of ± Input Amplifiers," National Semiconductor, Santa Clara, CA, 1972, AN-72, p 17–18.

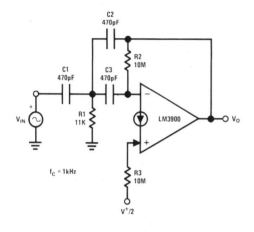

1-KHZ HIGH-PASS—Uses Norton current-differencing opamp. Report gives complete design procedure for obtaining desired passband gain and corner frequency. Values shown give unity gain.—T. M. Frederiksen, W. M. Howard, and R. S. Sleeth, "The LM3900—A New Current-Differencing Quad of ± Input Amplifiers," National Semiconductor, Santa Clara, CA, 1972, AN-72, p 14.

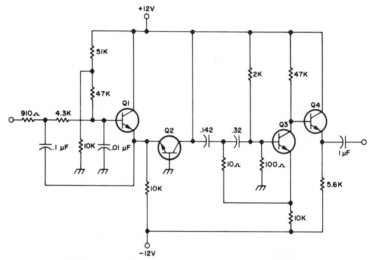

BANDPASS—Uses MPS6520 transistors. Combines characteristics of low-pass and high-pass filters having 2.5-kHz cutoffs. Requires some peaking at 2.5 kHz, achieved by adjusting Q of circuit. Response is reasonably flat for 2 octaves, with good slopes to —15 dB.—F. D. Williams, Active Filter Design and Use—Part II, 73, July 1972, p 97–106.

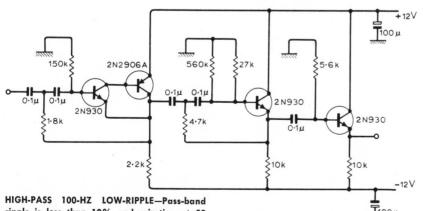

HIGH-PASS 100-HZ LOW-RIPPLE—Pass-band ripple is less than 10% and rejection at 50 Hz is better than 35 dB. Article gives design procedure and tables of coefficients for both low-pass and high-pass.—M. Bronzite, Simple Active Filters, Wireless World, March 1970, p 117–119.

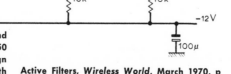

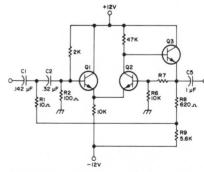

HIGH-PASS WITH 2.5-KHZ CUTOFF—Uses MPS6520 transistors. Article gives design equations for changing cutoff frequency and shape of response below 3-dB-down point.—F. D. Williams, Active Filter Design and Use—Part II, 73, July 1972, p 97–106.

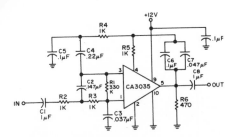

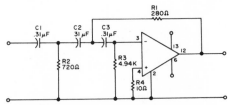

1.5-KHZ LOW-PASS 3-POLE—Provides sharper rate of attenuation beyond cutoff than 2-pole filter. Bias is obtained from 12-V line.—F. D. Williams, Active Filter Design and Use—Part III, 73, Sept. 1972, p 76–87.

1.5-KHZ HIGH-PASS—Uses RCA CA3030 IC linear amplifier. Insertion loss is about 70 dB. Overall response and attenuation slope are both good.—F. D. Williams, Active Filter Design and Use—Part III, 73, Sept. 1972, p 76–87.

1.5-KHZ HIGH-PASS 3-POLE—Arrangement of filter network resistors to form parallel circuits that do not interfere with biasing improves performance sufficiently to give insertion gain of 5 dB above cutoff.—F. D. Williams, Active Filter Design and Use—Part III, 73, Sept. 1972, p. 76–87.

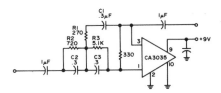

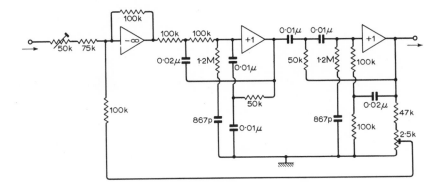

159-HZ 1/3RD-OCTAVE 2ND-ORDER—Uses type 741 opamps in balanced parallel-T network. Untuned input stage puts necessary sign reversal into loop and permits adjusting overall gain without affecting tuned stages. Peak gain is about 18.—F. E. J. Girling and E. F. Good, Active Filters, *Wireless World*, Oct. 1970, p 505–510.

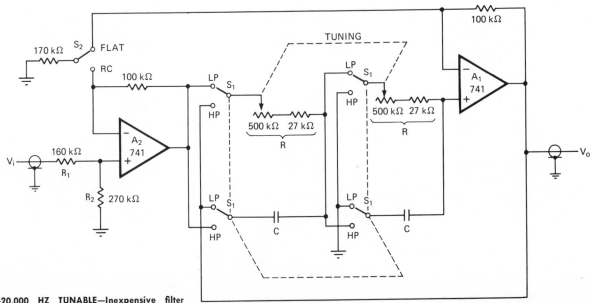

20–20,000 HZ TUNABLE—Inexpensive filter provides switch-selectable low-pass or high-pass responses with either Bessel or Butterworth characteristics. Cutoff is continuously tunable with R over entire frequency range.

Skirt rolloff is 40 dB per decade. Switch interchanges resistors and capacitors to change from low-pass to high-pass. S2 changes characteristics. Will cover 20–200 Hz if C is 0.02

μF, 200–2,000 Hz with 0.002 μF, and 2,000–20,000 Hz with 0.0002 μF.—P. Cushing, Tunable Active Filter Has Switchable Response, *Electronics*, Jan. 4, 1973, p 104.

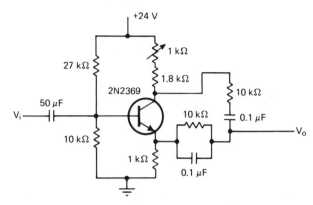

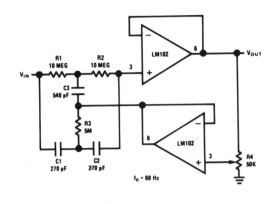

60-DB NOTCH—Combination of phase splitter and Wien-bridge network forms active notch filter that can be tuned with ganged capacitors or resistors to move notch from subaudio frequencies up to several hundred kHz. With standard two-gang variable capacitor for bridge, notch is greater than 45 dB over tuning range of 8 to 200 kHz.—D. DeKold, Wien Bridge in Notch Filter Gives 60 dB Rejection, *Electronics*, Aug. 28, 1972, p. 80.

60-HZ ADJUSTABLE-Q NOTCH—R4 at output of second voltage follower for twin-T R-C filter permits varying Q from 0.3 to 50 for varying width of 60-Hz notch. Notch depth depends on component match; use of 0.1% resistors and 1% capacitors minimizes trimming needed for 60-dB notch.—"High Q Notch Filter," National Semiconductor, Santa Clara, CA, 1969, LB-5.

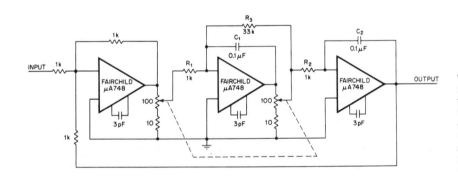

CONSTANT-Q TUNABLE BANDPASS—Ganged pots cover resonant frequency range of 150 to 1,500 Hz, with Q varying less than 5% from value of 30 over range. Operating frequency range can be shifted by changing values of C1 and C2 while still giving decade range of tuning.—R. Melen, Tunable Active Filter Maintains Constant Q, *Electronics*, July 19, 1971, p. 72.

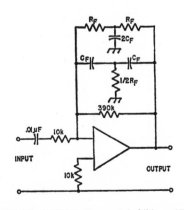

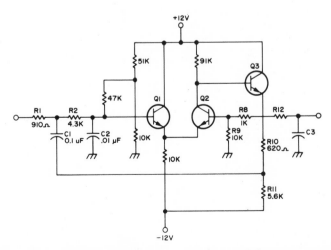

1,200-HZ TWIN-T—Uses Fairchild μA741 opamp. Voltage gain is about 40 at resonance, and gain drops to unity at 0.5 and 2.5 kHz. RF is 2.7 K and CF is 0.05 μF.—J. M. Pike, The Operational Amplifier, *QST*, Sept. 1970, p 54–57.

LOW-PASS WITH 2.5-KHZ CUTOFF—Provides Butterworth response. Article gives equations for changing C1, C2, R1, and R2 to get other cutoff frequencies. Transistors are MPS6520.—F. D. Williams, Active Filter Design and Use—Part II, *73*, July 1972, p 97–106.

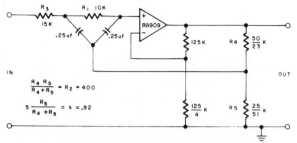

$$\theta(s) = 5[(s/2000)^2 + s/5000 + 1]/[(s/1000)^2 + s/1000 + 1]$$

ACTIVE R-C—Article covers synthesis of active filters, with mathematical analysis. In example given, dynamic input impedance exceeds 100 meg, while output impedance is completely negligible. Thermal stability depends on effective resistance of passive network. Circuit is realization of mathematical equation given. Designed for frequency band from d-c to a few kHz.—J. J. Padalino and A. Tuszynski, Note on the Analysis and Synthesis of Active RC Filters, *IEEE Trans. on Circuit Theory*, Dec. 1968, p 503–505.

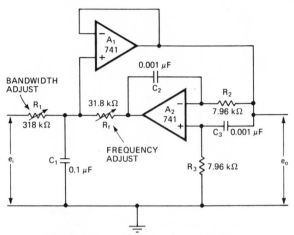

ADJUSTABLE 1–10 KHZ BANDPASS—Pots control bandwidths and center frequency independently in active filter having Q range of 2 to 200. Article give design equations. Maximum output voltage is 1 V p-p for values shown, having 5-Hz bandwidth at center frequency of 1 kHz.—J. Jenkins, Active Filter Has Separate Band and Frequency Controls, *Electronics*, Sept. 11, 1972, p. 110.

TUNABLE NOTCH—Designed for use between headphones and speaker output of receiver, to reduce heterodyne and noise interference. First section, between points marked X, is 500–2,000 Hz bandpass filter for reducing audio bandwidth to that required for speech. At right of this is Wien-bridge tunable filter whose notch can be moved anywhere in or out of passband, for removing heterodyne whistle without affecting speech intelligibility. Errors in tracking of ganged 10K pot and differences in 0.047-μF capacitors are balanced out with 1K pot.—M. Mann, Tunable Audio Filter, *73*, Jan. 1973, p 85–86.

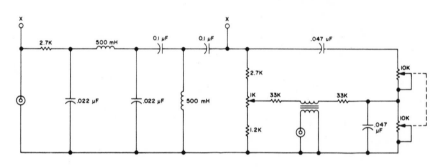

VOLTAGE-TUNED A-F FILTER—Center frequency changes linearly with control voltage of −4 V to +3 V over center-frequency range of 4.5 to 1. Center frequency is 1.46 kHz at zero control voltage, and can be tuned from 2.5 kHz to 0.5 kHz. D1 and D2 should be gold-doped to have low forward-voltage drop.—V. J. Georgiou, Voltage-Tuned Filter Varies Center Frequency Linearly, *Electronics*, Nov. 6, 1972, p 104.

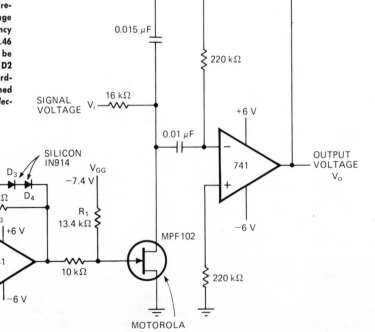

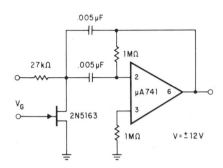

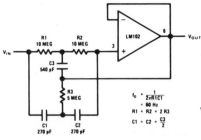

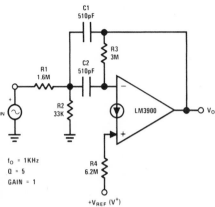

VOLTAGE-TUNED A-F BANDPASS—Center frequency is varied between 200 and 3,200 Hz when negative d-c voltage of 0 to 5 V is applied to terminal VG of fet. Bandwidth is fixed at 80 Hz, so Q of filter is varied from 2.5 at 200 Hz to 40 at 3.2 kHz. Filter gain is 26 dB, varying less than 1 dB over tuning range. Possible drawback is about 7% variation in center frequency at high end of tuning range for temperature change of 0–70 C. —N. Doyle, "Some Useful Signal Processing Circuits Using FET's and Operational Amplifiers," Fairchild Semiconductor, Mountain View, CA, No. 243, 1971, p 4–5.

60-HZ NOTCH—Combines LM102 voltage follower with twin-T R-C filter to give Q over 50. Junction of R3 and C3 is bootstrapped to output of follower, which raises Q in proportion to amount of signal fed back. High input resistance of opamp permits use of high resistance values in T, so only small capacitors are required even at low frequencies. Fast response of follower allows notch to be used at high frequencies.—"High Q Notch Filter," National Semiconductor, Santa Clara, CA, 1969, LB-5.

1-KHZ BANDPASS—Uses single Norton current-differencing opamp. Report gives complete design procedure for obtaining desired corner frequency and desired value of Q under 10.—T. M. Frederiksen, W. M. Howard, and R. S. Sleeth, "The LM3900—A New Current-Differencing Quad of ± Input Amplifiers," National Semiconductor, Santa Clara, CA, 1972, AN-72, p 15–16.

HIGH-PASS WITH 100-HZ CUTOFF—Uses IC voltage follower to give equivalent of two isolated R-C filter sections with buffered low-impedance output. Use metalized polycarbonate capacitors.—R. J. Widlar, "Integrated Voltage Follower," National Semiconductor, Santa Clara, CA, 1968, AN-5.

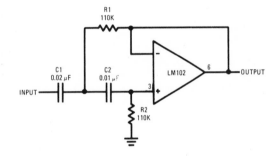

VOLTAGE-CONTROLLED LOW-PASS—Control voltage of 1 to 10 V adjusts cutoff of active low-pass filter from 100 to 1,000 Hz. Uses single IC multiplier and opamp plus only four additional components.—"AD530 Complete Monolithic MDSSR," Analog Devices, Inc., Norwood, MA, Technical Bulletin, July 1971.

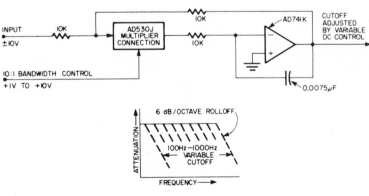

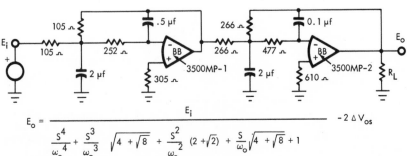

$$E_O = \frac{E_i}{\dfrac{s^4}{\omega_o{}^4} + \dfrac{s^3}{\omega_o{}^3}\sqrt{4+\sqrt{8}} + \dfrac{s^2}{\omega_o{}^2}(2+\sqrt{2}) + \dfrac{s}{\omega_o}\sqrt{4+\sqrt{8}} + 1} - 2\,\Delta V_{os}$$

1-KHZ LOW-PASS—Four-pole Butterworth filter uses inverting synthesis technique to realize each pole pair, so offset voltages of matched amplifiers cancel. Net offset error without trimming is less than 400 μV and total filter drift under 2 μV per deg C.— "Matched Low Bias Current Operational Amplifiers," Burr-Brown, Tucson, AZ, April 1972.

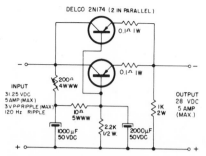

60:1 RIPPLE FILTER—Simple electronic filter reduces ripple 60 times below that of conventional L-C filter for 28-V power supply.—"Electronic Ripple Filter," Delco, Kokomo, IN, July 1971, Application Note 28.

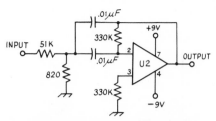

1-KHZ WITH 100-HZ BANDWIDTH—Uses Signetics N5741V opamp in circuit providing gain of 10, to provide additional selectivity after audio circuit of c-w receiver. Output will drive headphones having impedance of at least 600 ohms.—R. R. Knibb, A Simple Audio Filter, QST, Sept. 1971, p 46.

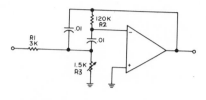

VARIABLE BANDPASS—Frequency can be varied without affecting bandwidth or gain. I3 controls center frequency while retaining constant bandwidth of about 260 Hz centered on about 2 kHz. Gain is over 26 dB. Opamp can be CA3030 or CA3035.—F. D. Williams, Active Filter Design and Use—Part IV, 73, Oct. 1972, p 43–44 and 46–47.

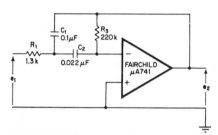

200-HZ BANDPASS—Article gives time-saving step-by-step technique for designing second-order active bandpass filters. Filter gain is 29.5 dB and Q is intermediate value of 5 at center frequency of 200 Hz.—F. G. Stremler, Simple Arithmetic: An Easy Way to Design Active Bandpass Filters, Electronics, June 7, 1971, p 86–89.

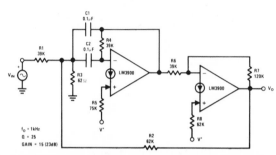

1-KHZ TWO-OPAMP BANDPASS—Use of two Norton current-differencing opamps gives Q between 10 and 50 along with higher gain than single opamp. Report gives complete design procedure for obtaining desired center frequency, Q value, and gain.—T. M. Frederiksen, W. M. Howard, and R. S. Sleeth, "The LM3900—A New Current-Differencing Quad of ± Input Amplifiers," National Semiconductor, Santa Clara, CA, 1972, AN-72, p 16–17.

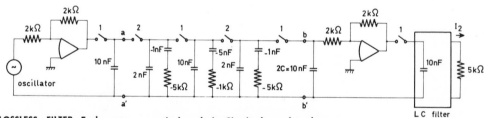

SYNTHESIZED LOSSLESS FILTER—Each capacitance in parallel with regenerating device constitutes, with two switches, basic building block of lossless filter which simulates behavior of transmission line. Article gives mathematical analysis. Circuit shown has three sections, with all switches operated at rate of 10 kHz. Filtering bandwidth of 0 to 5 kHz is selected by using L-C low-pass filter at output, with cutoff of 5 kHz.—R. Boite and J. P. V. Thiran, Synthesis of Filters with Capacitances, Switches, and Regenerating Devices, IEEE Trans. on Circuit Theory, Dec. 1968, p 447–454.

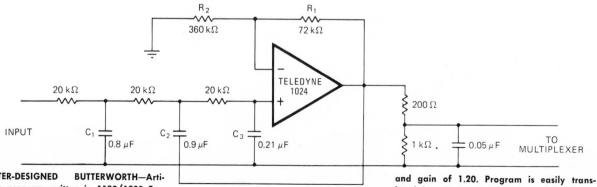

COMPUTER-DESIGNED BUTTERWORTH—Article gives program written in 1130/1800 Fortran that finds capacitor values and optimum-gain resistors for three-pole active low-pass Butterworth filters with gains between 1 and 2. Example shown provides 15-Hz cutoff and gain of 1.20. Program is easily translated into Basic.—I. P. Stapp, Jr., Brief Fortran Program for Active Low-Pass Filters, Electronics, Jan. 17, 1972, p 68–69.

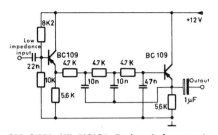

300–3,000 HZ VOICE—Designed for use in transmitter modulators and other applications where lower audio frequencies which do not contribute to intelligibility should fall off linearly 6 dB per octave. Can also be used in communication receivers to suppress noise outside voice band.—P. Zwart, The PM 5105 Sine-Square Generator for Service and Education, *Test and Measuring Notes*, Philips, Pub. Dept., Elcoma Div., Eindhoven, The Netherlands, 1972/3, p 5–7.

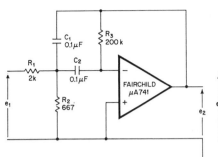

HIGH-Q BANDPASS AT 160 HZ—Article gives time-saving step-by-step technique for designing second-order active bandpass filters. Example shown has Q of 10, gain of 34 dB, and 16-Hz bandwidth centered on 160 Hz.—F. G. Stremler, Simple Arithmetic: An Easy Way to Design Active Bandpass Filters, *Electronics*, June 7, 1971, p 86–89.

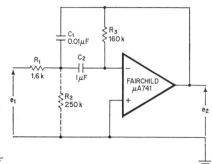

LOW-Q BANDPASS AT 100 HZ—Article gives time-saving step-by-step technique for designing second-order active bandpass filters. Example shown has Q of 1 and gain of 40 dB, holds phase shift constant within 10 deg about center frequency of 100 Hz, and rolls off at 20 dB per decade.—F. G. Stremler, Simple Arithmetic: An Easy Way to Design Active Bandpass Filters, *Electronics*, June 7, 1971, p 86–89.

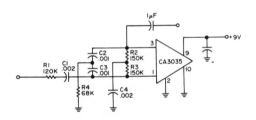

1-KHZ BANDPASS—Linear opamp is used in narrow bandpass version of twin-T configuration. Bandwidth is extremely narrow because Q is high, and can be made narrower by increasing value of R1. Decreasing R1 lowers Q and gives greater bandwidth.—F. D. Williams, Active Filter Design and Use—Part III, 73, Sept. 1972, p 76–87.

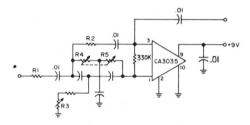

VARIABLE NOTCH FILTER—Variation of twin-T network allows active filter to be tuned over 10:1 range. Choice of R1 affects Q and gain. Choose value of R3 for fine-tuning Q. Choose R4 and R5 to give desired notch frequency. R2 limits feedback.—F. D. Williams, Active Filter Design and Use—Part IV, 73, Oct. 1972, p 43–44 and 46–47.

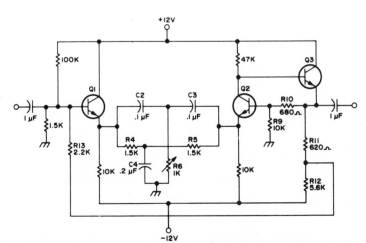

TWIN-T BANDSTOP FILTER—Values shown give maximum attenuation at 1.2 kHz. Article gives design equations for other frequencies. Uses MPS6520 transistors.—F. D. Williams, Active Filter Design and Use—Part II, 73, July 1972, p 97–106.

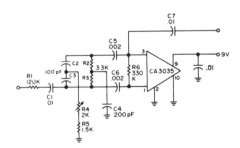

455-KHZ NARROW-BAND—Provides about 20 dB attenuation at 455 kHz with value of R1 shown, and 10 dB attenuation if R1 is increased to 270K. Useful as buffer amplifier following local oscillator, or for providing low attenuation over medium range such as might be desired in first i-f of scanning receiver. Uses twin-T network. Q is not high.—F. D. Williams, Active Filter Design and Use—Part IV, 73, Oct. 1972, p 43–44 and 46–47.

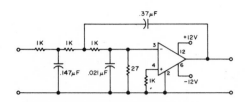

3-POLE LOW-PASS—Provides 3-dB cutoff at 1,500 Hz. Uses RCA CA3030 opamp. Provides Butterworth response beyond cutoff.—F. D. Williams, Active Filter Design and Use—Part III, 73, Sept. 1972, p 76—87.

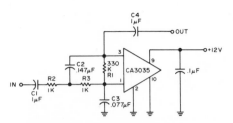

2-POLE LOW-PASS—Uses linear IC to eliminate high insertion losses of L-C filters. Network is normalized for frequency cutoff of 1 radian per second.—F. D. Williams, Active Filter Design and Use—Part III, 73, Sept. 1972, p 76—87.

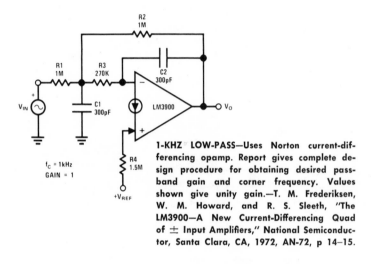

$f_C = 1kHz$
GAIN = 1

1-KHZ LOW-PASS—Uses Norton current-differencing opamp. Report gives complete design procedure for obtaining desired passband gain and corner frequency. Values shown give unity gain.—T. M. Frederiksen, W. M. Howard, and R. S. Sleeth, "The LM3900—A New Current-Differencing Quad of ± Input Amplifiers," National Semiconductor, Santa Clara, CA, 1972, AN-72, p 14—15.

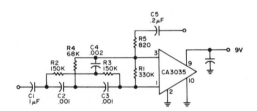

1-KHZ BANDSTOP—Parallel combination of low-pass and high-pass filter configurations gives bandstop response having attenuation of almost 40 dB at 1 kHz. Design resembles that of twin-T filter.—F. D. Williams, Active Filter Design and Use—Part III, 73, Sept. 1972, p 76—87.

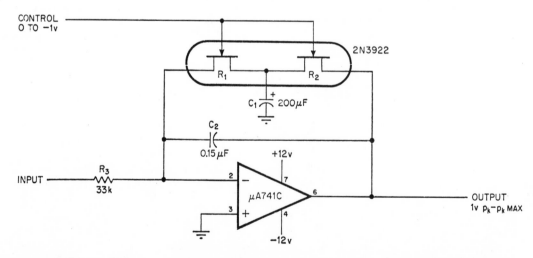

FET-TUNED NARROW-BAND—Replacement of resistors of conventional bridged-tee R-C network with on resistance of fet pair converts standard active bandpass filter into electronically tunable narrow-band filter. Frequency range is one octave, from about 70 to 140 Hz, for control voltage range of 1 V.—A. D. Delagrange, FETs in RC Network Tune Active Filter, Electronics, Dec. 7, 1970, p 76.

CHAPTER 34
Filter Circuits—Passive

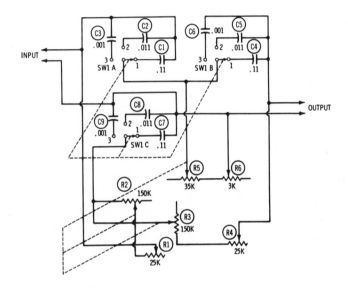

HUM AND RUMBLE KILLER—Notch filter can be used to suppress any desired frequency between range of 20 and 18,000 Hz. For 60-Hz hum suppression, set SW1 to position 1 and adjust ganged pot R2-R3-R5 until hum disappears. To kill 120-Hz rumble, adjust switch and pot similarly. Switch position 3 covers highest frequencies. Use R1, R4, and R6 for fine tuning, starting with 1K for R1 and R6 and 10K for R4.—R. Brown and T. Kneitel, "101 Easy Audio Projects," Howard W. Sams, Indianapolis, IN, 1971, p 119–120.

800-HZ FILTER IMPROVES C-W—Filter attenuates entire audio range except for very narrow band around 800 Hz, to suppress interference from other stations while copying c-w. Amplifier stage Q1 compensates for gain loss in filter. L1 and L2 are 88-mH toroids and T1 is transistor output transformer.—L. G. McCoy, A Solid-State Selectoroid, QST, May 1970, p 30–33.

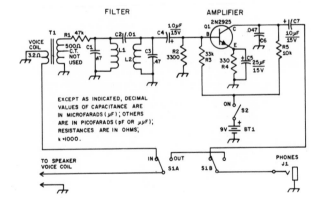

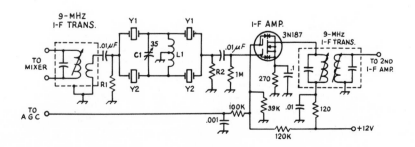

CASCADED HALF-LATTICE CRYSTAL FILTERS—Bandpass depends on frequency separation of vertical pairs of crystals; use about 200 Hz separation for c-w and 2 kHz for phone. Tune i-f strip about midway between crystal frequencies. Can be used equally well for 455-kHz and 9-MHz i-f strips.—"The Radio Amateur's Handbook," American Radio Relay League, Newington, CT, 49th Ed., 1972, p 247–248.

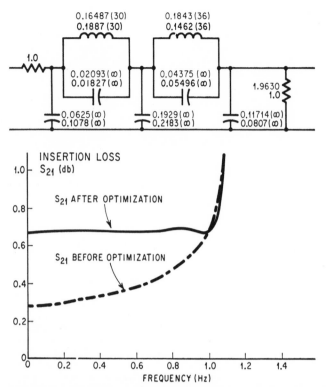

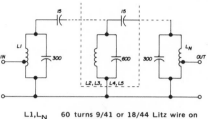

455-KHZ L-C—May have up to four identical center L-C sections, depending on selectivity required by detector of f-m receiver. Article covers tuning procedure.—G. K. Shubert, 455-kHz Filter for Amateur Fm, *Ham Radio*, March 1972, p 22–24.

OPTIMIZED CHEBYSHEV—Lower values, obtained from handbook for low-pass filter blocking above 1 Hz, give excessive parasitic losses in inductors. Filter optimization program for computer, described in article, gave upper values which provide more nearly ideal performance.—H. B. Lee, P. Carvey, R. Grabowski, and D. Evans, Program Refines Circuit from Rough Design Data, *Electronics*, Nov. 23, 1970, p 58–65.

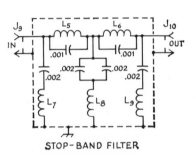

STOP-BAND FILTER

BC STOP-BAND FILTER—Offers sharp rejection to 500–1,600 kHz signals, to prevent interference when ham receiver is near high-power broadcast station. Use in 50 to 75-ohm lines. L5 and L6 are 33 μH, L7 and L9 are 10 μH, and L8 is 4.7 μH.—D. DeMaw, Rejecting Interference from Broadcast Stations, QST, Dec. 1967, p 35–38.

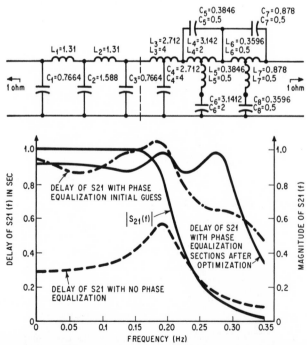

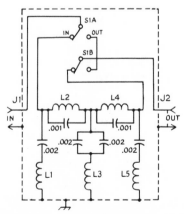

CHEBYSHEV ALL-PASS—Article describes use of Stanford Research Institute MATCH computer program for optimizing all-pass fifth-order Chebyshev filter (at left of vertical dashed line) by adding all-pass sections in cascade to provide phase equalization. Upper value for each component is obtained from computer after lower value is calculated by hand for first try.—H. B. Lee, P. Carvey, R. Grabowski, and D. Evans, Program Refines Circuit from Rough Design Data, *Electronics*, Nov. 23, 1970, p 58–65.

BROADCAST TRAP—Band-rejection filter having two constant-k sections in cascade is inserted in 50 or 75-ohm line between antenna and ham receiver, to provide sharp rejection to signals in 500 to 1,600-kHz range without impairing reception above and below broadcast band. Sometimes needed in receivers lacking front-end selectivity. L1 and L5 are 10 μH, L2 and L4 33 μH, and L3 4.7 μH, all on toroid cores.—"The Radio Amateur's Handbook," American Radio Relay League, Newington, CT, 49th Ed., 1972, p 252–253.

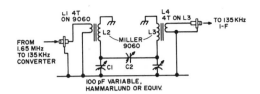

135-KHZ BANDPASS—Uses standard coils with four turns added on each, to give bandwidth of 3 to 4 kHz at output of converter changing 1.65 MHz to 135 kHz. Coupling method used is easy to adjust.—W. Hoisington, I-F Filter Converter AVC, 73, May 1970, p 62–64, 66–70, and 72.

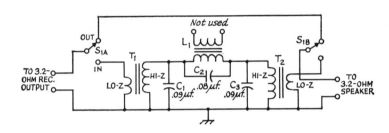

SPEECH FILTER—Uses m-derived pi-section filter having high-frequency cutoff of 2,100 Hz and maximum attenuation at 2,630 Hz. Lower 3-dB limit is 100 Hz. L1 is 44 mH (half of split-winding 88-mH toroid). Transformers have 500-ohm primaries, such as Stancor 8101. Used between audio amplifier and speaker of a-m or ssb receiver to suppress interference outside of speech intelligibility range.—J. H. Ellison, An Audio Filter for Speech Reception, QST, June 1967, p 45–46.

600–1,900 HZ SPEECH FILTER—Used between ham receiver and speaker to improve readability of weak a-m and ssb signals when interference is heavy. Response drops down 10 db at 500 Hz, affecting naturalness of voice but not impairing intelligibility. Insertion loss is about 3 dB.—The Torofil—A QRM Reducer for the Phone Man, QST, April 1967, p 28–29.

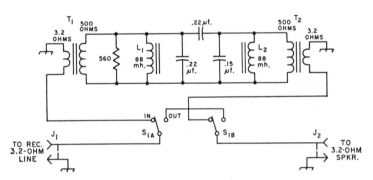

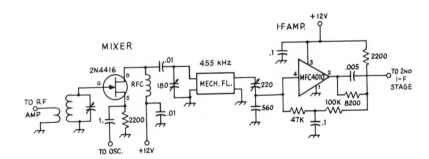

MECHANICAL I-F BANDPASS—Uses Collins 455-FB-21 mechanical filter having ssb bandpass of 2.1 kHz, to narrow i-f response for voice signals.—"The Radio Amateur's Handbook," American Radio Relay League, Newington, CT, 49th Ed., 1972, p 247–248.

455-KHZ TUNABLE SLOT FILTER—Q multiplier is followed by phase inverter to give sharp null instead of peak, tunable with C1 between about 449 and 461 kHz. Regeneration control R1 sets slot depth, which can be maximum of about 50 dB for suppression of heterodyne without appreciably affecting intelligence of phone signals. Used between i-f amplifier and second mixer in commercial fixed-station amateur receiver. Q1 is 2N3693 and Q2 is 2N3564. L1 is 150-μH adjustable.—Hammarlund HQ-215 Receiver, QST, Dec. 1968, p 50–54.

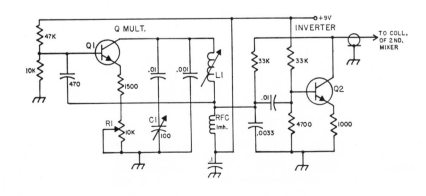

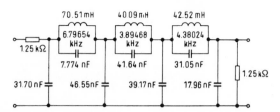

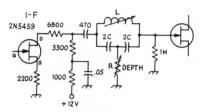

CAUER-PARAMETER LOW-PASS—Lossless L-C seventh-order low-pass filter has only 0.044-db ripple in passband of 0 to 3.4 kHz and 40.07 dB minimum attenuation in stopband from 3.85 kHz to infinity. Article tells how to convert design to active-C filter by replacing the three floating inductors with capacitance-loaded semi-floating gyrators using thick-film resistors.—W. H. Holmes, W. E. Heinlein, and S. Grutzmann, Sharp-Cutoff Low-Pass Filters Using Floating Gyrators, *IEEE Journal of Solid-State Circuits*, Feb. 1969, p 38–50.

T-NOTCH FILTER—Provides sharp tunable null at low intermediate frequencies, 50 to 100 kHz. For 50-kHz rejection notch, L is about 2.6 mH and C is 3,900 pF. R controls depth of notch. Not suitable for higher i-f values because notch broadens.—"The Radio Amateur's Handbook," American Radio Relay League, Newington, CT, 49th Ed., 1972, p 249.

HIGH-PASS FOR 300-OHM LINE—Installed at antenna terminals of tv receiver to attenuate interfering ham transmitter signals below about 40 MHz. Do not use direct ground on chassis of transformerless receiver; instead, ground through 0.001-μF mica capacitor.—"The Radio Amateur's Handbook," American Radio Relay League, Newington, CT, 49th Ed., 1972, p 499.

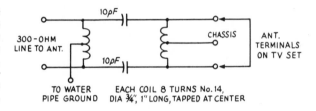

EACH COIL 8 TURNS No. 14, DIA ¾", 1" LONG, TAPPED AT CENTER

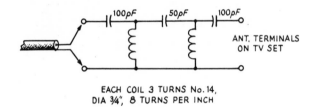

EACH COIL 3 TURNS No. 14, DIA ¾", 8 TURNS PER INCH

HIGH-PASS FOR 75-OHM COAX—Installed at antenna terminals of tv receiver to attenuate interfering ham transmitter signals below about 40 MHz. Do not use direct ground on chassis of transformerless receiver; instead, ground through 0.01-μF mica capacitor.—"The Radio Amateur's Handbook," American Radio Relay League, Newington, CT, 49th Ed., 1972, p 499.

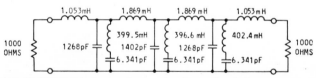

100-KHZ BAND-ELIMINATION—Uses three synthesized crystals (each is C in shunt with L-C), having series resonances at 100 kHz and at 350 Hz above and below this value, to suppress 800-Hz band around 100 kHz. Computed response is not too good, but remainder of article presents mathematical analysis of various methods for improving response enough to get at least 30 dB suppression in stopband. Parameters are given for final three-section network based on 360-deg constant-resistance delay in bridged-T sections, with stiff series resonant circuit across bridging arm. —G. Szentirmai, The Synthesis of Narrow-Band Crystal Band-Elimination Filters, *IEEE Trans. on Circuit Theory*, Dec. 1968, p 409–414.

300–2,000 HZ BANDPASS—Used for adapting hi-fi headphones to make short-wave listening more enjoyable. Battery is needed to provide polarizing voltage for maintaining chemical films on electrolytics; battery life equals shelf life because it delivers only tiny leakage current. R1 is 8.2 ohms, R2 18, and R3 15.—J. Ashe, How to Use Hi-Fi Headphones, *Popular Electronics*, July 1972, p 76–77.

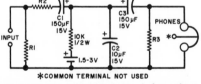

10-METER LOW-PASS—Based on modern design techniques that give superior harmonic attenuation with fewer components. Values are also given for 15-meter cutoff. Design is optimized for insertion in 52-ohm coaxial transmission line, as between exciter and linear amplifier or between amplifier and antenna. Both give true 50–70 dB harmonic attenuation. Useful for eliminating tv interference.—J. Schultz, Low-Pass Filters for 10 and 15 Meters, *Ham Radio*, Jan. 1972, p 42–45.

component C in pF L in μH	15-meter filter (22 MHz cutoff)	10-meter filter (30 MHz cutoff)
C1	68	50
C2	7	5
C3	200	100 & 39 parallel
C4	27	20
C5	180	100 & 33 parallel
C6	22	15
C7	57	47
L1	.41	.30
L2	.45	.33
L3	.35	.26

SWITCHED AUDIO FILTER—Used with amateur receiver to allow higher cutoff frequency for phone work and low cutoff for c-w.—B. Wildenhin, Inexpensive Audio Filters, *Ham Radio*, Aug. 1972, p 24–29.

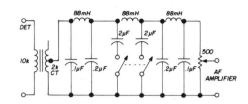

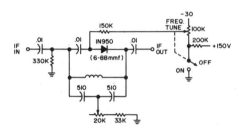

VARACTOR NOTCH FILTER—Used in i-f section of Collins ssb receiver to suppress interference over narrow band. Uses 100K pot to vary reverse bias on varactor for tuning notch filter to different frequency.—J. J. Schultz, IF Notch Filter, *73*, Nov. 1969, p 14–16.

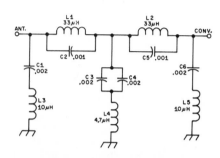

BC REJECTION—Designed for use with 2-meter f-m converter feeding broadcast-band auto radio. Prevents strong bc stations from feeding around or through converter. All coils are wound on toroid cores; article gives winding data.—L. G. McCoy, Low-Cost Hardware for 2-Meter FM Reception, *QST*, Sept. 1971, p 34–38.

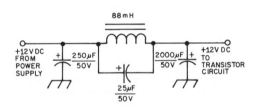

RIPPLE FILTER—Uses 88-mH toroid in parallel with 25 μF to form tuned choke having high impedance at 120 Hz, internal resistance of only 10 ohms, and 200-mA current rating. Reduces ripple by factor of 10 as compared to filter using 10-ohm series resistor. Ideal for miniature power supplies of transistor equipment.—R. M. Matteis, Reducing Ripple in Miniature Power Supplies, *QST*, May 1971, p 48.

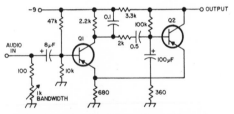

1,000 HZ—Uses modified Wien bridge in collector circuit of amplifier Q1, with constants chosen for center frequency of 1,000 Hz. Control varies bandwidth between 70 and 600 Hz. Developed for use in high-frequency receivers for separating stations in presence of heavy interference. Transistors can be 2N408, 2N2613, SK3004, GE-2, or HEP254.—J. Fisk, "73 Useful Transistor Circuits," 73 Inc., Peterborough, NH, 1967, p 15A.

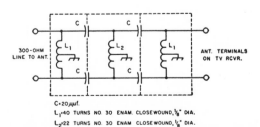

300-OHM HIGH-PASS TVI—Connected to antenna terminals of tv set to block interfering ham transmitter signals below about 40 MHz without affecting tv signals. Coils may be wound on ⅛" plastic knitting needles. Mount in three shielding boxes as shown by dashed lines, with tubular ceramic capacitors centered in partition holes. Ground shield through 0.001-μF mica capacitor.—"The Radio Amateur's Handbook," American Radio Relay League, Newington, CT, 49th Ed., 1972, p 499–500.

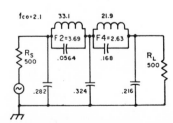

SPEECH FILTER—Dual-section elliptic-function low-pass filter has cutoff of 2,100 Hz, for attenuating upper noise frequencies. First frequency of maximum attenuation (50 dB) is 2,630 Hz.—E. E. Wetherhold, Modern Design Methods Applied to the Speech Filter, *QST*, Nov. 1967, p 51.

CHAPTER 35
Fire Alarm Circuits

MOTOR UNDERVOLTAGE—Circuit will not supply voltage to motor unless voltage is above safe threshold value, to prevent stall condition that could cause motor overheating and fire. Protective circuit is fail-safe so motor can operate even without protective circuit. With 26-V supply, power is always cut off below 20 V, may be cut off between 20 and 22.1 V, may be supplied between 22.2 and 24.3 V, and is always supplied above 24.3 V. Select Q3 to have higher switching voltage than Q6. Omit R7 and D1 if motor has built-in diode suppression.—W. H. Sahm, III, "A Highly Reliable, Fail Safe, Precision Undervoltage Protection Circuit," General Electric, Semiconductor Products Dept., Auburn, NY, No. 90.83, 1970, p 1—4.

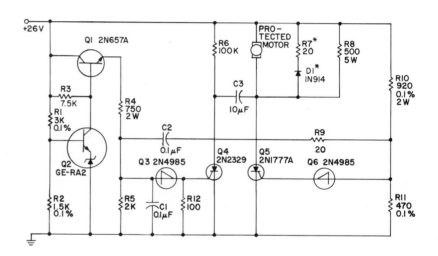

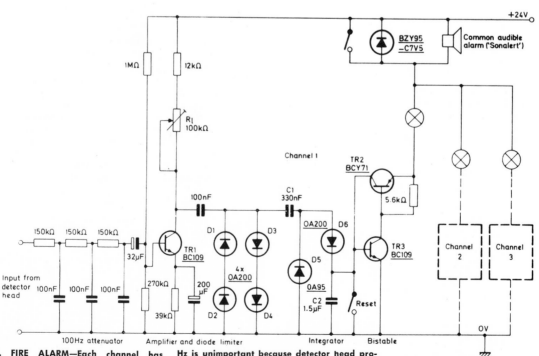

CENTRAL FIRE ALARM—Each channel has identical infrared detector, amplifier, and indicator lamp giving location of fire, with all channels driving audible alarm. Input R-C filter provides 60 dB attenuation of 100-Hz flicker from incandescent lamps on British 50-Hz power lines. Attenuation of 3 dB at 10 Hz is unimportant because detector head provides 40 dB of amplification. Circuit is also immune to large transient signals such as occur when lighting is switched on or sunshine level changes suddenly. With system set for minimum gain, about 8 cm of flame flicker can be detected at range of 6 meters within cone angle of 150 deg from each detector head. At maximum sensitivity, range is doubled.—"Applications of Infrared Detectors," Mullard, London, 1971, TP1201, p 51—54.

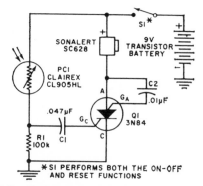

9-V ALARM—Battery-operated version of high-sensitivity burglar and fire alarm is triggered by sudden increase in light, such as lighting of paper match 6 feet away. If window of photocell is aimed at sky, can be used at camp to detect lightning flash and warn of approaching storm. Q1 is silicon controlled switch.—F. H. Tooker, Sensitive Burglar and Fire Alarm, *Electronics World*, June 1970, p 58–59.

FIRE ALARM—Used with infrared detector connected across input, as bistable circuit that activates indicator lamp and audible alarm whenever alarm pulse from detector is greater than threshold voltage of TR2. Transistors are connected to form equivalent of pnpn element. Chief drawback of this simple circuit is inability to discriminate between flame and any other spurious pulse.—"Applications of Infrared Detectors," Mullard, London, 1971, TP1201, p 50–51.

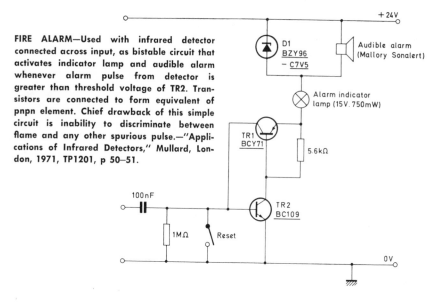

MULTIPLE DETECTORS—Uses any number of fire, smoke, or other detectors connected in parallel to basic relaxation oscillator using GE-X5 scr driving 8-ohm speaker to generate clicking sound. Connect speaker, 22½-V battery, 100K resistor, and on-off switch in series across scr, with 1-μF 25-V capacitor across scr-speaker combination. Use high-output ceramic or crystal mike, phono pickup, or ceramic contact mike to pick up sound or vibration of footsteps. Cadmium sulfide photocell will respond to flashlight beam of intruder. Choose other sensors as required for desired protection.—"Electronics Experimenters Circuit Manual," General Electric, Owensboro, KY, 1971, 3rd Ed., p 161–166.

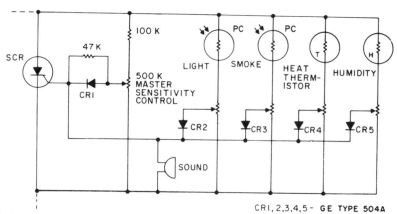

AMBIENT-IMMUNE HEAT ALARM—Pulls in relay to sound alarm or turn on warning lamp when sensing thermistor for high-power amplifier is reduced in resistance by overtemperature. Bridge connection with reference thermistor makes large changes in ambient temperature affect both thermistors equally and keep bridge balanced.—J. J. Schultz, Temperature Alarms for High-Power Amplifiers, *Ham Radio*, July 1970, p 48–50.

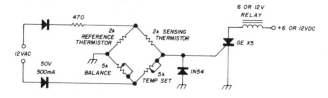

FIRE ALARM—Uses Radio Shack 276-035 or similar infrared detector kit, including reflector and filter, for PC1. Responds to objects over 75 C hotter than ambient temperature. Fet amplifier Q1 feeds threshold detector Q2-Q3-Q4, which is separated from output stage Q6-Q7 by buffer Q5. Settings of R2 and R3 determine infrared level at which alarm buzzer goes off. Can be used as burglar alarm if heat lamp is used as source and positioned so intruder interrupts broad infrared beam.—J. G. Sloat, Build Infrared Sensor, *Popular Electronics*, Jan. 1969, p 27–31 and 110.

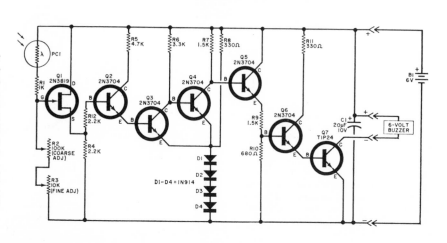

THIN-WIRE ALARM—Extremely thin wire stretched across doors, windows, and other potential intruder passageways, or around campsite for animal protection, breaks easily without knowledge of intruder and triggers Q1 and scr to apply power to Mallory SC628 Sonalert alarm. Use about No. 38 wire, which can be purchased on spools or salvaged from old transformers. Can also be used as fire alarm, by suspending wood block with rubber band in such a way that when heat makes rubber band stretch, wood block breaks wire.—G. M. Rawlings, The Prowler Howler, *Popular Electronics*, Nov. 1968, p 47–49.

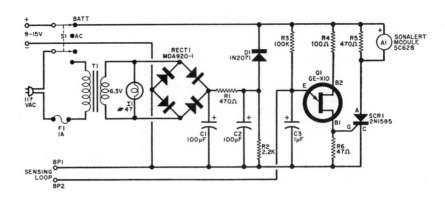

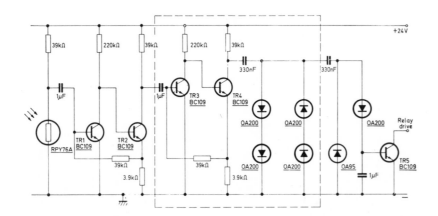

GAS FLAME-FAILURE—Used with gas furnace to provide holding current for main gas jet relay as long as flame is present. Intermittent radiation produced by flame flicker generates low-frequency signal in chemical lead sulfide cell. Signal is amplified and amplitude-limited to provide pulses for charging capacitor sufficiently to hold TR5 on and energize relay. With no flicker, and hence no pulses, capacitor discharges and TR5 switches off.—"Applications of Infrared Detectors," Mullard, London, 1971, TP1201, p 56–58.

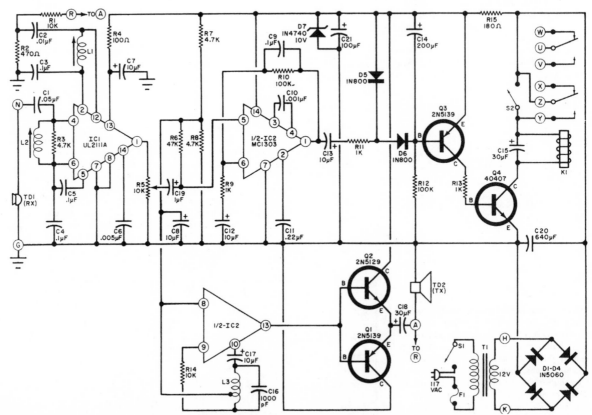

ULTRASONIC INTRUDER ALARM—Single step of intruder in protected area, in shape of 50-deg cone fanning out about 15 feet from detector, actuates alarm. Radiated 40-kHz signal is used as reference for comparing with slightly lower or higher frequency reflected from moving object. Will also serve as fire alarm, because air turbulence of flames will cause same Doppler shift. TD1 and TD2 are 40-kHz Massa MK-109 or similar transducers. Article covers construction and adjustment.—D. Meyer, One-Step Motion Detector, *Popular Electronics*, March 1970, p 57–61 and 104.

A-C BURGLAR AND FIRE—Can be triggered only by sudden increase in room illumination, such as sweep of burglar's flashlight, lighting of ordinary paper match within 6 feet in dark room, or fire in room. Not affected by dawn. Once triggered, alarm can be turned off only by operating concealed reset switch S1. Relay can be used to turn on 100-W lamp in room (plugged into S01) to scare off intruder, or to activate remote alarm. If S1 is mercury switch mounted vertically in housing so it is normally closed, box must be turned upside down to open switch and turn off alarm. Head should be held over window of photocell while turning box upright, then removed very slowly to prevent room illumination from turning alarm on again.—F. H. Tooker, Sensitive Burglar and Fire Alarm, *Electronics World*, June 1970, p 58–59.

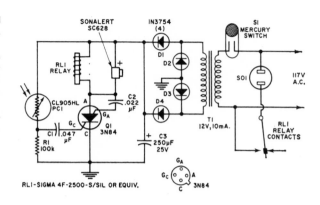

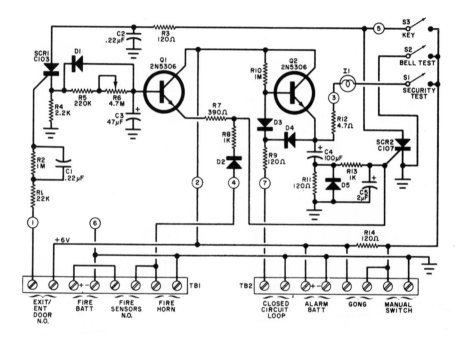

DELAYED FIRE & BURGLAR ALARM—Provides adjustable delay of 5 to 50 s to enter protected building legitimately without setting off gong. Battery life is almost shelf life on standby (gong not ringing). With appropriate thermal sensors, can also be used as fire alarm. R6 determines delay. Diodes are 1-A 600-V silicon. I1 is 6-V pilot lamp. Operates from 6-V lantern battery. Author recommends 10-inch 6-V gong and either 135 or 190 deg heat sensors.—G. Meyerle, Professional Intruder/Fire Alarm, *Popular Electronics*, Dec. 1971, p 61–65.

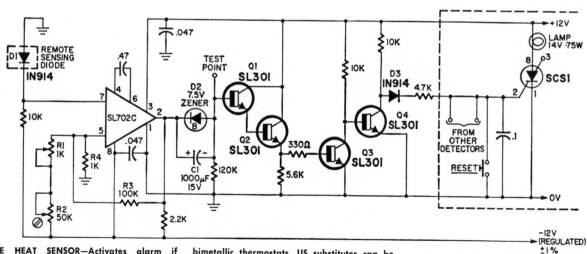

DIODE HEAT SENSOR—Activates alarm if rate of temperature rise becomes excessive or if ambient temperature exceeds preset level. Operation is more reliable than when using bimetallic thermostats. US substitutes can be used for European transistor types shown; try one of GE 3N80 series for SCS1. R1 and R2 set temperature trip level. Alarm must be reset after fire is extinguished.—R. F. Scott, Technical Topics, *Radio-Electronics*, Aug. 1969, p 41–42.

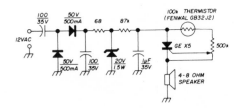

SPEAKER-DRIVING HEAT ALARM—Thermistor senses overtemperature in high-power amplifier and triggers scr, allowing 1-μF capacitor to discharge repeatedly through scr and speaker at audio rate determined by time constants. When temperature returns to normal, thermistor resistance increases and scr is cut off. Does not require resetting.—J. J. Schultz, Temperature Alarms for High-Power Amplifiers, *Ham Radio*, July 1970, p 48–50.

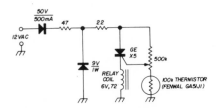

SELF-RESETTING HEAT ALARM—Uses scr to pull in relay and turn on buzzer or warning light when thermistor senses overtemperature in high-power amplifier. When thermistor returns to normal, scr current falls below holding value and relay drops out. 9-V zener stabilizes operating point of alarm against supply voltage changes.—J. J. Schultz, Temperature Alarms for High-Power Amplifiers, *Ham Radio*, July 1970, p 48–50.

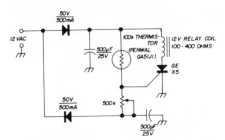

OVERTEMPERATURE ALARM—Thermistor senses overtemperature of high-power amplifier and triggers scr, pulling in relay that can turn on buzzer or warning light. Supply must be interrupted to turn off alarm after temperature returns to normal.—J. J. Schultz, Temperature Alarms for High-Power Amplifiers, *Ham Radio*, July 1970, p 48–50.

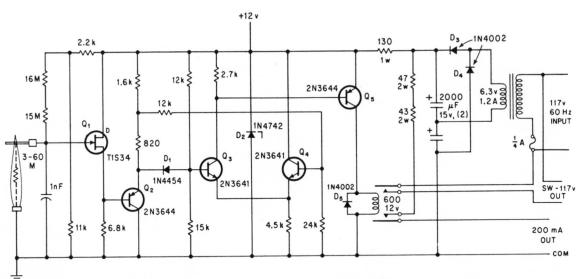

FLAME FAILURE—Operates over flame resistance range of 30 to 60 meg. Relay is energized as long as flame is burning. If flame goes out or if sensor rod is grounded, Q4 conducts and turns off Q3, Q5, and relay, to give fail-safe operation.—R. Wolfram, Fail-Safe Flame Sensor Provides Control Functions, *Electronics*, Aug. 31, 1970, p 68.

CHAPTER 36
Flasher Circuits

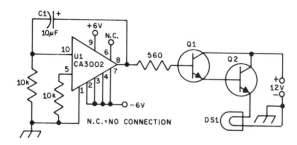

1-HZ SQUARE-WAVE FLASHER—RCA IC amplifier connected as square-wave generator delivers almost 8 V p-p to Darlington arrangement of transistors capable of furnishing 0.5 A to No. 76 pilot lamp or to relay. Q1 is silicon npn 150-mW transistor such as 2N333, and Q2 is 2N1485 npn silicon power transistor or equivalent. C1 determines frequency.—L. Nickel, Simple Integrated-Circuit Square-Wave Source, *QST*, Dec. 1971, p 33.

5-NEON BLINKER—Each neon lamp is relaxation oscillator. Connection as shown gives ring counter, with each neon in turn coming on. If mounted on styrofoam star, gives twinkling-star effect. Article includes simple voltage-doubling supply for operation from a-c line. Will also work from two 67½-V batteries. Any number of additional neons may be added.—W. F. Splichal, Jr., Twinkling Star, *Science and Electronics*, Jan. 1970, p 46—48 and 108.

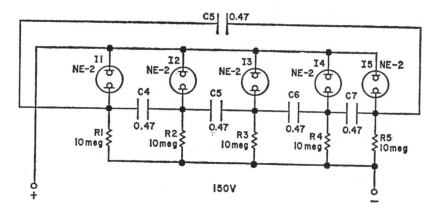

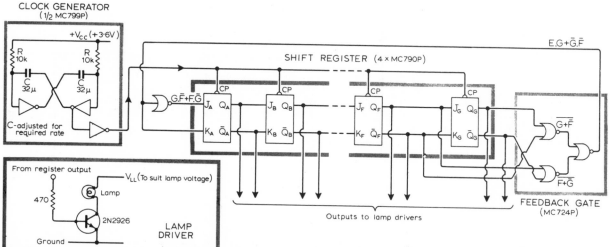

RANDOM-TWINKLING XMAS LIGHTS—Switches 14 lamps on and off in pseudo-random sequence having length of 127 clock periods. Uses 7-stage shift register to generate sequence, with exclusive OR gate providing feedback to input from stages 6 and 7. Model-railroad pea-type lamps of different colors, each driven by transistor, are arranged on Christmas tree to provide fascinating effect.—H. N. Griffiths, A Digital Christmas Tree, *Wireless World*, Jan. 1970, p 35.

150-W A-C TRIGGER—Two IC's and very few additional components form d-c power supply for flasher driving C106 triac flashing 117-V lamp load about once per second, with exact rate depending on connection shown on pin 11 of PD455 six-stage frequency divider. As shown, final output period is 64/60 s. Moving connection to any pin from 9 to 14 changes lines frequency division and output period. PA424 zero-voltage switch triggers triac every half-cycle while output of divider is in *on* state. For other lamp load, use appropriate triac.—R. W. Fox, "Flashers, Ring Counters and Chasers," General Electric, Semiconductor Products Dept., Auburn, NY, No. 200.48, 1969, p 3.

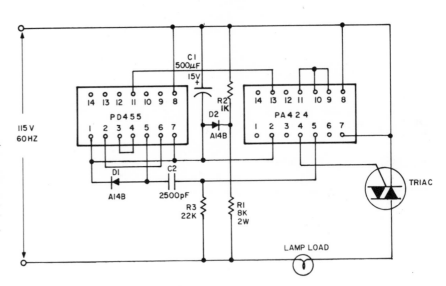

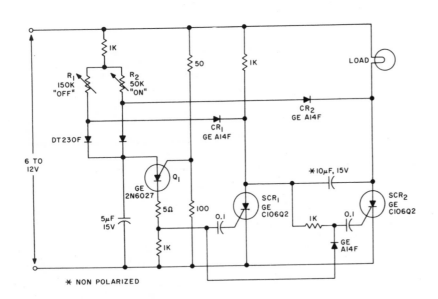

6–12 V ADJUSTABLE—Uses power flip-flop and put to obtain independently adjustable on and off times with R1 and R2. If desired, second lamp can be used in place of 1K resistor in anode circuit of SCR1. Will handle lamps drawing up to about 4 A.—"SCR Manual," General Electric, Syracuse, NY, 1972, 5th Ed., p 205–206.

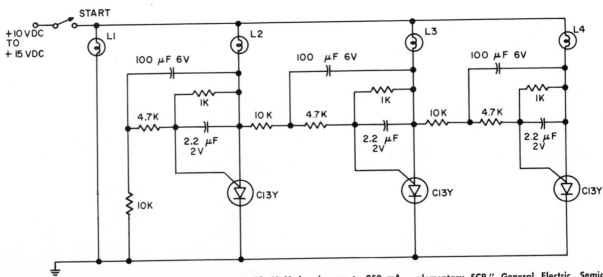

SEQUENTIAL DELAY—Lamp L1 lights when power is applied. L2 lights about ½ s later, and each succeeding lamp at ½-s intervals. Lamps are 12–40 V drawing up to 250 mA. Any number of stages may be cascaded.—W. H. Sahm, III, and F. M. Matteson, "The Complementary SCR," General Electric, Semiconductor Products Dept., Auburn, NY, No. 90.94, 1970, p 21–22.

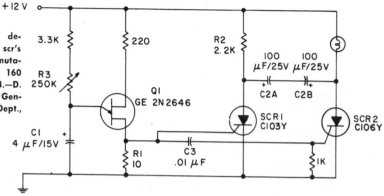

12-V FLASHER—Relaxation oscillator Q1 delivers train of trigger pulses to gates of scr's connected in circuit using C2 for commutation. R3 adjusts flash rate from 36 to 160 flashes per minute. Lamp is GE No. 1073.—D. R. Grafham, "Using Low Current SCR's," General Electric, Semiconductor Products Dept., Auburn, NY, No. 200.19, 1970, p 38.

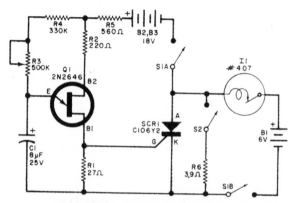

18-V FLASHER—Operates from two 9-V transistor batteries and 6-V lantern battery. Uses GE 407 flasher lamp mounted in red trailer clearance lamp assembly atop box containing circuit and batteries. Can be used as emergency flasher when changing tire at night, for guiding hikers back to campsite after dark, or warning of hazardous conditions. Electronic circuit extends off time of lamp up to 12 s to give maximum battery life, when S1 is closed. For faster flash, S2 is closed, making flash rate dependent on bimetallic strip built into flasher lamp.—L. E. Greenlee, Build a Blinky Blinker, *Popular Electronics*, Sept. 1969, p 51–54.

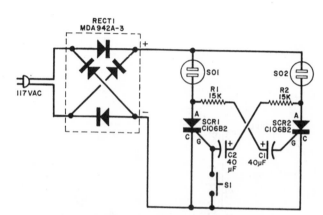

200-W TWO-LAMP BLINKER—Circuit is high-power flip-flop using scr's instead of transistors. If alternate blinking of lamps does not occur immediately, press S1 momentarily to start. Lamps do not need to be same wattage. Can be used for store-window displays, on piers to guide boaters home after dark, or as hazard warning. If 150-W and 25-W lamps are in same frosted-glass enclosure, effect of rotating beacon is obtained.—J. S. Simonton, Jr., Build 200-Watt Dual Flasher, *Popular Electronics*, Feb. 1969, p 50–52.

500-CP FLASHER—Too bright for indoor use. Ideal for marking pier at night or as emergency vehicle flasher. Draws 2 A from 12-V battery. Use ordinary 117-V 6-W D26 Christmas-tree lamp or 6S6. Oscillator Q1-Q2 converts battery supply to a-c for stepup by T1. Dual pot R2 determines duration of each flash, while dual pot R1 sets flashing rate. Despite overvoltage, lamp does not burn out because it cools between flashes.—T. Couch, Super Flash, *Popular Electronics*, May 1970, p 48–50 and 86.

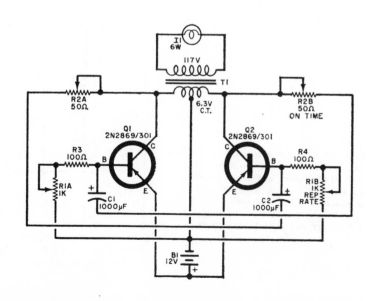

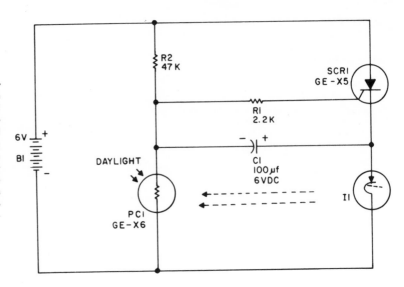

AUTOMATIC TURN-ON—Turns on automatically at night and shuts off at dawn, with short-duration flash to conserve life of GE No. 407 flasher lamp and 6-V lantern battery. Light from lamp falls on photocell and lowers its resistance, to make C1 reach firing voltage of scr about 5 s after bimetal switch of flasher lamp has reclosed, to spread out normal cycle of flasher as well as provide automatic turnoff by daylight.—"Electronics Experimenters Circuit Manual," General Electric, Owensboro, KY, 1971, 3rd Ed., p 105–107.

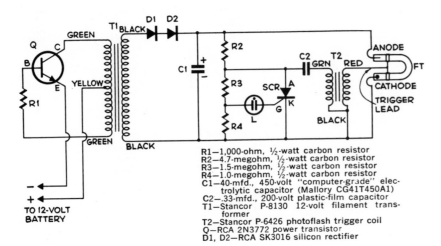

R1—1,000-ohm, ½-watt carbon resistor
R2—4.7-megohm, ½-watt carbon resistor
R3—1.5-megohm, ½-watt carbon resistor
R4—1.0-megohm, ½-watt carbon resistor
C1—40-mfd., 450-volt "computer-grade" electrolytic capacitor (Mallory CG41T450A1)
C2—.33-mfd., 200-volt plastic-film capacitor
T1—Stancor P-8130 12-volt filament transformer
T2—Stancor P-6426 photoflash trigger coil
Q—RCA 2N3772 power transistor
D1, D2—RCA SK3016 silicon rectifier

XENON-FLASH BEACON—Battery voltage is stepped up to about 450 V d-c by converter circuit, for charging C1. GE C106B1 scr acts with trigger coil T2 to deliver about 5,000-V pulse to GE FT-118 xenon flashtube about once per second to lower conductivity of flashtube so stored energy in C1 discharges and produces brilliant flash that is visible for miles on clear night. Can be used on auto, boat, or pier. L is NE-2 neon.—R. M. Benrey, *Flasher Beacon for Your Boat or Car, Popular Science*, Aug. 1970, p 86–87 and 116.

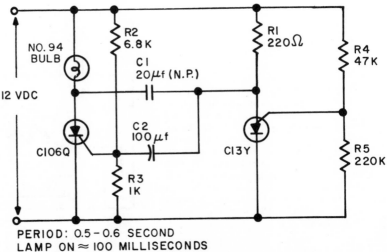

12-V SINGLE-LAMP—Economical circuit uses only two active devices. Off time is changed with R2-R3 divider, and on time with R4-R5 divider or R1.—W. H. Sahm, III, and F. M. Matteson, "The Complementary SCR," General Electric, Semiconductor Products Dept., Auburn, NY, No. 90.94, 1970, p 20–21.

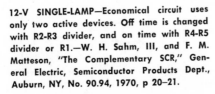

PERIOD: 0.5–0.6 SECOND
LAMP ON ≈ 100 MILLISECONDS

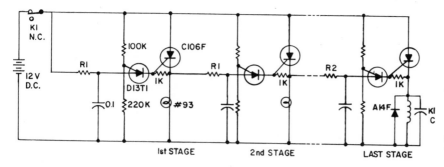

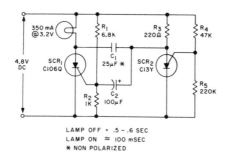

12-V CHASER—Used chiefly in advertising sign where string of lights is turned sequentially to produce effect of moving sign. Consists of series D13T relaxation oscillators each triggering an scr. One advantage of circuit is that delay of each stage is independent of all other stages. Energizing of one load allows following R-C circuit to charge. Final scr triggers relay that turns off system momentarily, to reset all scr's. Any number of identical stages can be used.—R. W. Fox, "Flashers, Ring Counters and Chasers," General Electric, Semiconductor Products Dept., Auburn, NY, No. 200.48, 1969, p 10.

LAMP OFF = .5 - .6 SEC
LAMP ON ≈ 100 mSEC
* NON POLARIZED

4.8-V D-C FLASHER—Simple low-cost low-voltage flasher for pilot lamp uses scr and complementary scr. Can be used for higher voltages if peak negative voltage on gate of SCR1 is limited to less than 6 V.—"SCR Manual," General Electric, Syracuse, NY, 1972, 5th Ed., p 206–207.

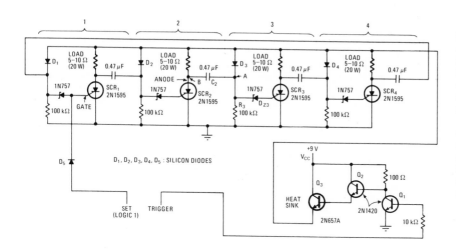

20-W SEQUENTIAL FLASHER—High-current ring counter handles 20-W lamp loads drawing up to 5 A each. Four stages are shown, but design can be extended to unlimited number of stages as required for advertising signs. Trigger input pulse should have minimum width of 200 μs, maximum frequency of 1 kHz, and amplitude of 6–9 V. Initially, all stages are nonconducting. To enable counter, apply set pulse to gate of first or any other scr; each trigger pulse then passes conduction to next stage.—C. Strangio, High-Power Counter Drives 20-Watt Loads, *Electronics*, March 1, 1973, p 84.

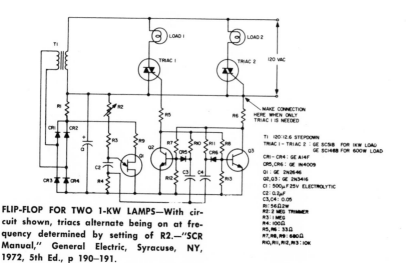

FLIP-FLOP FOR TWO 1-KW LAMPS—With circuit shown, triacs alternate being on at frequency determined by setting of R2.—"SCR Manual," General Electric, Syracuse, NY, 1972, 5th Ed., p 190–191.

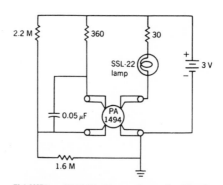

FLASHER SWITCH—Uses General Electric PA1494 Accu-Switch that provides logic function of Schmitt trigger but with better voltage and temperature stability.—Mass-Produced Integrated Circuits on Film Strip Speed Manufacturer's Use, *IEEE Spectrum*, Feb. 1971, p 100.

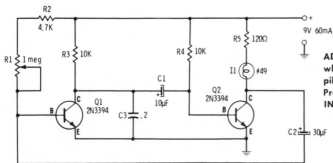

ADJUSTABLE FLASHER—R1 controls rate at which flip-flop mvbr Q1-Q2 makes No. 49 pilot lamp flash.—H. Friedman, "99 Electronic Projects," Howard W. Sams, Indianapolis, IN, 1971, p 54–55.

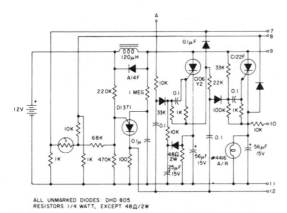

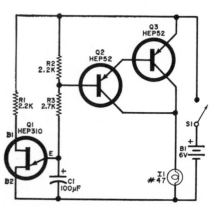

BARRICADE FLASHER—Contains oscillator, commutating inductor, and first two stages of ring counter. Use of dummy load for C106Y2 scr permits pwm of lamp load for brightness control and longer battery life. Photocell monitoring ambient light controls oscillator so duty cycle of flashing lamp varies with ambient. Also responds to car headlights by flashing warning.—E. K. Howell and R. W. Fox, "New Powerpac Thyristors and How to Use Them," General Electric, Semiconductor Products Dept., Auburn, NY, No. 671.18, 1969, p 5.

EMERGENCY FLASHER—Operates from 6-V lantern battery, for carrying in auto as emergency flasher. Can also be used on pier, driveway entrance, or store display. Ujt Q1 is relaxation oscillator driving Darlington pair. Changing C1 changes flash rate. For larger lamp, use 6-V d-c relay in place of I1.—V. X. Golden, Readers' Circuits, *Popular Electronics*, March 1971, p 85–86.

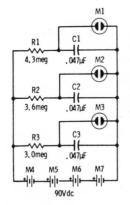

THREE-NEON FLASHER—Neons flash alternately in unusual patterns continuously for up to one year on one set of four 22½-V hearing-aid batteries. Can be built in empty cardboard cigarette pack. Use NE-2 bulbs.—R. M. Brown and R. Kneitel, "49 Easy Entertainment & Science Projects," Howard W. Sams, Indianapolis, IN, 1971, p 7–8.

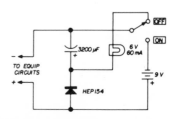

POWER-OFF FLASHER—Flasher circuit drawing less than 10 μA makes high-drain pilot lamp blink to indicate that switch is turned off. Designed for use with transistor equipment operating from 9-V battery.—S. Thomas, Switch-Off Flasher, *Ham Radio*, July 1971, p 64.

CHAPTER 37
Fluorescent Lamp Circuits

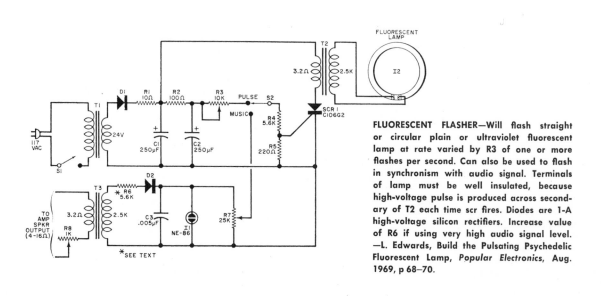

FLUORESCENT FLASHER—Will flash straight or circular plain or ultraviolet fluorescent lamp at rate varied by R3 of one or more flashes per second. Can also be used to flash in synchronism with audio signal. Terminals of lamp must be well insulated, because high-voltage pulse is produced across secondary of T2 each time scr fires. Diodes are 1-A high-voltage silicon rectifiers. Increase value of R6 if using very high audio signal level. —L. Edwards, Build the Pulsating Psychedelic Fluorescent Lamp, *Popular Electronics*, Aug. 1969, p 68—70.

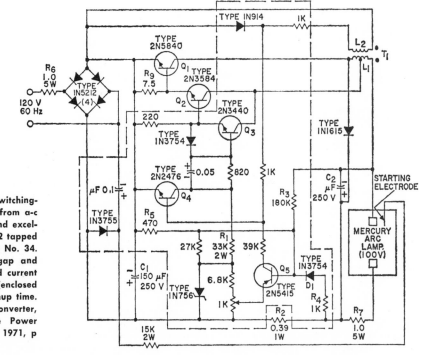

100-W MERCURY-ARC BALLAST—Switching-regulator ballasting circuit operates from a-c line with overall efficiency of 87% and excellent regulation. L1 is 120 turns No. 22 tapped 1 turn from collector. L2 is 18 turns No. 34. T1 is Arnold AH361 with 0.036" gap and 6.7:1 turns ratio. Output voltage and current are both sampled by control circuits (enclosed in dashed lines) to reduce bulb warmup time. Regulator is also known as down converter, with limited filtering.—"Solid State Power Circuits," RCA, Somerville, NJ, SP-52, 1971, p 416—419.

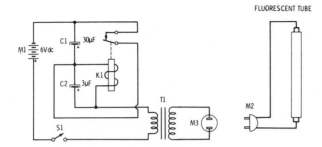

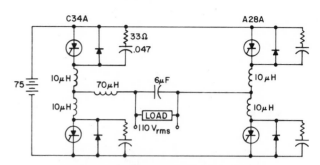

FLUORESCENT FLASHER—Uses standard audio output transformer in reverse to step up inductive surges produced by 120-ohm relay operating as flasher from hefty 6-V battery such as lattern battery. For values shown, flash is produced about every 2 s. With 2′ or 4′ red fluorescent lamp, can be used as long-range signal blinker, auto emergency light, or for pier marker at night.—R. M. Brown and T. Kneitel, "49 Easy Entertainment and Science Projects," Vol. 2, Howard W. Sams, Indianapolis, IN, 1969, p 54–55.

RESONANT BRIDGE—Uses low-cost inverter scr's capable of blocking up to 500 V both forward and reverse. Operates at 5 kHz while delivering 500 W to load. High-frequency output permits use of light-weight ballasts for fluorescent lamps. Will also operate on full-wave rectified 60-Hz a-c line voltage instead of d-c supply. Commutation of scr-diode switches (connected in inverse parallel) is accomplished by resonant L-C circuits. Diagonally opposite switches are turned on simultaneously to provide 110 V rms at 5 kHz for load.—A. P. Connolly, "Resonant Bridge Inverter," General Electric, Semiconductor Products Dept., Auburn, NY, No. 671.21, 1970, p 1–3.

180-V INPUT—High-speed converter allows use of small ferrite transformer with transistors rated only slightly higher than input voltage. Choose output turns to give desired load voltage from full-wave rectified 110-V or 220-V a-c line. Gives 4.6 W output power with 60% efficiency for 1,360-ohm load. Suitable for low-voltage unregulated supply, battery charger, small d-c motor, or (if d-c filter is omitted) fluorescent light. Inverter frequency is 14 kHz.—L. E. Donovan, "A High Input Voltage Converter," General Electric, Semiconductor Products Dept., Auburn, NY, No. 201.25.

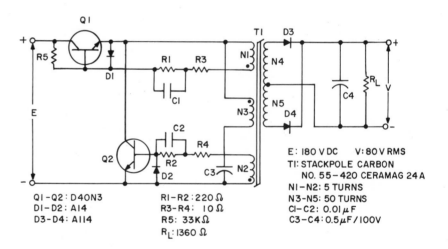

E: 180 V DC　　V: 80 V RMS
TI: STACKPOLE CARBON
　　NO. 55–420 CERAMAG 24 A
NI–N2: 5 TURNS
N3–N5: 50 TURNS
CI–C2: 0.01 μF
C3–C4: 0.5 μF/100V

Q1–Q2: D40N3
DI–D2: AI4
D3–D4: AII4

RI–R2: 220 Ω
R3–R4: 10 Ω
R5: 33 KΩ
R_L: 1360 Ω

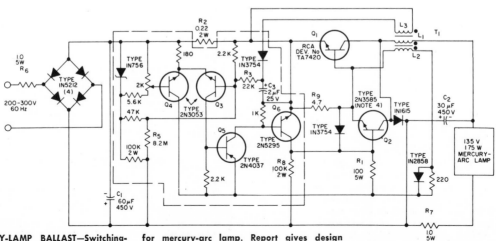

175-W MERCURY-LAMP BALLAST—Switching-regulator ballasting circuit operates from 200–300 V 60-Hz line and provides 135 V for mercury-arc lamp. Report gives design procedure and performance data, including transformer construction details.—"Power Transistors and Power Hybrid Circuits," RCA, Somerville, NJ, 1973 Ed., p 602–614.

DESK-LAMP FILTER—Will usually suppress rfi radiated over power line from two-bulb fluorescent desk lamp having pushbutton starting switch rather than starter. Requires good ground.—J. Fisk, Radio-Frequency Interference, *Ham Radio*, Dec. 1970, p 12–17.

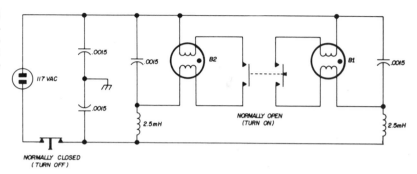

MUSIC CONTROL—Will drive up to ten 40-W fluorescent lamps. With S2 closed, brightness of lamps will vary smoothly with amplitude of audio signal fed into input jack J2 from connection to speaker terminals, to give music-controlled dancing lights keeping exact time because fluorescents do not have lag inherent in incandescent lamps. With S2 closed, for strobe effect, sensitivity is adjusted so light flashes only with higher sound levels. If desired, 100-W incandescent spotlight of any desired color can be plugged into J1.—C. L. Andrews, Dancing Fluorescent Strobe, *Radio-Electronic*, July 1971, p 46–47.

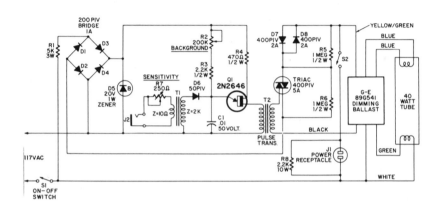

FLUORESCENT-LAMP FILTER—Can be built into base of lamp to suppress r-f interference that would otherwise go out over a-c power line. Does not cure directly radiated interference, but ¼-inch mesh screen across opening of reflector may reduce this.—J. Fisk, Radio-Frequency Interference, *Ham Radio*, Dec. 1970, p 12–17.

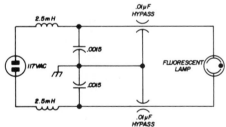

100-W MERCURY-LAMP BALLAST—Switching-regulator ballasting circuit is designed for 120-V a-c line applications in which both output voltage and current are sampled to reduce warm-up time of mercury-arc lamp. Overall efficiency is 87%. Voltage regulation is excellent.—"Power Transistors and Power Hybrid Circuits," RCA, Somerville, NJ, 1973 Ed., p 602–614.

L1 = 120 turns of No.22 wire tapped 1 turn from collector

L2 = 18 turns of No.34 wire

T1 = Arnold AH 361 (or equiv.) with 0.036" gap (6.7:1 turns ratio)

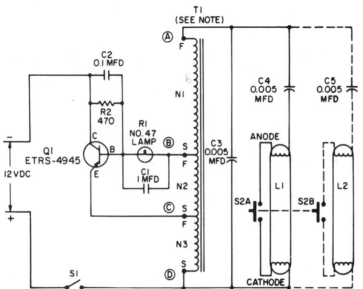

NOTE: AUTOTRANSFORMER WINDINGS—
N1 - 200 TURNS NO. 32 AWG
N2 -   6 TURNS NO. 32 AWG
N3 -   8 TURNS NO. 20 AWG
(WOUND ON ARNOLD A930157-2 CORE)

S - START OF WINDING
F - FINISH OF WINDING

16 W ON 12-V BATTERY—High-efficiency high-frequency inverter will drive two 8-W fluorescent lamps at current drain of 2 A from 12-V storage battery in trailer, motor home, truck, or boat. With one lamp, drain reduces to 1 A. Uses GE F8T5-CW lamps for L1 and L2. Construction details are given.—"Electronics Experimenters Circuit Manual," General Electric, Owensboro, KY, 1971, 3rd Ed., p 151–155.

BATTERY-POWERED 22-W FLUORESCENT—Operates from 12-V car, boat, or trailer battery. Provides large-area illumination without harsh glare, at high efficiency both for maximum brightness and for lower-level illumination at reduced battery drain. Uses pair of power transistors and inverter-ballast transformer EC-0501-LM from Milwaukee Electromagnetics, to form feedback-type power oscillator with frequency slightly above audible range. D1 is optional, preventing transistor damage if battery polarity is accidentally reversed. L1 and C1 minimize radio interference; wind L1 with 18.5 feet insulated No. 14 in four even layers on ⅝-inch dowel.—B. Richards, Battery-Powered Fluorescent Lamp, *Popular Electronics*, Dec. 1968, p 55–59.

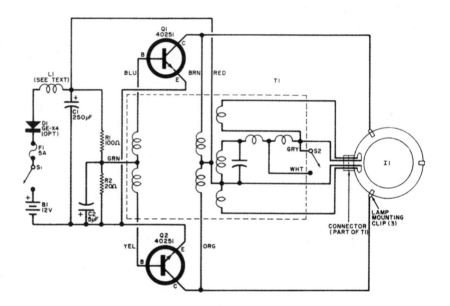

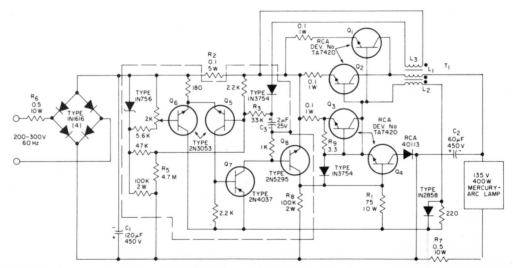

400-W MERCURY-LAMP BALLAST—Switching-regulator ballasting circuit operates from 200–300 V 60-Hz line and provides 135 V for mercury-arc lamp. Report gives design procedure and performance data, including transformer construction details.—"Power Transistors and Power Hybrid Circuits," RCA, Somerville, NJ, 1973 Ed., p 602–614.

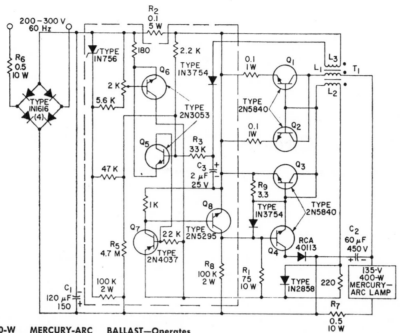

400-W MERCURY-ARC BALLAST—Operates from any 220-V a-c line and provides excellent voltage regulation for conventional 135-V mercury-arc lamp with switching-regulator ballasting circuit. Control circuit is shown within dashed lines. Efficiency of regulator alone is better than 90%. L1 is 98 turns No. 18, L2 is 10 turns No. 32, L3 is 6 turns No. 32, and T1 is Arnold AH223 with 0.125 air gap.—"Solid State Power Circuits," RCA, Somerville, NJ, SP-52, 1971, p 420–428.

CHAPTER 38
Frequency Counter Circuits

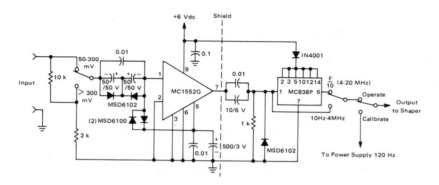

PRESCALER—Uses IC video amplifier as preamp having two input amplitude ranges for frequency counter, covering frequency range of 4 Hz to 42 MHz. Input impedance is 10K. IC decade counter increases frequency range to 20 MHz.—B. Botos, "A Frequency Counter Using Motorola RTL Integrated Circuits," Motorola Semiconductors, Phoenix, AZ, AN-451, 1969.

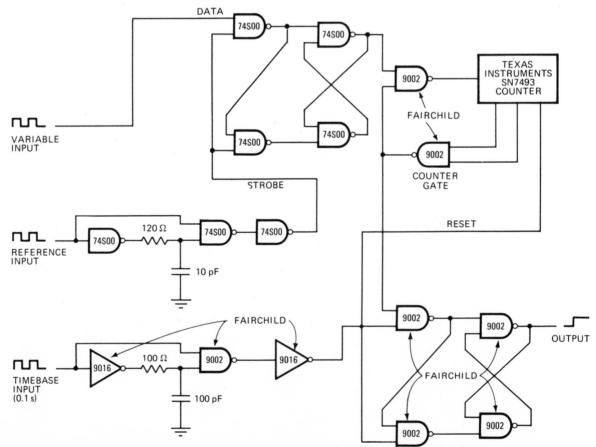

FREQUENCY-ERROR COUNTER—Reference square wave is differentiated, either with or without preparation by zero-crossover detector, and applied to strobe input of latch. Variable input goes to data input of latch. Number of positive or negative transitions made by latch output in given period equals number of times variable input has gained or lost a full cycle. Useful over reference range of 5 to 20 MHz, with error limit accuracy of 100 Hz for 0.1-s interval.—R. C. Rogers, Differentiate and Count to Find Frequency Error, *Electronics*, Feb. 28, 1972, p 81.

INPUT SIGNAL SHAPER—Used to amplify and clip signal up to 37 MHz whose frequency is to be determined by counting for accurate time interval, with resulting values shown on digital display. Article gives all circuits and covers construction.—T. Harper, Low Cost Frequency Counter, 73, Aug. 1973, p 57–62.

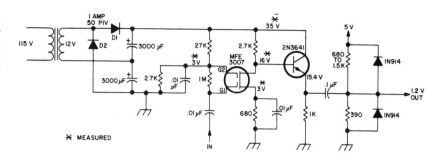

✳ MEASURED

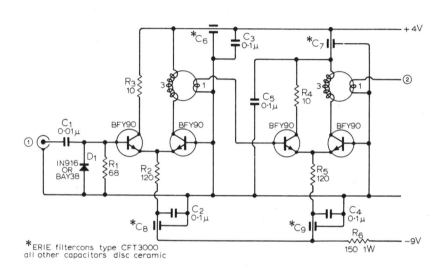

✳ERIE filtercons type CFT3000 all other capacitors disc ceramic

5–120 MHZ WIDEBAND—Uses Mullard type FX2249 ferrite cores having very low losses up to at least 100 MHz. Both transformers have 3 primary turns and 1 secondary turn of No. 24 s.w.g. enamel. Used with frequency divider to extend range of digital frequency meter.—D. R. Bowman, 100 MHz Frequency Divider, *Wireless World*, Aug. 1970, p 389–393.

DIGITAL DIAL—Simplified version of frequency counter uses subtractive mixing to cover 3.9–4.4 MHz vfo range of amateur radio transmitter or transceiver and give five-Nixie frequency readout. Uses 1-MHz crystal oscillator as reference. Article tells how to minimize digit blurring during tuning.—R. Factor, A Digital Readout for Your VFO, 73, Aug. 1971, p 56 and 58–65.

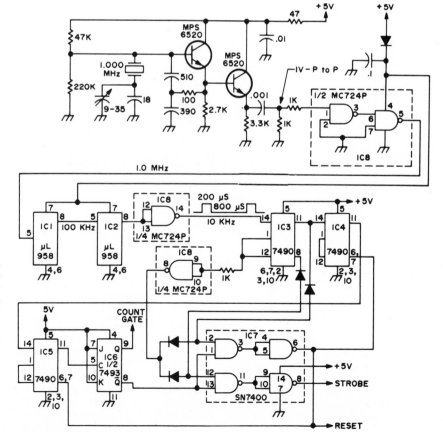

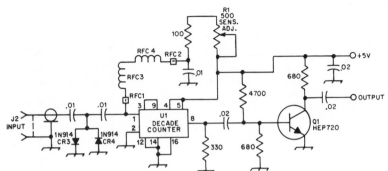

PRESCALER FOR FREQUENCY COUNTER—Uses Fairchild U6B95H9059X IC, rated for operation up to 300 MHz, to provide frequency division by 10 in system for providing digital display of f-m transmitter frequency. Requires ferrite beads RFC1 and RFC2, 0.82-μH choke RFC3, and 8.2-μH choke RFC4 for broadband decoupling to eliminate r-f from bias line.— D. A. Blakeslee, Notes on the Amateur Station Counter, QST, June 1972, p 31–35.

PULSE SHAPER—Used in frequency counter to condition incoming signal for J-K flip-flops. Uses half of hex inverter, connected as Schmitt trigger, to keep fall time of flip-flop pulse inputs from clock within range of 10–100 ns.—B. Botos, "A Frequency Counter Using Motorola RTL Integrated Circuits," Motorola Semiconductors, Phoenix, AZ, AN-451, 1969.

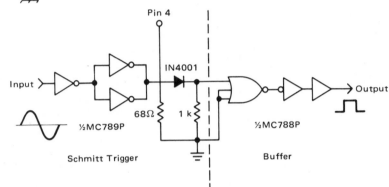

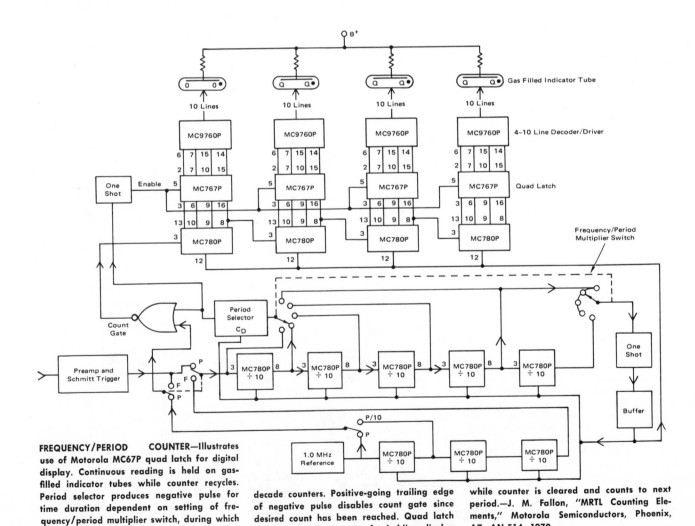

FREQUENCY/PERIOD COUNTER—Illustrates use of Motorola MC67P quad latch for digital display. Continuous reading is held on gas-filled indicator tubes while counter recycles. Period selector produces negative pulse for time duration dependent on setting of frequency/period multiplier switch, during which count gate transmits incoming pulses to four decade counters. Positive-going trailing edge of negative pulse disables count gate since desired count has been reached. Quad latch samples and stores count for holding display while counter is cleared and counts to next period.—J. M. Fallon, "MRTL Counting Elements," Motorola Semiconductors, Phoenix, AZ, AN-514, 1970.

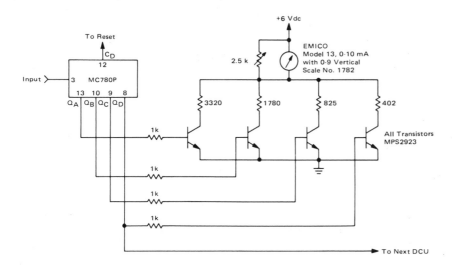

DECADE COUNTER—Used in frequency counter covering 10 Hz to 20 MHz. IC decade counter divides by 10 and delivers output in 1-2-4-8 binary-coded decimal form, for controlling on-off conditions of four transistors. Collector resistor values form sequence in which each is twice the preceding, to give binary weighted collector currents at summing junction driving meter.—B. Botos, "A Frequency Counter Using Motorola RTL Integrated Circuits," Motorola Semiconductors, Phoenix, AZ, AN-451, 1969.

VHF FREQUENCY COUNTER—Includes latches, decoders, and overflow provisions. If number of cycles counted during time period exceeds 99,999, IC18 sends pulse to pin 5 of IC21A to make it act as 1-bit latch which remembers this condition and turns on overflow lamp through Q3. IC5 and IC21 are reset by RESET A pulse when new count is started. Article gives all peripheral circuits.—P. A. Stark, A Modern VHF Frequency Counter, 73, July 1972, p 5–11 and 13.

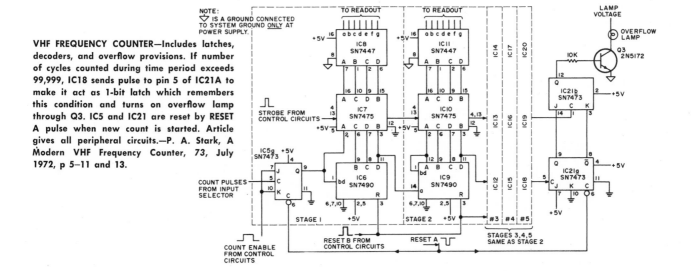

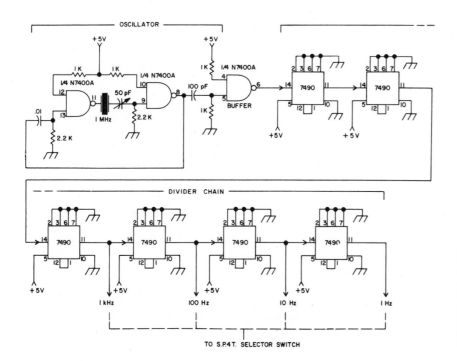

TIMING SIGNAL—Output of 1-MHz crystal frequency standard is divided down to four different frequencies for use as 1, 0.1, 0.01, and 0.001 s timing signals for frequency counter. Article gives all other circuits required to build counter that will count reliably to above 37 MHz, and give digital display.—T. Harper, Low Cost Frequency Counter, 73, Aug. 1973, p 57–62.

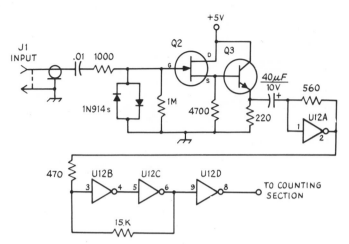

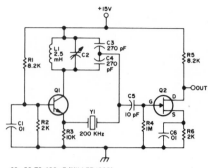

INPUT CONDITIONER—Used in frequency counter covering 20 Hz to 30 MHz. Q2 (MPF102) and Q3 (MPS3563) operate as unity-gain amplifiers for matching high input impedance to low impedance of remaining counter circuits. U12 (SN7404N TTL hex inverter) forms trigger circuit that switches at +1.35 V; for lower voltages, output remains low, and for higher voltage the output switches to high, with switching range of only a few mV.—HUA Electronics Frequency Counter Model 1BC-1a, QST, April 1972, p 60–62.

200-KHZ CRYSTAL STANDARD—Used as time base for frequency counter. Crystal is surplus FD241, having adequate stability for four-place accuracy of counter.—H. Olson, Addendum to the W1PLJ Counter, 73, March 1972, p 65–68.

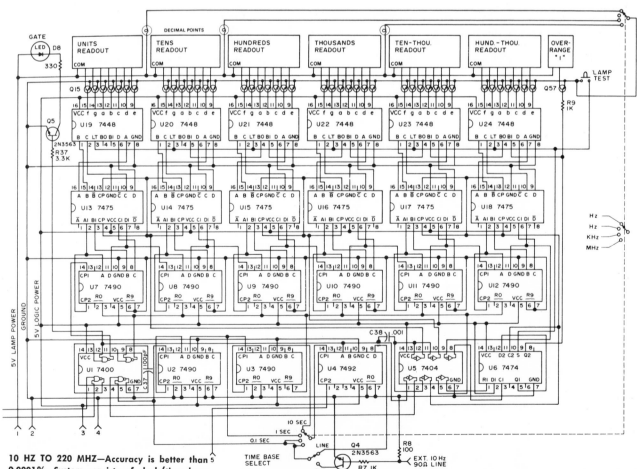

10 HZ TO 220 MHZ—Accuracy is better than 0.0001%. System consists of clock/time base unit (described in earlier issue) used with counter having 6½-digit 7-bar incandescent display compatible with accuracy of instrument. Very narrow trigger pulse eliminates one-count error which can be significant at lower frequencies. Article gives design theory along with operation and construction details. Two leads at lower left go to front end which processes and shapes input signal whose frequency is being counted. Numbered connections go to regulated power supply for which circuit is given in article.—M. McDonald, Time-Frequency Measuring System Part III, 73, March 1973, p 85–87 and 90–96.

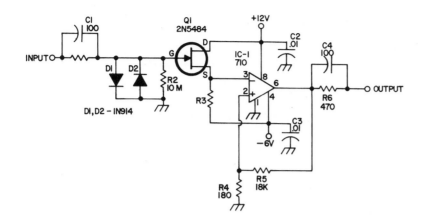

10-MHZ COUNTER INPUT WITH CLIPPER—Input signal as low as 32 mV provides reliable drive for trigger at frequencies up to 10 MHz. Diodes conduct for inputs above 2 V p-p and provide clipping, to eliminate need for input level control. IC-1 is Poly Pak 710. Use 100K in parallel with C1 at input, and 1.8K for R3. Article includes regulated power supply circuit required.—G. A. Powell, Frequency Counter Input Circuit, 73, Feb. 1973, p 89–90.

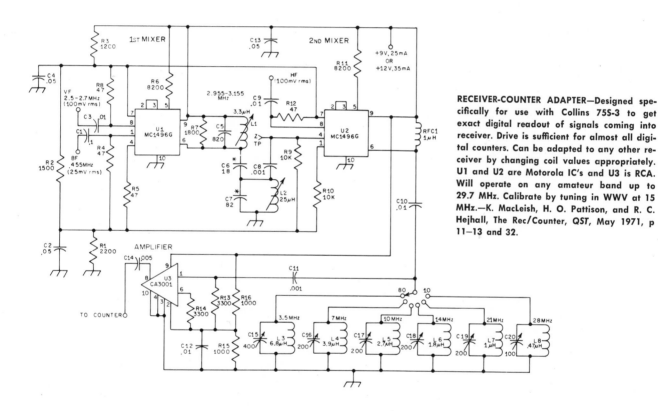

RECEIVER-COUNTER ADAPTER—Designed specifically for use with Collins 75S-3 to get exact digital readout of signals coming into receiver. Drive is sufficient for almost all digital counters. Can be adapted to any other receiver by changing coil values appropriately. U1 and U2 are Motorola IC's and U3 is RCA. Will operate on any amateur band up to 29.7 MHz. Calibrate by tuning in WWV at 15 MHz.—K. MacLeish, H. O. Pattison, and R. C. Hejhall, The Rec/Counter, QST, May 1971, p 11–13 and 32.

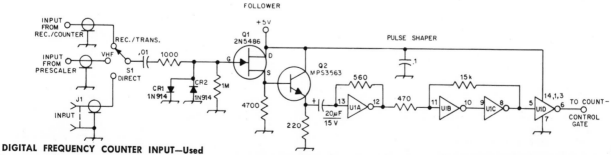

DIGITAL FREQUENCY COUNTER INPUT—Used in system for providing digital display of exact frequency value at desired point in f-m receiver or transmitter operating up to 250 MHz. With divide-by-10 prescaler at input, low-cost 30-MHz counter can be used to 300 MHz. Article discusses other prescaler arrangements. U1 is Signetics SN74H04.—D. A. Blakeslee, Notes on the Amateur Station Counter, QST, June 1972, p 31–35.

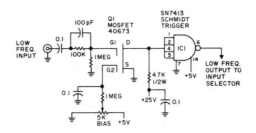

0–20 MHZ INPUT—Designed to convert variety of input signals, ranging up to 50 V with simple or complex waveforms, into digital pulse signals of correct voltage and speed for operating vhf frequency counter. Bias for gate G2 of Q1 is adjusted with 5K pot so Schmitt trigger operates in middle of its range, corresponding to about 1.3 V on fet drain with no signal.—P. A. Stark, A Modern VHF Frequency Counter, 73, July 1972, p 5–11 and 13.

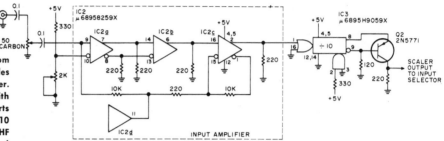

VHF PRESCALER—Accepts input signals from low r-f range up above 200 MHz and divides frequency by 10 for vhf frequency counter. Range can be extended to 300 MHz, with some loss of sensitivity, by eliminating parts in dashed box and jumper pins 2, 9, and 10 at IC2 socket.—P. A. Stark, A Modern VHF Frequency Counter, 73, July 1972, p 5–11 and 13.

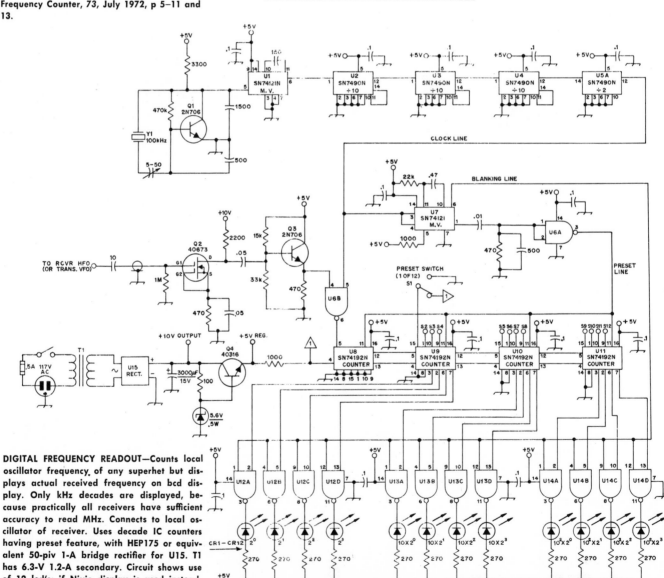

DIGITAL FREQUENCY READOUT—Counts local oscillator frequency of any superhet but displays actual received frequency on bcd display. Only kHz decades are displayed, because practically all receivers have sufficient accuracy to read MHz. Connects to local oscillator of receiver. Uses decade IC counters having preset feature, with HEP175 or equivalent 50-piv 1-A bridge rectifier for U15. T1 has 6.3-V 1.2-A secondary. Circuit shows use of 12 led's; if Nixie display is used instead, SN7400 blanking gates U12, U13, and U14 can be omitted since necessary decoders include blanking input. U6 is also SN7400. Arti-

cle tells how to set up counter by first tuning to WWV and adjusting 100-kHz oscillator, then tuning to known station and flipping preset switches to achieve correct display.—J. Hagen, A Simple Frequency Counter for Receivers, QST, Nov. 1972, p 11–13.

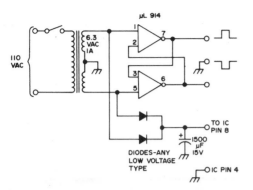

60-HZ SQUARE WAVES—Circuit is essentially R-S flip-flop using dual two-input gate which is triggered by 60 Hz from filament transformer. Used in demonstrating methods of frequency counting via cro patterns when using counting logic.—A. S. Joffe, IC Logic Doesn't Have to be Obscure, 73, 1972, 217–219.

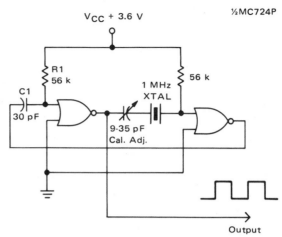

1-MHZ CRYSTAL OSCILLATOR—Used in frequency counter to turn count gate on for specific gate time so output of gate is burst of pulses whose number is directly proportional to original input frequency. Uses cross-coupled gates forming free-running mvbr whose square-wave output frequency is locked by crystal. Stability is adequate (frequency within 0.01%) without crystal oven.—B. Botos, "A Frequency Counter Using Motorola RTL Integrated Circuits," Motorola Semiconductors, Phoenix, AZ, AN-451, 1969.

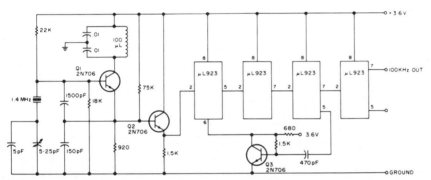

100-KHZ TIME BASE—Used in frequency counter reading up to 50 MHz. Output should be set exactly to 100 kHz either by comparison with commercial counter or with WWV, using scope rather than zero-beating by ear. Q2 is buffer driving first IC in divide-by-7 circuit. Third IC is connected to Q3 in half-mono mvbr resetting first IC to force division by 7. Last IC divides by 2 to give desired output.—T. Johnson, Another Integrated Circuit Frequency Counter, 73, Jan. 1973, p 39–42 and 44–45.

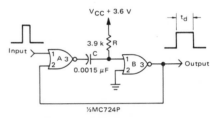

ONE-SHOT MVBR—Maintains reset pulse of frequency counter for 5 μs to ensure complete reset. Uses only two NOR gates, one resistor, and one capacitor. Circuit operation is analyzed.—B. Botos, "A Frequency Counter Using Motorola RTL Integrated Circuits," Motorola Semiconductors, Phoenix, AZ, AN-451, 1969.

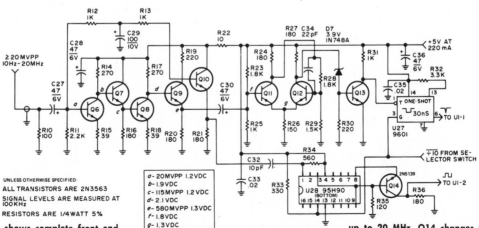

PRESCALER—Circuit shows complete front end of frequency counter providing accuracy better than 0.0001% from 10 Hz to 220 MHz, including preamp, squaring amplifier, one-shot, and prescaler providing division by 10. Squaring amplifier Q11-Q12 is Schmitt trigger feeding Q13 used as level shifter. U27 is type 9601 IC pulse generator connected for shortest possible pulse width, to allow operation up to 20 MHz. Q14 changes from ECL levels of 95H90 prescaler to TTL levels for counter gates.—M. McDonald, Time-Frequency Measuring System Part III, 73, March 1973, p 85–87 and 90–96.

CHAPTER 39
Frequency Divider Circuits

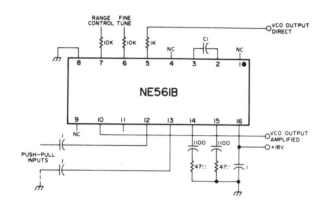

MULTIPLIER-DIVIDER—Uses Signetics IC phase-locked loop for multiplying input frequency up to ten times or for dividing by 3, 5, 7, or 9. Choose value for C1 to combine with fine-tuning adjustment for operation near desired output frequency. If input is within locking range, and of adequate strength, vco will lock to desired multiple or odd submultiple of input. For single-ended input, bypass one input lead to ground as shown by dotted line. Output is harmonic-rich square wave.—J. Kyle, The Phase-Locked Loop Comes of Age, 73, Oct. 1970, p 42, 44–46, 49–52, and 54–56.

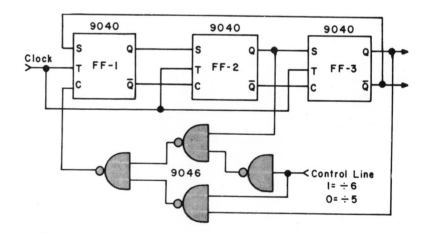

DIVISION CHOICE OF 5 OR 6—Addition of quad NAND gate to arrangement of three J-K flip-flops permits division by 6 when control line is high (logic 1) or division by 5 when control line is low. All IC's are Fairchild.—J. V. Sastry, Logic Levels Control Division Ratio, *The Electronic Engineer*, Dec. 1968, p 82.

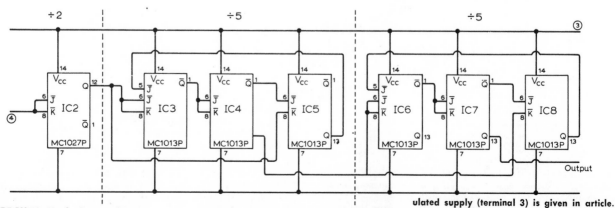

DIVIDE-BY-50—Used in combination with wideband amplifier, pulse shaper, and divide-by-two stage to extend range of digital frequency meter. Circuit for required 4-V regulated supply (terminal 3) is given in article.—D. R. Bowman, 100 MHz Frequency Divider, *Wireless World*, Aug. 1970, p 389–393.

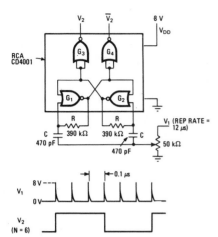

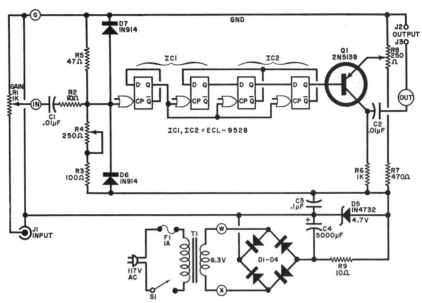

DIGITAL DIVIDER—Even-order frequency divider consists of two c-mos NOR gates G1 and G2 connected as simple latch. Pot adjusts division modulo N from 2 to 30. Gates G3 and G4 act as output pulse shapers. Generates complementary square waves at output frequency.—D. Newton, C-MOS Gate Package Forms Adjustable Divider, *Electronics*, March 1, 1973, p 104.

EXTENDING 20-MHZ COUNTER TO 175 MHZ —Divide-by-ten circuit used as front end for limited-range counter extends operating range to 175 MHz at cost of less than $35.

D1–D4 are 1-A silicon diodes. Secondary of T1 is 6.3 V at 600 mA.—D. Meyer, Build a 175-MHz Prescaler, *Popular Electronics*, April 1972, p 53–55.

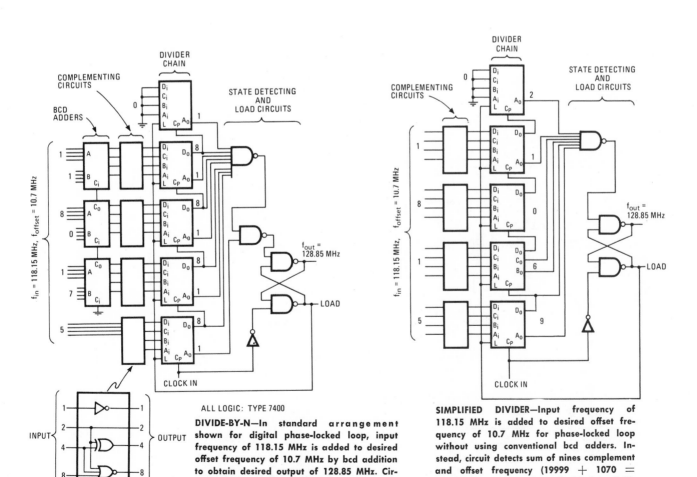

DIVIDE-BY-N—In standard arrangement shown for digital phase-locked loop, input frequency of 118.15 MHz is added to desired offset frequency of 10.7 MHz by bcd addition to obtain desired output of 128.85 MHz. Circuit takes nines complement of sum, then detects signal when counter outputs are all nines.—L. Martin, Circumventing BCD Addition in Digital Phase-Locked Loops, *Electronics*, May 22, 1972, p 100.

SIMPLIFIED DIVIDER—Input frequency of 118.15 MHz is added to desired offset frequency of 10.7 MHz for phase-locked loop without using conventional bcd adders. Instead, circuit detects sum of nines complement and offset frequency (19999 + 1070 = 21069). Correct count is reached before divider resets. All logic is type 7400.—L. Martin, Circumventing BCD Addition in Digital Phase-Locked Loops, *Electronics*, May 22, 1972, p 100.

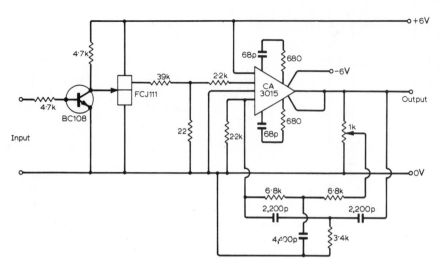

DIVIDER FOR 5 TO 20 KHZ—Modification of parallel T circuit which allows tuning with single 1K pot is connected in feedback path of opamp to give narrow-pass amplifier that can be tuned over range of input frequencies while preserving sinusoidal waveform.—F. C. Evans, *Frequency Divider with Variable Tuning*, *Wireless World*, July 1969, p 324.

INCREASING 10-MHZ COUNTER TO 40 MHZ—Frequency divider is triggered by sine or square waves or pulses having peaks between 0.5 and 50 V. Output is short-circuit-proof. Limiter-amplifier Q1 drives dual J-K flip-flop having Q2 as output buffer.—R. Van Dick, *Divide-by-Four at 40 MHz*, *The Electronic Engineer*, Nov. 1968, p 75.

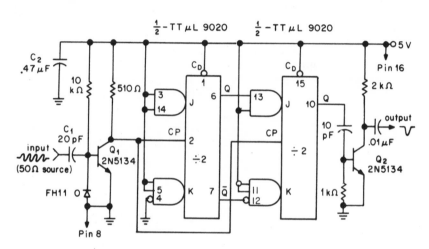

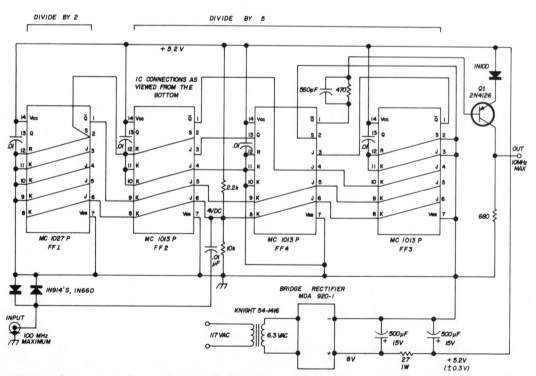

DIVIDE-BY-TEN SCALER—Delivers 4 V p-p for driving digital frequency counter, to increase range of counter by factor of 10. Uses inexpensive J-K flip-flops and unregulated four-diode bridge supply (all Motorola). Q1 increases scaler output voltage to 4 V and translates MECL logic level to saturated logic levels.—B. Kelley, *Divide-by-Ten Frequency Scaler*, *Ham Radio*, Aug. 1970, p 26–29.

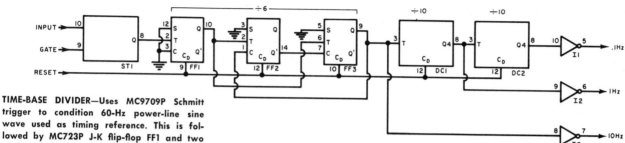

TIME-BASE DIVIDER—Uses MC9709P Schmitt trigger to condition 60-Hz power-line sine wave used as timing reference. This is followed by MC723P J-K flip-flop FF1 and two halves of MC790P dual J-K flip-flop which divide by 6 to give 10-Hz output. Two Motorola MC780P decade up counters each divide by 10 to give outputs of 1 and 0.1 Hz. Each output is through one-sixth of MC789P hex inverter. Used in complete 4-MHz electronic counter described in article.—D. L. Steinbach, Digital Instruments, *Electronics World*, Jan. 1971, p 35—38 and 56.

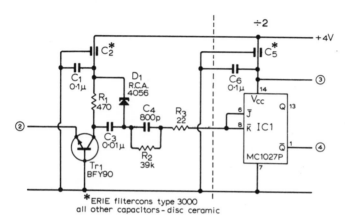

*ERIE filtercons type 3000 all other capacitors - disc ceramic

PULSE SHAPER AND DIVIDE-BY-TWO—Used after wideband amplifier to feed additional divider stages for extending range of digital frequency meter. Tunnel diode operates with rise time well within 2 ns required to drive IC logic stage.—D. R. Bowman, 100 MHz Frequency Divider, *Wireless World*, Aug. 1970, p 389—393.

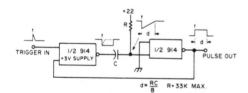

$d = \frac{RC}{8}$ R = 33K MAX.

IC ONE-SHOT—Will serve nicely as frequency divider in frequency standard. Duration of output pulse will be almost exactly ⅛th of RC time constant, with R in ohms and C in farads. R should not be larger than 33K.—J. Kyle, The Logical Approach to Surplus Buying, 73, March 1970, p 80—92.

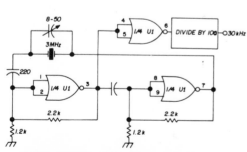

3-MHZ CRYSTAL WITH 100:1 DIVIDER—Uses three of four NAND gates in high-frequency 74H00 IC as crystal-controlled mvbr and buffer driving two 7490 decade dividers to give 30-kHz square-wave output.—E. Noll, Digital IC Oscillators and Dividers, *Ham Radio*, Aug. 1972, p 62—67.

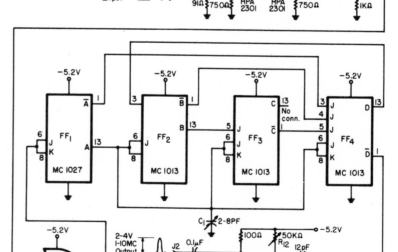

RANGE EXTENDER—IC dividers will multiply range of most 10-MHz counters by factor of 10. R1, which controls bias points of hot-carrier diodes, should be adjusted for —3.2 V d-c at anode of CR1. Limiter signal is shaped by MC1023 high-speed dual clock driver whose input is biased to 1.15 V d-c by setting of R2. Flip-flop FF1 divides signal by 2, and remaining flip-flops divide by 5.—R. L. Starliper, Inexpensive Divider Extends Range of 10 MHz Counter to 100 MHz, *The Electronic Engineer*, July 1971, p 56.

THREE-GATE DIVIDE-BY-30 ONE-SHOT—Pot Rt determines lockout time of gate G5 after one pulse of 1.1-MHz crystal reference passes to output. One-shot G3-G4 then resets to allow next crystal pulse to reach output. Provides precise, stable frequency divisions of any integral value between 2 and 30.—J. Snaper, Control One-Shot Divides Frequency by up to 30, *Electronics*, May 8, 1972, p 102.

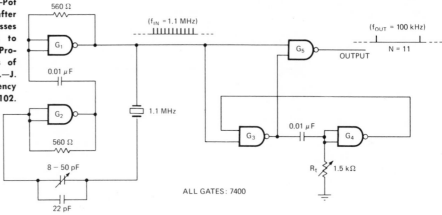

ALL GATES: 7400

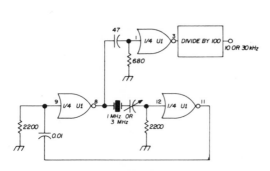

1 OR 3 MHZ CRYSTAL IC—Two 7400 NAND gates wired as mvbr feed two 7490 decade dividers through buffer gate to give 10-kHz or 30-kHz square-wave output with high frequency accuracy.—E. Noll, Digital IC Oscillators and Dividers, *Ham Radio*, Aug. 1972, p 62–67.

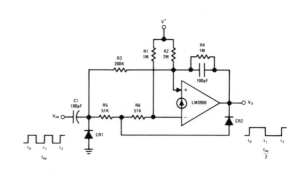

DIVIDING FLIP-FLOP—Uses only one Norton current-differencing opamp as asynchronous trigger for dividing input frequency. CR2 provides steering of differentiated positive input trigger.—T. M. Frederiksen, W. M. Howard, and R. S. Sleeth, "The LM3900—A New Current-Differencing Quad of ± Input Amplifiers," National Semiconductor, Santa Clara, CA, 1972, AN-72, p 27.

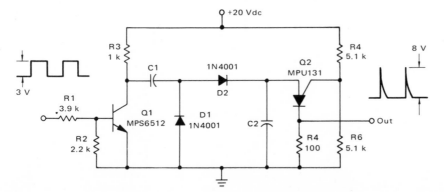

C₁	C₂	Division
0.01 μF	0.01 μF	2
0.01 μF	0.02 μF	3
0.01 μF	0.03 μF	4
0.01 μF	0.04 μF	5
0.01 μF	0.05 μF	6
0.01 μF	0.06 μF	7
0.01 μF	0.07 μF	8
0.01 μF	0.08 μF	9
0.01 μF	0.09 μF	10
0.01 μF	0.1 μF	11

LOW-FREQUENCY DIVIDER—Intended for square-wave input frequencies in audio range, with amplitude of about 3 V. Division ratio depends on capacitor values, as given in table. Circuit is fairly insensitive to amplitude, pulse width, rise time, and fall time of incoming pulses.—R. J. Haver and B. C. Shiner, "Theory, Characteristics and Applications of the Programmable Unijunction Transistor," Motorola Semiconductors, Phoenix, AZ, AN-527, 1971.

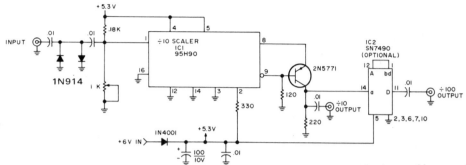

300-MHZ DIVIDE-BY-10—First IC is Fairchild μ6B95H9059X scaler, dividing input signal frequency by 10 and feeding transistor am-plifier. This will extend range of 15-MHz frequency counter to 150 MHz. If scaler is slower than 15 MHz, use second IC decade counter to provide second division by 10. Power supply can be four D cells.—P.A. Stark, 300 MHz Frequency Scaler, 73, June 1972, p 97 and 100.

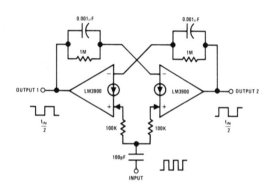

DIVIDING FLIP-FLOP—Uses pair of Norton current-differencing opamps to provide division of pulse input frequency with complementary outputs.—T. M. Frederiksen, W. M. Howard, and R. S. Sleeth, "The LM3900—A New Current-Differencing Quad of ± Input Amplifiers," National Semiconductor, Santa Clara, CA, 1972, AN-72, p 27.

CHAPTER 40
Frequency Modulation Circuits

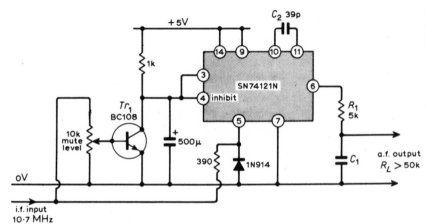

DEMODULATOR—Connecting TTL mono mvbr as shown gives high-linearity f-m demodulator with harmonic distortion less than 0.5% for up to 75-kHz deviation. Requires muting in absence of signal, to prevent random triggering of mvbr by noise; this is achieved by feeding inhibit terminals of Schmitt trigger in IC from collector of half-wave rectifier stage TR1.—P. Keenan, Low Distortion F.M. Demodulator, *Wireless World*, April 1972, p 185.

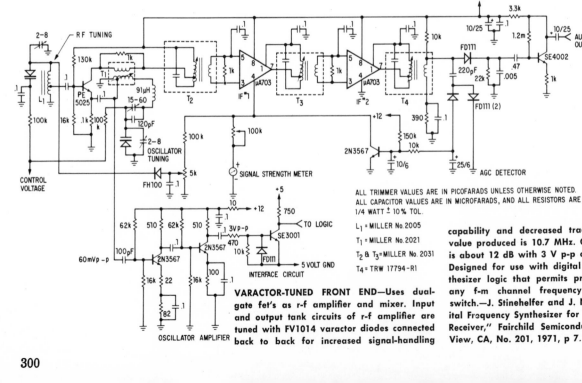

ALL TRIMMER VALUES ARE IN PICOFARADS UNLESS OTHERWISE NOTED.
ALL CAPACITOR VALUES ARE IN MICROFARADS, AND ALL RESISTORS ARE
1/4 WATT ± 10% TOL.

L_1 = MILLER No. 2005
T_1 = MILLER No. 2021
T_2 & T_3 = MILLER No. 2031
T_4 = TRW 17794-R1

VARACTOR-TUNED FRONT END—Uses dual-gate fet's as r-f amplifier and mixer. Input and output tank circuits of r-f amplifier are tuned with FV1014 varactor diodes connected back to back for increased signal-handling capability and decreased tracking error. I-f value produced is 10.7 MHz. Conversion gain is about 12 dB with 3 V p-p oscillator signal. Designed for use with digital frequency synthesizer logic that permits precise tuning to any f-m channel frequency with selector switch.—J. Stinehelfer and J. Nichols, "A Digital Frequency Synthesizer for an AM and FM Receiver," Fairchild Semiconductor, Mountain View, CA, No. 201, 1971, p 7.

VARACTOR QUINTUPLER FOR 220 MHZ—Requires driving frequency of 44 to 45 MHz. CR1 is Amperex H4A/1N4885 or equivalent varactor diode. Article gives coil-winding data, including construction of strip-line tank L6-L7. Circuit uses two idlers; L3-C4 is tuned to 89.6 MHz and L4-C5 is tuned to 134.4 MHz.—T. McMullen, An Easy Road to 220 MHz, QST, Jan. 1972, p 11–13.

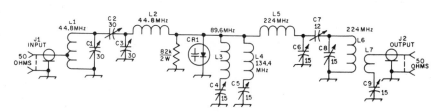

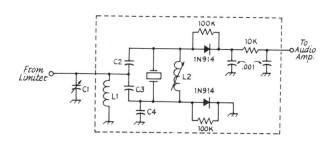

CRYSTAL DISCRIMINATOR—Adjustment-free f-m detector design uses quartz resonator shunted by coil to serve in place of secondary of conventional discriminator transformer. Unmarked parts are chosen to give desired bandwidth. C1 and L1 should resonate at i-f value. C2 equals C3. C4 corrects any circuit imbalance, so diodes receive equal amounts of signal.—"The Radio Amateur's Handbook," American Radio Relay League, Newington, CT, 49th Ed., 1972, p 443.

VFO FOR 2-METER RECEIVER—Tuning capacitors and tube provide flexibility for monitoring f-m repeaters and for scanning band, at much lower cost than solid-state circuit using crystals. Tuning range is continuous from 146 to 148 MHz, with vfo operating in 11-MHz region. Vfo plugs into crystal socket of one channel of commercial 2-meter receiver. CR1 is 33-pf HEP R2503 varactor diode. Article gives coil data.—D. DeMaw and G. Wilson, A High-Performance Tunable FM Receiver, QST, April 1972, p 44–48.

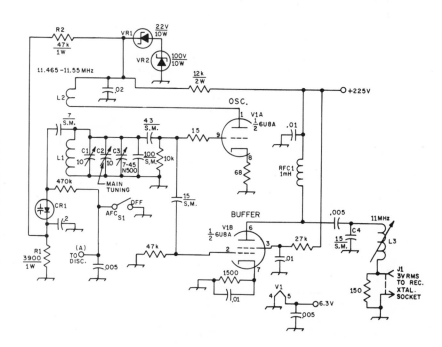

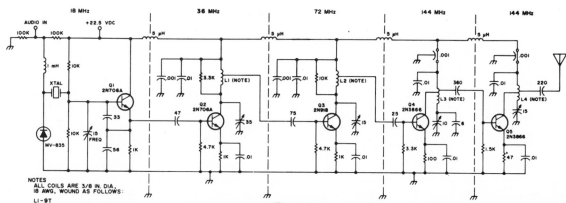

2-METER F-M TRANSMITTER—Operates from single 22.5-V battery and provides adequate power output at milliwatt level to activate f-m repeater within line-of-sight range. Crystal frequency of 18.29250 MHz is multiplied eight times to get 146.34-MHz output frequency. Oscillator is modulated with Motorola varicap. Article covers construction and tuning.—C. Klinert, A 2M Minitransmitter For Repeater Use, 73, Dec. 1970, p 32–34 and 36.

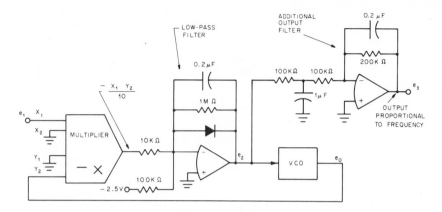

PHASE-LOCKED LOOP DEMODULATOR—Uses Burr-Brown opamps and analog function modules with any commercial voltage-controlled oscillator (VCO) which has output frequency of 1 kHz plus 1 kHz/0.5 V over 20:1 range. For values shown, frequency may vary over 20:1 range, and signal amplitudes may vary over 20:1 range without losing lock.—J. G. Graeme, G. E. Tobey, and L. P. Huelsman, "Operational Amplifiers," McGraw-Hill Book Co., New York, 1971, p 423–425.

50-W 2-METER AMPLIFIER—Uses single transistor operated Class C with zero bias, for f-m or c-w operation off auto battery. Not suited for a-m or ssb signals because amplifier is not linear. RFC3 is ferrite bead. Designed to match 50-ohm source and 50-ohm antenna system.—R. C. Hejhall, Some 2-Meter Solid-State Rf Power-Amplifier Circuits, QST, May 1972, p 40–45 and 68.

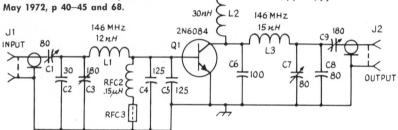

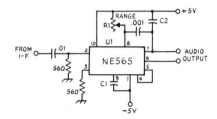

PLL DETECTOR—Signetics NE565 phase-locked loop IC gives sensitivity of 1 mV and bandwidth up to 10% of i-f for f-m detection. R1 and C1 are chosen to set vco of IC close to desired frequency. Loop filter capacitor C2 determines capture range for locking with input signal. Upper frequency limit is 500 kHz.—"The Radio Amateur's Handbook," American Radio Relay League, Newington, CT, 49th Ed., 1972, p 443.

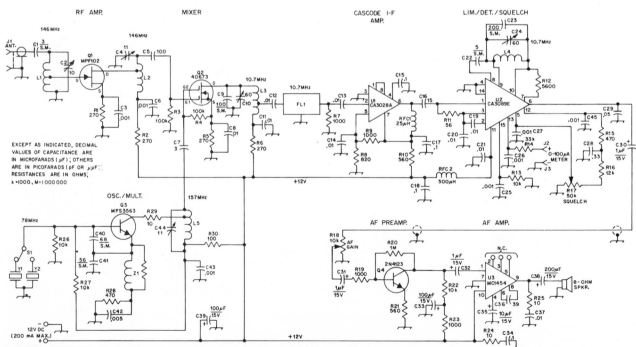

2-METER RECEIVER—Single-conversion solid-state receiver design is centered around RCA CA3089E IC serving multiple functions. Oscilator crystal frequency is 78 MHz, which is half the injection frequency. FL1 is any 10.7-MHz filter having bandwidth for amateur f-m, such as KVG XM 107S04 made by Spectrum International, Topsfield, MA.—D. DeMaw, A Single-Conversion 2-Meter FM Receiver, QST, Aug. 1972, p 11–15.

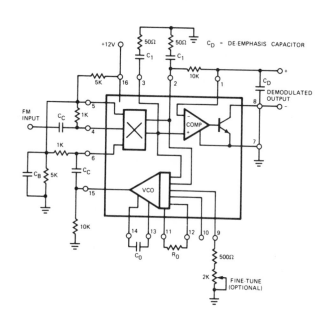

1–10 MHZ DEMODULATOR—Uses Exar XR-210 fsk modulator-demodulator for frequency-selective f-m demodulation. Value of C0 depends on carrier frequency, and low-pass filter C1 depends on selectivity requirements. For carriers between 1 and 10 MHz, C1 is in range of 10 to 30 C0. Demodulated output is about 5 R0 mV (rms) % of f-m deviation, where R0 is gain control resistance in kilohms. Harmonic distortion of output is under 0.5% with f-m deviations up to 10%.—"XR-210 FSK Modulator/Demodulator," Exar Integrated Systems Inc., Sunnyvale, CA, June 1972, p 8.

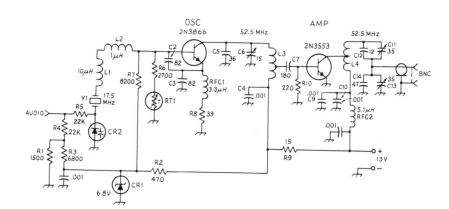

52.5-MHZ EXCITER—Uses varactor diode (4 pF at 4 V) to change series capacitance in Clapp oscillator adapted for crystal-controlled wide-band f-m transmitter. Article gives coil data. Is capable of at least 15-kHz deviation at 52.5 MHz, and is also usable for true wide-band f-m on 220 MHz with deviation of 75 kHz. Provides up to 0.8 W output into 50 or 70-ohm load. Article includes simple 2N2222 mike preamp for audio input.—C. F. Hadlock, Wide-Band FM with Crystal Control, QST, Oct. 1972, p 17–21 and 44.

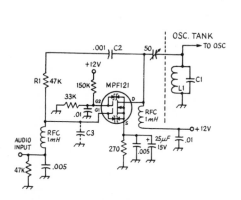

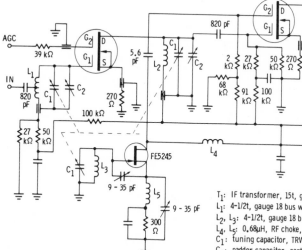

REACTANCE MODULATOR—Uses mosfet connected across oscillator tank and serving as variable inductance. Output frequency increases in proportion to amplitude of audio input. Developed for amateur f-m transmitters.—"The Radio Amateur's Handbook," American Radio Relay League, Newington, CT, 49th Ed., 1972, p 435.

TUNER—Uses conventional ganged tuning capacitor for fet oscillator, dual-gate mosfet r-f stage, and dual-gate mosfet mixer. Each mosfet is Fairchild FT0601. Report gives design

equations. L1 has ¼ inch coil form.—"RF Applications of the FT0601 Dual-Gate Mosfet," Fairchild Semiconductor, Mountain View, CA, No. 189, 1970, p 7–8.

T_1: IF transformer, 15t, gauge 32 wire, J-slug, adjustable
L_1: 4-1/2t, gauge 18 bus wire, tapped at 2-1/2t from ground
L_2, L_3: 4-1/2t, gauge 18 bus wire, on 1/4" ID coil form
L_4, L_5: 0.68μH, RF choke, Miller, encapsulated
C_1: tuning capacitor, TRW 57-3A
C_2: padder capacitor, part of C_1
All other capacitors: 1000 pF, FT

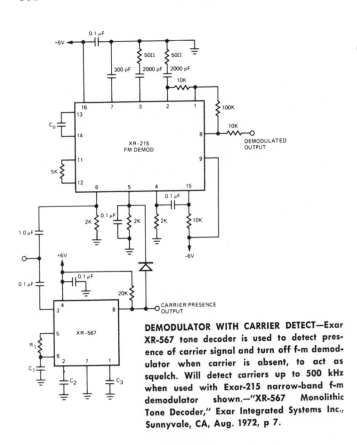

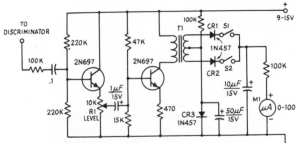

DEMODULATOR WITH CARRIER DETECT—Exar XR-567 tone decoder is used to detect presence of carrier signal and turn off f-m demodulator when carrier is absent, to act as squelch. Will detect carriers up to 500 kHz when used with Exar-215 narrow-band f-m demodulator shown.—"XR-567 Monolithic Tone Decoder," Exar Integrated Systems Inc., Sunnyvale, CA, Aug. 1972, p 7.

DEVIATION METER—Used for checking peak frequency deviation of f-m transmitter. T1 has 10K primary and 20K C-T secondary, such as Triad A31X. Used with output of discriminator in wideband receiver tuned to transmitter being checked. Calibration procedure is given, along with chart of audio frequencies that will produce carrier null for various deviation values. Detected noise gives meter reading with zero input, so accuracy is poor on weak signals.—"The Radio Amateur's Handbook," American Radio Relay League, Newington, CT, 49th Ed., 1972, p 437—438.

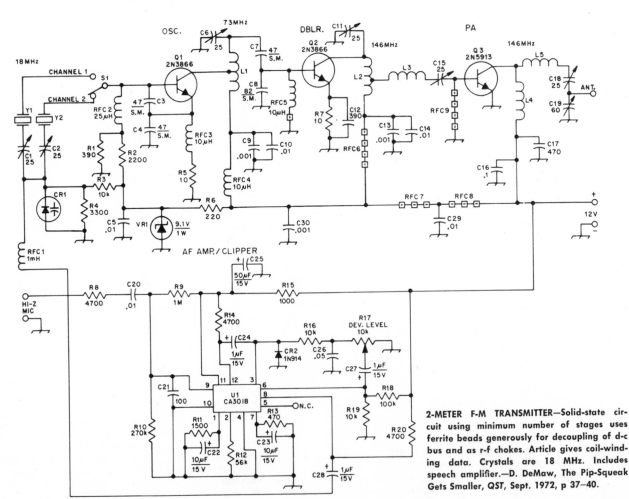

2-METER F-M TRANSMITTER—Solid-state circuit using minimum number of stages uses ferrite beads generously for decoupling of d-c bus and as r-f chokes. Article gives coil-winding data. Crystals are 18 MHz. Includes speech amplifier.—D. DeMaw, The Pip-Squeak Gets Smaller, QST, Sept. 1972, p 37—40.

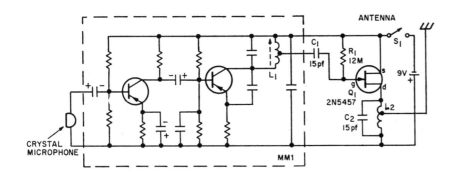

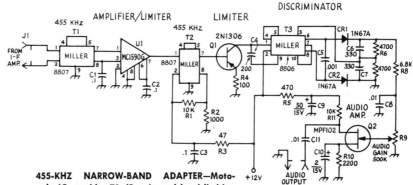

WIRELESS MIKE—Addition of fet buffer stage to output of commercial f-m wireless microphone prevents detuning by changing body capacitance when unit is worn like necklace by vocalist. Mike can be Allied Radio Shack 277-205 shown in dashed rectangle, or equivalent. L2 has 3½ turns No. 22, spread to 5/16 inch on 3/16-inch diameter, tapped at center. Use 18-inch antenna wire.—S. Daniels, Magic-Mike, *Elementary Electronics*, July-Aug. 1971, p 69–71.

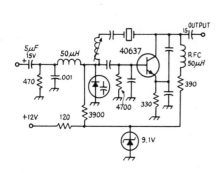

VARACTOR REACTANCE MODULATOR—May be used either with self-controlled or crystal oscillator in f-m transmitter. Since frequency deviation obtainable with crystal oscillator is quite small, resulting signal can be more phase-modulated than frequency-modulated. —"The Radio Amateur's Handbook," American Radio Relay League, Newington, CT, 49th Ed., 1972, p 435.

455-KHZ NARROW-BAND ADAPTER—Motorola IC provides 70 dB gain and hard limiting action. Bandwidth of miniature transformers T1 and T2 restricts adapter to narrow-band reception, but wideband performance can be obtained with Miller 8811 units. Used for f-m reception with older tube-type amateur receivers.—"The Radio Amateur's Handbook," American Radio Relay League, Newington, CT, 49th Ed., 1972, p 445-446.

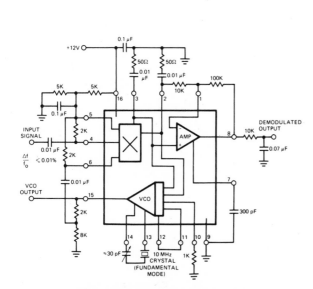

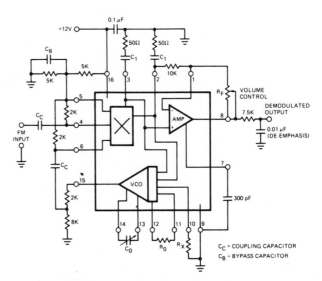

CRYSTAL-CONTROLLED PLL—Timing capacitor for Exar XR-215 phase-locked loop is replaced by crystal to improve stability of f-m demodulator. Typical pull-in range of circuit is 1 kHz at 10 MHz.—"XR-215 Monolithic Phase-Locked Loop," Exar Integrated Systems Inc., Sunnyvale, CA, July 1972, p 8.

DEMODULATOR—Uses Exar XR-215 phase-locked loop for frequency-selective demodulation. Value of C0 depends of f-m carrier frequency, ranging from about 10,000,000 pF for 100 Hz to 50 pF for 10 MHz. C1 is determined by selectivity requirements, and is in range of 10 to 30 C0. Report gives performance graphs.—"XR-215 Monolithic Phase-Locked Loop," Exar Integrated Systems Inc., Sunnyvale, CA, July 1972, p 6.

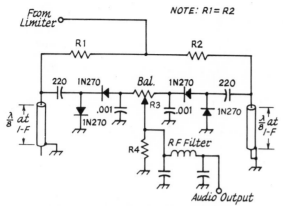

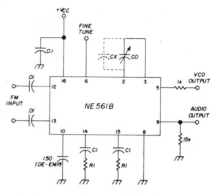

BRIDGE DISCRIMINATOR—Can provide nearly distortion-free demodulation of f-m signals. Detector consists of balanced transmission-line bridge. Choose values of R1 and R2 to give desired input impedance. Pi-section output filter removes residual i-f energy.—D. A. Blakeslee, Receiving FM, QST, Feb. 1971, p 16–23.

PHASE-LOCKED LOOP AS DETECTOR—Signetics IC requires only additional parts shown for use as efficient f-m detector. Capacitor between pins 2 and 3 determines frequency of vco. For fine tuning, small variable voltage may be applied to pin 6 or Cx added. Values of R1 and C1 in low-pass loop filter depend on i-f value.—E. Noll, Phase-Locked Loops, Ham Radio, Sept. 1971, p 54–60.

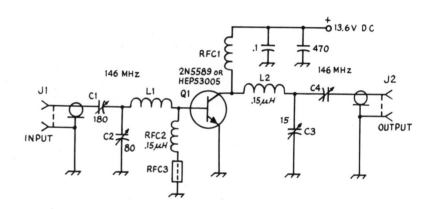

3-W 2-METER DRIVER—Designed for f-m or c-w operation off auto battery. L1 is 1 turn No. 18, 3/16 inch diameter. RFC1 is 10 turns number 20 wound over 510-ohm ½-W composition resistor. RFC3 is ferrite bead. Input and output impedances are 50 ohms.—R. C. Hejhall, Some 2-Meter Solid-State Rf Power-Amplifier Circuits, QST, May 1972, p 40–45 and 68.

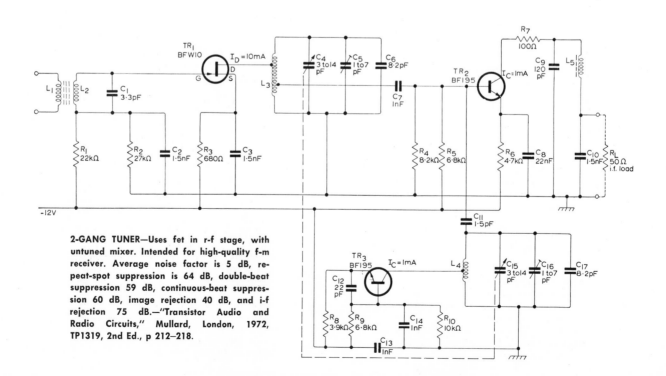

2-GANG TUNER—Uses fet in r-f stage, with untuned mixer. Intended for high-quality f-m receiver. Average noise factor is 5 dB, repeat-spot suppression is 64 dB, double-beat suppression 59 dB, continuous-beat suppression 60 dB, image rejection 40 dB, and i-f rejection 75 dB.—"Transistor Audio and Radio Circuits," Mullard, London, 1972, TP1319, 2nd Ed., p 212–218.

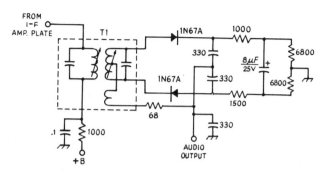

RATIO DETECTOR—Widely used in radio and tv receivers. Input signal may vary in strength over wide range without changing level of output voltage, for detection of f-m but not a-m. Audio output is taken from tertiary winding tightly coupled to primary of T1. For communication service, at least one limiter stage should be used ahead of ratio detector.—"The Radio Amateur's Handbook," American Radio Relay League, Newington, CT, 49th Ed., 1972, p 442.

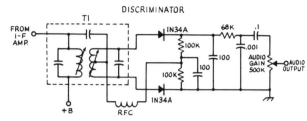

DISCRIMINATOR—T1 converts f-m signal to a-m for rectification by diodes. With no audio, rectified voltages are equal and opposite so output is zero. Shift in input frequency shifts phase of voltage components, increasing amplitude of one side of secondary of T1 and decreasing on other side. Resulting differences in rectified voltages constitute audio output.—"The Radio Amateur's Handbook," American Radio Relay League, Newington, CT, 49th Ed., 1972, p 442.

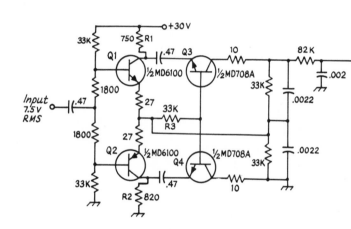

TRANSFORMERLESS DISCRIMINATOR—Operates up to 1 MHz. Uses active circuits to provide required phase shifts.—D. A. Blakeslee, Receiving FM, QST, Feb. 1971, p 16–23.

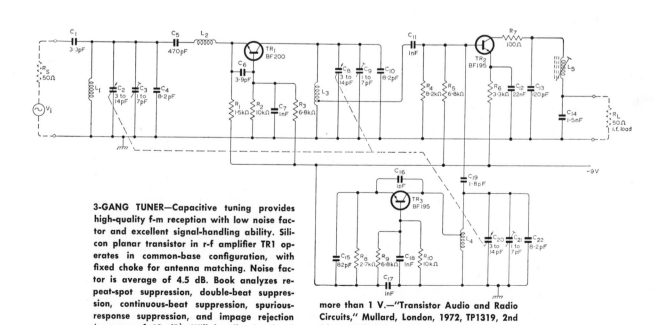

3-GANG TUNER—Capacitive tuning provides high-quality f-m reception with low noise factor and excellent signal-handling ability. Silicon planar transistor in r-f amplifier TR1 operates in common-base configuration, with fixed choke for antenna matching. Noise factor is average of 4.5 dB. Book analyzes repeat-spot suppression, double-beat suppression, continuous-beat suppression, spurious-response suppression, and impage rejection (average of 63 dB). Will handle signals of more than 1 V.—"Transistor Audio and Radio Circuits," Mullard, London, 1972, TP1319, 2nd Ed., p 202–211.

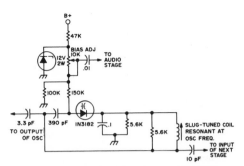

PHASE MODULATOR—Can be used with any type of transmitter to provide f-m operation, because circuit varies frequency indirectly by varying phase of tuned circuit. Uses 1N3182 varactor whose capacitance is changed by varying bias voltage.—M. Pecen, A Simple Varactor Modulator for Going FM, 73, April 1971, p 121.

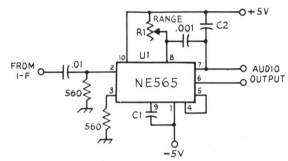

PLL AS DETECTOR—Sensitivity is about 1 mV with Signetics NE565 phase-locked loop IC. Bandwidth is 2% to 10% of i-f value for f-m detector. Upper frequency limit is 500 kHz, but NE561 is usable up to 30 MHz. Article gives design equations for determining values of C1 and C2 to give desired capture range.—D. A. Blakeslee, Receiving FM, QST, March 1971, p 29–34.

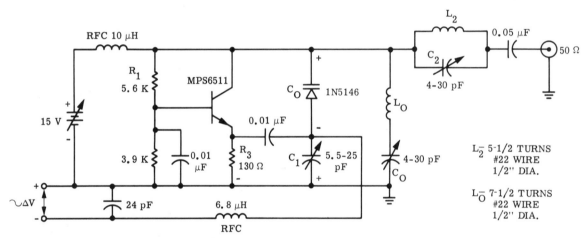

EPICAP 52-MHZ MODULATOR—Motorola 1N5146 voltage-variable capacitor, rated at 33 pF for reverse bias of —4 V, provides good linearity for full commercial f-m deviation of 75 kHz. Transistor is designed especially for oscillator use.—"FM Modulation Capabilities of EPICAP VVC's," Motorola Semiconductors, Phoenix, AZ, AN-210, 1972.

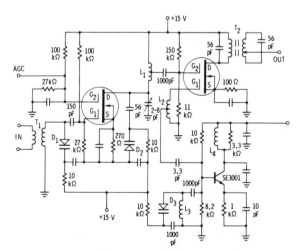

MOSFET TUNER—Fairchild RF401 voltage-variable capacitors D1, D2, and D3 serve in place of conventional air capacitors for tuning dual-gate mosfet r-f amplifier and mixer stages using FT0601 transistors. Performance can be improved by using two tuning diodes back to back in each tuning circuit. Report gives design equations and performance data. Gain is 18 to 22 dB. Frequency of local oscillator is 98.16 MHz at 0.4 V d-c control voltage and 117.96 MHz at 20 V, corresponding to varactor change over range of 3 to 15 pF.—"RF Applications of the FT0601 Dual-Gate Mosfet," Fairchild Semiconductor, Mountain View, CA, No. 189, 1970, p. 7.

T_1: (P) 22t, gauge 26 wire, (S) 6t, gauge 18 wire, on 15/64" ID coil form
L_1: 6t, gauge 18 wire, 15.64" ID, tap at 2t from cold side
L_2: 10t, gauge 18 wire, 0.2" ID, tap at 4t from Gate 1
L_3: 6t, gauge 18 wire, 15/64" ID
L_4: 8t, gauge 32 wire, wound on 3.3 kΩ (1/4 W) resistor
All other capacitors: 1000 pF, FT

CHAPTER 41
Frequency Multiplier Circuits

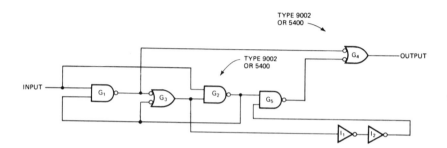

DIGITAL DOUBLER—Will operate at frequencies as high as 10 MHz, over temperature range of —55 C to +125 C, in doubling square-wave input signal frequency. Uses logic delay in place of R-C timing components. Leading and trailing edges of input signal are detained one gate delay by each logic element. Symmetry of doubled output signal matches that of input waveform.—C. H. Doeller III and A. Mall, IC Frequency Doubler Runs at Logic Speed, *Electronics*, Aug. 2, 1971, p 60.

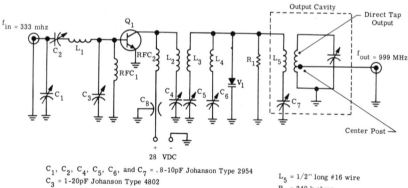

333 MHZ TO 999 MHZ—Transistor amplifier Q1 combined with varactor multiplier gives efficiency of almost 40% for 2.7-W output power when drive level is 1 W. Linearity is excellent and circuit is free from hysteresis (discontinuous mode jump in output power when input power is changed). Report covers construction of output cavity.—J. Cochran, "Transistor-Multiplier Versus Transistor-Varactor-Multiplier," Motorola Semiconductors, Phoenix, AZ, AN-243, 1971.

C_1, C_2, C_4, C_5, C_6, and C_7 = .8-10pF Johanson Type 2954
C_3 = 1-20pF Johanson Type 4802
C_8 = 330pF Feedthrough
L_1 = 4 turns 1/4" diameter 1/8" spacing #20 wire
L_2 and L_3 = 3 turns 1/4" diameter 1/8" spacing #20 wire
L_4 = 1 turn 1/4" diameter #16 wire

L_5 = 1/2" long #16 wire
R_1 = 240 k ohms
Q_1 = 2N3961
V_1 = 1N5149 (MV1806C)
RFC_1 and RFC_2 = 0.1 μh

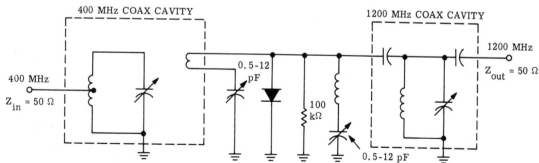

0.4 GHZ TO 1.2 GHZ—Tripler is achieved by adding shunt idler circuit to cavity-type doubler also covered in report. Coil and tunable capacitor following varactor (1N5149 or 1N5150) serve as idler circuit, mounted in same compartment with varactor and bias resistor. Report gives tuning procedure.—J. Cochran, "24 Watts at 1 GHz with Step Recovery Varactors," Motorola Semiconductors, Phoenix, AZ, AN-228, 1971.

PRR DOUBLER—Simple IC doubles pulse repetition rate of input square wave. Values of R and C will vary with type of IC used; those shown are for WC266. For quadrupling, cascade two sections and cut RC product in half for second section.—G. P. Scher, Simple Pulse Rep Rate Doubler Is Cascadable, *The Electronic Engineer*, Nov. 1968, p 74.

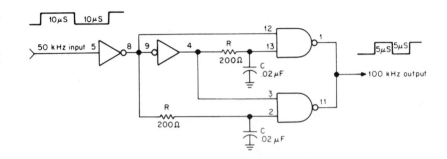

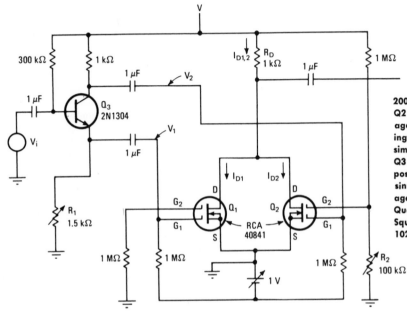

200 HZ–1 MHZ FILTERLESS—Mosfet's Q1 and Q2 operating in depletion mode square voltages applied to their main gates G1. Adjusting control gate G2 voltage forces mosfet's to simulate matched pair. Unity-gain transistor Q3 drives mosfet with equal voltages of opposite polarity. For sinusoidal input Vi, Vo is sinusoidal at twice input frequency, with voltage gain of about 3.5.—W. V. Subbarao, Quasi-Matched MOSFETs Form Filterless Squaring Circuit, *Electronics*, Oct. 9, 1972, p 102.

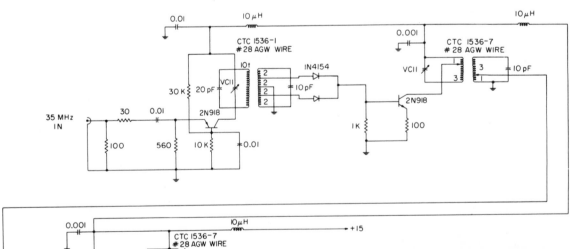

35–70 MHZ MULTIPLIER—Used in three-multiplier chain for boosting input from 5-MHz rubidium frequency standard to 140 MHz in phase-stable K-band radiometer suitable for very long baseline interferometric measurements.—G. D. Papadopoulos, Phase Stability of Two Independently Locked Local Oscillators at 22.235 GHz, *Proc. IEEE*, Nov. 1971, p 1620–1622.

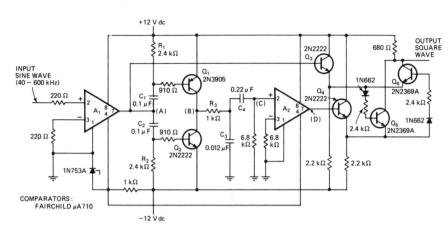

DOUBLER FOR 40–600 KHZ—Uses biconditional logic driven by quadrature square waves to double frequency of nearly any waveform and give square-wave output having duty cycle ranging from about 38% at 40 kHz to 65% at 600-kHz upper limit. Sine or other input having duty cycle of about 50% and at least 0.5 V p-p is converted into square wave by comparator A1, which acts on Q1-Q2 to produce triangular wave driving comparator A2. Resulting square-wave output of A2 is in quadrature with that of A1. Switches Q5 and Q6 conduct only when comparator states are different, to provide square wave at twice input frequency with amplitude of about 3.5 V p-p.—D. DeKold, *Frequency Doubler Accepts Any Waveshape,* Electronics, July 17, 1972, p 84–85.

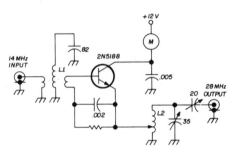

GROUNDED-COLLECTOR DOUBLER—Can be used in class-C r-f service up to about 150 MHz. Ideal for doubling 14 MHz to 28 MHz. —F. C. Jones, Transistor Frequency Multipliers, *Ham Radio,* June 1970, p 49–51.

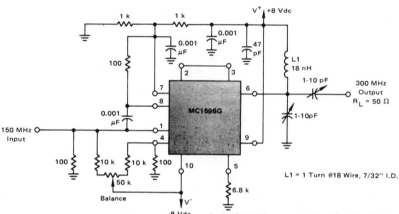

150 TO 300 MHZ—Uses IC balanced modulator. Spurious outputs are over 20 dB below desired 3-MHz output.—R. Hejhall, "MC1596 Balanced Modulator," Motorola Semiconductors, Phoenix, AZ, AN-531, 1971.

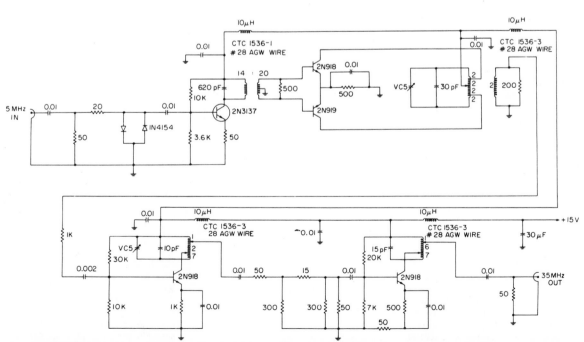

5–35 MHZ MULTIPLIER—Input from 5-MHz rubidium frequency standard is multiplied by 7 in first part of multiplier chain used to provide 140 MHz in phase-stable K-band radiometer suitable for very long baseline interferometric measurements.—G. D. Papadopoulos, Phase Stability of Two Independently Locked Local Oscillators at 22.235 GHz, *Proc. IEEE,* Nov. 1971, p 1620–1622.

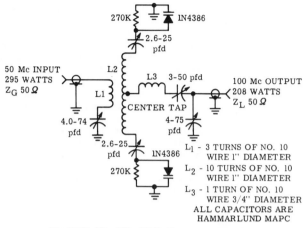

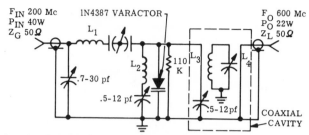

L₁ - 3 TURNS OF NO. 10 WIRE 1" DIAMETER
L₂ - 10 TURNS OF NO. 10 WIRE 1" DIAMETER
L₃ - 1 TURN OF NO. 10 WIRE 3/4" DIAMETER
ALL CAPACITORS ARE HAMMARLUND MAPC

L_1 - 3 TURNS OF 1/16" WIRE 1/2" DIA. x 1" LONG.
L_2 - 1 TURN 1/8" TUBING 3/8" O.D.
L_3 - STRAIGHT COUPLING LOOP 1/8" TUBING 2" LONG SPACED APPROX. 1/8" FROM CENTER CONDUCTOR.
L_4 - STRAIGHT COUPLING LOOP 1/16" WIRE 1-1/2" LONG, SPACED APPROX. 1/16" FROM CENTER CONDUCTOR.

50 MHZ TO 100 MHZ—Two varactors in push-push doubler provide output power of over 200 W, previously available only in vhf region from large vacuum tubes operating with high electrode voltages. Conversion loss is nominal 1.5 dB, corresponding to about 70% efficiency. Varactors are connected in phase opposition to input signal and in parallel to common load at even harmonics of output, to give odd-harmonic suppression with minimum spurious output of 26 dB.—G. Schaffner and J. Cochran, "Varactor Diodes and Circuits for High Power Output and Linear Response," Motorola Semiconductors, Phoenix, AZ, AN-191, 1971.

200 MHZ TO 600 MHZ—Harmonic tripler using Motorola 1N4387 varactor as harmonic tripler operates at about 50% efficiency. Output uses double-tuned cavity. Spurious output is minimum of 35 dB below desired output. For a-m signals, 55% modulated, distortion is negligible.—G. Schaffner and J. Cochran, "Varactor Diodes and Circuits for High Power Output and Linear Response," Motorola Semiconductors, Phoenix, AZ, AN-191, 1971.

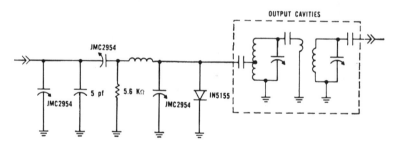

500 MHZ TO 4,000 MHZ—Uses Motorola 1N5155 step-recovery power varactor diode as one-step frequency multiplier. Efficiency is excellent, with power output up to several watts from cavity. Report compares performance formulas for various varactor harmonic generation theories.—G. Schaffner, "Selecting Varactor Diodes," Motorola Semiconductors, Phoenix, AZ, AN-260, 1968.

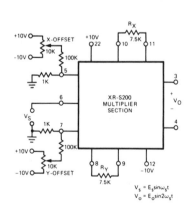

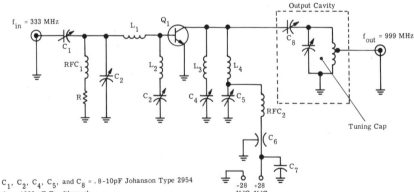

C_1, C_2, C_4, C_5, and C_8 = .8-10pF Johanson Type 2954
C_6 = 1000 pF Feedthrough
C_7 = 0.2 μF disc
L_1 = 2 turns 1/4" diameter #18 wire
L_2 = 1/'6" wide copper strap . 4" long
L_3 = 2 turns 3/8" diameter #18 wire
L_4 = 1 turn 1/4" diameter #18 wire
RFC_1 = 0.22 μh
RFC_2 = 0.33Ω W.W.
R = 2.4 ohms
Q_1 = 2N4012

DOUBLER—Uses multiplier section of Exar XR-S200 IC having sinusoidal input signal VS at frequency fs to produce low-distortion sine output at 2fs. Total harmonic distortion is under 0.6% with input of 4 V p-p at 10 kHz and output of 1 V p-p at 20 kHz. Offset controls are nulled to minimize harmonic content of output.—"XR-S200 Multi-Function Integrated Circuit," Exar Integrated Systems Inc., Sunnyvale, CA, June 1972, p 4.

333 MHZ TO 999 MHZ WITH TRANSISTOR—Uses lumped-constant input and idler circuits with coaxial-cavity output for which construction is covered. Cavity resonance is obtained by adjusting length of center post. With 1 W input drive, output is 2.7 W at collector efficiency of about 30%.—J. Cochran, "Transistor-Multiplier Versus Transistor-Varactor-Multiplier," Motorola Semiconductors, Phoenix, AZ, AN-243, 1971.

DOUBLER UP TO 1 MHZ—Motorola IC modulator shown (or lower-cost MC1496G) operates as frequency doubler from audio through vhf without any tuned circuits, when same signal frequency is injected at both IC inputs as shown. Sum-frequency output of modulator is then twice input frequency, and difference frequency is zero. All other frequencies are at least 30 dB down.—R. Hejhall, An Integrated-Circuit Balanced Modulator, *Ham Radio,* Sept. 1970, p 6–13.

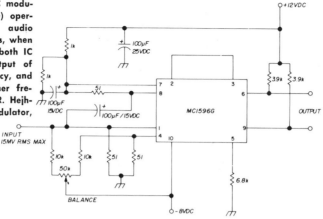

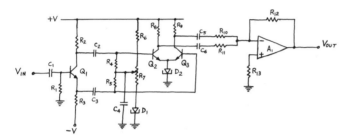

WIDEBAND DOUBLER—Utilizes approximately parabolic characteristic of tunnel-diode voltage-current response to achieve frequency-doubling over range of 10 Hz to 500 kHz. Circuit provides compensation for distortion and stabilizes bias voltage drift effects. Both supply voltages are 12 V. Q1 is 2N930, while Q2 and Q3 are µA726. Tunnel diodes are 1N3713. A1 is Motorola MC1539G opamp. All capacitors are 8 µF. R1–R6 are 1K, R7 is 100 ohms, R8 and R9 are 12K, and other resistors are adjusted to give desired output waveform.—K. K. Sum, A Wide-Band Tunnel-Diode Frequency Doubling Circuit, *Proc. IEEE,* Nov. 1969, p 2062–2063.

C_1-HammarLund MAPC-75	C_{11}-HammarLund MAPC-25	C_{21}-140 pf (Fixed)
C_2-HammarLund MAPC-75	C_{12}-Johanson-JMC-1801	C_{22}-15 pf (Fixed)
C_3-HammarLund MAPC-75	C_{13}-HammarLund MAPC-25	
C_4-HammarLund MAPC-75	C_{14}-HammarLund MAPC-25	L_1-3 Turns 1" dia. 1/8 Tubing
C_5-HammarLund MAPC-50	C_{15}-Johanson-JMC-1801	L_2-6 Turns 5/8" dia. 1/8 Tubing
C_6-HammarLund MAPC-50	C_{16}-Johanson-JMC-2951	L_3-2 Turns 13/16" dia. 1/8 Tubing
C_7-HammarLund MAPC-25	C_{17}-Johanson-JMC-1801	L_4-2 Turns 13/16" dia. 1/8 Tubing
C_8-E.F. Johnson Co.-Type "M"-160-110	C_{18}-20 pf (Fixed)	L_5-1 Turns 5/8" dia. 1/8 Tubing
C_9-E.F. Johnson Co.-Type "M"-160-110	C_{19}-32 pf (Fixed)	L_6-2 Turns 5/8" dia. #8 Wire
C_{10}-HammarLund MAPC-25	C_{20}-140 pf (Fixed)	L_7-3 Turns 1/4" dia. #8 Wire
		L_8-3 Turns 1/4" dia. #8 Wire

R_1 - 68KΩ	
R_2 - 68KΩ	
R_3 - 150KΩ	
R_4 - 150KΩ	

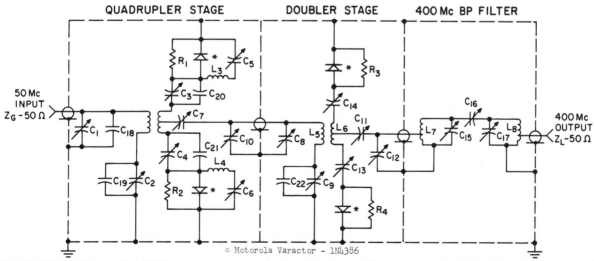

* Motorola Varactor - 1N4386

50 TO 400 MHZ—Provides nominal 40-W c-w output power at conversion efficiency of 30%. With pulse-modulated drive signals at input, will give 100-W peak pulse power at duty cycle of 0.0088. Uses two Motorola 1N4386 power varactors in push-push cascade circuits, with bandpass filter at output. Varactors are connected in phase opposition to input signal and in parallel to common load for even harmonics at output, with suppression of odd harmonics. Report gives construction and performance details.—J. Cochran, "Two-Stage Varactor Multiplier Provides High Power at 400 Mc," Motorola Semiconductors, Phoenix, AZ, AN-177, 1972.

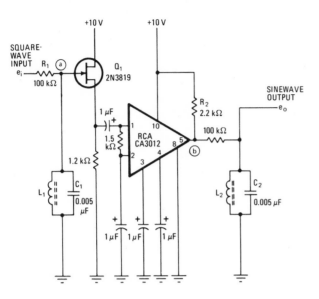

L_1, L_2: POWDERED, IRON-CORE, SLUG-TUNED INDUCTORS (MILLER 9005)

MULTIPLYING 625 HZ BY 40—Hard limiting of damped 625-Hz sine wave gives multiplication factor of 40 without need for low-Q a-f inductors. Circuit performs odd multiplication for true square wave, but even harmonics can be introduced by varying duty cycle. Both tuned circuits are adjusted to desired multiple of input frequency.—D. F. Dekold, Hard-Limited Sinusoid Eases Frequency Multiplication, *Electronics*, Nov. 22, 1971, p 75.

500 MHZ TO 1,000 MHZ—Uses coaxial cavities separated by varactor, which may be either 1N5149 or 1N5150. Serves for production testing of these varactors, to insure that they meet minimum efficiency requirements of 55% (for 1N5149) at 20 W input power. Report gives design and performance details, including cavity construction.—J. Cochran, "24 Watts at 1 GHz with Step Recovery Varactors," Motorola Semiconductors, Phoenix, AZ, AN-228, 1971.

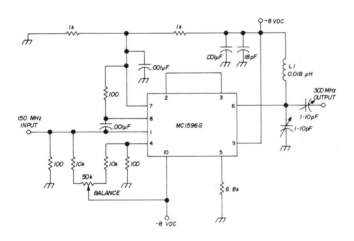

VHF DOUBLER—Uses Motorola IC modulator shown (or lower-cost MC1496G) for doubling any input frequency up to 200 MHz. Tank circuit gives cleaner output signal with all spurious outputs 20 dB down. L1 is 1 turn of No. 18, 7/32-inch ID.—R. Hejhall, An Integrated-Circuit Balanced Modulator, *Ham Radio*, Sept. 1970, p 6–13.

DOUBLER UP TO 1 MHZ—Uses IC balanced modulator with same signal injected in both inputs. Will double input frequency in range from 1 MHz down to audio. Maximum input signal for linear operation is 15 mV rms.—R. Hejhall, "MC1596 Balanced Modulator," Motorola Semiconductors, Phoenix, AZ, AN-531, 1971.

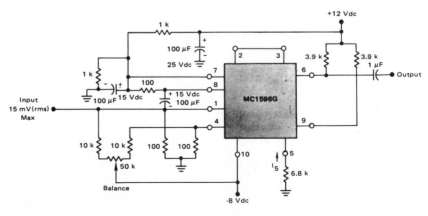

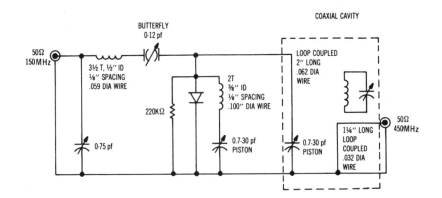

150 TO 450 MHZ VARACTOR TRIPLER—Uses 1N4387 varactor (shunted by 220K) in frequency-tripling circuit capable of handling up to 40 W input power at 150 MHz. Report covers development of circuit in detail and gives dimensions of coaxial cavity required at output.—"High-Power Varactor Diodes: Theory and Application," Motorola Semiconductors, Phoenix, AZ, AN-147, 1971.

500 MHZ TO 1,000 MHZ—Provides up to 15-W output from 25-W input, with output linear up to 11 W. Conversion efficiency is about 50%. Uses double-tuned input and output coaxial cavities.—G. Schaffner and J. Cochran, "Varactor Diodes and Circuits for High Power Output and Linear Response," Motorola Semiconductors, Phoenix, AZ, AN-191, 1971.

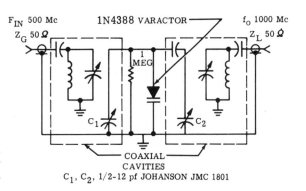

GATES MULTIPLY CRYSTAL RANGE—Provides accurate output in range of 150 to 250 MHz, which is well beyond practical overtone range of conventional crystal operation. ECL gates G1, G2, and G7 serve as oscillator, with L-C tank fine-tuning to 100-MHz crystal overtone. G2 is buffer. In frequency doubler, G3 and G4 are phase-shifters, while G5 and G6 operate as summers that combine four-phase 100-MHz signals to give 200-MHz output. G7 is biased generator for crystal stage. —W. Blood, ECL Gates Stretch Oscillator Range, *Electronics*, March 13, 1972, p 76.

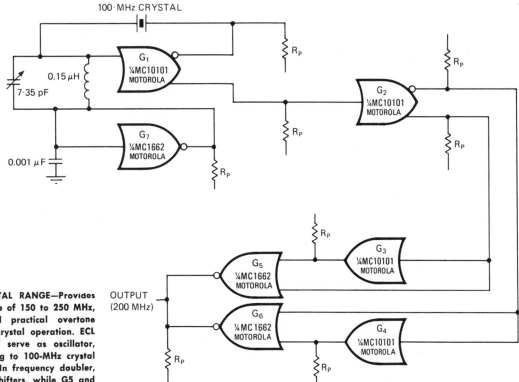

R_P: TYPICALLY 510 Ω TO V_{EE} OR 50 Ω TO −2 VDC

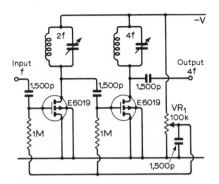

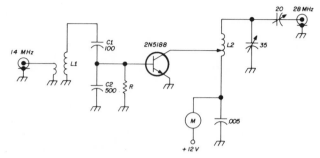

DOUBLER CHAIN—High impedance of mos transistors allows tuned frequency doublers to be cascaded without need for impedance transformation. Adjust pot to provide bias for maximum efficiency. Will operate up to 150 MHz. Does not require separate bias supply. —J. A. Roberts, A M.O.S.T. Frequency-Doubler Chain, *Wireless World*, Jan. 1970, p 12.

14 TO 28 MHZ—Provides excellent results as doubler, and with appropriate changes in L and C will serve as tripler or quadrupler at low or high power. Collector tap on L2 should be closer to grounded end of coil.—F. C. Jones, Transistor Frequency Multipliers, *Ham Radio*, June 1970, p 49–51.

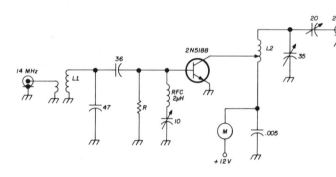

HIGH-OUTPUT DOUBLER—Gives twice as much output as conventional doublers because series-tuned circuit provides low-impedance path from base to emitter, eliminating degeneration. May also be used as tripler or quadrupler. Circuit requires careful adjustment.—F. C. Jones, Transistor Frequency Multipliers, *Ham Radio*, June 1970, p 49–51.

CHAPTER 42
Frequency Synthesizer Circuits

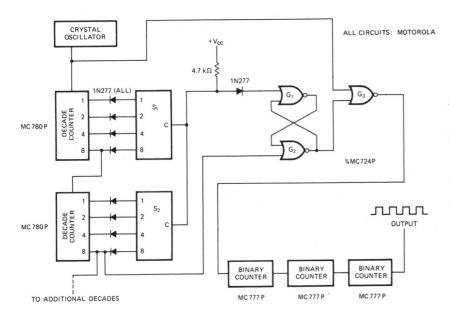

ALL CIRCUITS: MOTOROLA

10-MHZ SYNTHESIZER—Decade counters generate pulse train that is divided by binary counter chain to obtain output frequency set by two binary-coded thumbwheel switches. Accuracy and long-term stability of output frequency are same as that of crystal reference. Binary countdown string provides time-averaging for train of pulses and blank spaces coming from G3, to reduce pulse-to-pulse time variation in output.—J. L. Foote, Thumbwheel Switches Set Synthesizer Output Frequency, *Electronics*, June 5, 1972, p 100.

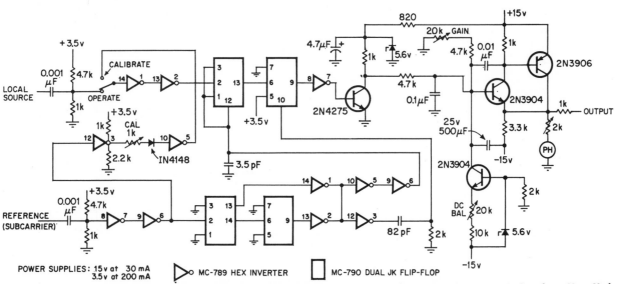

CALIBRATING 5 MHZ WITH 3.58-MHZ TV SIGNAL—Linear phase comparator using four IC's uses highly stable 3.58-MHz burst reference frequency of television network as low-cost local standard for calibrating 5-MHz signal. TV reference is generated by rubidium-controlled oscillator-synthesizer whose output is stable to within one part in 10^{12} per day. Local source is 3.58-MHz signal derived from 5 MHz by synthesizer. Be sure network program is live from New York.—D. D. Davis, Frequency Standard Hides in Every Color TV Set, *Electronics*, May 10, 1971, p 96–98.

1.6-MHZ REFERENCE—Uses Exar XR-210 fsk modulator-demodulator in combination with SN7493 or equivalent programmable counter or divide-by-N circuit connected between vco output and pin 6 of phase detector input. Counter divides down oscillator frequency by modulus N of divider. Will synthesize large number of discrete frequencies from given reference frequency. C1 is chosen to provide filter cutoff between 0.1% and 2% of signal frequency fs, which is 100 kHz when N is 16 and fo is 1.6 MHz.—"XR-210 FSK Modulator/Demodulator," Exar Integrated Systems Inc., Sunnyvale, CA, June 1972, p 8.

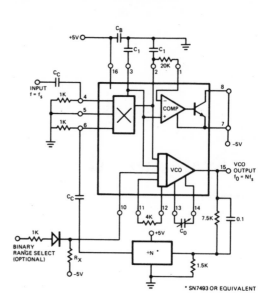

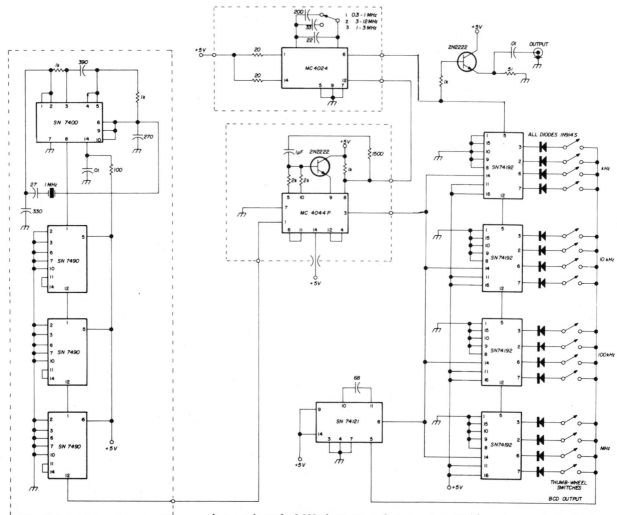

300 KHZ TO 10 MHZ IN 1-KHZ STEPS—Uses phase-locked loop synthesizer to read frequency counter connected to receiver and automatically adjust to desired frequency. Readjustment is made 1,000 times per s. System consists of vco, programmable divider, frequency discriminator, and reference frequency source. Discriminator provides d-c control voltage for vco. Output of synthesizer is square wave. Article covers construction and operation, including circuit for 5-V regulated power supply.—A. D. Helfrick, High-Frequency Frequency Synthesizer, *Ham Radio*, Oct. 1972, p 16–21.

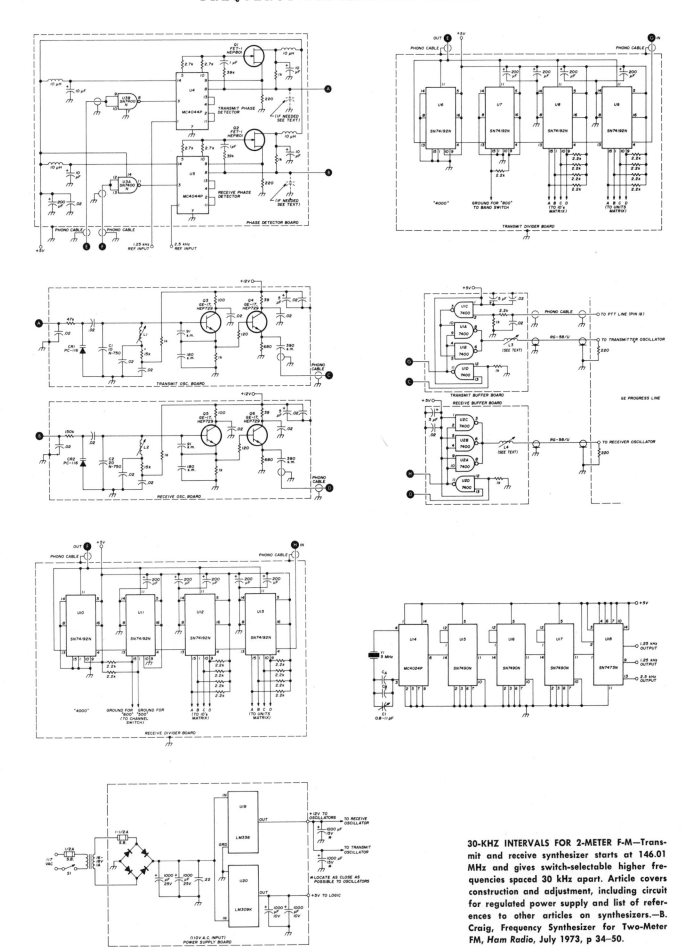

30-KHZ INTERVALS FOR 2-METER F-M—Transmit and receive synthesizer starts at 146.01 MHz and gives switch-selectable higher frequencies spaced 30 kHz apart. Article covers construction and adjustment, including circuit for regulated power supply and list of references to other articles on synthesizers.—B. Craig, Frequency Synthesizer for Two-Meter FM, *Ham Radio*, July 1973, p 34–50.

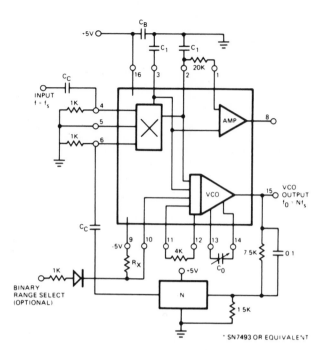

VCO WITH DIVIDER—Uses Exar XR-215 phase-locked loop with divide-by-N IC for which modulus N can be changed, for synthesizing large number of discrete frequencies from given reference input frequency. Designed to operate with programmable counters using TTL logic for divider action. Report shows typical waveforms for 100-kHz input and 1.6-MHz output when N is 16.—"XR-215 Monolithic Phase-Locked Loop," Exar Integrated Systems Inc., Sunnyvale, CA, July 1972, p 7–8.

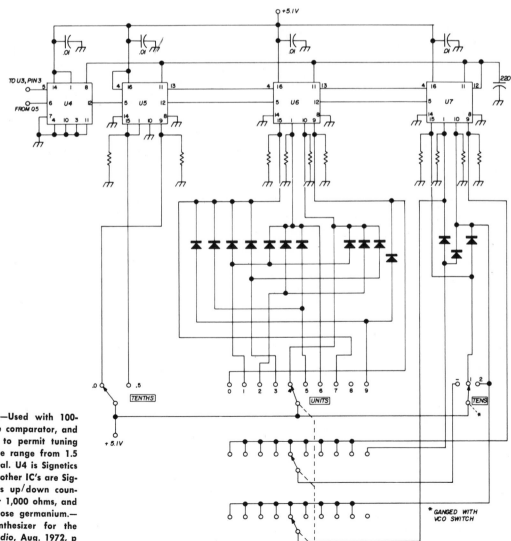

DIVIDER FOR SYNTHESIZER—Used with 100-kHz crystal oscillator, phase comparator, and vco (also given in article) to permit tuning amateur receiver over entire range from 1.5 to 30 MHz with single crystal. U4 is Signetics 8290A decade counter, and other IC's are Signetics N74192A synchronous up/down counters. All resistors are 910 or 1,000 ohms, and all diodes are general-purpose germanium.—R. S. Stein, Frequency Synthesizer for the Drake R-4 Receiver, *Ham Radio*, Aug. 1972, p 6–19.

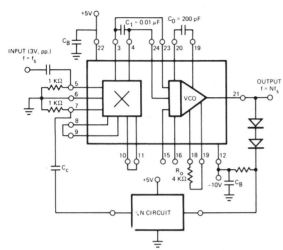

FLEXIBLE SYNTHESIZER—Uses Exar XR-S200 IC in combination with IC divider inserted in feedback loop of phase-locked loop, to synthesize large number of discrete frequencies from given reference input frequency fs. Output frequency is dependent on value of N for divider.—"XR-S200 Multi-Function Integrated Circuit," Exar Integrated Systems Inc., Sunnyvale, CA, June 1972, p 7.

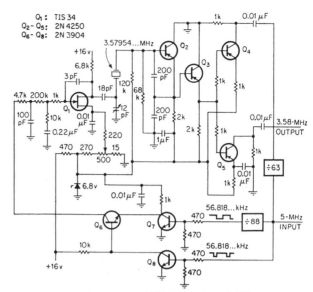

3.58-MHZ SYNTHESIZER—Converts 5-MHz local frequency reference to 3.58-MHz tv burst reference frequency, for comparison with high-accuracy tv network color subcarrier. One-day phase instability is less than 2 ns and short-term jitter under 1 ns. Used with linear phase comparator for calibrating other signal sources.—D. D. Davis, Frequency Standard Hides in Every Color TV Set, *Electronics*, May 10, 1971, p 96–98.

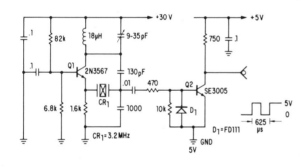

3.2-MHZ REFERENCE OSCILLATOR—Used with vco, frequency comparator, phase comparator, and programmable divider to permit precise tuning of a-m and f-m receivers to stations by means of selector switch. Fine tuning is not required, because accuracy depends on crystal. Q2 squares output of oscillator and provides proper signal level. Report covers entire system.—J. Stinehelfer and J. Nichols, "A Digital Frequency Synthesizer for an AM and FM Receiver," Fairchild Semiconductor, Mountain View, CA, No. 201, 1971, p 6.

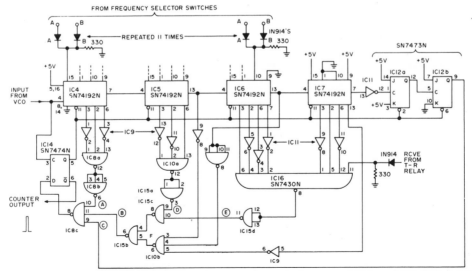

PROGRAMMABLE DIVIDER—Used in 6-MHz synthesizer covering entire 2-meter f-m band from 145 to 148 MHz in steps of 5 kHz with only single 10-kHz crystal. Article gives all circuits. Frequency selector switches set divider to divide by exactly correct value to give desired output frequency for transmitter or receiver with crystal stability. IC4–IC7 are programmable decade up-down counters which, along with IC12, do actual counting. Division number is between 29,000 and 29,599 during transmitting, and is 2,140 less during receiving. Designed for use with 10.7-MHz first i-f value in receiver.—P. A. Stark, Frequency Synthesizer for 2 Meter FM, 73, Oct. 1972, p 15–23.

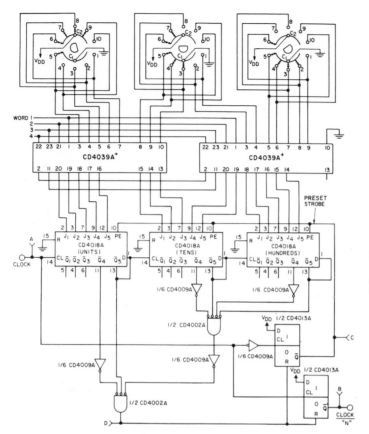

PROGRAMMABLE 2–999 COUNTER—Three-decade divide-by-N counter has four-channel preset memory settings for frequency synthesizer. Four counter-preset words, selected by rotary switches, can be stored in CD4039A devices and read into each CD4018A simply by addressing proper word. Switch settings shown read 5-3-2 from left to right, with equivalent N value of 3-1-0 or N = 13.— "COS/MOS Digital Integrated Circuits," RCA, Somerville, NJ, SSD-203A, 1973 Ed., p 186.

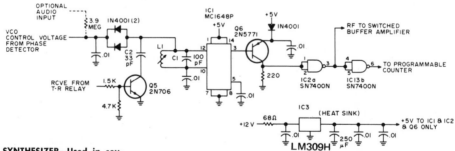

VCO FOR 6-MHZ SYNTHESIZER—Used in covering entire 2-meter f-m band from 145 to 148 MHz in steps of 5 kHz with only single 10-MHz crystal. Article gives all circuits, along with coil data. Crystal frequency is divided by 48,000 for phase and frequency comparator, while output of vco is divided by variable ratio in programmable divider and also fed to comparator. Comparator then adjusts vco frequency to give desired output for multipliers feeding f-m receiver and transmitter.—P. A. Stark, Frequency Synthesizer for 2 Meter FM, 73, Oct. 1972, p 15–23.

CHAPTER 43
Function Generator Circuits

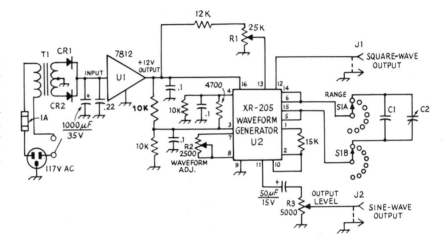

SINE-SQUARE-TRIANGLE A-F TO 1.2 MHZ— Uses Exar Systems XR-205 IC as function generator. Values of C1 range from 68 μF for 3.5–13 Hz range to 680 pF for 0.35–1.2 MHz range. Silicon rectifier diodes are 200-prv 500-mA, operating from 36-V 100-mA center-tapped secondary. Requires 12-V regulated supply.—D. A. Blakeslee, A Simple Function Generator, QST, Sept. 1972, p 11–13.

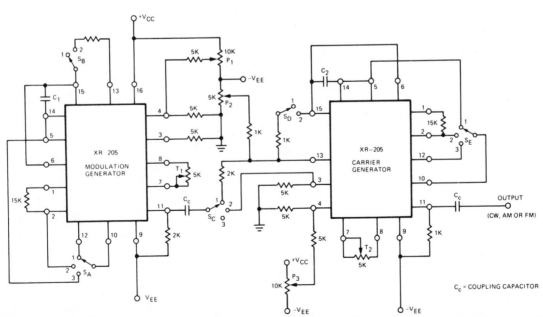

LAB A-M/F-M/C-W GENERATOR—Two interconnected Exar XR-205 waveform generators serve as self-contained a-m/f-m generator, with one unit providing modulation for the other. Can also generate c-w, fsk, and psk outputs over range of 1 Hz to 5 MHz. Requires only single 12-V supply. Controls are: P1—modulation level; P2—carrier output frequency; P3—carrier output level; T1—modulation waveform; T2—carrier waveform; SA—modulation waveform (1—square wave; 2—ramp; 3—sine or triangular); SB—modulation duty cycle offset (1—50% duty cycle; 2—20% duty cycle); SC—modulation mode select (1 —f-m; 2—a-m; 3—c-w); SD—output duty cycle offset (1—50% duty cycle; 2—20% duty cycle); SE—output waveform select (1—ramp; 2—sine, triangle, or sawtooth; 3—square or pulse).—"XR-205 Monolithic Waveform Generator," Exar Integrated Systems Inc., Sunnyvale, CA, March 1972.

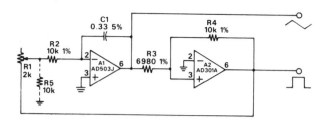

100-HZ TRIANGLE AND SQUARE—Oscillator using fet-input opamps simultaneously delivers ±13 V square waves and ±10 V triangular waves. Frequency is determined by values of R1, R2, and C1, and can be as low as 0.1 Hz if R2 is 10 meg. R1 serves for tuning. Integrator A1 feeds comparator A2 converted into Schmitt trigger by regenerative feedback through R4.—"AD503, AD506 I.C. Fet Input Operational Amplifiers," Analog Devices, Inc., Norwood, MA, Technical Bulletin, Aug. 1971.

0.002 TO 0.25 HZ—Gives choice of square or triangle waves. Second opamp is Philbrick/Nexus, having high-impedance fet input required for good linearity.—N. S. Nicola, Ultra-Slow Function Generator, *Wireless World*, March 1972, p 139–140.

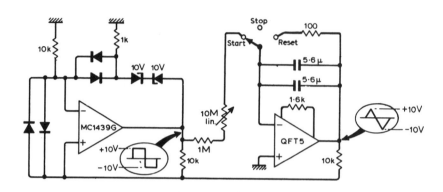

OPAMP TESTER—Provides three output functions along with range of supply voltages for testing opamp parameters over full range of common-mode and power-supply voltages. Report includes actual test circuit which is connected to terminals A, B, and C for testing with ±19 V square wave, −19 V to +19 V pulse with 1% duty cycle, and ±5 V triangular wave. Pulse is referenced to leading edge of square wave, while triangular wave is inverted and integrated square wave. Construction and operation are described in detail.—"A Simplified Test Set for Operational Amplifier Characterization," National Semiconductor, Santa Clara, CA, 1969, AN-24, p 6–7.

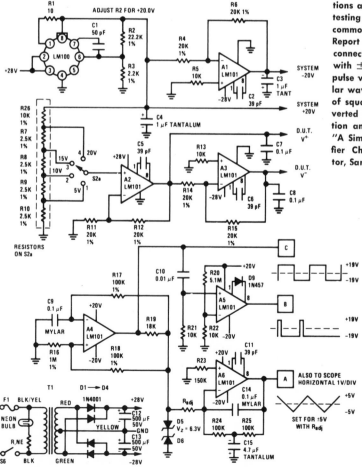

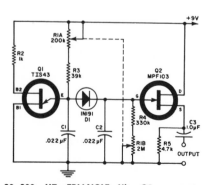

80–800 HZ TRIANGLE—Ujt Q1 generates sawtooth waveform at frequency determined by two-gang pot R1, and fet Q2 serves as source-follower. If several transistors are available for Q1, try all and choose the one that gives the most symmetrical output waveform.—F. H. Tooker, Triangular-Waveform Generator, *Electronics World*, Oct. 1970, p 79.

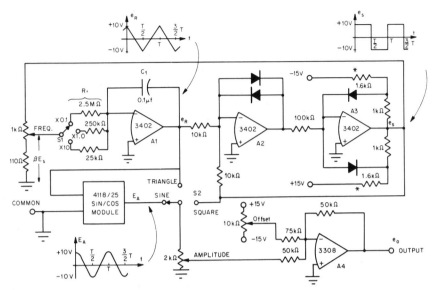

ULF MULTI-WAVEFORM—Provides square and triangle wave, and in addition uses Burr-Brown 4118/25 function generator to convert triangular wave to very good ultra-low-frequency sine wave. If voltage-controlled frequency is desired, use multiplier in place of 1K frequency pot.—J. G. Graeme, G. E. Tobey, and L. P. Huelsman, "Operational Amplifiers," McGraw-Hill Book Co., New York, 1971, p 377–378.

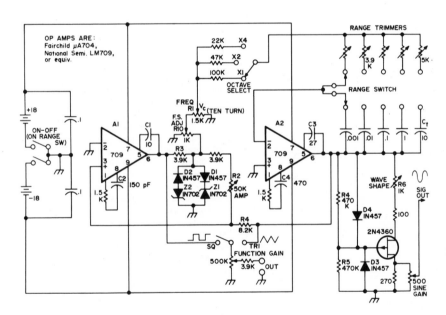

0.01–4,000 HZ TRIANGLE-SQUARE-SINE—Op-amps and fet generate triangle wave and shape it into low-distortion sine wave at very low frequencies for such applications as generation of slow-sweep displays, checking response of low-frequency filters, generating electronic music, and instrumentation circuits. R1 varies frequency over 500:1 range. Square-wave output is also available, directly from first opamp.—R. Factor, A Low Cost Function Generator for the Experimenter, 73, Oct. 1970, p 58–60 and 62–63.

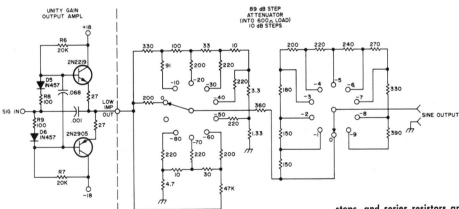

89-DB ATTENUATOR—Used with unity-gain output amplifier for 0.01–4,000-Hz function generator to provide attenuation in 1-dB steps into 600-ohm load when checking professional audio equipment. Constant-impedance 200-ohm T attenuator provides 10-dB steps, and series resistors are switched in for 1-dB steps.—R. Factor, A Low Cost Function Generator for the Experimenter, 73, Oct. 1970, p 58–60 and 62–63.

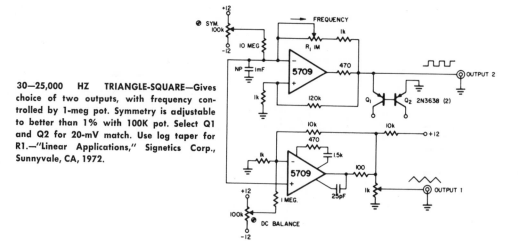

30—25,000 HZ TRIANGLE-SQUARE—Gives choice of two outputs, with frequency controlled by 1-meg pot. Symmetry is adjustable to better than 1% with 100K pot. Select Q1 and Q2 for 20-mV match. Use log taper for R1.—"Linear Applications," Signetics Corp., Sunnyvale, CA, 1972.

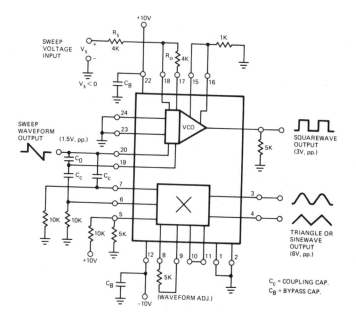

WAVEFORM GENERATOR—Connections shown for Exar XR-S200 IC give versatile waveform generator that can be voltage-tuned over 10:1 frequency range. Digital control inputs 15 and 16 can also be used for fsk applications, by removing grounded 1K disabling resistor. 5K pot between pins 8 and 9 can be used to round peaks of triangle waveform and convert it into low-distortion sine output. Terminals 3 and 4 can be used singly or differentially to provide both in-phase and out-of-phase output waveforms. For linear f-m modulation at deviations under 10%, modulation input can be applied across 23 and 24 after ungrounding. For larger deviations, apply negative-going sweep Vs to 18.—"XR-S200 Multi-Function Integrated Circuit," Exar Integrated Systems Inc., Sunnyvale, CA, June 1972, p 7.

SQUARE-TRIANGLE GENERATOR—Will generate both waveforms simultaneously. Design equations are given. Q1 acts as comparator and inverter for output of opamp. Q2 inverts signal and provides symmetrical signal output, while Q3 provides symmetrical output impedance. P1 controls frequency of triangular-wave output of opamp, P2 adjusts zero-crossing point of comparator, P3 adjusts time symmetry of triangular wave, P4 adjusts amplitude of square wave, and P5 provides additional control over frequency and triangular-wave amplitude.—J. G. Graeme, G. E. Tobey, and L. P. Huelsman, "Operational Amplifiers," McGraw-Hill Book Co., New York, 1971, p 375.

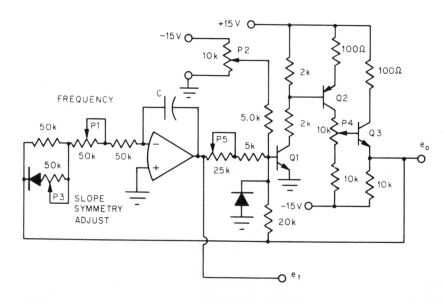

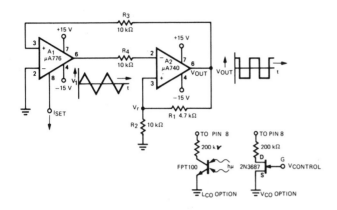

0.1 HZ–30 KHZ CURRENT-CONTROLLED—Provides over five decades of frequency range in response to similar range of set currents fed to pin 8 of first opamp. Second opamp acts as comparator with reference voltage Vr depending on values of R1 and R2. When output of A1 goes positive above reference value, comparator output flips to its negative extreme and circuit operates to generate square wave as shown. Use of fet in place of adjustable current supply gives vco action. Use of light-controlled phototransistor option for pin 8 makes frequency vary with light. Connect 4-μF capacitor from pin 8 to ground to eliminate unwanted a-c power-line flicker.—"The μA776, An Operational Amplifier with Programmable Gain, Bandwidth, Slew-Rate, and Power Dissipation," Fairchild Semiconductor, Mountain View, CA, No. 218, 1971, p 2–3.

SQUARE-TRIANGLE 1 HZ–100 KHZ—Amplitude of either output waveform is adjustable from 0.2 to 20 V p-p. Risetime is less than 450 ns at full amplitude, and under 200 ns at reduced amplitude. Will operate on supply voltage range of 10 to 20 V in either polarity. Triangle waveform is very linear, with symmetry adjusted by R6 which controls reference voltage of HA-2600 opamp which integrates output of square-wave generator. Output frequency is controlled by ramp rate of triangular wave, in five ranges selected by S1.—G. G. Miler, "A Simple Function Generator Using Operational Amplifiers," Harris Semiconductors, Melbourne, FL, No. 507/A, 1970.

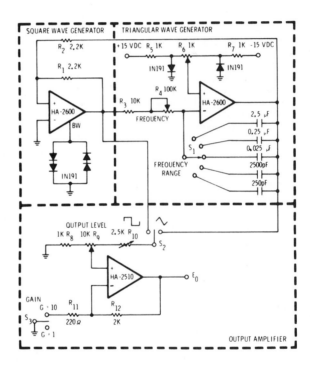

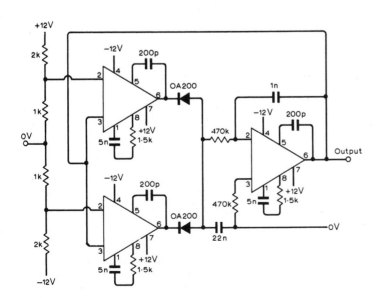

VARIABLE-AMPLITUDE TRIANGLE OUTPUT—Uses three MC709 opamps to generate triangular waveform, amplitude of which may be independently varied each side of zero by adjusting input voltages to first two opamps. Values shown give 1-kHz output at 8 V p-p, symmetrical about zero. Circuit has good amplitude stability and good linearity.—H. MacDonald, Triangular Waveform Generator, *Wireless World*, Feb. 1972, p 77.

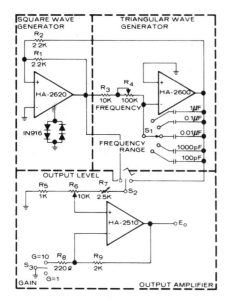

SQUARE-TRIANGLE 2.5 HZ–250 KHZ—Requires only three opamps. Amplitude of either output waveform is adjustable from 0.2 to 20 V p-p. Risetime of square wave is less than 100 ns. Supply voltage changes between 10 and 20 V have little effect on frequency, amplitude, or waveform.—G. G. Miler, "A Simple Square-Triangle Waveform Generator," Harris Semiconductor, Melbourne, FL, No. 510/A, 1970.

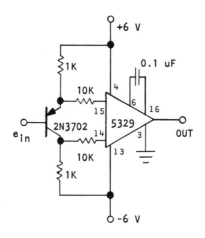

SINE-SQUARE-TRIANGLE—Uses OEI module operating up to 100 kHz in symmetrical connection giving three simultaneous outputs at frequency proportional to input voltage. Linearity is 0.9% of full scale over 50-dB dynamic frequency range. Output is 45 Hz at zero input and 725 Hz at 5 V. Use pin 12 for sine output, 16 for triangular, and 8 for square.—"Applications of the 5329," Optical Electronics, Tucson, AZ, Application Tip 10194, 1971.

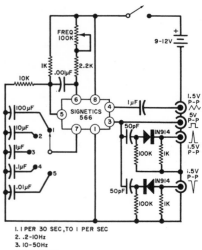

SQUARE-TRIANGLE-SPIKE—Versatile waveform generator using low-cost voltage-controlled oscillator IC provides wide choice of pulses at adjustable repetition rate for test purposes. For low-impedance loads, add transistor buffer stage.—Useful Circuits, Popular Electronics, Dec. 1971, p 94–95.

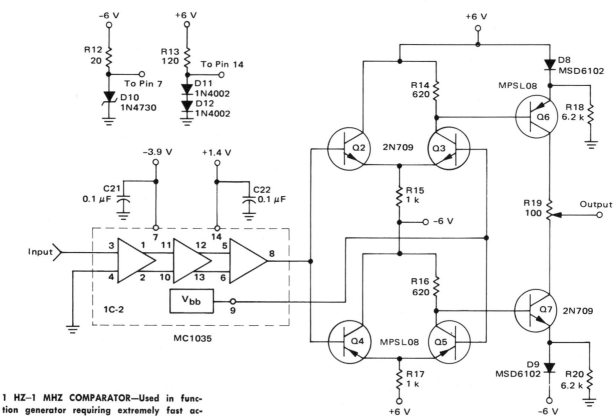

1 HZ–1 MHZ COMPARATOR—Used in function generator requiring extremely fast action, with square-wave output going from +5 V to −5 V. IC is triple differential amplifier, with level translation provided by differential pairs Q2 to Q5. Voltage gain is about 24,000 and delay time 20 ns. R9 should be composition pot to avoid inductance.—B. Botos, "A Low-Cost, Solid-State Function Generator," Motorola Semiconductors, Phoenix, AZ, AN-510A, 1971.

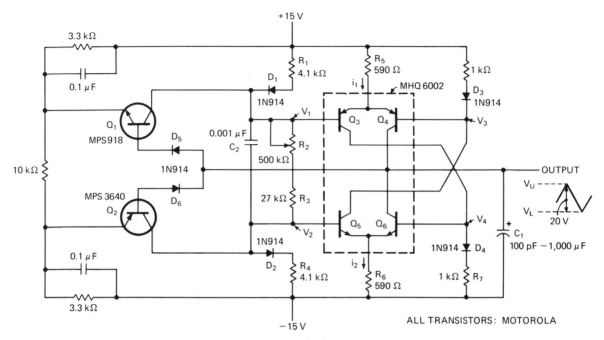

0.01-HZ–2-MHZ TRIANGLE—Outputs of constant-current sources Q1 and Q2 are switched to produce charging and discharging currents for output capacitor C1, whose value determines nominal output frequency. R2 sets upper limit and R3 lower limit.—W. S. Shaw, Triangular-Wave Generator Spans Eight Decades, *Electronics*, May 8, 1972, p 104.

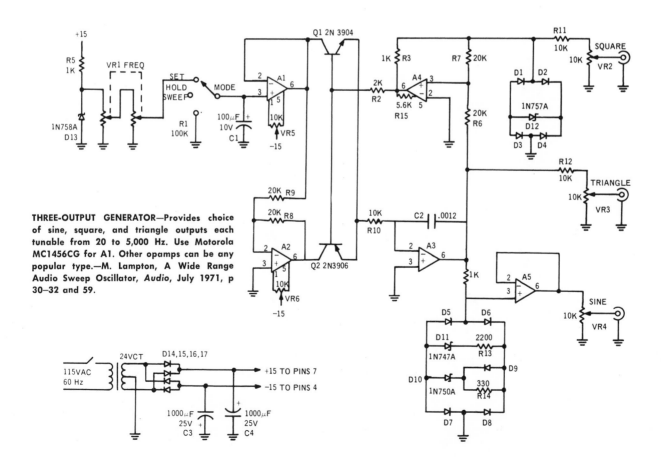

THREE-OUTPUT GENERATOR—Provides choice of sine, square, and triangle outputs each tunable from 20 to 5,000 Hz. Use Motorola MC1456CG for A1. Other opamps can be any popular type.—M. Lampton, A Wide Range Audio Sweep Oscillator, *Audio*, July 1971, p 30–32 and 59.

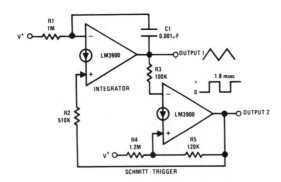

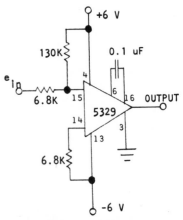

1.8-MS TRIANGLE—Use of two Norton current-differencing opamps allows generation of triangle waveform without requiring negative d-c input voltage. Requires only single supply voltage, which can be 4 to 36 V. One opamp does integration by operating first with current through R1 to produce negative output voltage slope; when output of second opamp (operating as Schmitt trigger) is high, current through R2 makes output voltage increase. For good symmetry, R1 equals 2R2. Half-period time is 1.8 ms for values shown. Schmitt circuit provides square-wave output also, of same frequency—T. M. Frederiksen, W. M. Howard, and R. S. Sleeth, "The LM3900—A New Current-Differencing Quad of ± Input Amplifiers," National Semiconductor, Santa Clara, CA, 1972, AN-72, p 21–22.

VOLTAGE-FREQUENCY CONVERTER—Module gives choice of sawtooth and pulse simultaneous outputs when operated as nonsymmetrical circuit. Nonlinearity is then about 6% of full-scale. Output is 4 Hz at zero input and 1,250 Hz at 5 V input. For sawtooth, use pin 16 for output, and pin 8 for pulse.—"Applications of the 5329," Optical Electronics, Tucson, AZ, Application Tip 10194, 1971.

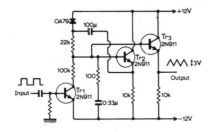

SQUARE TO TRIANGLE—Provides p-p amplitude of 3 V at 50 Hz.—S. Nagarajan, Triangular Waveform Generator, *Wireless World*, June 1970, p 292.

CHAPTER 44
Gate Circuits

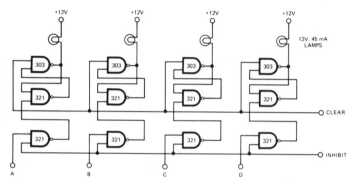

60-MA QUAD FLIP-FLOP—Uses type 303 quad 2-input power gates as lamp driver outputs on latches formed in combination with type 321 quad 2-input NAND gates. Circuit also provides inhibit and clear lines.—"HiNIL High Noise Immunity Logic," Teledyne Semiconductor, Mountain View, CA, 1972, p 13.

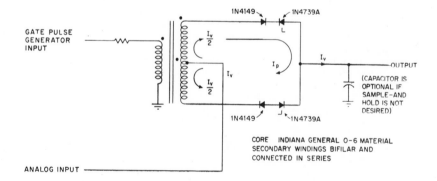

ZENER RESET—Input pulse to transformer primary reverse-biases zeners and forward-biases signal diodes, to give effect of connecting analog input to output for duration of sampling pulse. Use Indiana General 0-6 core material, with series bifilar secondaries.—R. W. Camp, Zener Diodes Reset Sampling Gate Automatically, *Electronics*, Nov. 9, 1970, p 80.

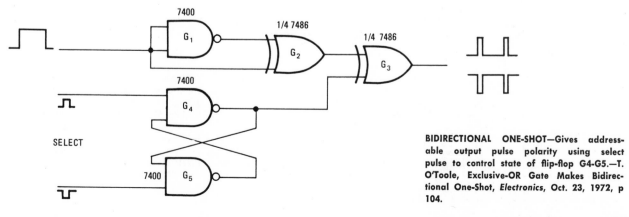

BIDIRECTIONAL ONE-SHOT—Gives addressable output pulse polarity using select pulse to control state of flip-flop G4-G5.—T. O'Toole, Exclusive-OR Gate Makes Bidirectional One-Shot, *Electronics*, Oct. 23, 1972, p 104.

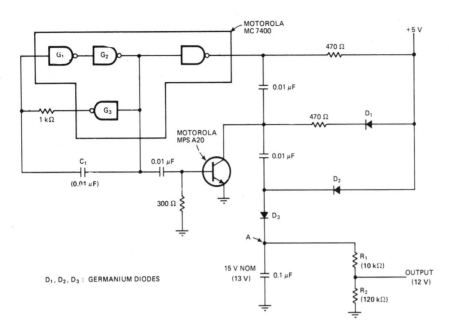

D₁, D₂, D₃ : GERMANIUM DIODES

12-V BIAS FOR N-MOS GATE—Consists of pulse generator formed by gates G1, G2, and G3 operating from 5-V supply and feeding voltage tripler serving with voltage divider R1-R2 to give required bias of 12 V at up to 10 μA for crt display system fed by character generator.—B. Broeker, Gate Bias Circuit for N-MOS Runs from 5-V TTL Supply, *Electronics*, March 15, 1973, p 106.

TTL-COMPATIBLE PULSES—Fast-switching scr in three-gate reset circuit converts steady or varying d-c input voltage into TTL-compatible pulses. Chopper-type opamp integrator accepts only negative voltages unless preceded by inverter. Error is only 1% for output pulse rate of 10,000 pps. Used to drive counters and other data processing devices.—M. W. Williams, SCR Reset for Integrator Provides High Speed, *Electronics*, Jan. 3, 1972, p 84.

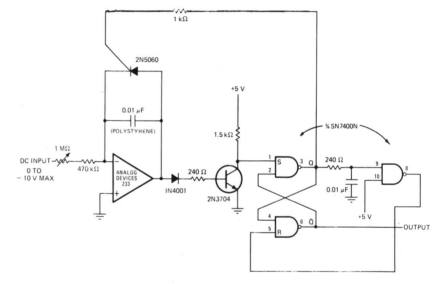

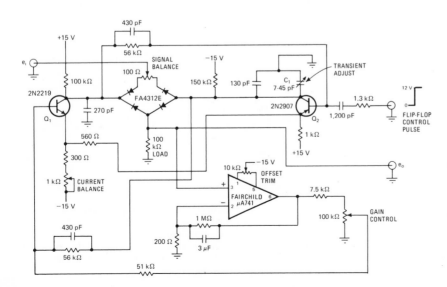

ZEROING GATE OUTPUT—Closed-loop balanced-diode-bridge gating circuit minimizes gating transients and d-c offset errors when tracking filter is changed from low-Q state to high-Q memory state, by using degenerative feedback loop to force d-c level at gate output to zero.—R. J. Turner, Feedback Zeros D Level of Diode Gating Circuit, *Electronics*, Aug. 30, 1971, p 49–50.

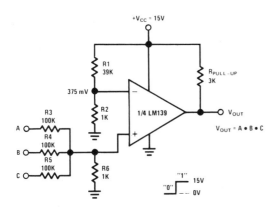

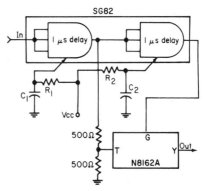

THREE-INPUT AND GATE—Divider R1-R2 establishes reference voltage at inverting input to comparator. Noninverting input is sum of voltages at inputs A, B, and C divided by voltage dividers R3-R4-R5-R6. Output goes high (+15 V) when all three inputs are high. —R. T. Smathers, T. M. Frederiksen, and W. M. Howard, "LM139/LM239/LM339 A Quad of Independently Functioning Comparators," National Semiconductor, Santa Clara, CA, AN-74, 1973, p 9.

PULSE WIDTH SORTER—Adjust delay of first gate with R1-C1 to be equal to minimum acceptable pulse width. Make delay of second gate equal to difference between minimum and maximum acceptable pulse widths. Delay times are about 0.35RC. When pulse width is acceptable, Signetics N8162A one-shot fires. —R. A. Anderson, Sort Pulse Widths with Two AND Gates and a One-Shot, The Electronic Engineer, Sept. 1968, p 79.

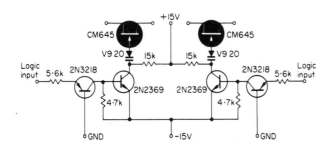

SPST WITH BREAK-BEFORE-MAKE—Dual-fet analog gate turns off faster than it turns on, to allow multiplexing without crosstalk.— Field-Effect Transistors, Design Electronics, May 1970, p 22–30.

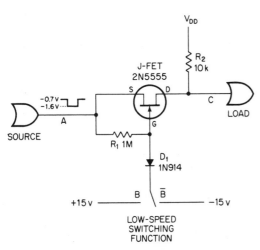

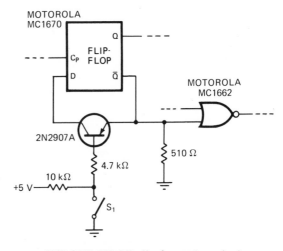

$1 JFET IS FAST DATA GATE—With positive supply, circuit is OR gate, and AND gate with negative supply. Low on resistance of jfet allows it to pass digital data with propagation delay under 1 ns.—E. S. Donn, Short-Delay J-FET Switch Gates High-Speed Data, Electronics, July 5, 1971, p 58.

FAST GATE FOR ECL—Simple transistor circuit cuts propagation delay to 100 ps for gating high-speed data signal by d-c control signal. Circuit shown minimizes signal delay between lower Q output of flip-flop and D input, to give fastest possible toggle rate.— A. J. Metz, Transistor Gating Circuit Cuts Signal Delay to 100 ps, Electronics, Nov. 20, 1972, p 116.

AND WITH DELAYED OR—Used at output of scs ring counter. When zero level is applied to D1 or D2 or D3, D4 is reverse-biased and has no effect. Operation is then same as synchronous AND gate. If positive 12 V is applied to diodes D1, D2, and D3, then D4 is forward-biased; negative-going input pulse at terminal 2 is then passed by D4 and absorbed by C2 so it has no effect. Network R2-C2 introduces delay ensuring that change of d-c level has no effect until next clock pulse appears.—D. J. G. Janssen, Circuit Logic with Silicon Controlled Switches, *Mullard Technical Communications*, March 1968, p 57–64.

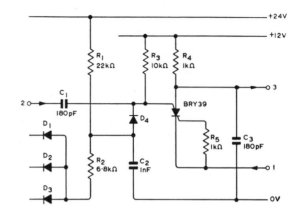

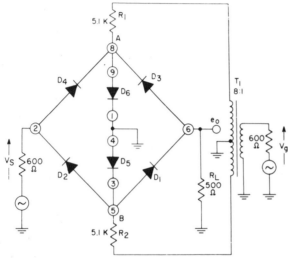

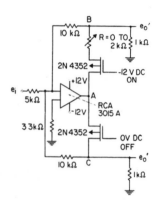

SERIES-SHUNT GATE—Uses CA3019 six-diode IC array to combine good on-to-off impedance ratio of shunt gate with low-output pedestal of series gate. Will operate with R1 and R2 shorted provided transformer centertap is ungrounded. No bias supply is needed for biasing gate diodes.—"Linear Integrated Circuits," RCA, Harrison, NJ, IC-41, p 304.

TRANSMISSION GATE—Negative feedback around electronic equivalent of spdt switch gives accurate transmission gate providing gain along with excellent isolation. Used with 1-kHz input sine wave for static measurements.—S. C. Kitsopoulos and R. W. Strauss, *IEEE Journal of Solid-State Circuits*, Dec. 1970, p 361–364.

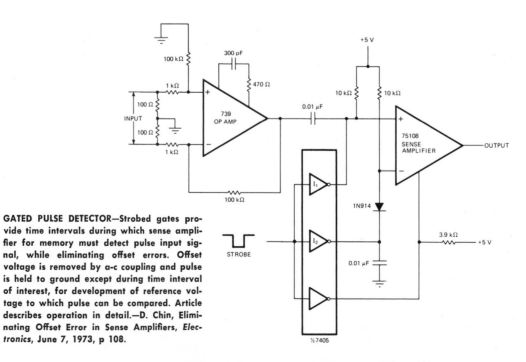

GATED PULSE DETECTOR—Strobed gates provide time intervals during which sense amplifier for memory must detect pulse input signal, while eliminating offset errors. Offset voltage is removed by a-c coupling and pulse is held to ground except during time interval of interest, for development of reference voltage to which pulse can be compared. Article describes operation in detail.—D. Chin, Eliminating Offset Error in Sense Amplifiers, *Electronics*, June 7, 1973, p 108.

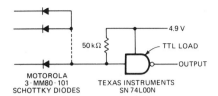

MOTOROLA
3 · MM80 · 101
SCHOTTKY DIODES

TEXAS INSTRUMENTS
SN 74L00N

SCHOTTKY GATES—Pull-up resistor and Schottky diodes replace NAND gates at input of low-power TTL circuits, improving gate response time and decreasing power consumption.—C. J. Huber, Schottky Diodes Eliminate Two-Level NAND Gating, *Electronics,* Jan. 31, 1972, p 59.

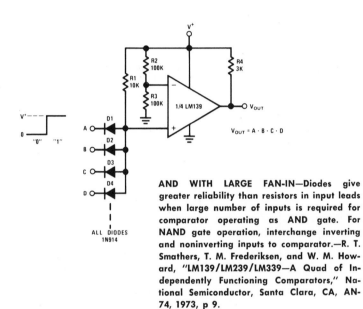

AND WITH LARGE FAN-IN—Diodes give greater reliability than resistors in input leads when large number of inputs is required for comparator operating as AND gate. For NAND gate operation, interchange inverting and noninverting inputs to comparator.—R. T. Smathers, T. M. Frederiksen, and W. M. Howard, "LM139/LM239/LM339—A Quad of Independently Functioning Comparators," National Semiconductor, Santa Clara, CA, AN-74, 1973, p 9.

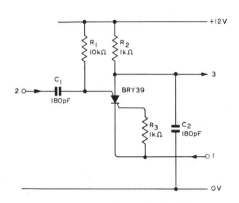

SYNCHRONOUS AND GATE—Negative 12-V input pulses from ring counter are applied to terminal 2 of scs AND gate and negative-going clock pulses are applied to terminal 1. Switching of scs occurs only if clock pulse is present at same time as input pulse. Resulting anode current through R2 gives negative-going square-wave output pulse S3 that persists until end of clock pulse.—D. J. G. Janssen, Circuit Logic with Silicon Controlled Switches, *Mullard Technical Communications,* March 1968, p 57–64.

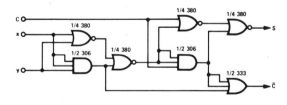

FULL ADDER—Gives binary sum of three binary digits—X, Y, and C (carry). One output indicates sum of all three inputs and other indicates value of resulting carry. Uses Signetics 380 quad 2-input NOR gates, Signetics 306 dual 3-input AND gates, and Signetics 333 dual and 3-input expandable OR gate.—"Signetics Digital Utilogic 2/600 TTL-DTL Data Book," Signetics Corp., Sunnyvale, CA, 1972, p 58.

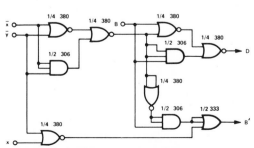

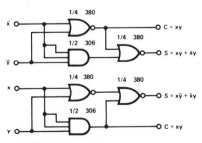

FULL SUBTRACTOR—Gives binary difference, in which output D is equal to X minus Y minus borrow, and B' output is new borrow. Similar to full adder except for requiring false input and change of output notation.— "Signetics Digital Utilogic 2/600 TTL-DTL Data Book," Signetics Corp., Sunnyvale, CA, 1972, p 59.

HALF-ADDER—Gives binary sum of X plus Y on two outputs, sum S and carry C. Uses Signetics 380 quad 2-input NOR gates and Signetics 306 dual 3-input AND gates.—"Signetics Digital Utilogic 2/600 TTL-DTL Data Book," Signetics Corp., Sunnyvale, CA, 1972, p 58.

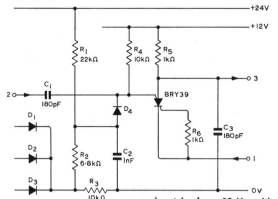

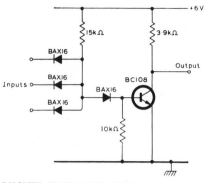

AND WITH DELAYED AND—Used at output of scs ring counter. Input pulses at terminal 2 can affect output only if all d-c inputs on diodes D1, D2, and D3 are at least zero and no input is above 12 V positive. Clock pulses go to terminal 1.—D. J. G. Jenssen, Circuit Logic with Silicon Controlled Switches, *Mullard Technical Communications*, March 1968, p 57–64.

DISCRETE NAND GATE—Will operate up to about 1 MHz. For higher frequencies, use BAX13 diodes. Used for decoding at outputs of counters.—"Electronic Counting," Mullard, London, 1967, TP874, p 154–155.

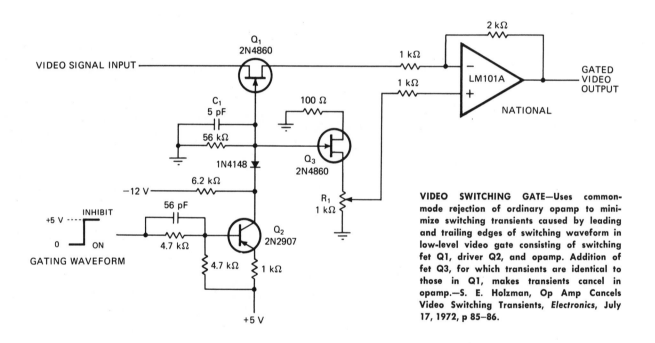

VIDEO SWITCHING GATE—Uses common-mode rejection of ordinary opamp to minimize switching transients caused by leading and trailing edges of switching waveform in low-level video gate consisting of switching fet Q1, driver Q2, and opamp. Addition of fet Q3, for which transients are identical to those in Q1, makes transients cancel in opamp.—S. E. Holzman, Op Amp Cancels Video Switching Transients, *Electronics*, July 17, 1972, p 85–86.

CHAPTER 45
High-Voltage Circuits

PHOTOMULTIPLIER SUPPLY—Developed for nine-dynode photomultiplier tube. Full-wave rectifier operating off one winding of power transformer T1 provides 15-V d-c bias voltage for IC voltage regulator. High-voltage output is obtained from voltage doubler operating from second winding. IC operates as current regulator, with cascode-connected transistors Q1–Q5 used as pass transistors. C1–C5 suppress and equalize transients across transistors, while clamp diodes D1–D5 protect transistors from voltage transients.—R. J. Widlar, "New Uses for the LM100 Regulator," National Semiconductor, Santa Clara, CA, 1968, AN-8.

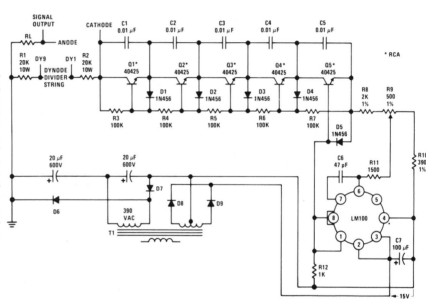

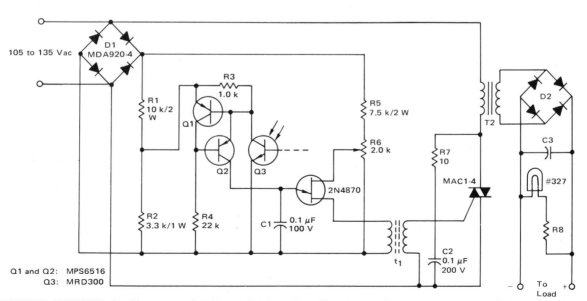

Q1 and Q2: MPS6516
Q3: MRD300

OPTOELECTRONIC REGULATOR—Provides regulation of extremely high-voltage power supply by using lamp across high-voltage rectifier to act on phototransistor Q3 which is in ujt oscillator circuit that changes firing angle of MAC1-4 triac in series with primary of high-voltage transformer C2. Output voltage determines specifications for high-voltage circuit. T1 is pulse transformer used for triggering triac by phase control of firing time. Arrangement provides high isolation between regulator and dangerous high-voltage circuitry.— J. Bliss, "Applications of Phototransistors in Electro-Optic Systems," Motorola Semiconductors, Phoenix, AZ, AN-508, 1971.

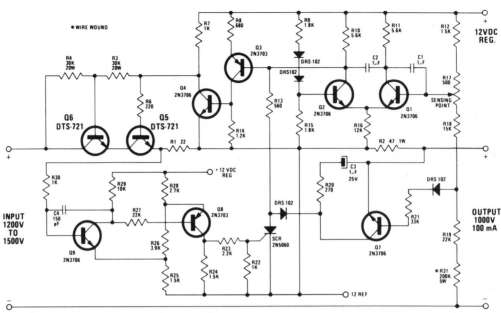

1,000-V D-C AT 100 W—Uses two Delco DTS-721 transistors in series as pass element. Regulation at full load is 0.1% for inputs from 1,200 V to 1,500 V. Circuit includes short-circuit protection. Q1 and Q2 form differential amplifier that compares sensing-point voltage (proportional to output voltage) to reference voltage at base of Q2. Difference signal is amplified by Q3-Q4 and coupled to Q5. Q6 tracks Q5 in linear region within 100 V. Regulated 12-V supply is referenced to high side of output voltage through R2, permitting direct drive of output stage.—"1,000-Volt Linear DC Regulator," Delco, Kokomo, IN, Feb. 1971, Application Note 45.

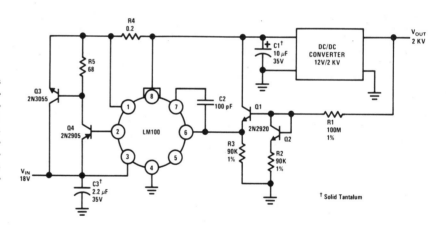

2-KV SUPPLY—IC voltage regulator senses output of high-voltage supply through resistive divider and varies input to d-c/d-c converter to hold output voltage constant. Q1 acts as buffer for high-impedance divider, permitting reduction of divider power dissipation to 40 mW. Q2 compensates for temperature drift in Q1.—R. J. Widlar, "New Uses for the LM100 Regulator," National Semiconductor, Santa Clara, CA, 1968, AN-8.

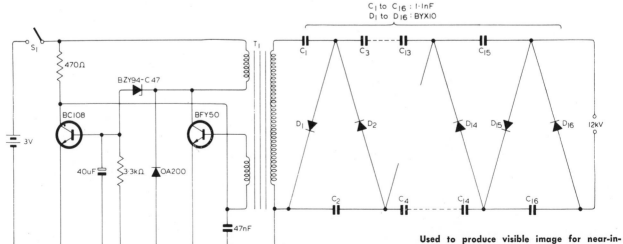

12-KV IMAGE CONVERTER SUPPLY—Simple blocking oscillator with high-ratio transformer operates from two flashlight cells in series and supplies 1.5-kV pulses to eight-stage Cockcroft-Walton bridge that delivers 12 kV for Mullard 6929 diode image converter tube. Used to produce visible image for near-infrared input.—K. A. Cook, Diode Image Converter Tubes and Their Applications, *Mullard Technical Communications*, Sept. 1967, p 256–260.

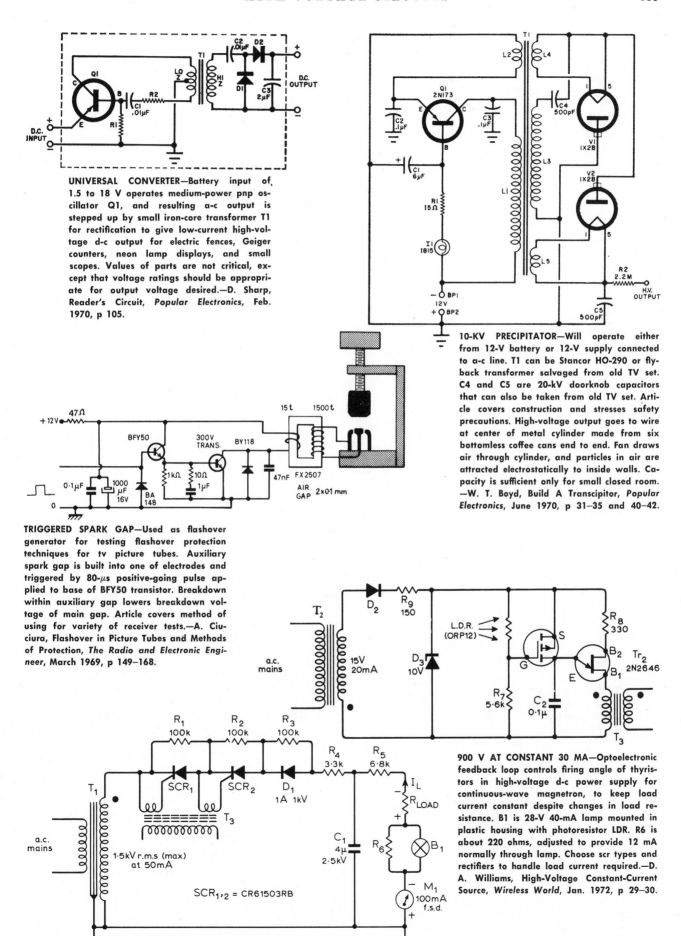

UNIVERSAL CONVERTER—Battery input of 1.5 to 18 V operates medium-power pnp oscillator Q1, and resulting a-c output is stepped up by small iron-core transformer T1 for rectification to give low-current high-voltage d-c output for electric fences, Geiger counters, neon lamp displays, and small scopes. Values of parts are not critical, except that voltage ratings should be appropriate for output voltage desired.—D. Sharp, Reader's Circuit, *Popular Electronics*, Feb. 1970, p 105.

10-KV PRECIPITATOR—Will operate either from 12-V battery or 12-V supply connected to a-c line. T1 can be Stancor HO-290 or flyback transformer salvaged from old TV set. C4 and C5 are 20-kV doorknob capacitors that can also be taken from old TV set. Article covers construction and stresses safety precautions. High-voltage output goes to wire at center of metal cylinder made from six bottomless coffee cans end to end. Fan draws air through cylinder, and particles in air are attracted electrostatically to inside walls. Capacity is sufficient only for small closed room.—W. T. Boyd, Build A Transcipitor, *Popular Electronics*, June 1970, p 31—35 and 40—42.

TRIGGERED SPARK GAP—Used as flashover generator for testing flashover protection techniques for tv picture tubes. Auxiliary spark gap is built into one of electrodes and triggered by 80-μs positive-going pulse applied to base of BFY50 transistor. Breakdown within auxiliary gap lowers breakdown voltage of main gap. Article covers method of using for variety of receiver tests.—A. Ciuciura, Flashover in Picture Tubes and Methods of Protection, *The Radio and Electronic Engineer*, March 1969, p 149—168.

900 V AT CONSTANT 30 MA—Optoelectronic feedback loop controls firing angle of thyristors in high-voltage d-c power supply for continuous-wave magnetron, to keep load current constant despite changes in load resistance. B1 is 28-V 40-mA lamp mounted in plastic housing with photoresistor LDR. R6 is about 220 ohms, adjusted to provide 12 mA normally through lamp. Choose scr types and rectifiers to handle load current required.—D. A. Williams, High-Voltage Constant-Current Source, *Wireless World*, Jan. 1972, p 29—30.

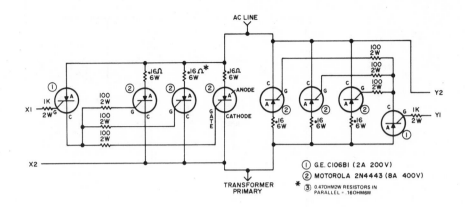

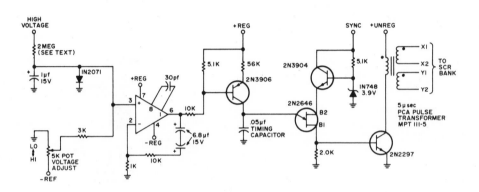

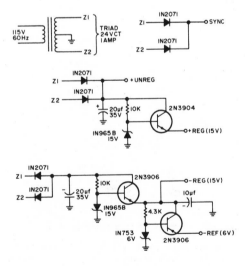

2.5 KV FOR 2-KW TRANSMITTER—Uses scr bank in series with primary of high-voltage transformer, controlled by regulator connected to d-c output voltage which can be varied from 100 V to over 3.5 kV with voltage adjust pot. Operates by manipulating waveform applied to transformer. Article covers construction and operation. IC used in error amplifier is LM201. Article gives equivalent three-transistor differential amplifier that can be built for fraction of cost of IC. Application of 600-W load gave no noticeable change in output voltmeter reading at voltages below 1.9 kV. Voltages in this circuit are dangerous.—R. Sebol, An SCR Regulator For Kilowatt Power Supplies, 73, Aug. 1972, p 107–113.

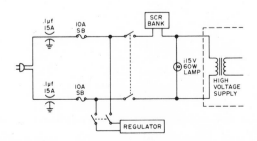

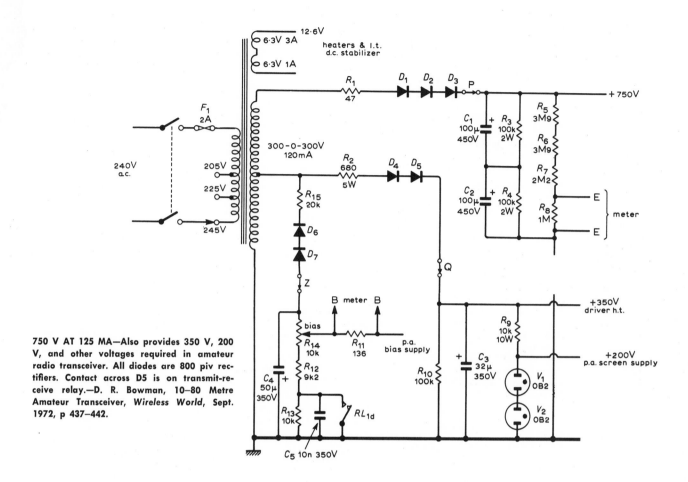

750 V AT 125 MA—Also provides 350 V, 200 V, and other voltages required in amateur radio transceiver. All diodes are 800 piv rectifiers. Contact across D5 is on transmit-receive relay.—D. R. Bowman, 10–80 Metre Amateur Transceiver, *Wireless World*, Sept. 1972, p 437–442.

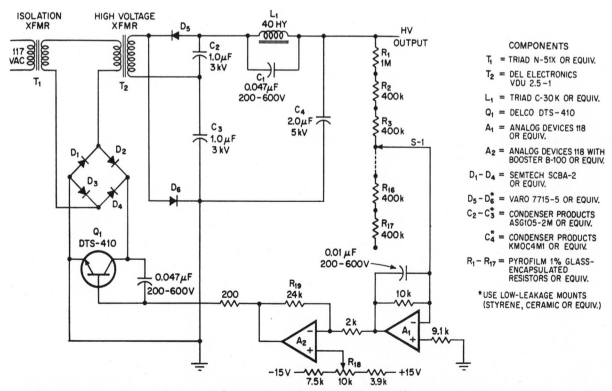

3.5 KV AT 1 MA—Regulator acts on primary side of high-voltage transformer, permitting use of lower-cost and less bulky low-voltage components. Regulation can be 0.001%. Used as photomultiplier tube supply requiring output ranges from 500 V to 3.5 kV.—R. J. Krusberg, Low-Voltage Feedback Loop Controls High-Voltage Supply, *Electronics*, March 29, 1971, p 69.

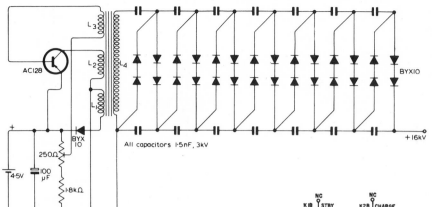

16-KV IMAGE CONVERTER SUPPLY—Operates from three flashlight cells in series. Designed for use with Mullard 6914 diode image converter tube which produces visible image for near-infrared input.—K. A. Cook, Diode Image Converter Tubes and Their Applications, *Mullard Technical Communications*, Sept. 1967, p 256–260.

3,000 V AT 700 MA—Designed for use with linear amplifier operating at legal maximum input power for amateur radio. Article includes modifications required for operation on 230-V a-c line to give improved regulation. Use string of ten 1-meg resistors for R2 to prevent arcover. C1–C5 are 0.01-μF disk. C6–C9 are 240-μF 450-V electrolytic. Diodes are 1,000-prv 2-A silicon rectifiers. R3–R7 are 470K and R8–R11 are 25K 20-W. Z1 is GE 20SP4B4 thyrector diode assembly.—C. Smith and D. DeMaw, A Power Supply for That Big Linear Amplifier, QST, Dec. 1969, p 41–43.

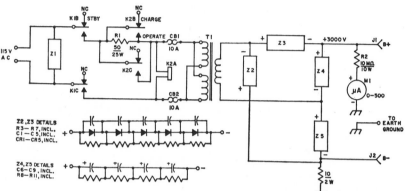

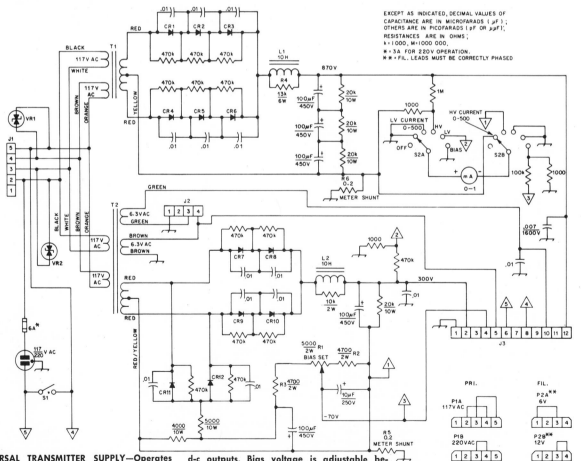

UNIVERSAL TRANSMITTER SUPPLY—Operates on 220 V, for handling transmitter having 2-kW amplifier without making house lights blink as exciter load is applied. Will also operate on 117 V. Choke-input filtering provides adequate regulation of both 300-V and 800-V d-c outputs. Bias voltage is adjustable between −40 and −80 V. All diodes are Mallory M2.5A 1,000-prv 2.5-A silicon. L1 is rated 200 mA, and L2 is 300 mA. T1 has 890 V each side of center tap at 300 mA, and T2 has 350 V each side of tap at 175 mA. VR1 and VR2 are GE 6RS20SP8B8 thyrector assembly. Cost of parts is about $100.—R. M. Myers and A. M. Wilson, Universal Power Supply for the Amateur Station, QST, Sept. 1972, p 42–43 and 63.

900-V TRANSFORMERLESS SUPPLY—Sextupler circuit actually multiplies a-c line voltage about 7.5 times because capacitors charge up to peak of applied a-c voltage before rectification. Provides adequate voltage regulation for use with tube-type ssb transceivers. Article covers construction and operation. Adjust R10 as required to give low voltage required for transceiver being used.—J. Bell, A Voltage Sextupler Power Supply, 73, Nov. 1969, p 62–64.

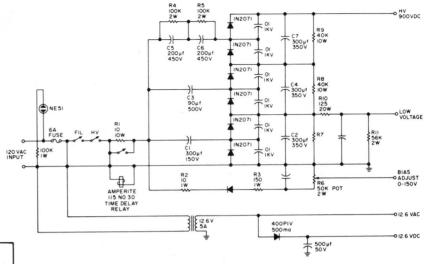

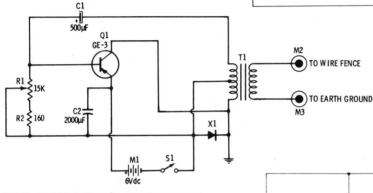

FENCE CHARGER—Transformer steps up output voltage of oscillator Q1 to several thousand volts for electric fence applications. Use at least 4' ground rod. R1 determines amount of shock. X1 is 1N540 diode. T1 is 6.3-V filament transformer used backward. Precaution: Avoid using where small children might be caught in fence with feet on ground. To use as portable shocker for warding off assailants, connect M2 and M3 to prongs at end of cane. Will discourage dogs from upsetting garbage cans (turn off on garbage pickup days).—R. M. Brown and T. Kneitel, "49 Easy Entertainment & Science Projects," Howard W. Sams, Indianapolis, IN, 1971, p 42–43.

LIGHT-OPERATED SERIES SWITCH—Xenon flash tube acts on phototransistors through fiber-output bundle to trigger series string of scr's simultaneously for applying 6 kV to load. Optoelectronic triggering eliminates inductive delays of conventional trigger wiring. Requires scr's having matched rise times, so slowest units will not be gated on by anode breakover. Useful in high-voltage pulse-forming networks and crowbar circuits.—J. Bliss, "Applications of Phototransistors in Electro-Optic Systems," Motorola Semiconductors, Phoenix, AZ, AN-508, 1971.

CHAPTER 46
Hobby Circuits

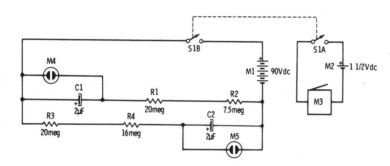

FISH BUZZER—Attracts fish by imitating sounds of wet bugs thrashing around on surface of water, while producing tiny intermittent flashes of light that penetrate water for hundreds of feet. Can be mounted in empty peanut-butter jar, with buzzer cemented to inside of cover. Lower into water by rope to desired depth, wait 15 minutes, then start pulling in fish. M4 and M5 are NE-2 neons. Use smallest 90-V battery available, or four 22½-V hearing-aid batteries in series.—R. M. Brown and R. Kneitel, "49 Easy Entertainment and Science Projects," Howard W. Sams, Indianapolis, IN, 1971, p 18—19.

TOOL MAGNETIZER—Screwdriver or other tool inserted in magnetizing coil L1 is magnetized. Isolation transformer T1, having 117-V secondary rated at 50 A, makes circuit safe. X1 is E 504A rectifier. M1 is NE-2E neon. Capacitors are 200-μF 150-V electrolytics connected in parallel. These charge when heavy-current knife switch S2 is at left, and discharge through L1 when S2 is at right. Make L1 with 8 turns No. 10 wire on ¾-inch form, spaced ⅛ inch between turns.—R. M. Brown and T. Kneitel, "101 Easy Test Instrument Projects," Howard W. Sams, Indianapolis, IN, 1968, p 84—85.

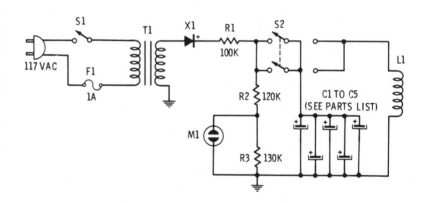

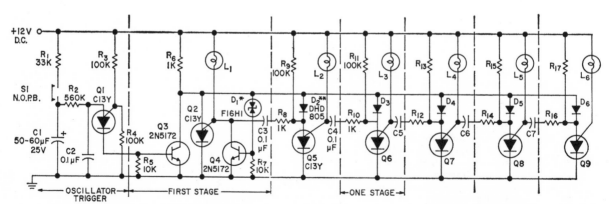

TO OPERATE, DEPRESS PUSHBUTTON FOR A SECOND OR SO AND RELEASE.
FOR ROULETTE, USE 36 LAMPS / STAGES.
✳ CAN BE ANY 6.2V ZENER   ✳✳ CAN BE 1N914   L₁ -L₆ - #53 LAMPS

RING COUNTER MAKES DICE—Q1, Q3, and Q4 form relaxation oscillator that feeds pulses to six-lamp ring counter when S1 is closed. When S1 is released, one of lamps stays lit. Probability of result is random because charging of capacitor C1 combines with 20-pps rate of oscillator to make control of on stage extremely difficult. Article includes optional encoder for energizing lamps to resemble spots on face of die.—R. W. Fox, Electronic Dice, Electronics World, Feb. 1970, p 34—35 and 79.

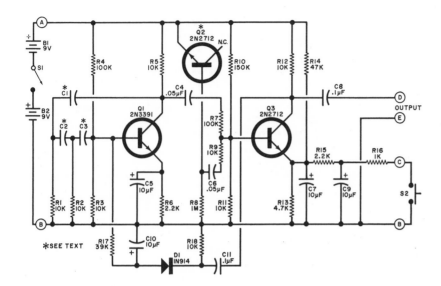

STEAM WHISTLE—Combination of tone generator Q1, white-noise source Q2, and gating amplifier Q3 provides signal simulating attack and decay characteristics of locomotive steam whistle. Try several 2N2712 transistors for Q2 if first one does not work as noise source. Values of C1, C2, and C3 should be 0.005 μF to give high-pitched screech of European trains, or 0.05 μF for throaty roar of American freight trains. Value of R9 determines amount of steam hiss. Use with any audio amplifier.—J. S. Simonton, Jr., Electronic Steam Whistle, *Popular Electronics*, May 1972, p 51–53.

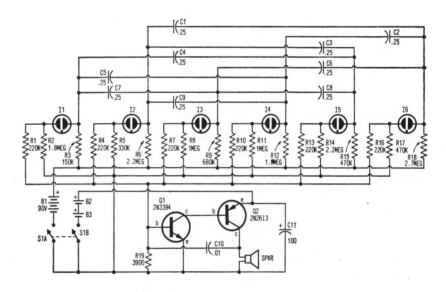

RANDOM BLINKS AND SQUEALS—Six NE-2 neons mounted on box blink on and off at random rate, with each turn-on accompanied by randomly changing soft glide tone. Current drain is so low that batteries should last several months if using 1.5-V AA batteries for B2 and B3, with RCA VS090 for B1. Use 2- or 3-inch 8-ohm miniature speaker. If necessary, increase or decrease R19 in 20% steps to make tone generator fire in step with lamps.—H. Friedman, Computomatic, *Elementary Electronics*, Jan.-Feb. 1969, p 53–55.

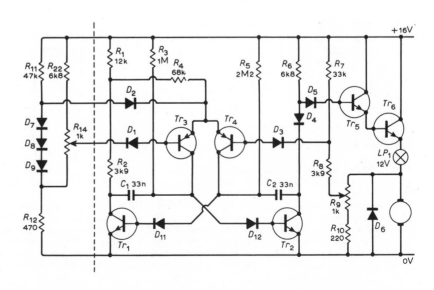

MODEL TRAIN SPEED CONTROL—R9 varies rate at which pulses are applied to small permanent-magnet motor to hold speed constant under varying loads. TR1, TR2, TR5, and TR9 should be very high gain silicon transistors with hfe above 200, such as 2N930; other transistors are general-purpose silicon. Wattage of lamp LP1 depends on maximum motor load, in range from 5 to 20 W for typical model locomotives. All diodes are small-signal silicon, such as 1N914.—P. L. Hollingberry, Towards Better Control of Small D.C. Motors, *Wireless World*, July 1972, p 333–335.

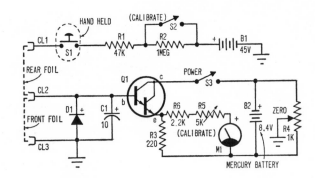

BASEBALL VELOCITY TIMER—If ball is thrown with sufficient accuracy to break two vertically mounted strips of tinfoil spaced 2 feet apart, time between interruptions of circuits is indicated as velocity on meter. Q1 is 2N5306. Use 1-mA d-c meter. D1 is 1N4742 zener. Article covers construction, calibration, and use.—R. Michaels, Build S/E's ... Baseball Velocity Timer, *Science and Electronics*, June-July 1969, p 59–63.

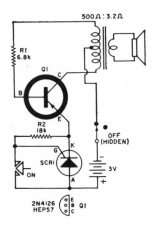

SQUEALING BOX—Circuit is mounted in box having holes for speaker, with only button of *on* switch projecting. *Off* switch is concealed inside, for operation with nail pushed through correct hole in housing. Button is labeled "DO NOT TOUCH." When pushed, oscillator generates ear-splitting screech that cannot be stopped because scr operates as latch. Penlight batteries are recommended, so they run down before box is smashed or thrown out of window if operated when owner is absent.—P. Franson, The Curiosity Box, *Electronics World*, Sept. 1970, p 82–83.

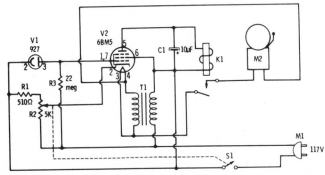

RIFLE RANGE—Bell sounds when flashlight on toy rifle hits bull's-eye phototube on target. T1 is 6-V filament transformer. K1 is 4,000-ohm relay, and M2 is 6-V a-c buzzer or bell.

—R. M. Brown and R. Kneitel, "49 Easy Entertainment and Science Projects," Howard W. Sams, Indianapolis, IN, 1971, p 16–17.

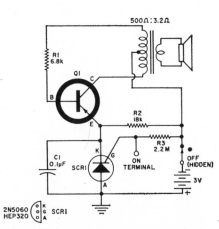

TOUCH-BUTTON SQUEALING BOX—Touching insulated *on* terminal marked "DO NOT TOUCH" latches scr *on*, energizing oscillator which generates ear-splitting screech that can be stopped only by poking nail through hole in housing to operate concealed *off* switch. Used as party novelty.—P. Franson, The Curiosity Box, *Electronics World*, Sept. 1970, p 82–83.

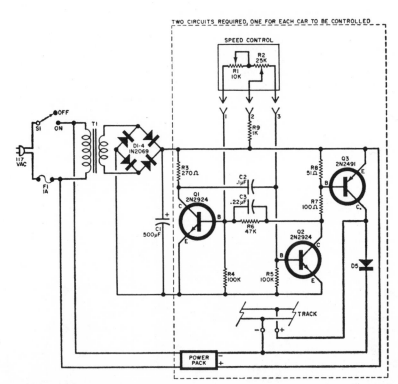

SLOT-CAR CONTROL—Provides complete control of slot car, at speeds ranging from slow crawl to all-out, by pulsing drive motor. Includes electronic braking. Provides optimum motor torque at all speeds. Power pack is conventional d-c supply sold for slot cars. T1 is power transformer with 6.3-V secondary at 0.6 A. All diodes are 1N2069. Will control any other small d-c motor requiring up to 30 V and not more than 4 A.—B. C. Snow, Modern Slot-Car Controller, *Popular Electronics*, Dec. 1967, p 41–44 and 92.

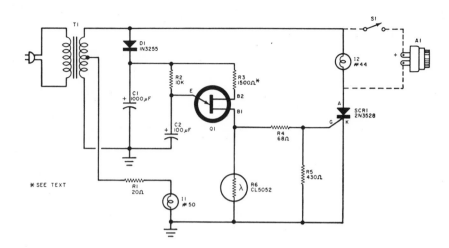

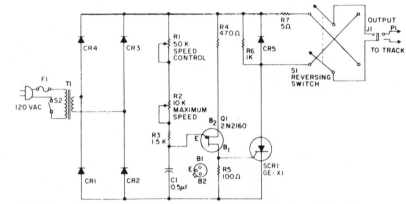

MESSAGE ALARM—When note for wife is inserted in box to block light from I1 to photoresistor R6, TIS43 or similar ujt Q1 makes lamp I2 flash about once per second as reminder. Optional Sonalert alarm A1 and switch S1 may be placed across lamp, for use when message is urgent.—R. Persing, Build Your Own Memo Minder, *Popular Electronics*, Oct. 1967, p 49–51 and 102.

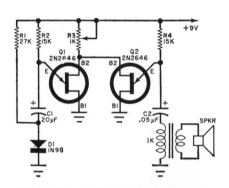

BELL - TONE OSCILLATOR — Low - frequency astable mvbr Q1 modulates 700-Hz relaxation oscillator Q2, to give bell-like sound with sharply rising and falling pitch. Use cheap 3-inch replacement speaker to get desired tinny sound. Adjust R3 for most pleasing effect.—F. H. Tooker, Getting to Know the UJT, *Popular Electronics*, April 1970, p 69–73.

MODEL-TRAIN CONTROL—Scr applies power to rails of toy train in pulses, with R1 varying pulse width to give smooth starting, stopping, and speed control. S1 reverses train motor.—"Electronics Experimenters Circuit Manual," General Electric, Owensboro, KY, 1971, 3rd Ed., p 121–124.

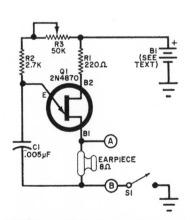

MOSQUITO REPELLER—Simple ujt relaxation oscillator operating from 22.5-V battery produces humming noise at about 2,000 Hz, which attracts male mosquitoes that don't bite while repelling the female of the species. Range is about 3 feet. Not guaranteed to repel all species.—L. E. Greenlee, Build the Bug Shoo, *Popular Electronics*, July 1970, p 27–30.

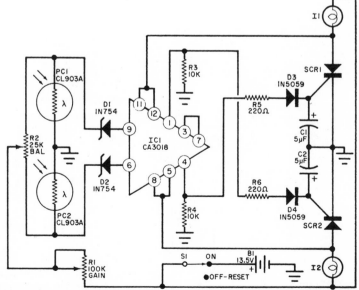

PHOTOELECTRIC FINISH-LINE—Determines which of two slot cars is winner even if cars are only 1/32 inch apart. Separate photoresistor circuits for each track are so arranged that when light is interrupted by winning car, other circuit is automatically deactivated and scr turns on pilot lamp for winning track. Use No. 330 14-V lamps and TIC-46 scr's.—W. T. Lemen, Build Slot-Car Win Detector, *Popular Electronics*, May 1969, p 41–45.

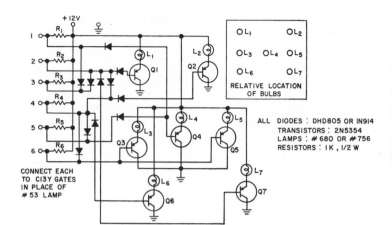

DICE DISPLAY—Diode-transistor encoder, when used with six-stage triggered ring counter driven by pushbutton-operated 20-pps oscillator, converts single numbered-lamp display of counter to equivalent numeric display of seven lamps positioned according to spots on face of die. Thus, scr triggered on between any one of input terminals 1–6 and ground energizes corresponding combination of lamps on display to represent same numeral.—R. W. Fox, Electronic Dice, *Electronics World*, Feb. 1970, p 34–35 and 79.

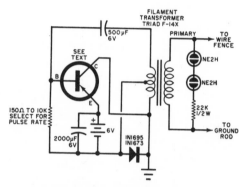

ELECTRIFIED FENCE—Battery-operated circuit operates with any good power transistor, such as 2N2869. Range of values shown gives 10 to 100 pulses per minute; adjust for optimum rate of 50 to keep the cows home. Can also be connected to garbage can, to discourage raids by dogs and cats; disconnect before garbage man comes.—C. J. Schauers, Electrified Fence, *Popular Electronics*, Feb. 1967, p 68–69.

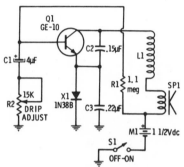

DRIP-DRIP SOUNDER—Applications include breaking apartment lease, annoying obnoxious neighbor, and preventing someone from falling asleep. Circuit is similar to that of metronome. Use higher battery voltage if louder dripping sound is required from 3.2-ohm p-m speaker. L1 is 100-mH choke.—R. M. Brown and T. Kneitel, "49 Easy Entertainment and Science Projects," Howard W. Sams, Indianapolis, IN, 1971, p 79.

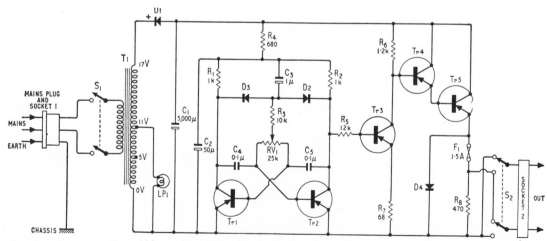

MODEL TRAIN CONTROL—Uses pulse width modulation at frequency of several hundred Hz to get good speed regulation as load changes on inclines. Transistor types are not critical. Tr1-Tr3 can be OC71. Tr4 can be OC81 or 2N381. Tr5 can be OC35 or 2N456. D1 is 50-piv 1.5-A rectifier. Other diodes are OA202. Designed for driving 12-V model train drawing up to 1.5 A. S2 is reversing switch. Tr1 and Tr2 form astable mvbr having diodes connecting back to back to insure that circuit will oscillate reliably at startup. RV1 is continuously variable control of mark-space ratio of pulses, serving as speed control.—T. E. Estaugh, Electronic Control for Model Locomotives, *Wireless World*, Oct. 1966, p 529–532.

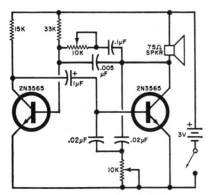

FISH CALLER—Based on hopeful theory that fish will be attracted by strange underwater noise. Consists of two-transistor tone generator and small speaker mounted in waterproof case for lowering into water. Changing settings of pots gives variety of tones.—A Fishy Project?, *Popular Electronics*, March 1971, p 84–85.

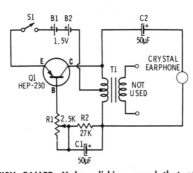

FISH CALLER—Makes clicking sound that attracts fish. Place in plastic bag or other waterproof container and lower over side of boat. T1 should be thumb-size subminiature output transformer for transistors, having 500-ohm center-tapped primary. Crystal earphone is transistor radio type. Use two 1.5-V AA or AAA cells. S1 is part of R1.—H. Friedman, "99 Electronic Projects," Howard W. Sams, Indianapolis, IN, 1971, p 120–121.

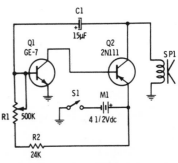

MOTORBOAT-SOUND GENERATOR—Designed to be installed in model boat powered by electric motor, to provide sound of motorboat engine for added realism. Use subminiature all-weather speaker. Adjust R1 to simulate revving up of engine.—R. M. Brown and T. Kneitel, "49 Easy Entertainment and Science Projects," Vol. 2, Howard W. Sams, Indianapolis, IN, 1969, p 17.

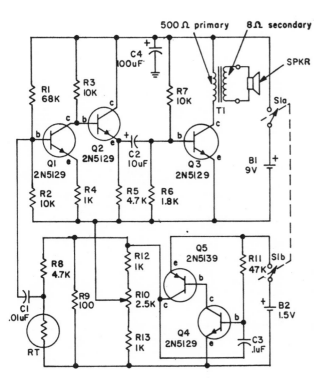

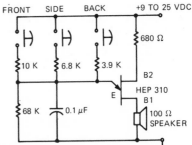

THREE-TONE DOORBELL—Produces three distinctively different audio tones from speaker, for identifying door at which button is pushed. Can be expanded to serve as toy organ.—"Home Handyman's Construction Projects," Motorola Semiconductors, Phoenix, AZ, HMA-37, 1972.

FISH FINDER—Fenwal GB32P8 2,000-ohm thermistor at end of 60-foot length of speaker wire is connected into one leg of a-c Wheatstone bridge powered by astable mvbr Q4-Q5. Three-transistor audio amplifier and speaker make unbalance of bridge audible as thermistor probe is moved through thermocline where water temperature drops markedly within few feet of depth. Fish tend to congregate at this depth. Article gives construction and calibration details.—C. E. Bryson, Build A Better Fish Finder, *Elementary Electronics*, Nov.-Dec. 1971, p 77–79 and 97.

CHAPTER 47
I-F Amplifier Circuits

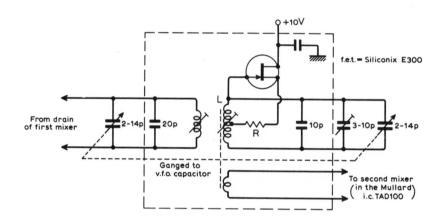

I-F IMAGE REJECTION—Replacement of traditional tube with Siliconix E300 fet in cathode-follower type of Q multiplier gives improved stability, for increasing image rejection of tunable i-f section (10.7–12.7 MHz, with second i-f of 470 kHz) of 2-meter amateur receiver. R is 5.6K, and L is 18 inches of No. 40 close-wound on 3/16-inch form, with 5-turn link.—D. A. Tong, A Stable Q-Multiplier, *Wireless World*, Dec. 1971, p 600.

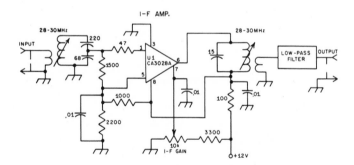

28—30 MHZ—Designed for use at output of 2-meter converter, for feeding into tunable i-f receiver. Provides additional 25 dB gain. Circuit also serves to isolate receiver from mixer of converter.—D. DeMaw, High Performance 2-Meter Converter, *QST*, June 1971, p 11-16 and 31.

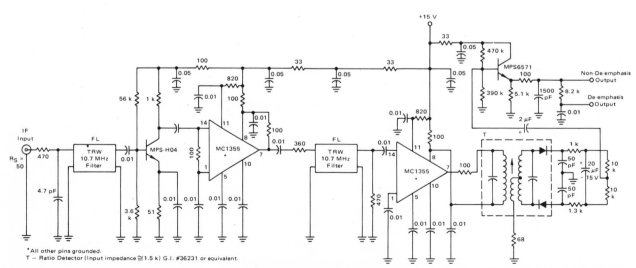

*All other pins grounded.
T – Ratio Detector (Input impedance ≅ 1.5 k) G.I. #36231 or equivalent.

I-F WITH RATIO DETECTOR—Designed for high-quality f-m system, with total harmonic distortion under 0.5%. Two TRW 10.7-MHz linear phase filters having combined bandwidth of 240 kHz provide excellent sensitivity. Transistor after first filter lowers noise figure.—B. Korth, "Integrated Circuit IF Amplifiers for AM/FM and FM Radios," Motorola Semiconductors, Phoenix, AZ, AN-543, 1971.

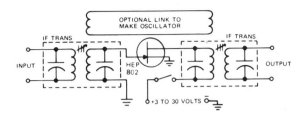

FET STAGE—Will operate equally well at 50 kHz, 455 kHz, or 10.7 MHz with appropriate i-f transformers. Addition of link having 6 turns of wire at each end, wound same diameter as i-f coils, converts circuit into stable tuned-drain tuned-gate oscillator. If it doesn't oscillate, reverse link at one end. —"Tips on Using FET's," Motorola Semiconductors, Phoenix, AZ, HMA-33, 1971.

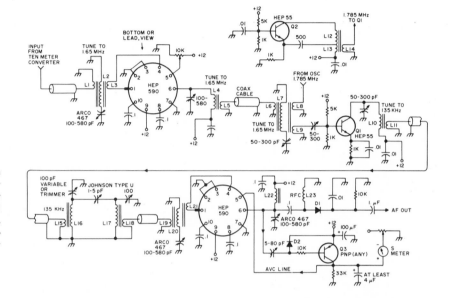

135-KHZ SOLID-STATE—Designed for use with converter changing 1.65 MHz to 135 kHz, for which 1.785-MHz local oscillator is shown at upper right. Article covers construction and adjustments for minimizing birdies, burbles, and hisses. All coil data is given.—W. Hoisington, I-F Filter Converter AVC, 73, May 1970, p 62–64, 66–70, and 72.

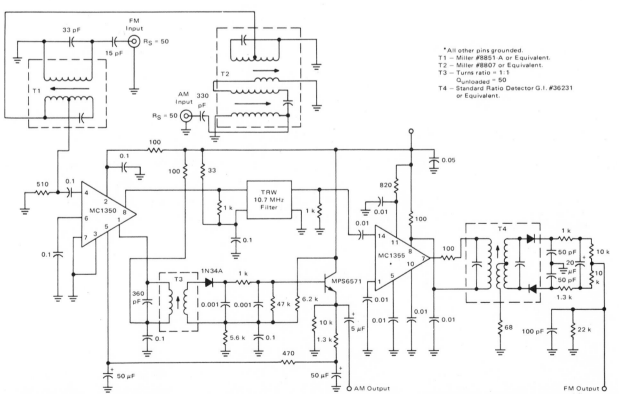

A-M AND F-M—Built-in agc capability of MC1350 IC i-f amplifier makes composite i-f strip possible. Double-tuned tank circuits used for input to MC1350 are wired in series because 10.7-MHz coil for T1 has low impedance at 455 kHz while 455-kHz coil for T2 has low impedance at 10.7 MHz. Uses standard a-m diode detector and standard f-m ratio detector.—B. Korth, "Integrated Circuit IF Amplifiers for AM/FM and FM Radios," Motorola Semiconductors, Phoenix, AZ, AN-543, 1971.

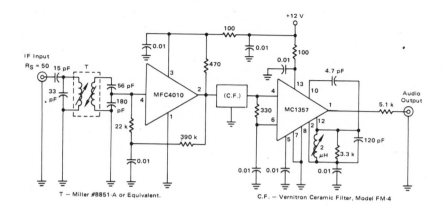

I-F WITH QUADRATURE DETECTOR—Gives excellent performance in f-m auto radio. Four-pole ceramic filter between IC i-f amplifiers provides most of filtering. Bandwidths are 240 kHz for i-f to give good stereo reproduction, and 500 kHz for low-distortion linear operation of quadrature detector. Q of quadrature coil must be about 20.—B. Korth, "Integrated Circuit IF Amplifiers for AM/FM and FM Radios," Motorola Semiconductors, Phoenix, AZ, AN-543, 1971.

T — Miller #8851-A or Equivalent.

C.F. — Vernitron Ceramic Filter, Model FM-4

9-MHZ PANORAMIC I-F—Two switched crystal filters give option of 500- or 5,000-Hz bandwidth. Provides high image rejection and gain of 46 dB. Input can be up to 0.5 V from fet mixer before cross-modulation starts. Although intended primarily for use in 2-meter panoramic receiver, can also be used in standard receiver. Large signal-handling capability eliminates need for agc.—J. Schuermann, A Universal IF Amplifier for a Standard or Panoramic Receiver, 73, Dec. 1972, p 38–42.

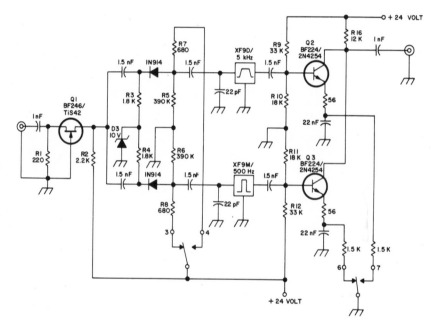

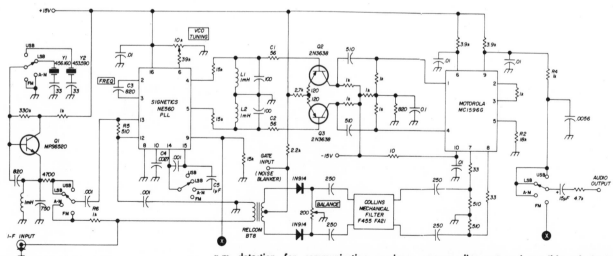

SOUPING UP THE I-F—Updates older receivers by providing ssb, f-m, and synchronous a-m detection for communication receiver. Uses phase-locked loop for synchronizing with carrier of incoming f-m signal. Article covers alignment and possible substitutions.—R. Factor, Multimode I-F System, Ham Radio, Sept. 1971, p 39–43.

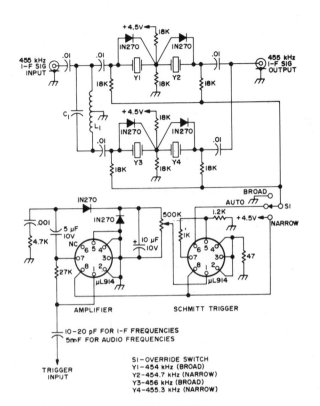

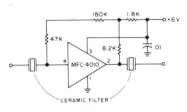

455-KHZ USING ACTIVE FILTER—Gives gain of 70 dB. Uses ceramic or crystal elements as frequency components.—F. D. Williams, Active Filter Design and Use—Part IV, 73, Oct. 1972, p 43–44 and 46–47.

AUTOMATIC BANDWIDTH CONTROL—Circuit automatically switches to narrower bandwidth when QRM (interference) develops while receiving a station. Provides two different i-f bandwidths which can be preselected for either phone or c-w service. Trigger input can be either a-f or i-f value up to about 1 MHz. Choose values for C1 and L1 to resonate at i-f being used. Choice of crystal frequencies depends on bandwidths desired. With circuit shown, bandwidths are 600 and 2,000 Hz. Article covers construction and adjustment.—J. J. Schultz, An Auto-Bandwidth Selector Unit, 73, April 1972, p 27–30.

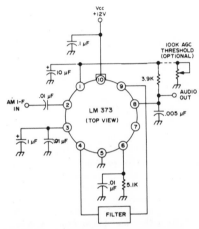

SINGLE-IC A-M I-F—Uses a-c coupling for signal transfer, to prevent possible damage to IC by direct current. Larger bypass capacitor at pin 3 is tantalum, for low frequencies. IC is by National Semiconductor.—R. Megirian, Using the LM 373, 73, April 1972, p 37–44.

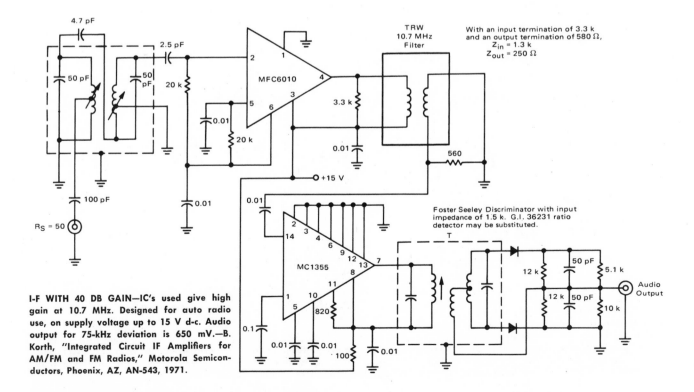

I-F WITH 40 DB GAIN—IC's used give high gain at 10.7 MHz. Designed for auto radio use, on supply voltage up to 15 V d-c. Audio output for 75-kHz deviation is 650 mV.—B. Korth, "Integrated Circuit IF Amplifiers for AM/FM and FM Radios," Motorola Semiconductors, Phoenix, AZ, AN-543, 1971.

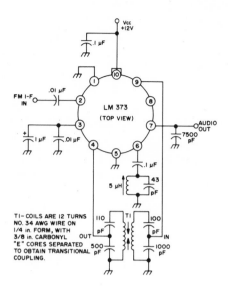

SINGLE-IC I-F—Uses a-c coupling and interstage transformer for signal transfer, to prevent possible damage to IC by direct current. Larger bypass capacitor at pin 3 is tantalum, for low frequencies. IC is by National Semiconductor.—R. Megirian, Using the LM 373, 73, April 1972, p 37–44.

60-MHZ WITH 1.5-MHZ BANDWIDTH—Two-stage IC amplifier provides power gain of about 80 dB. Differential-mode coupling is used for interstage and output networks for maximum output signal sweep capability.—B. Trout, "A High Gain Integrated Circuit RF-IF Amplifier with Wide Range AGC," Motorola Semiconductors, Phoenix, AZ, AN-513, 1969.

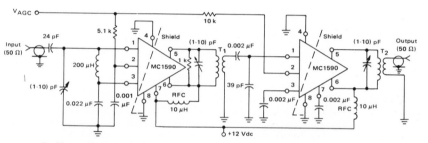

T_1: Primary — 15 turns, No. 22 AWG wire,
¼" I.D. Air Core Secondary — 4 turns, No. 22 AWG
wire, coef. of coupling ≈ 1.0

T_2: Primary — 10 turns, No. 22 AWG wire,
¼" I.D. Air Core Secondary — 2 turns, No. 22 AWG
wire, coef. of coupling ≈ 1.0

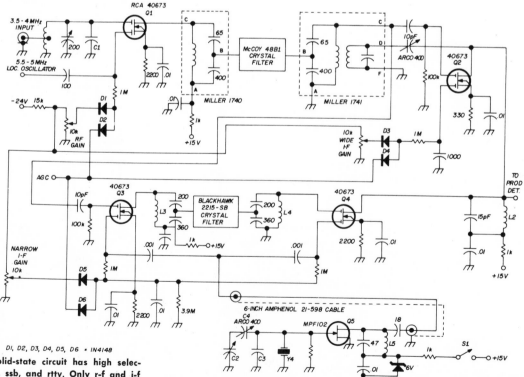

D1, D2, D3, D4, D5, D6 = 1N4148

9-MHZ I-F—Solid-state circuit has high selectivity for c-w, ssb, and rtty. Only r-f and i-f circuits are shown. Requires 80-meter tuner. Article gives coil data, along with product detector and a-f circuits. Wide filter passes only 8,998.6 to 9,001.4 kHz, making it neces- sary to convert slice of this passband to fre- quency of narrow filter; this is done with 11.215-MHz crystal oscillator feeding mixers Q3 and Q4.—G. W. Jones, A Versatile Solid-State Receiver, Ham Radio, July 1970, p 10–15.

10.7-MHZ TUNABLE—Uses active filter in unique circuit providing variable Q and means for tuning receiver i-f off slightly to get rid of interference. Fet acts as variable resistance in tuned-circuit network, to change input resistance of amplifier. Will increase circuit Q more than 2,000 while placing highest gain of response at desired frequency.—F. D. Williams, Active Filter Design and Use—Part IV, 73, Oct. 1972, p 43–44 and 46–47.

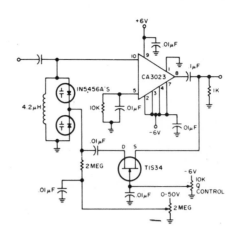

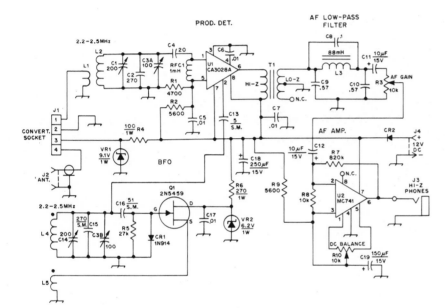

I-F AND A-F FOR THREE-BAND CONVERTER —Used with plug-in front ends to cover 80, 40, and 15 meters. Article includes converter circuits, along with coil data.—J. Kaufmann and D. DeMaw, A High-Performance Solid-State Receiver for the Novice or Beginner, QST, Oct. 1971, p 11–16 and 23.

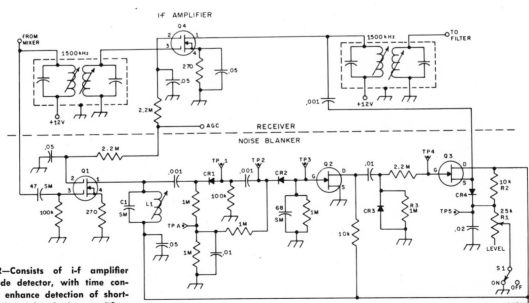

NOISE BLANKER—Consists of i-f amplifier followed by diode detector, with time constants chosen to enhance detection of short-duration noise pulses. After further amplification, noise pulses are applied to stage which is biased and threshold-controlled to act as switch which follows rise and fall time of noise pulses. By connecting switch to later i-f stage in receiver, noise pulses traveling in i-f chain can be blanked out. CR1–CR3 are 1N914 and CR4 is 1N34A. Q1 and Q4 are RCA 40673, while Q2 and Q3 are HEP802. Built for 1,500-kHz i-f. Article covers construction and adjustment.—F. N. Van Zant, A Solid-State Noise Blanker, QST, July 1971, p 20–22.

CHAPTER 48
Infrared Circuits

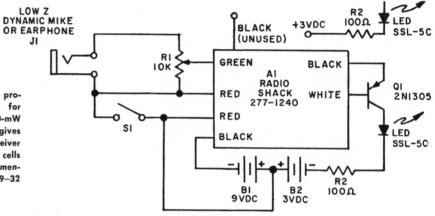

LED TRANSMITTER—Light-emitting diode produces invisible modulated light beam for voice communication when driven by 100-mW audio amplifier and microphone. Article gives construction details also for matching receiver using same amplifier with silicon solar cells and speaker.—F. Mims, Light-Comm, *Elementary Electronics*, May-June 1972, p 29–32 and 94.

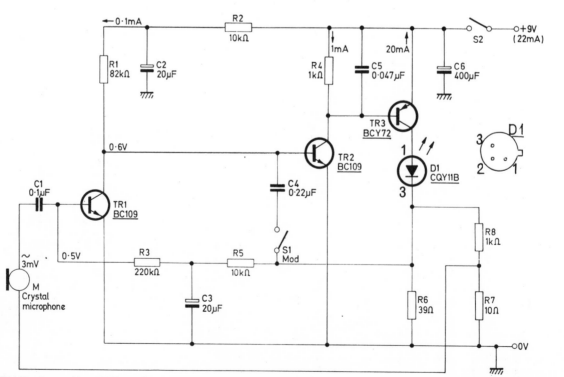

TRANSMITTER—Circuit provides 100% modulation of electroluminescent diode current (20 mA peak) for mike input of 3 mV peak. Will transmit over range of 450 feet to suitable receiver. Intended for voice communication when wire lines are not feasible, as at construction sites and sport events. Operates in near infrared region, at wavelength of about 0.9 μm.—"An Infra-Red Communication System," Educational Projects in Electronics, Mullard, London.

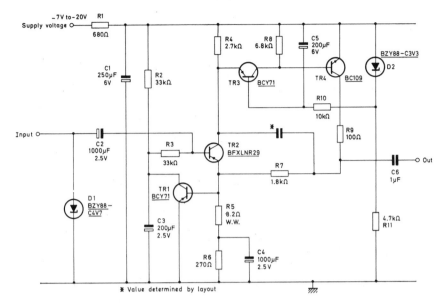

LOW-NOISE AMPLIFIER—Designed for use with mercury cadmium telluride detector cooled in liquid nitrogen to 77 K. Uses constant-current bias for detector connected to input. Voltage gain is 200, bandwidth above 1 MHz, and noise figure 1.5 dB with typical detector. Circuit uses two reference voltage sources to prevent supply voltage variations from affecting collector currents of TR2 and TR4.—"Applications of Infrared Detectors," Mullard, London, 1971, TP1201, p 36–38.

PHASE DETECTOR—Balanced full-wave phase-sensitive detector for infrared spectographic system uses RCA 7360 beam deflection tube having two deflector plates that serve to switch electron beam alternately from one electrode to the other. Switching action of tube is controlled by reference waveform in synchronism with input signal. Capacitor across output (8 or 40 μF) is charged on both half-cycles, to give multiplication of reference by input.—K. L. Smith, The Ubiquitous Phase Sensitive Detector, *Wireless World*, Aug. 1972, p 367–370.

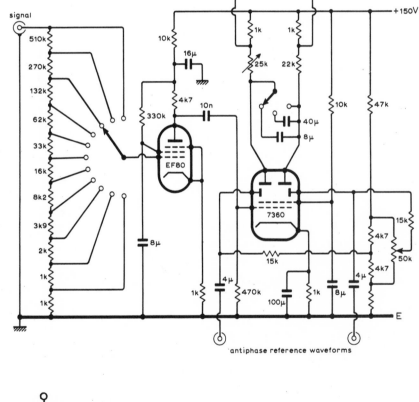

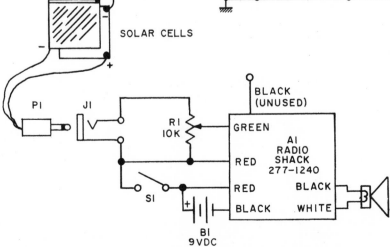

INFRARED RECEIVER—Used with led transmitter to provide short-distance voice communication over invisible infrared beam. Article gives construction details. Uses 100-mW IC audio amplifier driving 8- or 16-ohm speaker. Silicon solar cells, such as International Rectifier type S1M, are mounted in parabolic reflector.—F. Mims, Light-Comm, *Elementary Electronics*, May-June 1972, p 29–32 and 94.

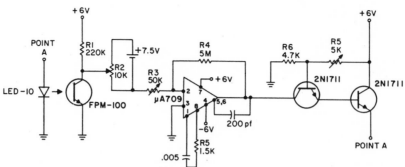

WEAK-IMAGE DETECTOR—Low-impedance combination of light-emitting diode and phototransistor feeds d-c opamp, current limiter, and emitter-follower amplifier. Operates from low-voltage d-c supply. Instead of using light bias to maintain system at operating point close to breakdown, R2 is adjusted so inverting input of opamp has exactly enough negative d-c signal to simulate desired light bias for maximum sensitivity at low levels of illumination encountered in application. When used as element of image-converting picture panel, positive-feedback circuit serves also for augmentation of contrast.—W. H. Knausenberger, Bistable Light-Detection Devices, *IEEE Journal of Solid-State Circuits*, April 1969, p 71–74.

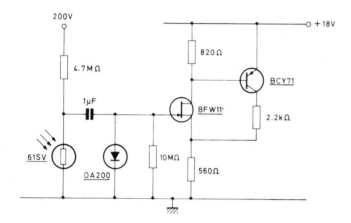

LEAD SULFIDE CELL—General-purpose amplifier for 61SV uncooled infrared detector uses fet as first stage to provide high input impedance for matching cell resistance of 1 to 4 meg. Diode protects fet input terminals from high-voltage transient occurring when cell bias is switched on.—"Applications of Infrared Detectors," Mullard, London, 1971, TP1201, p 40–42.

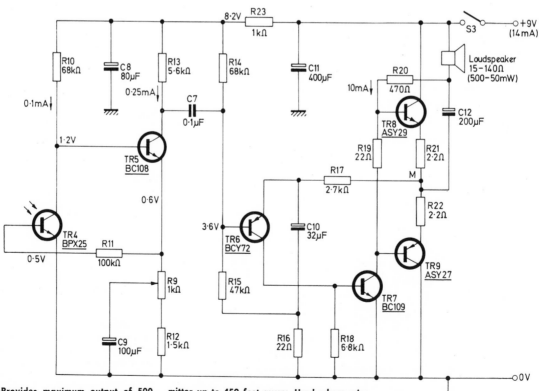

RECEIVER—Provides maximum output of 500 mW into 15-ohm speaker, when phototransistor is illuminated by voice-modulated electroluminescent diode of suitable infrared transmitter up to 450 feet away. Used where wire lines are not feasible, as at construction sites and sport events. Operates at wavelength of about 0.9 μm.—"An Infra-Red Communication System," Educational Projects in Electronics, Mullard, London.

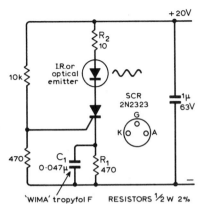

PULSING LED—Relaxation oscillator using scr generates short-duration 1-A pulses at 10 kHz for pulsing infrared or other light-emitting diode. With larger C1, pulses can be up to 10 A but at lower rate. R2 can be pot for controlling diode current.—D. M. Bussell, Pulse Generator for Diode Emitters, *Wireless World*, Jan. 1972, p 18.

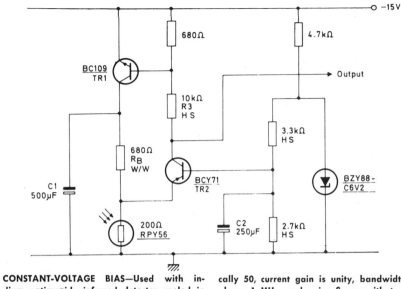

CONSTANT-VOLTAGE BIAS—Used with indium antimonide infrared detector cooled in liquid nitrogen to 77 K, placed in emitter of TR2. Zener provides regulated voltage for base of TR2. Voltage gain of circuit is typically 50, current gain is unity, bandwidth is above 1 MHz, and noise figure with typical detector is 3 dB.—"Applications of Infrared Detectors," Mullard, London, 1971, TP1201, p 34—35.

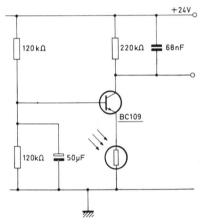

UNCOOLED PBS AMPLIFIER—Simple constant-voltage bias stage using single transistor serves for uncooled chemically deposited lead sulfide detector RPY75/76.—"Applications of Infrared Detectors," Mullard, London, 1971, TP1201, p 40—43.

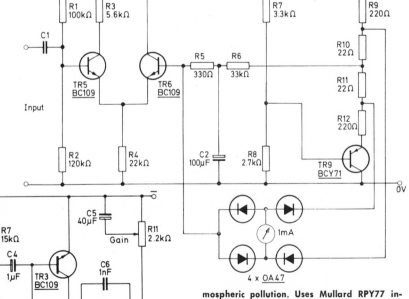

GAS ANALYZER—Developed as low-cost infrared spectrometer for checking carbon monoxide emissions of automobiles at testing stations and for monitoring other forms of atmospheric pollution. Uses Mullard RPY77 indium antimonide detector at room temperature in amplifier circuit and meter drive designed to operate satisfactorily over supply voltage range of 9 to 12 V. Low-noise input stage is more important than gain stability, because equipment is adjusted before each measurement. Detector is biased via 1-H choke for further reduction of noise. Bandwidth for 3-dB points is 600 Hz and gain is 50,000 to 500,000. Meter drive circuit uses d-c feedback through R5 and R6 to make output voltage level approximately equal to input d-c voltage set by R1 and R2 across supply.—"Applications of Infrared Detectors," Mullard, London, 1971, TP1201, p 129—142.

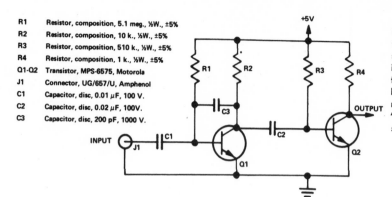

R1　Resistor, composition, 5.1 meg., ½W., ±5%
R2　Resistor, composition, 10 k., ½W., ±5%
R3　Resistor, composition, 510 k., ½W., ±5%
R4　Resistor, composition, 1 k., ½W., ±5%
Q1-Q2　Transistor, MPS-6575, Motorola
J1　Connector, UG/657/U, Amphenol
C1　Capacitor, disc, 0.01 μF, 100 V.
C2　Capacitor, disc, 0.02 μF, 100V.
C3　Capacitor, disc, 200 pF, 1000 V.

L-F AMPLIFIER—Input is Monsanto MD 2 photodiode used with appropriate lens and shield to pick up pulsed infrared beam from infrared led up to 30 feet away. Output feeds IC logic for detecting interruption of beam and using disturbance for resetting rather than triggering system.—R. C. Bach, "Modulated IR Beam Control System," *Monsanto GaAsLite Tips*, Vol. 1, 1970.

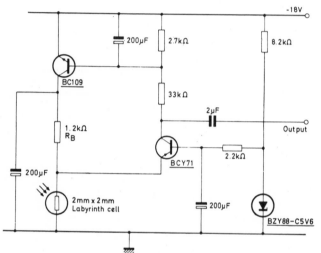

UNCOOLED-DETECTOR CONSTANT-VOLTAGE BIAS—Circuit is noise-optimized for RPY77 indium antimonide uncooled labyrinth cell, and provides voltage gain of 40 for bandwith greater than 1 MHz. Current gain is unity and noise figure is 2 dB.—"Applications of Infrared Detectors," Mullard, London, 1971, TP1201, p 38–41.

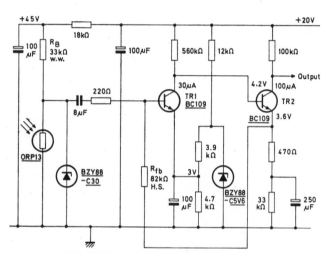

LOW-NOISE CONSTANT-POWER BIAS—Used with high-resistance indium antimonide infrared detector cooled in liquid nitrogen to 77 K. Noise figure with typical cell is only 0.8 dB, voltage gain is 400, and current 175. Bandwidth is above 0.5 MHz.—"Applications of Infrared Detectors," Mullard, London, 1971, TP1201, p 34–36.

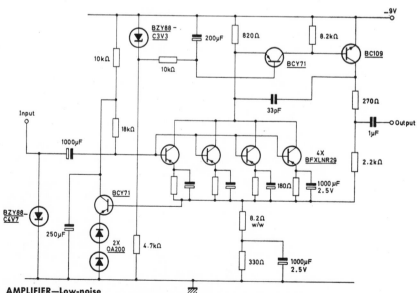

UNCOOLED-DETECTOR AMPLIFIER—Low-noise amplifier for ORP10 indium antimonide detector operating at room temperature uses four low-noise transistors in parallel as first stage, to match extremely low (30 ohms) cell resistance and noise level of detector connected to input. Effective noise figure is 3 dB, voltage gain is 220, and bandwidth is greater than 1 MHz.—"Applications of Infrared Detectors," Mullard, London, 1971, TP1201, p 38–40.

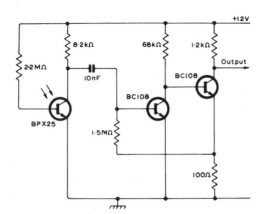

DETECTOR FOR MODULATED IR—Input of 1 lux peak on phototransistor will give 400-mV peak at output of 2-stage amplifier. Response at 4 kHz is 3 dB below that at 1 kHz. Used in high-secrecy communication system for distances up to 100 feet when f:2 collimating lenses or parabolic reflectors are used at infrared source and at detector.—"Mullard Silicon Planar Phototransistors BPX25 and BPX29," Mullard, London, 1968, TP1000A, p 9.

MISSING-PULSE LOCKOUT—Used in infrared receiver of pulsed infrared beamed system to count 60-Hz rectified pulses derived from a-c line. If allowed to count to nine, corresponding to nine infrared pulses pulsed every 16.7 ms, circuit delivers output pulse for triggering indicator relay driver of system. Counting feature makes system difficult to jam when used in burglar alarm application.—R. C. Bach, "Modulated IR Beam Control System," *Monsanto GaAsLite Tips*, Vol. 1, 1970.

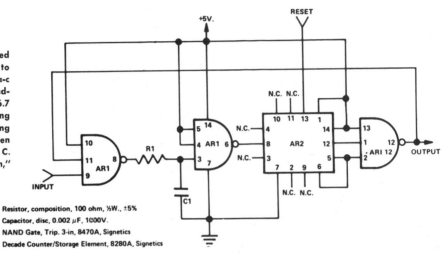

R1	Resistor, composition, 100 ohm, ½W., ±5%
C1	Capacitor, disc, 0.002 μF, 1000V.
AR1	NAND Gate, Trip. 3-in, 8470A, Signetics
AR2	Decade Counter/Storage Element, 8280A, Signetics

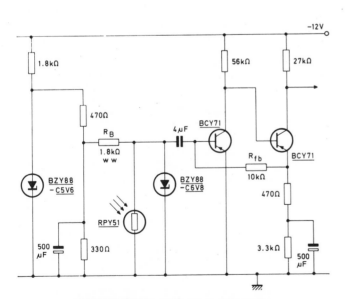

CONSTANT-POWER BIAS—Used with indium antimonide infrared detector cooled in liquid nitrogen to 77 K, to drive two-transistor current amplifier. Voltage gain of circuit is about 250 and current gain 21. Bandwidth is 1 MHz and noise figure with typical cell is 3 dB.—"Applications of Infrared Detectors," Mullard, London, 1971, TP1201, p 32–33.

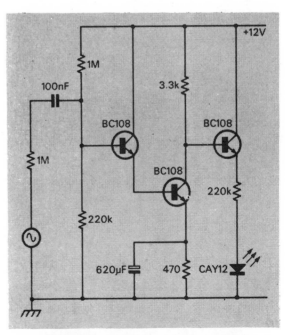

MODULATOR—Modulating signal, amplified by three transistors, drives gallium arsenide diode to produce modulated infrared radiation. Input of 150 mV peak gives diode current of 10 mA peak. Response at 80 kHz is 3 dB below that at 1 kHz.—Application of Photo Transistors in Communications and Control, *Electronic Engineering*, Nov. 1969, p 29–31.

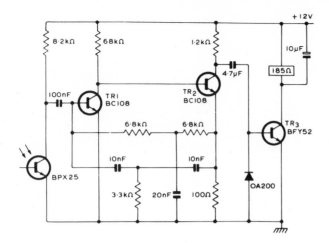

DETECTOR FOR CODED IR—Phototransistor is followed by tuned amplifier TR1-TR2 having twin-T network tuned to 2.7 kHz, at which network offers high impedance; at other frequencies, attenuation makes it low-impedance path for negative feedback, making signal passed on to TR3 at other frequencies too small to operate 185-ohm relay in collector circuit. Relay is energized each time phototransistor receives near-infrared signal modulated at 2.7 kHz. Light source can be gallium arsenide diode in collector circuit of mvbr transistor operating at 2.7 kHz. Although modulator gives square-pulse output, detector selects fundamental frequency.—"Mullard Silicon Planar Phototransistors BPX25 and BPX29" Mullard, London, 1968, TP1000A, p 9–10.

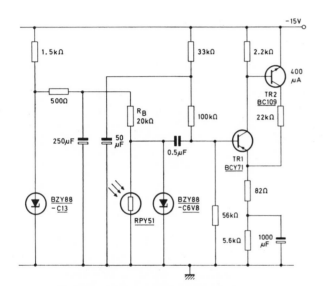

CONSTANT-CURRENT BIAS—Used with indium antimonide infrared detector cooled in liquid nitrogen to 77 K, to drive voltage-feedback amplifier. Voltage gain is about 250, bandwidth is above 1 MHz, and noise figure with typical cell is 0.8 dB.—"Applications of Infrared Detectors," Mullard, London, 1971, TP1201, p 32–33.

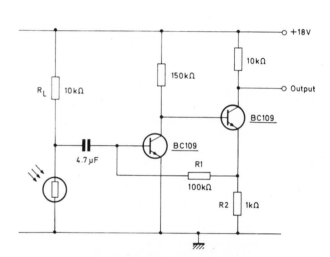

UNCOOLED PBS AMPLIFIER—Circuit provides constant-voltage bias by connecting lead sulfide cell to supply through 10K resistor. A-c coupling to amplifier prevents changes in d-c resistance of RPY75/76 cell from affecting amplifier performance. Signal current gain is 100.—"Applications of Infrared Detectors," Mullard, London, 1971, TP1201, p 42–43.

CHAPTER 49
Instrumentation Circuits

INSTRUMENTATION AMPLIFIER—Designed for strain gages, thermocouples, and other transducers requiring high input impedance to ground at differential input terminals and excellent common-mode rejection. Fixed gain of 10 can be increased with R5. Gain is 0 for common-mode signals. Temperature drift can be less than 1 μV per deg C. Frequency response at gain of 10 is down 3 dB at 500 kHz, with full output of $\pm$10 V attainable down to 1,800 Hz.—"Applying the AD504 Precision IC Operational Amplifier," Analog Devices, Inc., Norwood, MA, April 1972.

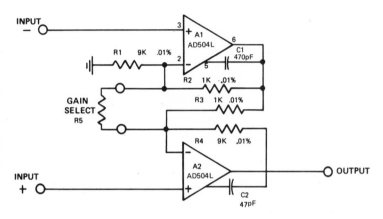

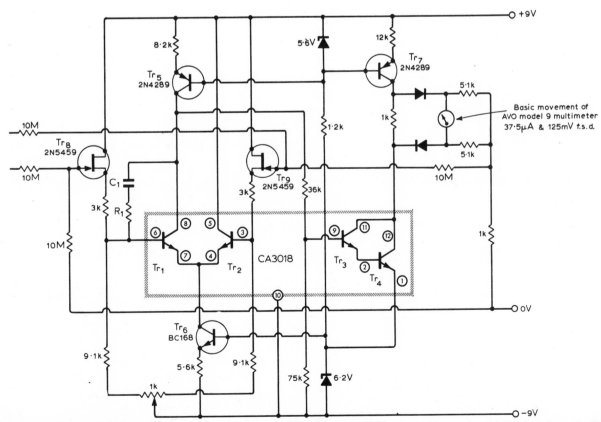

METER AMPLIFIER WITH FET'S—Use of fet's in source-follower configuration increases input impedance of universal meter amplifier by factor of 100 over that obtained with ordinary transistors in differential pair. Gives excellent scale linearity, though response falls off about 10% at 100-kHz limit of operating range. Article includes circuit for obtaining 10 voltage ranges. Input impedance is 20,000 meg per volt.—A. J. Ewins, Universal Meter Amplifier, *Wireless World*, Feb. 1972, p 61–65.

BRIDGE AMPLIFIER—Detects output of variable-resistance devices such as photoresistors and liquid resistance sensors. Delivers linear output of 2 V per 0.1% variation in R2 when bridge is connected across 30 V. R7 determines gain. At balance, power supply variations are cancelled.—"AD503, AD506 I.C. Fet Input Operational Amplifier," Analog Devices, Inc., Norwood, MA, Technical Bulletin, Aug. 1971.

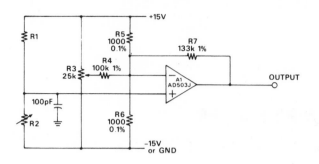

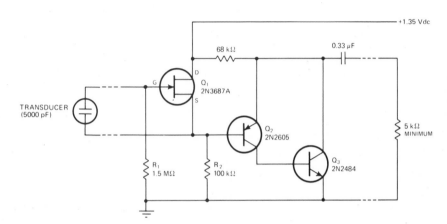

CAPACITIVE-TRANSDUCER PREAMP—Quiescent power dissipation is only 13 μW, corresponding to drain of 10 μA from 1.35-V supply. Equivalent input broadband noise is only about 33 μV from 140 Hz to 20 kHz. Voltage gain is 14 dB. Gate of fet Q1 is essentially biased at 0 V through R1. Q2 and Q3 serve as impedance converter.—R. F. Downs, Transducer Preamplifier Conserves Quiescent Power, *Electronics*, Feb. 28, 1972, p 80.

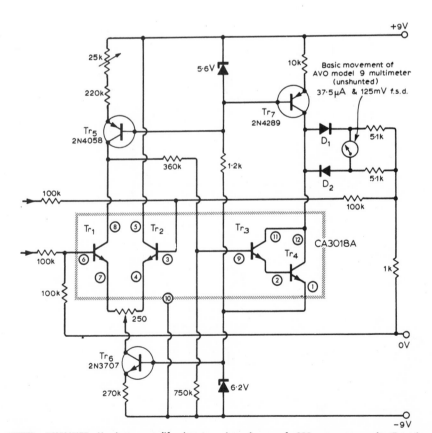

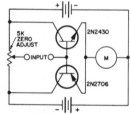

METER NULL AMPLIFIER—Uses complementary transistors in push-pull to provide current gain between 50 and 100 for zero-center meter. Suitable transistors include 2N2707 Amperex matched-pair or equivalent.—J. Fisk, Transistor Meter Amplifiers, 73, Jan. 1966, p 44–45.

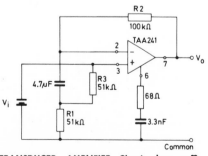

TRANSDUCER AMPLIFIER—Circuit has sufficiently high input impedance for amplifying output of piezoelectric transducer. Closed-loop gain is 3. Attenuation is −3 dB at 1 Hz.—"TAA241, TAA242 and TAA243 Differential Operational Amplifiers," Mullard, London, 1969, TP1086, p 20.

METER AMPLIFIER—Used to amplify inputs extending from d-c to 100 kHz before applying to moving-coil meter. Increases current sensitivity of 50-μA meter movement to 5 nA and voltage sensitivity to 100 mV, with input impedance of 200 meg per volt. For d-c measurements, meter gives positive indication regardless of input polarity. D1 and D2 are OA202.—A. J. Ewins, Universal Meter Amplifier, *Wireless World*, Feb. 1972, p 61–65.

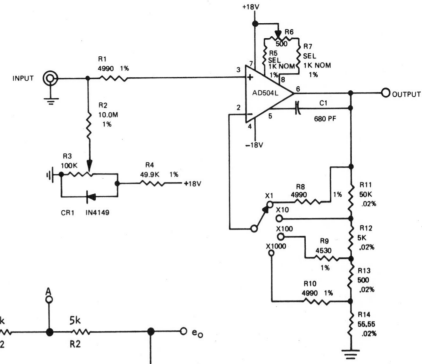

INPUT AMPLIFIER—Provides d-c gain from 1 to 1,000 in decade steps for d-c laboratory instruments and panel meters, with gain bandwidth of about 200 kHz. Input resistance of noninverting opamp is 10 meg, determined by R2. R3 is used to zero out input bias current. Gain switching is provided by X1 in feedback network. Will handle 50% over-range inputs.—"Applying the AD504 Precision IC Operational Amplifier," Analog Devices, Inc., Norwood, MA, April 1972.

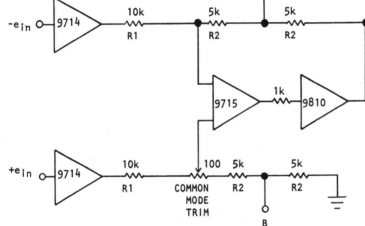

0–15 MHZ HIGH-ACCURACY—Will deliver up to 40 mA with either polarity, with unity gain and ±10-V full-scale output accurate to within 50 μV over 10 C temperature variation. Useful over full range of —55 C to 85 C. Report gives gain equation, and covers use of additional 9810 current booster to raise load current to 100 mA.—"Instrumentation Amplifier," Optical Electronics, Tucson, AZ, Application Tip 10240, 1971.

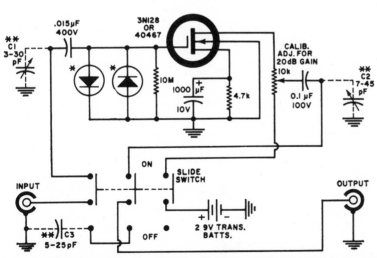

* IN914 OR ANY HIGH-FREQ., HIGH-BACK-RESIS. SILICON DIODES FOR GATE PROTECTION
** OPTIONAL ADJUSTMENTS FOR BEST SQUARE-WAVE RESPONSE

INSTRUMENT PREAMP—Used to extend range of scopes, a-c voltmeters, and 1.5-V a-c range of vtvm. Diodes in gate circuit protect against accidental overloads. Maximum input is 0.5 V p-p. Input impedance is above 1 meg at 1,-000 Hz, and frequency response is within 1 dB from 30 Hz to above 100 kHz. Calibration involves adjusting preamp for exactly 20 dB gain, which extends 1.5-V a-c scale to 0.15 V full-scale.—J. F. Sterner, MOSFET Utility Preamp for Test Equipment, *Electronics World*, Aug. 1970, p 62–63.

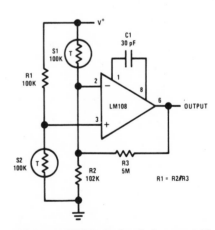

BRIDGE AMPLIFIER—Designed for use with high-impedance bridges such as those having high-resistance temperature sensors. Output of amplifier is 0 when bridge is balanced. When bridge goes off balance, opamp maintains voltage between input terminals at zero with current fed back from output through R3. Some nonlinearity in transfer function occurs for large imbalances, but this is usually not objectionable for small signal swings.—R. J. Widlar, "IC Op Amp Beats FETs on Input Current," National Semiconductor, Santa Clara, CA, 1969, AN-29, p 11–12.

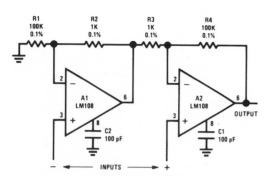

$R1 = R4; R2 = R3$

$A_V = 1 + \dfrac{R1}{R2}$

DIFFERENTIAL-INPUT AMPLIFIER—Provides input resistance above 10,000 meg without requiring large resistors in feedback circuit. Noninverting opamp A1 with gain of 1.01 feeds inverting opamp with gain of 100. With all resistors matched, circuit responds only to differential input signal and not to common-mode voltage.—R. J. Widlar, "IC Op Amp Beats FETs on Input Current," National Semiconductor, Santa Clara, CA, 1969, AN-29, p 12.

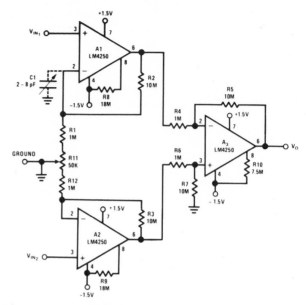

MICROPOWER 100-GAIN AMPLIFIER—Has full differential input center-tapped to ground. Input impedance is 100 meg with respect to ground and input bias current 0.2 nA for each input. Total current drain from pair of 1.5-V cells is 2.8 μA, having no effect on shelf life of most cells. R11 is adjusted to match gains of A1 and A2 for maximum d-c common-mode rejection ratio. Maximum output into 100K load is about 1.8 V p-p.—M. K. Vander Kooi and G. Cleveland, "Micropower Circuis Using the LM4250 Programmable Op Amp," National Semiconductor, Santa Clara, CA, 1972, AN-71, p 7.

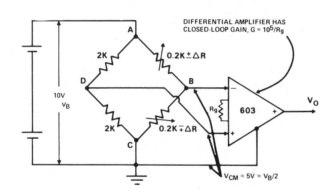

FET DIFFERENTIAL AMPLIFIER—Designed for data acquisition and instrumentation systems in which strain gages, differential transducers, physiological probes, and other signal sources are susceptible to ground-loop and common-mode interference. Gain is selected with single resistor Rg. Input impedance of fet opamp is 1 teraohm. Output can be offset up to 10 V with either polarity, for handling tare weight in load cells or barometric pressure in absolute-pressure measuring systems. Maximum output current is 5 mA.—"Model 603 J/K/L Low Cost Fet Differential Instrumentation Amplifier," Analog Devices, Inc., Norwood, MA, Sept. 1970.

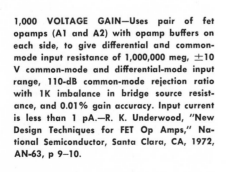

1,000 VOLTAGE GAIN—Uses pair of fet opamps (A1 and A2) with opamp buffers on each side, to give differential and common-mode input resistance of 1,000,000 meg, ±10 V common-mode and differential-mode input range, 110-dB common-mode rejection ratio with 1K imbalance in bridge source resistance, and 0.01% gain accuracy. Input current is less than 1 pA.—R. K. Underwood, "New Design Techniques for FET Op Amps," National Semiconductor, Santa Clara, CA, 1972, AN-63, p 9–10.

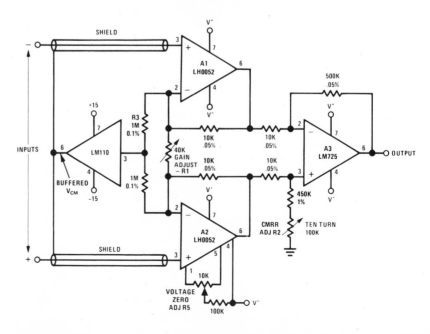

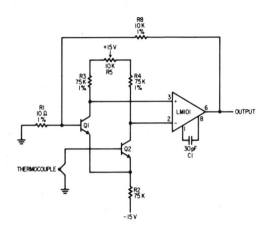

OFFSET VOLTAGE COMPENSATION—Uses IC transistor pair as preamp for opamp. Null pot, set for zero output with zero input, unbalances collector load resistors of transistor pair so collector currents are unbalanced for zero output, giving minimum drift. Circuit is relatively unaffected by supply voltage changes; 1-V change in either supply changes offset voltage only about 10 μV. Result is low-drift amplifier suitable for use with thermocouples, magnetometers, strain gages, and similar sensors requiring very low drift. Performance approaches that of chopper amplifiers.—R. J. Widlar, "Drift Compensation Techniques," National Semiconductor, Santa Clara, CA, 1967, AN-3.

ADJUSTABLE GAIN—Gain is linearly adjustable from 1 to 300 by changing single resistor R6 in differential-input instrumentation amplifier suitable for bridges, strain gages, and other low-level voltage-measuring applications. LM101A, connected as fast inverter, is used as active attenuator in feedback loop so common-mode rejection of 100 dB is unaffected by gain changes. Output is zeroed with R1. Two LM102 voltage followers buffer input signal to give 10,000-meg input impedance. LM107 is balanced differential amplifier.—B. Dobkin, "Instrumentation Amplifier," National Semiconductor, Santa Clara, CA, 1969, LB-1.

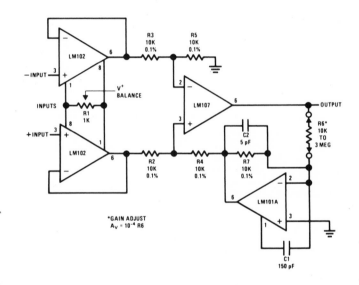

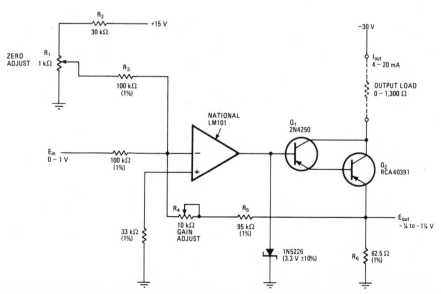

VOLTAGE-CURRENT CONVERTER—Used to limit output current to safe range of between 24 and 40 mA for process-control instrumentation. Converts input of 0–1 V to current of 4–20 mA for driving load of 0–1,300 ohms. Can be used with chart recorders.—H. L. Trietley, Jr., Voltage-to-Current Converter for Process-Control Systems, *Electronics*, April 26, 1973, p 102.

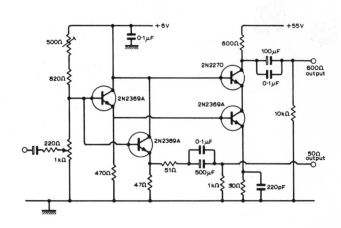

HIGH-LEVEL INSTRUMENT AMPLIFIER—Provides 45 V p-p into 600-ohm output impedance over bandwidth of 50 Hz to 2 MHz, with alternative low-level low-distortion 50-ohm output. Uses emitter-follower cascode arrangement. Absence of overall feedback eliminates high-frequency instability. Performance is adequate for majority of solid-state laboratory instruments.—J. A. Roberts and M. Taylor, High-Level Instrument Output Amplifier, *Design Electronics*, Jan. 1970, p 46–47.

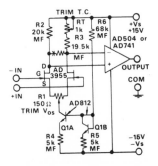

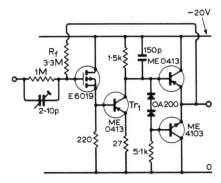

0.1 HZ TO 1 MHZ—Used as output amplifier for instrument requiring gain of only 10 dB. Has input resistance of 1 meg and output impedance of 5 ohms. Uses mos transistor connected in common drain mode to provide required high input impedance.—J. Roberts and H. C. Davies, Instrument Output Amplifier, *Wireless World*, March 1969, p 119.

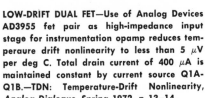

LOW-DRIFT DUAL FET—Use of Analog Devices AD3955 fet pair as high-impedance input stage for instrumentation opamp reduces temperaure drift nonlinearity to less than 5 μV per deg C. Total drain current of 400 μA is maintained constant by current source Q1A-Q1B.—TDN: Temperature-Drift Nonlinearity, *Analog Dialogue*, Spring 1972, p 13–14.

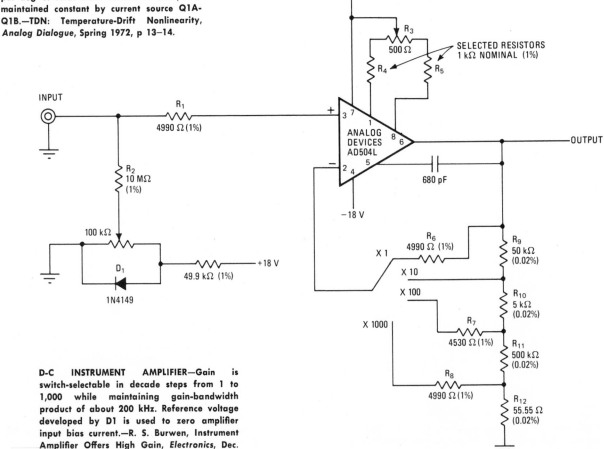

D-C INSTRUMENT AMPLIFIER—Gain is switch-selectable in decade steps from 1 to 1,000 while maintaining gain-bandwidth product of about 200 kHz. Reference voltage developed by D1 is used to zero amplifier input bias current.—R. S. Burwen, Instrument Amplifier Offers High Gain, *Electronics*, Dec. 20, 1971, p 56.

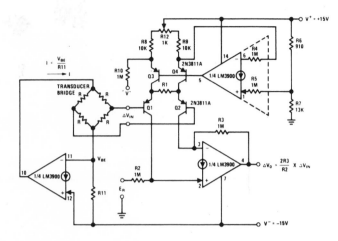

TRANSDUCER-BRIDGE AMPLIFIER—Uses LM-3900 quad current-differencing opamp for amplifying millivolt-level differential signals which may be riding on high common-mode voltage level. Report describes operation as basic instrumentation amplifier and gives design equations. Gain ranges from 72 dB when R1 is 0 to −34 dB when R1 is open. Small-signal response is down 3 dB at 1 MHz for gain of 1,000 and at 3 MHz for gain of 1.—H. H. Mortensen, "A Fully Differential Input Voltage Amplifier," National Semiconductor, Santa Clara, CA, 1972, LB-20.

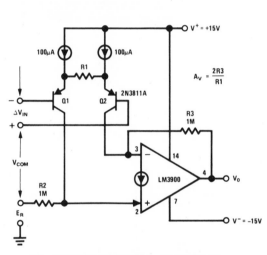

MV DIFFERENTIAL-SIGNAL AMPLIFIER—Uses LM3900 quad current-differencing opamp for amplifying millivolt-level differential signals which may be riding on high common-mode voltage level. Report describes operation as basic instrumentation amplifier and gives design equations.—H. H. Mortensen, "A Fully Differential Input Voltage Amplifier," National Semiconductor, Santa Clara, CA, 1972, LB-20.

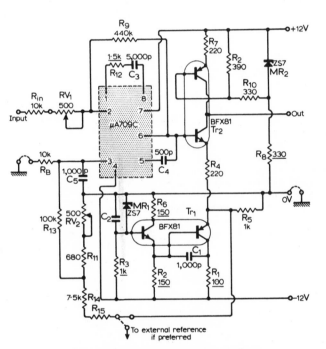

METER AMPLIFIER—Used in simple instrumentation systems where quantity to be monitored is represented by proportional d-c signal that has to be relayed over cable to meter at remote point. Maximum output swing is 10 mA positive or negative. Maximum error over temperature range of 0–60 C is 1%.—A. E. Crump, Instrumentation Amplifier, *Wireless World*, Feb. 1970, p 70–72.

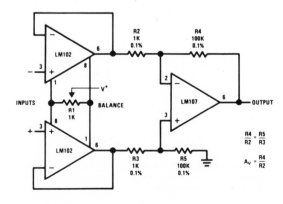

DIFFERENTIAL-INPUT AMPLIFIER—Two LM102 voltage followers buffer input signal to give 10,000-meg input impedance with 3-nA input currents. Followers drive LM107 balanced differential amplifier which provides gain and rejects greater than ±11 V of common-mode noise. Values shown give 10-V output for 100-mV input, for gain of 100.—B. Dobkin, "Instrumentation Amplifier," National Semiconductor, Santa Clara, CA, 1969, LB-1.

CHAPTER 50
Integrator Circuits

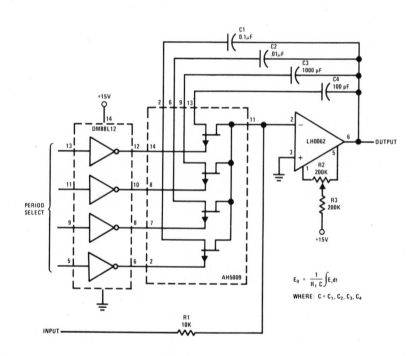

PROGRAMMABLE—Provides range in period of 1 μs to 1 ms. C1-C4 should be low-leakage polycarbonate or polystyrene. To adjust offset, insert 100K resistor temporarily between pins 2 and 6 of fet opamp and, with all switches of AH5009 off, set output to zero with R2.—B. Siegel, "Applications for a High Speed FET Op Amp," National Semiconductor, Santa Clara, CA, 1972, AN-75.

$$E_0 = \frac{1}{R_1 C} \int E \, dt$$

WHERE: $C = C_1, C_2, C_3, C_4$

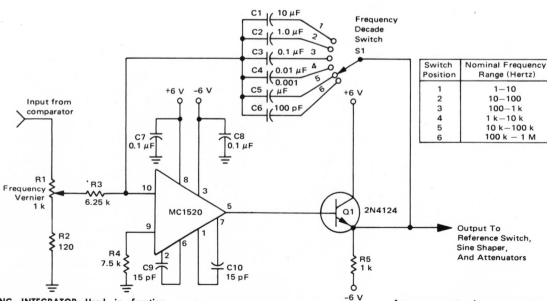

Switch Position	Nominal Frequency Range (Hertz)
1	1—10
2	10—100
3	100—1 k
4	1 k—10 k
5	10 k—100 k
6	100 k—1 M

INVERTING INTEGRATOR—Used in function generator for 1 Hz to 1 MHz. Output is positive when input from comparator of function generator is in high state. Produces ramp going from +2 V down to —2 V, by which time comparator output switches to negative bistable state and ramp goes positive. Input from comparator is square wave going from +5 V to —5 V.—B. Botos, "A Low-Cost, Solid-State Function Generator," Motorola Semiconductors, Phoenix, AZ, AN-510A, 1971.

OPAMP AS INTEGRATOR—Opamp is connected as integrator, so output is proportional to time integral of input signal. With symmetrical square-wave input having average value of 0 V, integrated output will be triangular as shown. R2 limits low-frequency gain, to minimize drift. Linearity of circuit is better than 1% at 1 kHz.—"Applications of the μA741 Operational Amplifier," Fairchild Semiconductor, Mountain View, CA, No. 289, 1972, p 1–2.

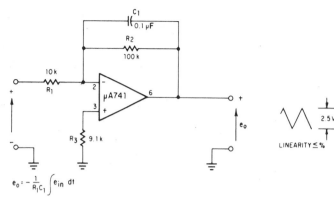

$$e_o = -\frac{1}{R_1 C_1} \int e_{in} \, dt$$

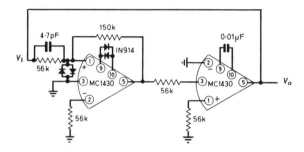

450-KHZ TRIANGLE WITH HYSTERESIS SWITCH—Output slew rate is 3 V per μs, and time for output of integrator to slew across hysteresis width of 3.6 V is 1.2 μs. Article gives detailed analysis of circuit operation.—R. N. Barnes, A Function Generator Using Hybrid Techniques, *The Radio and Electronic Engineer*, March 1970, p 153–159.

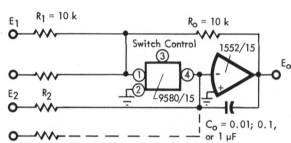

ANALOG WITH RESET—Uses Burr-Brown 9580/15 current-amplifying switch to switch between RESET mode (switching control 0 V) and COMPUTE mode (switching control +5 V) during which integration takes place. Switching time is about 200 ms. Error is typically less than 1 mV per s after amplifier has been balanced.—"Electronic Switches," Burr-Brown, Tucson, AZ, Sept. 1969, p 5.

QUADRATURE WITH LIMITING—Uses two Burr-Brown opamps as integrators, to give quadrature oscillator with nonlinear amplitude limiting. Book gives design equations. Values shown give frequency of about 16 Hz. —J. G. Graeme, G. E. Tobey, and L. P. Huelsman, "Operational Amplifiers," McGraw-Hill Book Co., New York, 1971, p 387–389.

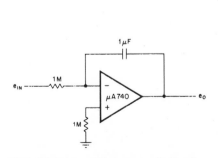

LONG-TERM—Use of jfet opamp having low bias and offset current allows source resistance of long-term integrator to be increased, so smaller-value capacitors with lower leakage can be used.—T. McCaffrey and R. Brandt, FET Input Reduces IC Op Amp's Bias and Offset, *Electronics*, Dec. 7, 1970, p 85–88.

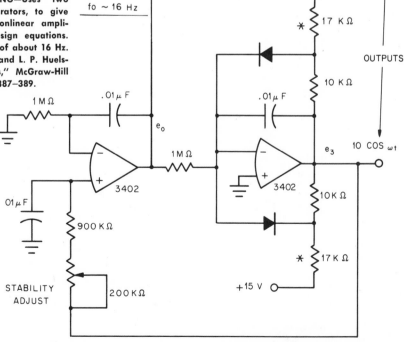

✳ TRIM FOR SYMMETRICAL ± 10V OUTPUT AND MINIMUM DISTORTION.

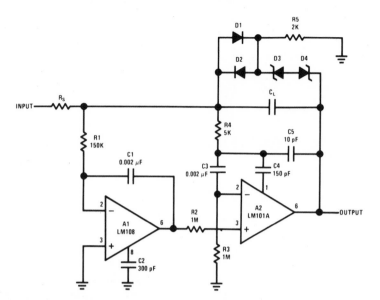

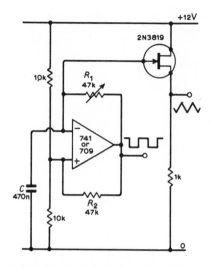

SPEED WITH LOW ERROR—Circuit is arranged so high-frequency gain is determined by opamp A2, while d-c and low-frequency gain is determined by A1 operating as integrator which goes through unity gain at 500 Hz. Above 750 Hz, feedback path is directly around A2, with A1 contributing little to gain. Report describes operation of circuit in detail. D1 and D2 serve simply to keep leakage currents of zeners from introducing errors in clamp circuit.—R. J. Widlar, "IC Op Amp Beats FETs on Input Current," National Semiconductor, Santa Clara, CA, 1969, AN-29.

CONSTANT-AMPLITUDE TRIANGLE—Instead of controlling frequency of square wave, integrator itself determines frequency of oscillation, while differential amplifier acts as Schmitt trigger between two voltages fixed by R2. Result is constant-amplitude triangular waveform at input, with frequency determined by R1C. For accurate integration, amplitude of triangular waveform should be limited to 1/10th of supply voltage. For values shown, circuit oscillates between about 100 and 3,000 Hz.—D. J. Price, Constant-Amplitude Triangular-Waveform Generator, Wireless World, July 1972, p 348.

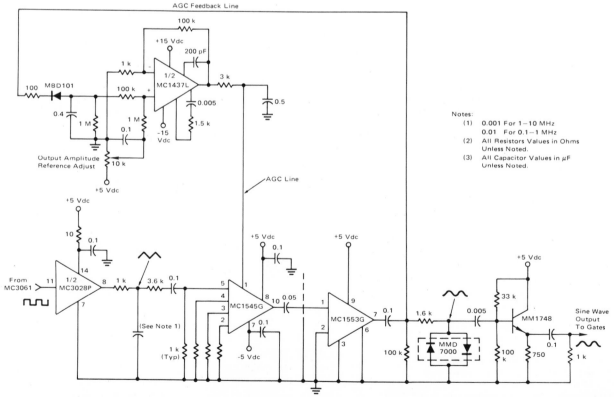

Notes:
(1) 0.001 For 1–10 MHz
     0.01 For 0.1–1 MHz
(2) All Resistors Values in Ohms Unless Noted.
(3) All Capacitor Values in μF Unless Noted.

0.1–10 MHZ INTEGRATOR-SHAPER-AGC—Used in 10 Hz–10 MHz lab sweep generator. Report gives all other circuits and covers operation of system in detail. Uses capacitive coupling, with MC1553 IC video amp to provide bandwidth required. Peak detecting diode is MBD101 hot-carrier diode having excellent rectification efficiency at highest frequencies used.—R. A. Botos and L. J. Newmire, "A Synchronously Gated N-Decade Sweep Oscillator," Motorola Semiconductors, Phoenix, AZ, AN-540, 1971.

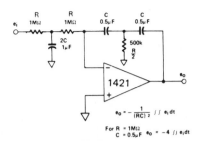

$$e_o = -\frac{1}{(RC)^2} \iint e_i\, dt$$

For $R = 1M\Omega$
$C = 0.5\mu F$ $e_o = -4 \iint e_i\, dt$

DOUBLE INTEGRATOR—Single Philbrick opamp is connected to generate second time-integral of input signal. Requires very close matching and tracking of resistance and capacitance values, for maximum accuracy.—"General Purpose FET Operational Amplifier 1421/01/02, Teledyne Philbrick, Dedham, MA, 1972.

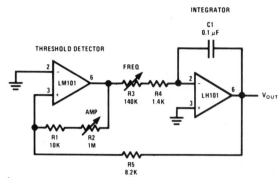

CONSTANT-AMPLITUDE TRIANGLE—Frequency change by R3 does not affect amplitude of triangular-wave output. Uses integrator as ramp generator, while threshold detector with hysteresis serves as reset circuit. Threshold detector is latch circuit with large dead zone, similar to Schmitt trigger. R2 changes amplitude of output waveform.—W. S. Routh, "Applications Guide for Op Amps," National Semiconductor, Santa Clara, CA, 1969, AN-20.

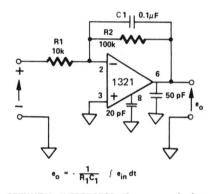

$$e_o = -\frac{1}{R_1 C_1} \int e_{in}\, dt$$

OPTIMIZED INTEGRATOR—Opamp used gives improved performance as compared to popular type 741. Report gives performance curves and detailed specifications for use as integrator and other applications. Will operate on voltage range of ± 8 V to ± 22 V, with ± 15 V nominal. Full-load current is ± 14 mA and quiescent current ± 4 mA.—"Wideband Operational Amplifier Series 1321," Teledyne Philbrick, Dedham, MA, 1972.

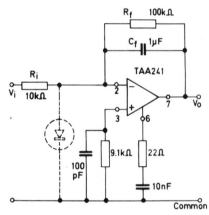

IC INTEGRATOR—Provides simple means of generating triangle waves from square-wave input. D-c gain is 10 for values shown. Operates satisfactorily above break frequency of 1.6 Hz, using Mullard IC.—"TAA241, TAA242 and TAA243 Differential Operational Amplifiers," Mullard, London, 1969, TP1086, p 11—12.

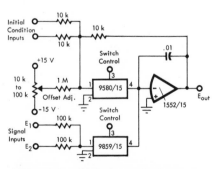

THREE-MODE ANALOG INTEGRATOR—Uses two current-amplifying switches, with upper switching between RESET and COMPUTE modes and lower providing HOLD mode. Switch control voltages are 0 V and +5 V. Report gives required offset balance procedure.—"Electronic Switches," Burr-Brown, Tucson, AZ, Sept. 1969, p 5.

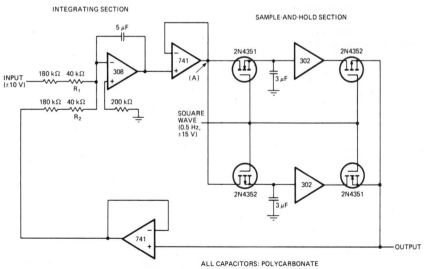

ALL CAPACITORS: POLYCARBONATE

CONTINUOUS SAMPLING—Self-resetting integrator with accuracy within 0.1% provides continuous integration of input. One sample-and-hold section holds previous integral and uses it to reset integrator, with sample-hold roles of sections reversing every integration. Switching is done by complementary mosfets.—D. J. Knowlton, Precision Integrator Resets as It Samples, *Electronics*, Aug. 28, 1972, p 78—79.

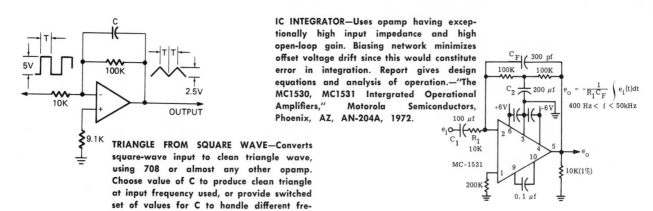

IC INTEGRATOR—Uses opamp having exceptionally high input impedance and high open-loop gain. Biasing network minimizes offset voltage drift since this would constitute error in integration. Report gives design equations and analysis of operation.—"The MC1530, MC1531 Intergrated Operational Amplifiers," Motorola Semiconductors, Phoenix, AZ, AN-204A, 1972.

$$e_o = -\frac{1}{R_1 C_F} \int e_i(t)dt$$

400 Hz < f < 50kHz

TRIANGLE FROM SQUARE WAVE—Converts square-wave input to clean triangle wave, using 708 or almost any other opamp. Choose value of C to produce clean triangle at input frequency used, or provide switched set of values for C to handle different frequencies.—B. R. Rogen, Experiments with Op-Amps, *Radio-Electronics*, June 1972, p 52–54.

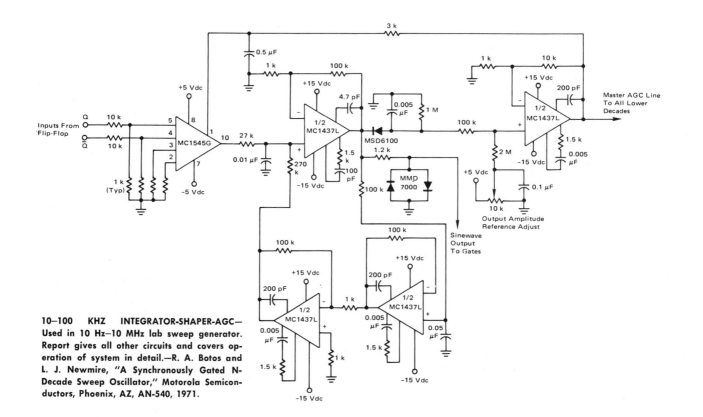

10–100 KHZ INTEGRATOR-SHAPER-AGC—Used in 10 Hz–10 MHz lab sweep generator. Report gives all other circuits and covers operation of system in detail.—R. A. Botos and L. J. Newmire, "A Synchronously Gated N-Decade Sweep Oscillator," Motorola Semiconductors, Phoenix, AZ, AN-540, 1971.

CHAPTER 51
Intercom Circuits

SIMPLE INTERCOM—Single IC audio power amplifier minimizes number of components needed for complete master-remote system having talk-listen switch only at master station. Each speaker acts as mike when switched for talk, because switch reverses role of master and remote speakers.—J. E. Byerly and M. K. Vander Kooi, "LM308 Power Audio Amplifier," National Semiconductor, Santa Clara, CA, 1972, AN-69, p 6.

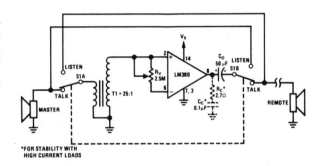

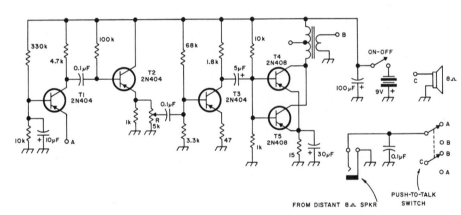

TWO-WAY—Can be used as communication link between basement workshop and kitchen or between hamshack and home. Many of parts used can be taken from discarded transistor pocket radio. Volume control R is optional.—C. R. MacCluer, The Shackcom, 73, Jan. 1967, p 87.

6-V INTERCOM—Battery power gives flexibility for portable use, as between tents in camps. Talk-listen switch S1, at main intercom location, is 4-pole double-throw spring-return switch such as Lafayette SW-68. R2 is sensitivity control for listening to call from remote speaker. Both speakers are p-m types that serve also as mikes.—R. Brown and T. Kneitel, "101 Easy Audio Projects," Howard W. Sams, Indianapolis, IN, 1971, p 19–21.

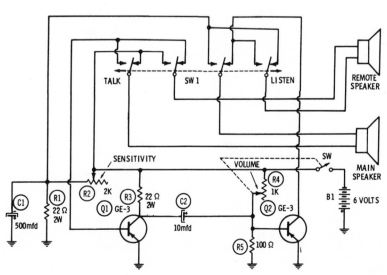

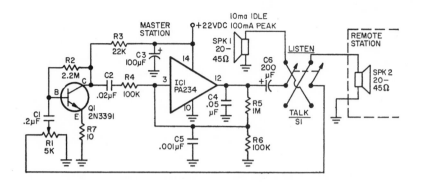

TINY INTERCOM—Can be built small enough to fit behind ordinary electric wall plate. Uses 1-W IC power amplifier driving speaker which is switched for use as mike. Only speaker is used at remote location, which is normally set for talk so master station can be called. Both IC and Q1 are GE units.—Eight Great IC Projects for Weekend Experimenters, *Elementary Electronics*, Sept.-Oct. 1971, p 73–74.

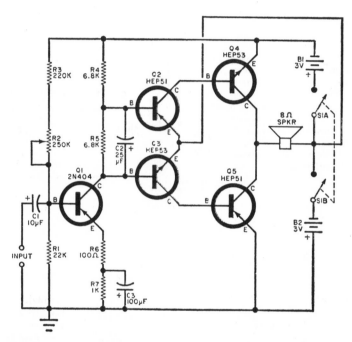

0.5 W ON FOUR D CELLS—Portable amplifier for field intercoms, record players, and receivers provides 67 dB gain from 16 Hz to 60 kHz and will operate 100 hours on one set of batteries.—R. J. Fajardo, Reader's Circuit, *Popular Electronics*, Feb. 1968, p 76–77.

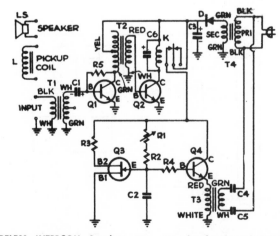

C1—10-mfd, 15v electrolytic capacitor
C2—300-mmfd dipped mica capacitor
C3—1,000-mfd, 15v electrolytic capacitor
C4,C5—0.1-mfd, 200v capacitor
C6—50-mfd, 25v electrolytic capacitor
R1—50,000-ohm trimmer potentiometer
R2—10,000-ohm, ½w carbon resistor
R3—330-ohm, ½w carbon resistor
R4—820,000-ohm, ½w carbon resistor
R5—220,000-ohm, ½w carbon resistor
D—silicon rectifier IN2070
Q1—transistor (RCA SK3020)
Q2,Q4—Darlington transistor (Motorola HEP S9100)
Q3—unijunction transistor (Motorola HEP 310)
T1,T3—miniature output transformer; 500-ohm CT to 3.2 ohms (Radio Shack 273-1379 or equal)
T2—miniature driver transformer; 2,000-ohm CT to 10,000 ohms (Radio Shack 273-1378)
T4—6.3v @ 0.6A filament transformer
K—miniature relay (Calectro D1-965)
LS—miniature loudspeaker (4- or 8-ohm)
L—telephone pickup coil (Radio Shack 44-533 or equal)
Misc.—line cord; al. minibox; perf. phenolic board; push-in terms.; twin-conductor cable

65-KHZ WIRELESS INTERCOM—Speaker acting as mike, or telephone pickup coil, serves as input for four-transistor transmitter that can be tuned by R1 to transmit signal over power lines at either 65, 55, or 45 kHz whenever signal at input is strong enough to energize relay K (Calectro D1-965). Used with receiver having three tuned circuits, each energizing distinctive lamp to indicate source of signal. Can be used on patio to pick up doorbell or phone by induction or baby cries in nursery.—R. M. Benrey, Patio Alert Lets You Relax Outdoors, *Popular Science*, April 1972, p 115–117.

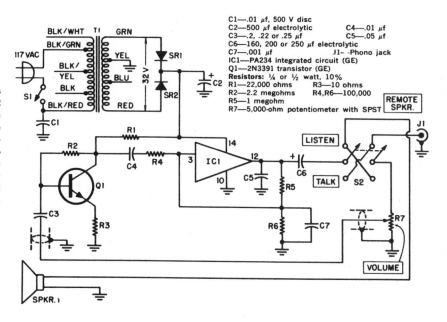

C1—.01 µf, 500 V disc
C2—500 µf electrolytic
C3—.2, .22 or .25 µf   C4—.01 µf
C5—.05 µf
C6—160, 200 or 250 µf electrolytic
C7—.001 µf   J1— -Phono jack
IC1—PA234 integrated circuit (GE)
Q1—2N3391 transistor (GE)
Resistors: ¼ or ½ watt, 10%
R1—22,000 ohms   R3—10 ohms
R2—2.2 megohms   R4,R6—100,000
R5—1 megohm
R7—5,000-ohm potentiometer with SPST

A-C INTERCOM—Master station operates from a-c power supply using Allied 54C4732 power transformer and 50-piv 10-mA silicon rectifiers. Feeds remote speaker over two-wire line. Use 32-ohm or 45-ohm intercom speakers at both stations. Article gives all construction details. Sensitivity is sufficient to pick up footsteps at 20 feet from either speaker. Only master has privacy, until talk-listen switch S2 is pressed.—V. Kell, *Tight-Budget Intercom, Electronics Illustrated,* Sept. 1969, p 64–67.

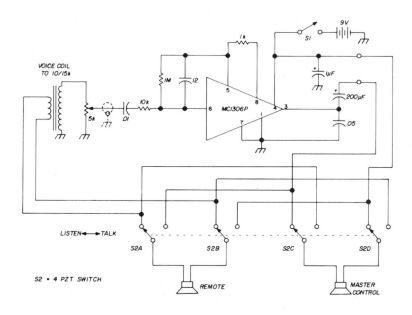

IC INTERCOM—Uses ½-W audio IC costing about $1, having high-impedance input and low-impedance output for driving 8-ohm speaker directly. Will operate from 9-V transistor radio battery or simple 9-V regulated power supply on a-c line. Remote is on at all times; pushing switch to *talk* changes circuit for master to talk to remote.—N. Stinnette, Simple Intercom, *Ham Radio,* July 1972, p 66–67.

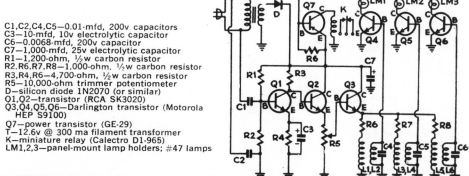

C1,C2,C4,C5—0.01-mfd, 200v capacitors
C3—10-mfd, 10v electrolytic capacitor
C6—0.0068-mfd, 200v capacitor
C7—1,000-mfd, 25v electrolytic capacitor
R1—1,200-ohm, ½w carbon resistor
R2,R6,R7,R8—1,000-ohm, ½w carbon resistor
R3,R4,R6—4,700-ohm, ½w carbon resistor
R5—10,000-ohm trimmer potentiometer
D—silicon diode 1N2070 (or similar)
Q1,Q2—transistor (RCA SK3020)
Q3,Q4,Q5,Q6—Darlington transistor (Motorola HEP S9100)
Q7—power transistor (GE-29)
T—12.6v @ 300 ma filament transformer
K—miniature relay (Calectro D1-965)
LM1,2,3—panel-mount lamp holders; #47 lamps

65-KHZ WIRELESS INTERCOM—Receiver for three-frequency power-line carrier intercom turns on one of three different lamps to indicate which of three transmitters in home is picking up signal. Article tells how to wind transformers L1–L6 for response to 65, 55, or 45 kHz. Can be used on patio to pick up doorbell or phone by induction or baby cries in nursery. Relay K is energized for all frequencies to turn on buzzer directing attention to lamp display.—R. M. Benrey, Patio Alert Lets You Relax Outdoors, *Popular Science,* April 1972, p 115–117.

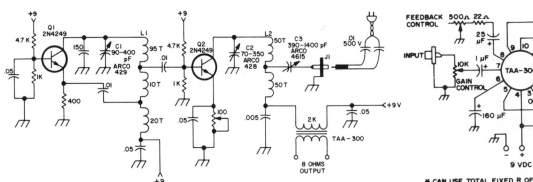

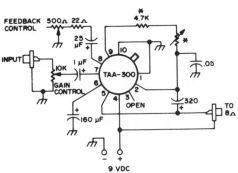

* CAN USE TOTAL FIXED R OF 6.2K, BUT BETTER TO USE 5K POT. AND SET FOR 8 mA TOTAL CURRENT

BC RADIO INTERCOM—Requires two portable a-m radios costing around $5 each, to provide two-way communication over a-c power lines of home. Circuit shows transmitter which operates on 1,600 kHz. To get second unit operating on 550 kHz, turns may have to be added to L1 and L2 while keeping tap ratios the same. Each location requires mike plugged into input of TAA-300 or equivalent IC/a-f amplifier serving as modulator for transmitter at that location, and bc receiver tuned to frequency of transmitter at other location. Article covers construction and adjustment.—B. Hoisington, The Amateur's Intercom, 73, Aug. 1973, p 39–41.

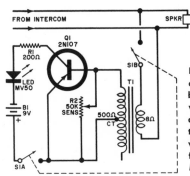

INTERCOM FLASHER—Audio signals on intercom line are stepped up by T1 and amplified by Q1 to activate light-emitting diode. Provides visual signal when noise of power saw drowns out intercom message or when listener is wearing earphones. Caller should whistle first or speak loudly to make led flash.—T. McTaggart, Useful Circuits, *Popular Electronics*, Jan. 1972, p 101–102.

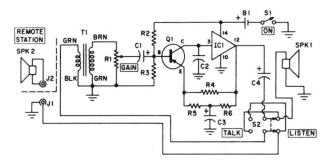

HOME INTERCOM—Requires only speaker at remote station. Operates from two 9-V transistor radio batteries in series for B1. Requires only two wires to remote speaker. T1 is Lafayette TR-99 a-f output transformer. Both speakers are 30- or 45-ohm intercom units. Q1 is GE 2N5355 and IC1 is GE PA-234. C1 is 10 μF, C2 500 μμF, C3 50 μF, C4 250 μF, R1 10K audio taper with switch, R2 100K, R3 47K, R4 10K, R5 15, and R6 1K. S2 is dpdt spring-return.—L. Powell, One-IC Home Intercom, *Electronics Illustrated*, Sept. 1972, p 34 and 98.

INTRUDER-INTERCOM NOISE SUPPRESSOR—Permits listening to sounds in store from house over special telephone line without continuous background noise. Noise input received from intercom is transformer-coupled to a-f amplifier Q1 which feeds rectifier-doubler supplying filtered positive voltage to pnp oscillator transistor Q3 to keep it reverse-biased to cutoff. Circuit is fail-safe; if noise input decreases, bias is removed from oscillator and alarm tone is heard from speaker in home. Alarm also sounds when intruders increase noise in protected building, when interconnecting wires are cut, power is turned off in store, or any part fails in intercom. Article covers construction and adjustment.—W. Lemons, Simple Silent Alarm, *Radio-Electronics*, Nov. 1966, p 42–43.

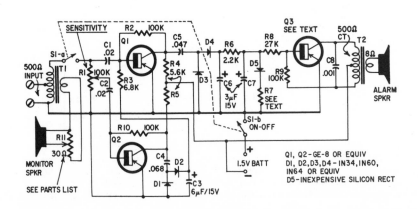

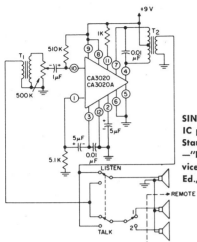

IC INTERCOM—Power amplifier feeds 250 mW into 16-ohm speaker yet is small enough for cementing to back of speaker that serves also as mike for intercom. S1 is on Mallory U-33 50K audio taper pot. Speaker at remote station plugs into J1. Uses Motorola IC.—H. Friedman, Piggyback Amp, *Elementary Electronics*, Nov.-Dec. 1970, p 41–44 and 106.

SINGLE IC—Requires only one RCA CA3020 IC power amplifier. Use 4-ohm speakers. T1 is Stancor A4744 and T2 is Thordarson TR-192. —"Linear Integrated Circuits and MOS Devices," RCA, Somerville, NJ, SSD-202A, 1973 Ed., p 191–192.

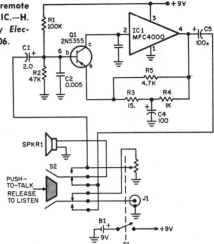

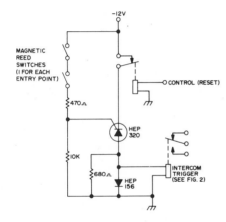

TRANSMITTER BREAK-IN ALARM—When door or window of unattended transmitter or repeater station is opened by intruder, switches connected as shown energize repeater intercom alarm so operator in nearby home can hear sounds and conversation of intruders and act accordingly. Can be used either with conventional intercom or with uhf repeater using speaker as mike. Article covers installation of switches and other intruder-detecting devices, including photodiodes.—K. W. Sessions, Jr., Repeater Site Break-In Alarm, *73*, April 1972, p 45–48.

CHAPTER 52
Inverter Circuits

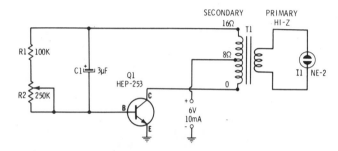

NEON ON 6 V—Circuit can serve as low-drain pilot lamp. Blocking oscillator Q1 generates audio frequency that is stepped up sufficiently by miniature transformer to light neon lamp. Primary of T1 is about 5,000 ohms.—H. Friedman, "99 Electronic Projects," Howard W. Sams, Indianapolis, IN, 1971, p 62–63.

117 V A-C IN CAR—Will provide about 117 V at about 60 Hz for electric shaver or other a-c device drawing up to 100 W, when plugged into 12-V lighter outlet in car, trailer, truck, or boat. T2 is conventional power transformer having many taps, such as Knight 54F2333, providing feedback for two-transistor oscillator. START position of S2 gives inverter chance to get started, after which S2 is switched to RUN.—J. Colt, Build a Power Inverter, *Popular Electronics*, May 1969, p 65–67 and 97.

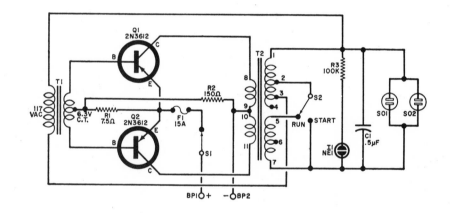

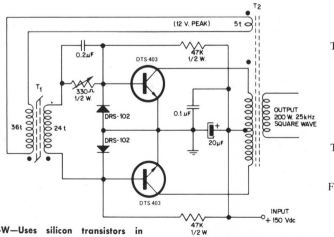

T1   Pri: 36 t #30 AWG
      Sec: 24 t #25 AWG
      Core: Ferroxcube
             266 T 125-3E2A
             Ferrite toroid

T2   Pri: 126 t tapped @ 63t,
             40 strands #38 AWG litz wire
      Sec: 2.38 V/t is used in this model
Feedback: 5t #25 AWG
      Core: Ferroxcube (ferrite toroid)
             528T500-3C5

25-KHZ 200-W—Uses silicon transistors in push-pull oscillator biased at 150 V d-c. Use 300-ohm pot to adjust transistor drive level for maximum efficiency at desired load. At 200 W, efficiency is 78%.—"25 kHz High Efficiency 200 Watt Inverter," Delco, Kokomo, IN, Feb. 1971, Application Note 47.

115-V 10-W MOBILE—Both output voltage and 400-Hz frequency are regulated. Base drive to output transistors is controlled by series regulator which senses output of additional reference winding. Typical efficiency is 72%.—"Regulated Inverter," Delco, Kokomo, IN, July 1971, Application Note 37.

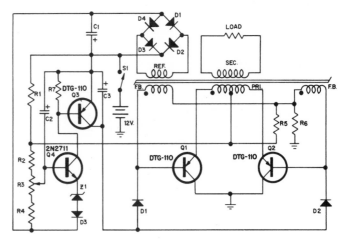

Q_1, Q_2, Q_3, — Delco DTG-110's
Q_4 — 2N2711
D_1, D_2, D_3, D_4, D_5, D_6, D_7
 Silicon, DRS-102
Z_1 — 1N3018 or Equiv., 8.2V
C_1, C_2 — 50μF, 25V
C_3 — 1000μF, 25V
R_1 — .25 Ω 2W
R_2, R_4 — 100 Ω 1W
R_3 — 400 Ω 2W
R_5 — 68 Ω
R_6 — 1.5K Ω
R_7 — 15 Ω

Toroid — Magnetics Inc.
51001-2A
Windings — Pri
                    60T #18 C.T.
Feedback
                    92T #30 C.T.
Reference
                    49T #29
Sec.
                    310 #24

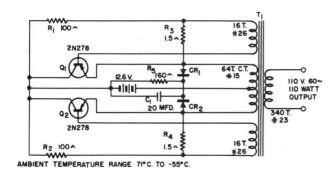

AMBIENT TEMPERATURE RANGE 71°C. TO -55°C.

12 V D-C TO 60 HZ AT 110 W—Developed for mobile applications requiring standard power-line voltage and frequency from auto storage battery. Transistors operate as switches, one being on while other is off, to give square-wave output whose frequency and amplitude are determined by supply voltage, primary turns, and saturation flux in 1 5/16-inch stack of 125 EI 0.014-inch silicon iron core. Frequency varies less than 5% from no load to full load.—"DC to AC Inverter," Delco, Kokomo, IN, Feb. 1972, Application Note 2-B.

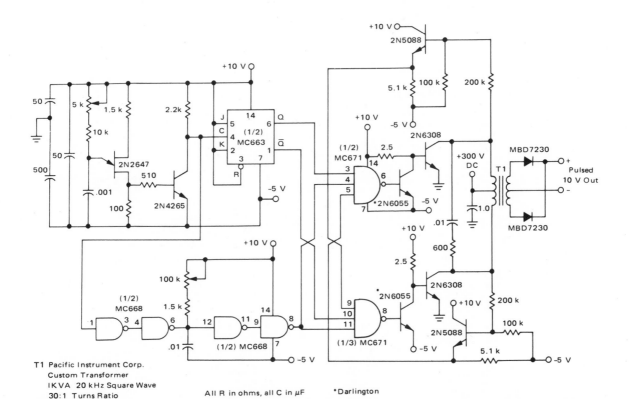

T1 Pacific Instrument Corp.
  Custom Transformer
  IKVA 20 kHz Square Wave
  30:1 Turns Ratio

All R in ohms, all C in μF   *Darlington

200 A AT PULSED 20 KHZ—Can serve as 1-kW main power supply for small computer. Use of ultrasonic inverter improves efficiency and minimizes size through elimination of 60-Hz power transformer. Required 300-V supply voltage would normally be obtained from three-phase 208-V power line through three-phase rectifier. Uses ujt relaxation oscillator to generate 40-kHz clock pulses which are shaped before going to frequency-dividing and phase-splitting J-K flip-flop. Report describes circuit operation in detail and gives waveforms.—R. J. Haver, "A 20 KHz, 1 KW Line Operated Inverter," Motorola Semiconductors, Phoenix, AZ, AN-588, 1972.

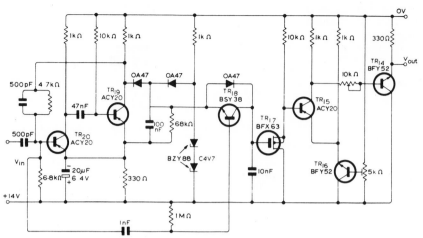

FREQUENCY-VOLTAGE CONVERTER—Provides smooth d-c output voltage proportional to output frequency of inverter, with fast response to changes, for a-c motor control over working range of 1 to 200 Hz. Circuit samples voltage across capacitor that is charged to fixed voltage at arrival of input pulse and then discharged through resistor. Sampled voltage is stored until next pulse arrives. Speed range can be 60 to 3,000 rpm, corresponding to pulse rate of 8 to 400 pps.—K. H. Williamson, Speed Control of Two-Phase A.C. Induction Motors, *Mullard Technical Communications*, Sept. 1967, p 214–238.

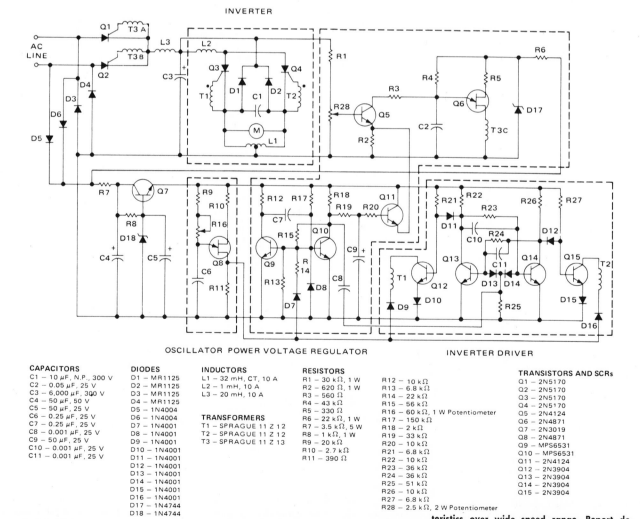

CAPACITORS	DIODES	INDUCTORS	RESISTORS		TRANSISTORS AND SCRs
C1 — 10 μF, N.P., 300 V	D1 — MR1125	L1 — 32 mH, CT, 10 A	R1 — 30 kΩ, 1 W	R12 — 10 kΩ	Q1 — 2N5170
C2 — 0.05 μF, 25 V	D2 — MR1125	L2 — 1 mH, 10 A	R2 — 620 Ω, 1 W	R13 — 6.8 kΩ	Q2 — 2N5170
C3 — 6,000 μF, 300 V	D3 — MR1125	L3 — 20 mH, 10 A	R3 — 560 Ω	R14 — 22 kΩ	Q3 — 2N5170
C4 — 50 μF, 50 V	D4 — MR1125		R4 — 43 kΩ	R15 — 56 kΩ	Q4 — 2N5170
C5 — 50 μF, 25 V	D5 — 1N4004		R5 — 330 Ω	R16 — 60 kΩ, 1 W Potentiometer	Q5 — 2N4124
C6 — 0.25 μF, 25 V	D6 — 1N4004	TRANSFORMERS	R6 — 22 kΩ, 1 W	R17 — 150 kΩ	Q6 — 2N4871
C7 — 0.25 μF, 25 V	D7 — 1N4001	T1 — SPRAGUE 11 Z 12	R7 — 3.5 kΩ, 5 W	R18 — 2 kΩ	Q7 — 2N3019
C8 — 0.001 μF, 25 V	D8 — 1N4001	T2 — SPRAGUE 11 Z 12	R8 — 1 kΩ, 1 W	R19 — 33 kΩ	Q8 — 2N4871
C9 — 50 μF, 25 V	D9 — 1N4001	T3 — SPRAGUE 11 Z 13	R9 — 20 kΩ	R20 — 10 kΩ	Q9 — MPS6531
C10 — 0.001 μF, 25 V	D10 — 1N4001		R10 — 2.7 kΩ	R21 — 6.8 kΩ	Q10 — MPS6531
C11 — 0.001 μF, 25 V	D11 — 1N4001		R11 — 390 Ω	R22 — 10 kΩ	Q11 — 2N4124
	D12 — 1N4001			R23 — 36 kΩ	Q12 — 2N3904
	D13 — 1N4001			R24 — 36 kΩ	Q13 — 2N3904
	D14 — 1N4001			R25 — 51 kΩ	Q14 — 2N3904
	D15 — 1N4001			R26 — 10 kΩ	Q15 — 2N3904
	D16 — 1N4001			R27 — 6.8 kΩ	
	D17 — 1N4744			R28 — 2.5 kΩ, 2 W Potentiometer	
	D18 — 1N4744				

1/6-HP INDUCTION—Inverter-driven circuit provides excellent speed control and regulation of 120-V 1,200-rpm a-c induction motor. Forced self-commutated inverter provides variable-frequency variable-voltage square-wave supply for motor for improved torque characteristics over wide speed range. Report describes operation in detail.—G. V. Fay, "Induction Motor Speed Control," Motorola Semiconductors, Phoenix, AZ, AN-450, 1972.

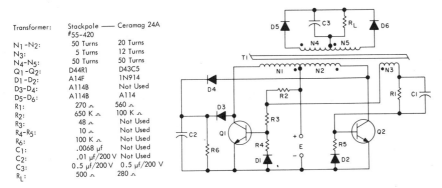

Transformer: Stackpole — Ceramag 24A
#55-420

N$_1$-N$_2$:	50 Turns	20 Turns
N$_3$:	5 Turns	12 Turns
N$_4$-N$_5$:	50 Turns	50 Turns
Q$_1$-Q$_2$:	D44R1	D43C5
D$_1$-D$_2$:	A14F	1N914
D$_3$-D$_4$:	A114B	Not Used
D$_5$-D$_6$:	A114B	A114
R$_1$:	270 Ω	560 Ω
R$_2$:	650 K Ω	100 K Ω
R$_3$:	48 Ω	Not Used
R$_4$-R$_5$:	10 Ω	Not Used
R$_6$:	100 K Ω	Not Used
C$_1$:	.0068 µf	Not Used
C$_2$:	.01 µf/200 V	Not Used
C$_3$:	0.5 µf/200 V	0.5 µf/200 V
R$_L$:	500 Ω	280 Ω

28 OR 180 V INPUT—With 180-V d-c input, gives 60 W at 172-V d-c output to load RL at efficiency of 88%, using inverter frequency of 28 kHz. With 28-V d-c input, gives 70 W at 69 V with efficiency of 81% and 10-kHz inverter action.—L. E. Donovan, "An Assortment of High Frequency, Transistor Inverters-/Converters Utilizing Saturating Core Transformers," General Electric, Semiconductor Products Dept., Auburn, NY, No. 200.57, 1970, p 8.

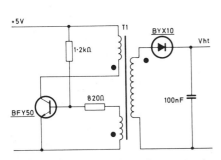

RINGING CHOKE—Operates from 5-V d-c supply and delivers above 180 V d-c for operating up to three numerical indicator tubes. Secondary winding on feedback transformer has 790 turns, primary 21 turns, and base winding 20 turns, all on 14-mm pot core.—"Numerical Indicator Tube Static Display Systems," Mullard, London, 1971, TP1203, p 19.

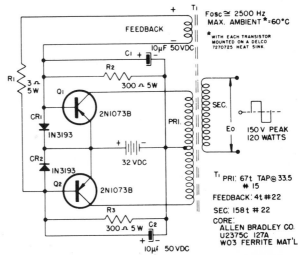

32 V D-C TO 2.5 KHZ AT 120 W—Switching frequency was selected for efficient use of inexpensive ferrite core. Circuit starts oscillating reliably under load from —60 C to +60 C.

Output may be fed into half-wave or bridge rectifier to get d-c output.—"DC to AC Inverter Using 2N1073B Transistors," Delco, Kokomo, IN, Feb. 1972, Application Note 18.

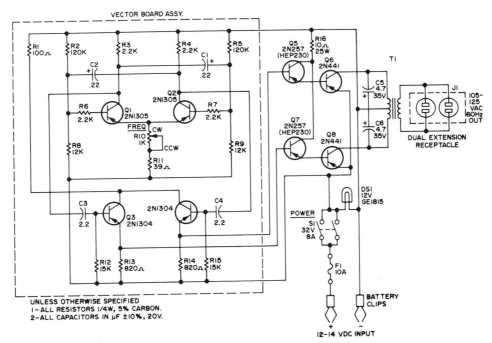

12 V D-C TO 110 V A-C AT 90 W—Developed to make small amateur station independent of a-c power-line failure. Symmetrical square-wave output has peak value of 110 V whereas 110-V a-c rms sine wave has peak of 155 V, but difference is not noticed in resistive loads. Circuit serves to switch 12-V d-c from one side of center-tapped transformer primary to other at predetermined rate. Q1 and Q2 form 60-Hz free-running mvbr for this purpose, with Q3 and Q4 serving as buffers and Q5 and Q7 as over-driven drivers for power-transistor switches Q6 and Q8.—W. F. Stortz, Portable House Power, 73, Oct. 1972, p 89—94.

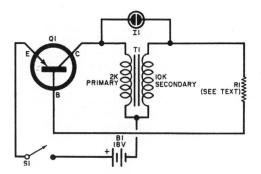

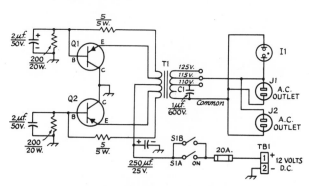

NEON ON 18 V D-C—Simple oscillator steps up battery voltage for NE-2 or other neon night light to over 60 V a-c for low-drain camp and other outdoor applications. Start with 10,000 ohms for R1, and increase until maximum light is obtained. Q1 can be any general-purpose audio transistor.—G. Richmond, Reader's Circuit, *Popular Electronics*, July 1967, p 76.

12 V D-C TO 115 V A-C AT 115 W—Permits mobile or portable operation of small a-c transceivers, receivers, and test equipment. Square-wave 60-Hz output may require use of brute-force line filter at a-c output to suppress hash noise. Transistors can be 2N278, 2N678, 2N1146, or 2N173. T1 is 12-V Triad TY-75A inverter transformer. Not suitable for motor-operated tape recorders or record players.—"The Mobile Manual For Radio Amateurs," The American Radio Relay League, Newington, CT, 4th Ed., 1968, p 158–159.

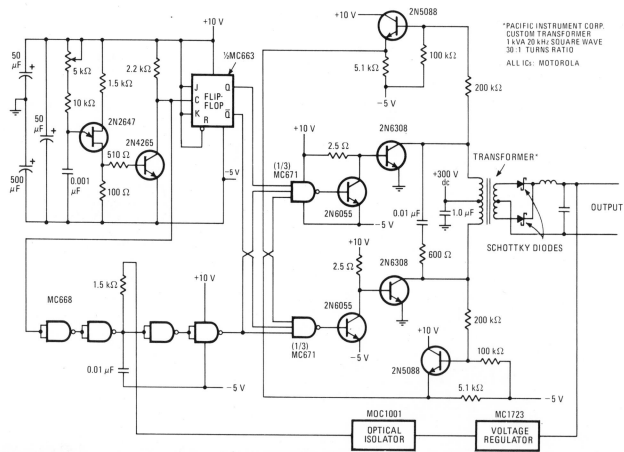

QUIET INVERTER—Uses switching power transistors operating at twice speed of older types, for doubling inverter switching frequency from audible 10 kHz to quiet 20 kHz. Eliminates need for bulky 60-Hz input power transformer. Over-all efficiency is increased by using Schottky diodes to rectify 20-kHz secondary voltage of 1-kVA transformer for applications requiring around 5 V. Article describes operation in detail.—L. Lehner, Doubling the Frequency of Switching Regulators, *Electronics*, April 12, 1973, p 123–124.

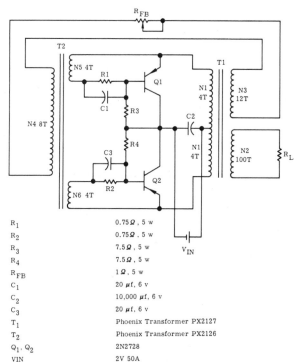

R₁   0.75 Ω, 5 w
R₂   0.75 Ω, 5 w
R₃   7.5 Ω, 5 w
R₄   7.5 Ω, 5 w
R_FB   1 Ω, 5 w
C₁   20 μf, 6 v
C₂   10,000 μf, 6 v
C₃   20 μf, 6 v
T₁   Phoenix Transformer PX2127
T₂   Phoenix Transformer PX2126
Q₁, Q₂   2N2728
V_IN   2V 50A

SOLAR-CELL VOLTAGE STEPUP—Uses high-performance power transistors for switching current as high as 50 A, for changing low-voltage high-current sources such as solar and fuel cells to useful higher a-c voltages. Can easily be changed over to converter giving desired high d-c output voltage, by rectifying and filtering a-c output across N2. Use of two transformers and two transistors minimizes core losses when switching high currents. Input voltage is 2 V at 50 A. Report covers construction of both transformers. Inverter frequency is 1 kHz.—J. Takesuye, "A Low Voltage High Current Converter," Motorola Semiconductors, Phoenix, AZ, AN-169.

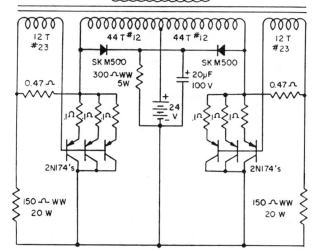

24 V D-C TO 60 HZ AT 400 W—Square-wave oscillator connected as inverter can replace vibrator and rotary power converters for 24-V battery supply. Secondary turns specified give 115 V at 60 Hz.—"Square Wave Oscillator Power Supplies," Delco, Kokomo, IN, Feb. 1972, Application Note 8-B.

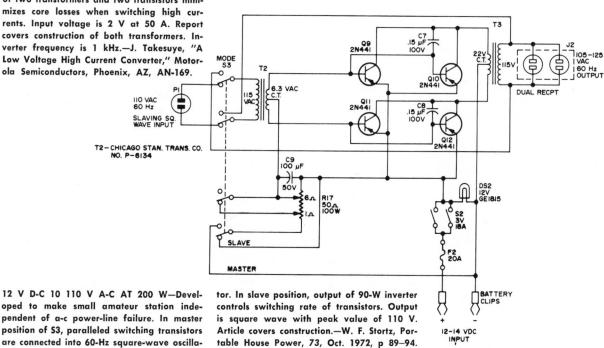

12 V D-C 10 110 V A-C AT 200 W—Developed to make small amateur station independent of a-c power-line failure. In master position of S3, paralleled switching transistors are connected into 60-Hz square-wave oscillator. In slave position, output of 90-W inverter controls switching rate of transistors. Output is square wave with peak value of 110 V. Article covers construction.—W. F. Stortz, Portable House Power, 73, Oct. 1972, p 89–94.

CHAPTER 53
Lamp Control Circuits

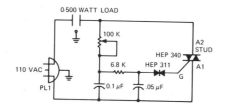

500-W MOOD-LIGHTING CONTROL—Pot provides phase control of triac to vary current through load up to 500 W, for changing brightness of lamps or speed of power tool plugged into output.—"Home Handyman's Construction Projects," Motorola Semiconductors, Phoenix, AZ, HMA-37, 1972.

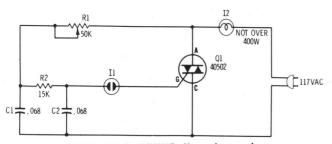

400-W TRIAC DIMMER—Uses inexpensive NE-2 or NE-83 neon lamp in place of trigger diode. As R1 is advanced, lamp turns on at about medium brilliance because neon must conduct before it can trip triac. R1 can then be backed down to give soft glow from lamp.—H. Friedman, "99 Electronic Projects," Howard W. Sams, Indianapolis, IN, 1971, p 61–62.

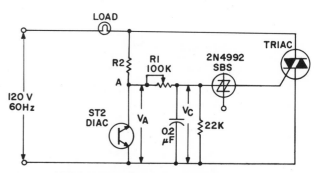

LINE COMPENSATION—Compensating circuit prevents lamp load from turning off if line voltage drops momentarily when dimmer control R1 is set for low light level.—R. W. Fox, "Solid State Incandescent Lighting Control," General Electric, Semiconductor Products Dept., Auburn, NY, No. 200.53, 1970, p 9.

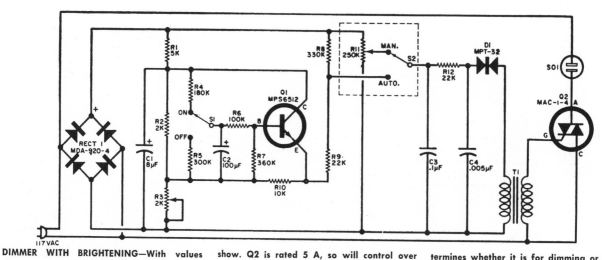

DIMMER WITH BRIGHTENING—With values shown, provides 1-min dimming cycle for room lights at start of home slide or movie show, and 20-s brightening cycle at end of show. Q2 is rated 5 A, so will control over 500 W. S2 gives choice of manual or automatic dimming. T1 is 1:1 pulse transformer (Sprague 11Z12). R3 sets interval, and S1 determines whether it is for dimming or brightening lights.—I. Gorgenyi, Build a "MALF", *Popular Electronics*, Sept. 1967, p 67–69 and 102.

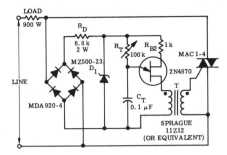

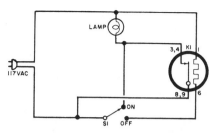

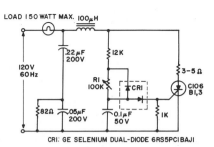

900-W FULL-WAVE—Basic ujt-triggered triac combined with bridge rectifier provides wide range of control with RT for resistive load. Pulse transformer T isolates triac gate from steady-state ujt current.—D. A. Zinder, "Unijunction Trigger Circuits for Gated Thyristors," Motorola Semiconductors, Phoenix, AZ, AN-413, 1972.

TURN-OFF DELAY—After lamp is switched off, light remains on long enough to get into bed or leave room. Delay is 30 s with Amperite 115C30T thermal relay, 60 s with 115C60T, and 120 s with 115C120T. Can usually be mounted in base of lamp.—J. Small, Mannerly Table Lamp, *Popular Electronics*, Nov. 1968, p 79 and 97.

LOW-COST 150-W—Dual-diode CR1 eliminates more expensive switching device usually used in phase controls. Provides half-wave control, with scr triggered through diode when line voltage goes positive.—R. W. Fox, "Solid State Incandescent Lighting Control," General Electric, Semiconductor Products Dept., Auburn, NY, No. 200.53, 1970, p 9–10.

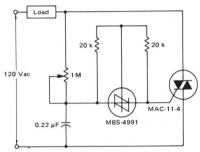

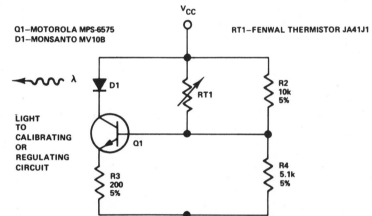

FLASH-ON SUPPRESSION—Shunting of sbs with two 20K resistors minimizes hysteresis effect wherein triac increases lamp brilliance suddenly as 1-meg control is turned up.—L. J. Newmire, "Theory, Characteristics and Applications of Silicon Unilateral and Bilateral Switches," Motorola Semiconductors, Phoenix, AZ, AN-526, 1970.

CONSTANT-BRIGHTNESS LED—Provides accurate and reliable brightness reference for industrial equipment, using thermistor R21 to compensate for temperature coefficient of led D1. Brightness is constant within 3% over temperature range of 10–50 C, with no need to compensate for aging or blackening as with incandescent lamps. Requires +12 V regulated d-c supply.—W. Otsuka, "Constant Brightness Light Source," *Monsanto GaAsLite Tips*, Vol. 1, 1970.

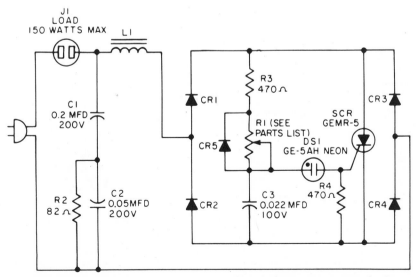

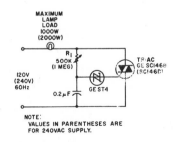

HIGH-INTENSITY LAMP—Designed for use with high-intensity study lamp having built-in transformer. Uses phase-control circuit acting on both halves of a-c cycle. Will also give 3:1 control of speed for shaded-pole fan motor.

Diodes are GE-504A. R1 is 100K for fan and 250K for lamp. L1 is 65 turns No. 18 on ¼" ferrite rod.—"Electronics Experimenters Circuit Manual," General Electric, Owensboro, KY, 1971, 3rd Ed., p 183–186.

FULL-WAVE ASBS-TRIAC—Use of ST4 asymmetrical silicon bilateral switch eliminates snap-on problem of diac-triac phase control, in which load current suddenly snaps from 0 to intermediate point at which smooth control starts. Operation of circuit is covered in detail. Waveform asymmetry of load current gives d-c components, precluding use with loads having transformers or fluorescent-lamp ballasts.—"SCR Manual," General Electric, Syracuse, NY, 1972, 5th Ed., p 253.

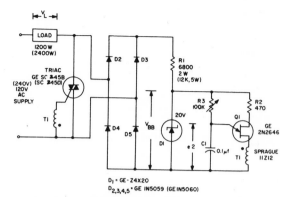

D1 = GE-Z4X20
D2,3,4,5 = GE 1N5059 (GE1N5060)

1.2-KW UJT-TRIAC—Uses zener to clamp control circuit voltage at fixed level of 20 V. R3 provides manual control of triac triggering over time-constant range of 0.3 to 8 ms.— "SCR Manual," General Electric, Syracuse, NY, 1972, 5th Ed., p 254–255.

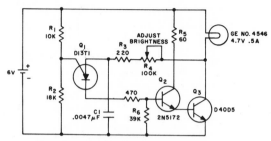

LOW-LOSS CONTROL—Provides almost continuous adjustment of light output between 0 and 100%, by changing average value of d-c supply voltage on 4.7-V tungsten lamp at high switching frequeny. Uses put Q1 as oscilator whose frequency is determined by R3, R4, and C1. Each time Q1 fires, Q2 drives Q3 into saturation and applies battery voltage to lamp.—"SCR Manual," General Electric, Syracuse, NY, 1972, 5th Ed., p 443–444.

TIME-DEPENDENT DIMMER—Designed for theater and other applications in which lamp load is to be turned on or off slowly. Uses full-wave ramp-and-pedestal control circuit. At maximum setting of R3, full transition from on to off takes about 20 min. Choose triac TR1 and fuse F1 to match load. (Use GE X12 for 500 W.)—R. W. Fox, "Solid State Incandescent Lighting Control," General Electric, Semiconductor Products Dept., Auburn, NY, No. 200.53, 1970, p 10–11.

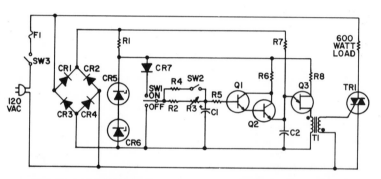

CRI THRU CR4 : G-E (1N5059) RECTIFIER DIODE
CR5, CR6 : G-E Z4XL7.5 ZENER DIODE
CR7 : G-E (1N5059) RECTIFIER DIODE
C1 : 100 µf, 15 WVDC ELECTROLYTIC CAPACITOR (G-E QT1-22)
C2 : 0.1µf, 15 WVDC CAPACITOR
Q1, Q2 : G-E (2N5172) n-p-n TRANSISTOR
Q3 : G-E 2N2647 UNIJUNCTION TRANSISTOR

R1 : 3.3 K OHM, 2 WATT RESISTOR
R2, R4 : 4.7 K OHM, 1/2 WATT RESISTOR
R3 : 5 MEGOHM, 1 WATT POTENTIOMETER
R5, R7 : 1 MEGOHM, 1/2 WATT RESISTOR
R6 : 2.2 K OHM, 1/2 WATT RESISTOR
R8 : 470 OHM, 1/2 WATT RESISTOR
SW1 : SPDT SWITCH
SW2 : SPST SWITCH
T1 : SPRAGUE 11Z12 PULSE TRANSFORMER

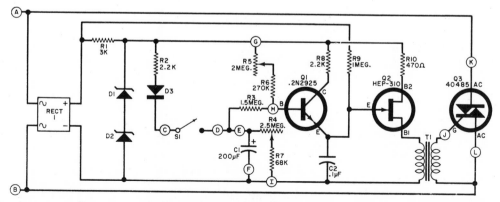

TIMER-DIMMER—Can be set to dim room lighting to any preset lower level, even full off, over period of 10 minutes so dimming is almost unnoticeable. Ideal for bachelor apartment. Closing S1 starts dimming cycle. R4 controls dimming time and R5 final level of light. D1 and D2 are GE Z4XL12B or similar 12-V 1-W zeners and D3 is HEP154. Use VS-248 bridge rectifier assembly. T1 is Sprague 11Z12 1:1 pulse transformer. Control is connected in series with lamp load being dimmed, via terminals A and B.—R. J. Bik, Build the Dynadim, *Popular Electronics*, Sept. 1968, p 71–74.

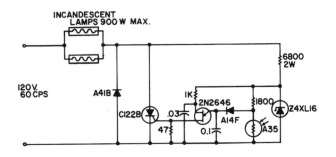

LIMITED-RANGE 900-W—For lamp loads in which power does not have to be reduced below half-power level. Rectifier supplies one half-cycle uncontrolled, and scr provides regulation of other half-cycle by phase control. Photocell used for automatic control is operated in low-resistance condition to achieve fast response time and eliminate hunting. Uses ramp-and-pedestal system.—R. W. Fox, "Solid State Incandescent Lighting Control," General Electric, Semiconductor Products Dept., Auburn, NY, No. 200.53 1970, p 9–10.

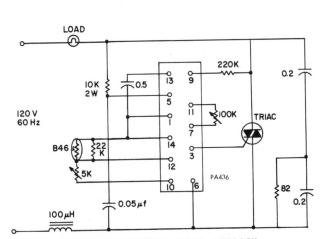

HIGH-GAIN CONTROL WITH FEEDBACK—Uses GE PA436 IC phase control, connected in simple light feedback arrangement with level set. Report gives optional soft-on circuit modification. Choose triac to match lamp load.—R. W. Fox, "Solid State Incandescent Lighting Control," General Electric, Semiconductor Products Dept., Auburn, NY, No. 200.53, 1970, p 14.

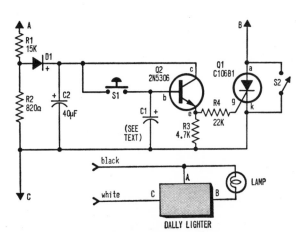

DELAYED TURN-OFF—Turns off garage, driveway, hall, or other light at any desired predetermined interval after pushbutton switch S1 is pressed, from 30 s to 15 min. S2 is conventional manual on-off switch used before. With 10 μF for C1, delay is 30 s; 100 μF gives 5 min and 300 μF 15 min. Use 12-V d-c electrolytic for C1. D1 is 200-V 50-mA or better silicon rectifier. B and C are house wiring leads to former switch and A goes to black wire.—R. Michaels, Dally Lighter, *Elementary Electronics*, May-June 1969, p 57–60.

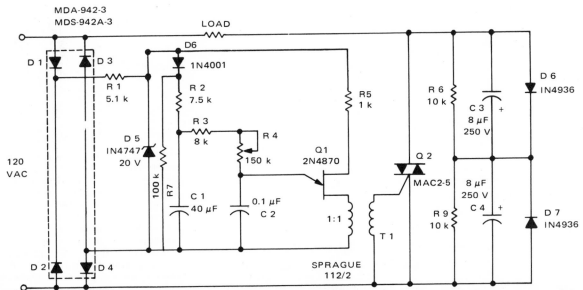

500-W SOFT-START—Minimizes high inrush current through very low resistance of cold filament, for protection of both lamp and dimmer circuit. Circuit is conservatively rated, because evaluation tests indicated triac was within ratings even for 1,000-W lamp load. Report describes operation of circuit in detail.—R. J. Haver and D. A. Zinder, "Conventional and Soft-Start Dimming of Incandescent Lights," Motorola Semiconductors, Phoenix, AZ, AN-436, 1972.

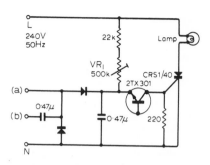

FIRING THYRISTOR—Uses npn silicon planar transistor (2TX301) connected to behave as zener diode for firing thyristor in simple half-wave lamp dimmer. Can be controlled manually with VR1, by d-c input of 0–10 V at (a), or by a-c input of 0–10 V p-p at (b).—J. A. H. Edwards, Negative Resistance of Transistor Junction, *Wireless World*, Jan. 1970, p 12.

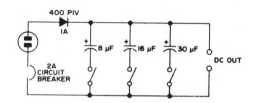

SIMPLE VOLTAGE REDUCER—Combination of three switched electrolytics and silicon rectifier gives seven different d-c output voltages for controlling lamps or other resistive loads of up to 50 W. Input at left is normal 120-V a-c line voltage. Theoretical maximum d-c output with all capacitors in use is 169 V, and without any capacitors is 60 V pulsating d-c.—N. Johnson, Simple Diode Controller, 73, March 1972, p 113–114.

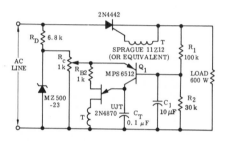

600-W HALF-WAVE AVERAGE-VOLTAGE FEEDBACK—Control responds to average load voltage. Pulse transformer is required for triggering of scr by ujt. Network R1-R2-C1 averages load voltage for comparison by Q1 with set point of Rc.—D. A. Zinder, "Unijunction Trigger Circuits for Gated Thyristors," Motorola Semiconductors, Phoenix, AZ, AN-413, 1972.

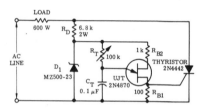

600-W HALF-WAVE—Basic ujt-triggered scr circuit is designed as two-terminal control replacing switch. If full-wave power is desired at upper limit of control, switch can be added that will short out thyristor (scr) when RT is set for maximum power. For inductive loads, two-position switch must be used to transfer load from scr to direct line.—D. A. Zinder, "Unijunction Trigger Circuits for Gated Thyristors," Motorola Semiconductors, Phoenix, AZ, AN-413, 1972.

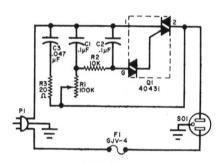

VARIABLE A-C VOLTAGE—Use of triac permits lowering a-c voltage of appliances up to 360 W for speed control, heat control, or dimming of lights.—R. C. Arp, Jr., Build a Solid-State Variable Transformer, *Popular Electronics*, Oct. 1969, p 42–44 and 117.

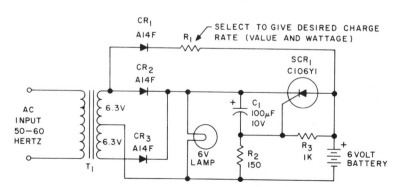

6-V STANDBY LIGHT—In event of a-c line failure, circuit switches automatically to 6-V lamp load. Turns off automatically when line voltage is restored, and converts to battery-charging mode that automatically keeps 6-V storage battery fully charged. Lamp rating should take into account battery capacity and maximum time of power failure.—"SCR Manual," General Electric, Syracuse, NY, 1972, 5th Ed., p 225–226.

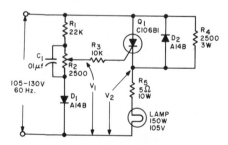

LAMP REGULATOR—Designed for use with specific lamps, for which circuit measures lamp power by measuring filament resistance. During negative half-cycle lamp receives full current through D2. At beginning of positive half-cycle, R4 provides bias current to develop voltage V2 that is dependent on lamp resistance. This voltage is compared with reference V1 by scr, to give phase-controlled triggering when V1 is more positive than V2. Possible drawback is 60-Hz component in lamp current, noticeable as flicker with peripheral vision.—R. W. Fox, "Solid State Incandescent Lighting Control," General Electric, Semiconductor Products Dept., Auburn, NY, No. 200.53, 1970, p 14–15.

CHAPTER 54
Laser Circuits

KRYTRON DRIVE FOR POCKELS CELL—Uses EG&G KN-6 gas tube to short d-c voltage between 3 and 7 kv to ground, in conjunction with lithium niobate Pockels cell, for Q-switching Nd:YAG laser. Bias voltage VB is typically —2 kV. Chief drawback is limited lifetime of tube.—W. R. Hook, R. P. Hilberg, and R. H. Dishington, High-Efficiency Electrooptic Q-Switching Using Slow Risetime SCR-Transformer Drive, Proc. IEEE, July 1971, p 1126–1128.

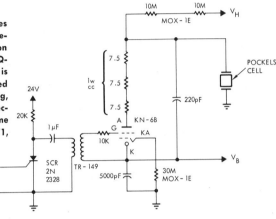

INJECTION LASER DRIVE—Paralleled avalanche transistors deliver pulses up to 50 A into gallium arsenide laser diode in series with 1-ohm sampling resistor at repetition rates up to 10 kHz. Rise time of pulses is 5 to 20 ms. Pulse current is varied by shutting off individual transistors and changing bias levels.—J. P. Hansen and W. A. Schmitt, A Fast Risetime Avalanche Transistor Pulse Generator for Driving Injection Lasers, Proc. IEEE, Feb. 1966, p 216–217.

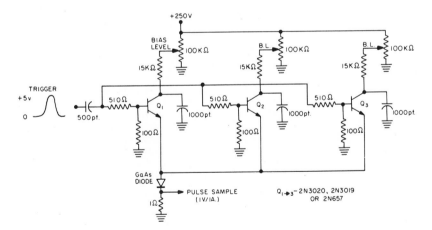

PFM VOICE TRANSMITTER—Provides voice communication range up to 3.5 miles with appropriate receiver. Q1 modulates frequency of relaxation oscillator using four-layer diode Q2, at rate proportional to amplitude of audio input. Q2 fires scr Q3 to give pulse-frequency modulation of RCA laser output. Q4 and Q5 provide sufficiently fast charging of scr pulser to permit modulation up to 8 kHz. —R. W. Campbell and F. M. Mims, III, "Semiconductor Diode Lasers," Howard W. Sams, Indianapolis, IN, 1972, p 104–106.

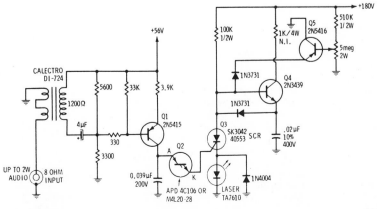

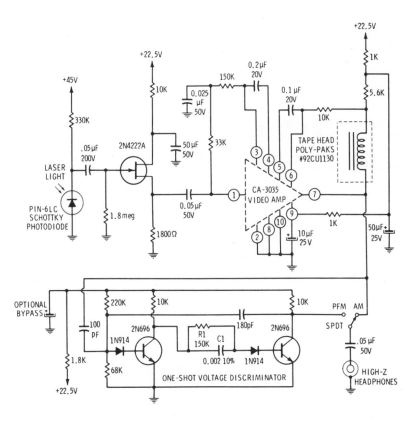

VOICE LASER RECEIVER—Provides voice communication range of about 3.5 miles using United Detector Technology photodiode shown, when modulating injection laser by pulse-frequency modulation for voice. Simple mono mvbr demodulates pulsed f-m signals. For tone transmission, use a-m position of switch. Experiment with values of R1 and C1 to obtain best voice-quality reception. Use lens with photodiode.—R. W. Campbell and F. M. Mims, III, "Semiconductor Diode Lasers," Howard W. Sams, Indianapolis, IN, 1972, p 130–131.

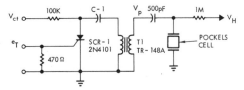

Q-SWITCHING WITH SCR—Uses EG&G TR148A transformer having 250-ns rise time and 1-cm lithium niobate Pockels-cell cube for Q-switching Nd:YAG laser. Requires fewer parts than conventional Krytron drive circuit but gives comparable lasing efficiency. Will operate at up to 60 pps. Instantaneous voltage across cell starts at 4.5 kV and swings down to —4.5 kV at peak of transformer output pulse.—W. R. Hook, R. P. Hilberg, and R. H. Dishington, High-Efficiency Electrooptic Q-Switching Using Slow Risetime SCR-Transformer Drive, Proc. IEEE, July 1971, p 1126–1128.

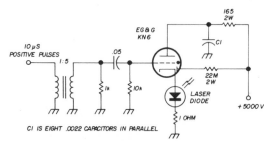

2,000-A DRIVER—Delivers pulses to injection laser when driven by positive-going input pulses. May be operated at pulse repetition rates up to 1,000 Hz. C1 consists of eight 0.002-μF capacitors, charged by 5,000-V supply until KN6 krytron is triggered on by positive pulse at grid. C1 then discharges through krytron and laser.—F. M. Mims, High Power Injection Lasers, Ham Radio, Sept. 1971, p 28–33.

PULSE DRIVER—GA201 nanosecond thyristor, capable of rising to 30 A in 20 ns, serves as reliable high-speed high-current switch for driving pulsed gallium arsenide laser diode D1. L-C lumped-constant delay line generates square current pulse that provides sharp turn-off for limiting excess power dissipation that would occur in laser diode if pulse fell exponentially. C1 is about 0.001 μF, to prevent false triggering. R2 is 50K to provide stable ground reference, and RGK is 100 ohms. Supply is 100 V.—"Nanosecond Thyristor for Laser Diode Pulse Driver," Unitrode Corp., Watertown, MA, New Design Idea No. 22.

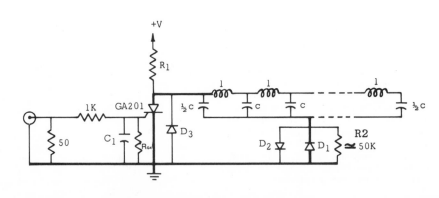

PULSED LASER POWER METER—Laser pulse is fired at photocell, reducing cell resistance and making C charge to voltage proportional to beam power. Voltage is then read by discharging C through transistor amplifier and meter. Time constant of R2 and C should be above 350 s; adjust R2 to keep meter reading on scale.—G. Bowman and T. K. Ishii, Simple Photocell Circuit Measures Pulsed Laser Power, *Electronics*, Jan. 4, 1971, p 69.

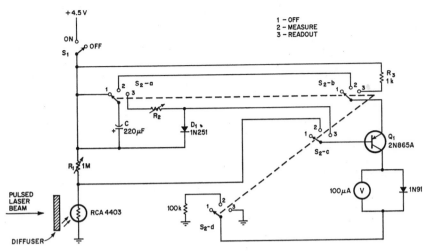

20-KV IMAGE-CONVERTER SUPPLY—Output voltage and current depend on battery supply used for two-transistor inverter. Image converter tube (RCA) serves for viewing invisible infrared light from injection laser, as required for optical alignment of laser system. Adjust supply voltage to give between 16 and 20 kV, then adjust 1-meg focus pot to give clear image on phosphor screen of tube. Objective lens is required to focus image on front photoemissive surface of tube.—R. W. Campbell and F. M. Mims, III, "Semiconductor Diode Lasers," Howard W. Sams, Indianapolis, IN, 1972, p 159–61.

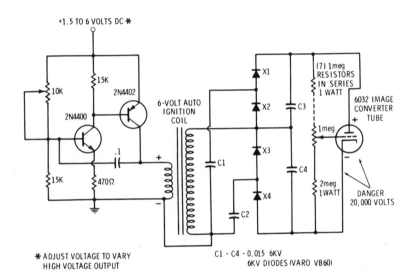

SCR PULSER—Pulse repetition rate is continuously adjustable from 1 Hz to 10 kHz, with 30-A peak current and 150-ns pulse width at half-current point. Suitable for testing variety of lasers. Uses ujt trigger, which can be 2N2647.—R. W. Campbell and F. M. Mims, III, "Semiconductor Diode Lasers," Howard W. Sams, Indianapolis, IN, 1972, p 91–93.

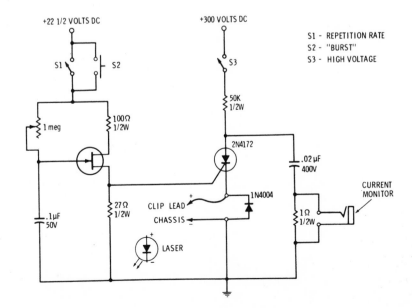

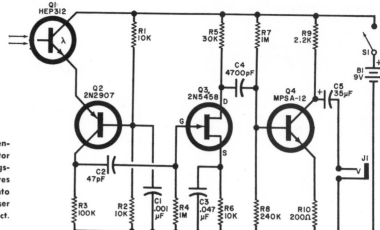

LASER RECEIVER—Circuit is simply conventional audio amplifier with phototransistor input capable of responding to 9,000-Angstrom output of laser diode. Article also gives transmitter circuits. Do not look directly into laser beam.—F. M. Mims, *Soild-State Laser for the Experimenter, Popular Electronics,* Oct. 1971, p 46—49 and 102—103.

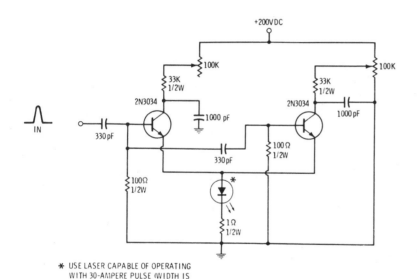

80-A PULSER—Paralleling of avalanche transistors increases current available for driving laser diode. Collector voltage of each transistor must be independently adjustable to insure coincidental switching. Circuit requires small input pulse for triggering transistors into avalanche for discharging capacitor through laser diode load.—R. W. Campbell and F. M. Mims, III, "Semiconductor Diode Lasers," Howard W. Sams, Indianapolis, IN, 1972, p 84—86.

＊ USE LASER CAPABLE OF OPERATING WITH 30-AMPERE PULSE (WIDTH IS LESS THAN 50 NANOSECONDS)

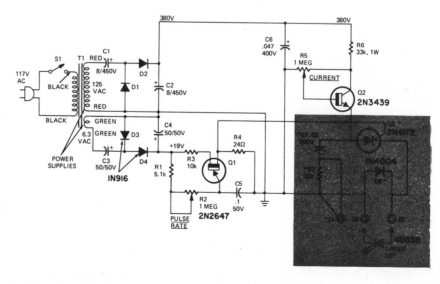

PULSED-DIODE LASER—With RCA laser shown, emits peak optical power of 3 W at 25 A. For more power, use RCA 40862 rated 10 W at 40 A. Can be used for intrusion alarm, optical communication, and optical radar. D1 and D2 are HEP58 and T1 is Calectro D1-760. Article gives construction details. Avoid looking directly into laser at close range even though its beam is invisible, and do not aim it at shiny surfaces that may reflect beam into eyes of person.—F. Mims, $32 Solid-State Laser, *Radio-Electronics,* June 1972, p 45 and 50—51.

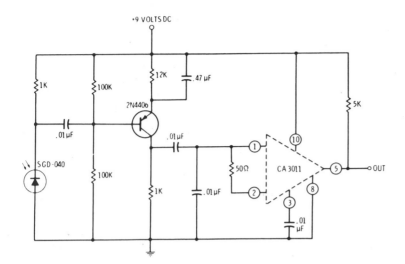

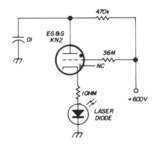

IC RECEIVER—Output of EG&G SGD-040 pin photodiode detector for injection laser pulses feeds RCA IC through transistor amplifier providing gain of about 5. IC provides gain of about 1,000. Will handle 40-ms pulses having frequencies in excess of 10 MHz. Communication range is up to 3.5 miles for pulse-frequency-modulated voice signals from laser transmitter.—R. W. Campbell and F. M. Mims, III, "Semiconductor Diode Lasers," Howard W. Sams, Indianapolis, IN, 1972, p 132–133.

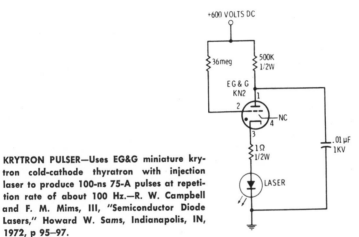

KRYTRON PULSER—Uses EG&G miniature krytron cold-cathode thyratron with injection laser to produce 100-ns 75-A pulses at repetition rate of about 100 Hz.—R. W. Campbell and F. M. Mims, III, "Semiconductor Diode Lasers," Howard W. Sams, Indianapolis, IN, 1972, p 95–97.

INJECTION LASER DRIVER—General-purpose circuit for use with variety of lasers is simple free-running relaxation oscillator producing 50-ns 75-A pulses at about 200 Hz. When voltage of charging capacitor is high enough, krytron fires and capacitor discharges through laser. 1-ohm resistor serves as current-monitoring point.—F. M. Mims, High Power Injection Lasers, *Ham Radio*, Sept. 1971, p 28-33.

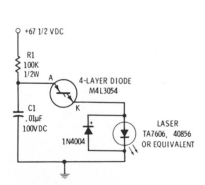

DIODE PULSER—Uses Shockley four-layer diode as switching element for discharging C1 through laser diodes when charging voltage across C1 reaches switching voltage of diode. Can be operated at repetition rates up to 20 kHz by adjusting value of R1.—R. W. Campbell and F. M. Mims, III, "Semiconductor Diode Lasers," Howard W. Sams, Indianapolis, IN, 1972, p 79–81.

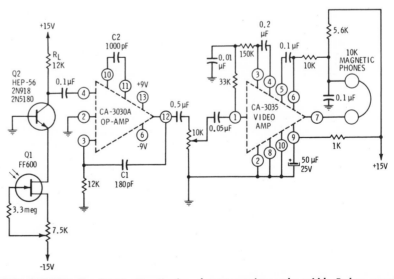

FOTOFET RECEIVER—Developed primarily for detecting 200-ns pulses of injection laser. Experiment with different values for RL to obtain maximum output signal amplitude as viewed on cro. Values of C1 and C2 should be similarly adjusted because they depend on laser transmitter pulse width. Reduce capacitor values for narrower pulse width. Use lens with Q1 to increase range.—R. W. Campbell and F. M. Mims, III, "Semiconductor Diode Lasers," Howard W. Sams, Indianapolis, IN, 1972, p 126–128.

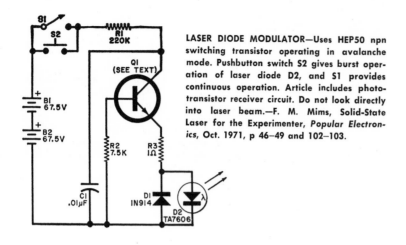

LASER DIODE MODULATOR—Uses HEP50 npn switching transistor operating in avalanche mode. Pushbutton switch S2 gives burst operation of laser diode D2, and S1 provides continuous operation. Article includes phototransistor receiver circuit. Do not look directly into laser beam.—F. M. Mims, Solid-State Laser for the Experimenter, *Popular Electronics*, Oct. 1971, p 46–49 and 102–103.

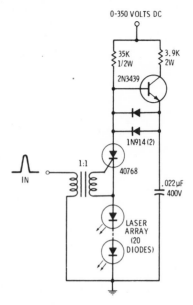

ARRAY PULSER—Provides peak currents of 30 A at 4 kHz for pulsing array of 20 laser diodes. Sprague 11Z12 pulse transformer provides impedance match between input trigger source and laser-array load.—R. W. Campbell and F. M. Mims, III, "Semiconductor Diode Lasers," Howard W. Sams, Indianapolis, IN, 1972, p 102–103.

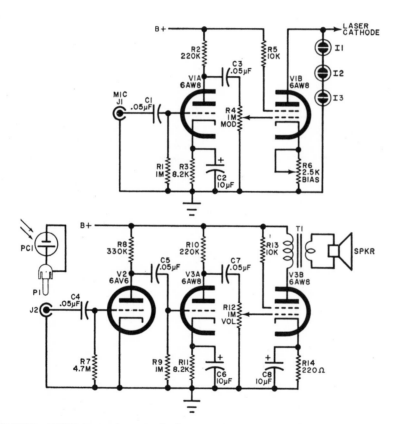

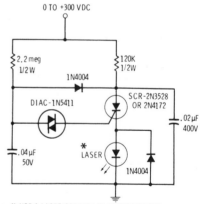

DIAC TRIGGER—When voltage across diac rises to certain point, diac becomes conductive and turns on scr for firing laser.—R. W. Campbell and F. M. Mims, III, "Semiconductor Diode Lasers," Howard W. Sams, Indianapolis, IN, 1972, p 89–90.

LASER-BEAM PHONE—Transmitter is basic pentode modulator V1B, driven by triode half of tube serving as microphone preamp. Neons I1–I3 are 200-V breakdown Signalite A-259. Requires 175 V d-c and 6.3 V a-c supplies. Neons chop off high-voltage spikes that trigger model 205 laser rated at 0.7 mW. Receiver is conventional three-stage audio amplifier driven by solar cell PC1 (Allied 60D7569 or similar). Article gives construction and adjustment details.—C. H. Knowles, Laser Beam Communicator, *Popular Electronics*, May 1970, p 27–35 and 40–42.

CHAPTER 55
Limiter Circuits

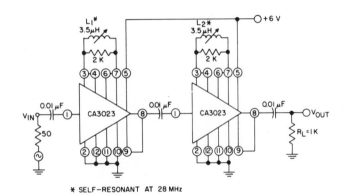

28-MHZ LIMITER-AMPLIFIER—Requires only two wideband IC amplifiers. Provides full limiting at input of 300 μV. Total gain is 61 dB for both stages, and power dissipation is 66 mW. Bandwidth before limiting is 3.8 MHz and effective Q is 7.35.—"Linear Integrated Circuits," RCA, Somerville, NJ, IC-42, 1970, p 208–211.

LIMITING AT MICROVOLTS—Below limiting level, input voltage V1 is not amplified enough by differential stage TR3-TR4 and opamp to turn on TR5 and TR6, so TR8 is off and TR9 saturated. Voltage fed back to TR1 is then high enough to saturate it, while TR2 is cut off. When input voltage exceeds about 200 μV for values shown, either TR5 or TR6 conducts (depending on polarity of input), TR8 saturates, TR9 turns off, TR1 cuts off, and TR2 saturates to provide limiting. First four transistors are p-channel enhancement-mode mosfet's.—A. Ivanov, High Impedance Input Protection at Microvolt Level, *Wireless World*, March 1972, p 140.

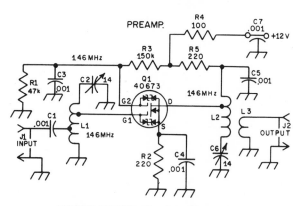

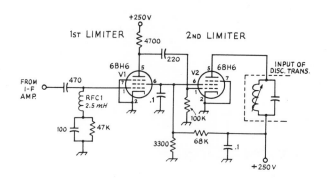

146-MHZ PREAMP—May be added to f-m receiver or transceiver to improve sensitivity and limiting. Uses dual-gate mosfet. Adjust all tuned circuits for maximum limiter current while receiving weak signal.—"The Radio Amateur's Handbook," American Radio Relay League, Newington, CT, 49th Ed., 1972, p 458.

PENTODE LIMITER—Uses sharp-cutoff tubes to clip both sides of input signal so as to chop off noise and amplitude modulation, for constant-amplitude output.—"The Radio Amateur's Handbook," American Radio Relay League, Newington, CT, 49th Ed., 1972, p 441.

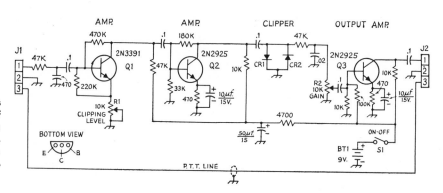

SPEECH CLIPPER—Includes amplifier stages as required with low-output mike to feed vhf transmitter. Silcon diodes (any type) begin clipping at audio level of about 0.6-V peak. Q3 makes up for gain loss by clipping.—C. Utz, A Handy Speech Amplifier-Clipper, QST, Sept. 1967, p 28–29.

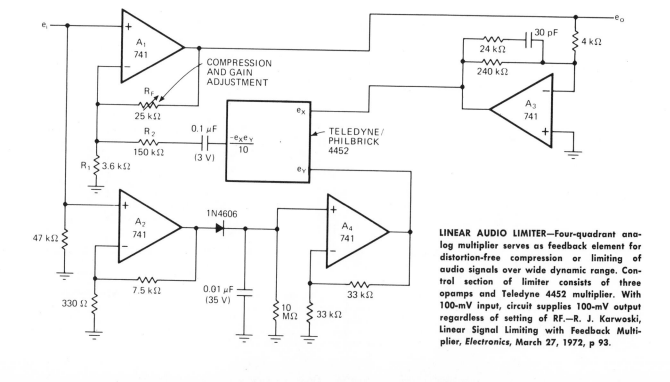

LINEAR AUDIO LIMITER—Four-quadrant analog multiplier serves as feedback element for distortion-free compression or limiting of audio signals over wide dynamic range. Control section of limiter consists of three opamps and Teledyne 4452 multiplier. With 100-mV input, circuit supplies 100-mV output regardless of setting of RF.—R. J. Karwoski, Linear Signal Limiting with Feedback Multiplier, Electronics, March 27, 1972, p 93.

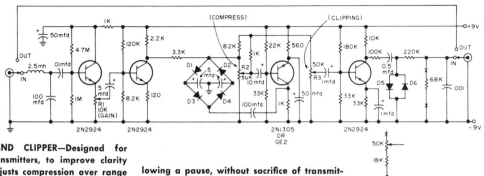

COMPRESSOR AND CLIPPER—Designed for use with ssb transmitters, to improve clarity of speech. R2 adjusts compression over range of 0 to 20 dB, and R3 adjusts clipping from 0 to 10 dB. Both input and output are high-impedance. Combination of clipping and compression prevents peaking of first syllable following a pause, without sacrifice of transmitter output capability. Output is about 150-mV peak, which is higher than average mike level. Article emphasizes importance of adjusting circuit properly.—P. Lovelock, A Versatile Premodulation Speech Processor, 73, Aug. 1972, p 28–30 and 31.

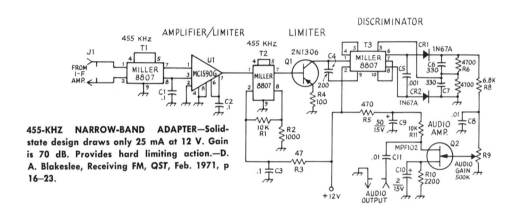

455-KHZ NARROW-BAND ADAPTER—Solid-state design draws only 25 mA at 12 V. Gain is 70 dB. Provides hard limiting action.—D. A. Blakeslee, Receiving FM, QST, Feb. 1971, p 16–23.

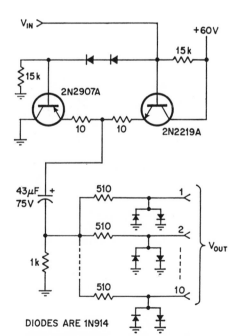

SHUNT LIMITER—Basic circuit for individual video channel requires large high-voltage coupling capacitor costing about $30 to block high voltage and maintain ground reference. Interfering signals above 400 mV drive limiting diodes to cutoff.—R. J. Turner, Series Limiter Tracks Signal, Finds Symmetry, *Electronics*, Feb. 15, 1971, p 87.

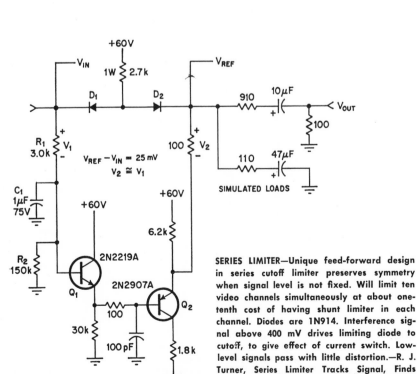

SERIES LIMITER—Unique feed-forward design in series cutoff limiter preserves symmetry when signal level is not fixed. Will limit ten video channels simultaneously at about one-tenth cost of having shunt limiter in each channel. Diodes are 1N914. Interference signal above 400 mV drives limiting diode to cutoff, to give effect of current switch. Low-level signals pass with little distortion.—R. J. Turner, Series Limiter Tracks Signal, Finds Symmetry, *Electronics*, Feb. 15, 1971, p 87.

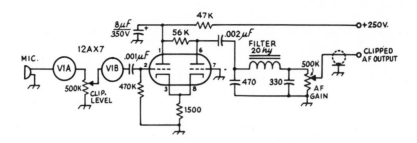

SPEECH CLIPPER—Used in ssb and other transmitters to raise average power level by clipping peaks of speech signal before modulation. Resulting high-order harmonics are removed by filter having high attenuation for all frequencies above 3,000 Hz. For drastic clipping, several times more amplification is required than for normal modulation, and noise must be lower. With moderate clipping levels of 6 to 12 dB, voice power is increased considerably with almost no change in voice quality. Clipper tube can be almost any a-f dual-triode, such as 12AX7.—"The Radio Amateur's Handbook," American Radio Relay League, Newington, CT, 49th Ed., 1972, p 403–404.

SPEECH CLIPPER—Developed for use in CB rig to pass only most intelligible voice frequencies on to modulator with amplification while clipping out highs and lows by filtering. Circuit requires no battery. Used to add audio power to CB or other transmitter.—R. M. Brown, "101 Easy CB Projects," Howard W. Sams, Indianapolis, IN, 1968, p 79–80.

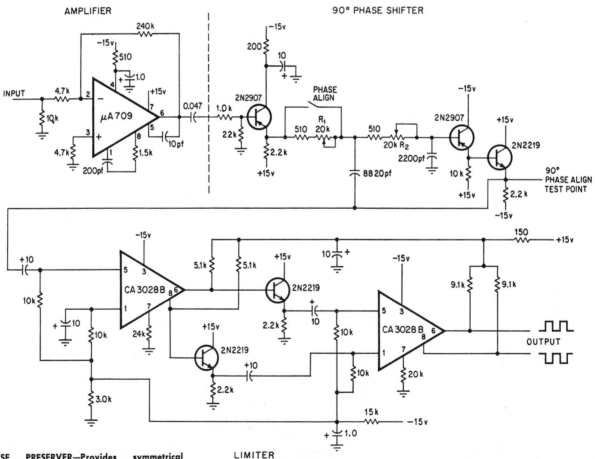

PHASE PRESERVER—Provides symmetrical complementary outputs in phase quadrature with input signal over 50-dB dynamic range, constant within 0.5 deg, using inexpensive CA3028B limiters. Circuit is insensitive to changes from 0.5 to 100 mV p-p in input amplitude, and provides 15-V p-p output swing between 2 and 60 kHz. Phase shifter uses active low-pass filter with Chebyshev response to give 90-deg shift at corner frequency where amplitude response is unity.—R. J. Turner, IC Limiter Preserves Phase over 50-DB Range, *Electronics*, Nov. 23, 1970, p 67–68.

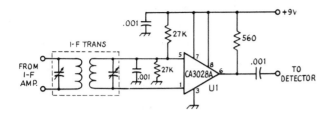

IC LIMITER—High-gain IC containing six to eight active transistor stages (MC1550G is alternate) has limiting knee of 100 mV and provides superior limiting action compared to pair of tubes or transistors. Provides constant-amplitude output, free of noise and amplitude modulation.—"The Radio Amateur's Handbook," American Radio Relay League, Newington, CT, 49th Ed., 1972, p 441.

500-KHZ LIMITER AMPLIFIER—Uses two CA3021 high-gain wideband amplifiers, each with 500-kHz self-resonant choke L1 in feedback path. Tuned circuit L2-C1 gives good sine-wave output. Voltage gain is at least 100 dB. Limited signal is apparent above noise at input of 1 μV, and good limiting occurs for input signals up to 3 V rms.—"Linear Integrated Circuits and MOS Devices," RCA, Somerville, NJ, SSD-202A, 1973 Ed., p 186.

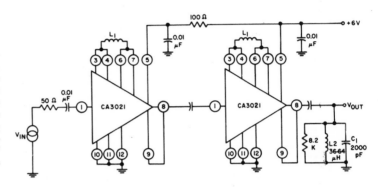

CHAPTER 56
Liquid Level Control Circuits

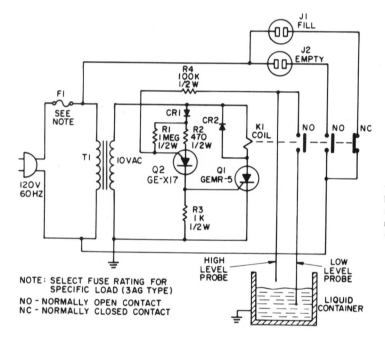

NOTE: SELECT FUSE RATING FOR SPECIFIC LOAD (3AG TYPE)
NO - NORMALLY OPEN CONTACT
NC - NORMALLY CLOSED CONTACT

LIQUID LEVEL—Keeps fluid level between two fixed points determined by probe wires. Loads can be either pump motors or solenoid-operated valves.—"Electronics Experimenters Circuit Manual," General Electric, Owensboro, KY, 1971, 3rd Ed., p 199–202.

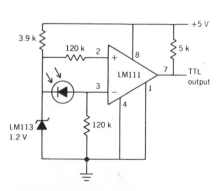

LEVEL DETECTOR—Transistors and resistors in National LM111 IC synthesize action of temperature-compensated low-voltage reference diode, for use with battery-powered photodiode of any type for controlling level of liquid or solid materials. Will operate over range of —55 to 125 C.—Synthesized Reference Diode Maintains Tight Regulation over Wide Current Range, *IEEE Spectrum*, June 1971, p 72.

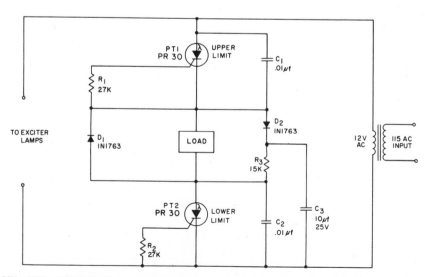

BIN FEED CONTROL—Uses two photoscr's and appropriately positioned exciter lamps to energize feed mechanism serving as load, for maintaining level of liquid or solid material between two limits in reservoir or bin. Load can be solenoid or relay drawing up to 300 mA d-c continuously, although scr's will handle surges up to 5 A.—"Automatic Bin Feed Control," Unitrode Corp., Watertown, MA, New Design Idea No. 12.

402

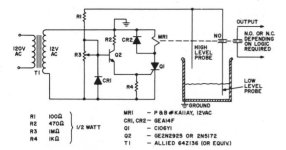

R1	100Ω	
R2	470Ω	1/2 WATT
R3	1MΩ	
R4	1KΩ	

MRI	— P&B #KA11AY, 12VAC
CRI, CR2	— GEA14F
Q1	— C106Y1
Q2	— GE2N2925 OR 2N5172
T1	— ALLIED 64Z136 (OR EQUIV.)

LIQUID LEVEL—Uses two wire probes, one sensing high level and the other low level. A-c voltage on probes eliminates electrolytic corrosion. Q2 amplifies probe signals sufficiently to trigger scr power switch Q1 that drives output relay MR1.—R. W. Fox and D. R. Grafham, "Two Automatic Liquid Level Controls," General Electric, Semiconductor Products Dept., Auburn, NY, No. 201.14, 1969.

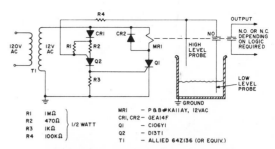

R1	1MΩ	
R2	470Ω	1/2 WATT
R3	1KΩ	
R4	100KΩ	

MRI	— P&B #KA11AY, 12VAC
CRI, CR2	— GEA14F
Q1	— C106Y1
Q2	— D13T1
T1	— ALLIED 64Z136 (OR EQUIV.)

HIGH-IMPEDANCE FLUID LEVEL—Designed for applications in which impedance between wire level-sensing probes may be as high as 20 meg. When liquid touches high-level probe, voltage drop across R1 triggers programmable unijunction transistor Q2; this energizes relay, which remains closed until liquid clears low-level probe.—R. W. Fox and D. R. Grafham, "Two Automatic Liquid Level Controls," General Electric, Semiconductor Products Dept., Auburn, NY, No. 201.14, 1969.

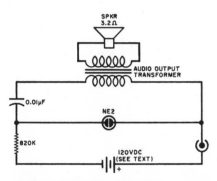

HELPING BLIND POUR COFFEE—Two stainless steel wire electrodes hooked over rim of coffee cup complete circuit of neon relaxation oscillator to produce audio tone indicating to blind person that cup is full enough. Use four small 30-V batteries in series. Get electrode wire from dental supply store, and mount on plug-in twin-lead connector to permit occasional washing. Will work on cold or hot liquids.—T. V. Cramer, Liquid Level Indicator for the Blind, *Popular Electronics*, May 1967, p 59–60.

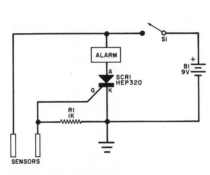

POOL SPLASH ALARM—When person falls into pool, water shorts sensors mounted just above surface, triggering scr on and energizing alarm. Alarm continues until S1 is opened to reset circuit. Alarm can be Mallory SC628 Sonalert or standard d-c buzzer or bell shunted by small resistor passing scr keep-alive anode current.—D. D. Mickle, Useful Circuits, *Popular Electronics*, Jan. 1972, p 101–102.

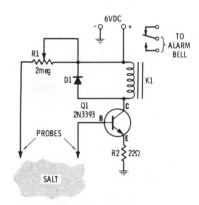

FLOOD ALARM—One teaspoon of dry salt placed between probe wires spaced 1 inch apart on basement floor will become conductive and sound alarm if water pipe breaks or outside water seeps into basement. Use sheet of plastic under salt to protect concrete from damage. To adjust, water salt and set R1 to make 300-ohm 6-V d-c relay K1 close. Diode is 1N60 or equivalent.—H. Friedman, "99 Electronic Projects," Howard W. Sams, Indianapolis, IN, 1971, p 30–31.

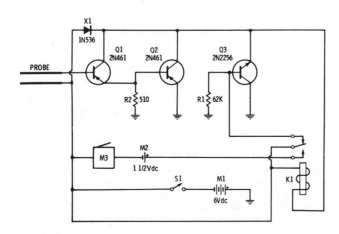

POOL SPLASH ALARM—Relay energizes and turns on buzzer or bell M3 whenever water is between probes. One probe is permanently under water, while other is positioned just enough above water level so wind-produced waves cannot touch it but splash of child falling in pool will reach it. Above-water probe can be stripe of conductive paint on wall around entire pool, provided water level is kept constant. Turn off when pool is in normal use. K1 is 6-V d-c relay.—R. M. Brown and T. Kneitel, "49 Easy Entertainment and Science Projects," Howard W. Sams, Indianapolis, IN, 1971, p 50–52.

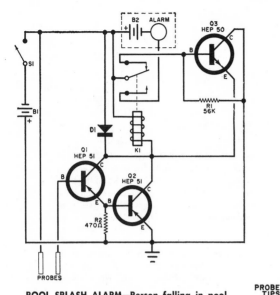

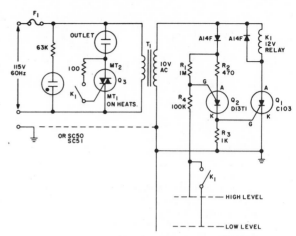

POOL SPLASH ALARM—Person falling in pool creates splash that places water resistance between probes positioned just above normal level, turning on Darlington pair Q1-Q2 and energizing relay that sounds alarm. One pair of relay contacts turns on Q3 to keep alarm sounding after splash has subsided, until person is rescued and S1 is opened to reset alarm. Probes must be repositioned each time pool is filled.—F. Maynard, Build a Swimming Pool Splash Alarm, *Popular Electronics*, July 1966 p 46—49 and 97.

MOISTURE SENSOR—Simple $3 circuit will detect leakage or overflow of liquid, such as water leaks in basements, aquariums, and boats. Can also be used to detect amount of moisture in any material, such as in lumber. Uses either GE C103Y or Motorola HEP320 scr. Serves as alarm if lamp or buzzer is connected in series with 6-V or 12-V battery between red and black leads. Relay can be used in place of alarm, for starting pump or other larger load. With d-c, battery circuit must be opened when probes are in clear after scr fires, to reset. With 6-V or 12-V a-c supply, reset is automatic. Probe tips can be simply ends of resistors.—H. St. Laurent, Build a Moisture Sensor, *Popular Electronics*, March 1971, p 67—69.

MAINTAINING LIQUID LEVEL—Controls either motor-driven pumps or solenoid-operated valves plugged in to outlet for operation on a-c line voltage. Uses metal-rod or wire probes to establish fixed points between which level of fluid is maintained. Reversing contact connections K1 changes from filling to emptying mode, to meet requirements of application. Gate of put Q2 forms detector that turns on Q1 for energizing relay when liquid rises to high probe level. Relay then turns on Q3 to activate load and simultaneously arm low-level probe which holds circuit on until level drops below this probe.—"SCR Manual," General Electric, Syracuse, NY, 1972, 5th Ed., p 226—227.

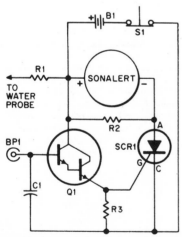

POOL LEVEL ALARM—Splash when person falls in unattended swimming pool raises water level momentarily enough to complete probe circuit of electrode connected to BP1 and positioned just above pool surface, other probe being immersed in water continuously. Mallory SE628 Sonalert alarm is then turned on by GE 2N5306 Darlington amplifier triggering HEP300 scr. Alarm continues until circuit is reset by opening S1. B1 is 6-V battery, R1 and R2 270 ohms, R3 100 ohms, and C1 0.02 μF. Adjust probe so wind-produced waves do not touch it. Raise probe or open S1 when pool is in use.—S. Daniels, Intrusion Detector for Backyard Pools, *Electronics Illustrated*, Sept. 1970, p 32—33.

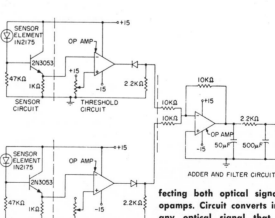

ADDER AND FILTER CIRCUIT

POSITION DETECTOR—Linearity of level detecting system using two photodiodes is improved by signal perturbation method in which extra higher-frequency signal is added to input of each receptor. Perturbation is introduced by using motor to vibrate mirror af-

fecting both optical signals. Uses high-gain opamps. Circuit converts into electrical signal any optical signal that exceeds threshold value of sensor as set by pot. Signals from each receptor are summed and sent through low-pass filter so only average value remains.—J. J. Stanaway, Jr., and G. Cook, Experimental Verification of Signal Perturbation Applied to the Accuracy of Position Detection, *IEEE Trans. on Industrial Electronic and Control Instrumentation*, Aug. 1970, p 369—374.

CHAPTER 57
Logarithmic Circuits

LOG CONVERTER—Gives accuracy over wide temperature range. Q1 is logging transistor, while Q2 provides fixed offset for temperature compensation. Both opamps are cascaded in overall feedback loop to reduce phase shift and improve stability for one-quadrant logarithmic conversion. R3 and R4 determine current for zero crossing on output, which is 10 μA for values shown.—R. J. Widlar, "IC Op Amp Beats FETs on Input Current," National Semiconductor, Santa Clara, CA, 1969, AN-29, p 12–13.

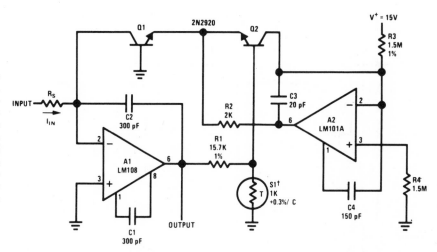

10 nA < I_IN < 1 mA
Sensitivity is 1V per decade

†Available from Tel Labs, Inc., Manchester, N.H., Type Q81.

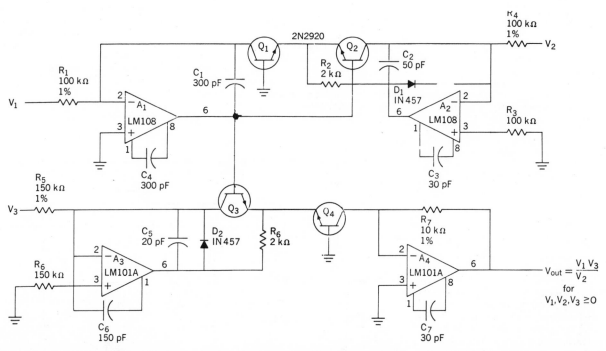

RECIPROCAL-FUNCTION GENERATOR—Complete one-quadrant multiplier-divider network is basically log generator driving antilog generator. Provides wide dynamic range with high accuracy, as required for such applications as measurement of transistor direct-current gain.—R. C. Dobkin, Logarithmic Converters, *IEEE Spectrum*, Nov. 1969, p 69–72.

APPROXIMATING ARCTANGENT—Single op-amp with Analog Devices Model 433 analog multiplier-divider module gives arctangent of VB/VA with maximum theoretical error less than 0.68 deg by using feedback techniques combined with arbitrary exponents for equation being solved.—D. Sheingold, Trigonometric Operations with the 433, *Analog Dialogue*, Vol. 6, No. 3, p 4–5.

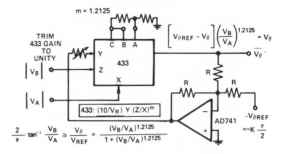

$$\frac{2}{\pi} \tan^{-1} \frac{V_B}{V_A} \cong \frac{V_\theta}{V_{REF}} = \frac{(V_B/V_A)^{1.2125}}{1 + (V_B/V_A)^{1.2125}}$$

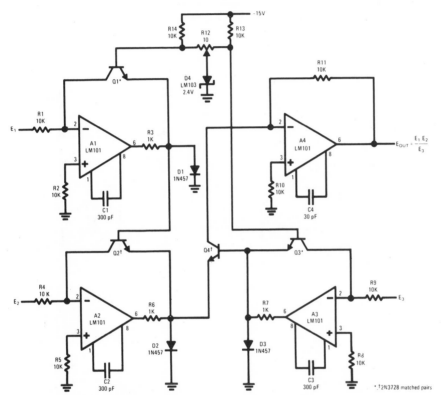

ANALOG MULTIPLIER-DIVIDER—Takes logs of input voltages, adds or subtracts as required, then takes antilog to give multiplication and division. Consists of three log converters using opamps A1–A3 and antilog transistor Q4. Collector current of Q4, proportional to

product of E1 and E2 divided by E3, is fed to summing amplifier A4 which gives desired output. Accuracy is 1% for inputs from 500 mV to 50 V.—R. J. Widlar, "Operational Amplifiers," National Semiconductor, Santa Clara, CA, 1968, AN-4.

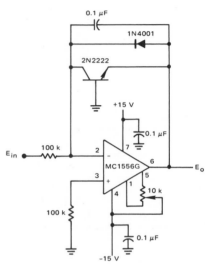

LOGAMP—Uses MC1556 opamp because input bias current is only 8 mA, permitting accurate operation down to millivolt input levels without bias current compensation. Input offset is adjusted with 10K pot. Requires positive input voltage and gives negative output voltage. For positive output with negative input, use 2N2907 transistor and turn diode around.—K. Huehne, "Transistor Logarithmic Conversion Using an Integrated Operational Amplifier," Motorola Semiconductors, Phoenix, AZ, AN-261A, 1971.

MULIPLIER-DIVIDER—Circuit is basically log generator driving antilog generator. Useful for measuring transistor current gains over wide range of operating currents. Output of A1 drives base of Q3 with voltage proportional to log of E1/E2. Q3 adds voltage proportional to log of E3 and drives antilog transistor Q4. Collector current of Q4 is then converted to output voltage by A4 and R7. Operates in one quadrant.—R. C. Dobkin, "Logarithmic Converters," National Semiconductor, Santa Clara, CA, 1969, AN-30.

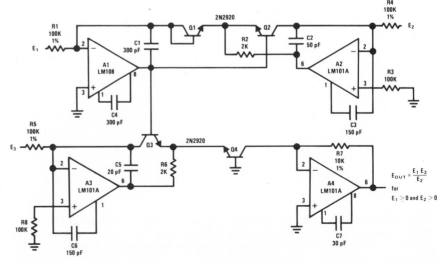

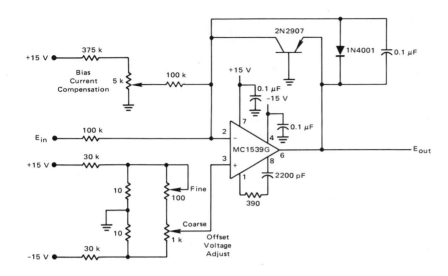

LOW-COST LOGAMP—Uses inexpensive MC1539G opamp. Requires external compensation, has about 200 nA bias current, and accomplishes offset adjustment at unused noninverting input. Pnp transistor serves as logarithmic element.—K. Huehne, "Transistor Logarithmic Conversion Using an Integrated Operational Amplifier," Motorola Semiconductors, Phoenix, AZ, AN-261A, 1971.

ONE-QUADRANT DIVIDER—Provides better than 2.8% of full-scale accuracy over dynamic range of 1,000:1. Scale factors are readily adjustable and circuit can easily be optimized for other signal ranges by using design equations given. Uses Burr-Brown Model 4116 modular d-c logamps rated 400 pA to 400 μA. Used to compute ratio of two positive analog input voltages, Z and X.— "Modular Logarithmic Amplifier," Burr-Brown, Tucson, AZ, April 1971, p 8.

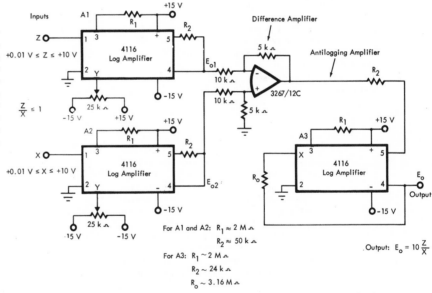

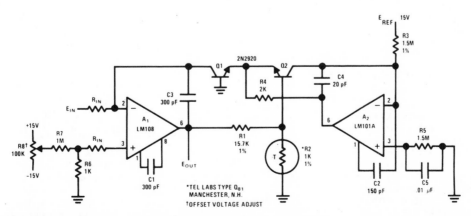

GENERATOR WITH 100-DB RANGE—Generates logarithmic output voltage for linear input current over dynamic range of over five decades. Q1 is nonlinear feedback element around LM108 opamp and Q2 is feedback element of other opamp. Circuit provides inverting operation; report gives modifications required for noninverting operation. Log output is accurate to 1% for inputs from 10 nA to 1 mA.—R. C. Dobkin, "Logarithmic Converters," National Semiconductor, Santa Clara, CA, 1969, AN-30.

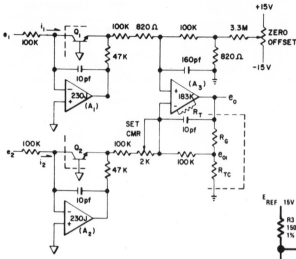

CURRENT OR VOLTAGE RATIO—Combination of Analog Devices 751N log module (dashed lines) with 230J chopper-stabilized amplifiers in each input provides log of ratio of two positive input currents with high precision. For negative inputs, use 751P log module. Third amplifier is high-impedance differential opamp.—W. Borlase and E. David, "Design of Temperature-Compensated Log Circuits Employing Transistors and Operational Amplifiers," Analog Devices, Inc., Norwood, MA, Sept. 1969, p 12.

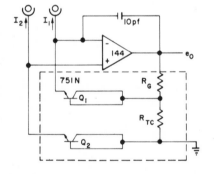

ANTILOG GENERATOR—Generates exponential output from linear input. Opamp A1 and Q1 drive emitter of Q2 in proportion to input voltage. Collector current of Q2, varying exponentially with emitter-base voltage, is converted to voltage by opamp A2.—R. C. Dobkin, "Logarithmic Converters," National Semiconductor, Santa Clara, CA, 1969, AN-30.

$$E_{OUT} = 10^{-[E_{IN}]}$$

*TEL LABS TYPE Q_{81}
MANCHESTER, N.H.

CURRENT RATIO—Simple combination of Analog Devices 751N logarithmic module and inexpensive Model 144 fet amplifier provides log of ratio of two input currents with high accuracy down to 1 nA. Pairs of phototubes or photomultipliers, one serving as reference, meet circuit requirement for "true" current sources. Overall accuracy is 0.1 dB.—W. Borlase and E. David, "Design of Temperature-Compensated Log Circuits Employing Transistors and Operational Amplifiers," Analog Devices, Inc., Norwood, MA, Sept. 1969, p 10.

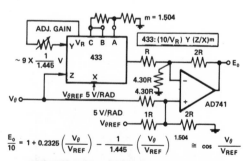

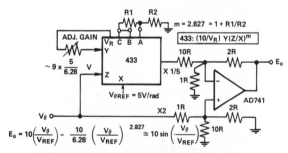

$$E_0 = 10\left(\frac{V_\theta}{V_{REF}}\right) - \frac{10}{6.28}\left(\frac{V_\theta}{V_{REF}}\right)^{2.827} \cong 10 \sin\left(\frac{V_\theta}{V_{REF}}\right)$$

APPROXIMATING SINES—Combination of opamp and Model 433 analog multiplier-divider module gives sine of input voltage value proportional to angle with accuracy of 0.25% in single quadrant. Based on infinite series for computing sine of angle. If coefficients are modified to allow truncation of series to three terms, error is less than 0.02%. By modifying exponents as well as coefficients the error can be less than 0.25% with just two terms.—D. Sheingold, Trigonometric Operations with the 433, Analog Dialogue, Vol. 6, No. 3, p 4–5.

$$\frac{E_0}{10} = 1 + 0.2325\left(\frac{V_\theta}{V_{REF}}\right) - \frac{1}{1.445}\left(\frac{V_\theta}{V_{REF}}\right)^{1.504} \cong \cos\frac{V_\theta}{V_{REF}}$$

APPROXIMATING COSINES—Single opamp and Model 433 analog multiplier-divider module gives cosine of input voltage value proportional to angle with better than 1% accuracy by using only single power term of infinite series but with arbitrary exponents and linear third term. Covers only single quadrant.—D. Sheingold, Trigonometric Operations with the 433, Analog Dialogue, Vol. 6, No. 3, p 4–5.

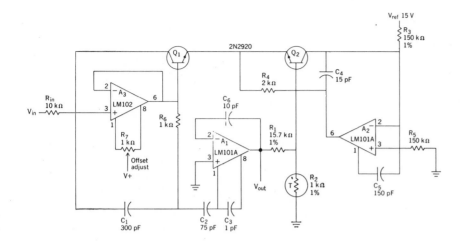

FAST LOG GENERATOR—Uses feed-forward compensation on LM101A opamp to extend bandwidth beyond 3.5 MHz and increase response rate of output voltage. Dynamic range is 80 dB. For input currents of 100 nA to 1 mA, output voltage varies logarithmically with input. T is Tel Labs type 981 compensator.—R. C. Dobkin, Logarithmic Converters, *IEEE Spectrum*, Nov. 1969, p 69–72.

LOG GENERATOR WITH 100-DB DYNAMIC RANGE—Generates logarithmic voltage for linear input, with Q1 as nonlinear feedback element around opamp A1. Q2 is feedback element for A2. Accuracy is 1% over current range of 10 nA to 1 mA. T is Tel Labs type 981 compensator, and R8 is offset-voltage adjusting resistor. Circuit is slow, taking up to 5 ms for output to settle to 1% of final value.—R. C. Dobkin, Logarithmic Converters, *IEEE Spectrum*, Nov. 1969, p 69–72.

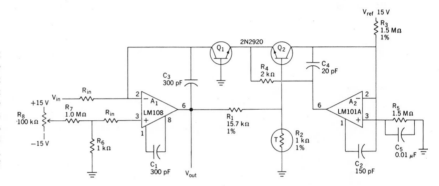

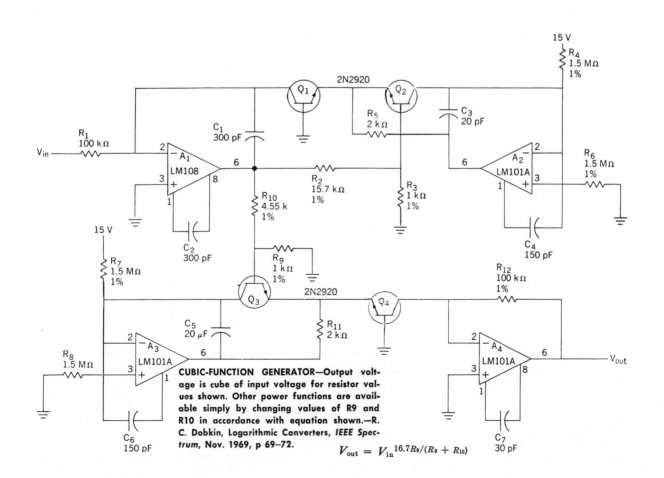

CUBIC-FUNCTION GENERATOR—Output voltage is cube of input voltage for resistor values shown. Other power functions are available simply by changing values of R9 and R10 in accordance with equation shown.—R. C. Dobkin, Logarithmic Converters, *IEEE Spectrum*, Nov. 1969, p 69–72.

$$V_{out} = V_{in}^{16.7R_9/(R_9 + R_{10})}$$

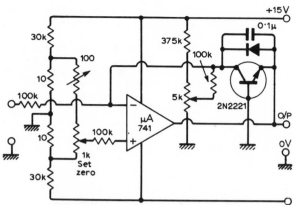

LOGAMP—Provides rapid change in gain around zero signal level and logarithmic fall-off in gain for higher signal levels. Can be used as null detector. Diode is 1N914.—T. D. Towers, Elements of Linear Microcircuits, *Wireless World*, Feb. 1971, p 76–80.

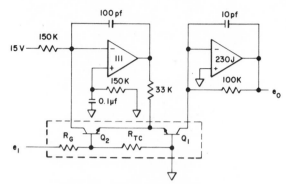

VOLTAGE ANTILOG—Combination of Analog Devices 751N log module with 230J chopper-stabilized amplifier and 111 opamp provides wide dynamic range for compression of positive input voltage. With input varying from 0 to 5 V, output varies from 10 V to 100 μV, representing one decade of output voltage change per volt of input voltage change.—W. Borlase and E. David, "Design of Temperature-Compensated Log Circuits Employing Transistors and Operational Amplifiers," Analog Devices, Inc., Norwood, MA, Sept. 1969, p 12.

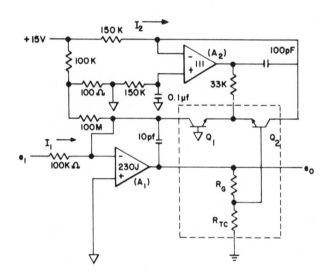

COMPRESSION—Combination of Analog Devices 751N log module with 230J chopper-stabilized amplifier and 111 opamp provides wide dynamic range, from 100 μA to 10 V, for logarithmic compression of input voltage. With 310 varactor-bridge amplifier in place of 230J and with 100K input resistor omitted, circuit gives logarithmic compression of current input over range of 10 pA to 100 μA. Use 751P for negative inputs.—W. Borlase and E. David, "Design of Temperature-Compensated Log Circuits Employing Transistors and Operational Amplifiers," Analog Devices, Inc., Norwood, MA, Sept. 1969, p 11.

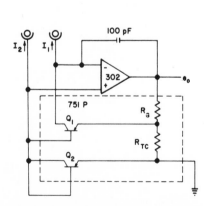

CURRENT RATIO—Combination of Analog Devices 751P logarithmic module and 302 opamp gives log of input current ratios over range from 3 pA to hundreds of nA. Designed for negative input currents, as from photomultipliers.—W. Borlase and E. David, "Design of Temperature-Compensated Log Circuits Employing Transistors and Operational Amplifiers," Analog Devices, Inc., Norwood, MA, Sept. 1969, p 11.

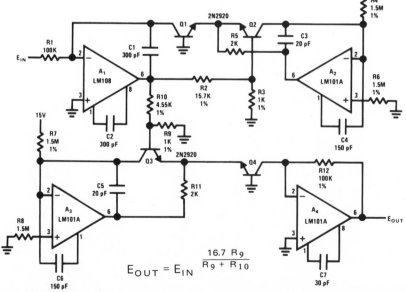

$$E_{OUT} = E_{IN}\frac{16.7\,R_9}{R_9 + R_{10}}$$

CUBE GENERATOR—Values shown make output voltage the cube of input voltage, but any other power function is available by changing values of R9 and R10.—R. C. Dobkin, "Logarithmic Converters," National Semiconductor, Santa Clara, CA, 1969, AN-30.

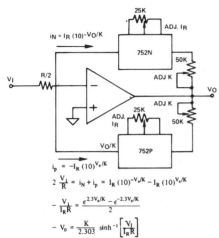

$$i_N = I_R (10)^{-V_O/K}$$

$$i_p = -I_R (10)^{V_o/K}$$

$$2\frac{V_1}{R} = i_N + i_p = I_R (10)^{-V_o/K} - I_R (10)^{V_o/K}$$

$$-\frac{V_1}{I_R R} = \frac{\epsilon^{2.3V_o/K} - \epsilon^{-2.3V_o/K}}{2}$$

$$-V_0 = \frac{K}{2.303} \sinh^{-1}\left[\frac{V_1}{I_R R}\right]$$

BIPOLAR SIGNAL COMPRESSION—Uses two complementary antilog transconductors (Analog Devices 752P and 752N) in feedback path of opamp to synthesize $\sinh^{-1}$ function. Resulting function is logarithmic for larger values of input, but passes through zero essentially linearly but slowly.—D. H. Sheingold, "Analog-Digital Conversion Handbook," Analog Devices, Norwood, MA, 1972, p I-37—I-40.

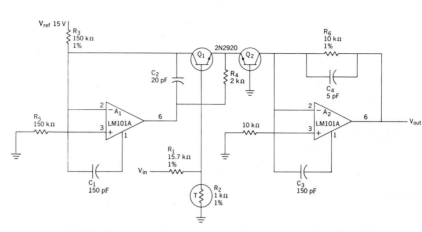

ANTILOG GENERATOR—Rearrangement of log converter circuit provides antilog or exponential generation of output voltage from linear input. T is Tel Labs type 981 compensator.—R. C. Dobkin, Logarithmic Converters, *IEEE Spectrum*, Nov. 1969, p 69–72.

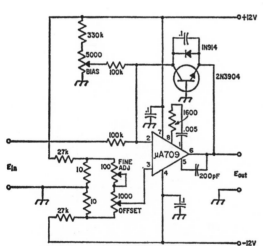

LOGAMP—Output voltage is proportional to log of input voltage over five-decade range, from 0.1 mV to 10,000 mV input. Adjust for zero output while temporarily grounding pin 2. Article tells how to adjust bias control for straight-line graph of d-c response on log paper.—J. M. Pike, The Operational Amplifier, QST, Sept. 1970, p 54–57.

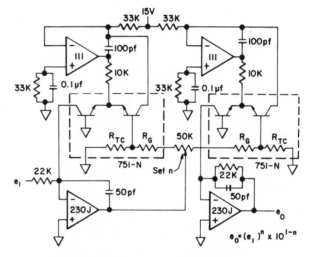

$$e_o = (e_1)^n \times 10^{1-n}$$

POWER OF VOLTAGE—Single 50K pot adjusts exponent between 0.25 and 4 for generation of exponential function. Uses Analog Devices 751N log modules, 230J chopper-stabilized amplifiers, and 111 opamps for positive input voltages.—W. Borlase and E. David, "Design of Temperature-Compensated Log Circuits Employing Transistors and Operational Amplifiers," Analog Devices, Inc., Norwood, MA, Sept. 1969, p 12.

FAST LOG GENERATOR—Converts linear input current to logarithmic output voltage over dynamic range of 80 dB, with additional opamp providing optimization for speed rather than range. Scale factor is 1 volt per decade for inputs from 100 nA to 1 mA.—R. C. Dobkin, "Logarithmic Converters," National Semiconductor, Santa Clara, CA, 1969, AN-30.

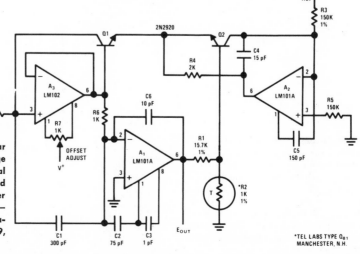

CHAPTER 58
Logic Circuits

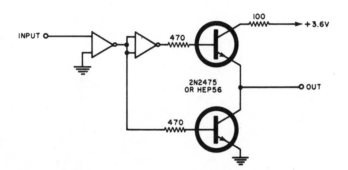

EMERGENCY NONINVERTING BUFFER—Can be assembled from MC799P or HEP571 IC and two transistors. Has fan-in of 3 and fan-out of at least 60. Used when complement of preceding-stage output is not available.—F. H. Tooker, Unusual, But Useful, Digital Circuits, *Electronics World*, April 1971, p 36–37 and 73.

PNP TTL-MOS INTERFACE—Used in translating bipolar logic levels of 5 or 6 V to 10 to 25 V levels associated with mos logic. Gate used must either have passive (resistive) pull-up output circuit or no load resistor in gate. Suitable only for low-speed operation. —"MOS Integrated Circuits and Their Applications," Mullard, London, 1970, TP11081, p 94–97.

POWER RESET—Turning on logic power supply switches on first buffer inverter. After interval determined by R1 and C1, 12-V supply reaches level sufficient to turn on Q1 and switch inverter off. Resulting delay produces required pulse at least 1-ms long for resetting logic correctly. Same reset action occurs after power interruption.—R. L. Wiker, Control Voltage Resets Logic at Power Turn-On, *Electronics*, Sept. 14, 1970, p 101.

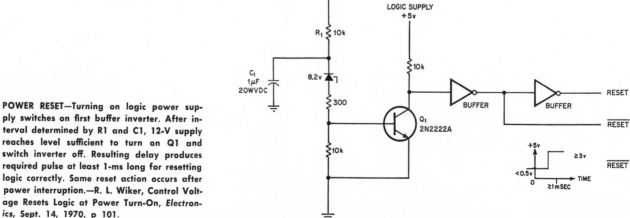

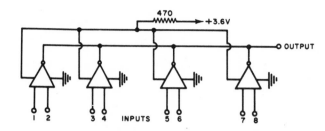

EMERGENCY 8-INPUT GATE—Can be easily assembled from quad 2-input gate such as MC724P or HEP570. Series resistor in supply lead is correct for collector resistors paralleled inside IC.—F. H. Tooker, Unusual, But Useful, Digital Circuts, *Electronics World*, April 1791, p 36–37 and 73.

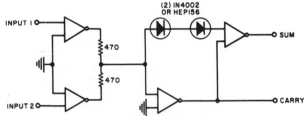

EMERGENCY EXCLUSIVE-OR—Provides sum and carry functions of positive-logic half-adder. Can be assembled with quad 2-input gate such as MC724P or HEP570. When both inputs are high, sum output is low and carry is high. When both inputs are low, both outputs are low. With power rectifiers shown, will operate up to 100 kHz.—F. H. Tooker, Unusual, But Useful, Digital Circuits, *Electronics World*, April 1971, p 36–37 and 73.

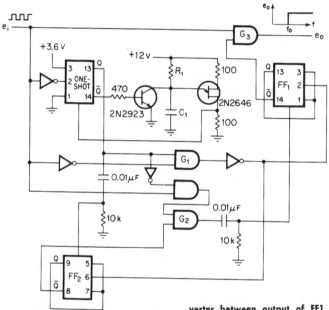

DIGITAL HIGH-PASS—R1 and C1 values set break frequency and gate time of MC790P one-shot. Flip-flops are Motorola MC790P. Other logic units are in MC789P hex inverter. Can be converted to low-pass by placing inverter between output of FF1 and input of G3. Article also covers conversions to bandpass and band-reject response.—R. J. McKinley, Versatile Digital Circuit Filters Highs, Lows, or Bands, *Electronics*, June 21, 1971, p 66.

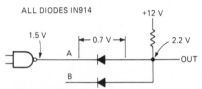

DIODE SWITCHING MATRIX—Improved noise immunity of IC logic permits use of diode switching matrices in logic systems, particularly for code conversion. With high-noise-immunity logic driving gate having guaranteed input threshold of 5 V (rather than 0.8 V of standard TTL), noise immunity is 2.8 V.—D. Guzeman, Diode Switching Matrices Make a Comeback, *Electronics*, Jan. 17 1972, p 76–77.

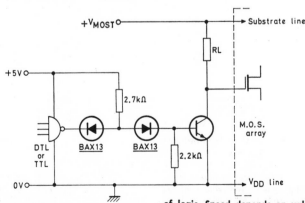

NPN TTL-MOS INTERFACE—Used in translating bipolar logic levels of 5 or 6 V to 10 to 25 V levels associated with mos logic. Uses grounded 0-V line as common to both types of logic. Speed depends on value of RL and transistor type used. Circuit does not reduce system noise immunity.—"MOS Integrated Circuits and Their Applications," Mullard, London, 1970, TP11081, p 98–99.

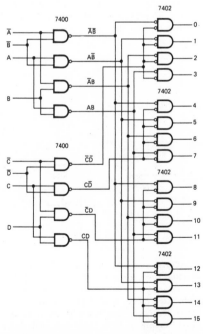

HEX DECODER—Mixing of positive and negative NAND gate packages reduces IC package count from eight to six for binary-to-hex decoder with active low output.—L. E. Frenzel, Jr., Positive and Negative Gates Trim Logic Package Count, *Electronics*, Oct. 25, 1971, p 78.

NPN TOTEM-POLE INTERFACE—Used in translating bipolar logic levels of 5 or 6 V to 10 to 25 V levels associated with mos logic. Uses diode-shifting input circuit and antisaturation diode clamps. If D2 is fast germanium type operating within its temperature range, D1 may be omitted. Will handle capacitive loads.—"MOS Integrated Circuits and Their Applications," Mullard, London, 1970, TP11081, p 99—100.

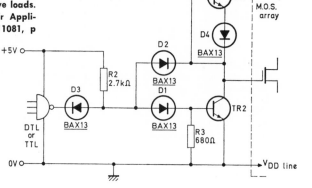

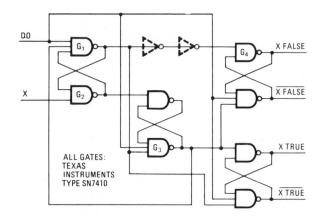

DECISIONS WITHOUT SYNC—Unique arrangement of gates makes logic decisions based on control signal at input DO, without requiring clock for synchronization. Input signal at X represents logic condition acted upon to give one of four possible outputs. Operating speed is above 10 MHz with input pulse widths of about 30 ns.—L. K. Torok, Unclocked Logic Element Makes Quick Decisions, Electronics, April 10, 1972, p 98.

LOGIC-POWERED 3-INPUT GATE—Gives high output when any one or all of inputs are high, but low output for all other conditions. Almost any low-power small-signal npn transistors can be used.—K. D. Dighe, Low-Cost Exclusive-OR Needs No Power Supply, Electronics, April 26, 1971, p 56.

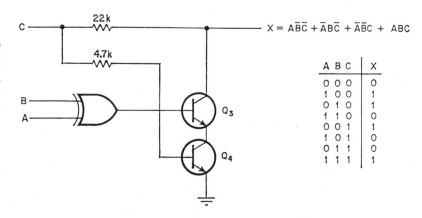

$$X = A\overline{B}\overline{C} + \overline{A}B\overline{C} + \overline{A}\overline{B}C + ABC$$

A	B	C	X
0	0	0	0
1	0	0	1
0	1	0	1
1	1	0	0
0	0	1	1
1	0	1	0
0	1	1	0
1	1	1	1

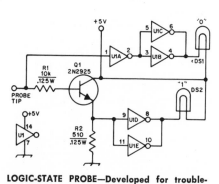

LOGIC-STATE PROBE—Developed for troubleshooting logic circuits and IC chips without cro. Do not exceed 5 V for power supply. Probe uses all sections of SN7404 TTL hex inverter IC. Power leads are attached to 5-V power supply of logic under test, and probe is touched in turn to terminals of circuit. Lamp DS2 will glow for high logic level, and DS1 for low level. For open circuit, both lamps stay off. Probe is compatible only with TTL logic levels.—E. H. Rogers, The Vest-Pocket Logic Probe, QST, Aug. 1972, p 46—47.

TOTEM-POLE INTERFACE—Used in translating bipolar logic levels of 5 or 6 V to 10 to 25 V levels associated with mos logic. Uses active pull-up stage and antisaturation diodes to increase operating speed even when driving highly capacitive loads such as are associated with clock lines of long shift registers or control lines of mos storage.—"MOS Integrated Circuits and Their Applications," Mullard, London, 1970, TP11081, p 97—99.

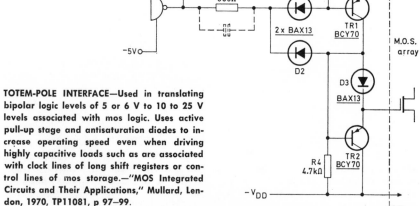

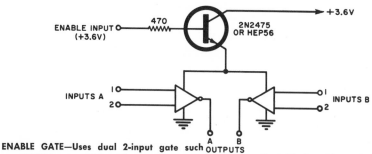

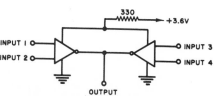

ENABLE GATE—Uses dual 2-input gate such as HEP584 or μL914 with transistor connected between pin 8 of IC and supply bus. Enable signal is fed directly to base of transistor. Outputs remain low, regardless of inputs, until enable input is made high.—F. H. Tooker, Unusual, But Useful, Digital Circuits, *Electronics World*, April 1971, p 36–37 and 73.

EMERGENCY 4-INPUT GATE—Can be easily assembled from dual 2-input gate such as HEP584 or μL914. Fan-out decreases only slightly.—F. H. Tooker, Unusual, But Useful, Digital Circuits, *Electronics World*, April 1971, p 36–37 and 73.

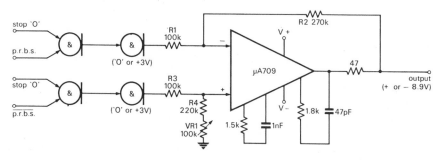

LEVEL CHANGER—Used in pseudo-random binary sequence generator to change logic levels of 0 and +3 V to new levels of ±8.9 V. Inverting and noninverting inputs of opamp are used alternately as inputs change state. Pot is used with R4 to make levels equal.—R. G. Young, P.R.B. Sequence Correlator Using Integrated Circuits, *Electronic Engineering*, Sept. 1969, p 41–45.

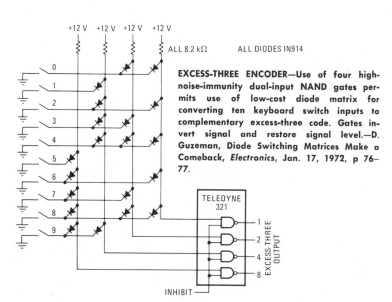

EXCESS-THREE ENCODER—Use of four high-noise-immunity dual-input NAND gates permits use of low-cost diode matrix for converting ten keyboard switch inputs to complementary excess-three code. Gates invert signal and restore signal level.—D. Guzeman, Diode Switching Matrices Make a Comeback, *Electronics*, Jan. 17, 1972, p 76–77.

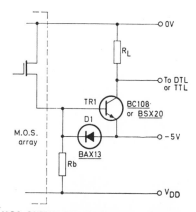

MOS OUTPUT INTERFACE—Uses substrate line as common for mos array and bipolar logic. Operating speed depends on transistor type, circuit values, and mos. Typical propagation delay is 70 ms with BSX20.—"MOS Integrated Circuits and Their Applications," Mullard, London, 1970, TP11081, p 103.

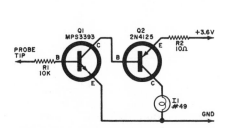

LOGIC LEVEL DETECTOR—When tip of probe is held on terminal of digital circuit, lamp comes on for any positive logic level corresponding to 1, and goes off for 0 level. Transistor types are not critical.—Logic Probe, *Popular Electronics*, Sept. 1969, p 60–61.

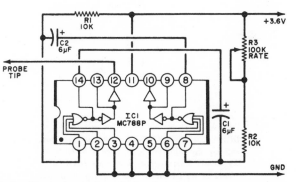

GATE DRIVER—Astable mvbr, using dual-buffer IC, is adjustable from about 1 pps to 10 pps. Will drive up to 80 low-power (mW) RTL gates or 26 medium-power gates for testing purposes. Power is obtained from circuit being tested.—Logic Pulser, *Popular Electronics*, Sept. 1969, p 61–63.

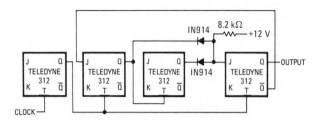

BCD DECODER-COUNTER—Simple diode AND gate here replaces more costly NAND gate and inverter in combinational logic. High noise immunity of IC units now makes substitution possible.—D. Guzeman, Diode Switching Matrices Make a Comeback, *Electronics*, Jan. 17, 1972, p 76–77.

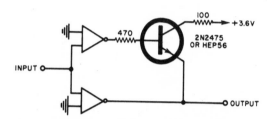

EMERGENCY INVERTING BUFFER—Can be readily assembled from MC799P or HEP571 IC. Has fan-in of 6 and will easily give fan-out of 60.—F. H. Tooker, Unusual, But Useful, Digital Circuits, *Electronics World*, April 1971, p 36–37 and 73.

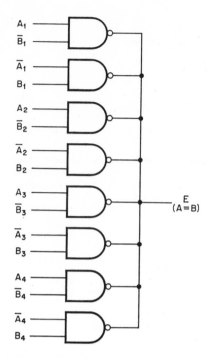

BINARY COMPARATOR—Comparing two multibit binary numbers is simpler and less expensive with wired OR circuit shown than with conventional use of exclusive NOR circuits and NAND gate. Circuit can use either 946 DTL or 7401 TTL quad two-input NAND gates.—L. E. Frenzel, Jr., Wired OR Circuit Simplifies Binary Number Comparison, *Electronics*, Nov. 23, 1970, p 66.

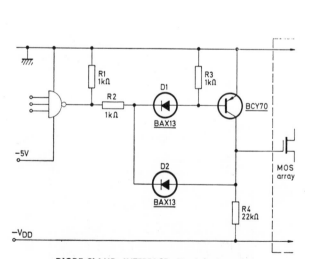

DIODE-CLAMP INTERFACE—Used in translating bipolar logic levels of 5 or 6 V to 10 to 25 V levels associated with mos logic. Uses diode clamp D1 and D2 to keep transistor out of saturation so as to increase operating speed.—"MOS Integrated Circuits and Their Applications," Mullard, London, 1970, TP11081, p 96–98.

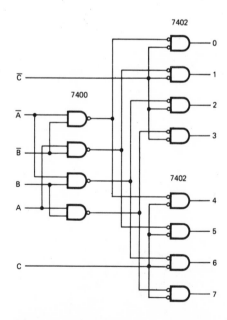

OCTAL DECODER—Mixing of positive and negative NAND gate packages reduces package count to three for binary-to-octal decoder with active high output, as compared to five normally required if using type 7410 positive NAND gates.—L. E. Frenzel, Jr., Positive and Negative Gates Trim Logic Package Count, *Electronics*, Oct. 25, 1971, p 78.

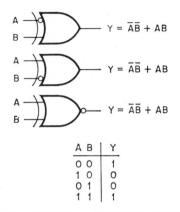

A	B	Y
0	0	1
1	0	0
0	1	0
1	1	1

SIGNAL-POWERED COINCIDENCE DETECTOR —Mixed positive and negative inputs or inversions change signal-powered exclusive-OR gate to coincidence detector.—K. D. Dighe, Low-Cost Exclusive-OR Needs No Power Supply, *Electronics*, April 26, 1971, p 56.

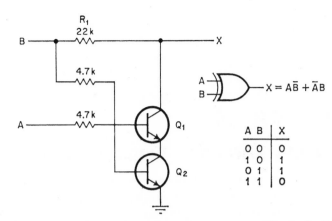

A	B	X
0	0	0
1	0	1
0	1	1
1	1	0

SIGNAL-POWERED EXCLUSIVE-OR—Requires only three resistors and two low-power small-signal npn transistors, and operates without power supply.—K. D. Dighe, Low-Cost Exclusive-OR Needs No Power Supply, *Electronics*, April 26, 1971, p 56.

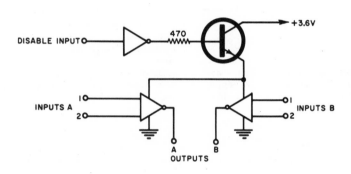

DISABLE GATE—Uses dual 2-input gate such as HEP584 or μL914 with transistor connected between pin 8 of IC and supply bus. Allows conventional inputs to control outputs in usual manner as long as disable input is low. When all conventional inputs are low and outputs are high, applying disable input makes outputs go low. Transistor is 2N2475 or HEP56.—F. H. Tooker, Unusual, But Useful, Digital Circuits, *Electronics World*, April 1971, p 36–37 and 73.

CHAPTER 59
Measuring Circuits—Capacitance

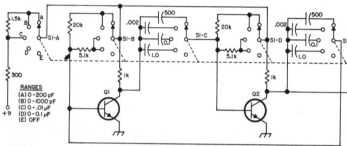

CAPACITANCE METER—Measures capacitors accurately from 4 pF to 0.1 μF in four ranges, and values down to 2 pF can be esti-mated. Uses 0–50 μA meter. Transistors can be 2N2926, SK3011, or HEP54. Capacitor under test is connected in series with meter across output of square-wave free-running mvbr Q1-Q2. Meter deflection is directly proportional to capacitance value, so calibration requires only precision capacitor for full-scale value of each range.—J. Fisk, "Useful Transistor Circuits," 73 Inc., Petersborough, NH, 1967, p 30A.

CAPACITOR LEAKAGE—Requires disconnecting one capacitor lead from its circuit, before connecting test leads across capacitor and throwing S1. Repeated blinking of neon NE51 indicates small amount of leakage. Continuous glow means capacitor is shorted. Single blink means capacitor is good. Uses voltage-doubling circuit with SK3016 rectifiers to apply about 250 V to capacitor being checked, so should be used only for capacitors rated at this voltage or higher.—R. M. Brown and T. Kneitel, "101 Easy Test Instrument Projects," Howard W. Sams, Indianapolis, IN, 1968, p 75–76.

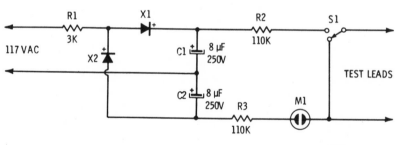

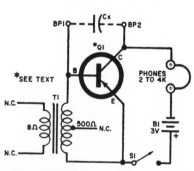

AUDIO-TONE CAPACITOR TESTER—Capacitor under test provides feedback in simple audio oscillator using general-purpose pnp power transistor such as 2N554, for checking 0.002 to 0.1 μF capacitors of any type for opens, shorts, and approximate value as determined by tone frequency heard in headphones. T1 is miniature output transformer used as choke. Higher pitch indicates smaller value.—K. Scharf, Reader's Circuits, *Popular Electronics*, April 1968, p 84–85.

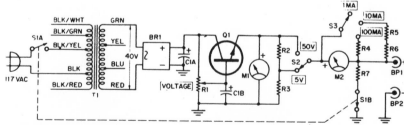

BP1,BP2—Insulated binding post
BR1—Full-wave bridge rectifier: 1 A, 200 PIV (Motorola HEP-176)
C1A,C1B—20/20 μf, 150 V dual electrolytic capacitor
M1—0-50 V DC voltmeter (Emico Model RF-2¼C, Allied 52 C 6097)
M2—0-1 ma DC milliammeter (Emico Model RF-2¼C, Allied 52 C 8012)
Q1—2N2405 transistor (RCA)
Resistors: ½ watt, 10% unless otherwise indicated
R1—50,000 ohm linear-taper potentiometer

R2—4,700 ohms
R3—470 ohms
R4—10 ohms, 5%
R5—39 ohms, 5%
R6—56 ohms, 5%
R7—22 ohms, 1 watt (see text)
S1—DPDT toggle or slide switch
S2—SPDT toggle or slide switch
S3—SR triple-throw rotary switch (see text)
T1—Low-voltage rectifier transformer; secondaries: 10-20 V center tapped and 40 V center tapped @ 35 ma (Allied 54 C 4731)
Misc.—7 x 5 x 3-in. Minibox, terminal strips

LEAKAGE IN ELECTROLYTICS—Designed primarily for checking electrolytics rated up to 50 V d-c, as used in solid-state circuits where excess leakage can be serious. Leakage current limits for capacitors from 1 to 3,000 μF range from 0.31 mA to 10 mA (given in table in article), but 100-mA range is provided for initial use to protect meter if capacitor is shorted. Capacitor under test is connected between BP1 and BP2, S2 is set for voltage range, and R1 adjusted to exact test voltage desired as indicated on M1. M2 reads leakage current after capacitor charges.—V. Kell, Electrolytic Leakage Checker, *Electronics Illustrated*, Nov. 1969, p 82–85.

418

CAPACITANCE METER—Range selector S1 provides four ranges, by switching different frequency-determining network into mvbr Q1-Q2. Output of mvbr is coupled through capacitor under test (at jacks J1-J2) to indicating meter having linear 0–100 scale for reading capacitance directly from a few pF to 0.1 μF. Zener D2 provides regulation for life of battery. Based on principle that square-wave alternating current passed by capacitor is directly proportional to capacitance value.—S. Sula, Build a Capacitance Meter, *Popular Electronics*, Oct. 1969, p 66–69 and 110.

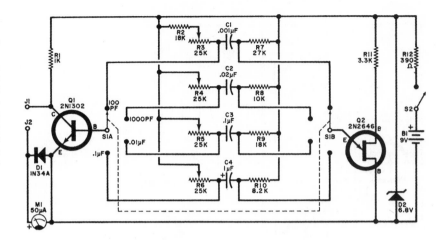

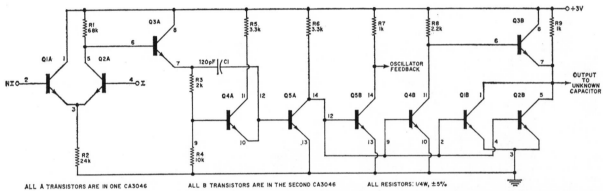

CAPACITANCE-METER OSCILLATOR—Used with switched frequency-determining R-C networks connected to terminal NI to generate seven different stable frequencies in range from 23 Hz to 460 kHz, as required for obtaining seven different capacitance ranges in meter. Article gives all other circuits and calibration procedure. Uses two RCA IC transistor arrays.—H. A. Wittlinger, IC Capacitance Meter, *Electronics World*, Sept. 1970, p 44–47 and 84.

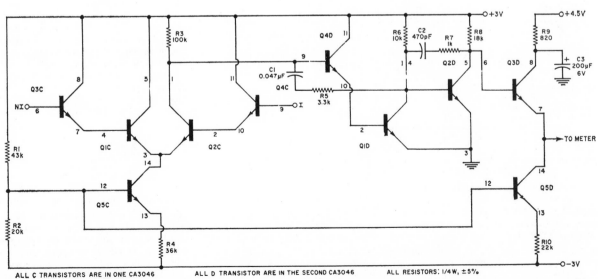

CAPACITANCE-METER AMPLIFIER—Uses two IC transistor arrays to form differential-input amplifier, two-stage d-c amplifier, and emitter-follower output for driving 1-mA meter of capacitance meter.—H. A. Wittlinger, IC Capacitance Meter, *Electronics World*, Sept. 1970, p 44–47 and 84.

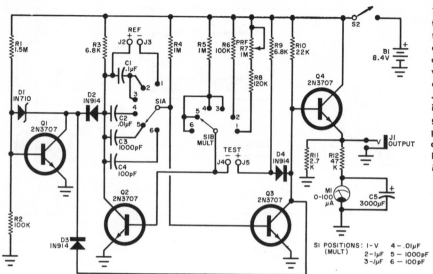

15 PF TO 10 μF—Test circuit is basically free-running mvbr which compares capacitances. Unknown capacitor connected to J4 and J5 becomes one of mvbr cross-coupling capacitors, while other is known precision value selected by S1A. At switch position 1, external reference capacitor is connected to J2 and J3 to determine pulse width, and R7 is used to adjust pulse repetition rate. Article gives construction and calibration details. For pulse generator applications, output pulses are available from J1.—D. Hileman, Direct-Reading Capacitance Meter, *Popular Electronics*, Feb. 1973, p 65–68.

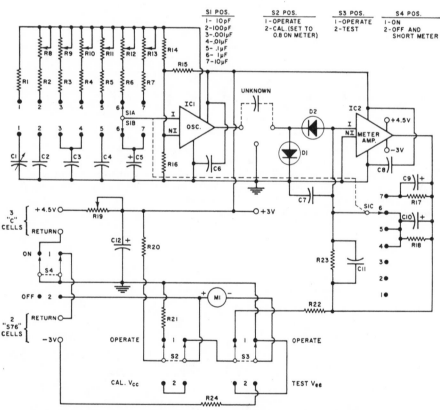

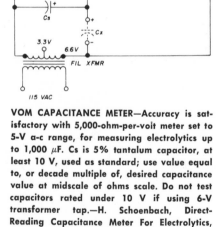

VOM CAPACITANCE METER—Accuracy is satisfactory with 5,000-ohm-per-volt meter set to 5-V a-c range, for measuring electrolytics up to 1,000 μF. Cs is 5% tantalum capacitor, at least 10 V, used as standard; use value equal to, or decade multiple of, desired capacitance value at midscale of ohms scale. Do not test capacitors rated under 10 V if using 6-V transformer tap.—H. Schoenbach, Direct-Reading Capacitance Meter For Electrolytics, *Ham Radio*, Oct. 1971, p 14–15.

R1, R2, R3, R4, R5, R6—5100 ohm, ¼ W res. ±5%
R7, R15, R16—10,000 ohm, ¼ W res. ±5%
R8, R9, R10, R11, R12—5000 ohm trimmer (Mallory MTC4-53L4)
R13—10,000 ohm trimmer (Mallory MTC4-14L4)
R14—15,000 ohm, ¼ W res. ±5%
R17—2000 ohm, ¼ W res. ±5%
R18—11,000 ohm, ¼ W res. ±5%
R19—500 ohm linear-taper pot (Mallory U-2)
R20—3600 ohm, ¼ W res. ±5%
R21—100 ohm, ¼ W res. ±10%
R22—910 ohm, ¼ W res. ±5%
R23—100,000 ohm, ¼ W res. ±5%
R24—5600 ohm, ¼ W res. ±5%
C1—7-100 pF trimmer capacitor (Elmenco 423)
C2—0.0027 μF polystyrene capacitor
C3—0.027 μF, 80 V capacitor
C4—0.27 μF, 80 V capacitor

C5—2.7 μF, 15 V tantalum capacitor
C6, C8—0.1 μF, 10 V disc ceramic capacitor
C7, C11—0.1 μF, 20 V disc ceramic capacitor
C9—150 μF, 6 V tantalum capacitor
C10—6.8 μF, 6 V tantalum capacitor
C12—1000 μF, 6 V elec. capacitor
D1, D2—1N914 diode
M1—0-1 mA meter
S1—3-pole, 11-pos. (7 pos. used) non-shorting rotary sw. (Centralab PA-1009)
S2—D.p.d.t. momenetary push-button sw. ("Cal")
S3—D.p.d.t. momentary push-button sw. ("Test")
S4—D.p.d.t. slide sw. ("On-Off")
IC1, IC2—Four RCA CA3046 integrated circuits
Three 1.5-volt "C" cells
Two 1.5-volt silver-oxide cells (Eveready S76 or Mallory MS76)

7-RANGE CAPACITANCE METER—Uses transistor arrays to form oscillator and meter amplifier, circuits for which are shown separately in article and in this chapter. Unknown capacitor is placed in series with square-wave oscillator, switching diode network D1-D2, and 1-mA indicating meter. Article covers construction and calibration.—H. A. Wittlinger, IC Capacitance Meter, *Electronics World*, Sept. 1970, p 44–47 and 84.

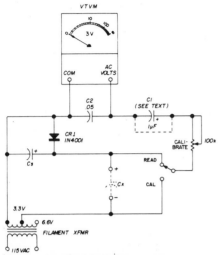

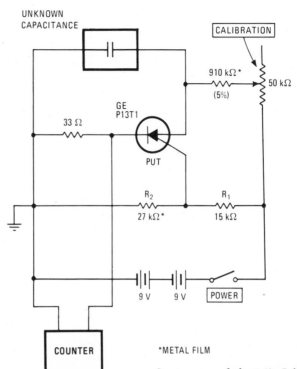

VTVM CAPACITANCE METER—Measures up to 1,000 μF for electrolytics, with value read directly on resistance scale. Use 3-V a-c volts range, and set calibration pot for 3 V full-scale. C1 is blocking capacitor; not needed if already in vtvm. Cs is 5% tantalum capacitor, at least 10 V, used as standard; its value should be equal to, or decade multiple of, desired capacitance value at midscale of ohms scale. If 100 μF is used, for vtvm reading 10 at midscale, range will be 10 to 1,000 μF for easy reading.—H. Schoenbach, Direct-Reading Capacitance Meter For Electrolytics, Ham Radio, Oct. 1971, p 14–15.

COUNTER AS CAPACITANCE METER—Relaxation oscillator converts any time-measuring counter into direct-reading capacitance meter in which seconds indicate μF, milliseconds nF, and microseconds pF values. Accuracy is better than 1% from 5,000 pF to over 10 μF.

Constant error of about 40 pF degrades accuracy of absolute-value measurements below 5,000 pF, but circuit can still be used for matching capacitors as small as 100 pF. Battery drain is only 0.5 mA.—M. J. Salvati, Oscillator Converts Counter to Capacitance Meter, Electronics, July 5, 1973, p 109.

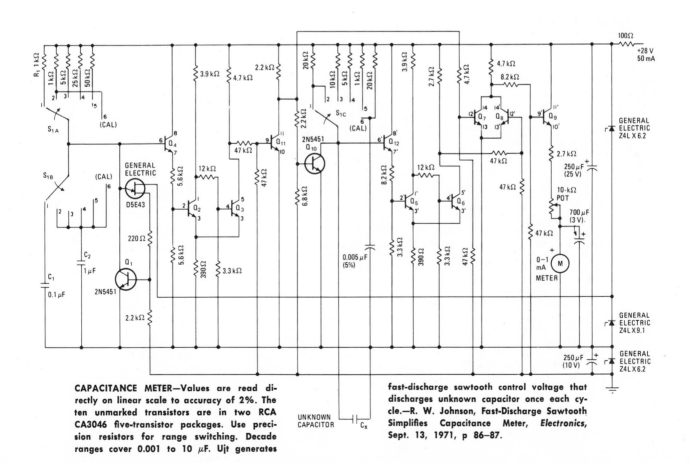

CAPACITANCE METER—Values are read directly on linear scale to accuracy of 2%. The ten unmarked transistors are in two RCA CA3046 five-transistor packages. Use precision resistors for range switching. Decade ranges cover 0.001 to 10 μF. Ujt generates fast-discharge sawtooth control voltage that discharges unknown capacitor once each cycle.—R. W. Johnson, Fast-Discharge Sawtooth Simplifies Capacitance Meter, Electronics, Sept. 13, 1971, p 86–87.

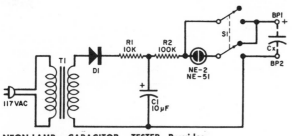

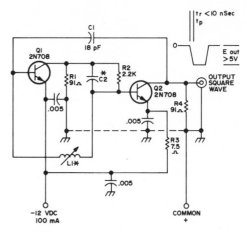

NEON-LAMP CAPACITOR TESTER—Provides visual indication for capacitors rated over 150 V, including electrolytics. With S1 as shown, neon flashes once with brightness and duration proportional to capacitor value for nonelectrolytics. If bulb stays dark, capacitor is open; flicker means intermittent, and dim continuous light means leakage. Use other position of S1 for electrolytics, observing polarity. Circuit now forms relaxation oscillator; the larger the capacitor, the lower the flash rate. If lamp stays on, capacitor is either open or so low in value that flashing is faster than eye can follow. No light means capacitor is leaky or shorted. T1 is 1:1 isolation transformer and D1 general-purpose silicon diode.—E. Richardson, Reader's Circuit, *Popular Electronics*, April 1968, p 84–86.

C BY FREQUENCY SHIFT—Simple 4,000-kHz Hartley oscillator with buffer amplifier is fed into antenna terminal of any amateur or other communication receiver covering 80-meter band, and oscillator is tuned for zero beat with receiver set at this frequency (by adjusting L1, which is Miller 4400 ⅜ inch slug-tuned form having 12 close-wound turns No. 26 tapped 3 turns from bottom). Unknown capacitor Cx is then connected, receiver retuned for zero beat at lower frequency, frequency shift is noted, and capacitance value is then read directly from computer-generated tabulation in article; range covered is from 2.574 pF for 5-kHz deviation to 313.829 pF for 499-kHz deviation. Accuracy is comparable to that of precision lab bridge. Article also gives equations. Transistors are HEP 1 or 2N964. RFC is 2.5 mH National R100S. CR1 is 1N34.—H. Lukoff, Capacitance Measurement by Frequency Shift, *73*, Feb. 1973, p 108–113.

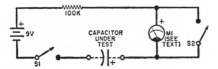

ELECTROLYTIC LEAKAGE TEST—Use 50-μA microammeter. Close S2 before closing S1, to prevent charging current of capacitor from damaging meter. Wait at least 3 minutes before opening S2 to read leakage, which should be less than 1 μA for high-quality capacitor. Switch to lower meter range if available. Large-value low-leakage low-voltage electrolytic connected across battery of transistor radio will give 30% longer play time from set of batteries by eliminating audio distortion, motorboating, and poor oscillator operation.—I. M. Gottlieb, Extending Battery Life, *Popular Electronics*, Feb. 1968, p 66.

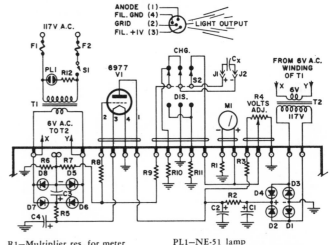

ELECTROLYTICS TESTER—Provides mondestructive test of miniature high-capacitance low-voltage electrolytic capacitors over range of 0.001 to 1,000 μF. Uses type 6977 triode indicator tube having phosphorescent coating on anode to give pale blue glow when grid voltage is zero and tube conducts. Tube should glow brightly in proportion to charging current of capacitor under test, and fade slowly as full charge is reached. S2 is then operated to float capacitor and observe its ability to hold charge. Choose value of R1 to produce 40-V full-scale reading on meter.—D. F. Fleshren, Electrolytic-Capacitor Tester, *Electronics World*, Oct. 1969, p 82–83.

R1—Multiplier res. for meter (optional, see text)
R2—150 ohm, 2 W res.
R3—1000 ohm, 2 W res.
R4—1000 ohm, 2 W linear-taper pot
R5—100 ohm, ½ W res.
R6—150 ohm, ½ W res.
R7—10,000 ohm, ½ W res.
R8—120,000 ohm, ½ W res.
R9—15,000 ohm, ½ W res.
R10, R11—680,000 ohm, ½ W res. ±5%
R12—22,000 ohm, ½ W res.
C1, C2—10 μF, 150 elec. capacitor
C3, C4—100 μF, 20 V elec. capacitor
S1—S.p.s.t. toggle switch
S2—3-pole, d.t., center-off, lever switch

PL1—NE-51 lamp
F1, F2—½ A, plug-mounted 3AG fuse
J1, J2—Banana terminal post
M1—0-40 V d. c. full-scale meter
D1, D2, D3, D4—1N2070 diode or equiv.
D5, D6, D7, D8—1N1693 diode or equiv.
T1, T2—117 V a. c./6 V a. c. at 1 A trans.
V1—6977 indicator tube (Sylvania, Amperex, or Tung-Sol)
1—3″ x 5″ x 7″ circuit box

CHAPTER 60
Measuring Circuits—Current

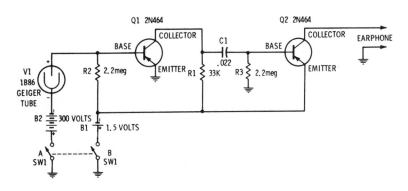

GEIGER COUNTER—Used for locating uranium ore, but will also respond to other radioactivity such as from luminous-dial wristwatch. Normally gives faint clicking sound, which becomes faster and louder in vicinity of radioactive material. Can be mounted in aluminum box since metal does not block radiation being measured. Protect Geiger tube with foam rubber.—R. M. Brown and T. Kneitel, "49 Easy Transistor Projects," Howard W. Sams, Indianapolis, IN, 1972, p 47—48.

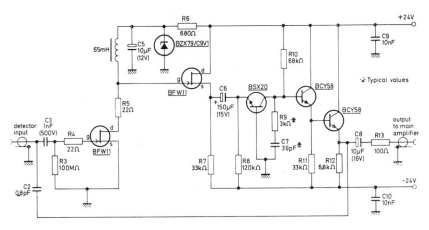

PREAMP FOR SOLID-STATE DETECTOR—Designed for use with silicon surface-barrier radiation detectors, which have 10 times better resolution than scintillation counter, even though requiring much higher gain. Uses low-noise fet's in charge-sensitive input section.—"Field Effect Transistors," Mullard, London, 1972, TP1318, p 75—78.

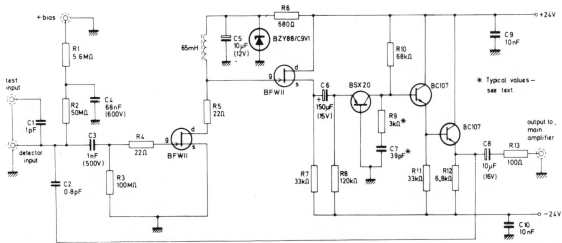

HIGH-GAIN PREAMP—Uses low-noise Mullard BFW11 fet's in charge-sensitive section. Used to convert into voltage the charge released by radiation detector connected to input. Test input is used to set up minimum preamp rise and fall times with minimum overshoot.—"Field-Effect Transistors in a Pre-Amplifier for Use with Solid-State Radiation Detectors," Mullard, London, 1969, TP1106, p 2.

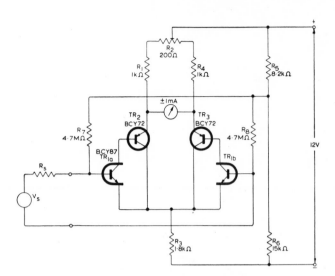

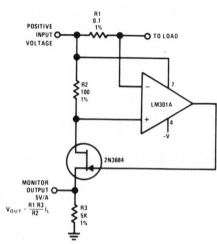

GALVANOMETER AMPLIFIER—Provides deflection from zero to full-scale value of 1 mA on rugged meter for input current change of 25 nA. Biasing is obtained by passing half the difference of input transistor base currents through signal source. Input current for zero output is typically between —10 and 10 nA.— "Circuits Using Low-Drift Transistor Pairs," Mullard, London, 1968, TP 994, p 16—17.

SUPPLY CURRENT MONITOR—R1 senses current flow of power supply. Combination of opamp and jfet buffer converts resulting voltage drop across R1 to output monitor voltage which accurately reflects current output of power supply.—"FET Circuit Applications," National Semiconductor, Santa Clara, CA, 1970, AN-32, p 11.

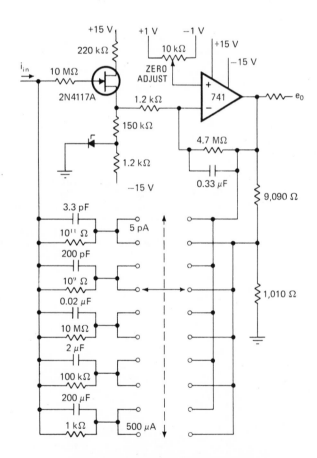

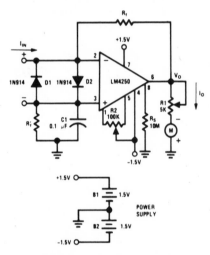

I FULL SCALE	$R_f\,[\Omega]$	$R_f'\,[\Omega]$
100 nA	1.5M	1.5M
500 nA	300k	300k
1 μA	300k	0
5 μA	60k	0
10 μA	30k	0
50 μA	6k	0
100 μA	3k	0

WIDE-RANGE ELECTROMETER—Measures low-frequency currents in ten ranges having full-scale values from 5 pA to 500 μA. Chief uses are in helium leak detectors, mass spectrometers, photomultipliers, and vacuum gages.—R. G. Weinberger, Solve Low-Current Measuring Woes by Designing Your Own Electrometer, Electronics, Aug. 30, 1971, p 58—62.

NANOAMMETER—Meter amplifier provides full-scale current ranges from 100 μA down to 100 nA by changing values of two resistors as shown in table. Circuit is built around programmable opamp, connected as differential current-to-voltage converter with input protection, zeroing and full-scale adjustments, and input resistor balancing for minimum offset voltage. Pair of D cells will serve minimum of 1 year even though operated continuously, corresponding to shelf life, so on-off switch is not needed.—M. K. Vander Kooi and G. Cleveland, "Micropower Circuits Using the LM4250 Programmable Op Amp," National Semiconductor, Santa Clara, CA, 1972, AN-71, p 5.

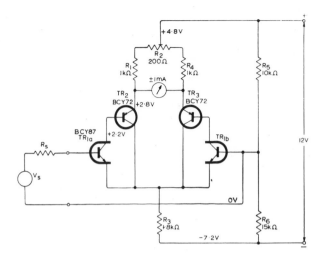

GALVANOMETER AMPLIFIER—Used in place of galvanometer to amplify small currents sufficiently for driving more rugged meter. Uses long-tailed pair directly coupled to two emitter-followers driving meter. Input current for zero output is typically 65 nA, and for full-scale deflection is 90 nA. Biasing is obtained by passing all base current of one input transistor through signal source.—"Circuits Using Low-Drift Transistor Pairs," Mullard, London, 1968, TP994, p 15–16.

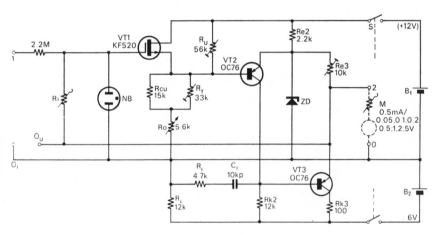

COMPENSATED MOSFET ELECTROMETER—Copper-wire resistor Rcu in parallel with adjustable zero-temperature-coefficient resistor RT are adjusted to offset positive temperature coefficient of circuit. RU and VT1 in parallel compensate for decrease in battery voltage during discharge. Neon NB protects input against overvoltage (Czech mosfet has 100-V breakdown).—M. Pacak, Simple MOSFET Electrometer Circuits, *Electronic Engineering*, Sept. 1969, p 24–27.

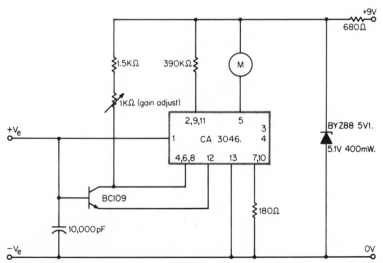

METER AMPLIFIER—Input of 1 µA gives full-scale deflection of 200-µA moving-coil meter. Accuracy and linearity are within 1% over temperature of 0 to 50 C and supply of 8 to 12 V. Warm-up drift is negligible.—H. MacDonald, Current Amplifier Circuit, *The Electronic Engineer*, Aug. 1971, p 70.

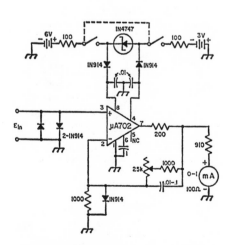

MICROAMMETER—Inputs as low as 40 mV will give full-scale reading. 25K control adjusts d-c gain of opamp from about 2 to 25. Uses 20-V zener for transient suppression.—J. M. Pike, The Operational Amplifier, *QST*, Sept. 1970, p 54–57.

CURRENT TO VOLTAGE—Values shown give bandwidth of 600 kHz and scale factor of 0.1 V per μA. Increasing R1 to 10 meg and reducing C2 to 1 pF, with C1 omitted, gives 25-kHz bandwidth and 10 V per μA. Useful for measuring small currents.—"AD513, AD516 I.C. Fet Input Operational Amplifiers," Analog Devices, Inc., Norwood, MA, Technical Bulletin, Aug. 1971.

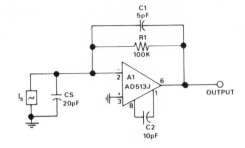

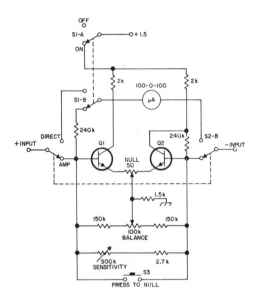

MICROAMMETER AMPLIFIER—Sensitivity can be adjusted with 500K pot so full-scale deflection on each side of zero is anywhere between 2 and 100 μA. Battery drain is only 1.5 mA. Uses differential amplifier with degenerative biasing and collector meter feed to give satisfactory compromise between sensitivity and stability. Transistors are 2N930, GE-10, or HEP50.—J. Fisk, "Useful Transistor Circuits," 73 Inc., Petersborough, NH, 1967, p 31A.

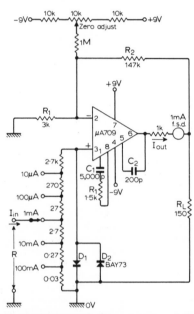

D-C MICROAMMETER—Gives accuracy of 1% at ambient temperature if meter is good quality and resistor values are accurate. Voltage drop is low.—D. C. Microammeter, *Wireless World*, Jan. 1970, p 16.

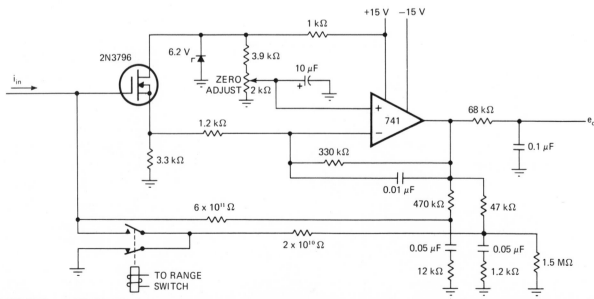

FAST-RESPONSE ELECTROMETER—Frequency-compensated feedback loop gives response time of 7 ms for ranges of 1 pA and 300 pA full-scale. Use of reed relay as range switch provides necessary isolation from ground.—R. G. Weinberger, Solve Low-Current Measuring Woes by Designing Your Own Electrometer, *Electronics*, Aug. 30, 1971, p 58–62.

I FULL SCALE	R_A [Ω]	R_B [Ω]	R_f [Ω]
1 mA	3.0	3k	300k
10 mA	.3	3k	300k
100 mA	.3	30k	300k
1A	.03	30k	300k
10A	.03	30k	30k

D-C AMMETER—Meter amplifier using programmable opamp provides full-scale ranges from 10 A down to 1 mA by changing resistor values as in table. Uses inverting amplifier configuration. Current drain from pair of flashlight D cells is so low that no on-off switch is needed.—M. K. Vander Kooi and G. Cleveland, "Micropower Circuits Using the LM4250 Programmable Op Amp," National Semiconductor, Santa Clara, CA, 1972, AN-71, p 5–6.

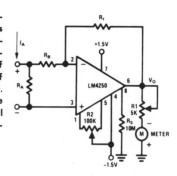

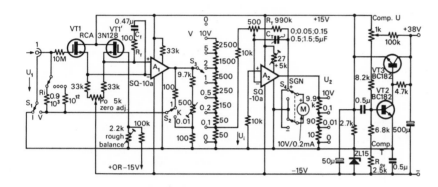

7-RANGE ELECTROMETER—Includes compensation for temperature and supply voltage. Opamp A1 supplies up to 10 V to range switch, and feedback-stabilized opamp A2 with gain of 100 boosts signal to 10 V full-scale on all ranges. Article covers construction and use.—M. Pacak, Simple MOSFET Electrometer Circuits, *Electronic Engineering*, Sept. 1969, p 24–27.

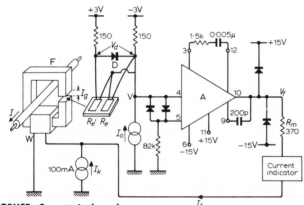

D-C TRANSFORMER—Current I through conductor is measured by means of magnetoresistance and Hall effects, using magnetoresistance Re in air gap of Ferroxcube core FX1795 having 500 turns of No. 24 s.w.g. wire. A is MC1709CP Motorola IC having gain of about 45,000. Indicator is 50-µA meter. Adjust offset current Io for zero meter deflection when I is zero.—B. E. Jones, Magnetoresistance and Its Application, *Wireless World*, Jan. 1970, p 17–19.

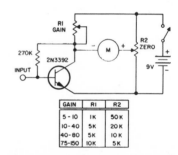

GAIN	RI	R2
5 - 10	1K	50K
10 - 40	5K	20K
40 - 80	5K	10K
75 - 150	10K	5K

METER AMPLIFIER—Resistor values used in bridge-type amplifier determine gain of circuit. With values for highest gain, 1-mA meter will read 10-µA full-scale. Requires high-quality linear meter. To calibrate, apply known desired full-scale current value to input and alternately adjust gain and zero controls until meter reads full scale.—J. Fisk, Transistor Meter Amplifiers, 73, Jan. 1966, p 44–45.

SIMPLE MOSFET ELECTROMETER—Uses feedback with medium loop gain of about 200. Requires far less space than comparable tube circuit.—M. Pacak, Simple MOSFET Electrometer Circuits, *Electronic Engineering*, Sept. 1969, p 24–27.

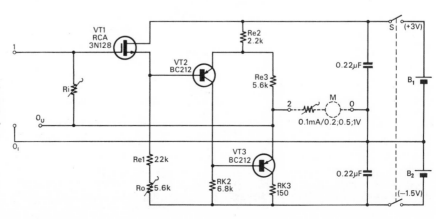

CURRENT-VOLTAGE CONVERTER—Current to be measured is injected directly into summing point of opamp connected in inverting configuration. Current is thus forced to flow through feedback resistor, for conversion into voltage with scaling factor of Rf volts per ampere. Input of 1 μA thus gives reading of 1 V on meter. Output voltage offset with zero input current is typically 0.2 V.—G. B. Clayton, Resistive Feedback Circuits, *Wireless World*, Aug. 1972, p 391–393.

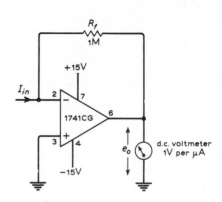

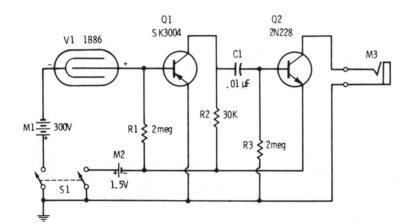

GEIGER COUNTER—Used in prospecting for uranium. Mount fragile Geiger tube V1 carefully in protective aluminum box, since gamma rays from uranium will pass through aluminum. Plug high-impedance headphones into jack M3, and check out circuit by holding radium-dial clock or watch near V1. Click rate of normal background radiation will then increase to that heard when V1 is close to samples of desired uranium ore.—R. M. Brown and T. Kneitel, "101 Easy Test Instrument Projects," Howard W. Sams, Indianapolis, IN, 1968, p 104–105.

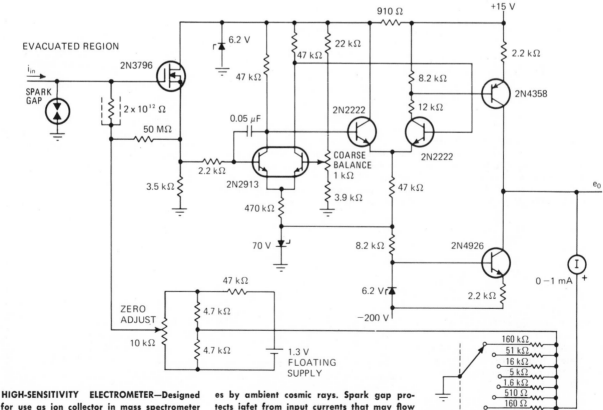

HIGH-SENSITIVITY ELECTROMETER—Designed for use as ion collector in mass spectrometer requiring noise level below 5×10^{-16} A. Input through 2N3796 must be in vacuum chamber to prevent production of noise pulses by ambient cosmic rays. Spark gap protects igfet from input currents that may flow when no power is applied. Measures maximum input current of 9×10^{-11} A in seven ranges covering 3½ decades.—R. G. Weinberger, Solve Low-Current Measuring Woes by Designing Your Own Electrometer, *Electronics*, Aug. 30, 1971, p 58–62.

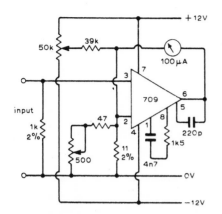

MICROAMMETER—Opamp increases sensitivity of moving-coil meter enough to give full-scale deflection for 1-μA input. Accuracy and linearity are within 1% for ambient range of 0—50 C. Range of 500—0—500 pA can be obtained by adjusting 50K pot. 500-ohm pot adjusts amplifier gain.—H. MacDonald, Opamp Meter Amplifier, *Wireless World*, Sept. 1972, p 443.

5-A A-C AMMETER—Current passing through 0.75-A r-f thermocouple (such as BC-442 antenna current indicator taken from surplus airplane command set) heats dissimilar metal junction, producing small current which is indicated on meter. R1 is made from 31-inch length of tv twin-lead shorted at one end. Article gives calibration table.—N. Johnson, Low-Cost A.C. Ammeter, *Popular Electronics*, May 1969, p 50—52.

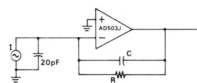

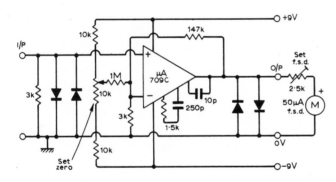

CURRENT TO VOLTAGE—Use of fet-input opamp permits measuring small currents such as are generated by photomultipliers and photodiodes. Bandwidth drops from 220 kHz to 12 kHz as R is increased from 100K to 10 meg and C is decreased from 100 pF to 1 pF.—"AD503, AD506 I.C. Fet Input Operational Amplifier," Analog Devices, Inc., Norwood, MA, Technical Bulletin, Aug. 1971.

DIRECT-CURRENT AMPLIFIER—Use with 50-μA meter for measuring 1-μA d-c full-scale. Diodes are 1N914.—T. D. Towers, Elements of Linear Microcircuits, *Wireless World*, Feb. 1971, p 76—80.

CHAPTER 61
Measuring Circuits—Frequency

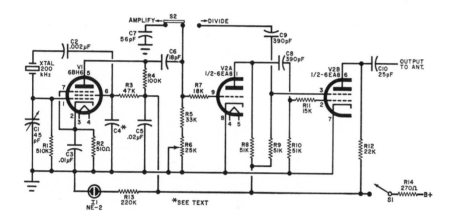

25-KHZ HAM-BAND CALIBRATOR—Absence of tuned circuits gives harmonic-rich output of V1 for feeding to frequency divider V2 controlled by R6. Designed to work with vacuum-tube receivers, from which between 150 and 200 V d-c for plate supply is obtained. C4 is between 50 and 150 pF, depending on crystal used. Article covers construction, calibration, and operation.—N. Johnson, Advanced Ham Frequency Calibrator, *Popular Electronics*, Jan. 1969, p 57—58.

50-KHZ MARKERS—Frequency standard for ham transmitter uses 100-kHz crystal oscillator Q1 to stabilize output of 50-kHz mvbr Q3-Q4. Q2 is pulse amplifier. Harmonics of mvbr serve to spot band edges in station receiver.—S. C. Creason, A Novice Frequency Standard, *QST*, Jan. 1967, p 22—23.

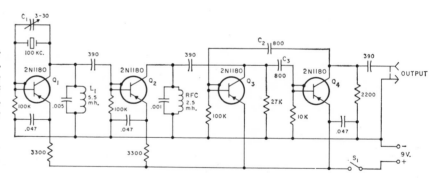

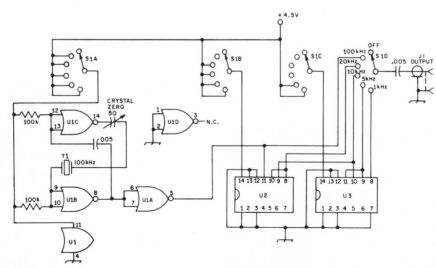

1-KHZ MARKERS—Uses low-cost Fairchild IC's with 100-kHz crystal mvbr to give strong marker signals at 100, 20, 10, 5, and 1 kHz intervals well into h-f range. U2 and U3 are MS19350 decade counters, and U1 is quad 2-input NOR gate such as International Rectifier 1C724-C. Power supply can be three penlight cells in series.—A. C. Beresford, An Inexpensive Secondary Frequency Standard, *QST*, May 1972, p 37—39.

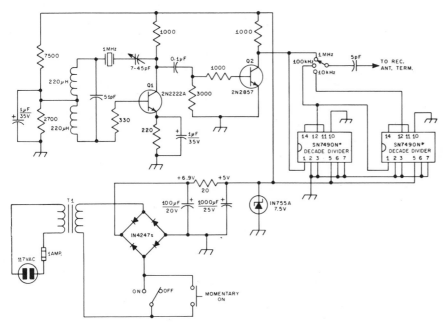

CRYSTAL CALIBRATOR—Used to find specific frequency on higher bands of all-wave receivers having fast tuning rate. With circuit shown, 1-MHz points can be spotted easily. Counting them gives desired 100-kHz marker after switching, then 10-kHz marker. Zener serves only to protect dividers in case one of them becomes defective and d-c bus rises above 8-V maximum.—G. D. Benskin, A Crystal Calibrator for General-Coverage Receivers, QST, Dec. 1970, p 51.

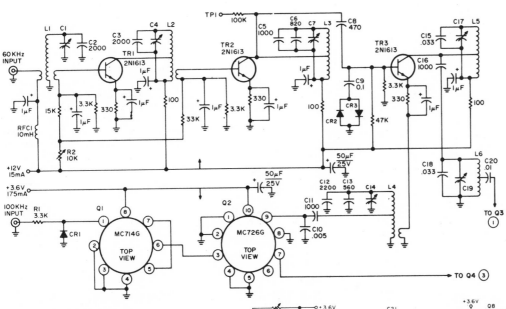

60-KHZ WWVB COMPARATOR—Uses heterodyning and integration to display error between standard frequency of WWVB as received at 60 kHz and local frequency standard. Display is linear plot of phase vs time. 60-kHz input is obtained from preamp using pair of HEP802 transistors fed by loop antenna. Article gives construction details and coil data. R-f amplifiers TR1 and TR2 feed mixer TR3 which also receives 50-kHz signal from IC squarer Q1 and IC divider Q2. Output of 10 kHz goes through Schmitt trigger Q3 to bistable flip-flop Q8. This also receives 10-kHz signal derived from divider Q2 by buffer Q4 and divide-by-five IC's Q5, Q6, and Q7. Phase difference at output of Q8 drives recorder through switching amplifier TR4. Resolution is one part in 10^{10}.—E. P. Manly, WWVB 60 kHz Frequency Comparator Receiver, 73, Sept. 1972, p 16–18, 20, 22–24, and 26.

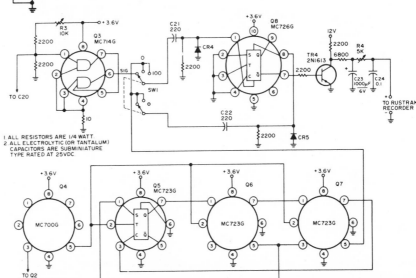

50 AND 500 KHZ STANDARDS—Used for calibrating short-wave receivers. V2 is multivibrator that runs free at about 50 kHz when R5 is at midrange. When 500-kHz output of crystal oscillator V1 is injected into mvbr, output is highly stable and accurate 50-kHz signal. Use HC-6U crystal with matching socket. D1 is 1N3195 and T1 is 100-mA power transformer with 125-V and 6.3-V secondaries. I1 is any neon, used to indicate when plate voltage is applied.—N. Johnson, 500/50-kHz Frequency Standard, *Popular Electronics*, May 1970, p 63–64 and 98.

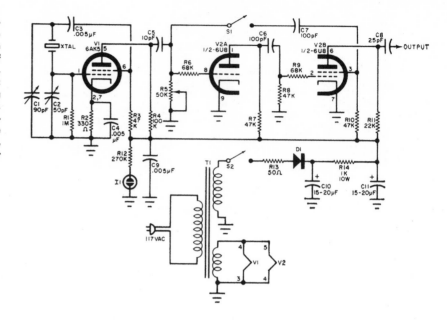

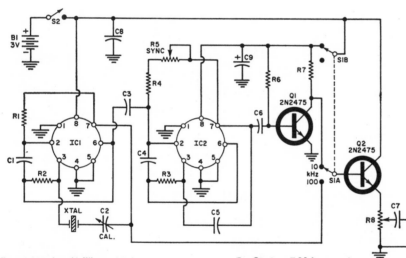

10 AND 100 KHZ—Uses 100-kHz crystal oscillator and 10-kHz IC synchronized mvbr to give choice of either frequency at output with negligible frequency drift and precision required for accurate frequency spotting of communication receivers. Can be calibrated by adjusting C2 for zero beat with signal of WWV.—F. H. Tooker, Portable Dual-Range IC Frequency Standard, *Electronics World*, Feb. 1970, p 76–77.

R1—15,000 ohm, ½ W res. ±5%
R2—22,000 ohm, ½ W res. ±5%
R3—2700 ohm, ½ W res. ±5%
R4—1200 ohm, ½ W res. ±5%
R5—2500 ohm linear-taper pot
R6—6800 ohm, ½ W res. ±5%
R7—3300 ohm, ½ W res. ±5%
R8—500 ohm linear-taper pot
C1—5000 pF polystyrene capacitor
C2—50 pF ceramic-insulated variable capacitor
C3—6.8 pF zero-temp coeff. ceramic capacitor
C4, C5, C6—0.022 µF Mylar capacitor

C7, C8—0.1 µF Mylar capacitor
C9—5 µF, 3 V elec. capacitor
S1—D.p.d.t. slide switch
S2—S.p.s.t. slide switch
Xtal—100-kHz frequency-standard crystal
B1—3-volt battery (2 "C" cells in series)
IC1, IC2—Dual two-input gate IC (µL914 or HEP584)
Q1, Q2—Silicon "n-p-n" transistor (2N2475)

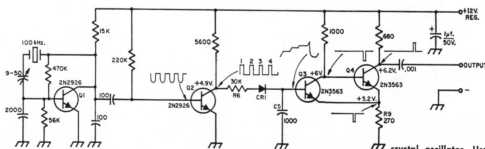

25-KHZ CRYSTAL CALIBRATOR—Q2 amplifies and squares 100-kHz output of crystal oscillator, to give reliable division by four in Q3-Q4. Waveforms are shown for four cycles of crystal oscillator. Used in commercial transceiver.—Galaxy GT-550 Transceiver, *QST*, June 1969, p 41–46.

1-MHZ REFERENCE—Provides square-wave output with amplitude of 5 V and about 1:1 on-off ratio, with long-term frequency accuracy of 1 part in 10⁸. Article includes temperature control circuit for crystal oven. Output of oscillator Tr1 drives two-transistor shaping circuit.—L. O. Nelson-Jones, Crystal Oven and Frequency Standard, *Wireless World*, June 1970, p 269–273.

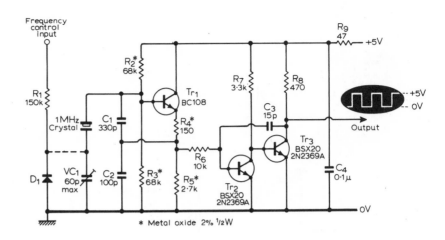

10-KHZ PIPS TO 30 MHZ—provides switch-selectable crystal-controlled marker pips at every 1,000, 500, 200, 100, 50, 20, and 10 kHz up to frequency limit of most commercial short-wave receivers, to simplify location of desired specific short-wave frequency. Closing S4 tone-modulates marker for easier spotting in crowded band. Use of IC flip-flops as frequency dividers makes one crystal do work of four. Different frequencies of square waves in circuit can be used for checking scope sweeps and testing audio amplifiers for ringing. XTAL1 is 100 kHz and XTAL2 is 1,000 kHz.—A. A. Mangieri, IC Frequency Spotter/Standard, *Popular Electronics*, Aug. 1969, p 27–32.

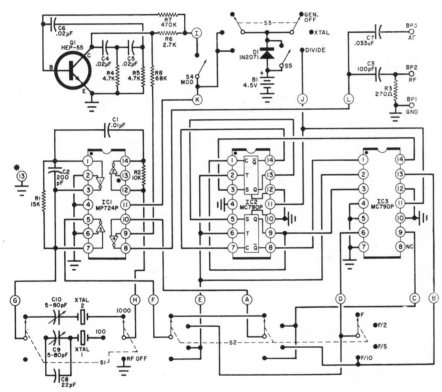

FET GRID-DIP METER—Light-weight battery-operated circuit provides greater sensitivity than comparable tube circuit, because fet is switched into operation as Q multiplier of absorption frequency meter. Covers 1.7 to 300 MHz. Developed by James Millen Mfg. Co.—Transistor Grid-Dip Frequency Meter Simplifies Portable Measurements, *IEEE Spectrum*, March 1971, p 93.

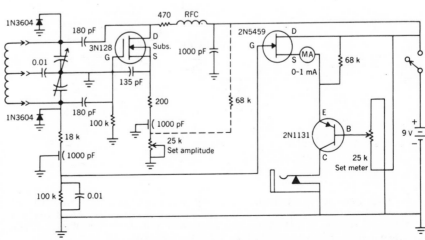

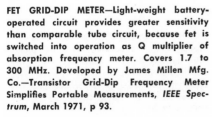

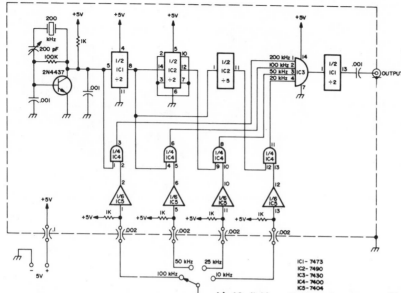

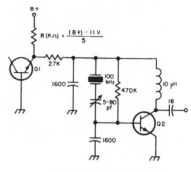

100-KHZ CRYSTAL CALIBRATOR—Provides usable harmonics up to about 150 MHz, has built-in voltage regulator Q1, and has padder in series with crystal for zero-beating with WWV. Regulated voltage provided by Q1 is about 11 V.—J. Fisk, "73 Useful Transistor Circuits," 73 Inc., Peterborough, NH, 1967, p 27A.

100, 50, 25, AND 10 KHZ—Switch gives choice of frequencies for symmetrical square waves derived from 200-kHz crystal oscillator with IC dividers. Used as frequency calibrator. Switch may be located remotely because lines to it have only d-c levels.—Circuits, 73, Aug. 1972, p 131.

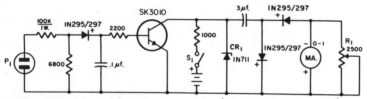

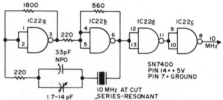

POWER-LINE FREQUENCY MONITOR—Used for checking frequency of emergency a-c generators. Calibrate by connecting P1 to regular 60-Hz power line after setting R1 for zero resistance. Readjust R1 for reading of 0.6 mA on meter. This will correspond to 60 Hz, and meter will indicate other frequencies with reasonable accuracy over upper half of scale regardless of voltage fluctuation or waveform.—"The Mobile Manual For Radio Amateurs," The American Radio Relay League, Newington, CT, 4th Ed., 1968, p 261.

10-MHZ CRYSTAL TIME REFERENCE—Simple circuit uses four gates from IC. Frequency chosen for crystal gives excellent frequency stability with minimum of temperature compensation, and permits zero-beating against 10-MHz signal of WWV.—P. A. Stark, A Modern VHF Frequency Counter, 73, July 1972, p 5–11 and 13.

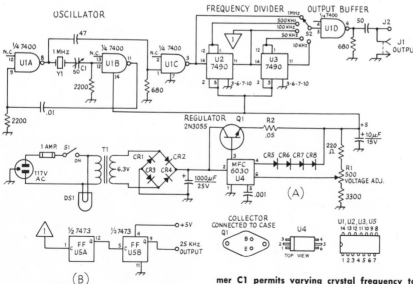

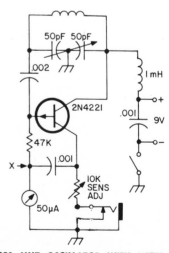

1-MHZ STANDARD—Output of crystal-controlled mvbr is divided to 500, 100, and 10 kHz by decade counters U2 and U3. For 25-kHz output, add two flip-flops as in (B). Trimmer C1 permits varying crystal frequency to align with WWV. Power supply uses Motorola IC voltage regulator U4. All diodes are HEP156 or equivalent.—D. A. Blakeslee, Double Standards, QST, April 1972, p 13–17.

1–250 MHZ OSCILLATOR WITH METER—Frequency range is covered with six plug-in coils, winding data for which is given in article. Used as grid-dip meter. 10K pot serves as sensitivity control. Jack is for headphones.—K. Brown, The Indicating Oscillator, 73, Sept. 1970, p 29–31.

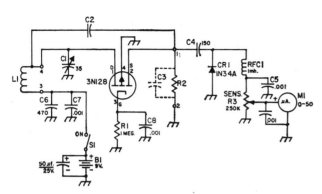

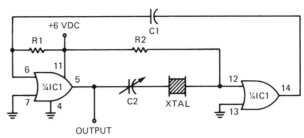

FET DIPPER—Single fet operating from 9-V battery gives portability and performance comparable to that of grid-dip meter. Article gives winding data for eight L1 plug-in coils and values of C2 and R2 for covering 1.8 to 150 MHz in eight ranges.—L. G. McCoy, A Field-Effect Transistor Dipper, QST, Feb. 1968, p 24–27.

CRYSTAL CALIBRATOR—Output is square wave with high harmonic content, at frequency determined by crystal. Only half of HEP570 IC is used; for dual-frequency calibrator, use both halves. R1 and R2 are 56K, C2 is 9–35 pF, and C1 is 1,000 pF for 100-kHz crystal, 430 pF for 500 kHz, and 39 pF for 1 MHz.—"Radio Amateur's IC Projects," Motorola Semiconductors, Phoenix, AZ, HMA-36, 1971.

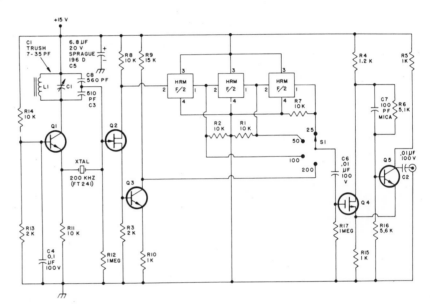

RECEIVER CALIBRATOR—Provides choice of 200, 100, 50, and 25 kHz calibration markers for use with general-coverage h-f receivers. 200-kHz crystal operates in series mode in Colpitts oscillator Q1. Q2 drives chain of three Hughes HRM-F/2 MOS binary dividers or equivalent, while Q3 drives Schmitt trigger Q4-Q5 when 200-kHz output is desired. Requires regulated supply. Q1, Q3, and Q5 are 2N3646 or HEP50; Q2 is NPF102 or HEP802; Q4 is 3N128. L1 is 2 mH having pot core.—H. Olson, The Ball of Wax—A Calibrator, 73, Nov. 1969, p 84–86.

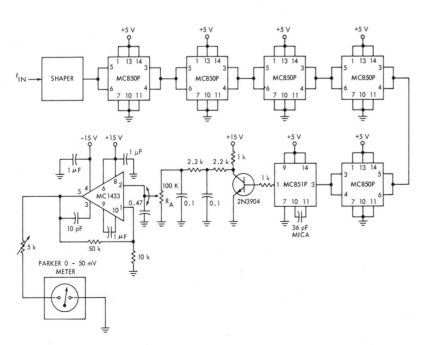

1–30 MHZ FREQUENCY METER—Uses seven IC's in linear frequency-voltage converter driving 0–50-mV meter having direct-reading frequency scale. Linearity is better than 5% over entire range. Minimum input voltage level is 2 V p-p. Report includes shaping circuit required at input, design equations, and detailed description of operation. Divide-by-32 binary uses five Motorola MC850 high-speed pulsed binary IC's feeding MC851P mono mvbr. Mono output is input to 2N3904 low-pass filter acting as averaging circuit that feeds MC1433 opamp biased to give d-c voltage gain of 5. Pot RA is adjusted to give 10-V d-c level at output of opamp for 30-MHz input, and 5K pot in series with meter is used to adjust current to 5 mA when opamp output is 10 V.—"Integrated Circuits for High Frequency to Voltage Conversion," Motorola Semiconductors, Phoenix, AZ, AN-297, 1971.

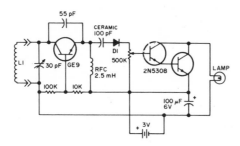

DIPPER—Simple oscillator and Darlington amplifier drive No. 48 pilot lamp that dims to indicate absorption of energy when oscillator is tuned to same frequency as circuit under test. Operates from pair of penlight cells. Oscillator will work up to 120 MHz. Vernier dial for tuning capacitor can be calibrated with any receiver having accurate tuning dial covering frequency range of input.—E. Babudro, The Dip Light, 73, March 1970, p 68–69.

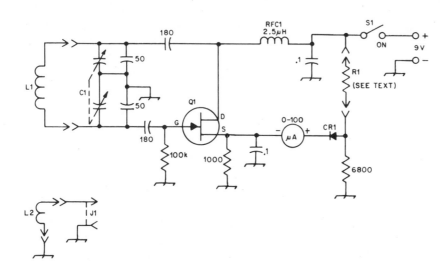

FET DIPPER—Will give indication of resonance down to 0.5 μA on meter when coupled to circuit 3 inches away. Article gives approximate winding data for eight plug-in coils covering frequency range of 1.5 to 30 MHz. With L2 added as pickup loop, jack J1 provides sampling of oscillator frequency for readout on frequency counter. Q1 is MPF102. CR1 is 1N34A. R1 is part of each plug-in coil, whose value is found experimentally by first using variable resistor and adjusting for about 95% of full-scale reading on meter; its value will be in range from 10K to 33K.—P. Lumb, High-Accuracy FET Dipper, QST, June 1972, p 46–47.

100-KHZ CRYSTAL CALIBRATOR—Provides usable signals well up into MHz range for receiver calibration. Temperature and frequency stability are excellent, especially if alkaline battery is used.—F. H. Tooker, Build FET Crystal Calibrator, Popular Electronics, March 1968, p 56–58.

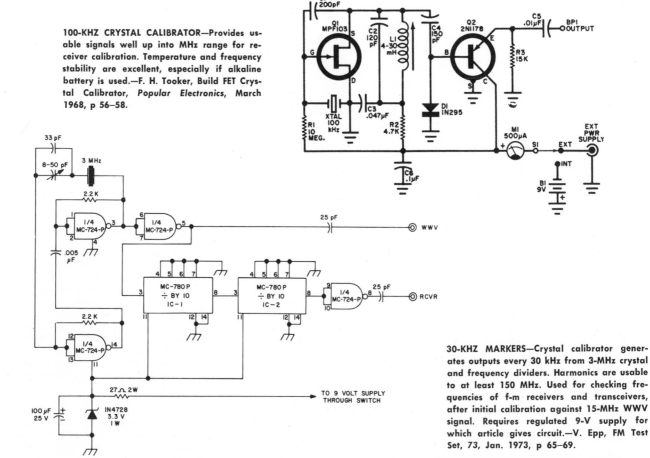

30-KHZ MARKERS—Crystal calibrator generates outputs every 30 kHz from 3-MHz crystal and frequency dividers. Harmonics are usable to at least 150 MHz. Used for checking frequencies of f-m receivers and transceivers, after initial calibration against 15-MHz WWV signal. Requires regulated 9-V supply for which article gives circuit.—V. Epp, FM Test Set, 73, Jan. 1973, p 65–69.

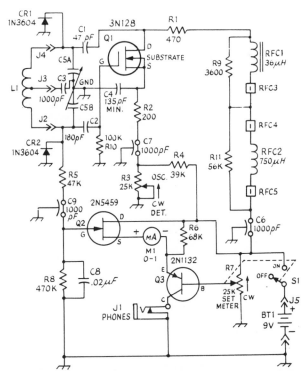

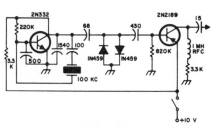

100-KHZ CALIBRATOR—Crystal oscillator and single amplifier stage provide markers 100 kHz apart in excess of 100 MHz.—D. J. Lynch, SB-33 Modification, 73, Nov. 1969, p 90 and 92—93.

1.7–300 MHZ DIP METER—Used in Millen 90652 solid-state dip meter. Operates from self-contained batteries and produces dip in output meter reading when power is absorbed by external circuit at a particular frequency as tuning dial is rotated. Uses seven plug-in coils to cover frequency range with two-gang tuning capacitor geared to drum dial. Article covers design problems encountered.—F. D. Lewis, The Anatomy of a Solid-State Dipper, QST, Dec. 1972, p 23—27.

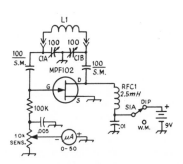

1.5–50 MHZ DIP METER—Checks resonant circuits by absorbing energy at specific frequency. Uses fet with meter connected to measure changes in source current. Use plug-in coils for L1. Calibrate tuning dial of C1 by monitoring output signal of dipper with calibrated receiver. For uhf, change transistor to 2N4416 or similar.—"The Radio Amateur's Handbook," American Radio Relay League, Newington, CT, 49th Ed., 1972, p 515—516.

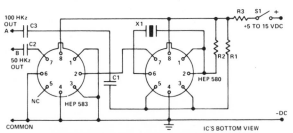

50 OR 100 KHZ—Can be used as source of accurate square-wave signals for test purposes or to generate markers for communications receivers. Output at A or B is 1 V p-p when using 9-V d-c supply. X1 is 100-kHz crystal. R1 and R2 are 100K, R3 is 1K, C1 is 0.001 μF, and C2 and C3 are 0.1 μF.—"Tips on Using IC's," Motorola Semiconductors, Phoenix, AZ, HMA-32, 1972.

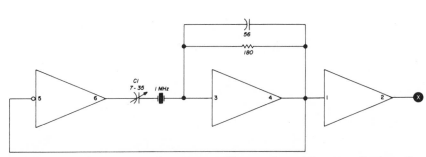

IC MARKER—Uses hex inverter, such as Motorola MC3008 or TI SN74H04, to generate harmonic-rich 1-MHz output for calibrating communication receiver to above 30 MHz.—G. Tillotson, General-Coverage Receiver Frequency Calibrator, Ham Radio, Dec. 1971, p 28—29.

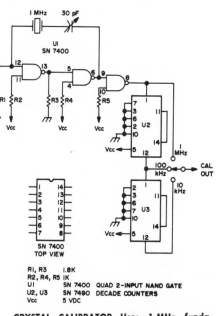

CRYSTAL CALIBRATOR—Uses 1-MHz fundamental crystal and provides direct output of 1 MHz along with divided outputs of 100 and 10 kHz.—Circuits, 73, April 1972, p 107.

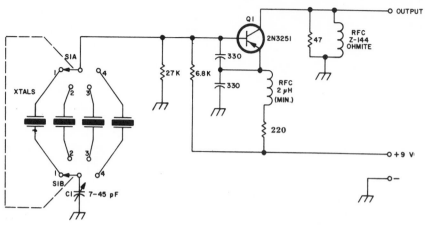

F-M TRANSMITTER OSCILLATOR—Single-transistor crystal oscillator Q1, with crystal switching, is rich in harmonics up to about 450 MHz. Used in conjunction with crystal calibrator of test set for f-m receivers and transmitters, deviation checker, and signal generator with calibrated output. C1 adjusts crystal to exact frequency desired. Output level can be adjusted with 15K linear-taper series rheostat.—V. Epp, FM Test Set, 73, Jan. 1973, p 65–69.

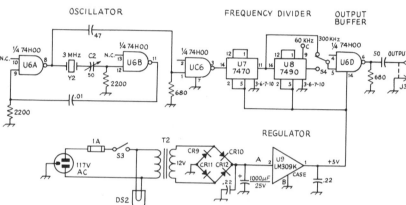

3-MHZ STANDARD—Uses high-speed gate package as oscillator and single IC as complete voltage regulator in power supply. Oscillator output is divided to 300 and 30 kHz, selected by S4, with optional 60-kHz output. All diodes are HEP156.—D. A. Blakeslee, Double Standards, QST, April 1972, p 13–17.

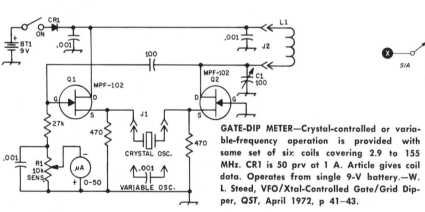

GATE-DIP METER—Crystal-controlled or variable-frequency operation is provided with same set of six coils covering 2.9 to 155 MHz. CR1 is 50 prv at 1 A. Article gives coil data. Operates from single 9-V battery.—W. L. Steed, VFO/Xtal-Controlled Gate/Grid Dipper, QST, April 1972, p 41–43.

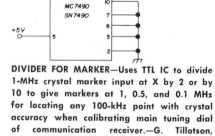

DIVIDER FOR MARKER—Uses TTL IC to divide 1-MHz crystal marker input at X by 2 or by 10 to give markers at 1, 0.5, and 0.1 MHz for locating any 100-kHz point with crystal accuracy when calibrating main tuning dial of communication receiver.—G. Tillotson, General-Coverage Receiver Frequency Calibrator, Ham Radio, Dec. 1971, p 28–29.

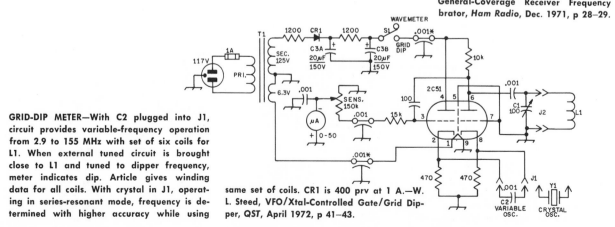

GRID-DIP METER—With C2 plugged into J1, circuit provides variable-frequency operation from 2.9 to 155 MHz with set of six coils for L1. When external tuned circuit is brought close to L1 and tuned to dipper frequency, meter indicates dip. Article gives winding data for all coils. With crystal in J1, operating in series-resonant mode, frequency is determined with higher accuracy while using same set of coils. CR1 is 400 prv at 1 A.—W. L. Steed, VFO/Xtal-Controlled Gate/Grid Dipper, QST, April 1972, p 41–43.

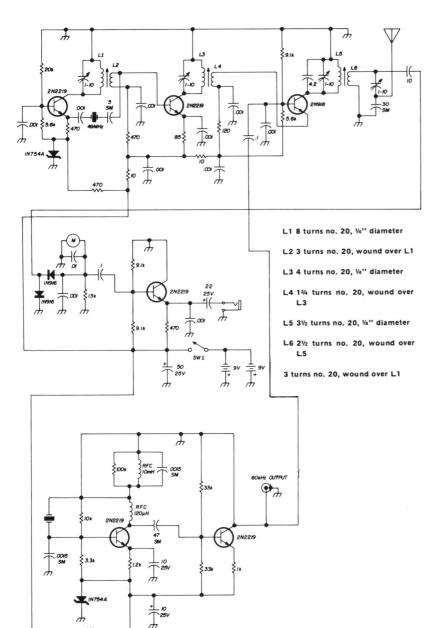

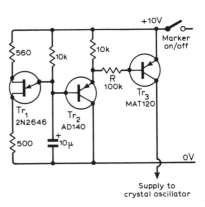

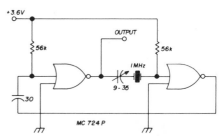

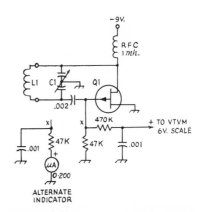

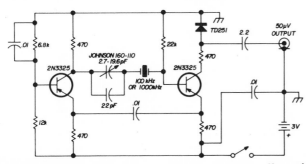

L1 8 turns no. 20, ¼" diameter

L2 3 turns no. 20, wound over L1

L3 4 turns no. 20, ¼" diameter

L4 1¾ turns no. 20, wound over L3

L5 3½ turns no. 20, ¼" diameter

L6 2½ turns no. 20, wound over L5

3 turns no. 20, wound over L1

MARKER IDENTIFIER—Supply to 500-kHz crystal oscillator is switched on and off by ujt at rate of around 8 Hz to make recognition of marker signals easier in crowded bands. Creates slight warble in addition when bfo is in use. Ideal for marker recognition between 18 and 30 MHz, where harmonics of crystal are quite weak.—G. Dann, Crystal Marker Identification, *Wireless World*, Feb. 1972, p 77–78.

1 MHZ WITH NOR GATES—Simple dual-gate IC arrangement provides square-wave output for receiver calibration.—E. Noll, Digital IC Oscillators and Dividers, *Ham Radio*, Aug. 1972, p 62–67.

2-METER F-M FREQUENCY METER—High-accuracy heterodyne-type meter provides crystal-controlled frequency markers. Covers range of 146.94 MHz to below 145 MHz with accuracy within 15 Hz immediately after calibration and within 100 Hz long-term. 49-MHz crystal oscillator feeds buffer-tripler providing 147 MHz, coupled to 2N918 mixer. Lower crystal oscillator, generating 60 kHz with high harmonic output, drives base of mixer through buffer, to modulate 147-MHz signal. First lower sideband of mixer is 146.94 MHz, and strong signals are available at 60-kHz intervals to below 144 MHz for 19-inch whip that radiates signals for receiver calibration. —C. A. Baldwin, Two-Meter FM Frequency Meter, *Ham Radio*, Jan. 1971, p 40–43.

GATE DIPPER—Uses Fairchild 2N4342, 2N4360, or equivalent p-channel jfet. C1 can be transistor radio type or dual-365-pF. L1 uses plug-in coils to cover range of 3 to 200 MHz. Used same as grid-dip oscillator for determining resonant frequency.—W. Hayward, Gate-Dip Oscillator, *QST*, Sept. 1967, p 45.

HARMONIC GENERATOR—With either 100 or 1,000 kHz crystal in reference oscillator, circuit will generate harmonics through 1,296 MHz.—C. Spurgeon, Harmonic Generator, *Ham Radio*, Oct. 1970, p 76.

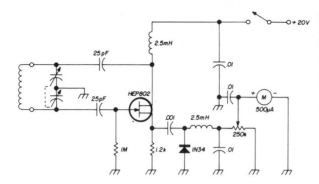

FET GRID-DIP OSCILLATOR—High input impedance of fet provides higher sensitivity for obtaining dip than is possible with tube or other transistor circuits. Supply can be two 9-V transistor batteries in series. Article covers other methods of modernizing tube-type gdo's—P. A. Lovelock, Solid-State Conversion of the GDO, *Ham Radio*, June 1970, p 20—23.

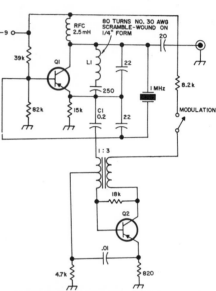

1—MHZ CRYSTAL—Provides distinctive markers up to 30 MHz, modulated at about 1,000 Hz. Particularly useful for band-edge marking. Modulation assists in identifying marker at higher frequencies where harmonics are quite weak.—J. Fisk, "73 Useful Transistor Circuits," 73 Inc., Peterborough, NH, 1967, p 28A.

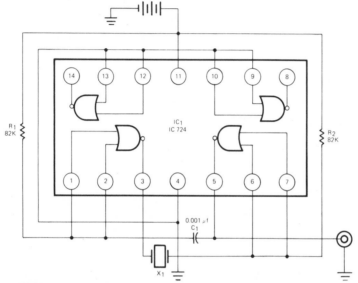

100-KHZ CRYSTAL CALIBRATOR—International Rectifier IC724 and 100-kHz crystal X1 serve with minimum of additional components as secondary frequency standard for calibrating receivers, transmitters, and other test equipment. Provides output of 3-V p-p at fundamental frequency, for use with cro. Battery voltage is critical and should be checked frequently; use two AA cells in series.—"Hobby Projects," International Rectifier, El Segundo, CA, Vol. II, p 62–63.

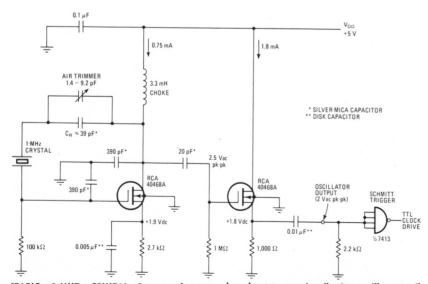

STABLE 1-MHZ CRYSTAL—Output changes less than 1 MHz over supply range of 3 to 9 V. Uses modified Pierce oscillator and isolating source-follower, with Schmitt trigger serving as TTL interface. Output is rich in r-f harmonics, for beating with 10-MHz WWV broadcast to permit adjusting oscillator easily within 0.1 Hz of nominal operating frequency. Circuit can then be used as secondary frequency standard.—T. King, Stable Crystal Oscillator Works over Wide Supply Range, *Electronics*, June 21, 1973, p 112.

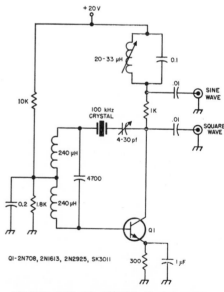

100-KHZ SINE OR SQUARE—Single-transistor Hartley crystal oscillator includes fine frequency adjustment control in series with crystal for zero-beating with WWV. Large amount of feedback drives collector of Q1 from cutoff to saturation for square-wave output, while sine-wave output is developed across tunable high-Q tank.—J. Fisk, "73 Useful Transistor Circuits," 73 Inc., Peterborough, NH, 1967, p 27A.

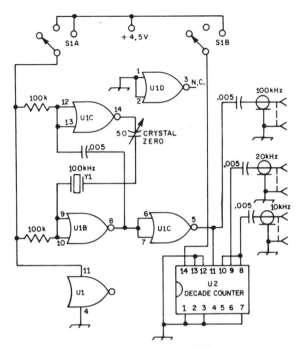

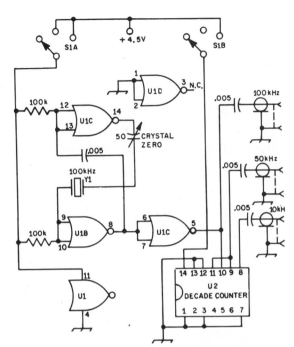

20-KHZ AND 10-KHZ MARKERS—Low-cost 100-kHz crystal mvbr provides 100, 20, and 10 kHz markers well into h-f range. U1 is quad 2-input NOR gate such as International Rectifier IC724-C, and U2 is Fairchild MS19350 decade counter.—A. C. Beresford, An Inexpensive Secondary Frequency Standard, QST, May 1972, p 37–39.

10-KHZ MARKERS—Low-cost 100-kHz crystal mvbr provides 100, 50, and 10-kHz markers well into h-f range. U1 is quad 2-input NOR gate such as International Rectifier IC724-C, and U2 is Fairchild MS19350 decade counter.—A. C. Beresford, An Inexpensive Secondary Frequency Standard, QST, May 1972, p 37–39.

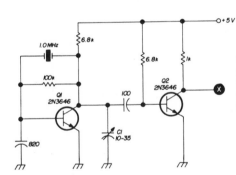

MARKER—Used to calibrate communication receiver. Output of 1-MHz crystal is rich in harmonics, to above 30 MHz.—G. Tillotson, General-Coverage Receiver Frequency Calibrator, Ham Radio, Dec. 1971, p 28–29.

CHAPTER 62
Measuring Circuits—General

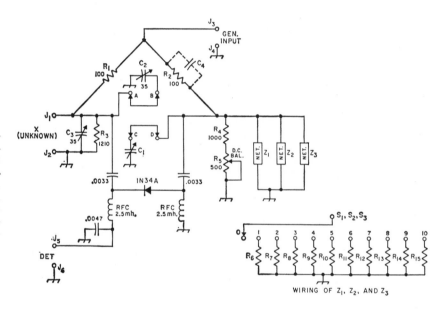

R-F ADMITTANCE BRIDGE—Can be used for measuring 10–1,000 ohm r-f resistance, 0–500 pF capacitance, 1–50 μH inductance, and complex impedances. Article covers construction and calibration. General input is transmitter or other 7,120-kHz frequency source. Meter or other detector detects null at bridge balance. C1 is about 500 pF. Table in article gives values of R6–R15 for three r-f conductance ranges.—F. Cherubini, An Admittance Bridge for R.F. Measurements, QST, Sept. 1967, p 30–33.

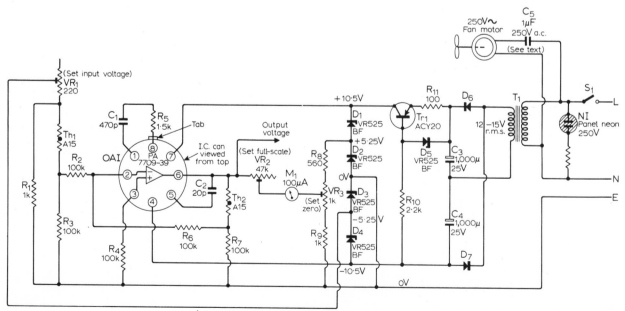

HYGROMETER—Uses Philco-Ford opamp as computing element. VR1 sets negative input voltage across type A15 wet thermistor Th1 and R3. Th2 is dry thermistor. Opamp output is capable of driving load up to 600 ohms to operate external humidity control switch. Capacitor is used in series with shaded-pole a-c motor providing continuous ventilation of wet thermistor, to reduce motor voltage and prevent overheating. Article covers construction, calibration, and use. Meter has nonlinear scale reading 0–100% humidity. Accuracy is better than 5%. D6 and D7 are A.E.I. SJ403-F diodes.—D. Bollen, A Thermistor Hygrometer, Wireless World, Dec. 1969, p 557–561.

AMPLIFIED VU METER—Will check signal level anywhere in audio system, amost down to microphone level. Sensitivity control R1 permits handling high power levels. Q1 presents high input impedance for meter circuit, and Q3 provides low output impedance for standard volume-unit meter. To use as signal tracer, replace M1 with 500 to 3,000 ohm headphones.—H. Friedman, "99 Electronic Projects," Howard W. Sams, Indianapolis, IN, 1971, p 101–102.

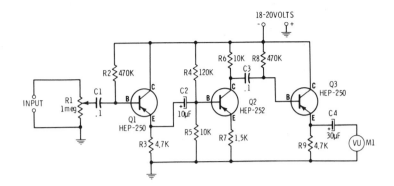

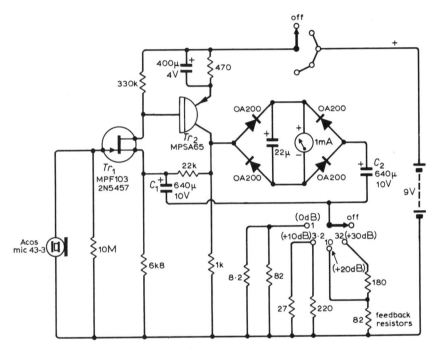

LOUDSPEAKER RESPONSE METER—Used to obtain frequency response curve of loudspeaker system while experimenting with acoustic loading, damping material, and input impedance. With crystal microphone cartridge, meter itself has flat response from 20 to 5,000 Hz. Use of pnp Darlington transistor TR2 as amplifier following input fet avoids shunting drain load resistor of fet.—J. L. Linsley Hood, Sound Pressure-Level Meter, *Wireless World*, April 1972, p 167–168.

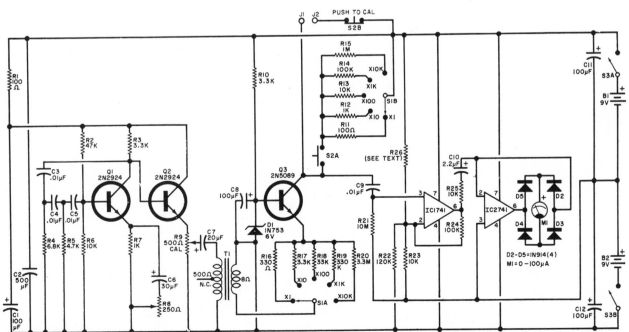

IMPEDANCE METER—Lowest impedance range is 0–100 ohms and highest is 0.1–1 meg. Measurements are made at 1 kHz, generated by oscillator Q1. Ranges can be calibrated with any values of precision resistors because meter has linear scale. Will also check R-L or R-C networks provided they have no series capacitors. Article covers construction and calibration.—C. D. Rakes, Build an Impedance Meter, *Popular Electronics*, Oct. 1972, p 68–71.

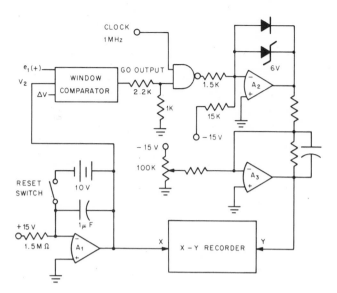

PROBABILITY DENSITY ANALYZER—Uses window comparator for which basic circuit is also given, though without values. GO output goes high each time input signal is inside window width centered on V2. If clock is also high, output of first NAND gate goes low. A2 is opamp used as comparator to generate precision voltage levels for R-C averaging filter used with remainder of circuit to drive recorder. Opamps can be Burr-Brown.—J. G. Graeme, G. E. Tobey, and L. P. Huelsman, "Operational Amplifiers," McGraw-Hill Book Co., New York, 1971, p 364–366.

ION CHAMBER—Measures polarity and amount of ionization present in air. Ions detected by ion chamber develop small voltage drop across very high resistance of R3, for amplification by Victoreen electrometer tube V1, which operates transistor bridge Q1-Q2 having polarity-indicating meter. Article covers adjustment and operation.—H. Burgess, Make Your Own Ion Chamber, *Popular Electronics*, Nov. 1969, p 31–35.

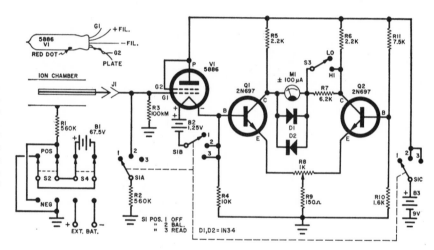

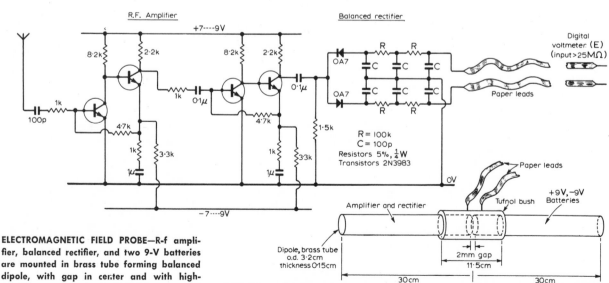

ELECTROMAGNETIC FIELD PROBE—R-f amplifier, balanced rectifier, and two 9-V batteries are mounted in brass tube forming balanced dipole, with gap in center and with high-impedance paper leads (Teledeltos pen recording paper) going from gap ends to high-impedance digital voltmeter. Used to measure near electromagnetic field in frequency range of 150 kHz to 30 MHz. Paper leads with resistance of about 25,000 ohms per foot avoid perturbation of field being measured. Has also been used to check electric field distribution in aperture of 11-dB pyramidal horn at 1,000 MHz.—J. Thickpenny, Electric Field Probe, *Wireless World*, June 1970, p 293–294.

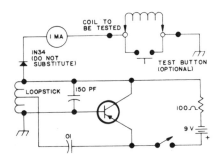

INDUCTANCE METER—After calibration with known inductance values, meter will indicate inductance values of chokes and coils accurately over range of 0.25 to 10 mH. Almost any pnp transistor may be used, but diode type is critical. To use, connect unknown coil, push test button and adjust control for 1 mA (corresponding to 0 mH since scale is like that of ohmmeter), then release button and read inductance value.—J. Clack, A Direct Reading Inductance Meter . . . The "Henry-ometer," 73, Dec. 1972, p 88.

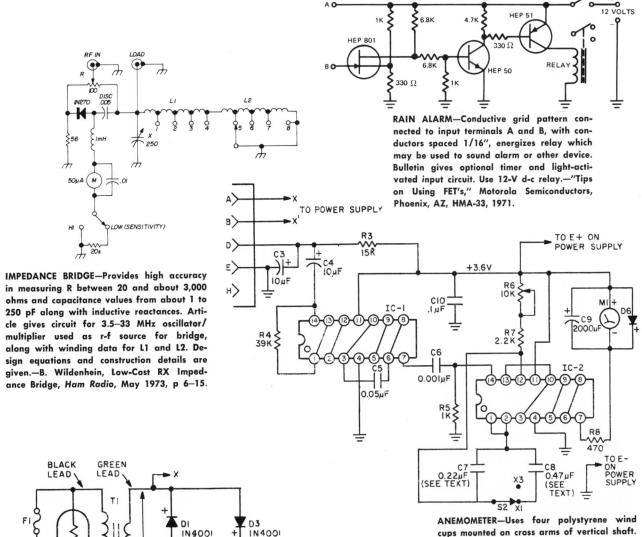

IMPEDANCE BRIDGE—Provides high accuracy in measuring R between 20 and about 3,000 ohms and capacitance values from about 1 to 250 pF along with inductive reactances. Article gives circuit for 3.5–33 MHz oscillator/multiplier used as r-f source for bridge, along with winding data for L1 and L2. Design equations and construction details are given.—B. Wildenhein, Low-Cost RX Impedance Bridge, Ham Radio, May 1973, p 6–15.

RAIN ALARM—Conductive grid pattern connected to input terminals A and B, with conductors spaced 1/16", energizes relay which may be used to sound alarm or other device. Bulletin gives optional timer and light-activated input circuit. Use 12-V d-c relay.—"Tips on Using FET's," Motorola Semiconductors, Phoenix, AZ, HMA-33, 1971.

ANEMOMETER—Uses four polystyrene wind cups mounted on cross arms of vertical shaft. Also on shaft is 2.25-inch-diameter disc having eight equally spaced holes around edge for interrupting light on Clairex C L-703L photocell connected to inputs D and E. Circuit converts resulting pulses into d-c voltage driving 1-mA meter calibrated to read wind velocity. Both IC's are Motorola MC789P hex inverters. Q1 is 2N697. Terminals X and X' are for 6.3-V pilot lamp near photocell. Change meter scale to read 0–30 mph. Meter reading for range X1 should be three times reading for range X3. Calibrate in auto having reasonably accurate speedometer.—E. A. Morris, "Windy" the Wind Gauge, Science and Electronics, April-May 1970, p 29–38 and 99–100.

IMPEDANCE BRIDGE—Will measure reactive and resistive components of unknown impedance throughout high-frquency range. Applications include measurement of antenna impedance, characteristic impedance and electrical length of coaxial lines, and input impedance of r-f components and amplifiers over range of 2 to 30 MHz. Operates from single 9-V transistor radio battery. Uses diode noise generator and 3-stage amplifier to feed r-f bridge. T1 is Micro-Metals T-37-10 toroid core having 9-mm OD and 8-turn trifilar windings. Receiver having S meter is connected to DETECTOR terminal and tuned to frequency of measurement. Bridge is balanced for noise null as indicated on S meter. —G. Pappot, Noise Bridge for Impedance Measurements, *Ham Radio*, Jan. 1973, p 62—64.

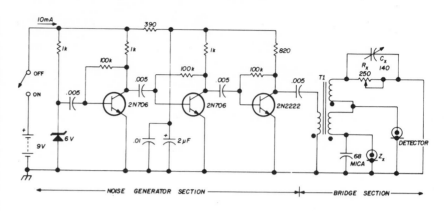

HYGROMETER—Uses 400-Hz twin-T oscillator as source for bridge having humidity sensor in one leg and transistor driving meter to indicate ambient relative humidity values up to 100%. Sensor construction details are given in article. Fiberglass cloth clamped between two pairs of brass plates is dipped in solution of lithium chloride to give sensor whose resistance varies with humidity. Article covers construction and calibration. Use 11K for R10, which should give about ¾-scale reading on meter for new battery when S2 is up for battery test.—J. Giannelli, Electro-Chemical Hygrometer, *Popular Electronics*, Oct. 1972, p 33—35.

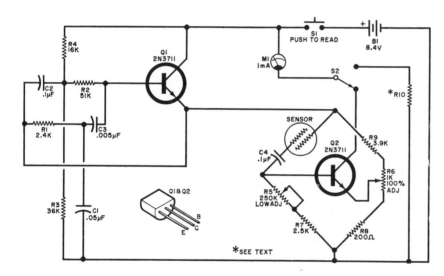

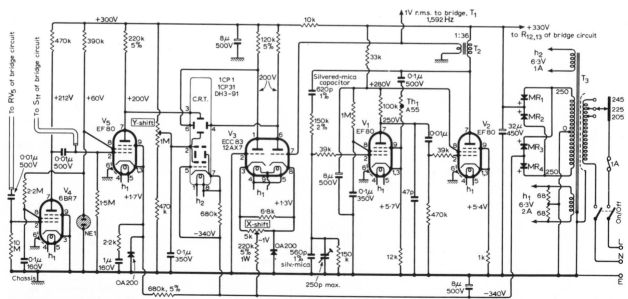

INDICATOR FOR R-L-C BRIDGE—Used with universal bridge to indicate amplitude and phase of bridge unbalance on screen of miniature crt. Y amplifier has voltage gain of about 8,700, making input of 2.5 mV rms produce 1-cm deflection of crt spot. Article gives construction details.—L. Nelson-Jones, Universal Component Bridge, *Wireless World*, Dec. 1968, p 434—436.

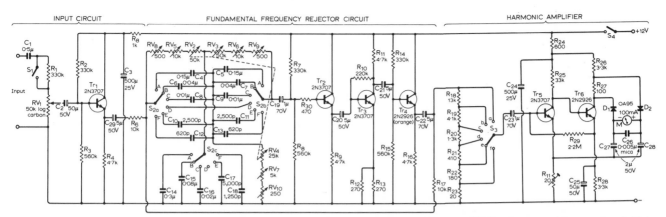

AUDIO DISTORTION METER—Covers 0.1 to 100% harmonic distortion in five ranges, with each range having frequency ratio of about 4:1 to cover audio spectrum from 20 to 20,-000 Hz. Uses twin-T filter in combination with negative feedback provided by transistor stages Tr2, Tr3, and Tr4 to give over 70 dB attenuation of fundamental frequency without significantly attenuating second and third harmonics. Article covers construction and adjustment.—L. Haigh, Distortion Factor Meter, *Wireless World*, July 1969, p 317–320.

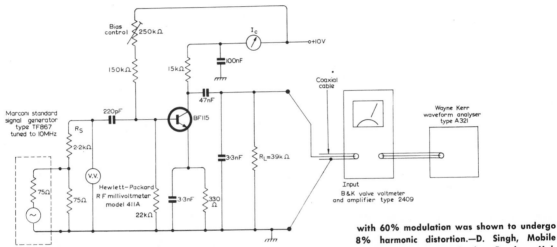

A-M DEMODULATION DISTORTION—Used in determining optimum i-f amplifier output for transistor used as detector for satisfactory demodulation with minimum percentage of harmonic distortion. With BF115 detector transistor shown in circuit, 50-mV rms signal with 60% modulation was shown to undergo 8% harmonic distortion.—D. Singh, Mobile 166 MHz A.M. Communications Receiver, *Mullard Technical Communications*, Jan. 1968, p 14–29.

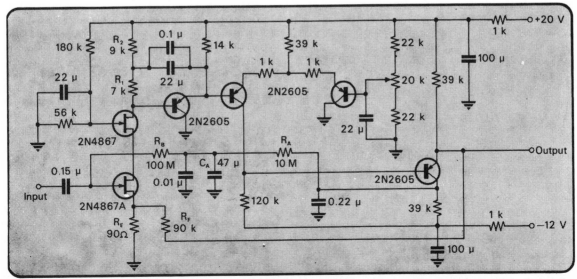

MAGNETOMETER AMPLIFIER—Low-noise fet-input amplifier is used with search coil on spacecraft to measure variations in interplanetary magnetic fields. Covers 1 Hz to 100 kHz. Power drain is 24 mW. Designed for operating temperature range of −30 to +60 C.—S. Cantarano and G. V. Pallottino, A Low-Noise FET Amplifier For A Spaceborne Magnetometer, *Electronic Engineering*, Sept. 1970, p 57–60.

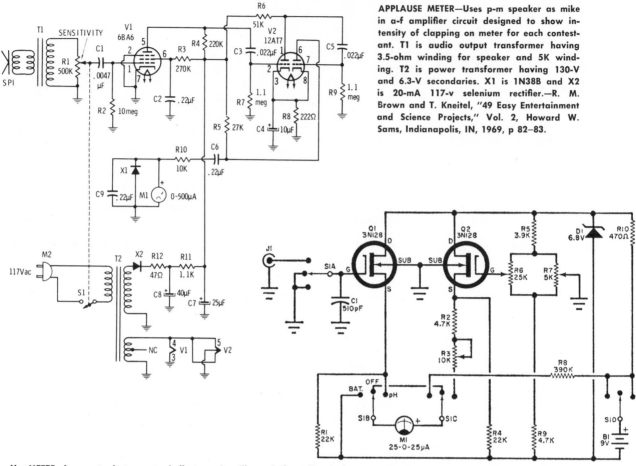

APPLAUSE METER—Uses p-m speaker as mike in a-f amplifier circuit designed to show intensity of clapping on meter for each contestant. T1 is audio output transformer having 3.5-ohm winding for speaker and 5K winding. T2 is power transformer having 130-V and 6.3-V secondaries. X1 is 1N38B and X2 is 20-mA 117-v selenium rectifier.—R. M. Brown and T. Kneitel, "49 Easy Entertainment and Science Projects," Vol. 2, Howard W. Sams, Indianapolis, IN, 1969, p 82–83.

pH METER—Low-cost instrument indicates with high accuracy amount of acidity or alkalinity in a solution. Includes battery-condition switch; with S1 in BAT position, battery is still good if reading is between pH 12.5 and 14 on scale. pH probe plugged into J1 is available at about $20 from Analytical Measurements, Chatham, NJ.—R. C. Dennison, Solid-State PH Meter, *Popular Electronics*, Nov. 1968, p 33–35 and 40–42.

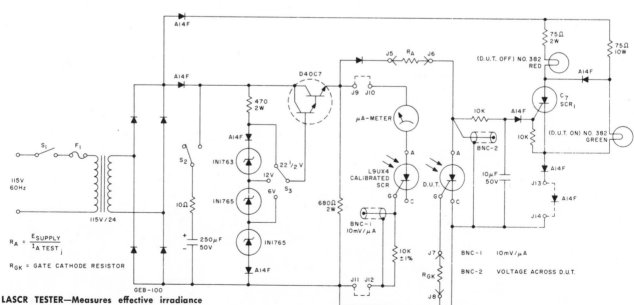

LASCR TESTER—Measures effective irradiance needed to trigger lascr or other light-sensitive device under test, by comparison with performance of L9UX4 calibrated lascr. Operates from regulated power supply. Current through light source (not shown) needed to trigger device under test is first established. Light emitter is then positioned at same distance and manner opposite calibrated lascr and effective irradiance is determined with microammeter and lascr calibration curve. If desired, meter scale can be calibrated to read directly in watts per square centimeter.—S. R. Korn, "How to Evaluate Light Emitters and Optical Systems for Light Sensitive Silicon Devices," General Electric, Semiconductor Products Dept., Auburn, NY, No. 200.59, p 11–12.

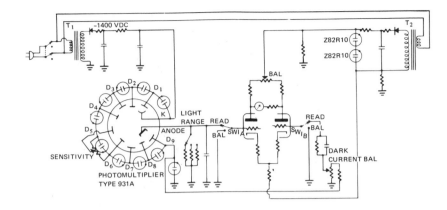

STARLIGHT PHOTOMETER—Uses Z84R2 Signalite voltage regulators D1–D9 for accurate voltage regulation of photomultiplier operating from unregulated d-c supply. Output indication is obtained with low-cost neon relaxation oscillator in which output current charges capacitor linearly. Accuracy approaches 0.01 starlight magnitude. Blink rate of neon is directly related to illumination level. Conventional vtvm circuit at right of SW1 may be used for meter indication if desired; use trial and error to find range switch resistor values that match vtvm in use.—M. I. Distefano, T. R. Cram, and T. E. Houck, Versatile, Low-Cost, Very Sensitive Photometer, *Electromechanical Design*, Nov. 1971, p 42–44.

LAWN-WATERING METER—Gives indication that lawn needs water before grass turns yellow, without wasting money by watering excessively. Probes are made from ⅛-inch brass brazing rods spaced ½ inch apart, held in block of insulating material. Probes are 7 inches long, with ends that go into soil sharpened. Choose value for R1 that gives full meter deflection when probes are shorted. In general, reading below half-scale when probes are inserted in lawn means watering may be needed. Experimentation is required because soil resistance varies with pH.—R. M. Brown and T. Kneitel, "101 Easy Test Instrument Projects," Howard W. Sams, Indianapolis, IN, 1968, p 23–24.

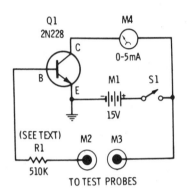

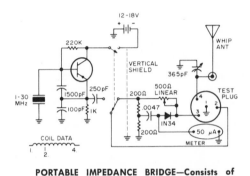

PORTABLE IMPEDANCE BRIDGE—Consists of common impedance bridge coupled to and driven by untuned crystal oscillator using any fundamental crystal between 1 and 30 MHz. Requires four plug-in coils to cover 3–30 MHz; article gives winding data. Transistor is GE9 or similar. Connect unknown impedance between pins 1 and 3 and read its value on calibrated dial. To check crystal activity, connect test crystal and read relative activity on meter. To use as field-strength meter, leave power off and connect whip antenna according to instructions given in article.—W. W. Pinner, The Third Hand . . . Tester, *73*, Nov. 1972, p 65 and 68.

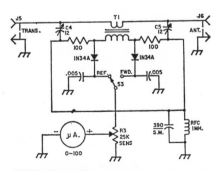

R-L-C BRIDGE—Amplitude and phase of bridge unbalance are indicated on cro. Covers resistances of 1 ohm to 1 meg and capacitances of 10 pF to 100 μF with better than 1% accuracy, and inductances of 10 μH to 100 H with almost same accuracy. Article includes circuit of cro using 1CP31 miniature crt which can be directly connected to bridge.—L. Nelson-Jones, Universal Component Bridge, *Wireless World*, Dec. 1968, p 434–436.

VSWR INDICATOR—Low-power circuit is designed especially for use with c-w transmitters operating at around 2 W input power to last stage. T1 primary is 2 turns No. 20 and secondary is 50 turns No. 30, wound on Amidon T-68-2 toroid core. Use 50-ohm dummy load during calibration.—D. DeMaw, The "QRP 80-40" C.W. Transmitter, *QST*, June 1969, p 11–16.

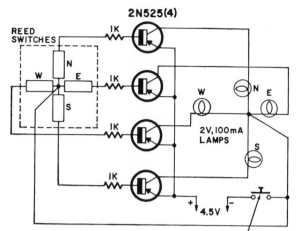

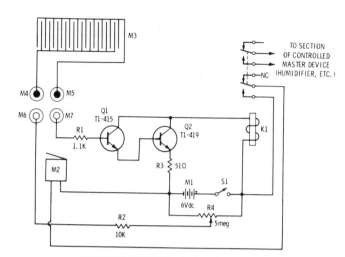

WIND INDICATOR—Bar magnet mounted on weather vane rotates over four reed switches which are normally open and close each time magnet passes over as wind fluctuates around average direction. Leads from switches are run through cable to transistors driving indicator lamps at remote point. For reading, switch is held down until lamp corresponding to N-E-S-W direction closest to average flickers on as wind varies. Hermetically sealed switches are not affected by atmospheric conditions.—R. F. Scott, Tech Topics, *Radio-Electronics*, Oct. 1969, p 63 and 68–69.

HUMIDITY CONTROL—Standard moisture-sensor plate M3 (not water sensor) completes connection between R1 and R2 to energize 6-V d-c relay K1 whenever excess of moisture builds up. Choose location for M3 at which dust accumulation is minimized. Relay can be connected to sound alarm or turn on electric heater or dehumidifier. M2 is 6-V d-c buzzer.—R. M. Brown and T. Kneitel, "49 Easy Entertainment and Science Projects," Vol. 2, Howard W. Sams, Indianapolis, IN, 1969, p 85–87.

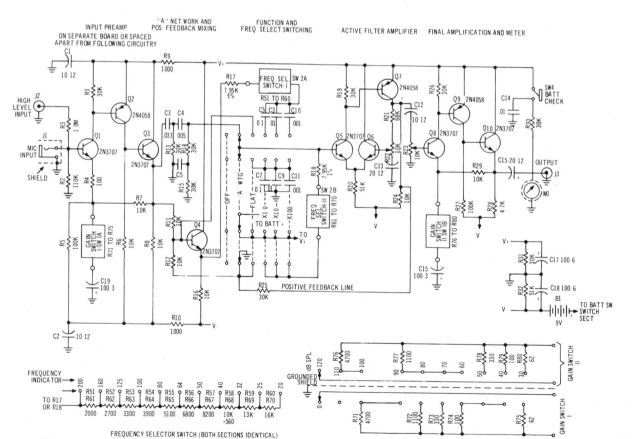

SOUND-LEVEL METER—Covers 20 to 20,000 Hz in three ranges, with frequency accuracy of 10%. Will measure sounds down to 20 dB SPL and third-octave noise down to 0 dB SPL when used with microphone sensitivity of —60 dB re 1 V/μbar. Includes bass filter to give standard "A" weightings. Ideal for measuring room acoustics. Article covers measuring techniques in detail. Preamp Q1-Q2-Q3 with overall gain adjustable from 1 to 200 feeds combination amplifier and active filter Q4. Remaining amplifier stages provide extra sensitivity for driving standard vu meter M1.—J. D. Griesinger, A Sound-Level Meter, *Audio*, Dec. 1970, p. 28, 30, 32, 34, 36, and 38.

INDUCTANCE CHECKER—Used with short-wave receiver to measure r-f inductance values from 0.3 μH to 7 mH. Consists of two-terminal oscillator to which unknown coil and known capacitor are connected in parallel. Receiver is placed near oscillator for use in determining frequency. Verify correct tuning by placing hand near tuned circuit and listening for frequency change. Oscillator range is 600 kHz to above 30 MHz, which can be covered with only two standard capacitors—10 pF and 100 pF. With 10 pF and 3 μH, frequency is 30 MHz, and 600 kHz with 7 mH. With 100 pF, 0.3 μH gives 30 MHz and 700 μH gives 600 kHz.—J. A. Rolf, Quickie Inductance Checker, *Popular Electronics*, Aug. 1973, p 98–99.

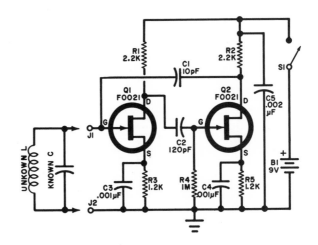

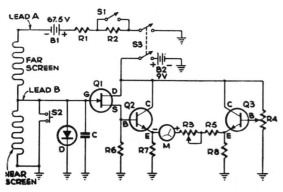

R1—10,000 ohm, ½-watt carbon resistor
R2—10 megohm, ½-watt carbon resistor
R3—3,000 ohm carbon potentiometer
R4—10,000 ohm carbon potentiometer
R5—470 ohm, ½-watt carbon resistor
R6—22,000 ohm, ½-watt carbon resistor
R7, R8—470 ohm, ½-watt carbon resistor
C—2ufd, 100-volt paper capacitor
D—Zener diode. Motorola HEP-104

BULLET TIMER—Circuit measures time required for bullet to pass between two screens placed exactly 1 ft apart. When bullet cuts first screen, B1 starts charging C through R1. When bullet cuts second screen, battery circuit is opened and charging stops; voltage across C, proportional to time of flight, is then measured with voltmeter circuit Q1-Q2-Q3. Article covers calibration of indicating meter. For bullets slower than 1,000 fps, place screens 6 inches apart. Use any 1-mA meter. B2 is 8.4-V VS146X mercury battery. Q1 is HEP 802 fet, and other transistors are 2N5172.—R. M. Benrey, Measure the Speed of a Bullet with This Electronic Stopwatch, *Popular Science*, July 1969, p 144–146 and 178.

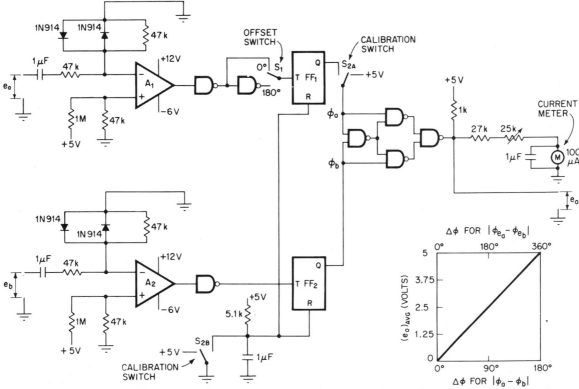

360-DEG DIGITAL PHASE METER—Detects average phase (time) difference between any two sine, square, triangle, or pulse inputs with same frequency in range from 100 Hz to above 1 MHz and amplitude range of 0.5 to 10 V p-p, using only five IC packages. A1 and A2 are type 710 comparators used for clipping and squaring input signal to give logic-level pulses that are buffered and inverted with gates from two type 7400 NAND gate packages before being applied to Texas Instruments SN7473 dual flip-flop. Flip-flops perform as binary counters, dividing input frequency by two so phase difference between inputs is also divided by two to extend measurement range to 360 deg.—C. A. Herbst, Detector Measures Phase over Full 360° Range, *Electronics*, July 19, 1971, p 73.

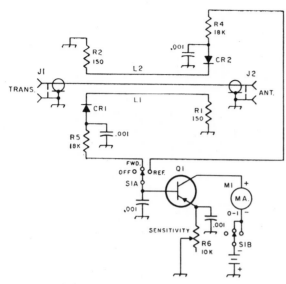

SWR METER—Designed for use with low-power transmitters having outputs of a few milliwatts, for tuneup and antenna adjustment. Insert in 50-ohm line. Q1 can be almost any pnp transistor, such as 2N705. Pickup wires L1 and L2 are 3⅜ inch lengths of No. 14 wire spaced ⅛ inch from coax. For 75-ohm coax line, use 100 ohms for R1 and R2. Diodes are 1N34A and battery is 1.5 V. Article covers construction and use.—L. G. McCoy, The Millimatch, QST, Aug. 1967, p 44–46.

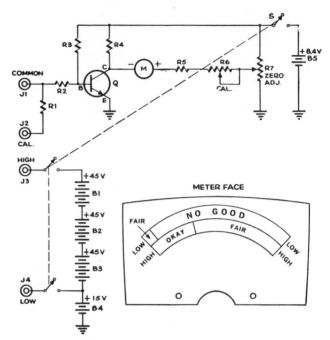

WOOD MOISTURE METER—Electronic ohmmeter circuit measures resistance between 6d 2-inch finishing nails soldered into Jones socket for convenience in pressing points firmly into wood being tested. Leads on socket go to banana plugs, one of which is always in J1; other goes into J3 or J4, depending on range desired. Since current drain from all batteries is insignificant, use smallest sizes available, and cement batteries in place. Meter is 50 μA and transistor is HEP54. R1 is 15 meg, R2 47K, R3 3.3 meg, R4 22K, R5 6.8K, R6 100K pot, and R7 5K pot. Article covers calibration and use.—R. M. Benrey, Is That Wood Dry Enough?, Popular Science, July 1971, 88–89 and 102.

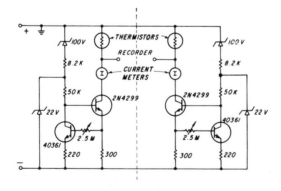

THERMAL-CONDUCTIVITY MONITOR FOR GAS—Drives recorder to provide information required for achieving constant carrier gas flow through evaporator of epitaxial reactor. Transistorized constant-current supply consists of two matched circuits, each serving as current source for one thermistor. Currents are adjusted to give 5 mA normally through each thermistor. Recorder measures difference in voltage drop across thermistors as function of time, as it changes with variations in thermal conductivity of gas mixture.—G. A. Riley and J. A. Amick, Monitoring Silicon Tetrachloride Concentration in Hydrogen Carrier Gas, RCA Review, June 1970, p 396–406.

CHAPTER 63
Measuring Circuits—Power

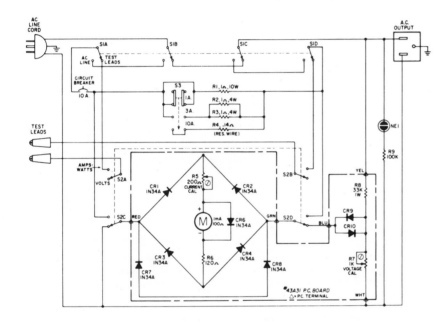

POWER DRAIN—Indicates amount of power drawn from a-c line by appliance or other device under test, on meter calibrated to read up to 345 W for line voltage of 115 V. Meter also has conventional voltage and current scales. Power conversion chart is required if line voltage differs significantly—P. Dahlen, Sencore's PM 157 Power Monitor, *Electronic Technican/Dealer*, Dec. 1970, p 57.

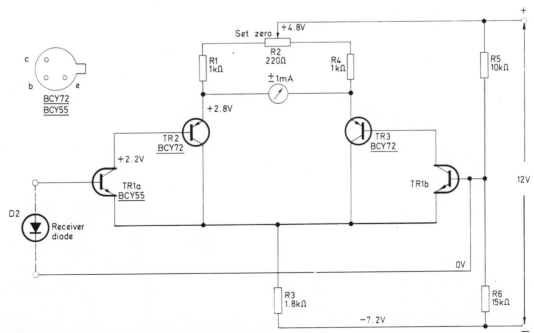

10-GHZ FIELD-STRENGTH METER—D-c amplifier used between 1N21B microwave diode D2 and indicating meter extends useful range for making field measurements at 10 GHz from Gunn-device transmitter to about 20 yards. Differential amplifier TR1 uses low-drift transistor pair.—"A Gunn Device Transmitter (10 GHz)," Educational Projects in Electronics, Mullard, London.

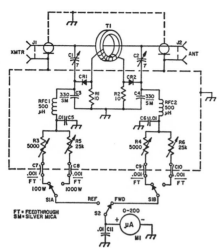

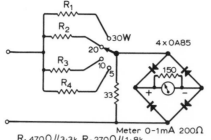

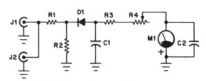

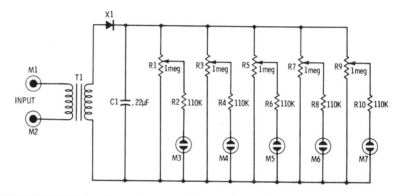

POWER METER—Square-law characteristic of bridge diodes up to about 1.4 V makes linear scale possible for meter. Provides four ranges between 5 and 30 W at impedance of 15 ohms.—K. D. James, Linear Scale Power Meter, *Wireless World*, June 1969, p 269.

CB WATTMETER—Used in series with coax line to antenna, for continuous check of output power of transmitter or transceiver. Will also indicate jump in standing-wave ratio of antenna by showing improbably high power reading on meter. C1 and C2 are 0.001 μF, D1 is 1N60, M1 is 0–1 mA d-c meter, R1 is 3.3K, R2 4.7K, R3 10K, and R4 10K trimmer. —H. Friedman, Inline RF Wattmeter for CBers, *Electronics Illustrated*, May 1972, p 59–61 and 100.

WATTS	M1	WATTS
100	200	1000
90	180	900
80	170	800
70	155	700
60	145	600
50	125	500
40	105	400
30	85	300
20	65	200
10	40	100
5	20	50

R-F WATTMETER—Provides two power ranges, 0–100 and 0–1,000 W, forward and reflected, for 3.5 to 30 MHz. Cable gives data for making calibrated meter scales. Toroidal transformer T1 has 35 turns No. 26 spread evenly to cover entire core of Amidon T-68-2 toroid. Diodes are 1N34A. Article covers construction, and gives circuits of comparable commercial versions.—D. DeMaw, In-Line RF Power Metering, *QST*, Dec. 1969, p 11–16.

NEON WATTMETER—Pot for each neon is adjusted during calibration so bulb just lights at known a-c or a-f wattage level. Can be calibrated with various audio amplifiers having known full-volume ratings, or with variable-wattage source such as 12.6-V a-c filament

transformer feeding 100-ohm pot. X1 is 100-mA 400-piv silicon diode and T1 is universal output transformer. All neons are NE-2A.—R. M. Brown and T. Kneitel, "49 Easy Entertainment & Science Projects," Howard W. Sams, Indianapolis, IN, 1971, p 29–30.

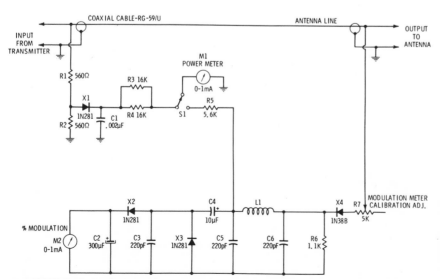

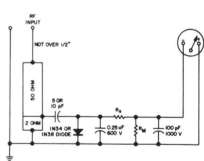

R-F WATTMETER—Reads 0–10 W on 0–1 mA meter M1. Lower portion of circuit gives percentage modulation reading for transmitter. Three-quarter swing of M2 to right corresponds to 75% modulation of a-m carrier. L1 is 24 μH.—R. M. Brown, "101 Easy CB Projects," Howard W. Sams, Indianapolis, IN, 1968, p 12–14.

500-MHZ POWER OUTPUT METER—Simple meter circuit is relatively nonreactive (vswr under 1.5:1 in operating range) and capable of dissipating 25 W for up to 30 s without overheating or 50 W for 5 s. Article covers construction of input resistor from five 10-ohm and one 2-ohm composition resistor discs. With 1-mA meter, 4K for Rs, and 180 for RM, full-scale reading is 400 W; article recommends higher-current meters and gives resistor values for them to provide more desirable lower full-scale readings.—J. A. Houser, An Improved UHF Power Output Meter, *73*, June 1973, p 37–38, 40, and 43.

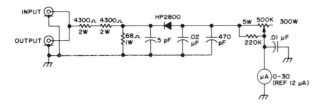

5–300 W R-F WATTMETER—Designed for use with large dummy antenna rated up to 500 MHz, for measuring transmitter output power over frequency range of 2 to 450 MHz. Article covers construction and calibration. Uses HP2800 hot carrier diode having rating of 75 piv.—F. C. Jones, RF Power Measurement with Hot Carrier Diodes, 73, Sept. 1971, p 42–44 and 46.

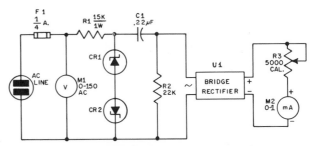

POWER FREQUENCY METER—Intended for use with portable a-c generator. To calibrate, connect to commercial power line and adjust R3 for reading of 0.6 mA on M2 to indicate 60 Hz. Readings are then accurate within 5% for 20 to 100 Hz. CR1 and CR2 are 1N5245 and U1 is HEP175.—J. Hall, A Field-Day AC-Power Monitor, QST, March 1971, p 40–41.

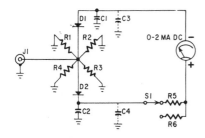

UHF WATTMETER—Provides full-scale ranges of 30 or 100 mW with acceptably flat response from 2 to 500 MHz. With separate 10-dB 50-ohm pad between transmitter and wattmeter, ranges are increased to 300 mW and 1 W. Diodes are 1N82A, R1–R4 are 200, R5 and R6 are meter multipliers selected to give desired full-scale ranges as described in text, R7 and R8 are 27, R9–R12 are 150, C1 and C2 are 1,000 pF, and C3–C4 are 0.01 μF.—H. Balyoz, An RF Wattmeter for UHF, Radio-Electronics, Sept. 1966, p 58.

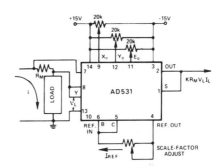

INSTANTANEOUS WATTMETER—Uses Analog Devices AD531 IC to reject common-mode interference by taking advantage of differential X input. Scale factor can be adjusted with 20K pot. Ideal for instrumentation, agc applications and for situations where two input signals must be subtracted before being subjected to further processing. NOTE: End terminals of 20K pot for Xo should be connected between +15 V and 42K resistor going to ground.—Monolithic Analog Multiplier-Dividers, Analog Dialogue, Vol. 6, No. 3, p 10.

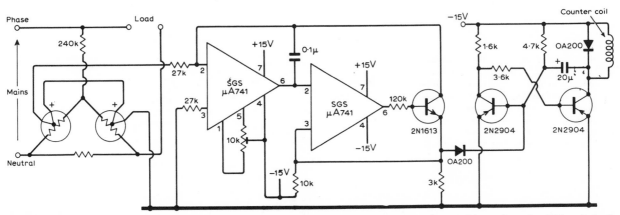

WATTHOUR METER—Electronic replacement for rotating-disc meter uses vacuum thermocouple multiplier operating on quarter-square principle to measure power consumption of loudspeaker, electric motor, or any other device drawing power from source in frequency range from d-c to 100 MHz. Integrator using opamps drives mono mvbr feeding electromagnetic counter, for converting sensed energy in watts to watthours. Heater acting on thermocouple has range of 5 mA and resistance of 400 ohms, while each couple is 2 ohms. Article analyzes feasibility of circuit.—L. A. Trinogga, Experimental Electronic Electricity Meter, Wireless World, Jan. 1972, p 33–35.

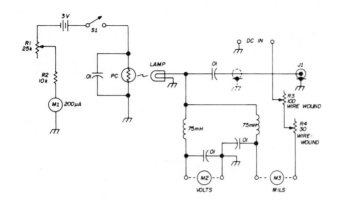

OPTOELECTRONIC R-F WATTMETER—Measures low power levels up to 148 MHz without elaborate calibration, based on fact that small lamp with short filament will reach given temperature for same amount of d-c, a-f, or r-f power. Suitable pilot lamp for transmitter power, as specified in article, is connected as transmitter load and light level is measured with cadmium sulfide photocell and meter. Direct current required through lamp to give same light reading is then measured for computing power.—H. F. Burgess, An Accurate RF Power Meter for Very Low Power Experiments, *Ham Radio*, Oct. 1972, p 58–61.

0.025–10 W R-F WATTMETER—Useful from low radio frequencies to above 450 MHz. Article covers construction and calibration. Uses HP2900 hot carrier diode having rating of 10 piv, limiting rms r-f voltage across it to 3 V for safe operation.—F. C. Jones, RF Power Measurement with Hot Carrier Diodes, *73*, Sept. 1971, p 42–44 and 46.

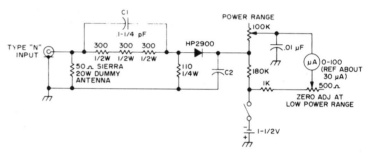

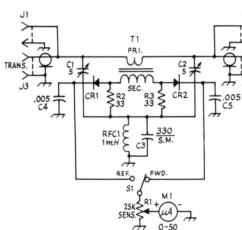

0–5 W FOR 10–80 METERS—Gives same deflection per watt on all bands in frequency range. With sensitivity control at maximum, requires only .1 W r-f energy to drive meter full-scale. Article gives construction and adjustment details. Diodes are 1N34A germanium. T1 has 60 turns No. 28 on Amidon T-68-2 toroid for secondary, with two turns over it for primary.—D. DeMaw, A QRP Man's RF Power Meter, *QST*, June 1973, p 13–15.

FIELD METER—Detector located as far as possible from antenna of CB, ham, or commercial transmitter is connected by ordinary two-wire line to indicator at operating position, for continuous check of field strength whenever transmitter is on air. Transistor can be RCA SK3009 or equivalent, and diode D is 1N34A.—J. B. White, Remote-Reading Field Meter, *Radio-Electronics*, Aug. 1967, p 60–61 and 66.

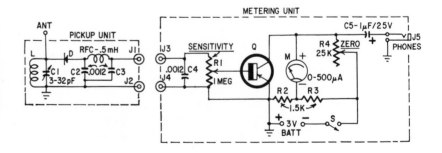

CHAPTER 64
Measuring Circuits—Resistance

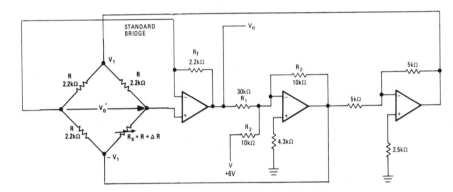

LINEAR RESISTANCE BRIDGE—Feedback circuit makes output voltage Vo vary linearly within 0.1% with value of bridge resistor Rx over range of 0 to 4.4K. Article gives design equations. Either a-c or d-c voltage can be used to excite bridge. All opamps are Burr-Brown 3024/15.—R. D. Buyton, Feedback Linearizes Resistance Bridge, *Electronics*, Oct. 23, 1972, p 102.

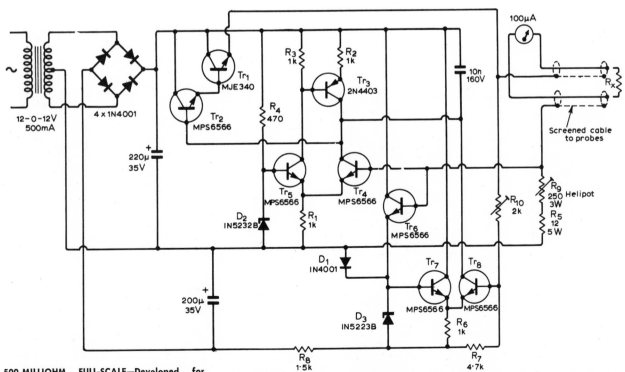

500-MILLIOHM FULL-SCALE—Developed for measuring resistance of contacts and soldered joints. R9 can be set to give either 500 milliohms or 5 ohms full-scale deflection. Output voltage is limited to under 1 V to protect meter and active devices in circuit under test. Supplies constant current to resistance under test and measures voltage drop. TR4 and TR5 form current-error amplifier controlling series pair TR1-TR2, with TR3 forming constant-current source for TR4.—J. Johnstone, Low-Range Ohmmeter, *Wireless World*, June 1971, p 294.

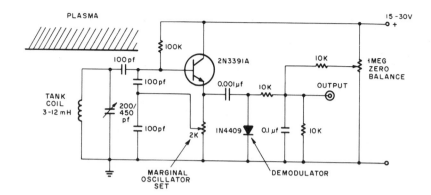

CONDUCTIVITY PROBE—Uses sensitive marginal oscillator to measure changes in Q and inductance of nonimmersive conductivity coil system because these changes are closely related to conductivity of plasma under study. Designed both for use in laboratory and under reentry flight conditions. Center-tapped capacitor divider in tank circuit is direct analog of center-tapped coil of Hartley oscillator. Basic oscillator frequency is 2 MHz.—S. Aisenberg and K. W. Chang, A Wide Range RF Coil System for the Measurement of Plasma Electrical Conductivity, *Proc. IEEE*, April 1971, p 710–712.

FOLLOWER VOLTMETER—Input impedance is 5 teraohms, as required for measuring insulation resistance of high-quality capacitors. Basic circuit accommodates input ranges of 0 to 0.5 V of either polarity for voltage measurement and 0 to +1.5 V for resistance measurement. Input range divider and function switch may be added if desired. Response is flat from 10 Hz to 100 kHz, and down 3 dB at 320 kHz for a-c voltage measurement. Voltage follower circuit with fet input gives inherently high input impedance that is further boosted by feedback. Use of balanced long-tailed pair to compare input and output voltages ensures low offset voltage.—L. E. MacHattie, Voltmeter Using F.E.T.s Measures Capacitor Insulation Resistance, *Wireless World*, Jan. 1971, p 13.

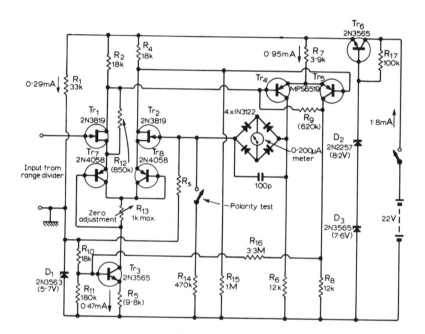

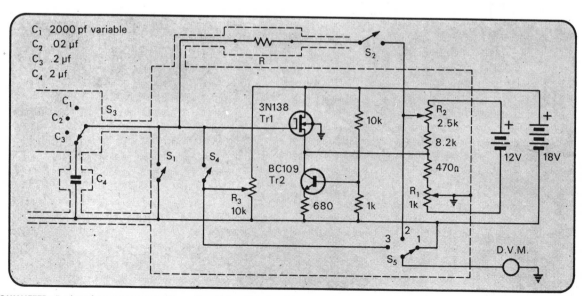

TERAOHMMETER—Designed to measure resistance values in range from 10^8 to 10^{12} ohms with precision of 1%. Based on timing the charging of an active analog integrator. Except for digital voltmeter and precision air variable capacitor, components are inexpensive. Article gives construction, operation, and performance details.—A. C. Corney, Simple, Medium-Precision High-Resistance Measuring Device, *Electronic Engineering*, Aug. 1970, p 46–47.

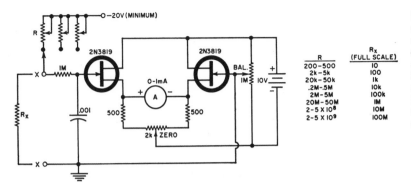

R	Rx (FULL SCALE)
200–500	10
2k–5k	100
20k–50k	1k
.2M–.5M	10k
2M–5M	100k
20M–50M	1M
2–5 × 10⁸	10M
2–5 × 10⁹	100M

LINEAR-SCALE OHMMETER—Requires no zero adjustment after initial setting, and has sufficient precision for matching resistors for critical circuits. Can be operated from batteries or from rectified transformer outputs. To calibrate, adjust R for full-scale deflection when desired value of Rx is connected to input.—R. L. Carroll, A Simple Linear Labratory-Quality Ohmmeter, *Electronics World*, May 1970, p 60.

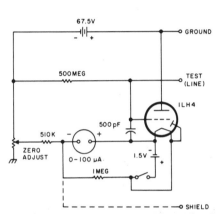

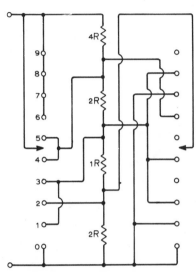

MEGOHMMETER—Simple portable circuit will measure resistances up to 1,000 meg, corresponding to reading of 20 μA on meter. Shield terminal should be used for accurate readings above 100 meg, as explained in article. Developed for checking antenna systems and transmission lines for leakage.—J J. Shultz, How To Megger Your Antenna, 73, April 1970, p 125–128.

4 RESISTORS PER DECADE—Requires only double-pole 10-position switch. Increase resistors by powers of 10 to get higher decades.—J. Johnstone, Decades of Resistance, *Wireless World*, Feb. 1971, p 66.

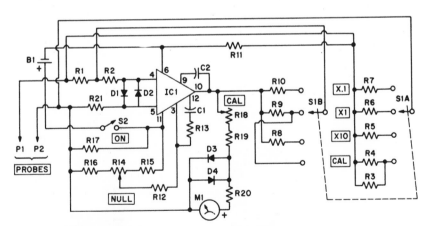

B1—Two 9-V transistor radio batteries wired in series
Capacitors: all 20% tolerance, 1000 VDC
C1—470 μμf disc ceramic
C2—22 μμf silvered mica
D1-D4—General-purpose germanium diode (IN34A or equiv.)
IC1—Integrated circuit (Motorola MC1709CL) see text
M1—0-1 ma. DC milliammeter (Lafayette 99R 50403)
P1,P2—Needle-point test prod (Calectro 33-384 or equiv.)
Resistors: ½-watt, 5% unless otherwise noted
R1,R5—100 ohms
R2,R21—1,000 ohms, ¼ watt
R3,R7—1 ohm, ¼ watt
R4,R6—10 ohms
R8,R11,R17—620 ohms, ¼ watt

R9—6,200 ohms
R10—62,000 ohms
R12—470,000 ohms, ¼ watt
R13—1,500 ohms
R14—100,000-ohms, linear-taper pot
R15—100,000 ohms
R16—470,000 ohms (see text)
R18—500-ohm, linear-taper pot
R19—750 ohms
R20—200 ohms
S1—Two-pole, six-position shorting-type rotary switch (Centralab PA-2002 or equiv.)
S2—SPDT miniature toggle switch
2—1-in. spacers
1—14-pin IC socket

Misc.—6 x 5 x 4-in. aluminum cabinet, perfboard, push-in terminals, rubber grommets, hardware, knobs, decals, etc.

LOW-RESISTANCE OHMMETER—Measures resistances between 10 milliohms and 40 ohms, as required for locating poor solder joints, bad connectors, and measuring resistance of coils. Article covers construction and calibration.—C. R. Lewart, A Lo-Lo-Ohm Ohmmeter, *Electronics Illustrated*, Sept. 1972, p 63–65 and 104.

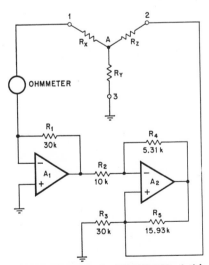

TROUBLESHOOTING T ATTENUATOR—Article tells how unity-gain current amplifier A1-A2 can be used to determine value of unknown resistor in encapsulated T attenuator or ladder network. For resistor values shown, ohmmeter reading is accurate for currents up to 0.3 mA. Resistor values can be scaled up or down for higher or lower currents.—W. J. Travis, Op Amps Find Values of Buried Resistors, *Electronics*, Sept. 28, 1970, p 80.

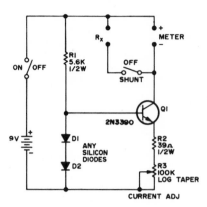

METER RESISTANCE—Practical high-accuracy circuit measures internal resistance of almost any d-c meter movement. Transistor serves as constant-current source, with R3 providing current range of about 8 μA to 13 mA. Rx is rheostat serving as shunt for meter, covering range of expected meter resistances. Connect meter and shunt, turn on circuit, turn off shunt switch, adjust R3 for full-scale meter reading, turn on shunt switch, and adjust Rx until meter reads half-scale. Rx is then exactly equal to meter resistance, and can be disconnected for measuring with ohmmeter. —Marovich, The Meter Evaluator, 73, April 1971, p 116–119.

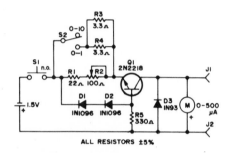

LOW-RANGE OHMMETER—Has two linear scales, 0–10 ohms and 0–1 ohm. Momentary pushbutton is used for test because of high current drain from single D cell. Q1 serves as constant-current generator providing known current through test resistance. Requires no zeroing once calibration control is set, by adjusting for full-scale reading on 10-ohm range while testing precision 10-ohm resistor. —A. Schecner, The Low-Ohm Meter, 73, Feb. 1971, p 87.

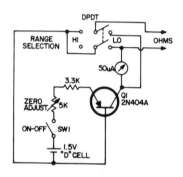

LOW-VOLTAGE OHMMETER—Can be used safely in solid-state circuits because maximum voltage across terminals is only 250 mV. Switch gives choice of 3,000-ohm low-resistance range with highest value at right end of scale, and high range with 150K near zero on microammeter scale. Article covers calibration.—I. M. Gottlieb, An Ohmmeter for Solid-State Circuits, 73, June 1973, p 91–93.

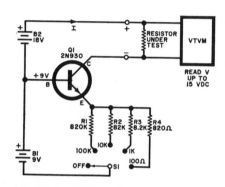

LINEARIZING VTVM SCALE—Permits reading ohmmeter scales of 100, 1K, 10K, and 100K ohms on linear scale of d-c voltage range of vtvm instead of on hard-to-read log-type resistance scale.—G. Beene, A Linear-Scale Ohmmeter, *Popular Electronics*, May 1972, p 45.

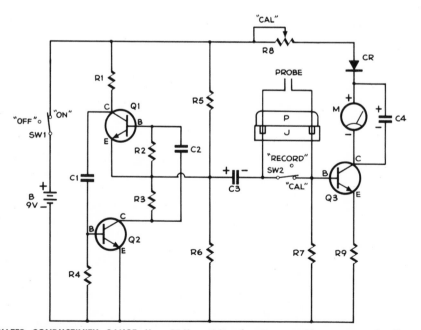

C1, C2—0.25uf, 100v paper capacitors	2
C3, C4—25uf, 16v electrolytic capacitors	2
CR—Diode, 1N34A	1
Holder, battery— Lafayette No. 34E50053	1
M—Meter, 0-1mA DC milliammeter, Lafayette No. 99E50528	1
J—Connector, chassis, Lafayette No. 34E20460	1
P—Connector, probe (see text), Lafayette No. 34E20015	1
Q1, Q2, Q3—Transistors 2N3704, MPS 3704, or equal	3
R1, R3, R5, R6—15,000-ohm, ½-watt carbon resistors	4
R2, R4—100,000-ohm, ½-w.tt carbon resistors	2
R7—5,100-ohm, ½-watt carbon resistor	1
R8—10,000-ohm linear-taper midget potentiometer	1

WATER CONDUCTIVITY GAUGE—Uses 55-Hz mvbr Q1-Q2 to convert battery voltage to about 1.5 V a-c between probes, to overcome polarization effects when measuring conductivity of water as guide to extent of pollution. Article tells how to calibrate tester with salt solution. Choose R9 between 0 and 100 ohms to make meter read 0.75 mA when 1K resistor is between probes and SW2 is open. —P. Emerson, Portable Water Tester, *Popular Science*, Sept. 1970, p 127–128, 132, and 140.

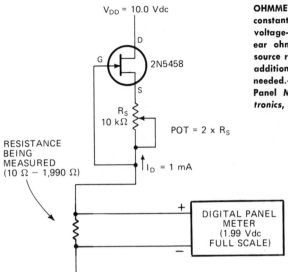

OHMMETER WITH DIGITAL VM—Simple fet constant-current source converts conventional voltage-measuring digital panel meter to linear ohmmeter having range determined by source resistance RS. Switch can be added for additional resistors if more than one range is needed.—J. L. Turino, Converting a Digital Panel Meter into a Linear Ohmmeter, *Electronics*, March 1, 1973, p 102.

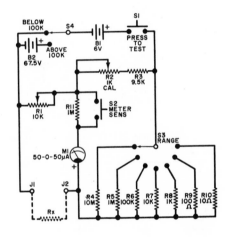

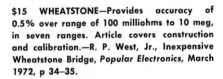

$15 WHEATSTONE—Provides accuracy of 0.5% over range of 100 milliohms to 10 meg, in seven ranges. Article covers construction and calibration.—R. P. West, Jr., Inexpensive Wheatstone Bridge, *Popular Electronics*, March 1972, p 34–35.

CHAPTER 65
Measuring Circuits—Temperature

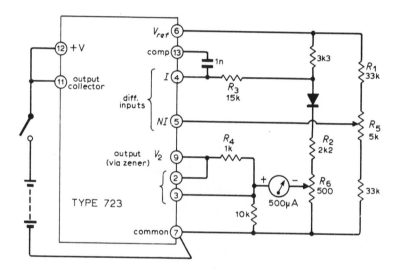

SURFACE TEMPERATURE—Single IC voltage regulator simplifies circuit for measuring forward voltage drop of silicon diode, which varies essentially linearly with temperature. Circuit is independent of meter resistance. Battery drain is only about 3 mA.—B. E. Kerley, Electronic Thermometer Uses I.C., *Wireless World*, Aug. 1972, p 366.

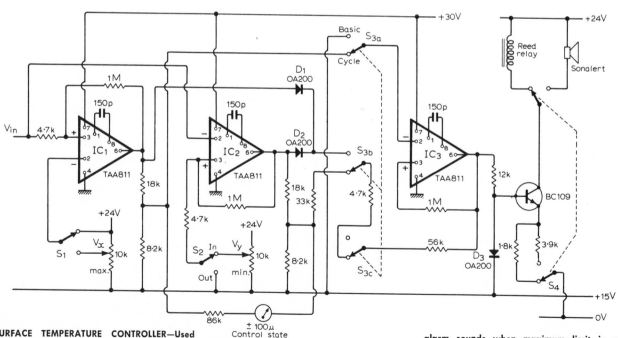

SURFACE TEMPERATURE CONTROLLER—Used with YIG radiometer for measuring and controlling temperatures in range of 100 to 250 C. Provides independent and adjustable maximum and minimum limits. In basic mode of operation, relay is energized or Sonalert alarm sounds when pointer of indicating meter crosses either limit, without differentiating between high and low. In cycle mode, alarm sounds when maximum limit is exceeded and remains on until lower limit is crossed.—I. J. Kampel, A YIG Radiometer and Temperature Controller, *Wireless World*, Oct. 1970, p 501–504.

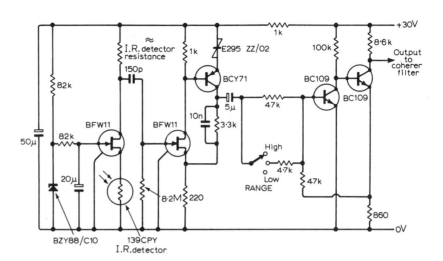

RADIOMETER PREAMP—Infrared detector responds to radiant energy chopped at 400 Hz by YIG modulator. Output goes to coherent filter along with feed from modulator, and this in turn feeds main amplifier driving meter calibrated to read surface temperatures in range of 100 to 250 C.—I. J. Kampel, A YIG Radiometer and Temperature Controller, *Wireless World*, Oct. 1970, p 501–504.

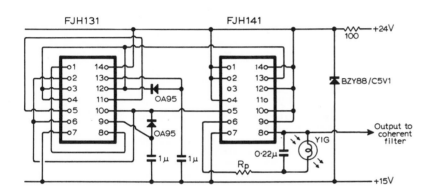

RADIOMETER MODULATOR—YIG solid-state optical modulator replaces bulky motorized chopper in radiometer for measuring surface temperatures of objects in range from 100 to 250 C. Uses quadruple NAND gate FJH131 connected as 400-Hz square-wave generator, feeding FJH141 dual buffer gate that provides required current drive to YIG coil. Article includes infrared detector circuit and coherent filter.—I. J. Kampel, A YIG Radiometer and Temperature Controller, *Wireless World*, Oct. 1970, p 501–504.

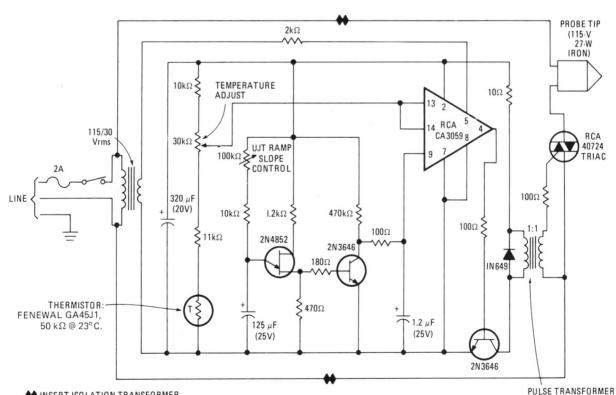

◆◆ INSERT ISOLATION TRANSFORMER

SOLDERING-IRON HEAT PROBE—Thermistor epoxy-cemented into drilled hole in tip of soldering iron actuates control circuit which maintains tip of soldering iron within few degrees of desired temperature setting from room ambient to 125 C. Used as selective heat source by holding against component for which temperature stability is being checked. Use silicone vacuum grease to improve heat transfer from probe to component.—M. J. Shah, Soldering Iron Converts to Constant-Temperature Probe, *Electronics*, Oct. 23, 1972, p 103.

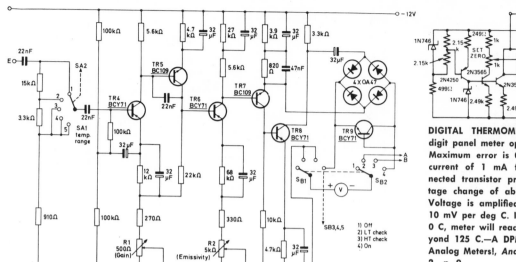

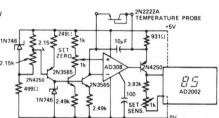

DIGITAL THERMOMETER—Circuit drives 2½-digit panel meter operating from 5-V supply. Maximum error is 0.5% ±1 digit. Constant current of 1 mA flows through diode-connected transistor probe which develops voltage change of about −2.2 V per deg C. Voltage is amplified and typically scaled to 10 mV per deg C. If output is set to zero at 0 C, meter will read directly from 0 C to beyond 125 C.—A DPM That Really Challenges Analog Meters!, *Analog Dialogue*, Vol. 6, No. 2, p 9.

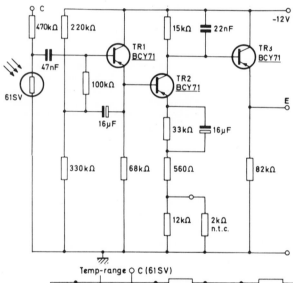

RADIATION THERMOMETER—Uses lead sulfide detector for measuring surface temperatures above 100 C. Three-stage amplifier shown is mounted in detector head. Input matches 300K impedance of cell, by bootstrapping of bias chain. All stages are biased for minimum noise. Uses 2K negative-temperature-coefficient thermistor to adjust gain to offset decrease in cell sensitivity with increasing ambient temperature. Final emitter-follower stage has low output impedance, for driving main amplifier and indicating meter. With input of 1 mV from head amplifier, for cell output of 0.3 mV, main amplifier delivers 20 mV to provide 10% deflection on 200-μA meter on most sensitive range. Will cover temperature range of 100 to 500 C in five ranges. Requires regulated 12-V supply.—"Applications of Infrared Detectors," Mullard, London, 1971, TP1201, p 71–81.

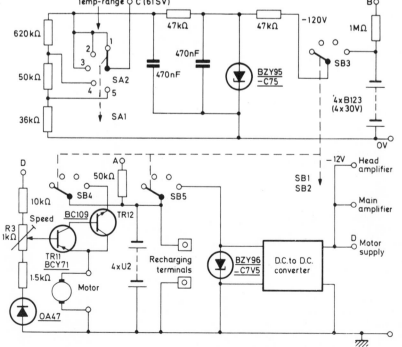

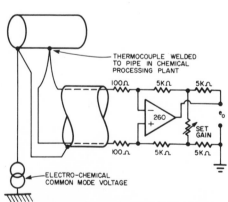

THERMOCOUPLE VOLTMETER—Chopper-stabilized Analog Devices Model 260 noninverting opamp operates off-ground, permitting differential connection for measuring push-pull signals developed symmetrically around common-mode level by thermocouple source. Circuit rejects common-mode voltages generated in soil by electrochemical effects, power-line leakage currents, and Peltier voltages developed by buried metal.—P. Zicko, "Designing With Chopper Stabilized Operational Amplifiers," Analog Devices, Inc., Norwood, MA, Sept. 1970.

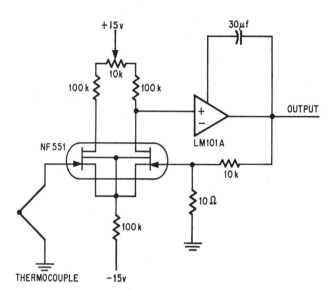

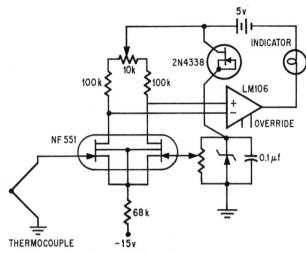

THERMOCOUPLE AMPLIFIER—Matched fet's present required high input impedance to thermocouple while amplifying its output.—R. Christensen and D. Wollesen, Matching FET's by Design Is Faster and Cheaper Than by Pick and Choose, *Electronics*, Dec. 8, 1969, p 114–116.

COMPARATOR—Matched fet's present required high input impedance to thermocouple while amplifying its output. Third fet combines with zener diode to give reference voltage.—R. Christensen and D. Wollesen, Matching FET's by Design Is Faster and Cheaper Than by Pick and Choose, *Electronics*, Dec. 8, 1969, p 114–116.

COHERENT FILTER—Complementary germanium transistors form dynamically synchronized narrow-band filter for YIG radiometer and surface-temperature controller. Gating waveform taken from YIG drive circuit switches transistors in anti-phase. Signal from preamp charges 125-μF capacitor through one of 10K resistors as transistors are switched by gating waveform, to build up d-c charge proportional to signal waveform, while noise or spurious signals at other frequencies average to zero.—I. J. Kampel, A YIG Radiometer and Temperature Controller, *Wireless World*, Oct. 1970, p 501–504.

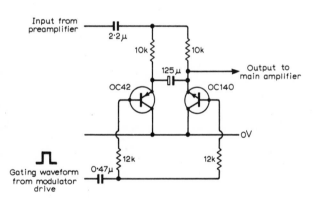

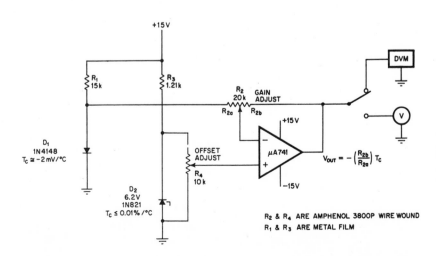

DVM READOUT—Opamp adjustments allow temperature of silicon-diode probe to be read directly in degrees on ordinary digital voltmeter. R4 calibrates circuit.—R. J. Battes, Any Voltmeter Reads Electronic Thermometer, *Electronics*, March 29, 1971, p 68.

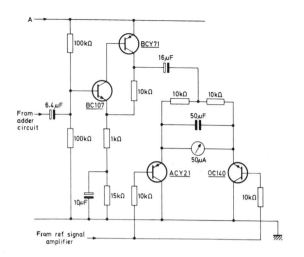

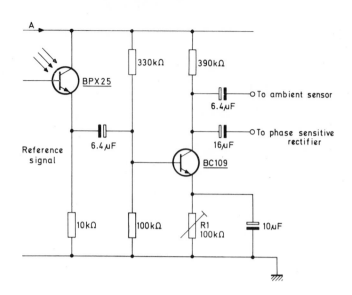

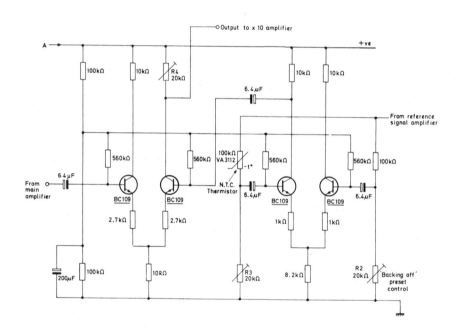

CLINICAL RADIATION THERMOMETER—Covers range of 30 to 40 C, for measuring temperature of external auditory canal. Radiation from ear is chopped at about 600 Hz by motor-driven disc having radial slots. Uses Mullard ORP10 uncooled indium antimonide cell to measure radiation from ear, and combination of CQY11A led and BPX25 detector on opposite sides of chopper disc to measure temperature of disc for reference signal. Circuits include ambient-temperature sensor using thermistor to produce signal that is added to radiation signal in differential amplifier. Output of adder goes to amplifier having gain of 10, driving phase-sensitive rectifier and indicating meter.—"Applications of Infrared Detectors," Mullard, London, 1971, TP1021, p 82–89.

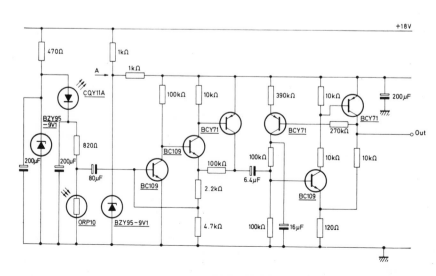

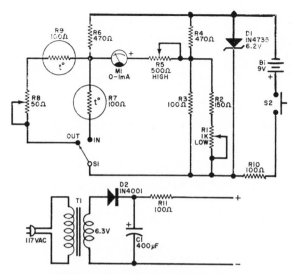

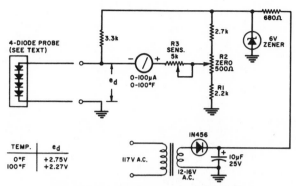

TEMP.	e_d
0°F	+2.75V
100°F	+2.27V

INDOOR-OUTDOOR THERMOMETER—Uses 100-ohm Texas Instruments TM¼ Sensitors selected by S1, with R9 outdoors and R7 indoors. Circuit is simple resistance bridge. Calibrate with boiling and freezing water, for use with 0–100F scale placed on milliammeter. Article tells how to install polarity-reversing switch at meter for reading outdoor temperatures below zero. Alternative a-c supply is shown.—C. P. Troemel, Indoor/Outdoor All-Electronic Thermometer, *Popular Electronics*, Oct. 1969, p 27–31.

DIODE-PROBE THERMOMETER—Probe uses practically any type of silicon junction diode for monitoring room, oven, solution, refrigerator, or other temperatures below about 200 F. Examples include 1N916, 1N456, 1N482, and 1N4383. Probe lead is Teflon-insulated subminiature coax which can be up to 100 feet long. Article covers calibration using conventional thermometer as standard in hot and cold water.—G. Gregg, Solid-State Probe Thermometer, *Electronics World*, March 1971, p 70–71.

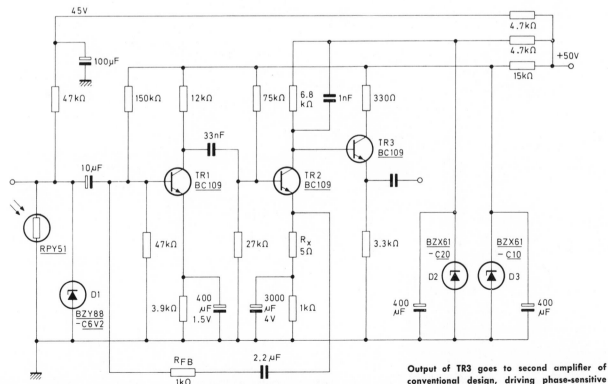

INFRARED MICROSCOPE—Developed to examine temperature distribution over surface of *transistors* and IC's. Uses RPY51 indium antimonide detector cooled in liquid nitrogen, with appropriate optical system and motor-driven chopper. Book gives all circuits required for driving temperature-indicating meter covering 0 to 200 C in four ranges. Output of TR3 goes to second amplifier of conventional design, driving phase-sensitive signal from conventional photoelectric amplifier also chopped by disc.—"Applications of Infrared Detectors," Mullard, London, 1971, TP1201, p 90–106.

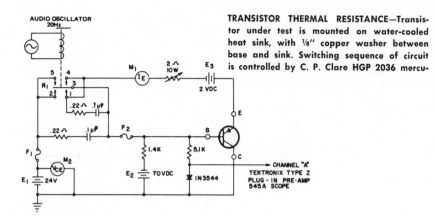

TRANSISTOR THERMAL RESISTANCE—Transistor under test is mounted on water-cooled heat sink, with ⅛″ copper washer between base and sink. Switching sequence of circuit is controlled by C. P. Clare HGP 2036 mercu-ry-wetted reed relay which is on 98% of each cycle. During reed transit time, diode is forward-biased to 50 mA by 70-V supply. Change in voltage across diode is measured and used as thermometer to indicate temperature of junction with given level of power. Thermocouple is inserted in copper washer to measure its stabilized temperature. Dividing junction-washer temperature differential by power dissipated gives thermal resistance, expressed as deg C per W. Fuses are 3-A fast-blow. M1 is 3 A d-c and M2 is 30-V d-c.—"Germanium Alloy Power Transistor Electrical Test Procedures," Delco, Kokomo, IN, April 1972, Application Note 6-A.

THERMOCOUPLE AMPLIFIER—Will give maximum indication at 1,000 C with R10 set at about 40 ohms when using a chromel-alumel junction having an output of 41 μV per deg C. Network R1—R6 is used to balance input voltage offset. With Mullard differential opamp IC shown, drift error of entire circuit is typically 0.1 deg C for each degree variation in amplifier ambient temperature.—"TAA241, TAA242 and TAA243 Differential Operational Amplifiers," Mullard, London, 1969, TP1086, p 10.

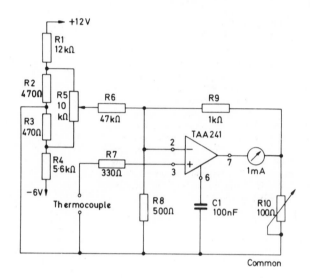

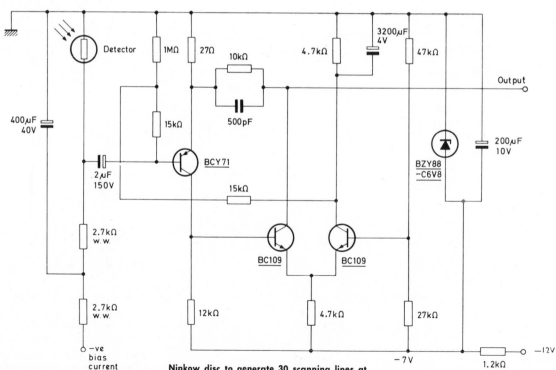

INFRARED TV—Circuit gives preamp for indium antimonide detector (Mullard RPY56) cooled in liquid nitrogen, for use in low-cost thermal scanning system using motor-driven Nipkow disc to generate 30 scanning lines at 17 frames per second covering area under observation. Book gives all other circuits for system, including motor drive. Will cover temperature range of —40 to +800 C. Used as closed-circuit tv system, with oscilloscope as monitor.—"Applications of Infrared Detectors," Mullard, London, 1971, TP1201, p 107–128.

CHAPTER 66
Measuring Circuits—Voltage

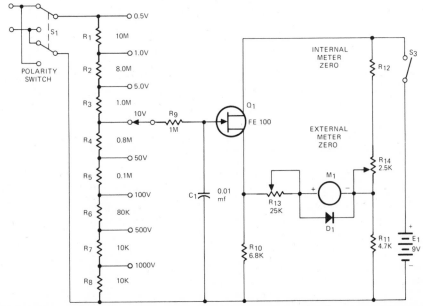

FET D-C VOLTMETER—Provides full-scale ranges of 0.5 V to 1,000 V for total of eight ranges, with 11-meg input impedance on all ranges. Silicon rectifier D1 (IR-804) protects 50-μA meter against overload. Construction and calibration are covered. Q1 (IR-FE100) and D1 are available from International Rectifier.—"Hobby Circuits," International Rectifier, El Segundo, CA, Vol. II, p 53—58.

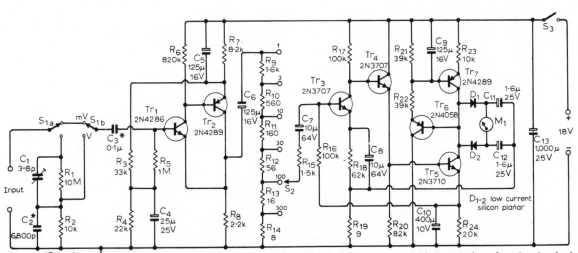

LINEAR A-C MILLIVOLTMETER—Use of constant-current load for output transistor makes current through d-c meter virtually independent of impedance of rectifying circuit. Meter is 50-μA, used with attenuator to provide desired ranges, such as for full-scale values of 1, 3, 10, 30, 100, and 300 mV. Current drain on battery is less than 2 mA.—A. J. Ewins, Linear Scale Millivoltmeter, *Wireless World*, Dec. 1970, p 592—595.

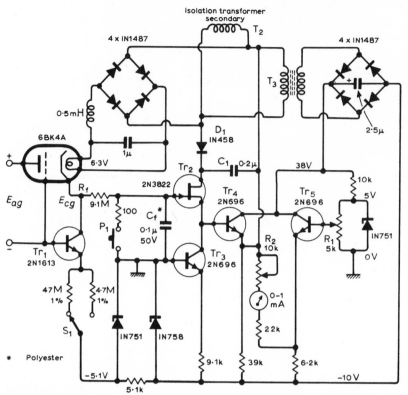

50-KV VOLTMETER—Measures a-c voltages of either polarity on same scale as peak value of a-c voltage. Uses 6BK4A high-voltage beam triode connected so anode voltage is independent variable and grid voltage is dependent variable. Five transistors form automatic balance circuit driving 1-mA meter having linear 50-kV scale. For safety, primary of isolation transformer is outside high-voltage compartment and energized by 50-kHz Col-pitts power oscillator operating from a-c line. Article gives construction details and stresses safety precautions.—A. M. Albisser and N. F. Moody, Electronic Voltmeter for 2 to 50 kV, *Wireless World*, March 1971, p 119–122.

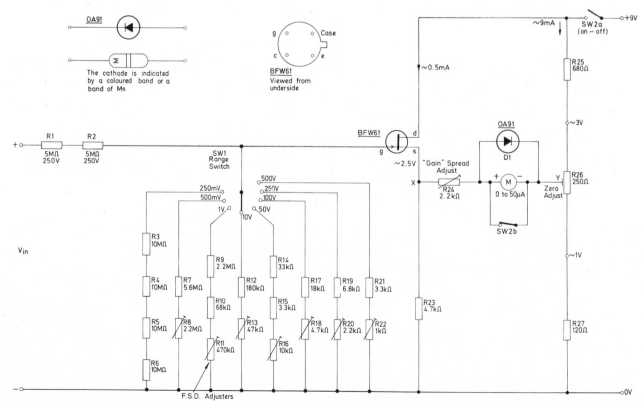

FET VOLTMETER—Voltage-dividing resistors for each range except lowest include pot that is adjusted initially to give full-scale deflection when voltmeter is calibrated against voltage standards providing full-scale voltages. On lowest range, R24 is used for full-scale adjustment. Report gives calibration procedure.—"A Simple F.E.T. Voltmeter," Educational Projects in Electronics, Mullard, London.

FET BOOSTS VOM SENSITIVITY—Provides effective input resistance of 10 meg when used with 20,000 ohms-per-volt voltohmmeter having 0.1-V d-c range. Article covers construction and operation.—S. D. Prensky, FET Rejuvenates VOM, *Popular Electronics*, Nov. 1968, p 72–73 and 114.

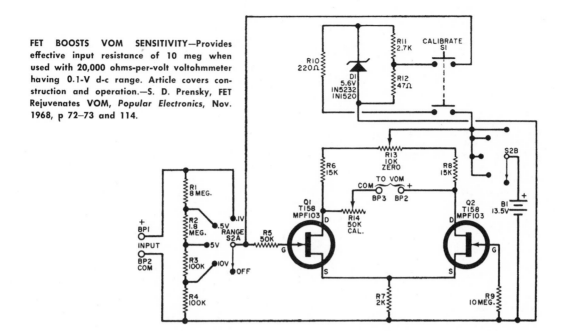

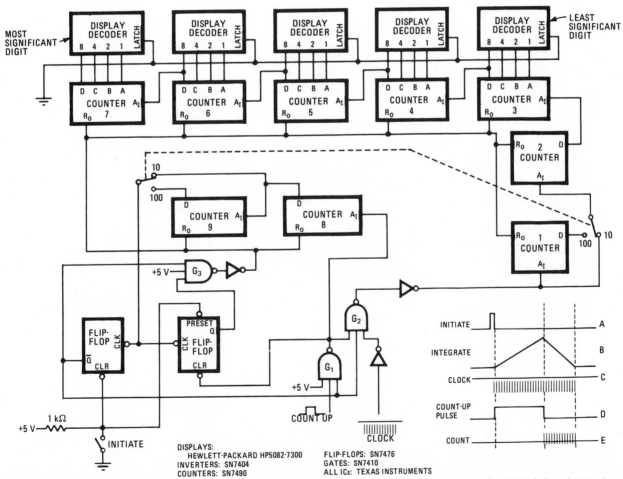

DISPLAYS:
HEWLETT-PACKARD HP5082-7300
INVERTERS: SN7404
COUNTERS: SN7490

FLIP-FLOPS: SN7476
GATES: SN7410
ALL ICs: TEXAS INSTRUMENTS

TIME-AVERAGING FOR DVM—Simple averaging circuit added to digital meter reduces measurement uncertainty of noisy signals. Circuit produces sum of 20 or 100 measurements, depending on switch position, to reduce data uncertainty by factor of 3.2 or 10. Clock pulse train, obtained from dual-slope converter of digital meter, is transmitted during integration period of analog signal and reference input. Article describes operation of circuit in detail.—G. Mitchell and R. D. Spencer, Data Averager for Panel Meter Operates from Meter's Clock, *Electronics*, April 26, 1973, p 103.

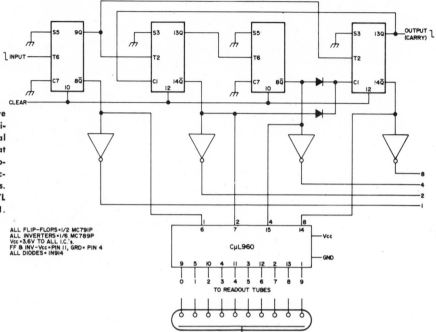

DISPLAY DRIVER FOR DVOM—Used to drive ten-digit numeric indicator tubes providing directly readable reading of value for digital volt-ohmmeter. Supply voltage required is at least 170 V d-c. Optional bcd output is provided for driving digital printer, with four sections of MC789P hex inverter used as buffers. Article covers construction.—R. Vaceluke, RTL Decade and Driver, 73, Oct. 1972, p 99–101.

ALL FLIP-FLOPS=1/2 MC791P
ALL INVERTERS=1/6 MC789P
Vcc=3.6V TO ALL I.C.'s.
FF & INV-Vcc=PIN 11, GRD=PIN 4
ALL DIODES=1N914

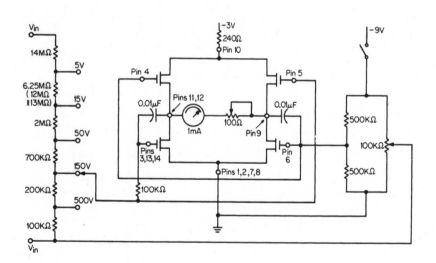

IC VOLTMETER—Uses mos transistors of either MC1155 or MC2255 general-purpose logic IC as voltage-controlled resistors in meter bridge configuration. Meter and 100K pot (adjust for full-scale deflection) are connected externally to IC. Design gives excellent linearity, with input resistance of voltmeter essentially constant for all ranges.—L. Parker, A Simple-to-Make Voltmeter, The Electronic Engineer, March 1972, p 52.

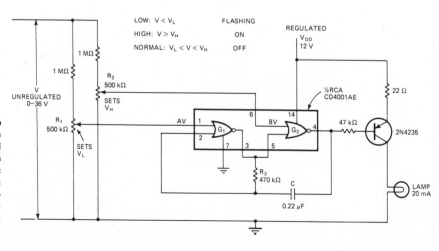

BATTERY-MONITORING LAMP—Single lamp stays off when unregulated supply voltage is normal, stays on when voltage is high, and flashes when voltage is low. Voltage limits and flashing frequency are adjustable. Logic is performed by two complementary-mos NOR gates. R3 and C determine flashing frequency.—N. D. Thai, One Lamp Can Monitor Battery Voltage, Electronics, April 12, 1973, p 106.

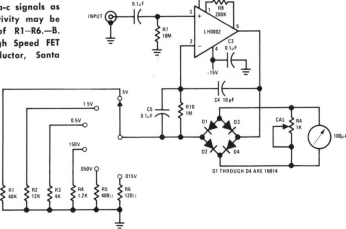

100 HZ–500 KHZ A-C VM—Wideband circuit using fet opamp will measure a-c signals as low as 15 mV. Full-scale sensitivity may be changed by altering values of R1–R6.—B. Siegel, "Applications for a High Speed FET Op Amp," National Semiconductor, Santa Clara, CA, 1972, AN-75.

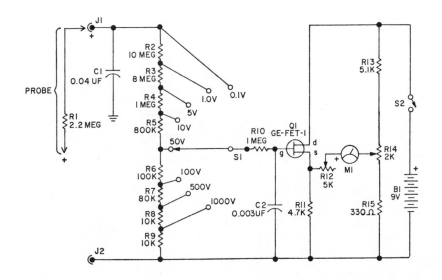

FET VTVM—Inexpensive equivalent of vacuum-tube voltmeter provides eight voltage ranges for accurate measurements in practically any circuits. Uses 0–50 μA microammeter. Resistors can be 5%, but 1% resistors improve accuracy. R1 is mounted in handle of probe.—S. Daniels, Debug With Hi-Fet, *Elementary Electronics*, Jan.-Feb. 1970, p 43–46.

VOLTAMMETER—Transistor circuit driving zero-center meter provides sensitivity of 1.33 meg per V on 0.25 to 25 V full-scale ranges and 33.3 meg per V on higher voltage ranges. Circuit also includes 0.75 and 7.5 μA current ranges. Can withstand most overloads without damage. Q1 and Q4 are 2N3906 or 3N3702, Q2 and Q3 are 2N3904 or 2N2925.—R. W. Fergus, The Indestructible (But Sensitive) Voltmeter, 73, Nov. 1972, p 134–136 and 138–140.

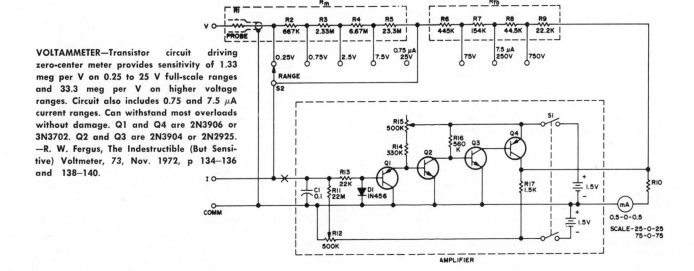

ROOT-MEAN SQUARING—Uses Burr-Brown squaring modules combined with inexpensive opamps to make rms voltmeter which does not require sine-wave inputs. Operates from 3-wire regulated ±15 V supply. Accuracy is about 0.5% of full scale for 10 Hz to 10 kHz. Gain switch is set so output is between 1 and 10 V.—"Squaring Modules," Burr-Brown, Tucson, AZ, April 1972, p 6.

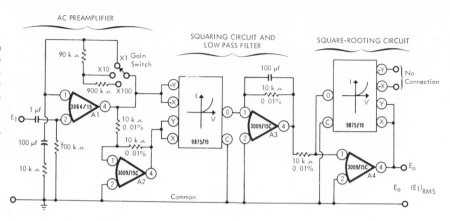

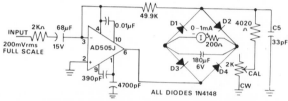

10 HZ–200 KHZ A-C VOLTMETER—Provides accuracy within 1% at 200-mV rms full-scale input. Higher voltages can be measured by switching in additional resistance in series with input and reducing input capacitor size proportionally. Opamp converts input voltage to current which is measured by meter in bridge containing diodes which serve only to route current according to polarity.—H. Krabbe and R. S. Burwen, Stable Monolithic Op Amp Slews at $130V/\mu s$, *Analog Dialogue*, Vol. 5, No. 4, p 3–5.

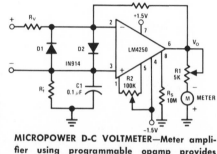

MICROPOWER D-C VOLTMETER—Meter amplifier using programmable opamp provides full-scale ranges from 100 V down to 10 mV by changing resistor values as in table. Diodes provide complete amplifier protection for input overvoltages up to 500 V on 10-mV range. Inverting amplifier configuration has gain varying from −30 for lowest range to −0.003 for 100-V range.—M. K. Vander Kooi and G. Cleveland, "Micropower Circuits Using the LM4250 Programmable Op Amp," National Semiconductor, Santa Clara, CA, 1972, AN-71, p 6.

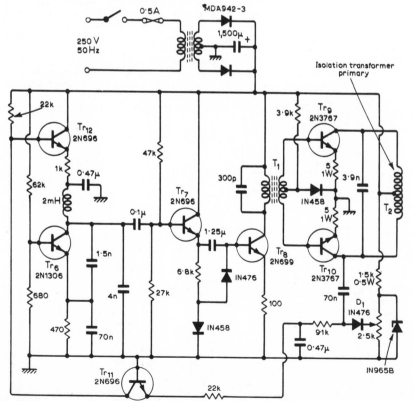

50-KHZ SUPPLY FOR 50-KV VOLTMETER—Used outside of high-voltage compartment of voltmeter to transfer supply energy through plastic housing to isolation transformer secondary inside compartment. Colpitts oscillator TR6 excites driver for push-pull class-B output stage through isolation transistor TR7. TR11 and TR12 are in negative d-c feedback circuit that holds amplitude of a-c output voltage across T2 constant despite changes in d-c supply voltage derived from a-c line.—A. M. Albisser and N. F. Moody, Electronic Voltmeter for 2 to 50 kV, *Wireless World*, March 1971, p 119–122.

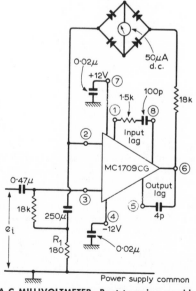

A-C MILLIVOLTMETER—Bootstrapping provides high input impedance. Input of 10 mV gives full-scale deflection of meter. Upper 3-dB limit is 150 kHz.—G. B. Clayton, Operational Amplifiers, *Wirelss World*, Oct. 1969, p 482–483.

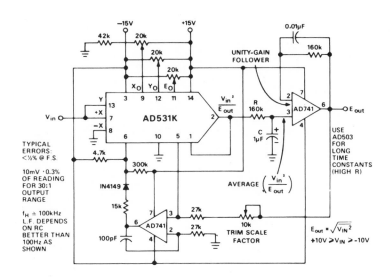

TRUE RMS TO D-C—Uses Analog Devices AD531K IC multiplier/divider with opamps and feedback loop to give true rms value of input voltage regardless of whether it is square, triangular, heavily distorted sine, random noise, or any arbitrary waveform.—L. Counts, *True RMS Measurement Using the AD531*, Analog Dialogue, Vol. 7, No. 1, p 13.

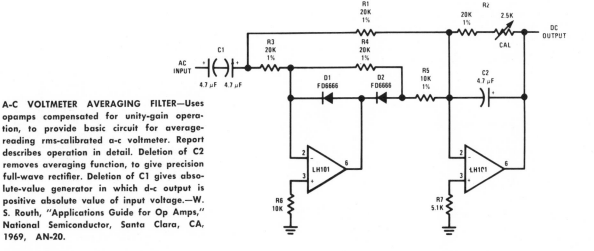

A-C VOLTMETER AVERAGING FILTER—Uses opamps compensated for unity-gain operation, to provide basic circuit for average-reading rms-calibrated a-c voltmeter. Report describes operation in detail. Deletion of C2 removes averaging function, to give precision full-wave rectifier. Deletion of C1 gives absolute-value generator in which d-c output is positive absolute value of input voltage.—W. S. Routh, "Applications Guide for Op Amps," National Semiconductor, Santa Clara, CA, 1969, AN-20.

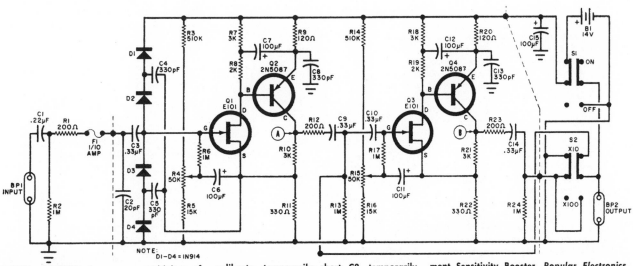

RANGE BOOSTER—Increases sensitivity of scope or vtvm by factor of 10 and factor of 100. Can be used up to 10 MHz. Scope or vtvm should have high input impedance. To calibrate, temporarily short C2, temporarily short R13, and adjust R4 until d-c voltmeter measures exactly 7.2 V between points A and B.—J. Bongiorno, Build the X10/X100 Instrument Sensitivity Booster, *Popular Electronics*, July 1970, p 43–47.

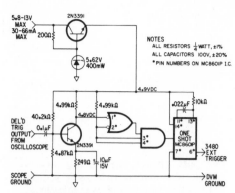

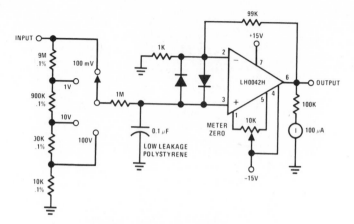

DVM TRIGGER—Designed for use with Hewlett Packard 3480 digital voltmeter when measuring amplitude of low-frequency pulses. Consists of single-transistor pulse amplifier followed by one-shot. Peak amplitude of delay trigger pulse from cro is about 1.5 V. When this pulse is absent, amplifier presents constant 4.9 V to input of one-shot. Arrival of delayed trigger pulse saturates transistor, causing one-shot to trigger on negative-going slope of collector voltage. Output of one-shot then falls from 5 V to ground for 75 μs due to external R-C timing network, for triggering 3480 dvm.—"Low Frequency Pulse Amplitude Measurements," Hewlett Packard, Palo Alto, CA, No. 133-1, 1971.

LOW-COST VTVM—Uses LH0042 fet opamp to replace all active circuitry. Supply can be eight flashlight cells, from which it draws only 20 mW.—R. K. Underwood, "New Design Techniques for FET Op Amps," National Semiconductor, Santa Clara, CA, 1972, AN-63, p 11.

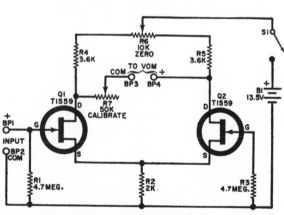

FET BOOSTS VOM SENSITIVITY—Provides effective input resistance of 5 meg when used with 1-V range of 1,000 ohms-per-volt volt-ohmmeter. Article covers construction and operation.—S. D. Prensky, Fet Rejuvenates Vom, *Popular Electronics*, Nov. 1968, p 72-73 and 114.

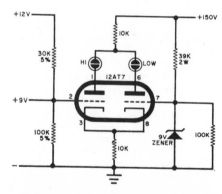

METERLESS VTVM—Indicates whether either 9-V or 12-V input is within 0.2 V of correct value. If Input is high, HI neon lamp glows brighter than other. If voltage is correct, both lamps are equally bright. Serves as low-cost vtvm for these two voltages.—J. Kyle, The Ubiquitous Neon Lamp, Popular Electronics, Nov. 1968, p 55–57.

VTVM ADAPTER FOR MULTIMETER—Draws only 1 μA from measured circuit for full-scale deflection of 60 or 100 μA meter of conventional multimeter, to give sensitivity of 1 meg per V for combination. Permits making low-voltage d-c measurements, (1-V or 10-V scale) at high impedance. To use, set multimeter to 60 or 100 μA position and plug adapter leads into corresponding jacks. Book gives calibration procedure.—"Electronics Experimenters Circuit Manual," General Electric, Owensboro, KY, 1971, 3rd Ed., p 239–242.

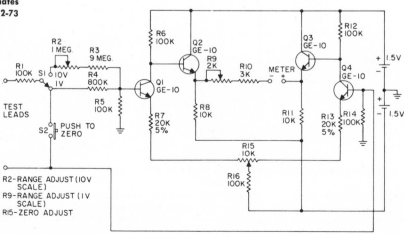

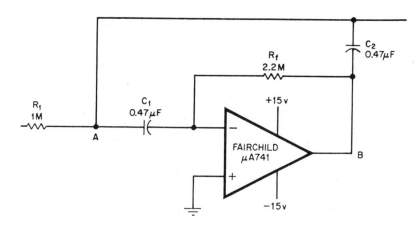

RIPPLE ATTENUATOR — Impedance-lowering opamp attenuates ripple in d-c signals of digital voltmeters by inverting, amplifying, and feeding back ripple to input. Gives 100-dB attenuation at 60 Hz. Will bring output within 1% of final value in only 5 s after step change in d-c level. Serves to multiply performance of small capacitors.—R. J. Battes, Impedance-Lowering Op Amp Speeds Filter Response, *Electronics*, May 10, 1971, p 82.

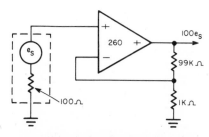

0.01% MICROVOLTMETER—Chopper stabilizing minimizes errors caused by thermal transients and holds drift sufficiently low for resolving microvolt signals. Inexpensive low-value resistors are adequate for meeting accuracy requirements. Total noise is only about 0.4 μV, and total drift only 0.1 μV per deg C. —P. Zicko, "Designing With Chopper Stabilized Operational Amplifiers," Analog Devices, Inc., Norwood, MA, Sept. 1970.

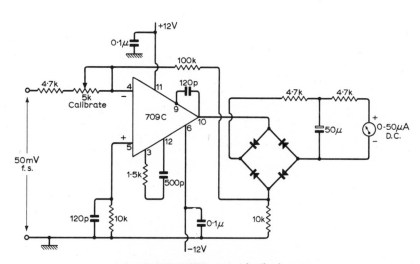

A-C MILLIVOLTMETER—Forced-feedback meter circuit using Fairchild opamp gives full-scale reading with 50-mV rms input down to 3 Hz. —M. V. Dromgoole, Op. Amp. A. C. Millivoltmeter, *Wireless World*, Feb. 1970, p 75.

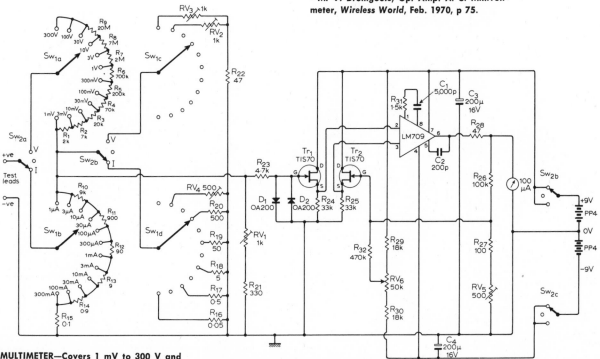

D-C MULTIMETER—Covers 1 mV to 300 V and 1 μA to 300 mA by switching in square-root-of-10 steps. Current drain is low enough to give battery life of 1 year in normal intermit-tent use. Input impedance is 1 meg per V up to 30 V and constant 30 meg on highest voltage ranges.—J. Johnstone, Direct Current Multimeter, *Wireless World*, Feb. 1971, p 87—88.

VOLTMETER AMPLIFIER—Permits measuring 2.5-mV d-c full-scale on 50-μA range of multimeter. Opamp brings measured voltage up to level at which it can be read accurately. Diodes are 1N914.—T. D. Towers, Elements of Linear Microcircuits, *Wireless World*, Feb. 1971, p 76-80.

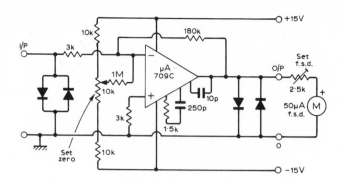

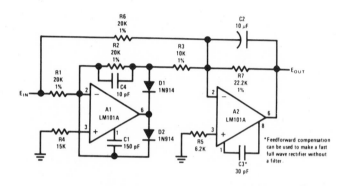

PRECISION 100-KHZ WEAK SIGNAL TO D-C—Provides accurate rectification of millivolt signals by placing silicon diode in feedback path of opamp and using feedforward compensation with inverting amplifier connection. Arrangement increases slew rate and reduces gain error at high frequencies. Opamp A2 converts half-wave rectifier to full-wave by summing half-wave rectified output with input signal. Report describes operation in detail. Conversion accuracy is better than 1% to above 100 kHz. Ripple at 20 Hz is less than 1%. Output is calibrated to read rms value of sine-wave input.—R. C. Dobkin, "Precision AC/DC Converters," National Semiconductor, Santa Clara, CA, 1969, LB-8.

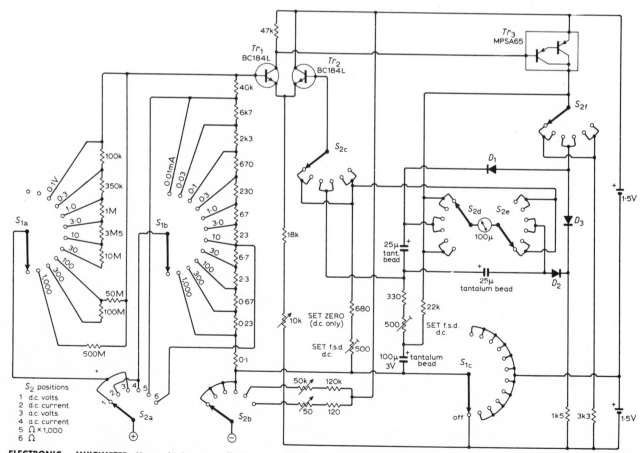

ELECTRONIC MULTIMETER—Uses high-gain silicon planar transistors in opamp voltage comparator serving as current multiplier for d-c mode. In a-c mode, modification of Waddington a-c millivoltmeter circuit is used. Current drain is only 300 μA on d-c ranges and 600 μA on a-c, from pair of 1.5-V pen-cells. Frequency range on a-c is 10 Hz to above 100 kHz. Use small-signal silicon diodes. Article covers construction.—J. L. Linsley Hood, Simple Electronic Multimeter, *Wireless World*, June 1972, p 279–281.

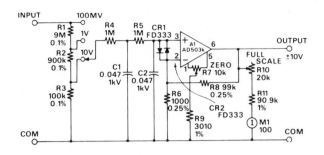

LAB D-C VOLTMETER—Attenuator at input provides full-scale input ranges of 10 V, 1 V, and 100 mV. Uses 100-μA microammeter at output. Optional output terminal provides up to ±10 V for feeding recorder. Full-scale accuracy is 0.5%. Low-pass filter R4-R5-C1-C2, having 1-Hz bandwith, permits reading d-c component without being overloaded by large a-c input signals.—"AD503, AD506 I.C. Fet Input Operational Amplifiers," Analog Devices, Inc., Norwood, MA, Technical Bulletin, Aug. 1971.

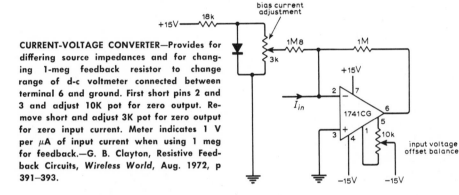

CURRENT-VOLTAGE CONVERTER—Provides for differing source impedances and for changing 1-meg feedback resistor to change range of d-c voltmeter connected between terminal 6 and ground. First short pins 2 and 3 and adjust 10K pot for zero output. Remove short and adjust 3K pot for zero output for zero input current. Meter indicates 1 V per μA of input current when using 1 meg for feedback.—G. B. Clayton, Resistive Feedback Circuits, *Wireless World*, Aug. 1972, p 391–393.

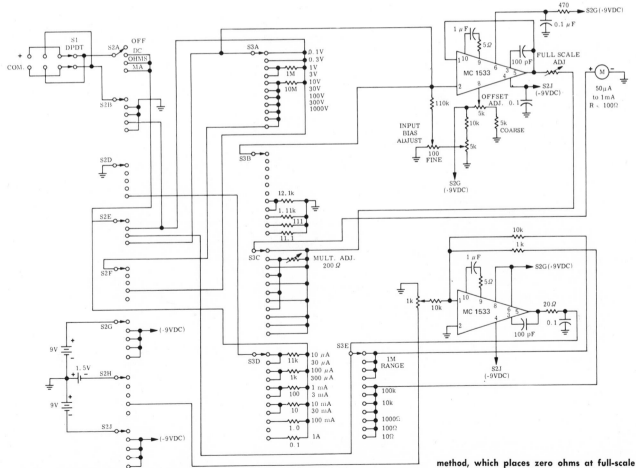

VOLTAGE FOLLOWER—Uses pair of MC1533 IC opamps featuring high open-loop gain, large common-mode input signal, and low drift, to give voltmeter having high input impedance on nine ranges up to 1,000 V d-c full-scale, along with current and resistance ranges. Report gives design equations. Ohmmeter section uses half-scale deflection method, which places zero ohms at full-scale deflection and minimizes meter zeroing between ranges.—L. L. Wisseman, "The MC1533 Monolithic Operational Amplifier," Motorola Semiconductors, Phoenix, AZ, AN-248, 1972.

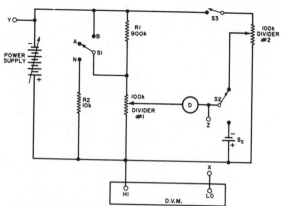

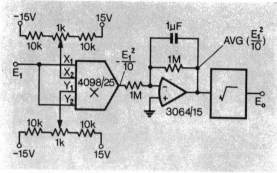

D-C TO 200-KHZ RMS VOLTMETER—Combines analog function modules with opamp. Input need not be sine wave.—B. Gledhill, Can You Afford to Ignore Analogue Modules?, *Electronic Engineering*, March 1970, p 63–65.

DVM CALIBRATOR—Used in precision metrology lab as voltage standard having at least 10 times better accuracy than manufacturer rating of 25 parts per million for high-quality digital voltmeters. Sc is saturated standard cell rated 1.017586 V, held at 35 C in hermetically sealed oven. Divider No. 1 is seven-decade Kelvin-Varley having terminal linearity of 0.1 ppm. Divider No. 2 is five-decade unit of any accuracy. R1 and R2 have short-term stability better than 1 ppm. Null detector D is electronic detector with variable sensitivity to 1 μV full-scale. Power supply is stable seven-digit source with manufacturer's stated accuracy of 100 ppm, maintained at accuracy of Sc by switching S2 and adjusting supply for balance. Article gives procedures for calibrating digital voltmeters up to 1,000 V to required accuracy.—D. L. Kierstead, Calibrating Digital Voltmeters, *Electronics World*, June 1971, p 36 and 62.

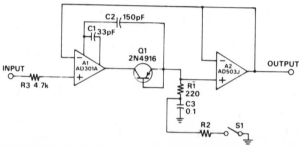

PEAK RECTIFIER—Precision half-wave rectifier delivers d-c output equal to positive peak value of input signal. Will also read negative peaks if polarity of Q1 (connected as diode) is reversed. A2 serves to eliminate loading on C3. Bleeder R2 can be used to return output to 0 after input falls to 0. For 7 V rms input, d-c output level is accurate within 2% for 30 to 1,000 Hz when R2 is 10 meg.—"AD503, AD506 I.C. Fet Input Operational Amplifier," Analog Devices, Inc., Norwood, MA, Technical Bulletin, Aug. 1971.

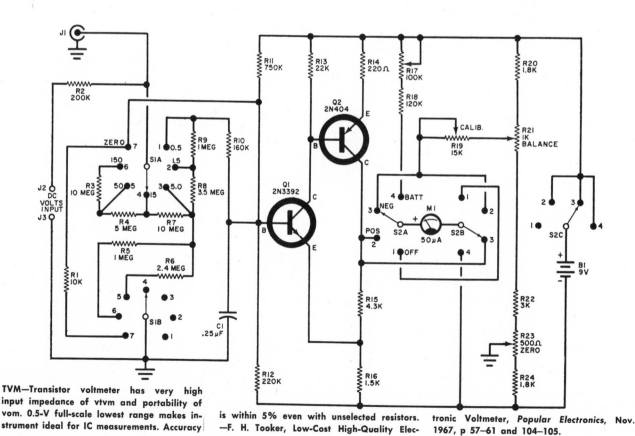

TVM—Transistor voltmeter has very high input impedance of vtvm and portability of vom. 0.5-V full-scale lowest range makes instrument ideal for IC measurements. Accuracy is within 5% even with unselected resistors.—F. H. Tooker, Low-Cost High-Quality Electronic Voltmeter, *Popular Electronics*, Nov. 1967, p 57–61 and 104–105.

CHAPTER 67
Medical Circuits

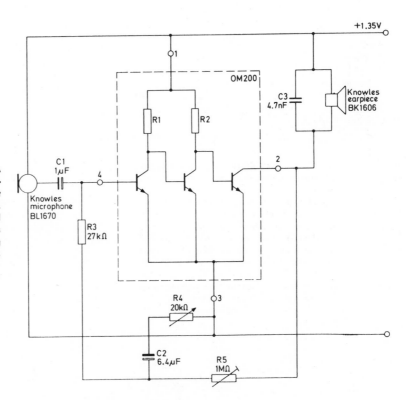

200-μW HEARING AID—Draws only 1.5 mA from 1.4-V mercury cell. Has good low-frequency response, being only 3 dB down at 350 Hz. Volume control range of R4 is 30 dB. Preset pot R5 in feedback circuit is adjusted so d-c voltage at collector of output stage is 1.1 V. Uses Mullard IC.—"Mullard Hearing Aid Amplifier Circuits," Mullard, London, 1969, TP1134, p 25–27.

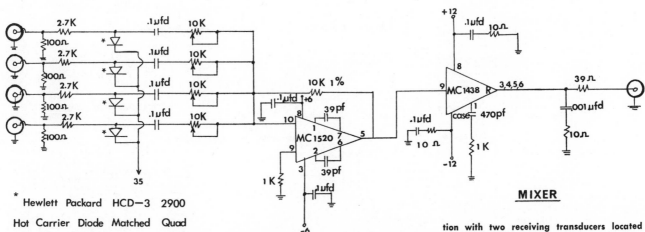

* Hewlett Packard HCD—3 2900

Hot Carrier Diode Matched Quad

MIXER

RECEIVER FOR CARDIAC CATHETER—Four-channel mixer with independent gain controls feeds single receiver using low-noise opamps. Used in connection with pulse-driven cathet-er-borne array of four transducers introduced into left ventricle of heart for sensing contours at various stages during cardiac cycle by echo-ranging techniques. Used in combination with two receiving transducers located on chest wall.—R. C. Eggleton et al, Ultrasonic Visualization of Left Ventricular Dynamics, *IEEE Trans. on Sonics and Ultrasonics,* July 1970, p 143–153.

RECORDING LEVEL INDICATOR FOR BLIND—
Tiny d-c motor with reduction drive is driven
by mono mvbr. Tr1 is normally conducting
and Tr2 cut off. With audio input, 100K control is adjusted so knob on output shaft of
motor just pulses when tape recorder magic
eye shows maximum recording level. Blind
person can feel movements of knob and reduce volume when making recording.—G. J.
Andriessen, Tactile Recording Level Indicator,
Wireless World, Feb. 1967, p 79–80.

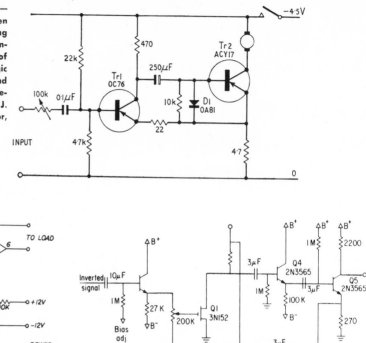

A-C CONSTANT-CURRENT PUMP—Used to excite bipolar electrode in tank of conductive
solution or in torso models. Furnishes about
200 μA p-p at 10 Hz to resistive impedance
of about 300 ohms in 0.45% saline solution
used. Constant current is obtained by placing
electrode in feedback loop of opamp. Will
produce excellent square wave of constant
p-p current for several days with less than
200 mA of offset and drift.—C. W. Brandon
et al, A Bipolar Electrode and Current Pump
for Volume Conductor Experiments, *IEEE
Trans. on Bio-Medical Engineering*, Jan. 1971,
p 70–71.

**ELECTROENCEPHALOPHONE AMPLITUDE MOD-
ULATOR—**Used in system for giving stereo
presentation of inaudible 8–13 HZ signals
from four quadrants of human skull. Input is f-m signal centered on 1,240 Hz for
one channel and on 700 Hz for other. Circuit
provides amplitude modulation by varying
channel resistance of fet Q1. To insure that
zero input produces null on outpt amplitude,
some of original non-a-m audio signal is subtracted from a-m audio signal.—S. B. Gine,
M. H. Graham, and C. Susskind, Processing
and Stereophonic Presentation of Physiological Signals, *IEEE Trans. on Bio-Medical Engineering*, Jan. 1971, p 9–15.

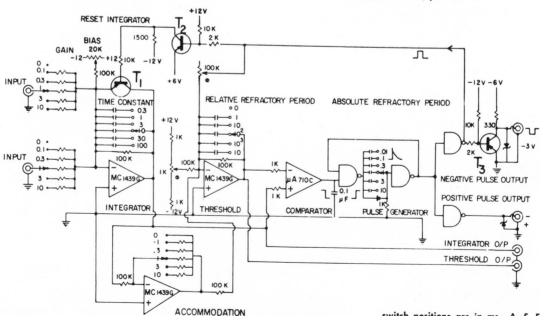

NEURAL ANALOG—Used to stimulate activity
of nerve cells. Includes circuitry for modeling
important neural properties of adaptation
and accommodation. T1 is 2N1305; T2 and T3
are 2N3906. All gates are in single Texas Instruments 7400 IC chip. All times indicated on
switch positions are in ms.—A. S. French and
R. B. Stein, A Flexible Neural Analog Using
Integrated Circuits, *IEEE Trans. on Bio-Medical
Engineering*, July 1970, p 248–253.

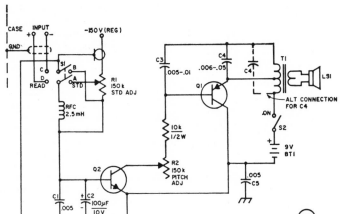

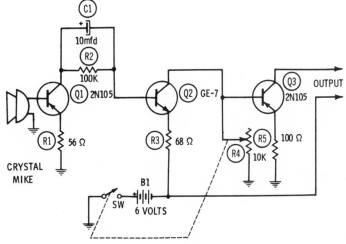

TRANSMITTER-TUNING TONE FOR BLIND—Voltage-sensitive tone generator uses voltage amplifier Q2 (any npn audio type) to change frequency of relaxation audio oscillator Q1 (any pnp audio type). Input is voltage normally used to drive tuning meter of transmitter, generally up to 100 mV d-c.—A. E. Schwaneke, Equipment Modification for the Blind, QST, Feb. 1970, p 11–19.

HEARING AID—Can easily be constructed in plastic box fitting in shirt pocket, with tiny high-output crystal mike mounted in hole drilled in box. Output can be fed through two-conductor fine insulated cord to transistor-radio earpiece. R4 adjusts volume.—R. Brown and T. Kneitel, "101 Easy Audio Projects," Howard W. Sams, Indianapolis, IN, 1971, p 158–159.

HAM-STATION CONTROL FOR BLIND—Completely integrated control unit can be used with antenna, transmitter, and receiver to make it possible for any sightless amateur radio operator to adjust equipment and go on the air. Tone comparator allows operator to estimate transmitter voltages and currents with high degree of accuracy by listening to frequency of tone generated by audio oscillator Q2. Circuit includes antenna changeover relay. Article covers construction, adjustment, and operation. Q1 is HEP641 and Q2 is HEP51. Construction of Z1 is covered in Oct. 1969 QST.—L. G. McCoy, A Station Control Unit for the Blind Amateur, QST, Nov. 1970, p 32–36.

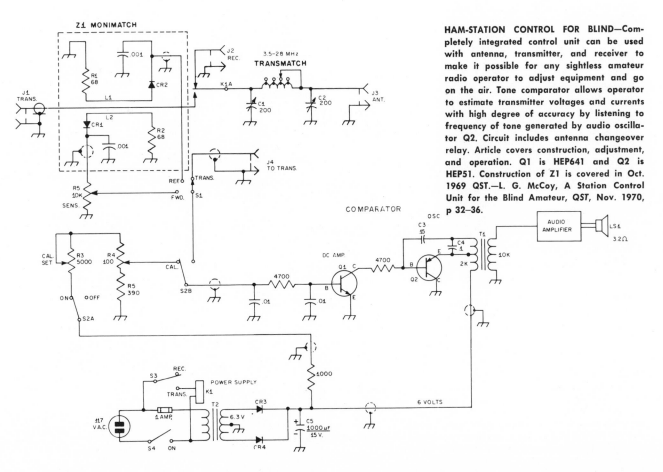

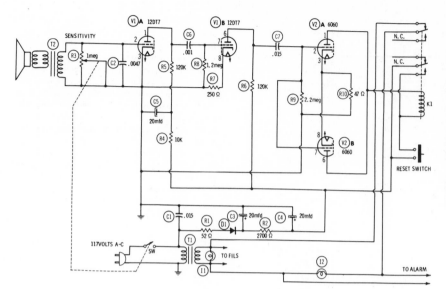

VOICE ALARM FOR DEAF MOTHER—P-m speaker serving as mike picks up sound of child calling or crying and energizes relay that can serve to turn on attention-getting red lamp, loud buzzer, or other device needed to alert deaf person. Can also be used to pick up splash of person falling in swimming pool, to wake up person when he starts snoring, or as intrusion alarm. K1 is 5K relay. D1 is 65-mA selenium rectifier. T1 is 6.3-V filament transformer and T2 is output transformer such as Stancor A-4744. Push reset switch to stop alarm and reset for next sound. I1 and I2 are No. 47 pilot lamps.—R. Brown and T. Kneitel, "101 Easy Audio Projects," Howard W. Sams, Indianpolis, IN, 1971, p 82–84.

ECG AMPLIFIER—Battery-powered amplifier drives zero-center moving-coil pen recorder of electrocardiograph to give full-scale deflection of 260 mA from d-c to 100 Hz. Balanced input from direct-coupled preamp drives long-tailed input pair of transistors. These in turn provide balanced drive to inverting inputs of GE opamps.—G. B. C. Harrop, Driver Amplifier for Pen Recorder, *Wireless World*, Aug. 1969, p 379.

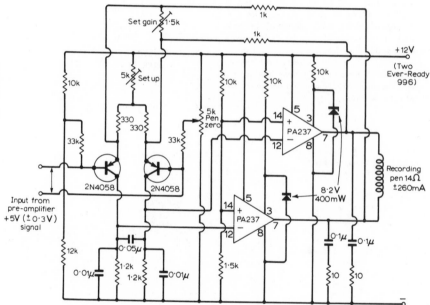

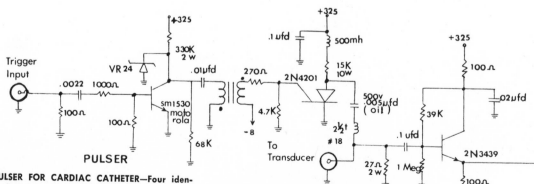

PULSER

PULSER FOR CARDIAC CATHETER—Four identical pulsers like that shown drive catheter-borne array of four transducers spaced 90 deg apart, introduced into left ventricle of heart for sensing contour of this ventricle at various stages during cardiac cycle. Transducers are pulsed sequentially at rate of 1,000 Hz. Used with receiver circuit and computer for sorting and storing resulting echo-ranging data.—R. C. Eggleton et al, Ultrasonic Visualization of Left Ventricular Dynamics, *IEEE Trans. on Sonics and Ultrasonics*, July 1970, p 143–153.

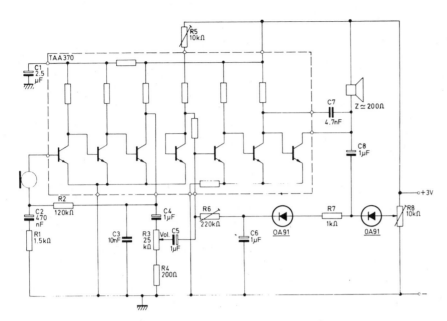

10-MW HEARING AID—Uses Mullard IC with power amplifier connected as class A sliding-bias output stage. With no signal, output transistor is biased only slightly on and draws very little current. With signal, reduction in collector voltage is rectified by voltage-doubling detector and resulting d-c is fed back to increase bias of output stage. Reduces battery drain, because power is taken from supply only while signal is being amplified. Adjust R5 for 0.5-mA no-signal quiescent current in output stage. Adjust R8 so output stage current starts to rise as soon as signal is applied. Adjust R6 for 12.5 mA in output stage at full output.—"Mullard Hearing Aid Amplifier Circuits," Mullard, London, 1969, TP1134, p 49–51.

BLOOD PRESSURE SIMULATOR—Opamps coupled to Schmitt triggers generate and combine triangular waveforms to simulate blood pressure wave in patient, for calibrating devices that record analog output of catheter. Heart rate is adjustable from 0.2 to 20 Hz, and respiration rate from 0.1 to 10 Hz.—M. E. Swinnen, Amplifiers and Triggers Simulate Blood Pressure, *Electronics*, Oct. 31, 1966, p 68–69.

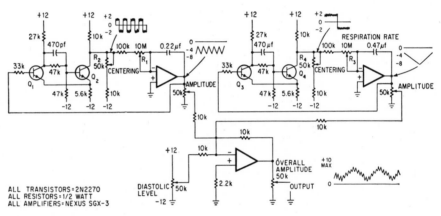

ALL TRANSISTORS = 2N2270
ALL RESISTORS = 1/2 WATT
ALL AMPLIFIERS = NEXUS SGX-3

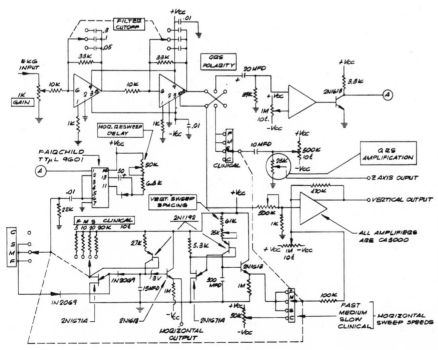

ECG CONTOUROGRAPH—Produces electrocardiographic display in contourographic format in real time on face of variable-persistence crt, to facilitate on-line comparison of successive ECG cycles. Also gives instantaneous heart rate readout from scale on crt face on beat-by-beat basis. All opamps are CA3000. Terminals A are connected together. Z-axis output, when positive, decreases brightness of trace on Hewlett-Packard H12-1208B scope. Circuit provides storage, continuous, and scalar modes of operation. Covers heart rate range of 30 to 200 beats per minute.—D. P. Golden, D. G. Mauldin, and R. A. Wolthuis, *IEEE Trans. on Bio-Medical Engineering*, Oct. 1970, p 296–302.

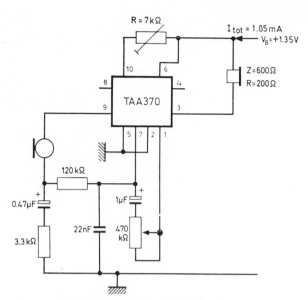

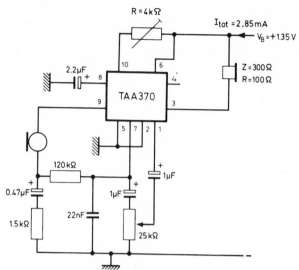

SIMPLIFIED 0.9-MW HEARING AID—Uses Mullard IC with minimum of additional components for acceptable performance. Uses mike having maximum of 4 mV rms output.—"Mullard Hearing Aid Amplifier Circuits," Mullard, London, 1969, TP1134, p 46.

1.5-MW HEARING AID—Uses Mullard IC amplifier built on 0.9-mm-square silicon chip. Designed for minimum power drain from 1.4-V mercury cell. Control range of volume control is 45 dB. Maximum output voltage of 5K mike used is about 4 mV rms. For values shown, response is down 3 dB at 30 and 20,000 Hz. Book also gives similar circuits for outputs of 0.9, 0.8, and 0.5 mW.—"Mullard Hearing Aid Amplifier Circuits," Mullard, London, 1969, TP1134, p 29–37.

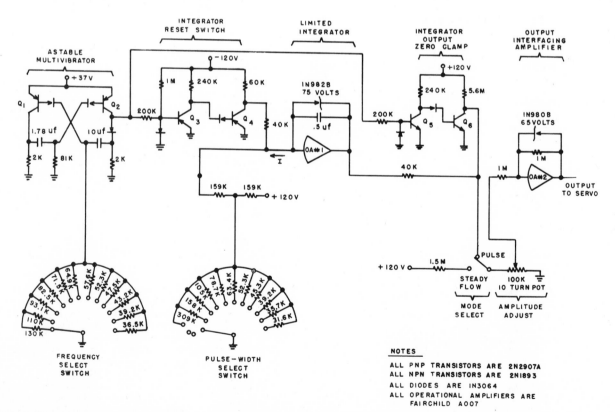

BLOOD-PUMP DRIVE—Pulse generator produces required triangular waveform for driving d-c torque motor which in turn drives roller pump in cardiac bypass line. Generator allows independent variation of pulse rate between 60 and 170 pulses per minute by varying astable mvbr frequency. Systolic duration may be varied from 0.1 to 0.35 s by changing d-c bias of integrator input. Peak flow of blood may be varied continuously from 0 to 25 liters per minute by attenuating integrator output with 100K pot. Output of tachometer on motor shaft is fed back to power amplifier to make shaft velocity follow input signal closely.—B. F. Hoffman et al, A Pulsatile Roller Pump for Cardiac Bypass, *IEEE Transactions on Bio-Medical Engineering*, Jan. 1970, p 78–80.

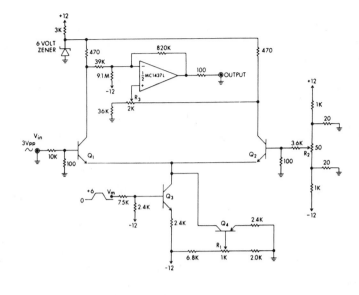

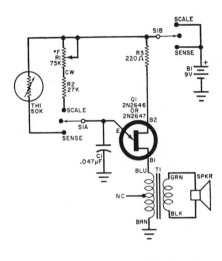

EAR STIMULATOR—Sinusoidal audio signals are linearly modulated to generate tone bursts of controlled duration and shape for audio research. Modulator uses transistor differential pair Q1-Q2 which with Q3 are in RCA CA3006 IC. Q4 is 2N3638. Article also gives circuit of trapezoidal generator providing modulating waveform. Response is flat to 50 kHz.—V. Klig, S. Beitler, and E. Stephenson, Sinusoidal Burst Generator for Auditory Evoked Response Research, *IEEE Trans. on Bio-Medical Engineering*, Jan. 1970, p 74–76.

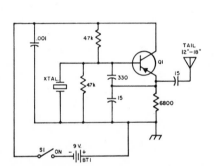

FREQUENCY SPOTTER FOR BLIND—Circuit using 2N1177 or eqivalent will oscillate with any FT-243 crystals from 3 to 30 MHz. When operator of blind receiver wants to know where specific frequency is on receiver tuning dial, he plugs crystal for that frequency into oscillator and tunes receiver until marker signal is heard. Crystals are identified with Braille labels. Antenna is insulated wire poked into ventilation hole in receiver or transceiver.—A. E. Schwaneke, Equipment Modification for the Blind, *QST*, Feb. 1970, p 11–19.

AUDIBLE THERMOMETER—Allows blind person to determine air temperature in range of 55 to 100 F, simply by comparing two tones. User notes tone with switch in SENSE position, then switches S1 to SCALE position and adjusts R1 until ujt oscillator Q1 gives same tone. Braille numbers on panel are then read at position of pointer for R1. T1 is transistor output transformer driving 3.2-ohm 2½-inch speaker. TH1 is Fenwal GA45P2 or similar 50K probe-type thermistor, which can be mounted outdoors.—F. H. Tooker, Build Sound-Signal Thermometer, *Popular Electronics*, p 59–61.

ELECTROENCEPHALOPHONE FREQUENCY MODULATOR—Used in system for presenting effective stereophonic display of four channels of electroencephalography (EEG), one from each quadrant of the human skull. EEG signals are 8 to 13 Hz, below audibility. Signals from right and left front of skull are amplified, combined, and applied to voltage-controlled relaxation oscillator shown, to create symmetrical triangular waves and synthesize sine waves from them in diode network, for frequency-modulating carrier having center frequency of 1,240 Hz. Same circuit is used for frequency-modulating 700-Hz carrier for rear signals. Resulting audible signals are fed to two speakers or stereo headphones.—S. B. Fine, M. H. Graham, and C. Susskind, Processing and Stereophonic Presentation of Physiological Signals, *IEEE Trans. on Bio-Medical Engineering*, Jan. 1971, p 9–15.

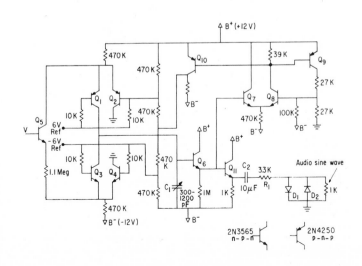

TRANSCEIVER TUNER FOR BLIND—Provides output tone whose frequency is proportional to deflection of transceiver tuning meter. Developed for HW-16, but can be adapted to almost any transmitter. Also provides reference frequency that is set to tone corresponding to plate current for 75-W input power for final stage of transmitter. Although output is square wave, its tone is acceptable. C4 sets output frequency range of oscillator, 1,000 to 4,000 Hz. Diodes are 1N914. Article covers calibration and use.—T. P. Riley, An IC Audio Tune-Up Device for the Blind Amateur, QST, June 1972, p 41–43.

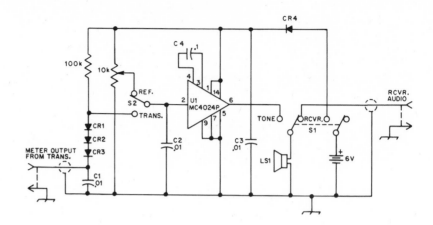

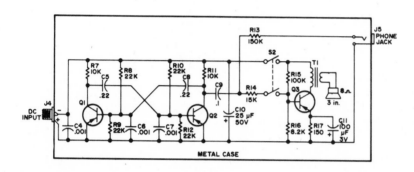

AUDIO-TONE TUNING FOR SSB—Requires no external power source. Final amplifier cannot overheat even when tuned circuits of ssb transmitter are far off resonance. Intended for use by blind or otherwise handicapped ham radio operators. For speaker operation, feed output to small IC a-f amplifier. When r-f voltage in antenna feedline increases during tuning, frequency and intensity of audible tone increases. Article covers construction and operation. Q1 and Q2 are 2N4354; Q3 is 2N270.—F. J. Fox, An Aural Transmitter Tuning Aid, 73, Nov. 1972, p 305–309.

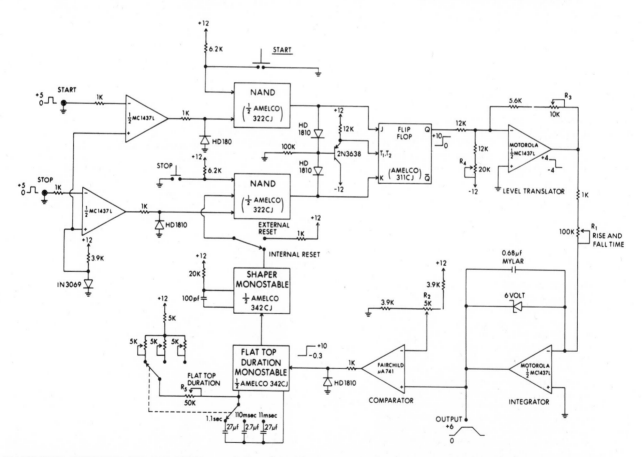

TRAPEZOIDAL GENERATOR FOR EAR STIMULATOR—Generates trapezoidal modulating voltage for stimulator producing sinusoidal signals in audio range, linearly modulated to generate tone bursts of controlled duration and shape for auditory research. Output has equal rise and fall times, independent of duration of flat top of trapezoid. Output terminal (lower right) provides 6-V p-p waveform for modulator.—V. Klig, S. Beitler, and E. Stephenson, Sinusoidal Burst Generator for Auditory Evoked Response Research, *IEEE Trans. on Bio-Medical Engineering*, Jan. 1970, p 74–76.

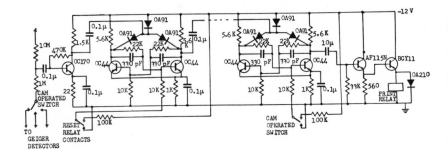

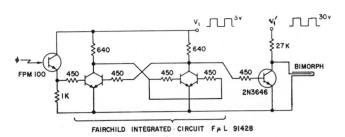

FAIRCHILD INTEGRATED CIRCUIT FµL 91428

COUNTER FOR GEIGER DETECTOR—Mica-window geiger tubes in contact with surface breast tumors are selectively switched to input of counter while radiophosphorus is administered to patient. After preset count of 2,048 is reached, print relay is energized on chart recorder, to print dot at time-proportional spacing from previous dot. Only first and last (12th) binaries are shown.—W. M. Burch, Measurement of Periodicity in 32-P Concentration in Breast Tumors with an Automated Integral Counting Technique, *IEEE Trans. on Bio-Medical Engineering*, Jan. 1970, p 66–69.

READING AID FOR BLIND—Multiplicity of optoelectronic circuits, all like that shown, each drive vibratory pin consisting of bimorph or piezoelectric reed. Reading material is focused on phototransistor array, and pins are gated on or off to give tactile image. Phototransistor at left sets IC flip-flop which determines whether or not switching transistor shunts driving current for bimorph pin. Chief drawback is slowness of reading rate achieved, about 35 words per minute.—R. C. Joy and J. G. Linvill, Optoelectronic Circuitry for a Reading Aid, *IEEE Journal of Solid-State Circuits*, Dec. 1968, p 452–453.

CHAPTER 68
Memory Circuits

SENSE AMPLIFIER—Motorola MC1510 high-gain broadband sense signal amplifier is gated off during application of digit pulse, to prevent saturation so circuit recovers in time to amplify bit signal. Will recover from 5-V digit pulse in 700 ns. R2 forms network with R1 to bias μA710 comparator beyond cutoff and keep output at zero when signal is applied. Circuit is effective for noise up to 0.5 mV on sense lines from laminated ferrite-core memories.—H. P. Brockman, Amplifier Recovery from Saturation Slowing You Down? Give It the Gate, *The Electronic Engineer*, May 1968, p 72.

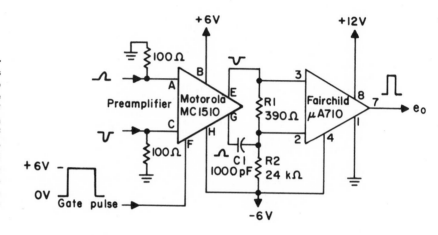

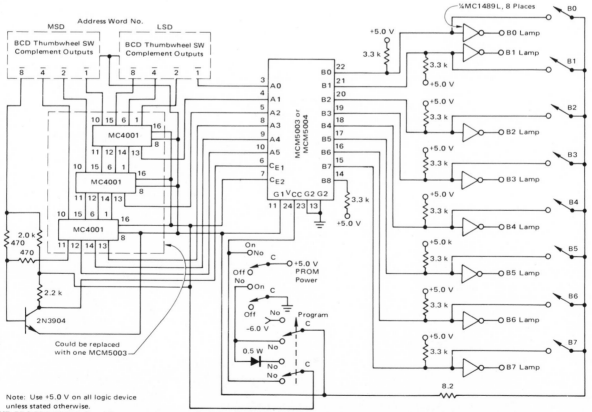

512-BIT PROGRAMMER—Provides current gain required to ensure high reliability when fusing link of programmable read-only memory. Address word number is selected by setting bcd thumbwheel switches, and three MC4001 TTL standard read-only memories are used to convert bcd code to code required at address inputs. Display verifies that bit has been fused. Report covers operation in detail.—J. E. Prioste, "Programming the MCM5003/5004 Programmable Read Only Memory," Motorola Semiconductors, Phoenix, AZ, AN-550, 1971.

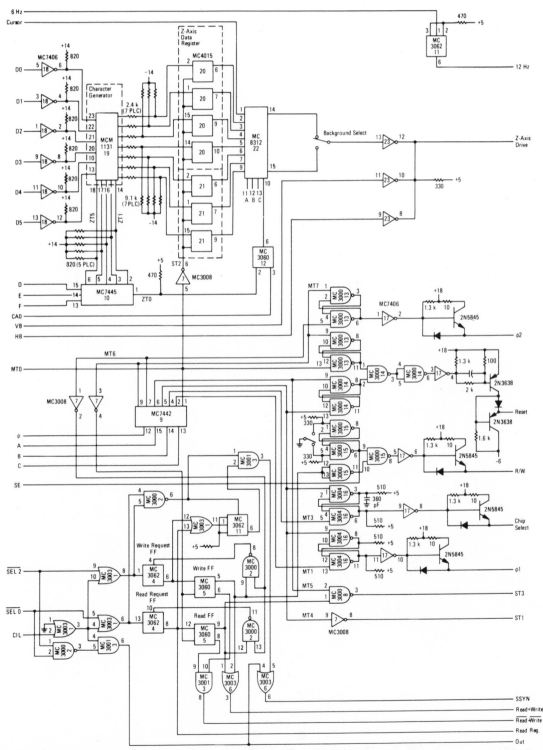

CRT DISPLAY DRIVE FOR RAM—Used to provide required timing and character generator signals for alphanumeric display having 16 horizontal rows of 64 characters each on crt, for dynamic read-only memory. Report describes operation in detail and gives all other circuits needed for system. Will provide either read or write operation.—B. Bratt, "CRT Display with Dynamic MOS RAM Storage," Motorola Semiconductors, Phoenix, AZ, AN-558, 1972.

CORE MEMORY SENSE AMPLIFIER—Uses IC dual differential comparator as interface between memory and logic elements of computer. Designed to detect small signals while distinguishing between signals differing only few mV in amplitude. When strobe input is grounded, output is clamped at about —0.5 V at amplifier input from falsely triggering memory data register.—R. Brunner, "A High Speed Dual Differential Comparator—The MC1514," Motorola Semiconductors, Phoenix, AZ, AN-547, 1971.

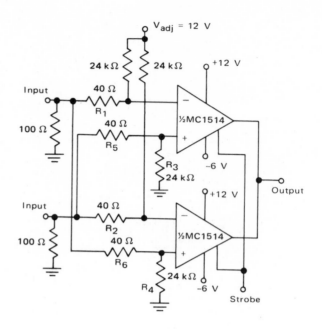

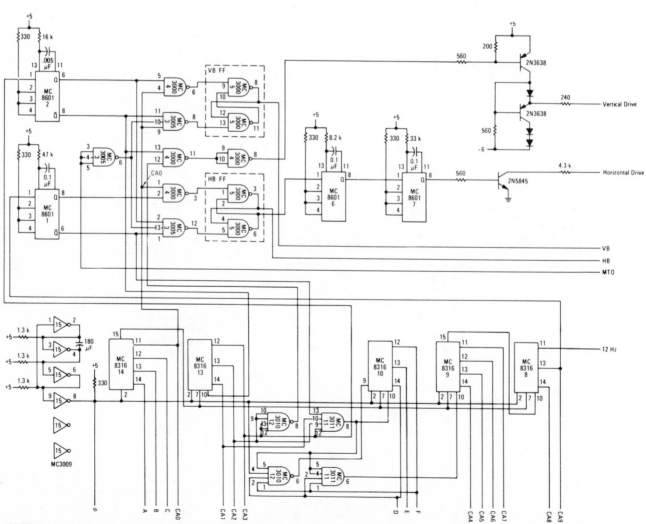

COUNTER AND RETRACE CONTROL—Used to generate interval, position, and scan rate signals required to provide alphanumeric crt display with random-access memory. IC15 hex-inverter forms ring oscillator providing 5.4- MHz clock signal driving 18-bit synchronous counter providing division of clock frequency by 196,608 to give display frequency of 51 Hz (allowing for retrace times), for freedom from flicker. Final output of 12 Hz generates blinking cursor. Report describes operation in detail and gives all other circuits needed for system.—B. Bratt, "CRT Display with Dynamic MOS RAM Storage," Motorola Semiconductors, Phoenix, AZ, AN-558, 1972.

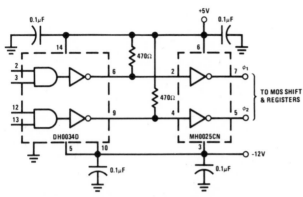

NEGATIVE-LEVEL TRANSLATOR—Permits direct coupling for National MH0025 monolithic clock driver when shifting to negative level is desired. Uses DH0034 dual high-speed TTL to negative level converter.—B. Siegel and M. Scott, "Applying Modern Clock Drivers to MOS Memories," National Semiconductor, Santa Clara, CA, AN-76, 1973, p 9.

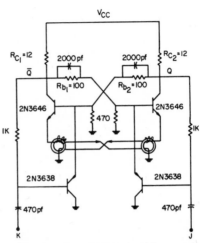

NONVOLATILE MEMORY—Provides nondestructive readout and requires no sensing amplifiers. Uses only conventional transistors, magnetic switching cores, resistors, and capacitors. Useful for systems requiring low memory capacity, from a few bits to about 100 bits. Cores ensure that flip-flop will return to state it was in prior to power removal. Uses Electronic Memories 50-mil linear-select cores. Will operate on supply between 3 and 5 V.—M. Michael and W. C. Lin, A Nonvolatile Memory Circuit—A Novel Approach, *IEEE Journal of Solid-State Circuits*, Oct. 1969, p 288–291.

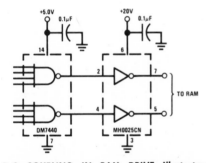

D-C COUPLING IN RAM DRIVE—Illustrates how National MH0025 monolithic clock driver may be direct-coupled in applications when level-shifting to positive value only, as with MM1103 random-access memory operating between ground and +20 V.—B. Siegel and M. Scott, "Applying Modern Clock Drivers to MOS Memories," National Semiconductor, Santa Clara, CA, AN-76, 1973, p 9.

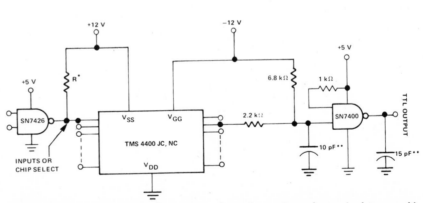

4,096-BIT STATIC RAM—Read-only memory has typical access time under 1 μs, when used with input switching circuit shown and TTL interface for output. Value of R depends on system speed and power requirements.

Lumped capacitors shown simulate parasitic capacitances that provide loading in practical application.—"MOS/LSI Standard Products Catalog," Texas Instruments, Dallas, TX, Catalog CC-402, 1971, p 218.

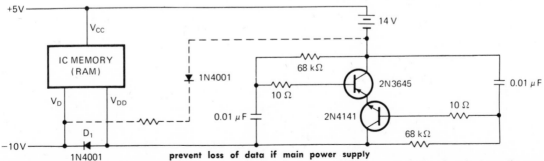

RAM DATA PROTECTION—Series mvbr provides 1-μs pulses at 1 kHz from 14-V standby battery to replenish mos gate charges and prevent loss of data if main power supply fails. Standby battery drain is comparable to shelf life. Dashed-line circuit recharges standby battery when line power is restored.

—K. C. Herrick, Pulsed Standby Battery Saves MOS Memory Data, *Electronics*, May 8, 1972, p 102–103.

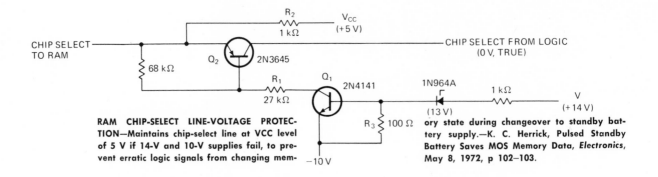

RAM CHIP-SELECT LINE-VOLTAGE PROTEC-TION—Maintains chip-select line at VCC level of 5 V if 14-V and 10-V supplies fail, to prevent erratic logic signals from changing mem-ory state during changeover to standby battery supply.—K. C. Herrick, Pulsed Standby Battery Saves MOS Memory Data, *Electronics*, May 8, 1972, p 102–103.

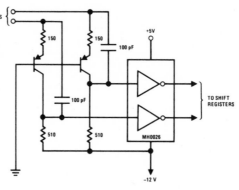

TRANSISTOR-COUPLED MOS CLOCK DRIVER—Provides level shifting along with direct coupling to National MH0026 monolithic clock driver feeding shift registers of random-access memory. Transistors can both be 2N3906.—B. Siegel and M. Scott, "Applying Modern Clock Drivers to MOS Memories," National Semiconductor, Santa Clara, CA, AN-76, 1973, p 11.

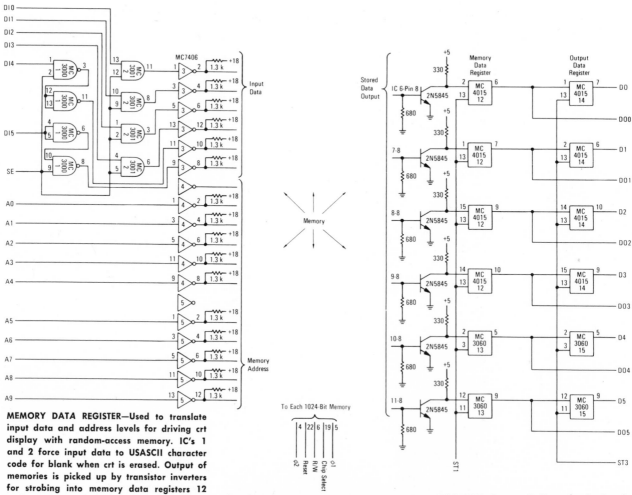

MEMORY DATA REGISTER—Used to translate input data and address levels for driving crt display with random-access memory. IC's 1 and 2 force input data to USASCII character code for blank when crt is erased. Output of memories is picked up by transistor inverters for strobing into memory data registers 12 and 13, from which desired character code goes to character generator. Memory consists of eight MCM1172 IC's, each holding 1,024 bits.—B. Bratt, "CRT Display with Dynamic MOS RAM Storage," Motorola Semiconductors, Phoenix, AZ, AN-558, 1972.

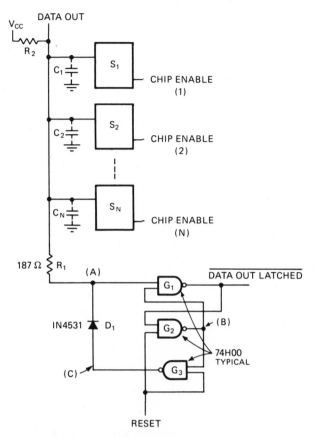

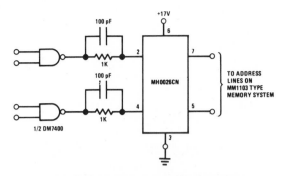

RAM DRIVER—National MH0026 monolithic clock driver with direct-coupled input drives address or pre-charge lines on MM1103 random-access memory. Requires only positive supply.—B. Siegel and M. Scott, "Applying Modern Clock Drivers to MOS Memories," National Semiconductor, Santa Clara, CA, AN-76, 1973, p 11.

SPEEDING UP WIRE-ORED MEMORY—Turn-off delay can be reduced from 70 ns to 10 ns by adding feedback gate to NAND gate latch, for 16 wire-ORed memory packages and 4.7K pull-up resistor R2.—J. McDowell and W. Moss, Feedback Latch Reduces Memory Recovery Time, *Electronics*, Feb. 28, 1972, p 82.

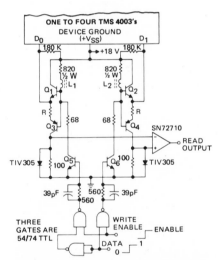

READ-WRITE FOR RAM—Provides high-speed read or write for up to four TMS 4003 256-bit random-access memories. Data can be sensed within 50 ns after being written. L1 and L2 are 2½ turns No. 30 on ferrite bead. R is 47 for four memories and 180 for one. Q3 and Q4 are 2N3829 and all other transistors 2N3014.—G. B. Hoffman, "MOS Random-Access 256-Bit Memory," Texas Instruments, Dallas, TX, Bulletin CA-127, 1969, p 6–7.

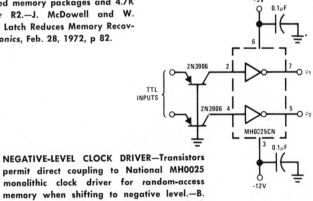

NEGATIVE-LEVEL CLOCK DRIVER—Transistors permit direct coupling to National MH0025 monolithic clock driver for random-access memory when shifting to negative level.—B. Siegel and M. Scott, "Applying Modern Clock Drivers to MOS Memories," National Semiconductor, Santa Clara, CA, AN-76, 1973, p 9.

LEVEL-SHIFTING RAM DRIVE—Arrangement provides direct coupling along with d-c level shift, for National MH0026 monolithic clock driver for random-access memory. Interface to clock is DH0034 dual high-speed TTL to negative-level converter.—B. Siegel and M. Scott, "Applying Modern Clock Drivers to MOS Memories," National Semiconductor, Santa Clara, CA, AN-76, 1973, p 11.

CHAPTER 69
Metal Detector Circuits

METAL SENSOR—Simple 3-transistor design can be used for locating nails in wall studs, buried pipes, and other buried or concealed objects. L1 is ferrite-rod loopstick antenna, and L2 is standard a-m loop antenna covered with metal window screen serving as shield. Loop antenna is mounted at one end of wood rod and rest of circuit on other end. To adjust, hold loop near metal object and adjust L1 for maximum buzz or tone in phones, then adjust C6 for maximum tone. Construction details are given.—R. M. Brown and R. Kneitel, "49 Easy Entertainment & Science Projects," Howard W. Sams, Indianapolis, IN, 1971, p 21–24.

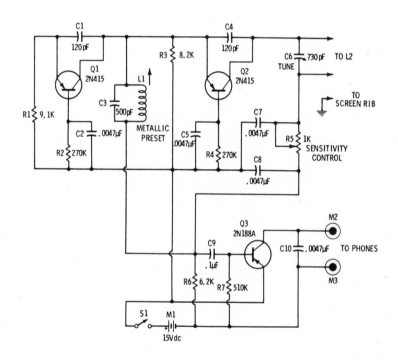

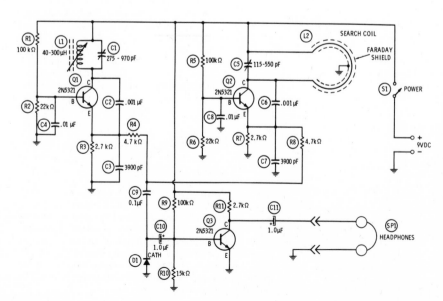

THREE - STAGE DETECTOR — Built in housing mounted on broom handle, with search coil on lower end of handle. Uses beat-frequency principle with Q2 as variable oscillator whose frequency is changed by metal near search coil and Q1 as fixed r-f oscillator tuned to get beat note when no metal is present. Q3 amplifies beat to level sufficient for headphones. Search coil is 12 turns No. 18 enamel on 12-inch diameter form, partly enclosed by Faraday shield for which construction details are given. Smaller-diameter coil will be more sensitive to coins but have less depth of penetration. L1 has adjustable ferrite core. Diode is 1N34A.—J. P. Shields, "How to Build Proximity Detectors and Metal Locators," Howard W. Sams, Indianapolis, IN, 1972, p 120–129.

TUNTED-LOOP OSCILLATOR WITH CRYSTAL FILTER—Simple, stable, and sensitive circuit for locating metal objects in ground is easy to build and operate. Will detect coins up to 8 inches from loop and larger objects up to several feet. Article gives construction and calibration details, including add-on a-f amplifier-speaker unit that can be plugged into J2 for audible indication supplementing meter reading.—C. D. Rakes, Build a Treasure Finder, *Radio-Electronics*, Nov. 1967, p 32–33.

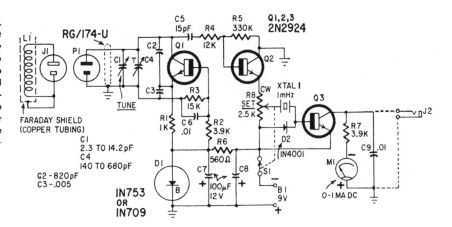

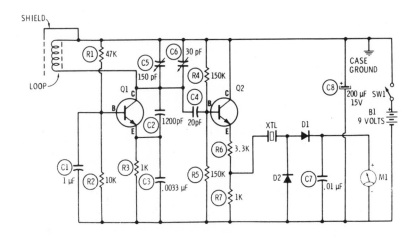

CRYSTAL-FILTER LOCATOR—Suitable for finding coins and other small objects at depths around 1 ft. Consists of Colpitts oscillator Q1 connected to 8-turn search loop mounted in 5-inch diameter Faraday shield (construction details are given), feeding buffer amplifier Q2. Both transistors are 2N2924. Oscillator is tuned to 1-MHz frequency of crystal operating in series-resonant mode as narrow-pass filter. Diodes (1N914 silicon) rectify r-f signal for driving 0–50 μA panel meter. Crystal can be any frequency from 700 kHz to 1.5 MHz as long as oscillator is tuned to it.—C. D. Rakes, "Solid State Electronic Projects," Howard W. Sams, Indianapolis, IN, 1972, p 67–73.

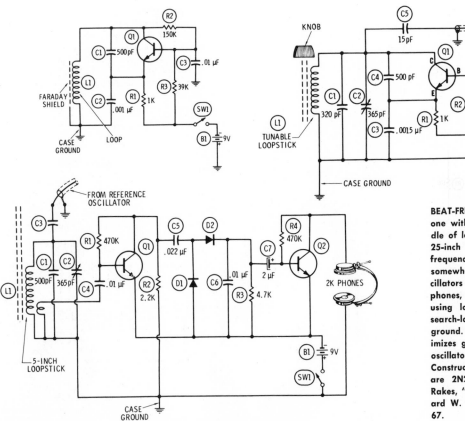

BEAT-FREQUENCY LOCATOR—Two oscillators, one with tunable loopstick mounted on handle of locator and other mounted on 11-turn 25-inch search loop, are adjusted for beat frequency of about 10 Hz while operating somewhere between 500 kHz and 2 MHz. Oscillators feed mixer amplifier driving headphones, mounted on handle of locator and using loopstick to pick up radiation from search-loop oscillator being moved over ground. Faraday shield over search loop minimizes ground effect, so metal changes loop oscillator frequency and beat note in phones. Construction details are given. All transistors are 2N2924 and diodes are 1N90.—C. D. Rakes, "Solid State Electronic Projects," Howard W. Sams, Indianapolis, IN, 1972, p 57–67.

TWO-OSCILLATOR LOCATOR—Reference oscillator mounted on handle and loop oscillator at lower end of handle are both radiation-coupled to i-f of ordinary transistor a-m radio strapped over reference oscillator. With no metal nearby, tuning capacitor of reference oscillator is adjusted to give about 1-kHz tone from radio speaker. Tune radio to spot between stations, preferably at low-frequency end of bc band. Best performance is obtained with reference oscillator initially about 1 kHz below that of loop oscillator. Ferrous or nonferrous objects near loop will then cause rise in pitch. Article covers construction and operation.—I. M. Gottlieb, New Approach for the Metal Locator, 73, Feb. 1971, p 10–14.

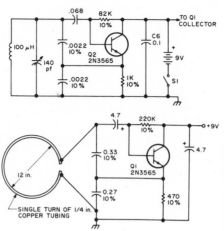

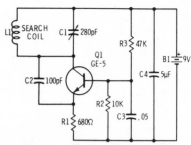

COIN FINDER—Will also locate bottle caps under beach sand. Search coil is 18 turns No. 22 enamel scramble-wound on 4-inch diameter form. Use with small transistor radio mounted on upper end of carrying handle, with search coil on lower end and its circuit just above. Tune radio to weak station, then adjust C1 for beat whistle with no metal near coil. Frequency of whistle will now change when coil is brought near buried metal.—H. Friedman, "99 Electronic Projects," Howard W. Sams, Indianapolis, IN, 1971, p 119–120.

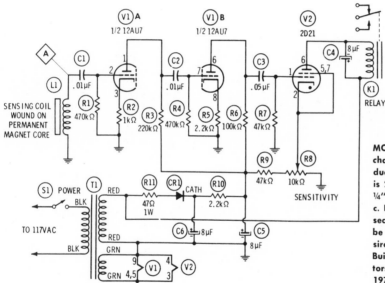

MOVING-IRON DETECTOR—Responds only to changing magnetic fields such as are produced by moving ferrous objects. Sensing coil is 250 turns No. 30 enamel wound on ¼" × ¼" × 2" permanent magnet. Relay is 5K d-c. Power transformer T1 has 125-V and 6.3-V secondaries. CR1 is 1N2484 silicon. Relay can be connected to counter, alarm, or any desired control circuit.—J. P. Shields, "How to Build Proximity Detectors and Metal Locators," Howard W. Sams, Indianapolis, IN, 1972, p 140–143.

PIPE FINDER—Single-coil portable unit will locate pipes or conduit inside walls or buried in ground. Operates from single 15-V battery. L2 is loop antenna of older broadcast radio, enclosed in Faraday shield to eliminate effect of external capacitance. Phones plugged into M2 should be about 2,000 ohms impedance. L1 is loopstick antenna. If no beat is heard when adjusting L1 through its range, with unit away from metal, try larger or smaller value for C8. When beat is obtained, adjust R5 for growl without metal. Tone in phones will then change in pitch when loop is brought near metal object.—R. M. Brown and T. Kneitel, "101 Easy Test Instrument Projects," Howard W. Sams, Indianapolis, IN, 1968, p 115–117.

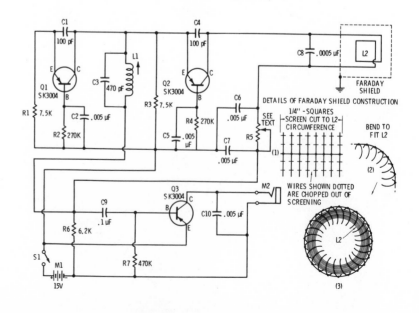

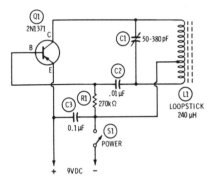

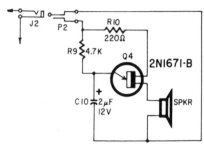

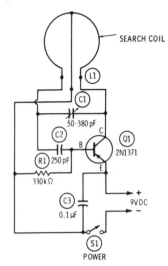

PIPE LOCATOR—Used for locating metal pipes or nails of studs concealed in walls or floors. Single-transistor oscillator using flat ferrite loop stick is mounted in flat box held against a-m transistor radio by tape or rubberbands. Radio and oscillator are adjusted until beat-frequency tone of about 1,000 Hz is heard from speaker, on side of zero beat that makes tone increase in frequency when loopstick is brought near metal object.—J. P. Shields, "How to Build Proximity Detectors and Metal Locators," Howard W. Sams, Indianapolis, IN, 1972, p 117–120.

AUDIBLE INDICATOR—Designed for use with tuned-loop oscillator type of metal detector, having crystal filter, to supplement meter reading with audible indication from speaker. Meter alone gives maximum sensitivity, so unplug ujt a-f oscillator and watch only meter when pinpointing exact location of buried metal. Article covers construction and calibration of complete locator.—C. D. Rakes, Build a Treasure Finder, *Radio-Electronics*, Nov. 1967, p 32–34.

SINGLE-TRANSISTOR USED WITH A-M RADIO—Circuit shown serves as variable-frequency oscillator for metal locator using ordinary portable a-m transistor radio as fixed-frequency oscillator. Search coil has 20 turns of No. 22 enamel on 6-inch diameter plastic hoop, tapped at tenth turn. Radio is tuned to local station between 600 and 900 kHz and C1 is adjusted until beat-frequency whistle is heard. Readjust C1 until whistle vanishes, for zero beat; metal object near search coil will bring whistle back. Construction details are given.—J. P. Shields, "How to Build Proximity Detectors and Metal Locators," Howard W. Sams, Indianapolis, IN, 1972, p 112–117.

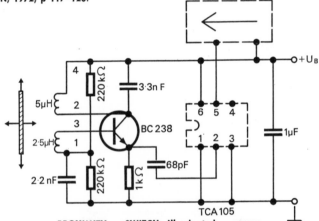

PROXIMITY SWITCH—Illuminated arrow lights up when metal strip moves between the two coils and stops oscillator. Can be used as limit switch for machine tools, elevator level control, and metal-object counter. Uses Siemens TCA105 threshold switch module. Circuit will operate on supply voltages as low as 1.8 V, but best range is from 4.5 to 30 V.—Integrated Circuits and Versatile Switches, *Electronic Engineering*, May 1973, p 21.

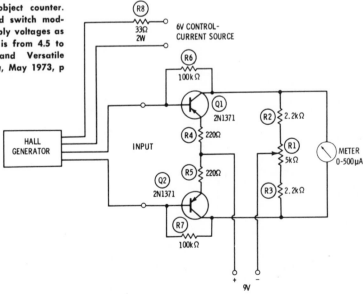

HALL-EFFECT DETECTOR—Used for detecting changes in magnetic field such as might be caused by large ferrous object. Sensor is Ohio Semitronics HR-33 or equivalent Hall generator, used with differential amplifier driving indicating meter. Construction and adjustment are covered, including use of small permanent magnet near Hall generator to null out residual Hall voltage.—J. P. Shields, "How to Build Proximity Detectors and Metal Locators," Howard W. Sams, Indianapolis, 1972, p 136–139.

CHAPTER 70
Modem Circuits

FSK WITH SLOPE AND VOLTAGE DETECTOR —Designed for 2,125 and 2,975 Hz at data rate of about 170 Hz. Output is alternately switched between a-f inputs f1 and f2 by applying signal to each differential amplifier input pair and changing gate voltage from one extreme to other. Report describes operation.—J. Reinert and E. Renschler, "Gated Video Amplifier Applications—The MC1545," Motorola Semiconductors, Phoenix, AZ, AN 491, 1969.

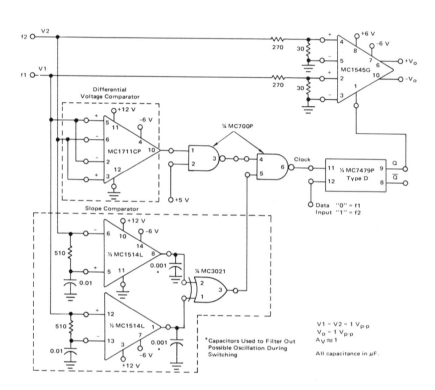

FSK FOR TWO TONES—Opamp arrangement minimizes component count in system for generating 1,070 and 1,270 Hz sine-wave audio tones representing logic 1 and logic 0 when transmitting digital data at low speed over telephone lines. First dual opamp is integrator and Schmitt trigger loop, feeding dual-opamp second-order Butterworth active filters that provide packing to equalize signal amplitude while giving maximum suppression of second harmonics, so as not to interfere with received 2,025 and 2,225 Hz signals.—D. Kesner, Circuit Basics For FSK Modems, *Digital Design*, April 1972, p 32–34.

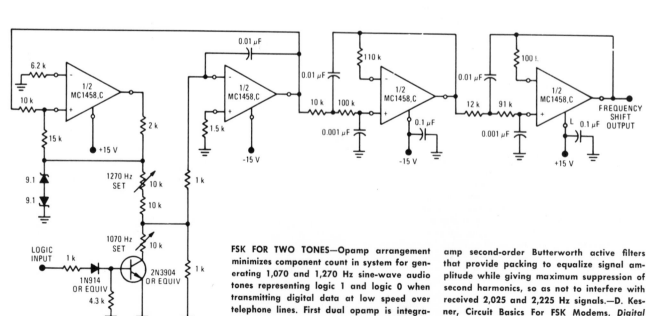

DEMODULATOR WITH D-C RESTORATION— Overcomes problem of d-c drift by using one LM111 opamp as accurate peak detector to provide d-c bias for one input to comparator in LM565 phase-locked loop. Circuit then acts as d-c restorer that tracks changes in drift, to make comparator self-compensating for changes in frequency or other effects of drift. Values shown are for 2,025-Hz mark and 2,225-Hz space frequencies. Will handle keying rate of 300 baud (150 Hz).—"The Phase Locked Loop IC As a Communication System Building Block," National Semiconductor, Santa Clara, CA, 1971, AN-46, p 10–11.

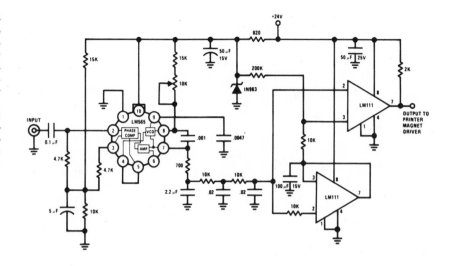

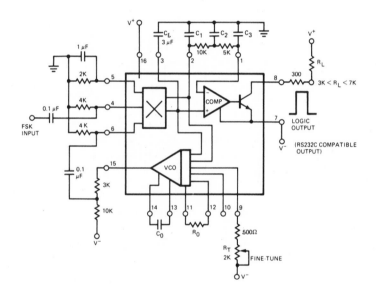

SPLIT-SUPPLY DEMODULATOR— Logic output is compatible with RS-232C because split supply is used with Exar XR-210 fsk modulator-demodulator. External components used with IC are same as for single-supply fsk demodulator, for which values are given in report both for 300-baud and 1,200-baud modem applications. Supply voltage can be 5 to 26 V.—"XR-210 FSK Modulator/Demodulator," Exar Integrated Systems Inc., Sunnyvale, CA, June 1972, p 6–7.

SELF-GENERATING FSK— Two gated oscillators combined with switching network provide self-generation of 2,125-Hz and 2,975-Hz keying signals. Negative feedback in each channel provides gain control. Switching transients are smaller than with separate oscillators because one oscillator is driven at frequency of other oscillator while first oscillator is off. Report describes operation of circuit in detail. Output voltage is about 1 V p-p.—J. Reinert and E. Renschler, "Gated Video Amplifier Applications—The MC1545," Motorola Semiconductors, Phoenix, AZ, AN-491, 1969.

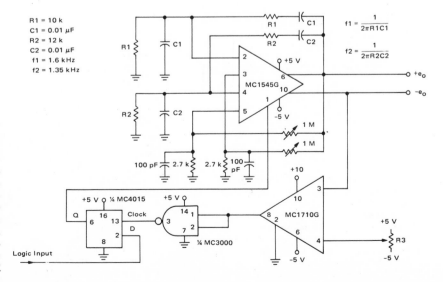

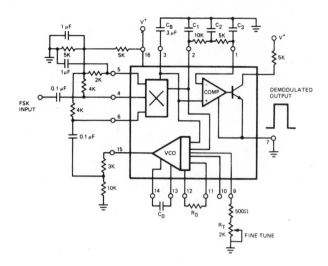

DEMODULATOR WITH SINGLE SUPPLY—Uses Exar XR-210 fsk modulator-demodulator connected as phase-locked loop system by using a-c coupling between vco output (pin 15) and pin 6 of phase detector. Fsk input is applied to pin 4. When input frequency is shifted, corresponding to data bit, polarity of d-c voltage across phase detector outputs (pins 2 and 3) is reversed. Voltage comparator and logic driver section convert level shift to binary pulse. C0 and fine-tune adjustments set vco midway between mark and space frequencies. For 1,200-baud modem, fsk values are 1,200 and 2,000 Hz, R0 is 2K, C0 0.14 μF, C1 0.033 μF, C2 0.01 μF, and C3 0.02 μF. Report also gives values for 300-baud operation. Single supply can be 5 to 26 V.—"XR-210 FSK Modulator/Demodulator," Exar Integrated Systems Inc., Sunnyvale, CA, June 1972, p 6–7.

FSK GENERATOR—Uses Exar XR-210 fsk modulator-demodulator IC for frequency-shift keying of carrier up to 100 kHz. C0 is chosen to give free-running frequency about 5% lower than space frequency. RT and RX are then set to give desired space and mark frequencies for data communication. Square-wave output is 2.5 V p-p.—"XR-210 FSK Modulator/Demodulator," Exar Integrated Systems Inc., Sunnyvale, CA, June 1972, p 7.

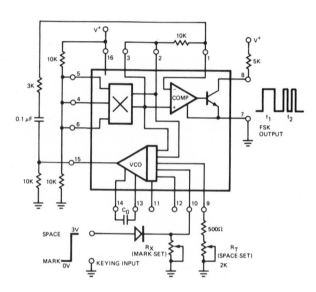

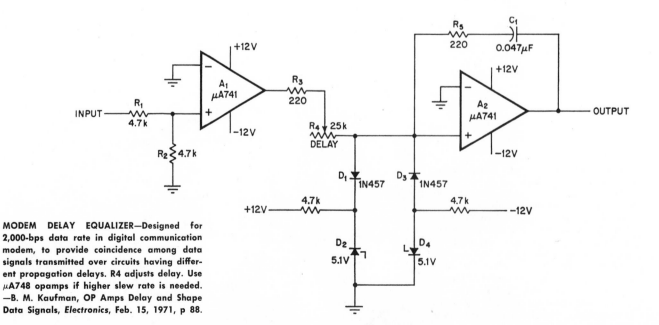

MODEM DELAY EQUALIZER—Designed for 2,000-bps data rate in digital communication modem, to provide coincidence among data signals transmitted over circuits having different propagation delays. R4 adjusts delay. Use μA748 opamps if higher slew rate is needed. —B. M. Kaufman, OP Amps Delay and Shape Data Signals, *Electronics*, Feb. 15, 1971, p 88.

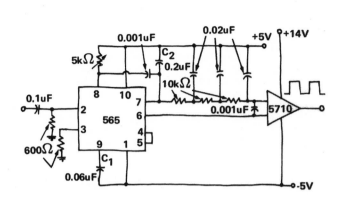

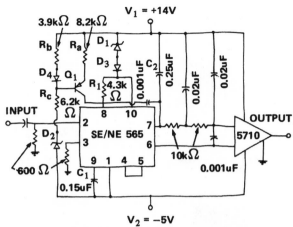

1,070 AND 1,270 HZ FSK DECODER—Three-stage R-C ladder filter removes carrier component from output of Signetics 565 phase-locked loop connected to data communication receiver, to provide shifting d-c voltage at output of opamp for driving printer. Maximum keying rate is 300 baud.—Frequency Shift Keying Demods with Phase-Locked Loop Devices, *Digital Design*, Feb. 1972, p 30–31.

FSK DECODER WITH VCO—Sophisticated fsk decoder for data communication receiver having narrow frequency deviation (1,070 and 1,270 Hz) requires adjusting free-running vco frequency in Signetics 565 phase-locked loop so output voltage swings equally above and below reference voltage on pin 6. Band edge of two-stage RC ladder network is about 800 Hz.—Frequency Shift Keying Demods with Phase-Locked Loop Devices, *Digital Design*, Feb. 1972, p 30–31.

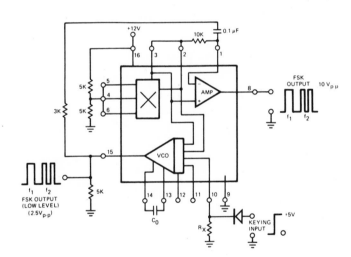

FSK GENERATOR—Utilizes digital programming capability of Exar XR-215 phase-locked loop, with control logic pulse applied to pin 10. Circuit will provide two different levels of fsk output, each with second harmonic content less than 0.3%.—"XR-215 Monolithic Phase-Locked Loop," Exar Integrated Systems Inc., Sunnyvale, CA, July 1972, p 7.

DEMODULATOR—Uses Exar XR-S200 IC with phase-locked loop connection shown, as modem suitable for Bell 103 or 202 data sets operating at up to 1,800 baud. Input frequency shift corresponding to data bit makes d-c output voltage of multiplier reverse polarity. D-c level is changed to binary output pulse by gain block connected as voltage comparator.—"XR-S200 Multi-Function Integrated Circuit," Exar Integrated Systems Inc., Sunnyvale, CA, June 1972, p 6.

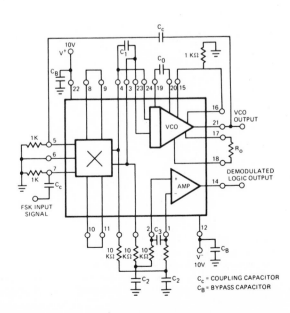

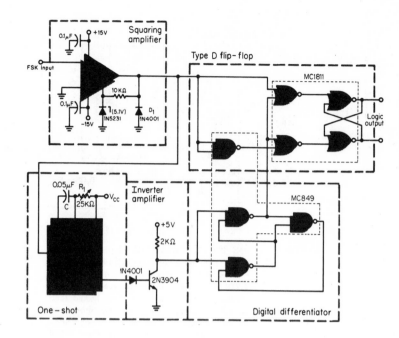

FSK RECEIVER—Converts two input audio tones to saturated logic levels. Input opamp changes sinusoidal inputs to series of square-wave pulses. D1 clips negative half. Article describes operation of circuit, which can be considered as sample-and-hold modem with one-shot providing timing signal.— D. Kesner, A Simple FSK Receiver, *The Electronic Engineer*, Dec. 1971, p 56.

FSK DECODER—Signetics 560B phase-locked loop is used as receiving converter to demodulate fsk audio tones from data communication receiver and provide shifting d-c voltage for driving printer. Will decode frequencies from near 0 to 500 kHz, giving 2-V p-p swing at up to 600-baud rate.—Frequency Shift Keying Demods with Phase-Locked Loop Devices, *Digital Design*, Feb. 1972, p 30–31.

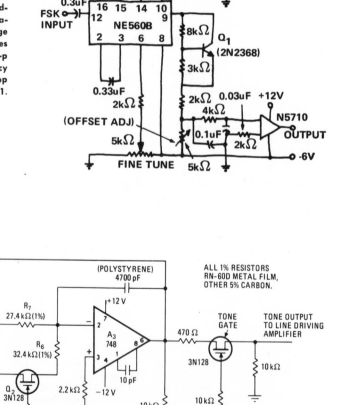

TONE OSCILLATOR—Supplies two tones, 1,-200 and 2,200 Hz, compatible with Bell type 202 modems. Uses frequency-shift tone oscillator delivering phase-continuous constant-amplitude signal that holds jitter distortion below 5% even for data rates of 1,800 bits per second, which corresponds to bit period approaching tone signal period. Circuit uses state-variable active bandpass filter A1-A2-A3 that changes its frequency when resistance is switched. Data voltage of −8 V cuts off Q1 and Q2, to give 1,200-Hz tone; +8 V unblocks these transistors to give 2,200 Hz.—B. M. Kaufman, Resistance Switching Cuts Tone Oscillator Jitter, *Electronics*, Sept. 13, 1971, p 87–88.

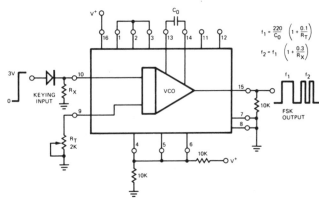

$$f_1 = \frac{220}{C_0}\left(1 + \frac{0.1}{R_T}\right)$$

$$f_2 = f_1\left(1 + \frac{0.3}{R_X}\right)$$

GENERATOR USING VCO—Generates data-communication mark and space frequencies for carrier up to 10 MHz by using vco output of Exar XR-210 fsk modulator-demodulator directly, with comparator and logic driver sections of IC removed from signal path. Supply can be anywhere between 5 and 26 V. —"XR-210 FSK Modulator/Demodulator," Exar Integrated Systems Inc., Sunnyvale, CA, June 1972, p 7–8.

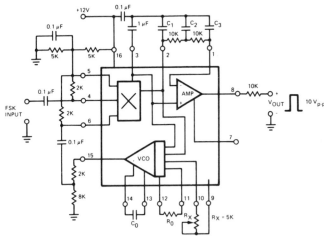

DEMODULATOR—Uses Exar XR-215 phase-locked loop IC. When input frequency is shifted for data bit, d-c voltage at phase comparator outputs 2 and 3 reverses polarity. Opamp converts d-c level shift to binary output pulse. RX sets vco frequency. Report gives typical component values both for 300-baud and 1,800-baud operation.—"XR-215 Monolithic Phase-Locked Loop," Exar Integrated Systems Inc., Sunnyvale, CA, July 1972, p 7.

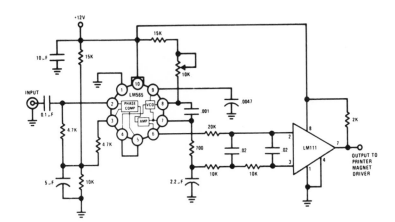

DEMODULATOR—Uses National LM565 phase-locked loop driving opamp. Values shown are for 2,025-Hz mark and 2,225-Hz space frequencies. Will handle keying rate of 300 baud (150 Hz). Chief problem is d-c drift, which may lock comparator in one state or other because demodulated output is only 150 mV; report includes circuit which overcomes drift with d-c restorer stage.—"The Phase Locked Loop IC As a Communication System Building Block," National Semiconductor, Santa Clara, CA, 1971, AN-46, p 9–11.

CHAPTER 71
Modulator Circuits

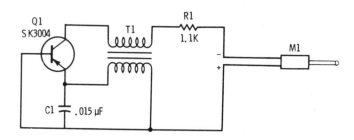

GRID-DIPPER MODULATOR—M1 is plugged into phone jack of grid-dip meter, to obtain power for feedback oscillator Q1 and add modulation to r-f generated by grid dipper. T1 can be Lafayette TR-98 transistor transformer.—R. M. Brown and T. Kneitel, "101 Easy Test Instrument Projects," Howard W. Sams, Indianapolis, IN, 1968, p 111.

AMPLITUDE MODULATOR—Uses MC1596 IC balanced modulator as amplitude modulator by unbalancing carrier null to insert proper amount of carrier into output signal. Provides excellent modulation from 0% to greater than 100%.—R. Hejhall, "MC1596 Balanced Modulator," Motorola Semiconductors, Phoenix, AZ, AN-531, 1971.

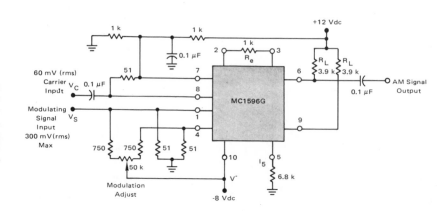

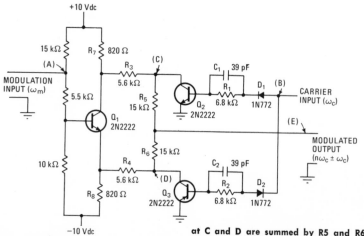

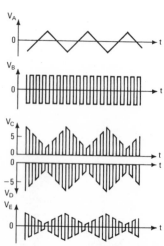

LINEAR CHOPPER—Q1 splits modulation input into equal signals having opposite polarity and phase. Q2 and Q3 pass alternate half-cycles, and chopped modulated signals at C and D are summed by R5 and R6. Provides good linearity at modulation levels up to 97.5% and modulation frequencies ranging from d-c to half of carrier frequency. Envelope shows distortion above 250 kHz and maximum linear modulation drops to 88% at 1 MHz.—D. DeKold, Amplitude Modulator Is Highly Linear, *Electronics*, June 5, 1972, p 101–102.

PWM WITH SWITCHED INTEGRATOR—Uses Burr-Brown 9580/15 switched current amplifier and standard opamps to serve as very linear and stable synchronized pulse width modulator. Pulse train provides clocking signal, so output is pulse train synchronized to clock pulse input. Output pulse width is linear function of input voltage. Values of VR, C1, and R1 depend on pulse repetition rate and desired dynamic range. For 1-kHz clock and 0.1–10 V input, C1 can be 0.01 μF and R1 90K.—J. G. Graeme, G. E. Tobey, and L. P. Huelsman, "Operational Amplifiers," McGraw-Hill Book Co., New York, 1971, p 413–414.

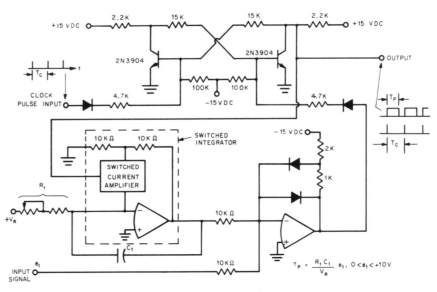

$$T_P = \frac{R_1 C_1}{V_R} e_1, \quad 0 < e_1 < +10V$$

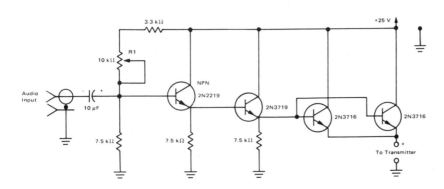

SERIES MODULATOR FOR 4-W A-M TRANSMITTER—Provides convenient and simple method of modulating 118–136 MHz broadband transmitter for light aircraft. R1 adjusts d-c voltage to transmitter. Harmonic distortion is less than 1%, and can be reduced by including external feedback.—D. Brubaker, "A Broadband 4-Watt Aircraft Transmitter," Motorola Semiconductors, Phoenix, AZ, AN-481, 1972.

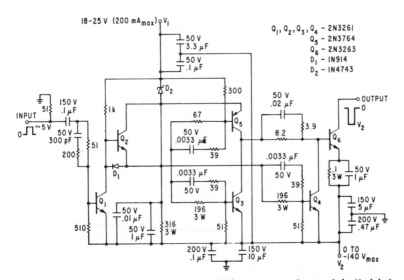

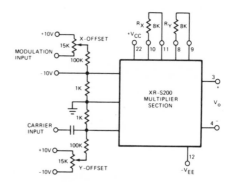

DRIVER FOR PULSE MODULATOR—Used in L-band transponder for biasing silicon diodes into avalanche breakdown. Modulator is capable of delivering up to 500 V at up to 10 A, with pulses having rise times of the order of 50 ms. Driver is in effect a high-voltage high-current transistor switch. Modulation efficiency is up to 90% for single-pulse operation.—J. F. Reynold, J. Assour, and A. Rosen, A Solid-State Transponder Source Using High-Efficiency Silicon Avalanche Oscillators, RCA Review, June 1972, p 344–356.

DOUBLE-SIDEBAND A-M—Multiplier section of Exar XR-S200 IC uses d-c offset adjustment on modulation input to set carrier output level, while d-c offset of carrier input governs symmetry of output waveform. Modulation input can also be used as linear agc to control amplification with respect to carrier input signals. Will operate up to 30 MHz.—"XR-S200 Multi-Function Integrated Circuit," Exar Integrated Systems Inc., Sunnyvale, CA, June 1972, p 4.

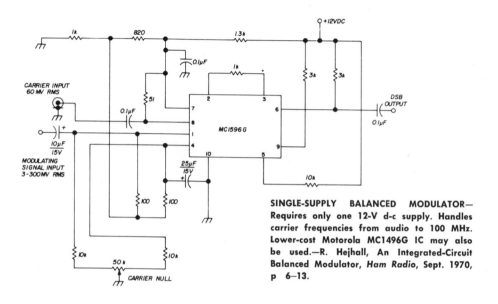

SINGLE-SUPPLY BALANCED MODULATOR— Requires only one 12-V d-c supply. Handles carrier frequencies from audio to 100 MHz. Lower-cost Motorola MC1496G IC may also be used.—R. Hejhall, An Integrated-Circuit Balanced Modulator, *Ham Radio*, Sept. 1970, p 6–13.

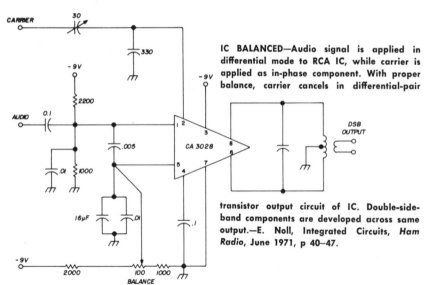

IC BALANCED—Audio signal is applied in differential mode to RCA IC, while carrier is applied as in-phase component. With proper balance, carrier cancels in differential-pair transistor output circuit of IC. Double-side-band components are developed across same output.—E. Noll, Integrated Circuits, *Ham Radio*, June 1971, p 40–47.

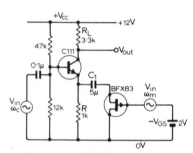

FET AMPLITUDE MODULATOR—Basic circuit provides linear control over wide range of carrier and modulation frequencies by using junction fet as variable resistance. Will give 1.5-V p-p output with modulation depth of at least 33%, up to at least 10 MHz for carrier and d-c to 25 kHz for modulation.—M. E. Cook, Amplitude Modulation Using an F.E.T., *Wireless World*, Feb. 1970, p 81–82.

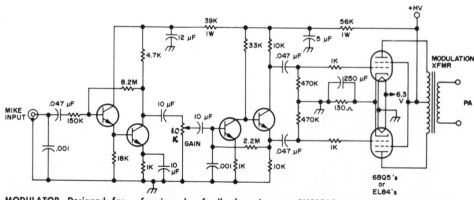

2-METER HYBRID MODULATOR—Designed for driving low-power vhf transmitter using 3630 tube in final stage. First two transistors, forming d-c feedback pair, are 2N2926G. Second pair may be any type having collector breakdown voltage of at least 80 V. Modulation transformer match to final is critical.—J. E. Kasser, The Minimod, *73*, Nov. 1972, p 202–203.

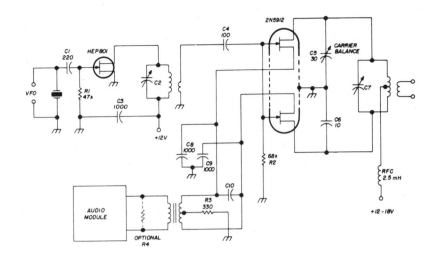

DUAL-FET BALANCED MODULATOR—Will operate on 80 through 10 meters using appropriate plug-in coils. Drains of dual-fet are connected in push-pull for carrier cancellation. Stable vfo may be used in place of crystal.—E. Noll, Experiments with Phase-Locked Loops, *Ham Radio*, Oct. 1971, p 58–63.

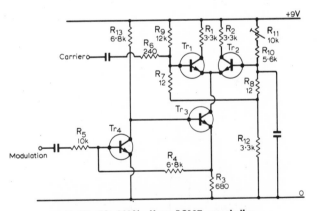

PWM DRIVE—Generates pwm waveform by using delay of R-C phase-shift network to control sharp-cutoff transistors in d-c converter for changing 275 V to 50 V at low power levels.—"One Kilowatt Regulated Power Converter with the 2N5157 Silicon Transistor," Delco, Kokomo, IN, April 1972, Application Note 44.

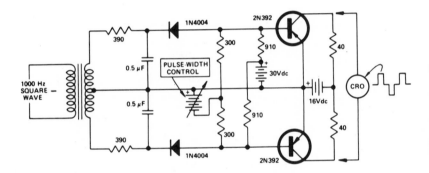

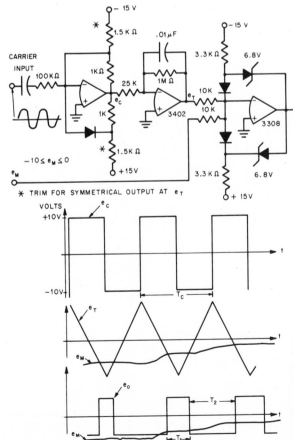

A-M UP TO 100%—Uses BC107 or similar transistors. Article includes mathematical analysis of circuit.—A. F. Newell, A Transistor Multiplier Circuit, *Wireless World*, June 1969, p 285–289.

PWM—Uses Burr-Brown opamps to amplify and clip sine-wave carrier input and then convert it to triangle wave with integrator. Modulation input biases triangle and thus modulates pulse width about 50% duty-cycle condition. Design equation is given.—J. G. Graeme, G. E. Tobey, and L. P. Huelsman, "Operational Amplifiers," McGraw-Hill Book Co., New York, 1971, p 411–413.

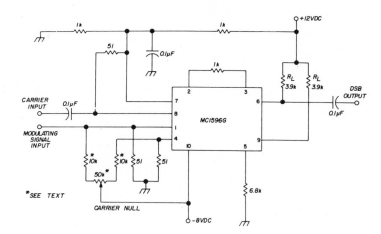

BALANCED MODULATOR—Simple circuit requires no transformers, handles carrier frequencies from audio to 100 MHz, provides 65 dB carrier suppression at 500 kHz, and provides 50 dB suppression at 9 MHz. Carrier null pot can be remotely located. Lower-cost Motorola MC1496G IC may also be used. To use as a-m modulator, change 10K resistors to 750 ohms and adjust pot for carrier insertion instead of carrier null (ssb).—R. Hejhall, An Integrated-Circuit Balanced Modulator, *Ham Radio,* Sept. 1970, p 6–13.

DOUBLY BALANCED MIXER FOR 9 MHZ—Has broadband inputs and 9-MHz tuned output, so r-f and local oscillator inputs may be any two frequencies with sum or difference of 9 MHz, in range from 160 meters up to 300 MHz.—R. Hejhall, An Integrated-Circuit Balanced Modulator, *Ham Radio,* Sept. 1970, p 6–13.

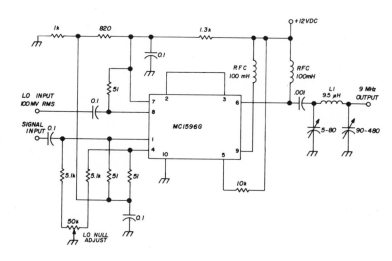

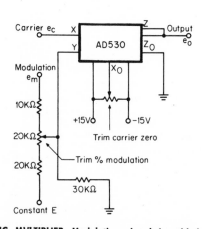

IC MULTIPLIER—Modulating signal is added to constant voltage derived from power supply through attenuator that provides for 50% scale factor correction to allow full positive swing at 100% modulation. Modulation adds ± 5 V to unmodulated carrier level of ± 5 V.—D. Sheingold, Amplitude Modulator Uses an IC Multiplier, *The Electronic Engineer,* Aug. 1971, p 68.

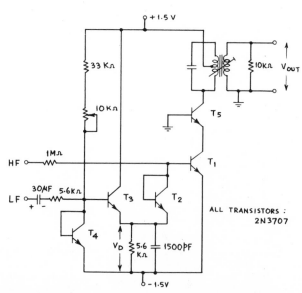

500-KHZ MICROPOWER A-M—Developed for miniature communication and instrumentation systems. Modulation with 400-Hz signal is satisfactory up to several megahertz because additional transistor T5 is used in cascode with current-gain block to separate functions of modulation and voltage gain.—V. Venkateswarlu and B. S. Sonde, A Micropower Amplitude Modulator, *Proc. IEEE,* July 1971, p 1114–1116.

BALANCED MODULATOR—Gives excellent gain and carrier suppression by operating upper (carrier) differential amplifiers at saturated level and lower differential amplifier of IC in linear mode. Recommended input levels are 60-mV rms for carrier and 300 mV rms for modulating signal. Suppression of spurious sidebands is 55 dB at 500 kHz.—R. Hejhall, "MC1596 Balanced Modulator," Motorola Semiconductors, Phoenix, AZ, AN-531, 1971.

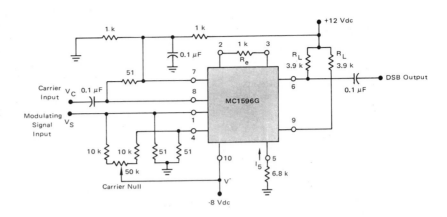

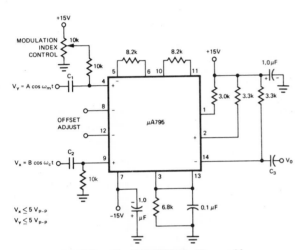

MULTIPLIER AS MODULATOR—IC provides both balanced modulation and amplitude modulation as function of position of modulation index control. Index is zero with pot wiper at ground, giving double-sideband suppressed-carrier modulation. Design equations are given.—"The μA795, A Low-Cost Monolithic Multiplier," Fairchild Semiconductor, Mountain View, CA, No. 211, 1971, p 2–3.

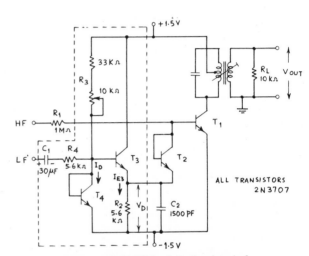

10-KHZ MICROPOWER A-M—Developed for miniature communication and instrumentation systems. Modulation is free from distortion, operation is class A, and overmodulation is prevented. Maximum modulation index realizable is less than 0.9. Chief drawback is effect of modulator input capacitance on performance at higher frequencies.—V. Venkateswarlu and B. S. Sonde, A Micropower Amplitude Modulator, Proc. IEEE, July 1971, p 1114–1116.

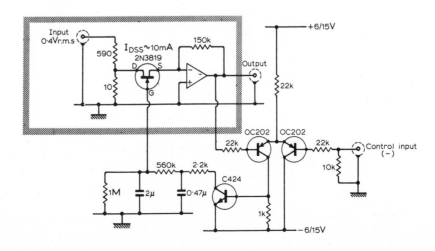

AMPLITUDE MODULATOR—Allows amplitude of repetitive input signal to be controlled by slowly varying control voltage. Any opamp having gain above 10,000 can be used. Peak negative value of output signal is very nearly equal to value of negative control voltage. Circuit was used to control 10-kHz sine wave.—J. Vanderkooy, Amplitude Modulator Using Operational Amplifier, Wireless World, March 1969, p 119.

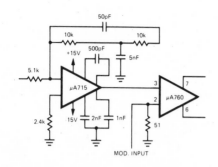

COMPARATOR FOR PWM—Ramp output of μA715 is fed into input 3 of μA760, with audio signal going to input 2. Outputs from 6 and 7 then change state each time audio input equals ramp voltage. Pulse width of output varies with instantaneous value of audio input.—P. Holtham, "The μA760—A High Speed Monolithic Voltage Comparator," Fairchild Semiconductor, Mountain View, CA, No. 311, 1972, p 7.

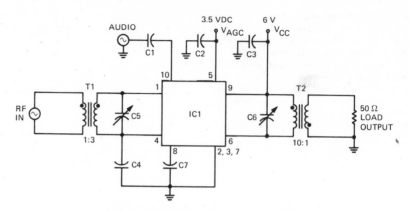

6-METER A-M—Single IC gives up to 90% modulation with very little distortion.—"Radio Amateur's IC Projects," Motorola Semiconductors, Phoenix, AZ, HMA-36, 1971.

FET BALANCED MODULATOR—Designed to operate with carrier inputs of 100–150 kHz and modulating frequencies from d-c to 10 kHz. Provides linear control by using junction fet as variable resistance. Tr4 and Tr5 form simple summing amplifier providing low output impedance.—M. E. Cook, Amplitude Modulation Using an F.E.T., *Wireless World*, Feb. 1970, p 81–82.

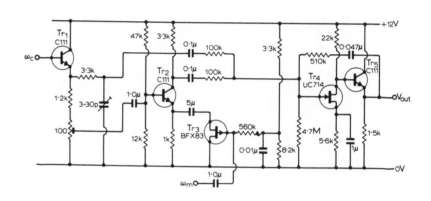

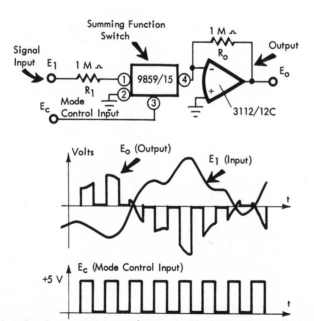

CHOPPER—Uses Burr-Brown Model 9859/15 summing junction switch and opamp to provide chopping action for analog input signal E1. 5-V pulses are fed into Ec as mode con-trol input for switch, so output voltage goes to zero when switch is gated OFF.—"Electronic Switches," Burr-Brown, Tucson, AZ, Sept. 1969, p 6.

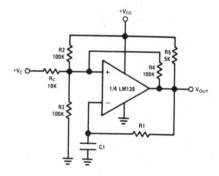

PWM DOWN TO 1 HZ—Simple connection of IC comparator gives square-wave output when input control voltage is equal to half of supply voltage. Increasing or decreasing control voltage about this value changes duty cycle. Report tells how to calculate upper and lower trip points. Center frequency depends on values of R1 and C1, with maximum being limited by value of supply voltage and output slew rate.—R. T. Smathers, T. M. Frederiksen, and W. M. Howard, "LM139/LM239/LM339 A Quad of Independently Functioning Comparators," National Semiconductor, Santa Clara, CA, AN-74, 1973, p 15.

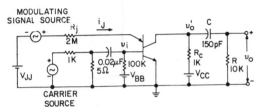

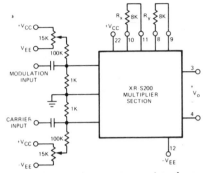

PNPN TETRODE MODULATOR—Transistor is operated in its linear mode to serve as amplitude modulator. Article gives design equations. With 100-kHz carrier and 60-Hz modulating signal, maximum percentage modulation with negligible signal distortion is about 50%.—N. C. Voulgaris and E. S. Yang, Linear Applications of a P-N-P-N Tetrode, *IEEE Journal of Solid-State Circuits*, Aug. 1970, p 146–150.

SUPPRESSED-CARRIER A-M—Multiplier section of Exar XR-S200 IC generates suppressed-carrier a-m signals, with about 60 dB carrier suppression at 500 kHz and 40 dB suppression at 10 MHz. Carrier and modulation inputs are interchangeable. 15K offset adjustments optimize carrier suppression.—"XR-S200 Multi-Function Integrated Circuit," Exar Integrated Systems Inc., Sunnyvale, CA, June 1972, p 3.

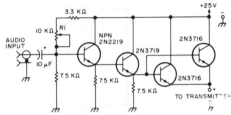

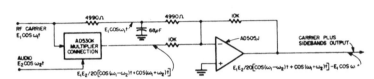

SERIES MODULATOR—Simple four-transistor circuit provides convenient means for modulating 4-W or other low-power transmitter for test purposes. Harmonic distortion is less than 1%.—D. Brubaker, A VHF AM Transmitter Using Low-Cost Transistors, *73*, Aug. 1970, p 54–59.

A-M USING ANALOG MULTIPLIER—Single-IC multiplier develops output proportional to product of carrier and audio input signals, shown as sum and difference sidebands. Process is known as suppressed-carrier modulation, since neither carrier nor modulation frequencies are present in output. Opamp serves as summing amplifier which reinstates carrier in output.—"AD530 Complete Monolithic MDSSR," Analog Devices, Inc., Norwood, MA, Technical Bulletin, July 1971.

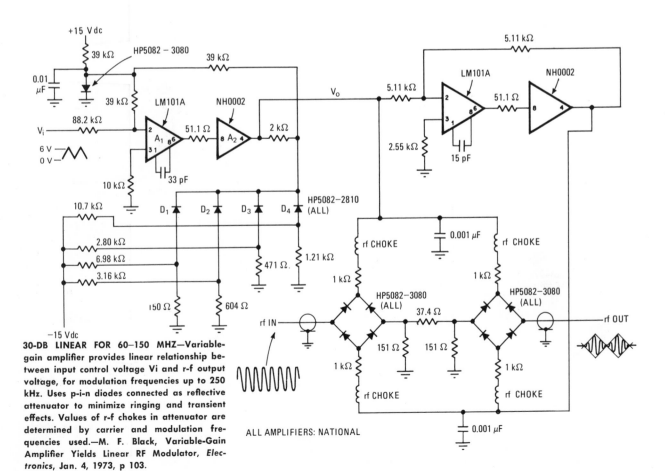

30-DB LINEAR FOR 60-150 MHZ—Variable-gain amplifier provides linear relationship between input control voltage Vi and r-f output voltage, for modulation frequencies up to 250 kHz. Uses p-i-n diodes connected as reflective attenuator to minimize ringing and transient effects. Values of r-f chokes in attenuator are determined by carrier and modulation frequencies used.—M. F. Black, Variable-Gain Amplifier Yields Linear RF Modulator, *Electronics*, Jan. 4, 1973, p 103.

ALL AMPLIFIERS: NATIONAL

CHAPTER 72
Motor Control Circuits

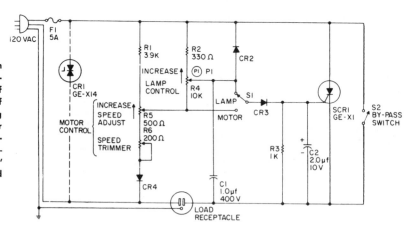

UNIVERSAL MOTOR OR LAMP—Combination half-wave motor speed and lamp control varies output voltage from 0 to about 70% of full a-c line voltage, with S1 giving choice of lamp or motor control circuit for controlling loads up to 500 W. CR1 is GE-X14 thyrector diode used as optional transient voltage suppressor, and other diodes are GE-504A.—"Electronics Experimenters Circuit Manual," General Electric, Owensboro, KY, 1971, 3rd Ed., p 179–182.

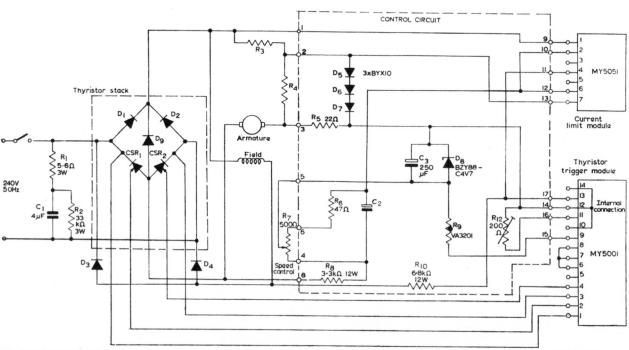

10:1 SPEED RANGE—Simple system uses Mullard modules and Mullard thyristor stack rated to handle 2 times full-load motor current. Choose D3 and D4 to handle field current. R4 is 1/7th of armature resistance and R3 in ohms is 1 ÷ full-load armature current in amperes. Speed droop at maximum speed is less than 10% when increasing from 25% to full load. Report gives regulation curve obtained for 5-hp d-c shunt motor.—"Simple System for D.C. Motors Supplied From Single-Phase A.C. Mains," Mullard, London, 1968, TP922/1.

514

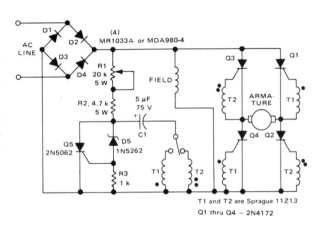

SHUNT-WOUND SPEED AND DIRECTION—
Will handle 1/15-hp 5,000-rpm motors, with
speed determined by R1 and direction by
switch setting. Field is across rectified supply
and armature is in scr bridge, making armature current reversible for changing rotation.
—G. V. Fay, "Direction and Speed Control
for Series, Universal and Shunt Motors," Motorola Semiconductors, Phoenix, AZ, AN-443,
1972.

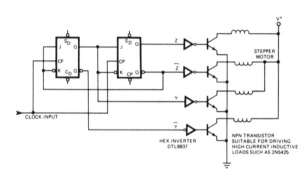

STEPPING MOTOR TRANSLATOR—Converts
electrical pulses from clock to correct number
of phases required for activating stepping
motor. Translator is 2-bit shift register or
Johnson counter. Reverse Z and Y connections
to get operation in opposite direction. Motor
shown requires 440 mA per winding at 11.6
V d-c.—"MSI Logic Simplifies Control of Mechanical Systems," Fairchild Semiconductor,
Mountain View, CA, No. 252, 1971, p 3.

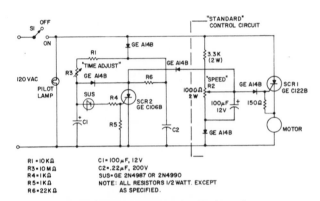

SPEED CONTROL WITH TIMER—R2 determines
speed of universal motor up to 2 A, and R3
in timer circuit determines interval up to 1
min after which motor is shut off automatically. Designed for large food mixers and
blenders. Circuit has instant-ready capability
and requires no large high-voltage electrolytic.—D. R. Grafham and S. R. Zimmer,
"Universal Motor Controls with Built-in Self-
Timer," General Electric, Semiconductor Products Dept., Auburn, NY, No. 201.13, 1970.

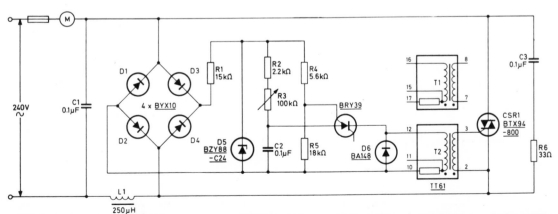

2-KW SQUIRREL-CAGE FAN MOTOR—Triac
triggered by BRY39 scs through pulse transformer provides full-wave a-c power control
of single-phase capacitor start-run fan motor.
—"Applications of the BRY39 Transistor,"
Mullard, London, 1973, TP1319, p 7.

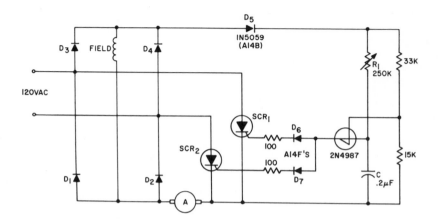

SCR FOR SHUNT OR P-M MOTOR—Scr's conduct on alternate half-cycles but are triggered from same sus. Chief advantage is that scr's can under no condition fail to turn off. Only D1, D2, and scr's need be stud-mounted for motors having under 4 A field current, because diodes D3 and D4 carry only field current. Choose diodes and scr's for motor rating, such as GE A15B rectifier for up to 4.5 A. Speed regulation is moderately good, on order of 10%.—"SCR Manual," General Electric, Syracuse, NY, 1972, 5th Ed., p 294–295.

PUT-TRIGGERED UNIVERSAL—Use of cosine-modified ramp and pedestal circuit enclosed in dashed lines makes phase control of universal series appliance motor independent of gate trigger requirements of scr used. Control uses motor counter emf as feedback signal. Provides good low-speed operation, controlled by R2.—"SCR Manual," General Electric, Syracuse, NY, 1972, 5th Ed., p 290–291.

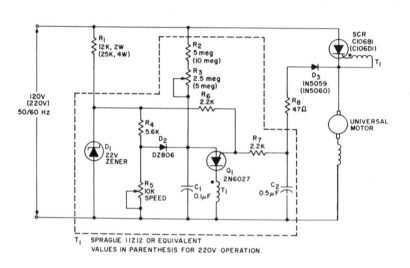

T_1 SPRAGUE 11Z12 OR EQUIVALENT
VALUES IN PARENTHESIS FOR 220V OPERATION.

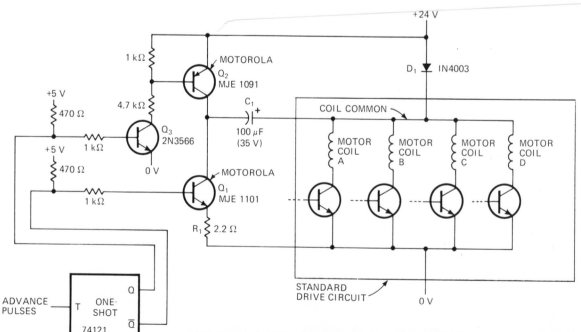

STEPPER DRIVE—Designed for four-phase 28-V stepping motor. Q1 and Q2 form complementary Darlington pair acting with Q3 to boost output power of motor when it is stepping while minimizing power dissipated during dwell or hold intervals. C1 charges to supply voltage during dwell, and discharges to double supply voltage under control of stepping pulse from one-shot.—E. Wolf, Stepper Drive Circuit Boosts Motor Torque, *Electronics*, Nov. 6, 1972, p 103.

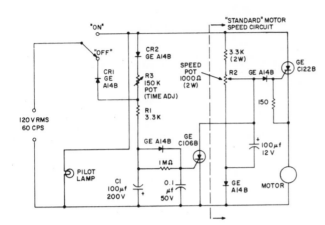

MIXER CONTROL WITH TIMER—Shuts off motor automatically after interval determined by setting of R3, up to 30 s. Speed-dependent feedback in circuit of universal motor gives excellent torque at low speed. Scr shown will handle motors rated up to 2 A.—D. R. Grafham and S. R. Zimmer, "Universal Motor Controls with Built-in Self-Timer," General Electric, Semiconductor Products Dept., Auburn, NY, No. 201.13, 1970.

200:1 SPEED RANGE—Designed for use with 2.5-V d-c motor having armature resistance of 40 ohms. Motor winding is connected in one arm of bridge made up of R1, R2, and R3. Motor drive voltage is made proportional to difference between control signal and back emf of motor.—V. B. Gerard, Motor Control Circuit with 200:1 Speed Range, *Wireless World*, Aug. 1969, p 379.

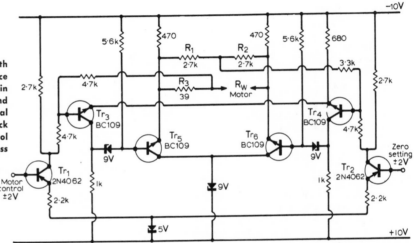

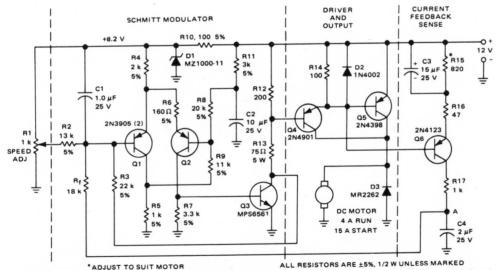

PWM FOR D-C MOTOR—Control circuit operates from 12-V d-c source and provides speed control for motors with inrush current up to 20 A. Maximum running current is less and depends on heatsinking of Q5. Schmitt trigger Q1-Q2 feeds pulse-width-modulated waveform to phase-inverter Q3 whose output is variable-width variable-frequency pulse having duty cycle and frequency dependent on d-c current sent through R2 by R1 and current through overall feedback resistor Rf. Report describes operation in detail and gives performance graph.—G. V. Fay and N. Freyling, "Pulse-Width Modulation for DC-Motor Speed Control," Motorola Semiconductors, Phoenix, AZ, AN-445, 1972.

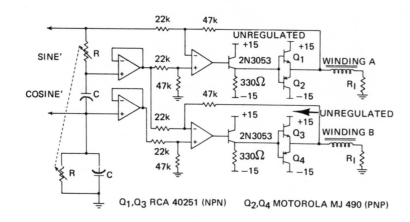

QUADRATURE-SIGNAL SPEED CONTROL—Op-amp feedback resistor connects to output of complementary-symmetry transistor pair and senses voltage directly at load, for precise speed control of a-c synchronous induction motor in which quadrature signals are re-established to same amplitude as oscillator output.—G. I. Johnston, Quadrature Signals Provide Precise Speed Regulation, *Electromechanical Design*, June 1971, p 16 and 19–21.

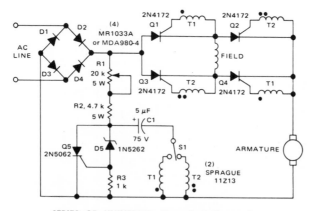

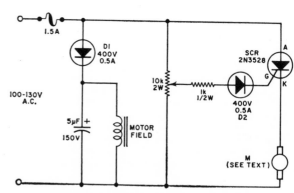

SERIES OR UNIVERSAL—Controls both speed and direction of 1/15-hp 5,000-rpm motors, with setting of R1 determining speed. T1 and T2 are Sprague 11Z13 coupling transformers. Armature current is always in same direction, with S1 reversing field current by selecting either Q2-Q3 or Q1-Q4 for conduction.—G. V. Fay, "Direction and Speed Control for Series Universal and Shunt Motors," Motorola Semiconductors, Phoenix, AZ, AN-443, 1972.

HALF-WAVE SCR FOR P-M AND SHUNT MOTORS—Compares back emf of motor with bias applied to gate of scr by motor speed pot. Will handle up to 1/6 hp motors. Choke in series with motor will give smoother operation but at sacrifice of torque at very low speeds.—L. Fleming, SCR Controls for Small Motors, *Electronics World*, June 1970, p 36 and 83.

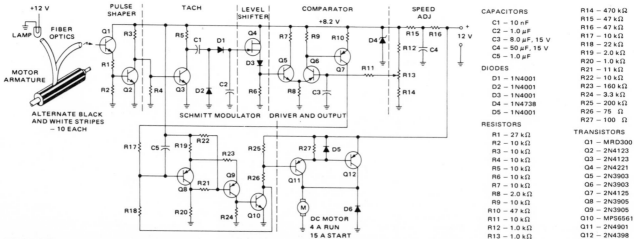

CAPACITORS		R14 — 470 kΩ
C1 — 10 nF		R15 — 47 kΩ
C2 — 1.0 μF		R16 — 47 kΩ
C3 — 8.0 μF, 15 V		R17 — 10 kΩ
C4 — 50 μF, 15 V		R18 — 22 kΩ
C5 — 1.0 μF		R19 — 2.0 kΩ
		R20 — 1.0 kΩ
DIODES		R21 — 11 kΩ
D1 — 1N4001		R22 — 10 kΩ
D2 — 1N4001		R23 — 160 kΩ
D3 — 1N4001		R24 — 3.3 kΩ
D4 — 1N4738		R25 — 200 kΩ
D5 — 1N4001		R26 — 75 Ω
		R27 — 100 Ω
RESISTORS		
R1 — 27 kΩ		TRANSISTORS
R2 — 10 kΩ		Q1 — MRD300
R3 — 10 kΩ		Q2 — 2N4123
R4 — 10 kΩ		Q3 — 2N4123
R5 — 10 kΩ		Q4 — 2N4221
R6 — 10 kΩ		Q5 — 2N3903
R7 — 10 kΩ		Q6 — 2N3903
R8 — 2.0 kΩ		Q7 — 2N4125
R9 — 10 kΩ		Q8 — 2N3905
R10 — 47 kΩ		Q9 — 2N3905
R11 — 10 kΩ		Q10 — MPS6561
R12 — 1.0 kΩ		Q11 — 2N4901
R13 — 1.0 kΩ		Q12 — 2N4398

OPTOELECTRONIC FEEDBACK—Alternate black and white stripes on armature of d-c motor reflect light to phototransistor Q1 to give chopping of feedback information for regulated d-c motor control at frequency determined by motor speed. Report describes operation in detail. Circuit uses pulse-width modulation of voltage applied to 12-V d-c motor.—G. V. Fay and N. Freyling, "Pulse-Width Modulation for DC-Motor Speed Control," Motorola Semiconductors, Phoenix, AZ, AN-445, 1972.

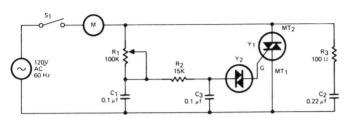

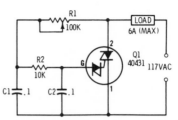

8-A A-C DRILL MOTOR SPEED—Can also be used on power saws. Uses double-time-constant circuit to provide delay required for triggering triac Y1 (IR-IRT82) at very low conduction angles. Provides practically full power to load at minimum-resistance position of speed control pot R1. Y2 is IR-IRD54 diac. Construction details are given.—"Hobby Projects," International Rectifier, El Segundo, CA, Vol. II, p 21–26.

750-W TRIAC CONTROL—Permits converting old appliance and other shaded-pole induction motors into adjustable slow-speed motors for drills, display turntables, and other applications. Triac Q1 has built-in diac trigger diode, and requires heatsink. Insulate triac from metal cabinet or other heatsink with epoxy.—H. Friedman, "99 Electronic Projects," Howard W. Sams, Indianapolis, IN, 1971, p 64–65.

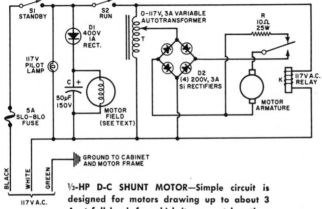

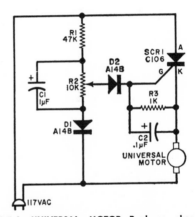

1/3-HP D-C SHUNT MOTOR—Simple circuit is designed for motors drawing up to about 3 A at full load, for which it can cost less than scr and triac speed controls. Will instantly stop, reverse, and change speed. Uses dynamic braking. Has no feedback loops to cause hunting. T controls armature voltage and hence speed. To stop motor, open S2 to release relay and connect dynamic braking resistor R across armature. S1 must be left closed during braking, to keep motor field energized.—L. Fleming, Speed Control for Large D.C. Motors, Electronics World, Jan. 1971, p 78.

1.5-A UNIVERSAL MOTOR—Replaces rheostat-type controller in variable-speed a-c/d-c motors used in sewing machines, mixers, and blenders. Speed-dependent feedback insures adequate torque at all speeds.—Manufacturer's Circuit, Popular Electronics, March 1971, p 89–91.

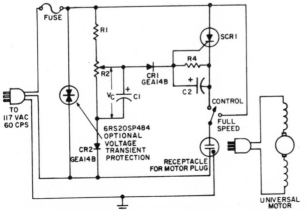

	LOW UP TO 1 AMP NAMEPLATE	MEDIUM UP TO 3 AMPS NAMEPLATE	HIGH UP TO 5 AMPS NAMEPLATE
P1	10K, 1W	1K, 2W	1K, 2W
R1	47K, 1/2W	3.3K, 2W	3.3K, 2W
R2	1K, 1/2W	150, 1/2 W OPTIONAL	150, 1/2 W OPTIONAL
C1	.5 μf, 10 V	10 μf, 10 V	10 μf, 10 V
C2	1 μf, 10 V	.1 μf, 10 V OPTIONAL	.1 μf, 10 V OPTIONAL
SCR1	GE C106 B	GE C22 B12 OR C122 B12	GE C33 B

5-A SPEED CONTROL—Simple plug-in halfwave scr phase control varies speed of series universal motors for drills, mixers, and other appliances. At given setting of R2, uses counter emf of motor armature as feedback signal to maintain essentially constant speed for varying torque requirements. Circuit values are given in table for three sizes of motors. —"A Plug-in Speed Control for Standard Portable Tools and Appliances," General Electric, Semiconductor Products Dept., Auburn, NY, No. 201.1, 1969.

⅓-HP D-C SHUNT MOTOR SPEED—Full-wave scr circuit provides speed control, reversing, and dynamic braking. Used on metal-working lathe requiring frequent stops and starts and wide range of loads. Construction requires good insulation because entire circuit is hot from power line. For p-m motor, omit field connections.—L. Fleming, Controlling DC Motors, *Popular Electronics*, March 1972, p 36–37.

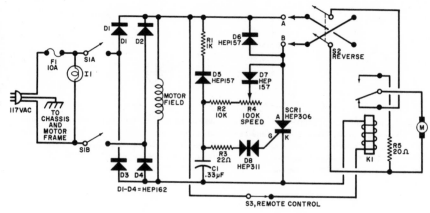

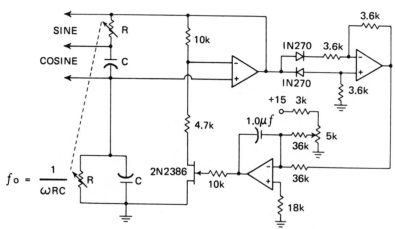

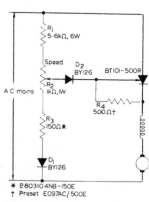

FHP SPEED CONTROL—Gives high degree of accuracy and regulation by using Wien-bridge oscillator with negative feedback (upper left opamp) to generate quadrature signals for controlling a-c synchronous induction motor or stepper. Full-wave rectified output of absolute-value opamp circuit is com-

pared with reference d-c voltage at summing junction of integrator opamp below, to generate error voltage applied through fet to bridge.—G. I. Johnston, Quadrature Signals Provide Precise Speed Regulation, *Electromechanical Design*, June 1971, p 16 and 19–21.

5:1 SPEED CONTROL—Back emf of electric drill or food mixer motor is used to change scr triggering angle, to hold speed constant at setting determined by R2 despite changes in line voltage or load. Half-wave circuit provides conduction only on positive half-cycles.—R. P. Gant, Electronic Speed Control for Electric Drills and Food Mixers, *Mullard Technical Communications*, Sept. 1967, p 252–255.

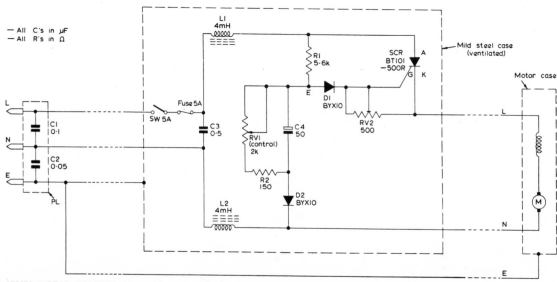

FHP A-C SERIES MOTOR CONTROL—Use with electric hand drills and food mixers up to ½ hp. Not suitable for induction motors. Motor speed is maintained despite changes in sup-

ply voltage or load by using back emf to change conduction angle of scr. RV2 adjusts minimum motor speed, while RV1 changes speed by changing conduction angle of scr

between about 0 and 160 deg.—"A Motor Speed Controller," Educational Electronic Experiments, No. 21, Mullard, London.

VOLTAGE FEEDBACK—Suitable for applications where quantity sensed can be obtained as varying d-c output voltage of tachometer (es). Pot Rc adjusts operating point. Circuit is used to trigger either half-wave or full-wave scr or triac phase control, depending on rectified line.—D. A. Zinder, "Unijunction Trigger Circuits for Gated Thyristors," Motorola Semiconductors, Phoenix, AZ, AN-413, 1972.

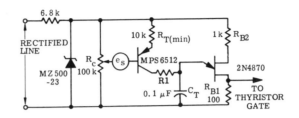

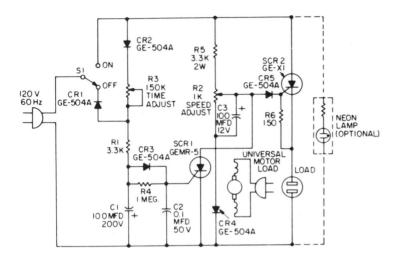

MIXER OR BLENDER—Combines half-wave scr phase control of speed with built-in adjustable 0–1 min timer for control of brush-type series universal motors for such appliances as shop tools, sewing machines, mixers, blenders, and fans. If motor is larger than 3-A rating for GE-X1, use larger C30B for SCR2. To increase time delay, increase value of C1.—"Electronics Experimenters Circuit Manual," General Electric, Owensboro, KY, 1971, 3rd Ed., p 203–204.

400-HZ SPINMOTOR SUPPLY—Transistors form long-tailed pair for damping-controlled oscillator serving as 400-Hz excitation supply. Two-section R-C filter using 100-μF capacitors is in amplitude control loop to give faster control-loop response time for equal ripple attenuation. Article includes mathematical analysis of circuit.—H. E. Jones, On a Damping-Controlled Oscillator Using a Long-Tail Pair, IEEE Journal of Solid-State Circuits, Dec. 1969, p 427–428.

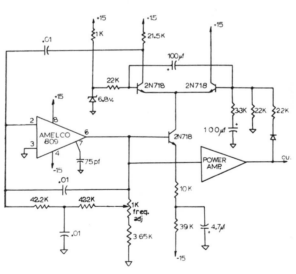

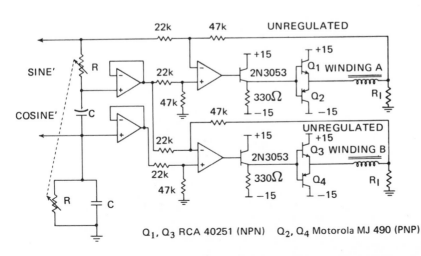

Q_1, Q_3 RCA 40251 (NPN) Q_2, Q_4 Motorola MJ 490 (PNP)

CONSTANT-CURRENT OPAMP CONTROL—Gives precise speed control of a-c synchronous induction motor having four leads, with each winding returned to ground through small current-sampling resistor. To prevent high-frequency oscillation, it may be necessary to shunt opamp with 100 pF. Article gives several variations of this quadrature-signal control.—G. I. Johnston, Quadrature Signals Provide Precise Speed Regulation, Electromechanical Design, June 1971, p 16 and 19–21.

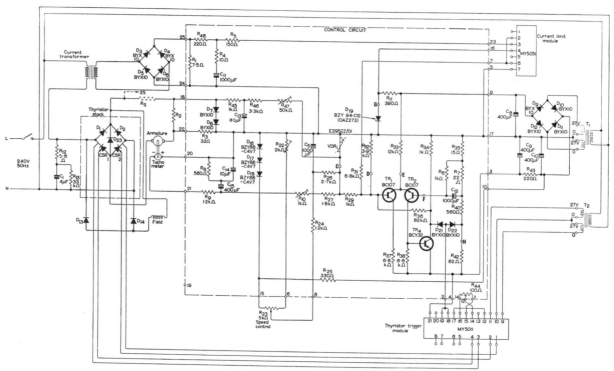

TACHOMETER FOR SPEED CONTROL—Used with thyristor system for motors from ¼ to 10 hp, to provide controlled speed range of 200:1 with less than 1% speed droop at maximum speed when load is increased from 20% to full load. All modules are Mullard. Thyristor stack should have mean current rating at least 1.5 times full-load current of motor. Transformers T1 and T2 are Mullard MY5201. Current transformer should deliver 6 V rms across 7.5 ohms when motor delivers full load. D13 and D14 are selected to handle field current. Tachometer should deliver 10 V per 1,000 rpm.—"Tachometer System For D.C. Motors Supplied From Single-Phase A.C. Mains," Mullard, London, 1967, TP920/1.

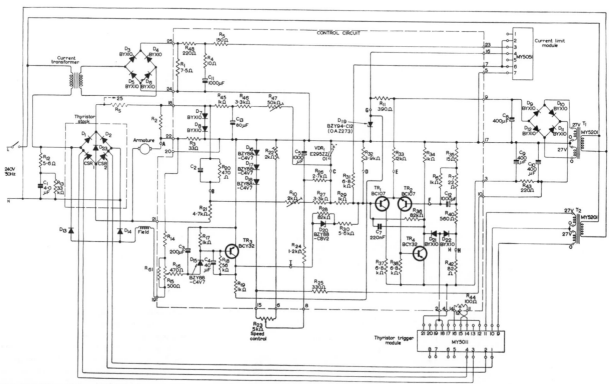

ARMATURE-VOLTAGE FEEDBACK—Used in thyristor speed control system for motors from ¼ to 10 hp, to give controlled speed range of 200:1 with speed droop of only 1% when increasing from 20% load to full load at maximum speed. Uses Mullard modules and Mullard thyristor stack rated at 1.5 times full-load motor current. D13 and D14 are chosen to handle motor field current. R2 is 1/10th of armature resistance.—"Armature-Voltage Feedback System for D.C. Motors Supplied From Single-Phase A.C. Mains," Mullard, London, 1968, TP921/1.

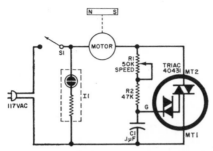

MAGNETIC STIRRER—Use of triac with built-in trigger provides smooth control of motor speed from creep to maximum. Motor drives plastic-coated stirring magnet placed in beaker, by means of similar magnet attached to shaft of motor and rotating under beaker. I1 is 117-V neon indicator lamp, optional. Use heat sink for triac. Motor is shaded-pole a-c such as R.M.S. M4RK.—R. C. Dennison, Build Variable-Speed Magnetic Stirrer, *Popular Electronics*, Nov. 1968, p 43–45 and 116.

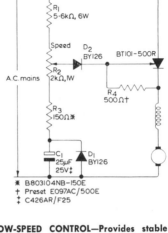

* B803104NB–150E
† Preset E097AC/500E
‡ C426AR/F25

LOW-SPEED CONTROL—Provides stable triggering at small conduction angles, corresponding to low speeds for electric drill and food mixer motors. Values of R2 and R3 depend on back emf of motor and minimum speed required, while voltage rating of C1 depends on motor back emf.—R. P. Gant, Electronic Speed Control for Electric Drills and Food Mixers, *Mullard Technical Communications*, Sept. 1967, p 252–255.

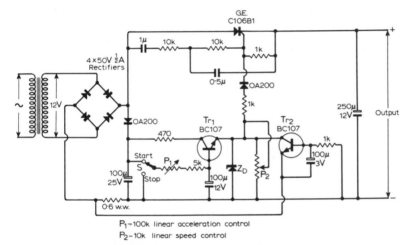

P_1—100k linear acceleration control
P_2—10k linear speed control

SMALL D-C MOTOR—Designed for speed control of small permanent-magnet or separately excited d-c motor. Circuit is heavily damped, so current falls to zero before next *on* period of rectifier. Zener reference that determines gating of scr is therefore measuring actual back emf of motor. Applied d-c voltage of motor rises as load increases, with speed remaining approximately constant. Can be used as electric drill control.—A. R. Bailey, Thyristor-Stabilized Power Supplies, *Wireless World*, Aug. 1969, p 388–390.

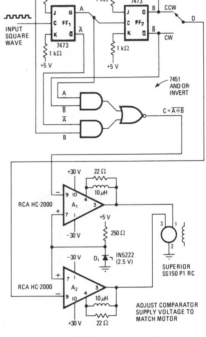

STEPPING SYNCHRONOUS MOTOR—Logic circuit combined with high-power comparators steps two-phase synchronous motor in either direction at speed determined by frequency of input square wave. Switch D controls direction of rotation. Motor is 115-V a-c having two coils. Each comparator has 75-V 100-W output.—M. D. Doering, Logic Circuit Converts Synchronous Motor to Stepper, *Electronics*, Sept. 25, 1972, p 108.

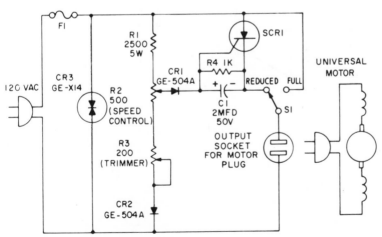

HALF-WAVE FOR UNIVERSAL MOTOR—Will handle 3-A motor if SCR1 is GE-X1, using 3-A fuse for F1. With GE-C30B scr and 5-A fuse, will handle 5-A motor. Ideal for electric drills, shop appliances, kitchen appliances, fans, and other tools using brush-type universal motors.—"Electronics Experimenters Circuit Manual," General Electric, Owensboro, KY, 1971, 3rd Ed., p 191–194.

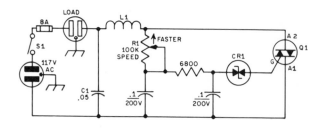

POWER-TOOL SPEED CONTROL—Triac Q1 provides full-wave control for electric drill or other small power tools. Use HEP340, Motorola MAC2-4, or other 8-A 200-V triac. L1 is 18 feet of No. 18 enamel scramble-wound on body of C1 to give about 70 μH. CR1 is HEP311 or equivalent 2-A 300-mW diac. L1-C1 form r-f hash-suppression filter.—J. Hall, Motor-Speed Control for Power Tools, QST, June 1971, p 34–35 and 55.

SUS-TRIGGERED UNIVERSAL — Half-wave phase-control circuit for universal series motors of power tools, blenders, and other small appliances uses low-voltage sus trigger as gate amplifier for scr chosen to match current rating of load. Control circuit is independent of trigger current requirements of scr.—"SCR Manual," General Electric, Syracuse, NY, 1972, 5th Ed., p 290.

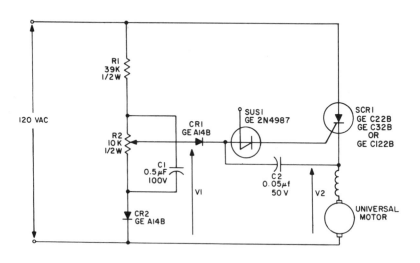

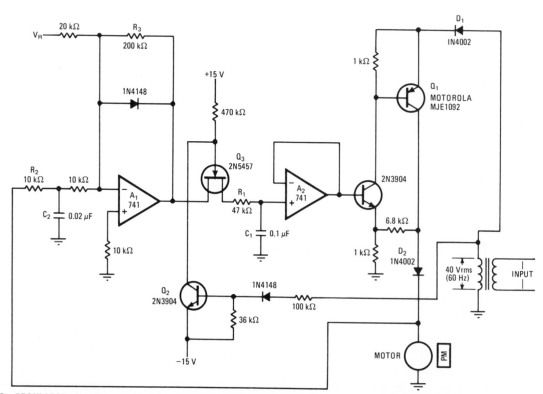

SAMPLING REGULATOR—Permits speed of permanent-magnet d-c motor to be varied over 20:1 range by making motor serve as its own tachometer for speed control. Drive power is removed from motor for portion of each negative half-cycle of input so back emf of motor can be compared with reference voltage VR. Any resulting error is stored in C1 until next positive half-cycle of input, then applied to motor for speed correction.—P. Dempster, Sampling Regulator Controls Motor Speed, Electronics, Aug. 2, 1973, p 97.

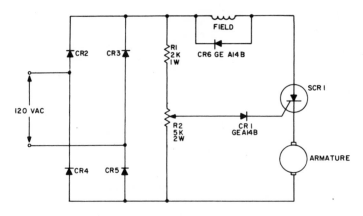

FULL-WAVE D-C CONTROL—Requires that separate connections be available for armature and field of series motor. Full-wave bridge supplies power, with counter emf of armature serving as feedback signal. R2 is speed control. One drawback is hunting at low speed settings because scr is not negative long enough to turn off because of back emf. For 6-A motor, use C22B or C122B scr and A15B rectifier diodes.—"SCR Manual," General Electric, Syracuse, NY, 1972, 5th Ed., p 291—292.

FAST STOP—Circuit uses extra contact of on-off switch to discharge capacitor bank through a-c motor to make it stop almost instantly or to provide faster reversing. Used with ⅓-hp motor of antenna rotator. Add electrolytics one by one as required for desired stopping action.—J. Vicente, Quick Stop and Reversing For Antenna Rotator, 73, Feb. 1970, p 137.

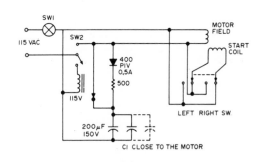

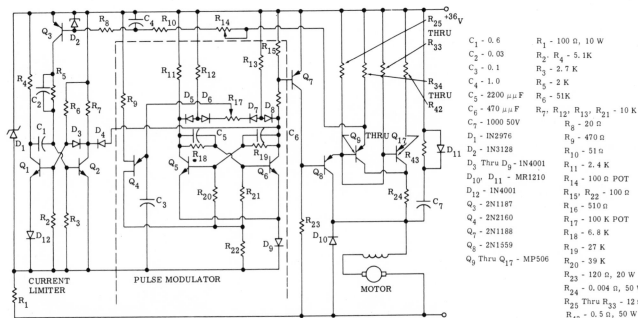

C₁ - 0.6	R₁ - 100 Ω, 10 W
C₂ - 0.03	R₂, R₄ - 5.1K
C₃ - 0.1	R₃ - 2.7 K
C₄ - 1.0	R₅ - 2 K
C₅ - 2200 μμF	R₆ - 51K
C₆ - 470 μμF	R₇, R₁₂, R₁₃, R₂₁ - 10 K
C₇ - 1000 50V	R₈ - 20 Ω
D₁ - 1N2976	R₉ - 470 Ω
D₂ - 1N3128	R₁₀ - 51 Ω
D₃ Thru D₉ - 1N4001	R₁₁ - 2.4 K
D₁₀, D₁₁ - MR1210	R₁₄ - 100 Ω POT
D₁₂ - 1N4001	R₁₅, R₂₂ - 100 Ω
Q₃ - 2N1187	R₁₆ - 510 Ω
Q₄ - 2N2160	R₁₇ - 100 K POT
Q₇ - 2N1188	R₁₈ - 6.8 K
Q₈ - 2N1559	R₁₉ - 27 K
Q₉ Thru Q₁₇ - MP506	R₂₀ - 39 K
	R₂₃ - 120 Ω, 20 W
	R₂₄ - 0.004 Ω, 50 W
	R₂₅ Thru R₃₃ - 12 Ω
	R₄₃ - 0.5 Ω, 50 W

Q₁, Q₂, Q₅, Q6 - 2N2218

R₃₄ Thru R₄₂ - 0.03 Ω, 10 W

36-V GOLF-CART CONTROL—Uses pulse-width modulation method of controlling d-c voltage to provide speed regulation of series d-c motor under varying torque conditions up to battery drain of 300 A under stall or starting conditions. Current limiting does not begin until above 200 A to provide sufficient torque for climbing steep inclines. Requires minimum of eight MP506 transistors in parallel for output stage. Report covers circuit operation in detail.—H. F. Weber, "Solid-State DC Motor Control for Traction Drive Vehicles," Motorola Semiconductors, Phoenix, AZ, AN-189, 1968.

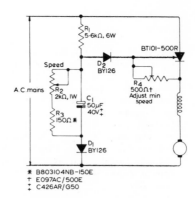

* B803104NB-150E
† E097AC/500E
‡ C426AR/G50

STABLE LOW-SPEED CONTROL—Improved half-wave scr speed control for drills and food mixers gives stable triggering point at low speeds and allows 160-deg conduction angle with minimum speed drop. Values of R2 and R3 depend on back emf of motor and minimum speed required.—R. P. Gant, Electronic Speed Control for Electric Drills and Food Mixers, *Mullard Technical Communications*, Sept. 1967, p 252–255.

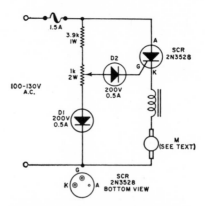

HALF-WAVE SCR FOR SERIES MOTOR—Designed for ordinary intermittent use with sewing-machine and other small appliance or portable tool motors rated around 1/15 hp at up to 5,000 rpm. Since scr operates in switching mode, heat is no problem. Residual magnetism in motor field core during *off* half-cycles provides required regulating action. Pot controls speed.—L. Fleming, SCR Controls for Small Motors, *Electronics World*, June 1970, p 36 and 83.

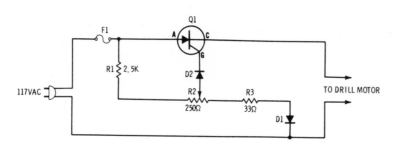

HIGH-TORQUE CONTROL—Compensating circuit increases torque of electric drill when motor speed is decreased by loading. Extra gate voltage of scr, developed by back emf across motor, increases firing angle of scr so as to provide additional current for holding motor speed constant. Use extra-heavy heatsink for scr, such as ¼-inch aluminum or copper at least 1 inch square. Diodes are 500-mA 200-piv silicon.—H. Friedman, "99 Electronic Projects," Howard W. Sams, Indianapolis, IN, 1971, p 65–66.

CHAPTER 73
Multiplexer Circuits

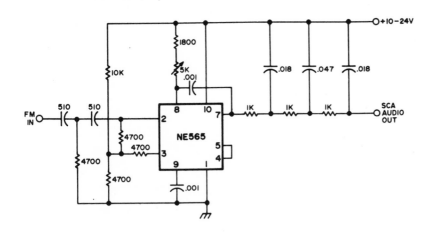

SCA ADAPTER—Signetics IC phase-locked loop serves as background music adapter for f-m receiver. Adjust 5K pot to 67 kHz to pick off subcarrier having background music of "storecast" services. Input connection should be made ahead of deemphasis filter in f-m receiver. High-pass filter at input of IC prevents overloading by much stronger audio of normal program or by stereo information on 38 kHz.—J. Kyle, The Phase-Locked Loop Comes of Age, 73, Oct. 1970, p 42, 44–46, 49–52, and 54–56.

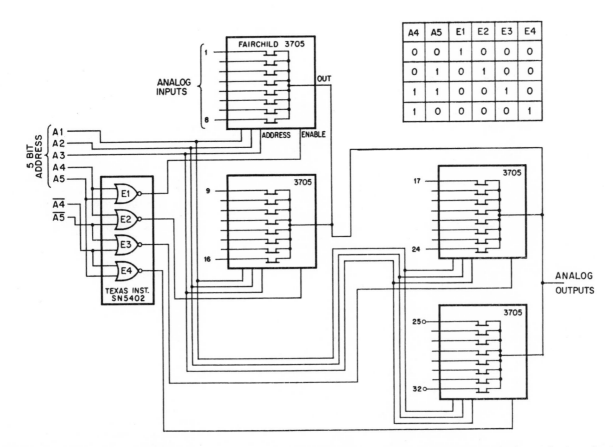

A4	A5	E1	E2	E3	E4
0	0	1	0	0	0
0	1	0	1	0	0
1	1	0	0	1	0
1	0	0	0	0	1

32-CHANNEL WITH 5-BIT CONTROL—Uses Fairchild 3705 multiplexer-commutator building block. NOR gates in TTL IC select package outputs, and first three address bits select one of eight channels in each package. Can also serve as commutator.—L. Accardi, Five Bits and Five ICs Switch 32 Analog Signals, *Electronics*, Dec. 7, 1970, p 75.

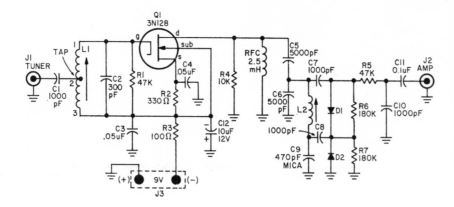

SCA ADAPTER—Simple amplifier-discriminator version of SCA (Subsidiary Communications Authorization) multiplex adapter can be connected between detector MPX (multiplex) output and audio amplifier input of f-m receiver, to hear commercial-free background music broadcast by many f-m stations on 67-kHz subcarrier as commercial service to stores and other businesses. Article gives construction details and shows how to modify receivers that do not have MPX jack. L1 is 12–40 mH J. W. Miller 9016 tapped adjustable coil, L2 is 4–30 mH Miller 6315, and matched germanium diodes are Sylvania ECG-110 or equivalent. —C. Green, Music-Music-Music, *Elementary Electronics*, March-April 1972, p 29–32.

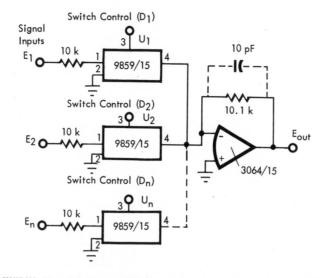

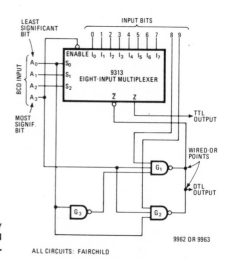

ALL CIRCUITS: FAIRCHILD

9962 OR 9963

ANALOG SWITCH—Uses Burr-Brown 9859/15 switched current amplifiers and opamp for analog switching and multiplexing applications. Properly sequenced control pulses to switches will multiplex inputs E1-En into opamp summing junction. With properly weighted input summing resistors, set of N switches can be connected as N-bit digital-analog converter.—"Electronic Switches," Burr-Brown, Tucson, AZ, Sept. 1969, p 4.

10-INPUT MULTIPLEXER—Achieved by adding only one DTL gate package to eight-input multiplexer IC having open-collector output. Bcd input controls selection of input bits to decade multiplexer. When most significant bit is low, multiplexer operates normally and accepts input bits 0 through 7, with gates G1 and G2 disabled. For bcd inputs 8 or 9, most significant bit is high, multiplexer is disabled, and bits 8 and 9 can pass to output since both G1 and G2 are enabled.—E. G. Breeze, Wired-OR DTL Gates Increase Multiplexer Input Capacity *Electronics*, Aug. 28, 1972, p 78.

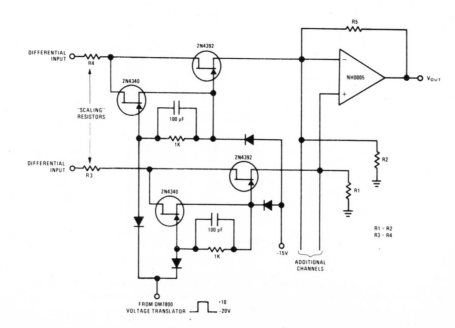

WIDE-BAND DIFFERENTIAL MULTIPLEXER—Permits simultaneous handling of signals up to several MHz at toggle rates up to 1 MHz. —"FET Circuit Applications," National Semiconductor, Santa Clara, CA, 1970, AN-32, p 10.

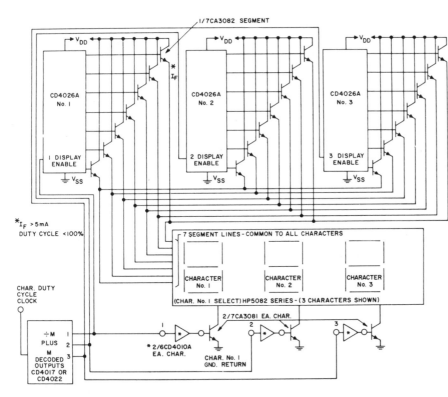

CHARACTER MULTIPLEXING—Used as interface between RCA CD4026A logic and HP5082 series multiple character display, to improve brightness and/or reduce power drain.—"COS/MOS Digital Integrated Circuits," RCA, Somerville, NJ, SSD-203A, 1973 Edition, p 440–441.

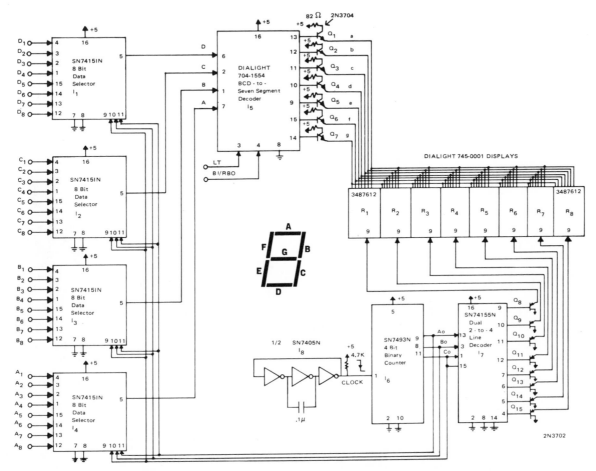

8 DIGITS WITH PARALLEL MULTIPLEXER—Three bits of binary counter I6 serve as scanner driving I7 wired as 1 out of 8 decoder. Each negative-going transition of clock results in three-bit code which selects one line from each of the four data selectors I1–I4 for presentation to decoder I5. Simultaneously, decoder I7 activates appropriate GaAsP display. Transistors Q1–Q7 provide drive currents required for strobed operation, while Q8–Q15 sink current which flows through segments of displays.—"LED Readout Displays," Dialight Corp., Brooklyn, NY, Application Note AN-7210, 1972, p 2–3.

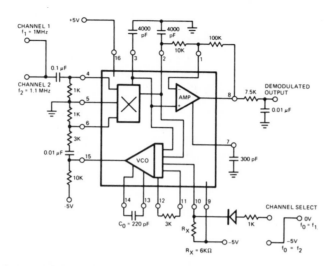

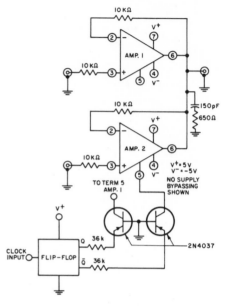

TIME-MULTIPLEXING OF F-M CHANNELS—A-c digital programming capability of Exar XR-215 phase-locked loop permits time-sharing or multiplexing between two f-m channels, at 1 and 1.1 MHz respectively. Channel-select logic signal is applied to pin 10. Inputs are selectively demodulated for output.—"XR-215 Monolithic Phase-Locked Loop," Exar Integrated Systems Inc., Sunnyvale, CA, July 1972, p 6.

TWO-CHANNEL—Clock input to flip-flop produces strobe that determines which input to RCA 3080 opamp appears at output of multiplexer. Requires bipolar supply in range of 2 to 15 V. Power consumption is adjustable from 10 μW to 30 mW.—"Linear Integrated Circuits and MOS Devices," RCA, Somerville, NJ, 1973 Edition, SSD-201A, p 465.

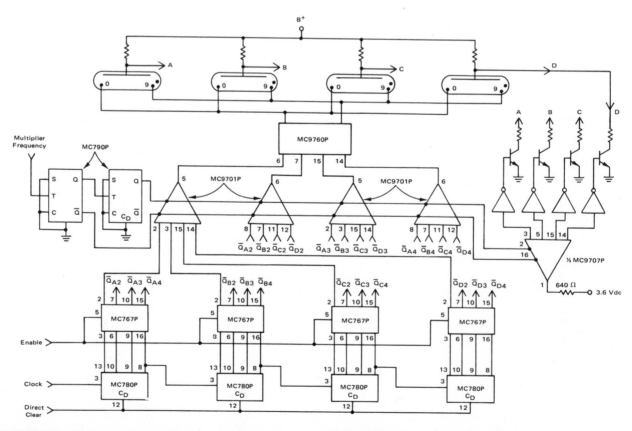

MULTIPLEXING—Used to reduce number of MC9760P decoder/drivers required for digital display. Supply voltage and display-tube resistors are chosen to meet display-tube requirement.—J. M. Fallon, "MRTL Counting Elements," Motorola Semiconductors, Phoenix, AZ, AN-514, 1970.

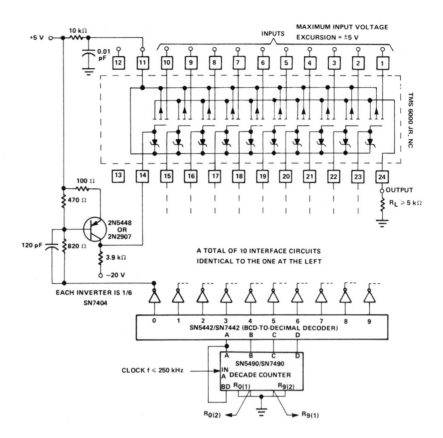

10-CHANNEL ANALOG SWITCH—Each input is sequentially connected through 10-channel mos switch to output circuit which can be represented by load resistance RL of at least 5K. Interface using pnp transistor translates TTL output voltage levels to those required by mos switch; +5 V turns switch off and —20 V turns it on. Simple R-C filter prevents supply noise from interfering with switch operation.—"MOS/LSI Standard Products Catalog," Texas Instruments, Dallas, TX, Catalog CC-402, 1971, p 303.

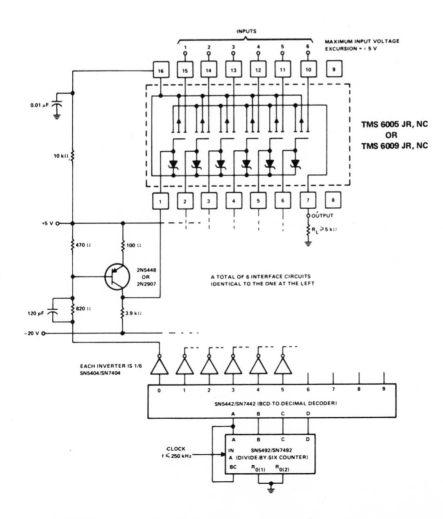

6-CHANNEL ANALOG SWITCH—Each input is sequentially connected through 6-channel mos switch to output circuit which can be represented by load resistance RL of at least 5K. Counter and decoder provide sequential drive from single clock having frequency up to about 250 kHz. Switch turns off with +5 V and on with —20 V on interface transistor stage. Transistor delay of about 300 ns allows previous mos switch to turn off completely before next one turns on.—"MOS/LSI Standard Products Catalog," Texas Instruments, Dallas, TX, Catalog CC-402, 1971, p 310.

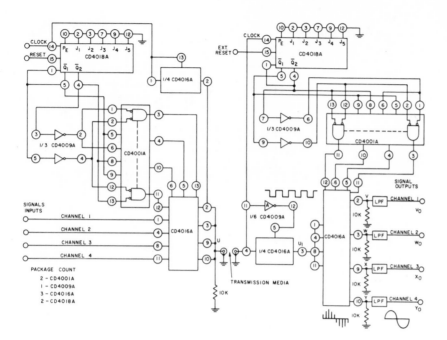

FOUR-CHANNEL PAM—Complete pulse amplitude modulation system for multiplexing and demultiplexing of four data transmission channels is shown. Each input signal is sampled sequentially and applied to single transmission line, but in different time slot. Signals are detected and reconstructed accurately at outputs if sampling rate is greater than twice maximum frequency component of input signal. Clock input is 80 kHz. Use of cos-mos devices keeps power dissipation low, for airborne instrumentation telemetry applications. —"COS/MOS Digital Integrated Circuits," RCA, Somerville, NJ, SSD-203A, 1973 Ed., p 383–385.

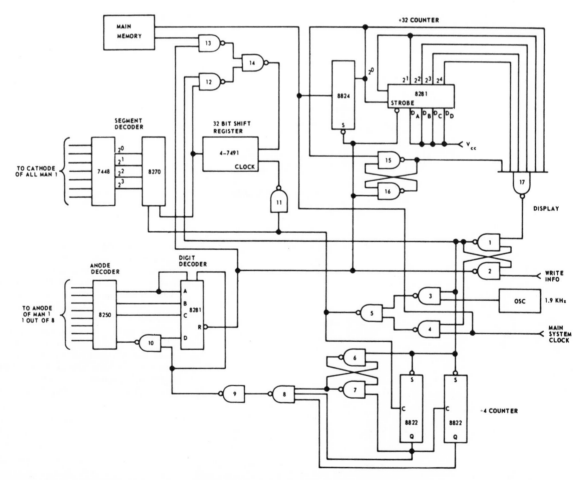

MULTIPLEXING ALPHANUMERIC—Will multiplex up to eight seven-segment Monsanto MAN 1 GaAsP alphanumeric displays, with connections shown, and can handle 16 by adding another anode decoder/driver. Economics depends on IC pricing, with multiplex system becoming more economical as number of digits increases beyond crossover point of six digits. Article describes operation in detail. IC's used are from Signetics 8000 series, equivalent to Fairchild 9000 series.—R. T. Gill, "Multiplexing and Individual Addressing the Man 1 Seven-Segment Display," *Monsanto GaAsLite Tips*, Vol. 1, 1970.

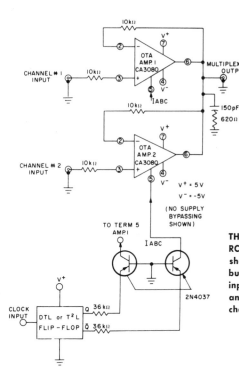

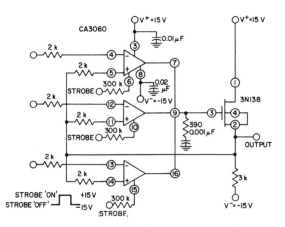

THREE-CHANNEL—Each channel uses single RCA CA3060 low-power opamp, with all sharing common 3N138 mosfet serving as buffer and power amplifier. When strobe input for channel is on, opamp is activated and output swings to level of input for that channel. Open-loop voltage gain for complete channel is then over 100 dB, assuring excellent accuracy in voltage follower mode with 100% feedback. Bandwidth of system is 1.5 MHz and slew rate 0.3 V/μs.—"Linear Integrated Circuits and MOS Devices," RCA, Somerville, NJ, 1973 Edition, SSD-201A, p 476.

TIME SHARING—High-performance IC operational transconductance amplifiers serve in two-channel linear time-shared multiplex circuit for data transmission. Maximum level shift from input to output is about 5 mV. Open-loop gain of system is typically 100 dB when output loading is low.—"Linear Integrated Circuits and MOS Devices," RCA, Somerville, NJ, SSD-202A, 1973 Edition, p 312–313.

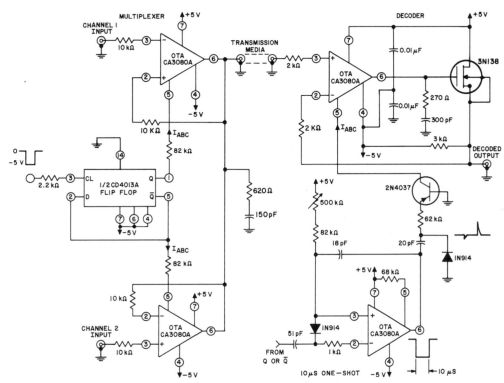

TWO-CHANNEL MULTIPLEXER-DECODER—Built around high-performance CA3080A operational transconductance amplifiers. Number of channels may be extended as desired. Delay of 10 μs in decoder insures that sampling can occur only after input signal has settled. Used for data transmission systems. —"Linear Integrated Circuits and MOS Devices," RCA, Somerville, NJ, SSD-202A, 1973 Edition, p 320–322.

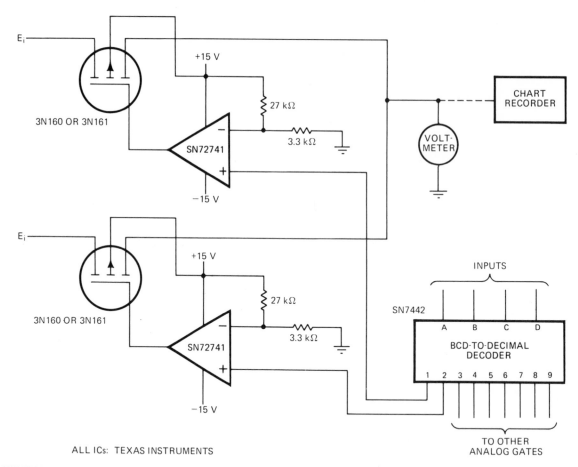

ALL ICs: TEXAS INSTRUMENTS

TIME-MULTIPLEXING — Four-terminal mosfet serves as series analog switch for time-multiplexing a-c or d-c signal voltages being monitored periodically by chart recorder or voltmeter. Decoder switches one analog gate at a time for given interval. Decoder outputs are high except for output being decoded. Maximum input voltage swing is ±10 V.—G. Coers, Gated MOSFET Acts as Multiplexing Switch, *Electronics*, Nov. 6, 1972, p 102.

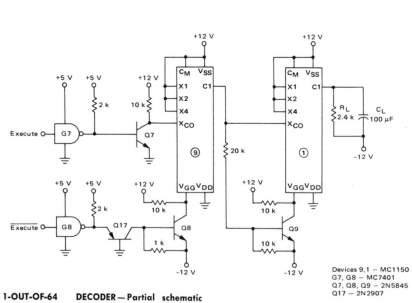

Devices 9,1 — MC1150
G7, G8 — MC7401
Q7, Q8, Q9 — 2N5845
Q17 — 2N2907

1-OUT-OF-64 DECODER—Partial schematic shows method of interconnecting MC1150 IC multiplex switches so bits 1 through 6 can be used to address one of 64 output lines. Selected output line changes from —12 V to +10 V whenever execute signal goes from 0 (0 V) to 1 (+5 V). Report describes operation and gives typical switching waveforms as well as complete schematic.—T. Reynolds, "MOS Multiplex Switches," Motorola Semiconductors, Phoenix, AZ, AN-523, 1970.

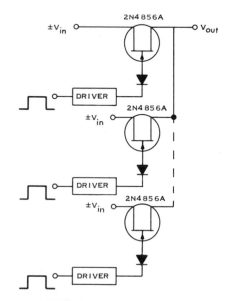

FET MULTIPLEX SWITCH—Formed by combining any number of single-fet analog switches, turned on or off by gate-drive signal acting through driver which can be made up from 2N3829 and 2N3013 transistors in two-stage amplifier.—L. Delhom, "Switching-Circuit Applications of the 2N4856A Series FET," Texas Instruments, Dallas, TX, Bulletin CA-109, 1969, p 8.

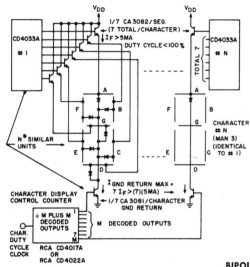

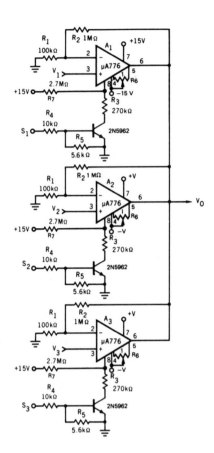

CHARACTER-DRIVE MULTIPLEXING—Gives light enhancement factor greater than one through multiplexing or, for same display brightness, reduces power dissipation of display. Developed for use with MAN-3 7-segment diffused planar GaAsP light-emitting diode display (Monsanto or equivalent). N represents number of characters in display, and M is number of character display control counter decoded outputs.—"COS/MOS Digital Integrated Circuits," RCA, Somerville, NJ, SSD-203A, 1973 Edition, p 440.

BIPOLAR MULTIPLEXING—Use of opamp with programmable gain eliminates need for fet switches when multiplexing bipolar analog signals. Voltage requirement for switches is logic signal referenced to ground, obtained from TTL or DTL device. Isolation between on and off amplifier is 80 dB at 50 kHz. Supply voltage for opamp can be 1.2 to 18 V. Circuit also provides signal conditioning.—"The μA776, An Operational Amplifier with Programmable Gain, Bandwidth, Slew-Rate, and Power Dissipation," Fairchild Semiconductor, Mountain View, CA, No. 218, 1971, p 3.

CHAPTER 74
Multiplier Circuits

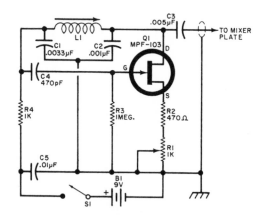

455-KHZ Q MULTIPLIER—Uses positive feedback to narrow passband of receiver and provide Q multiplication of 20 to 30 that has effect of sharply attenuating off - resonance signal frequencies. L1 is slug-tuned bc antenna coil. R1 controls regeneration. Circuit is connected to mixer plate of communication receiver having 455-kHz i-f value.—R. N. Tellefsen, Build a FET-QM, *Popular Electronics*, Dec. 1968, p 51–53.

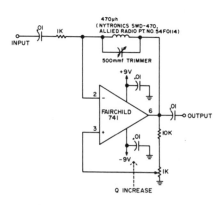

Q MULTIPLIER—Can be inserted after ssb filter in i-f amplifier of receiver or transceiver, to improve selectivity. Is particularly effective on c-w. Values shown are for 455-kHz i-f, for which this IC multiplier multiples Q of coil over 50 times. Same IC arrangement can be used to improve Q and selectivity of audio filters and tuned circuits of fsk converters.—J. J. Schultz, A Simple Integrated Circuit Q Multiplier, 73, Feb. 1970, p 134–137.

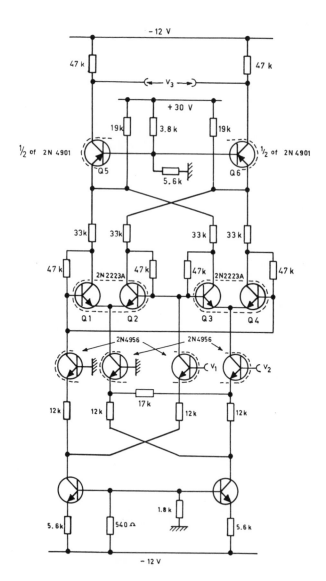

FOUR-QUADRANT MULTIPLIER—Single-stage analog multiplier provides output voltage V3 as product of input voltages V1 and V2. Circuit utilizes exact exponential characteristics of emitter-base junction of bipolar transistor to realize controlled current source. Article gives design equations. Developed for use in high-accuracy audiocorrelator.—H. Bruggemann, Feedback Stabilized Four-Quadrant Analog Multiplier, *IEEE Journal of Solid-State Circuits*, Aug. 1970, p 150–159.

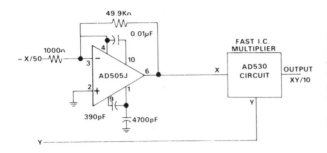

PREAMP FOR FAST MULTIPLIER—Opamp connected for gain of 50 will drive IC multiplier at up to 10 V with either polarity for frequencies up to 1 MHz. Response is down 3 dB at 1.6 MHz.—H. Krabbe and R. S. Burwen, Stable Monolithic Op Amp Slews at 130V/μs, *Analog Dialogue*, Vol. 5, No. 4, p 3–5.

ANALOG DIVIDER—With single IC multiplier connected as divider, numerator input signal is fed to Z input, and denominator to X input. Output is obtained from Y input via gain-trimming pot. To divide by positive values of X input signal, sign inversion stage is required for X. Accuracy is highest for maximum denominator input, with error increasing from 0.15% to 1.5% as X decreases to —1 V. Bandwidth is 700 kHz when X is —10 V and drops to 600 kHz at X = —1 V.— "AD530 Complete Monolithic MDSSR," Analog Devices, Inc., Norwood, MA, Technical Bulletin, July 1971.

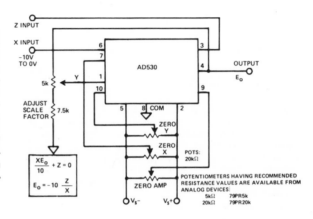

ANALOG DIVIDER—Produces analog quotient of A/B by multiplying analog dividend input A by reciprocal of divisor input B. Reciprocal circuit consists of Tr1, Tr2, Tr3, and Tr8, acting with feedback amplifier Tr4-Tr5 and constant-current sources Tr6-Tr7. Completely accurate for inputs from 1 to 10 V, with 1% error at 0.4 V and 2% error at minimum input of 1.2 V.—A. F. Newell, A Transistor Multiplier Circuit, *Wireless World*, June 1969, p 285–289.

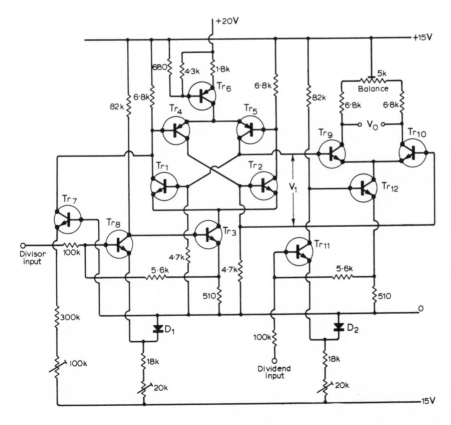

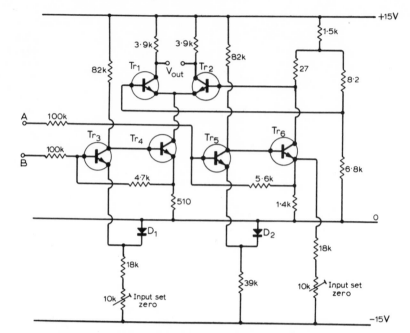

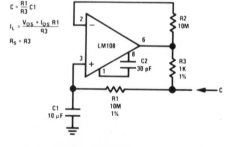

ADJUSTABLE SCALE FACTOR—Uses Analog Devices AD531 IC for computing function XY/Z, with any or all of the three input voltages variable. Scale factor (1/Z) is changed with 20K pot, or can be varied dynamically by applying externally controlled reference current. NOTE: End terminals of 20K pot for Xo should be connected between +15 V and 42K resistor going to ground.—Monolithic Analog Multiplier-Dividers, *Analog Dialogue*, Vol. 6, No. 3, p 10.

ANALOG MULTIPLIER—Linearity depends heavily on characteristic of long-tailed pair. With values shown, providing considerable negative feedback, maximum departure from linearity is about 3%. Diodes D1 and D2 compensate for base-emitter voltages of Tr3 and Tr5, so emitter current of Tr4 is zero for zero input and emitter current of Tr6 is that required to balance long-tailed pair. Transistors are not critical, and can be BC107 or similar.—A. F. Newell, A Transistor Multiplier Circuit, *Wireless World*, June 1969, p 285–289.

CAPACITANCE MULTIPLIER—Opamp eliminates need for large capacitance values by increasing effective capacitance of small capacitor and coupling it into low-impedance system. Circuit generates equivalent capacitance of 100,000 μF with worst-case leakage of 8 μA over −55 C to 125 C temperature range. Not suitable for tuned circuits and filters because Q is low, but satisfactory for timing circuit or servo compensation networks.—R. J. Widlar, "IC Op Amp Beats FETs on Input Current," National Semiconductor, Santa Clara, CA, 1969, AN-29, p 10–11.

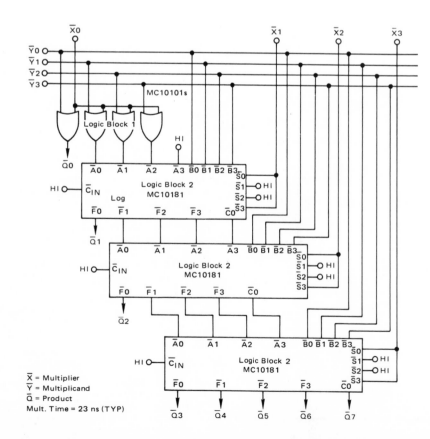

BINARY MULTIPLIER—Use of IC 10181 logic blocks makes ripple multiplier possible without large numbers of interconnects and parts, while giving typical multiplying time of only 23 ns for two 4-bit binary numbers. May be expanded easily to accommodate larger number of bits as covered in report. Based on use of one's complement of multiplicand and multiplier to provide one's complement of product. Product output is inverted.—T. Balph, "High Speed Binary Multiplication Using the MC10181," Motorola Semiconductors, Phoenix, AZ, AN-566, 1972.

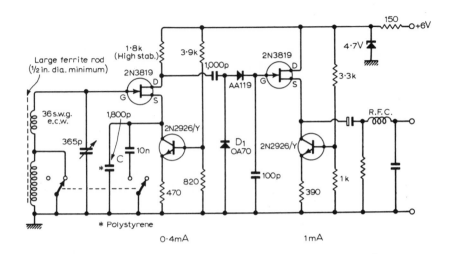

FET AS Q MULTIPLIER—Has inherent automatic bandwidth control up to 2 MHz. C determines no-signal Q; decreasing C moves stage toward oscillation. Circuit will replace two double-tuned 470-kHz i-f stages.—K. W. Mawson, High-Gain F.E.T. Tuned Amplifier, *Wireless World*, April 1970, p 182.

2-QUADRANT MULTIPLIER—Useful as gain controller for either a-c or d-c signals. Control input of 0 to 10 V at Y terminal will change gain of X-input signal in ±10-V range. Gain is unity when Y is +10 V. Linearity of control is 1% for Y input of 1 to 10 V. Bandwidth is 45 kHz at gain of 0.1 and decreases to 4.5 kHz at gain of 1.—Choosing and Using N-Channel Dual J-Fets, *Analog Dialogue*, Dec. 1970, p 4–9.

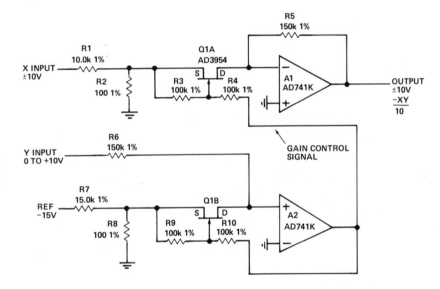

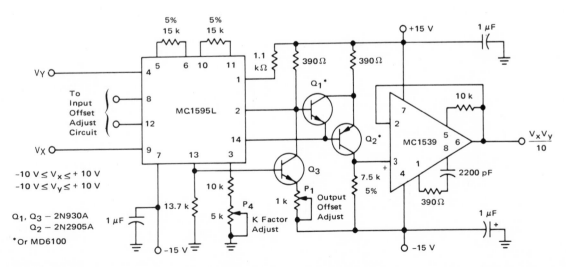

OPAMP BOOSTS CURRENT DRIVE—Uses discrete components for level-shifting of output to ground reference when multiplying two signal voltages, with opamp connected as source follower to increase current drive of single-ended output. Temperature problems are minimized by using MD60100 complementary-pair transistor package for Q1 and Q2.—E. Renschler, "Analysis and Basic Operation of the MC1595," Motorola Semiconductors, Phoenix, AZ, AN-489, 1970.

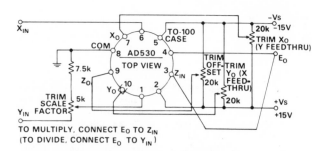

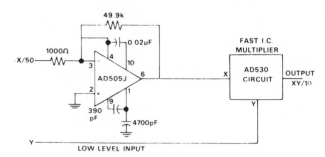

ANALOG MULTIPLIER—Single IC serves as complete multiplier or divider having transfer function of XY/10. X, Y, and Z input levels are ±10 V and output Eo is ±10 V at 5 mA. Accuracy can be 1%. Frequency response is down 3 dB at 1 MHz.—"AD530 Complete Monolithic MDSSR," Analog Devices, Inc., Norwood, MA, Technical Bulletin, July 1971.

PREAMP FOR FAST MULTIPLIER—Low-cost AD505J opamp has fast enough slew rate to drive AD530 multiplier-divider with ±10 V at up to 1 MHz. Preamp is connected for gain of 50. Response is 3 dB down at 1.6 MHz, with full output capability. Other input of multiplier can be run through preamp similarly. Multiplier accuracy is improved because both inputs and output can swing through full-scale range.—"Integrated Circuit High Speed Operational Amplifier," Analog Devices, Norwood, MA, Aug. 1972.

ANALOG MULTIPLIER—Uses Exar XR-S200 IC with external components connected to provide linear four-quadrant multiplication of input signal levels up to 6-V p-p without need for d-c level shifting between input and output.—"XR-S200 Multi-Function Integrated Circuit," Exar Integrated Systems Inc., Sunnyvale, CA, June 1972, p 3.

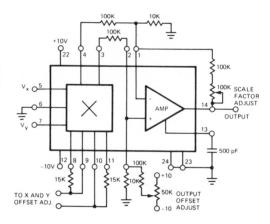

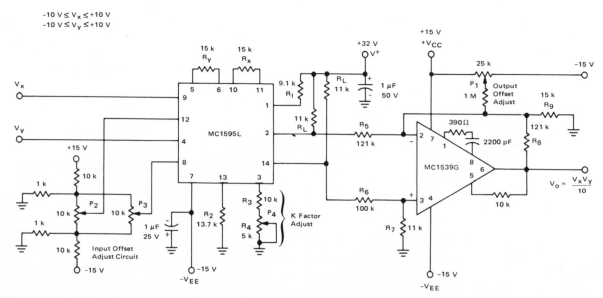

MULTIPLIER WITH OPAMP LEVEL SHIFT—Uses opamp for level-shifting multiplier output to ground reference while still giving product of input voltages (divided by 10) at single-ended output. Circuit is frequency-limited to about 50 kHz for signals swinging over full input range of ±10 V. Requires three power supplies.—E. Renschler, "Analysis and Basic Operation of the MC1595," Motorola Semiconductors, Phoenix, AZ, AN-489, 1970.

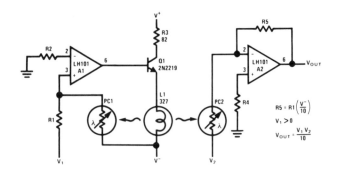

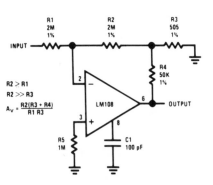

THREE-QUADRANT ANALOG MULTIPLIER— Optoelectronic circuit uses opamp A2 as controlled-gain amplifier for V2, with gain depending on ratio of PC2 resistance to R5. Control opamp A1 provides drive for lamp L1 until current to summing junction from negative supply through PC1 is equal to that from V1 through R1. With photoconductors matched, equal illumination makes their resistances equal. Mount cells in hole in aluminum block, with lamp midway between them. Circuit gives inverting output with magnitude equal to 1/10th product of two analog inputs. Time constant of lamp restricts use to low frequencies.—W. S. Routh, "Applications Guide for Op Amps," National Semiconductor, Santa Clara, CA, 1969, AN-20.

RESISTANCE MULTIPLIER—Inverting opamp with gain of 100 provides 2-meg input resistance using only 2-meg resistors, as contrasted to 200-meg feedback resistor required in conventional circuit. Possible drawback is multiplication of offset voltage by 200, instead of by 101 in conventional inverter.—R. J. Widlar, "IC Op Amp Beats FETs on Input Current," National Semiconductor, Santa Clara, CA, 1969, AN-29, p 14.

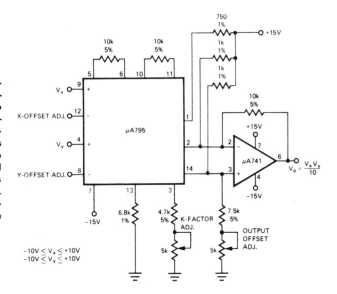

LEVEL SHIFT—IC opamp and multiplier together provide d-c level shift that causes output to vary in either direction with respect to ground. Applications include power measurement, wherein Vx input receives signal proportional to instantaneous voltage across load, and Vy receives voltage proportional to instantaneous load current. Power measured may be d-c, sinusoidal a-c, or combinations of frequencies up into low megahertz range. —"The μA795, A Low-Cost Monolithic Multiplier," Fairchild Semiconductor, Mountain View, CA, No. 211, 1971, p 3.

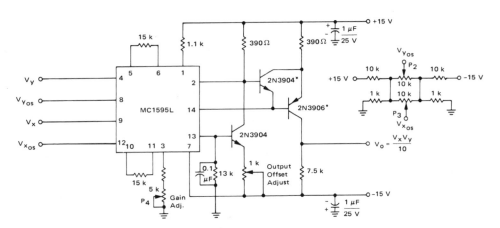

DISCRETE LEVEL-SHIFTING—Simple circuit using discrete components permits multiplying of two signal voltages, each up to 10 V with either polarity, at frequencies well above 50 kHz. Possible drawback is temperature sensitivity if base-emitter junctions of upper two transistors are not matched to track with temperature. Use of MD60100 complementary-pair transistors will minimize this problem.—E. Renschler, "Analysis and Basic Operation of the MC1595," Motorola Semiconductors, Phoenix, AZ, AN-489, 1970.

CHAPTER 75
Multivibrator Circuits

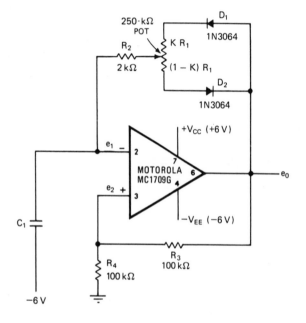

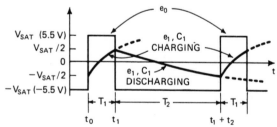

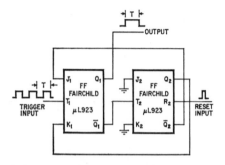

PULSE SEPARATOR—Two J-K flip-flops operate as pulse separator to align reset pulses with trigger input pulse train. In standby, both flip-flops are latched so Q1 output is low. Reset unlatches both, and trigger then makes Q1 go high and latches FF2. On time thus varies with trigger period, and off time with reset pulse period.—L. E. Baker, Flip-Flop Pair Synchronizes Pulses and Floats Clocks, *Electronics*, May 24, 1971, p 56.

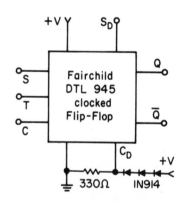

C_1 (µF)	$(T_1)_{min}$ (µs)	$T_1 + T_2$ (ms)	FREQUENCY (Hz)
0.0027	7	0.6	1,670
0.01	21.3	2.1	464
0.056	128	13	76.6
0.1	212	21.7	46.1
0.33	690	70.9	14.1
1	2,130	218	4.6

1–99% DUTY CYCLE—Simple circuit has good temperature stability, with duty cycle range nearly independent of repetition rate, yet requires only one timing capacitor C1.—

M. J. Shah, Feedback Pot Extends Multivibrator Duty Cycle, *Electronics*, Sept. 27, 1971, p 62.

INITIALIZER—Diode-resistor network is superior to R-C network for placing IC flip-flop in a particular state each time power is turned on. Whenever supply voltage V drops below 3.5 V, CD goes *low* and forces *low* Q output. When supply returns to above 3.5 V, CD goes *high* and restores flip-flop operation. Circuit is fast enough to respond to negative-going ine transients.—J. R. Giroux, Initialize Flip-Flops Fast, and Don't Worry about Time Constants, *The Electronic Engineer*, Aug. 1968, p 68.

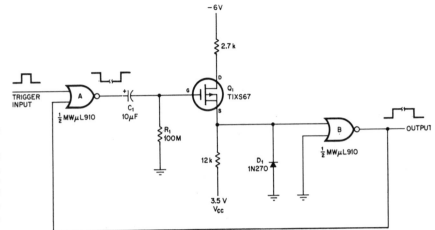

PULSE STRETCHER—Gives pulse duration of 260 s by maintaining output high until C1 discharges through R1 to point where Q1 is cut off. Trigger input, inverted by RTL gate A, drives Q1 into conduction initially. Conventional IC one-shot would give only about 0.4 s for value of C1 shown.—P. M. Salomon, FET Multiplies Pulse Time of IC One-Shot, *Electronics*, April 12, 1971, p 90.

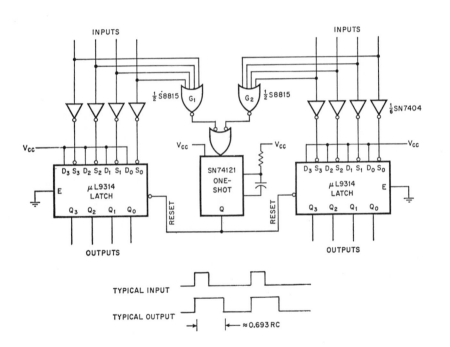

STRETCHER—Uses single one-shot to control latches that hold pulse level high for period of one-shot. Requires only single four-bit latch for each four channels, plus shared one-shot, one resistor, and one capacitor. When one-shot times out, latch is reset and all data outputs return to low.—K. Erickson, Shared One-Shot Simplifies Pulse Width Converter, *Electronics*, Sept. 14, 1970, p 102.

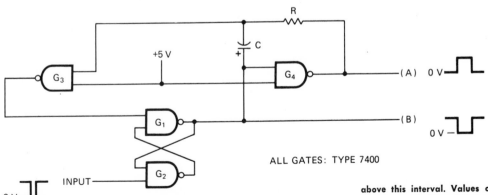

DUAL-PURPOSE—Provides complementary outputs when operating either as one-shot or as synchronous astable mvbr. Input can be ordinary grounding pushbutton switch. When input is low for less than three RC time constants, circuit is one-shot, but is astable above this interval. Values of R range from 330 ohms to 1.5K, and C from 0.001 to 1,000 μF.—E. Beach, Double-Duty Multivibrator Gives Complementary Outputs, *Electronics*, June 19, 1972, p 92.

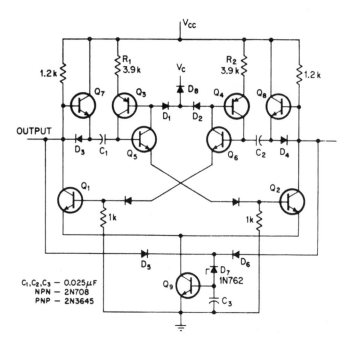

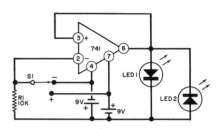

LED STATUS INDICATOR—With linear IC opamp connected as bistable mvbr (flip-flop), LED1 will be forward biased and on when opamp pin 6 is initially positive, and other led will be off. If S1 is momentarily switched to positive supply voltage, output at pin 6 becomes negative and LED2 comes on, with LED1 off. Use any led.—H. Garland, Flip-Flop with Op Amps, *Popular Electronics*, July 1972, p 58–59.

1–12,000 HZ VOLTAGE CONTROL—Control voltage Vc, between 2 V and supply voltage Vcc, provides broad frequency range for free-running mvbr. Pnp transistor Q3 acts with R1 as constant-current source for charging C1.—M. J. Fisher and J. Byrne, Voltage Changes Frequency of Multivibrator by 10,-000:1, *Electronics*, March 1, 1971, p 58.

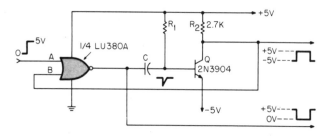

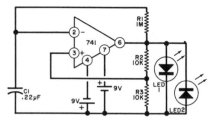

ASTABLE STATUS INDICATOR—When opamp output at pin 6 is positive, LED1 is on and C1 charges through R1. When C1 charge exceeds positive input of opamp, output swings negative and turns LED1 off; LED2 then comes on and charging process repeats. With values shown, led's flash back and forth about twice a second. Use any led.—H. Garland, Flip-Flop with Op Amps, *Popular Electronics*, July 1972, p 58–59.

BIPOLAR OUTPUT—Hybrid one-shot uses one NOR gate and common-emitter transistor operating between complementary supplies. Positive 5-V step at input A starts timing cycle by driving output of gate down and thereby turning Q off. C discharges until base voltage of Q rises enough to turn it on and reset circuit.—W. G. Jung, Hybrid One-Shot Has Logic Input and Bipolar Output, *The Electronic Engineer*, April 1971, p 66.

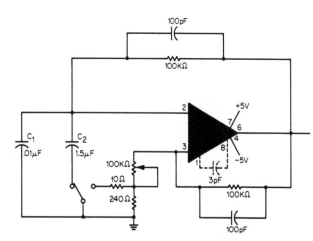

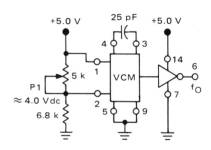

1 HZ TO 1 MHZ—Will operate up to 4 MHz with LM301C IC but is sensitive to stray capacitance above 1 MHz. Use of 100-pF capacitor across each feedback resistor of essentially standard mvbr extends control range with relatively small R-C network.—R. T. Scarpulla, Multivibrator Covers 1 Hz to 1 MHz, *The Electronic Engineer*, Nov. 1971, p 58.

10-MHZ ASTABLE—Uses Motorola MC4324/4024 IC consisting of two independent voltage-controlled mvbr's with output buffers. Input d-c control voltage of 1 to 5 V varies output frequency over range of 3.5 to 1. Maximum operating frequency with appropriate changes in external components is 25 MHz.—"MC4324/MC4024 Dual Voltage-Controlled Multivibrator," Motorola Semiconductors, Phoenix, AZ, Data Sheet 9108, 1972.

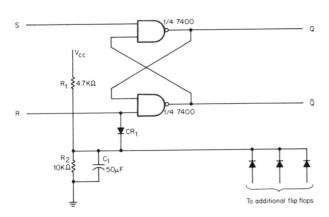

R-S RESET—Single diode makes action predictable with only two-input NAND gates, instead of three-input gates otherwise required.—A. H. Roshon, *Predictable R-S Flip-Flop, The Electronic Engineer,* May 1971, p 78.

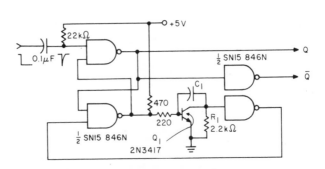

92% DUTY CYCLE—Combination of quad two-input DTL NAND gate and Miller integrator is stable one-shot providing long pulse durations, each equal in milliseconds to about $4.1C_1$, where C_1 is in μF.—R. C. Baskin and D. A. Esakov, *Very High Duty Cycle One-Shot, The Electronic Engineer,* Feb. 1969, p 86.

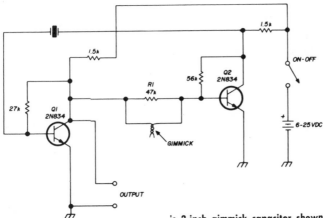

CRYSTAL CONTROL—Will oscillate at any frequency from 2.5 kHz to 15 MHz by changing crystal, without tuned circuits. Only reactance is 2-inch gimmick capacitor shown. Output is rich in harmonics, permitting use as marker generator.—M. Centore, *Crystal-Controlled Multivibrator, Ham Radio,* July 1971, p 65.

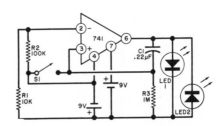

MONO STATUS INDICATOR—With connections shown, stable output state is positive so LED1 is on and LED2 off. Closing S1 momentarily triggers one-shot, making output go negative and turn off LED1 so LED2 comes on. After switch is opened, C1 charges through R3 until output goes back to stable positive state (about 0.5 s) and led's flash back to original condition. Use any led.—H. Garland, *Flip-Flop with Op Amps, Popular Electronics,* July 1972, p 58–59.

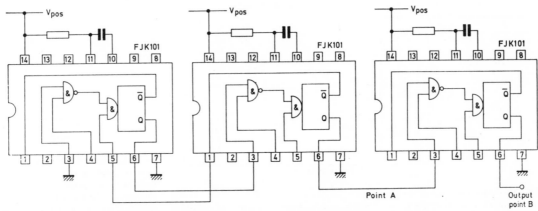

Point A

Output point B

1,000:1 DUTY CYCLE—Uses Mullard monostable IC's connected to avoid restriction on duty cycle normally existing because timing capacitors must recharge to initial conditions during off periods. First two monos have similar external timing components of 0.1 μF and 6.8K, cross-coupled to form free-running mvbr with mark-space ratio of 1:1. Third mono, triggered from Q output of either of others, has timing components chosen to give required pulse duration; 1,000 pF and 1.5K give 1,000:1 duty cycle.—"Mullard TTL Integrated Circuits Applications," Mullard, London, 2nd Ed., 1970, p 193–194.

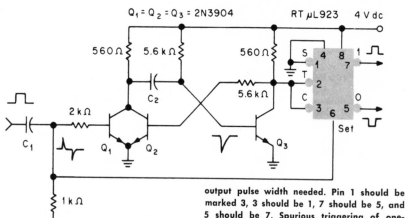

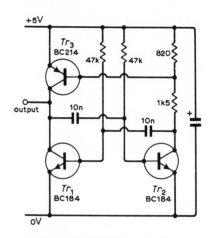

FLIP-FLOP KILLS NOISE—False triggering of one-shot Q1-Q2-Q3 by noise is avoided by adding IC flip-flop. Value of C2 depends on

output pulse width needed. Pin 1 should be marked 3, 3 should be 1, 7 should be 5, and 5 should be 7. Spurious triggering of one-shot does not affect output of flip-flop because one-shot would simply be trying to reset flip-flop again.—K. L. Stone, Bothered by Hiccup? Cure Them with an IC, *The Electronic Engineer*, Oct. 1968, p 90.

FAST RISE TIME—Collector load resistor of TR1 is replaced by pnp transistor TR3 which is switched on and off by collector current of TR2. With values shown, rise and fall times are 0.5 μs, increasing only to 2.5 μs with load of 0.05 μF.—C. R. Masson, Fast Rise-Time Multivibrator, *Wireless World*, May 1972, p 239.

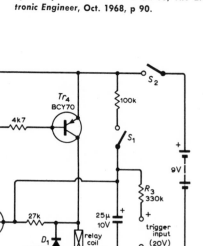

LONG-RECOVERY MONO—Developed for applications in which recovery time during which circuit cannot respond to second trigger pulse can be up to 1 hour with good-quality capacitor C1. Cross-coupled astable

mvbr TR1-TR2 feeds relay driver TR4, with TR3 used as gate.—D. A. Tong, Monostable with Long Adjustable Recovery, *Wireless World*, July 1972, p 349.

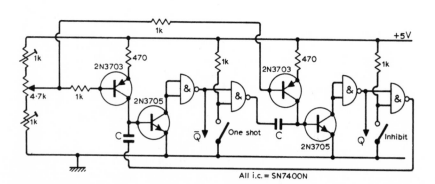

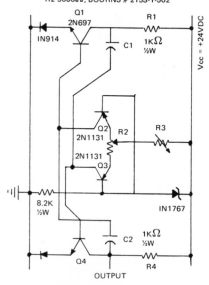

VARIABLE ASTABLE—Generates square waves from 1 to 30 Hz under control of single linear pot. Preset 1K resistors are adjusted to set desired frequency limits of range. Value of C determines frequency of oscillation; 33 μF gives 0.167 Hz, and 5,000 pF gives 350 kHz. Uses quad 2-input NAND as twin 360-deg in-

verter, to reduce rise and fall time. Closing inhibit switch stops oscillator. If one-shot switch is then pressed, single square wave is generated.—C. C. Ward, Variable Astable Multivibrator, *Wireless World*, Oct. 1971, p 512.

3:1 PULSE RATE RANGE—R3 controls current level supplied by constant-current sources Q2 and Q3 to conventional astable mvbr Q1-Q4. This current charges C1 and C2 alternately to control frequency over range of 1,500 to 4,-500 pps or other 3:1 range, depending on values used for C1 and C2. Q4 is 2N697.—G. Poggi, Pulse Repetition Rate Control with Two Pots, *Electromechanical Design*, Oct. 1971, p 38.

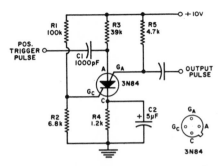

10-MS PULSE DURATION—Positive-going spike pulse at input triggers scs on for duration depending on value of C2, after which circuit idles until turned on by next trigger pulse. For shorter pulse duration, decrease value of C2.—F. H. Tooker, SCS Monostable Multivibrator, *Electronics World*, Dec. 1969, p 74.

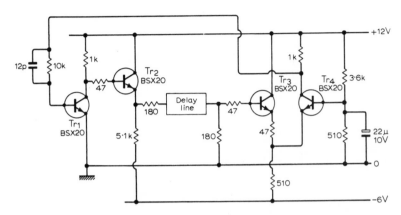

DELAY-LINE COUPLING—Use of delay line in place of conventional R-C coupling provides excellent frequency stability for square-wave generator. Total frequency drift is less than 0.1% for 0–45 C.—J. Heinzl, Delay-Line Coupled Multivibrator, *Wireless World*, Dec. 1969, p 556.

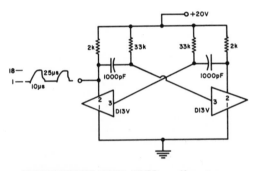

PROGRAMMABLE-ZENER MVBR — Uses two General Electric D13V programmable zeners in place of switching transistors. Capacitors determine frequency. Output is nonsymmetrical pulse.—R. L. Starliper, Programmable Zener Applications, *Electronics World*, June 1971, p 60–61.

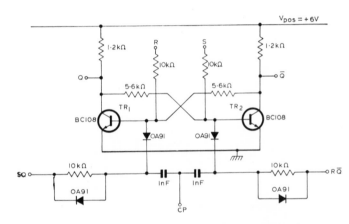

CLOCKED R-S BISTABLE—Levels (0 or 1) of inputs SQ and RQ determine state that bistable assumes after complete positive clock pulse at CP.—"Electronic Counting," Mullard, London, 1967, TP874, p 45.

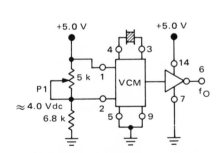

3–15 MHZ CRYSTAL—Uses Motorola MC4324/4024 IC consisting of two independent voltage-controlled mvbr's with output buffers. Exact frequency depends on crystal used. Crystal frequency can be pulled slightly by adjusting P1.—"MC4324/MC4024 Dual Voltage-Controlled Multivibrators," Motorola Semiconductors, Phoenix, AZ, Data Sheet 9108, 1972.

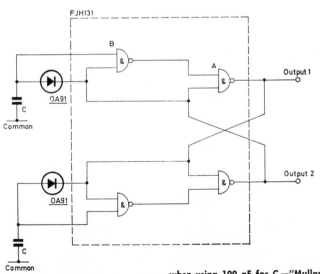

4-GATE MVBR—Uses Mullard four-gate IC to give square-wave output of 3.5 V at 3.3 kHz when using 100 nF for C.—"Mullard TTL Integrated Circuits Applications," Mullard, London, 2nd Ed., 1970, p 190.

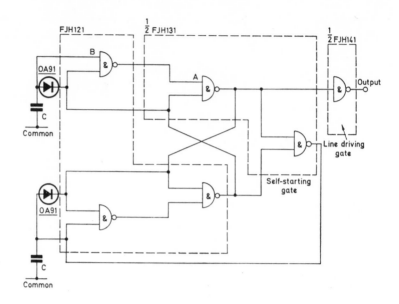

SELF-STARTER GATE—Mullard IC gates are combined to give conventional 4-gate mvbr with self-starter gate and line-driving gate to give square-wave output of 3.5 V at 3.3 kHz when using 100 nF for C.—"Mullard TTL Integrated Circuits Applications," Mullard, London, 2nd Ed., 1970, p 191.

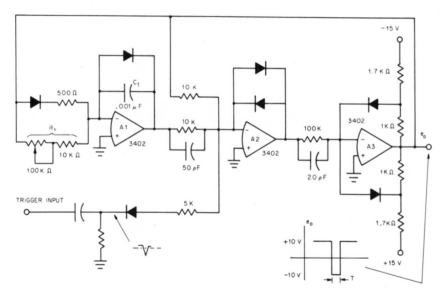

PRECISION OPAMP MONO—Provides adjustable trigger sensitivity, very wide range of pulse width adjustment (10–100 μs for values shown), and very flat well-controlled square-wave output. Uses regenerative switching action for completing single-shot pulse. Will work with almost any opamp, but fast slew rate and good settling time are desirable for very high-speed narrow pulses.—J. G. Graeme, G. E. Tobey, and L. P. Huelsman, "Operational Amplifiers," McGraw-Hill Book Co., New York, 1971, p 394–396.

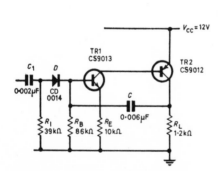

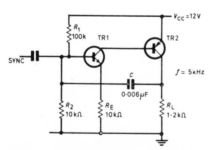

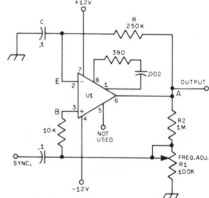

ZERO QUIESCENT POWER—Simple mono using complementary pair of transistors draws power only while in operation. Useful over wide range of frequency. With values shown, conduction period is 1 ms. Article gives mathematical analysis of circuit operation. Trigger voltage is about 2 V.—C. F. Ho, Zero Quiescent Current Monostable Multivibrator, *The Radio and Electronic Engineer*, Jan. 1969, p 22–24.

ASTABLE WITH LOW QUIESCENT POWER—Both transistors of complementary pair are either on or off simultaneously. While off, only current drain is that flowing through divider R1-R2, whose values can be made reasonably large to minimize quiescent power dissipation. Article includes mathematical analysis. TR1 is CS9013 and TR2 is CS9012.—C. F. Ho, Zero Quiescent Current Monostable Multivibrator, *The Radio and Electronic Engineer*, Jan. 1969, p 22–24.

120–3,000 HZ ASTABLE—Produces clean, symmetrical 24-V p-p square waves at frequency determined essentially by R and C. Triangular wave of same frequency is available at E. Opamp can be MC1439G, 709C, or equivalent. Triggering or synchronizing signal may be applied to sync input.—B. Balakrishnan, More on Operational Amplifiers, *QST*, Dec. 1970, p 45.

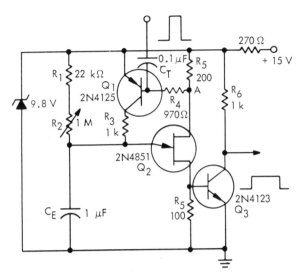

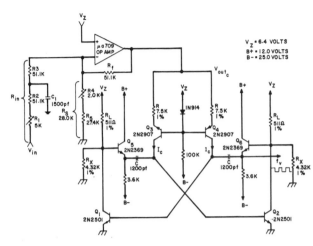

UJT MONO—Ujt Q2 is normally on and emitter saturation current is supplied by Q1 which is also on. Positive trigger at base of Q1 turns it off, turning off ujt also and starting timing cycle. When voltage on CE is high enough, ujt fires and turns on Q1 which supplies emitter current keeping ujt on. Zener provides voltage regulation because circuit is sensitive to supply voltage changes. Width of output pulse from Q3 is determined by setting of R2.—"Unijunction Transistor Timers and Oscillators," Motorola Semiconductors, Phoenix, AZ, AN-294, 1972.

VOLTAGE-CONTROLLED ASTABLE—Used in ultra-linear double-mix wideband voltage-controlled 30-MHz oscillator having crystal stability. Produces variable-frequency square wave by voltage-controlled variation of pulse width. Basic mvbr switching transistors Q1 and Q2 are driven by constant-current source Q3-Q4. Emitter-followers Q5 and Q6 reduce recovery time of timing circuit, to eliminate second-order nonlinearities. R1 adjusts tuning slope factor over 2.5% range. R4 tunes vfo over 3.2% of its center frequency of 120 kHz. Control voltage is 0–12 V for tuning range of 61 kHz.—W. A. Geckle, Jr., and R. C. Brackney, An Ultra-Linear Voltage Controlled Oscillator With Crystal Oscillator Stability, Frequency, Nov. 1968, p 18–21.

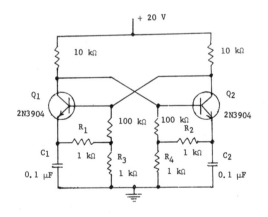

DIRECT-COUPLED ASTABLE—Unique arrangement of resistors completes d-c bias path between base and emitter of each transistor, permitting direct coupling of collectors and bases of both transistors to each other. Triangular waves are produced at emitters. Values of capacitors may be in range from 2,200 pF and 2,000 μF. Increasing resistor values lengthens time constant and reduces repetition frequency. Operating range is 3 Hz to 400 kHz.—S. Chang, A New Astable Direct-Coupled Multivibrator, Proc. IEEE, Aug. 1970, p 1278–1279.

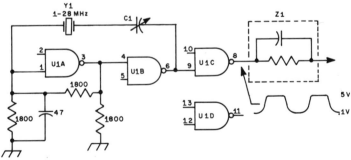

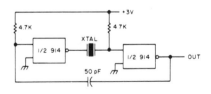

CRYSTAL MVBR—Uses quadruple 2-input positive NAND-gate IC (SN7400N or SG7400N) connected as astable mvbr, with crystal as frequency-determining element. Buffer stage enhances stability. Operates well with supply voltages of 3.5 to 6 V on pin 14 of IC and with any crystal from 1 to 28 MHz. Pin 7 should be grounded. Output is good approximation of square wave. With 1-MHz crystal, rich harmonics make output usable above 200 MHz. Omit 50-pF trimmer C1 if fine frequency adjustment is not needed. Network Z1, with 20 pF in parallel with 1K, is required only if driving base of npn power amplifier—J. W. Pollock, TTL Crystal Oscillator, QST, Oct. 1971, p 54–55.

CRYSTAL ONE-SHOT—Can serve as frequency standard, providing square-wave output rich in harmonics. Becomes free-running oscillator if capacitor is used in place of crystal.—J. Kyle, The Logical Approach to Surplus Buying, 73, March 1970, p 80–92.

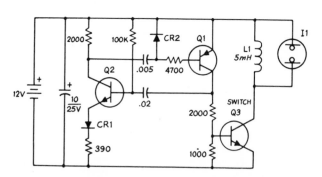

12-V NEON SUPPLY—Complementary astable mvbr Q1-Q2 turns on switching transistor Q3 when Q1 is on. Current then flows in L1. When Q2 comes on and turns off Q3, voltage spike across L1 fires NE-51H neon I1. Current drain is only 6 mA because transistors are off most of time. Will operate on 8 to 16 V, but lower voltage reduces neon brightness. Q1 is MPS 6523. Q2 is MPS 6521. Q3 is MJE 340. —J. H. Duncan, Neon-Bulb Lamp Driver, QST, May 1970, p 56.

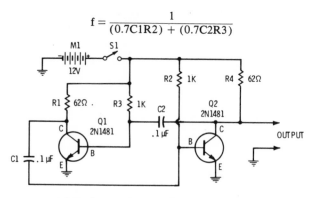

$$f = \frac{1}{(0.7C1R2) + (0.7C2R3)}$$

7-KHZ SQUARE-WAVE ASTABLE—Output frequency in kHz is determined by values of four components in equation, where capacitance values are in μF and resistance values in ohms.—R. M. Brown and T. Kneitel, "101 Easy Test Instrument Projects," Howard W. Sams, Indianapolis, IN, 1968, p 31.

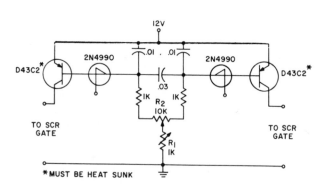

TRIGGER FOR SCR PAIR—Basic transistor flip-flop will drive gates of scr's directly as shown, or through transformer or pulse shaper, for use in inverters. Connection shown is free-running at about 1 kHz, but can be driven by ujt of put relaxation oscillator for precise timing. Frequency is trimmed by R1 and symmetry by R2.—"SCR Manual," General Electric, Syracuse, NY, 1972, 5th Ed., p 118.

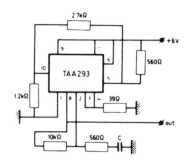

ASTABLE IC MVBR—Uses general-purpose Amperex three-transistor IC amplifier. Basic voltage-operated Schmitt trigger is made into astable mvbr by connecting output back to input and adding RC network to ground. Period is 16 μs and high-low ratio is 0.6 when C is 10 nF. Period varies proportionately with C down to 2 μs.—"The Amperex TAA 293 As a Schmitt Trigger and Multivibrator," Amperex Electronic Corp., Slatersville, RI, Application Report S-147.

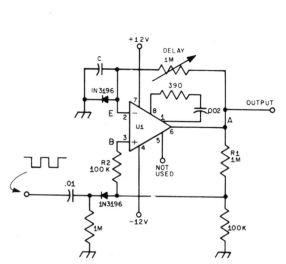

MONO IC—Circuit has stable state at which point A is at positive saturation. Incoming negative pulse drives A to negative saturation, where it stays for time interval determined by value of C and delay control, before flopping back to stable state. With 1 μF, time delay can be varied from 0.8 to 135 ms. —B. Balakrishnan, More on Operational Amplifiers, QST, Dec. 1970, p 45.

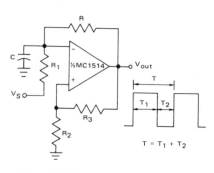

5-MHZ ASTABLE MVBR—Uses IC dual differential comparator designed for high-frequency operation. VS is —6 V, R 10K, R1 100K, R2 and R3 18K, and C 150 pF. Design equations are given.—R. Brunner, "A High Speed Dual Differential Comparator—The MC1514," Motorola Semiconductors, Phoenix, AZ, AN-547, 1971.

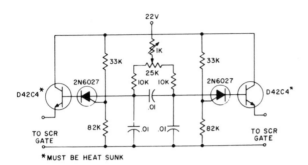

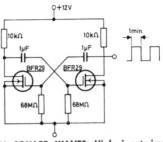

500-HZ TRIGGER FOR INVERTER—Provides alternate output pulses to scr gates at 500 Hz, by cross-coupling of two relaxation oscillator circuits, as required by many inverter circuits. Frequency is trimmed by 1K pot and symmetry by 25K pot. Upper frequency limit is 20 kHz.—"SCR Manual," General Electric, Syracuse, NY, 1972, 5th Ed., p 118–119.

2-MIN SQUARE WAVES—High input impedance of mosfets permits use of 68-meg resistors which, in combination with 1-μF capacitors, give output pulses with long time duration.—"Field Effect Transistors," Mullard, London, 1972, TP1318, p 115.

MONO IC—Uses general-purpose Amperex three-transistor IC amplifier. Negative edge of rectangular input wave makes output go to high state at 6 V, after which R-C cross-coupling makes circuit revert to low state after time delay. Recovery time for reliable retriggering is nominally equal to delay time, which is 100 μs for 10-nF capacitor. Times vary with capacitance in direct proportion down to about 1 μs. Report includes modification for delaying positive-going edge of input.—"The Amperex TAA 293 As a Schmitt Trigger and Multivibrator," Amperex Electronic Corp., Slatersville, RI, Application Report S-147.

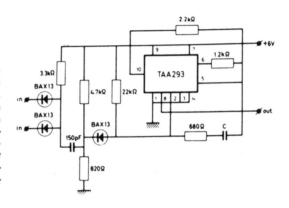

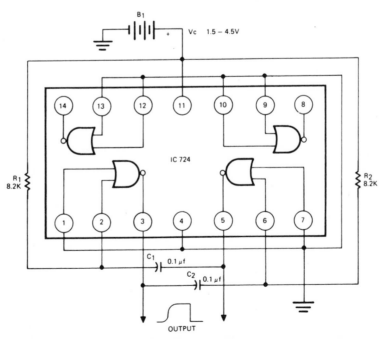

1-KHZ ASTABLE—Uses International Rectifier IC724 logic element and external frequency-determining components. Frequency can be increased to 10 kHz by reducing capacitor values to 0.01 μF. Supply voltage can be one to three dry cells in series. Possible drawback is rounded leading edge of output waveform. —"Hobby Projects," International Rectifier, El Segundo, CA, Vol. II, p 60–61.

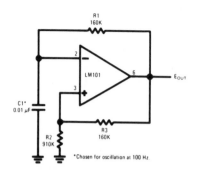

100-HZ FREE-RUNNING OPAMP—Operates at low frequencies with relatively small capacitors. Gives completely symmetrical square-wave output that is buffered, is always self-starting, and cannot hang up since there is more negative than positive feedback. Speed limitations of opamp make maximum output about 2 kHz.—R. J. Widlar, "Operational Amplifiers," National Semiconductor, Santa Clara, CA, 1968, AN-4.

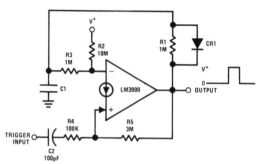

POSITIVE-PULSE ONE-SHOT—Requires only one Norton current-differencing opamp. R2 keeps output in low-voltage state. Differentiated positive trigger switches output to high-voltage state and R5 latches this state. When C1 charges to about 25% of supply voltage, circuit latches back to quiescent state, with CR1 allowing rapid retriggering. —T. M. Frederiksen, W. M. Howard, and R. S. Sleeth, "The LM3900—A New Current-Differencing Quad of ± Input Amplifiers," National Semiconductor, Santa Clara, CA, 1972, AN-72, p 28–29.

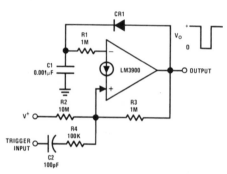

NEGATIVE-PULSE ONE-SHOT—Requires only one Norton current-differencing opamp. Quiescent output is high, and drops to 0 V for duration of output pulse after differentiated negative trigger input switches circuit. Large voltage across C1 keeps output low until C1 has discharged to about 10% of supply voltage.—T. M. Frederiksen, W. M. Howard, and R. S. Sleeth, "The LM3900—A New Current-Differencing Quad of ± Input Amplifiers," National Semiconductor, Santa Clara, CA, 1972, AN-72, p 29.

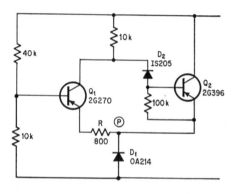

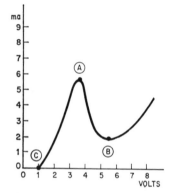

VOLTAGE-CONTROLLED NEGATIVE RESISTANCE—For applied voltages below 1 V (point C) neither transistor conducts. As voltage applied to circuit is increased to about 4 V, only Q2 conducts and current increases almost linearly. Between points A and B on characteristic, both transistors conduct and current decreases to give negative-resistance action. For voltages above point B, only Q1 conducts and circuit again acts as positive resistance. Can be used as flip-flop.—F. Broch-Tonolio, A Negative-Resistance Circuit Doubles for V or I Control, *Electronics*, April 28, 1969, p 78–79.

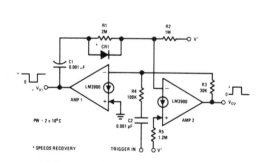

TWO-OPAMP ONE-SHOT—Uses pair of Norton current-differencing opamps. Circuit is rapidly retriggered, in about 10 mA.—T. M. Frederiksen, W. M. Howard, and R. S. Sleeth, "The LM3900—A New Current-Differencing Quad of ± Input Amplifiers," National Semiconductor, Santa Clara, CA, 1972, AN-72, p 28.

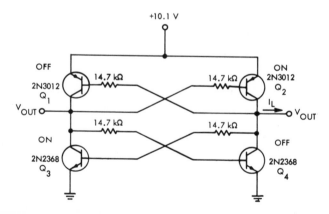

COMPLEMENTARY FLIP-FLOP—Use of complementary transistors makes power dissipation much lower than conventional flip-flop. Diagonally opposite transistors always have same conducting state. Q1 and Q4 are initially at cutoff, while Q2 and Q3 are in saturation. Total dissipation of stage is 12 mW, as compared to 432 mW with two-transistor flip-flop.—L. Delhom, "Circuit Applications of the PNP Transistor," Texas Instruments, Dallas, TX, Bulletin CA-89, 1969, p 8–9.

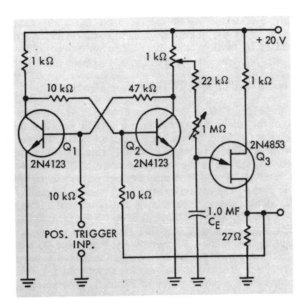

BIAS-INSENSITIVE UJT MONO—Circuit is relatively independent of changes in bias voltage. Positive input trigger turns on normally-off ujt relaxation oscillator Q3 after time interval required for charging CE as determined by setting of 1-meg pot. Positive voltage developed across 27-ohm resistor serves as output signal and as trigger for flip-flop to turn on Q2. Supply voltage range from 10 to 30 V causes only 2% change in timing.—"Unijunction Transistor Timers and Oscillators," Motorola Semiconductors, Phoenix, AZ, AN-294, 1972.

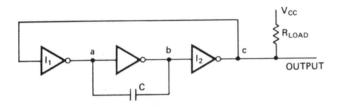

1 HZ TO 1 MHZ—Simple low-cost astable mvbr gives square-wave output frequencies from 1 Hz to 1 MHz for values of C from 300 μF to 300 pF, using only half of inexpensive TTL hex inverter with open-collector outputs such as type 7405. Ratio of on to off time is almost unity and independent of C.—M. Faiman, Widerange Multivibrator Costs Just 25¢ to Build, *Electronics*, Aug. 2, 1971, p 59.

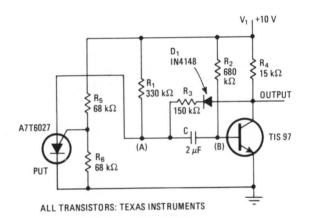

ALL TRANSISTORS: TEXAS INSTRUMENTS

1-HZ SINGLE-CAPACITOR—Use of programmable ujt makes circuit operate as astable mvbr with only one inexpensive Mylar capacitor. R1 controls width of negative output pulse and R2 controls width of positive output pulse, for adjusting output symmetry. Values of R1, R2, and C together determine output pulse duration.—G. Coers, Astable Multivibrator Needs Only One Capacitor, *Electronics*, Jan. 18, 1973, p 171.

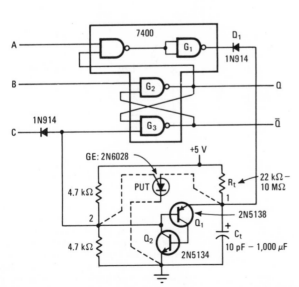

A	B	C	OPERATING MODE
LOW	SET	RESET	R·S LATCH: B = SET, C = CLEAR
HIGH	⎍	HIGH	MONOSTABLE: T = 1.3 $R_t C_t$, C = CLEAR
HIGH	LOW	HIGH	ASTABLE: $\overline{Q}$ = OUTPUT, C = CONTROL
A TIED TO B		HIGH	RETRIGGERABLE MONOSTABLE: C = CLEAR

FOUR-FUNCTION—Input signals and logic levels on lines A, B, and C determine whether circuit operates as latch, mono, astable, or retriggerable mono mvbr. Duration of output pulse can be adjusted from about 280 ns to several hours by changing Rt and Ct.—E. Beach, Programmable Multivibrator Is Four-in-One Circuit, *Electronics*, Feb. 15, 1973, p 105.

CHAPTER 76
Music-Controlled Circuits

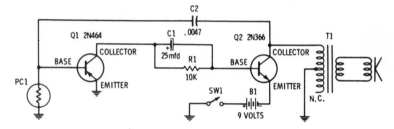

LIGHT-CONTROLLED MUSIC—Waving hand between photocell (International Rectifier B2M or equivalent) changes frequency of generated tone. The stronger the light, the higher the pitch. T1 is audio output transformer such as Argon AR-170.—R. M. Brown and T. Kneitel, "49 Easy Transistor Projects," Howard W. Sams, Indianapolis, IN, 1972, p 18–19.

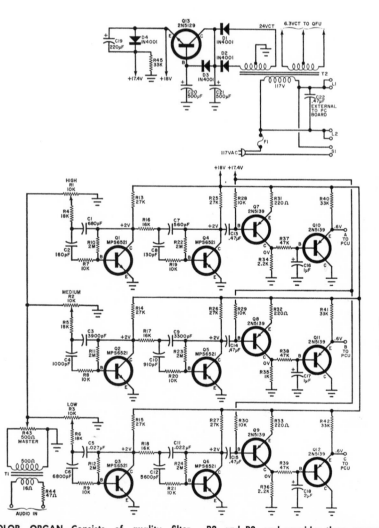

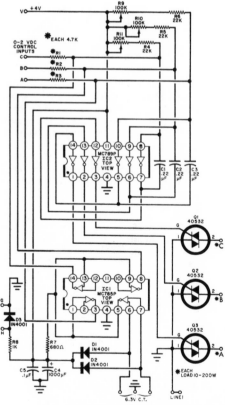

COLOR ORGAN—Consists of quality filter unit that divides low-level audio input of T1 into three isolated frequency bands by means of active filters controlled individually by R1, R2, and R3, and provides three proportional control voltages up to 2 V d-c for terminals A, B, and C of power control unit that uses triacs to control up to 200-W lamp per channel. Triac in each channel acts as series switch between supply and lamp load, turning off each time a-c supply passes through zero and staying off until ramp voltage drops below 0.6 and inverter supplies gate signal. The higher the input, the faster the ramp decays, the sooner the triac turns on during a cycle, and the brighter is the lamp for that channel. Article gives detailed construction and adjustment data.—D. Lancaster, Psychedelia 1, *Popular Electronics*, Sept. 1969, p 29–35 and 40–44.

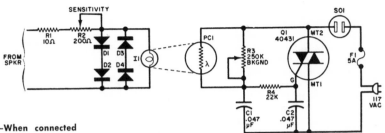

500-W LAMP MODULATOR—When connected to speaker terminals of hi-fi set or electronic musical instrument, will modulate up to 500-W lamp in proportion to audio level. With three units with suitable audio filters and different color filters, stage can be illuminated to give effect of color organ. Diodes are 50-V piv silicon and I1 is No. 49 pilot lamp. PC1 is RCA 2529 photoresistor.—D. Meyer, Build the "Sonolite", *Popular Electronics*, May 1968, p 27–30.

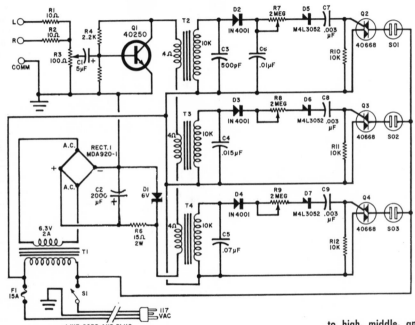

COLOR ORGAN—Simplified circuit uses only three channels for translating sound to light as adjunct of home stereo system. Use of triacs rather than scr's eliminates need for heavy-duty d-c power supply. Shunt capacitors tune audio transformers T2, T3, and T4 to high, middle, and low frequency ranges respectively. Triacs Q2, Q3, and Q4 drive color display lamps plugged into output sockets SO.—L. Garner, Solid State, *Popular Electronics*, Nov. 1971, p 82–89.

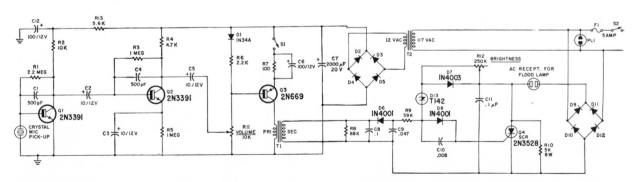

300-W RHYTHM LIGHT—Picks up live or recorded music with crystal mike and amplifies it sufficiently to vary brightness of two 150-W flood lamps which can be different colors. Can also be used with strings of Christmas-tree lights. PL1 is neon lamp with series resistor. T1 is interstage audio transformer with 10,000-ohm primary and 200-ohm secondary. D9–D12 are RCA 40267.—R. T. Montan'e, Build Rhythm Lights for Psychedelic Music, *Radio-Electronics*, July 1968, p 34–35 and 71.

SOUND-CONTROLLED LAMP—Small speaker serves as microphone in single-transistor audio amplifier used to trigger 5-A scr and make brightness of lamp load up to several hundred watts vary with sound level. Transformer is small voice-coil-to-500-ohm unit, such as Lafayette 99-6127. Use insulated housing, because circuit is not isolated from a-c line.—J. F. Kennedy, Sound-Operated Light Controllers, *Electronics World*, Nov. 1971, p 39.

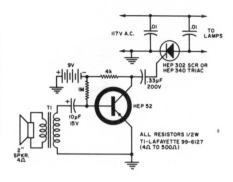

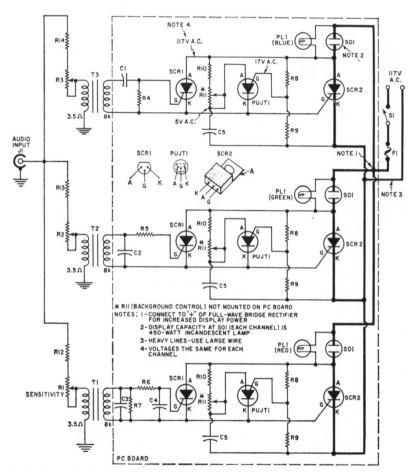

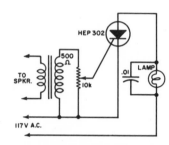

AUDIO-CONTROLLED LAMP—When circuit is connected to voice-coil leads of radio or hi-fi speaker, lamp up to 200 W varies in brightness with audio signal. HEP302 scr, rated at 5 A, requires only 0.15 V rms at speaker for triggering, corresponding to rather low volume of sound. Use small voice-coil-to-500-ohm transformer, such as Lafayette 99-6127. Should be mounted in plastic box because parts of circuit are hot and dangerous.—J. F. Kennedy, Sound-Operated Light Controllers, *Electronics World*, Nov. 1971, p 39.

R1, R2, R3—100 ohm pot ("Sensitivity")
R4—1000 ohm, 1/2 W res.
  (or 6800 ohm, 1/2 W res. See text.)
R5—220,000 ohm, 1/2 W res.
R6, R7—56,000 ohm, 1/2 W res.
R8—100,000 ohm, 1/2 W res. (3 required)
R9, R10—27,000 ohm, 1/2 W res. (6 required)
R11—250,000 ohm pot ("Background," 3 required)
R12, R13, R14—27 ohm, 1/2 W res.
C1—820 pF disc capacitor (or 0.002 μF, 100 V paper)
C2—0.01 μF, 100 V paper capacitor
C3—0.1 μF, 100 V paper capacitor
C4—0.033 μF, 100 V paper capacitor
C5—0.15 μF, 200 V paper capacitor (3 required)
T1, T2, T3—Audio output trans. 8000 ohm: 3.5 ohms (Stancor A3329 or equiv.)

SCR1—0.8 A (r.m.s.), 200 V SCR (GE C103B)
SCR2—8 A (r.m.s.), 200 V silicon controlled rectifier, (General Electric C-122B)
PUJT1—Programmable unijunction transistor (General Electric D13T1)
SO1—Receptacle (Cinch Jones 2R2 or equiv. 3 required)
PL1—Pilot light, 125 V a.c. (Dialco 0431-302, red; 0432-302, green; 0434-302, blue), with 6S6, 12.5 V lamp
S1—S.p.s.t. switch
F1—6 A fuse
J1—Audio input jack
The PC board (Part 408) is available at $6.00 while a kit of all semiconductors can be obtained for $10.00. Both from WWW Electronics, P.O. Box 5363, Charlottesville, Va. 22903.

450-W COLOR ORGAN—Three channels each have output scr rated to handle 450 W of lamp load. Simple audio filters using audio transformers divide audio input into three frequency channels: low (20-400 Hz) for red, middle (200-2,000 Hz) for green, and high (4,000 Hz and up) for blue. Hole in response between middle and high channels enhances response to solo instruments, while overlap between low and middle channels creates color blending for response to bass and rhythm instruments. Can be used for wall display made up of Christmas tree lights, spot or flood lights illuminating stage, or display under frosted-glass bar top.—R. A. Hertzler, High-Power Color Organ, *Electronics World*, Sept. 1970, p 78–80.

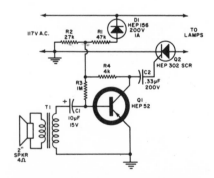

SOUND-CONTROLLED LAMP—Small speaker serves as microphone in single-transistor audio amplifier used to trigger 5-A scr and make brightness of lamp load up to several hundred watts vary with sound level. Half-wave rectifier provides supply voltage for transistor, eliminating need for 9-V battery. No filtering is needed. With good transistor, lights can be controlled halfway across room from sound source. If triac is used in place of scr, lights will be brighter because triac conducts on both halves of each a-c cycle.—J. F. Kennedy, Sound-Operated Light Controllers, *Electronics World*, Nov. 1971, p 39.

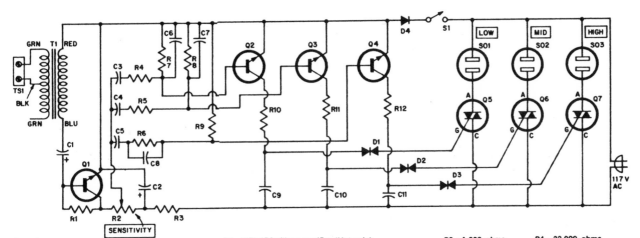

Capacitors:
C1—10 µf, 10-V electrolytic
C2—40 µf, 150-V electrolytic
C3—1 µf, 200-V mylar (not electrolytic)
C4,C5—.01 µf, 200 V paper or mylar
C6—.1 µf, 100-V mylar
C7—.01 µf, 100-V mylar
C8—.0039 µf, 500-V or higher ceramic disc
C9,C10,C11—.1 µf, 200 V paper or mylar
D1,D2,D3—HEP-311 bilaterial trigger diode
(Motorola)

D4—HEP-156 silicon rectifier (Motorola)
Q1,Q2,Q3,Q4—HEP-712 medium-power NPN
silicon transistor
Q5,Q6,Q7—Triac; minimum ratings: 200 V, 8 A
(RCA 2N5569 or HEP-340)

Resistors: ½ watt, 10% unless otherwise
indicated
R1—220,000 ohms
R2—5,000 ohm pot (2 watts or higher)

R3—1,000 ohms   R4—22,000 ohms
R5—4,700 ohms   R6—330,000 ohms
R7—47,000 ohms   R8—33,000 ohms
R9—5,600 ohms
R10,R11,R12—4,700 ohms, 1 watt
S1—SPST toggle or slide switch
SO1,SO2,SO3—Chassis-mount AC receptacle
(Calectro F3-100)
T1—Transistor driver transformer; primary:
100 ohms, secondary: 100 ohms center tapped
(Calectro D1-733)

COLOR ORGAN—When connected to terminals of one of speakers in hi-fi system, divides audio spectrum into three sections for varying brightness of three 500-W lamps in step with music. Most exciting light patterns are obtained with music having natural bounce. Use red, green, and blue floodlights aimed at different portions of large light-colored surface such as wall. Do not overlap, because light would then become white whenever amplitudes were equal in all three bands.—R. M. Benrey, Turn On with a Hi-Power Color Organ, *Electronics Illustrated*, July 1971, p 59–63, 100, and 102.

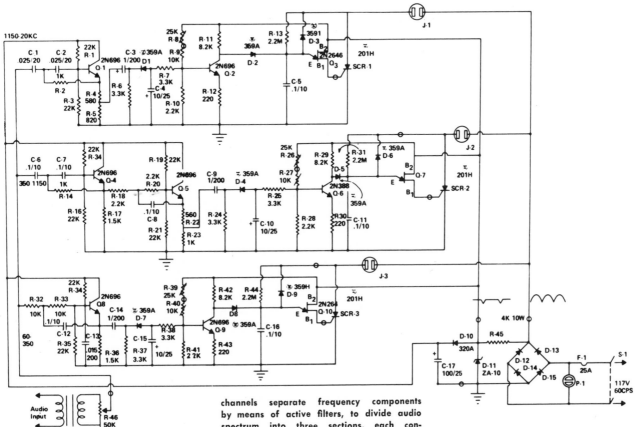

FREQUENCY-SELECTIVE LIGHT DISPLAY—When input is connected to audio amplifier, three channels separate frequency components by means of active filters, to divide audio spectrum into three sections, each controlling different color of light, such as red, green, and blue. Output of each channel modulates emitter voltage of ujt in accordance with signal level, to change delay angle of scr in lamp circuit.—"Silicon Controlled Rectifier Designers Handbook," Westinghouse, Youngwood, PA, 1970, 2nd Ed., p 7-63–7-65.

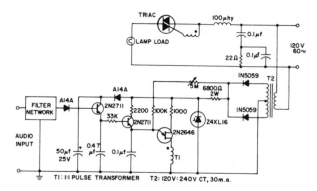

AUDIO-MODULATED LAMP—Input filter determines portion of audio spectrum for which lamp brightness will vary with audio amplitude. For color organ, use three systems, with low, medium, and high bandpass filters each controlling different color of lamp (red, blue, and green). Choose triac to match load. For red, which for equal power will dominate over blue and green lamps, reduce voltage by using scr in place of triac.—R. W. Fox, "Solid State Incandescent Lighting Control," General Electric, Semiconductor Products Dept., Auburn, NY, No. 200.53, 1970, p 11–12.

AUDIO-CONTROLLED LAMP—Circuit provides on-off control of lamp by triac triggered with transformer-isolated 1-V audio signal. Since switching action is fast compared to response times of lamp and human eye, effect of audio control is similar to proportional control. Adjust R1 for maximum resistance that will not turn on lamp with zero input. Choose triac to match lamp load.—J. H. Galloway, "Using the Triac for Control of AC Power," General Electric, Semiconductor Products Dept., Auburn, NY, No. 200.35, 1970, p 14.

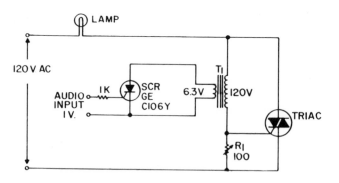

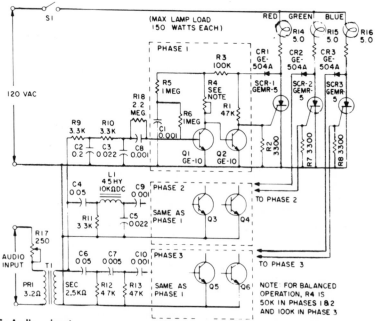

AUDIO DANCING LIGHTS—Audio input, taken directly from speaker terminals, is separated into three channels for low, medium, and high-frequency tones. Each channel signal is amplified separately and used to drive its own colored lamp up to 150 W. Lamps then vary in brightness as music changes. Best results are obtained with music having good highs and lows, such as modern combo groups. Construction details are given.—"Electronics Experimenters Circuit Manual," General Electric, Owensboro, KY, 1971, 3rd Ed., p 117–120.

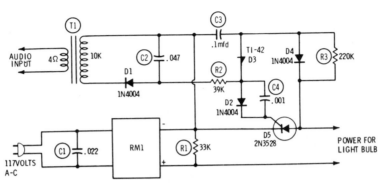

200-W SOUND-CONTROLLED LIGHT—Brilliance of lamp is maximum for full volume and it decreases as volume goes down. Input is connected to speaker leads of hi-fi, radio, or recorder. For stereo, use one circuit for each channel, with different colored lamps. If desired, three 60-W lamps of different colors can be used in place of one large lamp. If background noise turns on lights before music starts, experiment with different value for R2. RM1 is Motorola MDA942-3 or similar full-wave bridge rectifier module.—R. Brown and T. Kneitel, "101 Easy Audio Projects," Howard W. Sams, Indianapolis, IN, 1971, p 58—60.

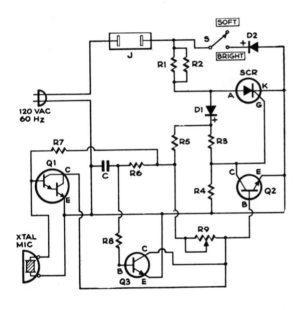

C—0.001 mfd 500VDC disc ceramic capacitor
R1, R2—10-ohm 2w resistor
R3—22k ohm ½w resistor
R4—3.3k ohm ½w resistor
R5—100k ohm ½w resistor
R6—1-megohm ½w resistor
R7—10-megohm ½w resistor
R8—470k ohm ½w resistor
R9—50k ohm potentiometer
D1—GE 1N5059 silicon diode
D2—GE A15F silicon diode
Q1—Sprague or GE 2N5308 Darlington amplifier
Q2, Q3—Sprague or GE 2N-5172 silicon transistor
SCR—GE type C105B1
S—SPST slide switch
J—AC receptacle
XTAL MIC—Lafayette 99T-4518

MUSIC-CONTROLLED XMAS LIGHTS—Phase-controlled scr circuit is gated at audio rate by external pulses picked up by crystal mike and amplified by Q1, Q2, and Q3. D1 furnishes rectified positive-going pulses of a-c line to all three transistors. Switch S and control R9 together provide various combinations of sound control for string of Christmas-tree lights plugged into J.—G. J. Whalen and R. F. Graf, Christmas Lights Keep Time to Music, *Popular Science*, Dec. 1969, p 145–146 and 188.

CHAPTER 77
Navigation Circuits

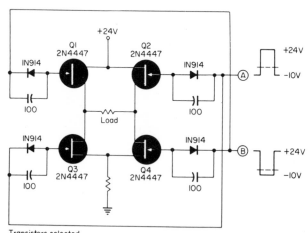

H SWITCH—Four fet's provide directional control for motor drives in inertial guidance systems. High current and power gain ratings of fet allow corrective torques to be controlled by low-power amplifiers, to reduce size and rate while increasing reliability.—Field-Effect Transistors, *Design Electronics*, May 1970, p 22–30.

Transistors selected
for BVDGO of 35V

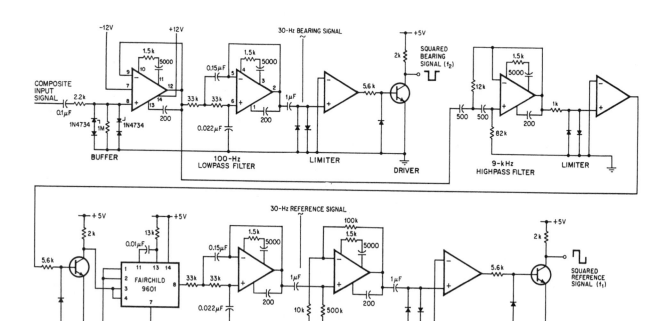

NOTES:
1. ALL TRANSISTORS 2N3904
2. ALL AMPLIFIERS ½ MC1437L
3. ALL RESISTORS ¼ W, 5% CARBON COMPOSITION
4. ALL DIODES 1N914 or EQUIVALENT
5. ALL CAPACITORS IN pF UNLESS INDICATED OTHERWISE

VOR SIGNAL CONDITIONER—Separates bearing and north-reference signals received from vhf omnidirectional range and shapes them in limiters to provide required steep leading and trailing edges for use as timing marks in air-borne vor converter-indicator developed for light planes. Article describes circuit operation in detail.—I. Breikss, Digital ICs + VOR = Simpler Navigation, *Electronics*, March 15, 1971, p 80–84.

MARKER-BEACON TONE FILTER—Opamp 1A acts as impedance transformer and line driver for audio output of 75-MHz airborne marker beacon receiver. Other three opamp sections are connected as multiple-feedback-path active bandpass filters. Opamp 1B is tuned to 400 Hz with gain of 6 and Q of 10, 2A is tuned to 1,300 Hz with gain of 5 and Q of 10, and 2B is tuned to 3,000 Hz with gain of 5 and Q of 10. Outputs of tone filters are capacitively coupled to drivers of three-lamp display.—J. Stinehelfer and P. Macdonald, "A Marker Beacon Receiver and Indicator Using Integrated Circuits," Fairchild Semiconductor, Mountain View, CA, No. 186, 1970, p 3–5.

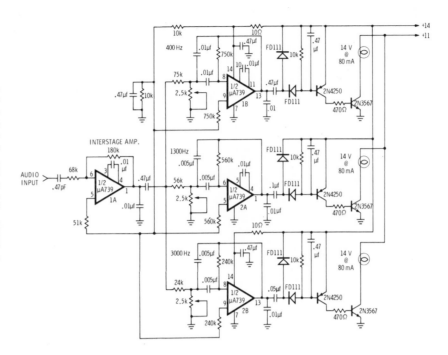

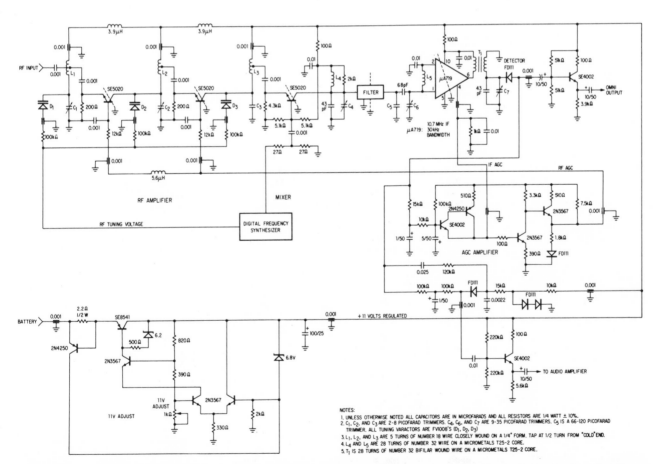

NOTES:
1. UNLESS OTHERWISE NOTED ALL CAPACITORS ARE IN MICROFARADS AND ALL RESISTORS ARE 1/4 WATT ± 10%.
2. C₁, C₂, AND C₃ ARE 2-8 PICOFARAD TRIMMERS. C₄, C₆, AND C₇ ARE 9-35 PICOFARAD TRIMMERS. C₅ IS A 66-120 PICOFARAD TRIMMER. ALL TUNING VARACTORS ARE FV1008'S (D₁, D₂, D₃)
3. L₁, L₂, AND L₃ ARE 5 TURNS OF NUMBER 18 WIRE CLOSELY WOUND ON A 1/4" FORM. TAP AT 1/2 TURN FROM "COLD" END.
4. L₄ AND L₅ ARE 28 TURNS OF NUMBER 32 WIRE ON A MICROMETALS T25-2 CORE.
5. T₁ IS 28 TURNS OF NUMBER 32 BIFILAR WOUND WIRE ON A MICROMETALS T25-2 CORE.

NAVIGATION RECEIVER—Single-conversion superhet uses digital frequency synthesizer to supply large number of precise crystal-controlled frequencies. Varactor tuning from 105 to 120 MHz, with 1-MHz bandwidth, is used in r-f amplifier. Includes 10.7-MHz crystal filter, IC i-f amplifier providing 80 dB voltage gain and —30 dB of agc, and total agc capability of 110 dB. Report covers receiver operation and gives synthesizer circuit.—J. Stine-helfer, "A Navigation Receiver That Uses a Digital Frequency Synthesizer," Fairchild Semiconductor, Mountain View, CA, No. 178, 1969, p 1–2.

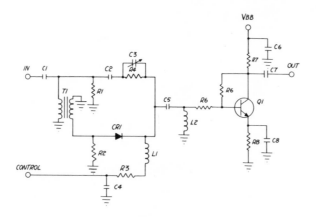

CR1 = AEL. P-432
C1 = C2 = C4 = C5 = C6 = C7 = 2200 pF
C3 = 0.8–10 pF
C8 = 470 pF
L1 = L2 = 10 μH
Q1 = 2N2857
R1 = 120 Ω

R2 = R4 = 270 Ω
R3 = 1000 Ω
R5 = 390 Ω
R6 = 1800 Ω
R7 = 560 Ω
R8 = 180 Ω

ATTENUATOR—Uses pin diode as control element of bridge circuit providing voltage attenuation over 50-dB dynamic range with phase-shift change of less than 2 deg. Used in optical tracking system which supplies range, velocity, and acceleration data by sensing Doppler shift. T1 provides two arms of bridge, with R4 and pin diode CR1 forming other two. T1 has 7 turns of No. 33 twisted pair on Micrometals T25-3 toroidal core, to provide accurate phase split. Source and load are both 50 ohms, and center frequency of circuit is 30 MHz.—G. B. Shelton, A Near-Constant-Phase Variable Attenuator, *Proc. IEEE*, July 1969, p 1345–1346.

R1—1,000-ohm°
R2—25,000-ohm°
R3—200-ohm°
R4,R7—100-ohm, screwdriver adjust°
R5,R9—50,000-ohm trimmer potentiometer
R6—10,000-ohm trimmer potentiometer
R8—10-ohm, ½w carbon resistor
R10—6.8-megohm, ½w carbon resistor
R11—2,200-ohm, ½w carbon resistor
C—120,000-mfd, 6-volt, electrolytic
D—6.2 volt zener diode (Motorola HEP 103)
Q—Darlington amplifier (Motorola HEP S9100)
M—0-50 μA DC meter (Calectro D1-910)
S1—4DPT miniature rotary switch
S2,S3,S4—DPDT mini. toggle (Calector E2-118)
S5—SPST mini. toggle (Calectro E2-116)
S6—SPST pushbutton switch
B1,B2—67½ v (RCA VS218)
B3—9v (Burgess D6)
B4—1.5v (AA)
°½w, wirebound potentiometers

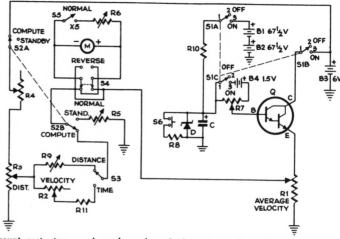

TRAVEL-TIME COMPUTER—Circuit uses small voltages to represent distance and time, for performing variety of time-distance-speed calculations and keeping track of predetermined schedule for long-distance driving while taking into account rest stops and unplanned side trips. All settings are entered manually, so no connections are required to auto circuits. For 600-mile trip, accuracy is about 10 miles or 15 minutes, for meter that indicates whether trip is ahead of or behind schedule near end of day. Article covers setup and use.—R. M. Benrey, The Travel Computer—Electronic Navigation for Your Car, *Popular Science*, May 1971, p 79–81.

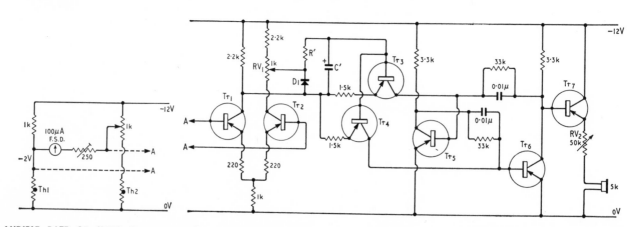

AUDIBLE RATE OF CLIMB—Generates audio tone whose pitch is proportional to reading of rate-of-climb meter, above preset threshold set by RV1. Used in gliders to relieve pilot of need to glance at meter frequently. Sensors are thermistors Th1 and Th2 mounted in Thermos bottle. Air flowing into or out of bottle, due to pressure change with rise or fall of glider, is directed by small jets over one thermistor or the other to give up or down reading of meter and corresponding change in tone. Oscillation range is about 300 to 10,000 Hz. All transistors are low-frequency small-signal pnp germanium. Inexpensive headphone is used as speaker.—J. M. Firth, Audible Rate-of-Climb Indicator, *Wireless World*, April 1965, p 204–205.

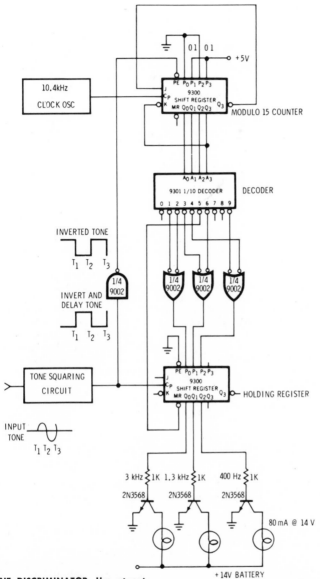

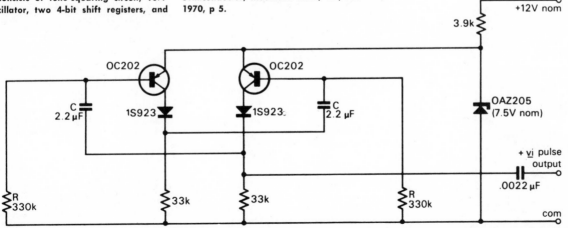

OCR AND STAR TRACKER—Silicon photodiode and opamp combination gives better linearity, frequency response, and stability than photomultiplier for optical character recognition, photometric instruments, earth-resource scanners, and celestial navigation. In typical circuit shown, fet opamp is used with UDT-500 photodiode to keep input bias current and photodiode dark current as small as possible, to hold offset voltage down.—P. H. Wendland, Solid State Combo Senses Light Well Enough to Vie with Tubes, *Electronics*, May 24, 1971, p 50–54.

DIGITAL TONE DISCRIMINATOR—Use at output of airborne marker beacon receiver to identify beacon frequencies (400, 1,300, and 3,000 Hz) and turn on appropriate display lamp. Consists of tone-squaring circuit, 10.4-kHz oscillator, two 4-bit shift registers, and 1-out-of-10 decoder.—J. Stinehelfer and P. Macdonald, "A Marker Beacon Receiver and Indicator Using Integrated Circuits," Fairchild Semiconductor, Mountain View, CA, No. 186, 1970, p 5.

1-PPS PINGER DRIVE—Developed for underwater acoustic beacons producing short pulses of 10-kHz sound, for position-fixing by ship carrying suitable hydrophone and display system. Stability is excellent and cost is low enough for expendable applications. Alloy-junction transistors are used because of higher reverse emitter-base voltage than most modern transistors.—M. J. Tucker and C. A. Hunter, A Stable Very Low-Frequency Multivibrator, *Electronic Engineering*, Sept. 1969, p 37–39.

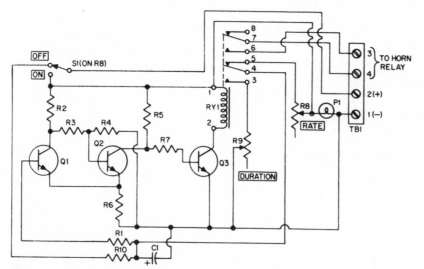

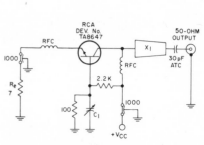

FOGHORN CONTROLLER—Used on boat in parallel with manual horn switch. Provides independent controls over length of each horn blast and rate at which blasts are repeated, when power boat is under way in fog. Coast Guard regulations prescribe duration of 4 to 6 s at not more than 1-min intervals. Circuit achieves this with Schmitt trigger Q1-Q2 turning on Q3 to energize relay whose contacts go to horn relay. Q1 and Q2 are 2N5172, Q3 is GE D27C1, RY1 is Potter & Brumfield KA11DY 12-V 120-ohm d-c relay, P1 is 12-V pilot lamp, and C1 is 350 μF. R1 is 10K, R2 4.7K, R3, R5, and R7 2.7K, R4 6.8K, R6 68, R8 and R9 1 meg linear-taper, and R10 100K.—R. F. Graf and G. J. Whalen, Horatio's Hornblower, *Electronics Illustrated*, March 1970, p 58–61.

1.68-GHZ RADIOSONDE OSCILLATOR—High-Q frequency-determining network is placed in base-ground circuit of high-power microwave oscillator transistor. Frequency remains constant over supply voltage range of supply voltages from 20 to 28 V. Power output at 24 V is 1.2 W, and 1.9 W at 28 V. C1 is 0.4–6 pF high-Q air piston capacitor. X1 is 5.5-ohm to 50-ohm tapered-line transformer.—G. Hodowanec, "High-Power Transistor Microwave Oscillators," RCA, Somerville, NJ, AN-6084, 1972.

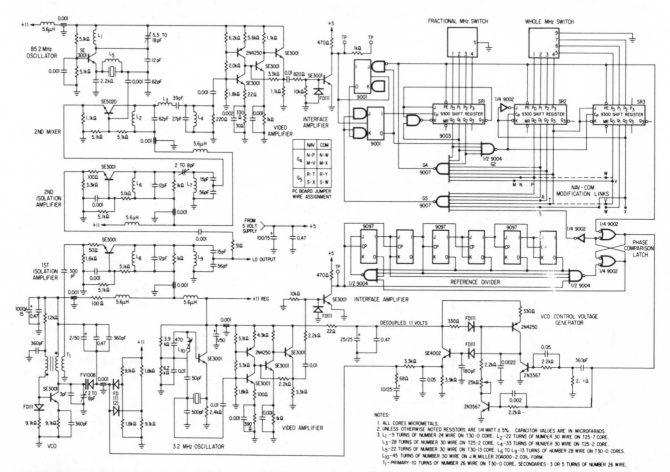

NAVIGATION RECEIVER CONTROL—Uses IC's as part of phase-locked loop for stabilizing vco to precision of reference crystal. With digital counter in loop, any frequency needed for navigation or communication receiver or transceiver can be generated. Output signal of vco is down-converted by 85.2-MHz crystal oscillator and mixer, to bring difference signal to 12.1–22.1 MHz, within frequency range of logic. For navigation receiver, division ratios of 242 to 442 are required.—J. Stinehelfer, "A Navigation Receiver That Uses a Digital Frequency Synthesizer," Fairchild Semiconductor, Mountain View, CA, No. 178, 1969, p 3–5.

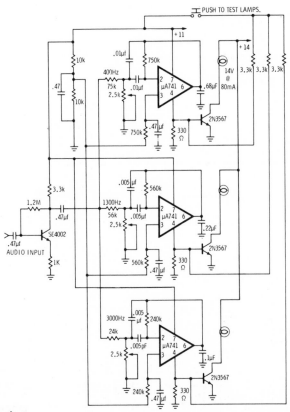

MARKER-BEACON TONE FILTER—Uses three standard opamps with minimum external components to form active filters responding to 400, 1,300, and 3,000-Hz tones at output of airborne marker beacon receiver, each driving appropriate indicator lamp.—J. Stinehelfer and P. Macdonald, "A Marker Beacon Receiver and Indicator Using Integrated Circuits," Fairchild Semiconductor, Mountain View, CA, No. 186, 1970, p 3–5.

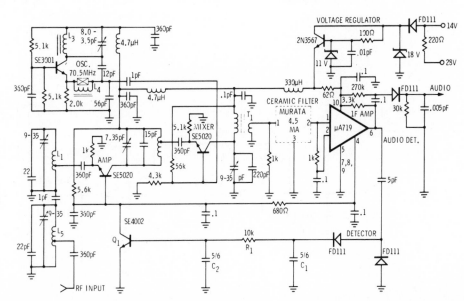

75-MHZ MARKER BEACON—Single-conversion airborne superhet receiver tunes only one frequency (75 MHz) and operates only at short range, typically 300 ft for middle marker, 1,600 ft for outer marker, and up to 40,000 ft for Z marker. Selectivity must be adequate enough to prevent triggering by tv signals close to 75 MHz. Crystal-controlled 70.5-MHz local oscillator gives i-f of 4.5 MHz, to place image frequency at 66 MHz, in quiet spot between tv channels 3 and 4.—J. Stinehelfer and P. Macdonald, "A Marker Beacon Receiver and Indicator Using Integrated Circuits," Fairchild Semiconductor, Mountain View, CA, No. 186, 1970, p 1–2.

CHAPTER 78
Noise Circuits

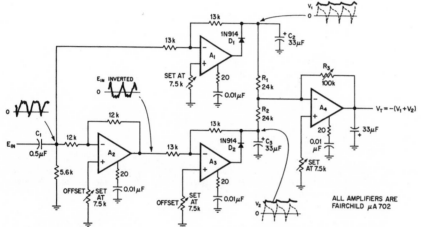

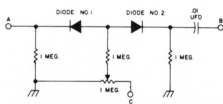

POWER SUPPLY TESTER—Measures absolute p-p amplitude of low-level a-c noise regardless of waveshape, for automatic production testing of d-c power supplies. Detects inductive-switching spikes as well as output ripple, by converting positive and negative peak values to two voltages that are summed in A4. —L. F. Caso and J. Fazio, Supply Tester Outdoes Scope as AC Noise Meter, *Electronics*, March 1, 1971, p 59.

DIODE NOISE LIMITER—Inserted in audio circuit between input of first a-f and blocking capacitor going to detector circuit, to provide limiting on top of both negative and positive halves of audio signal for suppressing interference produced by vacuum cleaners and other appliances. Almost any signal-type crystal diodes will work. Positive gating voltage required at terminal C is 5 V. Adjust 1-meg pot for this value if voltage on agc line or at cathode of audio output tube is higher than 5 V.—M. B. Crowley, Another Hedge Clipper, 73, Jan. 1973, p 113–114.

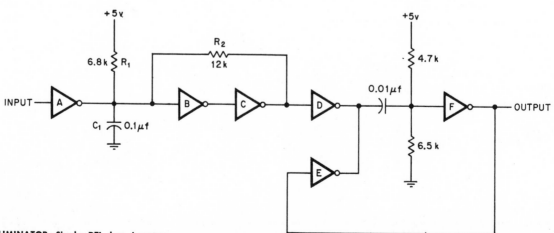

SPIKE ELIMINATOR—Single DTL hex inverter package and a few timing components give pulse width discriminator and pulse generator for determining if incoming pulses are too narrow to be data bits. Mono mvbr E-F cannot be triggered unless input inverter A sees pulse long enough to charge C1 to 1.5 V.—C. A. Herbst, Single Hex Inverter Picks Data Signals from Noise, *Electronics*, Aug. 31, 1970, p 69.

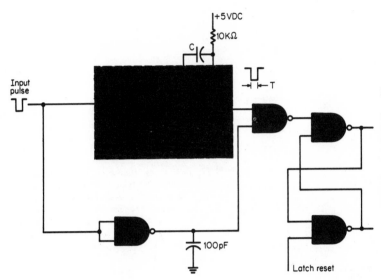

NOISE REJECTION—Circuit can be activated only by pulse longer than predetermined minimum length controlled by R-C values. When R is 10K as shown, divide desired minimum pulse length in ns by 3.424 to get value of C in pF. Rectangular IC block is 9601, with C connected between pins 11 and 13. Output from pin 6 feeds gates, all of which are 846.—S. R. Martin, Latch Circuit Provides Noise Immunity, *The Electronic Engineer*, Nov. 1971, p 58.

ENVIRONMENTAL NOISE MONITOR—S1 gives choice of three noise levels at which indicator lamp begins flashing—50, 70, and 85 dB. Can be used to check noise level when shopping for air conditioner. Lowest range is bedroom tester; if lamp flashes, room is too noisy for sleeping. Level of 70 dB affects concentration during study or desk work. Highest level of 85 dB, if continuous all day, can damage hearing permanently. M1 is crystal mike and indicator LED is light-emitting diode.—H. Cohen, Sound Pollution Tipster, *Elementary Electronics*, July-Aug. 1972, p 49–52.

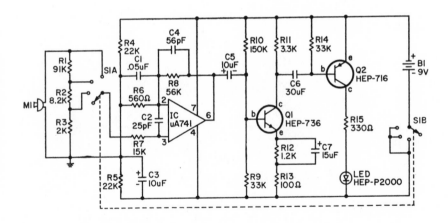

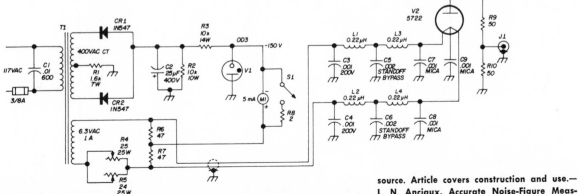

NOISE-FIGURE METER—Accuracy is within 0.5 dB up to 432 MHz and better than 0.1 dB below 220 MHz. Uses Sylvania type 5722 special diode as temperature-limited noise source. Article covers construction and use.—L. N. Anciaux, Accurate Noise-Figure Measurements for Vhf, *Ham Radio*, June 1972, p 36–40.

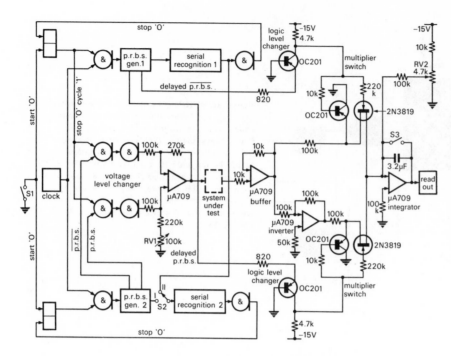

SUBSTITUTE FOR WHITE NOISE—Complete pseudo-random binary sequence correlator generates sequences approximating white noise for use in obtaining impulse response of system under test. Clock frequency of 800 Hz gives integration over a few seconds. Best suited for time constants under 3 ms. Digital voltmeter is used for readout. Article gives theory of operation and method of use.—R. G. Young, P.R.B. Sequence Correlator Using Integrated Circuits, *Electronic Engineering*, Sept. 1969, p 41–45.

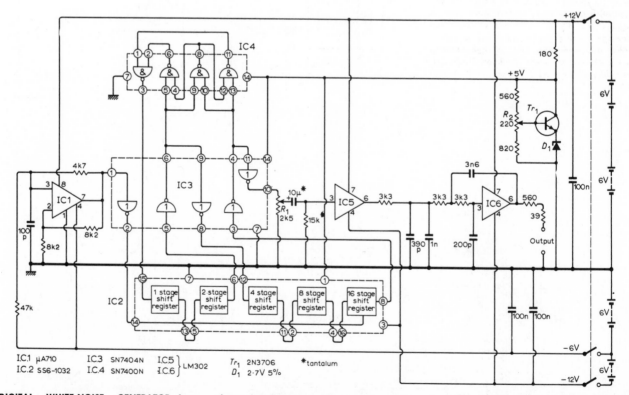

I.C.1 μA710
I.C.2 SS6-1032
I.C.3 SN7404N
I.C.4 SN7400N
I.C.5 }
I.C.6 } LM302
Tr_1 2N3706
D_1 2·7V 5%
*tantalum

DIGITAL WHITE-NOISE GENERATOR—Inexpensive digital version provides white noise with Gaussian amplitude probability distribution up to 50 kHz, with 18 dB/octave rolloff above this frequency, for measuring noise figure of a-f amplifier. Comparator IC1 is used as 900-kHz square-wave oscillator to provide clock pulses for shift register IC2. Binary noise signal is developed across R1 for feeding active filter IC6 through buffer IC5. —H. R. Beastall, White-Noise Generator, *Wireless World*, March 1972, p 127–128.

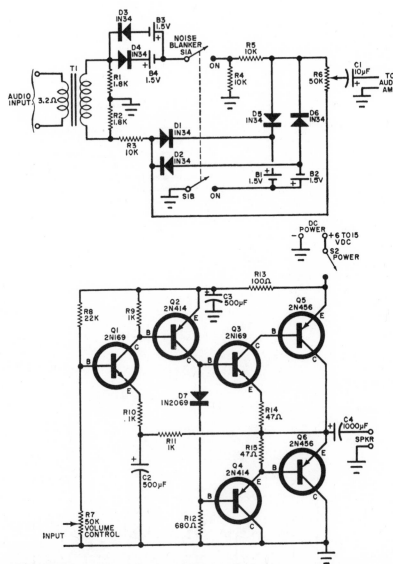

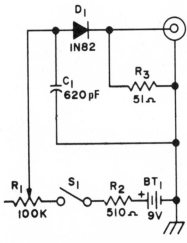

DIODE NOISE GENERATOR—Used as input for receiver or converter while adjusting circuit for minimum noise. Almost any uhf or microwave diode will serve for D1.—S. Goldstein, A $1.40 Noise Generator, *73*, Nov. 1970, p 97.

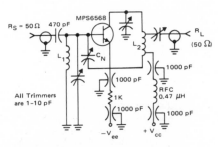

L_1 = 3 TURNS NO. 18 TINNED WIRE, 1/4 INCH I.D. AIR WOUND — LENGTH IS 3/16 INCH.

L_2 = 5 1/2 TURNS NO. 16 TINNED WIRE, 3/8 INCH I.D. AIR WOUND — LENGTH IS 1/2 INCH. V_{cc} FEEDS TAP 2 1/2 TURNS FROM COLLECTOR END —OUTPUT TAP, 1/2 TURN FROM COLLECTOR END.

200-MHZ NOISE-FIGURE TESTER—Common-emitter neutralized circuit requires two bias supplies. Noise diode is connected to test circuit through special coaxial cable which transforms diode impedance to higher value such as 100 ohms with negligible circuit loss. Report covers trimming of cable to exactly one-quarter of 200-MHz wavelength and gives design equations.—R. Brubaker, "Semiconductor Noise Figure Considerations," Motorola Semiconductors, Phoenix, AZ, AN-421, 1972.

BLANKER—When connected between low-impedance output of audio amplifier and speaker of short-wave, ham, or CB receiver, will accept signal that is nearly indistinguishable because of impulse noise and make it 90% readable. Additional audio amplifier, also shown, is required to raise low-level output of blanker. T1 can be audio output transformer for 50L6 tube. Based on peak clipping, in such a way that noise pulses exceeding blanking level are reduced to less power than that of desired signal which is below blanking level.—A. E. McGee, Jr., Build Noise Blanker, *Popular Electronics*, Oct. 1968, p 49–52.

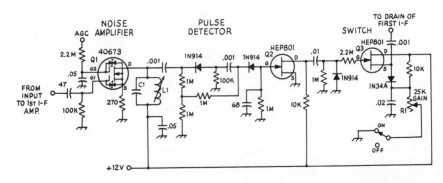

I-F NOISE SILENCER—Amplifies and rectifies noise pulses to provide negative-going d-c pulses for cutting off amplifier stage during noise. Clamp transistor Q3 shorts positive-going pulse overshoots. Choose L1 and C1 to resonate at desired i-f value. Construction details are given in July 1971 *QST*.—"The Radio Amateur's Handbook," American Radio Relay League, Newington, CT, 49th Ed., 1972, p 245.

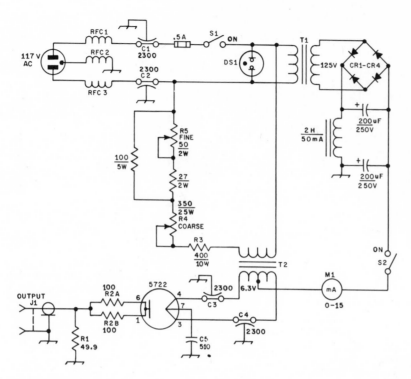

DIODE NOISE GENERATOR—Uses temperature-limited vacuum-tube diode as source of white and Gaussian noise for testing receivers. RCA R-6212A may be used in place of Sylvania diode shown, if mounted within 50-ohm coaxial line. Filament control of diode is applied to primary of T2 to get required very fine adjustment. Article tells how to minimize effects of stray inductance and capacitance.—R. E. Guentzler, Noise Generators, *QST*, March 1972, p 44–49.

NOISE BRIDGE—Wide-band signal source operating from 9-V battery serves for antenna and other r-f resonant frequency and radiation resistance measurements. Conventional receiver covering desired frequency range is used as frequency-selective detector. Transistors are Fairchild 2N3563, zener is Hoffman HW6.8A, and T1 is 4 quadrifilar turns No. 28 on ⅜-inch ferrite core.—R. T. Hart, The Antenna Noise Bridge, *QST*, Dec. 1967, p 39–41.

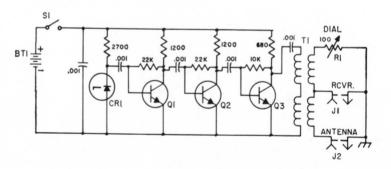

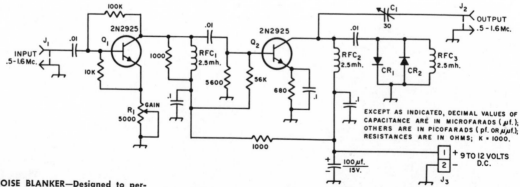

IGNITION NOISE BLANKER—Designed to permit using standard auto radio as i-f unit for h-f and vhf converters ahead of it for 10, 6, and 2-meter mobile reception without cutting into circuit of car radio. Output of converter is fed in at J1 for amplification by Q1 and Q2 before clipping of positive and negative noise-pulse peaks by germanium diodes, which can be 1N64 or similar. J2 feeds into auto radio.—"The Mobile Manual For Radio Amateurs," The American Radio Relay League, Newington, CT, 4th Ed., 1968, p 172–174.

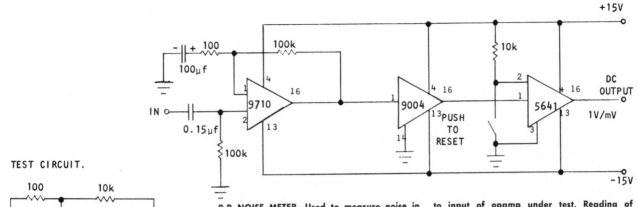

TEST CIRCUIT.

P-P NOISE METER—Used to measure noise in opamps and other high-gain devices. Uses standard OEI modules to provide preamplification, absolute value, and peak sensing over bandwidth of 10 to 10,000 Hz. Module 9004 produces output peak for sensing by 5641 having closed-loop gain of 100, so output is 1 V per 10 μV of peak noise referred to input of opamp under test. Reading of DVM connected to output must be doubled for p-p value. After connecting amplifier under test to input as shown, push reset button and wait 10 s before making reading, to allow for low-frequency noise.—"A Noise Tester," Optical Electronics, Tucson, AZ, Application Tip 10225, 1971.

RADIATION-DETECTING SWITCH—Circuit will open at onset of radiation but not noise, is controllable in reclosure, and generates no noise of its own. Circuit is equivalent of two-diode linear gate in which transistors are normally not conducting. Radiation turns both on, to present balanced-bridge low-impedance shunt to ground. Q1 and Q2 are their own radiation detectors so there is no inherent delay in response. Any radiation-induced noise at input to switch is not passed. Used in protecting integrity of information on fuse-arming circuit, memory input of computer, or other signal channel from radiation-induced noise.—J. W. Crowe, A Radiation-Sensitive Time-Selector Switch, *IEEE Journal of Solid-State Circuits*, April 1969, p 83–84.

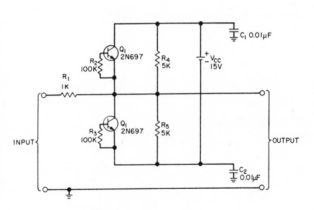

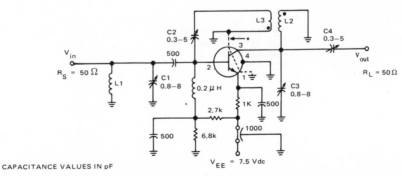

CAPACITANCE VALUES IN pF

L1, L2 — SILVER-PLATED BRASS ROD, 1-1/2″ LONG AND 1/4″ DIAMETER. INSTALL AT LEAST 1/2″ FROM NEAREST VERTICAL CHASSIS SURFACE.

L3 — 1/2 TURN NO. 16 AWG WIRE; LOCATED 1/4″ FROM AND PARALLEL TO L2.

* — EXTERNAL INTERLEAD SHIELD TO ISOLATE COLLECTOR LEAD FROM EMITTER AND BASE LEADS.

450-MHZ NOISE-FIGURE TESTER—Will measure noise figures as low as 2.1 dB. Report covers design considerations, including choice of transistor.—R. Brubaker, "Semiconductor Noise Figure Considerations," Motorola Semiconductors, Phoenix, AZ, AN-421, 1972.

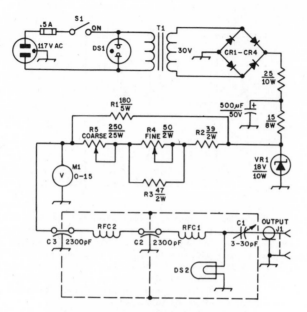

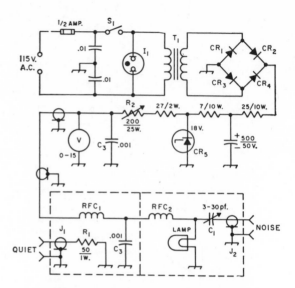

THERMAL NOISE GENERATOR—Generates white and Gaussian noise for testing receivers. Uses No. 12 incandescent pilot lamp as noise source, with R1–R5 providing adjustment of lamp voltage over range of 6 to 9 V. Keep lead length as short as possible. Rectifier diodes are 500-mA 100-piv silicon, such as 1N4002.—R. E. Guentzler, Noise Generators, QST, March 1972, p 44–49.

HOT-RESISTOR NOISE GENERATOR—Used as noise source for noise-figure measurements or as reference source for comparison with other noise source. Based on fact that noise output is known when temperature and resistance are known for tungsten filament of No. 12 pilot lamp heated by d-c source. R2 varies d-c voltage applied to lamp, to vary noise output. Diodes are 500-mA 100-piv silicon. CR5 is 1N1819 or equivalent. Secondary of T1 is 30 V at 2 A. Chokes are 1.8-μH 1,000-mA.—R. E. Guentzler, The "Monode" Noise Generator, QST, April 1967, p 30–33.

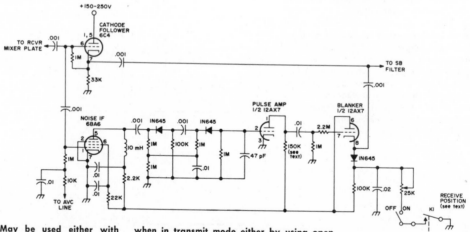

NOISE BLANKER—May be used either with receiver or with transceiver. Cathode follower is installed between mixer and first i-f filter. In transceiver, blanker must be deactivated when in transmit mode either by using open section of transmit-receive relay or installing extra relay. Blanker causes slight decrease (1 to 2 dB) in signal level while giving dramatic reduction in noise.—R. Grenell, Aftermath: Noise Blanker that Works, 73, April 1971, p 18–19.

CHAPTER 79
Operational Amplifier Circuits

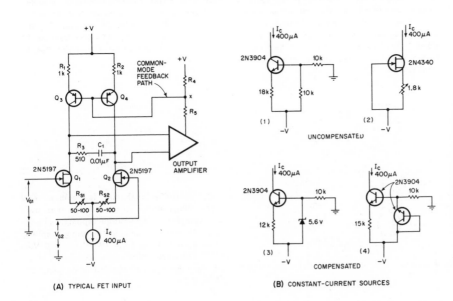

(A) TYPICAL FET INPUT

(B) CONSTANT-CURRENT SOURCES

UNCOMPENSATED

COMPENSATED

DIFFERENTIAL FET INPUT—Article covers design of fet input for opamp, using common-mode feedback to constant-current sources Q3 and Q4 (matched pair like 2N5255). Four choices for implementing constant-current source Ic are shown. Design optimizes thermal stability, sensitivity, noise, and common-mode characteristics. Supply is 15 V.—L. Diamond and A. V. Siefert, Designing Differential FET Inputs with Overall Performance in Mind, *Electronics*, June 21, 1971, p 76–80.

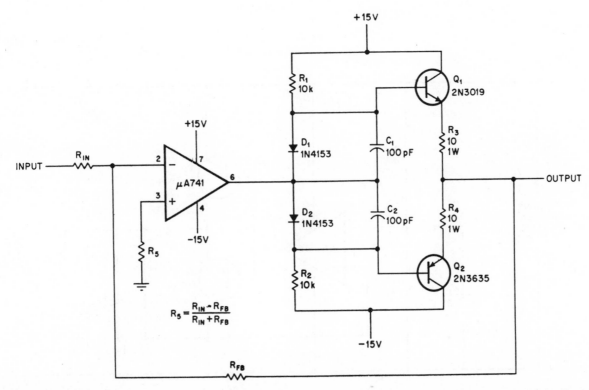

$$R_5 = \frac{R_{IN} \cdot R_{FB}}{R_{IN} + R_{FB}}$$

LINEAR CURRENT BOOSTER—Increases 25-mA opamp output to 500 mA, with no output crossover distortion up to 300 mA for ±3-V output swing. Diodes maintain voltage offset so transistors conduct early. Feedback from booster output maintains waveform.—J. R. Cox, Op Amp's Current Booster Ends Crossover Distortion, *Electronics*, March 29, 1971, p 70.

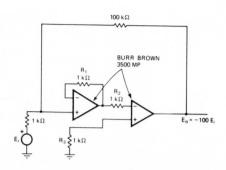

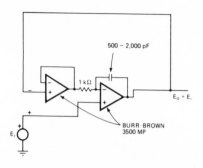

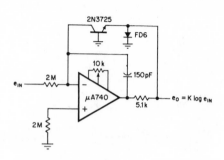

ZERO OFFSET—Second opamp is connected at inverting input of first to hold input offset voltage and offset voltage drift of differential opamp essentially to zero. First amplifier acts as floating voltage source. Output is inverting.—L. Choice, Series-Connected Op Amps Null Offset Voltage, *Electronics*, March 27, 1972, p 92.

ZERO-OFFSET BUFFER—Circuit shows compensation required with series-connected opamps used as unity-gain noninverting buffer having zero input offset voltage and offset voltage drift. Compensation does not reduce small-signal bandwidth.—L. Choice, Series-Connected Op Amps Null Offset Voltage, *Electronics*, March 27, 1972, p 92.

LOG-AMP—Use of jfet opamp having low bias and offset current eliminates complex nulling circuitry in logarithmic amplifier while covering full five-decade range.—T. McCaffrey and R. Brandt, FET Input Reduces IC Op Amp's Bias and Offset, *Electronics*, Dec. 7, 1970, p 85–88.

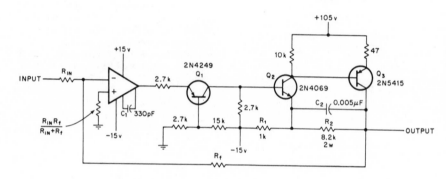

110-V SWING—Combining low-voltage and high-voltage stages costs much less than buying complete high-voltage opamp. Closed-loop operation of unipolar inverting amplifier gives output swing from 100 V to —10 V.—R. P. Patterson, With Some Discrete Aid IC Op Amp Swings 100 V, *Electronics*, Sept. 28, 1970, p 78.

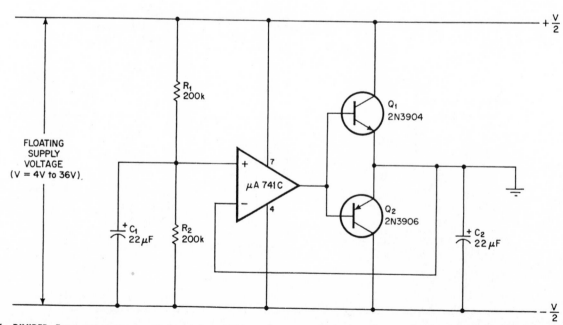

SUPPLY DIVIDER—Extra opamp connected across single supply provides balanced positive and negative voltages for opamp application requiring split power supply. Voltage regulation almost equals that of parent supply. Transistors act as current boosters in feedback loop of control circuits—R. D. Pierce, Op Amp Splits Supply for Other Op Amps, *Electronics*, March 1, 1971, p 57.

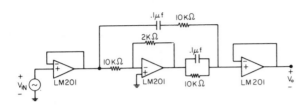

ALL-PASS TRANSFER FUNCTION—Values shown provide second-order all-pass operation without inductors. Phase response is 0 deg at about 160 Hz and increases almost linearly to 150 deg at 16 Hz and at 1,600 Hz. Article gives equation.—P. Aronhime and A. Budak, An Operational Amplifier All-Pass Network, *Proc. IEEE*, Sept. 1969, p 1677–1678.

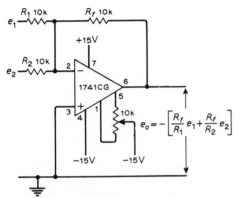

INVERTING ADDER—Both d-c and a-c input signals may be applied. Typical application is adding triangular and square waves. Input terminals should be grounded initially and 10K offset balance pot adjusted to give zero output voltage.—G. B. Clayton, Resistive Feedback Circuits, *Wireless World*, Aug. 1972, p 391–393.

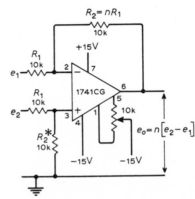

SUBTRACTING AMPLIFIER—Illustrates use of resistive feedback. Initially, both input terminals should be grounded and 10K offset balance pot adjusted for zero output voltage. Inputs should then be connected together and fed with sinusoidal signal; R2 is then trimmed to eliminate common-mode signal at output.—G. B. Clayton, Resistive Feedback Circuits, *Wireless World*, Aug. 1972, p 391–393.

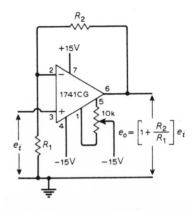

NONINVERTING AMPLIFIER—Illustrates use of resistive feedback with opamp. Bandwidth depends on closed-loop gain; 3-dB limit is 850 kHz when R1 and R2 are 10K for 6 dB gain. Bandwidth decreases to upper limit of 1 kHz when R1 is 100 and R2 is 100K for 60 dB gain. Ground input initially and adjust 10K pot for zero output.—G. B. Clayton, Resistive Feedback Circuits, *Wireless World*, Aug. 1972, p 391–393.

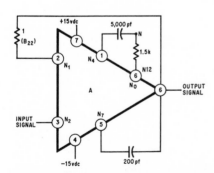

LINEAR VOLTAGE FOLLOWER—Pin connections shown are for 709 opamp, but article covers use of computer for worst-case and statistical analyses of any linear opamp. Values shown give closed-loop gain of 1 V/V.—R. Reid, Linear IC Model Takes to Analysis by Computer, *Electronics*, Aug. 31, 1970, p 78–82.

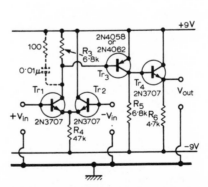

OPAMP FOR STUDENTS—Permits better understanding of IC opamp circuit action. Not recommended for actual use because of poor d-c response and limited high-frequency response. Noise figure is good for a-f applications.—D. Griffiths, A Simple Op. Amp., *Wireless World*, July 1970, p 337–338.

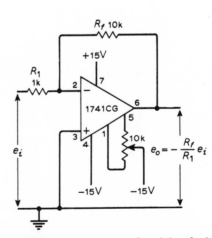

INVERTER—Illustrates use of resistive feedback in simple opamp inverter. Input terminals should be grounded initially and 10K offset balance pot adjusted to give zero output voltage.—G. B. Clayton, Resistive Feedback Circuits, *Wireless World*, Aug. 1972, p 391–393.

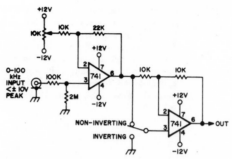

LEVEL CHANGER—Changes any input level to desired new value determined by setting of 10K pot, as required for low-speed counter, cro, or audio equipment. Switch gives choice of output polarity.—Circuits, 73, Aug. 1972, p 130.

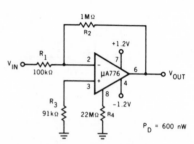

600-MW OPAMP—Provides gain of 20 dB from 1 to 55 Hz, dropping linearly to 0 dB at 500 Hz above turnover frequency. Power drain is so low that circuit can be operated continuously on two mercury cells for shelf-life time of batteries. Maximum output signal is 1.2 V p-p.—"The μA776, An Operational Amplifier with Programmable Gain, Bandwidth, Slew-Rate, and Power Dissipation," Fairchild Semiconductor, Mountain View, CA, No. 218, 1971, p 4.

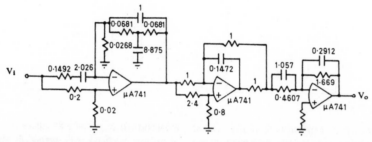

DELAY FUNCTION—Normalized values are given for network using three opamps in two cascaded stages. Article shows responses of denormalized network (1 meg and 0.1 s delay) to step and triangular waves, and gives mathematical analysis.—T. Deliyannis, Six New Delay Functions and Their Realization Using Active RC Networks, *The Radio and Electronic Engineer*, March 1970, p 139–159.

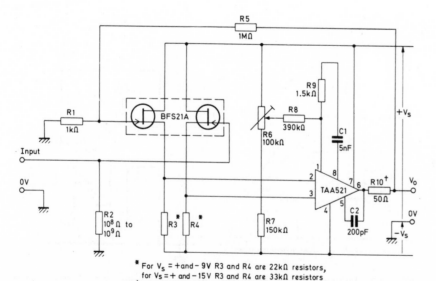

* For V_S = +and−9V R3 and R4 are 22kΩ resistors,
for V_S = + and −15V R3 and R4 are 33kΩ resistors
† R10 is required when the amplifier is used with capacitive loading

FET INPUT—Uses matched fet pair as front end for Mullard high-gain opamp to permit use with high source resistance. For 10K load, minimum output voltage swing is —12 V and +12 V for supply voltages of —15 V and +15 V. Closed-loop voltage gain is 60 dB up to 1 kHz. R6 balances out offset voltage.—"Fet Input Stage for an Integrated Operational Amplifier," Mullard, London, 1970, TP1149.

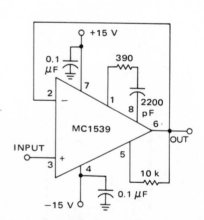

VOLTAGE FOLLOWER—Improved characteristics of second-generation version of 709 opamp make it excellent choice for voltage follower operating in noninverting unity-gain mode. Does not have common-mode latch-up problem of 709. Design equations are given.—E. Renschler, "The MC1539 Operational Amplifier and Its Applications," Motorola Semiconductors, Phoenix, AZ, AN-439, 1972.

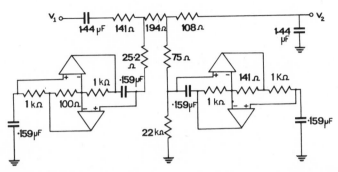

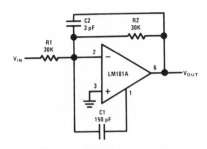

ELLIPTIC LOW-PASS LADDER FILTER TRANSFER FUNCTION—Voltage gain is —6 dB to 1 kHz, then drops sharply to —80 dB. Opamps are μA702C. Article gives design equations and response curve.—L. T. Bruton, Network Transfer Functions Using the Concept of Frequency-Dependent Negative Resistance, *IEEE Trans. on Circuit Theory*, Aug. 1969, p 406–408.

FAST COMPENSATION FOR SUMMING—Feedforward compensation shown improves high-frequency performance of standard opamp, giving slew rate of 10 V/μs, power bandwidth of 250 kHz, and small-signal bandwidth of 10 MHz. Feedforward works only when device is used as summing amplifier.—R. J. Widlar, Design Techniques for Monolithic Operational Amplifiers, *IEEE Journal of Solid-State Circuits*, Aug. 1969, p 184–191.

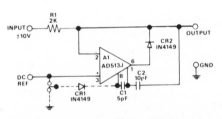

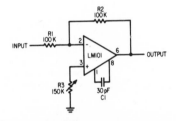

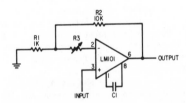

PRECISION CLIPPER—Sharply limits excursion of input voltage at level equal to d-c reference input. If reference is 0, circuit can be used as half-wave rectifier up to 100 kHz. Useful down to millivolt input levels at d-c and low frequencies. With CR1 omitted, input range is 70 mV to 7 V rms up to 10 kHz. With CR1 connected, useful input extends down to 0.3 V rms for up to 100 kHz.—"AD513, AD516 I.C. Fet Input Operational Amplifiers," Analog Devices, Inc., Norwood, MA, Technical Bulletin, Aug. 1971.

BIAS CURRENT COMPENSATION—Offset produced by bias current on inverting input of IC opamp is cancelled by offset voltage produced across variable resistor R3. Gives low drift over wide temperature range. Can be used as summing amplifier.—R. J. Widlar, "Drift Compensation Techniques," National Semiconductor, Santa Clara, CA, 1967, AN-3.

NONINVERTING BIAS CURRENT COMPENSATION—Offset voltage produced across d-c resistance of source due to input current is cancelled by drop across R3, which should have maximum value about three times source resistance. Gives low drift over wide temperature range. Source must have fixed d-c resistance.—R. J. Widlar, "Drift Compensation Techniques," National Semiconductor, Santa Clara, CA, 1967, AN-3.

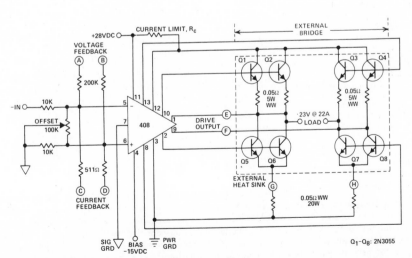

500-W POWER BOOSTER—Output of Model 408 power opamp is boosted with external bridge using eight inexpensive 2N3055 transistors. For voltage amplifier, connect A to E and B to F for gain of 20 V/V. For current amplifier, connect C to H and D to G for gain of 1 A/V.—"400 Series Power Operational Amplifiers," Analog Devices, Inc., Norwood, MA, July 1970.

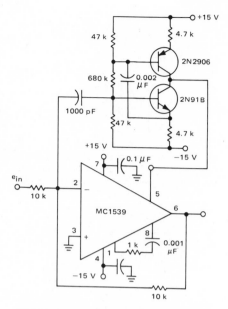

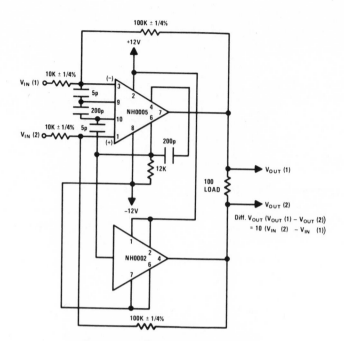

FEED-FORWARD UNITY-GAIN—Second-generation version of 709 opamp improves feed-forward method of extending power bandwidth of amplifier. Provides 10-V p-p output at unity gain for frequencies beyond 1 MHz. Gives fast-responding slow-settling response to step-function input. Report gives design considerations.—E. Renschler, "The MC1539 Operational Amplifier and Its Applications," Motorola Semiconductors, Phoenix, AZ, AN-439, 1972.

DIFFERENTIAL INPUT-OUTPUT—Combination of NH0002 current amplifier with NH0005 opamp gives differential inputs as well as outputs. Load must be floated between outputs of devices to provide complete loop of feedback. With compensation shown, 20-V p-p signal can be applied up to 80 kHz with about 33% efficiency.—C. M. Wittmer, "NH0002 Current Amplifier," National Semiconductor, Santa Clara, CA, 1968, AN-13.

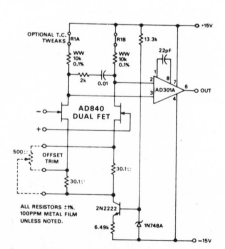

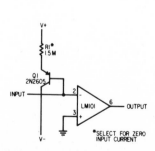

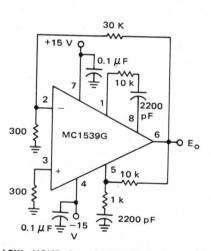

LOW DRIFT AND NOISE—With low-cost Analog Devices AD840 dual-fet input for opamp, noise is down to 1 μV p-p for 0.1 to 10 Hz and 8 μV rms for 5 Hz to 10 kHz. Small-signal unity-gain bandwidth is 700 kHz. Output is 10 V at 5 mA for either polarity.—A Low-Noise, Low-Drift Fet-Input Amplifier Design, *Analog Dialogue*, Vol. 7, No. 1, p 14.

BIAS-CURRENT COMPENSATION—Does not depend upon having fixed source resistance, so works well as summing amplifier. Base current of external pnp transistor balances out base current of npn input transistor of IC to give improved drift performance.— R. J. Widlar, "Drift Compensation Techniques," National Semiconductor, Santa Clara, CA, 1967, AN-3.

LOW NOISE—Second-generation version of 709 opamp operates in noninverting mode with gain of 100 down to reasonably low frequencies, with bandwidth of about 5 kHz and output noise less than 0.1-mV rms. Source impedance is about 300 ohms.—E. Renschler, "The MC1539 Operational Amplifier and Its Applications," Motorola Semiconductors, Phoenix, AZ, AN-439, 1972.

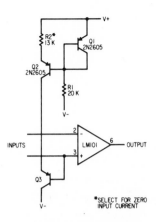

LARGE COMMON-MODE RANGE—Two transistors, serving as current source for Q3, improve bias-current compensation of opamp for changes in temperature and supply voltage, for noninverting operation.—R. J. Widlar, "Drift Compensation Techniques," National Semiconductor, Santa Clara, CA, 1967, AN-3.

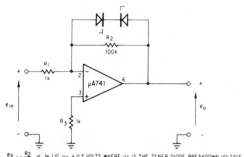

$$\frac{e_o}{e_{in}} = -\frac{R_2}{R_1} \quad \text{IF } |e_o| \leq v_Z + 0.7 \text{ VOLTS WHERE } v_Z \text{ IS THE ZENER DIODE BREAKDOWN VOLTAGE}$$

CLIPPING OPAMP—Output swing of opamp is held within specific limits by adding nonlinear elements to feedback network. Zeners quickly reduce gain if output tries to exceed zener voltage limits. Under these limits, zeners do not conduct and gain is determined by R1 and R2. Opamp shown meets clipper requirement that amplifier must be frequency-compensated for unity gain.—"Applications of the μA741 Operational Amplifier," Fairchild Semiconductor, Mountain View, CA, No. 289, 1972, p 3–4.

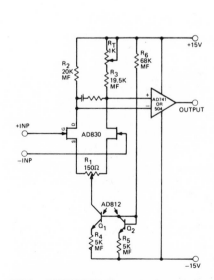

DRIFT CORRECTION—Uses Analog Devices AD830 dual-fet to enhance input characteristics of opamp, by reducing input current and offset drift. Changing ratio of fet currents with R1 can essentially eliminate any overall amplifier drift with temperature. Report recommends use of electrometer practices to get best performance.—"Trak-Fets—Dual Monolithic Field Effect Transistors," Analog Devices, Norwood, MA, Nov. 1972.

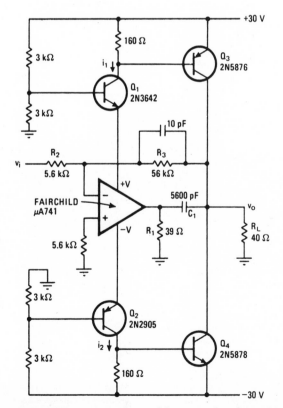

22 W PEAK AT 60-V SWING—Uses popular opamp with four transistors to give voltage gain of 10 with flat response from d-c to 30 kHz. Q1 and Q2 protect opamp by keeping voltage below power-pin rating of 36 V while generating base drive voltage for power output stage Q3-Q4.—P. P. Garza, Jr., Getting Power and Gain out of the 741-Type Op Amp, *Electronics*, Feb. 1, 1973, p 99.

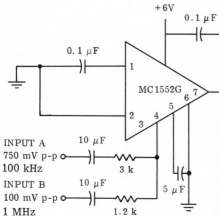

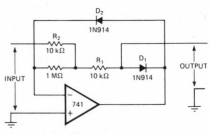

SUMMING-SCALING AMPLIFIER—Wide bandwidth and output swing capability of IC video amplifier permits use in noninverting configuration for summation of input signal current at pin 4. Scale factors are handled by adjusting values of input resistors.—"A Wide Band Monolithic Video Amplifier," Motorola Semiconductors, Phoenix, AZ, AN-404, 1971.

LOW-LEVEL SIGNAL RECTIFIER—Feed-forward resistive element makes one opamp do job of two for full-wave rectification in frequency-doubling and other small-signal rectifying applications. D1 conducts during positive inputs and D2 for negative input.—R. Knapp and R. Melen, Op Amp with Feedback Makes Full-Wave Rectifier, *Electronics*, Sept. 11, 1972, p 109.

DIFFERENTIAL-FET INPUT—Combination of discrete components and IC gives unconditional stability for all values of closed loop gain down to unity. Offset voltage is balanced out with R6. With supply of 9 V, R3 and R4 are 47K, and for 15 V are 82K.—"Field Effect Transistors," Mullard, London, 1972, TP1318, p 98.

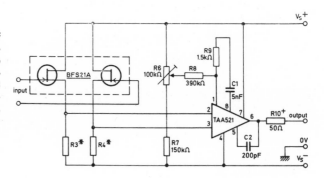

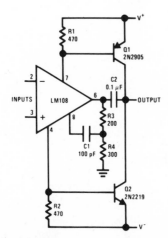

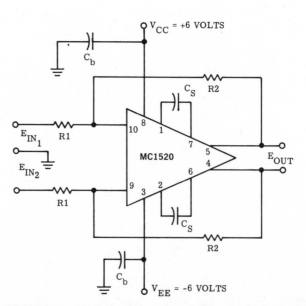

POWER BOOSTER—Increases output current of LM108 opamp to ±50 mA and swings output up to within fraction of volt of supply voltages. Output transistors are driven from supply leads of opamp. Bootstrapped shunt compensation works for all loads. Dead zone in open-loop transfer characteristic becomes noticeable around 1 kHz, but can be neglected at lower frequencies because gain there is high enough.—R. J. Widlar, "IC Op Amp Beats FETs on Input Current," National Semiconductor, Santa Clara, CA, 1969, AN-29, p 15–16.

BALANCED DIFFERENTIAL IC—Connections shown give differential output as well as differential input, to increase open-loop voltage gain of IC by 6 dB over single-ended connection having gain of about 70 dB. Closed-loop voltage gain, however, is reduced by same factor. Voltage gain is 40 from 1 kHz to about 3 MHz when R1 is 1K, R2 100K, R3 1K, and CS 2 pF. Ground reference is not critical; input may be left floating.—"A General Purpose I/C Differential Output Operational Amplifier," Motorla Semiconductors, Phoenix, AZ, AN-407, 1971.

LEVEL ISOLATION—Takes voltage referred to some d-c level and gives amplified output referred to ground without loading of input signal source. Uses fet to produce voltage drop across feedback resistor R1 which equals input voltage, so voltage across R2 is equal to input voltage multiplied by ratio of R2 to R1. Common-mode rejection is as good as that of opamp. Output voltage across R2 is buffered by LM102 voltage follower to give low output impedance. Will work with input voltages equal to positive supply voltages up to 15 V.—R. J. Widlar, "Operational Amplifiers," National Semiconductor, Santa Clara, CA, 1968, AN-4.

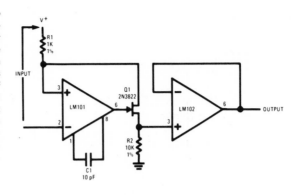

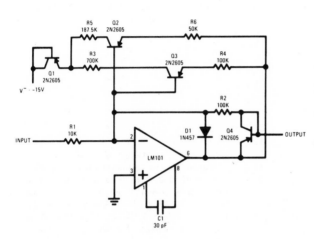

NONLINEAR WITH STABLE BREAKPOINTS—Provides very sharp temperature-stable breakpoints in nonlinear transfer function of opamp. Report describes operation in detail.—R. J. Widlar, "Operational Amplifiers," National Semiconductor, Santa Clara, CA, 1968, AN-4.

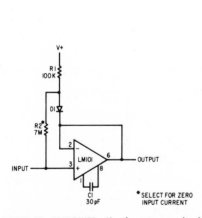

VOLTAGE FOLLOWER—Simple opamp circuit provides bias current compensation, through resistor connected across diode which is bootstrapped to output. Diode acts as regulator so compensating current does not change appreciably with signal level, giving input impedance of about 1,000 meg.—R. J. Widlar, "Drift Compensation Techniques," National Semiconductor, Santa Clara, CA, 1967, AN-3.

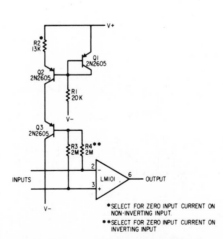

DIFFERENTIAL INPUTS WITH COMPENSATION—Both inputs of opamp are bias-current-compensated over full common-mode range as well as against power supply and temperature variations. Can be used either as summing or noninverting d-c amplifier.—R. J. Widlar, "Drift Compensation Techniques," National Semiconductor, Santa Clara, CA, 1967, AN-3.

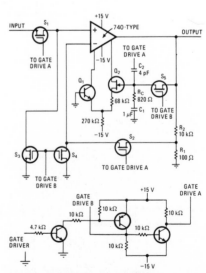

ZERO CORRECTION TO 3 μV—Uses mosfet switches to cycle opamp between error storage and amplification modes. Sample-and-hold circuit S5-Q1-Q2-C1 inserts correction current rather than correction voltage into intermediate state of opamp. Use p-channel mosfet's having floating substrates. Circuit also holds offset-voltage temperature drift to less than 0.05 μV per deg C.—R. C. Jaeger and G. A. Hellwarth, Dynamic Zero-Correction Method Suppresses Offset Error in Op Amps, *Electronics*, Dec. 4, 1972, p 109–110.

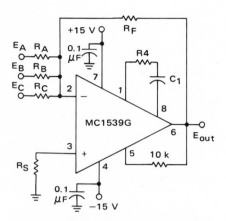

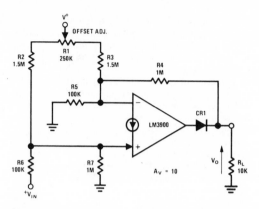

SUMMING AMPLIFIER—Second-generation version of 709 opamp functions with small amount of loop-gain error as closed-loop summing amplifier. Design equations are given.—E. Renschler, "The MC1539 Operational Amplifier and Its Applications," Motorola Semiconductors, Phoenix, AZ, AN-439, 1972.

NONINVERTING ZERO-OUTPUT—Uses diode with Norton current-differencing opamp to provide d-c level shift which allows output voltage to go to ground when input voltage is zero. Circuit gain is 10. Will operate on supply of 4 to 36 V.—T. M. Frederiksen, W. M. Howard, and R. S. Sleeth, "The LM3900—A New Current-Differencing Quad of ± Input Amplifiers," National Semiconductor, Santa Clara, CA, 1972, AN-72, p 9.

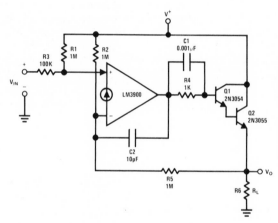

3-A POWER AMPLIFIER—Darlington pair at output of Norton current-differencing opamp delivers over 3 A to load when transistors are properly mounted on heatsinks. Supply can be 4 to 36 V.—T. M. Frederiksen, W. M. Howard, and R. S. Sleeth, "The LM3900—A New Current-Differencing Quad of ± Input Amplifiers," National Semiconductor, Santa Clara, CA, 1972, AN-72, p 9.

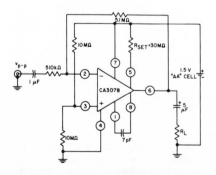

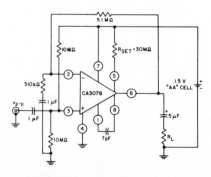

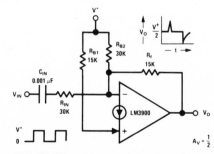

INVERTING 20-DB AMPLIFIER—Uses RCA opamp drawing standby power as low as 700 mW, operating from single supply having minimum of 1.5 V. Total power consumption is about 675 mW while delivering output voltage swing of 300 mV p-p into 20K load.—"Linear Integrated Circuits and MOS Devices," RCA, Somerville, NJ, 1973 Ed., SSD-201A, p 486–487.

NONINVERTING 20-DB AMPLIFIER—Uses RCA opamp drawing standby power as low as 700 mW, operating from single supply having minimum of 1.5 V. Total power consumption is about 675 mW while delivering output voltage swing of 300 mV p-p into 20K load.—"Linear Integrated Circuits and MOS Devices," RCA, Somerville, NJ, 1973 Ed., SSD-201A, p 486–487.

SWINGS ABOVE AND BELOW BIAS—Output voltage acts with RB1 and RB2 to bias at half of supply voltage, allowing positive and negative swings above and below bias point.—T. M. Frederiksen, W. M. Howard, and R. S. Sleeth, "The LM3900—A New Current-Differencing Quad of ± Input Amplifiers," National Semiconductor, Santa Clara, CA, 1972, AN-72, p 34.

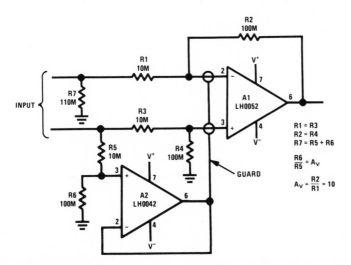

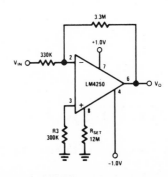

GUARDED FULL DIFFERENTIAL—Guard driver amplifier develops proper voltage for guard, to prevent deterioration of opamp performance by leakage caused by improper cleaning of circuit boards, improperly cured protective coatings, or socket leakage. Placing guard conductor in leakage path, operating at same potential as inputs, intercepts leakage current.—R. K. Underwood, "New Design Techniques for FET Op Amps," National Semiconductor, Santa Clara, CA, 1972, AN-63, p 6–7.

500-MW INVERTING AMPLIFIER—Provides gain of 10 while drawing only 260 nA from +1-V supply and 210 nA from —1-V supply. Uses programmable opamp capable of providing variety of circuit functions. If operated from 1.35-V mercury cells, power drain is only 1 μW per cell, having no effect on shelf life of cell. With mercury cells, output swing is 1.4 V p-p. Response is 3 dB down at 300 Hz.—M. K. Vander Kooi and G. Cleveland, "Micropower Circuits Using the LM4250 Programmable Op Amp," National Semiconductor, Santa Clara, CA, 1972, AN-71, p 4.

CHAPTER 80
Optoelectronic Circuits

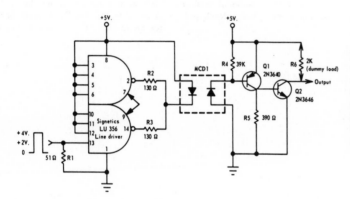

D-C AMPLIFIER FOR LED—Required with photodiode for driving DTL or TTL. When output of line driver is low, led emits light to detector, driving Q2 into saturation and making collector output low. When line driver output is high, collector output of Q2 is high.—B. Otsuka and R. Hunt, Light Emitting Diodes, *Electromechanical Design*, Jan.-Feb. 1971, p 16–19.

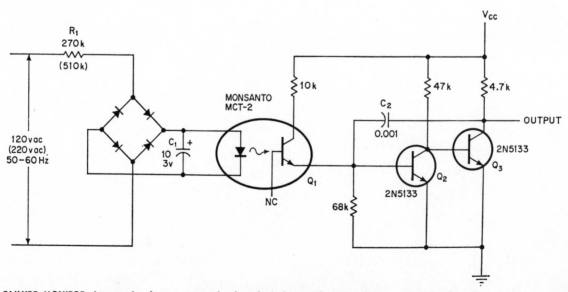

POWER-FAILURE MONITOR—Inexpensive low-level logic circuit, isolated from any a-c or d-c power supply by optical coupling, drops output level to logical zero if power fails, to set flip-flops in shut-down circuit. Response is 2 ms. Not affected by normal zero crossings of a-c line. Draws 50 mW.—J. van Zee, Optoelectronic Switch Monitors Line Power, *Electronics*, Nov. 23, 1970, p 68.

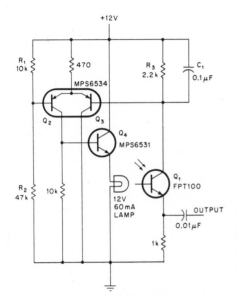

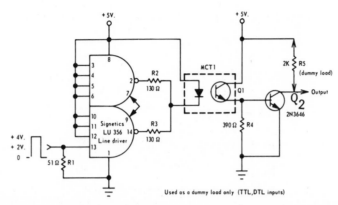

MEASURING LOW LIGHT LEVELS—Use of light to bias phototransistor improves response time and gives maximum sensitivity to changes in light.—D. Knowlton, Optical Biasing Maintains Phototransistor Sensitivity, *Electronics*, Oct. 26, 1970, p 87.

PHOTOTRANSISTOR DRIVES LOGIC—Will drive TTL or DTL directly, without amplification. With line driver output high, led is not conducting and Q1 is off, turning off Q2 and giving high output. With line driver output low, led conducts and turns on Q1, to drive Q2 into saturation and give low output.—B. Otsuka and R. Hunt, Light Emitting Diodes, *Electromechanical Design*, Jan.-Feb. 1971, p 16–19.

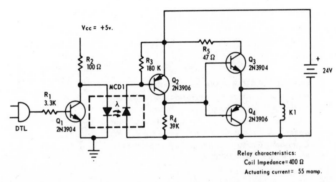

RELAY ISOLATOR FOR DTL—Coupled transistor pair eliminates ground currents and spikes originating in relay circuit. When DTL output is high, Q1 is turned on and no current flows in led. Q2 then cannot conduct and in turn prevents Q3 from conducting, so relay is not energized. When DTL output goes low, circuit response is so fast that K1 is energized in time depending on its mechanical response.—B. Otsuka and R. Hunt, Light Emitting Diodes, *Electromechanical Design*, Jan.-Feb. 1971, p 16–19.

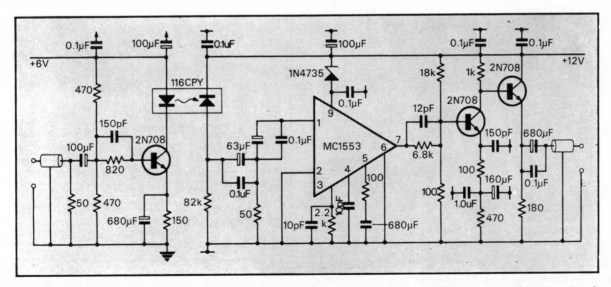

20 HZ–7 MHZ BANDWIDTH—Uses 116CPY photocoupler having silicon photodiode and GaAsP-base photoemitter. Developed in France. Provides complete isolation of input and output along with fast response and immunity to external magnetic fields. Used in plasma physics research. Separate battery supplies are required.—L. R. P. Symons, Small-Signal Opto-Electronic Transformer, *Electronic Engineering*, Nov. 1969, p 35–39.

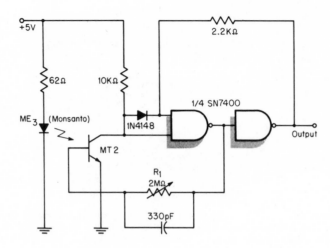

SLOW-MOTION DETECTOR—Punched card or any other object moving even at very slow speed between led and phototransistor actuates IC Schmitt trigger to generate single clean TTL-compatible pulse. R1 sets light threshold. Pulse rise and fall times are under 40 ns.—P. Obrda, Position Detector with Direct TTL Interface, *The Electronic Engineer*, Oct. 1971, p 69.

LED MODULATOR—Uses pnp overlay transistors, for improved performance between audio and video frequencies. Some shielding may be required. Used for modulating light beam in communications system.—R. W. Campbell, Gallium Arsenide LED Experiments, *Ham Radio*, June 1970, p 6–14.

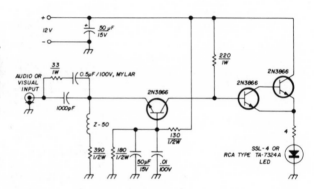

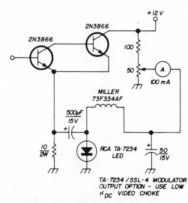

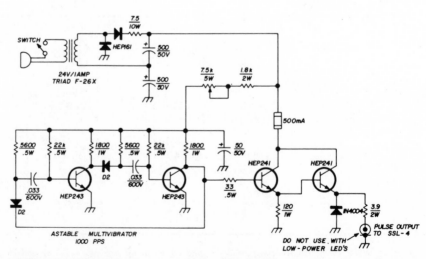

LED VOICE MODULATOR—With no modulating signal present, led is forward-biased to half of maximum c-w current value. With 500-Hz tone applied, average forward current is increased to 2 A, and tone sources replaced by speech amplifier. Led must have heat sink. Voice peaks will not cause lasing in led.—R. W. Campbell, Gallium Arsenide LED Experiments, *Ham Radio*, June 1970, p 6–14.

PULSED 1.25 W FOR LED—Operates at 50% duty cycle. Puts out square pulses having less than 200-ns rise time. For testing, use No. 248 miniature lamp in place of led.—R. W. Campbell, Gallium Arsenide LED Experiments, *Ham Radio*, June 1970, p 6–14.

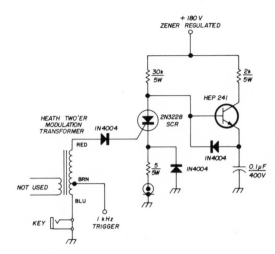

PULSED 900 MW FOR LED—Has 1% duty cycle. When 0.1-μF capacitor is fully charged, transistor switches pulses through scr, which acts as gate. Amplitude of resulting current pulse depends on dynamic response of scr. Good for 70-A peak high-threshold light-emitting diodes, but not for injection lasers.—R. W. Campbell, Gallium Arsenide LED Experiments, *Ham Radio*, June 1970, p 6–14.

50-MHZ PHOTODETECTOR—Parallel tuning is suitable for single GE photodiode shown. With up to five photodiodes in parallel, series tuning is required; 10 pF should then allow response up to about 400 MHz. T1 has 2-turn primary and 7-turn secondary on Permacor 57-6075 toroid.—R. W. Campbell, Gallium Arsenide LED Experiments, *Ham Radio*, June 1970, p 6–14.

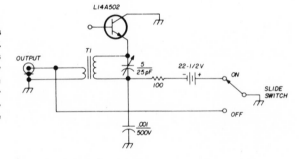

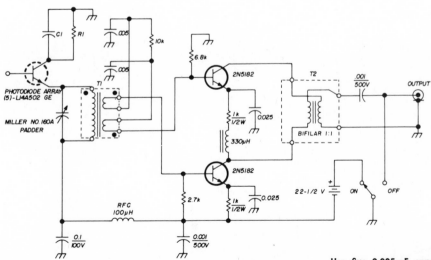

QUADRATURE DETECTOR FOR LED—Designed for receiving output of photodiode array illuminated by led of communication system, translated to broadcast band. Circuit is similar to 1:1 balun, with single-port input and 90-deg phasing between the two output ports. Gives reasonably high gain, needs no neutralization, and has low noise response. Use five 0.005-μF capacitors in parallel for C1, and five 560-ohm resistors in parallel for R1. Article gives transformer data.—R. W. Campbell, Gallium Arsenide LED Experiments, *Ham Radio*, June 1970, p 6–14.

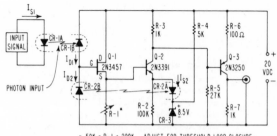

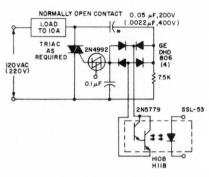

LOW-LEVEL PHOTO COUPLER—Uses one HPA 4310 photon-coupled isolator in feedback path of output amplifier for same device used as isolator, to provide required high power gain for applications which can not supply required several mA of drive to light-source diode of CR-1. Response of entire circuit is linear from under 100 μA to over 1 mA. Report covers circuit operation in detail.—"Electrical Isolation Using the HP 5082-4310," Hewlett Packard, Palo Alto, CA. No. 909, 1968.

NORMALLY OPEN 10-A RELAY—Triac matching load is triggered by 2N4992 which in turn is controlled by photodarlington acting through diode bridge.—S. R. Korn, "Photon Couplers," General Electric, Semiconductor Products Dept., Auburn, NY, No. 200.62, p 21.

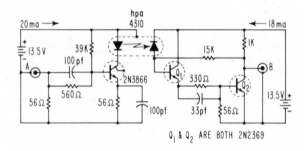

Q$_1$ & Q$_2$ ARE BOTH 2N2369

10-MHZ ISOLATOR—Preamp for photodiode of photon-coupled isolator has frequency response of about 3.5 MHz, corresponding to risetime under 0.1 μs, for high-speed performance. Response is further increased by using minority-carrier lifetime compensation in light-source drive circuit. If input pulse has fast rise, under 10 ns, photodiode preamp gives risetime of 15—20 ns.—"Electrical Isolation Using the HP 5082-4310," Hewlett Packard, Palo Alto, CA, No. 909, 1968.

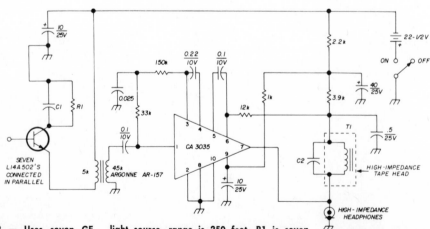

BROAD-AREA DETECTOR — Uses seven GE phototransistors connected as photodiodes. T1 is surplus tape head used as high impedance for coupling to headphones. With SSL-4 led light source, range is 250 feet. R1 is seven 680-ohm resistors in parallel, and C1 is seven 0.01 μF capacitors in parallel. Use from 0.01 to 0.033 for C2, depending on audio response desired.—R. W. Campbell, Gallium Arsenide LED Experiments, *Ham Radio*, June 1970, p 6—14.

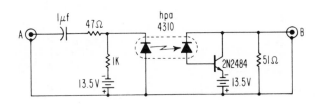

UNITY COUPLING WITH PHOTONS—Uses photon-coupled isolator to provide near-unity coupling coefficient between 50-ohm signal source and 50-ohm load, with complete isolation. Coupling will withstand 200-V difference between input and output supply voltages.— "Electrical Isolation Using the HP 5082-4310," Hewlett Packard, Palo Alto, CA, No. 909, 1968.

CLOCK-PULSE COUPLING — Uses Monsanto MCA2-55 gallium arsenide infrared emitter optically coupled to silicon planar photodarlington to pass 5-kHz clock pulse with high efficiency at maximum rated current and voltage while providing required isolation. Circuit uses base lead of photodarlington to improve release time with 400-ohm load.— "Photo-Darlington Relay," Monsanto Co., Cupertino, CA, No. MCA2-30, 1972.

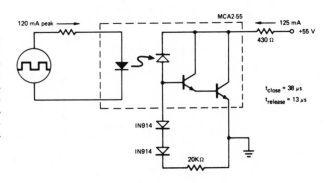

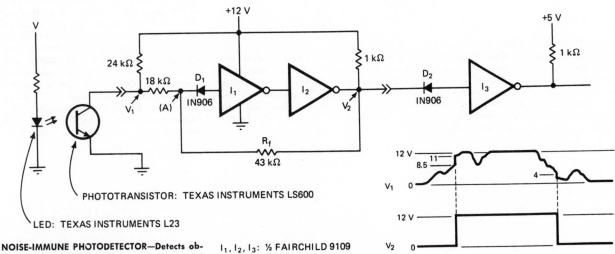

PHOTOTRANSISTOR: TEXAS INSTRUMENTS LS600

LED: TEXAS INSTRUMENTS L23

NOISE-IMMUNE PHOTODETECTOR—Detects objects breaking light beam between led and phototransistor. Operates reliably even with long cables between photodetector and amplifier in electrically noisy environment. Interruption of beam makes V1 increase and

I_1, I_2, I_3: ½ FAIRCHILD 9109

switch on inverters I1 and I2. Additional current through Rf makes V1 rise to give latching effect. V2 stays high until V1 drops to 4 V and makes V2 return to ground. Inverter I3

converts V2 to TTL-compatible output.—R. T. Laubach, Photodetector Senses Motion in Noisy Surroundings, *Electronics*, Oct. 9, 1972, p 101.

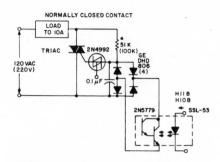

NORMALLY CLOSED 10-A RELAY—Triac matching load is triggered by 2N4992 which in turn is controlled by photodarlington acting through diode bridge.—S. R. Korn, "Photon Couplers," General Electric, Semiconductor Products Dept., Auburn, NY, No. 200.62, p 21.

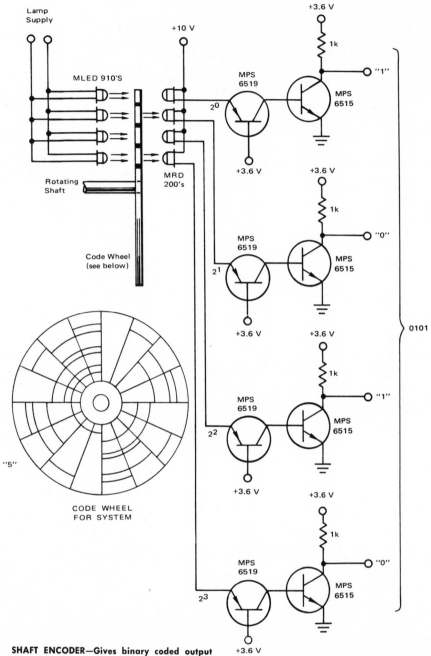

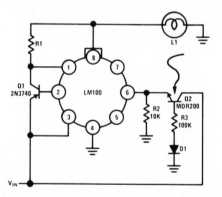

SHAFT ENCODER—Gives binary coded output corresponding to angular position of rotating shaft on which is mounted code disk having alternate transparent and opaque sections. When two phototransistors are illuminated as shown, output data word is 0101 or 5. Control system (not shown) would then recognize that shaft is in fifth of 16 possible positions.

If command code to control system then called for position 12, or 1100, comparator circuit would drive shaft until that position was reached.—J. Bliss, "Applications of Phototransistors in Electro-Optic Systems," Motorola Semiconductors, Phoenix, AZ, AN-508, 1971.

LIGHT-INTENSITY REGULATOR—Phototransistor Q2 senses light level and drives feedback terminal of IC voltage regulator to control current flow of incandescent lamp L1. Choose value of R1 to limit inrush current to lamp when circuit is first turned on. Current gain of Q2 is fixed at 10 by R3 and temperature-compensating diode D1 to make transistor less temperature-sensitive.—R. J. Widlar, "New Uses for the LM100 Regulator," National Semiconductor, Santa Clara, CA, 1968, AN-8.

CHAPTER 81
Oscillator Circuits—A-F

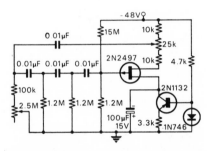

10-HZ PHASE-SHIFT—Junction fet in four-mesh phase-shift oscillator permits varying frequency several Hz above or below 10 Hz with 2.5-meg pot. Attenuation of feedback network is 18.—Some Applications of Field-Effect Transistors, *Electronic Engineering*, Sept. 1969, p 18–23.

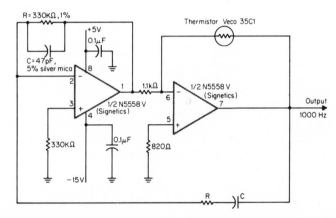

1-KHZ WIEN BRIDGE—Modification of basic bridge uses larger resistor values, simplifying generation of pure sine waves for laboratory audio work.—H. Olson, Dual Op Amp Makes Simple Sine Wave Generator, *The Electronic Engineer*, Aug. 1971, p 70.

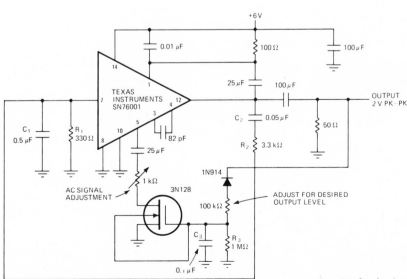

MOSFET FEEDBACK—Uses Wien-bridge type IC audio oscillator with R1, R2, C1, and C2 as frequency-determining elements, with IC also serving as 1-W amplifier. Mosfet feedback control keeps output level constant and prevents limiting. Will deliver several volts into 50-ohm load with total harmonic distortion under 1%.—G. Coers, MOSFET Network Minimizes Audio Oscillator Distortion, *Electronics*, Jan. 3, 1972, p 84–85.

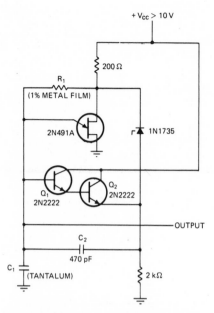

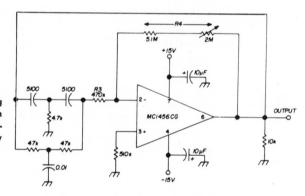

60-HZ STABILIZED—Bootstrapping voltage of timing capacitor C1 to base 2 of ujt in relaxation oscillator makes circuit independent of inter-base resistance, giving 0.05% frequency stability for 0–55 C range and 0.5% voltage stability for 100% change in supply. For 60 Hz, R1 is 10K to 50K and C1 must be above 0.001 μf.—M. J. Debronsky, Bootstrapped Capacitor Stabilizes UJT Oscillator, *Electronics*, Jan. 17, 1972, p 68.

1-HZ SINE-WAVE—Uses two opamps, with A1 connected as two-pole low-pass active filter and A2 as integrator. Zener clamps D1 and D2 stabilize output amplitude. Circuit virtually eliminates even-order harmonics, so dominant third harmonic is down about 40 dB at output of A1 and down 50 dB at output of A2. Opamps used will oscillate up to 1 kHz but higher frequencies require costly high-value resistors with low temperature coefficients.—R. J. Widlar, "IC Op Amp Beats FETs on Input Current," National Semiconductor, Santa Clara, CA, 1969, AN-29.

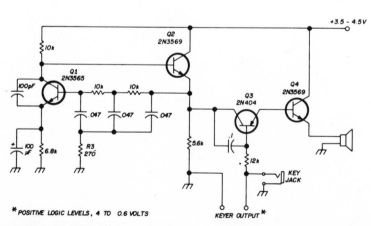

TWIN-T OPAMP—RC network has 180 deg phase-shift at frequency of oscillation, much as in phase-shift oscillator.—H. Olson, Resistance-Capacitance Oscillators, *Ham Radio*, July 1972, p 18–25.

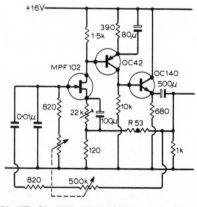

30 HZ—20 KHZ WIEN-BRIDGE—Uses fet to reduce damping on bridge and permit use of 500K twin pot for frequency control. Harmonic distortion is kept under 0.05% over entire tuning range by properly adjusting 22K preset resistor. Requires no capacitor switching to cover audio range.—C. A. Pye, Low-Distortion 30 Hz–20 kHz Oscillator, *Wireless World*, Jan. 1970, p 12.

1-KHZ SINE-WAVE—Will drive any speaker up to 40 ohms, connected directly to power amplifier Q4 as shown. For 3.2-ohm speaker, use 22 ohms in series. Used with keyer. Volume control may be connected between Q3 and Q4 if desired; use 10K pot between emitter of Q3 and ground.—N. J. Nicosia, Solid-State Audio Oscillator-Monitor, *Ham Radio*, Sept. 1970, p 48–49.

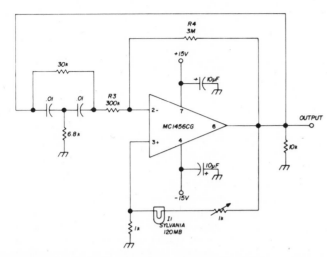

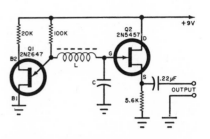

SINE-WAVE UJT—Generates essentially pure sine waves at output when charge-discharge period of C is equal to resonant frequency of L-C circuit. Will operate well up to 50 kHz.—F. H. Tooker, Getting to Know the UJT, *Popular Electronics*, April 1970, p 69–73.

BRIDGED-T OPAMP—Uses all-pass resistive voltage divider network in positive feedback path. Bridged-T network gives minimum negative feedback at its null frequency, corresponding to maximum gain and therefore giving oscillation at that frequency.—H. Olson, Resistance-Capacitance Oscillators, *Ham Radio*, July 1972, p 18–25.

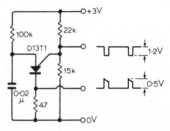

1 KHZ ON 3-V SUPPLY—Low-voltage relaxation oscillator has total current gain under 100 μA. Reliable triggering is achieved by holding anode gate at about 1.2 V and bringing anode up to about 2 V for triggering.—O. Greiter, PUT—The Programmable Unijunction, *Wireless World*, Sept. 1970, p 430–434.

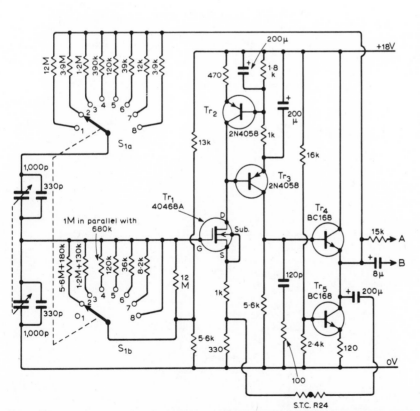

10 Hz TO 100 KHZ—Range is covered in eight square-root-of-10 steps, starting with 10—32 Hz for position 1 of S1, 32—100 Hz for position 2, etc. Uses conventional Wien-bridge circuit with frequency-selective positive feedback. Thermistor R24, with output of about 1.4 V rms, provides voltage-stabilizing negative feedback. Output A goes to frequency meter and square-wave shaper, while B goes to output attenuator.—A. J. Ewins, Wien-Bridge Audio Oscillator, *Wireless World*, March 1971, p 104–107.

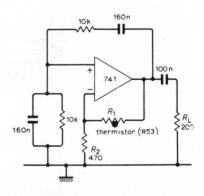

1-KHZ WIEN-BRIDGE—Thermistor in feedback loop stabilizes gain of opamp operating in noninverting mode. Output is 3-V p-p. Frequency and voltage change less than 0.1% for supply range of 4 to 15 V.—L. D. Thomas, Wien Bridge Oscillator, *Wireless World*, June 1972, p 275.

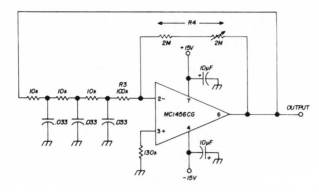

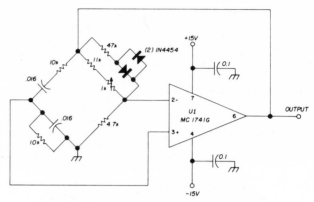

PHASE-SHIFT OPAMP—Uses three R-C sections, each with 60 deg of phase shift at operating frequency or 180 deg total, making input 360 deg out of phase with output so oscillation will start if opamp gain is adequate.—H. Olson, Resistance-Capacitance Oscillators, *Ham Radio*, July 1972, p 18–25.

WIEN-BRIDGE WITH DIODE CONTROL—Back-to-back diodes in negative feedback network operate instantaneously, with no time constant. Produces fairly pure sine-wave output.—H. Olson, Resistance-Capacitance Oscillators, *Ham Radio*, July 1972, p 18–25.

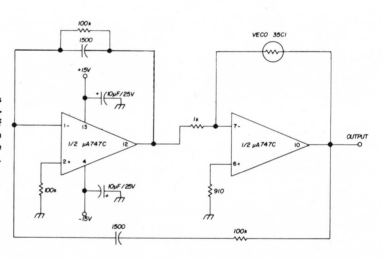

BAXANDALL WIEN-BRIDGE—Thermistor serves as nonlinear resistor in feedback control circuit. Circuit prevents input impedance of opamp from loading parallel R-C branch of bridge.—H. Olson, Resistance-Capacitance Oscillators, *Ham Radio*, July 1972, p 18–25.

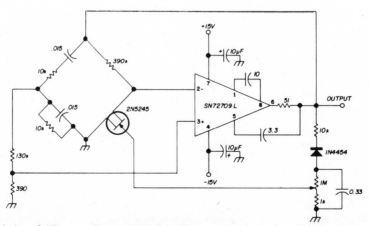

FET FEEDBACK CONTROL—Variation of Wien-bridge oscillator uses opamp as gain stage, with fet in one leg of bridge to serve as voltage-controlled resistance. Fet is operated with zero d-c voltage between source and drain, and very small a-c voltage.—H. Olson, Resistance-Capacitance Oscillators, *Ham Radio*, July 1972, p 18–25.

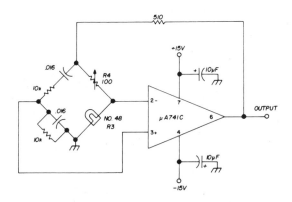

WIEN-BRIDGE OPAMP—RC network controls positive feedback, with negative feedback determined by bridge legs R3 and R4.—H. Olson, Resistance-Capacitance Oscillators, *Ham Radio*, July 1972, p 18–25.

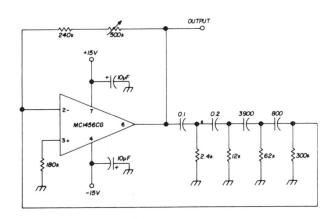

SERIES-C PHASE-SHIFT—Uses tapered multiple R-C sections to reduce loss in phase-shift network and give oscillation with less amplifier voltage gain. With four sections shown, gain can be less than 5, whereas three sections require gain of 29.—H. Olson, Resistance-Capacitance Oscillators, *Ham Radio*, July 1972, p 18–25.

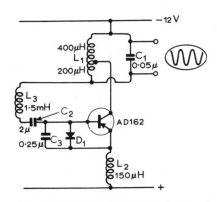

13 W AT 20 KHZ—Arrangement of coils allows transistor to swing above and below ground with virtually no limits, provided transistor limits are not exceeded during cutoff. Shorting output stops oscillation and current drops to zero until short is removed. D1 and C3 provide bias.—H. L. Armer, Sine-Wave Power Oscillator, *Wireless World*, Aug. 1970, p 402.

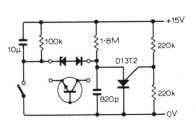

1-KHZ TONE PULSE—Combination of zener and diode can be replaced by emitter-collector path of small transistor. When switch is closed, 10-μF capacitor is changed, biasing diode off and causing oscillation at about 1 kHz until capacitor has discharged through 100K and zener conducts. Tone lasts about 1 s.—O. Greiter, PUT—The Programmable Unijunction, *Wireless World*, Sept. 1970, p 430–434.

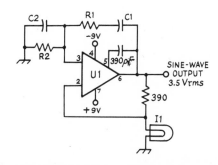

WIEN-BRIDGE IC—Uses μA709C opamp and No. 327 pilot lamp. If R1 and R2 are equal and C1 and C2 are equal, frequency in Hz is 1/6.28R1C1, where C1 is in farads and R1 in ohms.—G. Rothwell, A Simple and Inexpensive Audio Oscillator, *QST*, Sept. 1971, p 46.

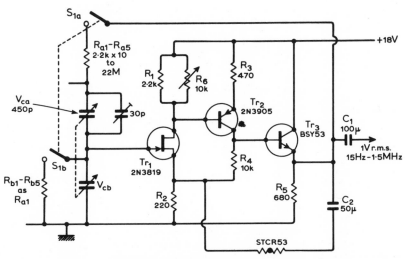

15 HZ–1.5 MHZ—Covered in five decade ranges, with output of 1 V and total harmonic distortion below 0.09%. Preset resistor R6, which determines quiescent operating bias conditons for direct-coupled amplifier, is initially adjusted to give half of supply voltage at emitter of TR3, and further adjusted for minimum distortion if Hewlett Packard 333A Distortion Analyzer is available.—V. R. Krause, F.E.T. Audio Oscillator, *Wireless World*, Sept. 1971, p 427.

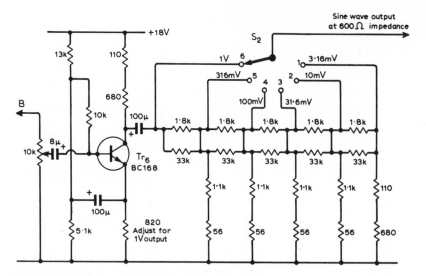

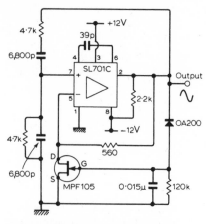

ATTENUATOR—Used at output of Wien-bridge audio oscillator (point B) to vary output voltage from 0 to 1 V in six square-root-of-10 steps with constant output impedance of 600 ohms.—A. J. Ewins, Wien-Bridge Audio Oscillator, *Wireless World*, March 1971, p 104–107.

5-KHZ CONSTANT-AMPLITUDE—Frequency variation is only 0.02% for 30% change in supply voltage. Output of Wien bridge oscillator is fed back to gate of fet after rectification and filtering, to change dynamic resistance of fet so as to hold output constant.—A. Basak, Constant Amplitude Oscillator, *Wireless World*, Nov. 1969, p 530.

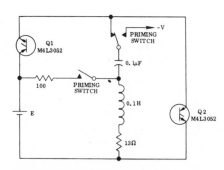

SINUSOIDAL RELAXATION—Simple series R-L-C circuit with single d-c source and two pnpn diodes produces sine-wave output voltage across capacitor even though it operates on relaxation principle. When voltage across Q1 reaches its forward breakdown value of 8 V, it conducts and R-C-L circuit oscillates for a half-period. At end of period, current in Q1 is zero and it blocks, reestablishing initial condition except that capacitor voltage is now reversed and higher. Q2 then breaks down similarly and R-L-C circuit oscillates for second half-period. Diodes continue conducting on alternate half-period to establish continuous oscillation. Capacitor can be charged initially with either priming switch. Values shown give about 150 Hz.—Z. D. Farkas, Avalanche Breakdown Sinusoidal Oscillator, *Proc. IEEE*, Jan. 1970, p 167–168.

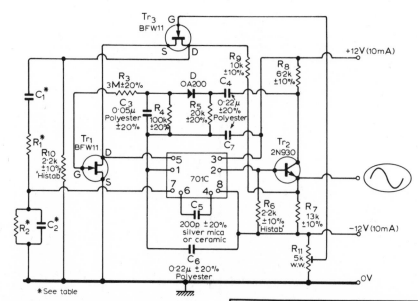

20 HZ—70 KHZ AT 70 C—Provides constant-amplitude sine-wave output at high ambient temperatures with good frequency stability. Uses Wien half-bridge with positive feedback from output of IC amplifier. Temperature range of 25–70 C produced amplitude and frequency changes of only 2%.—P. Williams, Sinusoidal Oscillator for High Temperatures, *Wireless World*, July 1970, p 332.

Operating frequency	Value of C_1, C_2	Value of R_1, R_2
20 Hz	0.068 μF polyester (with 0.022 μF shunting gate of Tr_1 to earth)	120 k Ω ± 5%
500 Hz	0.0027 μF silver mica	120 k Ω
5 kHz	0.0027 μF silver mica	12 k Ω
25 kHz	500pF silver mica	12 k Ω
70 kHz	120pF silver mica	12 k Ω (all 'Histab')

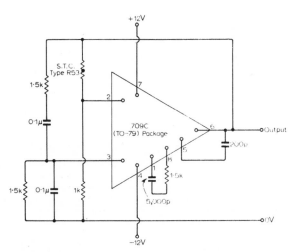

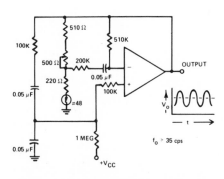

1-KHZ WIEN OPAMP—Bridge is connected between output of opamp and its noninverting input. Thermistor and resistor connected between output and inverting input limit output amplitude so as to make output sinusoidal. Delivers 3.5 V p-p to 100-ohm load. Entire audio range can be covered with switched capacitors and two-gang pot.—D. W. J. Bly, Oscillator Using Operational Amplifier, *Wireless World*, May 1969, p 234.

SINE-WAVE WIEN-BRIDGE—Uses current-differencing single-supply opamp operating over supply voltage range of 3.5 to 35 V, and providing p-p output voltage swing only 1 V less than supply voltage. Special Motorola opamp used has four amplifiers on single chip.—T. M. Frederiksen, W. F. Davis, and D. W. Zobel, A New Current-Differencing Single-Supply Operational Amplifier, *IEEE Journal of Solid-State Circuits*, Dec. 1971, p 340–347.

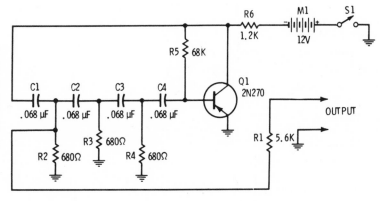

2.1-KHZ TONE GENERATOR—Basic phase-shift oscillator produces clean sine wave and starts reliably, even with battery voltage well below recommended 12 V. Can be used for testing audio equipment or incorporated in musical toy. To get 1.1-kHz tone, change all capacitors to 0.1 μF.—R. M. Brown and T. Kneitel, "101 Easy Test Instrument Projects," Howard W. Sams, Indianapolis, IN, 1968, p 112–113.

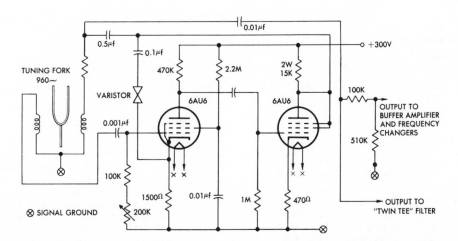

960-HZ TUNING-FORK OSCILLATOR—Used as time-base generator for missile guidance system. Book contains many other basic circuits for missiles but does not give component values. Available from Supt. of Documents at $4 (Stock No. 0270-0231).—"Guided Missiles Fundamentals," Dept. of the Air Force, Washington, DC, July 1972, AF Manual 52-31, p 6–54.

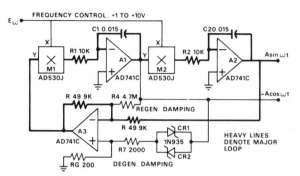

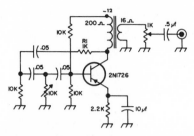

100–1,000 HZ VOLTAGE-CONTROLLED—D-c controlled voltage of 1 to 10 V changes two-phase sine-wave output over 10:1 frequency range. Zener reference diodes hold output voltage constant within 1 dB of 7 V rms over entire frequency range. Integrators A1 and A2 serve with unity-gain inverter A3 as negative feedback loop. Effective time constants of integrators are varied by multipliers M1 and M2, to increase conductance of R1 and R2 as control voltage is increased, thereby decreasing time constant and increasing natural frequency. Distortion for sine output is only 0.64% at 100 Hz and 0.18% at 1,000 Hz.—Frequency Modulator, *Analog Dialogue*, Vol. 5, No. 5, p 14–15.

1,000-HZ MODULATOR—Can be used for modulating r-f oscillators up to 1,300 MHz. Frequency can be trimmed with 10K pot, while 1K pot controls amount of modulation.—B. Hoisington, The 1215 Transistor Superhet, *73*, Jan. 1966, p 90–94.

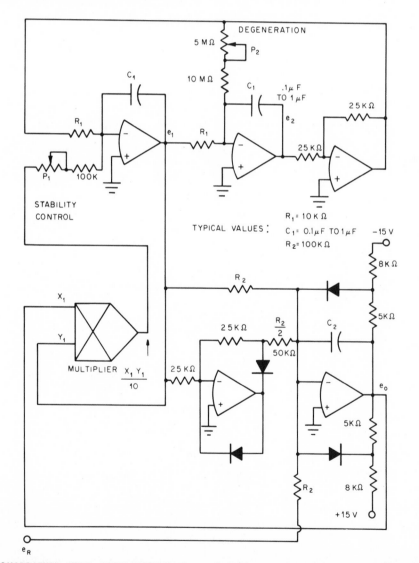

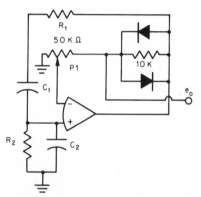

ONE-OPAMP WIEN BRIDGE—Requires only single opamp, keeping component cost at minimum, yet distortion can be kept in range of 1 to 5% by adjusting carefully. Circuit has high output impedance and changes in loading affect output amplitude, so should be used with fixed load or with buffer. R1 is same as R2 and C1 same as C2, with values determining frequency. Matching of diodes minimizes distortion.—J. G. Graeme, G. E. Tobey, and L. P. Huelsman, "Operational Amplifiers," McGraw-Hill Book Co., New York, 1971, p 383–385.

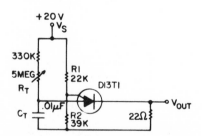

QUADRATURE WITH GAIN CONTROL—Uses Burr-Brown 3402 or equivalent opamps to form quadrature oscillator in which output amplitude is sensed and used to control loop damping. Book gives design equations. Value of C2 is critical; it must be large enough to provide adequate low-pass filtering for rectified current proportional to e1, but not so large as to make control loop go into slow limit-cycle oscillation. Designed for subaudio frequencies.—J. G. Graeme, G. E. Tobey, and L. P. Huelsman, "Operational Amplifiers," McGraw-Hill Book Co., New York, 1971, p 388–391.

PUT RELAXATION—Programmable unijunction transistor serving as trigger in low-frequency relaxation oscillator is simply small thyristor having anode gate. Device remains off until anode voltage exceeds gate voltage by one diode forward voltage drop, at which it turns on. Frequency is controlled by RT. Report covers design of circuit.—J. M. Reschovsky, "Design of Triggering Circuits for Power SCR'S," General Electric, Semiconductor Products Dept., Auburn, NY, No. 200.54, 1970, p 2–3.

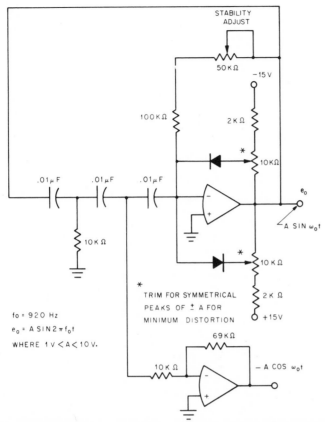

$f_0 = 920$ Hz

$e_o = A \sin 2\pi f_o t$

WHERE $1V < A < 10V$.

* TRIM FOR SYMMETRICAL PEAKS OF ± A FOR MINIMUM DISTORTION

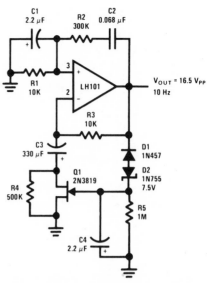

WIEN SINE-WAVE OSCILLATOR—Amplitude-stabilized bridge oscillator provides high-purity sine-wave output down to low frequencies. Wien bridge is used as positive feedback element around amplifier, so oscillation occurs at frequency at which phase shift is zero. Additional negative feedback is provided to make loop gain unity at this frequency.—W. S. Routh, "Applications Guide for Op Amps," National Semiconductor, Santa Clara, CA, 1969, AN-20.

920-HZ PHASE-SHIFT—Requires only two opamps, one for generating sine term and other for cosine term. Either single-ended input or differential-input opamps may be used. Three 0.01-μF capacitors must be matched, so circuit is best suited for fixed-frequency use. Design equations are given.—J. G. Graeme, G. E. Tobey, and L. P. Huelsman, "Operational Amplifiers," McGraw-Hill Book Co., New York, 1971, p 391–393.

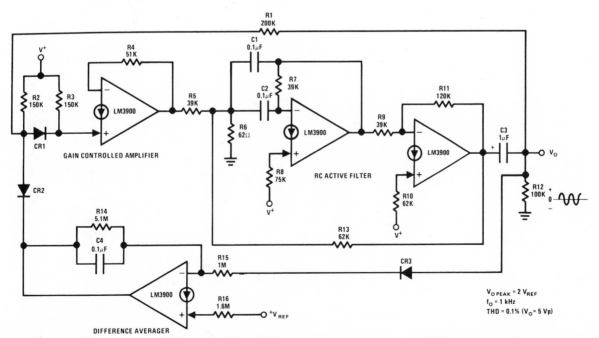

$V_{O PEAK} = 2 V_{REF}$
$f_o = 1$ kHz
THD = 0.1% ($V_O = 5$ Vp)

1-KHZ SINE-WAVE—Requires four Norton current-differencing opamps, with two forming R-C active filter providing overall noninverting phase characteristic, with noninverting gain-controlled opamp around filter to form oscillator. Resulting sine-wave output voltage is sensed and regulated as average value is compared to d-c reference voltage with differential averaging circuit of fourth opamp. Peak output voltage is twice reference voltage, essentially independent of temperature and supply voltage.—T. M. Frederiksen, W. M. Howard, and R. S. Sleeth, "The LM3900—A New Current-Differencing Quad of ± Input Amplifiers," National Semiconductor, Santa Clara, CA, 1972, AN-72, p 19.

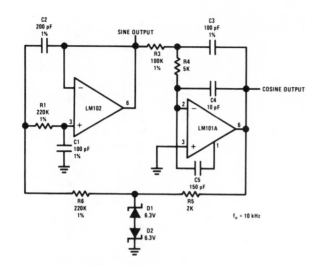

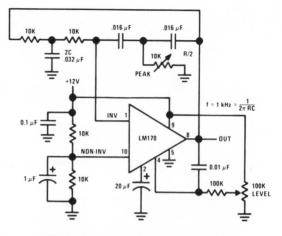

10-KHZ SINE-WAVE—With opamps shown, having improved high-frequency response, operating frequency of oscillator can be as high as 100 kHz hy changing resistor and capacitor values appropriately. LM102 is connected as two-pole low-pass active filter and LM101A as integrator. Zener clamps stabilize output amplitude.—R. J. Widlar, "IC Op Amp Beats FETs on Input Current," National Semiconductor, Santa Clara, CA, 1969, AN-29, p 10 and AN-31, p 5.

TWIN-T 1-KHZ WITH AGC—Output amplitude is regulated by feeding detected output voltage back to gain-control input of opamp. External pot sets agc threshold. Level may be controlled remotely.—R. A. Hirschfeld, "AGC/Squelch Amplifier," National Semiconductor, Santa Clara, CA, 1969, AN-11.

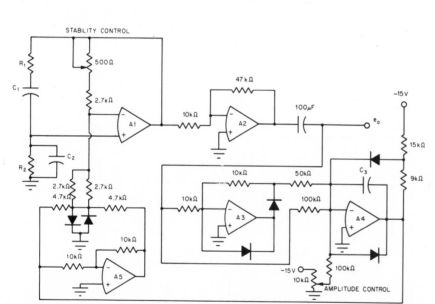

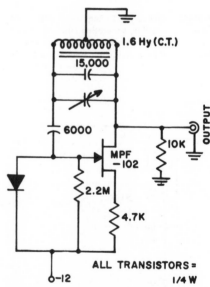

FIVE-OPAMP WIEN BRIDGE—R1, C1, R2, and C3 form Wien bridge with opamp A1 connected as oscillator, and determine its frequency. Amplified output of A2 is sensed by absolute-value circuit using next two opamps, with A4 acting as error integrator that stabilizes only when absolute value of input equals reference amplitude. Diode bridge used with A5 varies negative feedback of A1 to achieve high stability. Frequency range is 10 to 10,000 Hz.—J. G. Graeme, G. E. Tobey, and L. P. Huelsman, "Operational Amplifiers," McGraw-Hill Book Co., New York, 1971, p 383–384.

AUDIO FET—Provides sine-wave output that does not change more than 0.5 dB in amplitude or 1 Hz in frequency when supply voltage varies between 7 and 15 V. Uses Hartley configuration. Is not pure class A since junction fet is cut off during part of cycle. Operates at 1 kHz.—N. C. Hekimian, An Amplitude and Frequency Stable Sinusoidal Field Effect Transistor Oscillator, *Frequency Technology*, Oct. 1969, p 19–20.

CHAPTER 82
Oscillator Circuits—R-F

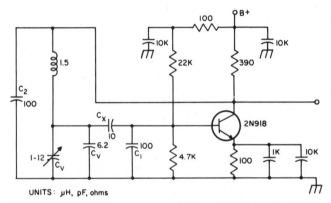

UNITS: μH, pF, ohms

30-MHZ VACKAR VFO—Provides tuning range of 2.5:1 with inherent stability and essentially constant output amplitude. Article analyzes design and performance of circuit originally developed in Czechoslovakia by J.

Vackar in 1949. Example shown was designed to operate as transmitter vfo.—G. B. Jordan, The Vackar VFO, A Design To Try, *The Electronic Engineer*, Feb. 1968, p 56–59.

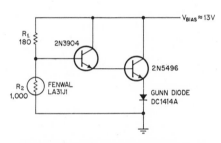

FAST WARMUP OF GUNN—Thermistor in voltage-biasing network speeds stabilization of 8-GHz Gunn-diode oscillator, and reduces frequency drift due to temperature changes. After 4-min warmup, stability is 4 ppm per deg C and drift down to —0.03 MHz per deg C.—T. V. Seling, Thermistor Stabilizes Gunn Oscillator, *Electronics*, Oct. 26, 1970, p 88.

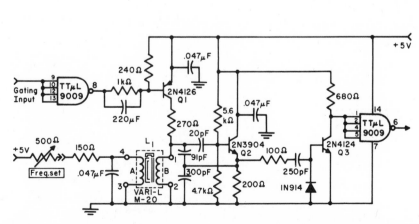

GATED 8.5 TO 12 MHZ—Output always starts in same phase with respect to gating signal. Has excellent frequency stability. Frequency can be changed remotely with 500-ohm pot. Q3 squares up signal and feeds it

to half of dual four-input buffer IC.—W. F. Harrick, Gated Oscillator Has Remote Frequency Control, *The Electronic Engineer*, Feb. 1969, p 87.

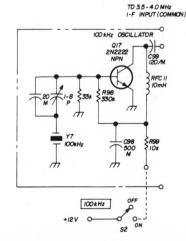

100-KHZ MARKER—Developed for amateur receiver. Article covers construction of entire set.—M. A. Chapman, Five-Band Solid-State Communications Receiver, *Ham Radio*, June 1972, p 6–21.

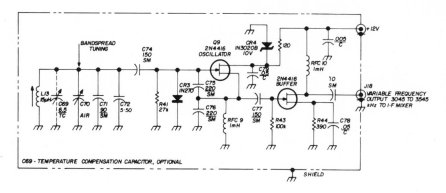

3.045 TO 3.545 MHZ—Has excellent stability, as required for amateur-band receiver. Article covers construction of entire set.—M. A. Chapman, Five-Band Solid-State Communications Receiver, *Ham Radio*, June 1972, p 6–21.

7–7.3 MHZ SEILER—Uses fet oscillator Q1, fet buffer-amplifier Q2, and emitter-follower Q3. Article also covers Vackar oscillator for same frequency.—D. R. Nesbitt, Miniature Solid-State Variable-Frequency Oscillator, *Ham Radio*, Dec. 1971, p 8–11.

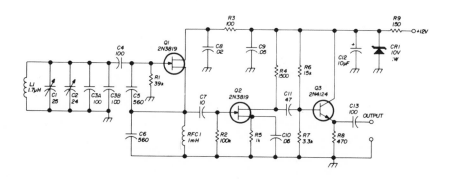

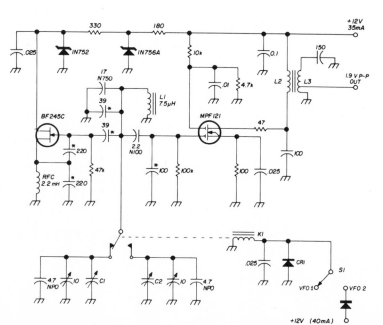

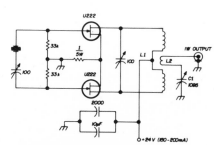

C1 1095 pF (three-gang broadcast variable)

L1 60 turns no. 26 enameled, closewound on 1¼" form, center-tapped. Leave space for L2 at center

L2 15 turns no. 26 enameled, closewound between two windings of L1

5–5.5 MHZ VFO—Uses Seiler oscillator in which transistor and tank circuit are very lightly loaded, giving excellent frequency stability. Relay is used to change bands, with 150-kHz overlap of bands. If relay is controlled by mvbr, two channels can be monitored simultaneously. C1 and C2 are three-section 3.2–18.5 pF variable, parallel-connected. Article gives coil-winding data. Mica capacitors have asterisks. Output is taken from frequency-determining circuit, at which point second harmonic is 50 dB down.—V. Aumala, High-Stability Variable Frequency Oscillator, *Ham Radio*, Jan. 1972, p 27–31.

160-METER PUSH-PULL FET—Output is better than 1 W. Series variable capacitor in gate circuit gives helpful frequency spread for c-w operation.—E. Noll, Power Fets, *Ham Radio*, April 1971, p 34–39.

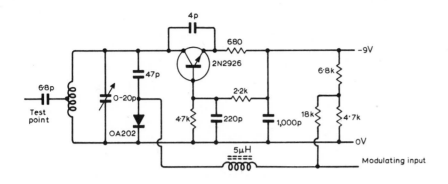

VHF F-M OSCILLATOR—Simple circuit oscillates readily. Tuning coil is 4 turns of No. 16, wound 0.4 inch long on 0.4-inch form.—W. H. H. Kelk, Simple V.H.F./F.M. Oscillator, *Wireless World*, Oct. 1971, p 512.

470 KHZ MODULATED WITH 220 HZ—Output is 60-mV p-p with about 90% modulation by pure sine wave. Uses ordinary i-f transformer from transistor radio.—K. E. Potter, A.M. Oscillator, *Wireless World*, Dec. 1969, p 556.

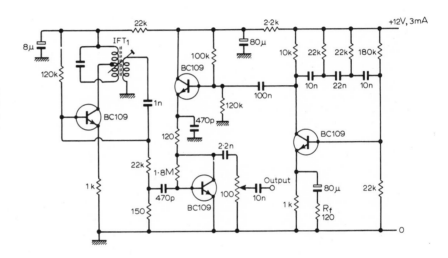

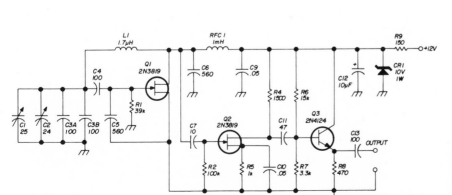

7–7.3 MHZ VACKAR—Uses fet oscillator Q1, fet buffer-amplifier Q2, and emitter-follower Q3. Article also covers Seiler oscillator for same frequency.—D. R. Nesbitt, Miniature Solid-State Variable-Frequency Oscillator, *Ham Radio*, Dec. 1971, p 8–11.

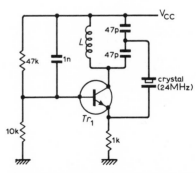

24 MHZ WITH LOW-ACTIVITY CRYSTAL—Simple circuit works well on either overtone or fundamental operation. Output may be taken directly from emitter resistor or by means of link coupling. Supply is 12 V. TR1 is 2N706 or similar. Choose L to resonate with series capacitors at desired output frequency.—L. V. Gibbs, Overtone Oscillator, *Wireless World*, April 1972, p 186.

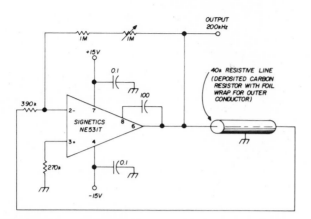

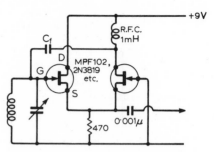

200-KHZ DISTRIBUTED R-C PHASE-SHIFT—Equivalent to length of coax with high-resistance center conductor. Frequency is about 200 kHz. Requires opamp with high slew rate.—H. Olson, Resistance-Capacitance Oscillators, *Ham Radio*, July 1972, p 18–25.

1 MHZ TO VHF—Circuit is basically twin-triode cathode-coupled oscillator that provides output over very wide frequency range merely by changing constants of tank circuit. Good compromise value for feedback capacitor Cf is 30 pF.—L. F. Heller, F.E.T. 'Two Terminal' Oscillator, *Wireless World*, Sept. 1969, p 409.

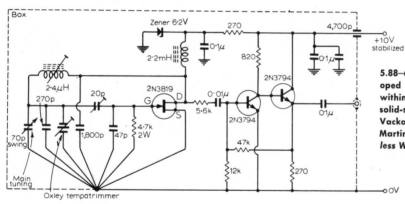

5.88–6.38 MHZ FET VACKAR—Circuit developed for amateur radio remains constant within 2 Hz after 1-min warmup. Circuit is solid-state version of highly popular Czech Vackar vacuum-tube oscillator circuit.—P. Martin, High-Stability F.E.T. Oscillators, *Wireless World*, Feb. 1970, p 92.

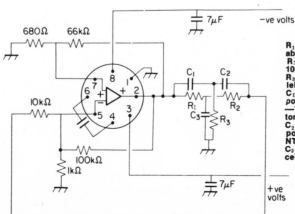

TWIN-T 100-KHZ—With SL701C opamp set up to give gain of 100, frequency stability is within 3 parts per million per deg C over range of 25 to 50 C. Elements of twin-T network should be matched within 2%.—J. R. Jones, High-Stability RC Oscillator for Thin-Film Circuits, *Design Electronics*, Feb. 1970, p 47.

R_1 4544Ω series with 250Ω variable resistor.
R_2 4622Ω (4.5kΩ nominal + 100Ω nominal).
R_3 2310Ω (4.7kΩ nominal paralleled with a 4.5kΩ nominal).
C_1 334.4 pF (322.8 pF NPO porcelain ceramic + 11.6 pF —1500 ppm/deg C NTC capacitor)
C_2 334.3 pF (322.9 pF NPO porcelain ceramic + 11.4 pF NTC).
C_3 669.1 pF (645.7 pF NPO porcelain ceramic + 23.7 pF NTC).

SCR RELAXATION—Provides damped sine-wave output having duration of 150 μs, repeated at frequency determined by value of R4. Will oscillate with supply voltages between 5 and 35 V.—R. D. Clement and R. L. Starliper, Relaxation Oscillators—Old and New, *Electronics World*, May 1971, p 36–37.

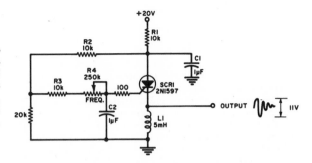

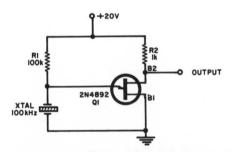

99.925-KHZ CRYSTAL RELAXATION—Output frequency of basic ujt relaxation oscillator is held constant within 1 Hz by replacing charging capacitor with 100-kHz crystal. Output is 2-V p-p distorted sine wave. With 1-MHz crystal, output is 999.663 kHz within 1 Hz.—R. D. Clement and R. L. Starliper, Relaxation Oscillators—Old and New, *Electronics World*, May 1971, p 36–37.

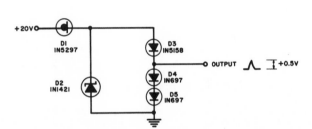

DIODE RELAXATION—Provides pulse output at about 40 kHz over supply range of 10 to 35 V, using only diodes. When power is applied, junction capacitance of zener D2 is charged with constant current of 1 mA from field-effect diode D1. Charging continues until forward breakover voltage of four-layer diode D3 is reached. D3 then fires, discharging D2 and producing small output pulse across conventional diodes D4 and D5.—R. D. Clement and R. L. Starliper, Relaxation Oscillators—Old and New, *Electronics World*, May 1971, p 36–37.

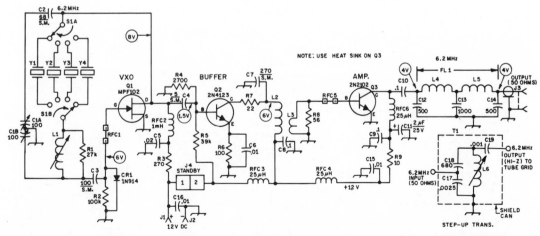

6.1-MHZ VARIABLE CRYSTAL—Approximates performance of frequency synthesizer at much lower cost. Suitable for either transmitter or receiver control. Based on fact that crystals are rubbery and can be changed in frequency over certain range with L1 and C1. With four crystals as shown, operating range was 146 to 147 MHz because some crystals provide swings in excess of 300 kHz although average swing is about 225 kHz. Article gives coil data. RFC1 and RFC5 are ferrite beads. Use crystals around 6.1 MHz to provide desired coverage at 146 MHz.—D. DeMaw, Some Practical Aspects of VXO Design, *QST*, May 1972, p 11–15 and 39.

400-KHZ 60-KW INDUCTION HEATER—Circuit using ceramic-envelope industrial triode provides high load efficiency and complete freedom from parasitics even at severe overloads, despite absence of chokes. Supply voltage is 6 to 8 kV, depending on which triode *is* used. Article gives coil construction data. All coils are water-cooled.—D. E. Nightingale, A 400 kHz Induction Heater of Advanced Design for Powers up to 60 kW, *Mullard Technical Communications*, July 1968, p 146–149.

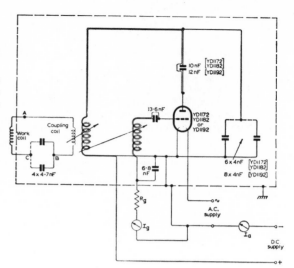

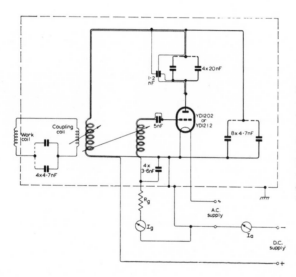

300-KHZ 120–240 KW INDUCTION HEATER—With YD1212 ceramic triode, circuit gives output power over 240 kW, and half this with YD1202. Has high load efficiency and no parasitics. Article covers construction of coils, all of which are water-cooled. Supply voltage is 14 kV.—D. E. Nightingale, A 300 kHz Induction Heater of Advanced Design for Powers up to 120 kW/240 kW, *Mullard Technical Communications*, July 1968, p 150–156.

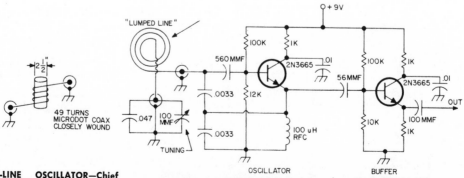

21-MHZ LUMPED-LINE OSCILLATOR—Chief feature is frequency stability achieved by using coax to simulate series-tuned L-C tank. Emitter-buffer follower operated as class A amplifier provides good isolation from following circuits. Power output is purposely kept around 100 mW to minimize thermal problems affecting frequency.—I. M. Gottlieb, High Stability "Lumped Line", *73*, Nov. 1972, p 69 and 72–74.

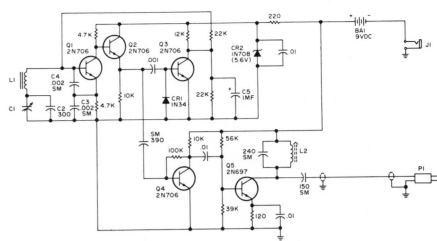

3.5-MHZ VFO—Uses Clapp circuit which is amplitude-limited to give low and constant output level. Q2 isolates oscillator Q1 from Q3 and output amplifiers. Q3 provides bias for oscillator transistor. Article gives coil data for 40-meter band. C1 is 100-pF maximum. Thermal drift is negligible.—C. Sondgeroth, A Transistorized VFO, 73, Dec. 1972, p 112–113, 115, and 120.

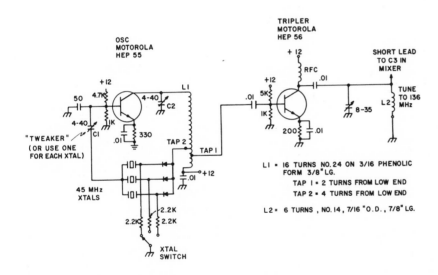

L1 = 16 TURNS NO. 24 ON 3/16 PHENOLIC FORM 3/8" LG.
TAP 1 = 2 TURNS FROM LOW END
TAP 2 = 4 TURNS FROM LOW END

L2 = 6 TURNS, NO. 14, 7/16 "O.D., 7/8" LG.

45-MHZ WITH TRIPLER—Used in narrow-band 2-meter amateur f-m receiver. Crystals are switched electronically by applying d-c voltage to appropriate series diode. Article includes circuits for fet r-f stage, fet mixer, i-f, audio, and squelch. Any number of crystals may be used.—B. Hoisington, Single Conversion Two Meter FM Receiver, 73, Dec. 1972, p 23–27.

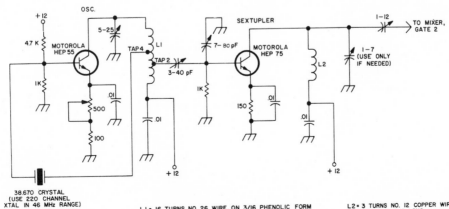

L1 = 16 TURNS NO. 26 WIRE, ON 3/16 PHENOLIC FORM
TAP AT 2 FOR OUTPUT
TAP AT 4 FOR FEEDBACK
(TUNE TO 38.6 OR 46 WITH CHANNEL XTAL)

L2 = 3 TURNS NO. 12 COPPER WIRE
5/16 O.D., 5/8 LONG

232-MHZ OSCILLATOR-SEXTUPLER—Used in 220-MHz receiver containing high-frequency 3N200 fet's in r-f stage and mixer. Article covers construction and shows parts layouts required for proper shielding.—B. Hoisington, 220-MHz: Front End Using Fet's, 73, Nov. 1972, p 45–46 and 48–50.

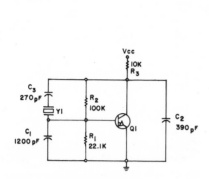

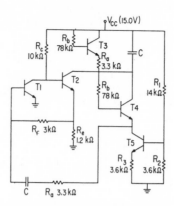

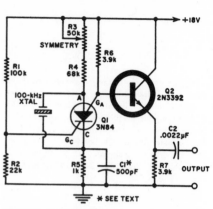

307-KHZ CRYSTAL—Article presents mathematical analysis of how changes in supply voltage affect frequency and phase shift. Q1 is 2N910. For 5% change in supply voltage, frequency changes only 0.0091%.—W. A. Magee, Crystal Oscillator Frequency Variations with Changes in Supply Voltage, *Frequency Technology*, Oct. 1969, p 24–26.

130-KHZ COMPENSATED WIEN OSCILLATOR—Circuit is current-amplifier Wien-type oscillator with series Miller-type compensation, presented as example of method developed in article to achieve temperature compensation in R-C oscillator using IC construction. Design equations are given. Output is 2-V p-p. C is 325 pF.—R. F. Adams and D. O. Pederson, Temperature Sensitivity of Frequency of Integrated Oscillators, *IEEE Journal of Solid-State Circuits*, Dec. 1968, p 391–396.

100-KHZ CRYSTAL SCS—Gives approximately rectangular output waveform. Draws about 5 mA. Choose value of C1 for optimum squareness.—F. H. Tooker, SCS Crystal-Controlled Oscillator, *Electronics World*, Oct. 1969, p 74.

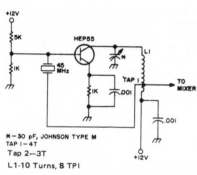

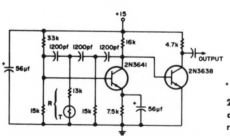

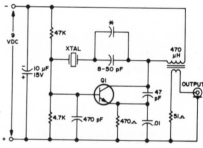

45-MHZ CRYSTAL—Used with 5–6 MHz vfo to produce 50–51 MHz at output of mixer without usual multiplication of frequency drift.—B. Hoisington, Solid-State 6 Meter Crystal-Het-VFO, *73*, Oct. 1972, p 29–30 and 32–35.

TEMPERATURE COMPENSATION—Circuit illustrates use of thermistor T for temperature compensation of phase-shift oscillator.—H. Olson, Nonlinear Resistors, *73*, March 1972, p 101–108.

2–30 METER CRYSTAL—Circuit is sufficiently active to handle crystals over wide frequency range. Output is 18-V p-p across 470-μH choke, so two turns wound around choke provide adequate loose-coupling to receiver when oscillator is used for alignment. Q1 is vhf pnp, such as GE-9 or ES19. Add fixed capacitance across 8–50 pF trimmer in 50-pF steps if trimmer will not zero-beat.—*Circuits*, *73*, May 1972, p 110.

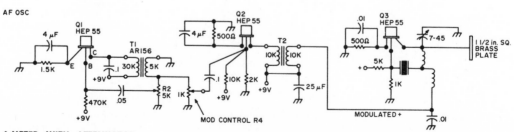

MODULATED 6-METER WITH ATTENUATOR—Serves as signal generator providing stable variable-strength signal having maximum of 1 mV and dropping gradually to true zero as circuit is moved inside 24-inch length of 4¼″ × 2⅛″ waveguide. Circuit consists of crys-tal-controlled 50-MHz crystal oscillator, a-f oscillator, and a-f modulator. Receiver under test is plugged into jack going to brass pickup plate inside closed end of waveguide. Wood or cardboard strip on which generator is mounted projects out of open end and can be calibrated to indicate receiver sensitivity in tenths of a microvolt.—B. Hoisington, Low Cost Oscillator and Infinite Attenuator, *73*, Sept. 1970, p 32–34.

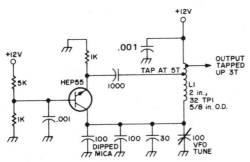

5–6 MHZ VFO—Used with 45-MHz crystal oscillator to produce 50–51 MHz at output of mixer without usual multiplication of frequency drift.—B. Hoisington, Solid-State 6 Meter Crystal-Het-VFO, 73, Oct. 1972, p 29–30 and 32–35.

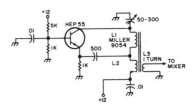

1.785 MHZ—Article gives instructions for adding L2 and L3 on standard oscillator coil L1 used as local oscillator for converter changing 1.65 MHz to 135 kHz.—W. Hoisington, I-F Filter Converter AVC, 73, May 1970, p 62–64, 66–70 and 72.

40- AND 80-METER VFO—Uses separate tuned circuits for each band. Designed primarily for c-w, so tuning range is intentionally restricted to give reasonable bandspread. L1 is slug-tuned 0.68–1.25 μH and L2 is 2.2–4.1 μH. Q1 and Q2 are MPS6514. RFC1 is three Amidon ferrite beads threaded on wire. Article covers construction and debugging.—D. DeMaw, Building a Simple Two-Band VFO, QST, June 1970, p 20–24.

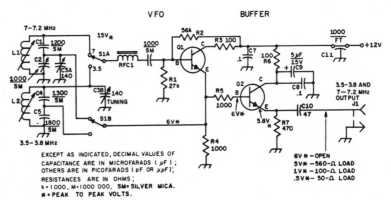

EXCEPT AS INDICATED, DECIMAL VALUES OF CAPACITANCE ARE IN MICROFARADS (μF); OTHERS ARE IN PICOFARADS (pF OR μμF); RESISTANCES ARE IN OHMS; k=1000, M=1000 000. SM=SILVER MICA. ∗=PEAK TO PEAK VOLTS.

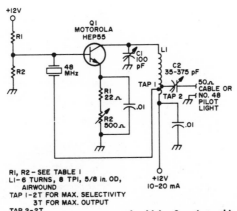

48-MHZ CRYSTAL—Provides over 150-mW r-f output with stability required for multiplying to 1,296 MHz. R2 can be 1K to 5K and R1 should be five times this value. Circuit is excellent for driving tripler.—B. Hoisington, The 1296'er Ideal Crystal Oscillator, 73, Aug. 1972, p 115–116.

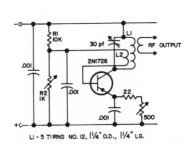

40–75 MHZ—Developed as local oscillator for 6-meter superhet receiver. L2 has 1 or 2 turns for feedback, wound within first turn at collector end of L1. Will oscillate on supply as low as 2 V.—B. Hoisington, The 1215 Transistor Superhet, 73, Jan. 1966, p 90–94.

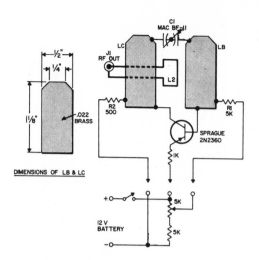

DIMENSIONS OF LB & LC

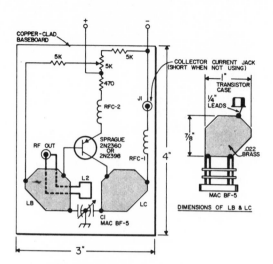

DIMENSIONS OF LB & LC

850–1,100 MHZ—Uses two half-wave strap lines, one for base and one for collector of transistor. Attach R1 and R2 to points having lowest r-f.—B. Hoisington, The 1215 Transistor Superhet, 73, Jan. 1966, p 90–94.

1,200–1,300 MHZ—Construction of r-f choke must be determined experimentally, using various lengths of wire or metal straps. Try 1¼-inch insulated wire for RFC-2. Can be used as test oscillator or for antenna testing. Article covers tuning procedure. Output jack should be mounted on insulating material, not on metal panel.—B. Hoisington, The 1215 Transistor Superhet, 73, Jan. 1966, p 90–94.

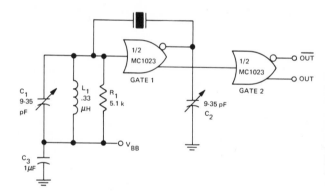

50–100 MHZ CRYSTAL—Output frequency depends on tuning of tank L1-C1 and choice of crystal. VBB is 1.2-V d-c. Overtone operation is accomplished by adjusting tank to desired crystal overtone. Circuit design insures that oscillator will always start at correct overtone. Provides square-wave output suitable for use as clock driver, frequency-marker generator, frequency synthesizer, or frequency comparator. Gives choice of inverted or noninverted outputs.—C. Byers and B. Blood, "IC Crystal Controlled Oscillators," Motorola Semiconductors, Phoenix, AZ, AN-417A, 1972.

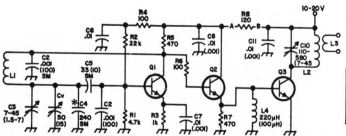

TO COLLECTOR-MODULATE FINAL:
OMIT R8
APPLY MODULATED VOLTAGE AT B
APPLY 9 TO 10V AT A

*C4 NOT USED IN 6M VERSION

COIL DATA		
COIL	80M	6M
L1	35T NO. 22 ON T68-2 CORE	9T NO. 22 ON T50-10 CORE
L2	SAME AS L1	SAME AS L1
L3	3 TO 4T NO.22 ON COLD END OF L2	SAME AS 80 M

6- AND 80-METER VFO—Simple circuit has excellent stability on both bands for which coil data is given, and will operate on other bands with coils determined by experimentation. Transistors can all be MPS706, but higher output is possible with 2N2270 or 2N3053 for Q3.—Circuits, 73, Feb. 1973, p 143.

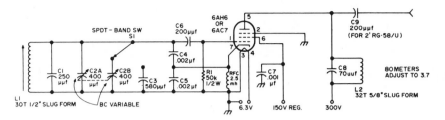

80- AND 40-METER VFO—Variable-frequency oscillator starts at 160 meters and doubles in plate circuit to give 80 meters. Switch S1 inserts C3 to give full bandspread when doubling in transmitter to 40 meters.—H. Presley, VFO Circuit, 73, June 1970, p 36 and 38.

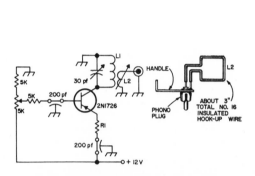

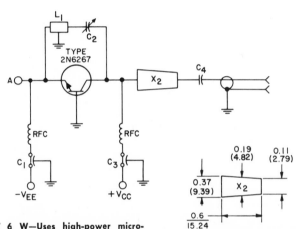

140–300 MHZ—Uses strap-line circuit. Higher output will be obtained at 300-MHz end of range with 2N2360 transistor. L1 is 1¼″ × 2⅜″ strip of brass mounted above L2, and R1 is 1K. Article includes dimensioned sketch of strap line.—B. Hoisington, The 1215 Transistor Superhet, 73, Jan. 1966, p 90–94.

1.7 GHZ AT 6 W—Uses high-power microwave transistor with resonant feedback loop placed in collector-emitter circuit. Power output is 6 W with 28-V supply, with frequency stability better than 0.1% for voltage or current excursions of up to 25%. C1 and C3 are Allen-Bradley SMFB-A1 Filtercons. C3 is 0.3–3.5 pF Johanson 4700. C4 is 300-pF ATC-100.

L1 is 1-inch miniature 50-ohm cable. RFC is three turns No. 32 3/16 inch long and 1/16th inch ID. X2 is exponentially tapered 13-mil thick Teflon-Kapton double-clad circuit board.—G. Hodowanec, "High-Power Transistor Microwave Oscillators," RCA, Somerville, NJ, AN-6084, 1972.

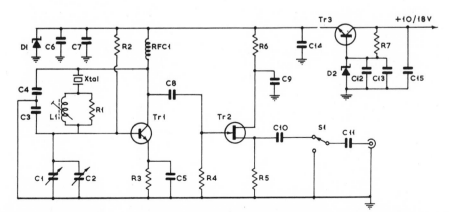

8-MHZ TUNABLE CRYSTAL—Tuning components L1 (40 turns No. 32 g. enamel on ⅜-inch slug-tuned form), C1, and C2 can make fundamental frequency of crystal swing down up to 40 kHz, giving change of 720 kHz after multiplication for 2-meter ham band. This covers practically all frequencies needed by operators in this band. Article includes circuit for extra pentode for amplifier stage needed to drive tube transmitter. Series regulator TR3 provides stabilized 9 V.—G. P. Adams, VXO for Two Metres, The Short Wave Magazine, Oct. 1972, p 484–485.

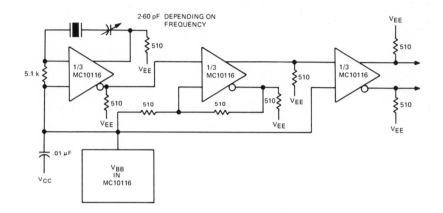

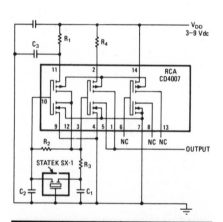

1—20 MHZ CRYSTAL—Operates on fundamental frequency of crystal selected, without tank circuit. Provides noninverting output. VBB is 1.2 V, available from IC. VEE is —5.2 V. Second section of IC is connected as Schmitt trigger driving third section connected as buffer, to give good square-wave output suitable for use as clock driver.—C. Byers and B. Blood, "IC Crystal Controlled Oscillators," Motorola Semiconductors, Phoenix, AZ, AN-417A, 1972.

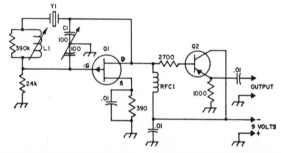

6-MHZ VARIABLE CRYSTAL OSCILLATOR—Designed for use with either 6 or 8 MHz crystal in vhf transmitter. Crystals cut especially for variable-frequency use will give greater frequency swing. L1 is 16–29 μH slug-tuned. Q1 is Siliconix U110 and Q2 is 2N3251. RFC1 is 2.5 mH.—W. M. Rowe, Jr., Transistor VXO for VHF Transmitters, QST, Jan. 1970, p 37.

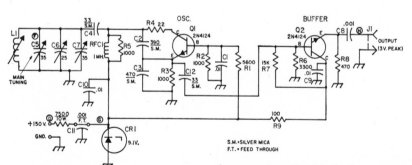

1.7–9 MHZ VFO—Range is covered in five bands, using Miller plug-in coils for L1 as specified in table in article. Output is 3-V p-p into low-impedance load. Use VR-150 or 0A2 when operating off 150-V d-c from receiver. Can also operate from 12-V source if 7,500-ohm resistor is changed to 100 ohms.—A General-Purpose V.F.O., QST, Sept. 1968, p 40–41 and 158.

TYPICAL COMPONENT VALUES						
Frequency (kHz)	C_1 (pF)	C_2 (pF)	R_1 (kΩ)	R_2 (MΩ)	R_3 (kΩ)	R_4 (kΩ)
10–40	20	5	100	22	500	10
40–100	10	0	100	22	50	10

10–100 KHZ CRYSTAL WITH C-MOS—Uses low-cost ultraminiature quartz crystal shaped like tuning fork and produced by photolithographic techniques (Statek Corp., Orange, CA). With complementary-mos IC shown, will draw as little as 10 μA from 5-V supply. Article includes circuit modification for operating from 9–15 V d-c supply.—S. S. Chuang, C-Mos Minimizes the Size of Crystal Oscillators, Electronics, June 7, 1973, p 117–118.

CHAPTER 83
Phase Control Circuits

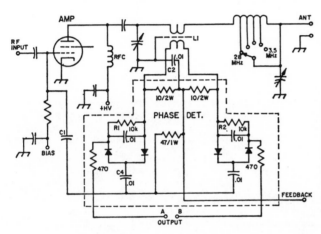

PHASE DETECTOR—Compares phase of grid-driving voltage of r-f amplifier with phase-opposed voltages induced in center-tapped link circuit coupled to amplifier plate tank to develop error signal if tank circuit is not at resonance. Error signal drives servo amplifier that controls rotation of motor-driven tank capacitor to give automatic tuning. Diodes are 1N34 or 1N38. Has sufficient range to cover both 40 and 80-meter amateur bands, when using 200-pF tuning capacitor in amplifier. Article gives coil data. C1 is 1½ inch length of RG-59/U coax.—F. Walsmith, Automatic Amplifier Tuning, *QST*, Sept. 1970, p 32–36.

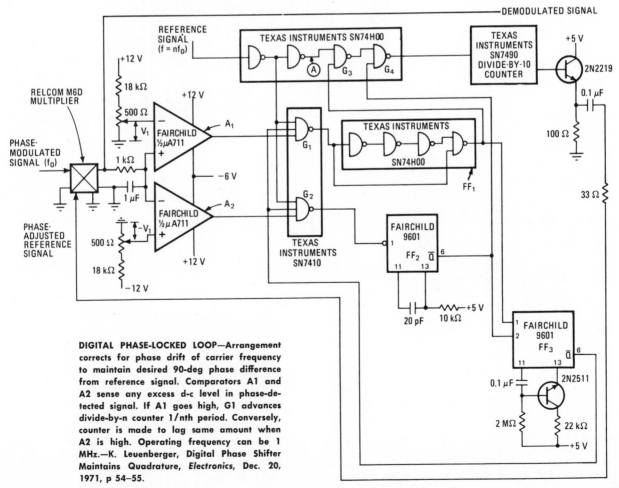

DIGITAL PHASE-LOCKED LOOP—Arrangement corrects for phase drift of carrier frequency to maintain desired 90-deg phase difference from reference signal. Comparators A1 and A2 sense any excess d-c level in phase-detected signal. If A1 goes high, G1 advances divide-by-n counter 1/nth period. Conversely, counter is made to lag same amount when A2 is high. Operating frequency can be 1 MHz.—K. Leuenberger, Digital Phase Shifter Maintains Quadrature, *Electronics*, Dec. 20, 1971, p 54–55.

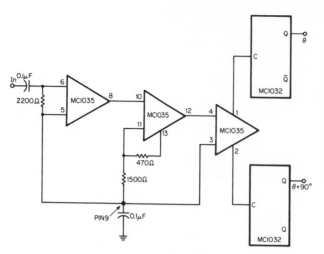

SSB PHASE SHIFTER—MC1035 triple differential amplifier provides differential output for driving 85-MHz MC1032 dual J-K flip-flop to give accurate 90-deg phase shift required for ssb or synchronous detection and for generation of ssb signals.—G. K. Shubert, A Digital 90° RF Phase Shifter, *The Electronic Engineer*, Aug. 1971, p 68.

30-MHZ PHASE SPLITTER—Used in r-f phase shifter having accuracy of 0.5 deg over 25% bandwidth and constant delay. Sine and cosine outputs go to multipliers for combining with outputs of sine-cosine potentiometer to give phase-shifted output signal.—R. H. Frater, Radio-Frequency Phase Shifter Has Constant Delay, *Electronic Engineering*, Dec. 1970, p 52–55.

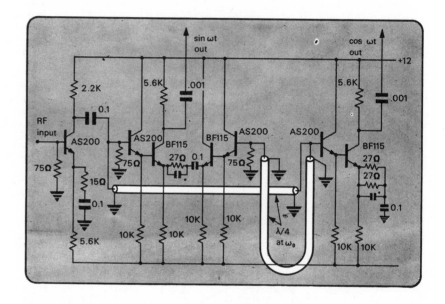

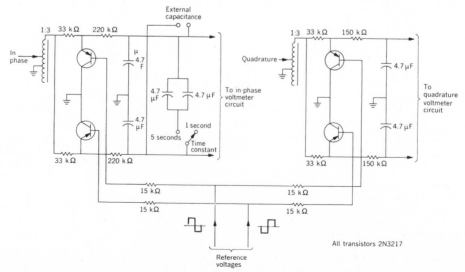

PHASE-SENSITIVE DETECTOR—Circuit receives two inputs, 90 deg apart, from phase-splitter circuit and 1,005-Hz square-wave reference from oscillator and squarer. Outputs go to voltmeters used for accurate balancing of low-frequency bridge. Frequency was set at 1,005 Hz to avoid harmonics of 50-Hz power-line frequency in Denmark.—J. M. Diamond, A Double Phase-Sensitive Detector for Bridge Balancing, *IEEE Spectrum*, June 1969, p 62–70.

PHASE SPLITTER—Consists of high-pass R-C filter in one channel and similar low-pass filter in other channel. With identical filter elements and filter loads, phase difference between channels is exactly 90 deg, independent of frequency and of values for R and C. Used in double phase-sensitive detector (PSD) for low-frequency bridge-balancing applications.—J. M. Diamond, A Double Phase-Sensitive Detector for Bridge Balancing, *IEEE Spectrum*, June 1969, p 62–70.

1–10 KHZ PHASE SHIFTER—Provides constant phase shift and constant output amplitude of sine wave over entire frequency range, by taking advantage of variable internal rolloff characteristic of opamp shown. Fet senses peak output voltage and begins cutting off current fed to master bias set terminal of opamp when peak exceeds predetermined value. Report gives relevant design equations. —"The μA776, An Operational Amplifier with Programmable Gain, Bandwidth, Slew-Rate, and Power Dissipation," Fairchild Semiconductor, Mountain View, CA, No. 218, 1971, p 3.

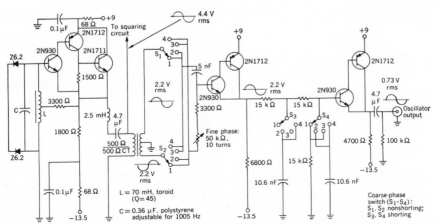

360-DEG PHASE SHIFTER—First three transistors form 1,005-Hz Hartley oscillator having White follower as active element. Remainder of circuit provides continuous full-range control of phase with constant amplitude. Used in conjunction with double phase-sensitive detector for precision balancing of low-frequency bridge.—J. M. Diamond, A Double Phase-Sensitive Detector for Bridge Balancing, *IEEE Spectrum*, June 1969, p 62–70.

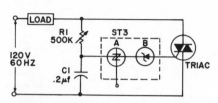

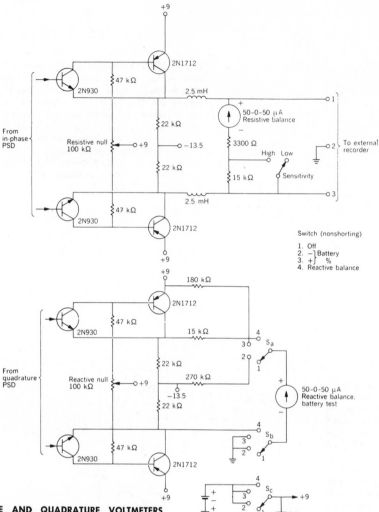

SBS-ZENER TRIGGER—Uses silicon bilateral switch A combined with zener B in GE ST3 device to achieve phase control of triac while eliminating hysteresis effect by introducing dissymmetry in trigger circuit. R1 provides control of lamp load up to capacity of triac in use, which for SC146 is 1,000 W. Report describes operation of circuit.—R. W. Fox, "Solid State Incandescent Lighting Control," General Electric, Semiconductor Products Dept., Auburn, NY, No. 200.53, 1970, p 6–7.

IN-PHASE AND QUADRATURE VOLTMETERS —Used with double phase-sensitive detector for low-frequency bridge balancing (1,005 Hz). Voltmeter circuits use pairs of White followers as simple differential meters. Input transistors are matched and tied thermally. In-phase channel has two sensitivities, 0.15- and 0.6-V rms at signal input terminal for full-scale deflection on zero-center meter. Sensitivity on quadrature channel is 0.48 V. Reactive balance meter also serves for checking batteries.—J. M. Diamond, A Double Phase-Sensitive Detector for Bridge Balancing, *IEEE Spectrum*, June 1969, p 62–70.

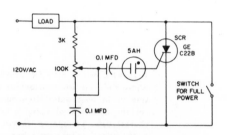

NEON TRIGGER FOR SCR—Half-wave a-c phase control is triggered when voltage across the two 0.1-μF capacitors reaches breakdown voltage of 5AH neon. Provides control from full off to 95% of half-wave rms output voltage, for load up to rating of scr. —"SCR Manual," General Electric, Syracuse, NY, 1972, 5th Ed., p 114.

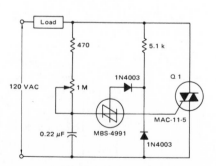

1,000-W TRIAC—Uses 1-meg pot to control conduction angle of triac Q1 over range of 0 to 170 deg, for better than 97% of full-power control. Report covers inrush current problems of lamps as related to surge-handling ability of triacs, and gives soft-start circuit which reduces initial surge.—R. J. Haver and D. A. Zinder, "Conventional and Soft-Start Dimming of Incandescent Lights," Motorola Semiconductors, Phoenix, AZ, AN-436, 1972.

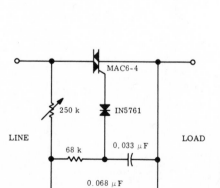

600-W FULL-RANGE—Uses double phase-shift network to obtain reliable triggering of scr (from MCR 2304 series) in both half-cycles at conduction angles as low as 5 deg. Ideal for incandescent lamps and some motors which require essentially full-range control. Uses bidirectional three-layer trigger.—D. Zinder, "SCR Power Control Fundamentals," Motorola Semiconductors, Phoenix, AZ, AN-240, 1972.

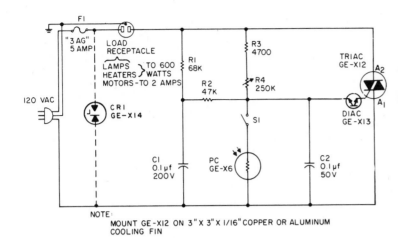

600-W FULL-WAVE—Provides full symmetrical voltage control from 0 to 100% for dimming lamps, controlling heaters, or controlling speed of shaded-pole, induction, or capacitor-start a-c motors. Gives speed range of about 4:1 for shaded-pole fan motor. Closing S1 places cadmium sulfide photocell across part of control circuit, for turning on lights automatically at darkness to discourage prowlers. CR1 is optional, for suppressing transient voltages.—"Electronics Experimenters Circuit Manual," General Electric, Owensboro, KY, 1971, 3rd Ed., p 179–182.

FULL-WAVE NEON TRIGGER—Transformer-coupled a-c phase control uses 5AH neon as trigger for two-terminal system, with separate secondaries allowing scr's to alternate in firing. Choose pulse transformer and value of R to give proper shape for pulses fed to scr gates. Provides control from full off to about 95% of total rms output voltage, for load up to rating of each scr.—"SCR Manual," General Electric, Syracuse, NY, 1972, 5th Ed., p 114–115.

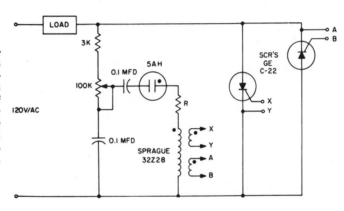

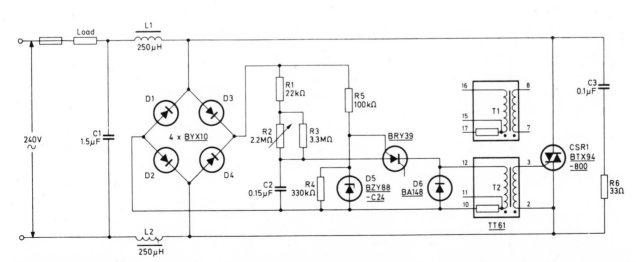

SCS TRIGGER FOR TRIAC—Uses BRY39 scs and pulse transformer TT61 to vary conduction angle of triac with R2 for full-wave a-c power control. Chokes L1 and L2 minimize r-f interference. R-C network across triac protects it from large voltage transients. With inductive load, bleeder resistor or R-C combination may be required across load to aid in latching on. Circuit requires adequate fusing to protect triac.—"Applications of the BRY39 Transistor," Mullard, London, 1973, TP1319, p 6–7.

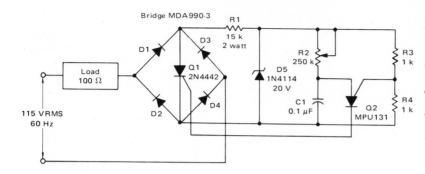

PUT CONTROLS SCR—Relaxation oscillator formed by put Q2 provides conduction control of scr in bridge circuit from 1 to 7.8 ms for 22 to 168 deg, equivalent to control over 97% of power available to load. Location of scr in bridge gives control over both positive and negative halves of sine wave.—R. J. Haver and B. C. Shiner, "Theory, Characteristics and Applications of the Programmable Unijunction Transistor," Motorola Semiconductors, Phoenix, AZ, AN-527, 1971.

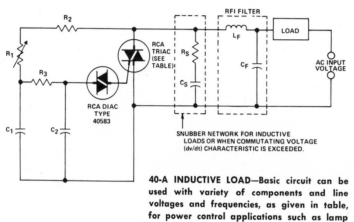

40-A INDUCTIVE LOAD—Basic circuit can be used with variety of components and line voltages and frequencies, as given in table, for power control applications such as lamp dimming, heat control, and speed control of universal motors. Rfi filter values are for lamp dimmer.—"40-A Silicon Triacs," RCA, Somerville, NJ, File No. 593, 1972.

AC INPUT VOLTAGE		120V 60Hz	240V 60Hz	240V 50Hz
C_1		0.1μF 200V	0.1μF 400V	0.1μF 400V
C_2		0.1μF 100V	0.1μF 100V	0.1μF 100V
R_1		100KΩ 1/2W	200KΩ 1W	250KΩ 1W
R_2		2.2KΩ 1/2W	3.3KΩ 1/2W	3.3KΩ 1/2W
R_3		15KΩ 1/2W	15KΩ 1/2W	15KΩ 1/2W
SNUBBER NETWORK FOR 40-A (RMS)●INDUCTIVE LOAD	C_S	0.18-0.22μF 200V	0.18-0.22μF 400V	0.18-0.22μF 400V
	R_S	330-390Ω 1/2W	330-390Ω 1/2W	330-390Ω 1/2W
RFI FILTER	C_F	0.1μF 200V	0.1μF 400V	0.1μF 400V
	L_F	100μH	200μH	200μH
RCA TRIACS		2N5441 2N5444 40688	2N5442 2N5445 40689	2N5442 2N5445 40689

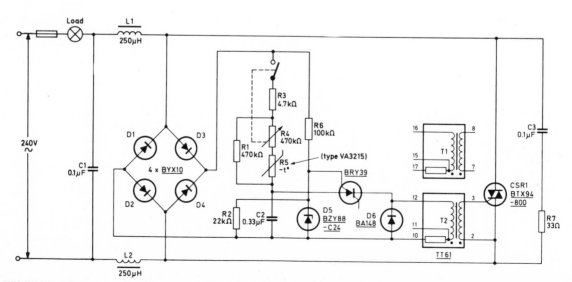

3-KW FULL-WAVE—BRY39 scs operating as put varies conduction angle of triac pulse transformer to provide full-wave a-c power control of incandescent-lamp load with R4. NTC thermistor protects triac by limiting amplitude and duration of current surge when lamps are switched on.—"Applications of the BRY39 Transistor," Mullard, London, 1973, TP1319, p 7—8.

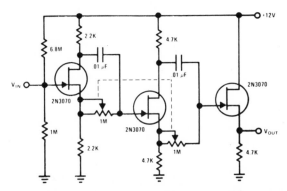

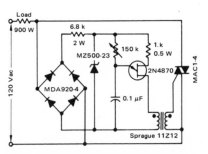

0–360 DEG PHASE SHIFTER—Cascading of two jfet stages, each providing 0–180 deg phase shift, gives maximum range, without loading phase-shift networks.—"FET Circuit Applications," National Semiconductor, Santa Clara, CA, 1970, AN-32, p 8.

900-W FULL-WAVE—Pulse transformer provides required oscillation between ujt and power line, required by presence of bridge rectifier. Pot determines point in each half-cycle at which load current begins to flow, for varying brightness of lamp load.—D. A. Zinder, "Electronic Speed Control for Appliance Motors," Motorola Semiconductors, Phoenix, AZ, AN-482, 1972.

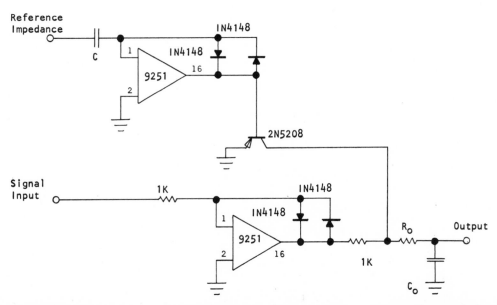

0–10 MHZ PHASE DETECTOR—Circuit shifts phase of reference 90 deg with coupling capacitor at input of OEI 9251 bipolar logamp whose voltage output is not sensitive to input level and has 90-deg phase shift relative to reference. Same logamp is in signal circuit where signal voltage is converted into current by resistor, to give output voltage 180 deg out of phase with signal input and therefore 90 deg out of phase wit reference output. High-speed bipolar switch chops signal amplifier output for integration by Ro and Co, chosen to produce smooth d-c output consistent with phase-modulation bandwidth.—"Very Wide Bandwidth Phase Detector Uses Operational Amplifier," Optical Electronics, Tucson, AZ, Application Tip 10189, 1971.

CHAPTER 84
Phonograph Circuits

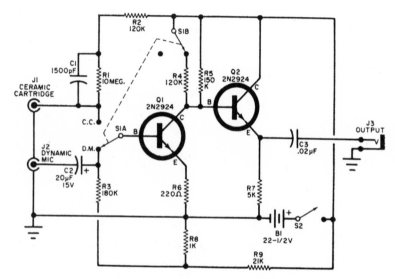

GENERAL-PURPOSE PREAMP—Designed for use with either ceramic phono cartridge or dynamic microphone.—L. Garner, Readers' Circuit, *Popular Electronics*, Sept. 1967, p 85–86.

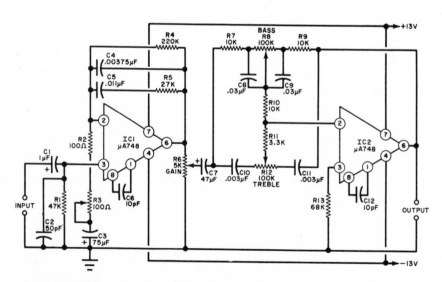

IC PREAMP—Contains total of 40 transistors. Provides adequate drive for power amplifier when used with high-quality phono cartridge. Separate bass, treble, and gain controls give maximum flexibility. Equalized for RIAA.—G. C. Sheatz, Reader's Circuit, *Popular Electronics*, July 1971, p 80–81.

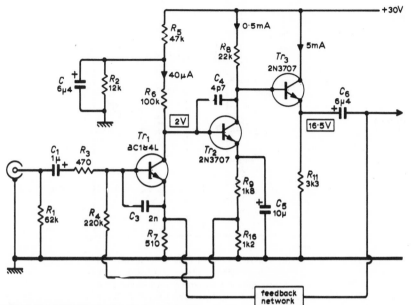

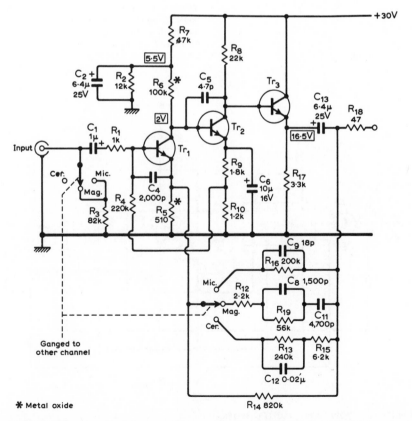

LOW-NOISE PREAMP—Designed for use with magnetic pickup. Uses series feedback and optimization of values to minimize noise. Signal-noise ratio referred to 2-mV input at 1 kHz is 70 dB.—H. P. Walker, Low-Noise Audio Amplifiers, *Wireless World*, May 1972, p 233–237.

PREAMP FOR CERAMIC—Gives RIAA equalized output with ceramic cartridges having anywhere from 800 to 10,000 pF capacitance. Output is equalized within 2 dB from 40 to 12,000 Hz. Article gives choice of two amplifier circuits and power supply. For stereo, other half of circuit is identical, flipped over below ground.—D. V. Jones, Constructing a 10-Watt Stereo Amplifier, *Audio*, May 1971, p 30 and 32–35.

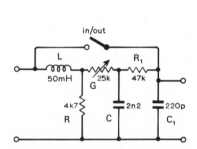

SCRATCH FILTER—Continuously variable passive filter serves in place of four or six ganged pots commonly used for stereo amplifiers. Values shown cover cutoff frequencies between 6 and 15 kHz. Rolloff is 18 dB per octave. Filter should be followed by circuit with at least 500K impedance.—A. E. Prinn, Scratch Filter, *Wireless World*, March 1972, p 140.

PHONO PREAMP—Uses bipolar input transistor to obtain good flicker noise performance with high-quality magnetic pickup. Provides bass boost required by majority of better ceramic cartridges, with values for C12, R15, and R4 providing turnover at about 1.5 kHz.

Mike input will match 50K high-impedance dynamic mike. TR1 is BC184LC or BC109C silicon. TR2 is 2N2707 or BC167. TR3 is BC107 or BC167.—H. P. Walker, Stereo Mixer, *Wireless World*, May 1971, p 221–225.

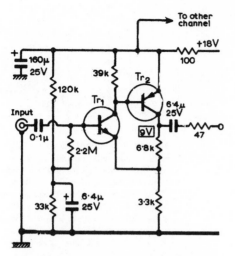

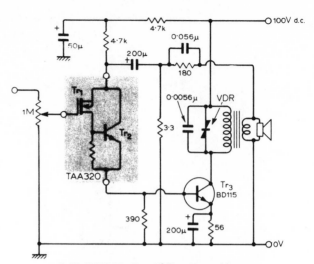

PREAMP FOR HIGH-OUTPUT CERAMIC—TR1 is BC184LC, BC169, or similar, connected to current-drive TR2 which is 2N4058, 2N4062, or similar. Distortion at 1 kHz is less than 0.02% for 600-mV rms input and does not exceed 0.1% for 1.5-V input.—H. P. Walker, Stereo Mixer, *Wireless World*, May 1971, p 221–225.

2 W WITH IC—Uses Philips IC to drive output transistor. Voltage-dependent resistor across output transformer suppresses voltage spikes that might damage transistor. Requires 85-mV input from crystal pickup for full output power.—T. D. Towers, Elements of Linear Microcircuits, *Wireless World*, March 1971, p 114–118.

COMMON-MODE VOLUME CONTROL—Eliminates signal attenuation of conventional voltage-divider volume control. Requires only single IC power audio amplifier.—J. E. Byerly and M. K. Vander Kooi, "LM380 Power Audio Amplifier," National Semiconductor, Santa Clara, CA, 1972, AN-69, p 4.

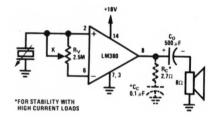

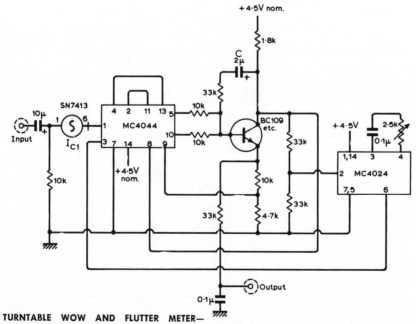

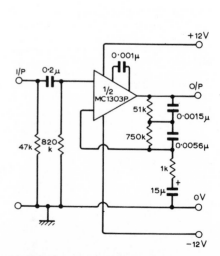

TURNTABLE WOW AND FLUTTER METER—Measures p-p deviation of recorded 3-kHz tone, by converting frequency modulations to changing d-c level with phase-locked loop and displaying result on cro. IC1 is Schmitt trigger squaring circuit, followed by IC phase detector and IC voltage-controlled oscillator. Article covers calibration and method of use for analyzing rumble, eccentricity, pinch, warp, and other turntable or record defects as well as wow and flutter.—R. Ockleshaw, Novel Wow and Flutter Meter, *Wireless World*, Dec. 1971, p 572–575.

MAGNETIC PICKUP PREAMP—Opamp circuit includes required compensation.—T. D. Towers, Elements of Linear Microcircuits, *Wireless World*, Feb. 1971, p 76–80.

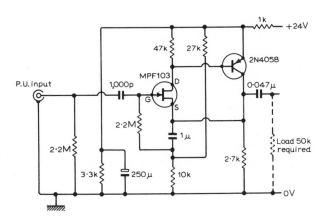

INPUT FOR CERAMIC PICKUP—Serves as first preamp stage when impedance conversion and equalization are required for ceramic cartridge. Response is flat from 35 Hz to above 200 kHz.—J. L. Linsley Hood, Modular Pre-amplifier Design, *Wireless World*, July 1969, p 306–310.

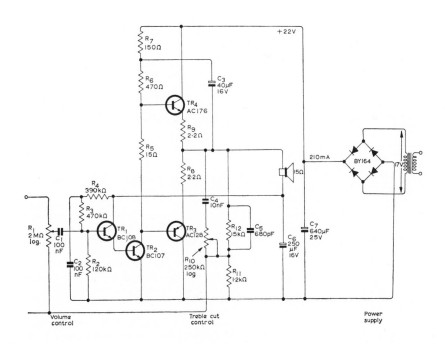

3-W RECORD PLAYER—Suitable for crystal pickups having 300-mV output. Feedback is applied in series with input to provide 2-meg input impedance required for adequate bass response. C4 and R10 form simple treble cut control. Distortion is lowest at maximum volume setting because feedback is high, being 0.4% for 2-W output.—"Transistor Audio and Radio Circuits," Mullard, London, 1972, TP1319, 2nd Ed., p 46–47.

MAGNETIC-PICKUP EQUALIZER—Equalization is incorporated in preamp to provide simultaneous low-distortion low-pass and high-pass filtering. Mid-point gain is 50 and filter slopes 18 dB per octave.—J. L. Linsley Hood, Combined Low-Pass and High-Pass Filter, *Wireless World*, March 1970, p 123.

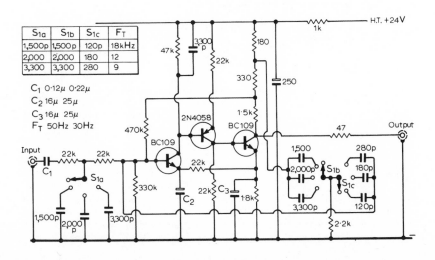

S1a	S1b	S1c	FT
1,500p	1,500p	120p	18kHz
2,000	2,000	180	12
3,300	3,300	280	9

C1 0.12μ 0.22μ
C2 16μ 25μ
C3 16μ 25μ
FT 50Hz 30Hz

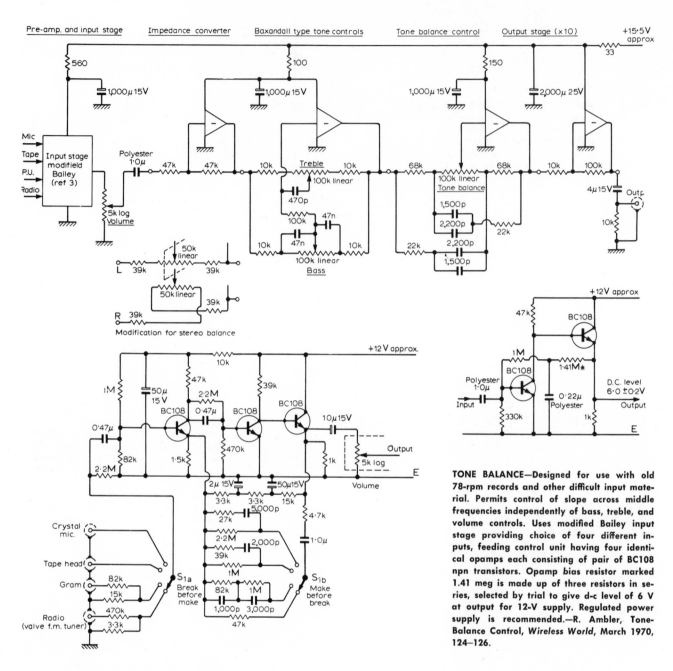

TONE BALANCE—Designed for use with old 78-rpm records and other difficult input material. Permits control of slope across middle frequencies independently of bass, treble, and volume controls. Uses modified Bailey input stage providing choice of four different inputs, feeding control unit having four identical opamps each consisting of pair of BC108 npn transistors. Opamp bias resistor marked 1.41 meg is made up of three resistors in series, selected by trial to give d-c level of 6 V at output for 12-V supply. Regulated power supply is recommended.—R. Ambler, Tone-Balance Control, *Wireless World*, March 1970, 124–126.

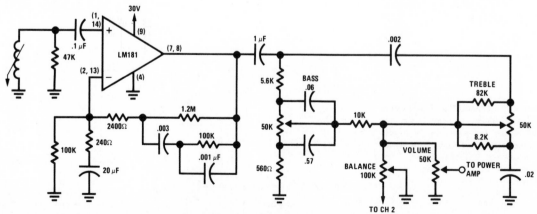

PHONO TONE CONTROLS—Complete single channel is shown for stereo phono preamp used with magnetic pickup driving power amplifier having 5-V rms input overload limit. Has separate bass, treble, balance, and volume controls. Includes RIAA equalization. Report gives design equations.—J. E. Byerly and E. L. Long, "LM381 Low Noise Dual Preamplifier," National Semiconductor, Santa Clara, CA, 1972, AN-64, p 10–11.

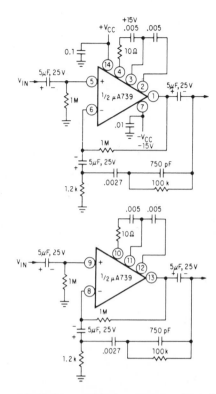

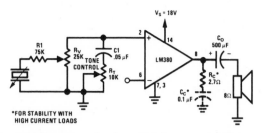

CRYSTAL PHONO AMPLIFIER—Uses only single IC power audio amplifier having fixed gain of 50 (34 dB) and output which automatically centers itself at half of 18-V supply voltage. Includes tone control providing high-frequency rolloff.—J. E. Byerly and M. K. Vander Kooi, "LM380 Power Audio Amplifier," National Semiconductor, Santa Clara, CA, 1972, AN-69, p 4.

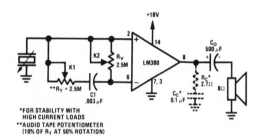

DUAL-SUPPLY PHONO PREAMP—Identical amplifiers on single chip each use +15 V and −15 V supply and provide RIAA-equalized gain of 40 dB at 1 kHz for stereo pickup. Noise level is 2 μV referred to input and input overload is 80 mV rms.—D. Campbell, W. Hoeft, and W. Votipka, "Applications of the μA739 and μA749 Dual Preamplifier Integrated Circuits in Home Entertainment Equipment," Fairchild Semiconductor, Mountain View, CA, No. 171, 1971, p 5.

COMMON-MODE VOLUME AND TONE CONTROLS—Used with single IC power audio amplifier when signal attenuation of conventional voltage-divider control is undesirable. Used with transducers having high source impedance, to realize full input impedance of amplifier. Report gives transfer function and response curves.—J. E. Byerly and M. K. Vander Kooi, "LM380 Power Audio Amplifier," National Semiconductor, Santa Clara, CA, 1972, AN-69, p 4.

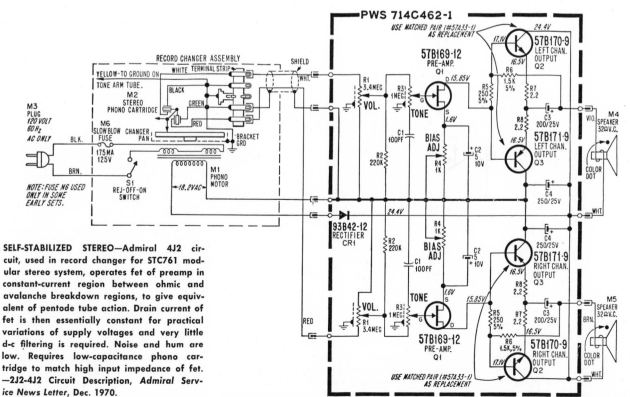

SELF-STABILIZED STEREO—Admiral 4J2 circuit, used in record changer for STC761 modular stereo system, operates fet of preamp in constant-current region between ohmic and avalanche breakdown regions, to give equivalent of pentode tube action. Drain current of fet is then essentially constant for practical variations of supply voltages and very little d-c filtering is required. Noise and hum are low. Requires low-capacitance phono cartridge to match high input impedance of fet. —2J2-4J2 Circuit Description, *Admiral Service News Letter*, Dec. 1970.

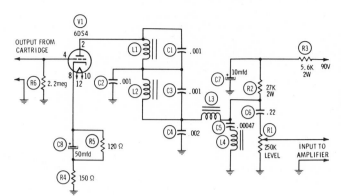

RECORD SCRATCH FILTER—Provides dramatic cutoff above 8,000 Hz, for suppressing scratches and other noise while listening to old 78-rpm records. Does not affect desired sounds on these records, since these are all below 8,000 Hz. L1 is 0.3-H filter choke, L2 0.4, L3 0.8, and L4 0.53. Required supply voltages can generally be obtained easily from tube-type phono amplifier.—R. Brown and T. Kneitel, "101 Easy Audio Projects," Howard W. Sams, Indianapolis, IN, 1971, p 51–52.

SCRATCH AND RUMBLE FILTER—Designed for use with professional sound systems to attenuate scratches on records at higher frequencies. Independent control SW2 serves for attenuating turntable rumble in range of 50 to 100 Hz. FLAT position of switches provides straight-through connection for use with good recordings. L1 and L2 are 1.5-H 10-mA filter chokes.—R. Brown and T. Kneitel, "101 Easy Audio Projects," Howard W. Sams, Indianapolis, IN, 1971, p 148–149.

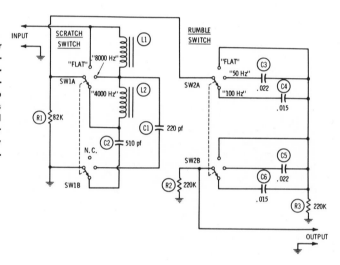

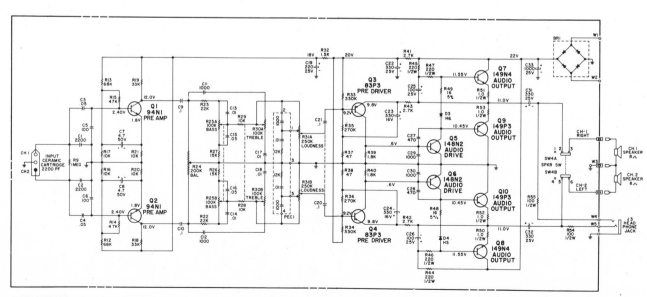

5-W PHONO—Uses ten transistors and two diodes to provide 5 W per channel of music power output from stereo ceramic cartridge in portable phonograph. Overwind on 117-V a-c phono motor supplies 18 V a-c to W1 and W2 of bridge rectifier delivering 22-V d-c for transistors. Output transistors are arranged in matched pairs.—"A511 Series Amplifier Chassis," Magnavox Co., Fort Wayne, IN, Service Manual No. 4343, 1970.

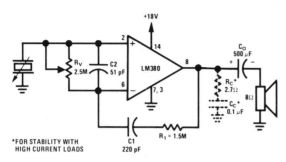

RIAA PHONO AMPLIFIER—Provides inverse frequency response to that used when recording phonograph records, wherein low frequencies are attenuated to prevent large undulations from breaking through groove walls and high frequencies are emphasized to improve signal-to-noise ratio. Uses single National IC power audio amplifier.—J. E. Byerly and M. K. Vander Kooi, "LM380 Power Audio Amplifier," National Semiconductor, Santa Clara, CA, 1972, AN-69, p 4–5.

PHONO PREAMP—Designed for use with ceramic or crystal pickup or microphone. Has excellent frequency response, and will operate on wide range of supply voltages. Use audio taper for 10K pot.—"Tips on Using FET's," Motorola Semiconductors, Phoenix, AZ, HMA-33, 1971.

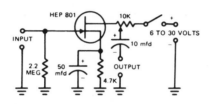

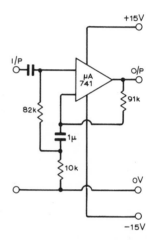

CRYSTAL PICKUP PREAMP—Opamp circuit has required high input impedance for use with any crystal pickup.—T. D. Towers, Elements of Linear Microcircuits, *Wireless World*, Feb. 1971, p 76–80.

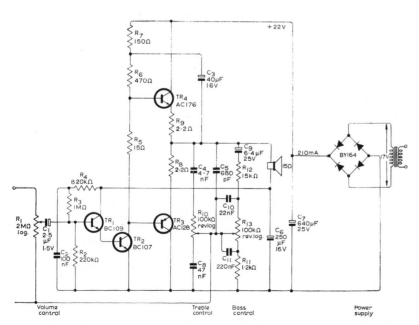

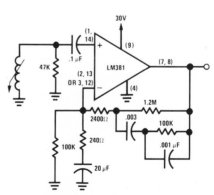

3-W PLAYER WITH TONE CONTROLS—Includes bass and treble cut and boost controls. Designed for crystal pickups having 300-mV output.—"Transistor Audio and Radio Circuits," Mullard, London, 1972, TP1319, 2nd Ed., p 49–50.

MAGNETIC-PICKUP PREAMP—Uses low-noise dual preamp designed for amplifying low-level signals. Standard RIAA cartridge load is provided by 47K input resistor. Designed for magnetic pickups having outputs of 3.5 to 8 mV at velocities of 5 cm/s. Will drive power amplifier having 5-V rms input overload limit. Report gives design equations.—J. E. Byerly and E. L. Long, "LM381 Low Noise Dual Preamplifier," National Semiconductor, Santa Clara, CA, 1972, AN-64, p 8–10.

CHAPTER 85
Photoelectric Circuits

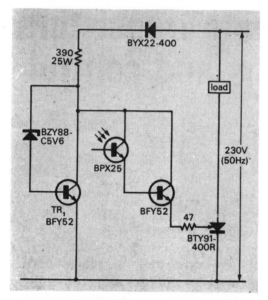

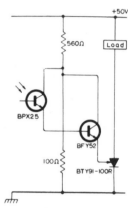

16-A THYRISTOR CONTROL—700 lux on BPX25 silicon phototransistor makes thyristor turn on about 6 deg away from zero voltage. Load power can be varied by synchronizing light pulses with line frequency, which may be 50 or 60 Hz.—Applications of Photo Transistors in Communication and Control, *Electronic Engineering*, Nov. 1969, p 29–31.

DARK-OFF SCR—Transformerless photoelectric trigger turns on scr with illumination, for applying power to d-c load.—"Mullard Silicon Planar Phototransistors BPX25 and BPX29," Mullard, London, 1968, TP1000A, p 7.

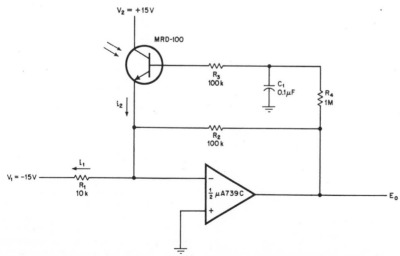

FAST RESPONSE—High-gain opamp improves response time of phototransistor by supplying current to keep transistor at more optimum operating point when handling extremely low-level modulated light signals.—M. L. McCartney, Feedback Amplifier Speeds Phototransistor's Response, *Electronics*, March 15, 1971, p 77.

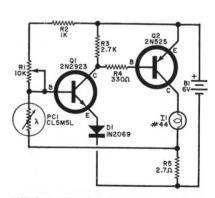

TWILIGHT TURN-ON—Automatic road-barrier lamp is turned on at dusk and off at dawn by Clairex photoconductive cell in complementary direct-coupled amplifier. Current drain in daylight is less than 1% of that at night. Can also be used as driveway or boat dock light. R1 determines sensitivity.—Manufacturer's Circuit, *Popular Electronics*, Feb. 1970, p 105 and 110.

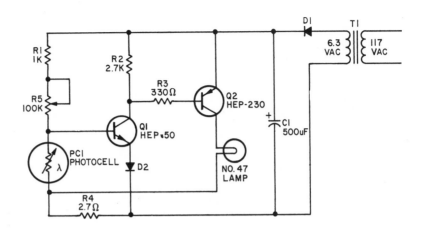

BEDSIDE LIGHT—Cadmium sulfide photocell in transistor control circuit turns on pilot lamp gradually as twilight falls, and dims lamp as sun rises in morning. No switch is needed. To control larger light, use Potter & Brumfield MR5D or equivalent 6-V relay in place of 6-V No. 47 pilot lamp. Can then be used to turn on porch light or other burglar-discouraging lamp at dusk when no one is home.—S. Daniels, Hobbyist's Night Light, *Elementary Electronics*, May-June 1972, p 43–44.

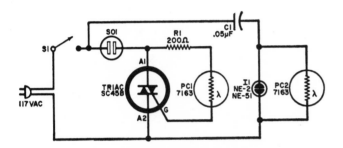

1,000-W SUNSET CONTROL—Provides greater power-handling capacity than commercial units for turning on lights automatically at twilight. Uses RCA 7163 cadmium sulfide photocells. Daylight on PC2 keeps its resistance so low that voltage drop across it is not enough to fire neon. PC1, in same housing with neon, is therefore dark. At darkness, PC2 turns on neon, which eliminates PC1 and switches triac into conduction to turn on lamps plugged into SO1.—G. L. Garvin, Reader's Circuit, *Popular Electronics*, Oct. 1968, p 85–86.

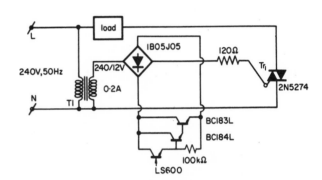

LIGHT FIRES TRIAC—Illumination of LS600 light sensor develops voltage across 100K resistor for turning on Darlington pair. These in turn short out bridge and fire triac Tr1 to energize load. To reverse operation, transpose 100K resistor and LS600.—J. Budek, Solid-State Switching Using Triacs and Thyristors, *Design Electronics*, May 1970, p 59–63 and 126.

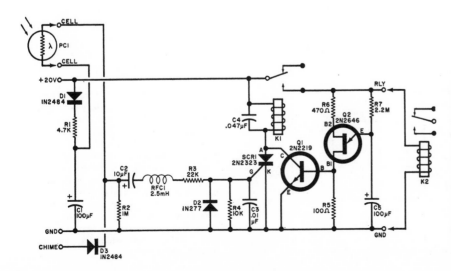

WELCOME LIGHT—Turns on porch light when hit by headlight beam of car turning into driveway of home and turns light off again automatically in about 5 min, to give guests sufficient time to get into house. Responds only to sudden increase in light, hence is not affected by dawn of new day. Uses cadmium sulfide photocell such as Lafayette 19T2101 and two 24-V d-c relays. Adjust value of R4 to give desired sensitivity for system, starting with value shown. Article also includes power supply circuit and modification for turning on lights for 5 min when either front or rear door chime button is pushed.—J. A. Archer, Your Own Private Owl, *Popular Electronics*, June 1969, p 51–53 and 58–59.

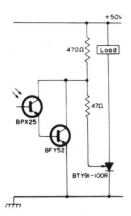

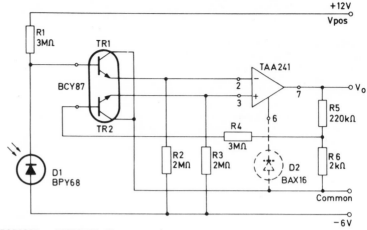

DARK-ON SCR—Transformerless photoelectric trigger turns on scr when illumination is removed from phototransistor, for applying power to d-c load.—"Mullard Silicon Planar Phototransistors BPX25 and BPX29," Mullard, London, 1968, TP1000A, p 7.

PHOTODIODE AMPLIFIER—Output voltage switches from negative to positive when voltage at inverting input of d-c amplifier falls below that at noninverting input, and switches back when voltage rises again. Once switching in either direction has been initiated, positive feedback from output to nonin- verting input drives IC into saturation. When photodiode is dark, inverting input is +12 V and higher than noninverting input.— "TAA241, TAA242 and TAA243 Differential Operational Amplifiers," Mullard, London, 1969, TP1086, p 16—18.

LOW DARK CURRENT—Designed for operation at extremely low light levels, where dark current of phototransistor can adversely affect circuit operation. R1 is adjusted so dark phototransistor is just cut off. Will operate with illumination of only 10 lux, with backlash of about 0.5 lux. Output for illumination is about 19.5 V. Useful for burglar alarm systems having long light beams.—"Mullard Silicon Planar Phototransistors BPX25 and BPX29," Mullard, London, 1968, TP1000A, p 7.

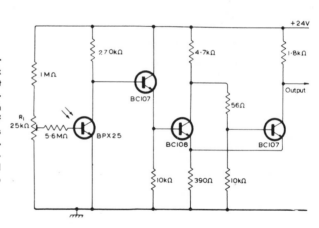

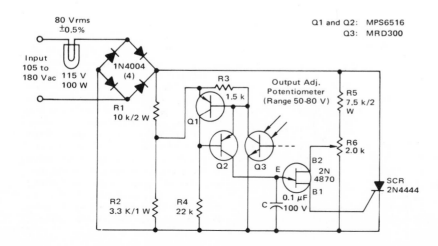

REGULATOR FOR LIGHT SOURCE—Used in light-beam projection system where it is desirable to maintain constant brightness level despite line voltage variations. Holds 100-W projection lamp voltage within 0.5% of 80 V by monitoring lamp brightness with phototransistor Q3 in ujt oscillator circuit that varies lamp current by changing firing angle of scr. R6 is set for desired brightness level. Light on Q3 from lamp must be attenuated with filter or small iris.—J. Bliss, "Applications of Phototransistors in Electro-Optic Systems," Motorola Semiconductors, Phoenix, AZ, AN-508, 1971.

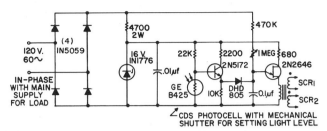

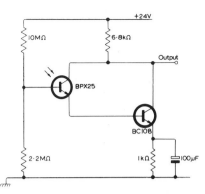

LAMP CONTROL—Cadmium sulfide photocell controls pedestal level of ramp-and-pedestal system for triggering scr's connected to regulate lamp load or illumination on surface. Circuit will accommodate wide range of impedance levels for photocell or other resistance-type sensor.—R. W. Fox, "Solid State Incandescent Lighting Control," General Electric, Semiconductor Products Dept., Auburn, NY, No. 200.53, 1970, p 11.

SOUND-TRACK READOUT—Required illumination of optical sound track on movie film can be achieved by focusing 1-W miniature lamp on 0.08 × 0.001-inch slot, corresponding to bandwidth of 6 kHz for sound. This gives equivalent of 15 lux on phototransistor at full volume, for which signal-noise ratio is 60 dB. Used with preamp and power amplifier.—"Mullard Silicon Planar Phototransistors BPX25 and BPX29," Mullard, London, 1968, TP1000A, p 8–9.

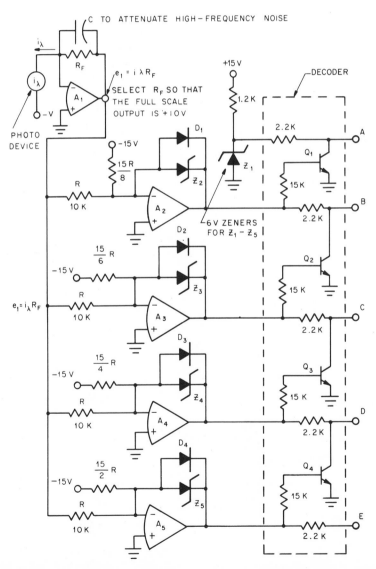

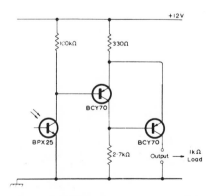

APPLE GRADER—Photodiode array, positioned for grading apples into five groups according to size, feeds opamp A1 operating as current-voltage converter. Opamps A2–A5 serve as biased comparators with single clamp circuit. Decoder ensures that only one logic output is high at a time. Decoder can drive TTL or DTL output system directly or control relays through npn switching transistor. Opamps can be Burr-Brown.—J. G. Graeme, G. E. Tobey, and L. P. Huelsman, "Operational Amplifiers," McGraw-Hill Book Co., New York, 1971, p 366–368.

LOW LIGHT LEVEL—Will operate at illumination of 50 lux. Can drive electronic counter at speeds up to 6,000 counts per second. Transistors form bistable circuit controlled by phototransistor so output is zero with illumination. When phototransistor is dark, circuit supplies 8 V at 8 mA for driving relay directly or through additional d-c amplifier. Can be used for elevator control, batch counter, edge detector, card reader, and burglar alarm.—"Mullard Silicon Planar Phototransistors BPX25 and BPX29," Mullard, London, 1968, TP1000A, p 6.

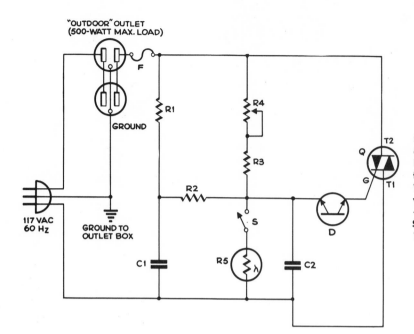

SUNSET GRADUAL-ON CONTROL—Brings outdoor lighting on gradually as darkness falls, by using photoresistor R5 with C2 to control frequency at which diac D triggers triac Q. With S open for manual control, lamp load will come on and R4 determines brightness. —J. G. Busse, Patio Control Brings Lights Up S-L-O-W-L-Y, *Popular Science*, Oct. 1970, p 123–124.

LIGHT COMPARATOR—Can be used as guide for adjusting two light sources for equal intensity. PC1 and PC2 are solar cells. To balance bridge initially, expose both cells to same light source and adjust R1 for zero reading on zero-center meter.—H. Friedman, "99 Electronic Projects," Howard W. Sams, Indianapolis, IN, 1971, p 55–56.

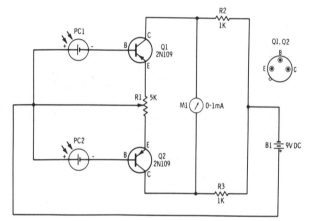

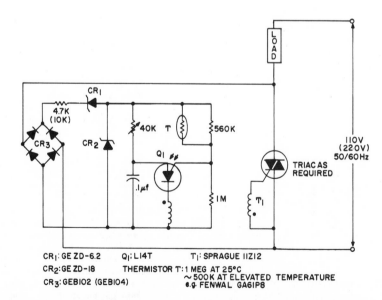

CR₁: GE ZD-6.2   Q₁: L14T   T₁: SPRAGUE 11Z12
CR₂: GE ZD-18   THERMISTOR T: 1 MEG AT 25°C
CR₃: GEB102 (GEB104)   ~500K AT ELEVATED TEMPERATURE
           e.g. FENWAL GA61P8

FULL-WAVE LIGHT CONTROL—Uses triac as switch to control load power over range of 0 to 90% in response to changes of light on L14T photoswitch Q1. Thermistor T provides temperature compensation to hold power constant within 3% over range of 25 to 50 C.— R. E. Locher, "Phase Control Voltage With Light," General Electric, Semiconductor Products Dept., Auburn, NY, No. 201.26.

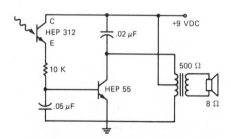

SUNLIGHT ALARM CLOCK—Operates from 9-V transistor radio battery, permitting use by hunters or fishermen who want to be out at dawn. Can also serve as fire alarm or burglar alarm triggered by flashlight or room lights. Audio alarm tone stops when light stops.—"Home Handyman's Construction Projects," Motorola Semiconductors, Phoenix, AZ, HMA-37, 1972.

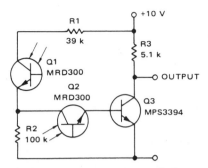

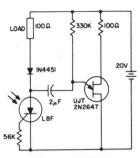

TRUTH TABLE	
Q1, Q2	OUTPUT = 0
Q1, $\bar{Q}$2	OUTPUT = 1
$\bar{Q}$1, Q2	OUTPUT = 1
$\bar{Q}$1, $\bar{Q}$2	OUTPUT = 1

PHOTOELECTRIC LOGIC—About 100 footcandles on phototransistors Q1 and Q2 gives zero output from Q3. Output is positive or 1 when either or both phototransistors are dark. Used as optical logic driver for power devices.—J. Bliss, "Applications of Phototransistors in Electro-Optic Systems," Motorola Semiconductors, Phoenix, AZ, AN-508, 1971.

LIGHT-TRIGGERED ONE-SHOT—Ujt is triggered 0.6 s after pulse of light turns on L8F lascr. If light pulse has ended then, lascr will turn off. If light is still on, ujt will operate as relaxation oscillator and turn lascr off the first time it fires after light is removed.—"SCR Manual," General Electric, Syracuse, NY, 1972, 5th Ed., p 439.

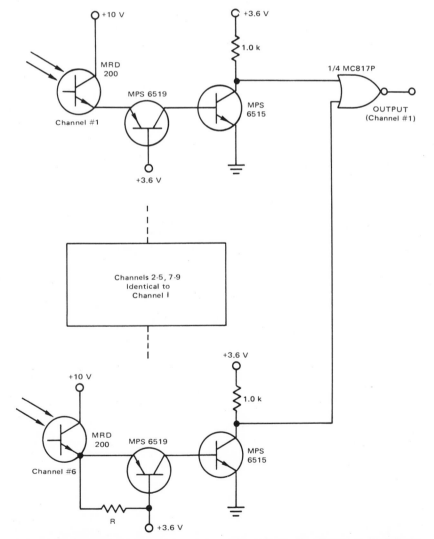

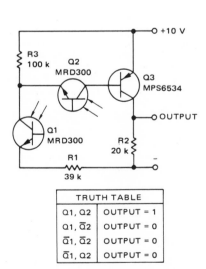

TRUTH TABLE	
Q1, Q2	OUTPUT = 1
Q1, $\bar{Q}$2	OUTPUT = 0
$\bar{Q}$1, $\bar{Q}$2	OUTPUT = 0
$\bar{Q}$1, Q2	OUTPUT = 0

PAPER-TAPE READER—Designed for use with RTL logic, with channel 6 of tape serving as clock input for logic. Hole in tape makes corresponding output line go low, to give "1" pulse at corresponding gate output. Choose value of R for best light-dark discrimination with color of paper tape used.—J. Bliss, "Applications of Phototransistors in Electro-Optic Systems," Motorola Semiconductors, Phoenix, AZ, AN-508, 1971.

PHOTOELECTRIC LOGIC—Output of Q3 is 0 at all times except when both phototransistors are illuminated with about 100 footcandles; light on either transistor alone will not give positive or 1 output.—J. Bliss, "Applications of Phototransistors in Electro-Optic Systems," Motorola Semiconductors, Phoenix, AZ, AN-508, 1971.

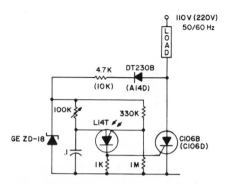

HALF-WAVE LIGHT CONTROL—Pot is set so L14T light-activated photoswitch triggers on at 170—180 deg. Then, as light increases, delay angle can be reduced to 10 deg, for varying power to load from 0 to 45%.—R. E. Locher, "Phase Control Voltage with Light," General Electric, Semiconductor Products Dept., Auburn, NY, No. 201.26.

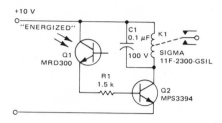

LIGHT-ENERGIZED RELAY—Required light level for turning on Q2 enough to pass 5-mA relay pull-up current is 125 footcandles, which can be supplied by flashlight using PR2 lamp or equivalent light source.—J. Bliss, "Applications of Phototransistors in Electro-Optic Systems," Motorola Semiconductors, Phoenix, AZ, AN-508, 1971.

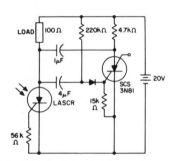

LIGHT-FLASH-WIDTH SENSOR—With no light, 3N81 silicon controlled switch conducts. Short pulse of light on L8F on other lascr turns it on and commutates scs off through 1-μF capacitor. After about 0.6 s, scs is turned on and lascr commutated off. Useful for detecting pulses of light lasting longer than some minimum time. If pulses last longer than 0.6 s, lascr stays on.—"SCR Manual," General Electric, Syracuse, NY, 1972, 5th Ed., p 439.

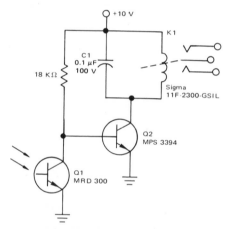

LIGHT-DEENERGIZED RELAY—Interruption of 125-footcandle minimum illumination on photodiode Q1 drops Q2 current below 5 mA and releases relay. Can be used as automatic door opener or moving-object counter.—J. Bliss, "Applications of Phototransistors in Electro-Optic Systems," Motorola Semiconductors, Phoenix, AZ, AN-508, 1971.

CHAPTER 86
Photography Circuits

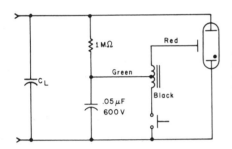

500-V TRIGGER—Photoflash firing circuit requires pushbutton switch with contacts rated at 500 V. Trigger transformer is Stancor P-6426. Flashtube can be General Electric FT52, FT118, or FT120.—L. T. Rees, Charging Energy-Storage Capacitors from Low-Voltage Sources, *The Electronic Engineer*, Jan. 1969, p 50—56.

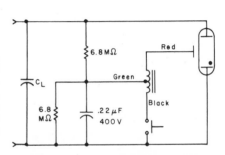

250-V TRIGGER—Voltage-dividing resistors across energy storage capacitor CL permit use of pushbutton switch with contacts rated at only 250 V when CL is charged to 500 V. Trigger transformer is Stancor P-6426. Flashtube can be General Electric FT52, FT118, or FT120.—L. T. Rees, Charging Energy-Storage Capacitors from Low-Voltage Sources, *The Electronic Engineer*, Jan. 1969, p 50—56.

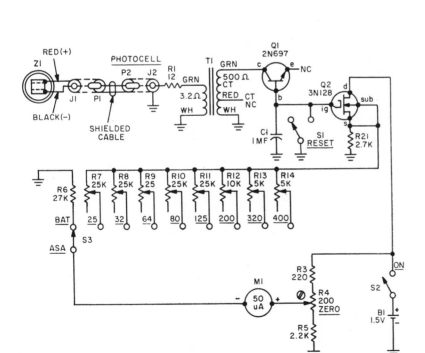

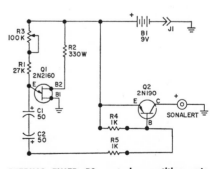

FLASH METER—Indicates correct f-stop for electronic flash and any of eight different ASA film speeds, without measuring distance to subject. With photocell at subject location, electronic flash on camera is set off without exposing film. Resulting generated pulse is stepped up in voltage by T1 (output transformer connected backward) and rectified by Q1 to charge C1. Resulting voltage across C1 is sensed by insulated-gate fet Q2, which drives meter calibrated to read directly in f-stops. Silicon photocell Z1 can be Calectro J4-800. Article gives construction and calibration details.—C. Green, Flash Master, *Elementary Electronics*, July-Aug. 1971, p 27—30 and 100—101.

BEEPING TIMER—R3 controls repetition rate of pleasant beep tone burst produced by relaxation oscillator Q1 when momentary-contact switch (plugged into jack J1) is held closed. R3 can be calibrated with stopwatch by counting beeps. Can serve as darkroom timer, especially if foot switch is used, or for timing sports events.—S. Daniels, SonoPulse Timer, *Elementary Electronics*, Jan.-Feb. 1970, p 47—49 and 108.

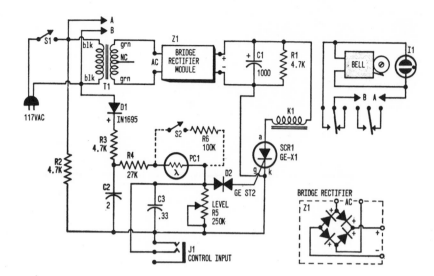

DARKROOM LIGHT ALARM—Photocell (Clairex CL5M3 or equivalent cadminum selenide) in ping-pong ball diffuser responds to unsafe white light in darkroom and triggers scr to pull in relay that sounds alarm and can also turn off enlarger or unsafe safelight. Z1 is 1-A bridge rectifier, such as IR 10DB6A. T1 is Stancor P-6465 6.3-V 0.6-A. K1 can be Guardian 200-6D. J1 permits plugging in normally closed switch on paper safe so it opens and activates circuit when paper is uncovered. S2 is optional manual circuit turn-off. —R. Michaels, Photomate, *Elementary Electronics*, March-April 1969, p 59–62.

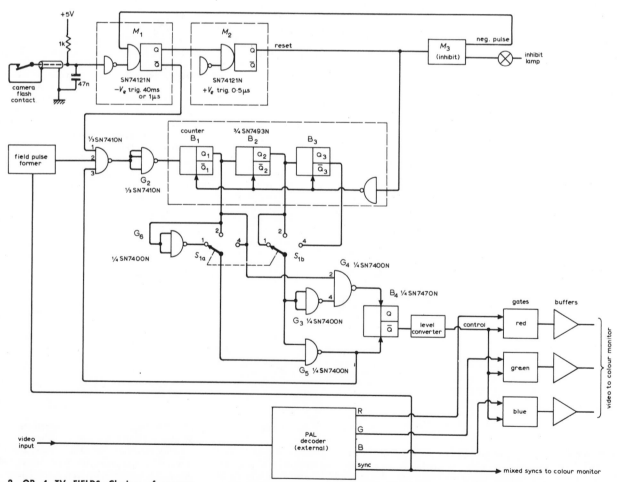

1, 2, OR 4 TV FIELDS—Closing of camera flash contact initiates electronic delay which in turn starts field counter having S1 set for desired number of fields to be exposed. Used in photographing tv coverage of Apollo 11 moon landing. Eliminates shutter bars usually present when approximate shutter speed of 1/25 s is used. Article includes inhibit mono circuit of M3, field sync pulse former, and level converter. Designed for 625-line British color tv using PAL decoder.—R. E. Knight and D. J. Bryan, Photographing Television Pictures, *Wireless World*, June 1972, p 290–293.

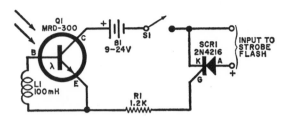

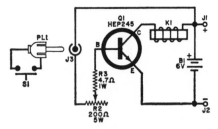

SLAVE FLASH—Phototransistor Q1 responds to sudden pulse of light from main probe flash and supplies trigger pulse to gate of scr to turn on remote flash. Range is up to 20 feet.—Manufacturer's Circuit, *Popular Electronics*, July 1970, p 86.

REMOTE SHUTTER TRIP—Provides true remote control over two-wire cable from several hundred feet, as required for wildlife or spy surveillance photography. Only bias current for Q1 goes through remote cable, permitting use of cheapest cable such as speaker wire. R2 permits varying actuating force of solenoid K1 that pushes in shutter release button. Article covers construction of required linkage for solenoid.—A. A. Mangieri, Remote Camera Shutter Release, *Popular Electronics*, July 1970, p 65–68.

ADJUSTABLE STROBE—Uses xenon flashtube such as Southwest Technical type 110. Can serve as auto timing light, slave flash for camera, stop-motion strobe for moving parts, or audio-controlled strobe lighting for parties. Caution: in darkened rooms, may induce hallucinations or undesired side effects, particularly in epileptics. Audio input is at B. For auto timing, ground one lead of B, connect 5-meg resistor to other lead, solder insulated wire to resistor, and wrap several turns of this wire around lead to No. 1 plug. Set R10 so flash just stops with engine dead. Strobe now fires each time No. 1 plug fires. For slave flash, use 1½-V cell in series with leads from audio input to camera switch. T1 is 2,000-ohm CT to 10,000-ohm interstage transformer.—J. Cuccia, Universal Strobe Goes Psychedelic, *Popular Electronics*, March 1968, p 45–48 and 98.

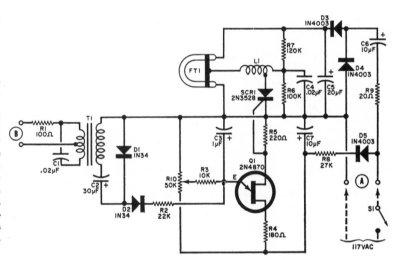

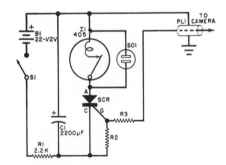

FILL-IN FLASH—Provides brief flash of incandescent light for close-ups, to avoid harsh shadows and washed-out effects of flashbulbs used too close. Gate of scr is triggered by pulse from internal flash battery of camera. I1 is GE 405 or similar 6.5-V flasher bulb that produces brilliant flash when operated on 22.5 V but cuts off current at critical value to avoid burnout. Use fairly slow shutter speed, to avoid sync problems. Will also work with flashbulbs plugged into socket SO1, if flasher is removed. Scr is GE X1, R2 1K, and R3 1.5K.—L. E. Greenlee, Build Li'l Winker, *Popular Electronics*, Dec. 1968, p 47–49.

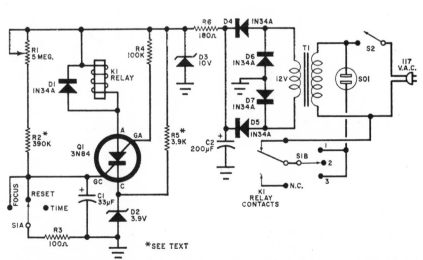

5–60 S TIMER—Use of silicon controlled switch improves accuracy because scs does not load R-C circuit that determines timing interval. Use Sigma 4F-2500S/SIL or similar 2,500-ohm relay with 2-A contacts for handling enlarger lamp. R1 has linear taper. Adjust value of R5 so lamp stays on 65 to 70 s with R1 set for maximum resistance.—F. H. Tooker, Build the SCS Darkroom Timer, *Popular Electronics*, Nov. 1969, p 52–54 and 117.

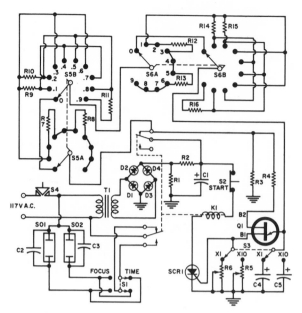

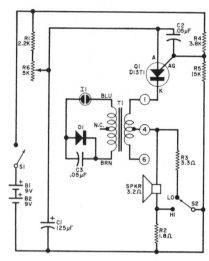

R1,R4—1000 ohm, ½ W res.
R2—190 ohm, 2 W res.
R3—330 ohm, ½ W res.
R5,R6—500-ohm miniature pot
R7,R8—150,000 ohm, ½ W res.
R9,R11—50,000 ohm, ½ W res.
R10—100,000 ohm, ½ W res.
R12,R13—1.5 megohm, ½ W res.
R14,R16—500,000 ohm, ½ W res.
R15—1 megohm, ½ W res.
C1—500 μF, 50 V elec. capacitor
C2,C3—0.005 μF, 1600 V capacitor
C4—1 μF, 50 V elec. capacitor
C5—10 μF, 50 V elec. capacitor
D1,D2,D3,D4—50 p.i.v. diode (or HEP 175 full-wave bridge)

K1—4-pole, d.t., 24 V d.c. relay (2 poles used) (Guardian 200-24D and 200-M5)
T1—117 V trans., 24 V sec. (Triad F45X)
S1,S3—D.p.d.t. bat-handle switch
S2—S.p. normally closed push-button switch
S4—S.p. push-on, push-off switch
S5,S6—D.p. 10-pos., two-deck rotary switch
SO1,SO2—A.c. receptacle
Q1—300 mW unijunction transistor (HEP 310)
SCR1—800 mA, 30 V thyristor (HEP 320)

0.1—99 S TIMER—Uses ujt to trigger scr. Provides repeatable accurate timing, as required for enlargers. Article covers construction and calibration. Extra outlet is provided to turn safelight off while enlarger is *on*, to facilitate focusing of enlarger, or for other purposes.—R. A. Walton, Solid-State Darkroom Timer, *Electronics World*, June 1971, p 66–67.

AUDIBLE TIMER—Provides loud click and flash of red light once per second, for timing printer and enlarger exposures in darkroom. Ujt relaxation oscillator rate is set at 1 Hz by R6. If speaker click is too loud, omit one 9-V battery. Use NE-51H high-intensity neon lamp for I1, and HEP156 1-A 200-V silicon diode for D1. T1 is universal 8-W speaker transformer, such as Allied Radio 54C2021.—A. A. Mangieri, One Second Metronome Timer, *Popular Electronics*, Feb. 1970, p 58–60.

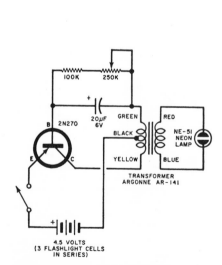

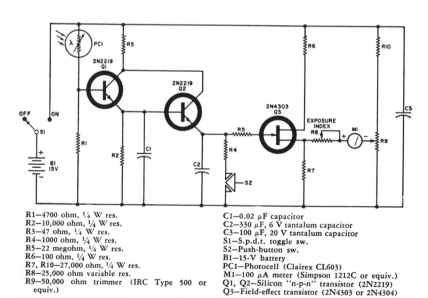

R1—4700 ohm, ¼ W res.
R2—10,000 ohm, ¼ W res.
R3—47 ohm, ¼ W res.
R4—1000 ohm, ¼ W res.
R5—22 megohm, ¼ W res.
R6—100 ohm, ¼ W res.
R7, R10—27,000 ohm, ¼ W res.
R8—25,000 ohm variable res.
R9—50,000 ohm trimmer (IRC Type 500 or equiv.)

C1—0.02 μF capacitor
C2—330 μF, 6 V tantalum capacitor
C3—100 μF, 20 V tantalum capacitor
S1—S.p.d.t. toggle sw.
S2—Push-button sw.
B1—15-V battery
PC1—Photocell (Clairex CL603)
M1—100 μA meter (Simpson 1212C or equiv.)
Q1, Q2—Silicon "n-p-n" transistor (2N2219)
Q3—Field-effect transistor (2N4303 or 2N4304)

BLINKING TIMER—Produces flashes of safe red neon lamp once per second, after calibration by adjusting 250K pot, for photographer in darkroom who is hard of hearing or likes to play radio loud while enlarging, printing, or developing.—F. H. Tooker, Blinking Darkroom Timer, *Popular Electronics*, Sept. 1967, p 87.

FLASH-READING METER—Meter is placed in front of or alongside subject and flash is triggered without exposing film. Meter holds reading long enough to make reading of correct f stop for film speed to which R8 is set. Pulse from photocell is amplified by Q1 and Q2 to charge C2. Charge is read by fet volt-meter circuit. Low leakage of fet makes it possible to hold reading long after flash has dissipated. To prepare for next flash reading, push reset button S2 to discharge C2. Article covers construction and calibration.—W. W. Schopp, Electronic Photoflash Meter, *Electronics World*, June 1970, p 62–63.

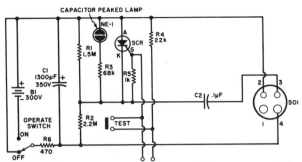

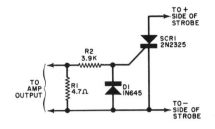

SHUTTER CONTACTS (TO CAMERA)

R1—1.5 megohm, ½ W res.
R2—2.2 megohm, ½ W res.
R3—68,000 ohm, ½ W res.
R4—22,000 ohm, ½ W res.
R5—1000 ohm, ½ W res.
R6—470 ohm, 2 W res.
C1—1300 μF, 350 V elec. capacitor
  (Mallory #CG 132 T 350 F 1)
C2—0.1 μF, 200 V capacitor (C-D 2P1)
Switch—S.p.d.t. switch (Cutler-
  Hammer 8810K15 or equiv.)
SO1—4-pin tube socket (Amphenol #77MIP4
  or equiv.)
NE1—Combination neon lamp / n.o. push-
  button (Grayhill #40-1)
B1—300-V battery (Burgess U-200)
SCR—2N2326 or C6B (General Electric)

3-S RECYCLING FLASH—Medium-power electronic flash produces 58 W-s for flashtube, giving sufficient light for most films and situations while still being small enough to carry comfortably. Circuit includes protection for flash contacts in camera because scr is used to trigger flashtube. Neon indicates when C1 is charged sufficiently for next flash, which is usually within time required to advance film. Article includes nomograph for determining correct f stop. Uses $22 Kemlite flashtube.—W. W. Schopp, Rapid-Flash, *Electronics World*, April 1971, p 68–69.

SOUND-FIRED STROBE—Used in photographing fast action such as hammer smashing light bulb or pin breaking balloon. Sound is picked up by carefully positioned microphone and fed to ordinary audio amplifier having strobe circuit connected in place of speaker. Allow about 1-ms delay for sound to travel between source and mike, to stop action at desired instant.—A. J. Lowe, Build the Sound Sync'er, *Popular Electronics*, April 1967, p 59–60.

FAST-RECHARGE FLASHGUN—Time to recharge for next picture is only 2 or 3 s because high-voltage photoflash battery is used (300-V Eveready 493). Uses MFT-110 flashtube and 6-kV trigger transformer TT-6. Guide number ranges from 50 to 110 depending on reflector used. Article covers construction and gives sources for parts. Diodes in charger are 300-piv 400-mA silicon.—H. Friedman, Speedy-Flash, *Science and Electronics*, Oct.-Nov. 1970, p 29–34.

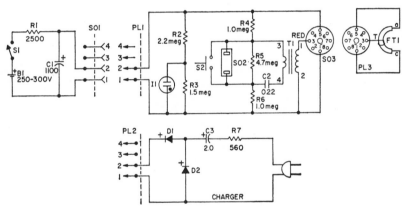

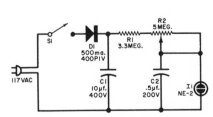

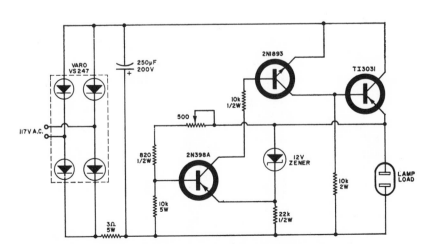

ENLARGER LAMP REGULATOR—Provides constant d-c voltage of 120 V for 75-W enlarger or photoflood lamp despite line voltage variations from 100 to 140 V a-c. Will handle up to 150-W enlarger lamps if normal exposure periods are not exceeded. Includes 500-ohm pot for reducing output voltage to adjust lamp brilliance. Circuit is series regulator using germanium pnp audio power transistor which should have heatsink, with two-stage error amplifier having 12-V zener reference.—R. A. Wolff, Voltage Regulator for Enlarger Lamp, *Electronics World*, Sept. 1969, p 72.

1-S FLASHING METRONOME—R2 is normally adjusted for one flash per second from neon relaxation oscillator. Eliminates need for resetting conventional timer. Calibrate by counting flashes for 30 s.—D. M. Gusdorf, Darkroom Metronome, *Popular Electronics*, Feb. 1965, p 84.

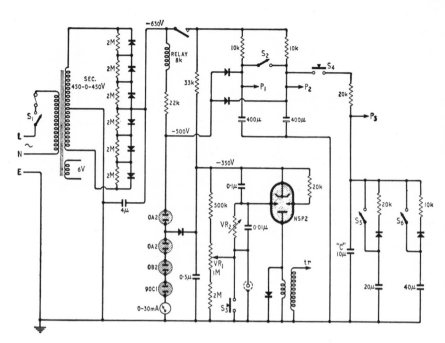

TWO-FLASH STROBE—Versatile circuit gives three choices of outputs—single 50-J or 100-J flashes, two simultaneous 50-J flashes from two flashtubes, or repeated flashes of 1 to 7 J at 1 to 80 per s. Two 400-μF electrolytics charged to 500-V form main energy storage for discharging through FA27 or similar flashtubes connected between ground and terminals P1 and P2. VR1 is preset control for trigger sensitivity, and VR2 controls strobe rate. All diodes are OY241 or similar. Trigger pulse generator is relaxation oscillator using Ferranti Neostron NSP2 (now EN10) cold-cathode trigger tube feeding high-ratio pulse transformer whose secondary goes to trigger electrodes of both flashtubes.—J. D. Pye, Simple Electronic Stroboscope, *Wireless World*, July 1964, p 339–342.

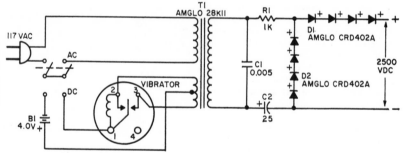

2,500-V FLASH SUPPLY—Power transformer operates either from line voltage or 4-V vibrator, with rectifiers providing 2,500-V d-c to capacitor having 75 watt-seconds capacity, to give flash speeds ranging from 90 to 700 μs.—J. Kyle, Quick as a Flash, *Science and Electronics*, Oct.-Nov. 1970, p 39–48.

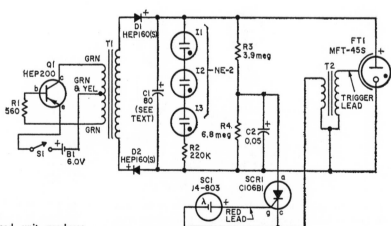

SLAVE FLASH—Self-contained unit produces auxiliary flash for secondary lighting when main flash hits Calectro silicon solar cell SC1. With 80 μF for C1 (450 V), output is 6 watt-seconds. Larger values of C1 give higher output, up to 30 watt-seconds for maximum permissible size of 400 μF. T1 is power transformer with 12.6-V CT secondary, used backward for stepping up a-c voltage generated by oscillator Q1. T2 is 4-kV Mura TR2 or equivalent trigger transformer.—S. Daniels, Li'l Blitzer, *Science and Electronics*, Oct.-Nov. 1970, p 35–38 and 98–99.

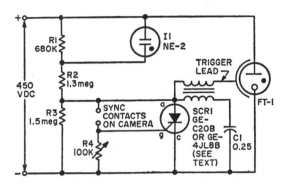

SCR TRIGGER FOR FLASH—Provides protection for camera sync contacts. Can be used as slave flash if GE 4JL8B is substituted for scr.—J. Kyle, Quick as a Flash, *Science and Electronics*, Oct.-Nov. 1970, p 39—48.

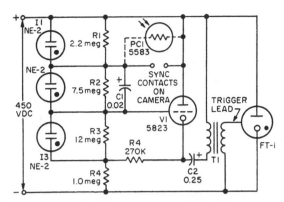

THYRATRON SWITCH FOR FLASH—Use of electronic switch in trigger circuit greatly minimizes possibility of damage to sync contacts of camera. Addition of photocell PC1 permits use of unit as slave flash. Neon I1 serves as ready indicator, while other neons regulate thyratron voltage.—J. Kyle, Quick as a Flash, *Science and Electronics*, Oct.-Nov. 1970, p 39—48.

PHOTORESIST EXPOSURE CONTROL—Light-sensitive ujt oscillator and 2N1306 logic driver are used to feed predetermined counter that energizes shutter-closing relay when total incident light flux on photoresist layer is precisely correct value. Article also gives logic and counter circuits. Type of light sensor is not critical. Choose value for timing capacitor CT to provide desirable count rate at light level used.—G. A. Riley, An Inexpensive Integrating Photoresist Exposure Control System, *RCA Review*, June 1970, p 407—413.

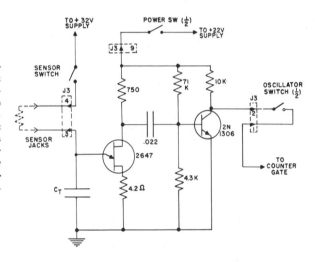

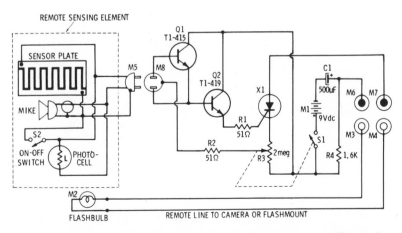

ANIMAL-TRIPPED FLASH—Circuit shows use of three different sensors, any one of which will fire scr X1 (2N3228) and flashbulb M2 aimed at area for which picture is desired at night. If camera is left open for flash, shutter must be closed before sunrise. Alternative is wiring shutter mechanism into flash circuit. Moisture-sensitive plate must be most sensitive available because feet of bird or animal are not normally as damp as drop of rain. Photocell requires light source positioned so beam is interrupted by animal. Place mike in cardboard tube or parabolic dish aimed at baited spot, so it will not respond to birds in flight or other sounds outside camera range.—R. M. Brown and T. Kneitel, "49 Easy Entertainment and Science Projects," Howard W. Sams, Indianapolis, IN, 1971, p 53—55.

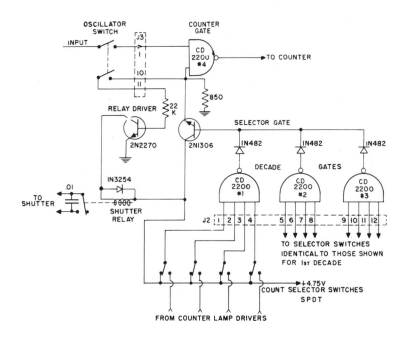

PHOTORESIST EXPOSURE CONTROL—Input from light-sensitive oscillator feeds predetermined counter through logic gates shown, to energize shutter-closing relay when total incident flux on photoresist layer is precisely correct value for exposure. When preselected count is reached, counter outputs connected to decade gates all become 1 simultaneously for first time since onset of counting. Since all other decade gate inputs were previously fixed at 1 by selector switches, every decade gate has a 1 at its input; all gate outputs are then 0, turning off 2N1306, turning off relay driver, and releasing relay to close shutter. Counter is then cleared with reset button to prepare for new cycle.—G. A. Riley, An Inexpensive Integrating Photoresist Exposure Control System, *RCA Review*, June 1970, p 407–413.

PHOTOFLOOD DIMMER—Triac Q1 and separate trigger diac D1 serve with R1 to provide continuous control of 500-W photoflood lamp from full off to essentially full on. Use type 8AG fast-action 5-A fuse in series with lamp to prevent damage to triac by surge current when lamp burns out.—H. Friedman, "99 Electronic Projects," Howard W. Sams, Indianapolis, IN, 1971, p 74.

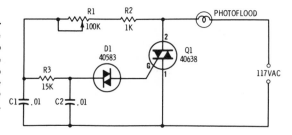

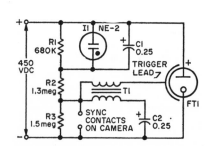

FLASH TRIGGER—Closing of camera sync contacts discharges C2 through primary of trigger transformer T1, producing trigger pulse for firing 450-V flash tube. Transformer secondary voltage depends on flashtube used. Neon lamp indicates when 450-V storage capacitor is charged and ready for next flash. —J. Kyle, Quick as a Flash, *Science and Electronics*, Oct.-Nov. 1970, p 39–48.

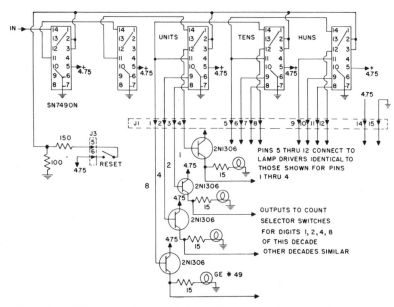

PHOTORESIST EXPOSURE CONTROL—Five-decade counter is used to count cycles of output of light-sensitive oscillator and, when preselected count is reached, make logic gates close shutter. Pressing reset button clears counters for start of next cycle. Exposure time is controlled with variation of less than 0.5% for mean exposure time of 93.7 s. —G. A. Riley, An Inexpensive Integrating Photoresist Exposure Control System, *RCA Review*, June 1970, p 407–413.

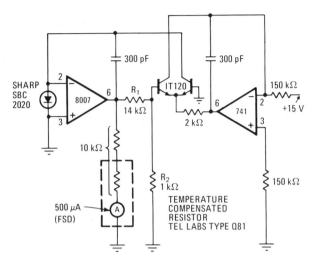

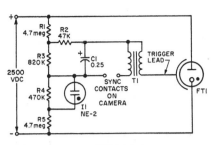

LIGHT METER—Meter gives log of light intensity as photographic exposure value, with each unit change corresponding to factor-of-two change in light intensity. Silicon photocell is operated at zero voltage to minimize leakage errors. Uses 8007 opamp designed for picoampere range, permitting measurements at low light levels where cell current may be only 20 or 30 pA.—D. Fullagar, Better Understanding of FET Operation Yields Viable Monolithic J-FET Op Amp, *Electronics*, Nov. 6, 1972, p 98–101.

2,500-V FLASH TRIGGER—Voltage rating of trigger transformer should match requirements of flashtube. Sync circuit is isolated from high-voltage circuit. Neon glows to indicate readiness for next flash.—J. Kyle, Quick as a Flash, *Science and Electronics*, Oct.-Nov. 1970, p 39–48.

PHOTOFLASH SLAVE—Flash from camera's own flashbulb or electronic flash turns on light-activated scr to trigger additional flashbulb without wire connections. Will operate from two 9-V transistor radio batteries in series. Be sure S1 is in load position. Use 1/10th-s shutter speed to allow for delay in slave.—"Electronics Experimenters Circuit Manual," General Electric, Owensboro, KY, 1971, 3rd Ed., p 135–136.

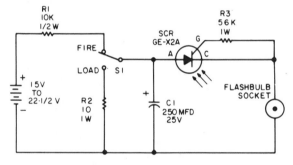

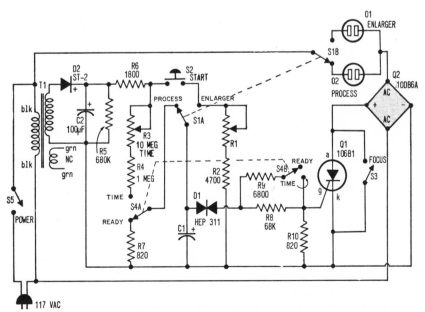

UNIVERSAL TIMER—Can be used either for exposure time or developing time, at flick of switch. With 50K for R1 and 200 μF for C1, exposure is 7 s and maximum developing time 7 min. For 15 s and 15 min, increase R1 to 100K and C1 to 400 μF. When enlarger is plugged into outlet O1, it comes on when S2 is pressed, and goes off automatically. For developing, bell or buzzer is plugged into O2.—R. Michaels, Universal Darkroom Timer, *Science and Electronics*, Oct.-Nov. 1969, p 35–39.

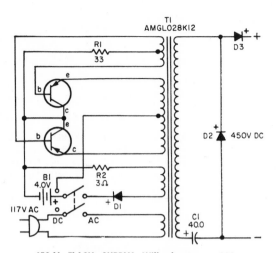

450-V FLASH SUPPLY—Will charge to 100 watt-seconds capacity. Provides automatic recharge of 4-V nickel-cadmium storage battery when operating from a-c line. Transistors can be 2N441 or 2N1168. Use low-voltage silicon rectifier diode for D1.—J. Kyle, Quick as a Flash, *Science and Electronics*, Oct.-Nov. 1970, p 39–48.

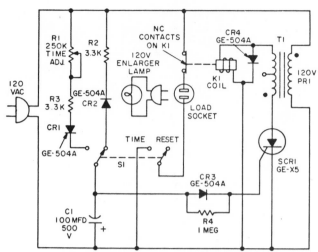

ENLARGER PHOTOTIMER—R1 gives range of about 0.01 to 60 s for automatic timing of enlarger lamp up to 500 W with repeatibility better than 2% if high-quality electrolytic is obtained for C1. K1 is 24-V a-c relay such as Potter & Brumfield No. MR5A. T1 is filament transformer with 12.6-V C-T secondary, such as Triad F25X.—"Electronics Experimenters Circuit Manual," General Electric, Owensboro, KY, 1971, 3rd Ed., p 131–133.

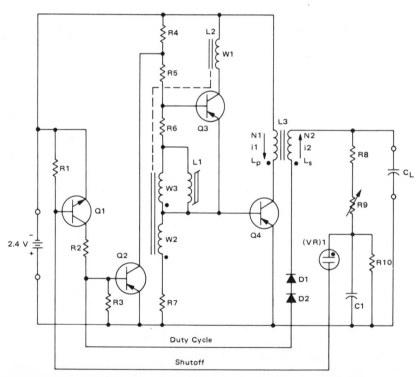

PHOTOFLASH CONVERTER—Steps up 2.4-V nickel-cadmium cell voltage (two cells in series) to 500 V for charging 480-μF energy-storage capacitor CL of camera-controlled flashgun. Report gives winding data for all three transformers and describes circuit operation in detail. Diodes are MR814. Q1 is MPS6520, Q2 MPS6563, Q3 MPS6562, and Q4 MP3613. Neon VR1 is selected 5AG. R1 is 39K, R2 100, R3 1K, R4 120, R5 150, R6 270, R7 725, R8 1 meg, R9 2 meg, and R10 390K.

Inverter frequency is about 2,000 Hz, which gives pleasant audible sound to indicate capacitor is charging. Circuit shuts off inverter automatically when capacitor is charged, so only standby current is battery drain through voltage divider. Charging time is about 15 s with AA cells.—L. T. Rees, "Designing DC-DC Converters for Capacitor Charging with Batteries," Motorola Semiconductors, Phoenix, AZ, AN-442, 1971.

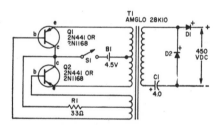

FLASH SUPPLY—Simple oscillator operating from three D cells feeds stepup transformer and rectifier to provide 450-V d-c for flashtube. Will charge storage capacitors of up to 60 watt-second capacity in 10 s.—J. Kyle, Quick as a Flash, *Science and Electronics*, Oct.-Nov. 1970, p 39–48.

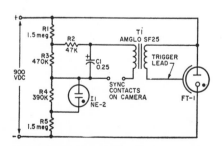

900-V FLASH TRIGGER—Neon glows when storage capacitor has been recharged to 900 V for next flash. Sync circuit is isolated from high-voltage circuit. Flashtube rating must be at least 900 V, and all insulation rated above 1,000 V.—J. Kyle, Quick as a Flash, *Science and Electronics*, Oct.-Nov. 1970, p 39–48.

CHAPTER 87
Power Supply Circuits

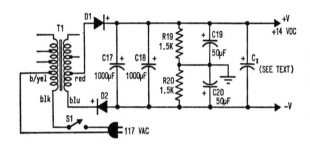

14-V BIPOLAR—Designed for use with stereo headphone amplifier to provide ±14 V with respect to ground, without regulation. Power transformer is Allied 54E4731 providing 10 V, 20-V CT, and 40-V CT at 0.035 A from secondaries. Cx is needed only if amplifier breaks into high-frequency oscillation; use 100-μF 15-V electrolytic. D1 and D2 are 750 mA 200-piv silicon diodes.—H. Friedman, The StereoFone, *Elementary Electronics*, Sept.-Oct. 1969, p 49—52 and 56.

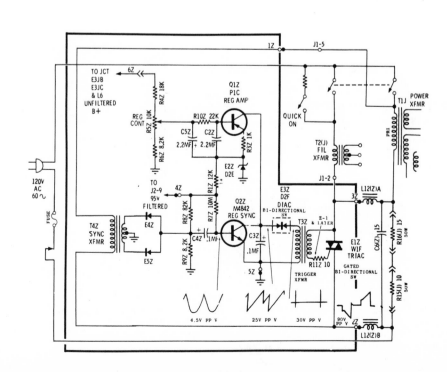

A-C LINE REGULATOR—Used in Motorola Quasar solid-state color tv receiver to provide constant 105-V a-c to power transformer of set over wide range of a-c line voltages.—What You Can Expect in 1970 Television Sets, *Electronic Technician/Dealer*, Oct. 1969, p 50—54.

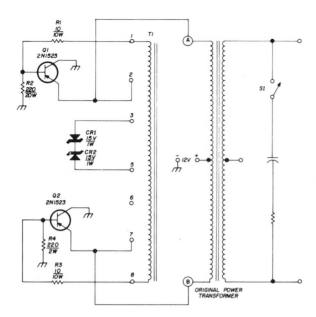

TWO-TRANSISTOR VIBRATOR—Solid-state switching circuit duplicates electrical function of vibrator in vhf mobile transmitter power supply. T1 is Osborne 2709 toroid, made by Osborne Transformer Co., 2823 Mitchell Ave., Detroit, Mich. 48207. Under load, switch frequency is about 1,000 Hz. Use heat sinks on transistors. Zeners protect transistors from transients.—R. Kashubosky, Solid-State Vibrator Replacement, *Ham Radio*, Aug. 1972, p 70–71.

ELECTRONIC VARIAC—Provides full range of voltage control from 0 to 120 V a-c at up to 10 A, operating from a-c line. Capacitors suppress r-f noise. Triac D2 should be mounted on good heatsink. If load is power transformer driving 24-V full-wave bridge, combination provides 0- to 30-V d-c for bench work.—J. Riff, Electronic Variac, *73*, Nov. 1969, p 88–89.

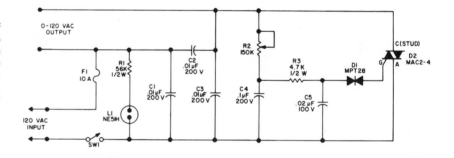

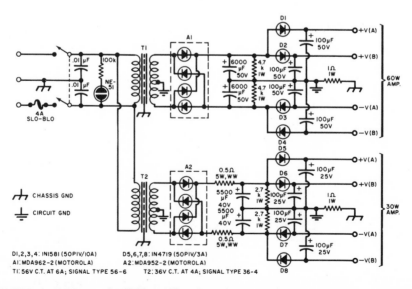

25 AND 36 V FOR 4-CHANNEL AMPLIFIER—Dual power supply provides 25 V with both polarities for two 30-W amplifiers feeding treble speakers and 36 V with both polarities for 60-W amplifiers feeding bass speakers of stereo system using 2-channel program material. Power transformers used have inherent regulation.—L. H. Garner, A 4-Channel Amplifier for Multi-Speaker Systems, *Electronics World*, Nov. 1970, p 28–30.

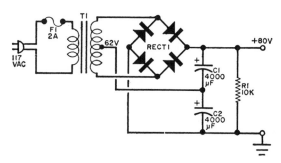

80 V AT 3 A—Designed to handle both channels of 75-W rms power amplifier for hi-fi stereo system, along with preamp. Uses 4-A bridge rectifier. Filters should be rated 50 V. Power transformer can be SW Tech P-3154 or similar, providing 62 V at 3 A.—D. Meyer, Tigers That Roar, *Popular Electronics*, July 1969, p 51–53, 58–63, and 99.

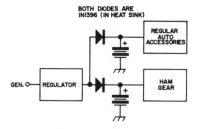

DUAL BATTERY SUPPLY—Circuit shows how to use separate storage battery for mobile transmitter without running down regular battery, while keeping both charged with regular auto generator. Diodes act as one-way switches that prevent power from flowing between batteries. Use silicon diodes.—P. Franson, "Diode Circuits Handbook," 73 Inc., Peterborough, NH, 1967, p 23A.

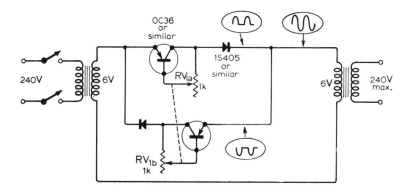

LINE VOLTAGE CONTROLLER—Ganged pot provides convenient and inexpensive means of reducing a-c line voltage. Output power depends on power ratings of transformers and transistors.—J. R. Harris, Transistor A.C. Mains Controller, *Wireless World*, May 1969, p 234.

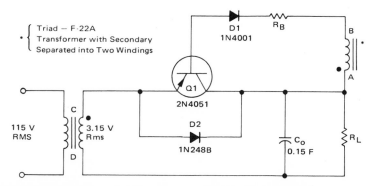

HALF-WAVE SYNCHRONOUS RECTIFIER—Uses germanium transistor Q1 having less than 0.3-V drop at 20 A collector current. Designed to provide high current at low voltage (3.2 V) for such systems as thermoelectric coolers and large digital IC systems.—B. C. Shiner, "Improving the Efficiency of Low Voltage, High-Current Rectification," Motorola Semiconductors, Phoenix, AZ, AN-517, 1971.

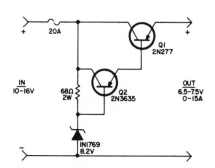

DROPPING 12 V TO 6 V—Permits operating 6-V mobile equipment in autos having 12-V storage battery. Use large heatsink for Q1. Alternatives for Q2 are HEP232, 2N2147, 2N4314, and 2N3616. Q1 can be HEP231.—Circuits, 73, Feb. 1973, p 143.

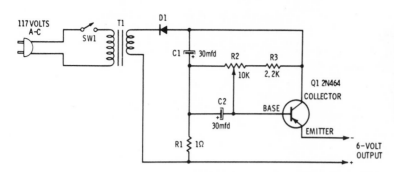

6-V BATTERY ELIMINATOR—Permits operating 6-V battery radio from a-c line. Many 9-V radios will also work satisfactorily on 6 V. D1 is 1N34A and T1 is 6.3-V filament transformer. R2 adjusts output voltage.—R. M. Brown and T. Kneitel, "49 Easy Transistor Projects," Howard W. Sams, Indianapolis, IN, 1972, p 39–40.

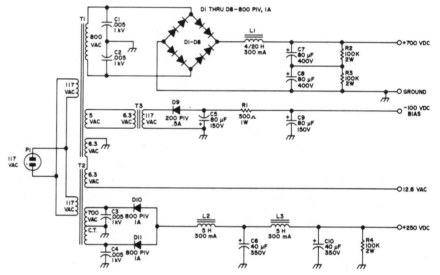

TRANSCEIVER SUPPLY—Provides 700 V at 200 mA, 250 V at 300 mA, and −100 V for bias as required by many types of transceivers.—D. Pongrance, Transceiver Power Supply, 73, Aug. 1972, p 126.

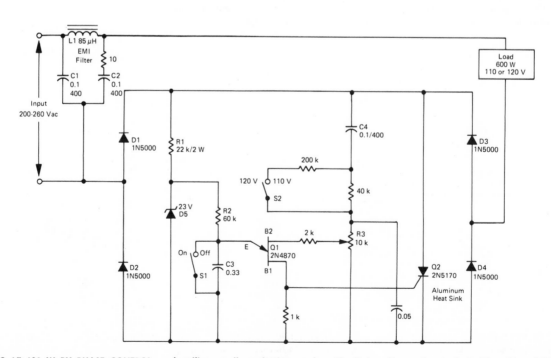

120-V A-C AT 600 W BY PHASE CONTROL— Single scr provides full-wave control, with simple open-loop compensator included for handling small conduction angles. Circuit is suitable primarily for conduction angles less than 90 deg, so requires at least 220-V a-c line as input.—D. Perkins, "True RMS Voltage Regulators," Motorola Semiconductors, Phoenix, AZ, AN-509, 1972.

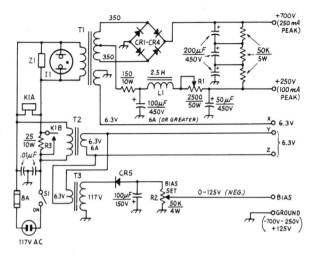

700 AND 250 V—Meets requirements of most medium-power ssb transceivers. Diodes are 1,000-prv 1-A silicon except CR5 which is 200-prv 500-mA silicon. Diode-protection relay K1 (115-V a-c with 10-A contacts) shorts out surge resistor R3 after capacitor bank at output of bridge rectifier is charged. Z1 is GE 6RS20SP4B4 thyrector for transient suppression.—"The Radio Amateur's Handbook," American Radio Relay League, Newington, CT, 49th Ed., 1972, p 123–124.

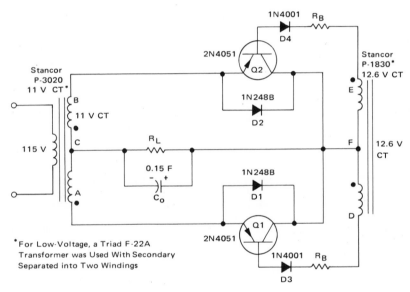

*For Low-Voltage, a Triad F-22A Transformer was Used With Secondary Separated into Two Windings

FULL-WAVE SYNCHRONOUS RECTIFIER—Uses germanium rectifiers having less than 0.3-V drop at 20 A collector current, for efficient supply of high current at low voltage for such systems as thermoelectric coolers and large digital IC systems. Silicon diodes D1 and D2 are placed across transistors to protect them when load is capacitive.—B. C. Shiner, "Improving the Efficiency of Low Voltage, High-Current Rectification," Motorola Semiconductors, Phoenix, AZ, AN-517, 1971.

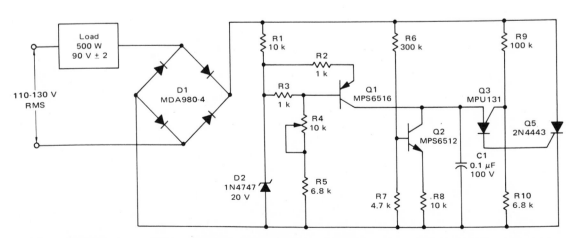

90-V 500-W RMS REGULATOR—Circuit operates by delaying firing of scr Q5 as input voltage increases. Q1 is in delay network that prevents circuit from latching up at beginning of each charging cycle. Report describes operation in detail.—R. J. Haver and B. C. Shiner, "Theory, Characteristics and Applications of the Programmable Unijunction Transistor," Motorola Semiconductors, Phoenix, AZ, AN-527, 1971.

CHAPTER 88
Protection Circuits

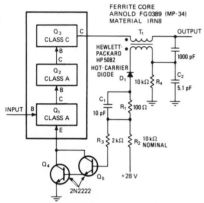

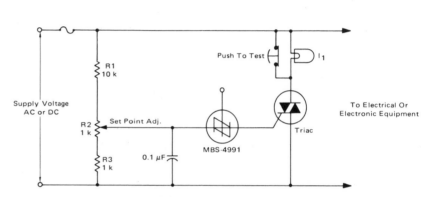

OPEN POWER AMPLIFIER—Output of class C r-f amplifier goes through transformer T1, which senses high vswr that occurs when load is removed. Resulting rectified voltage drives Q4 and Q5 into cutoff. This in turn acts through transistors Q1 and Q2 to remove drive to Q3 in power amplifier, shutting off amplifier before damage occurs to Q3. Protection circuit acts as integral notch filter whenever its secondary winding is tuned to second harmonic of amplifier.—F. A. Warren, Jr., VSWR Detector Protects Class C RF Amplifiers, *Electronics*, Feb. 14, 1972, p 78–79.

ADJUSTABLE CROWBAR—Operating point for turning on triac to short out load and protect it against overvoltage can be adjusted with R2 over range of 60- to 120-V d-c or 42- to 84-V a-c. Voltage rating of triac must be greater than highest operating point set by R2. I1 is low-wattage lamp rated for supply voltage. Mallory Sonalert or other alarm may be connected across fuse to provide audible indication that it has been blown by crowbar action.—L. J. Newmire, "Theory, Characteristics and Applications of Silicon Unilateral and Bilateral Switches," Motorola Semiconductors, Phoenix, AZ, AN-526, 1970.

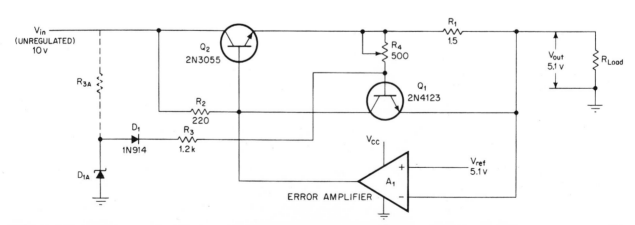

FOLDBACK FOR SHORTS—Prevents transistor burnout in voltage regulators during debugging, by providing foldback current-limiting that shuts off supply almost instantly when short occurs. R4 adjusts amount of foldback; maximum setting gives maximum foldback but regulator cannot restart until R4 is reduced. Anode of D1 should be connected to Vref (drawing has error).—B. Stopka, Shorted Load Folds Back Supply Current, *Electronics*, June 21, 1971, p 65.

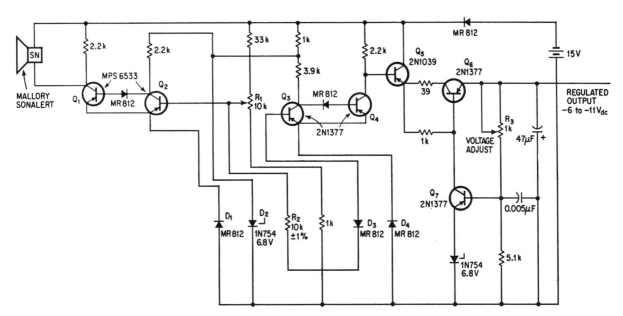

BATTERY-DISCHARGE ALARM—Protection circuit using two Schmitt triggers activates alarm when voltage of 15-V battery drops to trigger level determined by setting of R1, between —6 and —14 V. R2 controls second trigger point, about 0.5 V lower than first, at which voltage regulator is shut down. Entire circuit draws only 25 mA from fully charged battery.—D. Jeutter, Battery Discharge Triggers Alarm and Shuts Off Supply, *Electronics*, Jan. 4, 1971, p 68.

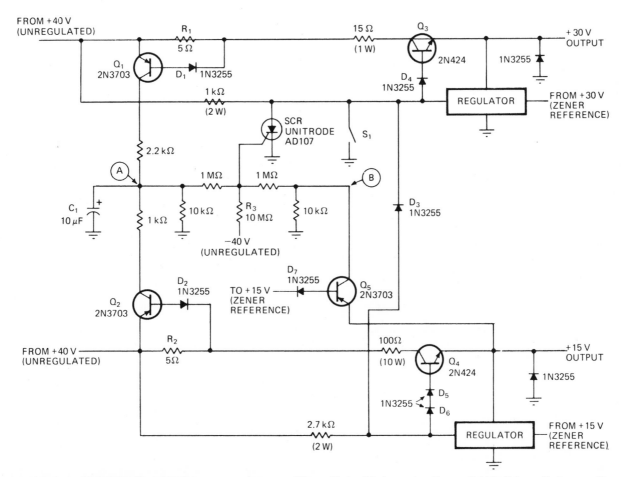

15-V AND 30-V OVERVOLTAGE MONITOR—Single low-power scr protects components in two different regulated power supplies from overvoltage conditions. Diode D3 interconnects supplies.—P. T. Uhler, Diode and SCR Protect Multiple-Voltage Equipment, *Electronics*, Oct. 11, 1971, p 75.

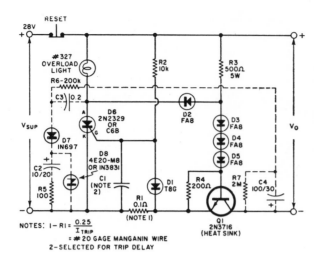

AUTOMATIC-RESET BREAKER—Manually reset solid-state circuit breaker shown in solid lines will, when modified by adding dashed-line connections and components, reset automatically within 3 s after circuit fault has cleared. R6-C2 combination determines time to first reset. With shorted output, consecutive reset attempts occur at about 3, 7, 12, 23, and 63 s after short. Fault current passes through R1 and triggers scr into conduction in original circuit.—S. W. Thomas, Automatic Reset For Circuit Breaker, *Electronics World*, Dec. 1969, p 92.

POWER SUPPLY—Prevents damage to 12-V regulated supply of commercial communication receiver in case of short-circuit or if battery supply is connected with wrong polarity. CB1 is 2-A thermal circuit breaker, I1 and I2 are 12-V pilot lamps, CR1 is 3-A 50-prv silicon, and CR2 is 9-V zener providing additional regulated output.—Hammarlund HQ-215 Receiver, *QST*, Dec. 1968, p 50–54.

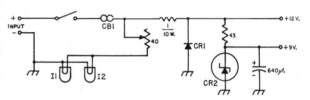

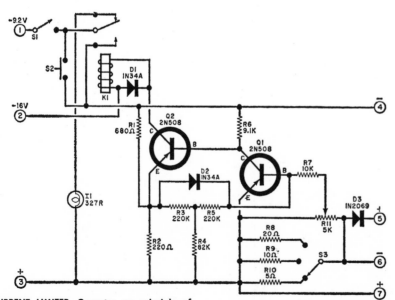

CURRENT LIMITER—Operates on principle of shunt current meter, to protect solid-state devices from overload. With overload, as determined by setting of R11, Q1 becomes forward-biased, Q2 bias is lowered, K1 is de-energized, and output circuit is disabled.

Pushing reset switch S2 momentarily restores operation only if overload has cleared. K1 is 5,500-ohm relay.—J. L. Keith, Overload Protection, *Popular Electronics*, March 1970, p 54–56.

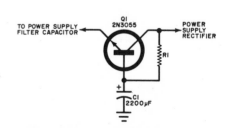

POWER DELAY—Protects solid-state rectifiers from current surges when power supply is switched on from cold start. The higher the values of R1 and C1, the slower will be the rate at which C1 charges to the point at which Q1 becomes conducting. Adjust R1 for delay of 0.5 to 1 s.—F. H. Tooker, Slow Turn-On Protects Power Supply, *Popular Electronics*, Sept. 1972, p 73.

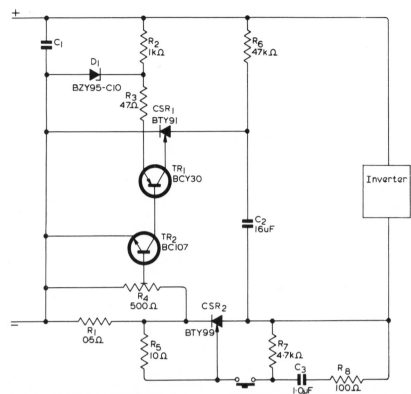

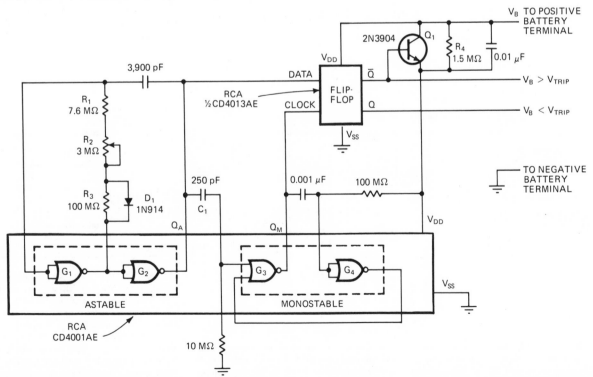

INVERTER OVERLOAD—Circuit protects thyristor of high-power high-frequency inverter against effects of commutation failure. Designed for inverter providing 15 A and requiring operation of safety circuit when fault increases load current to 20 A. When button at bottom is pressed, thyristor CSR2 conducts and completes circuits between inverter and supply. When fault increases current through R1, transistors turn on thyristor CSR1, making C2 discharge suddenly and reduce anode potential of CSR2 below conduction value, to disconnect inverter from supply.—"Fast Turn-Off Thyristors in High-Frequency Inverter—BTX64, BTX65," Mullard, London, 1967, TP895, p 5–8.

SAVING OUTPUT TRANSISTORS—Suitable for either single or dual polarity supplies of class B audio amplifiers. Dissipation in output transistors cannot appreciably exceed normal full-load value when load is reduced or becomes short-circuit.—J. R. I. Piper, Output Transistor Protection in Class B Amplifiers, *Wireless World*, Feb. 1972, p 77.

NI-CD VOLTAGE MONITOR—Turns off equipment operating from nickel-cadmium batteries when voltage falls below safe level, to prevent permanent damage to batteries. Circuit resets automatically when batteries are recharged. Monitor draws only about 0.5 μA. R2 is adjusted so periods of astable and mono are equal when battery voltage VB is at desired trip point, which can be 3.6 V for three batteries in series.—W. Wilke, C-Mos Voltage Monitor Protects Ni-Cd Batteries, *Electronics*, March 1, 1973, p 85–86.

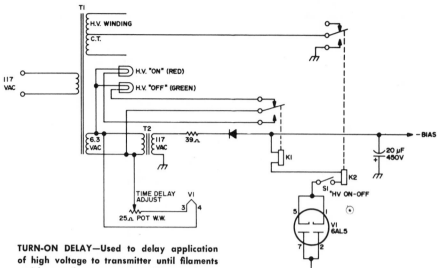

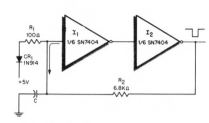

TURN-ON DELAY—Used to delay application of high voltage to transmitter until filaments and heaters have warmed up. Rheostat in series with heater varies heater voltage of 6AL5 tube, to adjust time for cathode to reach emitting temperature and make tube conduct sufficiently to energize coil of relay K2. Delay ranges from about 10 s for zero resistance to maximum practical value of 2

minutes, with greatest consistency around 30 s. K1 is optional, serving only to turn on red and green panel lights.—O. Wrench, Adjustable Time Delay Relay Circuit, 73, Oct. 1972, p 71–72.

AUTOMATIC RESET—Protective circuit uses part of IC package and only four discrete components to insure that latching circuits are in proper state when power is reapplied after turnoff. When power is applied, C is discharged, then recharged by current from I1 until logic-1 threshold is reached at which I1 makes output of I2 go high. C then charges through R2 to output voltage of I2, to give required direct clear input pulse to latching circuit. C is 0.1 to 1 μF.—J. O. Farmer, A Simple Power-On Reset Circuit, *The Electronic Engineer*, Oct. 1971, p 69.

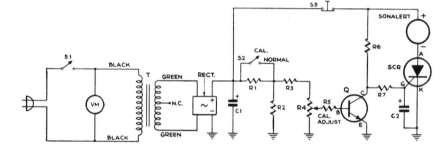

BROWNOUT ALARM—Uses scr to activate Sonalert alarm when a-c line voltage drops below predetermined value at which air conditioners and certain other home appliances may be damaged if not turned off. R4 determines voltage at which alarm sounds. Alarm stays on even after line voltage returns to normal, until reset button is pushed or S1 is opened. R1 is 7.5, R2 110, R3 3.3K, R4 500, R5 4.7K, R6 10K, R7 2.2K, T 6.3-V 0.6-A filament transformer, VM 150-V a-c voltmeter, RECT HEP175, Q HEP50, SCR HEP320, Sonalert SC628, C1 and C2 500 μF, and S3 normally closed pushbutton switch.—R. M. Benrey, Homemade BROWNOUT ALARM to Protect Your Appliances, *Popular Science*, Aug. 1971, p 90–91.

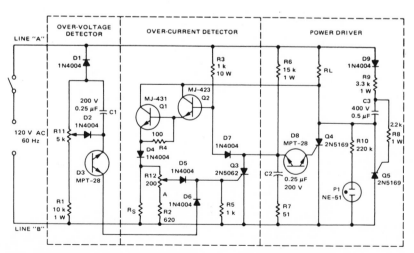

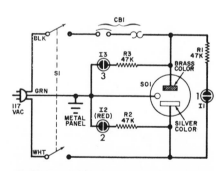

A-C OVERVOLTAGE AND OVERCURRENT—Protects resistive loads from excessive voltage and from currents over 20 A. Automatically resets when current returns to normal. Does not require current-sensing element in series with load RL. Voltage divider R2–R12 permits selection of trip point for overcurrent protec-

tion, and R11 serves same function for overvoltage. Report describes circuit operation in detail. RS is about 0.1 RL.—"AC Overvoltage and Overcurrent Protective Circuit with Automatic Reset," Motorola Semiconductors, Phoenix, AZ, AN-454, 1971.

GROUND TESTER—Must be plugged into three-terminal wall outlet, or ground (GRN) lead connected to positive ground. When two-lead or three-lead appliance is plugged into circuit outlet SO1, neon lamps I1 and I3 will light if appliance is safe. If I2 comes on, appliance is dangerous to human life because its "cold" lead is 117 V above ground. If all three lamps light when no appliance is plugged in, wall outlet has no ground or is improperly grounded.—L. E. Greenlee, Why Play Edison Roulette?, *Popular Electronics*, Aug. 1969, p 71–74.

CHAPTER 89
Public Address Circuits

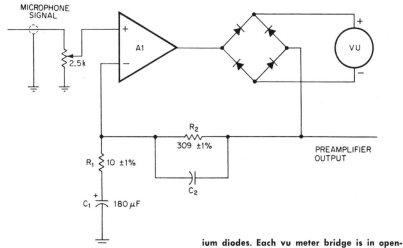

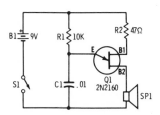

PREAMP WITH VU METER—Simplifies balancing of multi-input auditorium or other audio system, particularly for stereo reproduction, when used with each microphone. Amplifier A1 is built with differential input pair of 2N3391 transistors, and bridge with german-ium diodes. Each vu meter bridge is in open-loop gain path of its opamp, nearly eliminating diode offset voltages from output. Inexpensive IC opamp may be used.—G. R. Latham IV, Operational Preamps Help Balance Big Sound System, *Electronics*, Dec. 7, 1970, p 74.

MIKE BEEPER—Audio tone generator with miniature 3.2 or 8-ohm speaker and small transistor radio battery can be clamped to front of microphone with rubberband, for generating continuous tone signal that replaces human voice in checking out complete public address system.—H. Friedman, "99 Electronic Projects," Howard W. Sams, Indianapolis, IN, 1971, p 100–101.

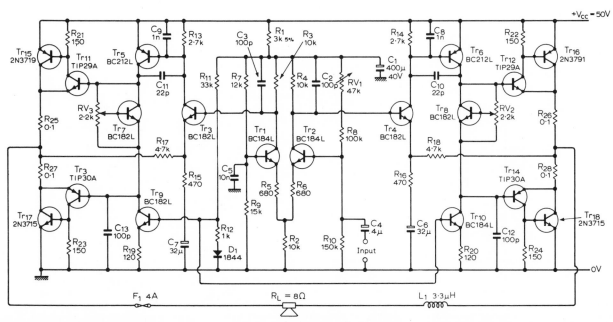

100 W WITH BRIDGE OUTPUT—Requires only 50-V regulated supply. Phase splitter is long-tailed pair TR1-TR2. Response is within 1 dB from 10 Hz to 20 kHz. Signal-noise ratio is 89 dB for 600-ohm source. Requires 180-mV input for full output power.—I. Hardcastle, High-Power Amplifier, *Wireless World*, Oct. 1970, p 477–481.

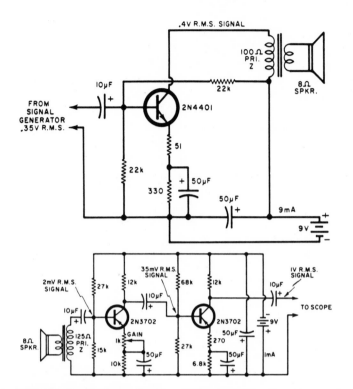

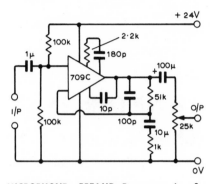

MICROPHONE PREAMP—Response is flat within 1 dB from 20 to 20,000 Hz, with bottom and top rolloff. Voltage gain is 50.—T. D. Towers, Elements of Linear Microcircuits, *Wireless World*, Feb. 1971, p 76—80.

PHASING SPEAKERS—Combination of a-f transmitter and a-f receiver is used to provide relative phasing of speakers in public-address systems for churches, halls, and auditoriums, for optimum placement as well as correct phasing. Signal generator is used to feed 200 Hz into audio amplifier driving speaker placed in front of one of microphones. All but one speaker are disconnected and leads terminated with resistors equal to speaker impedance. Two-stage receiver feeding cro is placed in listening area and amplitude changes in pattern are observed while reversing speaker leads and moving receiver around. Article gives step-by-step procedure for checking out other mikes and speakers one by one.—H. Stratman, Phasing P.A. Speakers, *Electronics World*, Dec. 1970, p 31 and 56.

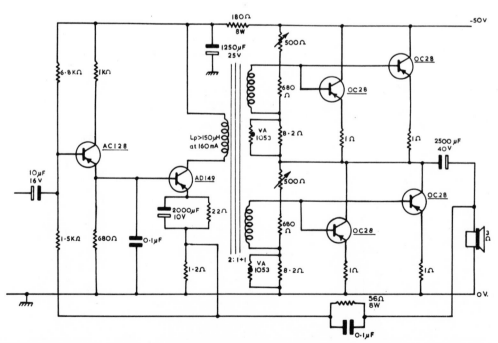

60-W PUSH-PULL CLASS B—Both driver and push-pull parallel output transistors require heatsinks. Response is 3 dB down at 25 and 16,500 Hz for 20-W output. Total harmonic distortion, predominantly 3rd, is 5% at full power and proportionally less at lower outputs. Input impedance is 6K. With higher-resistance speaker, output is reduced in inverse proportion.—"60 W Class B Push-Pull Transistor Power Amplifier," Mullard, London, 1967, TP866.

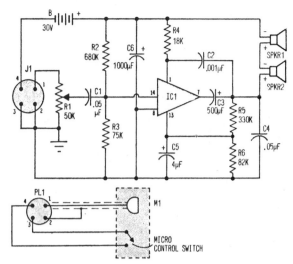

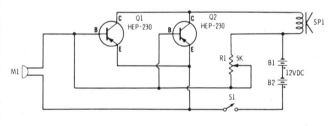

MEGAPHONE—Provides considerable volume from any p-m speaker with only single transistor and 4.5-V or 6-V (maximum) battery. Requires carbon mike; crystal or ceramic mikes will be damaged.—R. Brown and T. Kneitel, "101 Easy Audio Projects," Howard W. Sams, Indianapolis, IN, 1971, p 18—19.

BULLHORN—GE PA-246 IC and associated components are chosen for maximum amplification in intelligence-carrying mid-range voice frequencies, for maximum effectiveness of 5-W drive to pair of 5-W 8-ohm miniature reentrant horn speakers mounted on amplifier housing. To conserve life of 20 D cells in series (30 V), on-off switch is mounted alongside dynamic mike which can be Piezo DX-109. Can be used by football coach, cheerleader, for crowd control, or on boat. Article covers construction.—L. Jorgensen, Touchdown TwinHailer, *Elementary Electronics*, Sept.-Oct. 1969, p 31—36.

MEGAPHONE—Two transistors and 12-V battery combined with carbon mike provide ample speaker volume for use on boats and at sporting events. Transistors are connected in parallel to match 4-ohm impedance of speaker or horn. Use two 6-V lattern batteries or eight D cells in series. Adjust R1 for maximum sound with acceptable distortion. —H. Friedman, "99 Electronic Projects," Howard W. Sams, Indianapolis, IN, 1971, p 21—22.

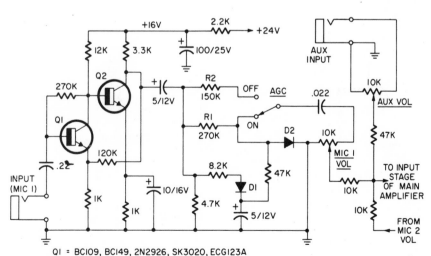

Q1 = BC109, BC149, 2N2926, SK3020, ECG123A
Q2 = BC108, BC148, 2N1566, SK3020, ECG123A
D1, D2 = BA100, IN914, OR EQUAL

AUDIO AGC—Can be used as preamp for mike in public address system. When amplified mike signal is high enough, D1 conducts and forward-biases D2, lowering its impedance and lowering signal voltage developed across it for transfer to input of main p-a amplifier. When agc circuit is not needed, R2 is cut into circuit in place of D1 so volume control settings do not have to be changed. —Add-On AGC for PA System, *Radio-Electronics*, March 1972, p 42.

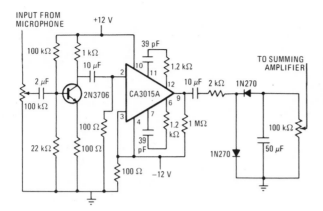

BACKGROUND-NOISE AMPLIFIER—Dynamic mike in auto, stadium, or room picks up background noise along with desired audio signal from speaker system, for combining with sample of audio signal to give d-c control voltage that increases output level of speaker system automatically as background noise increases. Article gives all circuits.—D. B. Hoisington, Automatic Control of Speaker Output Compensates for Noisy Background, *Electronics*, Nov. 20, 1972, p 118–121.

BACKGROUND NOISE CONTROL—Combines inputs from audio sampling circuit with audio-plus-noise amplifier to give d-c control voltage for changing output level of speaker system automatically by more than 45 dB to compensate for changing background noise in mobile receiver, public-address system, or lecture-hall sound system. Output voltage range is −3.6 to −5.3 V. Article gives all circuits. —D. B. Hoisington, Automatic Control of Speaker Output Compensates for Noisy Background, *Electronics*, Nov. 20, 1972, p 118–121.

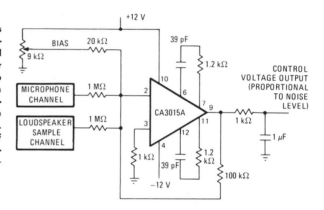

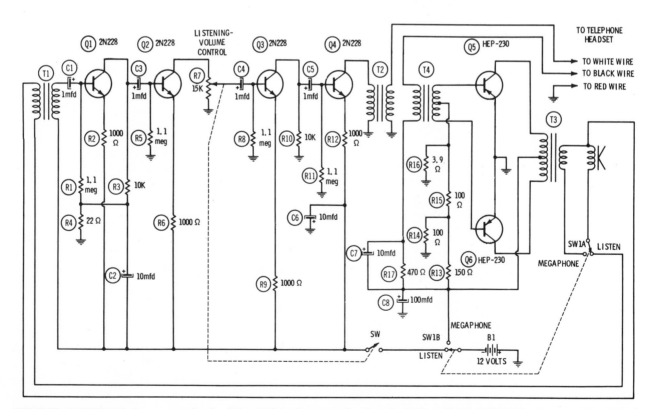

TWO-WAY MEGAPHONE—Uses conventional telephone handset as mike for megaphone and for listening to persons being shouted at. Since megaphone has considerable power and correspondingly heavy current drain, use hefty 12-V dry battery rather than flashlight cells. Speaker should preferably be outdoor-type p-m. Transformers recommended are Thordarson TR-36 for T1, Stancor TA-34 for T2, Triad TY-64X for T3, and Triad TY-61X for T4.—R. Brown and T. Kneitel, "101 Easy Audio Projects," Howard W. Sams, Indianapolis, IN, 1971, p 30–32.

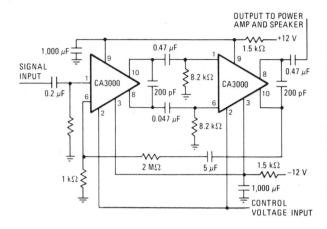

BACKGROUND NOISE CONTROL—D-c control voltage range of —4.8 to —5.2 V changes gain of audio control amplifier linearly over 30-dB range, to compensate automatically for changing background noise in mobile communication receiver or in public address system. Total control range of —3.6 to —5.3 V gives 45-dB gain variation. Article includes circuit for deriving control voltage from audio sample amplifier and audio-plus-noise amplifier.—D. B. Hoisington, Automatic Control of Speaker Output Compensates for Noisy Background, *Electronics*, Nov. 20, 1972, p 118—121.

SIGNAL SAMPLING FOR NOISE CONTROL—Two-stage transistor amplifier and peak detector are used in sampling audio signal in system which compensates automatically for changing background noise. Resulting output signal is added to that of audio-plus-noise amplifier fed by mike in room, auto, or stadium, to give d-c control voltage that boosts output level of speaker system as much as 30 dB as background noise increases. All circuits are given in article.—D. B. Hoisington, Automatic Control of Speaker Output Compensates for Noisy Background, *Electronics*, Nov. 20, 1972, p 118—121.

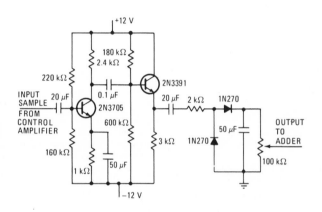

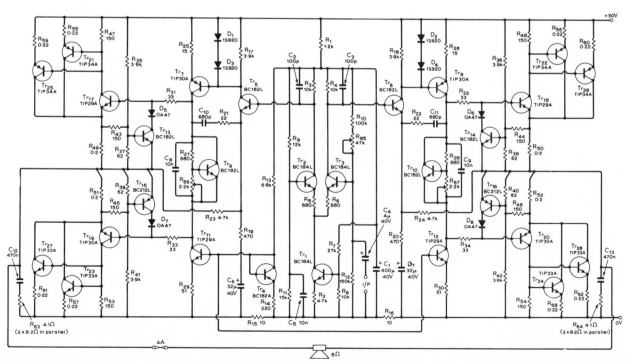

100-W CLASS B—Uses bridge arrangement of output stage to feed 80-V p-p at 5 A peak into 8-ohm load. Includes overload protection circuits for transistors. Both output lines should be fused at 4 A. If parasitic trouble occurs at higher power levels, connect 4,000 pF between collectors of TR7 and TR8. Article includes regulated power supply circuit and construction details.—R. B. H. Becker, High-Power Audio Amplifier Design, *Wireless World*, Feb. 1972, p 79—84.

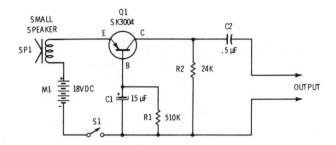

DYNAMIC MIKE—Small 8-ohm p-m speaker and single-transistor amplifier provide output impedance comparable to that of commercial dynamic mike, at low cost, for testing p-a and hi-fi systems.—R. M. Brown and T. Kneitel, "101 Easy Test Instrument Projects," Howard W. Sams, Indianapolis, IN, 1968, p 25—26.

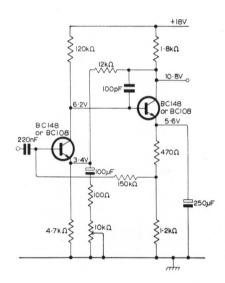

MIKE AMPLIFIER—Voltage gain is adjustable between 13 and 40 dB with 10K pot that varies feedback. Input impedance is about 125K and output impedance about 100 ohms. Frequency limits are 20 Hz and 20 kHz.—"Transistor Audio and Radio Circuits," Mullard, London, 1972, TP1319, 2nd Ed., p 170—171.

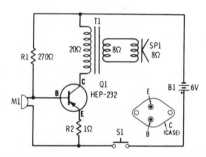

LOUDHAILER—Single-transistor megaphone delivers up to 5-W a-f output. Circuit can be mounted inside trumpet speaker if carbon mike is acoustically isolated to prevent feedback howling. T1 must be rated at least 5 W.—H. Friedman, "99 Electronic Projects," Howard W. Sams, Indianapolis, IN, 1971, p 22—23.

CHAPTER 90
Pulse Generator Circuits

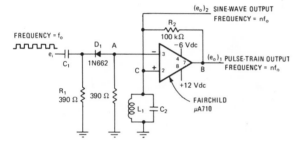

FREQUENCY = f_o

e_i

D_1 A 1N662

C_1

R_1 390 Ω

390 Ω

C

L_1 C_2

R_2 100 kΩ

$(e_o)_2$ SINE-WAVE OUTPUT
FREQUENCY = nf_o

−6 Vdc

$(e_o)_1$ PULSE-TRAIN OUTPUT
FREQUENCY = nf_o

+12 Vdc

FAIRCHILD
μA710

FOR nf_o = 10 kHz, f_o = 1 kHz:
C_1 = 0.1 μF
C_2 = 0.03 μF
$L_1 \cong$ 85 mH MILLER 9007
Q = 10 (LOADED)
E_C = 1.2 V pk·pk

FOR nf_o = 100 kHz, f_o = 10 kHz:
C_1 = 0.01 μF
C_2 = 510 pF
$L_1 \cong$ 5 mH MILLER 9004
Q = 16 (LOADED)
E_C = 1.6 V pk·pk

1–500 KHZ WITH SUBHARMONIC SYNC—Sine or square-wave output signal is kept in step with subharmonic pulse or square-wave input from low-impedance source. Differentiated trailing edge of input signal provides sync pulse. Tank L1-C1 values are given for two frequencies.—D. F. DeKold, Low-Frequency Oscillator Uses Subharmonic Sync, *Electronics*, Aug. 30, 1971, p 48–49.

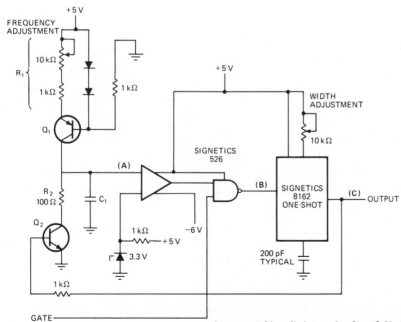

HZ TO MHZ RANGE—High-speed analog comparator and linear current source give stable generator that can oscillate from a few hertz to several megahertz. Silicon transistors, switching diodes, and value of C1 depend on frequency desired.—W. Standke, Stable Square-Wave Generator Provides Broad Bandwidth, *Electronics*, Feb. 14, 1972, p 77.

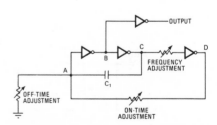

550 KHZ WITH FOUR INVERTERS—Inexpensive adjustable pulse generator using two-thirds Fairchild 936 DTL hex inverter IC package gives variable output frequency up to 500 kHz along with variable on and off times. Range can be extended to 10 MHz with Fairchild 9016 TTL inverter. Off-time rheostat can be 1.2K to infinity. Use 2K for on-time and for frequency, and 0.001 to 500 μF for C1.—M. L. Harvey, Variable Pulse Generator Consists of Four Inverters, *Electronics*, Aug. 30, 1971, p 48.

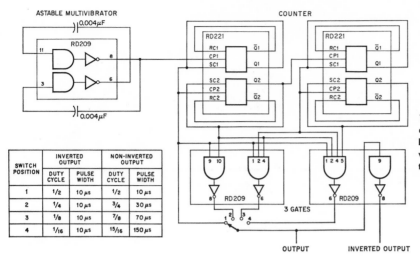

ASTABLE MULTIVIBRATOR

COUNTER

SWITCH POSITION	INVERTED OUTPUT		NON-INVERTED OUTPUT	
	DUTY CYCLE	PULSE WIDTH	DUTY CYCLE	PULSE WIDTH
1	1/2	10 μs	1/2	10 μs
2	1/4	10 μs	3/4	30 μs
3	1/8	10 μs	7/8	70 μs
4	1/16	10 μs	15/16	150 μs

100-KHZ SQUARE WAVES—Astable mvbr drives four-bit counter gated by RD209 IC's. Provides variety of duty cycles and pulse widths.—E. Lafko, Pulse Generator Uses Digital ICs, *Electronics*, Oct. 26, 1970, p 86.

NOISE-FREE STROBE—Can be made from inexpensive quadruple two-input gate such as DTμL9946. Has high noise immunity and is relatively insensitive to input pulse fall time. Pulse stretcher using gates G1 and G2 acts with gates G3 and G4 to generate strobe pulse at end of each high *true* input. Input pulses narrower than limit set by time constant R1C do not generate strobe output. For values shown, any positive input pulse wider than 10 μs gives negative output strobe about 10 μs wide.—F. D. Terry, This Simple Strobe Pulse Generator Has High Noise Immunity, *The Electronic Engineer*, Sept. 1968, p 80.

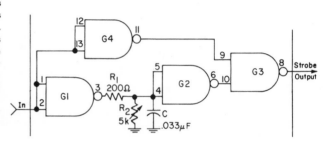

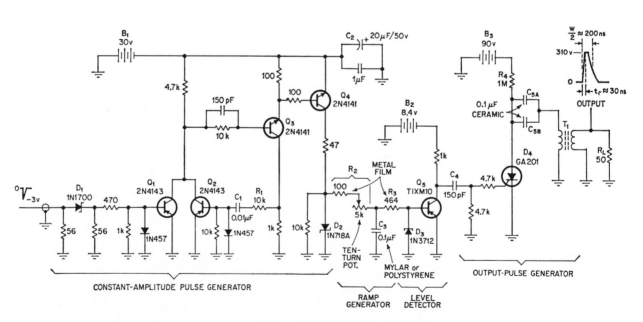

T_1: TOROID — O.D. = 23 mm ⎫
        I.D. = 13 mm ⎬ FERRITE 3E2
        HEIGHT = 7 mm ⎭

PRIMARY — 2 TURNS COPPER FOIL (6 mm WIDE) COVERING 1/3 CIRCUMFERENCE OF TOROID.

SECONDARY — 8 TURNS AWG #22 ENAMELLED, DISTRIBUTED OVER PRIMARY.

LOW-DRAIN PULSER—Gives 300-V pulses with 30-ns rise time yet draws only 2 μA on standby. Input pulse triggers sequence of charge transfers with variable delay between 0.5 and 20 μs while generating high-voltage output for laser and plasma physics research. Q1-Q4 form mono timed by R1 and C1. Q5 releases charge stored in C4, triggering thyristor D4 which can handle up to 30 A while discharging C5 through T1.—W. J. Orr, High-Voltage Pulser Spares Battery Supply, *Electronics*, May 24, 1971, p 58.

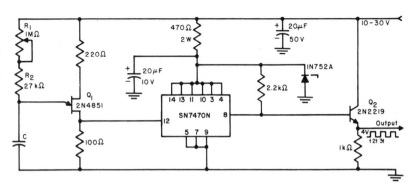

STABLE 50 KHZ—Output is stable to better than 100 Hz at 50 kHz, temperature coefficient is about 0.08%/°C, and voltage coefficient of frequency is 0.05%/V. Uses ujt oscillator Q1, IC flip-flop, and emitter-follower output driver.—A. J. Steinman, Stable Square-wave Generator, *The Electronic Engineer*, Feb. 1969, p 86.

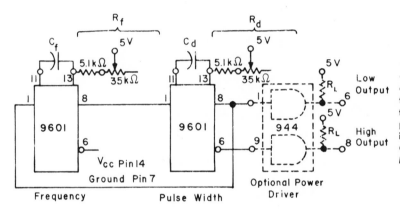

1 HZ TO 5 MHZ—First one-shot sets repetition rate, equal in seconds to 0.36RfCf. Second one-shot sets pulse width equal in seconds to 0.36RdCd. Pulse width must be set first, because frequency one-shot fires on trailing edge of pulse-width one-shot's output. Optional Fairchild DTµL944 power driver inverts high and low output.—R. G. Sullivan, One-Shots Generate Accurate Pulse Train, *The Electronic Engineer*, Dec. 1968, p 81.

STROBE UP TO 20 MHZ—Output strobe pulse width can be as narrow as 10 ns. To generate strobe during positive-going phase 1 of input signal, connect points A and C together; for strobe during negative-going phase 2 of input, connect B and C together. Value of C1 is chosen to give desired delay between start of phase and start of strobe. Value of C2 determines width of strobe.—D. A. Paris, Wide-Range Strobe Generator from Two ICs and Two Capacitors, *The Electronic Engineer*, July 1968, p 87.

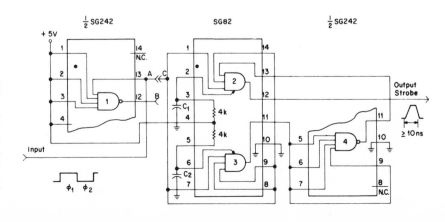

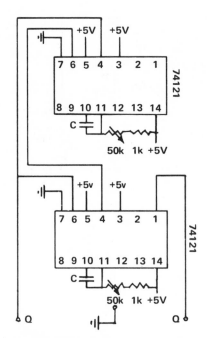

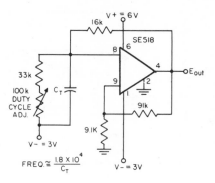

DIGITAL DRIVE—Linear opamp in oscillator circuit will operate down to very low frequencies, with adjustable duty cycle. Requires only one timing capacitor, which may be polarized.—R. K. Dahlem, Industrial Applications Of Linear ICs, *The Electronic Engineer*, June 1967, p 72–77.

VARIABLE WIDTH AND RATE—Motorola IC costing only about $1 can be cascaded for any number of stages to produce variety of pulse trains at low cost.—W. Lamb, Jr., More Square Waves, *Electromechanical Design*, March 1972, p 4.

$$FREQ \cong \frac{1.8 \times 10^4}{C_T}$$

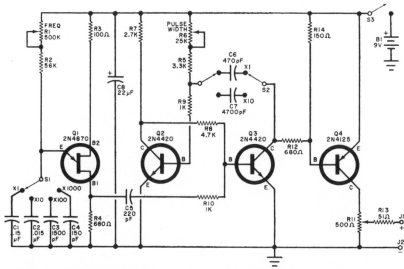

10 HZ TO 100 KHZ—Provides pulse widths of 1 to 100 μs (R6) at amplitudes up to 8 V (R11). Uses basic ujt relaxation oscillator Q1 having range multiplier S1 and fine frequency adjustment R1. One-shot mvbr Q2-Q3 generates pulses whose widths are independent of trigger from Q1, for amplification by Q4.—P. Harms, Portable Pulse Generator, *Popular Electronics*, Nov. 1969, p 61 and 66–68.

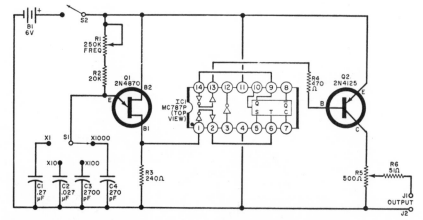

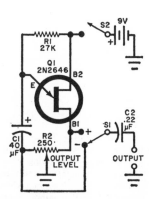

5 HZ TO 50 KHZ—Sharp spikes generated by ujt Q1 are shaped and used to trigger flip-flop whose output is sharp square wave at half the oscillator frequency. R5 adjusts output voltage over range of 0 to 6 V.—P. E. Harms, Squaring with an IC, *Popular Electronics*, July 1969, p 43–46.

BIPOLAR PULSES—Setting of S1 determines polarity of output pulses from ujt relaxation oscillator. Repetition rate for values shown is about one pulse every 2 s.—F. H. Tooker, Getting to Know the UJT, *Popular Electronics*, April 1970, p 69–73.

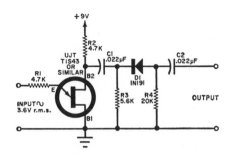

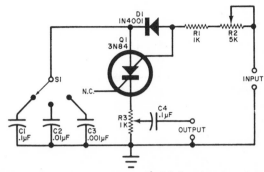

A-F SPIKE GENERATOR—Single ujt converts sine-wave input to very narrow negative-going spikes having amplitude of about 3 V for 3.6-V rms input at 60 Hz. Output frequency is same as input. Diode and R4 eliminate short positive-going spike.—F. H. Tooker, Sine-Wave To Pulse Converter, *Popular Electronics*, July 1969, p 50.

SPIKE GENERATOR—Produces one positive-going pulse for each cycle of sine-wave input signal that is between 1- and 4-V rms. Requires no power source other than input signal. Frequency range is 100 to 10,000 pps.

Useful for triggering digital logic. R3 is output level control. Adjust R2 and S1 for best spike waveform at desired input frequency.—A Self-Powered Pulse Generator, *Popular Electronics*, Sept. 1969, p 78.

COMPLEMENTARY PULSES—Provides equal-amplitude positive and negative pulses with time durations matching that of input.—A. Ivanov, Generating Fast Complementary Pulses, *Wireless World*, June 1970, p 292.

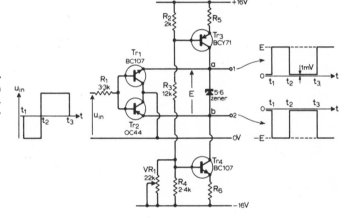

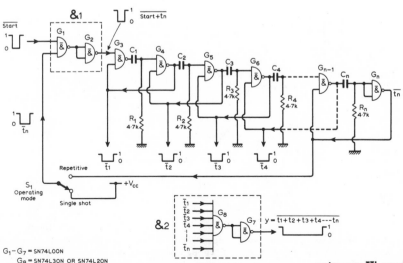

CASCADED MONOS—Input trigger causes chain of any number of pulses to be produced, each having different duration set by single R and C. S1 gives choice of single pulse train or selfsustaining operation after initial triggering. Optional circuit (below) provides pulse with duration equal to sum of durations of all other pulses in train. All gates are TTL quad two-input. The duration of each pulse is 1.3CR s, with C in farads and R in ohms.—H. A. Cole, T.T.L. Monostable Cascade, *Wireless World*, June 1971, p 301–302.

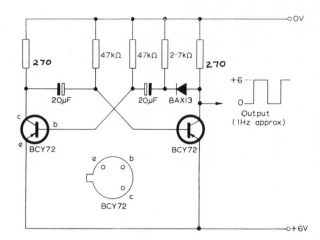

1-HZ SHIFT PULSE GENERATOR—Free-running mvbr may also be used as trigger generator.—"A Digital Integrated Circuit Training Aid," Educational Projects in Electronics, Mullard, London.

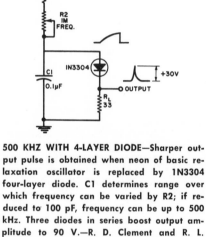

500 KHZ WITH 4-LAYER DIODE—Sharper output pulse is obtained when neon of basic relaxation oscillator is replaced by 1N3304 four-layer diode. C1 determines range over which frequency can be varied by R2; if reduced to 100 pF, frequency can be up to 500 kHz. Three diodes in series boost output amplitude to 90 V.—R. D. Clement and R. L. Starliper, Relaxation Oscillators—Old and New, Electronics World, May 1971, p 36–37.

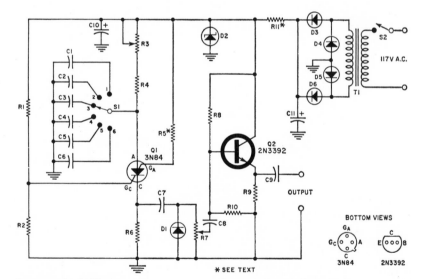

R1—82,000 ohm, ½ W res.
R2—15,000 ohm, ½ W res.
R3—350,000 ohm linear-taper pot
R4—56,000 ohm, ½ W res.
R5—6800 ohm, ½ W res. (see text)
R6—1000 ohm, ½ W res.
R7—10,000 ohm linear-taper pot
R8—220,000 ohm, ½ W res.
R9—3900 ohm, ½ W res.
R10—330,000 ohm, ½ W res.
R11—120 ohm, ½ W res. (see text)
C1—2.0 μF, 100 V Mylar capacitor
C2—0.47 μF, 100 V Mylar capacitor
C3—0.1 μF, 100 V Mylar capacitor
C4—0.022 μF, 100 V Mylar capacitor

C5—0.0047 μF, 100 V Mylar capacitor
C6—0.001 μF, 100 V Mylar capacitor
C7—0.01 μF, 100 V Mylar capacitor
C8—0.047 μF, 100 V Mylar capacitor
C9—0.22 μF, 100 V Mylar capacitor
C10—200 μF, 10 V elec. capacitor
C11—150 μF, 15 V elec. capacitor
S1—S.p. 6-pos. shorting-type miniature rotary sw.
S2—S.p.s.t. slide switch
T1—Miniature power trans. 12 V at 10 mA
D1—1N191 computer diode
D2—10 V, 400 mW zener diode
D3, D4, D5, D6—1N34A diode
Q1—3N84 silicon controlled switch
Q2—2N3392 transistor

2—25,000 PPS IN SIX RANGES—Produces sharp spike-like waveform with positive polarity, as required for counters, tachometers, and other applications. Lowest range is particularly valuable for troubleshooting, experimentation, and design work. Uses scs Q1 as oscillator and Q2 for decoupling to give rea-sonably low output resistance. Output is about 4 V at 1,000 pps. If transformer is rated higher than 10 mA at 12 V, adjust R11 to limit current to about 7.5 mA. Will operate from 12-V battery.—F. H. Tooker, SCS Positive-Pulse Generator, Electronics World, Jan. 1970, p 78–79.

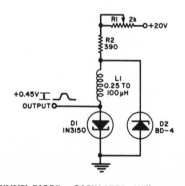

TUNNEL-DIODE OSCILLATOR—Will operate up to 10 MHz as relaxation oscillator if L1 is 0.25 μH. Uses backward diode D2 to bias D1 in negative-resistance region, so circuit returns to low-voltage state for start of new cycle after output pulse is produced.—R. D. Clement and R. L. Starliper, Relaxation Oscillators—Old and New, Electronics World, May 1971, p 36–37.

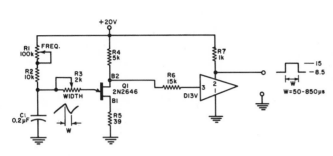

PROGRAMMABLE ZENER—Free-running relaxation oscillator using ujt Q1 is part of divider that determines output voltage of General Electric D13V programmable zener. When power is applied, C1 charges through R1 and R2 until Q1 fires. Sudden decrease in resistance of Q1 changes divider resistance of D13V, making output voltage across R7 rise. Width of output pulse is determined by *on* time of Q1, which in turn depends on discharge time of C1 through width control R3, Q1, and R5.—R. L. Starliper, Programmable Zener Applications, *Electronics World*, June 1971, p 60–61.

SINGLE-SUPPLY GENERATOR—Uses special Motorola current-differencing opamp operating over supply voltage range of 3.5 to 35 V. Provides output voltage swing only 1 V less than supply voltage.—T. M. Frederiksen, W. F. Davis, and D. W. Zobel, A New Current-Differencing Single-Supply Operational Amplifier, *IEEE Journal of Solid-State Circuits*, Dec. 1971, p 340–347.

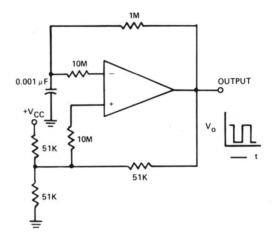

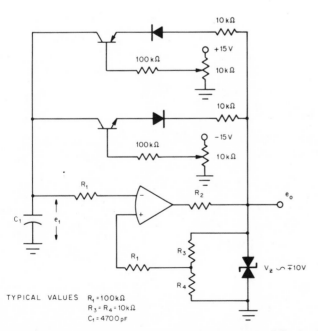

TYPICAL VALUES $R_1 = 100\,k\Omega$
$R_3 = R_4 = 10\,k\Omega$
$C_1 = 4700\,pF$

10–10,000 HZ—High-performance square-wave generator can use Burr-Brown 3401 or 3402 opamps having fast fet input. Book gives design equations. Range can be extended to 100 kHz by careful choice of opamps. Frequency drift is low.—J. G. Graeme, G. E. Tobey, and L. P. Huelsman, "Operational Amplifiers," McGraw-Hill Book Co., New York, 1971, p 373–374.

15-MS STROBE PULSE—Used with symmetric clock of 2-megabit digital memory module. Circuit is triggered by falling edge of clock pulse, and delivers 15-ms pulse at same frequency as clock.—High-Speed Digital Memory Modules are Usable in Low-Speed Applications at 1–2 Megabits, *IEEE Spectrum*, Dec. 1969, p 90.

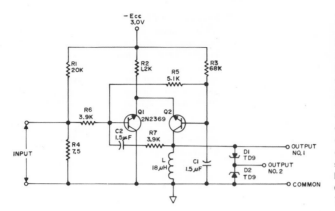

LOW DRAIN—Produces output pulse widths as narrow as 50 ns, with essentially flat tops, beyond duration of input triggering spike. Negative input spike turns off Q1, driving Q2 into conduction and starting pulse-generating action in tunnel diodes. Requires minimum input amplitude of 125 mV and gives 400-mV output pulses. May be adapted for use as delay, gated oscillator, frequency limiter, or discriminator for either leading or trailing edge of pulse. Once actuated, circuit is not affected by additional spikes until output pulses are completed.—C. A. Cancro, Versatile Low-Power-Drain Hybrid Tunnel Diode Pulse Generator, *Proc. IEEE*, Aug. 1968, p 1,389–1,390.

4–72 MHZ—Frequency of symmetrical square-wave output depends on values of L1 and C2, as given in article. Use 100 pF and 82 μH for 4 MHz; 47 pF and 15 μH for 10 MHz; 47 pF and 2.2 μH for 50 MHz. Trim L1 for best waveshape. Transistors are Fairchild. —Square Wave Source, *73*, Feb. 1973, p 139.

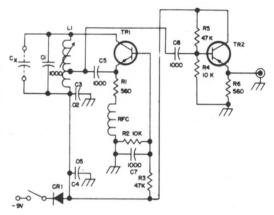

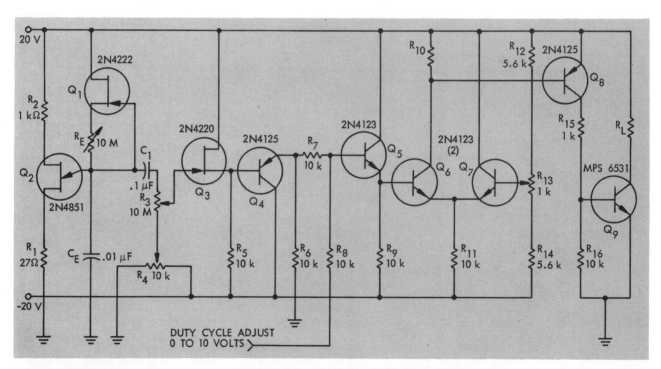

SPACE-MARK GENERATOR—Both frequency and duty cycle of output pulse are variable. Frequency is controlled by 10-meg pot in circuit of ujt relaxation oscillator Q1, while duty cycle is adjusted linearly over full range from 0 to 100% with 0–10 V d-c control voltage applied to base of Q5. Fet acts as constant-current source charging CE. Q2 generates linear ramp applied to gate of fet Q3, shifted up or down with R3 and R4 until ramp at emitter of Q4 goes from −10 V to 0. Report covers circuit operation in detail.—"Unijunction Transistor Timers and Oscillators," Motorola Semiconductors, Phoenix, AZ, AN-294, 1972.

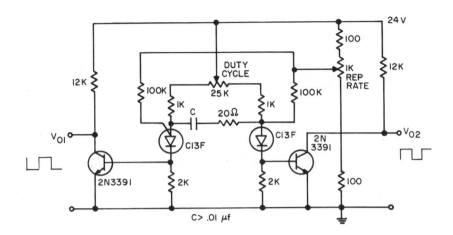

FLEXIBLE CONTROL—Circuit using scr pair provides independent control of repetition rate and duty cycle for square-wave output. Report gives design equations.—W. H. Sahm, III, and F. M. Matteson, "The Complementary SCR," General Electric, Semiconductor Products Dept., Auburn, NY, No. 90.94, 1970, p 19–20.

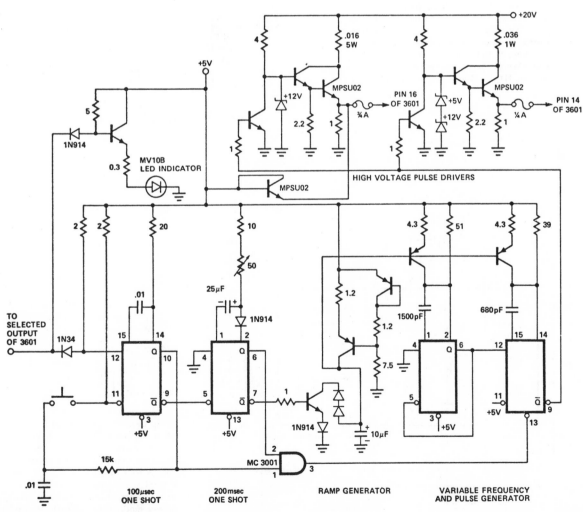

PROGRAMMING PULSE GENERATOR—Developed for use with Intel type 3601 high-speed electrically programmable 1,024-bit read-only memory. All npn transistors are 2N2369 and all pnp transistors are 2N4917, unless otherwise noted. Typical programming time is 2 ms per bit. All resistor values are in thousands of ohms. Requires four Fairchild 9602 one-shots.—"3601 Bipolar Programmable ROM," Intel Corp., Santa Clara, CA, 1972.

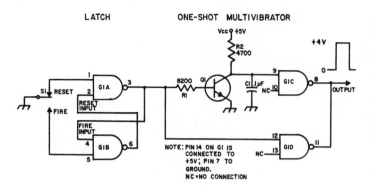

LATCH   ONE-SHOT MULTIVIBRATOR

NOTE: PIN 14 ON G1 IS CONNECTED TO +5V; PIN 7 TO GROUND.
NC = NO CONNECTION

DTL SINGLE-PULSE—Used for testing logic circuits, counters, shift registers, and frequency dividers when only one pulse is desired each time S1 is operated. Uses latch circuit mated with monostable mvbr to make circuit change state and remain that way no matter how many more positive pulses are applied to reset input. Circuit returns to original state only when positive pulse is applied to fire input by switching S1 to reset. G1 is MC846 or μL9946. Q1 is 2N3904. Pulse is 50 μs wide.—W. Simciak, Logic Pulser, QST, March 1970, p 39.

TIMED ONE-SHOT—Input triggers scr on, energizing load. Voltage across load energizes put timer. When put fires, pulse across R1 acts through C2 to lift scr cathode momentarily above supply and turn it off. Turn-off time depends on capacitor values and load.—D. R. Grafham, "Using Low Current SCR's," General Electric, Semiconductor Products Dept., Auburn, NY, No. 200.19, 1970, p 26–27.

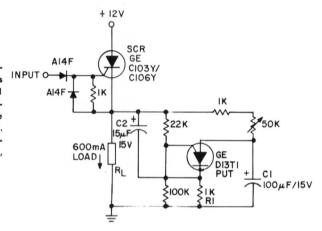

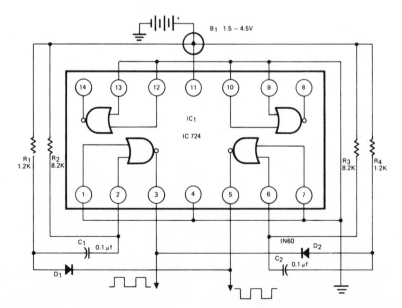

2 MHZ WITH IC—Values of C1 and C2 determine frequency of true square-wave output, up to maximum of 2 MHz. Both diodes are 1N60 germanium, and IC is International Rectifier IC724.—"Hobby Projects," International Rectifier, El Segundo, CA, Vol. II, p 64–65.

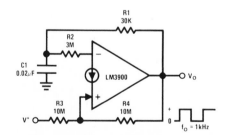

1-KHZ OPAMP—Uses Schmitt trigger configuration for Norton current-differencing opamp to give symmetrical square-wave output with amplitude essentially independent of supply voltage over range of 4 to 36 V. For unsymmetrical output, report gives design procedure for adjusting resistor values to produce any desired mark-space ratio.—T. M. Frederiksen, W. M. Howard, and R. S. Sleeth, "The LM3900—A New Current-Differencing Quad of ± Input Amplifiers," National Semiconductor, Santa Clara, CA, 1972, AN-72, p 19.

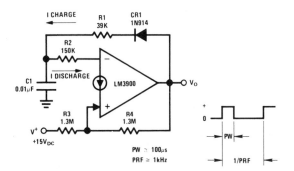

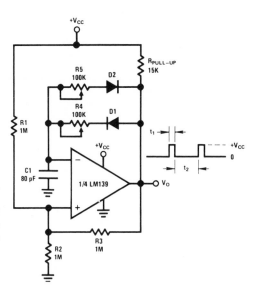

1-KHZ NARROW-PULSE—Uses single Norton current-differencing opamp, with diode serving to separate charge and discharge paths of C1. Report gives design equations and design example for circuit shown which produces 100-μs pulse every 1 ms.—T. M. Frederiksen, W. M. Howard, and R. S. Sleeth, "The LM3900—A New Current-Differencing Quad of $\pm$ Input Amplifiers," National Semiconductor, Santa Clara, CA, 1972, AN-72, p 20–21.

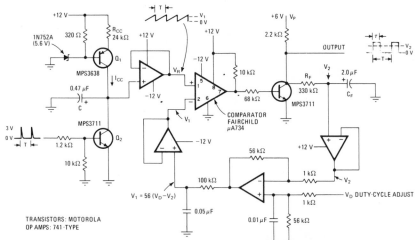

ADJUSTABLE DUTY CYCLE—Achieved by providing separate charge and discharge paths for C1. Path through R4 and D1 charges capacitor and sets pulse width, while path R5-D2 discharges capacitor and sets time between pulses. Adjustments are independent, although both change frequency. Design equations are given. Supply is 5 V.—R. T. Smathers, T. M. Frederiksen, and W. M. Howard, "LM139/LM239/LM339 A Quad of Independently Functioning Comparators," National Semiconductor, Santa Clara, CA, AN-74, 1973, p 7.

VOLTAGE-CONTROLLED DUTY CYCLE—Pulse generator with frequency range of 10 Hz to 10 kHz provides output duty cycle voltage-controllable from 5% to 95% independently of frequency. Can be used as constant-phase triggering source. Circuit has separate controls for output pulse period and width also; input frequency determines period, while pulse width depends on ramp input to comparator.—W. D. Harrington, Controlling Duty Cycle and Rep Rate Independently, Electronics, March 15, 1973, p 107.

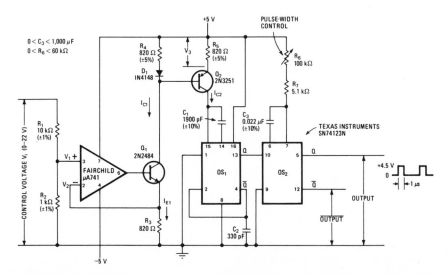

VARIABLE RATE AND WIDTH—Input control voltage of 0–22 V varies frequency of output pulse over range of 1 to 75 kHz, and R6 varies output pulse width from 100 ns to 18 s. Article describes operation and gives design equations.—M. Shah, Generator Independently Varies Pulse Rate and Width, Electronics, Oct. 9, 1972, p 100–101.

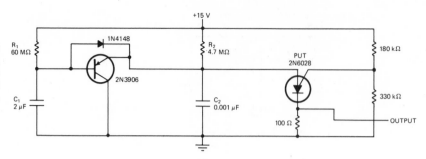

2.5-MIN PULSES—Uses programmable ujt fired by emitter-follower in high-resistance timing network to produce one output pulse every 2.5 min.—V. Hatch, Extending Time Delay with an Emitter-Follower, *Electronics*, March 29, 1973, p 82.

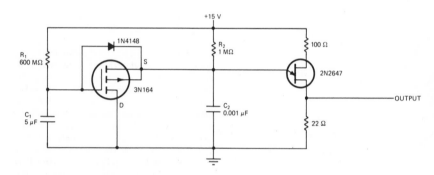

1 PULSE PER HOUR—Uses 2N2647 ujt with mosfet connected as source-follower in circuit permitting use of high value for timing resistor R1.—V. Hatch, Extending Time Delay with an Emitter-Follower, *Electronics*, March 29, 1973, p 82.

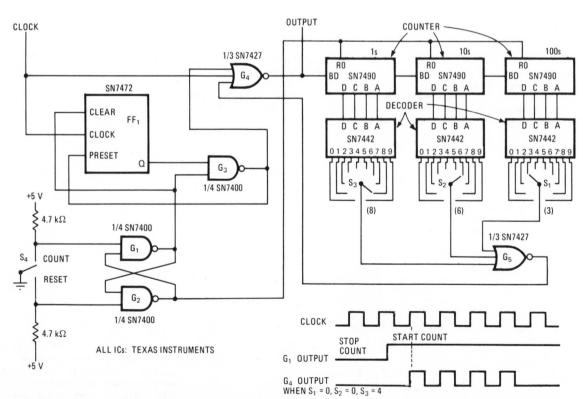

1–999 PULSE-COUNTING GENERATOR—Three ten-position switches are set to desired number of full-width pulses to be generated when S4 is set to count position. Gates G1–G4 eliminate count error caused by contact bounce of S4.—G. Coers, Preset Generator Produces Desired Number of Pulses, *Electronics*, July 3, 1972, p 88.

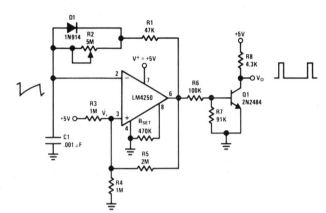

360–6,000 HZ MVBR—Micropower circuit is buffered-output free-running mvbr with constant-width output pulse having frequency determined by R2. Output buffer Q1 presents constant load to opamp, to prevent frequency variations caused by load voltage and current changes. Will interface with standard TTL or DTL logic. Power dissipation of mvbr is only 300 μW and that of buffer about 6 mW.—M. K. Vander Kooi and G. Cleveland, "Micropower Circuits Using the LM4250 Programmable Op Amp," National Semiconductor, Santa Clara, CA, 1972, AN-71, p 6–7.

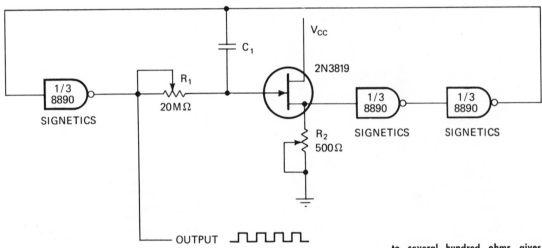

0.3 HZ–10 MHZ—Delivers 0.3-Hz square-wave output with timing capacitor as small as 0.1 μF, because high input impedance of fet and large timing resistance R1 keep time constant large. Operates over supply range of 4–6 V. Changing R1 from 20 meg down to several hundred ohms gives frequency change of over 50,000:1.—R. Siebert, Broadband Pulse Generator Uses Small Timing Capacitance, *Electronics*, Aug. 16, 1971, p 80.

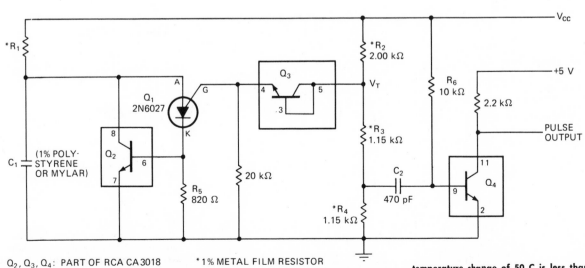

Q2, Q3, Q4: PART OF RCA CA3018   *1% METAL FILM RESISTOR

STABLE PRR UP TO 1 MHZ—Used for driving digital circuits. Value of timing resistor R1, which can be as low as 2K, is changed to vary output prr over 250:1 range for given value of C1 (500 pF to 1 μF). Worst-case variation in prr with aging of components or temperature change of 50 C is less than 1%. Article gives design equation.—F. Cicchiello, Pulse Generator Accuracy Is Immune to Aging, *Electronics*, July 17, 1972, p 86.

CHAPTER 91
Pulse Shaping Circuits

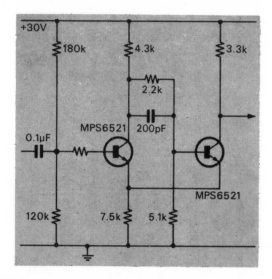

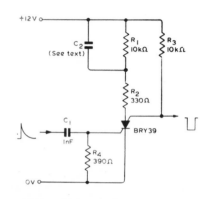

SINE TO SQUARE—Converts 2 to 9 V p-p sine wave into 5-V p-p square wave over frequency range of 0.1–500 kHz. Output square wave is symmetrical for 5-V p-p input. Used in f-m telemetry systems.—A. M. Kubo, Inexpensive F.M. Telemetry with Active Circuits, *Electronic Engineering*, July 1970, p 26–29.

COUNTER DRIVE—Positive-going pulses to be counted are fed to scs shaper for conversion to essentially square-wave pulses having fixed amplitude and duration, for driving scs ring counter. Conduction time of scs, corresponding to pulse duration, depends on value used for C2.—D. J. G. Janssen, Circuit Logic with Silicon Controlled Switches, *Mullard Technical Communications*, March 1968, p 57–64.

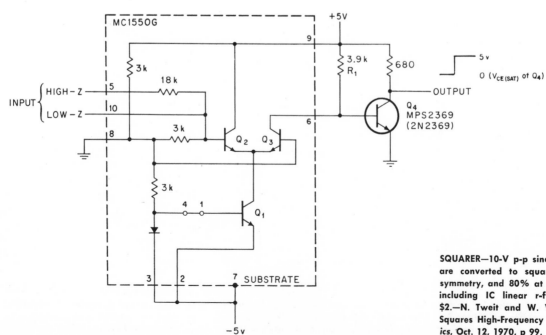

SQUARER—10-V p-p sine waves up to 1 MHz are converted to square waves with 90% symmetry, and 80% at 5 MHz. Cost of parts, including IC linear r-f amplifier, is about $2.—N. Tweit and W. Vincent, RF Linear IC Squares High-Frequency Sine Waves, *Electronics*, Oct. 12, 1970, p 99.

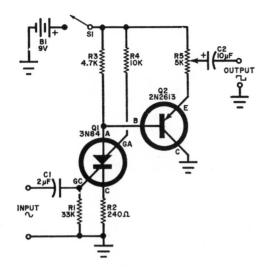

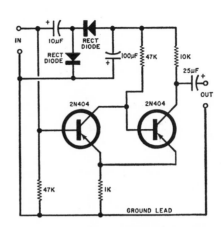

SQUARING SINE WAVES—Uses single silicon controlled switch Q1 which is triggered by input sine wave, followed by current amplifier Q2 for resulting square wave. With 1-V rms input, output swing is 7 V over range of 40 Hz to 20 kHz. Draws only 18 mW.—F. H. Tooker, SCS Signal-Squaring Adapter, *Popular Electronics*, Jan. 1970, p 65 and 89.

SINE TO SQUARE—In batteryless circuit, input sine waves betwen 50 and 15,000 Hz at 0.5 to 10-V rms supply power to rectifier diodes as well as trigger Schmitt circuit that produces square wave at same frequency as input. Rise time is excellent. Component values are not critical.—L. Solomon, Signal-Powered Signal Squarer, *Popular Electronics*, Oct. 1967, p 48.

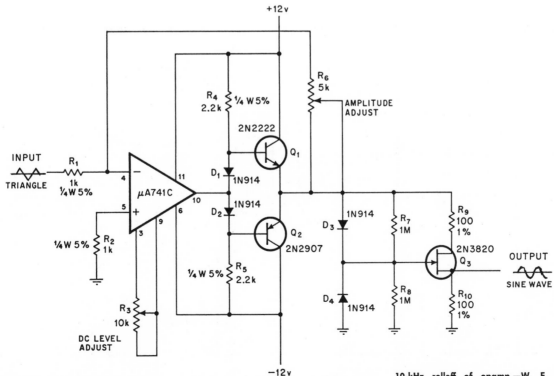

TRIANGLE-SINE CONVERTER—Transfer curve of fet gives sinusoidal output with less than 2% distortion, using only two adjustments. Circuit operates as nonreactive filter, up to 10-kHz rolloff of opamp.—W. E. Peterson, Field Effect Transistor Converts Triangles to Sines, *Electronics*, Aug. 31, 1970, p 69–70.

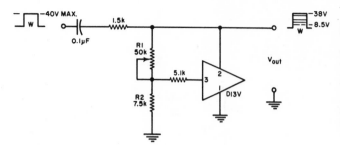

PROGRAMMABLE-ZENER CLIPPER—Will limit 40-V input pulses to value determined by setting of R1, in range of 8.5 to 38 V. Uses General Electric D13V programmable zener. Will operate up to 9,000 pps with pulse widths from 1 to 500 μs. At any setting of R1, output amplitude is unaffected by changes in input amplitude.—R. L. Starliper, Programmable Zener Applications, *Electronics World*, June 1971, p 60–61.

1,005-HZ SQUARER—Circuit converts sine-wave input to symmetrical square wave. Care is necessary to insure symmetrical clipping of 6.2-V peak input signal around its zero level. Stages, left to right, are preclipper using 1.5-V stabistors, differential-amplifier/clipper delivering 4-V semisquare wave, another differential clipper, and output amplifier. Used in conjunction with double phase-sensitive detector for precision balancing of low-frequency bridge.—J. M. Diamond, A Double Phase-Sensitive Detector for Bridge Balancing, *IEEE Spectrum*, June 1969, p 62–70.

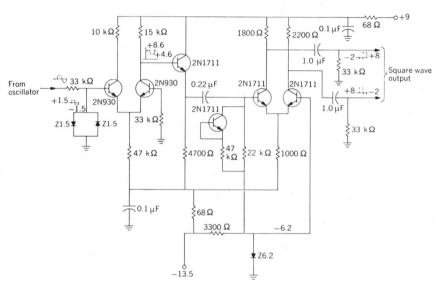

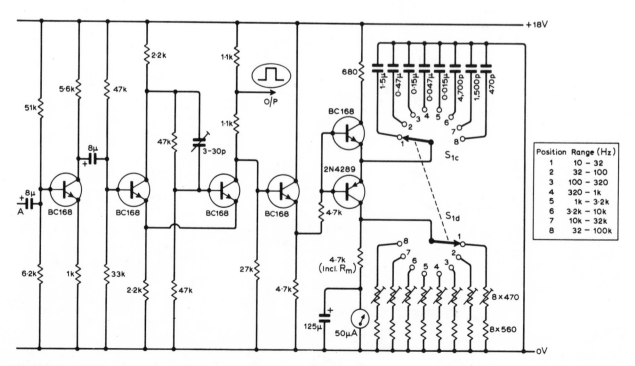

Position	Range (Hz)
1	10 – 32
2	32 – 100
3	100 – 320
4	320 – 1k
5	1k – 3·2k
6	3·2k – 10k
7	10k – 32k
8	32 – 100k

SQUARE-WAVE SHAPER—Used with 10 Hz–100 kHz eight-range Wien-bridge audio oscillator to provide optional 4-V p-p square-wave output from sine-wave output at A. Shaper also includes frequency meter, with range switch S1 ganged to range switch of audio oscillator.—A. J. Ewins, Wien-Bridge Audio Oscillator, *Wireless World*, March 1971, p 104–107.

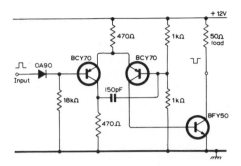

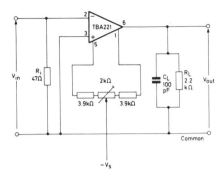

LONG-TAILED PAIR—Driven by at least 6-V positive pulse via diode gate, and gives 11.8-V output pulse having 150-ms rise time. Maximum repetition rate is 1 MHz and noise immunity is 5.5 V.—"P-N-P Practical Planar Transistors—BCY70 Family and BFX29 Family," Mullard, London, 1967, TP887, p 8.

10-HZ SQUARER—Mullard IC is operated open-loop to provide maximum gain, so output voltage is limited for small input signals. Offset voltage null circuit is adjusted so mean level of output voltage is zero, as required for symmetrical squaring. With 15-V supplies, maximum peak input voltage is 15 V.—"TBA221/222 Monolithic Internally Compensated Operational Amplifiers" Mullard, London, 1970, TP1238, p 9–10.

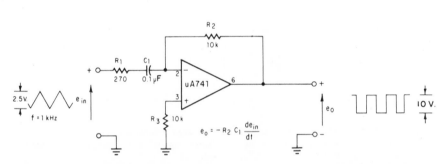

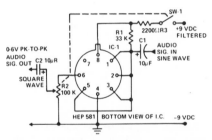

OPAMP AS DIFFERENTIATOR—Connection shown provides output proportional to derivative of input signal, so triangular input gives square wave at output. Report gives design equations.—"Applications of the µA741 Operational Amplifier," Fairchild Semiconductor, Mountain View, CA, No. 289, 1972, p 2–3.

SINE-SQUARE CONVERTER—Sine-wave input of 2–10 V produces symmetrical square-wave output having maximum amplitude of 2-V p-p. Frequency range is 0 to 1 MHz. Can be used with cro to check frequency response of hi-fi audio amplifier.—"Tips on Using IC's," Motorola Semiconductors, Phoenix, AZ, HMA-32, 1972.

$$e_o = - R_2 C_1 \frac{d e_{in}}{dt}$$

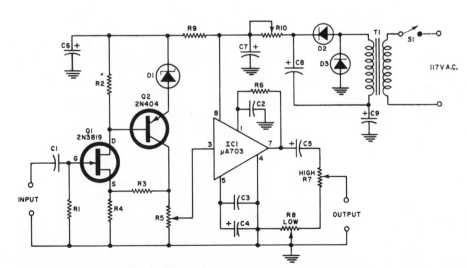

R1—1 megohm, ½ W res.
R2—5600 ohm, ½ W res.
R3—3300 ohm, ½ W res. (see text)
R4—2700 ohm, ½ W res.
R5—1000 ohm wirewound pot
R6—3900 ohm, ½ W res.
R7—10,000 ohm linear-taper pot
R8—1500 ohm linear-taper pot
R9—220 ohm, ½ W res.
R10—1500 ohm wirewound pot
C1—0.15 µF. 100 V Mylar capacitor
C2,C3—0.005 µF disc ceramic capacitor
C4—100 µF, 3 V elec. capacitor
C5—30 µF, 15 V elec. capacitor
C6—100 µF, 15 V elec. capacitor
C7,C8,C9—175 µF, 25 V elec. capacitor
S1—S.p.s.t. switch
T1—12 V at 10 mA
   miniature power transformer
D1—2.7 V, 1-watt zener diode
D2,D3—1N4002 diode
IC1—Integrated circuit (Fairchild µA703)
Q1—2N3819 FET
Q2—2N404 transistor

20–100,000 HZ SINE-WAVE CLIPPER—Input requires precision sine-wave audio signal generator. Clipper uses IC design for amplifying and limiting in i-f section of f-m receiver. Provides clean, symmetrical square wave with maximum amplitude of 8 V.—F. H. Tooker, IC Sine-Wave Clipper, *Electronics World*, Nov. 1969, p 96–97.

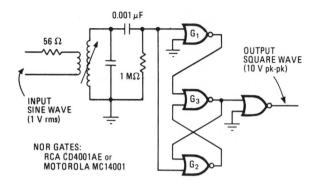

SQUARER—Link-coupled tuned circuit steps up 1-V rms sine-wave input sufficiently for squaring with c-mos gates to give 10-V square-wave output at megahertz rates. Gates are in quad two-input NOR package connected as forced latch. Rise and fall times of output are about 50 ns.—R. W. Mouritsen, Link-Coupled Tank Circuit Steps Up C-MOS Drive Voltage, *Electronics*, May 10, 1973, p 97–98.

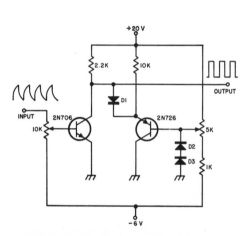

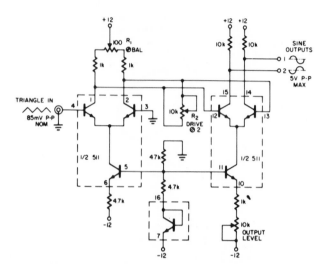

SQUARER—Designed for use with inexpensive signal generator in which square-wave output deteriorates badly at high frequencies. Diodes can be 1N914. Output amplitude is independent of frequency. Q1 is inverter, while Q2 and 5K pot provide variable clamp voltage for output.—J. Fisk, "Useful Transistor Circuits," 73 Inc., Peterborough, NH, 1967, p 30A.

TRIANGLE TO SINE WAVE—Gives two sine outputs at same frequency as triangle input, in range of 30 to 25,000 Hz. R1, R2, and input level can be adjusted to give less than 2% third harmonic distortion at output.— "Linear Applications," Signetics Corp., Sunnyvale, CA.

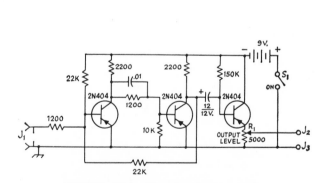

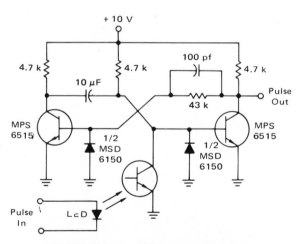

SQUARER—When driven by sine-wave audio generator, simple circuit produces square-wave output pulses having excellent symmetry and good rise time. Uses Schmitt trigger to produce positive output pulse for each positive swing of input signal. Pulse length, determined by input voltage, can be adjusted with output control of audio signal generator; although this is usually set for equal on and off times, duty cycle can be reduced to 20% if desired.—D. A. Blakeslee, The Squarer, QST, May 1967, p 36–37.

OPTOELECTRONIC PULSE STRETCHER—Used to stretch width of input voltage pulse (of order of 3 μs) to 55 ms with values shown, while providing isolation between input and output pulses. Phototransistor such as MRD300 picks up light pulse from light-emitting diode and triggers mono mvbr to initiate positive output pulse. Duration of output depends on time constant of R-C circuit.—J. Bliss, "Applications of Phototransistors in Electro-Optic Systems," Motorola Semiconductors, Phoenix, AZ, AN-508, 1971.

CHAPTER 92
Quadraphonic Circuits

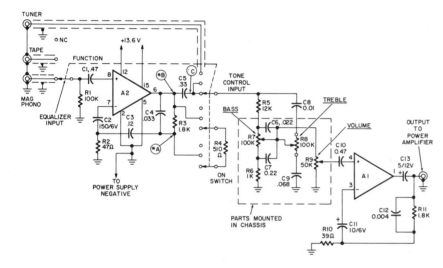

PREAMP FOR 4-CHANNEL STEREO—All four channels use identical two-IC preamp channel shown, which includes phono and tape-head equalization, input switching, and separate bass and treble tone controls. Has sufficient dynamic range for all modern phono cartridges. Gain per channel is 46 dB at 1 kHz, with flat response from 20 to 22,000 Hz. Article gives construction details and includes circuit for regulated power supply required. Both A1 and A2 are in RCA SK3071 IC, which has two of each so only two IC's are needed for entire 4-channel system.—L. Kaplan, Build R-E's 4-Channel IC Preamp, *Radio-Electronics*, Oct. 1971, p 41–45.

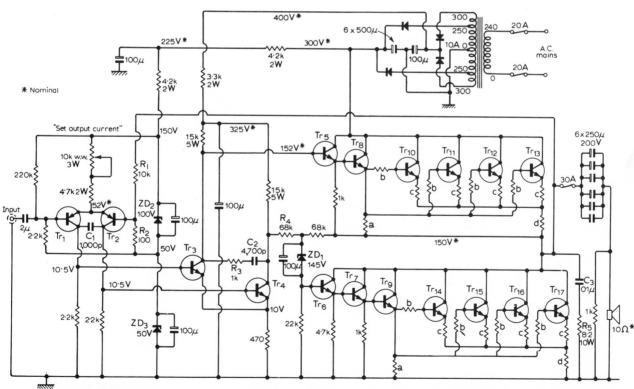

1,000-W POWER AMPLIFIER—Output stage uses relatively inexpensive npn high-voltage transistors in parallel for fully symmetrical push-pull class-B operation. (For simplicity, only four paralleled transistors are shown in each half of output stage; these are adequate only for intermittent full output, and six are actually required.) Tr1 and Tr2 are RCA 38496; Tr3–Tr7 MJE340; Tr8–Tr17 MJ413. Lettered resistor values in ohms: a = 22; b = 10; c = 0.5; d = 0.1. Output transistors require very generous heatsinks, and speaker leads must handle 30 A. Author used four 500-W 20-ohm metal-cone speakers, one in each corner of room, with opposite units connected in parallel but antiphase, mounted in sheet-steel column-loaded housings.—G. I. O'Veering, Dynamic Range Versus Ambient Noise, *Wireless World*, April 1970, p 189–190.

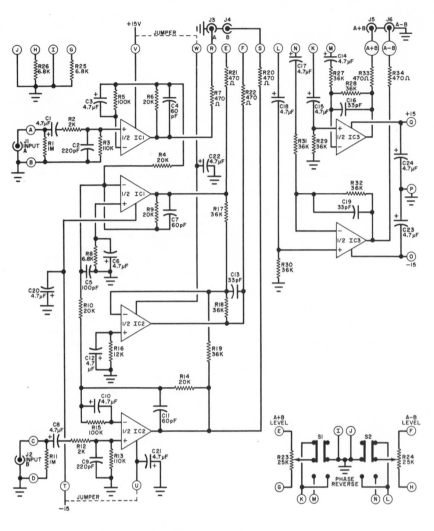

4 FROM 2—Uses 741 or equivalent opamps as active elements for converting two channels of stereo to four. Individual level controls are combined with phasing switches for new channels (lower right) so sound quality can be adjusted for listening taste and environment. Frequency response is flat within 0.5 dB from 20 Hz to 20 kHz for 1-V rms output. Uses sum and difference of original stereo channels to produce two new channels.—J. Bongiorno, Four-Channel Synthesizer, *Popular Electronics*, May 1972, p 32–35.

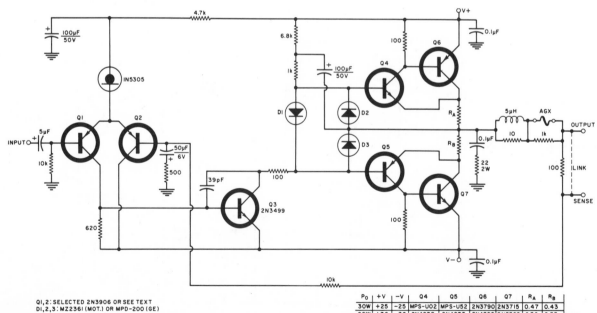

Q1,2: SELECTED 2N3906 OR SEE TEXT
D1,2,3: MZ2361 (MOT.) OR MPD-200 (GE)

P_O	+V	−V	Q4	Q5	Q6	Q7	R_A	R_B
30W	+25	−25	MPS-U02	MPS-U52	2N3790	2N3715	0.47	0.43
60W	+36	−36	2N4238	2N4235	2N4399	2N5302	0.36	0.33

30 OR 60 W FOR 4 CHANNELS—All four channels use circuit as shown. Choice of supply voltages and output transistors determines whether output per channel is 30 or 60 W. If desired, frequency-dividing network can be used at input to make two 60-W channels carry bass frequencies and two 30-W chan- nels carry treble if using two-channel program material. With four-channel programs, all channels should have same power. Designed for 8-ohm loads. Harmonic distortion is below 0.1% from 20 to 20,000 Hz. Instead of matching Q1 and Q2 for VBE of 1 mA, matched pairs such as 2N3350 or 2N4023 can be used. Use heatsinks for output transistors. Article includes power supply circuit, which is unregulated.—L. H. Garner, A 4-Channel Amplifier for Multi-Speaker Systems, *Electronics World*, Nov. 1970, p 28–30.

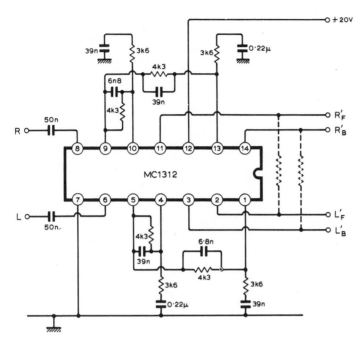

IC SQ DECODER—Single Motorola IC provides four-speaker outputs from conventional R and L stereo inputs. External components should be 5% tolerance to give maximum deviation of 8.5% from 90-deg norm between 100 and 10,000 Hz. To get 10% blending of front speakers, use 47K blend resistor between front outputs, as shown in broken lines. Similarly, use 7.5K between rear outputs for 40% blend of rear speakers. Serves as inexpensive method of simulating true quadraphonic effect.—G. Shorter, Surround-Sound Circuits, *Wireless World*, March 1973, p 114–117.

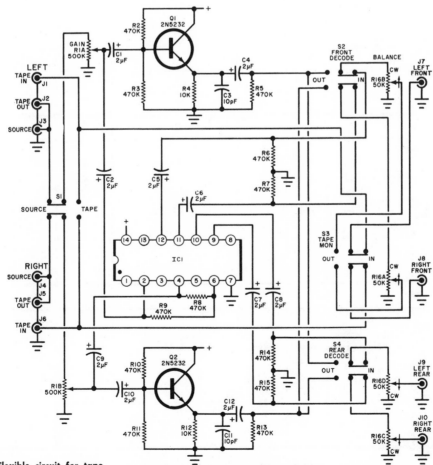

4-CHANNEL SOUND—Flexible circuit for tape or disc stereo sources permits choosing either regular or decoded stereo for front channels, with rear channels either in parallel with front ones or decoded to create four channels. Balance control permits changing ratio of front and rear signals without changing master gain control R1. IC is No. 701 from Metrotec Industries, Plainview, NY 11803 ($15). Article includes regulated power supply circuit using 20-V reference zener.—G. Meyerle, Four-Channel Stereo Decoder, *Popular Electronics*, July 1971, p 52–53 and 58–60.

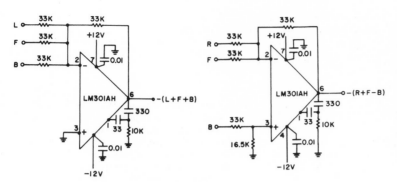

DYNAQUAD MIXER—Uses National IC having built-in short-circuit protection and no crossover distortion even with 600-ohm load. Harmonic distortion is less than 0.1% between 50 Hz and 15 kHz at +18 dBm. Based on adding additional information to two standard stereo channels to approximate multichannel reproduction.—E. Borbely, A Simple Matrix-Type Unit for the Dynaquad Four-Dimensional System, *Audio*, May 1972, p 42, 44, 46, and 48.

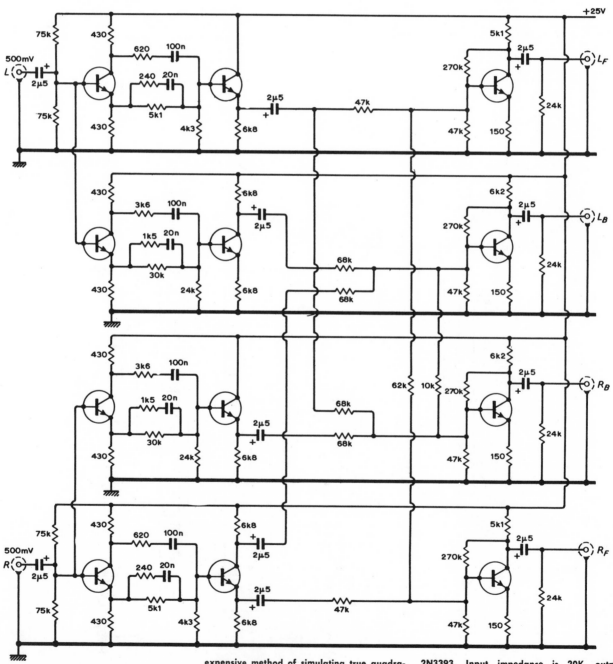

SQ DECODER—Provides 10% blending of front outputs and 40% between rear outputs of four-speaker quadraphonic system, as inexpensive method of simulating true quadraphonic effect. Uses networks providing required 90-deg phase difference to accuracy within 10% from 100 Hz to 10 kHz. Output transistors are 2N3390 and all others 2N3393. Input impedance is 20K, output impedance 1.8K, and nominal input level 500 mV rms. Circuit has unity gain.—G. Shorter, Surround-Sound Circuits, *Wireless World*, March 1973, p 114–117.

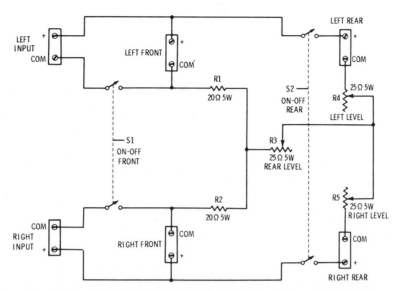

MATRIX ADAPTER—Passive circuit provides means for connecting two additional speakers to existing two-channel stereo setup to give fair approximation of four-channel effect. Circuit utilizes random out-of-phase information that is usually recorded deliberately or inadvertently on many stereo records, for feeding to rear speakers. With four-channel records, there is small amount of cross-blending in addition to full difference information from rear speakers, giving more natural rear ambience effect and some control over loudness of rear speakers.—L. Feldman, "Hi-Fi Projects for the Hobbyist," Howard W. Sams, Indianapolis, IN, 2nd Ed., 1972, p 51–53.

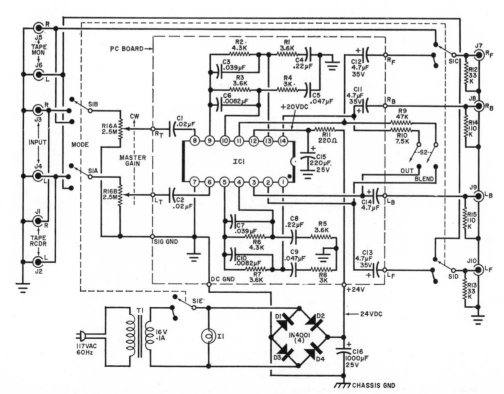

SQ DECODER—Uses Motorola MC1312P IC to provide outputs for four preamps of quadraphonic matrix system or two channels of standard stereo system. Article gives theory of operation and construction details. T1 is Stancor P-8611, with only half of secondary used to give 16 V at 0.1 A for bridge rectifier. I1 is No. 346 18-V 0.04-A pilot lamp. Two inputs, RT and LT, are expanded to four channels by phase shifters and matrix.—M. Esformes, SQ Four-Channel Decoder, *Popular Electronics*, July 1973, p 26–31.

50 W WITH ONE IC PER CHANNEL—Complete quadraphonic system requires only four RCA KD2131 audio power modules, each driven by preamp. May be used with electrostatic speakers. Includes overload protection for output stages. Half of system is shown; two identical units are required for four-channel system. Article covers construction and checkout. Requires input of 0.5 V rms per channel for rated output to 8-ohm speaker system. For 30 W output, response is down only 0.5 dB at 20 Hz and 20,000 Hz. T1 is Stancor TP-4.—G. D. Hanchett, Build R-E's 4-Channel IC Power Amplifier, *Radio-Electronics*, Oct. 1971, p 37–40.

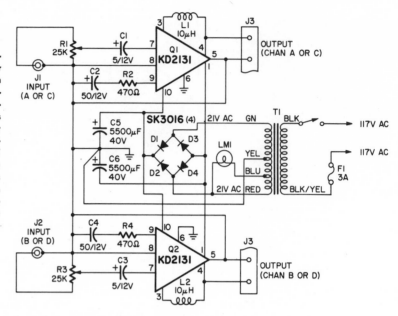

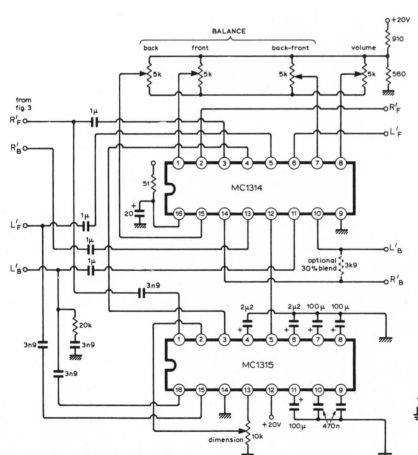

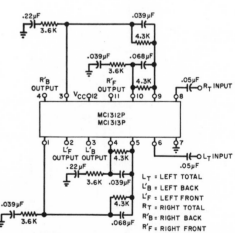

AUTOMATIC BLENDING—Motorola IC's provide automatic blending between front speaker outputs (commonly 10%) and between rear speaker outputs (commonly 40%) when fed by SQ decoder which does not have blending resistors. Blending serves to cancel antiphase components in rear signals when source appears at front center. Not advisable for use with recordings or programs having multiple sources. MC1314 includes voltage-controlled amplifiers, while MC1315 provides controlled voltages for improving front-center sounds and attenuating unwanted outputs for corner signals. Automatic action can be varied with 10K pot, normally set at 50% to give about 15 dB front-to-back crosstalk.—G. Shorter, Surround-Sound Circuits, *Wireless World*, March 1973, p 114–117.

SINGLE-IC DECODER—Uses Motorola IC chips designed for decoders of Columbia SQ encoded program material, with MC1312P operating from 20-V supply for home equipment and MC1313P from 12 V for auto installation. Both chips have zero gain typical signal-noise ratio of 80 dB, harmonic distortion of 0.1%, high input impedance, and low output impedance. Chips may not be available for home construction projects.—W. G. Jung, More New Devices & Applications, *Popular Electronics*, March 1973, p 96 and 98–99.

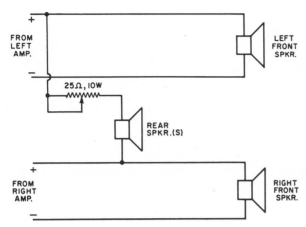

REAR SPEAKER—Extra speaker connected between high sides of left and right stereo speakers gives additional ambience effect. Speaker should be 8-ohm or higher, because added rear speaker can be considered to be in parallel with one of side speakers. Alternatively, two rear speakers can be used, connected either in series or in parallel. Series connection gives lower output from rear.—D. Hafler, Adding Extra Channel for Improved Hi-Fi Ambience, *Electronics World*, Oct. 1970, p 31.

SURROUND-SOUND FROM STEREO RECORDS—Toshiba TA7117P IC provides four-speaker simulated quadraphonic output from conventional stereo records and certain coded records such as early American Dynaco, Electro-Voice, some Japanese records, and coded QS/RM/Pye records. Chief drawbacks are increase in apparent width of stereo field, absence of precise back images, and some shifting to front speakers of sounds intended for back speakers. Experimentation with different values for 18K resistors may improve performance on certain records.—G. Shorter, Surround-Sound Circuits, *Wireless World*, March 1973, p 114–117.

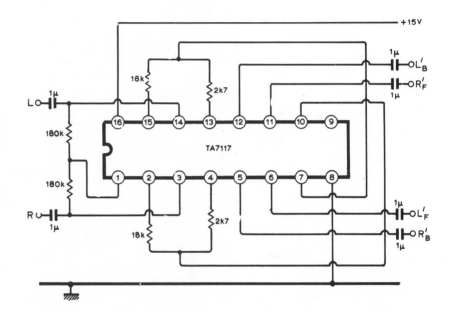

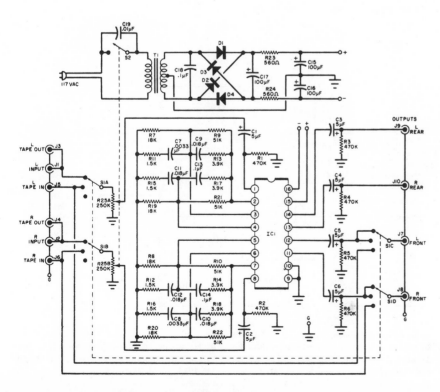

UNIVERSAL DECODER—Will handle CBS SQ, Sansui, Electro-Voice, and other matrix 4-channel systems. Designed for input signals in range of 0.1 to 1 V, with maximum of 4-V rms. Includes master volume control S1 for all four channels of preamps which feed four power amplifiers and 4-speaker system.—F. Nichols, Build a 4-Channel Universal Decoder, *Popular Electronics*, Dec. 1972, p 28–32.

CHAPTER 93
Radar Circuits

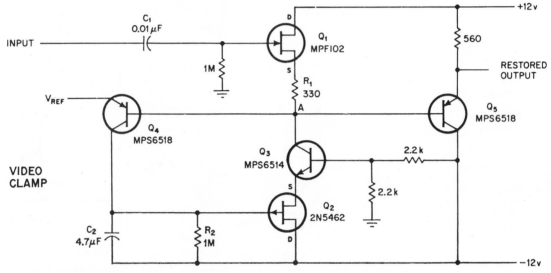

CLAMP AND D-C RESTORER—Locks level of output to reference voltage, following timing of randomly arriving video signals up to 10 MHz. Input impedance is high and output impedance low. Matching Q4 and Q5 makes temperature drift negligible.—T. E. Polcyn, Stable FET Clamp Operates at 10 MHz, *Electronics*, May 10, 1971, p 81.

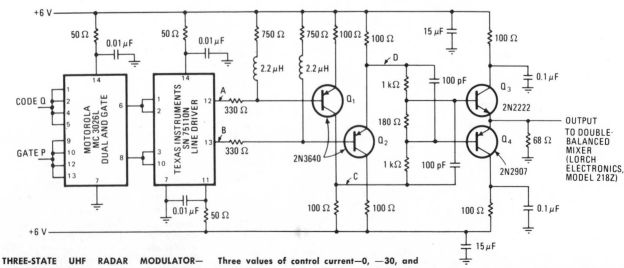

THREE-STATE UHF RADAR MODULATOR—Turns uhf signals on or off or reverses their phase between 0 and 180 deg in less than 10 ns. Off signal is attenuated at least 35 dB. Three values of control current—0, —30, and 30 mA—are generated by tri-state driver for double-balanced mixer that is driven by Q3 and Q4 with difference signal of B—A.—R. N. Assaly, Fast-Switching Modulator Reverses UHF Signal Phase, *Electronics*, March 13, 1972, p 74–75.

FLUX MONITOR FOR PHASE SHIFTER—Ferrite phase shifter control system minimizes effects of aging, temperature, power supply variations, and other parameter changes by monitoring magnetic flux and turning off driver pulse when desired flux level is reached. Circuit is covered by U.S. Patent 3,510,675. Values of R and C were not available.—H. C. Goodrich and R. C. Tomsic, Flux Monitoring Boosts Accuracy of Phased Array Radar Systems, *Electronics*, Nov. 23, 1970, p 77–80.

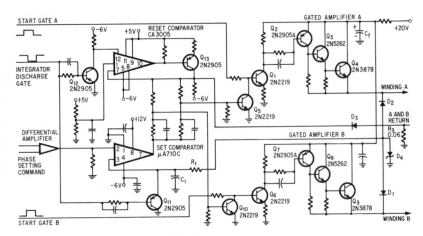

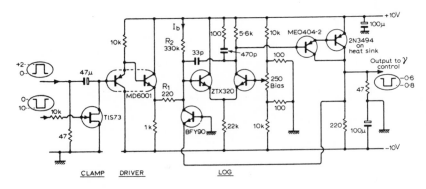

PPI GAMMA CONTROL—Logarithmic video amplifier accepts pulses with dynamic range up to 300:1 and compresses to 20:1 dynamic range or gamma required for intensity-modulated crt in ppi radar display. Can also be used for matching dynamic range of scene or recording being televised to that of crt in tv set.—A. M. Pardoe, 30-MHz Video Amplifier with Variable Power Law, *Wireless World*, Dec. 1970, p 597–598.

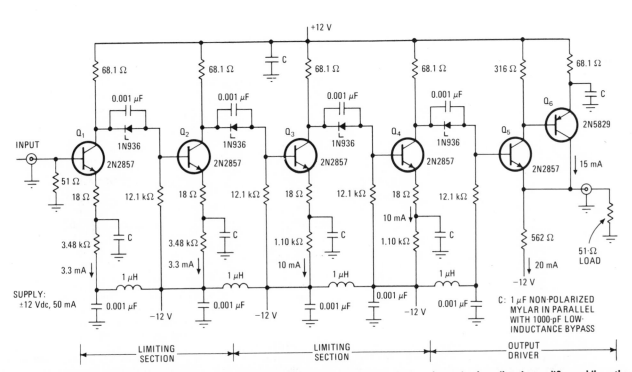

0–30 MHZ PHASE-TRANSPARENT—Does not alter zero crossings of input signal while providing gain of 20 dB over linear range. Used in radar and communication systems when information is transmitted in phase domain. Permits resolution of smaller targets in radar, by improving receiver sensitivity. Uses current-cutoff limiting over dynamic range of 40 dB for input. Depending on input signal polarity, one two-transistor limiter section is low-gain broadband amplifier while other acts as cutoff isolation amplifier.—R. J. Turner, Broadband Cutoff Limiter Is Phase-Transparent, *Electronics*, July 3, 1972, p 87.

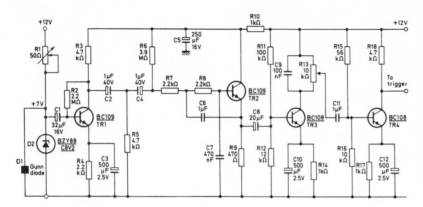

150-HZ CUTOFF—TR1 amplifies output of 10.69-GHz Mullard CL8630 Gunn diode modulated by signal reflections from two-blade propeller driven at speeds from 1,000 to 5,000 rpm, to give modulation frequencies of 30 to 150 Hz. TR2 is single-stage active low-pass filter with 150-Hz cutoff, falling 12 dB per octave above that. Resulting clean sine-wave modulation frequency is further amplified by TR3 and TR4 to give sufficient signal for operating Schmitt trigger driving counter. —T. G. Giles and J. E. Saw, Simple Doppler Radar Using the CL8630 Gunn Effect Oscillator for the Observation of Small Rotating Objects, *Mullard Technical Communications*, April 1972, p 114–119. Also in Mullard TP1303 Application Note.

SCHMITT DRIVE FOR COUNTER—Converts sine-wave inputs up to 150 Hz to negative-going pulses suitable for driving frequency tripler and counter. Used in doppler radar for monitoring speed of small rotating propeller. —T. G. Giles and J. E. Saw, Simple Doppler Radar Using the CL8630 Gunn Effect Oscillator for the Observation of Small Rotating Objects, *Mullard Technical Communications*, April 1972, p 114–119. Also in Mullard TP1303 Application Note.

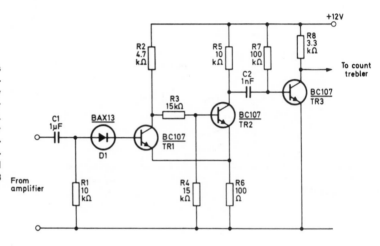

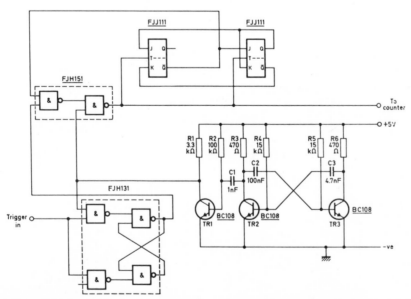

TRIPLER—Multiplies 150-Hz input frequency of negative-going pulses by three for driving counter in doppler radar system for monitoring speed of small rotating propeller. Consists of two Mullard FJJ111 J-K flip-flops, six gates, and three-transistor clock pulse generator. Each input pulse from Schmitt trigger opens gates long enough to allow three clock pulses to enter counter. FJH131 is used as bistable latch to make gates open in phase with clock pulses.—T. G. Giles and J. E. Saw, Simple Doppler Radar Using the CL8630 Gunn Effect Oscillator for the Observation of Small Rotating Objects, *Mullard Technical Communications*, April 1972, p 114–119. Also in Mullard TP1303 Application Note.

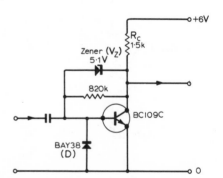

RADAR SPEED METER—Simple capacitance-coupled amplifier provides limiting when current from signal source exceeds value determined by input voltage, zener, and Rc, causing transistor to cut off. Limit is generally a few volts. Circuit will amplify small signals, around 1 mV, without blocking after receiving train of large signals. Two such stages were used together in microwave doppler radar speed meter.—F. Hibberd, A Non-blocking Limiting Amplifier, *Wireless World*, June 1969, p 269.

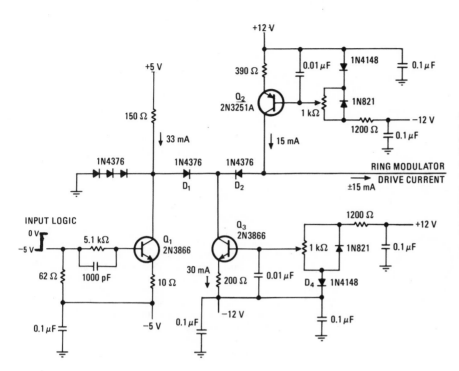

3-NS PHASE SWITCHER—Used in applying coded r-f phase modulation to interrogating radar. Can also serve for applying secure modulation to communication link, making system very difficult to jam. With diode-steered current source, binary r-f phase modulation is accomplished by translating TTL input logic levels to bidirectional current drive in less than 3 nF. Circuit switches Schottky diodes in ring modulator at high rates.—R. J. Turner, Binary RF Phase Modulator Switches in 3 Nanoseconds, *Electronics*, April 26, 1973, p 104.

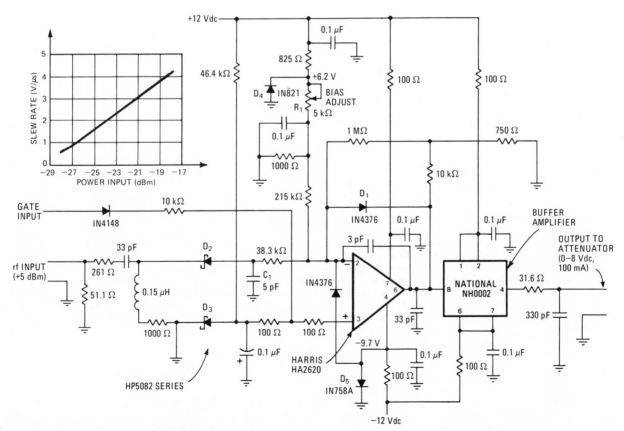

ENHANCED SIGNAL DETECTOR—R-f threshold detector for agc loop in radar and communication systems improves processing of signals from 1 MHz to 1 GHz at slew rates up to several V per μs. Signal is processed linearly in r-f section of receiver below detection threshold, but is rapidly leveled above threshold. Permits detection of low-level targets even in proximity of heavy clutter. Above detection threshold, output is 4.2 V from detector per dB of input power.—R. J. Turner, AGC RF Threshold Detector Provides Fast Slewing, *Electronics*, Dec. 18, 1972, p 107–108.

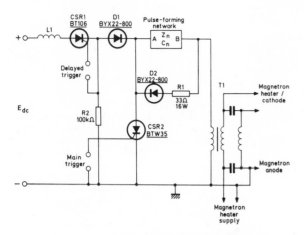

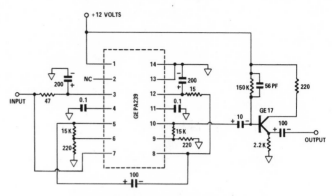

MAGNETRON MODULATOR—Direct-switching line-type modulator drives radar magnetrons having r-f pulse power output of up to 2-kW peak, such as Mullard YJ1390 and YJ1410. Report gives design criteria for pulse transformer T1.—M. Keohane, "Pulse-Modulator Thyristor BTW35 as a Radar Magnetron Driver," Mullard, London, 1971, TP1267, p 2–4.

SELF-DETECTING RADAR PREAMP—Uses nonlinear characteristic of oscillating diode as mixing detector in homodyne circuit to produce doppler output at sufficiently high level for operating alarm relay, driving counter, or performing other logic function. Can be used in radar speed detectors which must process doppler returns of about 300–5,000 Hz, corresponding to speeds from 10 to over 100 mph. Also suitable for intrusion alarm, which must respond to frequencies from a few Hz to less than 60 Hz. Carrier frequency is near 10.525 GHz.—Detection Range Measurements on a Simplified Doppler Radar, *The Microwave Journal*, Dec. 1971, p 20 and 22–23.

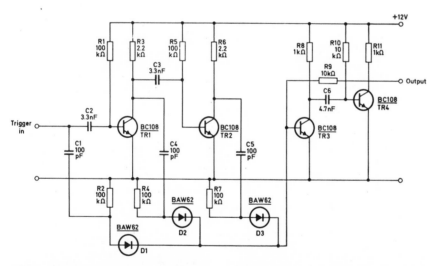

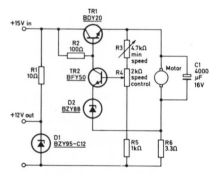

PULSE TRIPLER—Negative-going pulse from Schmitt trigger drives mono mvbr TR3-TR4 through pulse shapers TR1 and TR2. Diodes select positive-going edges of pulses at three points in circuit, to trigger mono which generates 20-μs pulses. Circuit thus serves as simple but precise frequency tripler for driving counter in doppler radar system for monitoring speed of small rotating propeller.—T. G. Giles and J. E. Saw, Simple Doppler Radar Using the CL8630 Gunn Effect Oscillator for the Observation of Small Rotating Objects, *Mullard Technical Communications*, April 1972, p 114–119. Also in Mullard TP1303 Application Note.

24-V D-C MOTOR SPEED—R4 adjusts speed of parallel-wound motor from about 1,000 to 5,000 rpm, while R3 sets minimum speed to prevent motor from being stopped and windings damaged by stall current. Regulator circuit uses small amount of positive feedback to stabilize motor speed. Used in testing doppler radar for observing speed of propeller mounted on shaft of motor.—T. G. Giles and J. E. Saw, Simple Doppler Radar Using the CL8630 Gunn Effect Oscillator for the Observation of Small Rotating Objects, *Mullard Technical Communications*, April 1972, p 114–119. Also in Mullard TP1303 Application Note.

CHAPTER 94
Receiver Circuits

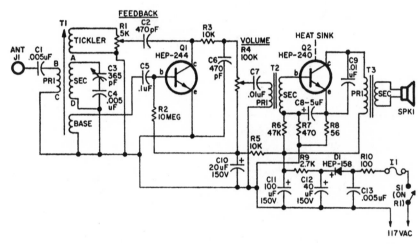

TRANSISTOR REGENERATIVE—Modern version of old circuit operates from a-c line, has high stability, and will drive 6½-inch 8-ohm speaker in cabinet. Q1 provides equivalent of grid-leak detection to give audio for driving Q2. Article gives coil-winding data.—C. Green, Mod Methuselah, *Elementary Electronics*, May-June 1971, p 29–34 and 96.

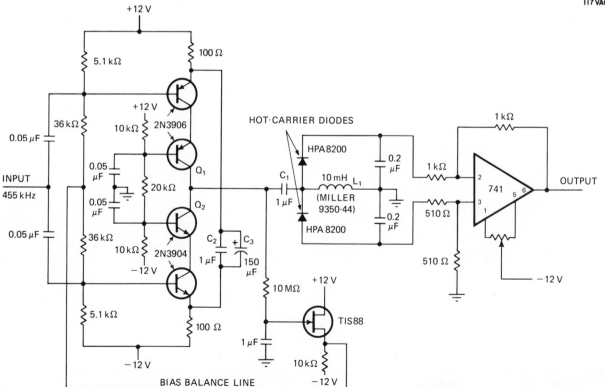

DETECTOR WITH 120-DB RANGE—Hot-carrier diodes in full-wave detector are driven by high-impedance 455-kHz current source to im-prove conduction of diodes below 0.5 V and thereby increase dynamic range to 120 dB. Ripple of 910 kHz in detected signal is smoothed by low-pass filter and differential-amplifier buffer.—R. J. Matheson, High-Impedance Driver Boosts Detector's Dynamic Range, *Electronics*, Oct. 25, 1971, p 79.

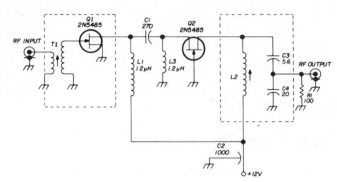

2-METER JFET PREAMP—Cascoding provides 16 to 25 dB of actual gain between terminals.—J. Vogt, Improved Two-Meter Preamplifier, *Ham Radio*, March 1972, p 25–31.

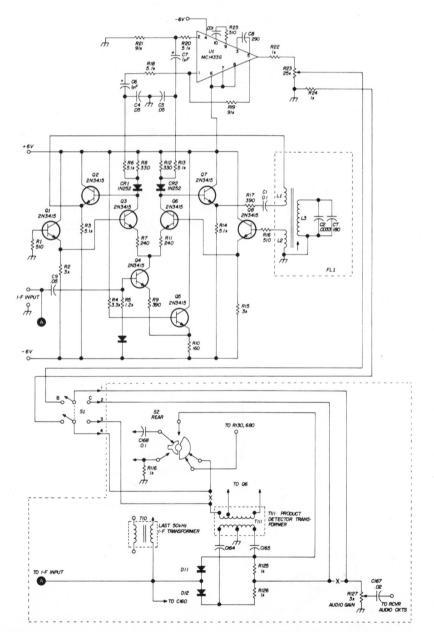

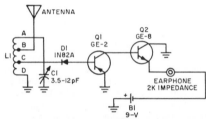

80–150 MHZ VHF—Picks up conversations between pilot and control tower, using simple tuned diode as receiver and two-stage audio amplifier. Does not interfere with flight operations. Antenna is 4 feet of stranded No. 22 hookup wire. L1 is 4 turns No. 16 enameled copper spaced out to ½ inch on ¾-inch-diameter form, with B ½ turn from A and C 2 turns from D. Can be concealed in pocket.—J. G. Busse, Flex-Jet VHF Rig, *Elementary Electronics*, July-Aug. 1971, p 65–68 and 101 –102.

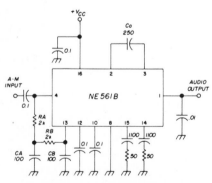

SYNCHRONOUS A-M DETECTOR—Uses Signetics phase-locked loop IC. Values are for broadcast band, up to 1.5 MHz. Phase-locked loop is locked to incoming a-m signal carrier frequency, and vco output of IC serves as local oscillator signal. External phase-shift network components RACA and RBCB establish quadrature relationship for synchronous demodulation.—E. Noll, Phase-Locked Loops, *Ham Radio*, Sept. 1971, p 54–60.

RECIPROCATING DETECTOR—Designed for double-sideband suppressed-carrier detection in communication receivers. Automatically adjusts bfo level in proportion to average signal level, to minimize noise on moon-bounce signals or weak c-w DX. Circuit is shown as incorporated in Drake R4A receiver, but may be used in any other receiver if filter FL1 is modified appropriately. Article covers theory of operation and method of use.—S. Olberg, Reciprocating Detector, *Ham Radio*, March 1972, p 32–35.

FET REGEN—Single field-effect transistor in regenerative circuit, used with any 100-mW IC audio amplifier, will give loudspeaker volume on local broadcast station with 25 feet of hookup wire as antenna. For more stations, use old-fashioned outdoor antenna and ground. Q1 can be Motorola MPF103. L is Miller A-5495-A with 8 turns No. 22 hookup wire wound over it for L1B.—D. J. Green, 1-FET BCBer, *Elementary Electronics*, Jan.-Feb. 1968, p 51–52 and 129.

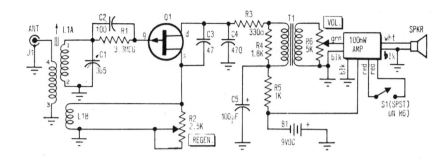

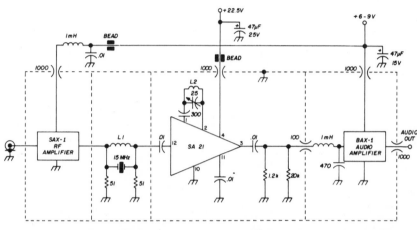

L1 10 to 18 µH (CTC X2060-4)

L2 3.3 µH. 30 turns no. 26 enameled on a T44-10 core

FIXED-TUNED WWV—Uses active i-f filter (SA21) for selectivity of about 6 kHz, more than enough for reliable WWV reception. R-f module, audio module, and IC are from International Crystal Co.—D. Pongrance, Simple WWV Receiver, *Ham Radio*, July 1970, p 68.

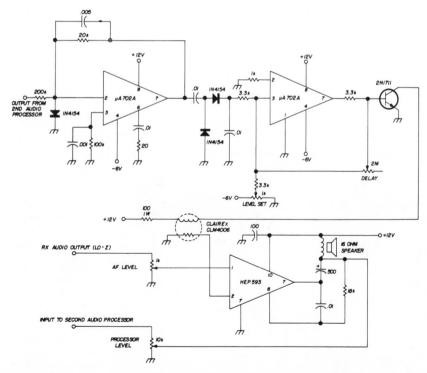

DIVERSITY COMBINER—Optoelectronic coupling provides effective, low-cost diversity receiving action. Used between audio output and speaker of receiver, with connections to identical audio processor of other receiver. Second IC stage, used as level detector, provides positive output when input level on terminal 3 exceeds that of level-set pot. This makes driver transistor turn lamp on and off in photocell module, to control gain of HEP593 and thereby provide smooth switching action to whichever receiver has the stronger audio signal.—J. J. Schultz, Diversity Receiving System, *Ham Radio*, Dec. 1971, p 12–17.

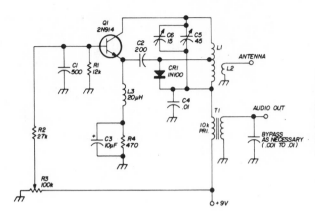

28-MHZ SUPERREGENERATIVE DETECTOR— Tuning range is about 21 to 40 MHz. Damping diode across part of tank coil eliminates hangover after oscillation burst, without preventing positive feedback.—C. L. Ring, Optimizing the Superregenerative Detector, *Ham Radio*, July 1972, p 32–35.

144-MHZ SUPERREGENERATIVE DETECTOR— Uses Colpitts oscillator to compensate for shunt capacitance. Time constant of R1-C1 should be as short as possible, but not down to point at which circuit tends to go into c-w oscillation.—C. L. Ring, Optimizing the Superregenerative Detector, *Ham Radio*, July 1972, p 32–35.

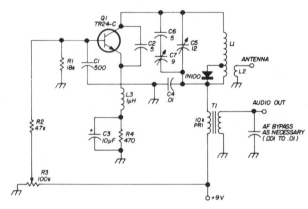

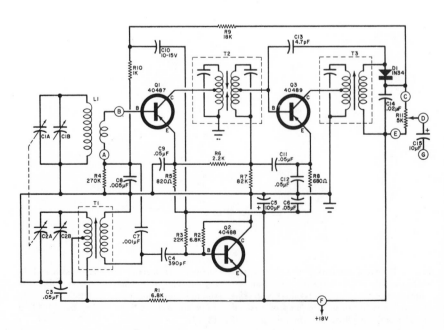

HI-FI A-M TUNER— Can be added to expensive hi-fi system. Image rejection is 48 dB, adjacent-channel attenuation 26 dB, agc figure of merit 30 dB, and (S + N)/N is 20 dB, comparable to performance of high-quality commercial a-m tuners. L1 is ferrite broadcast-band antenna with low-impedance secondary, feeding base of mixer Q1. Separate oscillator Q2 also feeds mixer. Resulting 455-kHz i-f signal is fed through standard double-tuned i-f transformer T2 to Q3 acting as high-gain neutralized amplifier providing power gain of 40 dB. T1 is standard local oscillator transformer. Article covers construction and alignment. Volume control R11 can be mounted remotely.—J. Cuccia, Top-Rated AM Tuner, *Popular Electronics*, Jan. 1969, p 43–46 and 112.

A-M REGENERATIVE—R-f detector-amplifier Q1 feeds two-stage audio amplifier Q2-Q3 driving speaker. Ferrite-core antenna coil L1-L2 is standard a-m coil such as Calectro D1-848, with Ct chosen to match. Feedback loop is single turn of insulated wire wrapped loosely around L1. R4 should have log taper.—Manufacturer's Circuit, *Popular Electronics*, Jan. 1970, p 84 and 86.

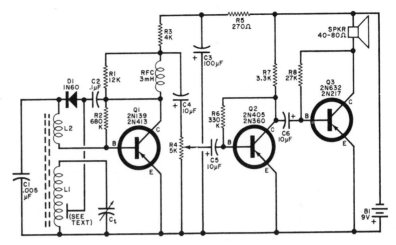

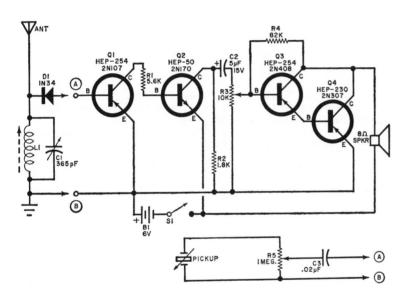

SIMPLE RADIO—Signal from long outdoor antenna is selected by broadcast-band tuned circuit L1-C1, detected by D1, and built up to speaker volume by four-transistor audio amplifier. R3 is volume control. High-output ceramic or crystal cartridge may be connected instead to points A and B, with additional volume control R5, for use as phono amplifier.—D. Hoff, Reader's Circuit, *Popular Electronics*, Aug. 1969, p 96.

VARICAP A-M TUNER—Regenerative broadcast-band tuner brings in local stations on headphones with only 3-foot antenna, and stations over 1,000 miles away with 20-foot antenna. R1 changes d-c voltage of varicap diode D1 to provide tuning. Q1 and Q2, connected to simulate single very-high-gain transistor, can be any general-purpose germanium r-f transistor. Do not use nonlinear converter-mixer transistors. L1 is tapped antenna loopstick such as Superex VLT 240, with feedback winding L2 close-wound over its center; determine number of turns by experiment.—A. A. Mangieri, A Varicap Front End AM Tuner, *Popular Electronics*, May 1969, p 76–77 and 96.

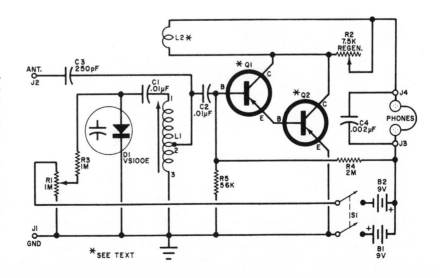

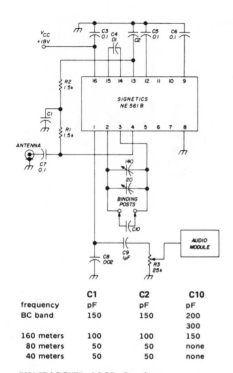

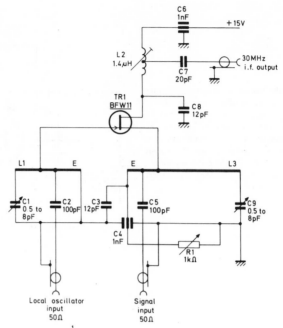

	C1 pF	C2 pF	C10 pF
frequency			
BC band	150	150	200
			300
160 meters	100	100	150
80 meters	50	50	none
40 meters	50	50	none

PHASE-LOCKED LOOP—Requires no resonant circuits for covering a-m broadcast band and amateur bands. Only the one signal is demodulated that has its carrier phase-locked to vco of IC. Frequency of oscillator is determined by R-C time constants. Audio module can be 1-W IC audio amplifier driving 4-inch p-m speaker.—E. Noll, Experiments with Phase-Locked Loops, *Ham Radio*, Oct. 1971, p 58–63.

L1 – 8.5cm of $\frac{1}{4}$ in o/d Cu, gate tap at 2.6cm from E, l.o. input at 1.5cm from E
L2 – 8.5cm of $\frac{1}{4}$ in o/d Cu, signal input tap at 0.9cm from E, source at 1.7cm from E

Characteristic impedance of line 70Ω

FET MIXER—Delivers 30-MHz signal to i-f amplifier of communication receivers. Typical noise figure is 9 dB. Conversion gain ranges from 8 to 11 dB for oscillator voltage of 0.2 to 1.2 V, and 3-dB bandwidth is 600 kHz.

Uses quarter-wave 70-ohm transmission line as tuning element for both inputs.—"R. F. Applications of FET's in Communications Receivers," Mullard, London, 1970, TP1175, p 5.

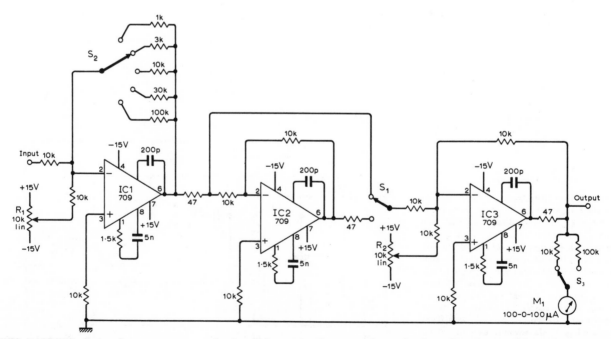

TUNER MATCHING—Designed to connect tuners with noncompatible i-f strips during experimentation, such as when tuner uses pnp transistors and i-f has npn. Also covers situation in which tuner requires agc voltage of opposite polarity to that available from i-f strip, or with different offset value than is required. With no signal input to tuner, set R1 for null on M1, set S2 for required gain, and set R2 to give required output with no input.—A. Clements, Matching Unit for Tuners and I.F. Strips, *Wireless World*, Jan. 1972, p 18.

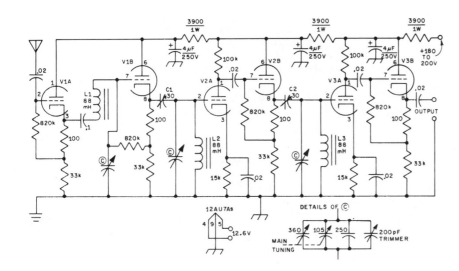

20-KHZ VLF FOR WWVL—Uses narrow-band amplifier for pulling in 2-kW standard-frequency signals while rejecting much more powerful NAA and NSS signals about 1 kHz away. All coils are toroids. Tuning range is roughly 18 to 27 kHz with padders shown. Tubes can be 12AU7A or 5963 dual triodes. Each tuning capacitor is a midget bc superhet unit with the two gangs connected in parallel.—E. Pearson, A WWVL Receiver, QST, Nov. 1971, p 34–37.

20-KHZ NAVY C-W RECEIVER—With 100-foot antenna and good ground, will pick up time signals, news flashes, and 5-letter cipher groups from NAA in Maine, NSS in Maryland, and NPG on West Coast. Simple circuit consists of regenerative detector tuning from 13 to 28 kHz, and one stage of audio for driving phones. Uses slug tuning in L1. S1 switches in C2 for covering lower end of tuning range. D1 is 400-piv 500-mA silicon, such as 1N2070. L1 is 16–42 mH tv horizontal oscillator coil.—H. B. Smith, Build Our 1-Tube Bottom Scraper, Science and Electronics, Jan. 1971, p 28–31.

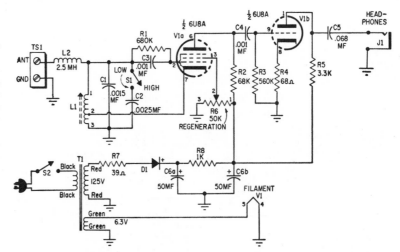

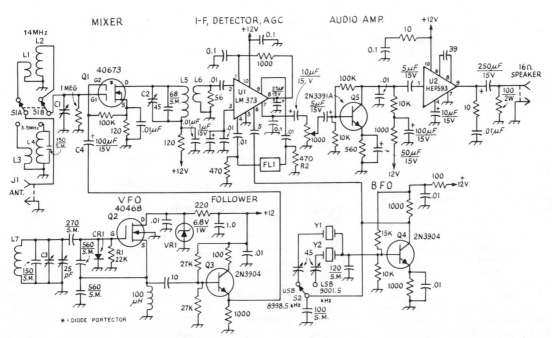

80 AND 20 METERS—Two IC's simplify construction of complete receiver using 5–5.5 MHz variable-frequency oscillator to provide 9-MHz i-f for either band. Includes beat-frequency oscillator for c-w reception. Article includes simple regulated power supply.—T. R. Sowden, The Super-Simple 80—20 Receiver, QST, April 1972, p 26–29.

32-GHZ GUNN DIODE MIXER—Simple afc circuit extends operating range of self-oscillating Gunn-mixer receiver by reducing frequency jitter. Gunn diode used was built at Research Center of Plessey Co. in England.—M. J. Lazarus and S. Novak, A Millimeter-Wave Gunn Mixer with −90-dBm Sensitivity, Using a MOSFET/Bipolar AFC Circuit, *Proc. IEEE*, June 1972, p 747−748.

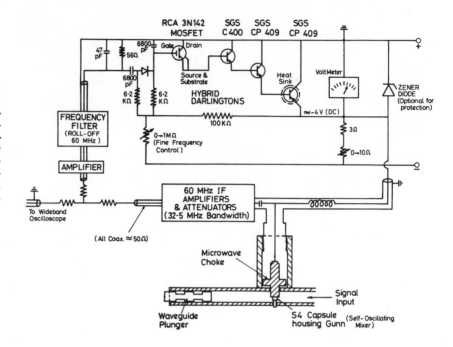

PHASE-LOCKED OSCILLATOR—Designed for use as first-conversion high-frequency oscillator in converter or communication receiver. Wide range of synchronizing capability allows locking to any integral harmonic of reference crystal oscillator Y1. Table gives crystal frequencies. Article gives coil-winding data, along with values of R2, C2, and C3 for different frequency ranges. Entire range from 7.5 to 45 MHz can be covered with nine sets of coils.—K. W. Robbins, Transistors and IC's in a Phase-Locked Local Oscillator, *QST*, Jan. 1972, p 43−47.

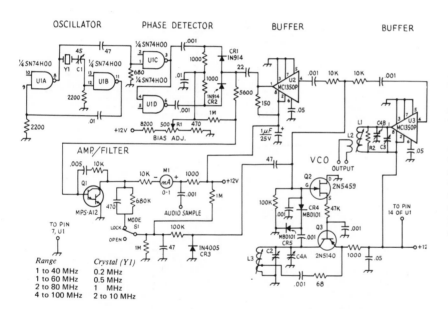

Range	Crystal (Y1)
1 to 40 MHz	0.2 MHz
1 to 60 MHz	0.5 MHz
2 to 80 MHz	1 MHz
4 to 100 MHz	2 to 10 MHz

AIRCRAFT BAND—Simple receiver is tunable over 117−150-MHz aircraft band as well as 2-meter amateur band. 9-V transistor radio battery may be connected between point B and ground, using spst switch to disconnect a-c supply at that point. Circuit is superregenerative receiver using GE-9 transistor. Use 43-inch folded dipole made from 300-ohm twinlead as antenna connected to L1. Article gives coil data. IC audio amplifier is 100 mW into 8-ohm speaker, such as Lafayette 99T90425. R1 may need changing to suit transistor, but should not be below 100K.—R. E. Kelland, Super Stable Receiver, *Science and Electronics*, Feb.-March 1970, p 39−43.

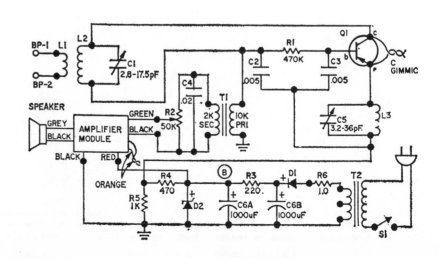

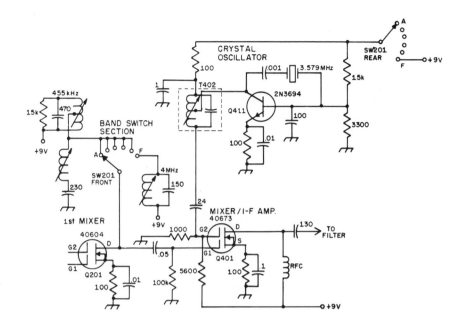

DUAL CONVERSION—Used in six-band receiver covering 200 kHz to 30 MHz. On all but highest-frequency band, 455-kz output of first mixer is fed directly to gate 1 of 40673 i-f amplifier. When band switch is set on F for 18–30 MHz, first h-f oscillator operates 4.034 MHz higher than desired incoming signal and 3.579-MHz crystal oscillator is energized. This feeds gate 2 of 40673, which then becomes second mixer instead of straight-through i-f amplifier. Technique reduces images.—The Heath GR-78 Receiver, QST, Oct. 1970, p 48–51.

IC TRF—Uses LM372 IC which provides complete gain, detection, and agc functions required in 455-kHz a-m i-f strips, plus a-f transistor driving speaker. Has useful sensitivity throughout a-m broadcast band, as well as 160-, 80-, and 40-meter bands when used to amplify pretuned a-m signals directly from antennas. IC is basically very high gain d-c amplifier preceded by agc attenuator stage which can cut incoming signals by as much as 80 dB. This is so effective that all broadcast-band stations come in with essentially same volume, making volume control unnecessary.—B. Hirschfeld, An Integrated Circuit TRF Receiver, 73, Nov. 1972, p 29–30 and 32.

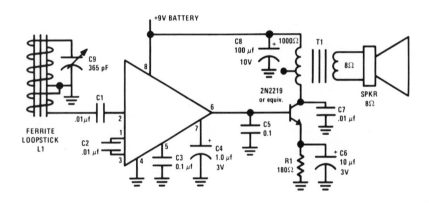

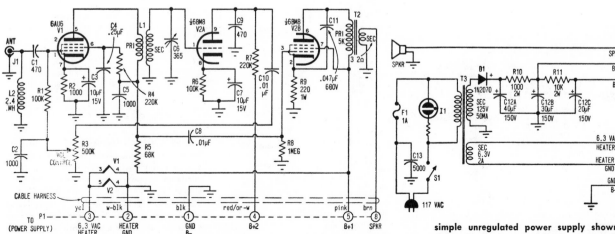

BROADCAST-BAND REFLEX—V1 serves for r-f and a-f amplification, to provide speaker volume with only two tubes. Antenna can be 6-foot length of hookup wire. Requires only simple unregulated power supply shown at right. L1 is bc r-f coil.—C. Green, Build TWOFER-FLEX, Science and Electronics, Jan. 1971, p 61–65.

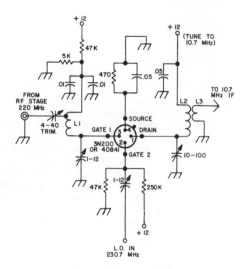

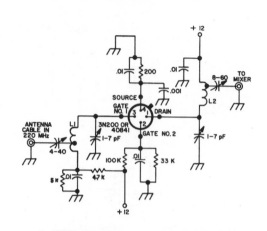

L1=3 TURNS NO. 12 BARE WIRE
7/16 O.D., 3/4 LONG, TAP AT
1/2 TURN FROM COLD END

L2=50 TURNS NO. 34 WIRE
ON 3/16 PHENOLIC FORM
(OR SIMILAR, FOR 10.7
MHz)
L3=2 TURNS ON L2

220-MHZ FET—Uses high-frequency fet with gate 1 fed by similar fet in r-f stage and gate 2 by local oscillator chain using 38.670-MHz crystal and sextupler. Use of dual-gate mixer eliminates self-oscillation.—B. Hoisington, 220 MHz: Front End Using Fet's, 73, Nov. 1972, p 45–46 and 48–50.

L1 = 3 TURNS NO. 12 WIRE
TAP AT 1/4 TO 1/2 TURN
FROM COLD END. OD=
7/16 , 3/4 LONG

L2 = 3 TURNS NO. 12 WIRE
TAP AT 1 TURN FROM
COLD END. OD=7/16 ,
3/4 LONG

220-MHZ R-F—Uses single high-frequency fet. Designed to feed same type fet connected as mixer. Article covers construction precautions. —B. Hoisington, 220 MHz: Front End Using Fet's, 73, Nov. 1972, p 45–46 and 48–50.

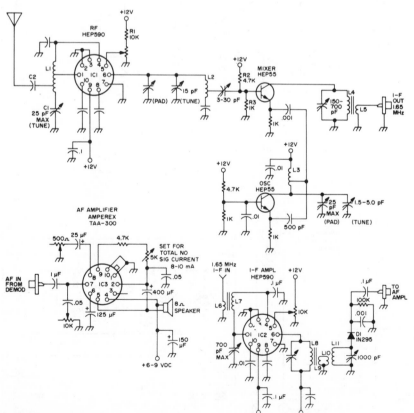

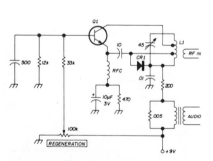

6-METER SUPERHET—Requires only three IC's and two transistors. A-f output is 1 W. Article gives all coil data. Broadband i-f stage contributes to excellent sensitivity. Uses diode demodulator.—B. Hoisington, IC Six Meter Receiver, 73, Sept. 1972, p 57–59 and 61–62.

SUPERREGENERATIVE—Use of germanium diode across tank circuit dissipates undesired energy immediately after oscillation burst, to eliminate hangover effects. Diode type is not critical. For 50 MHz, Q1 can be HEP55 or 2SC372; RFC is 10 μH and L1 7 turns No. 20 tapped at 1/2 and 2 1/2 turns from ground, 9 1/6 inch diameter and 3/4 inch long. For 144 MHz, use HEP56 or 2SC387 and smaller coils. —A. Iwakami, Improved Superregenerative Receiver, Ham Radio, Dec. 1970, p 48–49.

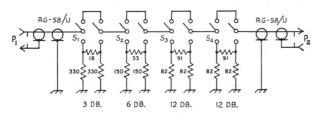

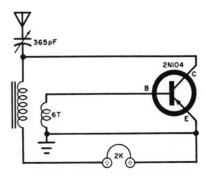

STEP ATTENUATOR—Use between antenna and receiver to prevent overloading by extremely strong signals. Attenuation is provided in 3-dB steps from 3 to 33 dB by closing various combinations of four slide switches. When all switches are up, direct connection between input and output gives 0 dB attenuation. Designed for insertion in 50-ohm line; will work with other impedances but attenuation values will be different.—B. Goodman, A Simple Step Attenuator, QST, Aug. 1967, p 24–25.

BATTERYLESS RECEIVER—Will give headphone reception of local a-m broadcast stations when used with outdoor antenna at least 100 feet long and good ground. Tuning coil is standard ferrite broadcast-band loopstick with 6 turns of No. 22 enamel wound on one end of coil.—C. J. Schauers, No Power Receiver, *Popular Electronics*, Oct. 1968, p 78.

LADDER ATTENUATOR—Used ahead of receiver, to prevent overloading of front end by extremely strong local signals when tuning across ham band. Switch gives five 10-dB steps of attenuation. Can be used in any 50 or 75-ohm line. Power rating is 0.5 W when using ½-W resistors. A Low-Z Ladder-Type Attenuator, QST, Nov. 1967, p 41, 150, and 152.

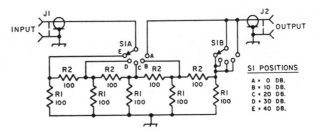

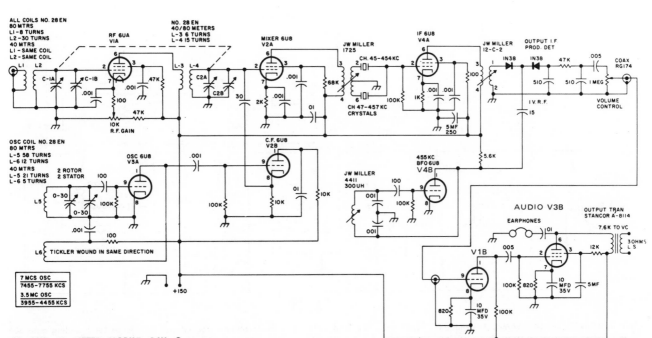

80 AND 40 METER MOBILE C-W—Operates from 12-V auto battery through simple two-transistor converter providing required 150 V unregulated for tubes, all four of which are identical 6U8 triode-pentodes. Crystal filters give selectivity of 3 kHz. Article covers construction, adjustment, and operation, including power supply circuit.—E. Marriner, Mobile CW Receiver, 73, July 1970, p 90–92 and 94–95.

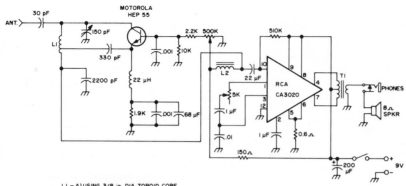

14–35 MHZ IC REGENERATIVE—Covers three ham bands, short-wave stations, and public service frequencies. Transistor serves as regenerative detector feeding IC audio amplifier. L2 is 100K primary of any small audio transformer, with secondary not used. Antenna is about 50 feet of wire. Use six D cells in series as supply.—S. Kelly, A Solid-State High Frequency Regenerative Receiver, 73, Feb. 1972, p 23–24.

L1 – A) USING 3/8 in. DIA TOROID CORE
(MICROMETALS T37-10)
12 TURNS 28 AWG ENAMEL
TAPPED 3 TURNS FROM COLD END.
B) USING 12 AWG SOLID WIRE – 8 TURNS
1/4 in. DIA TAPPED 2 TURNS FROM
COLD END.

L2 – AUDIO CHOKE (See Text)

T1 – AUDIO OUTPUT TRANSFORMER
500 mW, 250 Ω C.T. TO 8 Ω

DOUBLE-BALANCED WITH 9-MHZ OUTPUT—Uses IC balanced modulator as high-frequency mixer with broadband input and output tuned to 9 MHz. Bandwidth is 450 kHz for 3 dB down. Conversion gain is 13 dB for 30-MHz input signal and 39-MHz local oscillator. Performance curves are given.—R. Hejhall, "MC1596 Balanced Modulator," Motorola Semiconductors, Phoenix, AZ, AN-531, 1971.

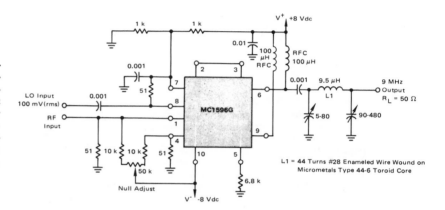

L1 = 44 Turns #28 Enameled Wire Wound on Micrometals Type 44-6 Toroid Core

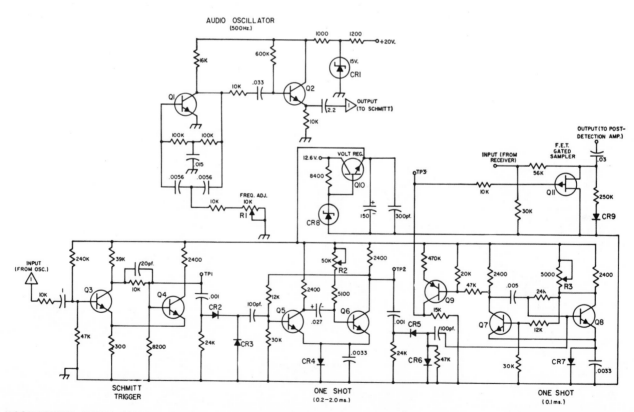

AUDIO OSCILLATOR
(500 Hz.)

SCHMITT TRIGGER

ONE SHOT
(0.2–2.0 ms.)

ONE SHOT
(0.1 ms.)

MOON-BOUNCE SIGNAL SAMPLER—Used in synchronous detection system for receiving 144-MHz signals transmitted to moon and reflected off its surface back to earth. Simple 500-Hz parallel twin-T oscillator Q1 and emitter-follower Q2 (both 2N1304) provide about 5 V drive for Schmitt trigger. Q9 is 2N508, Q11 is 2N3382, and other transistors are 2N914. CR2–CR7 are 1N4009, CR8 is 1N4737 zener, and CR9 is 1N270. Output pulse of second one-shot opens fet sampling gate 500 times per second. Article covers complete system.—W. R. Adey and R. T. Kado, Synchronous Weak Signal Detection with Real Time Averaging, QST, Dec. 1968, p 31–34.

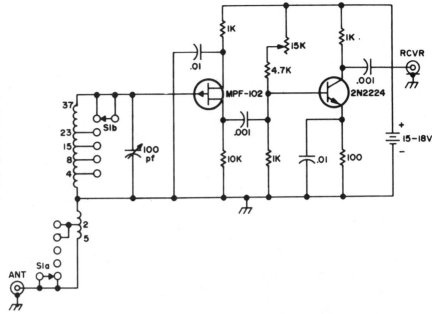

PRESELECTOR—Tapped coils permit use over entire high-frequency spectrum. High input impedance contributes to selectivity for image rejection. Used ahead of any ham receiver, with no modification of receiver itself. Has only single tuning control. Transistor types are not critical; HEP802 may be used for fet, and some other npn r-f transistor for 2N2224. —C. C. Drumeller, A Solid-State Preselector, 73, April 1971, p 128–130.

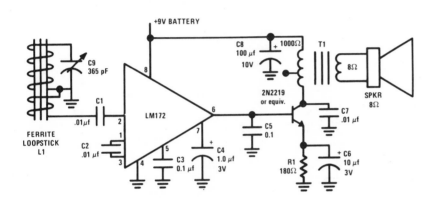

BROADBAND TRF—Single IC combined with inexpensive imported transistor radio components will amplify and detect signals below 2 MHz directly, with speaker volume. Agc action is so effective that volume control is not needed. Useful for monitoring loran (1.8-2 MHz), directional and informational channels below 550 kHz, and a-m broadcast band where sensitivity is sufficient for most urban reception areas. Although selectivity is poor with only single low-Q tuned circuit shown, selectivity can be improved by adding multisection tuning ahead of LM172.—R. A. Hirschfeld, "LM172 Monolithic AM/IF Strip," National Semiconductor, Santa Clara, CA, 1968, AN-15.

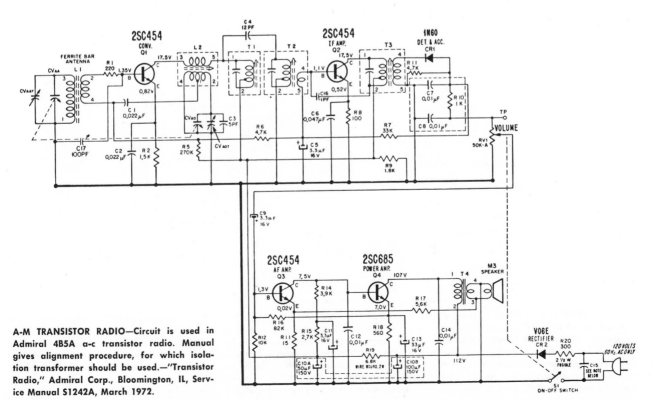

A-M TRANSISTOR RADIO—Circuit is used in Admiral 4B5A a-c transistor radio. Manual gives alignment procedure, for which isolation transformer should be used.—"Transistor Radio," Admiral Corp., Bloomington, IL, Service Manual S1242A, March 1972.

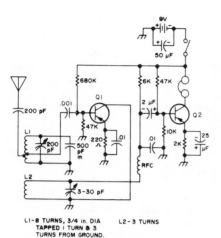

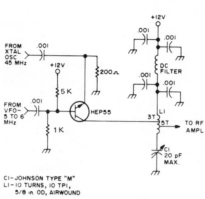

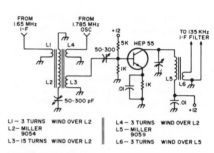

L1—8 TURNS, 3/4 in. DIA
TAPPED 1 TURN & 3
TURNS FROM GROUND.   L2—3 TURNS

C1—JOHNSON TYPE "M"
L1—10 TURNS, 10 TPI,
5/8 in. OD, AIRWOUND

L1— 3 TURNS   WIND OVER L2   L4— 3 TURNS   WIND OVER L2
L2—MILLER   L5—MILLER
      9054   9059
L3—15 TURNS   WIND OVER L2   L6— 3 TURNS   WIND OVER L5

7-MHZ REGENERATIVE—Portable receiver uses variable capacitor as regeneration control to provide reliable operation at 7 MHz with only a few Hz shift in received frequency as transistor slides in and out of regeneration. Q1 can be 2N370, 2N371, 2N372, or 3N2083. Q2 is OC70 or OC71.—J. H. Smith, Better Control for Transistor Regenerative Receivers, 73, Nov. 1972, p 157.

6-METER MIXER—Combines inputs from crystal oscillator and vfo to produce 50—51 MHz for r-f amplifier without usual multiplication of frequency drift.—B. Hoisington, Solid-State 6 Meter Crystal-Het-VFO, 73, Oct. 1972, p 29—30 and 32—35.

MIXER FOR 1.65 MHZ to 135 KHZ CONVERTER—Uses straight link coupling from oscillator coil to mixer base coil. To minimize birdies, keep oscillator power low and use adequate shielding. Article provides thorough coverage of solid-state design.—W. Hoisington, I-F Filter Converter AVC, 73, May 1970, p 62—64, 66—70, and 72.

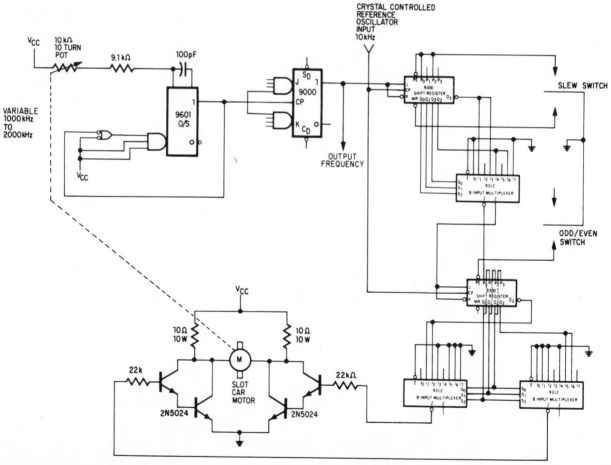

CRYSTAL-CONTROLLED DETENT TUNER—IC multiplexers and shift registers provide means for tuning communication receiver automatically and accurately to specified frequency even though no signal is present at that frequency. To select channel, slew switch is operated in proper direction until display indicates desired frequency range. Switch is then released, bringing into operation two-stage digital frequency discriminator consisting of two 9300 shift registers and three 9312 multiplexers. Two of multiplexers drive d-c tuning motor M in correct direction until local oscillator frequency generated by type 9000 IC flip-flop is synchronous with 10-kHz crystal reference. Tuning pot driven by motor has then brought vco to desired frequency.—"The 9309 and 9312 Multiplexers," Fairchild Semiconductor, Mountain View, CA, No. 181, 1971, p 15.

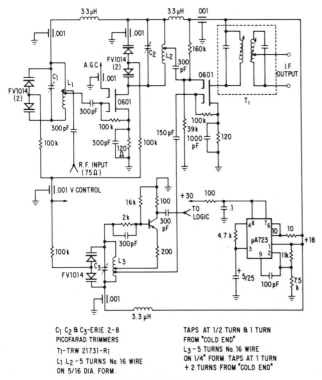

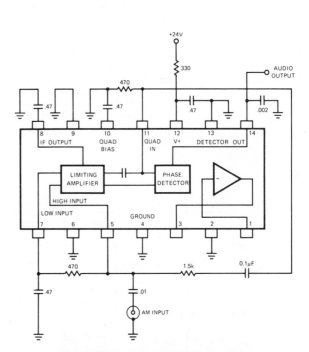

C₁ C₂ & C₃-ERIE 2-8
PICOFARAD TRIMMERS
T₁-TRW 21731-R₁
L₁ L₂-5 TURNS No. 16 WIRE
ON 5/16 DIA. FORM.

TAPS AT 1/2 TURN & 1 TURN
FROM "COLD END"
L₃-5 TURNS No. 16 WIRE
ON 1/4" FORM. TAPS AT 1 TURN
+ 2 TURNS FROM "COLD END"

SYNTHESIZER-TUNED A-M TUNER—Used with digital frequency synthesizer logic to permit precise selector-switch tuning to any assigned station frequency in broadcast band. Varactor tuning control voltage for r-f is 2–8 V. Audio output is 2 V p-p at 400 Hz for 10 mV per meter input.—J. Stinehelfer and J. Nichols, "A Digital Frequency Synthesizer for an AM and FM Receiver," Fairchild Semiconductor, Mountain View, CA, No. 201, 1971, p 8.

SYNCHRONOUS A-M DETECTOR—Uses Fairchild μA754 IC as synchronous a-m detector for receivers requiring greater sensitivity and linearity than is possible with conventional diode detector. Will operate at signal levels down to 10 mV, with wide dynamic range.—R. Smith, "Applications of the μA754 TV/FM Sound System," Fairchild Semiconductor, Mountain View, CA, No. 288, 1970.

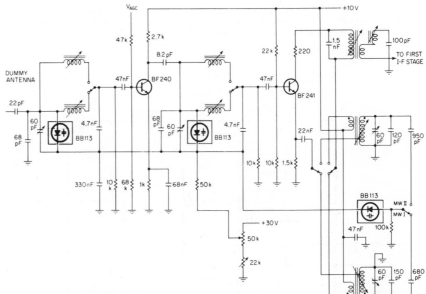

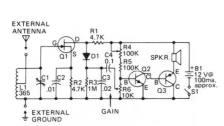

THREE-TRANSISTOR A-M—Single trf stage Q1 (HEP 802) feeds any signal diode D1 serving as detector, with two-stage audio amplifier consisting of npn transistor Q2 and pnp germanium power transistor Q3 driving 45-ohm speaker. Antenna can be any length of hookup wire. C1 is 365-pf variable capacitor.—"Calectro Handbook," GC Electronics, Rockford, IL, 1971, p 41.

VARACTOR-TUNED A-M AUTO RADIO—Triplet varactors on single chip tune a-m band by dividing into three frequency ranges, with each tuned stage using one diode section from each of three chips. Developed in West Germany. Permits locating auto radio in rear of car, far away from interference sources, with only remote tuning control within driver's reach.—G. Jonkuhn and C. H. Lembke, Matched-Varactor Chip Brings Electronic Tuning to A-M Radios, Electronics, July 19, 1971, p 60–65.

SUN-POWER RADIO—Photocell PC1 (International Rectifier B2M or equivalent) generates voltage in proportion to sunlight or artificial light, to make receiver much more sensitive than simple crystal set. L1 is transistor antenna coil; if variable loopstick antenna coil is used instead, C1 should be changed to fixed 360-pF capacitor. Antenna can be any long wire used alone or clipped to bedspring or metal window frame.—R. M. Brown and T. Kneitel, "49 Easy Transistor Projects," Howard W. Sams, Indianapolis, IN, 1972, p 37–38.

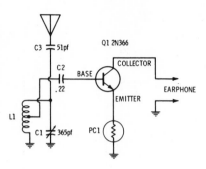

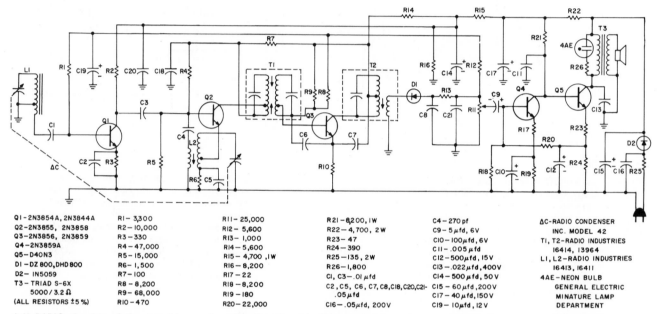

Q1 – 2N3854A, 2N3844A	R1 – 3,300	R11 – 25,000	R21 – 8,200, 1W
Q2 – 2N3855, 2N3858	R2 – 10,000	R12 – 5,600	R22 – 4,700, 2W
Q3 – 2N3856, 2N3859	R3 – 330	R13 – 1,000	R23 – 47
Q4 – 2N3859A	R4 – 47,000	R14 – 5,600	R24 – 390
Q5 – D40N3	R5 – 15,000	R15 – 4,700, 1W	R25 – 135, 2W
D1 – DZ 800, DHD 800	R6 – 1,500	R16 – 8,200	R26 – 1,800
D2 – 1N5059	R7 – 100	R17 – 22	C1, C3 – .01 μfd
T3 – TRIAD S-6X	R8 – 8,200	R18 – 8,200	C2, C5, C6, C7, C8, C18, C20, C21 – .05 μfd
5000 / 3.2 Ω	R9 – 68,000	R19 – 180	
(ALL RESISTORS ±5%)	R10 – 470	R20 – 22,000	

C4 – 270 pf	ΔC – RADIO CONDENSER INC. MODEL 42
C9 – 5 μfd, 6V	T1, T2 – RADIO INDUSTRIES 16414, 13964
C10 – 100 μfd, 6V	L1, L2 – RADIO INDUSTRIES 16413, 16411
C11 – .005 μfd	4AE – NEON BULB GENERAL ELECTRIC MINATURE LAMP DEPARTMENT
C12 – 500 μfd, 15V	
C13 – .022 μfd, 400V	
C14 – 500 μfd, 50V	
C15 – 60 μfd, 200V	
C16 – .05 μfd, 200V	
C17 – 40 μfd, 150V	
C19 – 10 μfd, 12V	

A-M RADIO—Consists of five transistors and rectifier diode, providing high performance at low cost. Untuned r-f stage Q1 improves performance without cost of adding extra gang to tuning capacitor. Sensitivity is 50–100 μV/m for 100-mW reference. Converter Q2 is autodyne circuit, Q3 is i-f amplifier, and detector diode D1 drives audio amplifier Q4-Q5 delivering maximum of 2 W to 3-ohm speaker.—E. L. Haas, "Five Transistor Line Operated AM Radio Receiver (with Untuned R.F. Stage)," General Electric, Semiconductor Products Dept., Auburn, NY, No. 90.69, 1970.

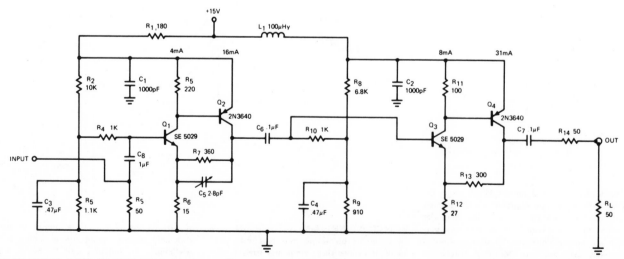

40 DB POWER GAIN AT LOW LEVEL—Bandwidth of 130 MHz permits use for boosting signal levels when measuring performance of oscillators, mixers, and r-f and i-f amplifiers. Draws less than 1 W from 15-V supply, while delivering 2 V p-p to 50-ohm load.—W. Cocke, "A Low-Level Wideband Video Amplifier," Fairchild Semiconductor, Mountain View, CA, No. 293, 1971, p 1–3.

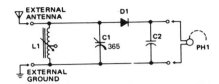

CRYSTAL DIODE SET—Requires no batteries. Use any signal diode and 2,000-ohm earphones. C2 is 0.001 μF. L1 is adjustable ferrite antenna coil used for tuning. Antenna can be about 50 ft of hookup wire.—"Calectro Handbook," GC Electronics, Rockford, IL, 1971, p 43.

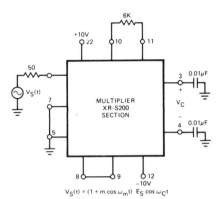

$$V_S(t) = (1 + m \cos \omega_m t) \, E_S \cos \omega_C t$$

SYNCHRONOUS A-M—Signal is applied to common input of multiplier section of Exar XR-S200 IC, and X and Y inputs are grounded. Output can be fed through low-pass filter to obtain demodulated output. Report shows cro waveforms for 1-kHz 30% modulation of 10-MHz carrier.—"XR-S200 Multi-Function Integrated Circuit," Exar Integrated Systems Inc., Sunnyvale, CA, June 1972, p 4.

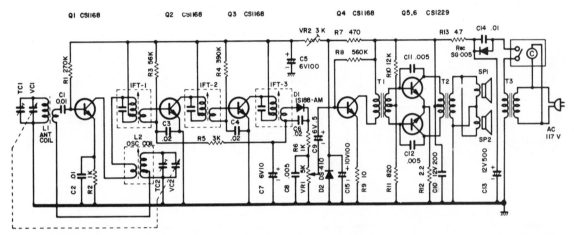

JAPANESE A-M RADIO—Circuit is typical of imported radios operating from a-c line and having two-transistor output stage driving two speakers through output transformer, and can therefore be used as rough guide for troubleshooting other makes of Japanese sets.—"Lloyd's Model 9J45-37A AM Receiver," Lloyd's Electronics Inc., Saddlebrook, NJ.

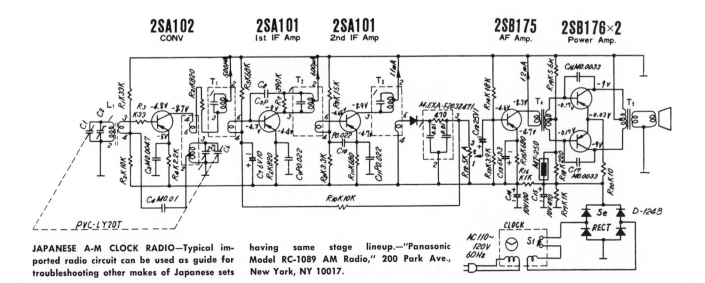

JAPANESE A-M CLOCK RADIO—Typical imported radio circuit can be used as guide for troubleshooting other makes of Japanese sets having same stage lineup.—"Panasonic Model RC-1089 AM Radio," 200 Park Ave., New York, NY 10017.

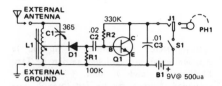

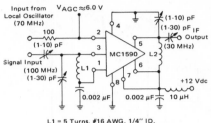

ONE-TRANSISTOR RADIO—Circuit is essentially crystal diode receiver with one stage of audio amplification for greater volume from 2,000-ohm earphone. Works best with long outdoor antenna, which can be 50' to 100' of insulated hookup wire running from house to tree. L1 is adjustable ferrite antenna coil. Diode and npn silicon transistor types are not critical.—"Calectro Handbook," GC Electronics, Rockford, IL, 1971, p 43.

L1 = 5 Turns, #16 AWG, 1/4" ID, 5/8" Long
L2 = 16 Turns, #20 AWG Wire on a Toroid Core, (T44-6 Micro Metal or Equiv)

100-MHZ IC—Input from local oscillator is 30 MHz below 100-MHz signal input. Circuit has good stability and good agc capability.—B. Trout, "A High Gain Integrated Circuit RF-IF Amplifier with Wide Range AGC," Motorola Semiconductors, Phoenix, AZ, AN-513, 1969.

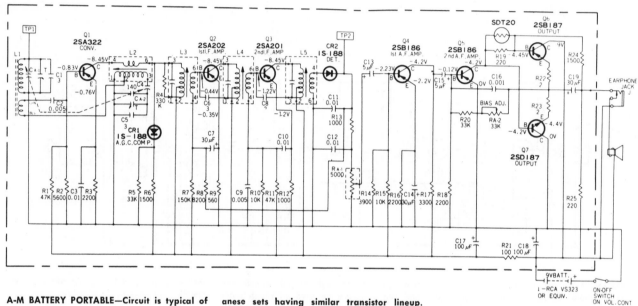

A-M BATTERY PORTABLE—Circuit is typical of this type of set as made in Japan for American manufacturer, and can therefore be used as rough guide for troubleshooting other Japanese sets having similar transistor lineup. Voltages shown are measured with vtvm with volume control at minimum and no signal. Q6 and Q7 are matched pair. Impedance of speaker is 60 ohms.—"RCA Model RCG 120 AM Portable Radio," RCA Sales Corp., Indianapolis, IN 46201.

CHAPTER 95
Regulated Power Supply Circuits

34-V PROTECTED—Electronic circuit breaker action is fast enough to protect IC amplifiers even when load is directly shorted at full power output. Power Darlington series regulator (D40C4) is controlled by IC voltage regulator IVR. Load current sensor R2 triggers scr C13F through R3, which in turn cuts off transistor serving as bias current source for Darlington. Must be reset after such overload shutdown, by opening S1 long enough for scr to turn off.—D. V. Jones, Constructing a 10-Watt Stereo Amplifier, *Audio*, May 1971, p 30 and 32–35.

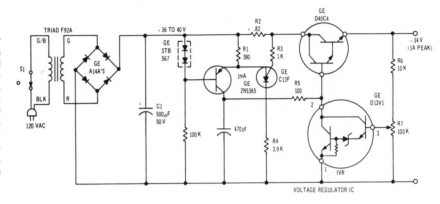

VOLTAGE REGULATOR IC

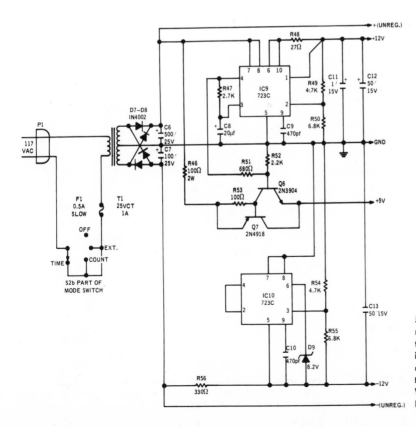

5 AND 12 V WITH IC PAIR—Internal 7.15-V reference available at pin 4 of series regulator IC9 serves for 5-V regulator Q6-Q7. IC10 is shunt regulator for negative leg, and acts as zener in absorbing current variations through R56. Used in tone burst generator.—W. G. Jung, IC Tone Burst Generator, *Audio*, Dec. 1971, p 30, 32, 34, 36, 38, and 40.

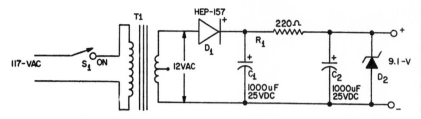

BATTERY SAVER—Transistor a-c power supply provides regulated and well-filtered 9-V d-c for use in place of 9-V batteries when transistor radio is used many hours in house by children. Zener D2 can be Motorola HEP-104. If 12 V is needed, change R1 to 100 ohms and change D2 to HEP-105.—D. A. Smith, TLPS . . . Power Supply, *Elementary Electronics*, May-June 1970, p 34–35.

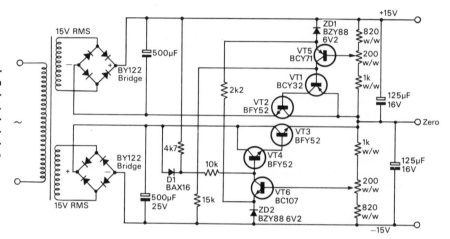

DUAL 15 V AT 100 MA—Uses compound emitter-followers as regulating elements. Main regulator transistors VT2 and VT3 have Redpoint 5F heatsinks for ambients up to 65 C. Output ripple is 0.1-mV p-p at full load. Developed at Mullard Ltd.—A. Gowthorpe, Economical Dual-Polarity Regulated Power Supplies, *Electronic Engineering*, March 1970, p 33–35.

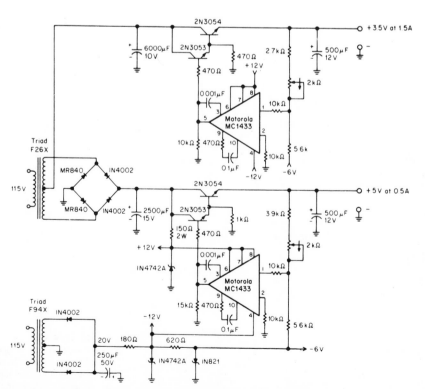

3.5 AND 5 V FOR COUNTERS—Uses inexpensive opamps to give two regulated outputs, one for high-speed DTL/TTL counting circuits and the other for lower-speed RTL circuits. Line and load regulations for both supplies are 0.1% and rms output ripple below 1 mV. —D. Purland, Regulated Supply Has Two Outputs, *The Electronic Engineer*, Dec. 1968, p 80.

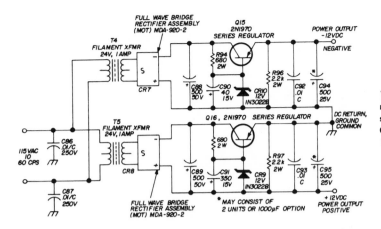

12-V BIPOLAR—Developed for amateur-band receiver. Article covers construction of entire set.—M. A. Chapman, Five-Band Solid-State Communications Receiver, *Ham Radio*, June 1972, p 6–21.

3.6 V—Reference voltage is provided by zener connected to base of series regulator Q3. Used with sstv sync generator.—D. R. Patterson, Sync Generator For Sstv, *Ham Radio*, June 1972, p 50–52.

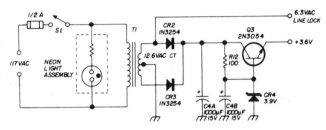

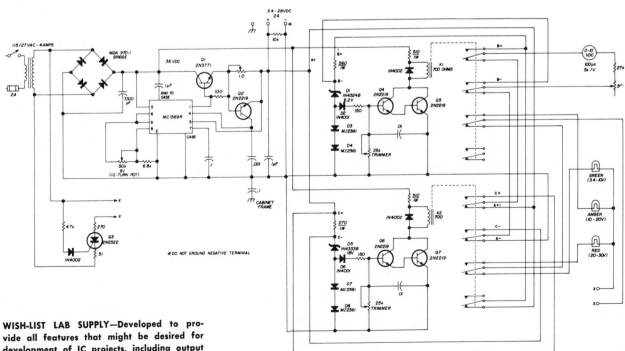

WISH-LIST LAB SUPPLY—Developed to provide all features that might be desired for development of IC projects, including output that can be varied from 3.4- to 28-V d-c at 2 A with regulation better than 0.01% and less than 50-mV ripple, current limiting regardless of voltage setting, three-lamp indication of voltage range being metered, and capability of operating continuously at full load. Relays are Advance/Hart 67DP-G-4C5 24-V. IC and Q1 have heat sinks.—A. Nusbaum, Precision Power Supply, *Ham Radio*, July 1971, p 26–31.

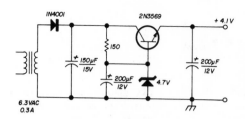

4.1 V AT 35 MA—Can be used in place of three D cells for transistor circuits.—N. J. Nicosia, Solid-State Audio Oscillator-Monitor, *Ham Radio*, Sept. 1970, p 48–49.

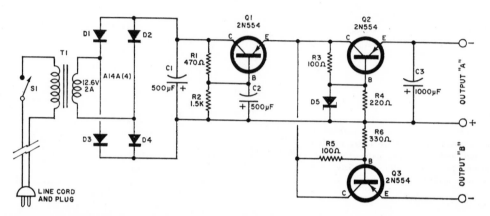

DUAL 12-V D-C—Provides equivalent of two 12-V batteries from a-c line, for driving auto radios, tape recorders, and other 12-V equipment requiring up to 1 A. D5 is 1-W zener such as HEP105, and other diodes are 100-piv 1-A rectifiers such as GE A14A. If only one supply is needed, omit Q3, R5, and R6. Use insulated heatsinks for all transistors.—A. H. Reichel, Jr., Reader's Circuit, *Popular Electronics*, Nov. 1969, p 78.

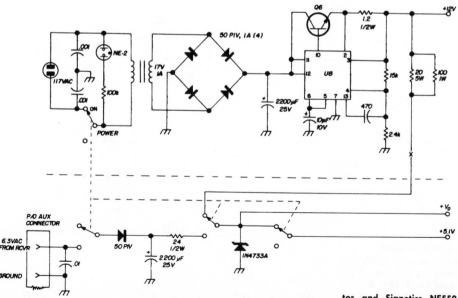

12 V AT 65 MA—Is filtered, to prevent frequency modulation of vco in 1.5–30 MHz frequency synthesizer. Also provides nominal 5 V at 360 mA. Uses 2N3054 power transistor and Signetics NE550A IC.—R. S. Stein, Frequency Synthesizer for the Drake R-4 Receiver, *Ham Radio*, Aug. 1972, p 6–19.

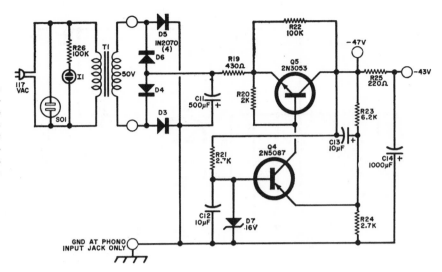

—43 V AT 100 MA—Designed for use with distortionless stereo preamp requiring essentially complete freedom from hum because preamp gain at 60 Hz is almost 60 dB. T1 is completely shielded toroid providing 50 V at 100 mA. Supply also provides —47 V regulated.—J. Bongiorno, Build a Distortionless Preamplifier, *Popular Electronics*, June 1972, p 58–62.

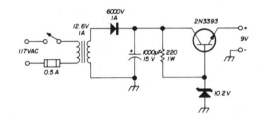

9 V FOR INTERCOM—Serves in place of 9-V transistor radio battery for intercom using 0.5-W IC audio amplifier.—N. Stinnette, Simple Intercom, *Ham Radio*, July 1972, p 66–67.

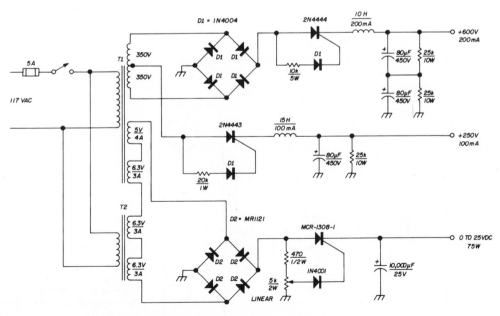

NOTE: T1 IS A STANDARD TELEVISION REPLACEMENT TRANSFORMER
T2 IS A STANDARD FILAMENT TRANSFORMER
ALL SEMICONDUCTORS ARE MOTOROLA TYPES

0–25, 250, AND 600 V—Provides two fixed and one adjustable voltage for radio transmitter and receiver applications. Each supply uses scr as series regulator. Diodes are used to prevent gate-to-cathode short-circuits.— W. E. Chapple, Scr Regulated Power Supplies, *Ham Radio*, July 1970, p 52–54.

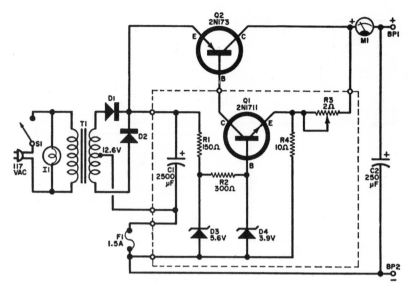

3.6 V AT 1.5 A—Has good regulation and very low ripple, for IC projects. 1.5-A d-c ammeter indicates load current drawn. D1 and D2 are 2-A silicon rectifier diodes.—F. H. Tooker, IC Experimenters' 3.6-Volt Power Supply, *Popular Electronics*, Aug. 1968, p 27–30.

LOW-RIPPLE 6 AND 12 V—Designed for use with f-m tuners and phase-locked-loop stereo decoders which require supply having very low ripple. Cascaded constant-current sources achieve ripple isolation. Shunt zener systems minimize likelihood of overvoltage failure. Is short-circuit-proof and provides reentrant overload protection. Hum and noise on output are below 0.03-mV rms. Twist d-c output leads together to prevent pickup of ripple from power transformer.—P. Lacey, Power Supply for F.M. Tuner and Decoder, *Wireless World*, Jan. 1972, p 10.

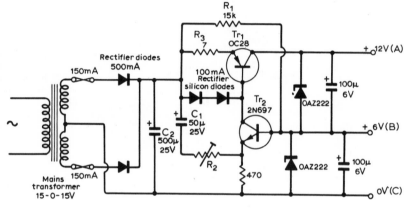

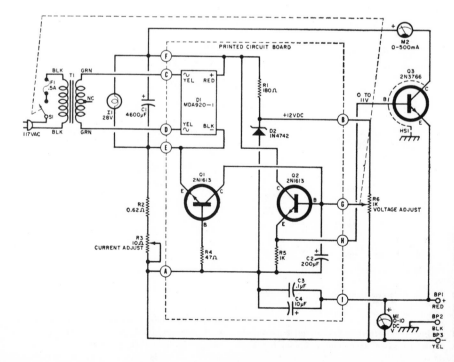

0—10 V AT 0—500 MA SHORT-PROOF—Output ripple is less than 1-mV rms. Regulation is better than 300 mV, no load to full load. Uses separate meters for voltage and current. Ideal for IC experiments, as battery eliminator for transistor radio servicing, and as general replacement for batteries.—D. Lancaster, Experimenter's "Professional" Power Supply, *Popular Electronics*, Nov. 1967, p 71–73, 105, and 110.

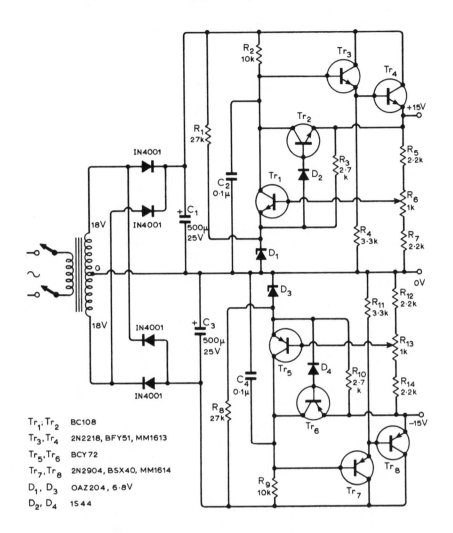

Tr_1, Tr_2 BC108

Tr_3, Tr_4 2N2218, BFY51, MM1613

Tr_5, Tr_6 BCY72

Tr_7, Tr_8 2N2904, BSX40, MM1614

D_1, D_3 OAZ204, 6·8V

D_2, D_4 1S44

15-V BIPOLAR PROTECTED—Addition of TR2 and D2 to conventional series regulator provides protection against short-circuit damage in regulated supply designed for use with fet tester.—D. E. O. Waddington, F.E.T. Tester, *Wireless World*, Dec. 1971, p 579–582.

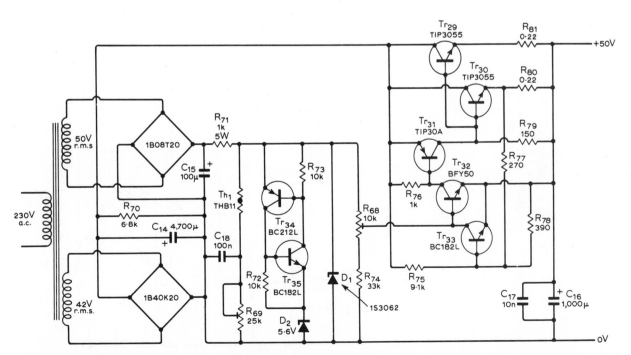

50-V FULLY-PROTECTED—Will trip out and supply almost zero current when output is shorted, without overheating. Amplifier load cannot be used again until power supply line is open for about 20 s. Paralleling of TR29 and TR30 eliminates need for costly 150-W regulator transistor. Developed for use with 100-W class B a-f amplifier.—R. B. H. Becker, High-Power Audio Amplifier Design, *Wireless World*, Feb. 1972, p 79–84.

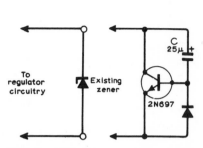

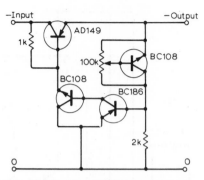

SLOW BUILDUP—Circuit at right, connected across existing zener in regulated supply, protects speaker by making voltage rise slowly at switch-on. Diode, which can be 1N480 germanium, discharges C at switch-off to ensure that runup is reinitiated after short-term power removal.—P. Lacey, Power Supply Modification, *Wireless World*, May 1971, p 234.

9 V FROM CAR BATTERY—Provides 9 V within 10 mV at 150 mA while battery voltage varies from 12 to 14.5 under normal driving conditions. Output voltage varies less than 15 mV when output current varies between 0 and 150 mA.—C. H. Banthorpe, Voltage Stabilizer, *Wireless World*, Jan. 1971, p 9.

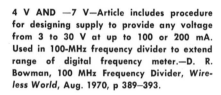

4 V AND —7 V—Article includes procedure for designing supply to provide any voltage from 3 to 30 V at up to 100 or 200 mA. Used in 100-MHz frequency divider to extend range of digital frequency meter.—D. R. Bowman, 100 MHz Frequency Divider, *Wireless World*, Aug. 1970, p 389–393.

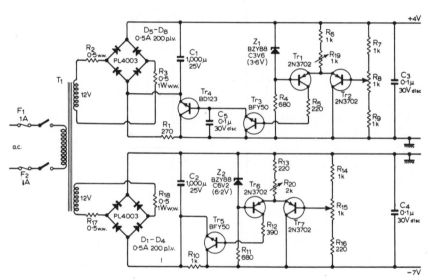

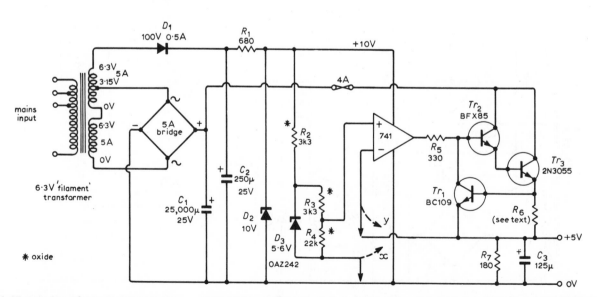

5 V AT 4 A—Use of opamp improves standard design. Output voltage is within few mV of reference voltage at all loads. R2 is adjusted to operate reference diode at its zero temperature coefficient point, to give outstanding temperature stability. TR1 provides current limiting when about 0.6 V is developed across sensing resistor R6, which for 4 A is 0.15 ohm. Adjust R3 or R4 as required to give exactly 5-V output.—J. Taylor, A Five-Volt Logic Power Supply, *Wireless World*, March 1972, p 139.

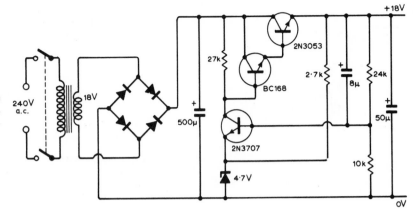

18 V AT 100 MA—Developed for use with Wien-bridge audio oscillator.—A. J. Ewins, *Wien-Bridge Audio Oscillator*, *Wireless World*, March 1971, p 104—107.

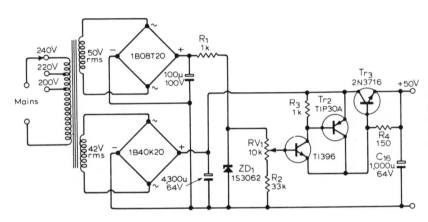

50 V AT 3.5 A—Designed for use with 100-W audio amplifier. R1 acts as constant-current source for zener.—I. Hardcastle, *High-Power Amplifier*, *Wireless World*, Oct. 1970, p 477—481.

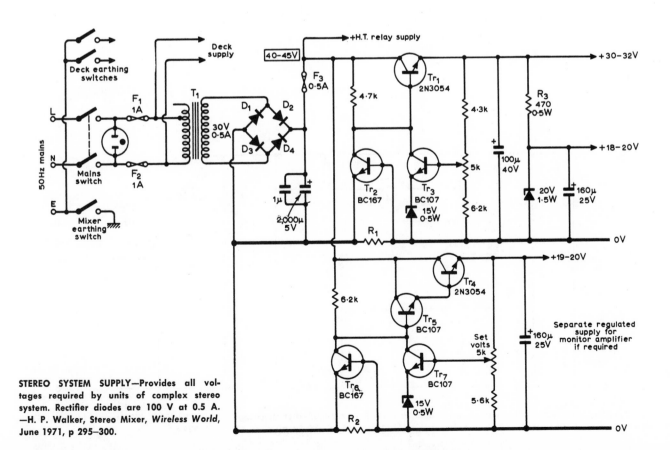

STEREO SYSTEM SUPPLY—Provides all voltages required by units of complex stereo system. Rectifier diodes are 100 V at 0.5 A.—H. P. Walker, *Stereo Mixer*, *Wireless World*, June 1971, p 295—300.

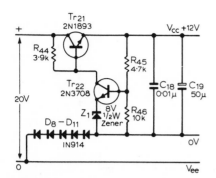

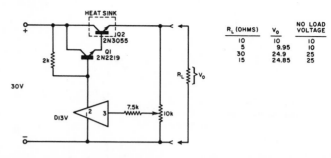

R_L (OHMS)	V_o	NO LOAD VOLTAGE
10	10	10
5	9.95	10
30	24.9	25
15	24.85	25

12 V—Output impedance is less than 1 ohm. Developed for use with digital gain control for tape recorder.—P. C. Grossi and C. Marcus, Digitally-Controlled Tape-Recorder Preamplifier, *Wireless World*, March 1970, p 127–129.

PROGRAMMABLE ZENER REFERENCE—Series regulator uses General Electric D13V programmable zener as adjustable reference that can be changed to make output voltage across RL any value between 10 and 25 V at 1.5 A. Regulation varies with load resistance, as shown in table.—R. L. Starliper, Programmable Zener Applications, *Electronics World*, June 1971, p 60–61.

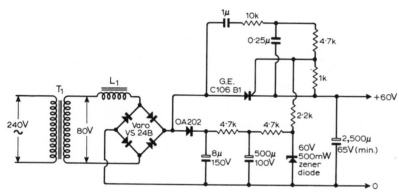

60 V AT 1.6 A—Developed for use with audio amplifier. Choke (180 turns) in series with secondary of T1 prevents uneven firing of scr.—A. R. Bailey, Thyristor-Stabilized Power Supplies, *Wireless World*, Aug. 1969, p 388–390.

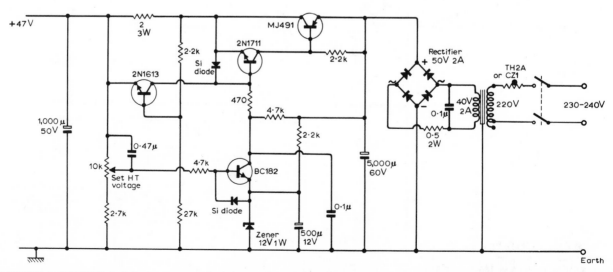

47 V WITH PROTECTION—Includes re-entrant short-circuit protection. Designed to handle both 20-W a-f power amplifiers of stereo system.—J. L. Linsley Hood, 15–20 W Class AB Audio Amplifier, *Wireless World*, July 1970, p 321–324.

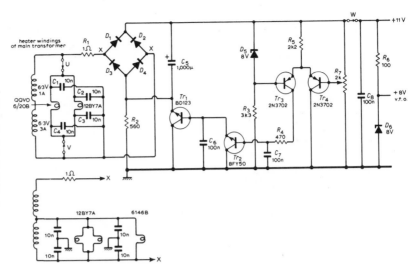

11-V SERIES-STABILIZED—Provides short-circuit protection. Used in transceiver designed for either a-c or mobile operation. If mobile is not required, heaters of transceiver tubes should be connected as in lower diagram. Diodes D1–D4 are 50-piv 250-mA rectifier types. Almost any npn power transistor can replace BD123, and any npn signal transistor can replace BFY50.—D. R. Bowman, 10–80 Metre Amateur Transceiver, *Wireless World*, Sept. 1972, p 437–442.

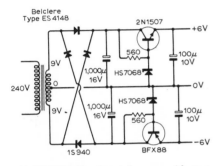

6-V BIPOLAR—Developed for use with transistor-IC decoder of f-m stereo receiver. Power transformer primary would be 117 V in U.S.—R. T. Portus and A. J. Haywood, Phase-Locked Stereo Decoder, *Wireless World*, Sept. 1970, p 418–422.

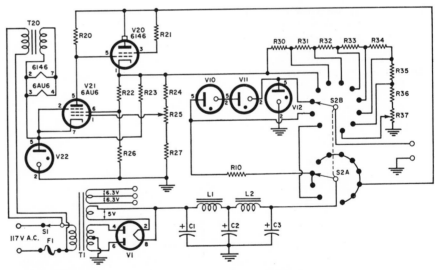

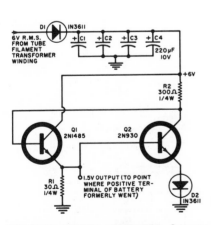

OHMMETER BATTERY ELIMINATOR—Converts 6 V a-c to regulated 1.5 V for ohmmeter of vtvm operating from a-c line. Will fit into space occupied by single C flashlight cell. Regulation is within 5% for ohmmeter loads from 10 ohms to infinity. If oscillation causes erratic behavior, connect 0.01 μF between collector of Q2 and ground. Filter capacitors can be replaced with single 900 μF unit.—W. G. Heller, V.T.V.M. Battery Eliminator, *Electronics World*, April 1971, p 72.

R10—2000 ohm res.
R20—2.2 megohm res.
R21—470 ohm res.
R22—100,000 ohm res.
R23—7500 ohm, 5 W wirewound res.
R24—47,000 ohm res.
R25—100,000 ohm pot
R26—150,000 ohm res.
R27—68,000 ohm res.
R30—1500 ohm, 25 W wirewound res.
R31—1000 ohm, 10 W wirewound res.
R32—2000 ohm, 5 W wirewound res.
R33, R34—10,000 ohm, 5 W wirewound res.
R35—20,000 ohm, 2 W res.
R36—68,000 ohm, 1 W res.
R37—50,000 ohm pot
C1—8 μF, 1000 V elec. capacitor (or 160-8 μF, 450 w. V d. c. min.)

C2—8 μF, 1000 V elec. capacitor (or 80-8 μF, 450 w. V d.c. min.)
C3—10 μF, 600 V elec. capacitor (or 40-10 μF, 450 w. V d.c. min.)
L1, L2—8 H choke (or 4-12 H)
F1—5 A fuse
S1—S.p.s.t., 2A, 110V switch
S2—11-pos., 2 deck rotary switch
T1—Power trans. 117 V: pri.: 610 V c.t.; 5 V, 12.6 V c.t. sec. (see text)
T20—Fil. trans. 117 V: 6.3 V
V1—5R4WGY (or 5AS4A, 5AV4, 5DB4, 5R4G 5V4G, 5V3, 5T4, 5Y3, 5Z3, 5931)
V10, V11—OA2 (or VR-150, 6073, 6626)
V12—OC2
V20—6146 (or 5933, 807, 6L6GB)
V21—6AU6 (or 6AU6A, 6BA6, 6136, 7543)
V22—OB2 (or 6074, 6627)

UNIVERSAL SUPPLY—Consists of four separate supplies tied together by S2, which determines voltage at output terminal. Provides regulated voltages of 200 to 300 V d-c continuously variable at up to 60 mA, 150 and 375 V d-c from voltage regulator tubes, and 0.1 to 150 V d-c for loads of 1 to 100 mA.—R. A. Walton, Universal Regulated Power Supply, *Electronics World*, Aug. 1970, p 60.

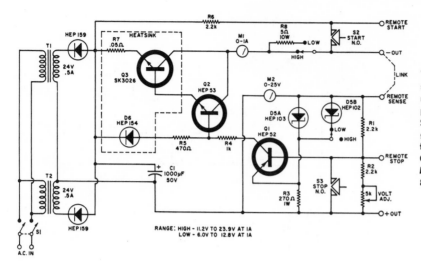

VARIABLE 6–24 V AT 1 A—Closely regulated supply is suitable for lab use. Contains terminals for connecting remote start and stop pushbutton switches in parallel with S2 and S3. Supply is floating, so either bus may be grounded. Q1 and D5 form differential amplifier of regulator.—R. H. Dutton, Design and Construction of Regulated Power Supplies, *Electronics World*, Feb. 1970, p 30–33 and 89.

RANGE: HIGH – 11.2V TO 23.9V AT 1A
LOW – 6.0V TO 12.8V AT 1A

0–250 V 0.1-A LAB SUPPLY—Single 300K pot provides full range of voltage control in circuit using Motorola MC1566L IC designed to serve as voltage and current control element in composite monolithic-discrete regulator. 500-ohm pot is current-adjusting control. Includes protective and frequency-compensating circuits.—D. Kesner, Monolithic Voltage Regulator, *IEEE Spectrum*, April 1970, p 24–32.

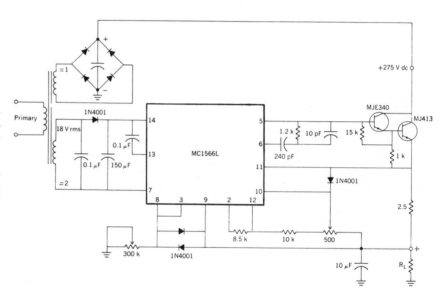

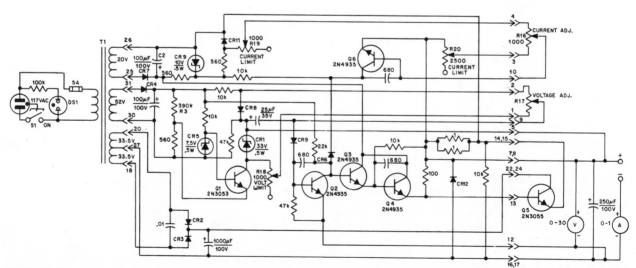

0.5–31 V AT 50 MA TO 1.1 A—Full-load hum level is below 1 mV. Q1 forms constant-current regulator which is independent of load. Mount Q5 on heatsink. Although 31-pin connector is shown, wires may be soldered directly. CR2, CR3, and CR12 are 150-prv 1-A silicon diodes; other diodes are 180-prv 0.25-A silicon.—H. Mauch, Current-Limited Power Supply, *QST*, June 1972, p 43–45.

12-V RECEIVER SUPPLY—Use of complete silicon-rectifier bridge U3 simplifies construction. Series regulator transistor Q5 comes mounted on heat sink. Provides excellent regulation for 80—20 meter receiver, with low output ripple. —T. R. Sowden, The Super-Simple 80—20 Receiver, QST, April 1972, p 26—29.

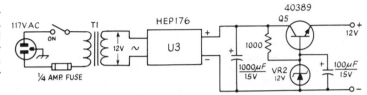

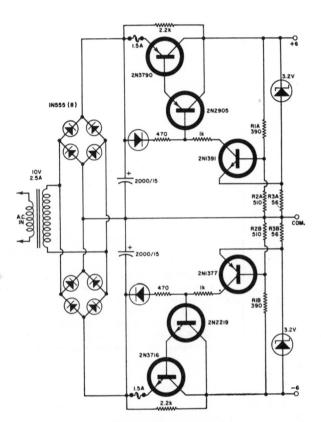

DUAL 6-V FOR SERVOS—Also suitable for other applications requiring symmetrical 6-V voltages, such as complementary-transistor audio amplifiers.—R. H. Dutton, Design and Construction of Regulated Power Supplies, *Electronics World*, Feb. 1970, p 30—33 and 89.

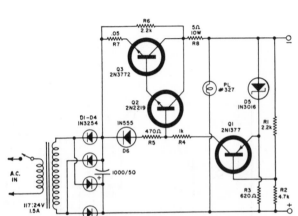

20 V AT 1.5 A—Protection against inadvertent thermal runaway is provided by D6, R7, and R8. Dynamic impedance is under 0.4 ohm and regulation better than 1%. Article covers design and construction.—R. H. Dutton, Design and Construction of Regulated Power Supplies, *Electronics World*, Feb. 1970, p 30—33 and 89.

12 AND 28 V FOR F-M REPEATER—Provides dual polarities at two voltages to meet practically all supply requirements. CR1—CR4 are 1,000-prv 2.5 A. Chokes are 2.8 μH. VR1 is 1N3023 zener (13 V at 1 W). VR2 is transient voltage suppressor rated 120 V (GE 6RS20SP4B4). Rated 1 A for continuous operation.—R. M. Myers, A Dual-Voltage Medium-Current Power Supply for Repeaters, QST, March 1972, p 20—21.

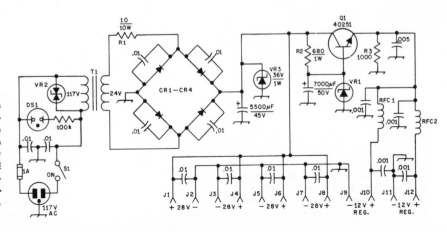

3.5—21 V AT 1 A—Designed as universal bench supply for transistor equipment. Uses RCA IC. R3 is 1.5-inch No. 30 copper wire wound on composition resistor. CR1—CR4 are 100 prv at 3 A, such as 1N4720. Pass transistor Q1 requires heatsink.—D. A. Blakeslee, AC-Operated Regulated DC Power Supplies for Transistorized Rigs, QST, Nov. 1971, p 11—14.

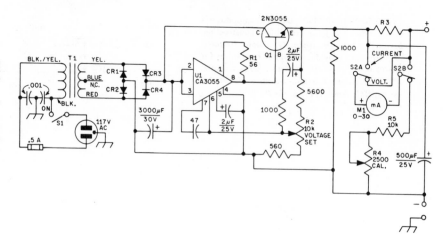

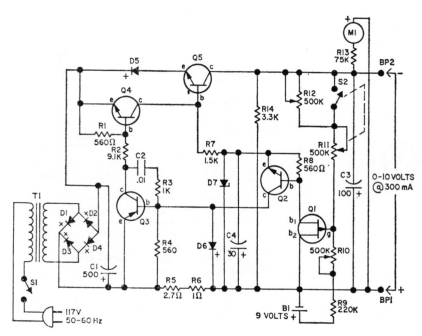

10 V AT 300 mA—Provides both current and voltage regulation. Current limiting starts at 250 mA, and output current will not exceed 300 mA even for dead short. Secondary of C1 is 12.6 V at 2 A. All six diodes are 750-mA 400-piv, and zener D7 is 5.6-V 250-mW such as HEP603. C1 is 500-μF at 25-V d-c electrolytic. Use 1-mA meter. Q1 is Motorola MPF155 fet. Q2 and Q4 are HEP54. Q3 is HEP57. Q5 is RCA 40316.—H. Cohen, Universal Regulated Power Supply, Science and Electronics, Feb.-March 1970, p 49—51 and 56.

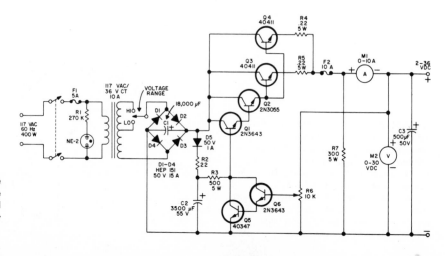

2—36 V AT 0—10 A—Ripple and noise are less than 10 mV at full load. Output voltage drops less than 0.5 V from no load to full load.—G. Schreyer, 10 Amp Variable Power Supply, 73, Dec. 1972, p 73—75.

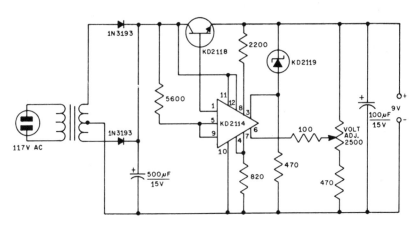

9 V AT 250 MA—Regulation at maximum load is 3%. Polarity is reversible. Available in kit form as RCA KC-4004.—New Apparatus, QST, Nov. 1971, p 39.

13 V FOR MOBILE TRANSCEIVER—Permits fixed-base operation of 12-V mobile equipment, such as 2-meter f-m transceiver, from a-c line. Output is 13 V on transmit and 13.4 V on receive.—J. P. Weir, Jr., Fixed Base Operation of the HR-2 Transceiver, 73, April 1972, p 51–53.

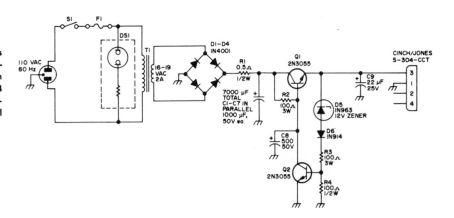

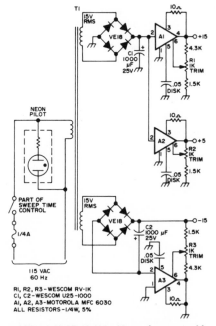

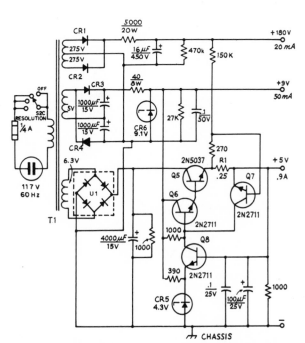

5 AND 15 V AT 40 MA—IC regulators provide 15 V with both polarities as well as 5 V for solid-state lab-type sweep generator and other applications. All three IC voltage regulators are overload-protected against accidental short-circuits. Power transformer can have 30-V CT secondary if center tap is uncovered and the leads separated.—R. Megirian, Lab Type IF/RF Sweep Generator Using IC's, 73, Nov. 1972, p 226–229, 232–234, 236–239, and 241–244.

5, 9, AND 180 V—Single power transformer provides all voltages required by frequency counter having digital display. 5-V supply for digital circuits and frequency standard uses series regulator, 9-V supply for mixers and linear IC's has simple shunt zener CR6, and 180 V for readout tubes is unregulated. At overload of about 1.2 A, 5-V supply changes to constant-current operation and can be short-circuited indefinitely without damage. U1 is 1.8-A silicon bridge. CR1 and CR2 are 800-prv 1-A silicon. CR3 and CR4 are 400-prv 1-A silicon. CR5 is 1N5229 zener and CR6 is 1N5239.—K. MacLeish, A Frequency Counter for the Amateur Station, QST, Oct. 1970, p 15–23 and 43.

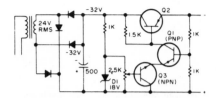

0.8–18 V—Provides regulated output over wide voltage range, for experimental low-voltage projects. Diode and transistor types are not critical.—A. G. Evans, Another Solid State Power Supply Article, 73, Sept. 1972, p 52 and 54.

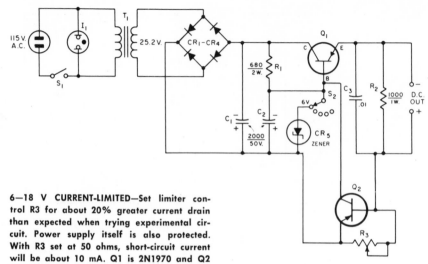

6–18 V CURRENT-LIMITED—Set limiter control R3 for about 20% greater current drain than expected when trying experimental circuit. Power supply itself is also protected. With R3 set at 50 ohms, short-circuit current will be about 10 mA. Q1 is 2N1970 and Q2 is 2N3906. 3 is 50-ohm linear-taper. Rectifiers are 50-prv 3-A silicon diodes. Different voltage ratings of zener CR5 are connected to terminals of S2.—J. M. Pike, Current Limiting For A Regulated Low-Voltage Power Supply, QST, Jan. 1971, p 46.

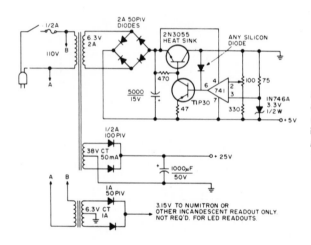

5 V AT 1.5 A—Provides 5% regulation as required for vhf frequency counter, along with unregulated voltages required. Is current-limiting and short-circuit-proof.—P. A. Stark, A Modern VHF Frequency Counter, 73, July 1972, p 5–11 and 13.

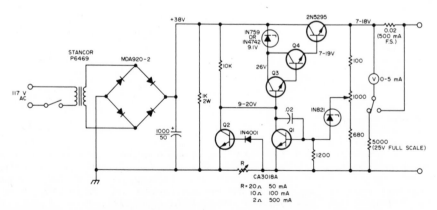

7–18 V AT UP TO 500 MA—Value of R is determined by load current requirement. Meter is switched to read either load voltage or load current on linear scale. Versatility makes supply ideal for experimental solid-state projects.—D. Nelson, Practical IC Regulator Circuits For Hams, 73, Oct. 1970, p 32, 34–36, and 38.

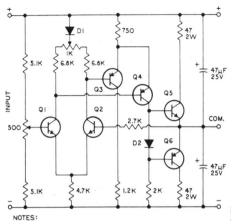

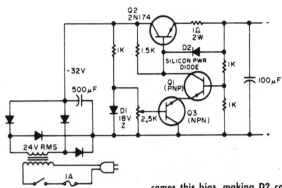

NOTES:

1. Q1, Q2 -- MOTOROLA HEP 243 OR FAIRCHILD 2N3641
2. Q3, Q4 -- MOTOROLA HEP 242 OR RCA 2N4037
3. Q5 ----- MOTOROLA HEP 245 OR MJE 520
4. Q6 ----- MOTOROLA HEP 246 OR MJE 370
5. D1, D2 -- MOTOROLA HEP 156 OR GE 1N4454

SPLITTER GIVES CENTERTAP—When plugged into any regulated power supply, circuit provides common terminal whose voltage stays midway between positive and negative terminals regardless of supply voltage and current drain. Developed for use with linear IC's requiring 6 or 15 V with both polarities. With 24-V input, output of splitter is ±12 V. Initial adjustment for voltage-splitting is made with 1K pot while sweeping input between 10 and 40 V, to find best compromise setting. —H. Olson, A Power-Supply Splitter for Linear IC's, 73, Oct. 1972, p 103 and 106–107.

1.4–32 V AT 420 MA—Provides current limiting along with wide output voltage range. D2 is normally reverse-biased by emitter-base voltage of 2N174. At output of about 420 mA, drop across 1-ohm resistor overcomes this bias, making D2 conduct and limit current flow to this level. For variable current limiting, use 1-ohm rheostat. Unmarked components are not critical.—A. G. Evans, Another Solid State Power Supply Article, 73, Sept. 1972, p 52 and 54.

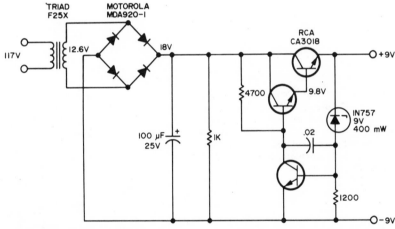

9 V AT 30 MA—Provides ample power for any portable a-m or f-m radio operating on 9-V battery.—D. Nelson, Practical IC Regulator Circuits For Hams, 73, Oct. 1970, p 32, 34–36, and 38.

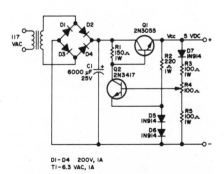

D1-D4 200V, 1A
T1-6.3 VAC, 1A

5 V FOR IC'S—Simple circuit provides commonest voltage required for IC projects.—Circuits, 73, May 1972, p 109.

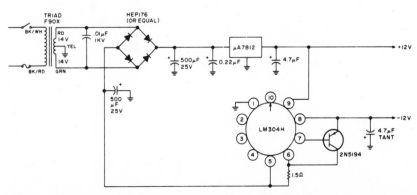

DUAL 12 V—Provides highly regulated voltages for two-tone test generator used in testing ssb amplifiers.—H. Olson, A Two-Tone Test Generator, 73, Jan. 1973, p 53–55.

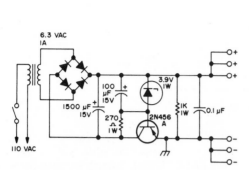

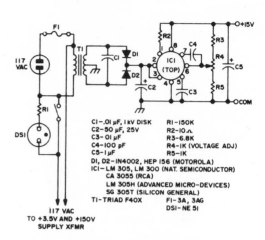

3.6 V AT 700 MA—Developed for experimentation with IC logic and use with dynamic logic demonstrators. Power transistor type is not critical.—A. S. Joffe, IC Logic Doesn't Have to Be Obscure, 73, Nov. 1972, 217–219.

C1—.01 μF, 1 kV DISK
C2—50 μF, 25V
C3—01 μF
C4—100 pF
C5—1 μF
D1, D2—1N4002, HEP 156 (MOTOROLA)
IC1—LM 305, LM 300 (NAT. SEMICONDUCTOR)
      CA 3055 (RCA)
      LM 305H (ADVANCED MICRO-DEVICES)
      SG 305T (SILICON GENERAL)
T1—TRIAD F40X
R1—150K
R2—10Ω
R3—6.8K
R4—1K (VOLTAGE ADJ)
R5—1K
F1—3A, 3AG
DS1—NE 51

15 V FOR CRYSTAL STANDARD—Single IC regulator provides adequate regulation for crystal oscillator and buffer of frequency counter.—H. Olson, Addendum to the W1PLJ Counter, 73, March 1972, p 65–68.

—6, —12, AND +6 V—General-purpose regulated supply provides all voltages required for automatic Morse-code generator driven by keyboard. Article gives example of each type of logic circuit required. Supply can be used for other logic applications as well.—F. E. Smith, The Button Box, 73, Feb. 1972, p 113–127.

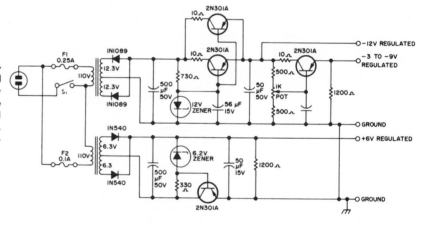

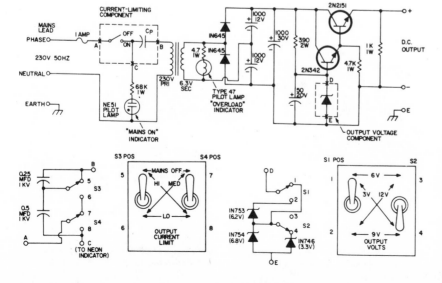

BENCH SUPPLY—Provides choice of 3, 6, 9, or 12 V by switching of zeners between terminals D and E, and choice of 100, 130, and 240 mA current limits by switching of capacitors at terminals A, B, and C. Article tells how to determine correct capacitor values for use with 117-V power transformer. Overload lamp is normally on, and extinguishes when output current is excessive.—F. Johnson, A Simple Bench Power Supply, 73, Jan. 1970, p 22–24 and 26–27.

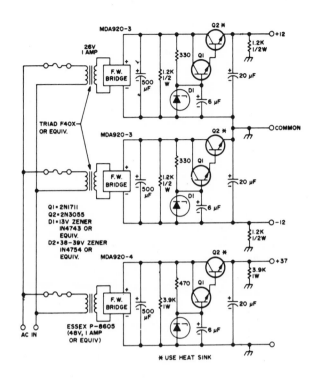

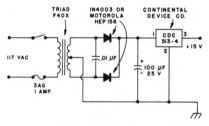

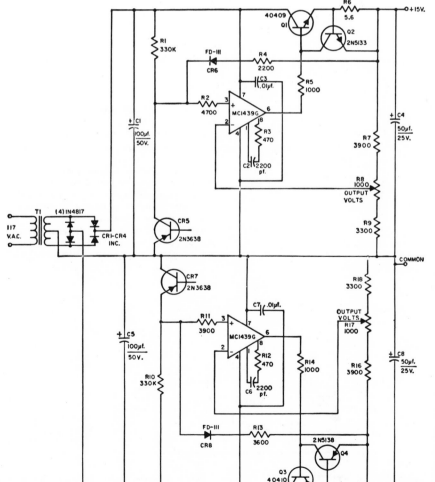

12 AND 37 V FOR 448-MHZ TRANSMITTER— Provides three regulated voltages for 5-W transmitter used at repeater site. Article gives all transmitter circuits.—W. Collier, 450 MHz Remote Site Transmitter, 73, May 1971, p 69—70 and 72—73.

15 V AT 100 MA—Designed for use with frequency calibrator providing markers at 200, 100, 50, and 25 kHz for h-f receivers.—H. Olson, The Ball of Wax—A Calibrator, 73, Nov. 1969, p 84—86.

9 V FOR F-M TEST SET—Output can be varied by changing ratio of R1 to R2.—V. Epp, FM Test Set, 73, Jan. 1973, p 65—69.

15-V BIPOLAR AT 100 MA—Voltage regulation is 0.05%, ripple under 2 mV, and temperature coefficient 0.01% per deg C. Provides short-circuit protection. If no part of circuit is grounded internally, outputs may be used in series to give 30 V. Output drops only 7 mV from no load to full load.—J. K. Gotwals, Regulated Dual Power Supply, QST, July 1969, p 38—41.

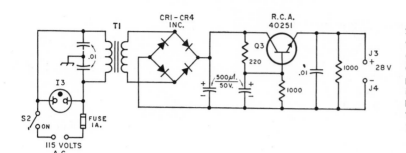

28 V AT 250 MA—Uses electronic filtering without voltage reference, to provide some degree of regulation. No-load output of 28-V drops to about 24 V when full load is applied. Chief advantage is simplicity. Filtering is adequate for use with 5-W c-w transmitter. Diodes are 750-mA 50-piv.—D. DeMaw, A Transistor 5-Watter for 80 And 40, QST, June 1967, p 11–14.

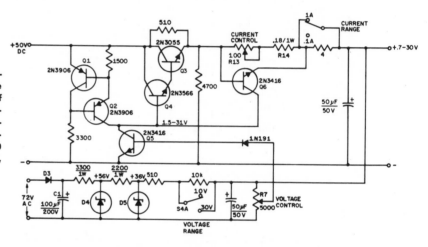

0.7–30 V AT 1 A—Includes voltage and current limiting, remote voltage sensing, and use of independent reference supply operating off a-c line through 72-V power transformer. Series regulator Q3-Q4 is Darlington pair. Output ripple is less than 4 mV, and shift in voltage from no load to full load is less than 50 mV.—Heath IP-28 Regulated DC Supply, QST, May 1970, p 49–50.

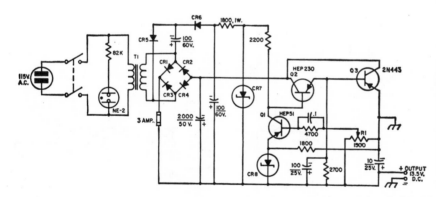

13.5 V AT 850 MA—Uses 21-V 3-A power transformer driving 15-A 50-piv bridge rectifier. CR5 and CR6 are 0.5-A 200-piv rectifiers, CR7 is 1N3030 zener, and CR8 is 1N754 zener. Voltage doubling with CR5 and CR6 improves regulation. Output voltage is sensed by Q1 and compared with 6.8-V emitter reference voltage maintained by zener CR8. Darlington output stage gives beta multiplication. Use good heatsink for Q3.—D. F. Becker, More Power on 144 MHz with Transistor, QST, Aug. 1969, p 11–16.

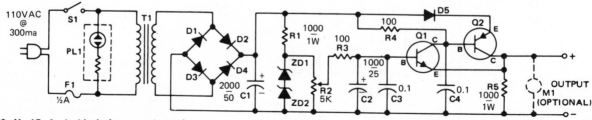

0–20 V AT 1 A—Ideal for experimental work with solid-state circuits. Output voltmeter is optional. R2 adjusts voltage. Q2 is HEP232 pnp power transistor connected as series-pass regulator. Q1 can be almost any npn transistor. Silicon diodes are 50 piv at 1 A, and zeners are 10-V HEP101. T1 is filament transformer having 24-V 2-A secondary.—"Calectro Handbook," GC Electronics, Rockford, IL, 1971, p 46.

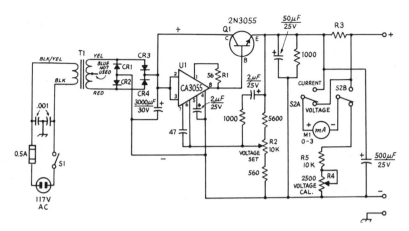

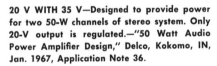

3.5–21 V AT 1.5 A—Includes short-circuit protection even at maximum output. Regulation is adequate for most solid-state equipment and for general-purpose bench supply. Diodes are 100 prv at 3 A, such as Motorola 1N4720. T1 has 21-V 2.5-A secondary.—"The Radio Amateur's Handbook," American Radio Relay League, Newington, CT, 49th Ed., 1972, p 121–122.

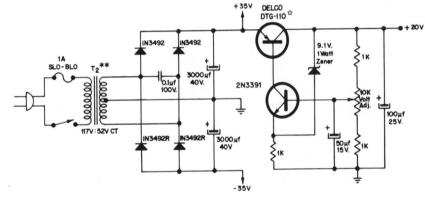

20 V WITH 35 V—Designed to provide power for two 50-W channels of stereo system. Only 20-V output is regulated.—"50 Watt Audio Power Amplifier Design," Delco, Kokomo, IN, Jan. 1967, Application Note 36.

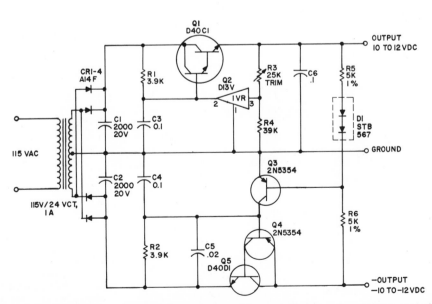

10–12 V AT 0.5 A—Gives better performance at lower cost than with equivalent zener supply. Output voltage is easily adjusted to required value while maintaining symmetry. Uses IC voltage regulator Q2, containing two transistors, reference diode, and resistor. Darlington power transistor Q1 serves as series-pass element.—W. H. Sahm, III, "A High Performance Symmetrical Power Supply Utilizing an Integrated Voltage Regulator (IVR)," General Electric, Semiconductor Products Dept., Auburn, NY, No. 90.85, 1970, p 2–3.

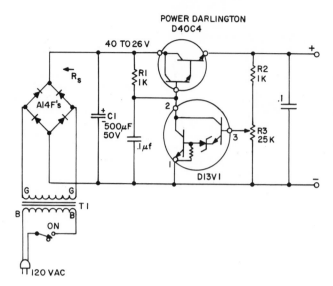

POWER DARLINGTON
D40C4

40 TO 26 V

10–28 V AT 300 MA—Uses IC power Darlington as series-pass regulator controlled by D13V1 IC voltage regulator containing two transistors, zener, and resistor. Increase size of C1 if supply impedance is above 15 ohms. T1 is Stancor P6469 with 25-V 1-A secondary. —D. V. Jones and M. L. Hyland, "General Purpose Power Supplies with Good Performance to Cost Ratio," General Electric, Semiconductor Products Dept., Auburn, NY, No. 90.95, 1970, p 1–3.

10–34 V AT 1 A—Uses IC power Darlington D40C4 as series-pass regulator controlled by IC voltage regulator, with current-sensing circuit breaker using C13F complementary scr. Regulation is better than 0.75% and ripple under 100-mV p-p.—D. V. Jones and M. L. Hyland, "General Purpose Power Supplies with Good Performance to Cost Ratio," General Electric, Semiconductor Products Dept., Auburn, NY, No. 90.95, 1970, p 4–6.

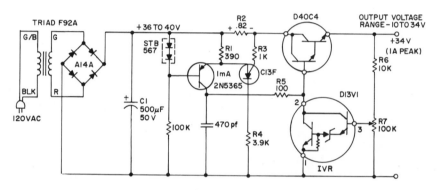

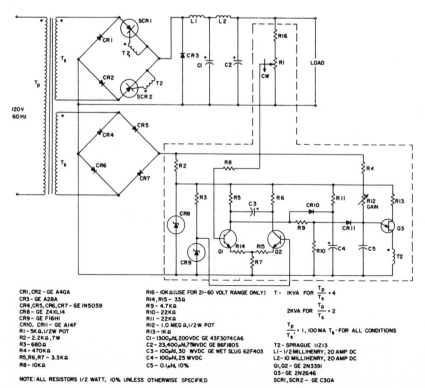

CR1, CR2 - GE A40A
CR3 - GE A28A
CR4, CR5, CR6, CR7 - GE 1N5059
CR8 - GE Z4XLI4
CR9 - GE F16HI
CR10, CR11 - GE AI4F
R1 - 5K Ω, 1/2W POT
R2 - 2.2K Ω, 7W
R3 - 680 Ω
R4 - 470 KΩ
R5, R6, R7 - 3.3 KΩ
R8 - 10 KΩ

R16 - 10K Ω (USE FOR 21-60 VOLT RANGE ONLY)
R14, R15 - 33 Ω
R9 - 4.7K Ω
R10 - 22 KΩ
R11 - 22 KΩ
R12 - 1.0 MEG Ω, 1/2 W POT
R13 - 1K Ω
C1 - 1300 μfd, 200 VDC GE 43F3074CA6
C2 - 23,400 μfd, 75 WVDC GE 86FI80S
C3 - 100 μfd, 30 WVDC GE WET SLUG 62F403
C4 - 100 μfd, 25 WVDC
C5 - 0.1 μfd, 10%

T - 1KVA FOR $\frac{T_p}{T_s}$ = 4

2KVA FOR $\frac{T_p}{T_s}$ = 2

$\frac{T_p}{T_s}$ = 1, 100 MA T_s - FOR ALL CONDITIONS

T2 - SPRAGUE 11Z13
L1 - 1/2 MILLIHENRY, 20 AMP DC
L2 - 10 MILLIHENRY, 20 AMP DC
Q1, Q2 - GE 2N3391
Q3 - GE 2N2646
SCR1, SCR2 - GE C30A

NOTE: ALL RESISTORS 1/2 WATT, 10% UNLESS OTHERWISE SPECIFIED

60 V AT 1.2 KW—Cosine-modified ramp and pedestal control allows use of gain pot R12 to adjust for maximum regulation and overshoot. Combination of CR10, R11, and C4 forms soft-start circuit to protect supply when starting under heavy load. Full-load 120-Hz ripple is 500-mV p-p. Load regulation is 1.5% for 12 to 20 A.—"SCR Manual," General Electric, Syracuse, NY, 1972, 5th Ed., p 274–275.

40–45 V AT 1 A—Current-limiting circuit and electronic circuit breaker together prevent overload pulses longer than 6 μs from destroying series-pass transistor Q1. Q3 is constant-current source for Q1, Q2, and IC voltage regulator D13V1. Will handle class-B amplifiers drawing up to 2.5-A peak sine-wave current.—D. V. Jones and M. L. Hyland, "General Purpose Power Supplies with Good Performance to Cost Ratio," General Electric, Semiconductor Products Dept., Auburn, NY, No. 90.95, 1970, p 6–7.

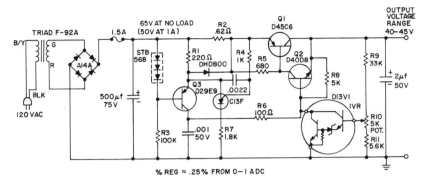

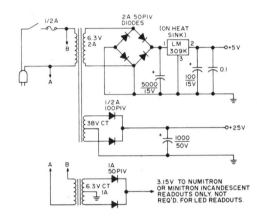

5 V AT 1.5 A—Uses inexpensive IC voltage regulator with bridge rectifier and only three filter capacitors to provide 5% regulation as required for vhf frequency counter. Other voltages are unregulated.—P. A. Stark, A Modern VHF Frequency Counter, 73, July 1972, p 5–11 and 13.

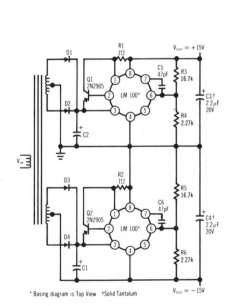

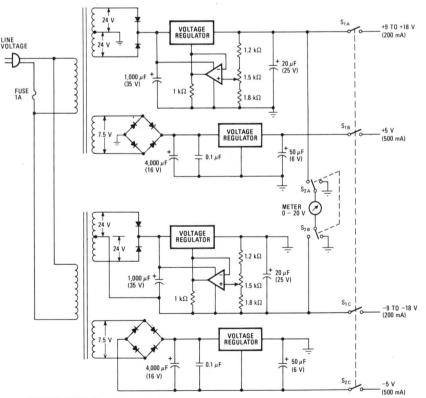

15-V POSITIVE-NEGATIVE—Split secondaries on power transformer create floating voltage source for negative National IC voltage regulator. Outputs have common ground. Rated maximum current is 250 mA. Each IC has current-boosting external transistor. Full-wave rectifier values are chosen to give desired 15-V outputs.—R. J. Widlar, "Monolithic Voltage Regulator," National Semiconductor, Santa Clara, CA, 1967, AN-1, p 9.

ALL DIODES: SILICON, 1A
RECTIFIER BRIDGES: INTERNATIONAL RECTIFIER 5B4
VOLTAGE REGULATORS: FAIRCHILD μA7805
OP AMPS: 741

FOUR-VOLTAGE BENCH SUPPLY—Wide choice of regulated voltages meets practically all supply requirements for breadboard checkout of new circuit designs using solid-state active elements.—M. J. Salvati, Versatile Breadboard Checks Out Designs Quickly, Electronics, March 29, 1973, p 94–95.

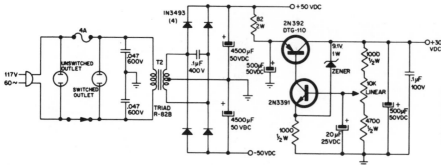

30 V WITH 50 V—Designed to provide power for two 80-W channels of stereo system. Only 30-V output is regulated.—"160 Watt Stereo Audio Power Amplifier Design," Delco, Kokomo, IN, Feb. 1972, Application Note 35.

9 V FOR SOLID-STATE PROJECTS—HEP245 transistor used as voltage regulator should be mounted on HEP500 heatsink for adequate heat dissipation.—"Home Handyman's Construction Projects," Motorola Semiconductors, Phoenix, AZ, HMA-37, 1972.

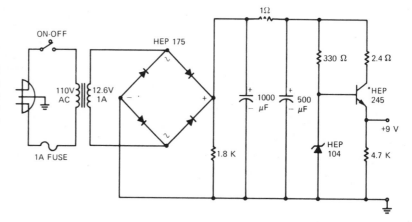

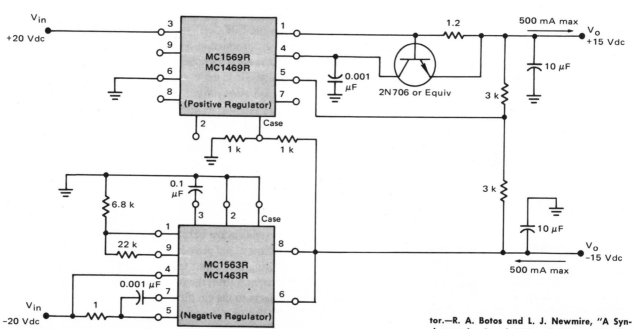

15-V BIPOLAR AT 400 MA—Uses separate complementary tracking IC voltage regulators for each output polarity. Used as one of supplies for 10 Hz—10 MHz lab sweep generator.—R. A. Botos and L. J. Newmire, "A Synchronously Gated N-Decade Sweep Oscillator," Motorola Semiconductors, Phoenix, AZ, AN-540, 1971.

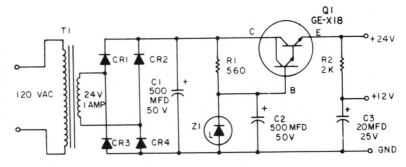

12 OR 24 V—Furnishes both output voltages at combined output current up to 500 mA. Ideal for bench experimentation and small stereo systems. Uses Darlington as voltage regulator.—"Electronics Experimenters Circuit Manual," General Electric, Owensboro, KY, 1971, 3rd Ed., p 205–208.

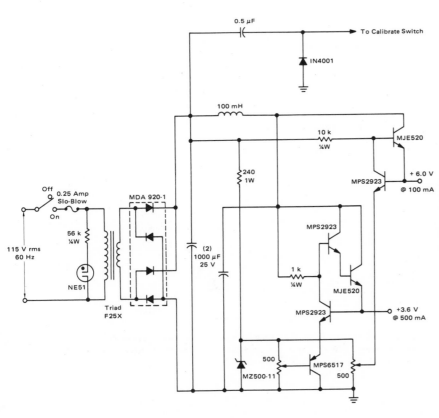

3.6 AND 6 V FOR FREQUENCY COUNTER—Series-regulated emitter-referenced circuit provides adjustment range of 1 to 6 V at 100 mA with 1% regulation for one output and 1.5 to 6 V at 500 mA with 2% regulation for other output.—B. Botos, "A Frequency Counter Using Motorola RTL Integrated Circuits," Motorola Semiconductors, Phoenix, AZ, AN-451, 1969.

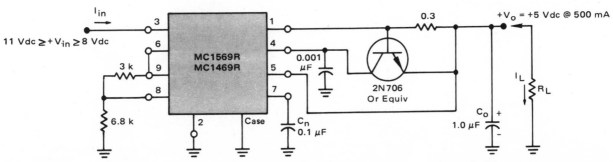

5 V AT 500 MA—Provides one of regulated voltages needed for 10 Hz–10 MHz lab sweep generator.—R. A. Botos and L. J. Newmire, "A Synchronously Gated N-Decade Sweep Oscillator," Motorola Semiconductors, Phoenix, AZ, AN-540, 1971.

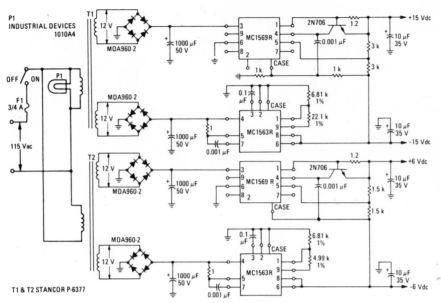

FOUR-VOLTAGE SUPPLY—Provides excellent performance in meeting requirements of 1 Hz—1 MHz solid-state function generator, at low cost. Requires little space.—B. Botos, "A Low-Cost, Solid-State Function Generator," Motorola Semiconductors, Phoenix, AZ, AN-510A, 1971.

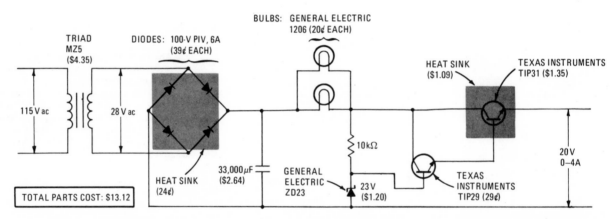

20 V AT 4 A—Cost is minimized by storing energy in 33,000-μF capacitor at voltage level of about 39 V, much higher than is normally required. Series regulator transistor drops level to desired 20-V output. Incandescent lamps serve for short-circuit protection.—J. Ennis, Four-Ampere Power Supply Costs Just $13 to Build, *Electronics*, Dec. 4, 1972, p 91.

CHAPTER 96
Regulator Circuits

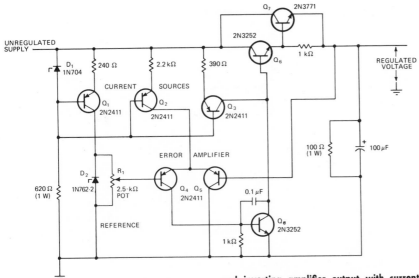

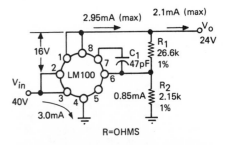

24-V DESIGN—Article tells how to determine that maximum output current is 2.1 mA when IC regulator shown is operated without heatsink under worst-case operating conditions of 40-V input and 125-C ambient temperature. —R. J. Widlar, Worst Case Power Dissipation in Linear Regulators, *Electromechanical Design*, Dec. 1969, p 40–41.

3.2 V AT 2 A—Provides better than 1% regulation if unregulated 14-V supply tolerance is 10%. Regulation is achieved at such low voltage by reversing error amplifier of regulator and inverting amplifier output with current-sink transistor Q8. Q6 and Q7 form series-pass Darlington.—C. H. Claasen, Low-Voltage Regulator Uses Reversed Error Amplifier, *Electronics*, Nov. 22, 1971, p 74.

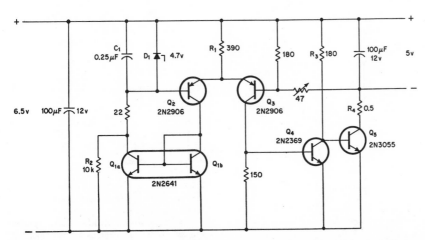

PROTECTED 5 V AT 1 A—Low-voltage regulator provides adequate protection against overvoltage damage even if pass transistor fails, and can operate with very small input-to-output voltage differential. If 3.6-V zener is used, output is 4 V.—T. K. Hemingway, Regulator Gives Overvoltage Protection for TTL, *Electronics*, Dec. 21, 1970, p 54.

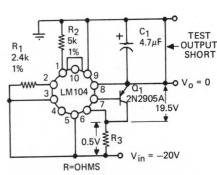

WORST-CASE DESIGN—Article tells how to compute maximum allowable short-circuit current for LM104 IC regulator used with series-pass transistor without heatsink for worst-case input of 20 V at 85 C; limit is 20.2 mA. —R. J. Widlar, Worst Case Power Dissipation in Linear Regulators, *Electromechanical Design*, Dec. 1969, p 40–41.

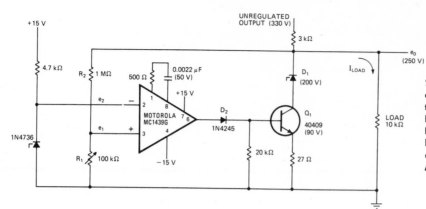

250-V REGULATOR—Zener D1 handles most of output voltage, permitting use of 90-V transistor as shunt regulator. Output changes less than 0.04%, from no-load to 25-mA full-load. Concept can be extended to regulating kilovolts.—M. J. Shah, Regulating High Voltage with Low-Voltage Transistors, *Electronics*, April 24, 1972, p 102.

FOLDBACK CURRENT LIMITER—Article tells how to determine worst-case dissipation in pnp driver Q1 under full load and 24-V input when using LM104 negative regulator to limit foldback current. Dissipation limit of Q1 is about 3.5 W here.—R. J. Widlar, Worst Case Power Dissipation in Linear Regulators, *Electromechanical Design*, Dec. 1969, p 40–41.

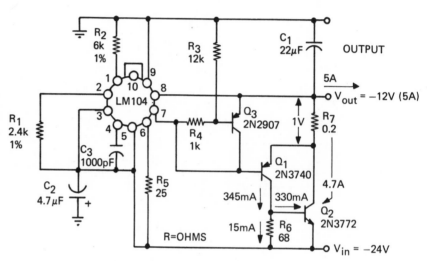

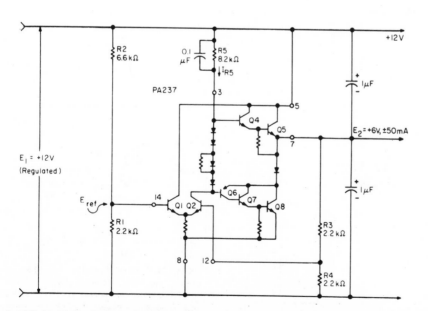

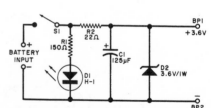

3.6-V ADAPTER—Can be mounted on top of Mallory M915 or equivalent 6-V lantern battery, to provide stable regulated 3.6-V output for IC experiments. Article covers optimization of values to minimize battery drain. D1 is light-emitting diode (optional) serving as low-drain pilot lamp.—J. A. Fred, 3.6-Volt IC Power Supply Regulator, *Popular Electronics*, April 1972, p 42–43.

6 AND 12 V—Single General Electric PA237 audio IC solves problem of providing two voltages from one regulated source while letting current flow in either direction. Regulation of second (6-V) source is 0.1% from 0 to ± 50 mA.—D. Lubarsky, Load Current Flows into or out of Voltage Regulator, *The Electronic Engineer*, Jan. 1969, p 94.

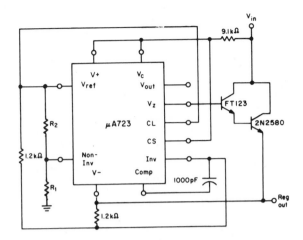

HIGH-VOLTAGE FLOATER—Base-emitter junction of current-limit transistor is zenered and added to Vref to form a pre-stabilized +14-V supply, eliminating need for extra floating supply. Output is 100 V.—J. D. Lieux and R. D. Ricks, IC Regulator Removes Restrictions, *The Electronic Engineer*, March 1969, p 53–57.

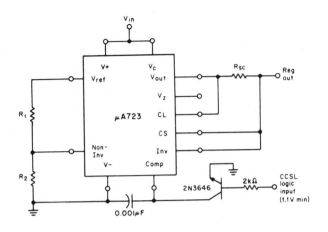

REMOTE-SHUTDOWN REGULATOR—Complementary current-sinking logic from external digital elements lets regulator be turned off remotely. Regulated output is 5 V within 1.5 mV for input of 9.5 to 40 V.—J. D. Lieux and R. D. Ricks, IC Regulator Removes Restrictions, *The Electronic Engineer*, March 1969, p 53–57.

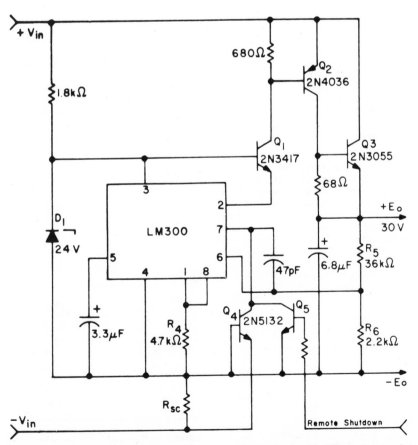

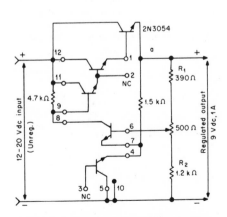

30-V REMOTE SHUTDOWN—Provides better than 0.05% load regulation and 0.002%/V for 1-A load, with short-circuit protection. Q4 saturates when load current through Rsc is too large, clamping pin 7 and switching regulator into current mode that protects IC. Q5 is collector-ORed with Q4 to shut regulator down with external d-c signal.—W. C. Jung, Voltage Regulator Has Extended Range, Remote Shutdown, *The Electronic Engineer*, March 1969, p 90.

9 V AT 1 A—Uses one of transistors in IC as reference element (pins 3, 4, and 5) in place of zener. External transistor can be eliminated if only 50-mA output is needed. Use RCA CA3018 IC.—L. Toth, Low Cost Voltage Regulator from One IC, *The Electronic Engineer*, Oct. 1968, p 89.

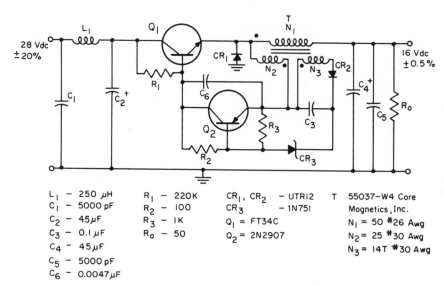

L_1 – 250 μH	R_1 – 220K	CR_1, CR_2 – UTR12	T	55037–W4 Core	
C_1 – 5000 pF	R_2 – 100	CR_3 – 1N751		Magnetics, Inc.	
C_2 – 45 μF	R_3 – 1K	Q_1 = FT34C	N_1 = 50 #26 Awg	
C_3 – 0.1 μF	R_0 – 50	Q_2 = 2N2907	N_2 = 25 #30 Awg	
C_4 – 45 μF			N_3 = 14T #30 Awg	
C_5 – 5000 pF				
C_6 – 0.0047 μF				

PDM F-M REGULATOR—Combination of pulse duration modulation and frequency modulation minimizes number of parts required to reduce 28-V d-c to 16 V with 0.5% regulation, for S-band and L-band transmitters. Efficiency is 92% for 5-W version shown. Article gives theory of operation.—J. S. Reece, *Power Supply Uses PDM-FM for Regulation, The Electronic Engineer,* Feb. 1967, p 50–53.

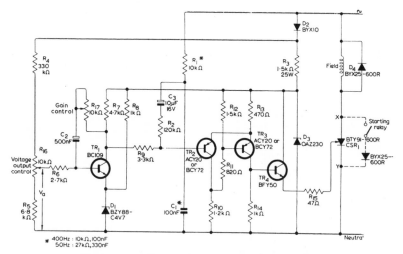

400-HZ 4-KVA ALTERNATOR CONTROL—Will also control 50-Hz or 60-Hz alternators, if values of R1 and C1 are changed. Regulation is better than 1% even for inductive or complex loads if gain control R17 is properly adjusted. Changes in output voltage of alternator are amplified by TR1 and fed into Schmitt trigger TR2-TR3, which in turn drives gate of thyristor through buffer TR4. Conduction angle of thyristor controls field current, and is determined by adding inverted half-sine-wave from first transistor to displaced waveform from phase-shifting network.—J. Lavallee, *Voltage Regulation of Lower-Powered Alternators, Mullard Technical Communications,* March 1968, p 34–39.

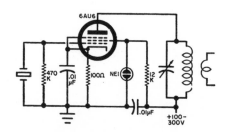

SCREEN-GRID VOLTAGE REGULATOR—Neon diode between screen and ground of pentode crystal oscillator provides improved frequency stability by regulating screen voltage.—J. Kyle, *The Ubiquitous Neon Lamp, Popular Electronics,* Nov. 1968, p 55–57.

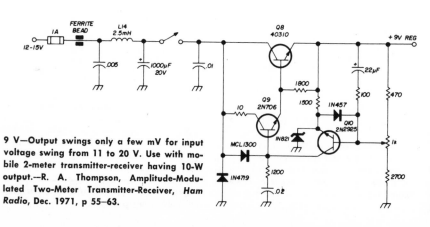

9 V—Output swings only a few mV for input voltage swing from 11 to 20 V. Use with mobile 2-meter transmitter-receiver having 10-W output.—R. A. Thompson, *Amplitude-Modulated Two-Meter Transmitter-Receiver, Ham Radio,* Dec. 1971, p 55–63.

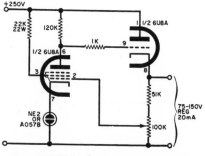

NEON REFERENCE—Neon lamp holds voltage constant on anode of first stage in 20-mA regulated supply. Output is adjustable from 75 to 150 V by changing feedback. Regulation is 0.5 V at 75 V.—J. Kyle, *The Ubiquitous Neon Lamp, Popular Electronics,* Nov. 1968, p 55–57.

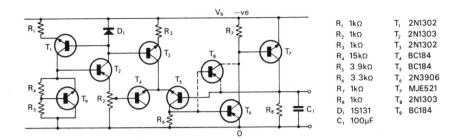

R_1	1kΩ	T_1	2N1302
R_2	1kΩ	T_2	2N1303
R_3	1kΩ	T_3	2N1302
R_4	15kΩ	T_4	BC184
R_5	3.9kΩ	T_5	BC184
R_6	3.3kΩ	T_6	2N3906
R_7	1kΩ	T_7	MJE521
R_8	1kΩ	T_8	2N1303
D_1	1S131	T_9	BC184
C_1	100μF		

1.6-V REGULATOR—Uses long-tailed-pair connection of transistors to permit operation from supply voltages down to 1.6 V with load currents up to 400 mA. Output voltage can be within 200 mV of supply or as low as 100 mv. Requires no additional bias supplies. Article gives theory of operation.—P. Williams, A D.C. Regulator for Low Voltages, *Electronic Engineering*, March 1970, p 41–43.

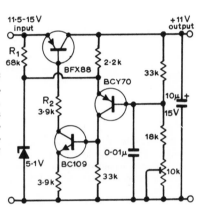

BATTERY REGULATOR—Gives stabilization of about 1:1,000, with very little difference between input voltage and regulated output. Circuit has good immunity to short-circuits and overloads, as compared to emitter-follower series regulators.—T. R. E. Owen, Battery Supply Regulator, *Wireless World*, May 1971, p 234.

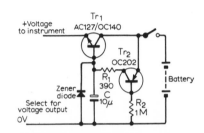

BATTERY REGULATOR—Drop in battery voltage from 18 to 12 V gives less than 0.2-V drop in output voltage when using OAZ206 zener, and current drain changes less than 5 mA. TR1 is germanium and TR2 silicon.—P. Lacey, Low-Drain Battery Regulator, *Wireless World*, Nov. 1970, p 548.

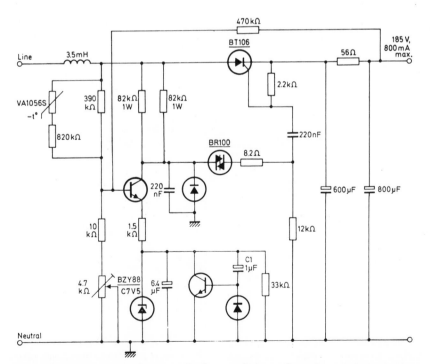

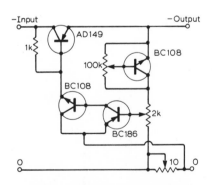

185-V SLOW-START—Active slow-start circuit using ordinary transistors has effect of multiplying value of C1 by hFE when receiver is switched on. When receiver is switched off, C1 discharges through 33K resistor and diode, protecting scr. Circuit has excellent regulation and transient response over range of 180 to 270 V line voltage and 400 to 1,000 mA load current.—R. E. F. Bugg, Thyristor Power Supplies for Television Receivers: Design Considerations, *Mullard Technical Communications*, April 1972, p 129–144.

9 V WITH NEGATIVE RESISTANCE—Output voltage varies less than 15 mV for load currents from 0 to 150 mA and battery input voltages of 12 to 14.5 V. Effective output resistance can be reduced to zero or made negative by adjusting 2K and 10-ohm pots, as desired for maintaining nearly constant speed in battery-operated tape recorder or record player having brush-type d-c motor. Pots can also be used for speed control.—T. D. Towers, Elements of Linear Microcircuits, *Wireless World*, Feb. 1971, p 76–80.

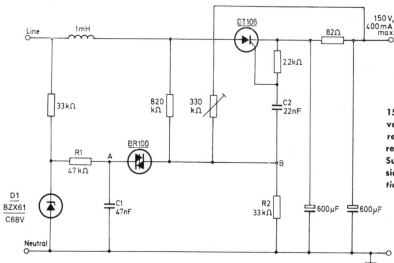

150 V AT 400 MA—Feedback to diac trigger varies mean d-c voltage at point B to provide regulation in circuit using scr as control series regulator.—R. E. F. Bugg, Thyristor Power Supplies for Television Receivers: Design Considerations, *Mullard Technical Communications*, April 1972, p 129–144.

20 V AT 500 MA—Designed for use with tape recorder. Input may be as high as 60 V if R40 is altered as indicated.—J. R. Stuart, High-Quality Tape Recorder, *Wireless World*, Dec. 1970, p 587–591.

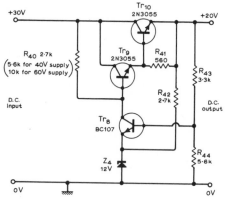

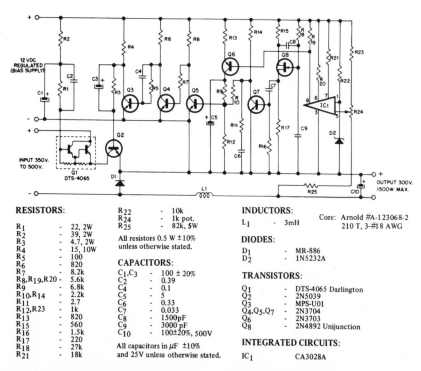

RESISTORS:

R₁	- 22, 2W
R₂	- 39, 2W
R₃	- 4.7, 2W
R₄	- 15, 10W
R₅	- 100
R₆	- 820
R₇	- 8.2k
R₈,R₁₉,R₂₀	- 5.6k
R₉	- 6.8k
R₁₀,R₁₄	- 2.2k
R₁₁	- 2.7
R₁₂,R₂₃	- 1k
R₁₃	- 820
R₁₅	- 560
R₁₆	- 1.5k
R₁₇	- 220
R₁₈	- 27k
R₂₁	- 18k

R₂₂	- 10k
R₂₄	- 1k pot.
R₂₅	- 82k, 5W

All resistors 0.5 W ±10% unless otherwise stated.

CAPACITORS:

C₁,C₃	- 100 ± 20%
C₂	- 0.39
C₄	- 0.1
C₅	- 5
C₆	- 0.33
C₇	- 0.033
C₈	- 1500pF
C₉	- 3000 pF
C₁₀	- 100±20%, 500V

All capacitors in µF ±10% and 25V unless otherwise stated.

INDUCTORS:

L₁	- 3mH

Core: Arnold #A-123068-2
210 T, 3-#18 AWG

DIODES:

D₁	- MR-886
D₂	- 1N5232A

TRANSISTORS:

Q₁	- DTS-4065 Darlington
Q₂	- 2N5039
Q₃	- MPS-U01
Q₄,Q₅,Q₇	- 2N3704
Q₆	- 2N3703
Q₈	- 2N4892 Unijunction

INTEGRATED CIRCUITS:

IC₁	- CA3028A

300-V SWITCHING REGULATOR—Single Delco DTS-4065 silicon Darlington transistor regulates 300 V at 5 A by pulse width modulation with accuracy better than 3%. Switching frequency is 10 kHz, maximum efficiency 85%, and output voltage ripple under 2%. Q1 and Q2 are series pass elements. Frequency is determined by ujt Q8 with R18 and C9. Q8 triggers sawtooth generator Q7 which is d-c modulated by output of differential amplifier IC1. Pulse width is directly proportional to error signal, so duty cycle of Darlington determines average output voltage.— "1.5 kW, 300 V, 10 kHz Darlington Switching Regulator," Delco, Kokomo, IN, Nov. 1972, Application Note 52.

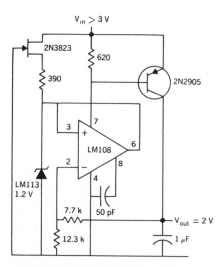

2-V IC—Uses National LM108 IC with transistors and zener to provide closely regulated low voltage. Temperature stability is typically 1% from −55 to 125 C.—Synthesized Reference Diode Maintains Tight Regulation over Wide Current Range, *IEEE Spectrum*, June 1971, p 72.

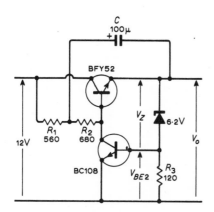

6.8 V AT 100 MA—Provides output regulation of 0.3% from no load to full load, using minimum number of components in series regulator with R2 bootstrap for ripple reduction. Output ripple is only 0.5 mV for 1-V p-p input ripple.—P. S. Ewer, Simple Voltage Stabilizer, *Wireless World*, Aug. 1972, p 366.

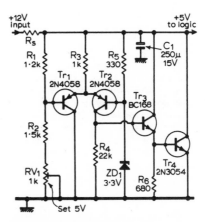

5-V SHUNT STABILIZER—Used as supply for oscillator of frequency standard in crystal oven and for DTL or TTL of counter-timer. Rs is about 15 ohms.—L. Nelson-Jones, Crystal Oven and Frequency Standard, *Wireless World*, June 1970, p 269–273.

185 V AT 800 MA—Transistor and reference diode provide feedback to diac trigger of series scr regulator by varying charging rate of C1.—R. E. F. Bugg, Thyristor Power Supplies for Television Receivers: Design Considerations, *Mullard Technical Communications*, April 1972, p 129–144.

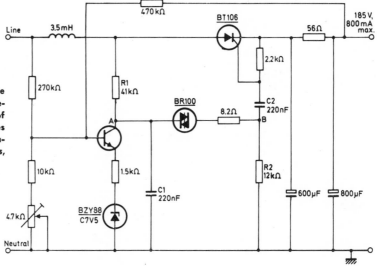

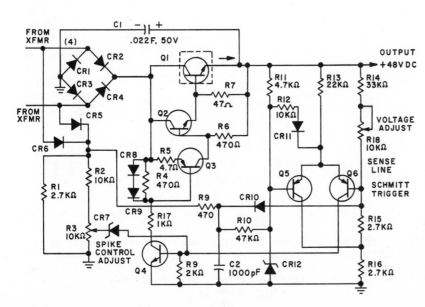

48 V AT 3 A—Will deliver power peaks up to 600 W. Efficiency is better than 90%. Uses solid-state switching techniques, with no inductor required for filtering. Q1, Q2, and Q3 need only 35-V breakdown ratings, but series switching transistor Q1 must handle peak repetitive currents of 15 A. Article gives design equations. Peak input voltage ranges from 75 to 83 V.—C. G. Keeney and M. McWhorter, A Highly Efficient Inductorless Voltage Regulator, *IEEE Journal of Solid-State Circuits*, Aug. 1969, p 192–195.

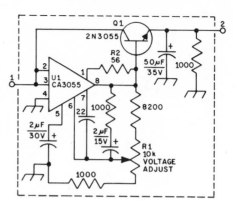

12.6 V AT 1 A—Input is from power transformer with 20-V secondary feeding silicon-diode bridge rectifier rated 100 prv at 1 A. Designed for operating solid-state f-m transceiver from a-c line.—D. A. Blakeslee, AC-Operated Regulated DC Power Supplies for Transistorized Rigs, *QST*, Nov. 1971, p 11–14.

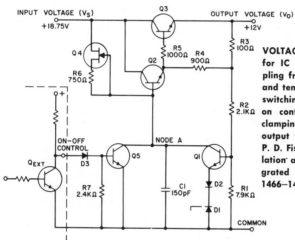

VOLTAGE REGULATOR AND SWITCH—Design for IC construction features excellent decoupling from power source, insensitivity to load and temperature variations, and provision for switching of power to load. Positive voltage on control lead (lower left) turns on Q5, clamping node A to ground and dropping output voltage to low value.—M. L. Embree, P. D. Fisher, and B. H. Hamilton, Power Regulation and Control Using Multifunctional Integrated Circuits, *Proc. IEEE*, Aug. 1967, p 1466–1476.

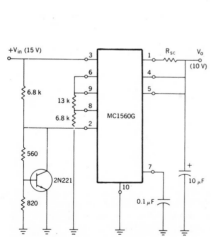

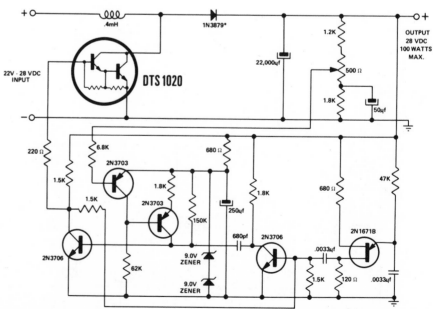

10 V AT 500 MA—Uses Motorola monolithic regulator which provides thermal shutdown automatically if junction temperature in regulator exceeds 140 C. Transistor network provides bias for shutdown control. Load regulation is 0.005%.—D. Kesner, Monolithic Voltage Regulator, *IEEE Spectrum*, April 1970, p 24–32.

28-V SWITCHING REGULATOR—Uses Delco DTS-1020 Darlington silicon power transistor. Efficient regulation by pulse width modulation is achieved at outputs up to 6 V above input of 22–28 V. Switching rate is 9 kHz. Regulation and ripple at full output of 100 W are less than 1%.—"28 Volt Darlington Switching Regulator," Delco, Kokomo, IN, Dec. 1971, Application Note 49.

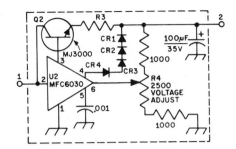

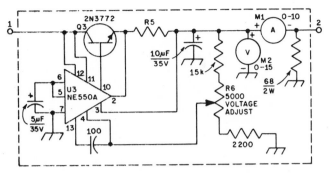

12.6 V AT 2.5 A—Input is from power transformer with 20-V secondary, feeding silicon-diode bridge rectifier rated 100 prv at 2 A. Designed for operating solid-state f-m transceiver from a-c line. Q2 is Motorola Darlington power transistor.—D. A. Blakeslee, AC-Operated Regulated DC Power Supplies for Transistorized Rigs, QST, Nov. 1971, p 11–14.

12.6 V AT 5 A—Input is from power transformer with 20-V secondary, feeding silicon-diode bridge rectifier rated 100 prv at 4 A. Designed for operating solid-state f-m transceiver from a-c line.—D. A. Blakeslee, AC-Operated Regulated DC Power Supplies for Transistorized Rigs, QST, Nov. 1971, p 11–14.

5 AND 12 V—Provides 5 V at about 0.6 A for logic circuits, and somewhat less regulated 12 V for other circuits of 2-meter f-m frequency synthesizer. IC should be mounted on heatsink or chassis. Designed specifically for mobile radio operation.—P. A. Stark, Frequency Synthesizer for 2-Meter FM, 73, Oct. 1972, p 15–23.

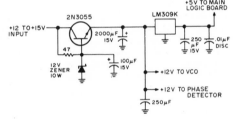

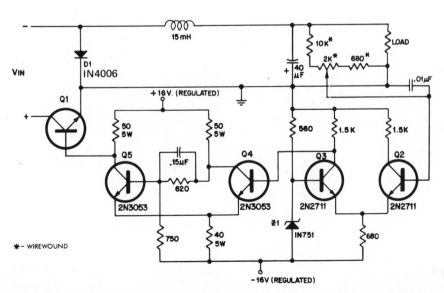

PWM SWITCHING REGULATOR—Uses single high-voltage silicon transistor Q1 as series element. Regulation is by pulse width modulation. Pot in error voltage divider adjusts output voltage. Various combinations of output voltages and currents are obtained by choice of transistor Q1; DTS-410 gives 150 V at 2 A for 200-V maximum input, DTS-431 gives 250 V at 3 A for 325-V maximum input, and DTS-411, DTS-423, and DTS-430 give intermediate outputs. Full-load efficiency is 92%, regulation less than 0.6%, and ripple 0.75-V peak at full load.—"Pulse Width Modulated Switching Regulator," Delco, Kokomo, IN, Feb. 1972, Application Note 39.

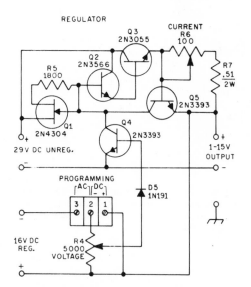

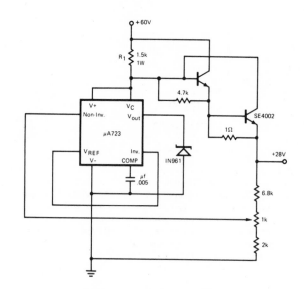

1–15 V AT 500 MA—Only regulator section of supply is shown. Power transformer has two independent secondary windings, with rectified output of lower-voltage winding regulated at 16 V with zener. Q1 is constant-current source, and Darlington-pair Q2-Q3 act as pass-current elements. Q4 provides reference voltage, and Q5 is current-sensing transistor. For external d-c programming, jumper is removed between terminals 2 and 3 and external source connected between 1 and 2. Voltage regulation at 15 V is 0.167% from no load to full load.—Heath IP-18 Regulated Power Supply, QST, Dec. 1971, p 45–46.

28 V FROM 60 V—Use of external transistor pair with zener allows input voltages exceeding 40-V maximum rating of IC regulator.— B. Ricks, "Voltage Regulator Applications Using μA723," Fairchild Semiconductor, Mountain View, CA, No. 283, 1971.

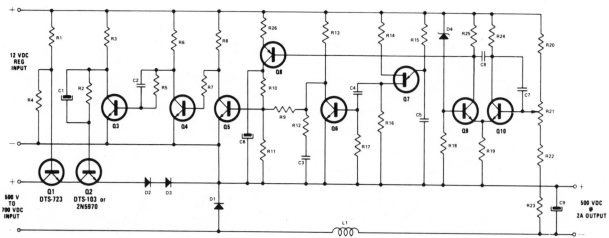

Q₁ - DTS-723	R₆, R₁₇ - 1.5k	R₂₃ - 200k, 5W
Q₂ - DTS-103, 2N5970	R₇ - 8.2k	R₂₆ - 820
Q₃, Q₄, Q₅, Q₆, Q₉, Q₁₀ - 2N3706	R₈, R₂₄, R₂₅ - 5.6k	C₁ - 100 μF
Q₇ - TIS 43	R₉, R₁₃, R₂₀ - 2.2k	C₂ - 0.082 μF
Q₈ - 2N3703	R₁₀ - 6.8k	C₃ - 0.33 μF
D₁ - See Text	R₁₁ - 1k	C₄ - 0.033 μF
D₂, D₃ - Delco DTS-102	R₁₄ - 620	C₅ - 1500 pF
D₄ - 5.6V zener	R₁₅ - 47k	C₆ - 5 μF
R₁ - 6.8, 25W (non-inductive)	R₁₆ - 220	C₇, C₈ - 0.1 μF
R₂, R₁₂ - 2.7	R₁₈ - 10k	C₉ - 2 μF
R₃ - 47, 2W	R₁₉ - 15k	L₁ - 7 mH
R₄ - 100, 1W	R₂₁ - 2k pot	
R₅ - 560	R₂₂ - 22k	

500-V SWITCHING REGULATOR—Uses single Delco DTS-723 silicon transistor as series switching element. Regulation is 0.4% at 500 V and 1 A, and better than 1% at 2 A. Regulates by pwm at constant switching frequency of 13 kHz, with efficiency of 90%.— "One Kilowatt DC Switching Regulator with the DTS-723 Transistor," Delco, Kokomo, IN, May 1971, Application Note 46.

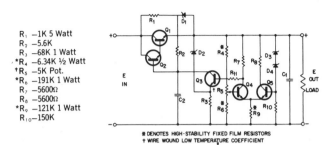

R₁ —1K 5 Watt
R₂ —5.6K
R₃ —68K 1 Watt
*R₄ —6.34K ½ Watt
†R₅ —5K Pot.
*R₆ —191K 1 Watt
R₇ —5600Ω
R₈ —5600Ω
*R₉ —121K 1 Watt
R₁₀ —150K

R₁₁ —120K
D₁ —IN1767
D₂ —IN4735A
D₃ —IN4735A
D₄ —IN4734A
Q₁ —DTS413
Q₂ —2N3439
Q₃ —2N2711
Q₄ Q₅ —2N2712
C₁ —5 μF 450V
C₂ —0.01 μF 1 KV

# DENOTES HIGH-STABILITY FIXED FILM RESISTORS
† WIRE WOUND LOW TEMPERATURE COEFFICIENT

290 V AT 600 MA—Regulation is better than 0.5% for 15% change in input voltage and load change of 50 to 600 mA. Uses high-voltage silicon transistor as series control element Q1.—"DC Voltage Regulator With High Voltage Silicon Transistors," Delco, Kokomo, IN, May 1972, Application Note 38.

12 V AT 2 A—Proportional voltage regulator using pwm gives up to 80% efficiency. Voltage proportional to output voltage is compared to reference diode CR4, differential signal is amplified by Q3, and resulting output of Q3 controls current of series transistor Q1 so voltage across load remains constant.—"Pulse Width Modulated Voltage Regulator," Delco, Kokomo, IN, Dec. 1971, Application Note 9B.

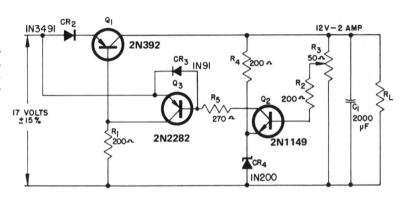

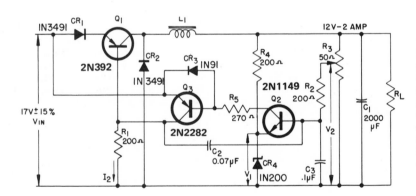

12 V AT 2 A IN SWITCHING MODE—Adding L1, C2, and CR2 to circuit of proportional voltage regulator gives operation in switching mode. L1 is 120 turns No. 20 on ½-inch stack of E1 21 core having 0.03-inch air gap. Q2 and Q3 form free-running mvbr that turns Q1 on and off as required to keep voltage constant across RL. Report gives design analysis.—"Pulse Width Modulated Voltage Regulator," Delco, Kokomo, IN, Dec. 1971, Application Note 9B.

12 V AT 2 A—Based on use of voltage difference between V1 and V2 to bias base of Q3. When voltage across Q2 rises above preset value, voltage-sensitive 90-ohm 5-V relay operates; this protects regulator from sudden increase in input voltage, overload, and short across output. Contacts of K1 short emitter of Q2 to base, to turn off collector current of Q2 much faster than is possible with fuse or circuit-breaker.—"Voltage Regulator," Delco, Kokomo, IN, July 1971, Application Note 3-B.

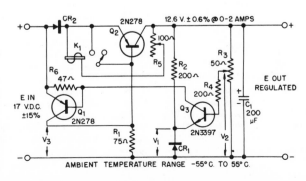

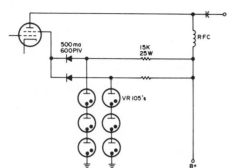

PARALLELING VR'S—Diodes act as switches that make both branches ignite when operating two strings of voltage regulator tubes in parallel to get higher output current along with higher regulated output voltage. Used in 4X150 linear amplifier stage.—C. Goodson, More Current From VR Tubes, 73, Nov. 1969, p 125.

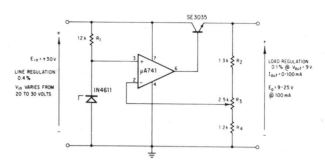

VARIABLE 9–25 V AT 100 MA—Uses opamp as reference amplifier connected to isolate zener voltage reference from changes in loading at supply output. Output impedance is less than 0.1 ohm and line regulation is 0.4% for inputs from 20 to 30 V.—"Applications of the μA741 Operational Amplifier," Fairchild Semiconductor, Mountain View, CA, No. 289, 1972, p 3.

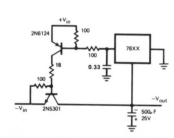

NEGATIVE REGULATOR—Combination of transistors and voltage regulator in Fairchild μA7800 series is regulated negative voltage in range of 5 to 24 V, depending on input voltage value. If base current requirements of 2N5301 series-pass transistor cannot be provided by positive voltage source, use Darlington npn series-pass pair instead.—J. W. Chu and R. D. Ricks, "The μA7800 Series, Three-Terminal Positive Voltage Regulators," Fairchild Semiconductor, Mountain View, CA, No. 312, 1971, p 6.

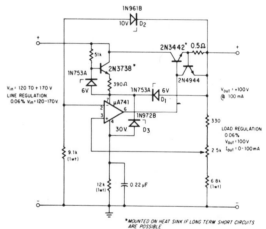

100 V AT 100 MA—Zeners D1–D3 supply proper operating voltages to opamps serving as control amplifier. Uses 2N4944 transistor and 5-ohm resistor for short-circuit protection. Maximum output current for dead short is about 100 mA. Line and load regulation are 0.6% for inputs from 120 to 170 V and output of 100 V at 0 to 100 mA.—"Applications of the μA741 Operational Amplifier," Fairchild Semiconductor, Mountain View, CA, No. 289, 1972, p 3–4.

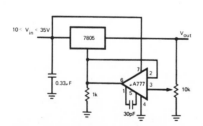

7–30 V AT 15 W—Opamp is active element for μA7805 series voltage regulator. Current limiting and thermal shutdown provide protection over full adjustment range.—J. W. Chu and R. D. Ricks, "The μA7800 Series, Three-Terminal Positive Voltage Regulators," Fairchild Semiconductor, Mountain View, CA, No. 312, 1971, p 5.

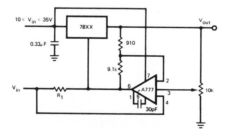

0.5–7 V AT 15 W—Opamp is active element for IC voltage regulator in Fairchild μA7800 series. Requires unregulated negative supply to extend output voltage range down to 0.5 V. Regulator provides current limiting and thermal shutdown to protect circuit over full adjustment range.—J. W. Chu and R. D. Ricks, "The μA7800 Series, Three-Terminal Positive Voltage Regulators," Fairchild Semiconductor, Mountain View, CA, No. 312, 1971, p 5.

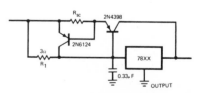

HIGH CURRENT AT 15 W—Circuit provides output currents exceeding rating of voltage regulator in Fairchild μA7800 series, along with protection against short-circuits, by adding pnp transistor and current-sensing resistor RSC. Value of R determines point at which 2N6124 begins conducting.—J. W. Chu and R. D. Ricks, "The μA7800 Series, Three-Terminal Positive Voltage Regulators," Fairchild Semiconductor, Mountain View, CA, No. 312, 1971, p 6.

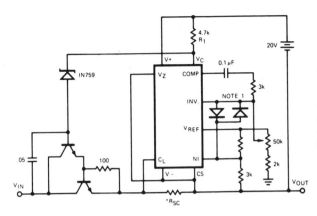

FLOATING REGULATOR—Permits use of input voltages exceeding 40-V maximum of μA723 IC voltage regulator, while achieving load regulation better than 0.005% of output voltage. Add diodes only if input voltage is above 40 V. RSC is equal to 0.51 divided by current limit in amperes.—B. Ricks, "Voltage Regulator Applications Using μA723," Fairchild Semiconductor, Mountain View, CA, No. 283, 1971.

3–25 V TRACKING PREREGULATOR—Uses basic μA723 IC voltage regulator in grounded configuration. Minimum input-output differential is set by 1N749 zener. Output is current-limited to 0.8 A. Line regulation is less than 1 mV and load regulation is 0.04%. Ripple and noise are less than 2-mV p-p over full output voltage range determined by setting of 20K pot.—R. Ricks, "More Voltage Regulator Applications Using the μA723," Fairchild Semiconductor, Mountain View, CA, No. 276, 1969, p 3.

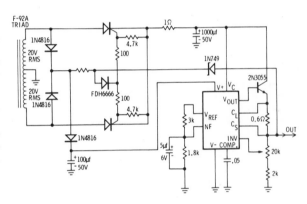

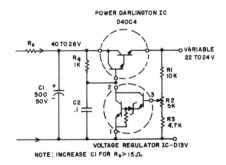

22–24 V FOR COLOR TV—Will safely furnish up to 500-mA d-c and 1-A peak. Output voltage changes less than 1% when output current changes from 0 to 300 mA, and ripple at 300 mA is less than 80-mV p-p. Unregulated input must be 2 V above desired output voltage.—D. V. Jones and M. L. Hyland, "Low Voltage Regulated Supply for Color TV," General Electric, Semiconductor Products Dept., Auburn, NY, No. 90.92, 1970.

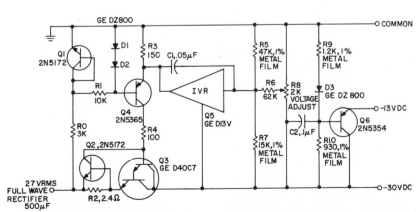

13 AND 30 V FOR ORGAN—Provides maximum regulation required for electronic organ at minimum cost, while eliminating ripple and noise in and near audio spectrum, to offset susceptibility of relaxation oscillators to power-supply noise triggering. Uses basic series-regulated supply referenced to D13V IC voltage regulator.—W. H. Sahm, III, "A Practical R-C Tone Generator System for Electronic Organs," General Electric, Semiconductor Products Dept., Auburn, NY, No. 90.90, 1970, p 5–6.

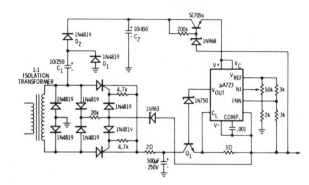

0–150 V—Uses basic μA723 IC voltage regulator in floating configuration to provide controlled voltage across series pass device. Output current is limited to 170 mA. Line regulation is 0.04% of output voltage and load regulation 0.1%. Ripple is less than 0.05% of output voltage. Choose diodes and Q1 to meet output power requirement.—R. Ricks, "More Voltage Regulator Applications Using the μA723," Fairchild Semiconductor, Mountain View, CA, No. 276, 1969, p 3.

15-V BIPOLAR AT 1.5 A—Uses Motorola IC dual-polarity tracking regulator designed to provide positive and negative output voltages balanced within 1%, line and load regulation of 0.06%, and 1% maximum output variation due to temperature changes. With proper heatsinking, circuit provides short-circuit protection.—"Dual ±15-Volt Regulator," Motorola Semiconductors, Phoenix, AZ, Data Sheet 9213, 1972.

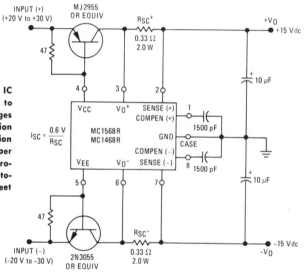

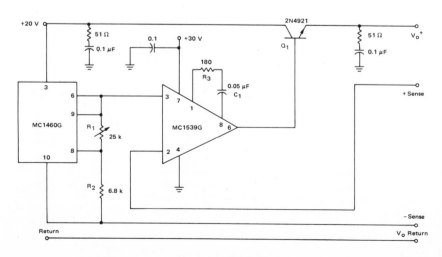

15-V REGULATOR—Uses MC1460G IC as voltage reference for MC1539G IC opamp serving as voltage regulator. Output voltage of 15-V drops only about 0.2 mV for change from no load to full load of 300 mA. Short-term regulation can be as good as 0.001%. Uses series-pass transistor with minimum gain of about 20.—D. Kesner, "Regulators Using Operational Amplifiers," Motorola Semiconductors, Phoenix, AZ, AN-480, 1971.

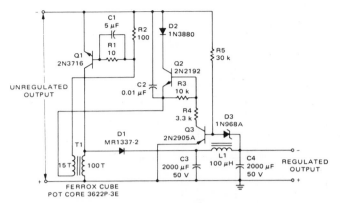

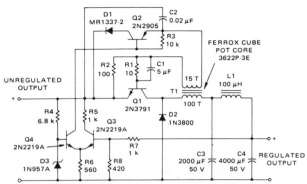

INVERTED-OUTPUT SWITCHING REGULATOR —Efficiency is about 60% for output range of 37 to 44 V, much higher than for series-pass regulator. Q1 and T1 form 6-kHz blocking oscillator for which feedback drive is controlled by Q2 which in turn is driven by voltage-sensing circuit of Q3 for which zener D3 provides voltage reference. Report gives design considerations.—J. Takesuye and H. Weber, "Silicon Power Transistors Provide New Solutions to Voltage Control Problems," Motorola Semiconductors, Phoenix, AZ, AN-147, 1971, p 4–6.

NONINVERTED-OUTPUT SWITCHING REGULATOR —Efficiency is about 85%, much higher than for series-pass or inverted-output switching regulators. Uses 6-kHz blocking oscillator Q1-Q2 for which feedback drive is controlled by reference voltage circuit. Report gives design considerations and performance graphs. —J. Takesuye and H. Weber, "Silicon Power Transistors Provide New Solutions to Voltage Control Problems," Motorola Semiconductors, Phoenix, AZ, AN-147, 1971, p 4–6.

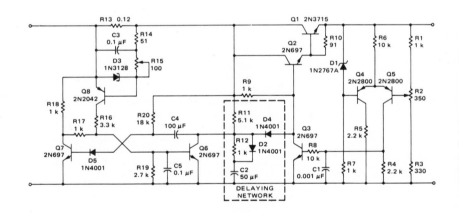

OVERLOAD PROTECTION WITH DELAY —Addition of delay network between overload-protecting mvbr Q6-Q7 and series regulator improves reliability of protection and prevents low-frequency oscillation when using large capacitive loads. Delay serves to apply drive slowly to series-pass transistor Q1, minimizing turn-on surge current. Report analyzes circuit in detail. Adjust R15 so regulator is turned off when load current exceeds 3.5 A, to keep within safe limit of Q1.—J. Takesuye and H. Weber, "Silicon Power Transistors Provide New Solutions to Voltage Control Problems," Motorola Semiconductors, Phoenix, AZ, AN-147, 1971, p 3–4.

SERIES-PASS WITH OVERLOAD PROTECTION —Overload-sensing circuit triggers mono mvbr Q6-Q7, which in turn removes drive from series-pass driver transistor Q2, to turn off regulator circuit until mvbr resets. If overload still exists, regulator is turned off again. Protection is adequate for resistive load, but may give trouble for large capacitive loads. Report analyzes circuit in detail.—J. Takesuye and H. Weber, "Silicon Power Transistors Provide New Solutions to Voltage Control Problems," Motorola Semiconductors, Phoenix, AZ, AN-147, 1971, p 3.

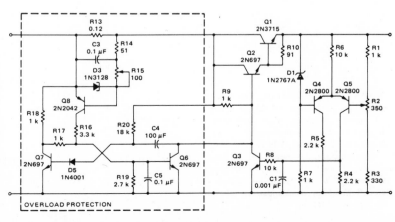

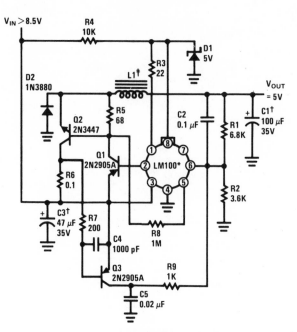

CURRENT-LIMITING SWITCHING—Circuit keeps regulator switching while providing protection by limiting current during overloads or output short-circuit. For input of 28 V, regulated output of 5 V begins dropping as output exceeds 2.5 A, and is 0 V at 4 A. Protects switching transistors for at least short-term overloads, but may not be adequate for continuous worst-case short-circuit. —R. J. Widlar, "Designing Switching Regulators," National Semiconductor, Santa Clara, CA, 1968, AN-2, p 8–9.

*Basing diagram is Top View

†Solid tantalum

‡70 turns #20 on Arnold Engineering A930157-2 molybdenum permalloy core

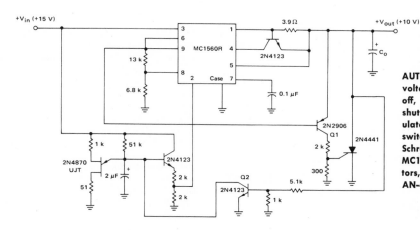

AUTOMATIC SHUTDOWN—If output of IC voltage regulator is short-circuited, Q1 turns off, voltage at pin 2 rises, and regulator is shut down. After short-circuit is removed, regulator must be restarted manually by pushing switch momentarily.—E. Renschler and D. Schrock, "Shutdown Techniques for the MC1560/61/69 Monolithic Voltage Regulators," Motorola Semiconductors, Phoenix, AZ, AN-499, 1972.

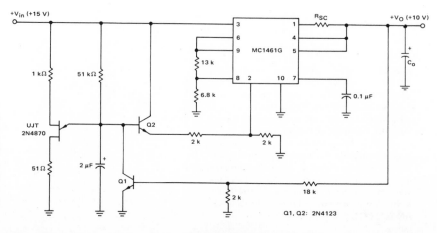

PROTECTION WITH AUTOMATIC RESTART—Short-circuited output turns off Q1, to make Q2 shut down IC regulator. When short is removed, circuit returns automatically to full regulation. Recycling duty cycle is controlled by R-C time constant, 51K and 2 μF.—E. Renschler and D. Schrock, "Shutdown Techniques for the MC1560/61/69 Monolithic Voltage Regulators," Motorola Semiconductors, Phoenix, AZ, AN-499, 1972.

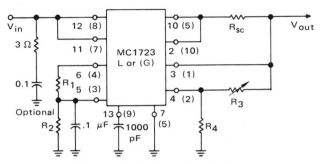

Numbers in parenthesis refer to G package.

R_{sc}	10 Ω	½W	10%
R_1	5 kΩ	1%	{Accurately Match T_C to within better
R_2	2 kΩ	1%	than 20 ppm/°C.
R_3	0-50 kΩ		50 ppm/°C potentiometer
R_4	2 kΩ	1%	20 ppm/°C

2–37 V HIGH-STABILITY—Designed for temperature range of —50 C to +125 C. Input is 40-V d-c, with R3 adjusting output.—D. Johnson and R. Hejhall, "Tuning Diode Design Techniques," Motorola Semiconductors, Phoenix, AZ, AN-551, 1972.

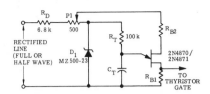

LINE-VOLTAGE COMPENSATOR—Used in trigger circuit for scr or triac control to provide constant output voltage to load regardless of line voltage changes. Pot P1 is adjusted to provide reasonably constant output over desired range of line voltage. Increase in line voltage introduces delay that reduces thyristor conduction angle and thus maintains average voltage reasonably constant.—D. A. Zinder, "Unijunction Trigger Circuits for Gated Thyristors," Motorola Semiconductors, Phoenix, AZ, AN-413, 1972.

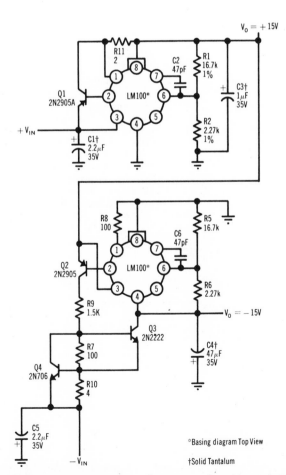

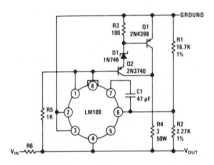

3-A NEGATIVE SHUNT—Output of LM100 voltage regulator drives compound emitter-follower which conducts excess input current. Zener provides level shift for output transistors of IC. Output voltage is determined by ratio of R1 to R2, in range of 2 to 30 V for inputs up to 40 V.—R. J. Widlar, "New Uses for the LM100 Regulator," National Semiconductor, Santa Clara, CA, 1968, AN-8.

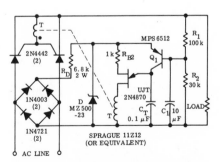

FULL-WAVE AVERAGE-VOLTAGE FEEDBACK—Control responds to average load voltage as derived by network R1-R2-C1. Requires pulse transformer T for triggering of scr pair by ujt. Both load and control circuit require d-c in this type of circuit. Voltage comparison is made directly with zener reference, so circuit acts as regulator without provision for manual control.—D. A. Zinder, "Unijunction Trigger Circuits for Gated Thyristors," Motorola Semiconductors, Phoenix, AZ, AN-413, 1972.

15-V POSITIVE-NEGATIVE—Uses same National IC as both positive and negative regulator, with both inputs and outputs having common ground. Q4 provides current limiting for outputs beyond rated 250 mA.—R. J. Widlar, "Monolithic Voltage Regulator," National Semiconductor, Santa Clara, CA, 1967, AN-1, p 9.

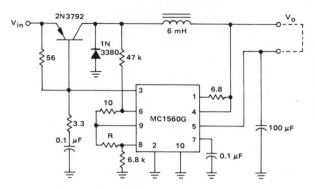

SELF-OSCILLATING SWITCHING REGULATOR—Output is held within 60 mV of 10 V for loads up to 1 A for input voltages within 1% of 28 V. Efficiency approaches 90%. Oscillating frequency of IC is 7 kHz. Report describes operation in detail and gives design equations.—M. Gienger and D. Kesner, "Voltage and Current Boost Techniques Using the MC1560-61," Motorola Semiconductors, Phoenix, AZ, AN-498, 1969.

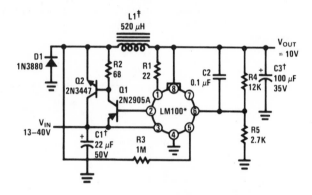

*Basing diagram is Top View
†Solid tantalum
‡60 turns #20 on Arnold Engineering A930157-2 molybdenum permalloy core

10-V SWITCHING—Addition of npn power switching transistor Q2 to basic switching regulator combination of LM100 IC and Q1 increases output current capability above 500 mA while still giving efficiency above 80% even at continuous output of 3 A. Will handle 5 A at reduced efficiency if heatsink is sufficiently large.—R. J. Widlar, "Designing Switching Regulators," National Semiconductor, Santa Clara, CA, 1968, AN-2, p 6.

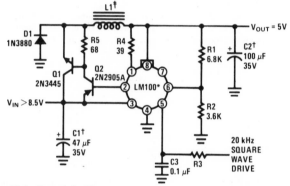

*Basing diagram is Top View
†Solid tantalum
‡100 turns #22 on Arnold Engineering A930157-2 molybdenum permalloy core

5-V DRIVEN SWITCHING—Square-wave 20-kHz drive signal is integrated in one section of IC voltage regulator, and resulting 40-mV p-p triangular wave is applied to reference bypass terminal of IC to control switching rate of regulator circuit. Report describes operation in detail. Arrangement gives lower output ripple and better transient response, along with better frequency stability.—R. J. Widlar, "Designing Switching Regulators," National Semiconductor, Santa Clara, CA, 1968, AN-2, p 8.

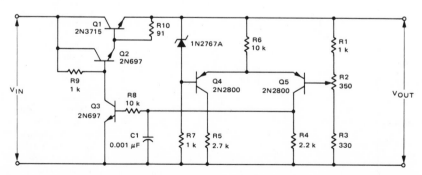

BASIC SERIES-PASS—Output voltage is regulated by series-pass transistor Q1, whose drive is derived by sampling output with divider R1-R2-R3 and comparing with reference voltage of 1N2767A zener. With proper heat-sinking, Q1 can pass 5 A at 30 V continuously without damage. Circuit is not protected from load short-circuit. Load regulation is 0.15% from no load to full load.—J. Takesuye and H. Weber, "Silicon Power Transistors Provide New Solutions to Voltage Control Problems," Motorola Semiconductors, Phoenix, AZ, AN-163, 1968, p 1–2.

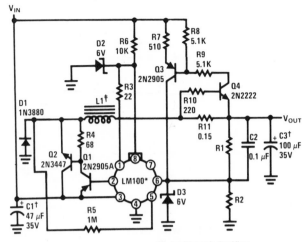

*Basing diagram is Top View

†Solid tantalum

‡70 turns #20 on Arnold Engineering
A930157-2 molybdenum permalloy core

SWITCHING WITH SHORT-CIRCUIT PROTECTION—Provides protection for switching transistors under worst-case continuous short-circuits. Rated 3-A output does not exceed 3.6 A on dead-short. Report covers circuit operation in detail and gives performance graphs.—R. J. Widlar, "Designing Switching Regulators," National Semiconductor, Santa Clara, CA, 1968, AN-2, p 9—10.

5-V HIGH-CURRENT SWITCHING—Efficiency approaches 100% although output is only fraction of 18-V unregulated input. Uses National IC voltage regulator connected as switching regulator, with external npn and pnp transistors connected in cascade to handle output current. Regulator is made to oscillate by applying positive feedback to reference terminal through R3. Report describes operation in detail.—R. J. Sidlar, "Monolithic Voltage Regulator," National Semiconductor, Santa Clara, CA, 1967, AN-1, p 10—11.

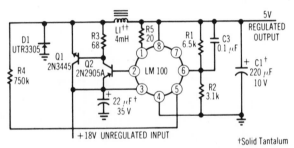

†Solid Tantalum

††160 turns #20 on Arnold Engineering
A-548127-2 Molybdenum Permalloy Core

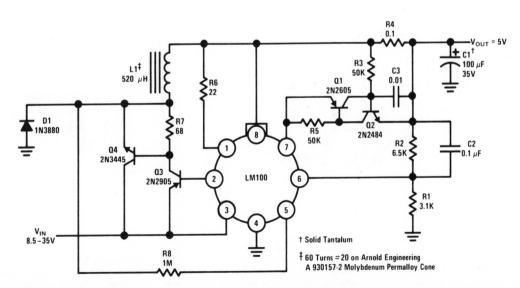

† Solid Tantalum

‡ 60 Turns #20 on Arnold Engineering
A 930157-2 Molybdenum Permalloy Cone

3-A SWITCHING WITH OVERLOAD CUTOFF—When output current becomes excessive, voltage drop across R4 rises to about 0.7 V and turns on Q2. This in turn removes base drive to output transistors in IC voltage regulator. Q1 then latches Q2 to hold regulator off until input voltage is removed. Circuit will restart automatically if overload is removed before power is applied again.—R. J. Widlar, "New Uses for the LM100 Regulator," National Semiconductor, Santa Clara, CA, 1968, AN-8.

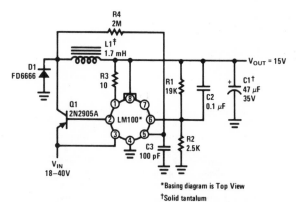

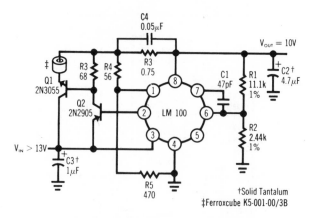

2–30 V SWITCHING—Values shown for resistive divider R1-R2 are adjusted if output other than 15 V is desired. C2 minimizes output ripple by applying full ripple to feedback terminal. Suitable for outputs up to 500-mA limit of switching transistor Q1. Switching frequency is in range of 20 to 100 kHz. Report gives design equations. Efficiency is 80 to 90% depending on input voltage and output current.—R. J. Widlar, "Designing Switching Regulators," National Semiconductor, Santa Clara, CA, 1968, AN-2, p 3–4.

10 V WITH SWITCHBACK CURRENT LIMITING—Reduces short-circuit current to values substantially less than full-load current of 2.1 A. R4 and R5 supply voltage which bucks out voltage drop across current-limit sense resistor R3, to increase maximum load current from 0.5 A to 2.1 A. When output is shorted, bucking voltage vanishes and short-circuit current is only 0.5 A. R4 and R5 also give 20-mA preload on regulator so it can operate without load.—R. J. Widlar, "Monolithic Voltage Regulator," National Semiconductor, Santa Clara, CA, 1967, AN-1, p 8–9.

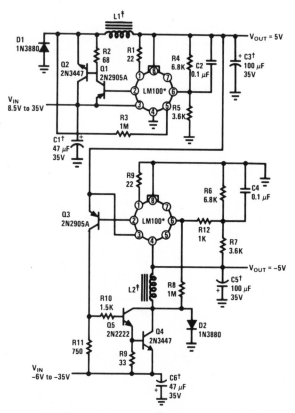

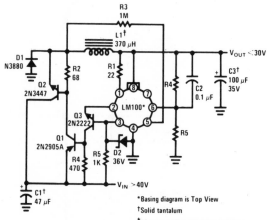

5-V POSITIVE-NEGATIVE—Permits common ground for both positive and negative unregulated inputs and regulated outputs. Only limitation is that positive output must be above 3 V in order to bias negative regulator properly. Pnp booster transistor Q3 drives Darlington-connected npn switch Q4-Q5 for switching regulator action giving negative output.—R. J. Widlar, "Designing Switching Regulators," National Semiconductor, Santa Clara, CA, 1968, AN-2, p 10–11.

SWITCHING FOR INPUTS ABOVE 40 V—Isolation of IC voltage regulator from unregulated supply permits use with input voltages limited only by switching transistors and catch diode. Rating of zener D2 must be at least 3 V above output voltage.—R. J. Widlar, "Designing Switching Regulators," National Semiconductor, Santa Clara, CA, 1968, AN-2, p 11.

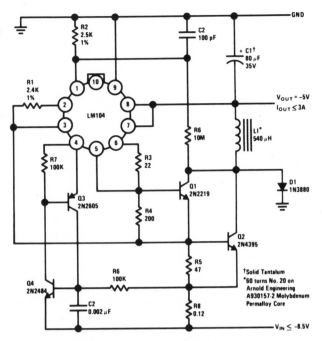

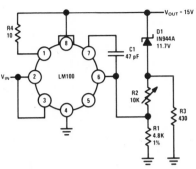

15-V HIGH-STABILITY—Addition of external temperature-compensated reference diode to IC voltage regulator improves performance considerably for outputs above 10 V. Adjust R3 to compensate for combined drift of D1 and IC's. Circuit improves regulation at least 4.5 times.—R. J. Widlar, "New Uses for the LM100 Regulator," National Semiconductor, Santa Clara, CA, 1968, AN-8.

—5 V WITH OVERLOAD SHUTOFF—Excessive output current makes Q3 latch Q4, to hold regulator off until input voltage is removed. Shutoff current is 1.5 times full-load current, so output capacitor charging current will not cause shutoff when regulator is switched on with full load.—R. J. Widlar, "Designs for Negative Regulators," National Semiconductor, Santa Clara, CA, 1968, AN-21.

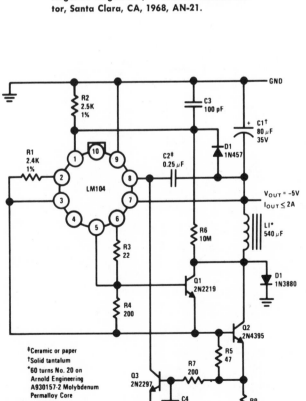

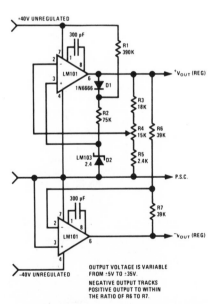

—5 V WITH CURRENT LIMITING—Provides regulator protection against short-circuits. Current limiting starts when switch transistor produces large enough voltage drop across R9 to turn on Q3.—R. J. Widlar, "Designs for Negative Regulators," National Semiconductor, Santa Clara, CA, 1968, AN-21.

TRACKING SUPPLY—Ideal for opamp supply because positive and negative voltages track, eliminating common-mode signals originating in supply voltage. Requires only one voltage reference and minimum number of passive components. Opamps are compensated for unity gain. Injection of current into wiper of R4 provides modulation giving equal and opposite voltage variations at positive and negative outputs.—W. S. Routh, "Applications Guide for Op Amps," National Semiconductor, Santa Clara, CA, 1969, AN-20.

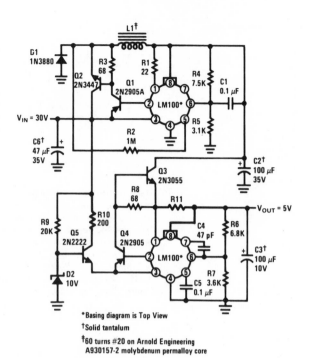

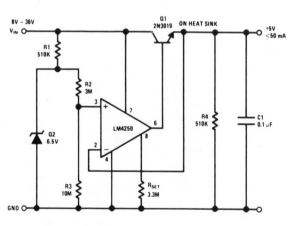

50–250 V WITH 24-V ZENER—Useful when low-voltage positive and negative supplies for conventional opamp regulators are not readily available. Regulation is typically 0.5%, for load currents up to 100 mA. Article gives design equations.—G. Coers, Regulating High Voltages with Low-Voltage Zeners, *Electronics*, June 21, 1973, p 113–114.

SWITCHING-LINEAR COMBINATION—Gives very low ripple and fast transient response for high input-output voltage differential. Uses switching regulator to reduce input voltage with high efficiency and regulate it, so linear regulator operates with fixed voltage differential that minimizes dissipation. Zener preregulator D2-R9-Q5 isolates linear IC regulator from noise on unregulated supply.—R. J. Widlar, "Designing Switching Regulators," National Semiconductor, Santa Clara, CA, 1968, AN-2, p 11.

5-V MICROPOWER—With 10-V input, regulator dissipates only 350 μW in standby mode but will deliver up to 50 mA. Circuit is basically boosted-output voltage-follower referenced to low-current special zener having good regulation in 2–60 μA reverse-current region, such as 2N3252 small-signal npn transistor connected as diode.—M. K. Vander Kooi and G. Cleveland, "Micropower Circuits Using the LM4250 Programmable Op Amp," National Semiconductor, Santa Clara, CA, 1972, AN-71, p 7–8.

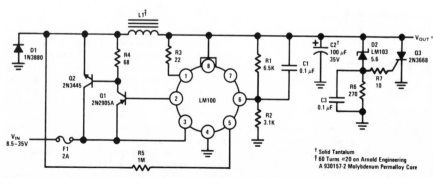

OVERVOLTAGE CROWBAR—If output voltage rises significantly above 6 V, zener D2 breaks down and fires scr Q3, shorting output and blowing fuse F1 on input line. C3 prevents voltage transients from firing scr.—R. J. Widlar, "New Uses for the LM100 Regulator," National Semiconductor, Santa Clara, CA, 1968, AN-8.

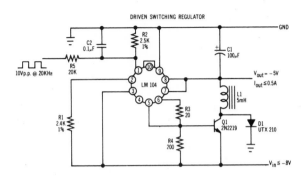

DRIVEN SWITCHING REGULATOR

DRIVEN SWITCHING—Square-wave drive signal permits synchronizing number of switching regulators operating from common power source, to distribute switched current waveforms more uniformly in input lines. Synchronous operation is also beneficial when switching regulator is operated with power converter. Triangular 25-mV p-p wave, obtained by integration of square-wave sync signal, is applied to noninverting input of error amplifier to produce switching action on duty cycle controlled by fed-back output voltage. Control of duty cycle produces desired —5 V output.—R. J. Widlar, "Designs for Negative Regulators," National Semiconductor, Santa Clara, CA, 1968, AN-21.

—50 V WITH FLOATING BIAS SUPPLY—Permits regulating output voltages higher than rating of IC. Bias is preregulated by D1. Second npn booster transistor can be used in compound connection with Q1 to increase output current of regulator.—R. J. Widlar, "Designs for Negative Regulators," National Semiconductor, Santa Clara, CA, 1968, AN-21.

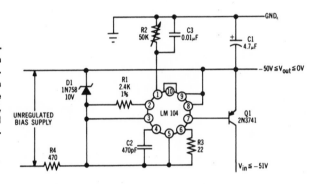

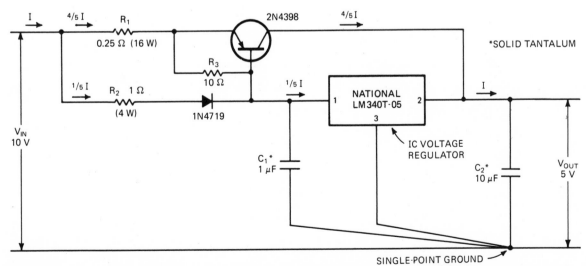

CURRENT BOOSTER—Input current is divided in 4:1 ratio between IC voltage regulator and external pass transistor. Current-sharing scheme preserves short-circuit, overload, and thermal-shutdown safety features of IC. Output is 5 V at 5 A, with 1.4% load regulation.—M. Vander Kooi, Current-Sharing Design Boosts Regulator Output, *Electronics*, March 29, 1973, p 83.

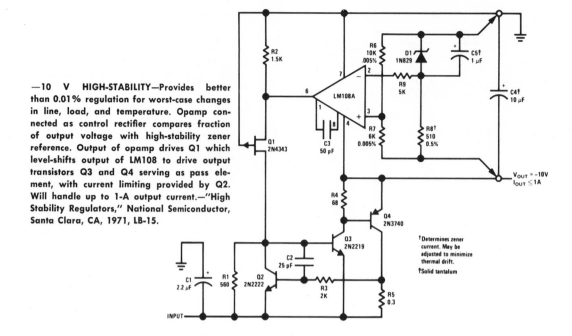

—10 V HIGH-STABILITY—Provides better than 0.01% regulation for worst-case changes in line, load, and temperature. Opamp connected as control rectifier compares fraction of output voltage with high-stability zener reference. Output of opamp drives Q1 which level-shifts output of LM108 to drive output transistors Q3 and Q4 serving as pass element, with current limiting provided by Q2. Will handle up to 1-A output current.—"High Stability Regulators," National Semiconductor, Santa Clara, CA, 1971, LB-15.

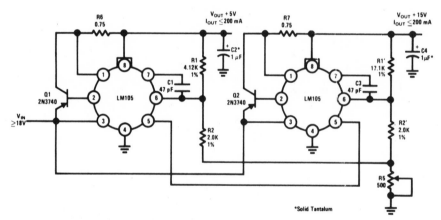

5 AND 15 V TRACKING—Single pot R5 adjusts output voltages of both IC voltage regulators simultaneously with accuracy under 2%. Internal reference voltages of regulators, available at pin 5, are tied together so both operate with same reference. Report gives design procedure.—"Tracking Voltage Regulators," National Semiconductor, Santa Clara, CA, 1969, LB-7.

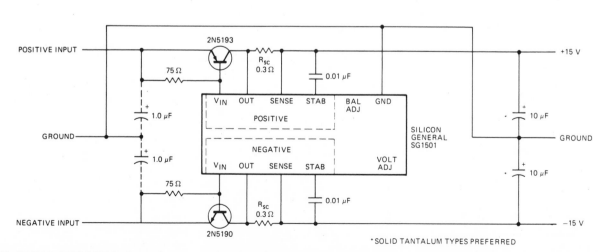

15-V BIPOLAR HIGH-POWER—External pass transistors supplement power-dissipating capacity of dual-polarity IC voltage regulator under short-circuit conditions. Low-frequency transistors reduce risk of oscillation.—R. A. Mammano, Dual-Polarity IC Regulators Aid Design and Packaging, *Electronics*, Feb. 15, 1973, p 108–111.

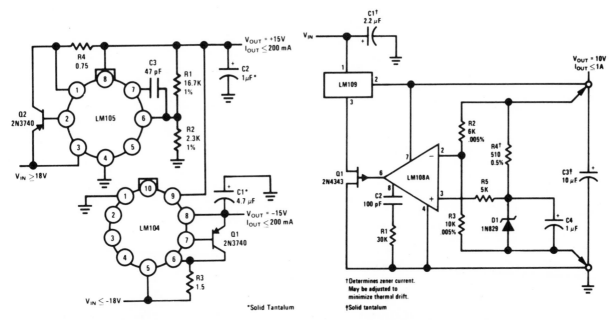

+15 AND −15 V TRACKING—LM104 negative regulator is adapted for use as inverting amplifier with LM105 positive regulator, with interconnections serving to make negative regulator track positive regulator. Positive voltage in LM104 is used as reference.— "Tracking Voltage Regulators," National Semiconductor, Santa Clara, CA, 1969, LB-7.

10-V HIGH-STABILITY—Provides better than 0.01% regulation for worst-case changes in line, load, and temperature. Opamp connected as control rectifier compares fraction of output voltage with high-stability zener reference. Output of opamp controls ground terminal of LM109 voltage regulator through source follower Q1. Will handle up to 1-A output current.—"High Stability Regulators," National Semiconductor, Santa Clara, CA, 1971, LB-15.

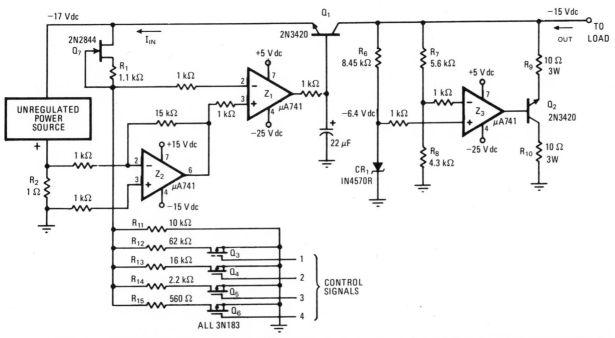

−15 V SERIES-SHUNT—Provides high output-to-input isolation of shunt design and efficiency of series design. Control signals corresponding to predetermined system load requirements set maximum current level through series regulator Q1. Shunt regulator Q2 then compensates for changes in load requirements within that preset current limit. Has built-in short-circuit protection. Developed for secure cryptographic radio communication system.—J. B. Denker and D. A. Johnson, *Electronics*, Aug. 2, 1973, p 91–94.

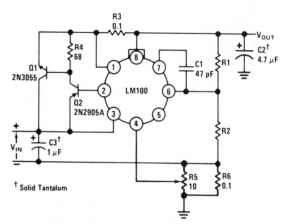

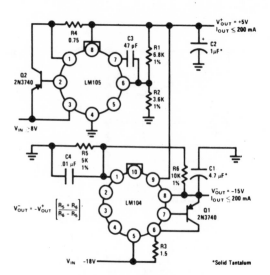

LINE-RESISTANCE COMPENSATOR—Effects of line resistance when sensing load voltage remotely are eliminated by connecting IC voltage regulator so negative-going voltage proportional to load current is produced across R6. Divider resistor R2 is returned to this voltage so output voltage will increase with increasing load current. Adjust R5 to cancel line resistance.—R. J. Widlar, "New Uses for the LM100 Regulator," National Semiconductor, Santa Clara, CA, 1968, AN-8.

+5 V AND —15 V TRACKING—Provides tracking of two voltages when negative output voltage is greater than positive output. Report describes operation in detail. Many negative regulators may be slaved to single positive regulator by connection shown.—"Tracking Voltage Regulators," National Semiconductor, Santa Clara, CA, 1969, LB-7.

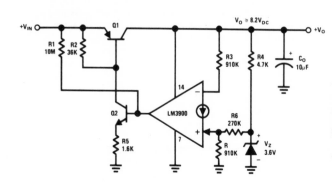

INPUT - VOLTAGE PROTECTION — Transistors Q1 and Q2 absorb excessively high input voltages so as to protect Norton current-differencing opamp. Transistors must be rated for higest voltage expected, but are otherwise not critical as to type. Report covers regulator design and performance.—T. M. Frederiksen, W. M. Howard, and R. S. Sleeth, "The LM3900—A New Current-Differencing Quad of ± Input Amplifiers," National Semiconductor, Santa Clara, CA, 1972, AN-72, p 12.

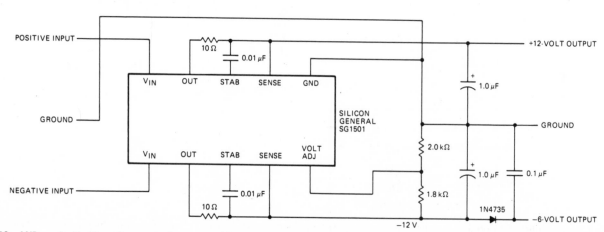

+12 AND —6 V—IC voltage regulator shown provides supply voltages required for 710 and 711 IC voltage comparators widely used in analog circuits. 6-V zener in negative output line drops negative output voltage to half-value.—R. A. Mammano, Dual-Polarity IC Regulators Aid Design and Packaging, *Electronics*, Feb. 15, 1973, p 108–111.

CHAPTER 97
Remote Control Circuits

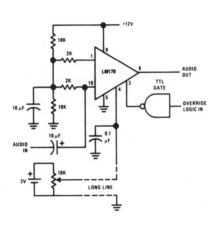

REMOTE GAIN CONTROL—Output voltage of National LM170 audio IC varies as logarithmic function of d-c voltage on pin 4; range of 2.0 to 2.55 V d-c changes level from +40 dB to −35 dB. Voltage can be derived from pot and 3-V supply through long line. TTL logic gate can turn off amplifier independently of d-c control level.—R. A. Hirschfeld, Monolithic Amplifier Has AGC and Squelch, The Electronic Engineer, Aug. 1968, p 60–66.

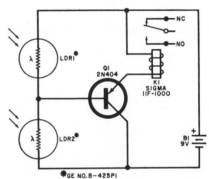

BISTABLE SWITCH—Can be operated by flashlight up to 30 feet away, if separation between light-dependent resistors is at least 7 inches. Light on LDR2 makes Q1 conduct and pull in relay K1. Relay stays energized after light is removed, because solenoid of relay is biased near its pull-in point. Light on LDR1 makes relay drop out. Not affected by ambient light range from bright sunlight to full darkness because this light affects both ldr's equally.—D. C. Conner, Light-Operated Bistable Switch, Popular Electronics, March 1971, p 70.

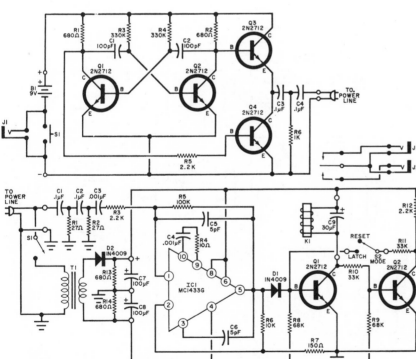

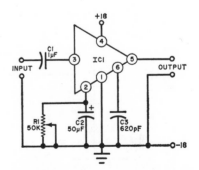

POWER-LINE CARRIER CONTROL—Transmitter plugged into a-c power line at one location operates relay of receiver plugged into power line at another location, by means of 500-kHz signal traveling over power line when S1 is closed or J1 shorted. Used when installation of wiring is not feasible. Burglar alarm in garage or storage building can trip alarm in house. 500-kHz output of astable mvbr Q1-Q2 is matched to low impedance of power line by buffer stage Q3-Q4. In receiver, nega-tive feedback around IC opamp is designed to amplify only over very narrow frequency range centered on 500 kHz. Output of opamp is detected by relay driver Q1. With S2 in latch position, Q1 drives Q2 to form bistable flip-flop that changes state to keep relay energized after input signal is removed.—J. S. Simonton, Jr., Build WWRC Super-Sensitive Wired Wireless Remote Control, Popular Electronics, June 1969, p 27–32.

ELECTRONIC ATTENUATOR—Uses Motorola MFC6040 IC containing ten npn transistors and three diodes, connected to serve as remote volume control supplying voltage gain up to 13 dB while providing attenuation range up to 90 dB. Maximum input signal is 0.5 V. Since control R1 handles only d-c, it can be located some distance from public address system or audio amplifier and connected with ordinary two-conductor cable.—Manufacturer's Circuits, Popular Electronics, March 1971, p 89–91.

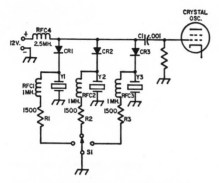

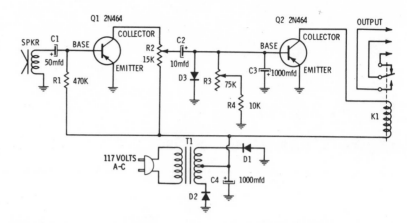

REMOTE-CONTROL CRYSTAL SWITCHING— Germanium diodes, such as 1N270 or 1N34A, provide quiet switching of as many as four crystals in fixed-frequency transmitters and receivers. Grounding through switch forward-biases diode, effectively connecting associated crystal to oscillator grid for r-f. Not suitable for f-m gear having trimming capacitor across each crystal.—H. D. Johnson, Diode Switching for V.H.F. F.M. Channel Selection, QST, Oct. 1969, p 16–17 and 122.

SOUND-ACTUATED CONTROL—Relay K1 (4,000-ohm) is energized when p-m speaker used as mike picks up sufficiently loud voice or horn sound. Can be used with car horn to turn on garage-door opener. R3 determines time that relay stays energized, in range of 1 to 20 s. All diodes ore 1N38B. T1 is 12.6-V C-T filament transformer.—R. M. Brown and T. Kneitel, "49 Easy Transistor Projects," Howard W. Sams, Indianapolis, IN, 1972, p 61–62.

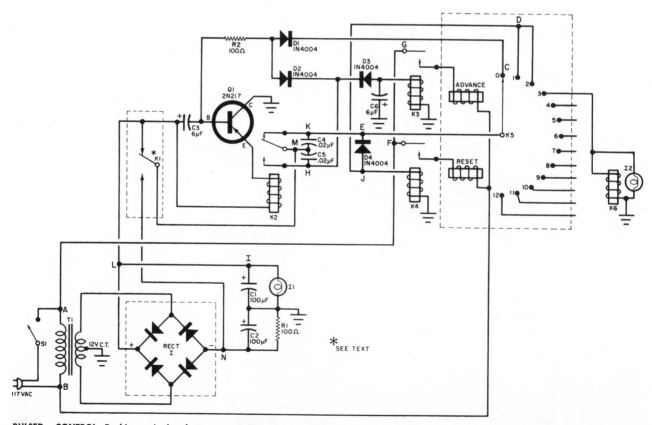

PULSED CONTROL—Pushing single button turns any one of twelve different loads on or off individually at distances up to several hundred feet. Control pulses can come from radio or carrier-current receiver or by direct wire, to actuate relay K1 for which only contacts are shown. Pulses trigger electronic switching circuit which in turn controls stepping relay K5. Number of pulses on first round determines stopping position of K5, each terminal of which may have its own power relay like K6 if needed. On second round, one or more pulses produce reset to 0. Response is foolproof, in that motor on step 5 can be started without affecting other steps, and subsequently any other step may be activated without affecting step 5.—E. C. Maynard, Build the Pulse Command Responder, Popular Electronics, July 1967, p 33–37.

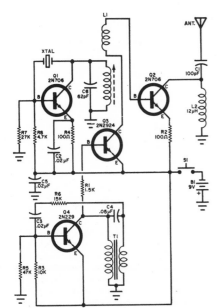

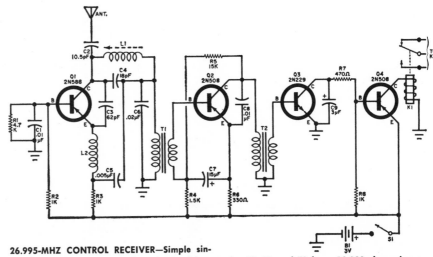

26.995-MHZ CONTROL TRANSMITTER—Can be used to open garage doors, turn tv or radio sets on or off, or turn lights off by radio. Uses 26.995-MHz crystal (to meet FCC specs) modulated with 800-Hz tone. Antenna can be telescoping tv or auto radio types. T1 has 10,000-ohm primary and 1,000-ohm secondary. L2 is 12 μH. L1 has 3½ and 10¾ turns No. 24 enamel on ¼-inch form with adjustable ferrite core.—E. C. Maynard, Remote Commander, *Popular Electronics*, Aug. 1967, p 42—46 and 100.

26.995-MHZ CONTROL RECEIVER—Simple single-channel superregenerative circuit drives 50-ohm subminiature relay for controlling power relay of garage-door opener or other load to be turned on or off by radio. Incoming modulated signal is detected by K1, and 800-Hz tone is amplified by Q2–Q4 to energize K1. T1 and T2 have 10,000-ohm primary and 1,000-ohm secondary. L2 is 12 μH. L1 is 10 turns No. 30 on ¼-inch form having adjustable ferrite core.—E. C. Maynard, Remote Commander, *Popular Electronics*, Aug. 1967, p 42—46 and 100.

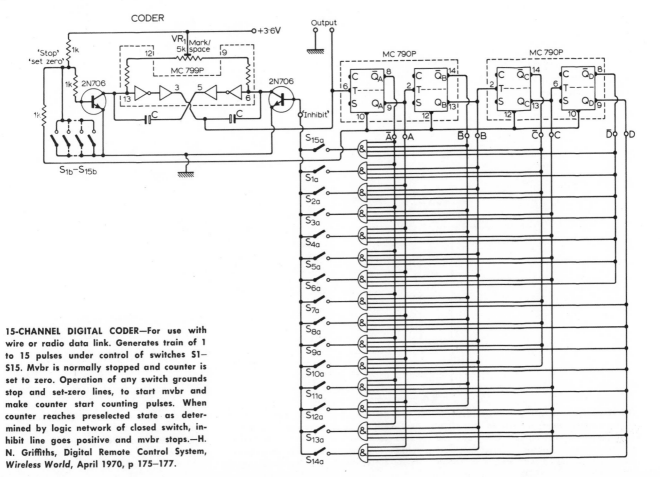

15-CHANNEL DIGITAL CODER—For use with wire or radio data link. Generates train of 1 to 15 pulses under control of switches S1—S15. Mvbr is normally stopped and counter is set to zero. Operation of any switch grounds stop and set-zero lines, to start mvbr and make counter start counting pulses. When counter reaches preselected state as determined by logic network of closed switch, inhibit line goes positive and mvbr stops.—H. N. Griffiths, Digital Remote Control System, *Wireless World*, April 1970, p 175—177.

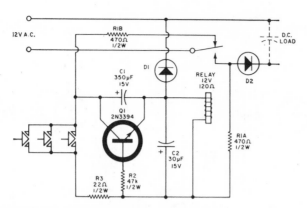

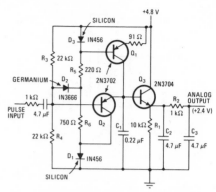

A-C PUSH ON AND PUSH OFF—Pushing any button once makes C1 discharge through relay and pull it in, to apply power to load and latch in relay. Pushing any button again releases relay. Uses rectifier diodes.—H. R. Mallory, Multiple-Function Remote-Control Relay Circuits, *Electronics World*, Aug. 1969, p 36 and 78.

PWM DECODER—Developed for proportional remote radio control system. Produces presettable fail-safe analog output if input control signal is interrupted for any reason. Converts time-modulated pulse input to analog output by starting ramp output of sawtooth generator Q1-Q2 when input goes positive, and resetting when input goes negative. Q3 acts as peak-voltage detector. With values shown, input pulse width of 1.25 ms gives analog output of 2.4-V d-c.—H. R. Beurrier, Temperature-Stable Decoder for Modulated Pulse Widths, *Electronics*, Nov. 20, 1972, p 114.

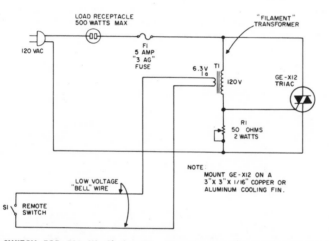

BELL-WIRE SWITCH FOR 500 W—Closing S1 shorts secondary of small filament transformer, triggering triac and energizing lamp or appliance load up to 500 W.—"Electronics Experimenters Circuit Manual," General Electric, Owensboro, KY, 1971, 3rd Ed., p 157–160.

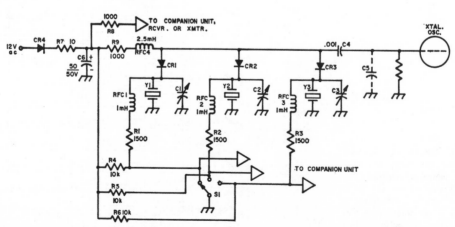

DIODE SWITCHING OF CRYSTALS—Used in f-m receivers or transmitters in which crystals have paralleled "netting" capacitors and one side of crystal is grounded. Up to four crystals can be switched with grounding switch S1. Diodes can be 1N270 or similar germanium. When switch is set to ground one diode, its current drain of about 5 mA produces voltage drop across R9 that reverse-biases diodes not selected, so as to place only desired crystal in oscillator grid circuit.—H. D. Johnson, Diode Switching for V.H.F. F.M. Channel Selection, *QST*, Oct. 1969, p 16–17 and 122.

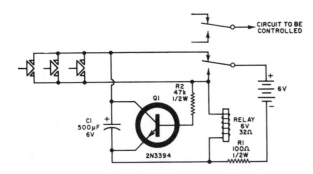

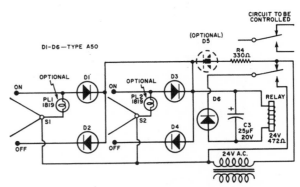

PUSH ON AND PUSH OFF—Completely safe battery-operated circuit energizes relay when any button is pushed, because energy stored in C1 is applied to relay coil. Normally open contacts of relay then provide latching action to keep relay energized. Pushing any of switches again applies discharged capacitor's voltage across relay and decreases relay voltage to zero, to release it and open control and power circuits.—H. R. Mallory, Multiple-Function Remote-Control Relay Circuits, *Electronics World*, Aug. 1969, p 36 and 78.

REMOTE SENSING WITH PUSH ON AND PUSH OFF—Variation of momentary-pushbutton remote control indicates by position of rocker switch whether power circuit is on or off. Such indication is possible only on a-c when using only two wires to remote control station, because diodes serve to selectively polarize control circuit and provide sensed signal. Since diodes at switches are used only momentarily, they can be almost any size and type. If D5 is inserted as shown, pilot lamps PL1 and PL2 may be used at remote stations; lamps glow when relay is not energized.—H. R. Mallory, Multiple-Function Remote-Control Relay Circuits, *Electronics World*, Aug. 1969, p 36 and 78.

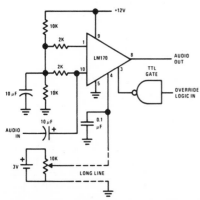

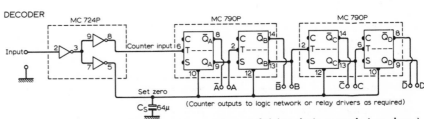

REMOTE GAIN CONTROL—Permits controlling gain of audio amplifier manually and noiselessly with d-c voltage from remote location. TTL gate provides override for turning off amplifier regardless of manual control voltage. Common-mode input voltage may be set anywhere between +4.5 V and +12 V. —R. A. Hirschfeld, "AGC/Squelch Amplifier," National Semiconductor, Santa Clara, CA, 1969, AN-11.

DIGITAL DECODER—Used with wire or radio data link fed by 15-channel digital coder for remote control. Train of 1 to 15 pulses from data link goes through inverting buffer amplifiers to counter input. Eight outputs of counter are fed into logic network (not shown) that provides required 15 control outputs. System uses 20-Hz pulse rate, with mark-space ratio of 1:2.—H. N. Griffiths, Digital Remote Control System, *Wireless World*, April 1970, p 175–177.

REMOTE CONTROL WITH PRESETS—For theatrical lighting, where it is necessary to return to given light level. Low-voltage wires to remote control sections can be several hundred feet long. Uses optoelectronic coupling. Choose triac to match load. Any number of presets may be used, selected with SW1.—R. W. Fox, "Solid State Incandescent Lighting Control," General Electric, Semiconductor Products Dept., Auburn, NY, No. 200.53, 1970 p 13–14.

R1: LEVEL SET.
R2: RATE SET.
R3: ADJUST WITH R1 AT HIGH SIDE TO POINT WHERE LOAD BEGINS TO DIM.

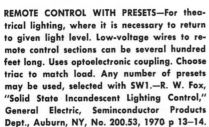

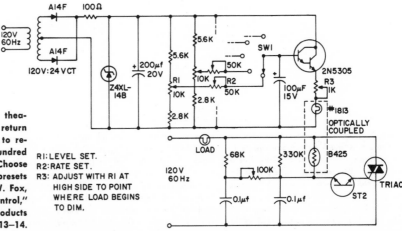

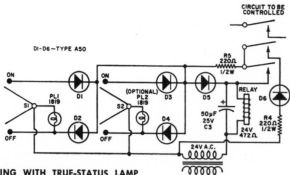

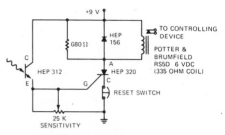

REMOTE SENSING WITH TRUE-STATUS LAMP—Running three wires to each momentary-push remote-control station makes indicator lamp come on when relay is energized, for true indication of status of circuit being controlled. Relay shown may be difficult to obtain, but general-purpose 3-pole double-throw

24-V d-c relay may be used instead. Value of R4 is for one pilot lamp; reduce R4 if two or more lamps are used, to insure that relay will latch in.—H. R. Mallory, Multiple-Function Remote-Control Relay Circuits, *Electronics World*, Aug. 1969, p 36 and 78.

GARAGE-DOOR OPENER—When car headlights hit HEP312 photodiode mounted on frame of garage door, relay is energized for actuating motor-driven door opener. May also be used to turn on any other device that is to be controlled by action of light. Relay stays energized until reset switch is pushed momentarily after light goes out. Adjust sensitivity control to prevent false triggering. Can also serve as burglar alarm that goes on when beam of flashlight hits photodiode mounted near wall safe.—"Home Handyman's Construction Projects," Motorola Semiconductors, Phoenix, AZ, HMA-37, 1972.

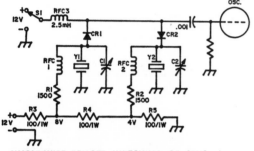

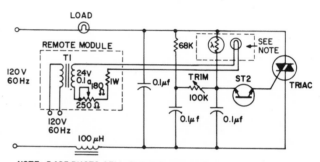

SINGLE-WIRE REMOTE SWITCHING OF CRYSTALS—Requires 12-V d-c supply at control position having switch S1. With S1 at +12 V, CR2 is forward-biased and Y2 is in circuit. Diodes can be 1N270 or similar germanium. Used in f-m receivers or transmitters in which crystals have paralleled "netting" capacitor and one side of crystal is grounded.—H. D. Johnson, Diode Switching of V.H.F. F.M. Channel Selection, QST, Oct. 1969, p 16–17 and 122.

NOTE: B425 PHOTO CELL AND 1813 MINATURE LAMP WRAPPED TOGETHER WITH ELECTRICAL TAPE.
T1: STANCOR P-8361 OR EQUIVALENT

LOW-COST REMOTE CONTROL—Optoelectronic coupling permits running of low-voltage wires from line-voltage control to triac dimmer. Voltage-current characteristic of photocell must be linear at 170 v, to prevent

snap-on effect instead of dimming action. Choose triac to match load.—R. W. Fox, "Solid State Incandescent Lighting Control," General Electric, Semiconductor Products Dept., Auburn, NY, No. 200.53, 1970, p 13.

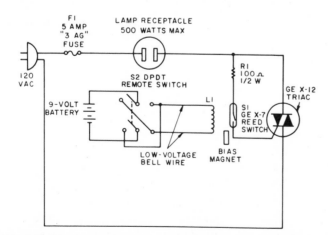

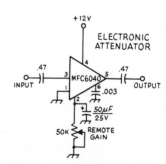

9-V CONTROL FOR 500 W—Since battery at remote location is used only momentarily to energize latching reed switch L1, long life is obtained even with smallest transistor radio size. L1 is 1,000 turns No. 36 wound on form supplied with GE-X7 reed switch having bias magnet. S2 has center-off position; when closed in one direction, magnetic field of L1

reinforces that of magnet so S1 closes and triggers triac on. Spring-loaded switch is desirable, so it opens automatically as soon as lamp comes on. In other position of S2, L1 opposes magnet field, so switch opens and turns off triac and load.—"Electronics Experimenters Circuit Manual," General Electric, Owensboro, KY, 1971, 3rd Ed., p 157–160.

ELECTRONIC ATTENUATOR—Gain of Motorola IC is controlled by adjusting d-c voltage with 50K pot, which may be remotely located. Range of control is from +6 dB to —85 dB, making shielded leads unnecessary for volume control.—"The Radio Amateur's Handbook," American Radio Relay League, Newington, CT, 49th Ed., 1972, p 399–400.

CHAPTER 98
Repeater Circuits

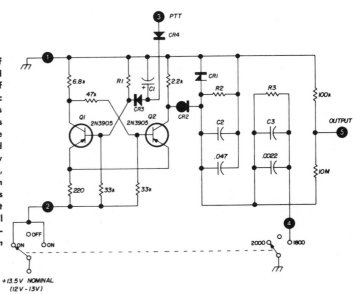

TONE - BURST KEYER — PROVIDES choice of 1,800- and 2,000-Hz tones that can be keyed automatically with push-to-talk circuit of transceiver, to eliminate accidental sporadic keying of 2-meter f-m repeater by stations using same frequency in another area. CR2 is Motorola field-effect diode 1N5299. Choose R1 and C1 for desired burst duration, R2 and C2 for desired high tone, and R3 for low tone; C3 is usually unnecessary. At start-up, oscillator is on; closing push-to-talk switch grounds terminal 3, to discharge C1; if it is 30 μF and R1 100K, discharge time and burst duration are ¾ s. Oscillator stays off until PTT switch is released.—R. B. Shreve, Compact Tone-Burst Keyer for Fm Repeaters, *Ham Radio*, Jan. 1972, p 36—39.

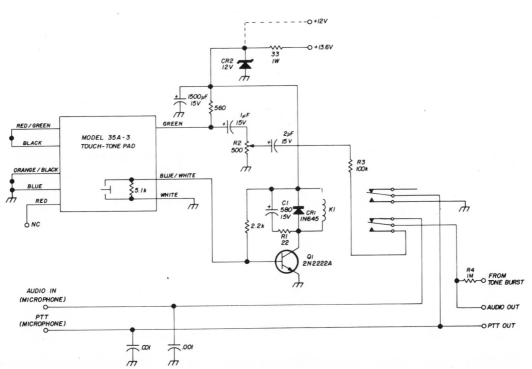

TOUCH-TONE CONTROL—Uses Western Electric 35A3 or 35Y3 12-button pad to control mobile f-m repeater system. If auto installation is noisy, add additional 1,500 μF of filter across zener CR2. Article covers construction, adjustment, and operation.—W. P. Lambing, Mobile Operation with the Touch-Tone Pad, *Ham Radio*, Aug. 1972, p 58—61.

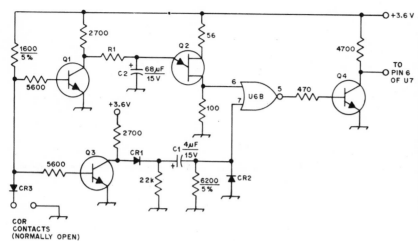

REPEATER-IDENTIFYING TIMER—Generates 1 pulse per minute as required for repetitive identification of f-m repeater with automatic digital Morse-code generator (Hall, "A Digital-Morse-code Message Generator," QST, June 1970). Identification occurs at beginning of transmission and every minute thereafter, as long as carrier-operated relay (COR) remains energized. Output of Q4 goes to code generator. Diodes are 1N914. Q2 is 2N2646. Other transistors are 2N3565. U6 is part of MC724P or HEP 570. Determine R1 experimentally, starting around 459K, to get 1-minute intervals.—C. Rowe, A Repeater Identifier, QST, Nov. 1972, p 44–45.

QUADRUPLER FOR 448-MHZ TRANSMITTER—Input comes from four doubler stages fed by 7-MHz crystal oscillator. Circuit consists of two transistor amplifier stages, varactor multiplier, and bandpass filter. Article gives all circuits for transmitter.—W. Collier, 450 MHz Remote Site Transmitter, 73, May 1971, p 69–70 and 72–73.

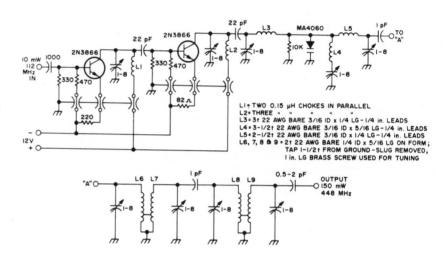

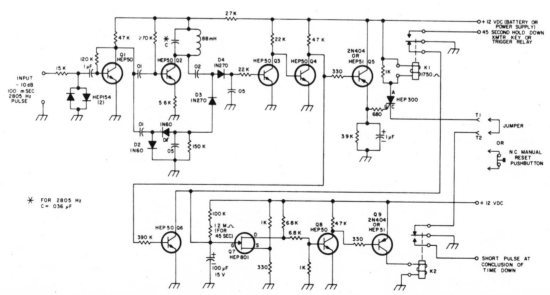

REPEATER TONE-BURST DECODER—Although originally designed for paging applications, circuit provides accurate control of f-m repeaters. Keying bandwith is less than 50 Hz at 2,805 Hz. Circuit holds relays on until 45 s after last tone burst is received. Fail-safe features are included. Article covers operation and construction, and gives circuit for simple 12-V zener-regulated power supply.—J. S. Hollar, Jr., A Short Tone Burst Decoder, 73, Dec. 1972, p 33–36.

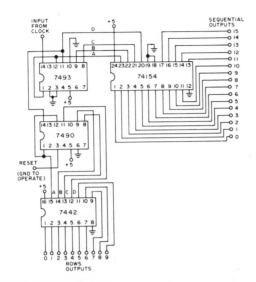

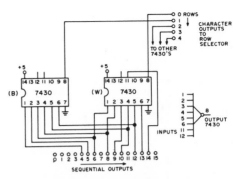

ID GENERATOR FOR REPEATER—Used to generate call letters automatically for f-m repeaters at regular intervals as required by law. Initiating switch must be bounceless-contact circuit as given in article. Operation for encoding desired character sequence is described in detail and all circuits are given.—C. Klinert, A Simple and Inexpensive ID, 73, Oct. 1972, p 64, 66–67, and 69–70.

CODE SEQUENCE GENERATOR—Circuit illustrates how counters of automatic call-letter generator are wired for letters W and B. Article gives all other logic circuits required and describes operation. Designed for use with f-m repeater.—C. Klinert, A Simple and Inexpensive ID, 73, Oct. 1972, p 64, 66–67, and 69–70.

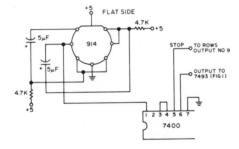

CLOCK FOR ID CODE GENERATOR—Used in automatic call-letter generator for f-m repeater. Astable mvbr generates square-wave train of pulses for driving counter. Article gives all circuits required and describes procedure for encoding desired character sequence.—C. Klinert, A Simple and Inexpensive ID, 73, Oct. 1972, p 64, 66–67, and 69–70.

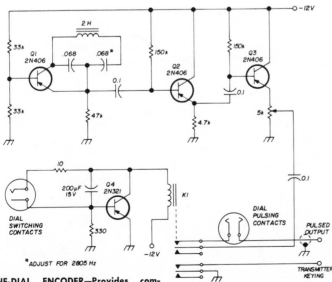

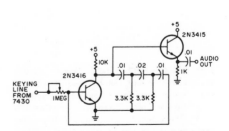

TELEPHONE-DIAL ENCODER—Provides complete dial control of f-m repeater from mobile transmitter. Consists of 2,805-Hz oscillator Q1, buffer Q2, and emitter-follower Q3. Circuit is open (no tone) until it is dialed and Q4 pulls in relay K1. Q4 is 3-s hold circuit that provides 3-s carrier and tone at end of pulsing sequence, for selective calling. Surplus 88-mH toroid may be used in place of 2-H unit shown, if capacitor values are juggled to get desired tone.—J. S. Hollar, Mobile Fm Sequential Encoder, Ham Radio, Sept. 1971, p 34–35.

TONE FOR ID CODE GENERATOR—Phase-shift oscillator is keyed by biasing transistor on from high output of last 7430 counter in automatic call-letter generator for f-m repeater. Pot is required to set bias exactly so collector voltage is about half of supply voltage. Will operate reliably on 5 V only at lower audio frequencies; for higher tone, increase supply voltage.—C. Klinert, A Simple and Inexpensive ID, 73, Oct. 1972, p 64, 66–67, and 69–70.

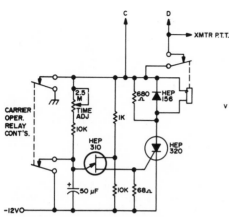

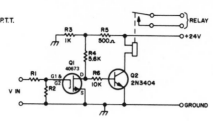

CARRIER-OPERATED RELAY—High input impedance of mosfet permits use with almost any receiver. Requires negative-going voltage of more than —3 V, normally obtainable in receiver squelch circuit, to energize relay which can control transmitter of repeater or connect speaker of standby receiver. To operate from 12-V supply, use 12-V relay, change R6 to 5.6K, and omit R3 and R5.—J. G. Oehlenschlager, Solid-State Carrier Operated Relay, 73, May 1972, p 108.

TIMER—Developed to turn off f-m repeater automatically after predetermined time limit in range of 2 s to 4.5 min, to prevent thoughtless ragchewer from tieing it up and preventing others from using the service. Will also handle situation where mobile f-m operator puts mike on seat and accidentally sits on push-to-talk switch. Reasonable transmission limit is 2 minutes. Use 6-V 335-ohm relay. Relay remains energized as long as offending carrier remains after timer acts, but resets automatically when carrier drops out. Article tells how to connect terminals C and D to circuit for defeating timer if manual reset is desired at times.—K. W. Sessions, Jr., Transmission-Limit Timer for Repeaters, 73, Oct. 1972, p 49—51.

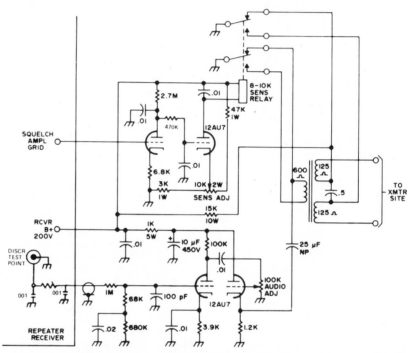

SPLIT-SITE F-M REPEATER—Use of separate sites up to half a mile apart for transmitter and receiver of repeater permits close-spacing of transmit and receive frequencies, improved receiver performance, and minimum shielding. Circuit provides audio conditioning and carrier switching, as required for sending both audio and d-c control signals over simple two-wire line from receiver to transmitter. Use of balanced line eliminates hum problems. Receiver squelch keeps audio off line until carrier appears and pulls in sensitive relay. All grounds are to receiver chassis.—K. W. Sessions, Jr., The Split Site Repeater, 73, Nov. 1971, p 39 and 42—44.

5 W ON 448 MHZ—Article gives all circuits for transmitter that can be used at repeater site. Power amplifier stages shown require only 150-mW input. Two-transistor protection circuit connected to directional coupler prevents damage to output stage Q3 if antenna is accidentally opened or short-circuited. Tune power amplifier with only 20 V applied at first. Amplifier feeds antenna through directional coupler.—W. Collier, 450 MHz Remote Site Transmitter, 73, May 1971, p 69—70 and 72—73.

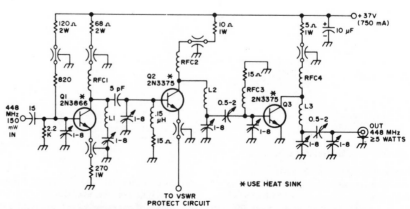

ALL FEED-THRU CAPACITORS = 1000 pF
RFC1, RFC2 = 12t 28 AWG ON 100K, 1/4W RESISTOR
RFC4 = 9t 28 AWG ON 100K, 1/4W RESISTOR
RFC3 = 12t 26 AWG ON 100K, 1/2W RESISTOR
L1 = 3t 18 AWG 3/16 ID, 3/8 LG, 1/8 in. LEADS

L2 = 2-1/2t 18 AWG 3/16 ID, 3/8 LG, 1/8 in. LEADS
L3 = 2t 18 AWG 3/16 ID, 3/8 LG, 1/8 in. LEADS

NOTE: EMITTER LEAD OF Q3 IS SOLDER LUG SOLDERED TO TRANSISTOR CASE

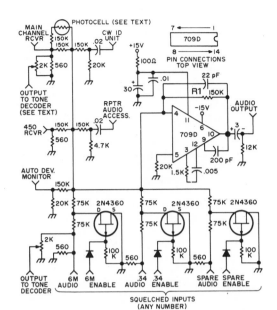

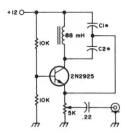

ENCODER—Tone frequency depends on capacitor values; C1 should be ten times C2. With standard 88-mH toroid, 2 μF and 0.22 μF give about 1,000 Hz. Transistor can be almost any npn. Connect output to mike input of 2-meter f-m transmitter and adjust 5K pot to give 5 to 10 kHz deviation. Article includes decoder circuit.—B. Kertesz, Simple Tone Units For Repeater Use, *73*, Feb. 1971, p 36 and 38.

AUDIO MIXER—Provides high degree of isolation (over 40 dB) between individual inputs, as required so tone command information on one channel is not affected by other inputs. Arrangement shown has three squelched inputs and five continuously on. Audio output can swing up to 10 V, for driving as many as ten high-impedance inputs. Photocell-lamp assembly for changing gain on main-channel input is optional. Diodes are 1N457.—P. Hoffman and R. Pichulo, Repeater Audio Mixer, *73*, March 1971, p 57–58.

DECODER—Uses twin-T circuit followed by d-c amplifier energizing relay for turning on f-m repeater. Input should be connected to discriminator output of receiver, and 15K pot adjusted so relay closes reliably each time encoder of transmitter is keyed while within desired range of repeater. Article includes circuit for 1,000-Hz encoder.—B. Kertesz, Simple Tone Units For Repeater Use, *73*, Feb. 1971, p 36 and 38.

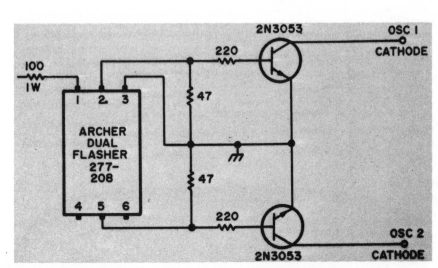

MONITOR SWITCHING—Permits monitoring two 2-meter f-m repeater frequencies simultaneously with single receiver. Uses dual flasher, which is essentially 6-V mvbr, to energize transistors that switch between channels by grounding receiver crystal oscillator cathodes of channels alternately. Article tells how to modify flasher module to give desired sampling rate of 2 s per channel. Eliminates tiresome manual flipping of channel selector switch.—L. Waggoner, Semiautomatic FM Channel Scanning, *73*, Nov. 1970, p 38–40.

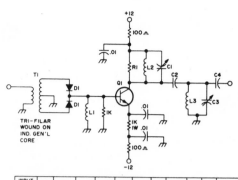

BASIC DOUBLER—Used with values given in table to double output frequency of 7-MHz crystal oscillator four times, to 112 MHz, for feeding quadrupler of 448-MHz transmitter at repeater site. Circuit is push-push, and requires tuning only for maximum output. Article gives all transmitter circuits.—W. Collier, 450 MHz Remote Site Transmitter, 73, May 1971, p 69–70 and 72–73.

INPUT FREQ	T1 *	D1	L1	L2	R1	C1	C2	L3	C3	C4	Q1
7 MHz	6t-Q1 CORE	1N914	150 μH	15 μH	2.0K	33 pF FIXED	3 pF	6.8 μH	5-18	5 pF	2N3904
14 MHz	6t-Q2 CORE	HP2800	47 μH	2.2 μH	2.2K	10 pF FIXED	3 pF	1.8 μH	5-18	5 pF	2N3904
28 MHz	6t-Q3 CORE	HP2800	6.8 μH	0.47 μH	OMIT	8-25	1 pF	1.0 μH	2-8	3 pF	MPS918
56 MHz	6t-Q3 CORE	HP2800	3.3 μH	0.33 μH	OMIT	2-8	1 pF	0.15 μH	5-18	3 pF	MPS918

* INDIANA GENERAL CORE MATERIALS SHOWN

CONTINUOUS-TONE SQUELCH—Phase-shift oscillator generates subaudible tone in range of 87 to 112 Hz for modulating transmitter to obtain access to f-m repeater. Article also gives component values for generating single 1,750-Hz tone with same circuit for tone-burst access to repeater. Stability is comparable to that obtained with vibrating-reed encoders.—J. Gallegos, Encoders for Subaudible, Tone-Burst, or Whistle-On Use, 73, Feb. 1970, p 74–79.

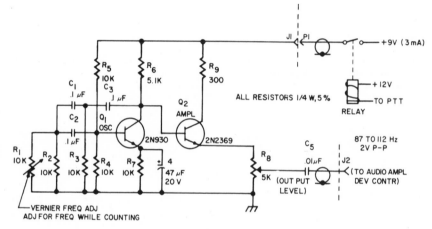

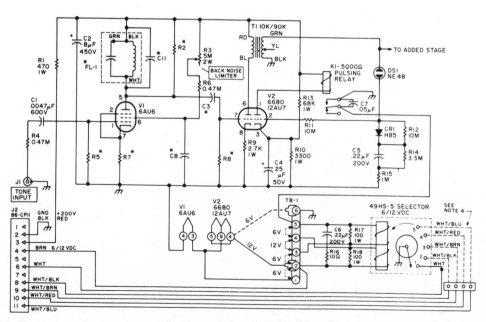

DIGITAL DECODER—Circuit is modification of Secode RPD-612 decoder for use on channel-A 2-meter f-m repeater (146.34 to 146.94 MHz). Pulsing relay operates each time tone is transmitted, for dial pulsing of phone line being fed by repeater for telephone patch. Not intended for simultaneous duplex operation. Article covers all connections needed for making patch.—D. E. Chase, The Wichita Autopatch, 73, May 1970, p 118–124.

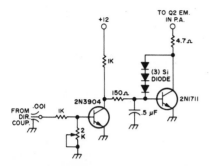

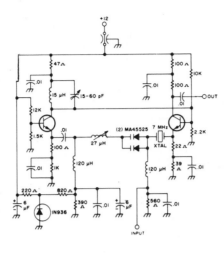

VSWR PROTECTION—Obtains d-c voltage from directional coupler in antenna feed line, for application to emitter of driver in solid-state 5-W power amplifier of 448-MHz transmitter to protect output stage if antenna is accidentally opened or short-circuited. Article gives all other circuits of transmitter.—W. Collier, 450 MHz Remote Site Transmitter, 73, May 1971, p 69–70 and 72–73.

7–MHZ CRYSTAL OSCILLATOR—Use with doublers and quadruplers to provide 448 MHz for input of 5-W power amplifier of transmitter used at remote site. Transistors are 2N3866 and 2N3904 in emitter-feedback circuit using crystal operating in series resonance mode. Article gives all transmitter circuits.—W. Collier, 450 MHz Remote Site Transmitter, 73, May 1971, p 69–70 and 72–73.

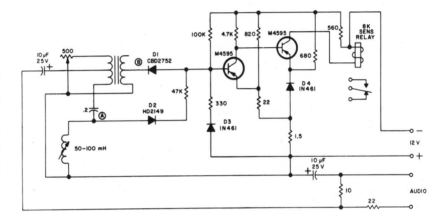

FAST DECODER—Responds to tone pulses having repetition rate of 10 per s, in range from 2,200 to 2,900 Hz. Frequency-sensitive elements can be adjusted to pinpoint bandwidth of 50 to 75 Hz in that range. Used for autopatch control of f-m repeater when call is to be initiated by telephone. Article covers alignment.—K. Session, Jr., For Emergencies: The Super Autopatch, 73, July 1970, p 26–32 and 34–35.

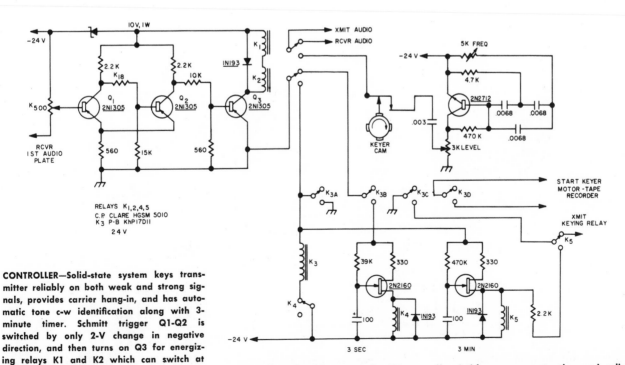

CONTROLLER—Solid-state system keys transmitter reliably on both weak and strong signals, provides carrier hang-in, and has automatic tone c-w identification along with 3-minute timer. Schmitt trigger Q1-Q2 is switched by only 2-V change in negative direction, and then turns on Q3 for energizing relays K1 and K2 which can switch at speeds up to 20 kHz. K2 makes K3 latch itself closed, so transmitter is keyed through K3C and K5. When K2 opens, it applies voltage to 3-s timer through K3C, to energize K4 momentarily at end of timing interval and allow K3 to reset. Tone oscillator frequency is 1,800 Hz. Article covers construction and adjustment.—W. B. Kincaid, A Repeater Controller, 73, April 1970, p 34–36 and 38–39.

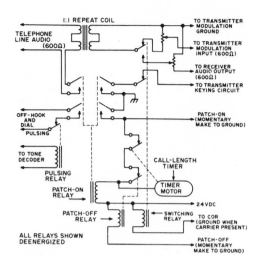

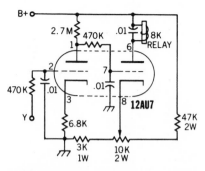

F-M REPEATER AUTOPATCH—Includes pulsing relay, switching relay, patch-on relay, and patch-off relay, along with call-length timer. Used for making telephone calls automatically from transmitters equipped to produce proper sequence of pulses for which repeater is designed.—D. E. Chase, The Wichita Autopatch, 73, May 1970, p 118–124.

CARRIER-OPERATED RELAY—Tube circuit is connected to squelch of f-m receiver in repeater and receives plate voltage from power supply of receiver. Relay is sensitive type, used to control heavy-duty relay that keys push-to-talk circuit of repeater.—K. W. Sessions, Jr., Understanding The Carrier-Operated Repeater, 73, April 1970, p 41–44.

2,300-HZ DECODER—Used at site of unattended f-m repeater to turn on its transmitter when receiver of repeater picks up signal having continuous tone of correct frequency for access. C1 determines exact frequency, generally chosen high enough to make access by whistling difficult. Speaker audio of repeater receiver is fed into input jack of decoder. —K. Sessions, Jr., Tone Decoder for Remote Switching Applications, 73, Feb. 1970, p 80–81.

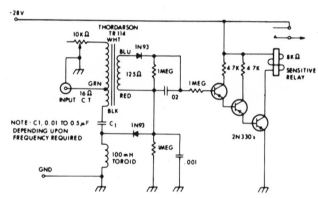

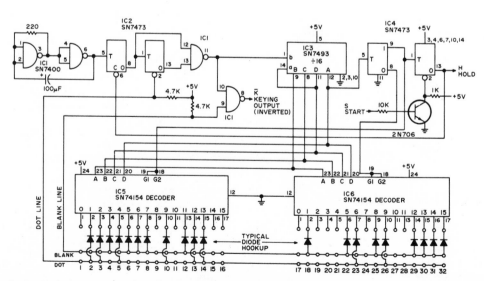

C-W IDENTIFIER—Designed for use with 2-meter repeater, for generating required station-identifying call letters automatically. Uses simple 32-position diode matrix for storing call; for each dot, diode is connected between corresponding IC output and dot line. For each blank, diode goes to blank line. For dash, no diode is used. Connections shown are for generating DE K2OAW. Used as part of solid-state repeater control system.—P. A. Stark, ATTL Logic CW ID Generator, 73, Feb. 1973, p 27–30.

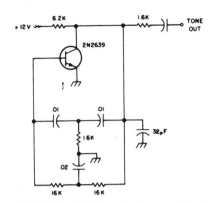

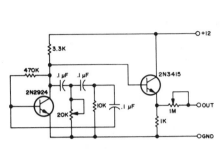

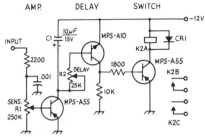

TONE-BURST CONTROL—Generates 1,750 Hz at sufficient amplitude for most transmitters, for turning on repeater. Oscillator uses twin-T feedback network. Circuit is turned on by pressing momentary-contact switch on control head of mobile transmitter.—J. Gallegos, Encoders for Subaudible, Tone-Burst, or Whistle-On Use, 73, Feb. 1970, p 74—79.

107-HZ ENCODER—Uses phase-shift oscillator followed by emitter-follower for generating low-deviation continuous-tone modulation for transmitter when access to repeater is desired. Article includes decoder circuit.—C. Klinert, Phase Locked Loop Decoder, 73, Feb. 1973, p 36—38.

CARRIER-OPERATED RELAY—Input is obtained either from grid current of tube limiter or from d-c output of noise detector in solid-state receiver of f-m repeater. R2 adjusts time that relay stays closed after input voltage disappears. CR1 is silicon diode.—"The Radio Amateur's Handbook," American Radio Relay League, Newington, CT, 49th Ed., 1972, p 450.

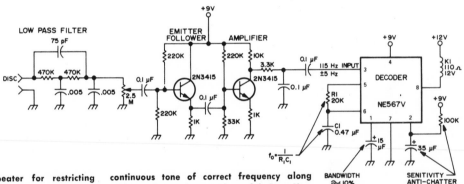

DECODER—Used at repeater for restricting use to certain individuals or for selecting one repeater from several sharing same channel. Relay K1 is energized only when accessing transmitter sends low-frequency low-deviation continuous tone of correct frequency along with carrier and normal modulation. Uses Signetics phase-locked loop decoder IC which responds consistently to tone as much as 6 dB below wideband noise level. Article also gives encoder circuit. Tone is 107 Hz, inaudible in receiver.—C. Klinert, Phase Locked Loop Decoder, 73, Feb. 1973, p 36—38.

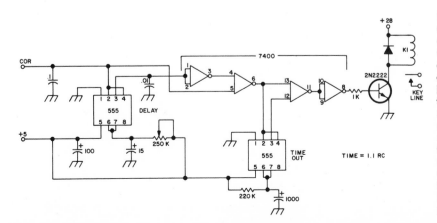

DROPOUT DELAY—Provides delay of about 2 s between release of carrier-operated relay and dropout of transmitter at repeater site, to prevent transmitter keying relays from chattering when signal is cluttering in and out of repeater receiver. Delay time is adjusted with 250K pot. Circuit also includes time-out timer for turning off transmitter of repeater if someone accidentally or deliberately leaves his transmitter keyed up on input channel of repeater. Values shown for R and C with Signetics NE555 at right give actual delay of about 10 min because leakage resistance of capacitor counteracts that of timing resistor (theoretical is 4 min).—C. Klinert, Repeater Keying Line Control, 73, Feb. 1973, p 60—61.

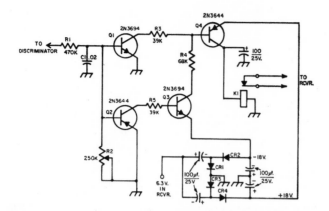

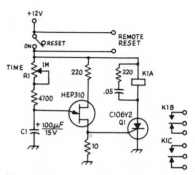

30-KHZ WINDOW—Circuit energizes relay that cuts off receiver of vhf f-m repeater when incoming signal is off frequency or has excessive deviation. Circuit operates when incoming signal is more than 3 kHz off intended center frequency or is deviating more than 15 kHz. Diodes are Motorola 1N4001. Relay is 12-V 576-ohm d-c. With Dunco MRR-1B-M miniature reed relay having operating time of milliseconds, less than 1 s of offending signal is heard through repeater.—G. J. Kowols, Interference Prevention for V.H.F. Repeaters, QST, July 1969, p 42–43.

3-MIN TIMER—R1 in ujt timer sets timing cycle. Range is up to 10 min. At end of cycle, scr turns on and energizes relay for turning on f-m repeater identification, restricting transmission to 3 min, or controlling logging functions. If desired, flip-flop IC can be used in place of relay. Supply voltage must be interrupted momentarily to reset timer.—"The Radio Amateur's Handbook," American Radio Relay League, Newington, CT, 49th Ed., 1972, p 450.

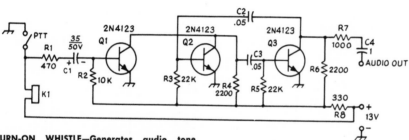

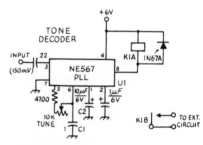

TURN-ON WHISTLE—Generates audio tone burst of about 650 Hz for modulating mobile f-m transceiver to obtain access to tone-controlled f-m repeater. Astable mvbr Q2-Q3 is triggered by one-shot when push-to-talk switch is closed to energize transmitter relay K1 of transceiver.—"The Radio Amateur's Handbook," American Radio Relay League, Newington, CT, 49th Ed., 1972, p 451.

SINGLE-TONE DECODER—Used at f-m repeater for tone-burst entry control. IC is Signetics phase-locked loop. C2 establishes bandwidth of decoder in range between 1 and 14% of operating frequency. Output of 100 mA will key relay directly or drive TTL logic.—"The Radio Amateur's Handbook," American Radio Relay League, Newington, CT, 49th Ed., 1972, p 451–452.

CHAPTER 99
Sampling Circuits

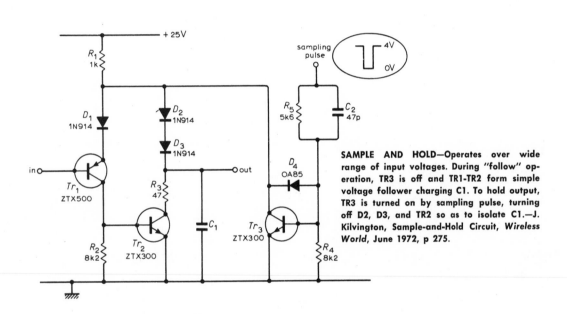

SAMPLE AND HOLD—Operates over wide range of input voltages. During "follow" operation, TR3 is off and TR1-TR2 form simple voltage follower charging C1. To hold output, TR3 is turned on by sampling pulse, turning off D2, D3, and TR2 so as to isolate C1.—J. Kilvington, Sample-and-Hold Circuit, *Wireless World*, June 1972, p 275.

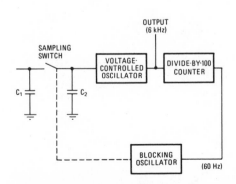

SAMPLED SINE WAVE CONTROLS VCO—Reference 60-Hz voltage applied to C1 is transferred at given time to C2 to serve as control voltage for voltage-controlled oscillator. Sampling switch shown provides lock-in in only about a dozen cycles of reference, without attendant delay of usual integrating circuit.

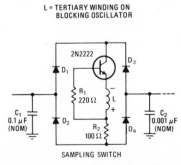

Block diagram shows how 6-kHz output of vco goes through divide-by-100 counter and blocking oscillator to provide required feedback for locking output to reference.—C. Deming, Divide-and-Sample Loop Cuts Phase-Locked VCO Slippage, *Electronics*, Dec. 20, 1971, p 54.

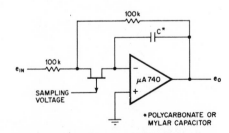

SAMPLE AND HOLD—Use of jfet opamp having low bias and offset current reduces capacitance leakage, therby extending holding time. With 0.1-μF charging capacitor, error due to bias current is only 0.1% per μs.—T. McCaffrey and R. Brandt, FET Input Reduces IC Op Amp's Bias and Offset, *Electronics*, Dec. 7, 1970, p 85–88.

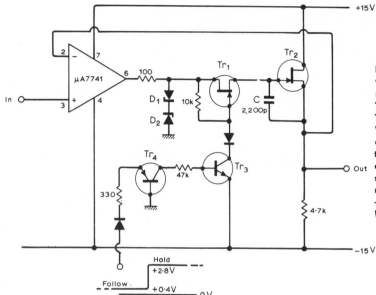

FOLLOW AND HOLD—Circuit follows input voltage until 2.8-V hold pulse is applied as shown, then holds input level of that instant. TR1 and TR2 are jfet's such as MPF102, while TR3 and TR4 are general-purpose transistors. When TR1 is on, very high open-loop gain of opamp ensures that output voltage is equal to input within input offset of 5 mV for opamp. When hold pulse turns TR1 off, source follower TR2 is isolated and output remains at voltage stored on C at that instant.—J. F. Roulston, Follow-and-Hold Circuit, *Wireless World*, Feb. 1971, p 66.

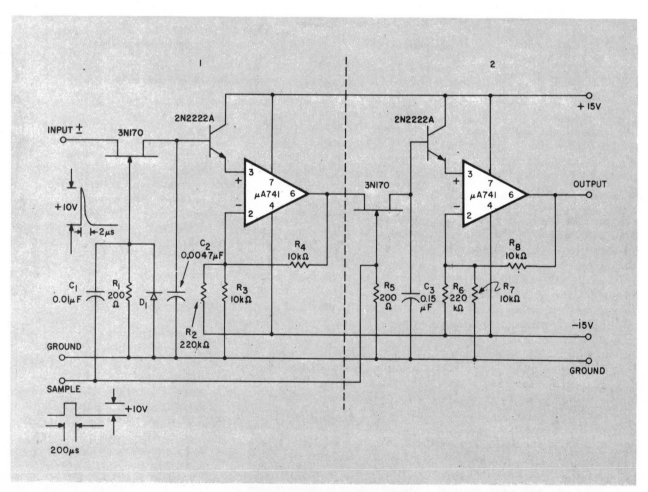

HIGH HOLD-SAMPLE RATIO—Uses two sample-and-hold circuits in series, each with ratio well below 500, to achieve net hold-sample ratio exceeding 40,000 with accuracy better than 1%. Only eight components are required for stage. Even higher ratios can be obtained by adding more stages. Input circuit samples analog signal and holds desired amplitude long enough so next stage can sample output of first.—Simple Circuit Maintains High Hold/Sample Ratio, *Electro-Technology*, Jan. 1970, p 11.

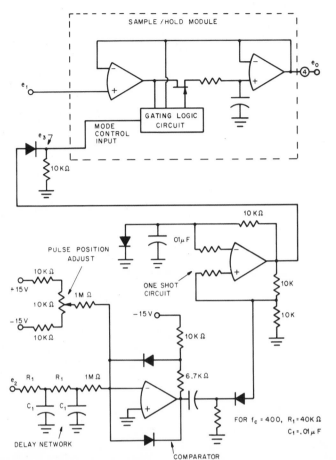

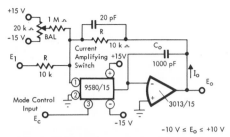

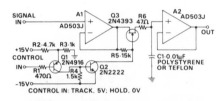

FAST SAMPLE-HOLD SWITCH—Current amplification of Burr-Brown 9580/15 switch increases sample-hold rate to 50 times that of conventional circuits, with gain accuracy of 0.01% if one of resistors R is trimmed for better match with other. Time required for circuit to slew to 20-V difference between samples is only about 1 μs, and depends chiefly on feedback capacitor Co. Circuit switches to HOLD mode in less than 400 ms.—"Electronic Switches," Burr-Brown, Tucson, AZ, Sept. 1969, p 4.

PULSE SAMPLE DEMODULATOR—With 400-Hz carrier, pulse width of about 70 μs centered on sine wave will sample voltage within 5 deg of peak. Based on sampling and holding peak values of a-c suppressed-carrier signal, so output is sequence of steps. Fundamental of output will be desired low-frequency signal, delayed in time by half a cycle of carrier because of sampling process.—J. G. Graeme, G. E. Tobey, and L. P. Huelsman, "Operational Amplifiers," McGraw-Hill Book Co., New York, 1971, p 417–418.

SAMPLE-HOLD AMPLIFIER—Utilizes low input current and high slew rate of fet-input opamps for tracking ±10 V input signal at up to 4 kHz. When control input changes from track ($+5$V) to hold (0 V), series fet switch Q3 opens and input signal voltage is retained on C1. Output opamp A2 provides high input impedance to keep C1 from discharging too rapidly.—D. H. Sheingold, "Analog-Digital Conversion Handbook," Analog Devices, Norwood, MA, 1972, III-85–III-86.

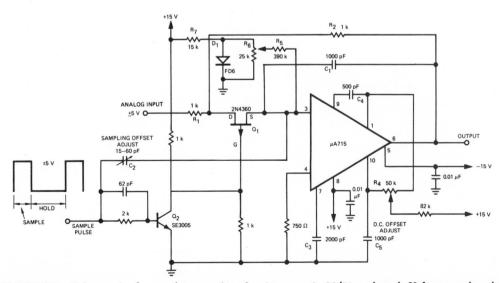

OPAMP SAMPLE-AND-HOLD—High speed of opamp used permits sampling of analog inputs and holding value constant long enough for a-c conversion. In sample mode, sampling jfet switch Q1 is turned on; circuit then functions as inverting opamp with voltage gain equal to negative of resistance ratio R2/R1. When Q1 is switched off, circuit enters hold mode and output voltage at that instant is retained across holding capacitor C1. A-c offset, independent of analog signal level, is removed by opposing signal obtained through C2 from sample pulse input to holding capacitor. With input waveform shown, settling time to 0.05% is 10 μs.—"A High Speed Sample and Hold Using the μA715," Fairchild Semiconductor, Mountain View, CA, No. 271, 1971.

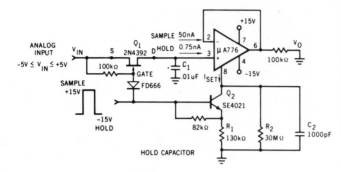

LOW-COST SAMPLE-HOLD—Uses low-cost programmable opamp which allows switching from required high slew rate during sample period to holding mode requiring only 750-pA input bias current. Although slew rate decreases proportionally with bias current, this has no effect on performance during hold period of cycle.—"The μA776, An Operational Amplifier with Programmable Gain, Bandwidth, Slew-Rate, and Power Dissipation," Fairchild Semiconductor, Mountain View, CA, No. 218, 1971, p 4–5.

OFFSET COMPENSATION—Compensating signal is bootstrapped to output of buffer amplifier of basic sample and hold, to compensate for variation in offset with input voltage and with nonlinearity of jfet used.—M. J. English, "An Improved Sample-and-Hold Circuit Using the μA740," Fairchild Semiconductor, Mountain View, CA, No. 165, 1971.

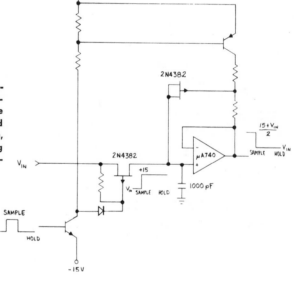

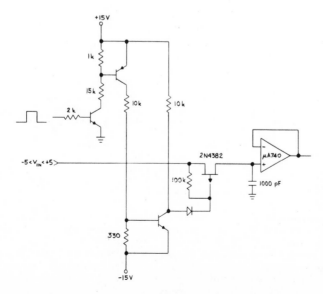

100-μS HOLD TIME—Uses 2N4382 jfet as sampling switch having maximum on resistance of only 350 ohms and maximum leakage of 1 mA. Signal range is ±5 V, for which accuracy is better than 0.1% of full-scale input and acquisition time better than 10 μs. Report covers circuit operation in detail and gives design equations.—M. J. English, "An Improved Sample-and-Hold Circuit Using the μA740," Fairchild Semiconductor, Mountain View, CA, No. 165, 1971.

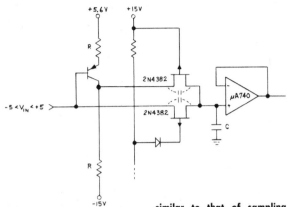

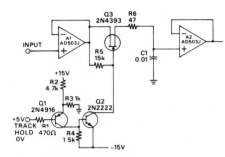

TRANSIENT COMPENSATION—Input signal is fed through unity-gain inverter and resulting inverted signal is coupled onto hold capacitor through fet that is always in off state. This coupling method provides capacitance very similar to that of sampling switch. Circuit also helps to equalize effects of stray capacitances.—M. J. English, "An Improved Sample-and-Hold Circuit Using the μA740," Fairchild Semiconductor, Mountain View, CA, No. 165, 1971.

TRACK-HOLD AMPLIFIER—Will track ±10 V input signals up to 4 kHz. When track-hold gate changes from +5 V to 0 V, fet Q3 opens and input voltage at that instant is retained on C1.—"AD503, AD506 I.C. Fet Input Operational Amplifiers," Analog Devices, Inc., Norwood, MA, Technical Bulletin, Aug. 1971.

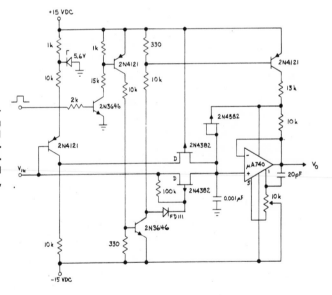

OFFSET AND TRANSIENT COMPENSATION—Provides complete compensation for sampling offset and for input transients during hold mode. Report covers circuit operation in detail.—M. J. English, "An Improved Sample-and-Hold Circuit Using the μA740," Fairchild Semiconductor, Mountain View, CA, No. 165, 1971.

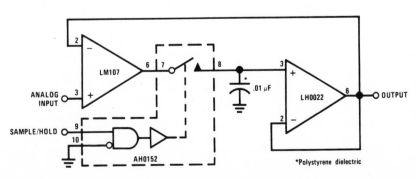

PRECISION SAMPLE-HOLD—Uses LH0022 fet opamp as buffer amplifier in long hold-time sampling. Circuit also uses conventional opamp as input buffer for analog signal.—R. K. Underwood, "New Design Techniques for FET Op Amps," National Semiconductor, Santa Clara, CA, 1972, AN-63, p 7–8.

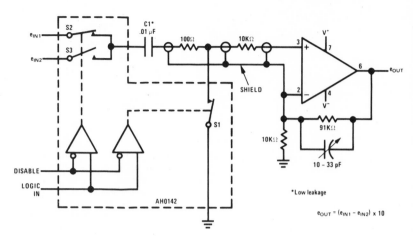

*Low leakage

$$e_{OUT} = (e_{IN1} - e_{IN2}) \times 10$$

VOLTAGE SUBTRACTOR—Used in automatic testing of linear circuits to take difference between two voltage readings occurring at different times. Initially, switches are in positions shown and logic input is in TTL state of 1, allowing C1 to charge to same voltage as signal on input 1. When logic input is changed to 0, S3 closes while S1 and S2 open, causing difference between stored value of voltage on input 1 and present voltage on input 2 to appear at noninverting input of LH0022 fet opamp. Low leakage and high input impedance of this opamp allow use of reasonable size of hold capacitor while providing gain for scaling if needed. Disable input is used to open all switches, such as for ignoring transient.—R. K. Underwood, "New Design Techniques for FET Op Amps," National Semiconductor, Santa Clara, CA, 1972, AN-63, p 10–11.

SAMPLE-HOLD WITH COMPARATOR—Clamping transistors Q1 and Q2 put Norton current-differencing opamp circuit in hold mode when driven on. When transistors are off, output voltage of opamp 1 can ramp either up or down as required to make output voltage of opamp 1 equal to d-c input voltage of opamp 3. R1 provides fixed down-ramp current which is balanced or controlled by R4 and comparator opamp 3. While stored voltage appears at output, continued comparison is made between output and input voltages by opamp 3, with switching based on comparison to keep input and output voltages equal.—T. M. Frederiksen, W. M. Howard, and R. S. Sleeth, "The LM3900—A New Current-Differencing Quad of ± Input Amplifiers," National Semiconductor, Santa Clara, CA, 1972, AN-72, p 35.

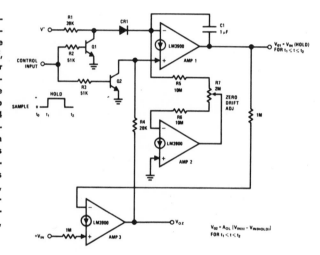

$$V_{O2} = A_{OL} [V_{IN(IN)} - V_{IN(HOLD)}]$$
FOR $t_1 < t < t_2$

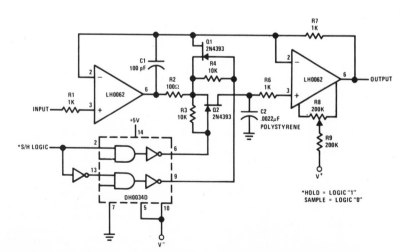

*HOLD = LOGIC "1"
SAMPLE = LOGIC "0"

10-MICROSECOND ACQUISITION TIME—Two fet opamps serve in high-speed sample-hold having 0.1% accuracy and about 25-ns aperture time. R6 limits input current during power-on and power-off transients.—B. Siegel, "Applications for a High Speed FET Op Amp," National Semiconductor, Santa Clara, CA, 1972, AN-75.

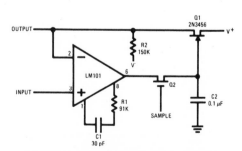

LOW-DRIFT SAMPLE-HOLD—Uses unity-gain opamp and fet source follower. Turning on sample switch Q2 closes feedback loop to make output differ from input only by offset voltage of opamp. When switch is open, charge stored on C2 holds output at last value of input voltage. Use polycarbonate-dielectric capacitor for C2. Circuit does not load input signal.—R. J. Widlar, "Operational Amplifiers," National Semiconductor, Santa Clara, CA, 1968, AN-4.

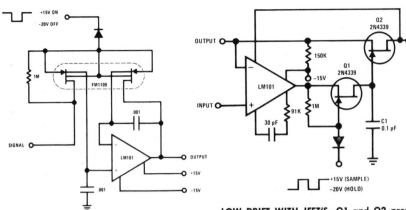

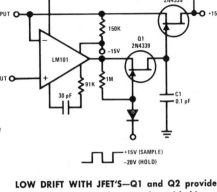

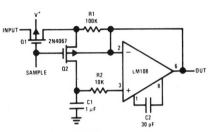

JFET SAMPLE AND HOLD—Logic voltage is applied simultaneously to both jfet's in package. Matched characteristics of jfet's greatly improve circuit performance and minimize errors.—"FET Circuit Applications," National Semiconductor, Santa Clara, CA, 1970, AN-32, p 7.

LOW DRIFT WITH JFET'S—Q1 and Q2 provide complete buffering for sample-and-hold capacitor C1. During sample, Q1 is on and provides charging path for C1. During hold, Q1 is off and the only two leakage paths for discharging C1 are less than 50 pA and less than 100 pA.—"FET Circuit Applications," National Semiconductor, Santa Clara, CA, 1970, AN-32, p 6.

PREVENTING LEAKAGE—Leakage is eliminated in fet switches of sample-and-hold opamp circuit by using fet Q1 as main sample switch, with Q2 isolating hold capacitor from leakage of Q1. When sample pulse is applied, both fet's turn on and C1 is charged to input voltage. Removing pulse shuts off fet's and output leakage of Q1 goes through R1 to output. Drop across R1 is less than 10 mV, so substrate of Q2 can be bootstrapped to output of opamp. Resulting cancellation of voltages reduces leakage of fet about two orders of magnitude.—R. J. Widlar, "IC Op Amp Beats FETs on Input Current," National Semiconductor, Santa Clara, CA, 1969, AN-29.

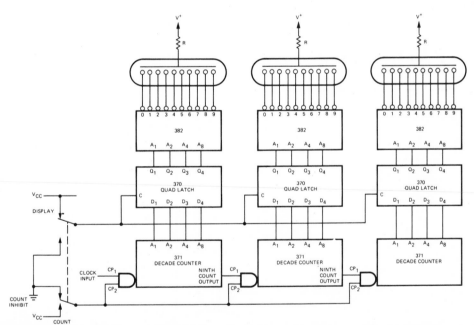

DIGITAL COUNT DISPLAY—When switch is in count position, decade counters count clock pulses and indicator tubes display count being held in quad latches. With switch in display position, counters stop, number counted is transferred to storage, and display tubes change to new number.—"HiNIL High Noise Immunity Logic," Teledyne Semiconductor, Mountain View, CA, 1972, p 65.

CHAPTER 100
Science Fair Circuits

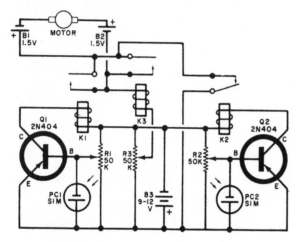

FLASHLIGHT-CONTROLLED MOTOR—Ideal science fair project. Flashlight on PC1 turns on motor by pulling in K1, which in turn energizes latching relay K3 to keep motor running after light is removed. Light on solar cell PC2 turns off motor.—B. Koval, Photocell Motor Control Demonstrator, *Popular Electronics*, June 1970, p 46—48.

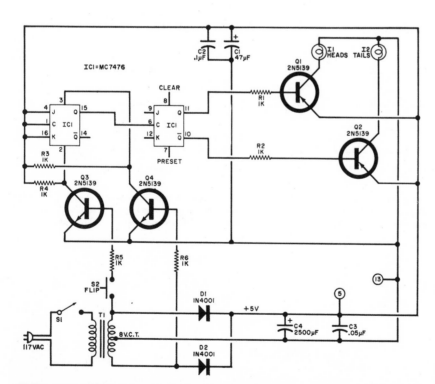

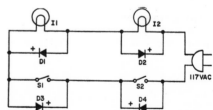

HONEST COIN-FLIPPER—When flip button is pressed, flip-flops count power-line frequency. When released, either heads or tails lamp lights to indicate final state of flip-flop. Length of time pushbutton is down and phase of power at instant it closes combine to provide truly random honest coin toss. Use 5-V 50-mA lamps.—J. Crawford, Heads'n'-Tails, *Popular Electronics*, Jan. 1972, p 35—37.

DIODE TRICKS—With 1N2070 or equivalent silicon diodes across lamps and switches in series circuit, but with diodes concealed in cabinet, switching actions are mysterious enough for any science fair exhibit. Diagram on housing should show only two lamps and two switches in series across a-c line. With switches closed, one lamp can be removed without affecting other lamp. Use ordinary 25-W lamps.—R. E. Pafenberg, Magic Lamp, *Elementary Electronics*, Sept.-Oct. 1970, p 67—68, 72 and 96—97.

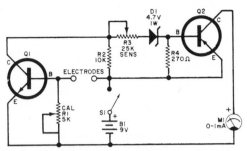

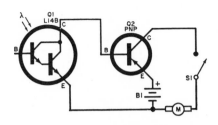

LIE DETECTOR—Designed for maximum response to changes in resistance from a preset normal value for a given person. Based on lowering of skin resistance during emotional stress of telling a lie. Consists of direct-coupled complementary amplifier Q1-Q2 using general-purpose small-signal npn and pnp transistors such as International Rectifier TR-09 and TR-05, reference diode, and milliammeter. Electrodes are attached to fingers of subject. R1 is calibration control and R3 sensitivity control adjusted for normal skin resistance of person. Electrodes can be copper pennies, taped on moistened index and middle fingers of one hand. With calm subject, adjust R1 for about 1/3 full-scale, adjust R3 until meter swings full-scale for emotion-arousing questions, then proceed with examination.—M. Dudley, Reader's Circuit, *Popular Electronics*, Oct. 1969, p 87 and 92.

FLASHLIGHT-CONTROLLED TOY—Two-cell flashlight about 5 feet from phototransistor Q1 will start motor of toy sports car, and keep it running for range of about 15 feet in average artificial light level of home. Use any pnp power transistor that will carry motor current of toy, and battery already in toy.—W. S. Gohl, Add Light Control to Battery-Powered Toys, *Popular Electronics*, May 1968, p 45—46.

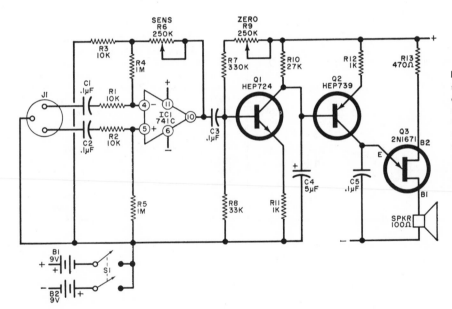

EMG-CONTROLLED WHISTLE—Electrodes on skin pick up electromyograph voltages generated by muscle activity, and change frequency of ujt audio oscillator Q3. The tenser the muscle, the higher the frequency of the tone heard from the speaker. High-gain differential opamp, Q1, and Q2 amplify emg voltage. Circuit has no known practical use. —H. Garland and R. Melen, Build the Muscle Whistler, *Popular Electronics*, Nov. 1971, p 60—61 and 103.

GALVANIC SKIN RESISTANCE—Makes skin resistance measurement used as one of tests in Keeler polygraph lie detector, but intended primarily for entertainment because only experts can ask meaningful questions. Electrodes can be 20-centavo Mexican coins or inch-square pieces of copper kept polished with sandpaper and attached to palm of each hand with bicycle-pants spring clips. Meter can be 2.5-V or 3-V voltmeter, or 1-mA ammeter in series with 2,500 or 3,000-ohm resistor. Circuit consists of measurement bridge with electrodes in one leg, IC d-c amplifier, and output indicator.—R. E. Devine, Build a Psych-Analyzer, *Popular Electronics*, Feb. 1969, p 27—32 and 116.

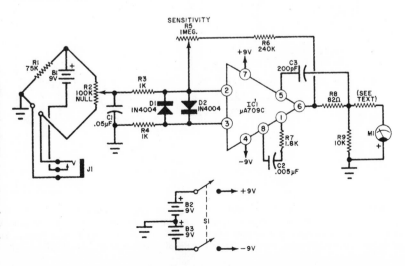

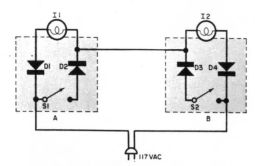

ENGINEERING PUZZLE—Two ordinary 117-V lamps and two switches are mounted on separate transparent boxes, with single power lead to each box and single wire running between boxes A and B. When S1 is on, I2 lights. When S2 is on, I1 lights. When both switches are on, both lamps light. When both are off, both lamps go out. Secret lies in diodes concealed in spaghetti tubing running from switch terminals to lamps; diodes should be rated 200 piv and 1 A or higher.—L. Vicens, The Impossible Circuit, *Popular Electronics*, Oct. 1966, p 72 and 79.

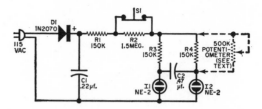

ELECTRONIC COIN-TOSSER—Neons normally flash alternately at rate too fast to be stopped at will. When S1 is pressed, R2 lowers supply voltage to point too low to ionize either of lamps, but high enough to keep lit whichever one is ionized at that instant. Bulbs are labeled YES and NO, or HEADS and TAILS.—R. C. Apperson, Jr., The Procrastinator's Companion, *Popular Electronics*, April 1964, p 88.

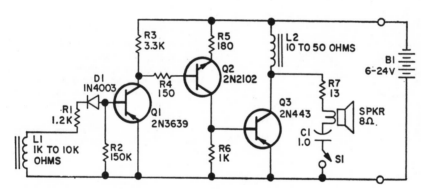

SCIENCE-FAIR MOTOR—Circuit generates pulses which energize electromagnets L1 and L2 (coils from relays) in sequence. Rotor of motor is plain disc on which four permanent magnets are mounted. Rotor is given a push to start rotation, causing one of permanent magnets to pass L1 and induce in it a voltage that turns on Q1, Q2, and Q3 in sequence to energize L2 and attract next permanent magnet. One permanent magnet is always approaching L2 while another is passing L1 and creating new pulse that keeps motor in motion. Speaker merely produces attention-getting clicks for exhibit. Article covers construction.—C. D. Rakes, The BOTDC Motor, *Science and Electronics*, Jan. 1970, p 31–35 and 102.

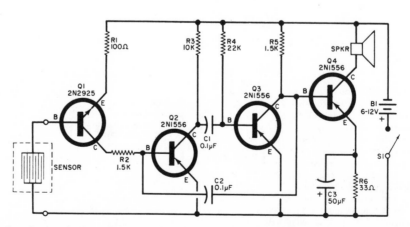

LIE DETECTOR—Based on fact that conductivity changes under emotional stress of lying. Sensor is pattern of separated parallel wires or foil strips attached to palm of subject tightly enough to make good contact without impairing circulation. After output tone has stabilized, proceed with questioning. Emotional response of lie should give easily noticed change in pitch of audio tone generated by mvbr Q2-Q3.—G. J. Beck, Reader's Circuit, *Popular Electronics*, June 1967, p 76–77.

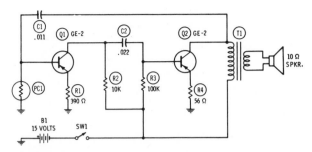

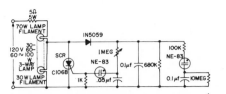

AUDIBLE LIGHT METER—Pitch of tone from speaker increases with intensity of tone from speaker. Intended primarily as science-fair exhibit, although theoretically it could serve as photographic exposure meter for person able to recognize tones. Output transformer has 2K primary and 10-ohm secondary. Photocell can be International Rectifier B2M.—R. Brown and T. Kneitel, "101 Easy Audio Projects," Howard W. Sams, Indianapolis, IN, 1971, p 153–155.

FLICKERING FLAME EFFECT—Simple circuit synthesizes flame effect by varying brightness of sections of three-way lamp. One neon relaxation oscillator triggers scr every other half-cycle at phase angle that is modulated by other neon oscillator operating at much slower rate.—R. W. Fox, "Solid State Incandescent Lighting Control," General Electric, Semiconductor Products Dept., Auburn, NY, No. 200.53, 1970, p 12.

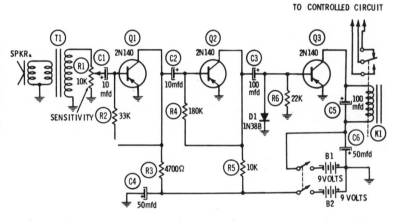

VOICE-OPERATED RELAY—Any p-m speaker serves as mike for energizing 300-ohm relay K1 by voice for controlling model train, lamp, or any other equipment within ratings of relay contacts. T1 is output transformer with 3.2-ohm secondary going to speaker and 500-ohm primary. Increase C3 to make relay stay closed for longer period after you stop speaking.—R. Brown and T. Kneitel, "101 Easy Audio Projects," Howard W. Sams, Indianapolis, IN, 1971, p 68–70.

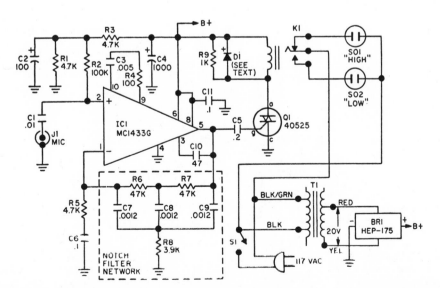

SOUND-DIMMED LAMP—Snap of finger or whistle up to 30 feet away, at frequency around 5 kHz, is amplified about 100 times by opamp in notch filter network for triggering triac Q1 and pulling in 24-V relay K1 for switching from 100-W lamp to very low-wattage lamp at appropriate moment for creating more romantic atmosphere. May also be used as noise-triggered burglar alarm for office or home. D1 is 100-piv silicon diode for suppressing inductive kickback voltage across relay. R9 provides holding current to keep triac energized so lamp stays dim or alarm stays on.—C. Jameson, Lover's Lamp, *Science and Electronics*, Feb.-March 1970, p 57–61.

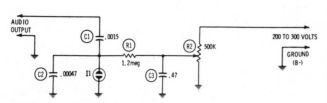

SUNLIGHT SCREAMER—Changes in amount of sunlight change tone of shrill screaming sound produced by relaxation oscillator using NE-2H neon bulb, because neon is highly sensitive to light falling on it. Sunlight streaming through branches of trees produces interesting sound effect on windy day. Use shielding on leads. Adjust triggering with R2.—R. Brown and T. Kneitel, "101 Easy Audio Projects," Howard W. Sams, Indianapolis, IN, 1971, p 149–150.

DANCING LIGHTS—Produces slow changes in brightness of lamp plugged into outlet next to R2. Operates directly from a-c line, using self-modulating half-wave phase control in which d-c feedback modifies both reference and pedestal levels in trigger circuit of T1. Difference in time constants for these levels makes triggering angle sweep slowly back and forth between 10 and 170 deg to achieve unique lighting control effect. R2 limits current surge when lamp burns out.—E. K. Howell and R. W. Fox, "New Powerpac Thyristors and How to Use Them," General Electric, Semiconductor Products Dept., Auburn, NY, No. 671.18, 1969, p 6.

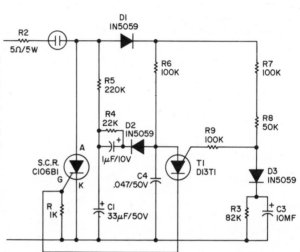

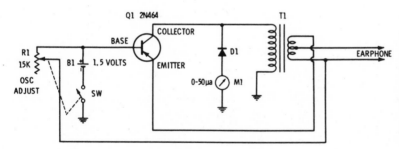

SCIENCE-FAIR BEEPER—Operating from single flashlight cell, produces beep tone in 1K to 2K magnetic earphone and deflection of meter automatically about once per second, as attention-getter. Has no practical use. T1 is transistor output transformer, such as Argonne AR-103, and D1 is 1N38B. Battery will last several months.—R. M. Brown and T. Kneitel, "49 Easy Transistor Projects," Howard W. Sams, Indianapolis, IN, 1972, p 17–18.

LIE DETECTOR—Uses bicycle trouser clips as probes to make firm contact with moist palms of subject, while opamp circuit amplifies changes in body resistance for driving meter. Scale can be calibrated over lie-truth range by asking questions for which answers are known; with 0-50 μA meter, range from 0 to 25 μA will indicate increased sweating and lowered body resistance associated with lie. —C. R. Lewart, Electronic "Lie Detector" Tells It Like It Is, *Popular Science*, May 1973, p 118–119.

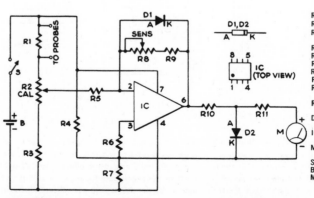

R1—47,000-ohm, ½w resistor
R3—68,000-ohm, ½w resistor
R5,6—3,300-ohm, ½w resistor
R4—10,000-ohm, ½w resistor
R7—20,000-ohm, ½w resistor
R9—15,000-ohm, ½w resistor
R10—22,000-ohm, ½w resistor
R11—5,600-ohm, ½w resistor
R2—50,000-ohm potentiometer
R8—100,000-ohm potentiometer
D1,2—silicon diode (1N914 or Motorola HEP 154)
IC—operational amplifier (Motorola HEP C6052P)
M—0-50 microammeter (Lafayette 99P50429)
S—SPDT mini toggle switch
B—9-volt battery
Misc.—bicycle clips, cabinet (Hazelton Scientific)

CHAPTER 101
Servo Circuits

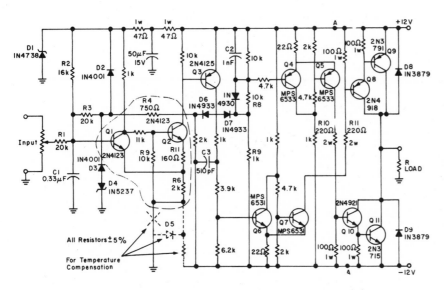

45-W PWM AMPLIFIER—Use of switching elements in direct-coupled servo amplifier gives higher efficiency than can be obtained with conventional continuously conducting class-B control amplifiers. Bidirectional output is changed simply by changing duty cycle. Operation is controlled by switching frequency of Schmitt-trigger input stage Q1-Q2; frequency is about 3 kHz. Will drive 5-A load. For 100-W output, use 25-V supplies, change R10 and R11, and add zeners as shown in article.—G. V. Fay and N. Freyling, Switching Circuit Improves Servo Efficiency, *The Electronic Engineer*, Dec. 1968, p 83–85.

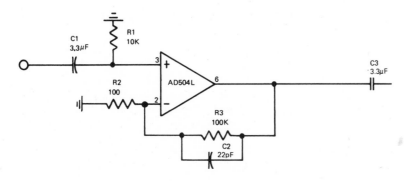

HIGH-GAIN A-C OPAMP—Use of low-offset opamp in circuit having fixed gain of 1,000 improves a-c preamplification. Uses 15-V supply voltages. Overdrive capability makes circuits suited for servos and other applications where amplifier may be overloaded. Bandwidth is 80 kHz.—"Applying the AD504 Precision IC Operational Amplifier," Analog Devices, Inc., Norwood, MA, April 1972.

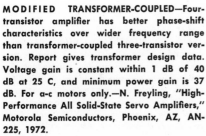

MODIFIED TRANSFORMER-COUPLED—Four-transistor amplifier has better phase-shift characteristics over wider frequency range than transformer-coupled three-transistor version. Report gives transformer design data. Voltage gain is constant within 1 dB of 40 dB at 25 C, and minimum power gain is 37 dB. For a-c motors only.—N. Freyling, "High-Performance All Solid-State Servo Amplifiers," Motorola Semiconductors, Phoenix, AZ, AN-225, 1972.

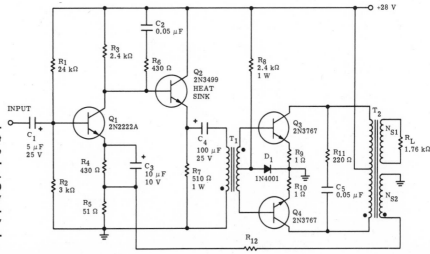

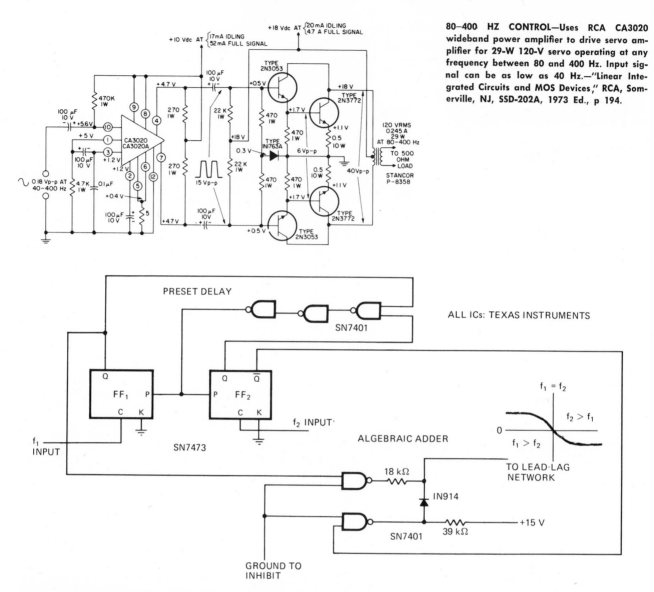

80–400 HZ CONTROL—Uses RCA CA3020 wideband power amplifier to drive servo amplifier for 29-W 120-V servo operating at any frequency between 80 and 400 Hz. Input signal can be as low as 40 Hz.—"Linear Integrated Circuits and MOS Devices," RCA, Somerville, NJ, SSD-202A, 1973 Ed., p 194.

ALL ICs: TEXAS INSTRUMENTS

PHASE COMPARATOR—Produces output proportional to difference between frequencies of pulses at inputs f1 and f2. Output current is positive and proportional to phase error when f1 is greater than f2, negative and proportional to phase error when f2 is greater, and zero when f1 equals f2.—F. E. Adams, *Phase Comparator for Servo Loops, Electronics*, Dec. 18, 1972, p 108.

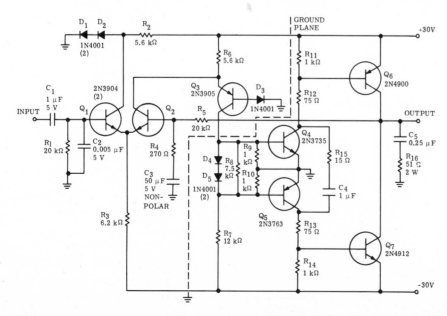

10-W COMPLEMENTARY AMPLIFIER—Chief advantages are elimination of transformers and use of direct coupling throughout, for driving both a-c and d-c loads. Report gives design procedure and describes operation in detail.—N. Freyling, "High-Performance All Solid-State Servo Amplifiers." Motorola Semiconductors, Phoenix, AZ, AN-225, 1972.

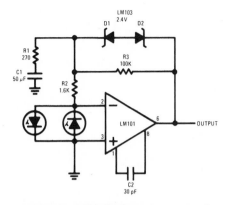

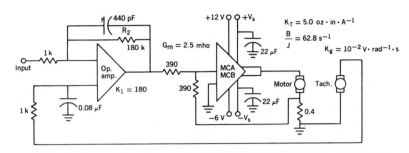

SATURATING PREAMP—Opamp operates in linear mode until output voltage reaches about 3 V with 30-μA output current from solar-cell sensors. Breakdown diodes in feedback loop then begin to conduct, drastically reducing gain. Rate signal will still be developed, however, so amplifier will not actually saturate until error current reaches 6 mA, corresponding to linear opamp with ±600 V output swing. Gives servo amplifier operation over extremely wide dynamic range.—R. J. Widlar, "Operational Amplifiers," National Semiconductor, Santa Clara, CA, 1968, AN-4.

VELOCITY SERVO LOOP—Closed-loop velocity system uses tachometer feedback. Power amplifier IC is selected from TRW MCA or MCB series of hybrid class-D power amplifiers using switching techniques with pwm operation. Choice of IC amplifier depends on supply voltage, required load current, and switching frequency. Switching is at either 20 or 40 kHz.—Semiconductor Power-Switching Amplifiers Are Designed to Actuate Motors and Solenoids. *IEEE Spectrum*, Jan. 1971, p 76.

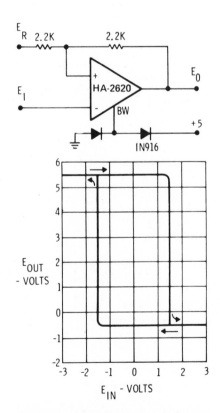

HYSTERESIS AMPLIFIER—Meets requirements of Schmitt triggers, analog simulators, differential comparators, and servomechanisms by having input impedance of 100 meg. Output voltage remains constant within several mV unless input is near threshold, as shown in graph. Report gives design equations.—G. G. Miler, "A High Impedance Hysteresis Circuit," Harris Semiconductor, Melbourne, FL, No. 505/A, 1970.

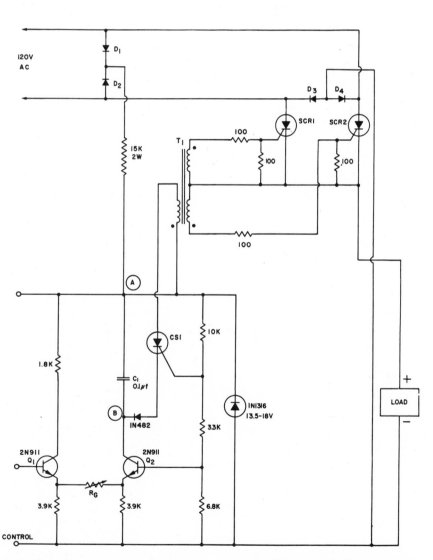

PROPORTIONING SCR CONTROL—Permits varying power continuously from essentially zero to maximum of 400 W for servo motor drive, with d-c input voltage of 4.9 to 5.1 V providing full range of control. Output voltage is 0 to 100 V d-c, at 0 to 4 A. All rectifiers and diodes are 200-V silicon. T1 is Aladdin 94-1400, scr's are 3B3200, and CSI is SSPI AA100. Report describes operation in detail.—"High Gain Proportioning SCR Power Controller," Unitrode Corp., Watertown, MA, New Design Idea, No. 16.

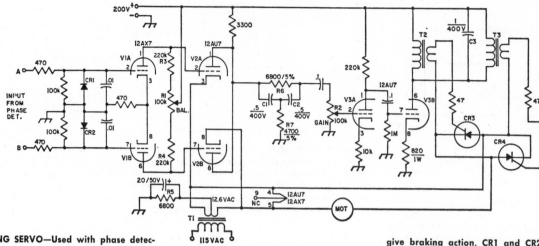

A-C TUNING SERVO—Used with phase detector in final r-f amplifier of transmitter to drive motor-driven tank capacitor that brings tank circuit back into resonance in response to error signal. Scr's in output permit use of low-cost d-c drive motor. Bridged-T network

C1-C2-R6-R7 suppresses hunting at zero error signal, as also does firing of both scr's during one a-c cycle of motor as error signal goes through crossover point at resonance, to

give braking action. CR1 and CR2 are 1N34 or SK3017A. CR3 and CR4 are International Rectifier SCR-01-C. T2 and T3 are audio output transformers, 8,000 ohms to 4 ohms. Motor is 12-V d-c.—F. Walsmith, Automatic Amplifier Tuning, *QST*, Sept. 1970, p 32–36.

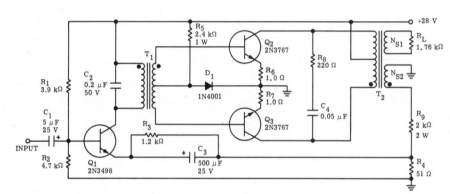

7.5-W TRANSFORMER-COUPLED—Simple circuit uses only three transistors, but transformers limit use to a-c motor loads. Stable voltage gain is 100. Feedback for amplifier is derived from separate winding NS2 rather than across load or motor winding. Report gives transformer design data.—N. Freyling, "High-Performance All Solid-State Servo Amplifiers," Motorola Semiconductors, Phoenix, AZ, AN-225, 1972.

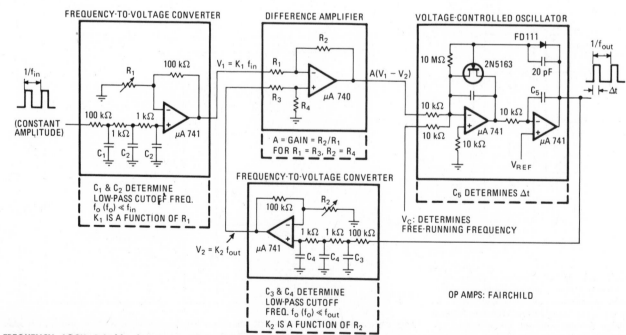

FREQUENCY LOCK—Suitable for servo systems in which two different frequencies are to be locked together. D-c voltages V1 and V2 are proportional to frequencies to be locked. Values shown cover range of 30 to

10,000 Hz for either frequency. Varying constants K1 and K2 multiplies or divides frequencies as required to achieve locking in single operation. Can also be used in phase-

locked loops and in tv receivers.—M. Hamaoui, Analog Multiplier/Divider Simplifies Frequency Locking, *Electronics*, July 5, 1973, p 99–100.

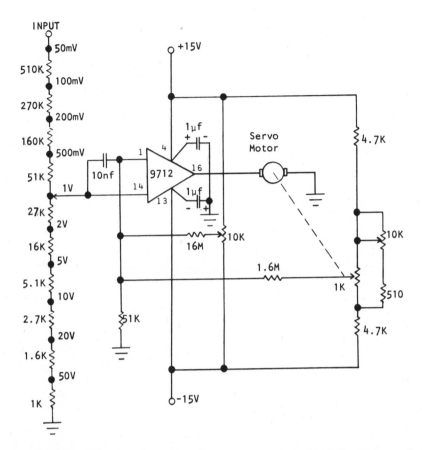

X-Y POSITION SERVO—Drives 10-V 0.2-A or similar servo motor in X-Y recorder. Uses wideband fet-input opamp permitting use of 1-meg input attenuator. Requires highly regulated power supplies.—"The 9712 as a Servo Motor Amplifier," Optical Electronics, Tucson, AZ, Application Tip 10206, 1971.

10-W A-C DRIVE—Will handle sizes 11, 15, or 18 a-c servo motors in 440-Hz closed-loop carrier system, providing gain of about 5.8 and clipping level of about 29-V rms. Amplifier is d-c coupled and will operate down to 40 Hz.—R. J. Wallace and J. M. Clarke, Power Amplifier for A.C. Servomotors, *Wireless World*, April 1971, p 201–202.

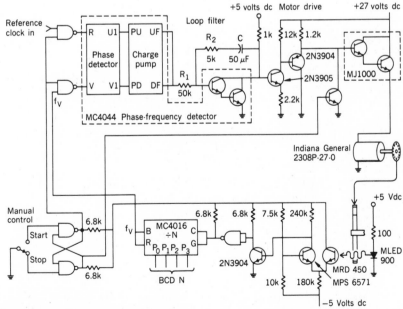

DIGITAL SERVO CONTROL—Phase-locked loop design using inexpensive digital IC's will regulate servo speeds to within 0.002% by comparison with reference clock signal. Optical encoder on shaft of motor produces 36 pulses per revolution for feedback to phase comparator which also receives reference frequency from clock. Frequency difference is converted to error signal for correcting servo speed. Article covers theory and design equations.—A. W. Moore, Phase-Locked Loops for Motor-Speed Control, *IEEE Spectrum*, April 1973, p 61–67.

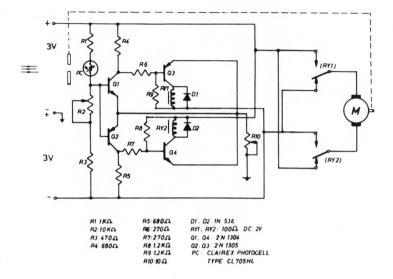

R1: 1KΩ	R5: 680Ω	D1, D2: 1N 536
R2: 10KΩ	R6: 270Ω	RY1, RY2: 100Ω, DC. 2V
R3: 470Ω	R7: 270Ω	Q1, Q4: 2N 1304
R4: 680Ω	R8: 1.2KΩ	Q2, Q3: 2N 1305
	R9: 1.2KΩ	PC: CLAIREX PHOTOCELL
	R10: 10Ω	TYPE CL 705HL

PHOTOELECTRIC SERVO—Motor which controls aperture drives illumination on photocell and cell resistance to value preset by R2. When level is correct, all four transistors are cut off and power drain is limited to current drawn by two high-resistance dividers. Can be operated off dry cells. R10 controls deadband.—"Application Notes," Clairex Electronics, Mount Vernon, NY, Data Sheet 101.1TD, 1970.

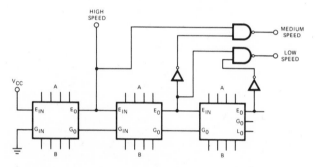

OVERSHOOT CONTROL—Three Teledyne 343 4-bit digital comparators in cascade provide speed information to servo positioning system. Most significant digit of control word is applied to comparator at left. If device being controlled is long way from new position, only high-speed output will be low. As new position is approached, this output goes high and medium-speed output goes low. When very close, only low-speed output goes low, thereby limiting overshoot. Simplified version is shown; in practical circuit, provision must be made for possibility that controlled device may be on other side of desired position.—"HiNIL High Noise Immunity Logic," Teledyne Semiconductor, Mountain View, CA, 1972, p 37.

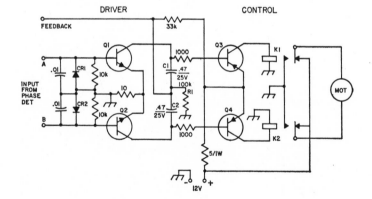

D-C TUNING SERVO—Used with phase detector in r-f amplifier to drive motor-driven tank capacitor that brings tank circuit back into resonance in response to error signal. Diodes are 1N34 or 1N38. Q1 and Q2 are ECG123 or equivalent; Q3 and Q4 are ECG104 or equivalent. Both relays are 100-ohm 12-V, and motor is 12-V reversible d-c. Feedback terminal goes to center tap of link circuit coupled to amplifier plate tank.—F. Walsmith, Automatic Amplifier Tuning, QST, Sept. 1970, p 32–36.

CHAPTER 102
Signal Generator Circuits

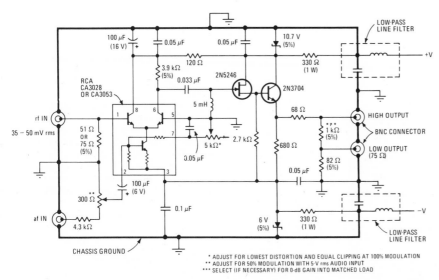

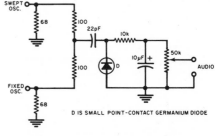

AUDIO BEAT DETECTOR—Used in place of swept audio oscillator. Swept r-f signal and fixed r-f signal are mixed in simple voltage-adding network feeding diode that provides nonlinear response required to bring out swept audio. Designed for low-impedance inputs. For high-impedance inputs, remove grounded 68-ohm resistors and change each 100-ohm resistor to 12K.—J. Ashe, Sine-Sweep Generator for R.F./I.F. Testing, *Electronics World*, Jan. 1971, p 72–75.

HI-FI TUNER TESTER—Amplitude modulator supplies a-m r-f signals at distortion level below 0.15%, as required for testing frequency response and distortion characteristics of a-m tuners in modern hi-fi equipment. Used in conjunction with standard r-f and a-f signal sources. Modulator gain can be set at 0 dB into matched load, so absolute value of output level can be read out from step attenuator and r-f generator settings. Requires only one 30-V power supply.—M. J. Salvati, Low-Distortion Modulator Tests Hi-Fi A-M Tuners, *Electronics*, Feb. 15, 1973, p 104–105.

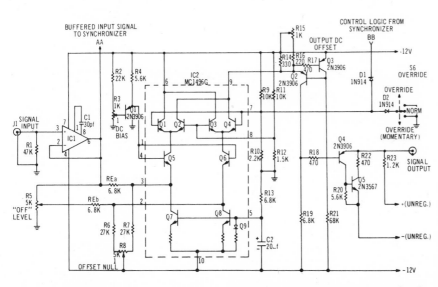

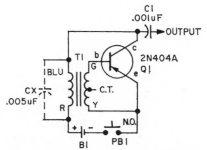

SIGNAL INJECTOR—Simultaneously generates a-f and r-f signals that can be injected into each stage of defective radio or amplifier, working back from speaker toward input or antenna, to locate stage that is not passing signal. Circuit is 1-kHz blocking oscillator with output waveform so distorted that harmonics go up to 1,500 kHz at top of a-m broadcast band. Article covers construction and use. T1 is subminiature transistor transformer having 10K primary and 2K C-T secondary. B1 is Mallory RM625 mercury cell. If it won't beep, reverse T1 secondary leads or add CX.—F. Markette, The Beeper, *Elementary Electronics*, July-Aug. 1971, p 45–46 and 98.

BALANCED-MODULATOR SWITCH—Used in tone burst generator to control signal from synchronizer. Article describes operation in detail and gives all circuits, including that for power supply.—W. G. Jung, IC Tone Burst Generator, *Audio*, Dec. 1971, p 30, 32, 34, 36, 38, and 40.

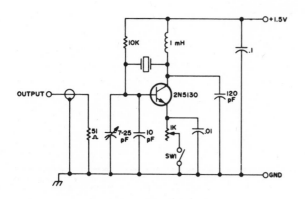

1.8–450 MHZ FOR F-M TUNEUP—Output is variable from about 80 nV to 50 mV. Almost any crystal from 1.8 to 12 MHz will oscillate in circuit and give rich harmonics up to 450 MHz, for tuning up any receiver from 160 to ¾ meters. Output impedance is 51 ohms. Maximum output level depends on lead lengths of 51-ohm resistor, and is ample with ½-inch leads. If transmitter is accidentally connected to output, only 51-ohm resistor is damaged.—H. S. White, Low Cost Signal Source, 73, April 1971, p 124–125.

455-KHZ MODULATED I-F GENERATOR—Uses high-Q 455-kHz ceramic filters in feedback and output of National LM170 IC amplifier to give a-m i-f signal with regulated output for alignment purposes.—R. A. Hirschfeld, Monolithic Amplifier Has AGC and Squelch, The Electronic Engineer, Aug. 1968, p 60–66.

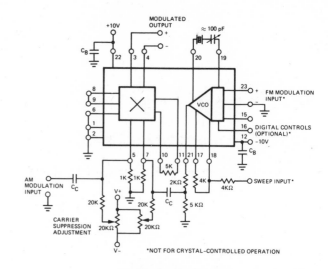

A-M/F-M/SWEEP—Multiplier and vco sections of Exar XR-S200 IC are interconnected as shown to form general-purpose r-f signal generator covering 0.1 Hz to 30 MHz and having a-m, f-m, sweep, and crystal-controlled a-m capability. Provides either suppressed-carrier or double-sideband a-m. Typical carrier suppression is over 40 dB up to 10 MHz. Digital control terminals of vco are used for fsk.—"XR-S200 Multi-Function Integrated Circuit," Exar Integrated Systems Inc., Sunnyvale, CA, June 1972, p 7.

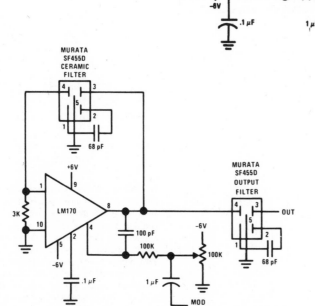

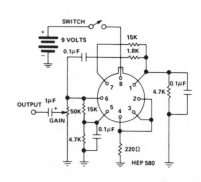

MODULATED 455-KHZ—Simple circuit serves as regulated-output a-m i-f alignment generator. If agc threshold voltage, which determines stabilized output, is varied at audio rate, output will have audio modulation, such as at 100 Hz.—R. A. Hirschfeld, "AGC/Squelch Amplifier," National Semiconductor, Santa Clara, CA, 1969, AN-11.

1-KHZ FIXED—For general audio testing. Provides sine-wave output adjustable from 0-V to 2-V p-p. Uses Motorola HEP580 dual 2-input gate IC. Operates from 9-V transistor radio battery.—"Tips on Using IC's," Motorola Semiconductors, Phoenix, AZ, HMA-32, 1972.

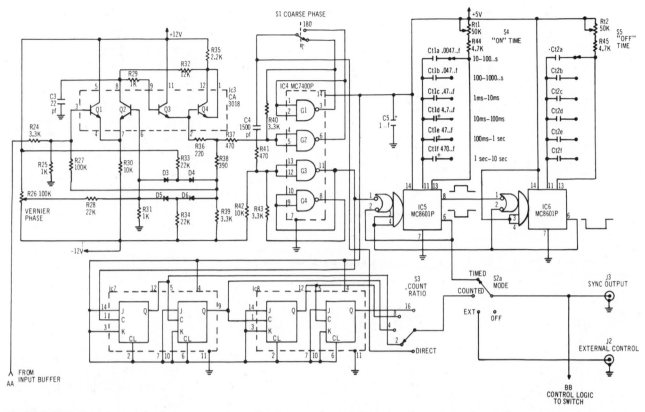

SYNCHRONIZER—Used in tone burst generator. Article describes operation in detail and gives all other circuits, including that for power supply.—W. G. Jung, IC Tone Burst Generator, *Audio*, Dec. 1971, p 30, 32, 34, 36, 38, and 40.

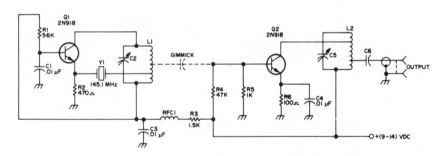

144-MHZ SOURCE FOR TESTS—Developed for checking 2-meter amplifier stages of transmitters and transceivers. Uses overtone crystal. Output is several mW.—C. Sondgeroth, Experimental Solid State VHF Amplifier, 73, July 1972, p 15–17.

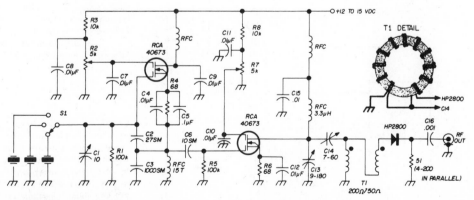

WEAK-SIGNAL SOURCE—Crystal-controlled signal generator with adjustable output provides required frequency stability for receiver alignment, frequency standard, and neutralization of transmitter or receiver r-f amplifiers. Suggested crystals are 1, 3.5, and 8 MHz. Use of dual-gate mosfets and Hewlett Packard 2800 hot-carrier diode multiplier gives usable harmonic energy (output level of −82.3 dBm) on 162nd harmonic of 1,296 MHz with 8-MHz crystal. T1 is 7 turns No. 28 bifilar-wound on Ferroxcube 3E2A toroid form. R-f chokes are also on toroids, with 10 turns of No. 24 unless otherwise specified.—B. Clark, A Stable Variable-Output Weak-Signal Source, *Ham Radio*, Sept. 1971, p 36–38.

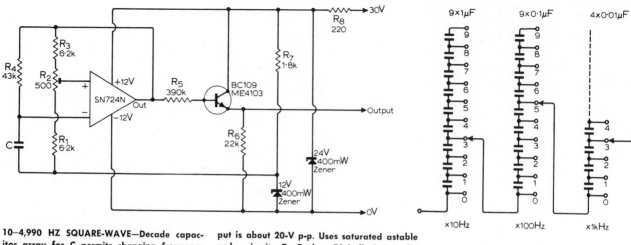

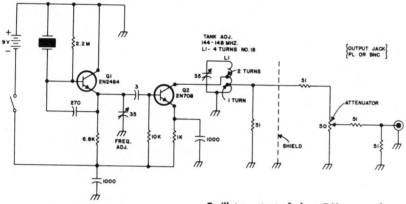

10–4,990 HZ SQUARE-WAVE—Decade capacitor array for C permits changing frequency in increments of 10, 100, and 1,000 Hz. Output is about 20-V p-p. Uses saturated astable mvbr circuit.—D. Taylor, Digitally-Set Audio Oscillator, *Wireless World*, Feb. 1970, p 73.

Set to 3·53kHz

F-M RECEIVER ALIGNER—Uses 6-MHz crystal in oscillator Q1, which is loosely coupled to class C multiplier Q2 for multiplying 24 times. Oscillator output of about 7-V p-p can be attenuated down to about 30 μV.—M. Oakes, FM Receiver Tweeker, *73*, Nov. 1969, p 80–81.

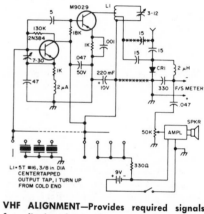

VHF ALIGNMENT—Provides required signals for aligning f-m or a-m transmitter to receiver frequency or to any preset crystal frequency simply by tuning transmitter oscillator for zero beat in speaker. Uses crystal oscillator and multiplier as peaking generator, followed by diode mixer-detector and any IC audio amplifier having about 1-W output for driving small speaker. Generator provides small signal for receiver alignment, while feeding enough signal to diode to heterodyne with transmitter circuit. Diode type is not critical.—E. Goldsby, FM-AM Transmitter-Receiver Aligner, *73*, May 1970, p 34.

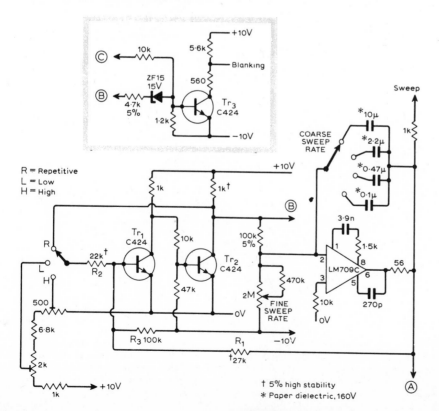

SWEPT A-F OSCILLATOR—Used in producing amplitude-frequency response on cro of a-f tuned circuits, equalizers, filters, and selective amplifiers. Uses bistable mvbr and integrator to give sweep times as low as 20 s. Frequency range is 100 Hz to 10 kHz, with output amplitude adjustable up to 3-V rms in six ranges with other circuits given in article. TR1 and TR2 form bistable mvbr. Positive blanking pulse of 10 V is available at terminal B during flyback.—R. J. Ward, Sweep-Frequency Audio Oscillator, *Wireless World*, Sept. 1971, p 412–417.

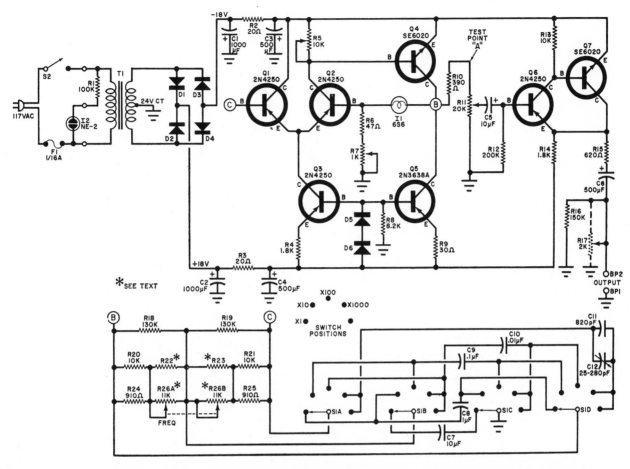

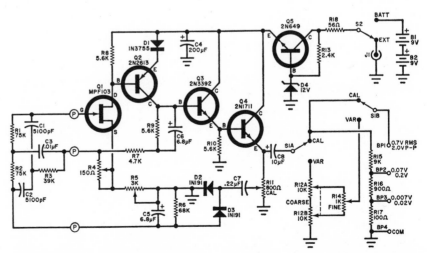

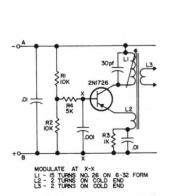

5–60,000 HZ SINE-WAVE—Low-distortion signal generator goes well beyond limits of audio and ultrasonic ranges. Residual distortion is less than 0.02% across entire range. Q1-Q2 form differential amplifier using Q3 as constant-current source. Q4 is output driver having Q5 as constant-current load. Does not require regulated supply. Frequency-determining network (below) connects to points B and C of main circuit. Article covers construction and calibration. Has four decade ranges. R26 must be calibrated with ohmmeter.—J. Bongiorno, Sine Wave Generator, *Popular Electronics*, Oct. 1969, p 55–61.

28 MHZ—Developed for use as signal generator for aligning 28-MHz i-f amplifier. Either A or B, but not both, can be connected to metal housing or panel for use as ground. L1 has powdered iron core at cold end.—B. Hoisington, The 1215 Transistor Superhet, *73*, Jan. 1966, p 90–94.

AUDIO SIGNAL GENERATOR—Stability is within 0.5% at 400 Hz. Uses twin-T feedback oscillator having both positive and negative feedback for stability and waveform purity. Article covers construction and calibration. Audio oscillator Q1-Q2 feeds output current amplifier through buffer current amplifier Q3.

Q5 is series voltage regulator for battery supply. Portion of output is rectified in voltage-doubler D2-D3 and applied as reverse bias to gate of Q1 to control operating level at calibrated setting.—T. Gilbert, Precision Audio Voltage Standard, *Popular Electronics*, Jan. 1969, p 71–76.

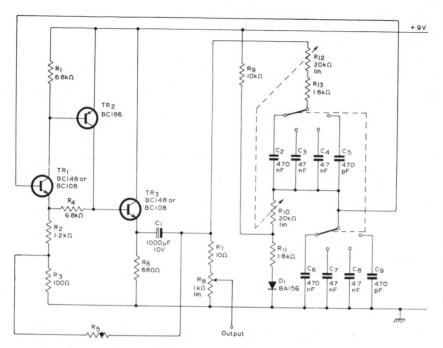

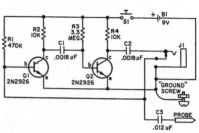

SIGNAL-GENERATOR PROBE—Battery-operated two-stage a-f oscillator is sufficiently rich in harmonics to inject r-f signals as well, for troubleshooting radios. Built into plastic pill box, with probe nail going through bottom. Switch and jack are mounted on cap. Can be used as signal tracer when crystal earphone is plugged into jack and radio is receiving signal from broadcast station or signal generator. Circuit then acts as amplifier because jack disconnects oscillator feedback capacitor C2.—R. Graf and G. Whalen, Sig-Prob, *Elementary Electronics*, Jan.-Feb. 1971, p 60–62 and 102.

A-F WIEN BRIDGE—Provides output of 1-V rms in four ranges: 15–200 Hz; 150–2,000 Hz; 1.5–20 kHz; 15–200 kHz. Uses thermistor R5 (680-ohm STC type R53) as amplitude control device for making output essentially independent of small changes in supply voltage and ambient temperature. Will operate with supplies of 6 to 10 V.—"Transistor Audio and Radio Circuits," Mullard, London, 1972, TP1319, 2nd Ed., p 253–254.

A-F AND R-F INJECTOR—Can be built in cigar-size housing with probe tip at one end and pushbutton switch at other end. Transistor types are not critical; with npn silicon and germanium, use polarity as shown for penlight cell. With pnp, reverse battery polarity. Circuit is 1,000-Hz astable square-wave mvbr having rich harmonics up to about 200 Hz.—W. F. Splichal, Jr., Cigar Tube GE Audio-RF Signal Generator, 73, Sept. 1972, p 27–28.

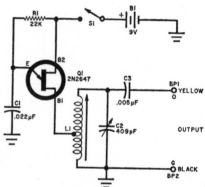

R-F SIGNAL GENERATOR—Provides modulated signal that can be tuned to any spot in a-m broadcast band, for aligning r-f circuit of superhet. Audio modulation is 750 Hz, determined by R1 and C1. L1 is high-Q a-m loopstick with adjustable slug.—F. H. Tooker, Beginner's Signal Generator, *Popular Electronics*, March 1970, p 47–50.

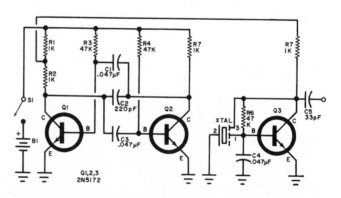

A-F AND R-F SIGNAL INJECTOR—Battery-powered 1,000-Hz mvbr generates square waves that can be varied in frequency over a-f and r-f ranges, with sufficient amplitude capability for testing or driving speaker. Ideal for testing a-m broadcast receivers.—D. Lancaster, Build a Signal Injector, *Popular Electronics*, June 1970, p 43–45.

BROADCAST RADIO ALIGNMENT—Low-cost signal source operating from 9-V transistor battery generates 455-kHz output modulated at 500 Hz, 500-Hz audio tone, and modulated signals at 910-kHz and 1,365-kHz harmonics, for aligning a-m receivers and for troubleshooting. Uses Clevite Corp. TO-01A crystal transfilter. A stable mvbr Q1-Q2 generates 500-Hz audio tone.—M. S. Robbins, Micro'lign Generator, *Popular Electronics*, April 1970, p 48–51:

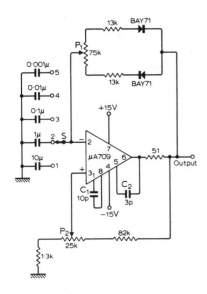

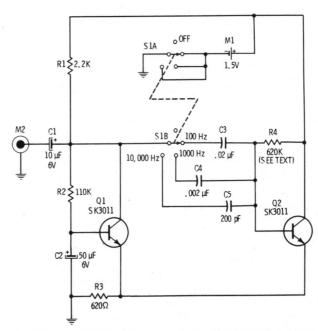

FIVE RANGES—Switch S in opamp oscillator changes capacitors to give ranges of 2–20 Hz, 20–200 Hz, 200–2,000 Hz, 2–20 kHz, and above 20 kHz. P2 is coarse frequency control and P1 is fine frequency control which operates by varying hysteresis cycle.—Square-Wave Generator, *Wireless World*, Jan. 1970, p 16.

AMPLIFIER LINEARITY CHECKER—Provides choice of three audio frequencies for checking linearity of a-f amplifiers. Output from jack M2 is fed into amplifier, and output of amplifier is monitored with cro. Use 1-meg pot initially for R4 and adjust for optimum square-wave output, then replace with corresponding fixed resistor. Amplifier should have resistance load corresponding to speaker impedance.—R. M. Brown and T. Kneitel, "101 Easy Test Instrument Projects," Howard W. Sams, Indianapolis, IN, 1968, p 88–90.

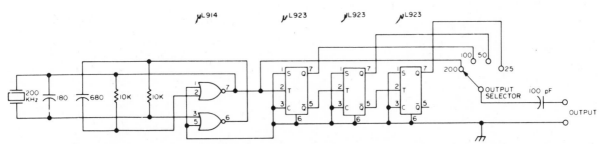

BAND-LIMIT MARKER GENERATOR—Three IC's serve as frequency dividers for μL914 200-kHz crystal-controlled mvbr, to provide calibration signals at 200, 100, 50, and 25 kHz intervals usable up through 6-meter band. Rotary switch selects desired output, for indicating edge of restricted ham band segment.—D. A. Poole, An IC Marker Generator, *73*, Oct. 1970, p 75–76.

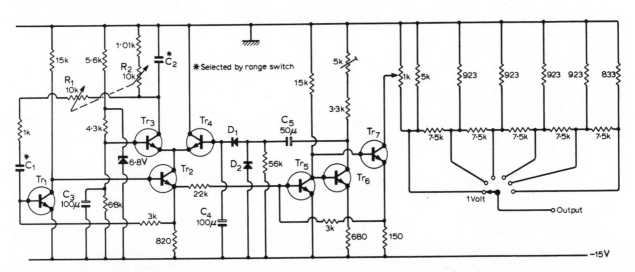

WIEN OSCILLATOR WITH AAC—Wien-network signal generator includes automatic amplitude control as part of oscillator, to provide constant-amplitude output by controlling gain, without distortion that would result from agc limiting action on amplifier itself. Circuit also includes agc to keep output approximately constant for changing input. Tr1 and Tr2 form oscillator. Transistors can be BC107 or similar.—A. F. Newell, A Transistor Multiplier Circuit, *Wireless World*, June 1969, p 285–289.

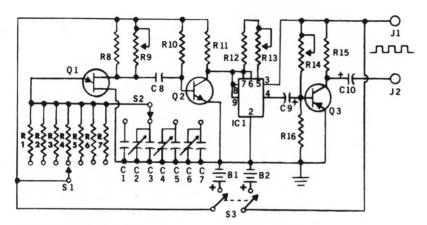

R1—4700 ohm, ½ W res.
R2, R3, R4, R5, R6, R7—See text*
R8—1000 ohm, ½ W res.
R9—5000 ohm pot (Master Osc. Adjust.)
R10, R11—5600 ohm, ½ W res. ±5%
R12—390 ohm, ½ W res.
R13—1000 ohm pot (Feedback Adjust.)
R14—10,000 ohm pot (Output Amplitude)
R15—750 ohm, ½ W res. ±5%
R16—10,000 ohm, ½ W res. ±10%
All resistors should be carbon, do not use wirewounds.
C1—1 µF, 50 V capacitor
C2, C4, C6—Trimmer or padder in anywhere near 100 to 800 pF
C3, C8—0.1 µF, 50 V (min.) capacitor
C5—0.01 µF, 50 V (min.) capacitor
C7—0.001 µF, 50 V (min.) capacitor
C9, C10—500 µF, 15 V elec. capacitor
S1—S.p. 11-pos (use 7 pos.) switch (Frequency)
S2—D.p. 5-pos. (use s.p. 4 pos.) switch (Multiplier)
S3—D.p.s.t. switch (On-Off)
J1, J2—5 way binding post
B1—9-volt transistor battery
B2—6-volt lantern battery
Q1—Unijunction transistor (HEP 310)
Q2—GE 10 (or HEP 55) transistor
Q3—2N404 transistor (or HEP 739)

6 HZ TO 60 KHZ—Range is covered in four multiplied steps (1, 10, 100, and 1,000 with S2) and seven increments (6, 10, 15, 20, 30, 40, and 60 Hz with S1) with accuracy of 1%. Uses HEP558 JK flip-flop as IC1. Ujt Q1 operates as relaxation oscillator feeding buffer Q2 whose output is divided by 2 in IC. Output stage Q3 provides isolation. Article covers construction and calibration. R14 controls amplitude of output, which can be up to 1 V. R2–R7 must be accurate within 1 ohm of 7.2K, 9.7K, 12K, 19K, 28K, and 48K, respectively, if R1 is 4.7K; use combinations of resistors in parallel to get required values.—R. A. Walton, Precision Square-Wave Audio Generator, *Electronics World*, May 1971, p 54–55.

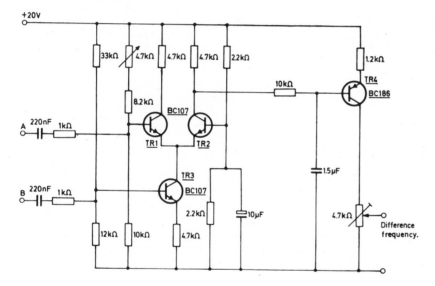

0.1–100 HZ BEAT NOTE—Simple mixer combines outputs of two oscillators set at about 1 kHz, to provide variable low-frequency beat signal required for loop gain measurements in regulated power supply using diac-triggered scr as series regulator. Input A should be overdriven, or supplied with square waves having sufficient amplitude to switch TR3 current between TR1 and TR2. TR4 provides some level shifting and gain.—R. E. F. Bugg, Thyristor Power Supplies for Television Receivers: Design Considerations, *Mullard Technical Communications*, April 1972, p 129–144.

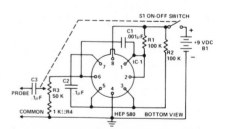

A-F SIGNAL INJECTOR—Ideal for testing audio amplifiers or audio sections of receivers. Generates 1,300-Hz tone that can be adjusted from 1-V to 7-V p-p with R3. Changing size of C1 changes frequency of tone. Uses Motorola HEP580 IC.—"Tips on Using IC's," Motorola Semiconductors, Phoenix, AZ, HMA-32, 1972.

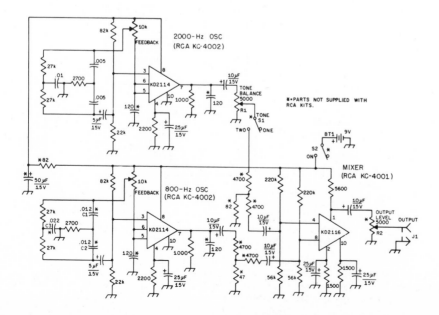

TWO-TONE GENERATOR—Applications include checking transmitter distortion. Harmonic-free 800 and 2,000-Hz oscillator outputs are combined in distortion-free mixer.—D. A. Blakeslee, The IC-TT Generator, *QST*, May 1970, p 21–25.

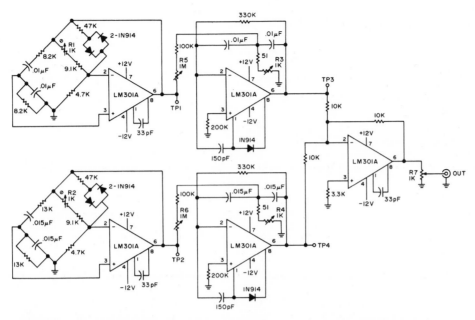

TWO-TONE GENERATOR—Separate Wien-bridge oscillators generate 800-Hz and 2,000-Hz sine waves for true algebraic summing in final IC opamp. Used in testing ssb amplifiers. Article includes circuit for required dual-voltage regulated supply.—H. Olson, A Two-Tone Test Generator, 73, Jan. 1973, p 53–55.

CHAPTER 103
Single-Sideband Circuits

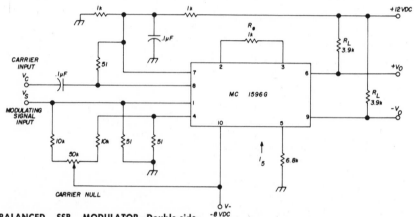

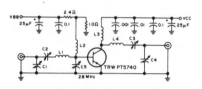

C1,   ARCO No.467, 110-580 pF
C2,3,4, ARCO No.466, 80-480 pF
C5   ARCO No.469, 170-780 pF
L2,3   10 TURNS No.18 AWG, 0.5"
         MEAN DIAMETER
L1,4   5 TURNS No.14 T.C., 0.5"
         MEAN DIAMETER, L1 EQUALS 1.0"

BALANCED SSB MODULATOR—Double-sideband suppressed-carrier generator uses single Motorola IC. Carrier is balanced out by adjusting bias on differential pair of transistors in IC with 50K pot. Pins 6 and 9 give balanced output; single-ended output can be taken between either pin and common.—E. Noll, Integrated Circuits, *Ham Radio*, June 1971, p 40–47.

15-W LINEAR FOR 10 METERS—Single transistor delivers full 15 W when operating from 12-V supply. Ideal for portable ssb transmitters and transceivers, for which collector efficiency averages 45%.—100 Watts Solid State Linear!, *73*, Nov. 1972, p 8–10.

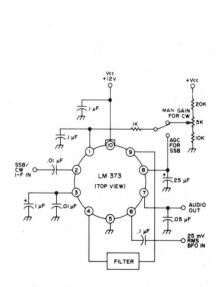

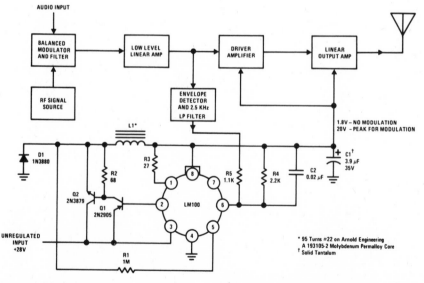

SINGLE-IC SSB/C-W I-F—Uses a-c coupling for signal transfer, to prevent possible damage to IC by direct current. Larger bypass capacitor at pin 3 is tantalum, for low frequencies. IC is by National Semiconductor.—R. Megirian, Using the LM 373, *73*, April 1972, p 37–44.

SWITCHING REGULATOR AS MODULATOR—IC voltage regulator operates linear output amplifiers of conventional ssb transmitter at voltage just higher than that required to accommodate envelope of r-f output signal, to improve efficiency. With no modulating signal, driver and output amplifiers are operated at 1.8-V reference voltage of IC. With modulation, envelope of r-f waveform is detected for driving regulator so its output voltage follows shape of envelope. Report also tells how to use with a-m transmitter.—R. J. Widlar, "New Uses for the LM100 Regulator," National Semiconductor, Santa Clara, CA, 1968, AN-8.

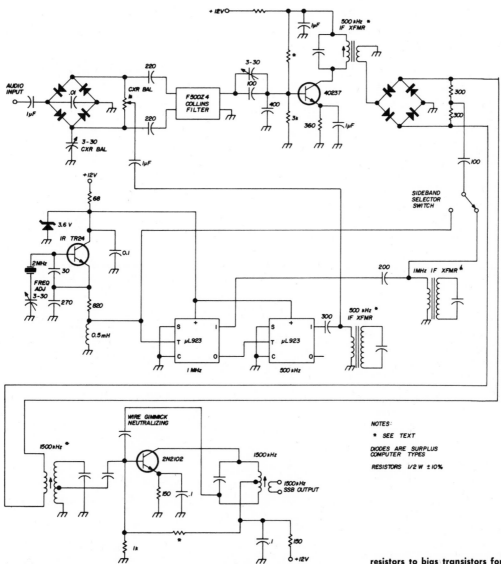

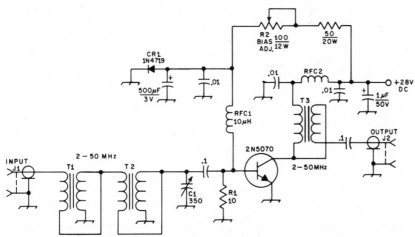

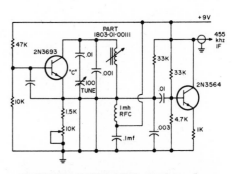

SSB EXCITER—Provides 1,500-kHz ssb output signal by using IC frequency dividers with 2-MHz crystal, for combining with 5-MHz to 6-MHz vfo to produce tunable ssb signal on 40-meter and 80-meter bands. Construction details are given in article. Choose asterisked resistors to bias transistors for class A operation. I-f transformers were taken from Japanese transistor radio.—K. Stone, Frequency Dividers for SSB Generators, *Ham Radio*, Dec. 1971, p 24–26.

2—50 MHZ 25-W LINEAR—Circuit uses transmission-line broadband toroidal transformers to achieve wide frequency response. Article gives winding data. Gain varies with frequency because no feedback is used. Should be followed by filter to reduce level of harmonics.—R. C. Hejhall, Broadband Solid-State Power Amplifiers for SSB Service, *QST*, March 1972, p 36–43.

ADJUSTABLE NOTCH FILTER—Used in ssb receiver to vary notch frequency up to 6 kHz above and below 455-kHz i-f value, for eliminating interference without losing voice intelligibility.—J. J. Schultz, IF Notch Filter, *73*, Nov. 1969, p 14–16.

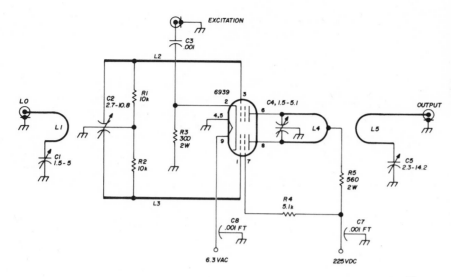

432-MHZ MIXER—High-efficiency circuit using Amperex 6939 tube can be driven with only 100-mW local oscillator power. Delivers 150-mW pep output for excitation of 240-mW pep at 50-MHz ssb. Article covers construction.—F. Telewski, A Practical Approach to 432-MHz SSB, *Ham Radio*, June 1971, p 6–21.

432-MHZ LINEAR—Developed for ssb equipment. Uses Amperex 6939 tube. Stage efficiency is 43%. Article gives construction details. Anode voltage is 225-V d-c and grid bias 3.5-V d-c.—F. Telewski, A Practical Approach to 432-MHz SSB, *Ham Radio*, June 1971, p 6–21.

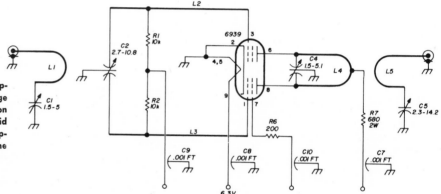

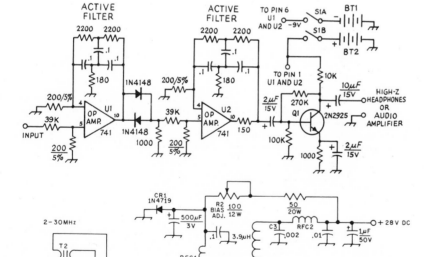

62-HZ BANDPASS C-W—Designed for use with ssb transceiver on c-w when bandwidth of i-f is not narrow enough. Uses two filter sections, each consisting of R-C notch network and high-gain opamp. Input is a-f from second detector. Filter sections are coupled by diodes that pass only peak voltages higher than about 0.6 V. Operates from 9-V transistor radio batteries. Opamps may also be Motorola MC1709.—"The Radio Amateur's Handbook," American Radio Relay League, Newington, CT, 49th Ed., 1972, p 261–262.

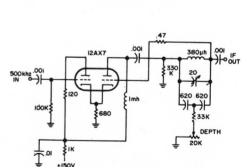

2–30 MHZ 80-W LINEAR—Circuit uses transmission-line broadband toroidal transformers to achieve wide frequency response. Article gives winding data. T1 and T2 are 4:1 series-connected to give 16:1 step down. Efficiency at 30 MHz is 43%. Gain varies with frequency somewhat, because no feedback is used.—R. C. Hejhall, Broadband Solid-State Power Amplifiers for SSB Service, *QST*, March 1972, p 36–43.

Q-MULTIPLIER NOTCH FILTER—Used in i-f section of Collins ssb receiver to increase Q of series bridged-T network in feedback amplifier. Will multiply circuit Q by factor of 2,500, for suppressing interfering signals over narrow band.—J. J. Schultz, IF Notch Filter, *73*, Nov. 1969, p 14–16.

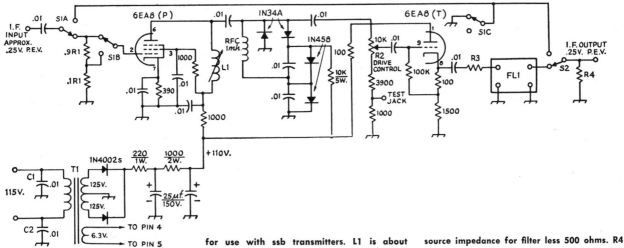

R-F SPEECH PROCESSOR—Diode clipper gives flat clipping, good symmetry, and freedom from rectification effects that can spoil symmetry during transient conditions. Designed for use with ssb transmitters. L1 is about 15 μH for 9 MHz or 500 μH for 455 kHz, adjusted to resonate at i-f of exciter, with slug tuning. FL1 is sideband filter of same type as in exciter. R1 is load resistance recommended for exciter filter, and R3 is recommended source impedance for filter less 500 ohms. R4 is recommended filter load resistance (0.1 meg for Collins mechanical filter).—"The Radio Amateur's Handbook," American Radio Relay League, Newington, CT, 49th Ed., 1972, p 407–408.

SSB TRANSCEIVER MONITOR—Used for listening to output of own transceiver, to check intelligibility. Crystal is 3.55 MHz, but value is not critical. Output can go to tape recorder, a-f amplifier, or phones. Tune receiver to monitor frequency, switch to transmit, then make necessary transmitter adjustments.—F. G. Rayer, Sideband Sniffer, 73, Dec. 1972, p 65–66.

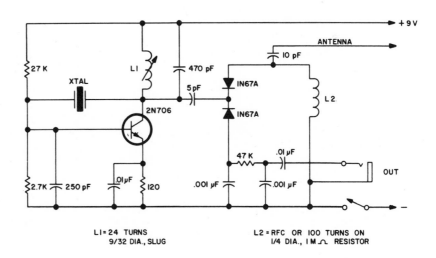

L1= 24 TURNS 9/32 DIA., SLUG

L2 = RFC OR 100 TURNS ON 1/4 DIA., 1 M Ω RESISTOR

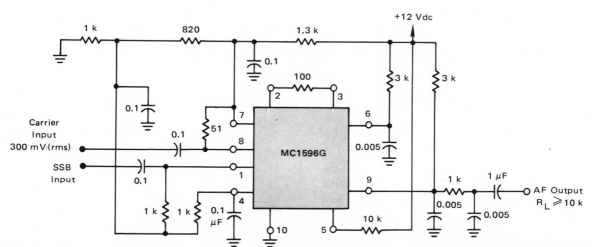

PRODUCT DETECTOR—Uses IC balanced modulator, with all frequencies except desired demodulated audio filtered out in output. Performs well with carrier input level of 100- to 500-mV rms. Will operate from very low frequencies up to 100 MHz. Sensitivity for 9-MHz ssb signal input and 10-dB ratio of signal plus noise to noise at output is 3 μV. Dual outputs are available, from pins 6 and 9, one for a-f amplifier drive and the other for agc system.—R. Hejhall, "MC1596 Balanced Modulator," Motorola Semiconductors, Phoenix, AZ, AN-531, 1971.

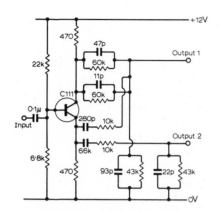

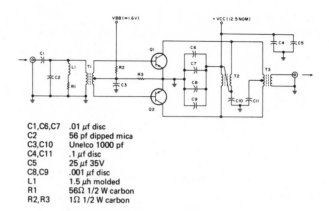

C1,C6,C7   .01 µf disc
C2   56 pf dipped mica
C3,C10   Unelco 1000 pf
C4,C11   .1 µf disc
C5   25 µf 35V
C8,C9   .001 µf disc
L1   1.5 µh molded
R1   56Ω 1/2 W carbon
R2,R3   1Ω 1/2 W carbon

SSB PHASE SHIFT—Dome 90-deg phase shift circuit for carrier of 100–150 kHz is designed for use with two balanced amplitude modulators feeding common summing amplifier, to provide required quadrature signals for single-sideband generator. Provides two outputs whose phase varies logarithmically with respect to input while phase difference between outputs remains constant at 90 deg and amplitudes of outputs also stay constant. Operates with modulation frequencies from 0.1 to 3,600 Hz.—M. E. Cook, Amplitude Modulation Using an F.E.T., *Wireless World*, Feb. 1970, p 81–82.

25 W FOR 1.5–30 MHZ—Two TRW PT5740 transistors in push-pull class AB will operate anywhere in frequency range without bandswitching. Article covers construction of T1 and T3. Ideal for compact low-power portable ssb transceivers.—100 Watts Solid State Linear!, *73*, Nov. 1972, p 8–10.

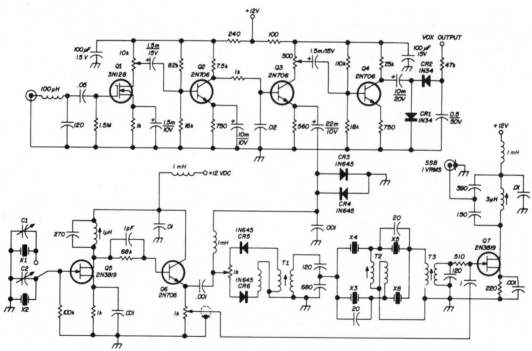

C1, C2 2 — 20 pF (variable or selected)

CR5, CR6 matched for forward resistance

T1 4 turns bifilar primary, 3 uH secondary*

T2 6 turns bifilar wound on 1/4" form

T3 5:1 ratio, 3 uH secondary*

X1 8.9995 MHz (parallel resonant with 32 pF)

X2 9.0015 MHz (parallel resonant with 32 pF)

X3, X5 9.0020 MHz (series resonant)

X4, X6 9.0000 MHz (series resonant)

*3 uH coils are J. W. Miller 40A336CBI

9-MHZ SSB GENERATOR—Mike input to audio amplifier Q1-Q2 drives buffer-limiter Q3, vox amplifier Q4, and rectifier-filter to give 6-V d-c vox output. Buffer Q3 output also goes to balanced diode modulator that is also fed by crystal oscillator Q5 through Q6. Pot in emitter of Q6 provides variable-level carrier for tuneup, c-w, and a-m operation. Balanced modulator feeds output amplifier Q7 through four-crystal ssb filter.—R. Bain, A Filter-Type Ssb Generator, *Ham Radio*, Dec. 1970, p 6–11.

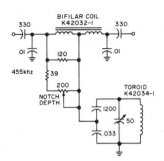

I-F NOTCH FILTER—Used in Hammarlund HQ180 ssb receiver to suppress interference over narrow band without affecting voice intelligibility. Bifilar coils provide close coupling. Notch frequency can be shifted up to several kHz in either direction from i-f center frequency. Circuit also includes notch depth adjustment. Can be placed anywhere in i-f after main crystal filter.—J. J. Schultz, IF Notch Filter, 73, Nov. 1969, p 14–16.

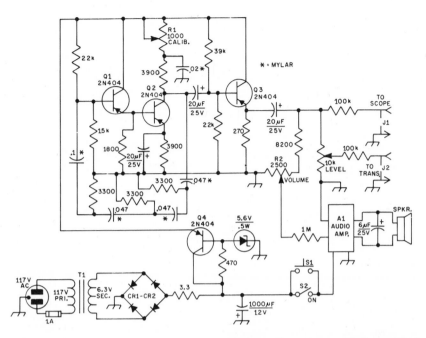

440-HZ TONE GENERATOR—Two-stage phase-shift oscillator generates sine wave that can be calibrated accurately against musical instrument or WWV transmissions. Used in adjusting ssb stations to same frequency for net operation. When received audio tone is same as 440-Hz standard, transmitted carrier signal will be same as carrier received. A1 can be RCA KD2115 or equivalent. CR1–CR4 are 100-prv 1-A silicon diodes.—S. Oehmen, A Tone Generator for Netting of SSB Stations, QST, Dec. 1971, p 38–39.

SSB CRYSTAL LATTICE—Uses three half-lattice sections to provide 6-dB bandwidth of 2,750 Hz extending downward from alignment frequency of 5,505 kHz. Y1–Y3 are same frequency near 5,500 kHz and Y4–Y6 are same frequency but 1,500 to 1,700 Hz different from Y1–Y3. L1 and L4 are 50 turns No. 38 and L2-L3 60 turns No. 38, close-wound on 17/64-inch ceramic form, with slug tuning for L1 and L4.—"The Radio Amateur's Handbook," American Radio Relay League, Newington, CT, 49th Ed., 1972, p 394–395.

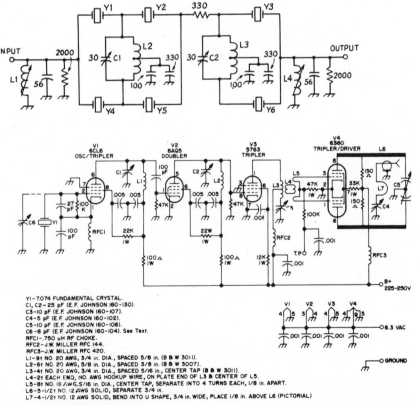

Y1 – 7.074 FUNDAMENTAL CRYSTAL.
C1, C2 – 25 pF (E.F. JOHNSON 160-130).
C3 – 10 pF (E.F. JOHNSON 160-107).
C4 – 5 pF (E.F. JOHNSON 160-102).
C5 – 10 pF (E.F. JOHNSON 160-108).
C6 – 8 pF (E.F. JOHNSON 160-104). See Text.
RFC1 – .750 uH RF CHOKE.
RFC2 – J.W. MILLER RFC 144.
RFC3 – J.W. MILLER RFC 420.
L1 – 9t NO. 20 AWG, 3/4 in. DIA., SPACED 5/8 in. (B & W 3011).
L2 – 6t NO. 20 AWG, 5/8 in. DIA., SPACED 3/8 in. (B & W 3007).
L3 – 4t NO. 20 AWG, 3/4 in. DIA., SPACED 5/16 in., CENTER TAP (B & W 3011).
L4 – 2t EACH END, NO. AWG HOOKUP WIRE, ON PLATE END OF L3 & CENTER OF L5.
L5 – 8t NO. 18 AWG, 5/16 in. DIA., CENTER TAP, SEPARATE INTO 4 TURNS EACH, 1/8 in. APART.
L6 – 5-1/2t NO. 12 AWG SOLID, SEPARATE 3/4 in.
L7 – 4-1/2t NO. 12 AWG SOLID, BEND INTO U SHAPE, 3/4 in. WIDE, PLACE 1/8 in. ABOVE L6 (PICTORIAL)

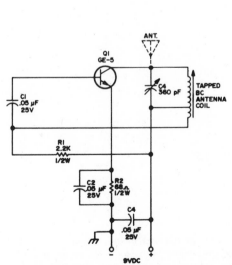

BEAT-FREQUENCY OSCILLATOR—Single-transistor circuit is placed near a-m receiver and C4 tuned until ssb signals become intelligible. Can also be used for c-w signals.—Circuits, 73, June 1972, p 111.

382-MHZ OSCILLATOR CHAIN—Output of crystal oscillator stage V1 is multiplied 54 times to get desired frequency for operation of transmitter on 432-MHz amateur band with ssb. C6 is optional, but will give up to 100-kHz change in output frequency to correct for crystal error or for moving to desired frequency. Use grid dip meter for tuning each multiplier stage in turn. Article also gives 432-MHz mixer circuit.—B. D. Ripley, SSB on 432 MHZ, 73, Nov. 1972, p 163–164 and 166–167.

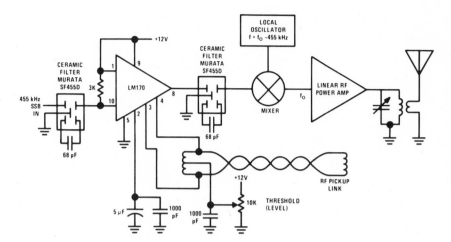

AUTOMATIC LOAD CONTROL—Opamp responds to both positive and negative peaks of transmitter output envelope and provides automatic load control to compensate for variations in power supply, load impedance, and tuning of final amplifier. Circuit provides fast attack time constant for speech and releases more slowly. Threshold pot may also be used as carrier level control in absence of modulation. Bandwidth of 2 MHz for opamp allows effective amplification of 455-kHz low-level ssb signal.—R. A. Hirschfeld, "AGC/Squelch Amplifier," National Semiconductor, Santa Clara, CA, 1969, AN-11.

PRODUCT DETECTOR—May also be used with lower-cost Motorola MC1496G IC. Circuit in effect acts as mixer, with audio output frequency equal to difference between ssb and carrier inputs. Low-pass filter at output cuts off above 3 kHz. Dynamic range is 90 dB. Has high sensitivity (3 μV for 9-MHz ssb i-f input).—R. Hejhall, An Integrated-Circuit Balanced Modulator, *Ham Radio*, Sept. 1970, p 6–13.

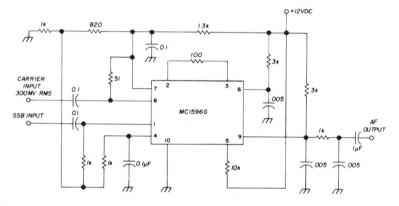

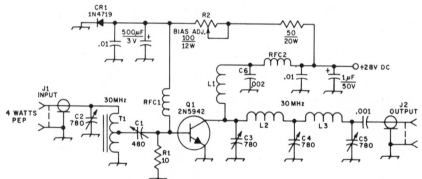

30-MHZ 80-W LINEAR—Output network is double pi section designed for 80-W pep output into 50 ohms. Article gives coil data. Power gain is 13 dB and intermodulation distortion is —34 dB.—R. C. Hejhall, Broadband Solid-State Power Amplifiers for SSB Service, *QST*, March 1972, p 36–43.

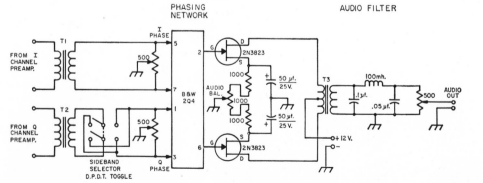

AUDIO PHASING NETWORK—Used in 14-MHz direct-conversion ssb receiver. Inputs to transformers are 0–3 kHz, obtained from balance to mixers operating at received signal frequency, with local oscillator signals 90 deg out of phase applied to mixers so one incoming sideband can be rejected and the other enhanced. Barker and Williamson 2Q4 phase-shift network is used in audio combiner section. Sharp-cutoff audio filter follows push-pull combiner. All audio transformers have 22,000-ohm primary and 600-ohm secondary.—R. S. Taylor, A Direct-Conversion S.S.B. Receiver, *QST*, Sept. 1969, p 11–14.

I-F NOTCH FILTER—Uses tightly coupled i-f transformer between filter terminals, with common point of transformer coupled to bridged-T network. Sharpness of notch depends largely on Q of ferrite-core transformer. Used in Davco DR-30 ssb receiver.—J. J. Schultz, IF Notch Filter, 73, Nov. 1969, p 14–16.

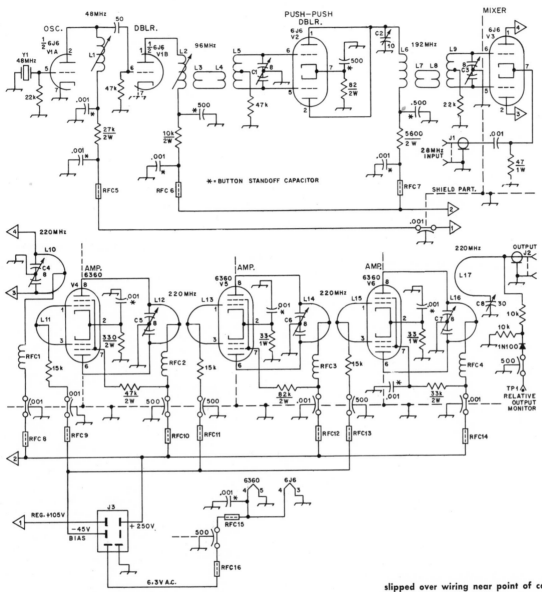

220-MHZ CONVERTER—Mixes 28-MHz injection signal from ssb exciter with output of crystal-controlled 192-MHz multiplier chain to produce sum frequency of 220 MHz. Uses 48-MHz third-overtone crystal oscillator and two doublers to get 192 MHz at high output. RFC1–RFC4 are Ohmite Z-220, and each other choke consists of two ferrite beads slipped over wiring near point of connection. Article gives coil data. Designed to drive final amplifier using tube from 4X150/4CX250 family.—D. V. Watters, An SSB and CW Transmitting Converter for 220 MHz, QST, March 1972, p 11–13.

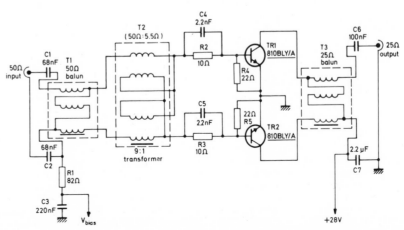

50-W PEP CLASS AB—Response is flat within 1 dB for 1.6–30 MHz. Report gives construction and performance details, including transformer winding data. Method of connecting two 50-W amplifiers in parallel with hybrid couplers to form 100-W amplifier is also given.—"Transistors for Single-Sideband Linear Amplifiers," Mullard, London, 1972, TP1337, p 11–17.

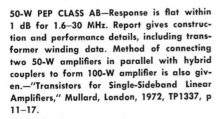

127.333-MHZ CRYSTAL—Uses International Crystal OE1 overtone oscillator module followed by two-stage transistor oscillator, with Amperex transistors, to amplify 1-mW output of crystal oscillator to 1 W. Developed for 432-MHz ssb equipment.—F. Telewski, A Practical Approach to 432-MHz SSB, *Ham Radio*, June 1971, p 6–21.

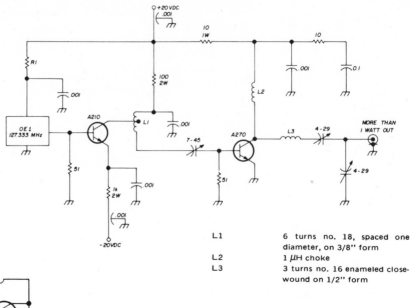

L1	6 turns no. 18, spaced one diameter, on 3/8" form
L2	1 μH choke
L3	3 turns no. 16 enameled close-wound on 1/2" form

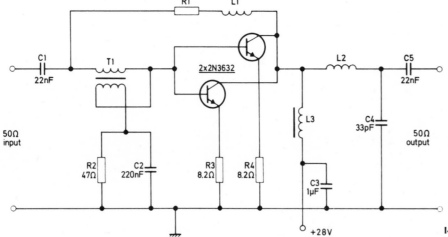

6-W PEP CLASS A BROADBAND DRIVER—Provides voltage gain of 14 over frequency range of 1.6 to 30 MHz. Intermodulation distortion level is better than −30 dB. Input transformer (50:12.5 ohms) is constructed as transmission-line type wound on Ferroxcube twin-bead core. L1 is 900 nH, L2 has 4 turns, and L3 60 turns.—"Transistors for Single-Sideband Linear Amplifiers," Mullard, London, 1972, TP1337, p 7–8.

I-F NOTCH FILTER—Uses bridged-T network in feedback path to eliminate narrow frequency band at which interference is present, without appreciably affecting voice intelligibility in ssb receiver. Attenuation at notch frequency can be as high as 60 dB, depending on Q of components used.—J. J. Schultz, IF Notch Filter, 73, Nov. 1969, p 14–16.

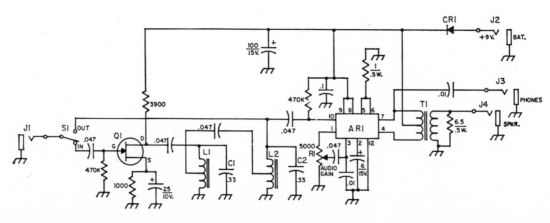

C-W FILTER—Provides about 15-dB improvement in signal-noise ratio by decreasing bandwidth of ssb receiver from normal 2.5-kHz value down to 80 Hz which is adequate for c-w. Can be plugged into phone jack of any receiver. Fet amplifier Q1 (2N3819) provides high-impedance input and gain of about 12 dB to overcome filter losses. L1 and L2 in filter circuits are 88-mH telephone toroids. IC is RCA CA3020 or any other high-gain a-f amplifier providing about 0.5-W push-pull output. T1 is transistor a-f output transformer. CR1 (almost any diode) protects transistors if battery polarity is accidentally reversed. Adjust C1 and C2 to tune coils to desired audio frequency, around 900 Hz.—L. N. Anciaux, A Solid-State Audio Filter, QST, Dec. 1968, p 35–37.

CHAPTER 104
Siren Circuits

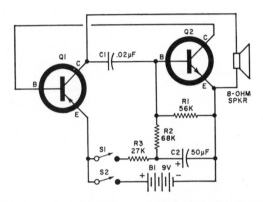

BICYCLE SIREN—Battery-operated siren can replace bell or horn on bicycle, doorbell in home, or just enliven parties. Uses complementary direct-coupled relaxation oscillator, with speaker as collector load for Q1 which can be any general-purpose pnp power transistor such as 2N554. Q2 is any low-power npn transistor. Use pushbutton switch for S1 to simulate siren.—J. Ramsey, Reader's Circuit, *Popular Electronics*, Oct. 1967, p 79.

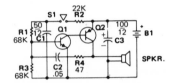

SIREN—Produces rising and falling wailing sound from 45-ohm 4-inch speaker. Transistor types are not critical, except that Q1 must be npn and Q2 pnp. For greater volume, feed into high-power audio amplifier. Uses mvbr circuit. Battery is 9 V.—"Calectro Handbook," GC Electronics, Rockford, IL, 1971, p 42.

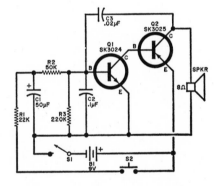

POLICE SIREN—Produces short, high-pitch scream when button is pressed, followed by drooping wail typical of police siren.—T. Dilger, Reader's Circuit, *Popular Electronics*, July 1970, p 85—86.

SIREN—Operates from 9-V transistor radio battery. Will drive speaker as shown, for use as toy siren on bicycles, or can be fed into fixed or mobile public-address amplifier when more volume is needed. Can serve as output device for burglar alarm in home or in car. Pots give wide variety of siren sound effects. —"Tips on Using IC's," Motorola Semiconductors, Phoenix, AZ, HMA-32, 1972.

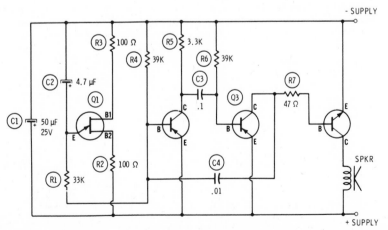

WAA-WAA SOUNDER—Ujt oscillator Q1 (2N2646) modulates basic oscillator using 2N3638 transistors driving MJE3055 power transistor to produce interrupted frequency-modulated tone from 8-ohm 20-30 W hi-fi tweeter that is ear-splitting at close range and distinctly attention-getting at greater distances than continuous alarms. Relay contacts of alarm circuit are connected to close power supply circuit of sounder.—C. D. Rakes, "Solid State Electronic Projects," Howard W. Sams, Indianapolis, IN, 1972, p 52—54.

813

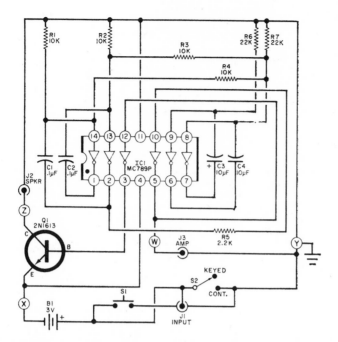

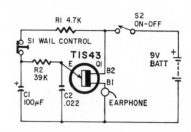

WAILING SIREN—Normally closed pushbutton switch S1 makes ujt operate as relaxation oscillator to produce wailing sound in earphone at frequency determined by R2, C2, and supply voltage that charges C1. Pressing S1 makes C1 discharge slowly to lower frequency gradually. If button is pressed and released at about 2-s intervals, realistic imitation of fire-engine siren is obtained while child plays with toys, without disturbing rest of family. Constant tone when siren is abandoned reminds child to turn it off by opening S2.—F. H. Tooker, Siren for Toy Ambulance or Fire Engine, *Radio-Electronics*, Aug. 1969, p 74.

TWO-TONE ALARM—Pair of audio oscillators interacts to produce attention-getting twee-dell twee-dell sound by switching audible output of speaker between 500 and 1,000 Hz about 5 times per second. Resulting sound cannot be ignored. IC is hex inverter; first two stages form astable mvbr operating at either 500 or 1,000 Hz depending on state of 5-Hz mvbr using last two stages. Center pair of inverters provides load oscillation, and Q1 boosts power enough to drive speaker.—D. Lancaster, Build the Two-Tone "Waverly" Alarm, *Popular Electronics*, Feb. 1970, p 29–31.

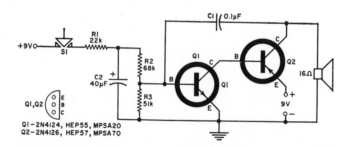

PUSHBUTTON SIREN—When S1 is pushed, C2 is charged at rate dependent on values of R1 and C2, making base bias of Q1 rise slowly and increase audio frequency of oscillator Q1-Q2. This creates slowly rising note like that of siren. When button is released, C2 discharges slowly and audio frequency drops correspondingly and finally stops unless button is pressed again. Loudest output is obtained with high-impedance speaker (up to 40 ohms).—P. Franson, Panic Button, *Electronics World*, May 1970, p 41 and 62.

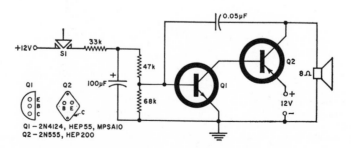

LOUD SIREN—Designed for operation from 12-V auto storage battery. Uses germanium power transistor for Q2, with lower impedance for matching 8-ohm speaker. Pressing button S1 starts slowly rising note of siren, while releasing button makes frequency drop slowly and finally stop unless button is pressed again.—P. Franson, Panic Button, *Electronics World*, May 1970, p 41 and 62.

TOY SOUNDBOX—Includes speaker. Pushing S2 charges C3 and turns on Q2 and Q3 to give sound that wails up and down scale like real siren. Closing S1 energizes Q1, which may be any general-purpose ujt, to give ticking sound at rate controlled by R3.—J. J. Tashetta, Gadget Box, *Popular Electronics,* Dec. 1968, p 64–65 and 109.

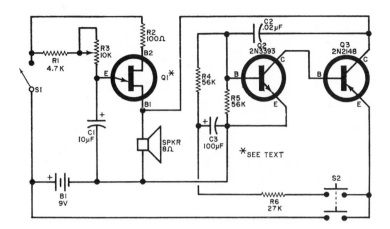

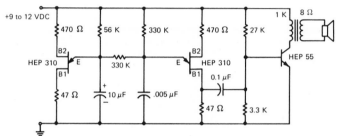

UJT SIREN—Produces attention-getting automatic rising and wailing output from speaker, to serve as attention-getting burglar alarm or warning device.—"Home Handyman's Construction Projects," Motorola Semiconductors, Phoenix, AZ, HMA-37, 1972.

REPEATING SIREN—Output of triangular-wave generator Q1 is applied to input of audio oscillator to make output frequency rise and fall slowly like siren without having to push and release button. Designed for operation from 12-V auto storage battery. Switch should be used in battery supply lead.—P. Franson, Panic Button, *Electronics World,* May 1970, p 41 and 62.

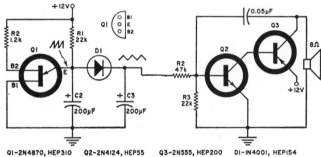

Q1-2N4870, HEP310   Q2-2N4124, HEP55   Q3-2N555, HEP200   D1-IN4001, HEP154

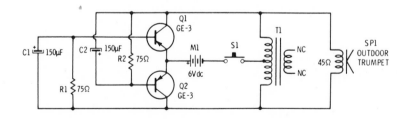

GOO-HAH HORN—Completely portable oscillator with 45-ohm p-a speaker imitates weird sound of old European auto horns. Can be used in modern sports car, for signaling touchdowns at football games, or for breaking apartment lease. T1 is Argonne AR-503 transformer.—R. M. Brown and T. Kneitel, "49 Easy Entertainment and Science Projects," Vol. 2, Howard W. Sams, Indianapolis, IN, 1969, p 34–35.

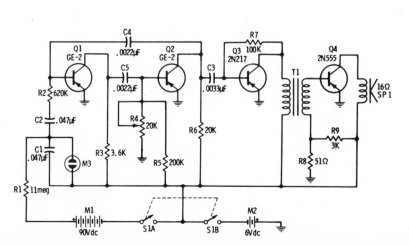

CHIRPER—Provides imitation of electronic siren, with setting of R4 determining pitch in range from 1,000 to 10,000 Hz for free-running mvbr. Chirping effect is achieved with low-frequency oscillator using NE-2 neon M3, which provides base bias for Q1 of mvbr. Q3 is driver for output transistor Q4. T1 can be Stancor TA-2 having 100-ohm primary and 10-ohm secondary. Can be used as sounder for burglar alarm.—R. M. Brown and R. Kneitel, "49 Easy Entertainment and Science Projects," Howard W. Sams, Indianapolis, IN, 1971, p 14–16.

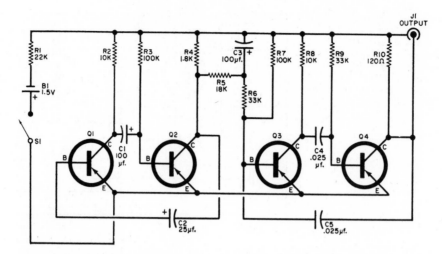

WAILING SIREN—When low-frequency mvbr Q1-Q2 switches at low-frequency rate, it makes frequency of high-frequency mvbr vary to produce wailing sound. Use any audio transistors, such as 2N1436 or 2N1754. Can be checked out with 8-ohm speaker plugged into J1, but would normally be used to drive audio amplifier.—T. Hillard, Reader's Circuit, *Popular Electronics*, March 1966, p 79–80.

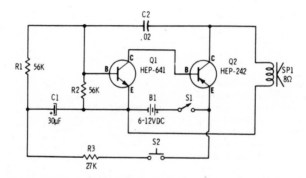

PORTABLE SIREN—To use as screamer for burglar protection, place jumper across S2 and use normally open door or window switch in place of S1 to sound alarm. Can be used as portable siren by pumping momentary-contact switch S2 to vary pitch.—H. Friedman, "99 Electronic Projects," Howard W. Sams, Indianapolis, IN, 1971, p 29–30.

CHAPTER 105
Square-Law Circuits

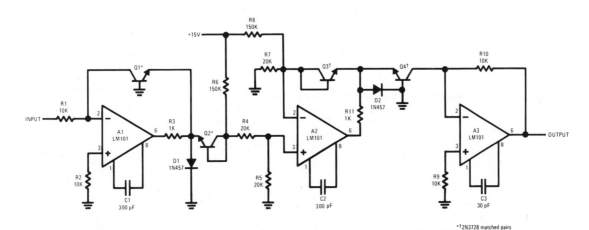

ROOT EXTRACTOR—Takes square root of number by taking log of voltage in log converter A1-Q1, dividing result by ½ for square root, then taking antilog with A3-Q4 combination. Q2 and Q3 are connected as diodes to simplify circuit. Opamp A2 is buffer, with Q3 serving as level-shifting diode so output is 1 V for 1 V input (square root of 1 is 1).

Accuracy is 1% for inputs from 0.5 to 50 V.—R. J. Widlar, "Operational Amplifiers," National Semiconductor, Santa Clara, CA, 1968, AN-4.

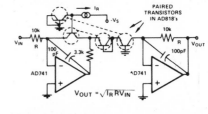

SQUARE ROOT—Circuit takes log of input voltage, divides by 2, then takes antilog to give square root. Scale factor is set by current source IR. Uses Analog Devices AD818 dual npn transistor IC's with two low-cost opamps. Gives excellent log conformance over eight decades.—Monolithic AD818 Has Excellent Log Conformity, *Analog Dialogue*, Vol. 6, No. 3, p 11.

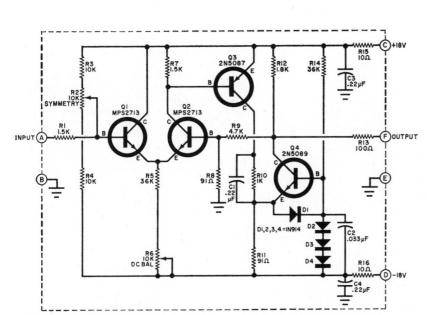

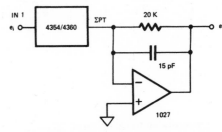

SQUARER—With any sine-wave input up to about 1 MHz, will deliver 15-V p-p symmetrical square-wave output into high-impedance load, with loading having no effect on quality of waveform as amplitude decreases. Rise and fall times are only 70 ns. Circuit is basically d-c opamp having positive rather than negative feedback. Ideal for testing hi-fi audio equipment.—J. Bongiorno, Build the Add-On Squarer, *Popular Electronics*, May 1970, p 51—53 and 58.

SINGLE-QUADRANT SQUARER—Uses Philbrick square-law module with inverting opamp. Will handle inputs up to 10 V with either polarity.—"Square Law Elements 4353/4354/4359/4360." Teledyne Philbrick, Dedham, MA, 1972.

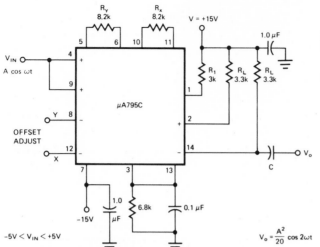

$$V_o = \frac{A^2}{20} \cos 2\omega t$$

SQUARING AND FREQUENCY DOUBLING—
Tying both inputs of IC multiplier together
provides squaring function. A-c term coupled
through output capacitor C is sine wave at
double frequency of input sine waveform.

Input may have any waveform; if triangular,
squared output is continuous set of unipolar
parabolas.—"The µA795, A Low-Cost Monolithic Multiplier," Fairchild Semiconductor,
Mountain View, CA, No. 211, 1971, p 4–5.

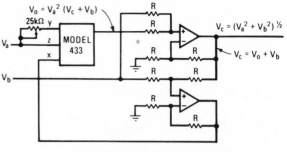

VECTOR LENGTH—Uses Analog Device's
model 433 log module with only two external
opamps to compute length of two-dimen-
sional vector. After each input is squared,
squared values are added and square root of
sum is taken. Accuracy is 0.1%. Typical value
of R is 10K. Choice of opamp is not critical.
—L. Counts and F. Pouliot, Computational
Module Stresses Applications Versatility, *Electronics*, July 17, 1972, p 87–88.

INPUT WORD

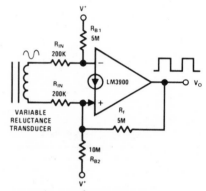

SQUARING WITH HYSTERESIS—Uses Norton
current-differencing opamp connected to have
symmetrical hysteresis above and below
zero-output state for noise immunity, to amplify low-level signals of variable-reluctance
transducers. Includes high-frequency rolloff
characteristic for filtering out high-frequency
input noise. Trip voltages are about 150 mV
each side of 0 V.—T. M. Frederiksen, W. M.
Howard, and R. S. Sleeth, "The LM3900—A
New Current-Differencing Quad of ± Input
Amplifiers," National Semiconductor, Santa
Clara, CA, 1972, AN-72, p 33–34.

SQUARE OF BINARY WORD INPUT

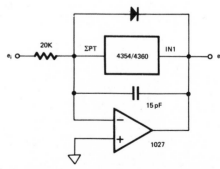

SQUARING 4-BIT BINARY—Uses three type
9304 dual full adders and two 9002 gate
packages to generate appropriate logic polarities and provide required AND functions for
generating square of 4-bit binary number.
Generating time is typically 80 ns. Can be
extended to larger word lengths by adding
logic levels. Each level increases number of

inputs by one and requires additional full adders. Generating square of 8-bit binary word
would require 28 full adders and have delay
time of 160 ns.—C. Ghest, "Generating the
Square of a Binary Number," Fairchild Semiconductor, Mountain View, CA, No. 142,
1970.

SQUARE ROOT—Uses Philbrick square-law
module in feedback circuit of opamp. Negative input up to 10 V provides positive output
with either of modules specified on diagram.
—"Square Law Elements 4353/4354/4359/
4360," Teledyne Philbrick, Dedham, MA, 1972.

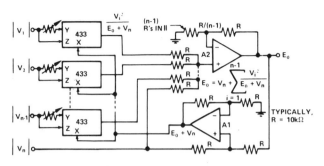

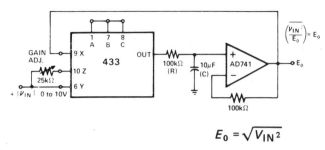

$$E_0 = \sqrt{V_{IN}^2}$$

ROOT SUM OF N SQUARES—Uses Analog Devices Model 433 programmable multifunction analog modules to compute square root of sum of squares of any number of input voltages, without cascading operations on pairs of inputs. Input values are limited by specified maximum output voltage and number of inputs. With four inputs as shown, maximum value of any input is 3.33 V.—D. Sheingold, Root-Sum-Of-Squares Circuits With the 433, *Analog Dialogue*, Vol. 6, No. 3, p 3.

TRUE RMS—Uses Analog Devices Model 433 programmable multi-function module, with simple unit-lag averaging filter connected between its output and square-root feedback point. Output of filter is then constrained to be square-root of average squared input. Averaging time constant is determined by values of R and C.—F. Pouliot and L. Counts, Versatile New Module: $Y(Z/X)^m$ at Low Cost, *Analog Dialogue*, Vol. 6, No. 2, p 3–6.

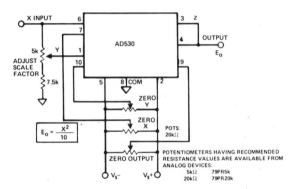

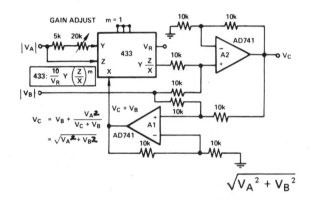

$$\sqrt{V_A^2 + V_B^2}$$

ANALOG SQUARER—Single IC multiplier connected as shown provides squaring by multiplying X input by itself. Output is connected to Z input of IC and gain is trimmed at Y input.—"AD530 Complete Monolithic MDSSR," Analog Devices, Inc., Norwood, MA, Technical Bulletin, July 1971.

ROOT SUM OF SQUARES—Uses Analog Devices Model 433 programmable multifunction analog module to compute square root of sum of squares of two input voltages, VA and VB. Circuit gives excellent stability and low noise, even for very small input. Inputs must be scaled to keep all amplifier outputs within specified 10-V limits, to avoid saturation; if inputs are equal, they should not exceed 4.142 V, which gives output of 5.858 V. Errors are less than 0.25%.—D. Sheingold, Root-Sum-Of-Squares Circuits With the 433, *Analog Dialogue*, Vol. 6, No. 3, p 3.

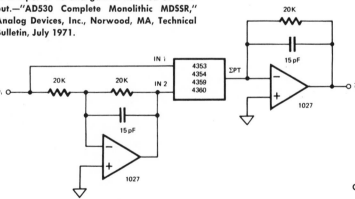

TWO-QUADRANT SQUARER—Uses opamps at input and output of Philbrick square-law module. Will handle inputs up to 10 V with either polarity.—"Square Law Elements 4353/4354/4359/4360," Teledyne Philbrick, Dedham, MA, 1972.

SQUARE-ROOTER—With single IC multiplier connected as shown, divider input is divided by output, so output is proportional to square root of input. Restricted to positive Z inputs.—"AD530 Complete Monolithic MDSSR," Analog Devices, Inc., Norwood, MA, Technical Bulletin, July 1971.

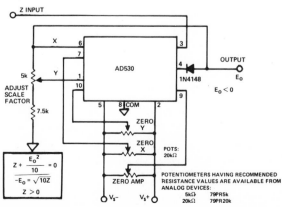

VECTOR DIFFERENCE—Based on square-rooting properties of Analog Devices AD531K three-variable analog multiplier-divider. Requires only one external opamp to convert fed-back output from voltage to current. Gives square root of difference between squares of input voltages Va and Vb, with less than 100-mV error. Bandwidth for 3-dB down is d-c to 600 kHz.—Vector Difference with the AD531, *Analog Dialogue*, Vol. 7, No. 1, p 14.

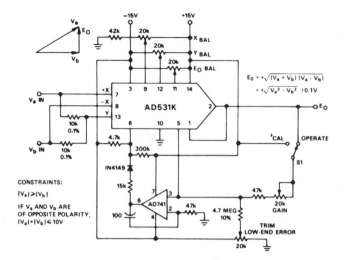

$$E_0 = +\sqrt{(V_a + V_b)(V_a - V_b)}$$
$$= +\sqrt{V_a^2 - V_b^2} \pm 0.1V$$

CONSTRAINTS:

$|V_a| > |V_b|$

IF V_a AND V_b ARE OF OPPOSITE POLARITY, $|V_a| + |V_b| < 10V$

VECTOR SUMMATION—Uses Analog Devices Model 433 programmable multifunction module, to give square root of sum of squares of two input voltages. Maximum tolerable value of (EW + VU), in any combination, is 10 V; for example, if VU is 4.1 V and VV is 4.1 V, EW is 5.8 V. Accuracy is typically 0.1%.—F. Pouliot and L. Counts, Versatile New Module: $Y(Z/X)^m$ at Low Cost, *Analog Dialogue*, Vol. 6, No. 2, p 3–6.

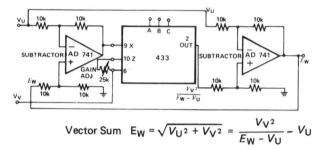

$$\text{Vector Sum} \quad E_W = \sqrt{V_U{}^2 + V_V{}^2} = \frac{V_V{}^2}{E_W - V_U} - V_U$$

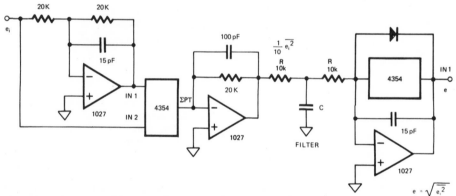

RMS OPERATION—Three opamps and two identical Philbrick consistent-polarity square-law modules give output signal which is true root-mean-square value of input voltage which may be either d-c or a-c. Circuit can be calibrated at d-c.—"Square Law Elements 4353/4354/4359/4360," Teledyne Philbrick, Dedham, MA, 1972.

RMS CONVERTER—Converts analog input voltage into equivalent rms value by using squaring operation followed by integration and square-rooting. Since OEI 5904 module performs squaring function in amplitude only and only on positive inputs, bipolar input signal must be converted first into linear absolute value with 9004 module. Two pot adjustments are used to give 10-V full-scale level. —"Simple RMS Converter," Optical Electronics, Tucson, AZ, Application Tip 10246, 1971.

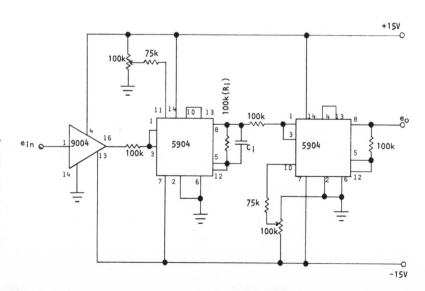

CHAPTER 106
Squelch Circuits

NOISE-OPERATED SQUELCH—Operates reliably on 0.1-μV r-f signal. Only noise component of audio signal, above 5 kHz, is passed by L1-C1 to noise rectifier. When noise is present, Q4 and Q5 are held on by output from Q3. When receiver quiets, Q4 and Q5 shut off, opening audio gate.—D. A. Blakeslee, Receiving FM, QST, March 1971, p 29–34.

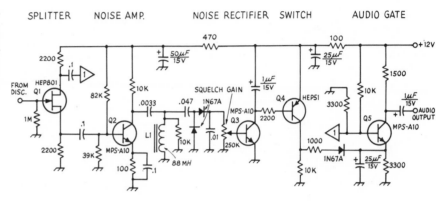

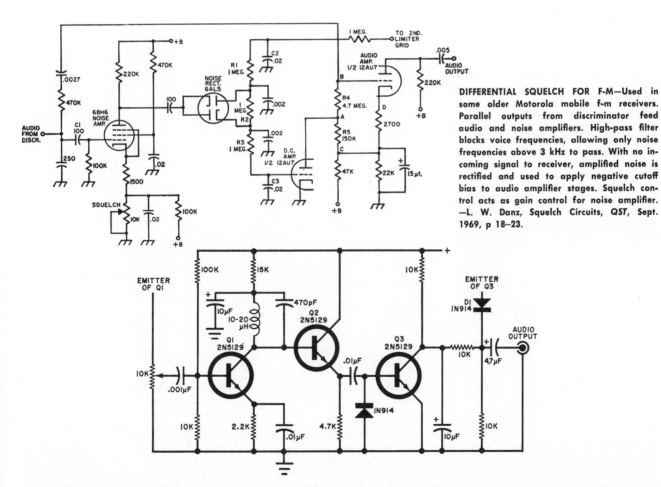

DIFFERENTIAL SQUELCH FOR F-M—Used in some older Motorola mobile f-m receivers. Parallel outputs from discriminator feed audio and noise amplifiers. High-pass filter blocks voice frequencies, allowing only noise frequencies above 3 kHz to pass. With no incoming signal to receiver, amplified noise is rectified and used to apply negative cutoff bias to audio amplifier stages. Squelch control acts as gain control for noise amplifier.—L. W. Danz, Squelch Circuits, QST, Sept. 1969, p 18–23.

SCA SQUELCH—Removes background noise when SCA carrier is dropped out at transmitter during portions of f-m program. Input of squelch is connected to emitter of first transistor in an SCA adapter. Output of SCA adapter, at emitter of third transistor, is fed to squelch circuit through D1, and new audio output is taken from squelch circuit itself. Adjust input pot of squelch so audio signal is blocked when 67-kHz subcarrier disappears.—L. Garner, Solid State, *Popular Electronics*, March 1971, p 84.

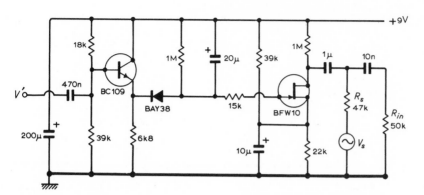

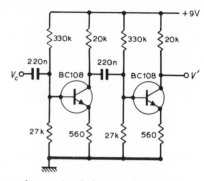

SQUELCH—Junction fet serves as switch in agc system providing automatic squelch for audio voice amplifier in communication receiver. Squelch stage can be controlled either by received carrier or by audio signal.

Threshold level is set just above receiver noise level, so only desired audio signal is amplified. Squelch operates about 2 s after carrier or audio ceases, to suppress noise. Vs is output of detector stage. Amplifier provid-

ing voltage gain of about 55 dB may be necessary for control voltage fed to V' of squelch; suitable circuit is shown at right.—R. M. Lea, The Junction F.E.T. as a Voltage-Controlled Resistance, *Wireless World*, Aug. 1972, p 394–396.

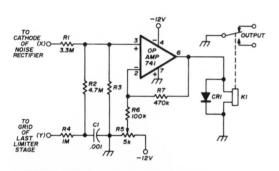

SIGNAL-OPERATED OPAMP RELAY—R1 is connected to squelch circuit of receiver and R4 receives limiter voltage of receiver, for silencing audio system to suppress noise when no carrier signal is present. K1 is Sigma 65FP1A and CR1 is silicon diode. R5 is set nominally to about 95 V. Opamp is used as d-c amplifier with almost infinite gain, for sensing difference between two input voltages and energizing relay at desired signal level. Article covers operation of squelch systems for f-m communication receivers in detail.—J. J. O'Brien, Operational-Amplifier Relay for Motorola Receivers, *Ham Radio*, July 1973, p 16–21.

SQUELCH FOR PAGER—Addition of noise-actuated squelch to Motorola f-m pocket paging receiver operating in 136–174 MHz band permits automatic triggering on by r-f signal at frequency to which pager is tuned. Set can then be used for monitoring 2-meter ham band or 150-MHz public-service band, without need for tone to turn it on. Article covers construction and installation in pager.—B. Mengel, Squelch Addition for the Pocket Pager, 73, June 1971, p 34 and 36–38.

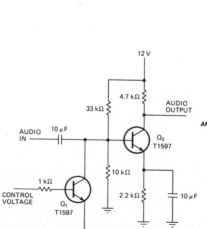

FET SQUELCH—Provides about 60-dB attenuation of audio signal when gate of Q1 is at zero level. Squelch is turned off, to pass audio input, by applying negative voltage to gate of Q1. Circuit is free of switching transients.—G. Coers, FETs Remove Transients from Audio Squelch Circuit, *Electronics*, June 5, 1972, p 102.

BIPOLAR SQUELCH—Q1 is control for amplifier Q2 in basic audio squelch circuit. Negative control voltage at Q1 turns squelch off. Chief drawback is that when Q2 is turned off by Q1, base voltage of Q2 switches from d-c level of 2.8 V to ground, creating large transient output voltage spike.—G. Coers, FETs Remove Transients from Audio Squelch Circuit, *Electronics*, June 5, 1972, p 102.

CARRIER-OPERATED RELAY—Tube-type circuit for communication receiver energizes relay for turning on audio amplifier only when carrier is received. D-c amplifier input is obtained from f-m receiver through noise-pass filter, noise amplifier, and noise rectifier that together serve to produce required d-c output for blocking audio when only noise is present. K1 is 8,000-ohm relay which picks up at 39 V and drops out at 27 V. Article covers squelch system in detail.—J. J. O'Brien, Operational-Amplifier Relay for Motorola Receivers, *Ham Radio*, July 1973, p 16–21.

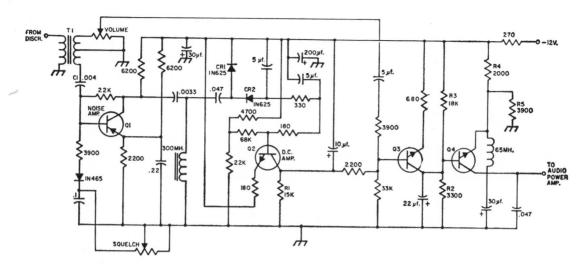

F-M RECEIVER SQUELCH—Used in General Electric type ER-31 mobile f-m receivers. Half of T1 supplies normal audio to volume control and a-f amplifier stages Q3-Q4, using 2N169 transistors. Other half feeds 2N450 noise amplifier Q1 through high-pass filter C1. Squelch control sets gain of Q1, whose rectified output determines base bias for GE A4003-428-1 d-c amplifier. With incoming signal, output of noise rectifiers goes to zero, Q2 conducts, Q3 is partially conducting, and resulting reduced bias on Q4 allows it to conduct and pass audio signal.—L. W. Danz, Squelch Circuits, QST, Sept. 1969, p 18–23.

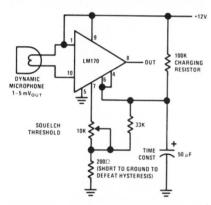

SQUELCHED PREAMP—Suppresses background microphone noises in audio systems. Also useful in receiving systems for suppressing unused transmission channel until useful information arrives. Charging resistor and capacitor values may be chosen to give squelch release times up to several seconds. Large current-sinking capability assures fast attack, so first speech syllables are not lost. Hysteresis action makes circuits smooth-acting and easy on ears. Entire circuit can be fitted in microphone case, even with batteries if operated from 4.5 or 6 V.—R. A. Hirschfeld, "AGC/Squelch Amplifier," National Semiconductor, Santa Clara, CA, 1969, AN-11.

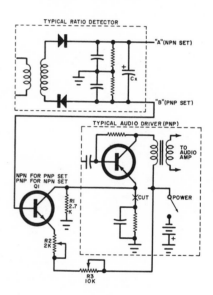

ADD-ON SQUELCH—Can be added to any solid-state vhf f-m receiver using pnp transistors. For set having npn transistors, change Q1 and change connection to ratio detector, as indicated on diagram. For pnp receivers, Q1 can be 2N5129, 2N706, or HEP55; for other sets, use 2N5139, 2N3638, or HEP52. Emitter resistor of audio driver in set is cut out and replaced by R1 as shown. With no signal, Q1 is on and audio stage is off. Received signal turns off Q1 and R1 completes normal path for audio driver stage. Article covers operation, construction, and adjustment.—J. G. Ramsey, Low-Cost Squelch Circuit, Popular Electronics, Feb. 1973, p 110–111.

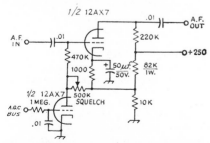

AGC-CONTROLLED SQUELCH—When no signal is received, there is no agc voltage so plate current through squelch tube produces voltage drop across squelch control, cutting off audio amplifier tube.—L. W. Danz, Squelch Circuits, QST, Sept. 1969, p 18–23.

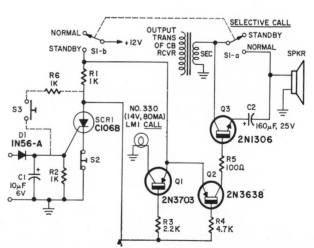

CB SOLID-STATE SQUELCH—Replaces electromechanical relay normally used in selective calling systems of CB receivers. Input to D1 is taken from base of transistor in receiver which energizes squelch relay. Article tells how connections are made to receiver for silencing speaker except when received carrier signal has pair of audio tones assigned to receiver for selective calling.—G. Neal, New Squelch for CB Receiver, Radio-Electronics, Nov. 1971, p 37.

CHAPTER 107
Staircase Generator Circuits

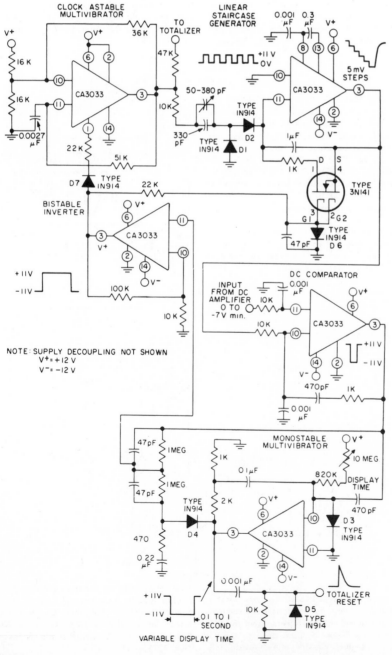

DIGITAL VOLTMETER—Uses RCA CA3033 opamp as linear staircase generator for digital voltmeter in which same IC is connected as squelchable mvbr serving as clock. Clock output drives staircase generator. Staircase output is applied to comparator that compares staircase with voltage to be measured. Book covers operation of circuit in detail.—"Linear Integrated Circuits," RCA, Somerville, NJ, IC-42, 1970, p 173–178.

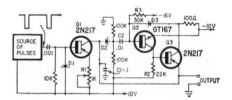

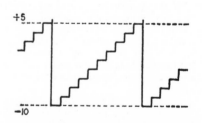

FREQUENCY-DIVIDING STAIRCASE—Provides accurate division of input pulses by 10. Each 5-V negative input pulse serves to make Q1 pass current during pulse intervals and charge C1. When charge on C1 has risen to 5 V, D2 conducts and applies positive pulse to base of Q2, giving current flow through R2 to initiate regenerative action of Q2 and Q3 to make Q3 saturate quickly to give output pulse while discharging C1 to —10 V for start of next staircase. Circuit is described in US patent No. 3,105,158.—B. B. Nichols, Transistor Counter Decade, *Radio-Electronics*, Aug. 1969, p 45.

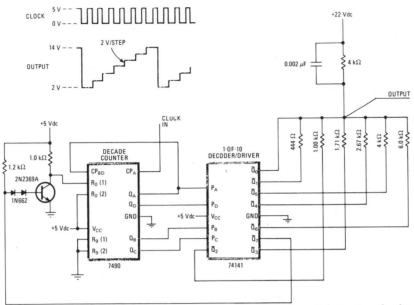

10 STEPS WITH DECADE COUNTER—Steps can be up to 65 V d-c before exceeding output breakdown limitations of IC decoder-driver used with decade counter. Useful in curve tracer, low-resolution a-d converter, and in control applications requiring sequential stepping of voltages. Connections shown give seven 2-V steps. Will accept clock frequencies up to 10 kHz.—D. F. DeKold, Counter and Decoder/Driver Produces Staircase Voltage, *Electronics*, Aug. 2, 1973, p 99.

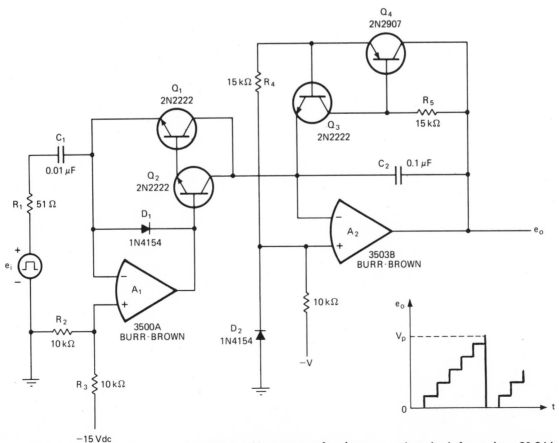

PRECISION STAIRCASE—Opamp A1 differentiates and rectifies input square wave. Negative-going transitions are transferred to second opamp A2 through Q1 and Q2. Each negative transition creates output step until trigger level of reset clamp Q3-Q4 is reached. —J. Graeme, Op Amps Generate Precision Staircase, *Electronics*, Jan. 31, 1972, p 58–59.

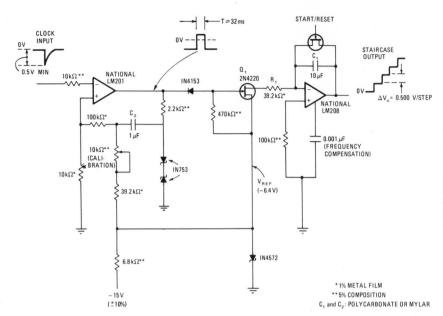

DRIFT-FREE ANALOG—Uses one-shot opamp triggered by clock to drive integrate-and-hold circuit. Tracking capacitors mutually cancel temperature drift. Staircase rise time is proportional to C2, while integrator slope is proportional to C1, so period of one-shot output pulse is directly proportional to ratio C2/C1. Step amplitude drift is held to 0.2%. When one-shot is off, integrator section holds step height constant.—M. Strange, Staircase Generator Resists Output Drift, *Electronics*, Dec. 4, 1972, p 90.

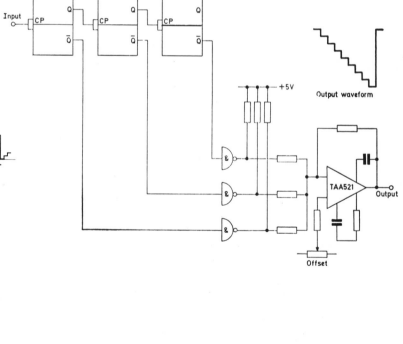

EIGHT STEPS—Uses 1½ Mullard IC's to provide J-K bistables feeding Mullard TAA521 summing amplifier to give staircase output having eight steps.—"Mullard TTL Integrated Circuits Applications," Mullard, London, 2nd Ed., 1970, p 203.

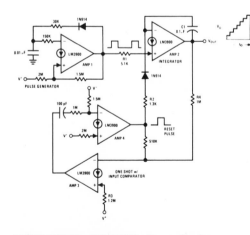

FREE-RUNNING STAIRCASE—Uses all four Norton current-differencing opamps in LM3900 package. Opamp 1 provides input pulses which pump up staircase via R1. Opamp 2 provides integrate and hold function along with staircase output. Opamps 3 and 4 provide both compare and one-shot mvbr function for sampling staircase output and comparing with power supply voltage via R3. When output exceeds about 80% of supply voltage, circuit generates 100-μs reset pulse that drives staircase output down to essentially 0 V.—T. M. Frederiksen, W. M. Howard, and R. S. Sleeth, "The LM3900—A New Current-Differencing Quad of ± Input Amplifiers," National Semiconductor, Santa Clara, CA, 1972, AN-72, p 23.

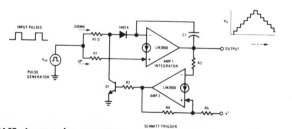

UP-DOWN STAIRCASE—Input pulse generator provides pulses which cause output of Norton current-differencing opamp to step up or down depending on conduction of clamp transistor Q1. When this is on, down pulse is diverted to ground so staircase steps up. Stepping is done between trip voltages of second opamp connected as Schmitt trigger.—T. M. Frederiksen, W. M. Howard, and R. S. Sleeth, "The LM3900—A New Current-Differencing Quad of ± Input Amplifiers," National Semiconductor, Santa Clara, CA, 1972, AN-72, p 24.

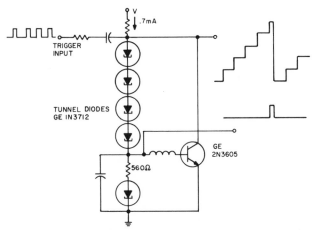

SERIES STRING OF TD'S USED AS 5:1
PULSE FREQUENCY DIVIDER OR STAIRCASE
WAVE GENERATOR

5:1 DIVIDER FOR STAIRCASE—Lowest tunnel diode is chosen to have highest peak-point current of string. All diodes are initially in low-voltage state, with 0.7 mA through string. Diode with lowest peak current is turned on first, creating voltage jump at output. Successive trigger pulses turn on each other diode in turn, bottom one last, after which transistor turns on and drops voltage across entire string to zero for start of next staircase sequence.—W. R. Spofford, Jr., "Applications for the IN3712 Series Tunnel Diodes," General Electric, Semiconductor Products Dept., Auburn, NY, No. 90.66, 1967, p 8–9.

CHAPTER 108
Stereo Circuits

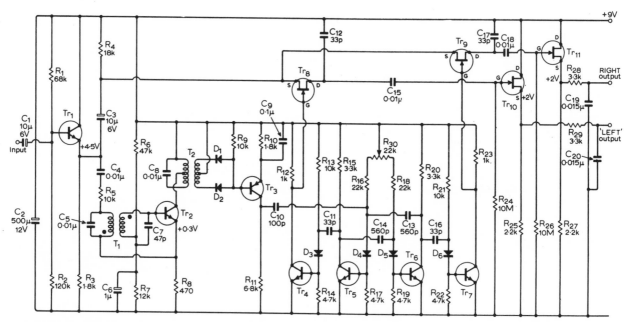

DECODER USING SAMPLING—Provides good channel separation, low distortion, and low f-m subcarrier breakthrough. Based on extraction of 19-kHz pilot tone from composite f-m stereo signal, generation of sampling pulses synchronized with and having correct phase relationship to pilot tone, sampling of multiplex signal, filtering out of unwanted components, and deemphasis. Sampling pulses are generated by continuously-running 38-kHz oscillator feeding sample-and-hold gates for right and left channels. Article covers construction and adjustment. TR1 is BC108 or 2N929. TR2 is 2N930. TR3 is 2N3702. TR4–TR7 are 2N2369. TR8–TR11 are 2N3819. —D. E. O. Waddington, Stereo Decoder Using Sampling, *Wireless World*, Feb. 1971, p 71–75.

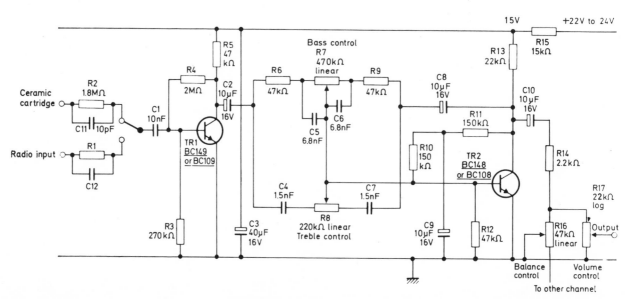

LOW-NOISE PREAMP—For 80-mV output, sensitivity is 180 mV for phono input. With 470K for R1, radio input sensitivity is 45 mV. Comprehensive tone control provides 10-dB boost and cut at 100 Hz and 10 kHz. Simple balance control is included for stereo use.— "Transistor Audio and Radio Circuits," Mullard, London, 1972, TP1319, 2nd Ed., p 44–45.

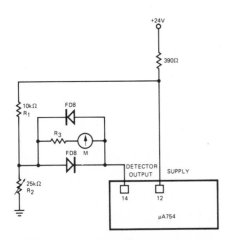

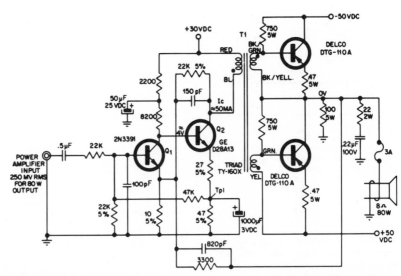

STEREO TUNING METER—Circuit shows how zero-center tuning meter can be connected to IC serving as four-stage limiting i-f amplifier, quadrature f-m detector, and audio driver for f-m stereo receiver. Value of R3 is chosen to give full-scale deflection of meter. R2 is adjusted only at setup, to accommodate variations found in d-c level of IC.—R. Smith, "Applications of the μA754 TV/FM Sound System," Fairchild Semiconductor, Mountain View, CA, No. 288, 1970.

80-W HI-FI—Designed for use in pairs to provide 160 W for stereo system using 8-ohm speakers. Response is flat within 1 dB from 20 to 20,000 Hz at 20-W output. Consists of two-stage driver Q1-Q2 transformer-coupled to half-bridge push-pull output stage direct-coupled to 8-ohm load. Bias circuit for output

transistors is thermally stable. Requires balanced 50-V unregulated supply and 30 V regulated. Report includes power supply, preamp, and tone control circuits.—"160 Watt Stereo Audio Power Amplifier Design," Delco, Kokomo, IN, Feb. 1972, Application Note 35.

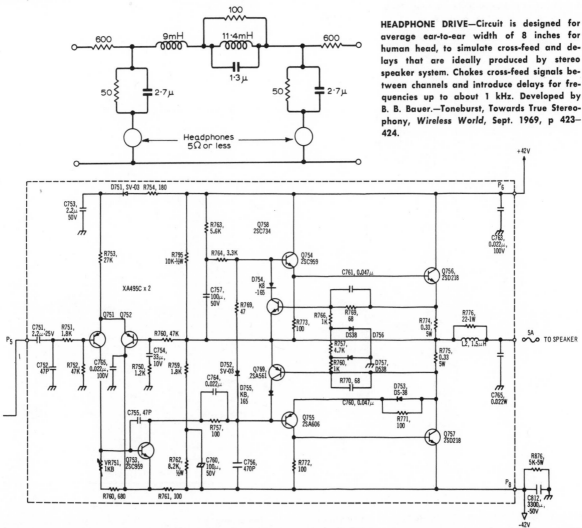

HEADPHONE DRIVE—Circuit is designed for average ear-to-ear width of 8 inches for human head, to simulate cross-feed and delays that are ideally produced by stereo speaker system. Chokes cross-feed signals between channels and introduce delays for frequencies up to about 1 kHz. Developed by B. B. Bauer.—Toneburst, Towards True Stereophony, Wireless World, Sept. 1969, p 423–424.

50-W MODULE—Used in foreign stereo receiver. Each channel (one shown) will deliver 50 W into 8-ohm load or 60 W into 4-ohm load. Power bandwidth is 15 Hz to 30 kHz. Performance of circuit is praised highly in this new-equipment review.—L. Feldman, BIC/LUX 71/3R AM/FM Stereo Receiver, Audio, Feb. 1972, p 50, 52, and 54.

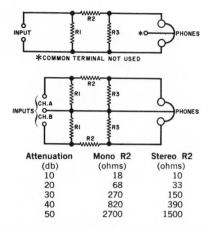

Attenuation (db)	Mono R2 (ohms)	Stereo R2 (ohms)
10	18	10
20	68	33
30	270	150
40	820	390
50	2700	1500

HEADPHONE ATTENUATOR—Used when hi-fi low-impedance headphones with 7-mW sensitivity for full volume are connected directly across speaker terminals of high-power amplifier. (Such amplifiers do not normally have milliwatt controls, and hum can become intolerable.) Both mono and stereo versions are shown. All resistors are ½ W except that R1 should be 1 W for 30 dB, 2 W for 40 dB, and 25 W for 50 dB. R3 is 15 for mono and 8.2 for stereo, and R1 is always 8.2 ohms.—J. Ashe, How to Use Hi-Fi Headphones, *Popular Electronics*, July 1972, p 76–77.

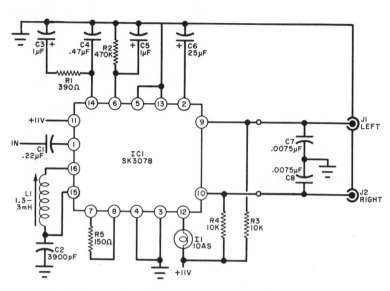

F-M MULTIPLEX DECODER—New RCA IC can be used to update older f-m receiver for multiplex stereo, or for incorporating latest circuits in present stereo f-m set. Includes stereo-on lamp. Decoder includes automatic switching from mono to stereo. Requires no alignment except for one adjustment that can be made with off-the-air signal. IC has built-in phase-locked loop to maintain accuracy of oscillator adjustment.—S. Reich, Unique Stereo Decoder, *Popular Electronics*, Oct. 1971, p 53–55 and 60.

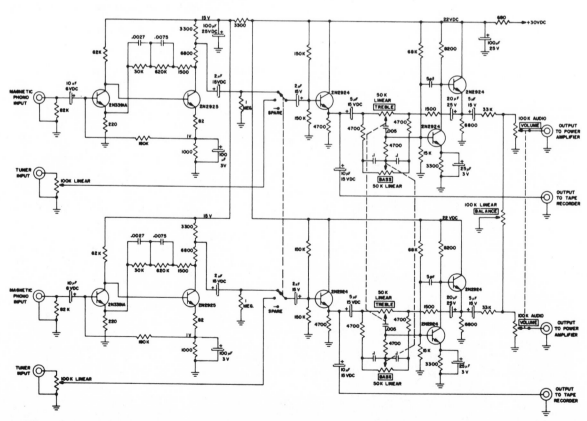

PREAMP WITH TONE CONTROLS—Although designed primarily for use with transistor audio amplifier rated 80 W per channel, can be used with any other stereo system. To minimize hum, insulate phono input jack from chassis. Use shielded cable and ground shields to input jacks and to ground point near preamp. Requires regulated 30-V supply. —"160 Watt Stereo Audio Power Amplifier Design," Delco, Kokomo, IN, Feb. 1972, Application Note 35.

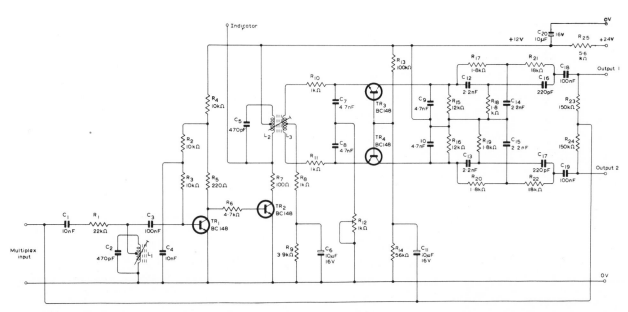

F-M DECODER—Connects to output of discriminator in f-m stereo receiver. Filter L1-C2 extracts 19-kHz pilot tone for amplification by TR1. Only positive half-cycles of amplified tone are further amplified by class B stage TR2, and second harmonic (38 kHz) is extracted by L2-C5. Secondary winding L3 applies anti-phase switching signals to TR3 and TR4 for synchronous detection to obtain difference signal along with matrixing of sum and difference signals. Resulting outputs 1 and 2 are fed to identical audio amplifiers of stereo system.—"Decoder for F.M. Stereo Transmissions," Mullard, London, 1968, TP989, p 2–4.

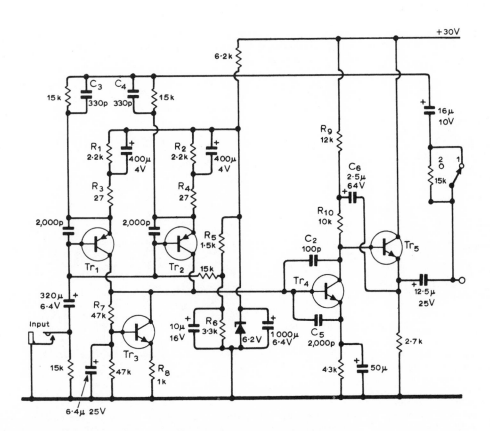

MIKE PREAMP—For use with medium-impedance mikes, between 200 and 1,500 ohms. Response is down 3 dB at 20 Hz and 25 kHz, and s/n ratio is not less than 60 dB. Paralleled TR1 and TR2 can be 2N4058, 2N4126, or 2N4289. TR3 is BC109 or similar. TR4 is 2N3707 or similar. TR5 is BC167 or similar. Switched sensitivity control in feedback loop makes circuit suitable for medium-impedance version of ribbon and moving-coil mikes as well as more sensitive capacitor mikes having about 1-mV output.—H. P. Walker, Stereo Mixer, *Wireless World*, May 1971, p 221–225.

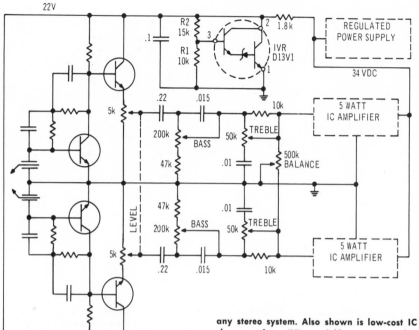

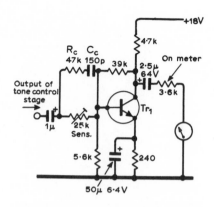

VU METER—Used with stereo mixer as guide for setting up system, checking pickup cartridges, checking channel balance, and monitoring output level. May be switched to other locations in stereo system. Quality of vu meter determines consistency of readings, particularly on transient signals. Transistor is BC109 or similar.—H. P. Walker, Stereo Mixer, *Wireless World*, June 1971, p 295–300.

STEREO CONTROLS—Bass, treble, and balance controls shown can be used between preamps and power amplifiers of practically any stereo system. Also shown is low-cost IC shunt regulator IVR, used like programmable zener in which ratio R2:R1 programs desired voltage of 22 V for phono preamp.—D. V. Jones, Constructing a 10-Watt Stereo Amplifier, *Audio*, May 1971, p 30 and 32–35.

DARLINGTON PREAMP—Designed for use with ceramic cartridge for driving load of at least 50K. Input of 100 mV will give 490-mV output at about 0.1% total harmonic distortion. Noise is 88 dB below maximum output signal. Only one channel is shown.—D. V. Jones, "Monolithic Darlington Preamplifier," General Electric, Semiconductor Products Dept., Auburn, NY, No. 90.98, 1971, p 3–4.

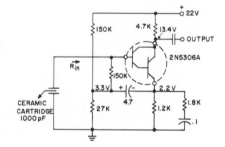

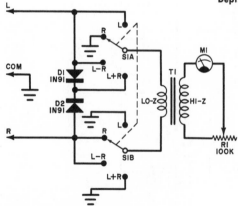

BALANCE METER—Passive circuit requires only three connections, to hot outputs of audio amplifier and to ground. M1 is Calectro D1-930 vu meter having two scales, one reading −20 to +3 volume units and the other voltage percentages, with 100% coinciding with 0 vu. Diodes provide sum and difference signals, with meter connected to read sum signal. Use only if amplifier outputs share common reference line. Article covers method of use. T1 is Stancor TA35.—J. R. Laughlin, Build a Signal Difference Stereo Balance Meter, *Popular Electronics*, April 1972, p 50–52.

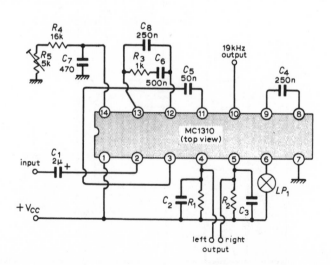

DECODER—Output of 76-kHz RC oscillator in Motorola IC (R5 sets frequency) is divided by four in two flip-flops to give 19-kHz output for multiplying incoming stereo pilot tone to produce d-c component extracted by low-pass filter R3-C6-C8 to control 76-kHz oscillator. Subsequent action energizes stereo 75-mA indicator LP1 and feeds switching decoder that produces left and right a-f outputs. With 8-V supply, use 2.7K for R1 and R2 and 18 nF for C2 and C3. Provides 40 dB stereo separation, and 485-mV output for 560-mV input.—Phase-Locked-Loop Stereo Decoder, *Wireless World*, July 1972, p 315.

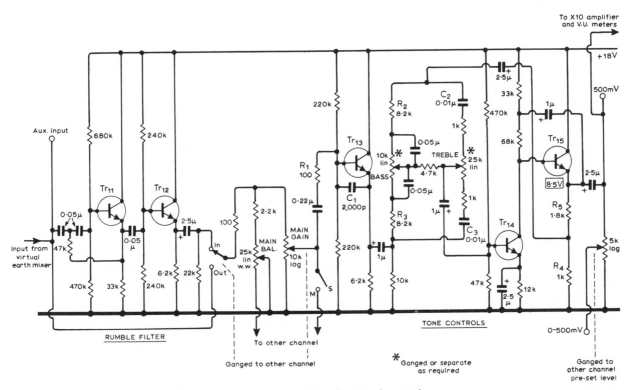

TONE CONTROL AND FILTER—Switchable rumble filter ahead of main gain and balance controls attenuates frequencies below 25 Hz at about 24 dB per octave. Tone control circuit has low interaction between bass and treble controls, low distortion even at maximum boost, and good square-wave response when controls are in flat position. Transistors are BC109 or similar.—H. P. Walker, Stereo Mixer, *Wireless World*, June 1971, p 295–300.

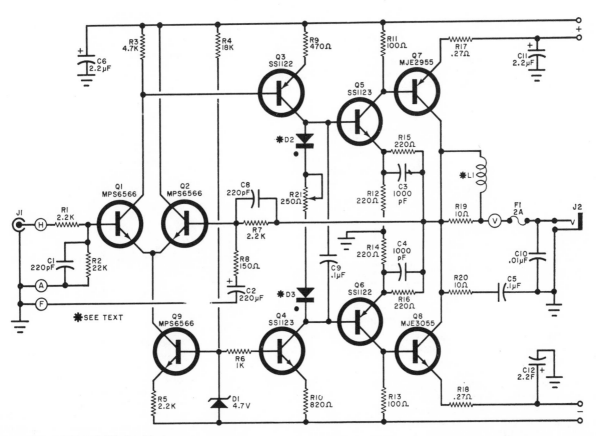

30 W FOR 15 TO 50,000 HZ—Provides outstanding performance as power amplifier for stereo system. Harmonic distortion is below 0.09% within power rating. Article covers construction and tells how to choose compensating diodes D2 and D3. L1 is single layer of No. 26 enamel close-wound on body of R19. —D. Meyer, The Plastic Tiger Audio Power Amplifier, *Popular Electronics*, Oct. 1971, p 27–34 and 100.

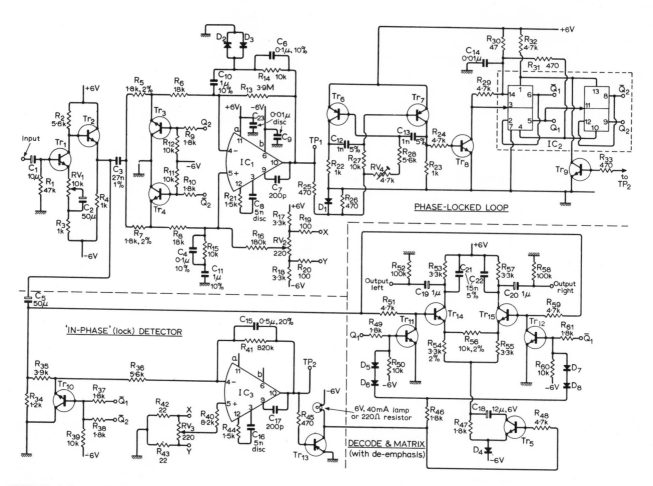

DECODER—Built around phase-locked loop in inductorless circuit providing improved channel separation and low distortion. IC1 and IC3 are type U6E770 9393 or equivalent, and IC2 is SN7474N. Transistors can be BC108 or equivalent for npn like Tr3, and ATX500 or equivalent for pnp. Diodes are 1S44 or equivalent. Article covers operation and performance of circuit in detail. Labeled terminals go to correspondingly labeled terminals, such as Q2 to Q2. Provides 45-dB separation of left and right channels at 1 kHz.—R. T. Portus and A. J. Haywood, Phase-Locked Stereo Decoder, *Wireless World*, Sept. 1970, p 418–422.

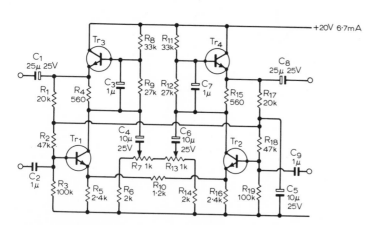

IMAGE WIDTH CONTROL—Permits changing apparent image width or spread of older stereo records. Circuit provides gain and gives fairly linear control of image width from 0% (mono) to about 165% simply by changing ganged controls R7 and R13. Operates at signal level of −10 dBm (250-mV rms), with overload factor of 18 dB so clipping occurs at about 1.75-V rms. Transistors are 2N930.—A. Roberts, Stereo Image Width Control, *Wireless World*, Dec. 1969, p 579–581.

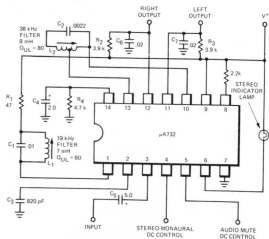

MULTIPLEX DECODER—IC contains functions of stereo demodulator, 19-kHz amplifier and doubler, stereo-monaural switch, stereo indicator lamp driver, and audio mute switch. External connections require only one 19-kHz tuned circuit. If stereo lamp gives false indications when f-m receiver is tuned off center carrier frequency, increase Q of 19-kHz filter. Supply is 12 V.—"The μA729, μA732 and μA767 Integrated Circuit Stereo Multiplex Decoders," Fairchild Semiconductor, Mountain View, CA, No. 286, 1971, p 6–7.

SOURCE WIDTH CONTROL—Ganged pots in both stereo channels permit varying apparent distance between sound sources of stereophonic system continuously, by adding part of signal voltage of one channel to other channel. Range limits are from 100% in-phase crosstalk (corresponding to mono operation) to 24% anti-phase crosstalk; higher anti-phase makes sound impression fall apart. Voltage gain of circuit is 0.5, input and output impedances are 750K and 47 ohms, and 3-dB response is 20 Hz to 20 kHz.—"Transistor Audio and Radio Circuits," Mullard, London, 1972, TP1319, 2nd Ed., p 180–181.

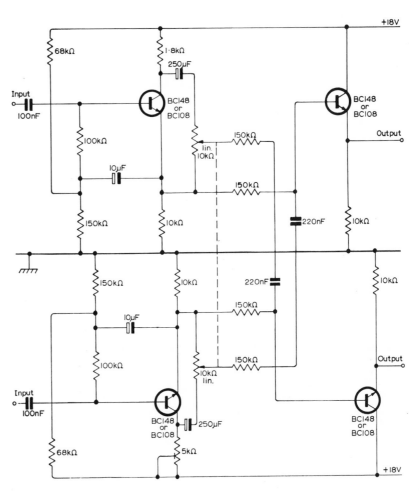

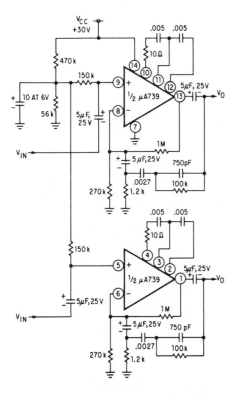

SINGLE-SUPPLY PREAMP—Identical amplifiers on single chip each provide RIAA-equalized gain of 40 dB at 1 kHz, for use with stereo phono pickup. Noise level is 2 μV referred to input and input overload point is 80-mV rms. —D. Campbell, W. Hoeft, and W. Votipka, "Applications of the μA739 and μA749 Dual Preamplifier Integrated Circuits in Home Entertainment Equipment," Fairchild Semiconductor, Mountain View, CA, No. 171, 1971, p 4.

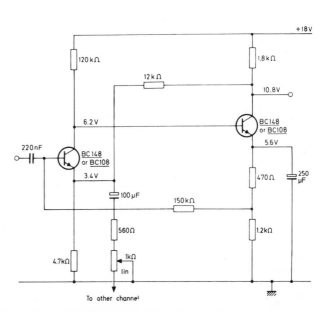

BALANCE CONTROL—Shown for one channel only. Permits varying voltage gain in both channels by 6 dB in opposite directions, with variable resistor inserted in feedback circuit. Average gain is 23.4 dB and 3-dB response is 20 Hz to 20 kHz.—"Transistor Audio and Radio Circuits," Mullard, London, 1972, TP1319, 2nd Ed., p 175–176.

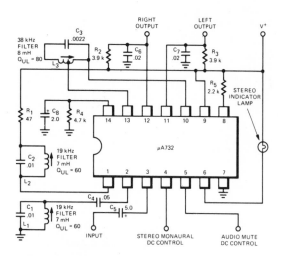

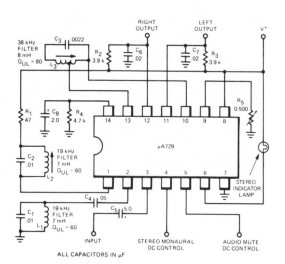

ALL CAPACITORS IN μF

MULTIPLEX DECODER—IC contains functions of stereo demodulator, 19-kHz amplifier and doubler, stereo-monaural switch, stereo indicator lamp driver, and audio mute switch. Uses fully balanced differential synchronous demodulator. Requires 19-kHz pilot level of 12-mV rms to turn stereo lamp on, and 8-mV rms to turn it off. Pin 4 requires 1.25-V d-c for stereo on and 0.85 V for stereo off. Pin 5 requires 1.2-V d-c for audio on and 0.85 V for audio mute. Supply is 12 V.—"The μA729, μA732 and μA767 Integrated Circuit Stereo Multiplex Decoders," Fairchild Semiconductor, Mountain View, CA, No. 286, 1971, p 5.

SIMPLE MULTIPLEX DECODER—Includes stereo demodulator, 19-kHz amplifier and doubler, stereo lamp driver, and audio mute, but requires external control R5 for separation adjustment. Supply is 12 V. Requires 19-kHz pilot level of 12-mV rms to turn stereo lamp on, and 8-mV rms to turn it off. Pin 4 requires 1.25-V d-c for stereo on and 0.85 V for stereo off. Pin 5 requires 1.2-V d-c for audio on and 0.85 V for audio mute.—"The μA729, μA732 and μA767 Integrated Circuit Stereo Multiplex Decoders," Fairchild Semiconductor, Mountain View, CA, No. 286, 1971, p 5.

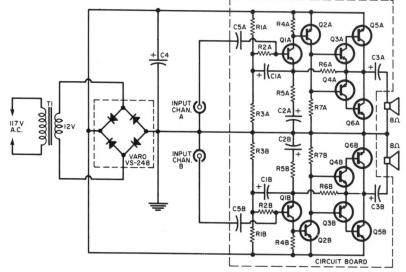

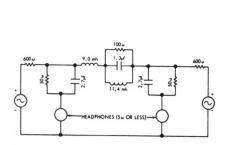

HEADPHONE BALANCE—Bauer cross-feed network for stereo phones permits introduction of crosstalk with appropriate phase and amplitude between channels to restore correct balance comparable to that heard from stereo speakers, even though channels are completely isolated by headphones.—C. G. McProud, Stereo Headphone Review, *Audio*, Dec. 1970, p 88.

R1—33,000 ohm, ¼ W res.
R2—39,000 ohm, ¼ W res.
R3—56,000 ohm, ¼ W res.
R4—12,000 ohm, ¼ W res.
R5—470 ohm, ¼ W res.
R6—22,000 ohm, ¼ W res.
R7—15,000 ohm, ¼ W res.
C1, C2—20 μF, 15 V elec. capacitor (Sprague TE1157)
C3—1000 μF, 15 V elec. capacitor (C-D CDE-1000-15)
C4—2000 μF, 25 V elec. capacitor (C-D CDE-2000-25)

C5—0.1 μF, 20 V elec. capacitor (Centralab UK-20-104)
T1—12 V at 2 A power trans. (Stancor P-8130)
Q1—2N3704, 2N3705, or 2N3706
Q2—2N3702 or 2N3703
Q3—2N1302, 2N1304, 2N1306, or 2N1308
Q4—2N1303, 2N1305, 2N1307, or 2N1309
Q5—2N5322 or 2N5323
Q6—2N5320 or 2N5321

3.7 W WITHOUT HEATSINKS—Provides clean 3.7 W per channel without overheating, with flat response from 15 Hz to 30 kHz. High-impedance inputs will take crystal stereo cartridge if volume control is added across input. Uses low-cost easy-to-get components. With no input, collector voltages of Q2A and Q2B should each be half of supply voltage. Voltage gain is 50.—W. W. Schopp, Small-Size Hi-Fi Stereo Amplifier, *Electronics World*, Jan. 1971, p 34.

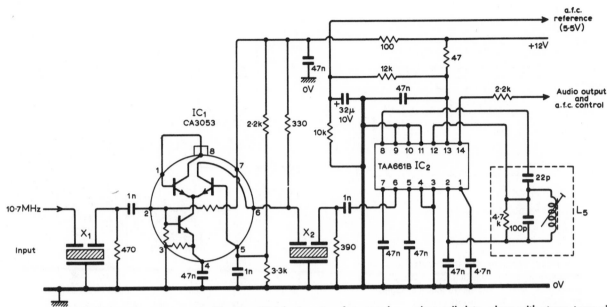

CERAMIC-FILTER I-F—Uses IC buffer amplifier between ceramic filters X1 and X2, which can be Vernitron FM-4 having same frequency between 10.625 and 10.775 MHz. IC2 is Siemens unit. Article covers construction and gives coil data, along with stereo tuner circuit.—L. Nelson-Jones, F.M. Stereo Tuner, *Wireless World*, April 1971, p 175–180.

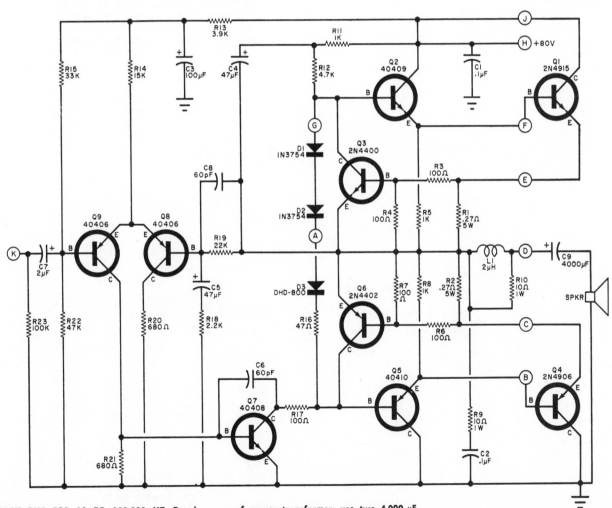

75-W RMS FOR 10 TO 100,000 HZ—Developed for use with stereo preamp and 4-ohm or 8-ohm hi-fi speaker system. Power supply can be unregulated, delivering 3 A at 80 V from 4-A bridge rectifier fed by 62-V secondary of power transformer; use two 4,000-μF filter capacitors and 10K bleeder. Other channel is identical and is operated from same supply. Distortion is under 0.5% total harmonic at 30-W output and 1 kHz. Article covers construction.—D. Meyer, Tigers That Roar, *Popular Electronics*, July 1969, p 51–53, 58–63, and 99.

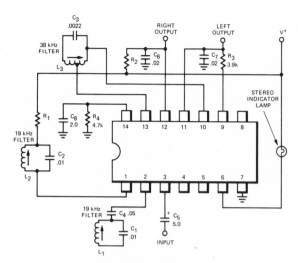

MULTIPLEX DECODER WITHOUT MUTE—Uses IC serving as stereo demodulator, 19-kHz amplifier and doubler, and stereo indicator lamp driver. Supply is 12 V. Report gives IC circuit and describes operation in detail.—"The μA729, μA732, and μA767 Integrated Circuit Stereo Multiplex Decoders," Fairchild Semiconductor, Mountain View, CA, No. 286, 1971, p 8.

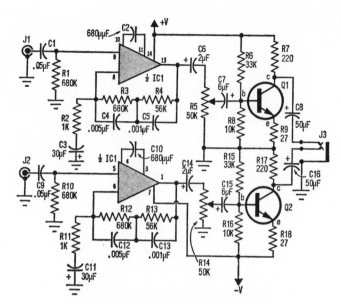

STEREO HEADPHONE AMPLIFIER—Preamp in each channel uses half of Motorola MC1303L IC, with input equalization for 3-mV magnetic cartridge. Amplifiers Q1 and Q2 are RCA 2N3242A silicon transistors for driving stereo headphones that can be 75 to 1,000 ohms each. Article gives circuit of simple unregulated a-c power supply that provides required ±14 V.—H. Friedman, The StereoFone, *Elementary Electronics*, Sept.-Oct. 1969, p 49–52 and 56.

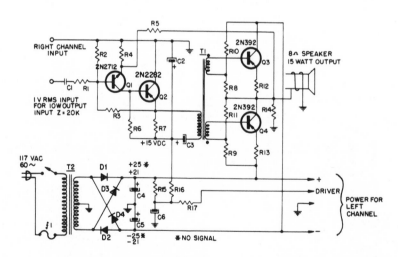

PARTS LIST

R_1	10K, ½W	R_{14}	100Ω, 1W	
R_2	33K, ½W	R_{15}	22Ω, 2W	
R_3	Select for Q_2	$R_{16,17}$	22Ω, 1W	
	V_{CE}= 8VDC	C_1	.47µF,12V	
	(100K is typical)	$C_{2,3}$	500µF, 15VDC	
R_4	75Ω, ½W	$C_{4,5,6}$	1000µF, 25VDC	
R_5	1.5K, ½W	$D_{1,2,3,4}$	Delco 1N2070 or 1N4004	
R_6	2.7K, ½W	f_1	1A, 3AG Fuse	
R_7	82Ω, 2W	Q_1	2N2712	
$R_{8,9}$	3.3Ω, ½W	Q_2	2N2282	
$R_{10,11}$	750Ω, 1W	$Q_{3,4}$	Delco 2N392	
$R_{12,13}$	.47 or .5Ω, ½W		Matched pair (Use Delco DTG-110 in high volume applications)	

T_1 Chicago Stancor TA-7 Core: EI-21, Radio Grade VI, ½" Stack, EI Laminations. Coil Form: Use nylon bobbin for EI-21. Winding Data: Simultaneously wind 2 #32 wires and 2 #28 wires to fill the nylon bobbin (approximately 250 turns). Interleave the EI Laminations in the bobbin. Connect the 2 #32 wires in series aiding to form the primary.

T_2 Triad F-92A Core: EI-12, M-19 Grade, 1" Stack, EI Laminations. Winding Data: Pri: 660 t #26 AWG Sec: 2 windings 120 T each #21 AWG Bifilar wind the secondary windings.

LOW-COST 15-W—Simple circuit has no noticeable distortion when used with good speaker system, even though limited to 6 W output at 20 kHz. T1 is wound multifilar to keep phase shift low at high frequency so feedback can be used without high-frequency oscillation when speaker impedance is much higher than nominal value. For greater power output down to 20 Hz, increase values of power supply filter capacitors.—"15 Watt Stereo Amplifier," Delco, Kokomo, IN, Nov. 1971, Application Note 31.

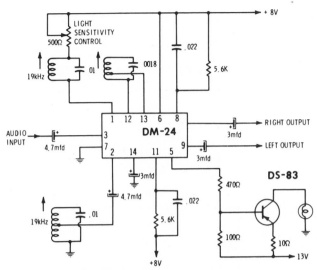

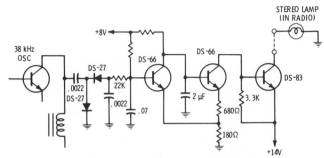

DEMODULATOR—Single IC contains most of stereo multiplex demodulator circuitry in 1971 Delco f-m stereo radios. Includes sensitivity control for stereo light.—J. J. Carr, 1971 Auto Radio Design—Changes That Affect Servicing, *Electronic Servicing*, April 1971, p 42–44 and 46–48.

STEREO LAMP DRIVE—Used in 1970 Delco f-m stereo radios to indicate that set is tuned to stereo station. Sample of signal at collector of 38-kHz oscillator is applied to half-wave diode bridge rectifier, to give negative output voltage whenever 38-kHz stereo signal is present. This turns off first transistor and turns on second transistor so as to drive lamp transistor into conduction.—J. J. Carr, A Look at Today's Stereo FM Auto Radio, *Electronic Servicing*, Dec. 1969, p 44–46 and 48–49.

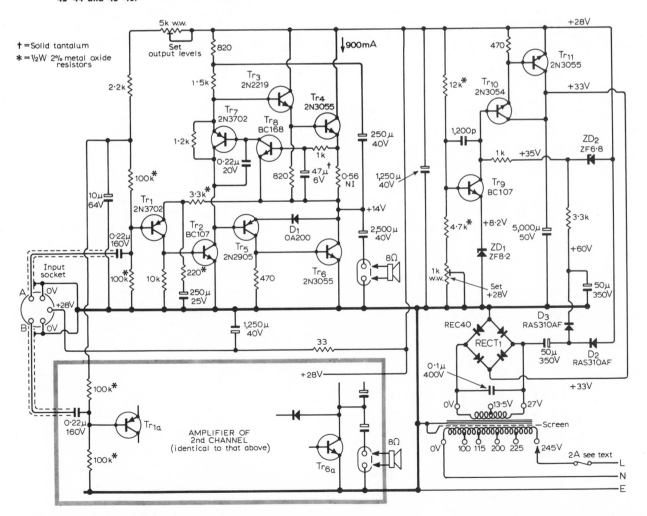

10-W CLASS A—Response is flat within 3 dB from 15 Hz to 92 kHz. Distortion is 0.015% at 1 kHz and full output. Channel separation is —43 dB at 20 Hz, rising to —60 dB at 1 kHz and above. Will deliver full power over entire bandwidth of 15 Hz to 30 kHz. Article gives construction details.—L. Nelson-Jones, Ultra-Low Distortion Class-A Amplifier, *Wireless World*, March 1970, p 98–103.

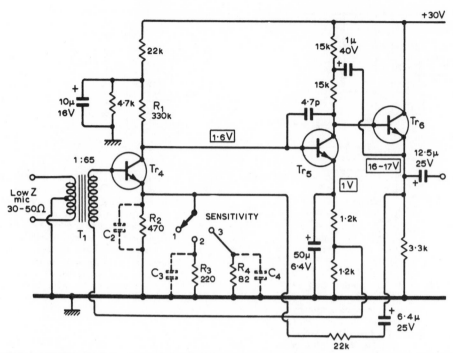

LOW-IMPEDANCE MIKE PREAMP—Bipolar input transistor TR4 is operated at very low collector current to provide good noise figure. Transformer secondary acts as d-c feedback loop, to avoid unnecessary attenuation of input signals. Bootstrapping improves linearity and contributes to excellent overload margin of 30 dB for 0.1% distortion. Switched feedback sensitivity control covers most mikes. Add C2, C3, and C4 as required to compensate for transformer losses. TR6 is BC107, BC167, or 2N3707. TR5 is 2N3707. TR4 is BC109, BC169, or BC1841C.—H. P. Walker, Stereo Mixer, *Wireless World*, May 1971, p 221–225.

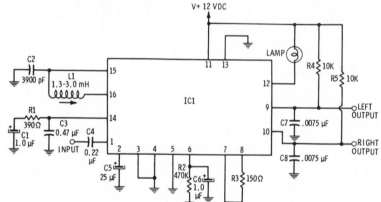

STEREO F-M ADAPTER—Single RCA CA3090Q IC containing equivalent of 128 transistors takes detector output signal of standard f-m receiver and separates stereo components for feeding to left and right stereo amplifier and speaker channels. Stereo indicator lamp is 10-V 14-mA, such as type 1869. Book gives construction and adjustment details.—L. Feldman, "Hi-Fi Projects for the Hobbyist," Howard W. Sams, Indianapolis, IN, 2nd Ed., 1972, p 106–113.

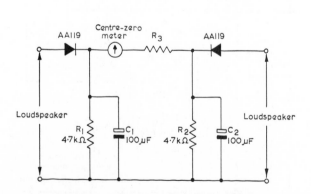

BALANCE METER—Connected between speakers of stereo system to balance outputs. Capacitors damp meter movement. R3 is typically 10K for 1-mA movement.—"Transistor Audio and Radio Circuits," Mullard, London, 1972, TP1319, 2nd Ed., p 258.

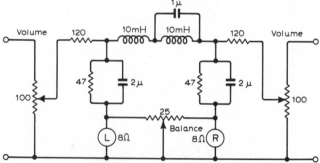

8-OHM HEADPHONE DRIVE—Developed to improve naturalness of stereo headphones spaced 8 inches apart on human head. Controls are adjusted until sound from phones creates same aural illusion as when listening to stereo speakers. Works best with high-quality phones having flat frequency response.—Toneburst, Towards True Stereophony, *Wireless World*, Sept. 1969, p 423–424.

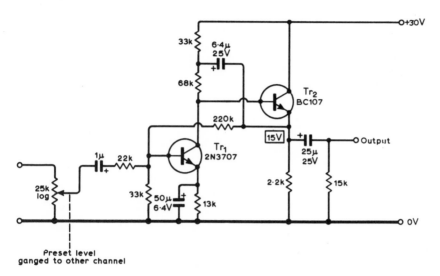

SINGLE-ENDED LINE AMPLIFIER—Provides gain of about 10, for use when stereo mixer must feed into 600-ohm termination at power level of several mW. Distortion is only 0.1% for output voltage swing of 2-V rms, and drops to 0.05% for 5-V swing with load above 10K.—H. P. Walker, Stereo Mixer, *Wireless World*, June 1971, p 295–300.

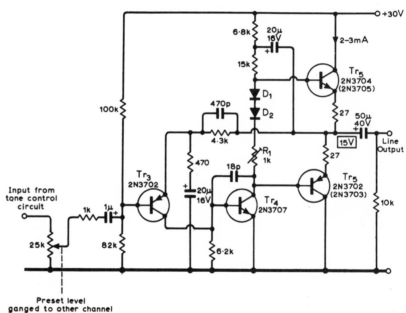

PUSH-PULL LINE AMPLIFIER—Provides gain of about 10, with voltage swing of 8-V rms into 600 ohms, for use when stereo mixer must feed into line at power level of several mW. Distortion is only 0.03% at +20 dBm at 1 and 10 kHz. Diodes are 1N914. Residual noise is less than —95 dB. Uses both a-c and d-c feedback loops.—H. P. Walker, Stereo Mixer, *Wireless World*, June 1971, p 295–300.

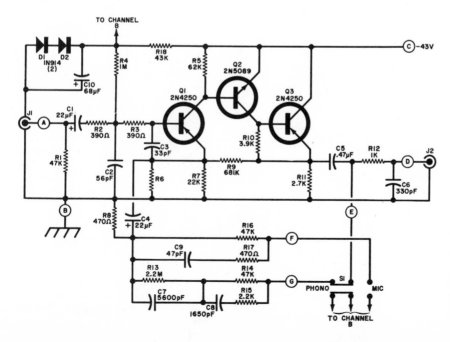

LOW-NOISE PREAMP—Cannot be overloaded at any frequency by any cartridge. Noise level is only 0.7 μV referred to input, or —83 dB below 10 mV. Distortion is only 0.1% at maximum output of 12 V rms, and just about unmeasurable below 4 V. Equalized for RIAA curve. Article covers construction, including power supply. R6 is 50K to 80K, determined experimentally. For stereo, second channel is identical.—J. Bongiorno, Build a Distortionless Preamplifier, *Popular Electronics*, June 1972, p 58–62.

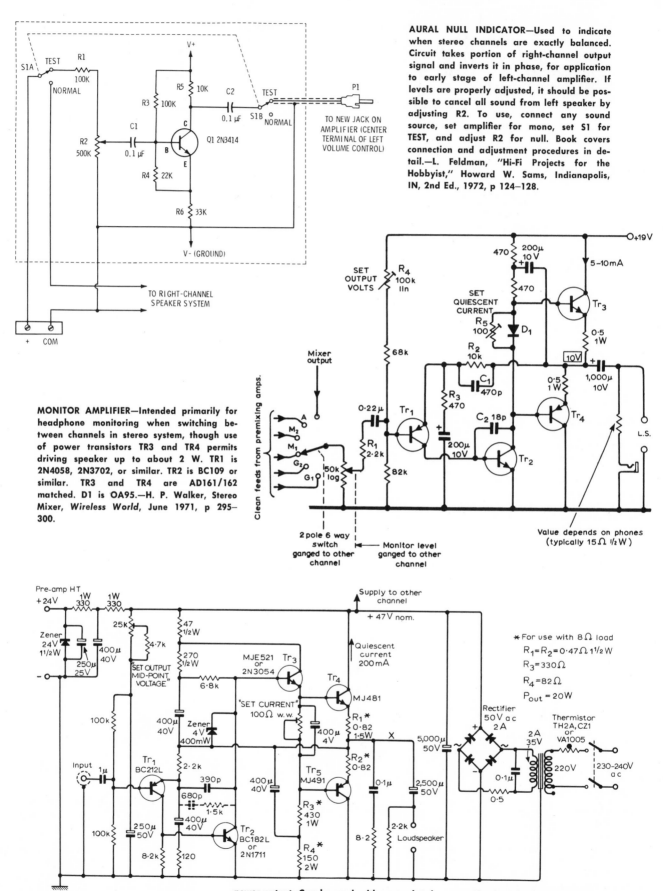

AURAL NULL INDICATOR—Used to indicate when stereo channels are exactly balanced. Circuit takes portion of right-channel output signal and inverts it in phase, for application to early stage of left-channel amplifier. If levels are properly adjusted, it should be possible to cancel all sound from left speaker by adjusting R2. To use, connect any sound source, set amplifier for mono, set S1 for TEST, and adjust R2 for null. Book covers connection and adjustment procedures in detail.—L. Feldman, "Hi-Fi Projects for the Hobbyist," Howard W. Sams, Indianapolis, IN, 2nd Ed., 1972, p 124–128.

MONITOR AMPLIFIER—Intended primarily for headphone monitoring when switching between channels in stereo system, though use of power transistors TR3 and TR4 permits driving speaker up to about 2 W. TR1 is 2N4058, 2N3702, or similar. TR2 is BC109 or similar. TR3 and TR4 are AD161/162 matched. D1 is OA95.—H. P. Walker, Stereo Mixer, *Wireless World*, June 1971, p 295–300.

20-W CLASS AB—Response is flat within 0.5 dB between 20 Hz and 50 kHz at maximum power output. Can be used with unregulated supply, but stabilized supply shown in article provides small improvement in already low hum and noise levels. Add dashed components if using electrostatic speakers. Has low thermal dissipation.—J. L. Linsley Hood, 15–20 W Class AB Audio Amplifier, *Wireless World*, July 1970, p 321–324.

BALANCE AND MONO SWITCH—Used ahead of mixer in five-channel stereo system. Wire-wound balance pot avoids crosstalk that would occur with high contact resistance of slider on carbon track. Use of 4.7K resistors ahead of mono-stereo switch provides proper mixing of stereo channels for mono.—H. P. Walker, Stereo Mixer, *Wireless World*, May 1971, p 221–225.

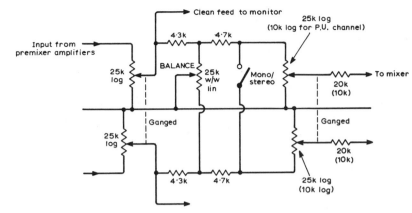

STEREO INDICATOR—5 K relay K1 is energized whenever f-m receiver is tuned to stereo broadcast, for turning on indicator lamp or audible buzzer. With buzzer, turn off SW after desired stereo station is located. L1 is 19-kHz multiplex oscillator coil (J. W. Miller 1354), adjusted along with R7 to make K1 pull in when input is stereo. T1 is Triad F-14X or other 6.3-V filament transformer.—R. Brown and T. Kneitel, "101 Easy Audio Projects," Howard W. Sams, Indianapolis, IN, 1971, p 66–67.

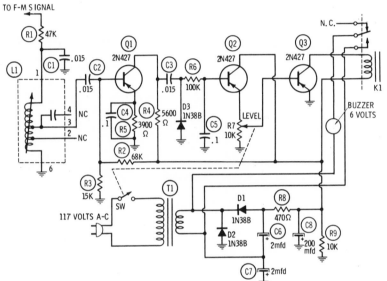

LOW-IMPEDANCE MIKE PREAMP—Includes high-frequency compensation to offset degrading effect of high secondary impedance, leakage inductance, and winding capacitance of stepup transformer. Fet at input has high input impedance, giving good noise figure and negligible attenuation of mike signals. R5 must be adjusted for each fet. Overload margin is 30 dB for 0.1% distortion. For 30-ohm mike, set sensitivity switch at 1 for 80–100 μV, 2 for 25–30 μV, and 3 for 12–20 μV. TR2 is 2N3707. TR3 is BC107, BC167, or 2N3707. Add C2, C3, and C4 as required to compensate for transformer losses. Change R3 to 220 ohms.—H. P. Walker, Stereo Mixer, *Wireless World*, May 1971, p 221–225.

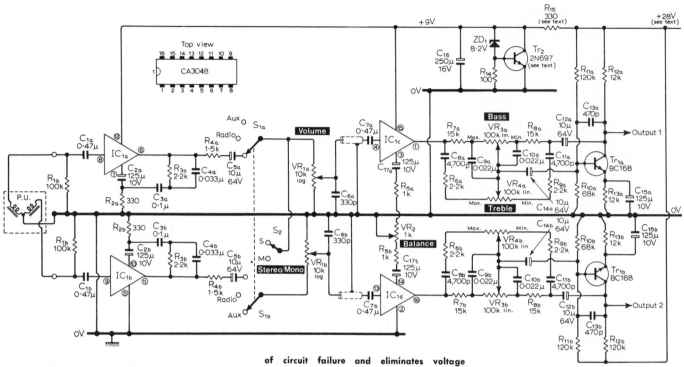

LOW-NOISE PREAMP—Uses single RCA IC containing four identical low-noise a-f amplifiers. Shunt regulator Tr2 protects IC in event of circuit failure and eliminates voltage surges at switch-on and switch-off. Response is flat within 1 dB between 30 Hz and 20 kHz. Uses active tone control to reduce noise.—L. Nelson-Jones, Integrated Circuit Stereo Pre-amplifier, *Wireless World*, July 1970, p 312–314.

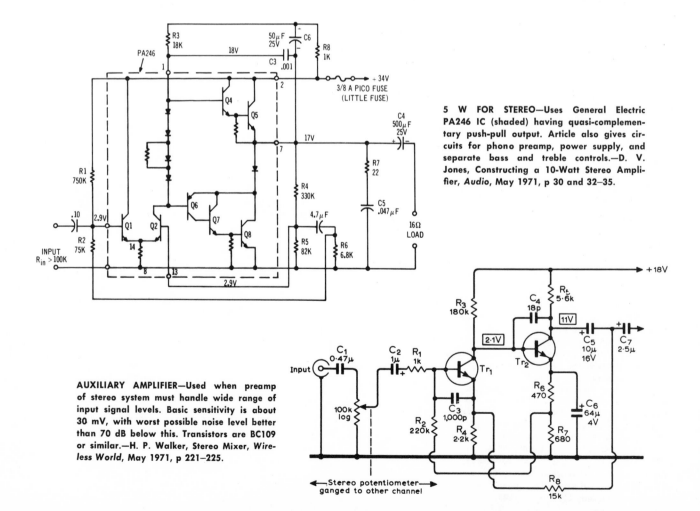

5 W FOR STEREO—Uses General Electric PA246 IC (shaded) having quasi-complementary push-pull output. Article also gives circuits for phono preamp, power supply, and separate bass and treble controls.—D. V. Jones, Constructing a 10-Watt Stereo Amplifier, *Audio*, May 1971, p 30 and 32–35.

AUXILIARY AMPLIFIER—Used when preamp of stereo system must handle wide range of input signal levels. Basic sensitivity is about 30 mV, with worst possible noise level better than 70 dB below this. Transistors are BC109 or similar.—H. P. Walker, Stereo Mixer, *Wireless World*, May 1971, p 221–225.

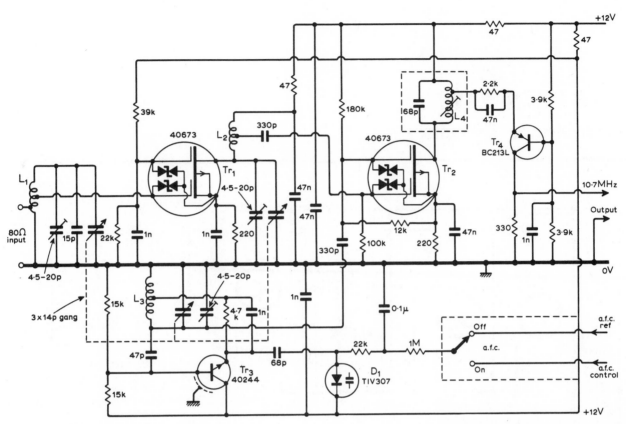

F-M TUNER—Both r-f amplifier and mixer stages use dual-gate fet's with gate protection diodes. Second gate of mixer is injection point for local oscillator voltage. Signal-noise ratio is better than 60 dB. Sensitivity for —3 dB limiting is 0.18 μV, and 0.75 μV for 20 dB quieting. Article also gives circuit for ceramic-filter i-f amplifier and quadrature demodulator.—L. Nelson-Jones, *F.M. Stereo Tuner, Wireless World*, April 1971, p 175—180.

CHAPTER 109
Surveillance Circuits

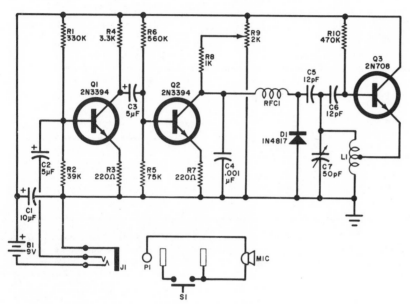

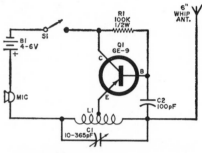

F-M WIRELESS MIKE—Gives range of 50 feet with short antenna and typical f-m broadcast receiver, using carbon mike. L1 is 36 turns No. 28 enamel close-wound on ¼-inch paper or plastic tubing, center-tapped. Adjust C1 so signal comes in at dead spot on f-m dial.— G. P. Golio, Reader's Circuit, *Popular Electronics*, May 1969, p 80–82.

F-M WIRELESS MIKE—Can be used with any standard f-m receiver. Uses as modulator a standard rectifier diode D1 operating as varactor across tuned circuit of modified Hartley oscillator Q3. L1 is 4 turns No. 18, ¼ inch diameter and 1 inch long, tapped at second turn from ground.—T. Duncan, Useful Circuits, *Popular Electronics*, Oct. 1971, p 86–88.

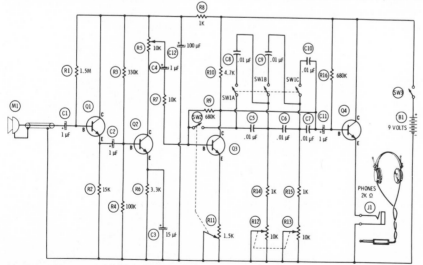

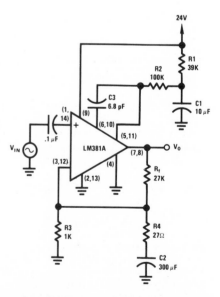

EAVESDROPPING AMPLIFIER—High-gain portable a-f amplifier is designed for use with crystal mike mounted in parabolic 12-inch metal disc that serves to concentrate sounds picked up at a distance. Selective filter, inserted with SW1 and SW2, is adjusted with R11-R12-R13 for optimum suppression of background noise. Q1 matches high impedance of crystal mike to input impedance of first amplifier Q2, without providing gain. Q2 is 2N5089 and other transistors are MPS6520. Construction details are given. Rifle scope is mounted on amplifier and aimed through hole drilled in plastic or metal reflector.— C. D. Rakes, "Solid State Electronic Projects," Howard W. Sams, Indianapolis, IN, 1972, p 109–115.

ULTRA-LOW-NOISE PREAMP—Designed for amplifying low-level signals in noise-critical applications such as surveillance tape recorders, instrumentation, hydrophones, and studio sound equipment. Wideband noise figure is 2.83 dB for 10 Hz to 10 kHz. Gain is 112 dB. Will operate on supplies from 9 to 40 V. Includes short-circuit protection.—J. E. Byerly, "LM381A Dual Preamplifier for Ultra-Low Noise Applications," National Semiconductor, Santa Clara, CA, 1972, AN-70.

TELEPHONE MONITOR—When 10K telephone pickup coil is connected to input, can be used for remote monitoring or recording both sides of telephone conversation. Gain is 20 dB. For conference recording, use 10K mike instead. Supply is two 9-V transistor radio batteries in series. Uses Fairchild IC.—Gabber Grabber, *Elementary Electronics*, Nov.-Dec. 1971, p 80.

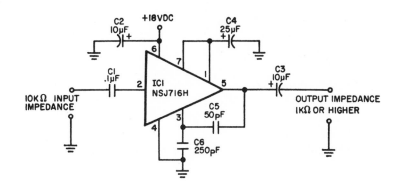

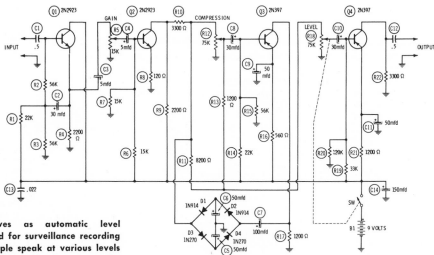

COMPRESSOR—Serves as automatic level control, as required for surveillance recording in room when people speak at various levels from various locations. Use between mike and input of recorder. Adjust gain control R18 for proper recording of softest-speaking person in back of room, then adjust compression control so loudest person is clipped down to normal level, and adjust level control for distortionless recording. Repeat all adjustments once if desired.—R. Brown and T. Kneitel, "101 Easy Audio Projects," Howard W. Sams, Indianapolis, IN, 1971, p 44—46.

A—Miniature transistor amplifier (Lafayette 99T9037)
B1, B2—9v alkaline battery (Mallory MN-1604)
C1—0.1 UFD, 200v
C2—1 UFD, 50v electrolytic
C3, C5—30 UFD, 16v electrolytic
C4—390 UFD, 12v electrolytic
C6—0.01 UFD, 150v ceramic

Q—PNP silicon transistor (GE D29A4)
R1—22k ½w ± 10%
R2—68k ½w ± 10%
R3—1.8k ½w ± 10%
R4—1k ½w ± 10%
R5—560-ohm ½w ± 10%
R6, R7—5,000-ohm linear taper potentiometer (Mallory U-12)
R8—33-ohm ½w ± 10%

NONMEDICAL STETHOSCOPE—Audio amplifier module with transistor preamp is designed for use with variety of surveillance pickups and for driving either speaker or phones. Crystal phono cartridge can be used as high-impedance vibration pickup held at end of rod "feeler" for picking up sounds through walls or vibrations at various points on auto or other engines or motors. Also high-impedance is capacitance pickup made from short length of shielded cable, used for detecting arcing, sparking, corona, and insulation breakdown in switches and other sparking devices. Ordinary telephone induction coil, mounted on suction cup for attaching to telephone handset when tape-recording phone conversations, serves as low-impedance pickup. With this, entire family group can hear phone conversation. Dynamic microphone cartridge serves as low-impedance acoustical pickup.—R. F. Graf and G. J. Whalen, Electronic Stethoscope Probes the Unhearable, *Popular Science*, Feb. 1970, p 99—101.

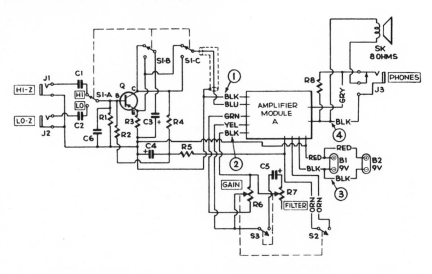

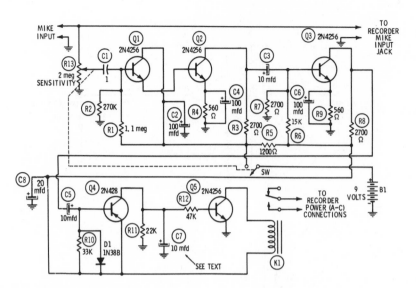

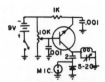

2-METER F-M TRANSMITTER—Range without antenna is at least 100 yards when used as miniature f-m transmitter for wireless mike. Audio is excellent even with 89¢ crystal lapel mike. Can be mounted in 2-inch-square metal box having end cut out to allow signal to radiate directly from tank circuit, without frequency shift due to hand capacitance. Transistor is Motorola MPS6512 and varactor is from Poly Paks. For 147 MHz, tank is two turns No. 16 ½ inch diameter and ½ inch long. For 432 MHz, use ½" × ½" hairpin loop. Can also be used testing 2-meter f-m transmitter, alignment, and r-f or i-f peaking.—W. Pinner, A Two Buck Signal Generator, 73, Aug. 1972, p 33.

VOICE TURN-ON—Use between mike and input of recorder to turn on recorder as soon as anyone begins speaking, and hold it on for about 0.5 s after last sound. Ideal for surveillance recording, to conserve tape and eliminate subsequent listening to blank tape run off while monitored room is vacant. K1 is 7-mA d-c current relay such as Sigma 11F-1000-G/SIL.—R. Brown and T. Kneitel, "101 Easy Audio Projects," Howard W. Sams, Indianapolis, IN, 1971, p 55–56.

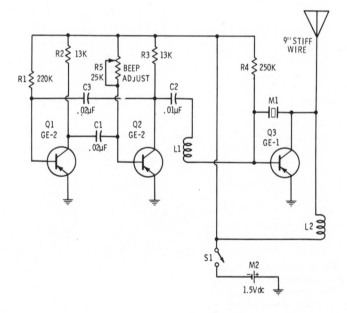

BEEPER TRANSMITTER—With antenna not over 9 inches long, puts out substantial signal legally on CB frequency determined by crystal used for M1, modulated with self-generated beeps. Can be used for following car under which it is concealed. Use third-overtone crystal with frequency between 26.980 and 27.260 MHz. Both r-f chokes are 1 mH.—R. M. Brown, "101 Easy CB Projects," Howard W. Sams, Indianapolis, IN, 1968, p 59–60.

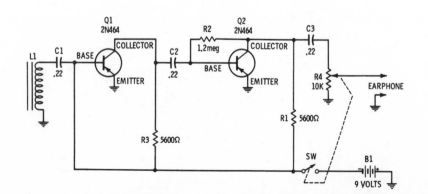

INDUCTION RECEIVER—Can be carried in pocket, for listening with transistor radio earpiece to commentary while studying exhibits in museum or art gallery wired for ultrasonic transmission. Each room has tape recorder and audio amplifier feeding loop antenna that completely encircles room. Each room has its own complete transmitting system, with appropriate recordings for its exhibits. L1 is standard telephone pickup coil like that used under handset to monitor telephone conversations.—R. M. Brown and T. Kneitel, "49 Easy Transistor Projects," Howard W. Sams, Indianapolis, IN, 1972, p 28–29.

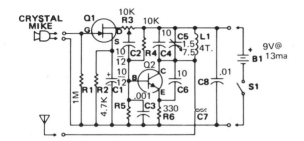

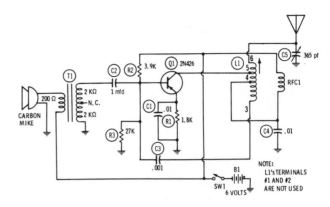

F-M WIRELESS MIKE—With short telescoping tv antenna, will transmit signals of crystal mike over distance of several blocks to f-m receiver tuned to quiet spot, with C5 adjusted until signal is picked up. Q1 is HEP802 n-channel fet transistor and Q2 is HEP55 npn silicon. C7 is gimmick made by twisting 2-inch lengths of insulated wire together.—"Calectro Handbook," GC Electronics, Rockford, IL, 1971, p 45.

A-M WIRELESS MIKE—Use about 22-inch length of stiff wire as antenna and adjust L1 and C5 to clear spot on nearby a-m radio. Input transformer T1 can be Argonne AR-123 and oscillator coil L1 is Miller 2020. RFC1 is 2.5 mH. Use four 1.5-V D cells in series for B1.—R. Brown and T. Kneitel, "101 Easy Audio Projects," Howard W. Sams, Indianapolis, IN, 1971, p 38–39.

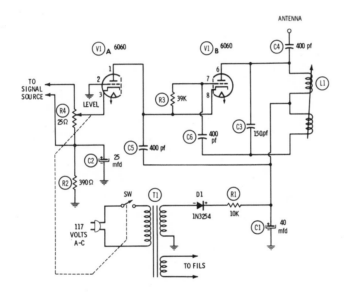

A-M TRANSMITTER—Can be used legally to broadcast tape recordings, records, or voice to any nearby a-m receiver. Operates from a-c line, using Olson Radio T-173 or equivalent isolation transformer with 117-V and 6.3-V secondary windings. L1 is Miller 71-OSC or equivalent, tuned to unoccupied frequency on a-m band of radio. To be legal, antenna should not be longer than 10 ft. Adjust modulation level with R4.—R. Brown and T. Kneitel, "101 Easy Audio Projects," Howard W. Sams, Indianapolis, IN, 1971, p 60–61.

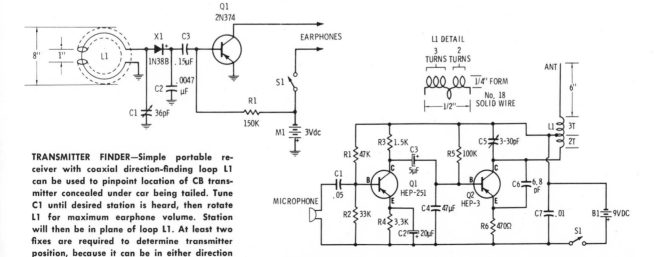

TRANSMITTER FINDER—Simple portable receiver with coaxial direction-finding loop L1 can be used to pinpoint location of CB transmitter concealed under car being tailed. Tune C1 until desired station is heard, then rotate L1 for maximum earphone volume. Station will then be in plane of loop L1. At least two fixes are required to determine transmitter position, because it can be in either direction from loop.—R. M. Brown, "101 Easy CB Projects," Howard W. Sams, Indianapolis, IN, 1968, p 113–114.

F-M WIRELESS MIKE—Will broadcast to f-m receiver up to 90 ft away. For voice, use crystal or ceramic mike; for music, use dynamic mike.—H. Friedman, "99 Electronic Projects," Howard W. Sams, Indianapolis, IN, 1971, p 113–114.

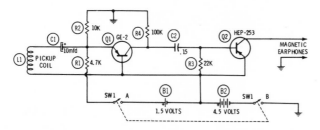

PICKUP-COIL AMPLIFIER—L1 is 650 turns No. 30 enamel scramble-wound on 1½-inch diameter form, concealed under phone near internal side-tone coil if used for surveillance. Both sides of conversation are heard in earphones of amplifier when telephone is in use. Shielded cable up to 12 ft long can be used between L1 and amplifier.—R. Brown and T. Kneitel, "101 Easy Audio Projects," Howard W. Sams, Indianapolis, IN, 1971, p 106–107.

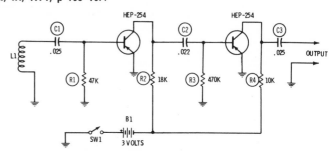

INDUCTION-COIL AMPLIFIER—L1 is standard telephone pickup coil placed as close as possible to telephone near internal side-tone coil. Amplifier can be up to 12 inches away from coil if shielded cable is used. Output will drive headphones or additional amplifier for speaker.—R. Brown and T. Kneitel, "101 Easy Audio Projects," Howard W. Sams, Indianapolis, IN, 1971, p 81–82.

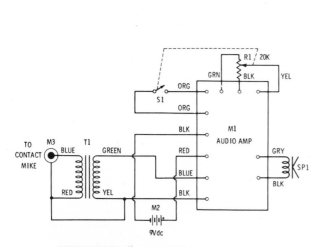

STETHOSCOPE—Uses standard contact mike and almost any a-f amplifier module, feeding 8-ohm speaker or low-impedance headphones. R1 adjusts volume. Designed for hard-of-hearing doctor, but can be used also for eavesdropping by taping mike to wall or floor. T1 is Lafayette 99C6034 audio transformer or equivalent.—R. M. Brown and T. Kneitel, "49 Easy Entertainment and Science Projects," Howard W. Sams, Indianapolis, IN, 1971, p 36–37.

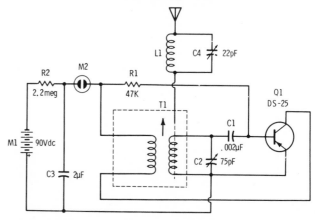

PILLBOX TRANSMITTER—Designed as beacon transmitter for sending out modulated tones continuously. Adjust C2 for oscillation and desired frequency in citizens band, then adjust C4 for optimum power transfer to antenna wire as indicated on nearby field-strength meter. T1 is J. W. Miller D-5495-C coil assembly, with adjustable slug serving same purpose as C2. L1 is 1.72 mH and M2 is NE-2 neon. Range is several blocks.—R. M. Brown, "101 Easy CB Projects," Howard W. Sams, Indianapolis, IN, 1968, p 103–104.

CHAPTER 110
Sweep Circuits

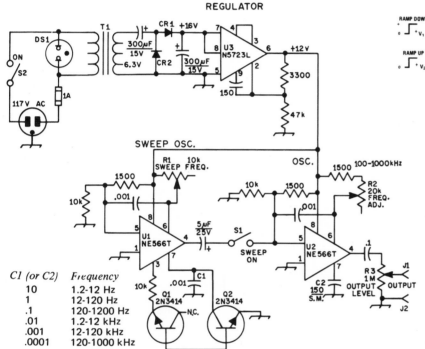

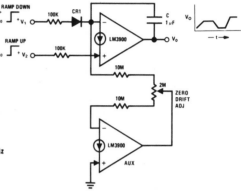

C1 (or C2)	Frequency
10	1.2-12 Hz
1	12-120 Hz
.1	120-1200 Hz
.01	1.2-12 kHz
.001	12-120 kHz
.0001	120-1000 kHz

RAMP AND HOLD—Simple low-drift circuit using Norton current-differencing opamps can be ramped up or down or allowed to remain in hold mode at any desired d-c output level. When both inputs are 0 V, circuit is in hold mode. Raising either input makes d-c output voltage to ramp go either up or down, depending on which input goes positive, with slope depending on input magnitude.—T. M. Frederiksen, W. M. Howard, and R. S. Sleeth, "The LM3900—A New Current-Differencing Quad of ± Input Amplifiers," National Semiconductor, Santa Clara, CA, 1972, AN-72, p 34–35.

SWEEP GENERATOR—Covers 1.2 Hz to 1,000 kHz in six ranges by using values shown in table for C1 and C2. Can be used for aligning f-m receiver i-f strips. Will also determine response characteristics of bandpass tuned circuits. Uses Signetics NE566T IC voltage-controlled oscillator (U1), which produces triangular and square waves simultaneously. R1 can shift frequency linearly over 10:1 range having upper limit of 1 MHz. Triangular-wave output is changed to sawtooth by Q1 and Q2. Lowest frequency ranges serve for rtty work.—A. E. Fury, A Simple Sweep Generator for FM Receiver Alignment, QST, Jan. 1972, p 48–49.

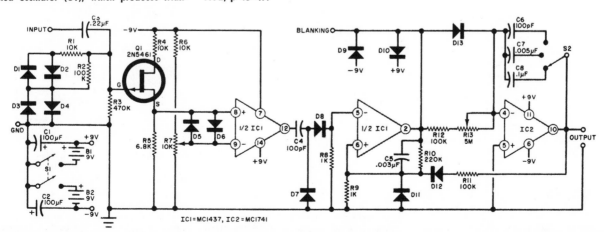

TRIGGERED SWEEP—Can be added to inexpensive cro to make waveforms stand still on screen. Also eliminates erratic multitriggering and permits accurate time and frequency measurements to be made on horizontal axis. Input can be either from existing scope input or from scope sync leads. Blanking output may be connected to scope blanking circuit if desired. All diodes are 1N914.—H. Garland and R. Melen, Add Triggered Sweep to Your Scope, Popular Electronics, July 1971, p 61–66.

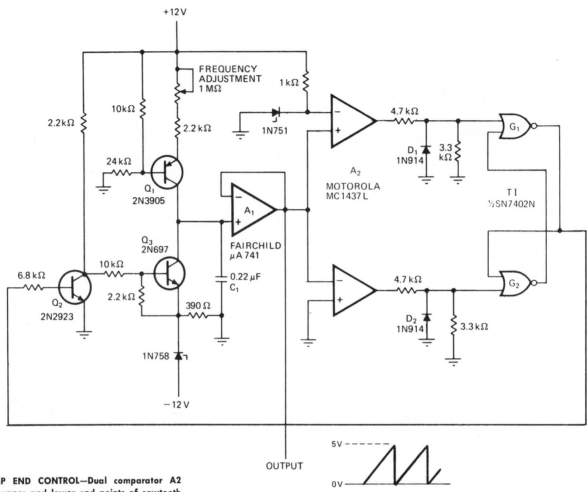

OUTPUT

RAMP END CONTROL—Dual comparator A2 sets upper and lower end points of sawtooth generator's ramp output, to make output independent of temperature. Frequency range is 0.33 to 1,000 Hz. Ramp limits are 0 and 5 V.—R. J. McKinley, Dual Op Amp Comparator Controls Ramp Reference, *Electronics*, Oct. 11, 1971, p 76.

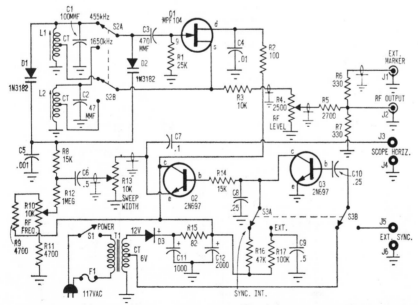

I-F SWEEP GENERATOR—Simple varactor-diode tuning of fet r-f oscillator Q1, under control of sawtooth generator Q2-Q3, sweeps 455 kHz over wide enough band to give i-f response curve of a-m superhet broadcast receiver on cro. Optional 1,650-kHz sweep required for double-conversion receivers is available at other setting of S2. L1 is Miller X-5496-C and L2 is A-5496-C.—C. Green, Dio-Tracer, *Elementary Electronics*, Jan.-Feb. 1969, p 35–39 and 109.

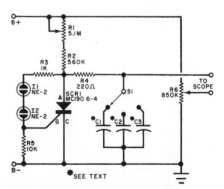

SCR SWEEP—Unique combination of neons with scr gives linear sawtooth signal having sufficient amplitude for direct horizontal drive of typical cathode-ray tube used in scopes. R6 provides d-c reference for centering beams. S1 gives coarse control of repetition rate and R1 serves as fine control. Choose capacitors experimentally to give desired operating frequency ranges, starting with values from 0.01 to 0.25 μF.—N. A. Steiner, Reader's Circuit, *Popular Electronics*, Sept. 1969, p 98.

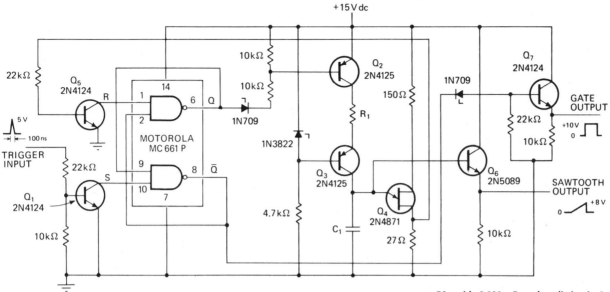

TRIGGERED SWEEP UP TO 10 S—Using dual high-threshold-logic NAND gate as set-reset flip-flop gives sweep generator providing 2% linear ramp from 10 μs to 10 s when triggered with 5-V 100-ns pulse. R1 can be in range from 100 ohms to 100K. C1 depends on R1, with 1,000 pF as low limit.—L. Sperling, Triggered Sweep Generator Responds to 100-ns Spike, *Electronics*, Aug. 16, 1971, p 79–80.

RAMP—Based on reversing bias of silicon planar transistor to make it behave as zener having typical breakdown voltage of 7–10 V. Transistor has negative resistance characteristic when connected as shown, permitting use as relaxation oscillator for generating ramp or sawtooth waveform.—J. A. H. Edwards, Negative Resistance of Transistor Junction, *Wireless World*, Jan. 1970, p 12.

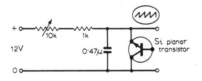

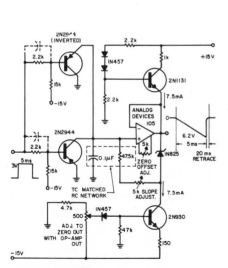

CHOPPER DISCHARGE—Precision ramp generator uses two 2N2944 chopper transistors in parallel to provide fast and complete discharge of feedback capacitor during retrace. Output is 5-ms negative-going ramp with linearity within 0.01%. Uses integrator opamp.—G. Richwell, An Op Amp and a Chopper Give Precision Ramp Generator, *Analog Dialogue*, March 1969, p 4.

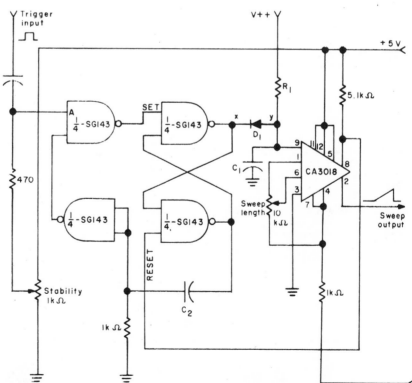

LABSCOPE SWEEP—Consists of control flip-flop, capacitor charging circuit, voltage comparator, and hold-off mvbr to prevent triggering of sweep during recovery time. C1 and R1 determine sweep rate, and C2 should be 0.1C1. Sweep linearity is excellent, and depends on values of V++ and R1, both of which can be very high.—C. Ulrick, Sweep Circuit Has Triggered, Free-Run Modes, *The Electronic Engineer*, April 1969, p 80.

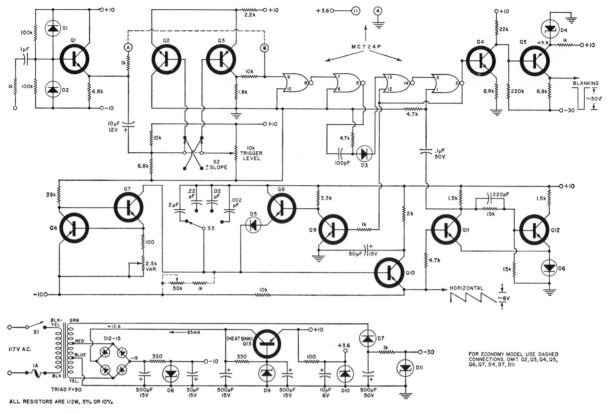

Q1, Q4, Q6, Q7, Q8, Q9, Q10, Q11, Q12—
  MPS6512, MPS3394, MPS2711 (or HEP729)
Q2, Q3, Q5—MPS6516 (or HEP52)
Q13—MJE520, MPS6552 (or HEP245)
D1, D2, D6, D7—1N4001
D3, D5—1N34 (or equiv.)
D4—6.2 V zener diode (1N4734 or 1N5235A)

D8, D9—11 V zener diode (MZ500-17, 1N4741,
  or 1N5241A)
D10—3.6 V zener diode (MZ500-5, 1N4728, or
  1N5227A)
D11—30 V zener diode (MZ500-27, 1N4751, or
  1N5256A)
D12, D13, D14, D15—MDA-920-1 bridge rectifier

(or HEP175)
Integrated Circuit—MC724P (or HEP570)

OUTBOARD SWEEP—Designed for use with low-cost general-purpose or service cro to provide triggered sweep requiring only 50-mV p-p trigger. Will work up to 1 MHz. Combination of 2.5K control and timing-capacitor switch S3 gives sweep rate range of 10 Hz to 100 kHz for locking desired number of cycles of pattern on screen. Beam is blanked automatically when there is no input.—I. Gorgenyi, Triggered Sweep For Any Scope, *Electronics World*, Nov. 1969, p 76–78.

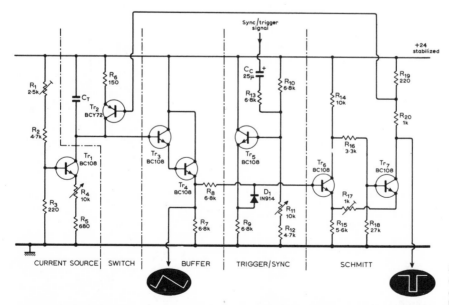

CURRENT SOURCE   SWITCH   BUFFER   TRIGGER/SYNC   SCHMITT

LINEAR RAMP—Designed for use as cro time base. Five-stage generator provides choice of free-running, synchronized, and trigger modes of operation. With 470 pF for CT, maximum ramp slope is 20,000 V/s. Article includes modification for further increasing ramp linearity.—J. B. F. Cairns, Linear Ramp Generator, *Wireless World*, Dec. 1971, p 604–605.

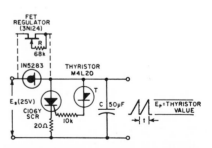

5-S RAMP—Simple circuit gives peak output of 20 V with values shown. Uses field-effect current-regulator diode in series with supply. Lower-cost optional fet regulator may be used in place of diode; choice of value for R then determines level of constant current.—A. Burns, Repetitive Ramp Generator, *Electronics World*, Oct. 1969, p 87.

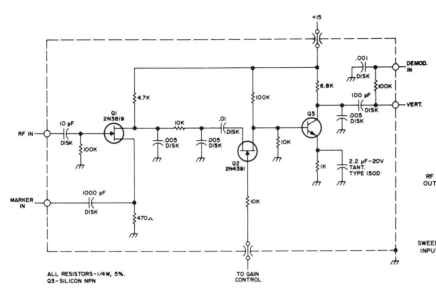

ALL RESISTORS-1/4W, 5%.
Q3-SILICON NPN.

POST-INJECTION MARKER—Used in 400 kHz—30 MHz lab sweep generator to provide control functions and permit use of external marker. Q2 is used as variable resistor to control signal input to amplifier stage Q3, which in turn determines height of marker on cro display. Q1 is mixer. Article includes all other circuits for sweep generator including regulated power supply.—R. Megirian, Lab Type IF/RF Sweep Generator Using IC's, 73, Nov. 1972, p 226–229, 232–234, 236–239, and 241–244.

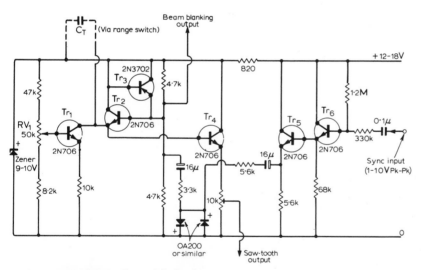

1 HZ—150 KHZ LINEAR—Five switched values of CT provide five frequency ranges for cro timebase generator having perfectly linear sawtooth output. Sync input, with impedance of 1 meg, is easily locked to low-level Y amplifier signals. Tr1 is constant-current source. RV1 changes frequency in each range.—R. M. Marston, Synchronized Oscilloscope Timebase Generator, Wireless World, June 1969, p 269.

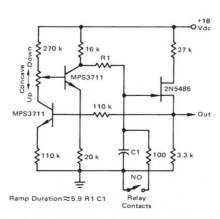

Ramp Duration ≈ 5.9 R1 C1

VARIABLE LINEARITY—Used in 10 Hz—10 MHz lab sweep generator to provide control over concave shape of vco output response curve between 2 and 24 MHz, when linearity far better than nominal 1.5% is desired. Circuit provides complementary signal, concave downward, for improving system frequency linearity.—R. A. Botos and L. J. Newmire, "A Synchronously Gated N-Decade Sweep Oscillator," Motorola Semiconductors, Phoenix, AZ, AN-540, 1971.

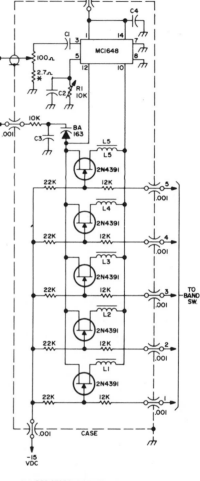

ALL RESISTORS-1/4W, 5%
L1-28 TURNS NO. 30 AWG ENAMEL.
L2-12 TURNS NO. 28 AWG ENAMEL.
  (L1 & L2 MOUNTED IN CUP CORE—See Text)
L3-18 µH MINIATURE MOLDED RFC.
L4-4.7 µH MINIATURE MOLDED RFC.
L5-1.1 µH MINIATURE MOLDED RFC.
C1, C2—TYPE 150D TANTALUM-2.2 µF/20V.
R1-BECKMAN 89PRIOK 15-T TRIMMER.

I-F/R-F SWEEP GENERATOR—R-f assembly, shown, uses d-c band-switching to cover 400 kHz to 30 MHz. When sweeping, return trace may or may not be blanked. Two calibrated dials set start and stop frequencies of sweep. Sweep time is variable between 20 ms and 6 s. Maximum output is 350 mV p-p across 50 ohms. Can also be operated in c-w mode as ordinary signal generator. Article gives all circuits and covers construction. Each of the five jfet's is in series with tank circuit, and serves as switch that is turned on by applying positive voltage to gate.—R. Megirian, Lab Type IF/RF Sweep Generator Using IC's, 73, Nov. 1972, p 226–229, 232–234, 236–239, and 241–244.

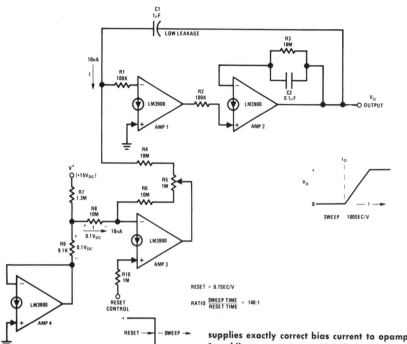

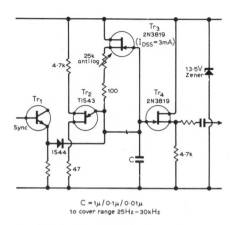

100 S/V SWEEP—Uses four Norton current-differencing opamps to generate very slow sawtooth waveform for producing long time delay intervals. Opamps 1 and 2 are cascaded to increase gain of integrator while giving desired output waveform. Opamp 3 supplies exactly correct bias current to opamp 1, while opamp 4 provides bias reference which equals d-c voltage at negative input of opamp 3. Values shown give sweep rate of 100 s/V and reset rate of 0.7 s/V.—T. M. Frederiksen, W. M. Howard, and R. S. Sleeth, "The LM3900—A New Current-Differencing Quad of ± Input Amplifiers," National Semiconductor, Santa Clara, CA, 1972, AN-72, p 22–23.

25 HZ TO 30 KHZ—Gives 10:1 frequency range for given value of C, with exceptionally good linearity. Uses fet as current source for timing circuit of ujt relaxation oscillator. —S. F. Weber, Linear Sawtooth Generator, *Wireless World*, Dec. 1970, p 580.

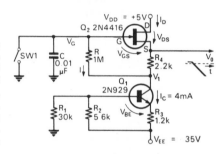

FET LINEAR RAMP—Constant-current source Q1 and source follower Q2 are combined with RC network coupling to give simple linear ramp output when switch is closed.— Some Applications of Field-Effect Transistors, *Electronic Engineering*, Sept. 1969, p 18–23.

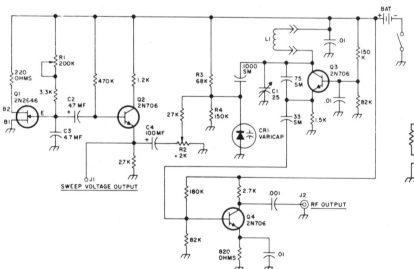

COIL TABLE		
FREQUENCY	DIAMETER	WINDING
7 MHz	7/8"	26 TURNS – 1 1/8" LONG
14 MHz	7/8"	7 TURNS – 1/2" LONG

ADJUSTABLE SWEEP—Easily assembled circuit will sweep over almost all of 40-meter band and over entire 20-meter and shorter-wavelength bands. Minimum sweep width is between 10 and 20 kHz. Can be used with cro to display frequency response of r-f tuned circuits. Uses 47-pF V12E Pacific Semiconductors varicap. Supply voltage can be 12 to 18 V. Ujt relaxation oscillator Q1 generates sawtooth used to provide voltage variation across varicap, with R1 setting sweep rate over range of 1 to 50 sweeps per second. R2 adjusts varicap voltage variation to control amount of frequency shift or sweep width.— C. Sondgeroth, An Experimental Sweep Oscillator, *73*, Feb. 1972, p 49–52.

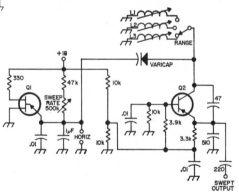

SWEEP FOR CRO—May be used at any spot frequency between 100 kHz and 60 MHz. Used for aligning communication receivers and vhf converters, plotting response curves, and checking bandwidth. Single ujt sawtooth generator Q1 (2N2646, 2N3480, or HEP310) provides sweep signal for horizontal input of cro and for fixed tuned r-f oscillator Q2 (2N1747, 2N2188, GE-9, or HEP2). Sweep rate can be varied between 5 and 30 sweeps per s with 500K control. Use Miller coils having ranges covering desired spot frequencies. If only 455 kHz is needed, use coil No. 9004 and omit range switch. Varicap is 56 pF, such as 1N955.—J. Fisk, "73 Useful Transistor Circuits," 73 Inc., Peterborough, NH, 1967, p 29A.

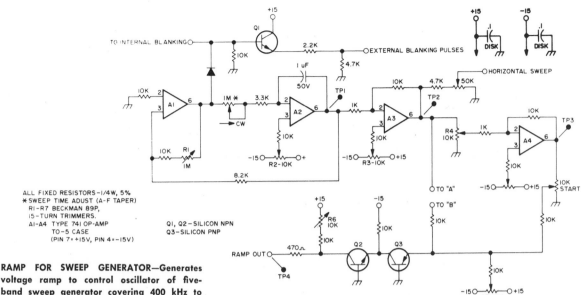

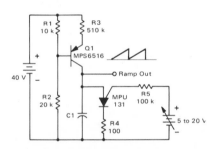

RAMP FOR SWEEP GENERATOR—Generates voltage ramp to control oscillator of five-band sweep generator covering 400 kHz to 30 MHz. Starting point of ramp can be adjusted independently of stopping point. Sweep time is also adjustable, between 20 ms and 6 s. Opamps A1 and A2 form triangle wave generator. Inverting opamp A3 gives positive-going ramp during sweep period, to serve as stop signal and horizontal sweep for cro. Inverting opamp A4 gives negative-going ramp for use as start signal. Article gives all other circuits and construction details.—R. Megirian, Lab Type IF/RF Sweep Generator Using IC's, 73, Nov. 1972, p 226–229, 232–234, 236–239, and 241–244.

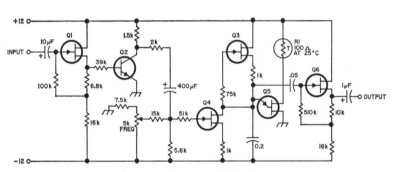

1–3 KHZ—Output waveform is linear within 2%. Q2 is 2N3391 or HEP54, Q5 is 2N3480 or HEP310, and other transistors are 2N3820 or HEP801.—J. Fisk, "Useful Transistor Circuits," 73 Inc., Peterborough, NH, 1967, p 29A.

VOLTAGE-CONTROLLED RAMP—Generates positive-going ramp having duration of 3 ms when C1 is 0.0047 μF and control voltage of put is changed from 5 to 20 V, and 5.4-ms ramp for 0.01 μF. Q1 is current source.—R. J. Haver and B. C. Shiner, "Theory, Characteristics and Applications of the Programmable Unijunction Transistor," Motorola Semiconductors, Phoenix, AZ, AN-527, 1971.

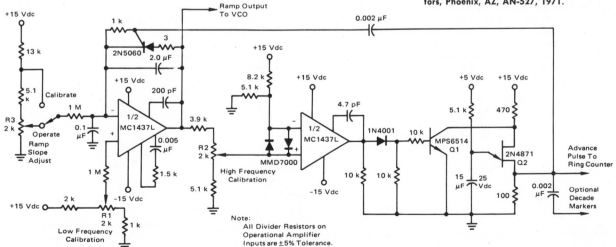

VCO RAMP—Used in 10 Hz–10 MHz lab sweep generator to generate ramp function of 6–10.5 V required for sweeping vco through frequency range. Sweep rate is 1 per 10 s, long enough to permit ample number of cycles of 10-Hz lowest frequency to be present at output. Report covers operation of ramp circuit and gives all other circuits used in sweep generator.—R. A. Botos and L. J. Newmire, "A Synchronously Gated N-Decade Sweep Oscillator," Motorola Semiconductors, Phoenix, AZ, AN-540, 1971.

CHAPTER 111
Switching Circuits

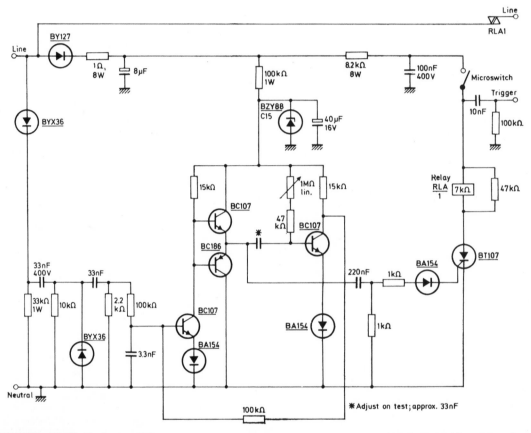

PHASE-CONTROLLED RELAY—Used to switch on line voltage at given phase angle of a-c waveform, as required when testing for dv/dt effects, overshoot, and current surges at switch-on of scr regulated power supply. Uses zero-crossing detector feeding mono mvbr that can be adjusted with 1-meg pot to make its output pulse occur at any point in a-c cycle, for turning on BT107 scr. Once scr is triggered, relay contacts begin to close. Phase angle at which relay contacts close is thus determined by setting of mono. Contacts remain closed until line microswitch is opened.—R. E. F. Bugg, Thyristor Power Supplies for Television Receivers: Design Considerations, *Mullard Technical Communications*, April 1972, p 129—144.

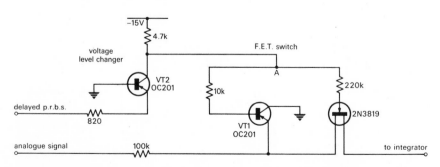

FET MULTIPLIER SWITCH—Used to pass either direct or inverted analog signal to integrator in pseudorandom binary sequence test signal generator. Voltage level changer provides re- quired logic level for operation of switch.— R. G. Young, P.R.B. Sequence Correlator Using Integrated Circuits, *Electronic Engineering*, Sept. 1969, p 41—45.

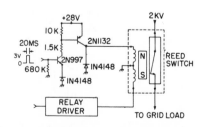

LOW-POWER REED DRIVE—Input pulse shown turns on transistors and energizes field coil of 2-kV reed switch. Resulting flux change makes switch hold in new state after pulse stops.—W. H. Holcombe, 17th Annual NARM Relay Conference, *Electromechanical Design*, June 1969, p 1—23.

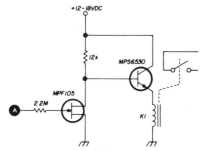

CARRIER-OPERATED RELAY—Works best when input terminal A is connected to grid of last 455-kHz i-f stage. Negative input voltage greater than about 3.5 V stops fet from conducting, allowing transistor to conduct and pull in relay for turning on tape recorder or monitor.—M. Ronald, Solid-State Carrier-Operated Relay and Call Monitor, *Ham Radio*, June 1971, p 22–23.

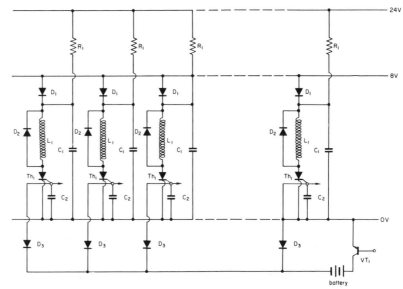

SWITCHING SOLENOIDS—Solenoids L1 are energized and turned off individually by applying control signals to gates of thyristors Th1 (TIC 145AO). Single signal applied to base of VT1 (2N1507) switches all solenoids off. D1 is 1S100. D2 and D3 are 1S130. C1 is 2,000-μF 25-V d-c. C2 is 0.1 μF. R1 is 500 ohms.—J. Budek, Solid-State Switching Using Triacs and Thyristors, *Design Electronics*, May 1970, p 59–63 and 126.

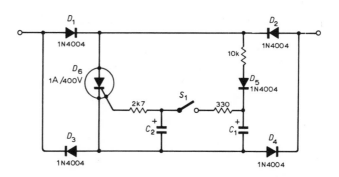

CONTACT TIME EXTENDER—When momentary switch S1 is released, thyristor D6 remains on for interval dependent on value of C1, which discharges into C2 when S1 is closed. Delay is about 1 s if C1 is 64 μF. Thyristor is triggered through 2.7K.—A. C. Grillet, Prolonging Effective Switch Contact Time, *Wireless World*, June 1972, p 274.

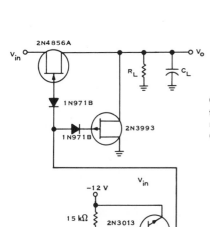

SERIES-SHUNT FET ANALOG SWITCH—Same gate drive signal switches both fet's, turning one off when other is on because they are of opposite polarity. Thus, n-channel fet allows CL to charge to Vin level, while p-channel fet quickly discharges CL to ground potential.—L. Delhom, "Switching-Circuit Applications of the 2N4856A Series FET," Texas Instruments, Dallas, TX, Bulletin CA-109, 1969, p 8.

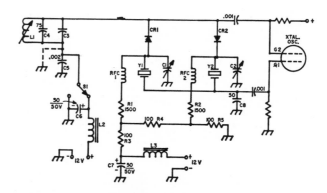

REMOTE SWITCHING OF UNGROUNDED CRYSTALS—Designed for use in Motorola 5V transmitter having L1 and C4. C3 is added to tune down into 2-meter band, using 30 to 40 pF, and netting capacitors C1 and C2 are added for individual frequency adjustment of crystals. S1 and L2-C6 are at remote-control location with separate 12-V supply. L2-C6 may be omitted if there is no hum on d-c leads. Requires only single control wire.—H. D. Johnson, Diode Switching for V.H.F. F.M. Channel Selection, *QST*, Oct. 1969, p 16–17 and 122.

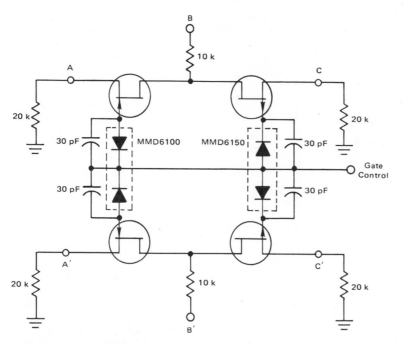

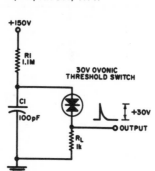

DPDT FET—Requires two MFE2012 n-type fet's (left pair) and two 2N3993 p-type fet's (right pair) to give solid-state equivalent of double-pole double-throw switch. Gate control voltage serves to connect terminal B to A or C and terminal B' to terminal A' or C'. On and off times are under 3 μs but vary with transistors. Switch takes longer to turn off than to turn on for either position.—B. Botos, "Low Frequency Applications of Field-Effect Transistors," Motorola Semiconductors, Phoenix, AZ, AN-511, 1971.

MEMORY CIRCUIT—Both sus's are initially off since supply voltage is only 5 V. Momentary closing of either switch will gate its sus on and keep it on until other sus is gated on. First sus then is commutated off through capacitor. When sus is on, output voltage at its anode is less than 1-V positive.—L. J. Newmire, "Theory, Characteristics and Applications of Silicon Unilateral and Bilateral Switches," Motorola Semiconductors, Phoenix, AZ, AN-526, 1970.

OVONIC SWITCH—Rise time of 30-V output pulse is reduced to less than 8 ns when neon of basic relaxation oscillator is replaced by 30-V ovonic threshold switch. Since this is bilateral, polarity of supply may be changed to give negative output pulse. Characteristics of switch are similar to those of Motorola MPT32 three-layer diode except for switching time.—R. D. Clement and R. L. Starliper, Relaxation Oscillators—Old and New, *Electronics World*, May 1971, p 36–37.

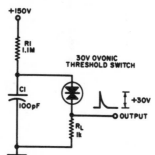

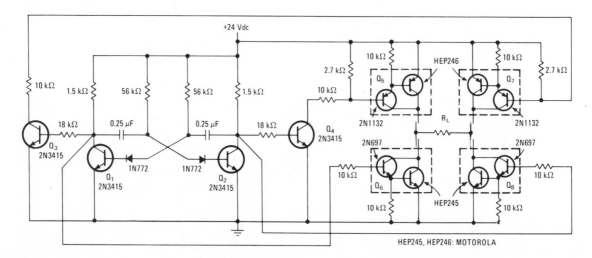

HEP245, HEP246: MOTOROLA

57-HZ DPDT—Used to reverse load current periodically to make square-wave load voltage equal to twice supply voltage. Astable mvbr Q1-Q2 determines switching frequency. Output stage has four Darlington pairs, with diagonally opposite pairs on while others are off. Q3 and Q4 alternate between saturation and cutoff 180 deg out of phase with Q1 and Q2. Gives high efficiency without requiring heavy power transformer of conventional inverter. Will supply 1.6-A a-c to 24-V load, but heavier load can be driven since HEP245 and HEP246 transistors are rated 3 A.—D. DeKold, Solid-State DPDT Switch Provides Current Reversal, *Electronics*, Feb. 1, 1973, p 100.

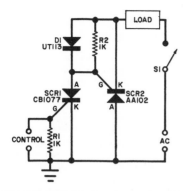

HIGH-SPEED SWITCH—Can be used as low-level replacement for triac static switch. With suitable pulse generator, will also serve as a-c proportional control. Furnishes full-wave a-c to load. Use 117-V 60-Hz a-c power, though circuit will operate up to 20 kHz. Requires positive control signal of only about 3 mA. Will drive loads of 1 mA to 1 A.—Manufacturer's Circuits, *Popular Electronics*, July 1971, p 81, 87, and 90–91.

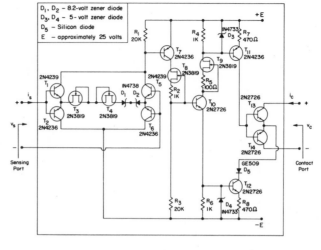

NORMALLY OPEN ELECTRONIC RELAY—Bilateral voltage-sensing circuit consists of isolation box combined with analog-to-binary converter using T1—T6 and zeners D1-D2, driving switching transistors T13-T14 through remainder of circuit which serves to provide sufficiently large turn-on current. T8 and T9 are current limiter used to compensate for supply voltage variations.—L. O. Chua, Theory and Design of Electronic Relays, *Proc. IEEE*, Nov. 1970, p 1818–1828.

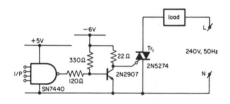

NAND CONTROL OF TRIAC—When all inputs of NAND gate are at positive level, pnp transistor is turned *on*, and triac is off. When any input of gate drops to low level, transistor is turned off and triac is turned *on* to energize load. Triac shown is rated at 25-A rms. Neutral of supply is not grounded.—J. Budek, Solid-State Switching Using Triacs and Thyristors, *Design Electronics*, May 1970, p 59–63 and 126.

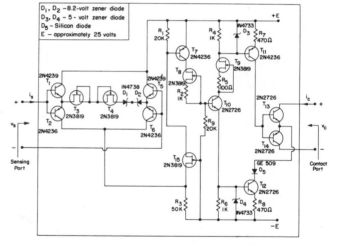

NORMALLY CLOSED ELECTRONIC RELAY—Bilateral voltage-sensing circuit drives switching transistors T13-T14 to provide turn-on or turn-off within about 20 μs, far exceeding speed of fastest mechanical relay. Except for addition of signal-inverting fet T15, circuit is essentially same as for normally open version.—L. O. Chua, Theory and Design of Electronic Relays, *Proc. IEEE*, Nov. 1970, p 1818–1828.

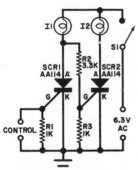

SWITCHING BETWEEN TWO LOADS—Uses silicon controlled switches to turn one lamp on when other is off, and vice versa. Used for driving indicator lamp, process solenoid valves, and fail-safe controls. Lamps are 6 V, such as No. 44. Control pulse should be about 1 V at 1 mA. For full lamp brightness, use 9-V a-c source.—Manufacturer's Circuits, *Popular Electronics*, July 1971, p 81, 87, and 90–91.

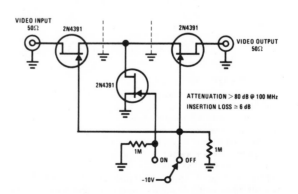

HIGH-FREQUENCY SWITCH—Uses transistors providing on resistance of only 30 ohms and extremely high impedance (less than 0.2 pF) when switch is off.—"FET Circuit Applications," National Semiconductor, Santa Clara, CA, 1970, AN-32, p 12.

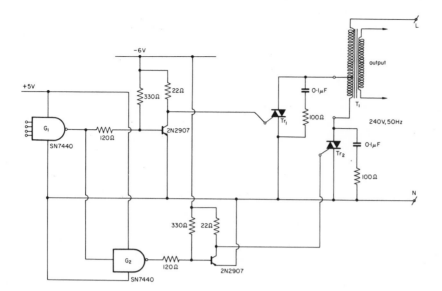

TRANSFORMER TAP-CHANGER—Two triacs, each controlling one tap of power transformer, are triggered by logic NAND circuit that eliminates possibility of triggering both triacs at same time and shorting out part of transformer. When all inputs to NAND gate G1 are at positive level, triac Tr2 is on and output of power transformer is lower. If any input of G1 drops to low level, Tr1 is turned on and transformer output is maximum voltage. Choose triacs to meet power requirements of application. G2, on same chip with G1, inverts output signal from G1 so one triac is off when other is on. Open —6 V supply to turn off both triacs.—J. Budek, Solid-State Switching Using Triacs and Thyristors, Design Electronics, May 1970, p 59—63 and 126.

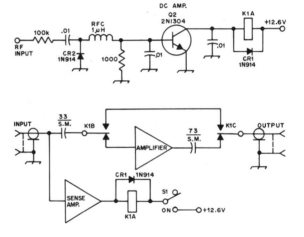

R-F POWERED RELAY—Used for switching r-f amplifier in and out of circuit. Circuit is designed to tune out reactance introduced by relay, to prevent mismatch. K1 is 4pdt relay with 10-A contacts and 12-V coil. Block diagram shows relay contact connections.—J. H. Johnson and R. Artigo, Fundamentals of Solid-State Power-Amplifier Design, QST, Nov. 1972, p 16—20.

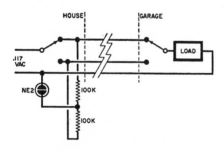

THREE-WAY SWITCH INDICATOR—Neon connected as shown lights only when switches at both ends are connected to same side of power line. Serves to determine off position of circuit being controlled, which depends entirely upon position of remote second switch. Circuit may be duplicated at other switch if desired. If neon is on and stays on when house switch is operated, load is open. If neon is on but goes off when switch is operated, load circuit is on. If neon is off, circuit is complete and load is off.—J. Kyle, The Ubiquitous Neon Lamp, Popular Electronics, Nov. 1968, p 55—57.

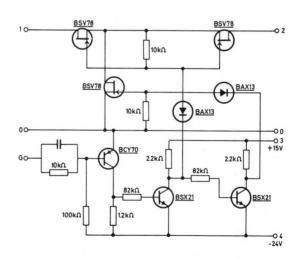

BIDIRECTIONAL ANALOG SWITCH—Used in multiplex switching circuits requiring high switching speed of fet's along with high off-resistance, low on-resistance, and low capacitance. Can handle voltage up to 30 V at 100 mA in either direction, opening at zero gate voltage and closing at 6 V. Series resistance of closed switch is 50 ohms, and leakage of open switch over 10,000 meg, with switching time less than 100 ns.—"Field Effect Transistors," Mullard, London, 1972, TP1318, p 84—85.

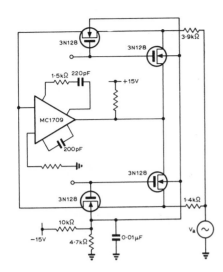

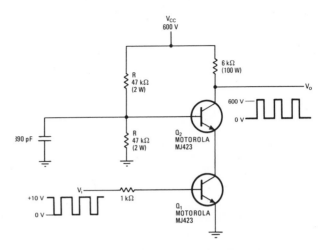

SHUNT SWITCH—Used as 10-bit digitally set potentiometer for hybrid computing techniques. Designed to operate for signal levels of 5 V with either polarity. Exhibits on resistance of less than 1 ohm for signals up to 10 kHz. Used with resistive ladder network.—S. Marjanovic and D. R. Noaks, A High Speed, High Accuracy, Digitally-Set Potentiometer, *The Radio and Electronic Engineer*, Dec. 1969, p 345–351.

SATURATING CASCODE SWITCH—Doubles transistor breakdown voltage without use of costly high-voltage zeners. Circuit generates 600-V rectangular positive-going output pulses while using only 350-V transistors.—P. T. Uhler, Doubling Breakdown Voltage with Cascoded Transistors, *Electronics*, Jan. 4, 1973, p 102.

CHAPTER 112
Tape Recorder Circuits

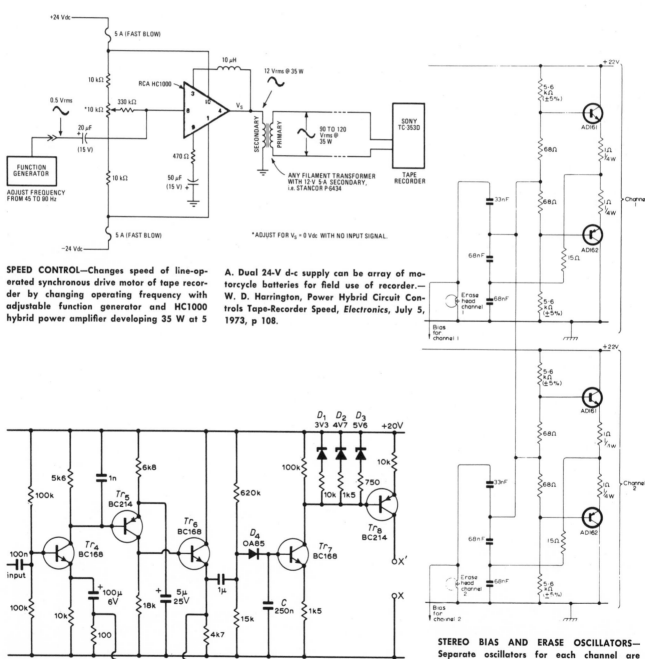

SPEED CONTROL—Changes speed of line-operated synchronous drive motor of tape recorder by changing operating frequency with adjustable function generator and HC1000 hybrid power amplifier developing 35 W at 5

A. Dual 24-V d-c supply can be array of motorcycle batteries for field use of recorder.— W. D. Harrington, Power Hybrid Circuit Controls Tape-Recorder Speed, *Electronics*, July 5, 1973, p 108.

LOGARITHMIC DETECTOR—Used in tape recorder to minimize noise. Provides log response over range of 100 μV to 10 mV for input (−60 dB to −20 dB).—J. R. Stuart, Tape Noise Reduction, *Wireless World*, March 1972, p 104—110.

STEREO BIAS AND ERASE OSCILLATORS— Separate oscillators for each channel are locked in frequency by connecting feedback paths together. Bias for each recording head is derived conventionally from voltage across its erase head.—"4W Tape Recorder for Mains Operation," Mullard, London, 1968, TP987, p 1—3.

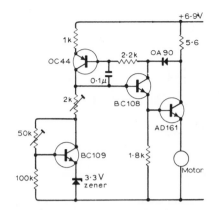

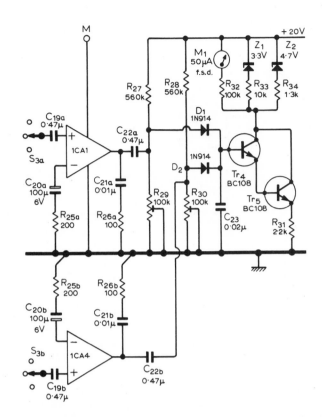

GOVERNOR FOR TAPE RECORDER—Circuit delivers constant output of about 3 V at 180 mA for motor of battery-operated recorder. Negative output resistance of circuit is adjusted to be slightly less than motor winding resistance of 10 ohms, to keep back emf of motor constant. 50K pot is fine speed control. Set 2K pot just short of hunting. Circuit is practically independent of battery and temperature changes.—D. Williams, Motor Speed Control, *Wireless World*, Dec. 1969, p 556.

LEVEL INDICATOR—Meter indicates on log scale positive peak value of whichever channel is greater at any instant in stereo tape recorder. Inputs to S3 come from ganged recording gain control at 0-dB level, and are boosted to 1 V rms by opamps for charging C23 to peak positive value through D1 or D2. Attack time depends on amplitude difference between successive peaks and is minimum of about 20 s. Decay time depends on rate at which C23 is discharged by high input impedance of amplifier TR4-TR5. Alternatively, meter can use A-B switching for checking individual channels.—J. R. Stuart, High-Quality Tape Recorder, *Wireless World*, Dec. 1970, p 587–591.

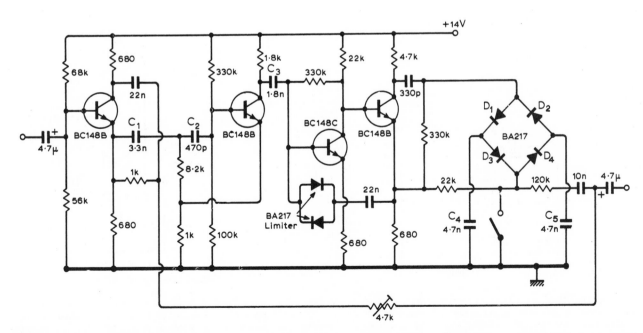

CASSETTE NOISE LIMITER—Developed by Philips to reduce noise 10 dB at 6 kHz and 20 dB at 10 kHz when signal on tape is zero at these frequencies. C1, C2, and C3 form part of high-pass filter. D1, D2, C4, and C5 form peak detector providing control voltage for attenuator diodes D3 and D4.—London Audio Fair, *Wireless World*, Dec. 1971, p 585–586.

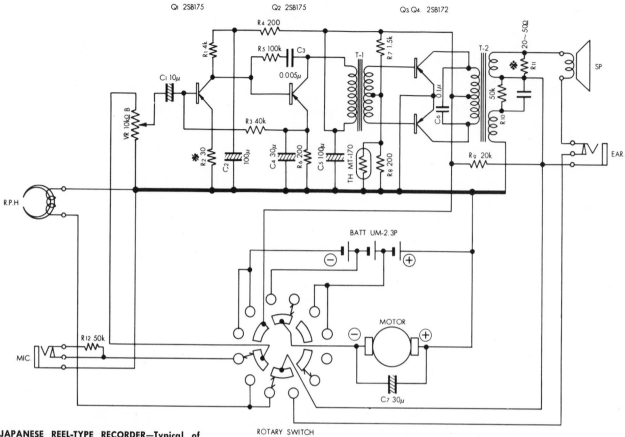

JAPANESE REEL-TYPE RECORDER—Typical of imported circuits for recording on 3-inch reels of tape, with same head used also for playback. Circuit can be used as troubleshooting guide for similar imported sets having same transistor lineup.—"Penncrest Model 6315 3-inch Reel-Type Recorder," J. C. Penney, 1301 Sixth Ave., New York, NY.

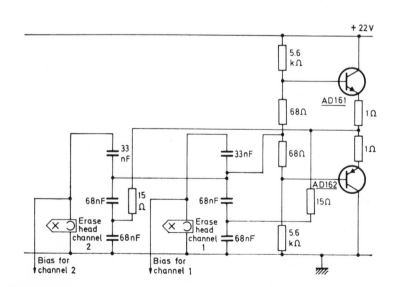

SINGLE-OSCILLATOR BIAS-ERASE—Two-transistor common oscillator serves for stereo recording operation. Tuned erase heads, which form load and feedback circuits of one of complementary pairs of transistors, are in parallel. Bias for recording heads is derived from voltages across erase heads just as in mono design. This permits use of other pair of complementary transistors as monitoring facility for output stage during playback.—"Transistor Audio and Radio Circuits," Mullard, London, 1972, TP1319, 2nd Ed., p 98.

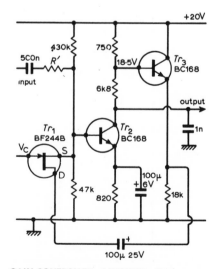

GAIN-CONTROLLED AMPLIFIER—Used to reduce noise in tape recordings. Will handle 0-dB level of 100 mV while gain-control device operates at 10 mV. R' is chosen for unity gain, in range of 120K. TR1 can be either diode or fet.—J. R. Stuart, Tape Noise Reduction, *Wireless World*, March 1972, p 104–110.

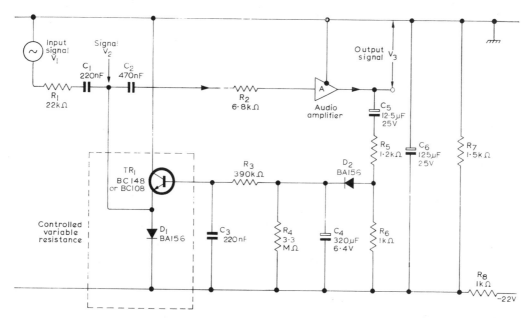

AGC—Used to prevent current in recording head from exceeding value at which tape saturation occurs, to prevent severe distortion. Circuit automatically attenuates signal before it reaches conventional or IC audio amplifier when optimum recording level is exceeded. Responds rapidly to overload signal, but has long recovery time so as to avoid volume compression. Can provide 40-dB attenuation within 150 ms.—"Transistor Audio and Radio Circuits," Mullard, London, 1972, TP1319, 2nd Ed., p 102–107.

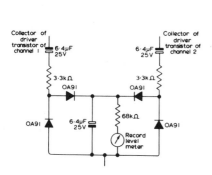

STEREO RECORDING LEVEL INDICATOR—Gives reading proportional to p-p voltage at collector of driver transistor in tape recorder channel having highest level. For input of 4 V rms, direct current through meter is 95 μA.—"4W Tape Recorder for Mains Operation," Mullard, London, 1968, TP987, p 1–3.

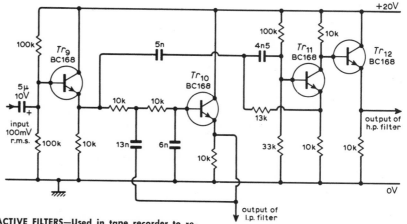

ACTIVE FILTERS—Used in tape recorder to reduce noise by compressing dynamic range of signal above 1.7 kHz. Both high-pass and low-pass filters have —3 dB point at 1.75 kHz. TR9 is input buffer. Used in system providing up to 20 dB compression above 1.7 kHz, for maximum noise reduction of 10 dB.—J. R. Stuart, Tape Noise Reduction, *Wireless World*, March 1972, p 104–110.

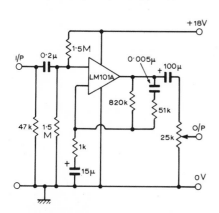

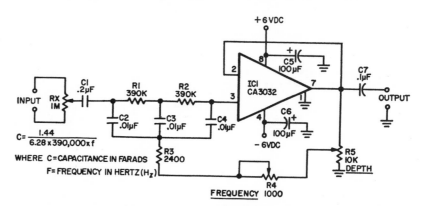

PLAYBACK PREAMP—Provides voltage gain of 50 at 1 kHz, along with 17-dB bass boost and 10-dB treble cut.—T. D. Towers, Elements of Linear Microcircuits, *Wireless World*, Feb. 1971, p 76–80.

60-HZ HUM FILTER—Used between playback recorder and amplifier when 60-Hz hum level is objectionable on recorded tape. For 50-Hz or other hum frequency, use formula to compute new values for C2, C3, and C4. R4 provides slight tuning, and R5 controls degree of attenuation. RX is not needed if playback recorder has output level control.—Active Hum Filter, *Elementary Electronics*, May-June 1971, p 52.

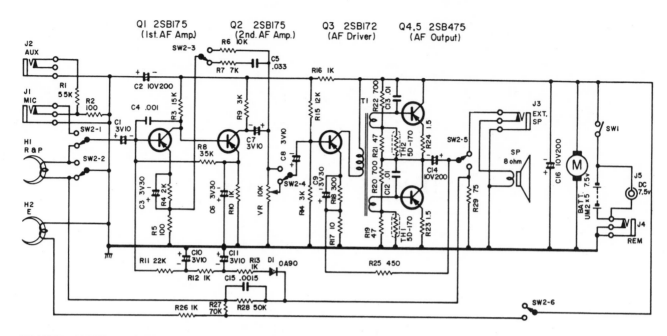

JAPANESE CASSETTE RECORDER—Circuit is typical of imported cassette tape recorders operating from five dry cells in series, and can therefore be used as rough guide for other Japanese recorders having similar transistor lineup.—"Lloyd's Model 9V95-114A Portable Cassette Tape Recorder," Lloyd's Electronics Inc., Saddlebrook, NJ.

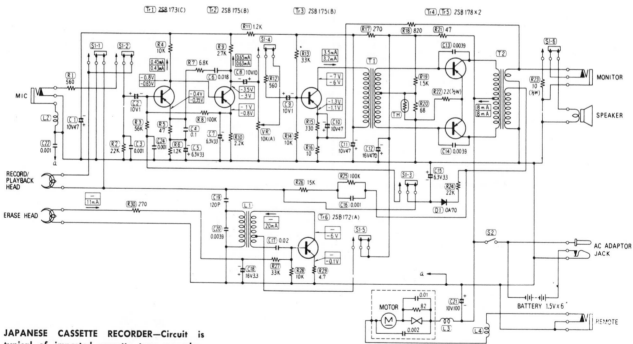

JAPANESE CASSETTE RECORDER—Circuit is typical of imported cassette tape recorders operating from six dry cells in series, and can therefore be used as rough guide for other Japanese recorders having similar transistor lineup. S1 is in playback position. D-c voltages are to chassis ground with no signal; upper values are for playback and lower for recording.—"Panasonic Model RQ-204S Portable Cassette Recorder," Matsushita Electric Corp. of America, 200 Park Ave., New York, NY 10017.

DELCO TAPE PLAYER—Centrifugal speed regulator contacts in motor turn on transistor connected to maintain relatively constant speed in Delco T-400 series tape players.—J. J. Carr, Troubleshooting Motor Circuits in Auto Tape Players, *Electronic Servicing*, June 1971, p 30—36.

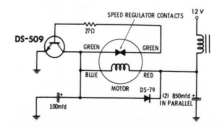

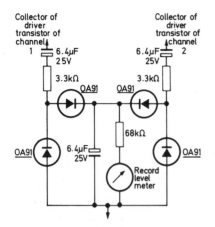

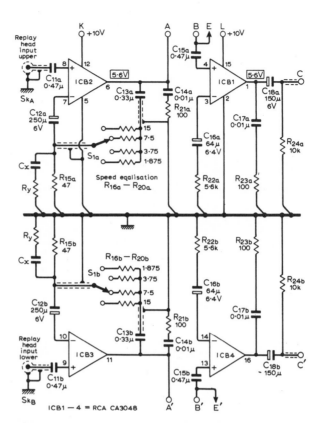

STEREO RECORDING LEVEL METER—Meter gives reading proportional to p-p voltage at collector of driver transistor in channel having highest recording level. For input of 4-V rms, at which distortion due to tape saturation becomes significant in most tape recorders, d-c meter current is 95 µA.—"Transistor Audio and Radio Circuits," Mullard, London, 1972, TP1319, 2nd Ed., p 100.

STEREO PLAYBACK—Designed as equalization stage having 20-dB gain to raise output level on each channel from 2 mV to 250 mV. Measured noise is 66 dB below 0 dB with 7.5-ips CCIR replay equalization in 20-kHz band. Cx and Ry are optional, for boosting high-frequency response.—J. R. Stuart, High-Quality Tape Recorder, *Wireless World*, Nov. 1970, p 524–529.

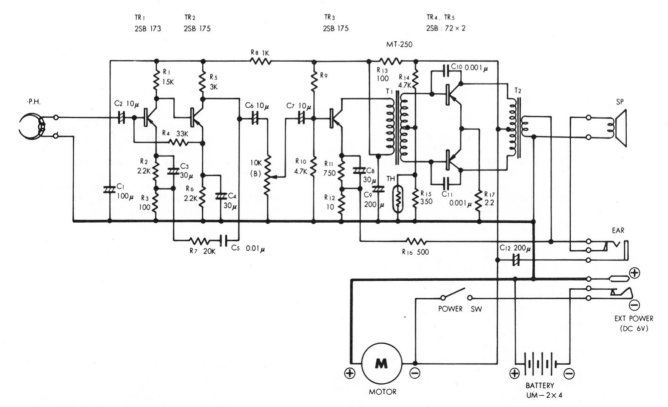

JAPANESE TAPE PLAYER—Typical of imported portable units which play prerecorded tapes but do not record, so circuit can be used as guide for troubleshooting similar sets having same transistor lineup.—"Penncrest Model 3110 Portable Tape Player," J. C. Penney, 1301 Sixth Ave., New York, NY.

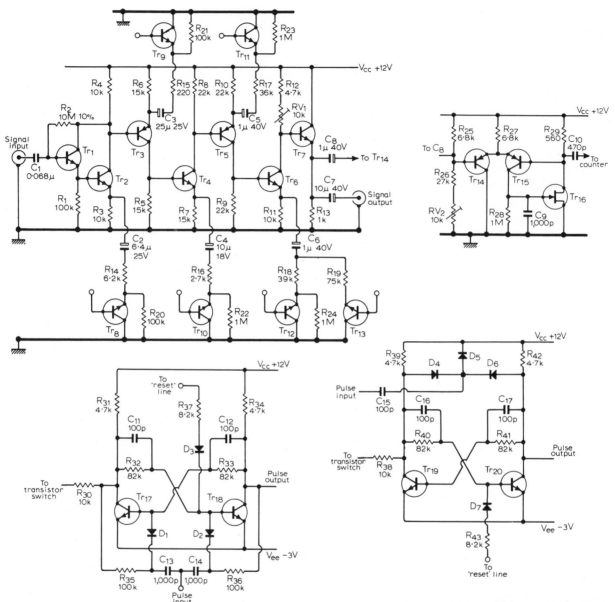

DIGITAL GAIN CONTROL—Automatically optimizes recording level to obtain maximum dynamic range without overmodulation. Input signal is fed to variable-gain amplifier having six cascaded stages. Voltage gain of each stage may have either of two preset values, selected by transistor switch. Output signal is fed to peak-level sensor Tr14–Tr16 which generates pulses whenever output exceeds preset level. These pulses are counted by six cascaded bistables which determine state of transistor switches. Tr17 and Tr18 in first bistable are 2N3708 and diodes are 1N914; other five bistables use same diodes and transistors in circuit of Tr19–Tr20. Tr14 and Tr15 are 2N4289 and Tr16 is 2N2646. Tr1 and Tr2 are 2N3707; Tr3 is 2N4286; Tr4 is 2N2925; Tr5 is 2N4289; Tr6–Tr13 are 2N2926. Power supply should be regulated. —P. C. Grossi and C. Marcus, Digitally-Controlled Tape-Recorder Pre-amplifier, *Wireless World*, March 1970, p 127–129.

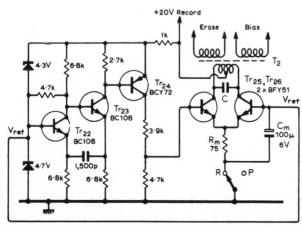

ERASE-BIAS OSCILLATOR—Uses current-switching design, with TR22 and TR23 forming 93-kHz mvbr. TR24 is buffer. Magnitude of current switched alternately between TR25 and TR26 must be such that transistors nearly saturate at required output level.—J. R. Stuart, High-Quality Tape Recorder, *Wireless World*, Jan. 1971, p 19–21.

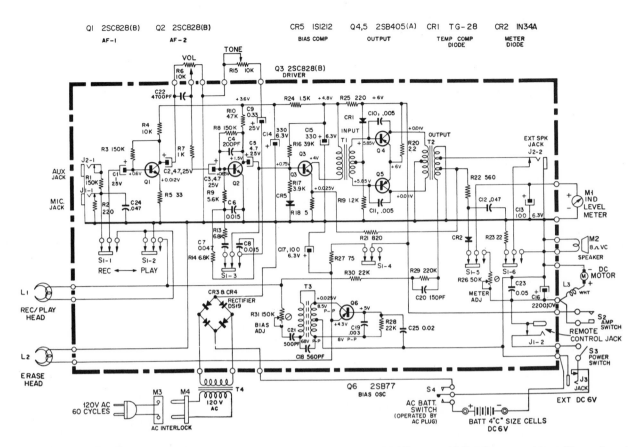

A-C OR BATTERY RECORDER—Circuit is Admiral 8A5 chassis, having power output of 600 mW. Designed for 1⅞ ips cassette tape cartridges. On recording, a-c bias is 35 kHz. Uses dynamic mike having remote-control switch. Manual gives maintenance and troubleshooting procedures.—"Tape Recorder," Admiral Corp., Bloomington, IL, Service Manual S1251, March 1971.

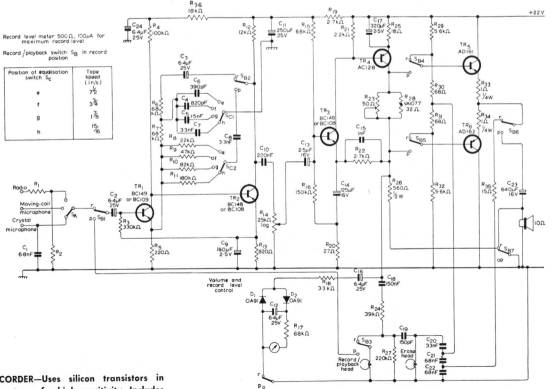

Position of equalisation switch S_C	Tape speed (in/s)
e	7½
f	3¾
g	1⅞
h	15⁄16

4-W RECORDER—Uses silicon transistors in first three stages for high sensitivity. Includes equalization for four tape speeds. Recording level indicator gives reading proportional to p-p voltage at collector of TR4. Full recording level of 4-V rms produces 95 μA through meter and 110 μA through recording head. Requires only simple bridge-rectifier power supply with single 640-μF 25-V filter capacitor. Frequency response depends on tape speed and is 55 Hz to 20 kHz for 7½ ips.—"Transistor Audio and Radio Circuits," Mullard, London, 1972, TP1319, 2nd Ed., p 88–94.

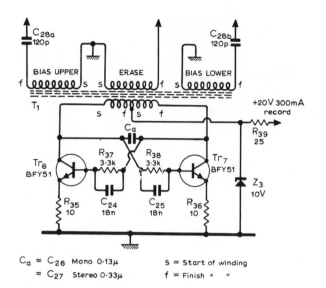

$C_a = C_{26}$ Mono 0.13μ
     $= C_{27}$ Stereo 0.33μ

S = Start of winding
f = Finish " "

MONO-STEREO ERASE-BIAS OSCILLATOR—Uses 107-kHz mvbr to provide current switching, with value of current depending on reflected load on transformer primary and on supply voltage. Designed for erase heads requiring 70-V rms at 45 mA, with recording head requiring 24-V rms at 9.7 mA for each channel.—J. R. Stuart, High-Quality Tape Recorder, *Wireless World*, Dec. 1970, p 587-591.

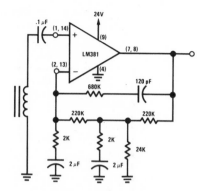

FAST TURN-ON PREAMP—Requires only 0.1 s to turn on for playback at 3¾ ips from head having sensitivity of 800 μV at 1 kHz. Uses NAB equalization. Report gives design equations.—J. E. Byerly and E. L. Long, "LM381 Low Noise Dual Preamplifier," National Semiconductor, Santa Clara, CA, 1972, AN-64, p 6–7.

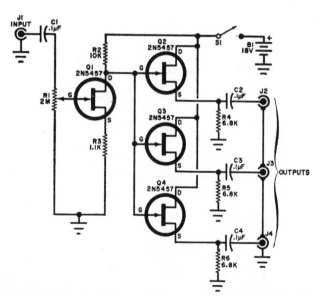

ECHO-CHAMBER EFFECT—Uses extra recording head mounted between existing head and takeup reel. If heads are as close together as possible, echo is short; head spacing of 2½ inches to 6 inches gives long echo. R5 controls volume of echo. Extra head can be Lafayette 99R6194. Supply voltage can be obtained from regular battery and recorder.—R. Brown and T. Kneitel, "101 Easy Audio Projects," Howard W. Sams, Indianapolis, IN, 1971, p 57–58.

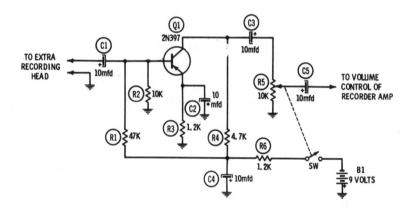

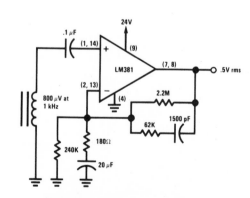

MULTICOUPLER FOR RECORDERS—Feeds signals from single source independently to several tape recorders. Provides about 10 dB of gain. Response is flat within 1 dB from 12 to 100,000 Hz. Failure of any one stage does not affect others.—D. M. Wherry, Build an Audio Multicoupler, *Popular Electronics*, July 1970, p 31–34.

PLAYBACK PREAMP—Uses LM381 low-noise dual preamp in differential input configuration of 3¾-ips playback speed of head having 800-μV sensitivity at 1 kHz. Includes NAB equalization. With 24-V supply and gain of 56 dB, turn-on time is about 5 s. Design equations are given.—J. E. Byerly and E. L. Long, "LM381 Low Noise Dual Preamplifier," National Semiconductor, Santa Clara, CA, 1972, AN-64, p 4–5.

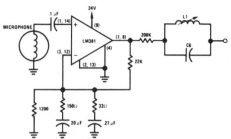

RECORDING PREAMP—Provides complement in frequency response of NAB playback equalization. Uses mike having 10-mV peak output and feeds 30-μA a-c drive current to recording head. High-frequency cutoff is 16 kHz. Report gives design equations. L1 and C6 form parallel resonant bias trap to present high impedance at bias frequency being used and prevent intermodulation distortion. Mid-band gain is 43.6 dB or 150.—J. E. Byerly and E. L. Long, "LM381 Low Noise Dual Preamplifier," National Semiconductor, Santa Clara, CA, 1972, AN-64, p 7–8.

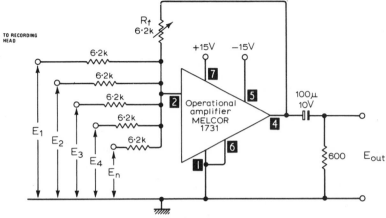

LOSS-FREE MIXER—Provides 114 dB isolation between inputs at 20 kHz and 134-dB isolation at 1 kHz. Distortion is 0.25% from 20 Hz to 20 kHz at full output of +20 dBm.—S. Feldman, Progress in Tape-recording Techniques, *Wireless World*, Jan. 1970, p 33.

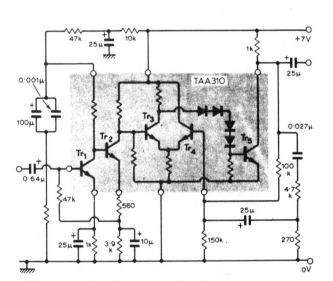

HIGH-GAIN PLAYBACK PREAMP—Uses four-stage Philips IC having d-c coupled input feedback pair TR1-TR2, long-tailed pair TR3-TR4, and four diodes at input of TR5 to carry out level-shifting required to set output at half the supply voltage. Includes compensation network.—T. D. Towers, Elements of Linear Microcircuits, *Wireless World*, March 1971, p 114–118.

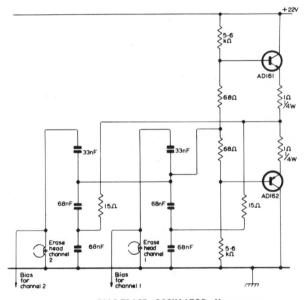

BIAS-ERASE OSCILLATOR—Uses common oscillator for two tuned erase heads of stereo tape recorder. Bias for recording heads is derived conventionally from voltages across erase heads.—"4W Tape Recorder for Mains Operation," Mullard, London, 1968, TP987, p 1–3.

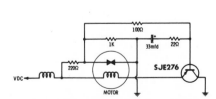

TAPE PLAYER—Speed regulator contacts, closed by centrifugal force, act on transistor to maintain relatively constant speed in Chrysler tape player made by Philips in Canada.—J. J. Carr, Troubleshooting Motor Circuits in Auto Tape Players, *Electronic Servicing*, June 1971, p 30–36.

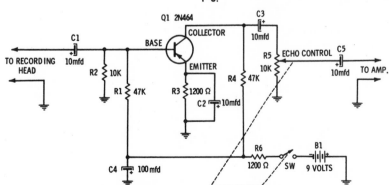

ECHO EFFECT—Input is connected to inexpensive replacement-type recording head mounted between existing tape head and takeup reel of recorder. Output goes to hot lead of volume control in recorder, for mixing with original recorded sound. Designed for recorders having positive ground; if ground is negative, reverse battery and capacitor polarities and change transistor to npn 2N366. Echo effect varies with distance between heads.—R. M. Brown and T. Kneitel, "49 Easy Transistor Projects," Howard W. Sams, Indianapolis, IN, 1972, p 12–13.

CHAPTER 113
Telemetry Circuits

IRIG DEMODULATOR—Used in missile telemetry when many channels of narrow-band data are sent via radio link. Data is frequency-modulated on set of subcarriers having center frequencies in range of 400 Hz to 200 kHz. Values shown are for IRIG channel 13 having center frequency of 14.5 kHz, maximum deviation of 7.5%, frequency response of 220 Hz, and deviation ratio of 5. Uses National LM565 phase-locked loop to provide output of 225-mV p-p, with LM107 opamp providing additional gain and level-shifting signal to ground so plus and minus output voltages are obtained for frequency shifts above and below center frequency. Report covers design of filter network and gives design equations.—"The Phase Locked Loop IC As a Communication System Building Block," National Semiconductor, Santa Clara, CA, 1971, AN-46, p 8–9.

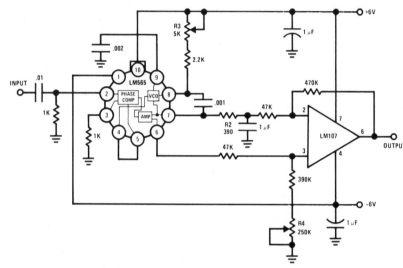

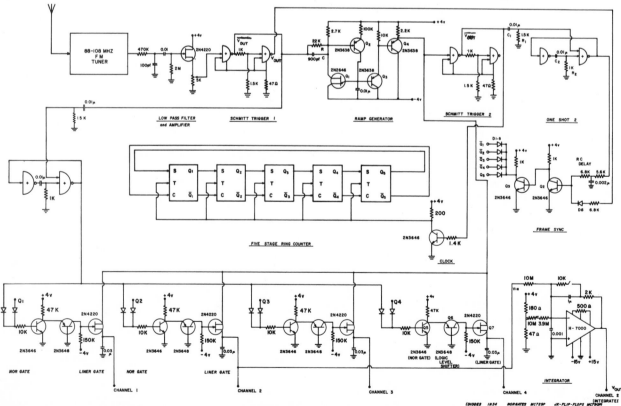

RECEIVER FOR ATHLETE TELEMETER—Picks up signal from telemeter strapped on athlete up to 100 meters away, extracts four channels of sequentially sampled pwm physiological data under control of ring counter, and converts to four outputs suitable for driving four-channel pen recorder. Article describes operation of system in detail.—H. R. Skutt, R. B. Fell, and R. Kertzer, A Multichannel Telemetry System for Use in Exercise Physiology, *IEEE Trans. on Bio-Medical Engineering*, Oct. 1970, p 339–348.

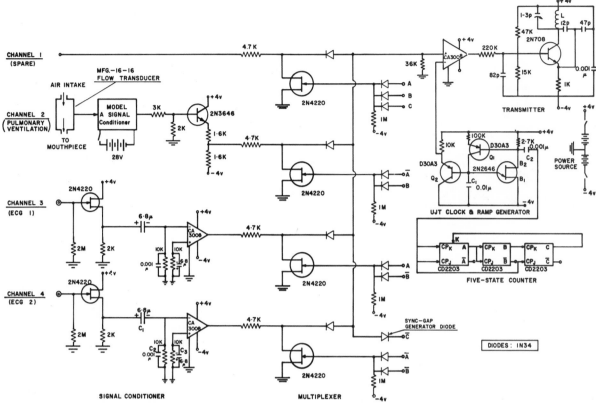

TELEMETER FOR ATHLETES—Transmits four channels of physiological data by time-division multiplexing on carrier in 88–108 MHz f-m band. Circuit samples each channel sequentially and generates pulse with width proportional to signal amplitude. Sampling rate is 200 times per second. Upper cutoff of 30 Hz is used for ecg channel to reduce interference from artifacts. With d-c input power of 10 mW to Colpitts oscillator with radiating tank coil in transmitter, usable range is over 100 meters. Transmitter weighs about 100 grams, constructed in two boxes that can be strapped on persons participating in squash, handball, track, and tennis.—H. R. Skutt, R. B. Fell, and R. Kertzer, A Multichannel Telemetry System for Use in Exercise Physiology, *IEEE Trans. on Bio-Medical Engineering*, Oct. 1970, p 339–348.

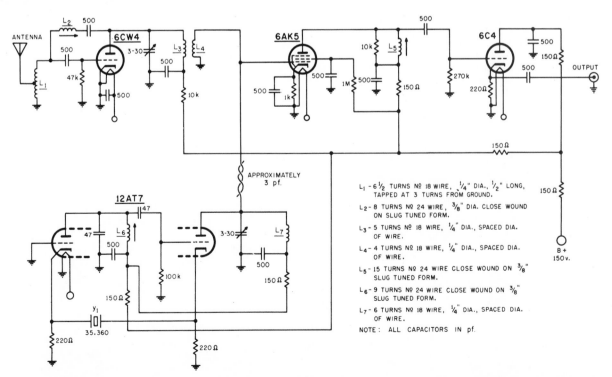

136 MHZ TO 30 MHZ—Uses 6CW4 Nuvistor front end and 106-MHz crystal-controlled oscillator. Noise figure of converter is at least 3 dB. Used in simple telemetry system for obtaining ionograms directly from ionosphere topside sounder in Alouette I satellite. Output at 30 MHz goes to commercial a-m/f-m receiver feeding tape recorder and scope with camera.—E. E. Ferguson and R. G. Green, A Simple Receiving and Display System for Alouette I Ionograms, *Proc. IEEE*, June 1969, p 945–947.

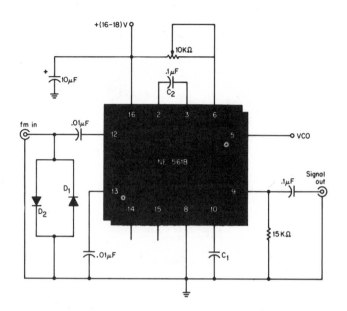

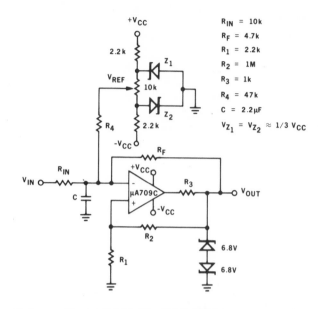

$R_{IN} = 10k$
$R_F = 4.7k$
$R_1 = 2.2k$
$R_2 = 1M$
$R_3 = 1k$
$R_4 = 47k$
$C = 2.2\mu F$
$V_{Z_1} = V_{Z_2} \approx 1/3\ V_{CC}$

15-MHZ F-M DEMODULATOR—Vco in IC phase-locked loop is set at transmitter quiescent frequency by C2 and 10K pot, for demodulating f-m conditioned signals such as those representing temperature, pressure, ecg, and other physiological data. Diodes provide input amplitude protection, for use with wide range of receivers and recorders.—R. L. Wilbur, FM Demodulator, *The Electronic Engineer*, Sept. 1971, p 92.

ANALOG TO PULSE WIDTH—Accuracy and temperature stability are excellent even though only one IC is required. Used to change analog signals from temperature, pressure, humidity, and other sensors or transducers into pulse width or time-ratio-modulated square wave. Design equations are given.—"Analog to Pulse Width Converter," Fairchild Semiconductor, Mountain View, CA, No. 229, 1968.

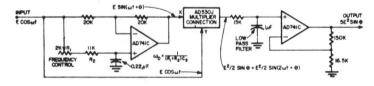

60-HZ DISCRIMINATOR—Develops d-c output proportional to deviation of input signal from tuned 60-Hz value. Applications include shaft speed control in which a-c tachometer develops frequency signal, alternator frequency control, and frequency deviation meters. Can also be used as demodulator for f-m telemetry data transmission or f-m tape recording and reproducing. Single-IC multiplier develops d-c output in response to phase differences between its X and Y inputs. Opamp shifts phase of input signal through 90 deg at 60-Hz center frequency determined by setting of R1. With 7-V rms input, sensitivity is 0.5-V output per percent of input frequency change.—"AD530 Complete Monolithic MDSSR," Analog Devices, Inc., Norwood, MA, Technical Bulletin, July 1971.

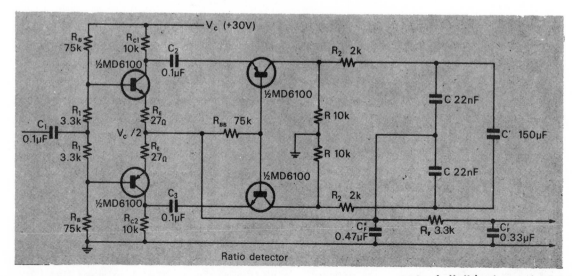

Ratio detector

RATIO DETECTOR—Transformerless circuit was developed for use in multiplexed f-m telemetry systems. Eliminates need for bulky inductances and expensive miniature components.—A. M. Kubo, Inexpensive F.M. Telemetry with Active Circuits, *Electronic Engineering*, July 1970, p 26–29.

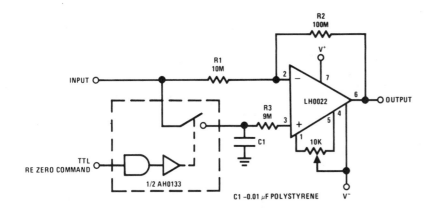

C1 –0.01 μF POLYSTYRENE

REZEROING AMPLIFIER—Used in telemetry applications where signal has unknown and variable d-c offset. Rezero command line is enabled while ground reference signal is applied to input, to make C1 charge to level proportional to system d-c offset. Amplifier then behaves like conventional inverting stage, subtracting off the system offset and giving true ground-referenced output. For 10-V full-scale system requiring accuracy of 0.1% or 10 mV, amplifier needs rezeroing every 100 ms.—R. K. Underwood, "New Design Techniques for FET Op Amps," National Semiconductor, Santa Clara, CA, 1972, AN-63, p 8.

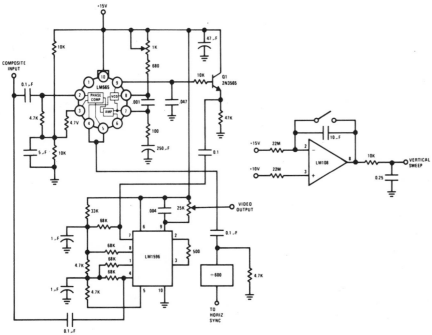

WEATHER SATELLITE PICTURE DEMODULATOR—Used in converting tape-recorded signals from weather satellite to photograph of earth taken from orbits of 100 to 800 miles. Signal recorded is detector output of f-m receiver, consisting of 2.4-kHz tone containing a-m video information. 2.4-kHz subcarrier is divided by 600 to obtain horizontal sync frequency of 4 Hz. Circuit uses LM565 phase-locked loop to track flutter and instability in recorder and filter out noise, in addition to providing large enough signal for digital frequency divider. In-phase component of 2,400-Hz vco signal is used to drive LM1596 synchronous demodulator for detecting video information. Camera is used to photograph 200-s picture produced on screen of readout cro.—"The Phase Locked Loop IC As a Communication System Building Block," National Semiconductor, Santa Clara, CA, 1971, AN-46, p 11–12.

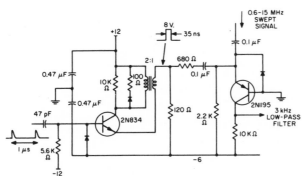

MARKERS SPACED 1 MHZ—Used in sweep calibrator of Alouette I satellite for echo sounding of ionosphere from above. Sampling pulse, about 35 ms wide, turns on 2N1195 gate transistor and allows short sample of swept-frequency input through to low-pass filter. Input to 2N834 blocking oscillator consists of 1-MHz pulses from crystal oscillator. Article includes other circuits used in system.—C. A. Franklin and M. A. Maclean, The Design of Swept-Frequency Topside Sounders, *Proc. IEEE,* June 1969, p 897–929.

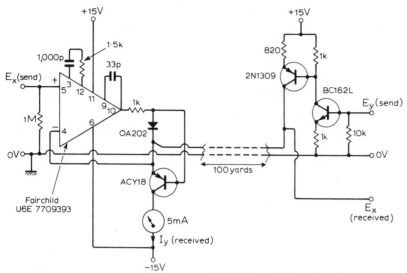

TWO-WAY D-C—Permits transmitting two d-c signals in opposite directions along each wire of 100-yard cable pair. Each wire has control signal of 0 to 5 V terminated into 1-meg resistance, and return signal of 0 to 5 mA.—R. C. Alcindor, Two-Way D.C. along Single Wire, *Wireless World*, Jan. 1971, p 9.

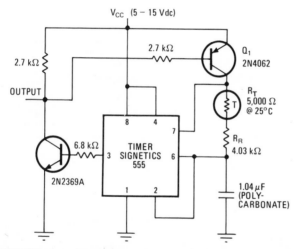

TEMPERATURE-FREQUENCY CONVERTER—IC timer connected as astable mvbr generates square-wave output frequency that varies nearly linearly from 38 to 114 Hz as temperature changes from 37 F to 115 F. Serves as low-cost temperature transducer for telemetry applications. Maximum error is 1 Hz. Design equations are given. Requires power-supply bypassing.—D. DeKold, IC Timer Converts Temperature to Frequency, *Electronics*, June 21, 1973, p 131–132.

CHAPTER 114
Telephone Circuits

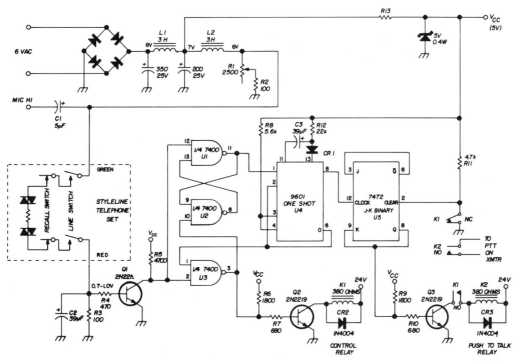

PUSH-TO-TALK WITH TOUCH-TONE—When connected to switches in Styleline telephone set used as mike, radio is switched to receive mode when handset is placed in cradle. When it is picked up and recall switch is pushed momentarily, circuit switches over to transmitter. Includes contact bounce suppression.—J. C. Tirrell, Push-to-Talk for a Styleline Telephone, *Ham Radio*, Dec. 1971, p 18–22.

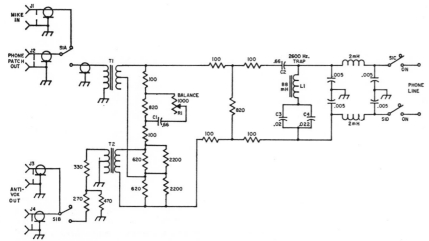

PHONE PATCH—Simple resistive bridge can be used with any amateur radio equipment, for two-way conversation between someone at amateur radio station overseas and any telephone. T1 and T2 are audio transformers with 1,500-ohm primaries and 500-ohm CT secondary. Article covers construction and null adjustment, along with method of use.—D. A. Blakeslee, A Phone Patch for the Collins S Line, *QST*, Nov. 1969, p 31–33.

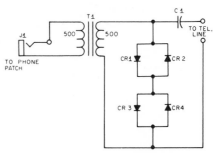

COUPLER—Designed for use in areas where telephone company is not part of Bell System. Used as ham phone patch. C1 is 2 to 4 μF nonpolarized, rated at least 200 V. Diodes are 100-prv 1-A silicon.—J. B. Berry, Jr., A Homemade Telephone Coupler. *QST*. Dec. 1970, p 51.

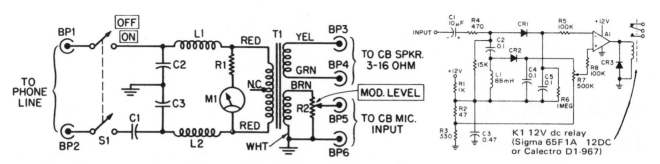

CB PHONE PATCH—Simple circuit makes connections automatically to CB transceiver at base station, for relaying telephone conversation by radio to CB transceiver in car. Driver can then talk directly by radio and telephone lines to person who placed phone call. Article covers construction and approved method of using phone patch without annoying tele-

phone company. C1 is 0.5, C2 and C3 0.001, L1 and L2 4.7 μH, R1 3.6K supplied as part of VU meter M1, R2 5K pot, and T1 modulation and audio output transformer with 500-ohm C-T primary and 8 and 3,000 ohm secondaries.—H. Friedman, CB Phone Patch, *Electronics Illustrated*, Jan. 1971, p 56–58.

1,800-HZ DECODER—L1 and C2 in series are tuned to 1,800 Hz. Audio signal across them is rectified and filtered by CR1 and C5. For application to negative input of 741 IC opamp which is used as voltage comparator and relay driver. Decoder is triggered only by pure 1,800-Hz tone. Article covers use as Touchtone decoder.—P. A. Stark, 741 Op-Amp COR and Tone Decoder Circuits, 73, July 1972, p 83–88.

PUSH-TO-TALK PATCH—Circuit is much simpler than hybrid types and does not require null, balance, or other adjustments. Designed for use with Bell System voice coupler QKT. R5 is 4-ohm L or T pad.—J. B. Berry, Jr., Phone Patches, *QST*, April 1970, p 54.

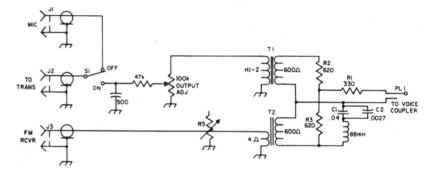

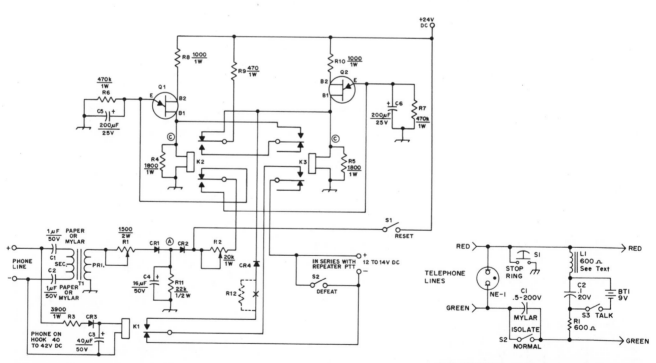

FAIL-SAFE REPEATER-PHONE INTERCONNEC-TION—With circuit as shown, f-m repeater is not on air and K1 is energized by constant 40-V d-c on telephone line. K3 is energized by repeater 24-V d-c power supply through normally closed upper contacts of K2. When phone rings specified number of times, ujt triggers and energizes K2 to release K3 and

place repeater on air for keying by user station. Q1 and Q2 are 2N2646. CR4 is 1N4002 and other diodes are 1N4004. K1 is 24-V 6,500-ohm and other relays are 12-V types. T1 has 115-V primary and 360-V 20-mA secondary. Article covers operation of system.—FM Repeater News, *QST*, March 1973, p 96–97 and 102.

D-C ISOLATOR—Keeps unwanted direct current off telephone lines when making amateur phone patches to radio equipment. Neon provides indication of ringing. Closing S1 stops ringing. Neon does not require current-limiting resistor. L1 can be half of transistor-type audio transformer, giving inductance of 30 to 160 mH.—M. B. Weinstein, A DC Isolator for Phones, 73, Sept. 1971, p 117–118.

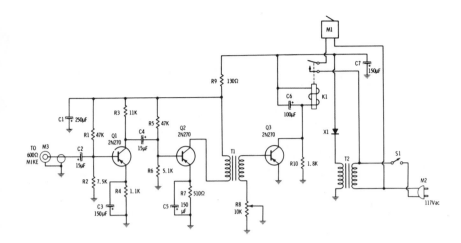

BELL AMPLIFIER—Mike picks up sound of phone ringer and amplifies resulting a-f signal sufficiently to energize 6-V relay K1 for turning on 117-V buzzer or bell M1 located in basement, garage, or other remote location. T1 is Lafayette 99H6124 a-f transformer and T2 is 6.3-V filament transformer. X1 is 1N2069 silicon diode. Circuit can be used for other sound-actuated applications as well.— R. M. Brown and T. Kneitel, "49 Easy Entertainment and Science Projects," Vol. 2, Howard W. Sams, Indianapolis, IN, 1969, p 9–10.

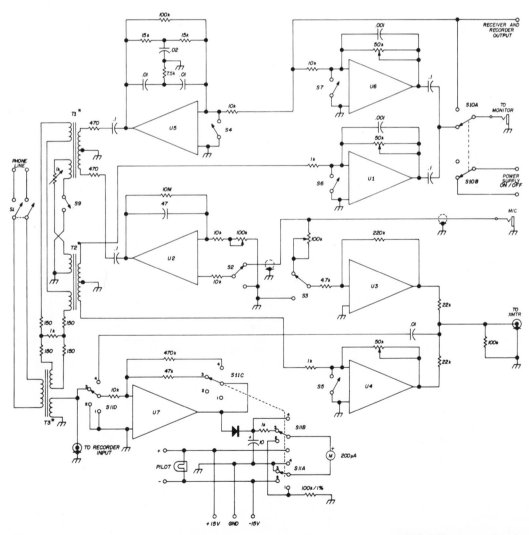

LEGAL PHONE PATCH—Permits using mike of amateur station and its speaker when interconnecting station with commercial telephone circuit. Eliminates overdrive of transmitter by telephone handset. Conversations of all three parties can be recorded and played back to telephone, radio, or local monitor. All opamps are Motorola MC1439G. T1 and T2 are UTC A-22, connected to form hybrid between telephone and patch. Circuit is nearly ideal for purpose, lacking only agc for opamps.—R. B. Shreve, Superior Phone Patch, Ham Radio, July 1971, p 20–24.

SCRAMBLER—Battery-powered unit requires no connections to telephone. User at each end must have audio sine-wave generator tunable between 1 and 3 kHz and delivering about 1 V to carrier input jack. For more difficult unscrambling, each user can feed audio output of ordinary transistor radio to J1, provided both radios are tuned to same station. Uses same balanced ring demodulator for both coding and decoding. Project phone can be any telephone handset having conventional carbon mike and dynamic earphone. All diodes are 1N34A or similar silicon, and all transformers 500-500 ohm CT Lafayette Argonne AR162 or similar.—J. Pina, Build Security 1, *Popular Electronics*, March 1970, p 27–33.

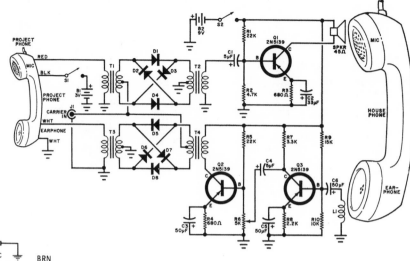

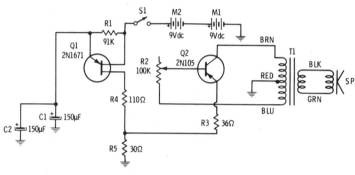

OBSCENE-CALL SQUELCHER—Serves in place of police whistle blown into phone, for discouraging unwanted telephone calls. When such a call comes in, speaker is held against mouthpiece and S1 closed to produce piercing audio blast in ear of person making call. Use 8-ohm 3-inch p-m speaker (such as Lafayette 99T6032) and matching transistor output transformer (such as Lafayette 33T7501). R2 adjusts tone. Volume control is not needed because maximum volume is desired.—R. M. Brown and T. Kneitel, "49 Easy Entertainment and Science Projects," Vol. 2, Howard W. Sams, Indianapolis, IN, 1969, p 8–9.

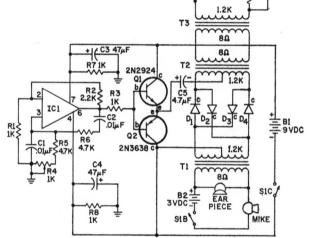

VOICE SCRAMBLER—Can be used to scramble messages on ordinary phone lines or for radiotelephone communication. May also serve for unscrambling conversations sometimes heard on 148–176 MHz vhf band. Provides single inversion only, and works best between 2,000 and 3,500 Hz for clarity. Only one unit is needed for receiver decoding, but two are required for telephone conversations.

Article covers construction, installation, adjustment, and use. Designed for installation in surplus telephone, using phone hook switch for S1. Diodes are 1N914 or HEP156 and IC is Signetics N5741K. Choose R9 to limit line current to 25 mA.—C. D. Rakes, Scramble Phone, *Elementary Electronics*, Sept.-Oct. 1972, p 75–78 and 96–97.

SIMPLE PHONE PATCH—Bridged-T 2,600-Hz filter prevents low-impedance loading of line when connecting amateur transmitter and receiver to telephone. Patch has intrinsic impedance of 900 ohms over entire voice band. L1 is toroid with Q of 63, rated 88 mH. For C1, use 0.04 and 0.0027 μF in parallel. T1 is output transformer.—J. B. Berry, Jr., An Improved Phone Patch, *QST*, Nov. 1970, p 51.

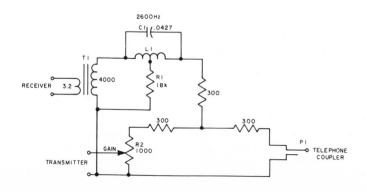

HAM PATCH—Low-pass filter cuts off frequencies below 400 Hz. Article covers construction, correct method of connecting to phone line, and operation. C1 and C2 are 0.05 µF, C3 0.22 µF, R1 10K audio taper, R2 10K, T1 Stancor A52C, and T2 Stancor A8101.—H. S. Brier, Simple Phone Patch for Hams, *Electronics Illustrated*, Nov. 1972, p 59—61 and 98.

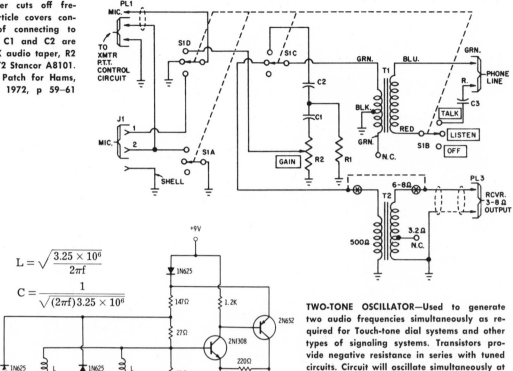

$$L = \sqrt{\frac{3.25 \times 10^6}{2\pi f}}$$

$$C = \frac{1}{\sqrt{(2\pi f)\, 3.25 \times 10^6}}$$

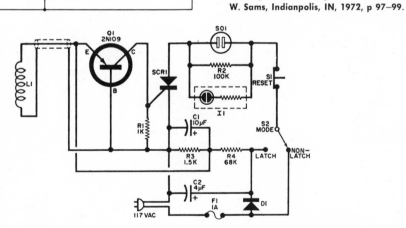

TWO-TONE OSCILLATOR—Used to generate two audio frequencies simultaneously as required for Touch-tone dial systems and other types of signaling systems. Transistors provide negative resistance in series with tuned circuits. Circuit will oscillate simultaneously at resonant frequencies of both tuned circuits. Equations give values of L and C, where L is in H, C in f, and f is frequency in Hz.—J. E. Cunningham, "Security Electronics," Howard W. Sams, Indianpolis, IN, 1972, p 97—99.

RINGER LIGHT—When phone rings, telephone pickup coil under handset picks up a-c ringing current for amplification by Q1 to turn on GE C106-B1 1-A scr. This turns on 117-V neon assembly I1 and applies power to socket that can be used to energize remote bell or buzzer. In LATCH position of S2, external circuit remains energized after phone stops ringing, until RESET S1 is pushed. When used in shop, without remote alarm, lamp indicates answering service should be called to get message. D1 is 1-A 400-piv silicon.—H. St. Laurent, Telephone Monitor Tells If Someone Called, *Popular Electronics*, Aug. 1968, p 31—33.

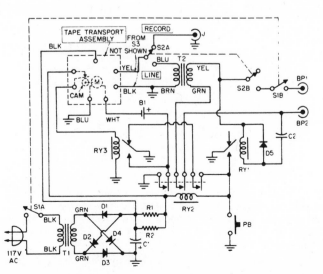

MESSAGE RECORDER—Developed for applications in which answer is same for all calls, such as telling when and where meeting is scheduled, giving prayer for day, or simply announcing when you will be back from lunch or vacation. Each time phone rings, message repeater answers phone and plays message. Built around continuous-tape-loop mechanism such as Burstein Applebee No. 18A1509. Mike for recording message is plugged into J1. Operates from 1.5-V D cell. D1–D4 are 25-piv silicon and D5 is 100-piv silicon. C1 is 1,000 µF at 10 V, C2 2 µF, R1 and R2 31 to 39 ohms, RY1 Potter & Brumfield RS5D-25,000, RY2 Magnecraft W88X-10, RY3 Potter & Brumfield RS5D, T1 6.3-V 1-A filament transformer, and T2 transistor transformer with both windings 500 ohms.—L. Powell, Build an Automatic Message Repeater for Your Phone, *Electronics Illustrated*, March 1972, p 40—43 and 100—101.

CHAPTER 115
Teleprinter Circuits

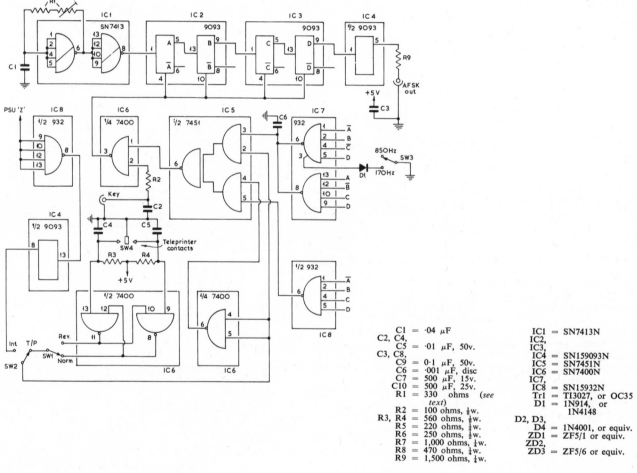

C1 = ·04 μF
C2, C4,
C5 = ·01 μF, 50v.
C3, C8,
C9 = 0·1 μF, 50v.
C6 = ·001 μF, disc
C7 = 500 μF, 15v.
C10 = 500 μF, 25v.
R1 = 330 ohms (see text)
R2 = 100 ohms, ¼w.
R3, R4 = 560 ohms, ¼w.
R5 = 220 ohms, ¼w.
R6 = 250 ohms, ¼w.
R7 = 1,000 ohms, ¼w.
R8 = 470 ohms, ¼w.
R9 = 1,500 ohms, ¼w.

IC1 = SN7413N
IC2,
IC3,
IC4 = SN159093N
IC5 = SN7451N
IC6 = SN7400N
IC7,
IC8 = SN15932N
Tr1 = TI3027, or OC35
D1 = 1N914, or 1N4148
D2, D3,
D4 = 1N4001, or equiv.
ZD1 = ZF5/1 or equiv.
ZD2,
ZD3 = ZF5/6 or equiv.

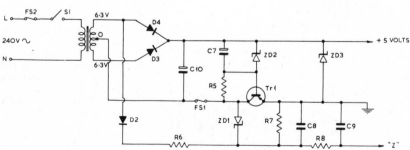

IC AFSK FOR RTTY—Efficient use of low-cost IC units holds cost of components down to about $15 including power supply. Stability is better than 100 Hz for 850-Hz shift and 45 Hz for 170-Hz shift. High-level output is free of switching transients. IC1 is dual-NAND TTL Schmitt trigger connected as 59.5-kHz oscillator and shaper. Other IC's are connected as dividers under control of teleprinter contacts. Little or no alignment is required, but article includes crystal-control modification using three additional IC's to eliminate all alignment.—A. D. Dickson and P. J. Perkins, Digital AFSK Oscillator and Combined RTTY Signal Generator, *The Short Wave Magazine*, Jan. 1972, p 670–675.

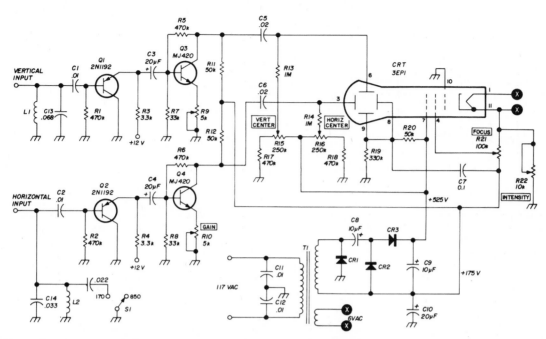

RTTY MONITOR CRO—Can be used with any radioteletype terminal unit as long as H and V amplifiers are not overdriven, for observing mark and space pulse waveforms. Toroid L1 is tuned to 2,125 Hz and L2 to 2,975 Hz with S1 open. With S2 closed, pad 0.022-μF capacitor to tune L2 to 2,295 Hz.—A. Sperduti, Solid-State RTTY Monitor Scope, Ham Radio, Oct. 1971, p 33–35.

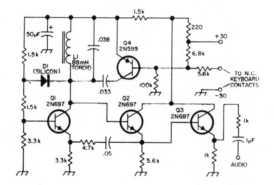

ENCODER FOR AFSK—Consists of audio oscillator with transistor switch, for generating 2,125-Hz mark tone and 2,975-Hz space tone for modulating carrier when transmitting rtty signals. Output is adequate for driving any transmitter. Beta value of Q4 is critical, so associated resistor values may have to be adjusted if either mark or space tone fails to come on.—Autostart Teletype Encoder and Decoder, 73, Jan. 1967, p 34–37.

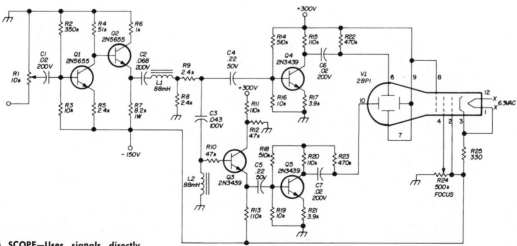

RTTY TUNING SCOPE—Uses signals directly from output of radioteletype receiver, and produces rotating-line display on face of crt. Angle of rotation is measure of frequency, and length of trace depends on amplitude of signal. Pattern aids in adjusting receiver to produce usable copy from marginal signals in presence of heavy interference. Article covers construction and adjustment.—E. E. Mooring, Phase-Shift RTTY Monitor Scope, Ham Radio, Aug. 1972, p 36–41.

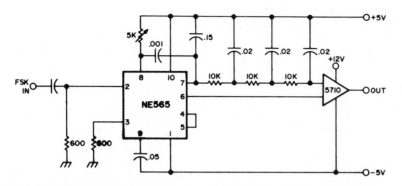

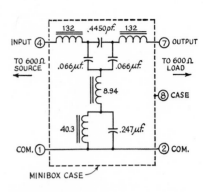

FSK CONVERTER—Uses Signetics NE565 phase-locked loop combined with IC voltage comparator that changes output levels to values compatible with digital IC chips. Center frequency is adjusted with 5K pot to produce slightly positive voltage at output when input is at low frequency.—J. Kyle, The Phase-Locked Loop Comes of Age, 73, Oct. 1970, p 42, 44–46, 49–52, and 54–56.

1,275–2,125 HZ BANDPASS—Used for rtty reception of 850-Hz fsk. Maximum pass-band attenuation is 0.5 dB and minimum stop-band attenuation is 40 dB. Inductances can be obtained with 44 and 88 mH toroids used alone, in combinations, or modified, as covered in article.—E. E. Wetherhold, An RTTY Bandpass Filter for 1,275/2,125 C.P.S., QST, Aug. 1967, p 21–23.

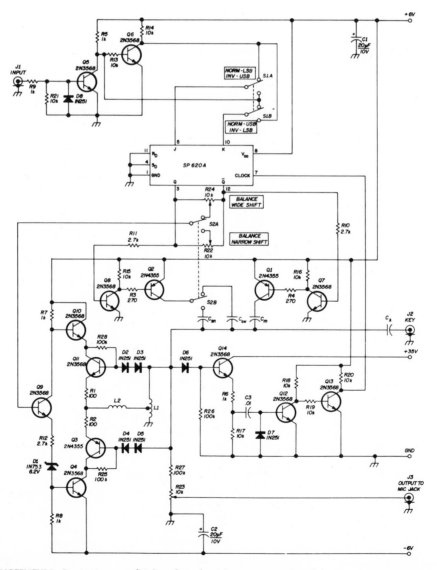

AUDIO FSK FOR RADIOTELETYPE—Generates stable sinusoidal audio tones required with ssb exciter for transmitting rtty carrier-shift signals on high-frequency bands. Ratio of amplitudes of mark and space tones is adjustable. Transition between tones is accomplished without phase discontinuity, to minimize generation of spurious signals. Uses Signetics IC flip-flop.—D. H. Phillips, A Synchronous-Phase AFSK Oscillator for RTTY, Ham Radio, Dec. 1970, p 30–36.

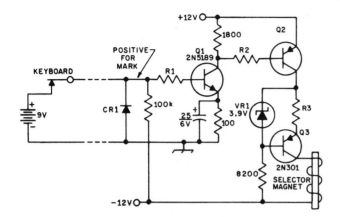

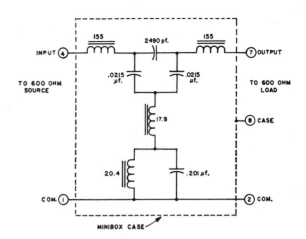

MARK-POSITIVE DRIVE—Uses positive and negative supply voltages for maximum flexibility in connecting selector-magnet driver to other sections of teleprinter. Q3 is in 60-mA constant-current drive circuit, with R3 chosen to give this loop current. Try 18K for R2, and use 25K pot to find value of R1 for which voltage across 100-ohm emitter resistor fails to increase as R1 is decreased. CR1 is any small-signal silicon diode. Q2 is 2N5142.—F. Merritt, The Modern Teleprinter Local Loop, QST, Jan. 1972, p 40–42.

BANDPASS FILTER—Elliptic-function passive filter passes 2,125 to 2,975 Hz for rtty. All inductance values are achieved with 88-mH toroids modified as per article. Article compares response with that of two other types of filters.—E. E. Wetherhold, An RTTY Bandpass Filter for 2,125-2,975 C.P.S., QST, April 1968, p 19–21.

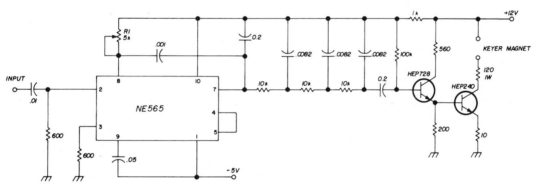

RITTY DEMODULATOR—Use of Signetics phase-locked loop IC as fsk demodulator for radio-teletype eliminates need for brandpass filters. Adjust R1 so vco of IC is at input frequency. Filter values are for 100 wpm maximum and 455-kHz i-f strip.—E. Noll, Phase-Locked Loops, Ham Radio, Sept. 1971, p 54–60.

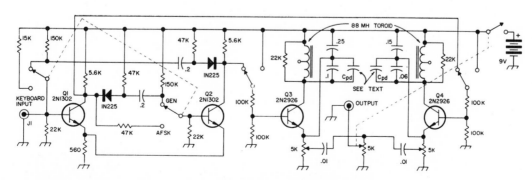

RTTY SIGNAL GENERATOR—Consists of free-running mvbr Q1-Q2 modified to produce true square waves on its collectors. When keyboard opens, collector of Q1 drops to ground and collector of Q2 rises to supply voltage, for turning on 2,975-Hz oscillator that generates space signal. When keyboard circuit closes, 2,125-Hz mark oscillator is similarly turned on. Used for making rtty transmitter modulation checks. With switch in AFSK position, circuit becomes automatic frequency-shift keyer because mvbr is then slaved to keyboard switch. Add small fixed capacitors Cpd as required to bring oscillator tuned circuits to correct frequencies.—J. Sakellakis, A Signal Generator for the RTTY Man, 73, Jan. 1966, p 34–36.

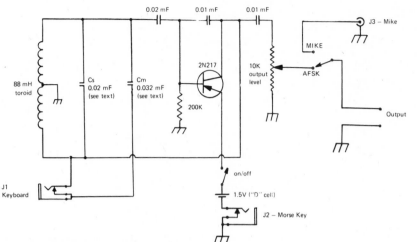

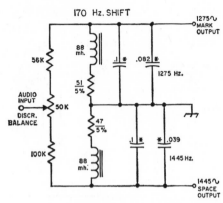

1,275–1,445 HZ DISCRIMINATOR FILTER— Provides linear response for rtty reception of 170-Hz fsk signals. To adjust, resonate individual L-C filter sections for frequencies shown on diagram while resistors going to chassis are temporarily shorted out.—J. Hall, 1,275/2,125-Hz Filters for the TT/L-2 F.S.K. Demodulator, QST, June 1969, p 29–31.

AFSK GENERATOR—Gives reasonably good sine-wave output with single transistor. Audio shift is achieved by switching additional capacitance in parallel with toroid to lower tone on "mark." On "space," the switch plugged into J1 is open and a higher tone is generated. Article covers measurement and adjustment of tone frequencies to required values of 2,125 and 2,975 Hz. Magnetic reed relay is used with Teletype keyer.—M. I. Leavey, AFSK Revisited, 73, Jan. 1972, p 35–37.

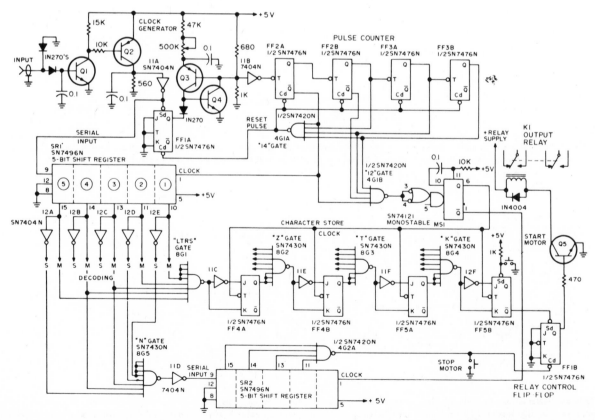

RTTY CALL-LETTER MONITOR — Automatically starts up printer for printing of message whenever call letters of station (or other start-up coded sequence) are received. Also includes provisions for stopping unattended printer with turn-off sequence. Article de- scribes operation of logic and setup for desired character recognition.—C. Sondgeroth, An RTTY Selcal with TTL Logic, 73, Nov. 1972, p 20–22, 24–26, and 28.

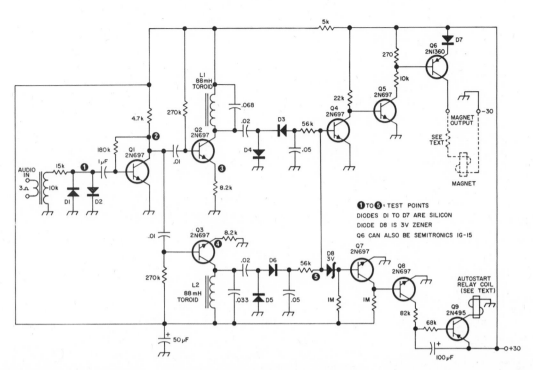

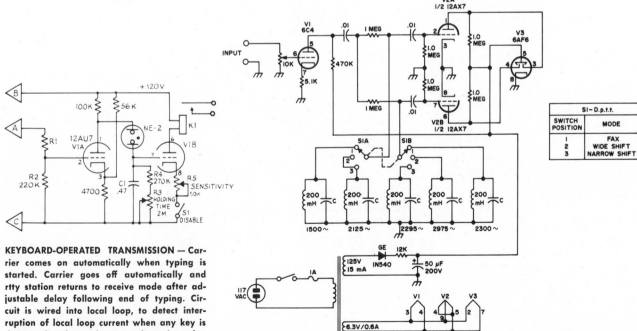

DECODER FOR AFSK—Permits automatic starting of unattended Teletype machine, with automatic shutoff when message is completed. Audio input is limited, amplified, and fed separately to 2,125-Hz mark filter Q2 and 2,975-Hz space filter Q3. Resulting audio outputs are rectified, filtered, and fed to pair of 56K resistors in divider network. Junction of resistors swings positive and negative on mark and space tones. R-C network between Q8 and Q9 gives about 1-s delay between arrival of mark tone and pull-in of autostart relay, and about 3-s dropout delay after tone is released. This prevents machine from shutting off during transmissions having high space content.—Autostart Teletype Encoder and Decoder, 73, Jan. 1967, p 34–37.

KEYBOARD-OPERATED TRANSMISSION — Carrier comes on automatically when typing is started. Carrier goes off automatically and rtty station returns to receive mode after adjustable delay following end of typing. Circuit is wired into local loop, to detect interruption of local loop current when any key is depressed and energize transmit-receive relay K1. Point A connects to keyboard-printer junction, point B to positive terminal of loop power supply, and point C to negative of supply. K1 is 10,000-ohm plate relay such as Potter and Brumfield LM5.—J. Hall, KOX—Keyboard-Operated Transmission on RTTY, QST, Nov. 1970, p 37–39.

TUNING INDICATOR—Circuit is connected so that when rtty station is received and set is tuned to mark (2,975 Hz), one of shadows on tuning eye V3 will close. Space signal will then close other shadow. Switch permits similar use for facsimile reception. Use 0.047 μF for C in 1,500-Hz filter, 0.15 μF for 2,975 Hz, and 0.022 μF in other three filters. Filters should be tuned accurately within a few Hz by varying tuning slug in each 200-mH tv width control.—D. Kadish, A Tuning Indicator for RTTY and FAX, 73, Feb. 1971, p 34–35.

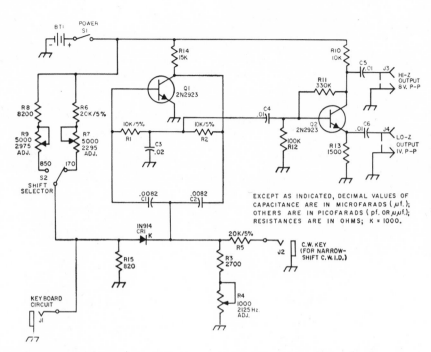

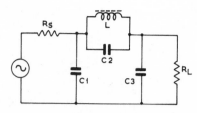

6.375-KHZ RTTY NOTCH FILTER—Three-pole elliptic low-pass audio filter has 1-dB cutoff of 3 kHz and deep notch at 6.375 kHz, as required for cleaning up output of audio frequency-shift oscillator for vhf rtty. Source impedance Rs and load impedance RL are 2,000 ohms. C1 and C3 are 0.05 μF, C2 is 0.007 μF, and L is 88-mH telephone-line loading toroid.—J. B. Hodgson, Low Pass Filter for Audio, *The Short Wave Magazine*, May 1972, p 158.

AFSK GENERATOR—Will drive either radioteletype transmitter or local loop, using either 170- or 850-Hz shift, and provide narrow-shift c-w identification. Uses R-C twin-T oscillator Q1 to give narrow-shift space tone of 2,295 Hz when S2 is in 170-Hz position and normal-shift 2,975-Hz space tone when S2 is in 850-Hz position (key open). Article covers construction and adjustment. Operates from 9-V transistor radio battery.—B. Antanaitis, Jr., A Simple Two-Transistor A.F.S.K. Generator, QST, Sept. 1969, p 36–39.

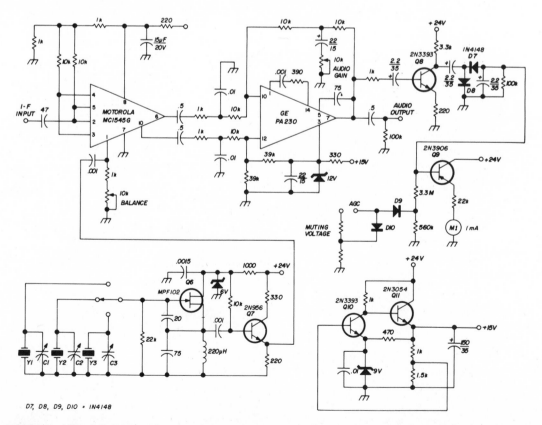

D7, D8, D9, D10 = 1N4148

PRODUCT DETECTOR AND A-F—9-MHz i-f output of solid-state receiver is mixed in MC1545G double-balanced modulator with one of three outputs of fet crystal oscillator. For ssb, crystals are 9.0015 and 8.9985 MHz as included with McCoy 48B1 filter for sideband selection. Third crystal is 8.99745 or 9.00255 MHz, to give standard audio tones of 2,125 and 2,975 Hz from rtty signal centered in 9-MHz filter passband. Output of product detector goes to IC opamp providing most of receiver gain. Regulated power supply is shown at lower right.—G. W. Jones, A Versatile Solid-State Receiver, *Ham Radio*, July 1970, p 10–15.

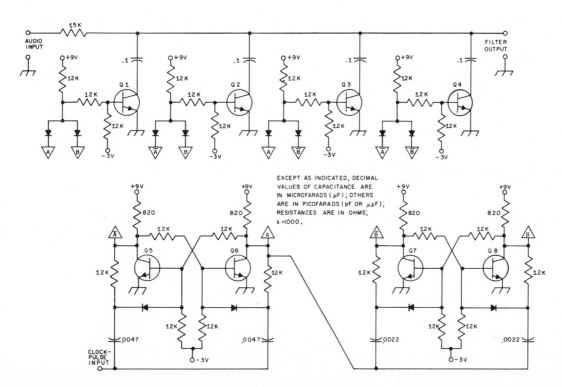

DIGITAL FILTER—R-C active device produces frequency response of L-C circuit having very high Q. Center frequency of response curve can be controlled with signal generator. Although circuit distorts input waveform, this is unimportant when detecting presence of signal as in rtty converters. All transistors are 2N706 or equivalent, and all diodes 1N914. Designed to pass range of 2,000 to 3,000 Hz. Filter output is limited to 1-V p-p to prevent overloading of transistors. Article explains operation of circuit.—L. V. Gibbs, Digital Filters, *QST*, April 1971, p 16—20.

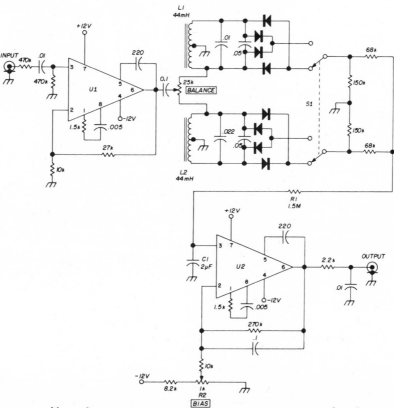

RTTY AFC—Developed to cure problem of drifting radioteletype signals in unstable communications receiver when copying Weather Bureau's rtty station on 14.395 MHz. Uses low-cost μL709C or equivalent opamps and surplus 44-mH toroids. Discriminator diodes can be 1N34A. Uses 2,975-Hz rtty mark tone as standard for frequency control. L1 demodulator is tuned about 150 Hz above 2,975, and L2 circuit 150 Hz below, so rectified d-c output voltages cancel for 2,975 Hz. If high or low, d-c control voltage is applied through opamp U2 to varactor or tuning diode installed in tuning oscillator of receiver, to correct drift.—E. R. Lamprecht, Automatic Frequency Control for Receiving RTTY, *Ham Radio*, Sept. 1971, p 50—52.

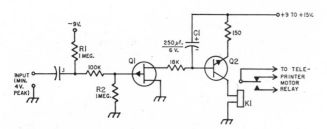

AUTOSTART—Simple circuit is designed for unattended vhf rtty operation. Will permit machine-speed keying without dropout. To change noise immunity, change ratio of R1 to R2 while keeping total of both at 2 meg. K1 can be 100 to 300 ohms, closing normally at 10 to 30 mA. Q1 is MPF-102 or TID-34, and Q2 is 2N644 or 2N4037.—C. Buttschardt, RTTY Autostart, QST, Dec. 1968, p 48.

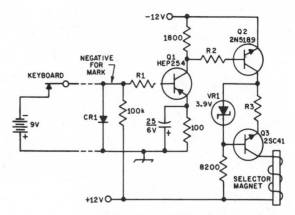

MARK-NEGATIVE DRIVE—Start with 100-ohm resistor for R3 and bridge it with higher values until Q3 passes desired constant selector-magnet current of 60 mA. Decrease R1 from 25K with pot until voltage across 100-ohm emitter resistor fails to increase. CR1 is any small-signal silicon diode.—F. Merritt, The Modern Teleprinter Local Loop, QST, Jan. 1972, p 40–42.

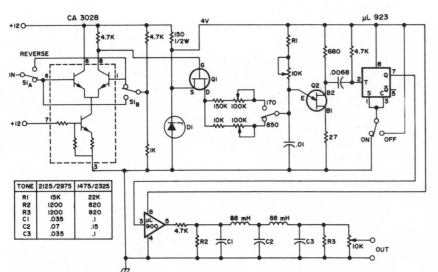

TONE	2125/2975	1475/2325
R1	15K	22K
R2	1200	820
R3	1200	820
C1	.035	.1
C2	.07	.15
C3	.035	.1

CLEAN AFSK—Can be used with two different pairs of tones, by changing values as per table. Lower-frequency pair minimizes spurious signal generation in transmitter having 2.1-kHz bandpass filter. Free-running ujt Q2 (2N4891) is in mvbr used as frequency-shifted pulse generator running at twice desired frequency. Output pulse train is divided by two in μL923 IC flip-flop to give constant-amplitude square wave at desired frequency. Five-pole Butterworth low-pass filter suppresses all odd harmonics above fundamental. Oscillator frequency is shifted by shifting resistance from supply voltage to emitter of ujt, using 2N3820 fet Q1 as switch.—J. Lovallo, A Clean AFSK Unit, 73, Feb. 1971, p 22 and 24–25.

1,275–1,445 HZ BANDPASS FILTER—Provides Butterworth response for rtty reception of 170-Hz fsk signals. Tune rtty signal as if it were lower-sideband ssb signal. Center frequency for all sections of filter is 1,350 Hz.—J. Hall, 1,275/2,125-Hz Filters for the TT/L-2 F.S.K. Demodulator, QST, June 1969, p 29–31.

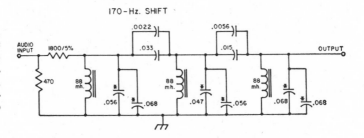

170-Hz. SHIFT

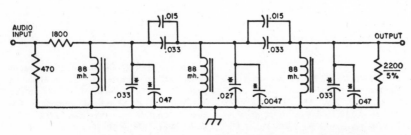

850-Hz. SHIFT

1,275–2,125 HZ BANDPASS FILTER—Provides Butterworth response for rtty reception of 850-Hz fsk signals. Tune rtty signal as if it were lower-sideband ssb signal. Center frequency for all sections of filter is 1,500 Hz.—J. Hall, 1,275/2,125-Hz Filters for the TT/L-2 F.S.K. Demodulator, QST, June 1969, p 29–31.

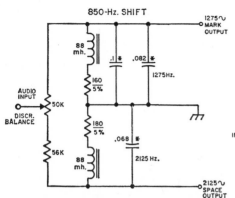

1,275–2,125-HZ DISCRIMINATOR FILTER—Provides linear response for rtty reception of 850-Hz fsk signals. To adjust, resonate individual L-C filter sections for frequencies shown on diagram while resistors going to chassis are temporarily shorted out.—J. Hall, 1,275/2,125-Hz Filters for the TT/L-2 F.S.K. Demodulator, QST, June 1969, p 29–31.

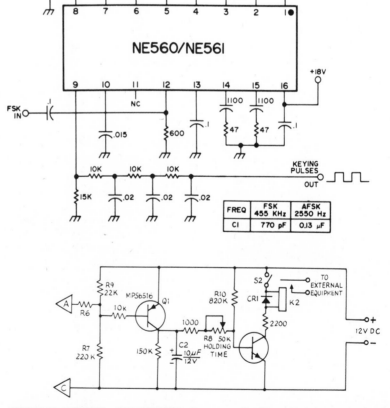

PHASE-SHIFT NETWORK—Provides improved selectivity for improving accuracy of rtty reception. Two filters are required, one for mark and one for space. For 2.125 kHz, Rb and Rc are 200K. For 2.975 kHz, Rb and Rc are 140K. For both, Cx is 370 pF, gain is 45 dB, and 3-dB bandwidth is 350 Hz. Article gives values of components for other passbands up to 1 MHz.—S. M. Olberg, Applications for an Active Filter, 73, Feb. 1973, p 45–47.

FSK CONVERTER FOR RTTY—Uses Signetics IC phase-locked loop for converting frequency-shift input at either 455 kHz or 2,550 Hz to keying pulses for driving teleprinter. Three-stage ladder filter is required with afsk to remove carrier component from output. Center frequency is adjusted to produce about 12-V output when input frequency is at its lower figure. Output then rises maximum of 4 V for higher input frequency.—J. Kyle, The Phase-Locked Loop Comes of Age, 73, Oct. 1970, p 42, 44–46, 49–52, and 54–56.

FREQ	FSK 455 KHz	AFSK 2550 Hz
C1	770 pF	0.13 μF

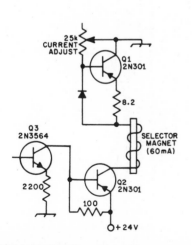

CONSTANT-CURRENT LOCAL LOOP—Q2 and Q3 form sophisticated switch in keying circuit of teleprinter demodulator, for increasing range of operation.—F. Merritt, The Modern Teleprinter Local Loop, QST, Jan. 1972, p 40–42.

KEYBOARD-OPERATED TRANSMISSION—Carrier comes on automatically when typing is started. Carrier goes off automatically and rtty station returns to receive mode after adjustable delay following end of typing. Circuit is wired into local loop, to detect interruption of local loop current when any key is depressed and energize transmit-receive 1,000-ohm sensitive relay K2. CR1 is any small silicon diode. Q1 can be HEP57 or MPS6516, and other transistor HEP50 or MPS3394. R6 is between 1.5 and 10 meg; article tells how to determine correct value experimentally. Point A goes to keyboard printer junction and point C to negative terminal of loop power supply.—J. Hall, KOX—Keyboard-Operated Transmission on RTTY, QST, Nov. 1970, p 37–39.

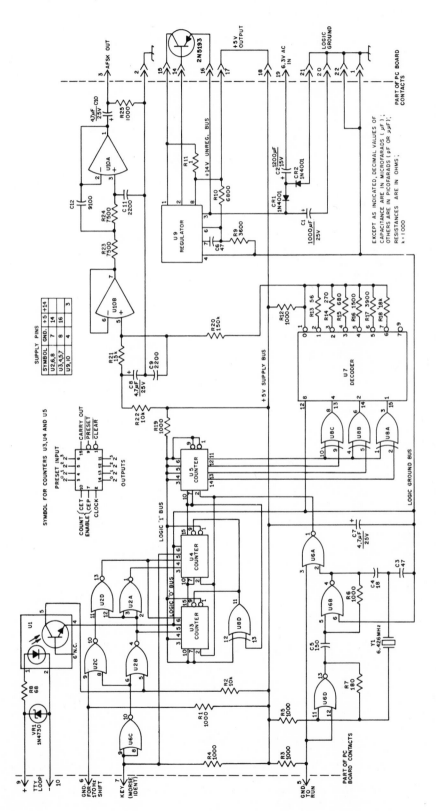

AUDIO FSK–IC circuits generate rtty tones with crystal-controlled accuracy. All four outputs (2,018.216, 2,125.000, 2,295.000, and 2,975.000 Hz) have same percentage accuracy as 6.426-MHz crystal. Photon-coupled isolator senses loop current without affecting teleprinter loop circuit. U1 is TIXL112; U2 and U6 are SN7402N; U3, U4, and U5 are SN7416N; U7 is SN74145N; U8 is SN7486N; U9 is LM376; U10 is LM1458N. Counters U3 and U4 serve in variable-modulus mode, dividing by 199, 189, 175, or 135. U2 places divider chain in 2,125-Hz mode when teleprinter is marking; for spacing, it changes counters to either 2,295 or 2,975-Hz mode depending on whether shift-select lead is open or grounded. Requires regulated high-V supply.—H. Drake, Jr., An Audio Synthesizer, QST, April 1972, p 35–39.

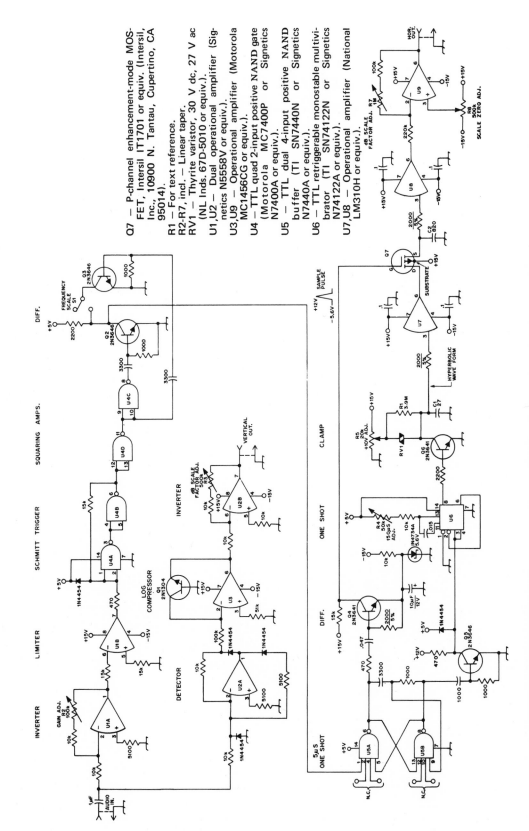

Q7 — P-channel enhancement-mode MOS-FET, Intersil IT1701 or equiv. (Intersil, Inc., 10900 N. Tantau, Cupertino, CA 95014).
R1 — For text reference.
R2-R7, incl. — Linear taper.
RV1 — Thyrite varistor, 30 V dc, 27 V ac (NL Inds. 67D-5010 or equiv.).
U1,U2 — Dual operational amplifier (Signetics N5558V or equiv.).
U3,U9 — Operational amplifier (Motorola MC1456CG or equiv.).
U4 — TTL quad 2-input positive NAND gate (Motorola MC7400P or Signetics N7400A or equiv.).
U5 — TTL dual 4-input positive NAND buffer (TI SN7440N or Signetics N7440A or equiv.).
U6 — TTL retriggerable monostable multivibrator (TI SN74122N or Signetics N74122A or equiv.).
U7,U8 — Operational amplifier (National LM310H or equiv.).

operation and use in detail.—H. Olson and J. Van Geen, The FVGT Box, QST, Nov. 1972, p 21–26 and 31.

RTTY TRANSFORM FOR CRO—Performs equivalent of fast Fourier transform by computer for spectrum analysis, using much simpler type of transform system since only one tone is present at a time in radioteletype. Scope requires high-persistence screen. Article covers

CHAPTER 116
Television Circuits—Color

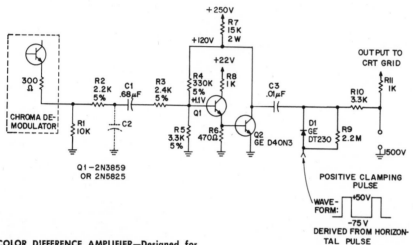

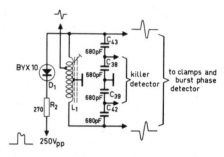

Q1—2N3859
OR 2N5825

POSITIVE CLAMPING
PULSE

WAVE-
FORM:
+50V

−75 V

DERIVED FROM HORIZON-
TAL PULSE

COLOR DIFFERENCE AMPLIFIER—Designed for use with IC chroma demodulator and with clamped output. Only R — Y amplifier is shown; G — Y and B — Y circuits are identical, all being connected between output of chroma demodulator and control grids of crt. Input is 6–10 V p-p. Does not require regulated supplies. Maximum output voltage between cutoff and saturation is 240 V p-p. Response is 3 dB down at 45 Hz and 500 kHz. Voltage gain is 36 dB.—R. C. Thielking, "TV Color Difference Amplifiers Using High Voltage Transistors," General Electric, Semiconductor Products Dept., Auburn, NY, No. 90.81, 1970, p 6.

AUXILIARY PULSE GENERATOR—Circuit delivers sine-wave voltage to clamps, burst phase detector, and color killer detector during flyback. Frequency of circuit is adjusted to give one cycle of oscillation for each flyback.—A. Boekhorst and W. Graat, "Chrominance Circuits for N. T. S. C. Colour Television Receivers," Philips, Pub. Dept., Elcoma Div., Eindhoven, Netherlands, No. 234, 1966.

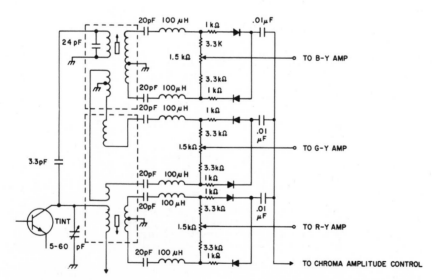

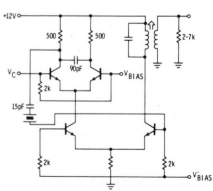

MATRIXING—Uses demodulation transformer to perform matrixing function, to provide required three-color difference signals at outputs of synchronous diode detectors. Each output requires one transistor stage for driving one of color grids in crt.—O. A. Kolody, "Simplified Transistor Color TV Processing Circuitry," General Electric, Semiconductor Products Dept., Auburn, NY, No. 90.63, 1969, p 8–9.

RENNICK SUBCARRIER OSCILLATOR—Uses crystal and transistors in configuration suitable for IC production. Report gives design equations.—N. P. Doyle, "A Comparison of Solid-State Subcarrier Oscillators for Color TV Receivers," Fairchild Semiconductor, Mountain View, CA, No. 203, 1971, p 2–3.

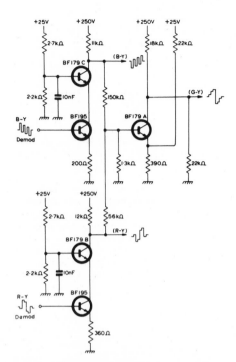

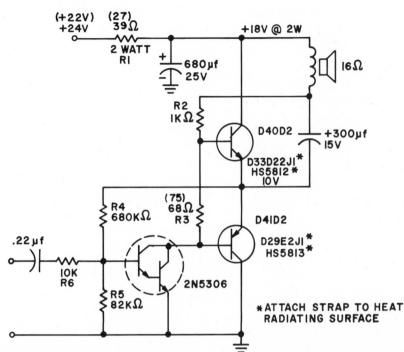

COLOR DIFFERENCE AMPLIFIER—Uses common-base output stage fed by cascaded stages. Matrix circuit is connected to collectors of B—Y and R—Y output stages.—" Silicon Planar Transistor Package—40822," Mullard, London, 1967, TP893, p 9.

2-W AUDIO—Low-cost complete audio amplifier operates from ratio detector or higher a-f level of sound IC. Circuit includes power-supply decoupling to prevent sound modulation of picture. Sensitivity is 115 mV for full power output, with frequency response down 3 dB at 80 Hz and 55 kHz. Output transistors are protected against intermittent load shorts. Report covers design procedure.—D. V. Jones, "TV Audio Amplifier," General Electric, Semiconductor Products Dept., Auburn, NY, No. 90.91, 1970, p 1–3.

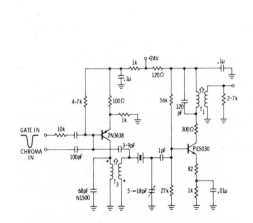

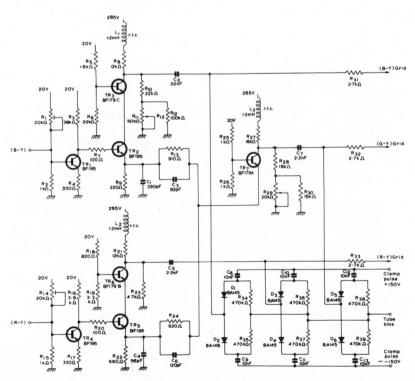

CRYSTAL RINGING—Low-cost method of regenerating color subcarrier. Burst is injected into crystal circuit, which then rings in phase and frequency with burst. Requires no color killer. Oscillator output may be used to derive automatic color control signal, since amplitude of ringing signal is proportional to that of injected burst. Requires well balanced demodulator, and is subject to dropout if signal input is poor.—N. P. Doyle, "A Comparison of Solid-State Subcarrier Oscillators for Color TV Receivers," Fairchild Semiconductor, Mountain View, CA, No. 293, 1971, p 5.

COMPLETE COLOR DIFFERENCE AMPLIFIER—Provides p-p voltages of 100 V for G—Y, 180 V for R—Y, and 200 V for B—Y, to drive grids of 25-inch shadow-mask color tv tube. Linearity is better than 80%. Bandwidth is 1 MHz for 3 dB down. Bandwidth compensation is achieved with 1.2-mH shunt compensation coils in collector circuits of output stages and capacitors in emitter circuits of earlier stages.—"Silicon Planar Transistor Package—40822," Mullard, London, 1967, TP893, p 10.

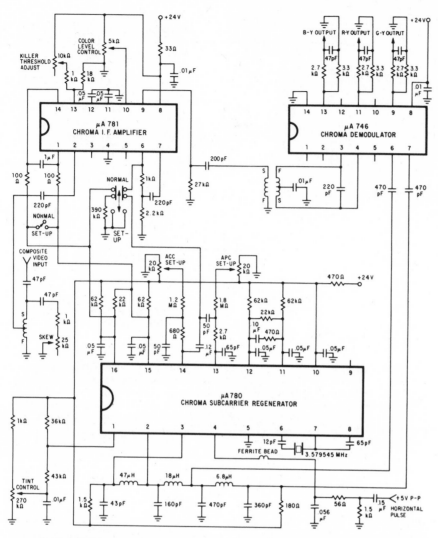

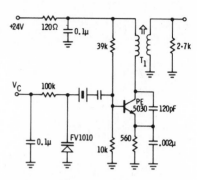

COLPITTS SUBCARRIER OSCILLATOR—Uses collector-emitter feedback. Crystal determines frequency, and sensitivity is function of how crystal is tuned by reactance device. Report gives design equations.—N. P. Doyle, "A Comparison of Solid-State Subcarrier Oscillators for Color TV Receivers," Fairchild Semiconductor, Mountain View, CA, No. 203, 1971, p 2–3.

CHROMA PROCESSOR—Completely integrated chroma processing system using three IC's converts composite video input above 0.4-V p-p to B—Y, R—Y, and G—Y outputs at required values of 4.5, 3.75, and 0.5 V p-p, respectively. Generates 3.58-MHz signal locked in phase and frequency to color burst, using phase-locked loop which is essentially vco, phase detector, and low-pass filter. Report covers operation of each IC and its interconnecting phase-shift and chroma output networks in detail.—J. W. Chu, "The µA780, µA781 and µA746 Integrated Circuit Color TV Chroma Processing System," Fairchild Semiconductor, Mountain View, CA, No. 210, 1972, p 2–6.

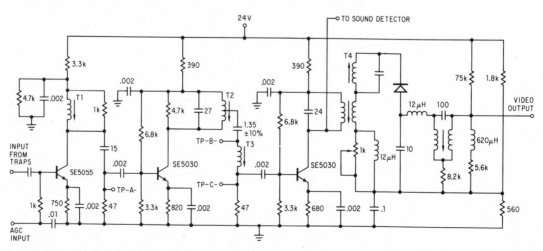

VIDEO I-F—Designed for nominal detected video output of 2 V, and capable of providing 4-V output before compression occurs. Gain control is provided only for first stage. All coil-winding data is given in report.—N. Doyle and D. Smith, "A Low-Cost Hybrid Color TV Receiver," Fairchild Semiconductor, Mountain View, CA, No. 174, 1969, p 1–2.

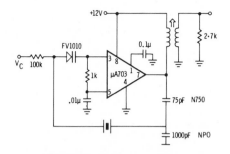

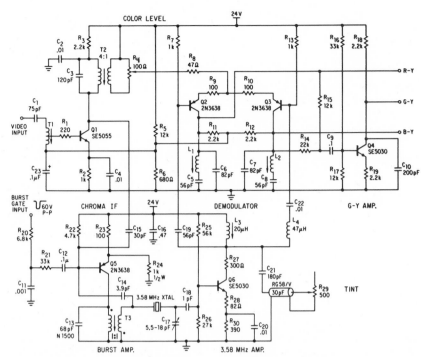

IC SUBCARRIER OSCILLATOR—Uses Colpitts circuit. Crystal determines frequency, and sensitivity is function of how crystal is tuned by reactance device. Report gives design equations.—N. P. Doyle, "A Comparison of Solid-State Subcarrier Oscillators for Color TV Receivers," Fairchild Semiconductor, Mountain View, CA, No. 203, 1971, p 2–3.

CHROMA SYSTEM—Low-cost transistor circuit uses only one stage of chroma amplification, with input and output transformers broadly tuned to 3.58 MHz. Input for burst amplifier Q5 is taken from collector of chroma i-f amplifier. Q6 is amplifier for 3.58-MHz ringing circuit serving as oscillator. Q2 and Q3 demodulate 3.58-MHz chroma subcarrier. Includes unique automatic color control for chroma i-f amplifier. During absence of color burst, there is no drive to demodulator and hence no output, so color killer is not needed. Winding data is given.—N. Doyle and D. Smith, "A Low-Cost Hybrid Color TV Receiver," Fairchild Semiconductor, Mountain View, CA, No. 174, 1969, p 4–5.

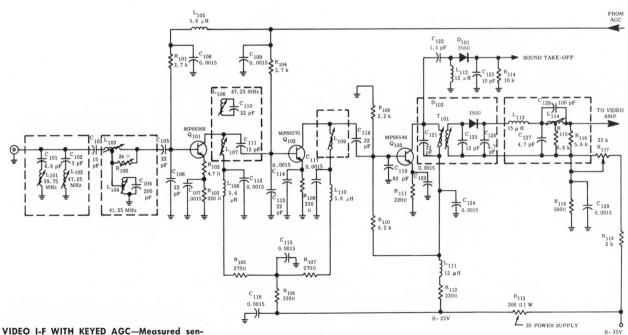

VIDEO I-F WITH KEYED AGC—Measured sensitivity of solid-state circuit is 1 V d-c across detector load for 70-μV input at full gain, using 44-MHz c-w test signal. Will deliver 5-V p-p undistorted audio across detector load at full gain when input is 80%-modulated a-m signal (400 Hz) at 44 MHz. Report covers design procedure of entire circuit in detail and gives agc circuit.—"Color IF Amplifier and AGC Circuit," Motorola Semiconductors, Phoenix, AZ, AN-287, 1971.

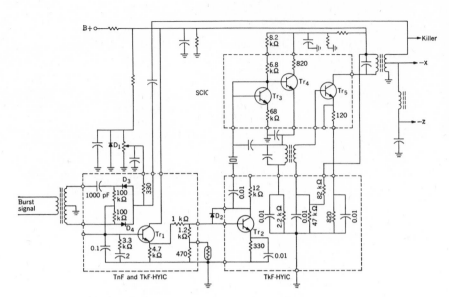

COLOR REFERENCE OSCILLATOR—Hybrid circuit uses IC's developed in Japan to prove feasibility of incorporating integrated circuits in commercial color tv sets. Color-reference oscillator and phase detector serve to demodulate and separate two color-difference signals while keeping phase error within 5 deg. —E. Sugata and T. Namekawa, Integrated Circuits for Television Receivers, *IEEE Spectrum*, May 1969, p 64–74.

SYNC—Signal processor is driven by video buffer Q1, which also feeds chroma and luminance circuits of color receiver. Sync separator Q3 provides 50 V of pulse output to horizontal and vertical oscillator circuits. Q2 is conventional keyed agc stage. Report gives winding data for T1.—N. Doyle and D. Smith, "A Low-Cost Hybrid Color TV Receiver," Fairchild Semiconductor, Mountain View, CA, No. 174, 1969, p 3.

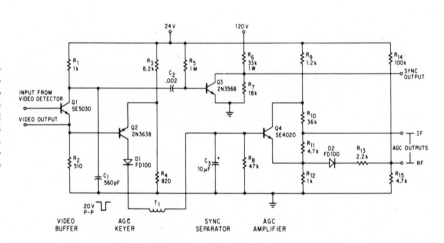

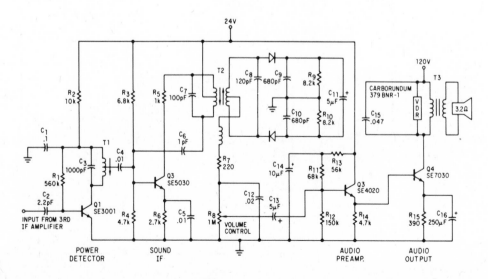

1-W TV SOUND—Use of power detector in place of diode detector and limiter gives enough conversion gain so only one 4.5-MHz sound i-f amplifier stage (Q3) is required. Coupling transformer T1 matches power detector to sound i-f amplifier, which drives standard ratio detector feeding audio preamp Q3 through volume control R8. Report gives transformer winding data.—N. Doyle and D. Smith, "A Low-Cost Hybrid Color TV Receiver," Fairchild Semiconductor, Mountain View, CA, No. 174, 1969, p 2–3.

LUMINANCE DRIVE WITH TRAP—In addition to providing compensation needed to give rising frequency response, amplifier includes 3.8-MHz resonant trap for suppressing undesired response at 3.58 MHz. Response of output amplifier peaks at about 900 kHz. Report covers operation of circuit in detail.—R. E. Williams, "Video Output Design Considerations Using a High Voltage Transistor," General Electric, Semiconductor Products Dept., Auburn, NY, No. 90.82, 1970, p 10–14.

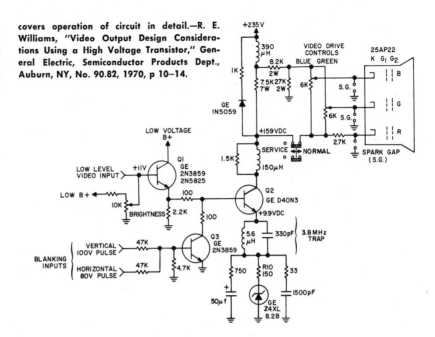

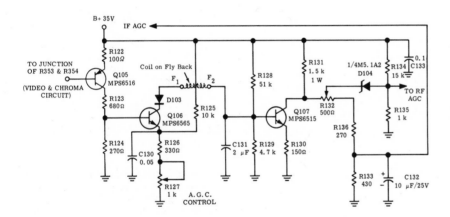

AGC KEYER-AMPLIFIER—Designed for use with solid-state video i-f amplifier. Uses height of horizontal pulses as reference, with R127 varying threshold level required for triggering agc keyer Q106. Report gives i-f circuit and covers design procedure in detail for both circuits.—"Color IF Amplifier and AGC Circuit," Motorola Semiconductors, Phoenix, AZ, AN-287, 1971.

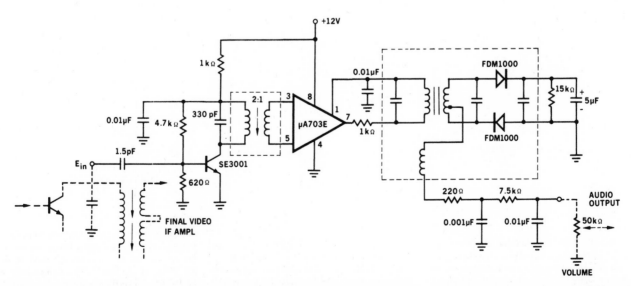

SOUND I-F—Opamp circuit provides audio output of 0.45 V for 600-mV rms input from video i-f amplifier, along with a-m rejection of 37 dB.—"Fairchild Semiconductor Integrated Circuit Data Catalog," Fairchild Semiconductor, Mountain View, CA, 1970, p 6–24.

INDICATOR
(V106) **6HU6**

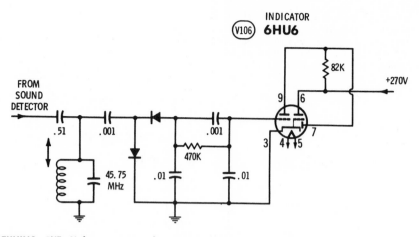

TUNING EYE—Makes accurate fine tuning easier in Andrea VCX325 color tv by indicating when picture carrier is tuned to 45.75 MHz.—C. Babcoke, New in Color TV For 1970, *Electronic Servicing*, Dec. 1969, p 54–66.

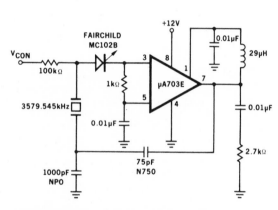

3.58-MHZ VCO—Varactor-tuned crystal oscillator can be increased up to 230 Hz below crystal frequency with negative 1-V d-c control voltage, and reduced up to 360 Hz with positive 1-V control voltage.—"Fairchild Semiconductor Integrated Circuit Data Catalog," Fairchild Semiconductor, Mountain View, CA, 1970, p 6–24.

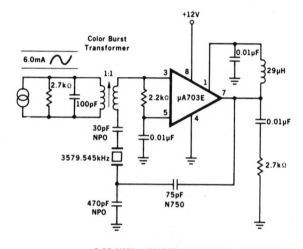

3.58-MHZ INJECTION-LOCKED OSCILLATOR —Phase shift is varied from −25 deg when oscillator is 300 Hz below normalized frequency to +35 deg when oscillator is 300 Hz above frequency. Uses linear IC.—"Fairchild Semiconductor Integrated Circuit Data Catalog," Fairchild Semiconductor, Mountain View, CA, 1970, p 6–25.

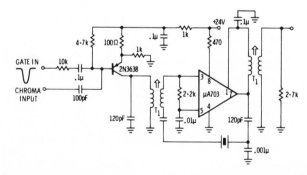

INJECTION-LOCKED OSCILLATOR—Color burst is injected into oscillator feedback circuit and presents impedance whose magnitude and sign vary as magnitude and sign of difference between burst and oscillator frequencies. Oscillator changes frequency to return loop to zero phase shift and lock on burst. Chief problem is susceptibility to noise. —N. P. Doyle, "A Comparison of Solid-State Subcarrier Oscillators for Color TV Receivers," Fairchild Semiconductor, Mountain View, CA, No. 293, 1971, p 4–5.

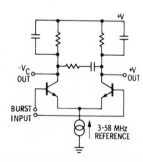

PRODUCT-TYPE PHASE DETECTOR—Circuit is suitable for IC production, hence values are not given. Typical sensitivity is about 10 mV per deg.—N. P. Doyle, "A Comparison of Solid-State Subcarrier Oscillators for Color TV Receivers," Fairchild Semiconductor, Mountain View, CA, No. 293, 1971, p 4.

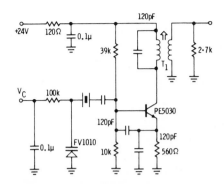

SUBCARRIER OSCILLATOR—Uses emitter-base feedback in circuit sometimes called Colpitts-Pierce or Gourier-Clapp. Crystal determines frequency, and sensitivity is function of how crystal is tuned by reactance device. Report gives design equations.—N. P. Doyle, "A Comparison of Solid-State Subcarrier Oscillators for Color TV Receivers," Fairchild Semiconductor, Mountain View, CA, No. 203, 1971, p 2–3.

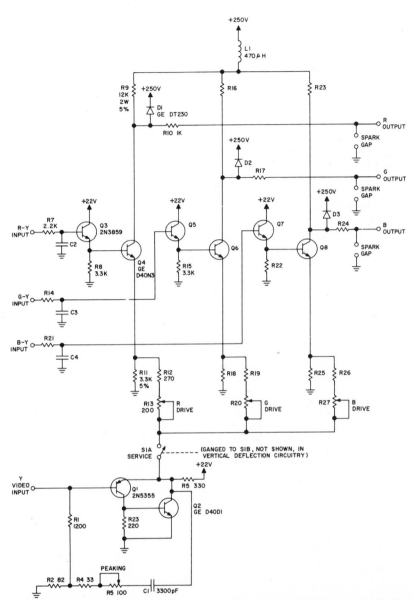

RGB AMPLIFIER—Provides tracking of chroma and luminance drives because chroma and luminance are matrixed ahead of gray-scale crt drive pots. Full d-c coupling of chroma information is achieved without clamping or other costly circuitry. Stages Q5-Q6 and Q7-Y8 are same as Q3-Q4.—R. C. Thielking, "RGB Video Amplifiers for Color TV Offer High Performance," General Electric, Semiconductor Products Dept., Auburn, NY, No. 90.88, 1970, p 3–10.

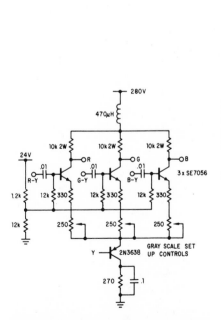

RGB OUTPUT—Use of SE7056 transistors in place of conventional 6MN8 triode tube sections provides much higher gain and permits placing drive controls in emitter circuits for better performance.—N. Doyle and D. Smith, "A Low-Cost Hybrid Color TV Receiver," Fairchild Semiconductor, Mountain View, CA, No. 174, 1969, p 6.

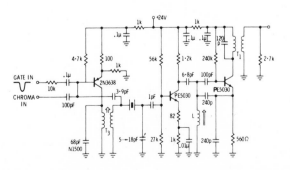

RINGER AND INJECTION LOCK—Combination of injection-lock and crystal ringing subcarrier oscillators overcomes many of disadvantages of each type, although at cost of increased complexity.—N. P. Doyle, "A Comparison of Solid-State Subcarrier Oscillators for Color TV Receivers," Fairchild Semiconductor, Mountain View, CA, No. 293, 1971, p 5.

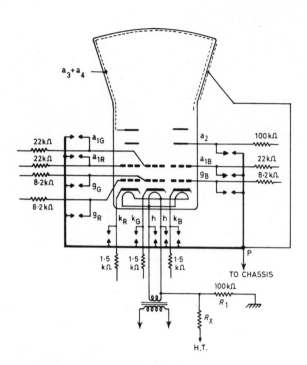

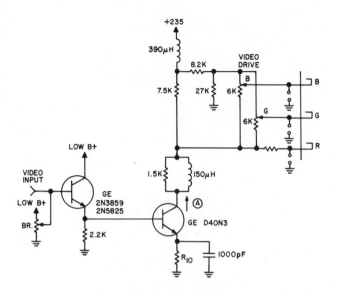

FLASHOVER PROTECTION—Combination of spark gaps and resistors provides full flashover protection for transistor color tv sets. Parts should be as close to pins of tube as possible and common wiring (shown as heavy line) to point P should be as short as possible.—A. Ciuciura, Flashover in Picture Tubes and Methods of Protection, *The Radio and Electronic Engineer*, March 1969, p 149–168.

COMPENSATED LUMINANCE DRIVE—Uses series-shunt peaking in collector of output amplifier and R-C peaking in emitter circuit, to provide required rising frequency response.—R. E. Williams, "Video Output Design Considerations Using a High Voltage Transistor," General Electric, Semiconductor Products Dept., Auburn, NY, No. 90.82, 1970, p 7–8.

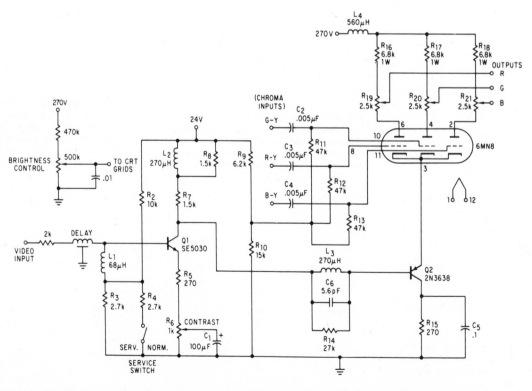

LUMINANCE AMPLIFIER—Signal from video buffer is coupled through luminance delay line to luminance amplifier Q1. Q2 delivers low-impedance voltage drive to 6MN8 triple-triode output tube. Hybrid design minimizes cost.—N. Doyle and D. Smith, "A Low-Cost Hybrid Color TV Receiver," Fairchild Semiconductor, Mountain View, CA, No. 174, 1969, p 3–4.

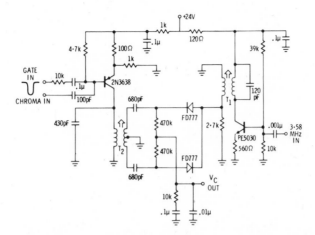

DISCRIMINATOR-TYPE PHASE DETECTOR— Commonly used in automatic phase control systems for color tv receivers. Requires high drive levels for efficient operation. Sensitivity is about 100 mV per deg.—N. P. Doyle, "A Comparison of Solid-State Subcarrier Oscillators for Color TV Receivers," Fairchild Semiconductor, Mountain View, CA, No. 203, 1971, p 4.

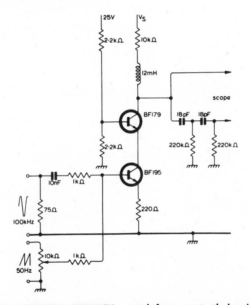

CHECKING COLOR TV AMPLIFIER LINEARITY —Cascaded two-transistor color difference amplifier stage, shown connected to scope for linearity measurements, is fed as shown with 50-Hz sawtooth having sufficient amplitude for driving output stage from cutoff to knee, and 100-kHz sine wave with amplitude of 10-mV p-p. High-frequency signal is separated from sawtooth by differentiating network at output of amplifier, and high-frequency signal is applied to dual-trace scope. Voltage swing at collector of output stage is also applied to scope. Report shows examples of curves obtained for various degrees of linearity.—"Silicon Planar Transistor Package— 40822," Mullard, London, 1967, TP893, p 15.

CHAPTER 117
Television Circuits—General

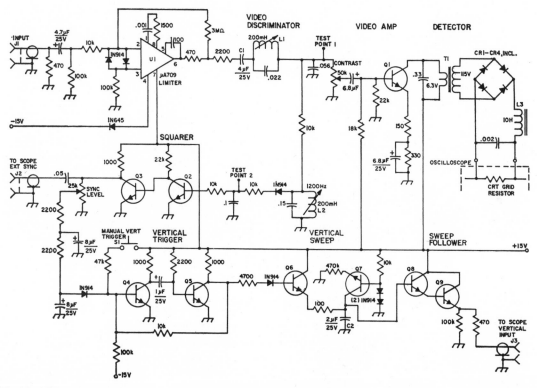

USING CRO AS SSTV MONITOR—Requires scope that will synchronize from 15-Hz external trigger and accept 10-V d-c at vertical input. Crt must be changed to tube having long-persistence P7 phosphor such as 5UP7. Slow-scan signal from audio output of communications receiver is fed to input. Article gives construction and adjustment details. All transistors are 2N2222, 2N697, 2N718, or 2N3641-3.—B. Briles and R. Gervenack, Slow-Scan TV Viewing Adapter for Oscilloscopes, QST, June 1970, p 46–49.

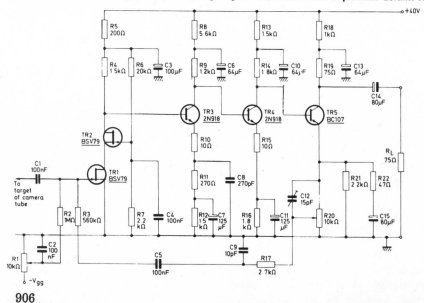

CAMERA PREAMP—Developed for use with Mullard Plumbicon camera tube. Uses cascoded fet's in input stage to keep noise down. Signal-noise ratio is about 46 dB for entire frequency range from 40 Hz to 5.5 MHz.—"Television Camera Preamplifier Using FET's in Cascode," Mullard, London, 1970, TP1219, p 1–3.

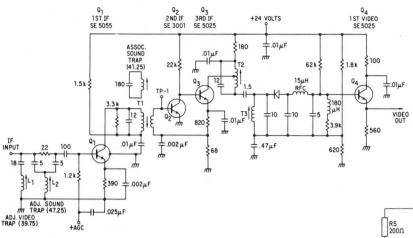

VIDEO I-F WITH AGC—Designed for use in low-cost solid-state monochrome tv receiver. With 2-V agc bias, requires 125 μV of 80% modulated 45.75 MHz for 1-V p-p detected video, with value increasing to 12,500 μV for 5-V agc bias.—T. B. Mills and H. S. Suzuki, "Design Concepts for Low-Cost Transistor AGC Systems," Fairchild Semiconductor, Mountain View, CA, No. 199, 1971, p 2–3.

CAMERA PREAMP—Uses fet's in cascode at input for high impedance matching that of camera tube and for very good signal-to-noise ratio. Developed for Plumbicon 55875 camera tube. Signal-to-noise ratio is about 46 dB throughout frequency range of 40 Hz to 5.5 MHz. Adjust R1 for 28 V at collector of TR5 to avoid output clipping. R20 adjusts voltage gain.—"Field Effect Transistors," Mullard, London, 1972, TP1318, p 57–60.

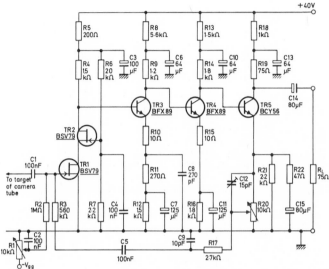

25-MHZ PREAMP—Serves as line driver for vidicon or image orthicon pickup tube providing modulating current output. Fet-input 9524 opamp has input compensation which can be utilized for aperture correction. T-type feedback network minimizes effects of stray capacitance. Capacitor coupling to 9412 driver passes desired beam-current modulation information while blocking d-c output of 9524. Input of 1 μA produces output of 1V.—"A 25 MHz Video Preamplifier—Line Driver," Optical Electronics, Tucson, AZ, Application Tip 10195, 1971.

C_1, C_2, C_5 = 1000 pF, feedthrough
C_3 = 470 pF, feedthrough
C_4 = 10 pF, NPO, ceramic disc, leadless
C_6 = 1000 pF, NPO, ceramic disc, leadless
C_8, C_{10} = 4.5 pF, ceramic disc, leadless
C_7, C_{12} = 1.5 – 18 pF, Arco 402
C_9 = 82 pF, feedthrough
C_{11} = 180 pF, ceramic disc
C_{13}, C_{14}, C_{15} = tuning capacitance, app. 1·5 – 18 pF
C_{16} = gimmick, approximately 1 pF
R_1 = 2.2 K
R_2 = 330 Ω
R_3 = 10 K
R_4 = 330 Ω
R_5 = 2.2 K
R_6 = 220 Ω
L_1 = RFC, 20 T, 1/8 ID, #38 wire
L_2 = Pick-up loop, 3/4 long, #16 bus wire
L_3 = -8 pF at 100 MHz
L_4 = -1 pF at 100 MHz
L_5 = -6 pF at 100 MHz
L_6, L_7, L_8 = tuning lines.

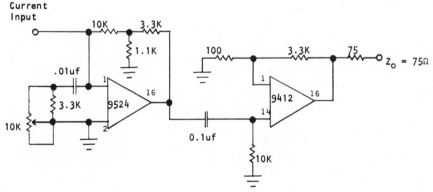

UHF AUTODYNE TUNER—Modification of standard Sickles tuner uses r-f amplifier to boost r-f signal and local-oscillator transistor in oscillator-mixer combination to reduce noise figure of tuner. Q1 is MT1061 used as r-f amplifier and Q2 is SE3005 local oscilla- tor. Gain ranges from 19 to 24 dB over uhf band, and noise figure is 7 to 8.5.—S. Sir, "A New Silicon Transistor for UHF Application," Fairchild Semiconductor, Mountain View, CA, No. 185, 1970, p 3–6.

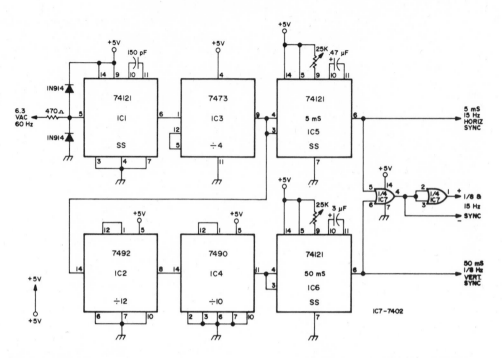

SSTV SYNC PULSE GENERATOR—Provides pulses needed for slow-scan tv cameras, flying-spot scanners, and pattern generators. Outputs of 15 Hz and ⅛ Hz are derived from 60-Hz power line by IC frequency dividers.—Circuits, 73, Aug. 1972, p 133.

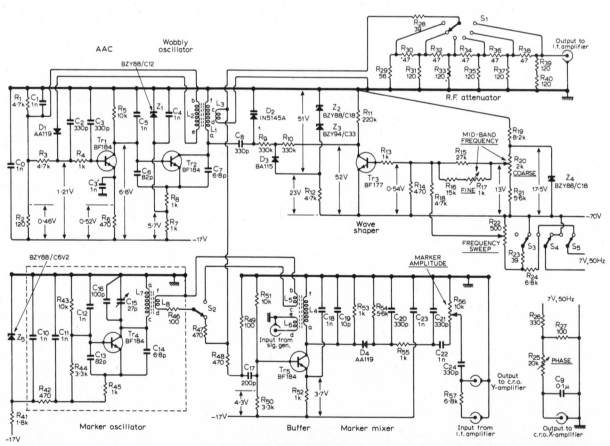

I-F ALIGNMENT WOBBULATOR—Tr2 operating as Colpitts oscillator generates signal varying between 30.5 and 42.5 MHz for feeding to amplifier under test through attenuator controlled by S1. Developed for British tv standards and 50-Hz power-line frequency, and will require some design changes for U.S. tv. Provides linear relation between frequency and displacement on cro. Traces for increasing and decreasing frequency superpose to give single trace only if X and Y channels have same phase shift and i-f amplifier under test gives same response to increasing and decreasing frequencies. Marker oscillator can be set to simulate video carrier frequency.—W. T. Cocking, Television Wobbulator, Wireless World, Sept. 1970, p 423–426.

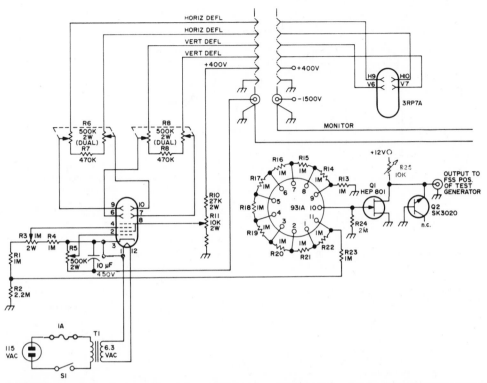

CRT FLYING-SPOT SCANNER—Scanning crt should be same type used in monitor, which here is 3RP7A. Type 931A photomultiplier delivers 6.5 to 12 V d-c to test generator that provides sweep triggers for monitor and scanning module. Use piece of cobalt glass between crt and 931A to filter out long-persistence yellow component of P7 phosphor. Article gives all circuits and covers construction and adjustment.—R. E. Taggart, A Simple Solid-State Flying Spot Scanner for Slow-Scan Television, 73, July 1972, p 89–95.

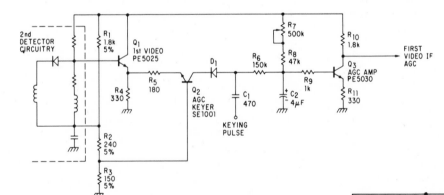

AGC KEYER—Keying current is diverted from horizontal stage through C1, away from R6 into keyer transistor Q2. As signal strength increases and sync pulse at keyer input becomes more negative, Q2 conducts and makes collector voltage of Q3 rise, reducing gain of first i-f stage.—T. B. Mills and H. S. Suzuki, "Design Concepts for Low-Cost Transistor AGC Systems," Fairchild Semiconductor, Mountain View, CA, No. 199, 1971, p 4–5.

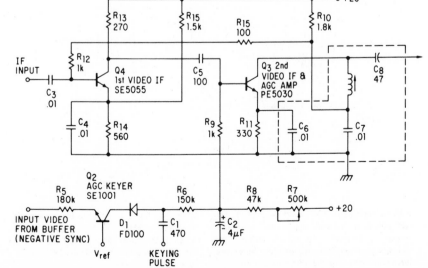

GAIN-CONTROLLED VIDEO I-F—Addition of components within dashed lines to agc amplifier makes it serve also as second video i-f amplifier, thereby reducing component count.—T. B. Mills and H. S. Suzuki, "Design Concepts for Low-Cost Transistor AGC Systems," Fairchild Semiconductor, Mountain View, CA, No. 199, 1971, p 4–5.

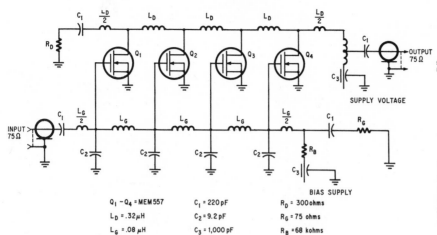

Q1 – Q4 = MEM 557   C1 = 220 pF   RD = 300 ohms
LD = .32 μH   C2 = 9.2 pF   RG = 75 ohms
LG = .08 μH   C3 = 1,000 pF   RB = 68 kohms

CATV DISTRIBUTED AMPLIFIER—Uses mosfet's in parallel distributed along artificial transmission line. Produces 11-dB gain and 6-dB noise figure over vhf band. Power output at 1-dB gain compression is 150 mW.—D. V. Lee, MOSFETs Rejuvenate Old Design for CATV Broadband Amplifiers, *Electronics*, March 15, 1971, p 72–75.

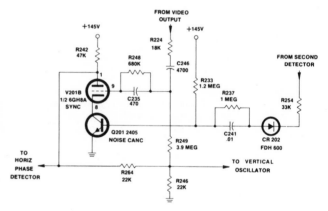

⅛TH-HZ RAMP Provides extremely linear ramp going from —10 V to +10 V, for sstv monitors, cameras, and flying-spot scanners. Positive-going pulse of 2 to 5 V resets ramp for next sweep.—Circuits, 73, June 1972, p 112.

NOISE CANCELLER—Noise spike that drives base of transistor Q201 more negative than peak of sync signal will cut off Q201, opening conduction path of sync amplifier tube V201B. This kills output from sync separator for duration of noise pulse, thereby maintaining stable vertical and horizontal synchronization. Used in RCA tv chassis KC5179/KCS183.—Sync and Noise Cancellation Circuit, *Electronic Technician/Dealer*, Feb. 1971, p 48.

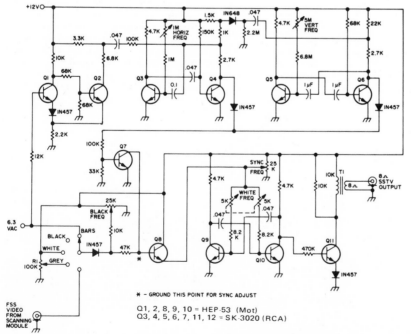

* – GROUND THIS POINT FOR SYNC ADJUST

Q1, 2, 8, 9, 10 = HEP-53 (Mot)
Q3, 4, 5, 6, 7, 11, 12 = SK-3020 (RCA)

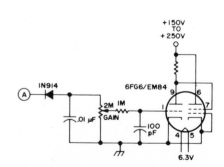

SSTV TUNING INDICATOR—Uses light-bar type of indicator tube as guide for tuning sstv signal to provide correct audio output tones. Point A is connected to high side of sync gain control in sstv receiver, while required high voltage is same as that supplied to 12AX7 limiter and receiver. Adjust gain control until light bars of display almost close for 1,200-Hz test signal connected to input of sstv monitor in which indicator is used. Maximum deflection of 15-Hz flicker in display then corresponds to optimum sstv tuning.—L. I. Hutton, Tuning Indicators for SSTV Monitors, 73, Jan. 1972, p 39–42.

TEST GENERATOR FOR FLYING-SPOT SCANNER—Used to furnish signal for triggering sweep in monitor and in crt scanner for slides. Generator will also produce black raster, white raster, raster with continuously variable gray level, and stable bar pattern. Article covers construction of complete system.—R. E. Taggart, A Simple Solid-State Flying Spot Scanner for Slow-Scan Television, 73, July 1972, p 89–95.

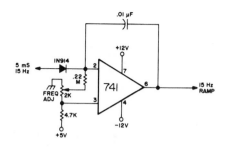

15-HZ SSTV RAMP—Provides extremely linear ramp going from −10 V to +10 V, for sstv monitors, cameras, and flying-spot scanners. Positive-going pulse of 2 to 5 V resets ramp for next sweep.—Circuits, 73, June 1972, p 112.

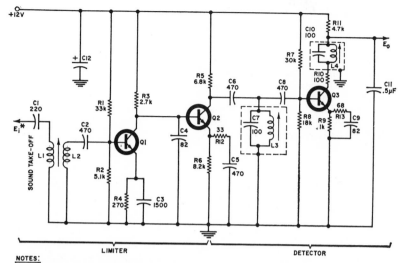

NOTES:
CAPACITORS ARE IN pF UNLESS OTHERWISE SPECIFIED
Q1, Q2—FAIRCHILD S1062 OR 2N3394
✻ FOR HIGH-IMPEDANCE SOURCE USE PARALLEL RESONANCE
L3,L4—12µH, Q–50
Q3—2N2369 OR 2N5027

LOCKED-OSCILLATOR DETECTOR—Solid-state version provides better limiting, more audio recovery, and higher sensitivity than ratio detector in U.S. tv sets, which use intercarrier sound having 4.5-MHz f-m sound carrier.—E. J. Jarrold, Solid-State Locked-Oscillator FM Limiter/Detector, Electronics World, July 1970, p 62–64.

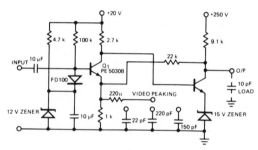

TV CAMERA AMPLIFIER—Developed for use in camera view finders and picture monitors. Output transistor is SE7056, capable of providing over 100-V p-p when driven by transistor shown for Q1. Bandwidth is over 6 MHz.—B. L. Jones, "Applications of the PE5030B," Fairchild Semiconductor, Mountain View, CA, No. 256, 1969.

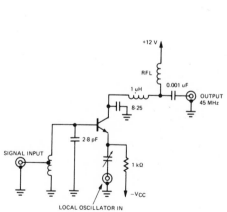

VHF MIXER—Local oscillator signal at 258 MHz is injected into emitter of PE5030B high-frequency npn transistor, while input signal at 213 MHz (channel 13) is applied to base to give i-f output at 45 MHz. Used for testing mixer performance. Step-up transformer is needed to transform 50-ohm signal source impedance to about 200 ohms. Noise figure is 7.5 dB and conversion gain is 12 dB.—B. L. Jones, "Applications of the PE5030B," Fairchild Semiconductor, Mountain View, CA, No. 256, 1969.

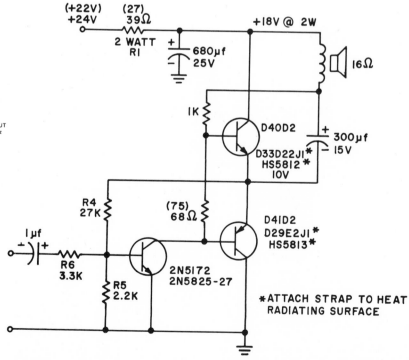

2-W AUDIO—Input resistance is compatible with audio output of most IC tv sound processors having audio preamps. Frequency response is down 3 dB at 60 Hz and 55 kHz. Sensitivity is 1-V rms for full output.—D. V. Jones, "TV Audio Amplifier," General Electric, Semiconductor Products Dept., Auburn, NY, No. 90.91, 1970 p 2–4.

SSTV SYNC—RTL flip-flops and gates provide stable horizontal and vertical sync pulses for slow scan amateur television. Uses 15-Hz line-scanning rate and 8-s frame rate to give 120-line raster. Referencing to 60-Hz power line provides stability and reduces hum in picture. Uses binary ripple counter to count in powers of 2, with gates to sense desired count and restart first counter.—D. R. Patterson, Sync Generator For Sstv, *Ham Radio*, June 1972, p 50–52.

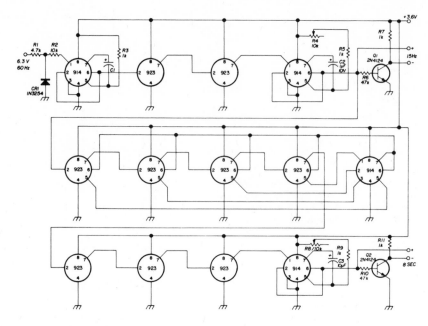

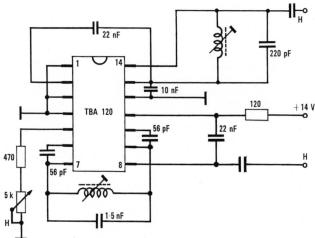

F-M TV I-F IC—Uses IC package developed by Intermetall in West Germany to serve as sound I-F amplifier and demodulator for tv sets using frequency modulation for sound, as in Germany. Provides i-f voltage gain of 60 dB, a-m rejection ratio of 55 dB, and a-f output of 0.6 V. Limiting starts at 70 μV.—Integrated FM/IF Amplifier and Demodulator TBA 120, *Electrical Communication*, Vol. 46, No. 3, 1971, p 221.

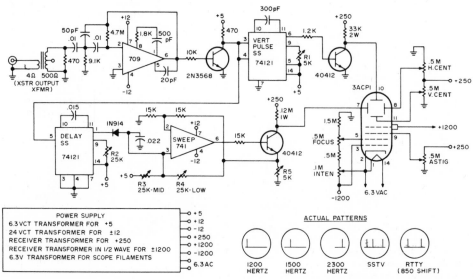

SSTV ANALYZER—Used primarily for monitoring on cro tube the most critical slow-scan tv receiving frequency of 1,200 Hz and the critical 1,500–2,300 Hz transmitting frequencies. Circuit analyzes every hertz generated and displays signals comparatively on screen.

Sweep is triggered when incoming signal crosses zero and goes positive, and vertical pulse is generated 180 deg later for opposite zero crossing. Display also shows gray scale,

gamma correction, and transients. Analyzer can also be used to display rtty signals for analysis of instantaneous shifting errors. Article covers construction and operation.—R. T. Suding, The AFSA IV SSTV Analyzer, *73*, Dec. 1972, p 17–21.

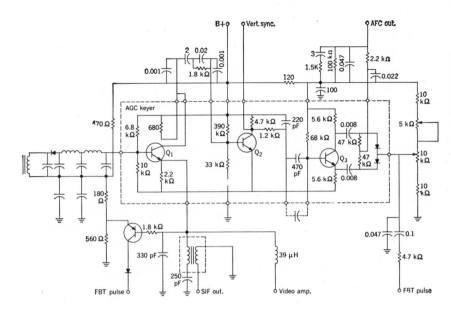

THIN-FILM HYBRID VIDEO I-F—Single series feedback stage Q1 serves as video amplifier, with base driven by video detector and with emitter delivering video driving signal and sound i-f signal. Collector provides positive video signal for synchronous pulse separator Q2. Q3 is in balanced sawtooth-type afc circuit.—E. Sugata and T. Namekawa, Integrated Circuits for Television Receivers, *IEEE Spectrum*, May 1969, p 64–74.

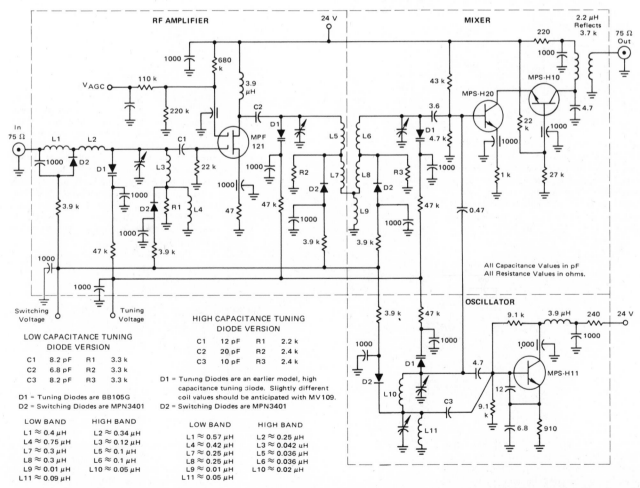

LOW CAPACITANCE TUNING
DIODE VERSION

C1	8.2 pF	R1	3.3 k
C2	6.8 pF	R2	3.3 k
C3	8.2 pF	R3	3.3 k

D1 = Tuning Diodes are BB105G
D2 = Switching Diodes are MPN3401

LOW BAND	HIGH BAND
L1 ≈ 0.4 µH	L2 ≈ 0.34 µH
L4 ≈ 0.75 µH	L3 ≈ 0.12 µH
L7 ≈ 0.3 µH	L5 ≈ 0.1 µH
L8 ≈ 0.3 µH	L6 ≈ 0.1 µH
L9 ≈ 0.01 µH	L10 ≈ 0.05 µH
L11 ≈ 0.09 µH	

HIGH CAPACITANCE TUNING
DIODE VERSION

C1	12 pF	R1	2.2 k
C2	20 pF	R2	2.4 k
C3	10 pF	R3	2.4 k

D1 = Tuning Diodes are an earlier model, high
capacitance tuning diode. Slightly different
coil values should be anticipated with MV109.
D2 = Switching Diodes are MPN3401

LOW BAND	HIGH BAND
L1 ≈ 0.57 µH	L2 ≈ 0.25 µH
L4 ≈ 0.42 µH	L3 ≈ 0.042 µH
L7 ≈ 0.25 µH	L5 ≈ 0.036 µH
L8 ≈ 0.25 µH	L6 ≈ 0.036 µH
L9 ≈ 0.01 µH	L10 ≈ 0.02 µH
L11 ≈ 0.05 µH	

VHF VARACTOR TUNER—Uses mosfet mixer with varactor-tuned dual-mosfet r-f amplifier and 101–257 MHz transistor oscillator which is also varactor-tuned. Designed for printed-circuit construction. Report gives performance on each vhf channel for two types of varactors, one having lower capacitance than the other.—J. Hopkins, "Printed Circuit VHF TV Tuners Using Tuning Diodes," Motorola Semiconductors, Phoenix, AZ, AN-544A, 1972.

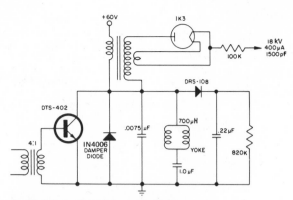

HORIZONTAL OUTPUT—Developed for use in large-screen monochrome tv receivers requiring up to 2,000 V-A and operating from 60-V d-c supply. Uses npn triple-diffused silicon power transistor having maximum dissipation of 100 W.—"Video Horizontal and Vertical Output Transistor Pair," Delco, Kokomo, IN, May 1972, Engineering Data Sheet DTS-401 DTS-402.

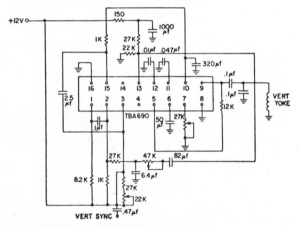

IC VERTICAL SYSTEM—Single Amperex TBA690 IC serves as complete vertical deflection system adequate for up to 12-inch tv set, coupled directly to yoke. For larger screens, external vertical amplifier is required.—"A High Performance Multifunction Novel IC," Amperex Electronic Corp., Slatersville, RI, Application Report S-150, July 1971.

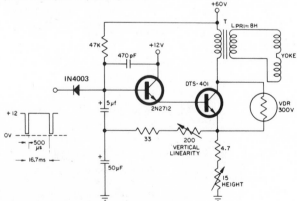

VERTICAL OUTPUT—Uses DTS-401 npn triple-diffused silicon power transistor in circuit operating from 60-V d-c supply. Has good gain linearity, as required for large-screen monochrome tv receivers.—"Video Horizontal and Vertical Output Transistor Pair," Delco, Kokomo, IN, May 1972, Engineering Data Sheet DTS-401 DTS-402.

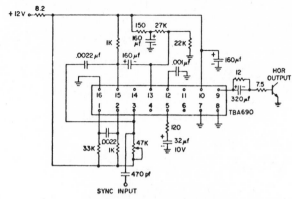

IC HORIZONTAL SYSTEM—Single Amperex TBA690 IC provides complete horizontal drive for external horizontal output transistor.—"A High Performance Multifunction Novel IC," Amperex Electronic Corp., Slatersville, RI, Application Report S-150, July 1971.

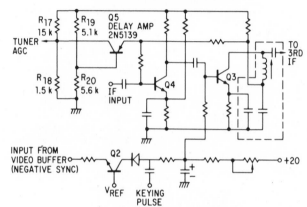

TUNER AGC DELAY—Addition of r-f delay amplifier to tv tuner reduces gain on strong signals. When signal reaches desired maximum level, Q5 base-emitter forward-bias makes lowered input impedance of Q5 load base of Q4, so very little additional change in agc voltage occurs at Q4. D-c loop gain of system thus remains fairly constant over wide range of input signal amplitudes.—T. B. Mills and H. S. Suzuki, "Design Concepts for Low-Cost Transistor AGC Systems," Fairchild Semiconductor, Mountain View, CA, No. 199, 1971, p 5.

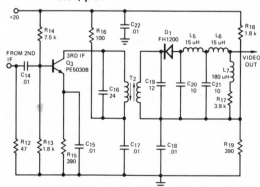

FINAL VIDEO I-F—Low feedback capacitance of transistor Q3 permits high gain without neutralization, while delivering large output currents with little distortion to hot-carrier diode D1 serving as detector.—B. L. Jones, "Applications of the PE5030B," Fairchild Semiconductor, Mountain View, CA, No. 256, 1969.

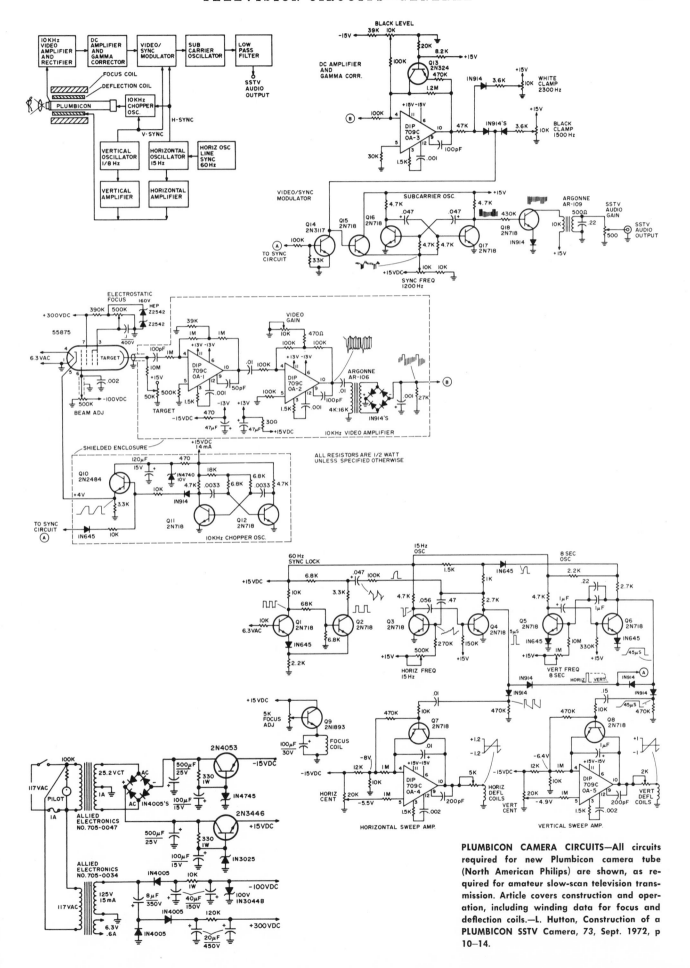

PLUMBICON CAMERA CIRCUITS—All circuits required for new Plumbicon camera tube (North American Philips) are shown, as required for amateur slow-scan television transmission. Article covers construction and operation, including winding data for focus and deflection coils.—L. Hutton, Construction of a PLUMBICON SSTV Camera, 73, Sept. 1972, p 10–14.

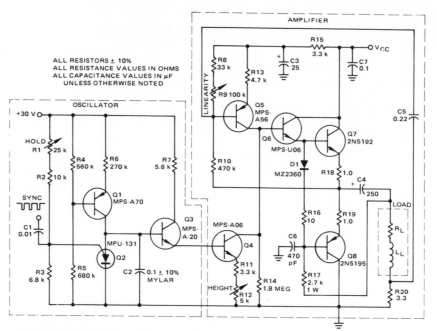

VERTICAL DEFLECTION—Complementary transistors permit use of current drive. Oscillator generates voltage sawtooth which is converted into current sawtooth for scan coils by amplifier. Use of d-c coupling throughout eliminates need for large electrolytic coupling capacitors. Will deliver 1.2-A p-p to standard yoke.—P. E. Crouse, "A Vertical Deflection Circuit Using Complementary Transistors," Motorola Semiconductors, Phoenix, AZ, AN-549, 1971.

CHAPTER 118
Television Circuits—Remote Control

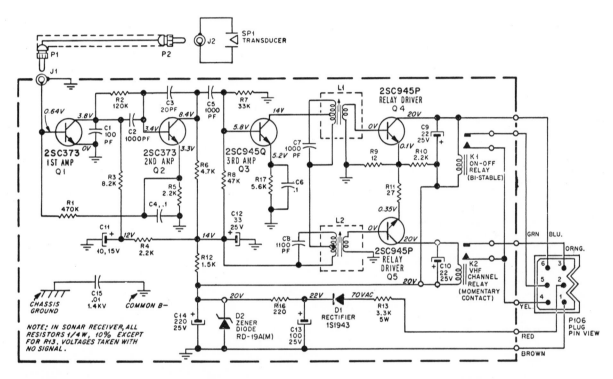

REMOTE - CONTROL RECEIVER — Circuit of R200R sonar receiver uses crystal transducer SP1 to pick up either 38.5-kHz or 41.5-kHz c-w signal produced when hammer strikes one of two resonator bars in hand-held remote control for turning tv set on or off and changing vhf channels. Both frequencies are amplified by common amplifier Q1–Q3, then separated by frequency-selective transformers L1 and L2 driving separate relays through Q4 and Q5.—Sonar Remote Control, Admiral Corp., Bloomington, IL, Service Manual S1291A, May 1972.

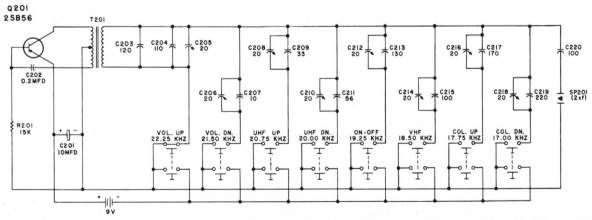

COLOR TV CONTROL TRANSMITTER—Single-transistor oscillator with eight pushbutton-switched capacitors generates eight different frequencies from 17.00 to 22.25 kHz to provide control functions shown on diagram. Transmitter feeds SP201 ultrasonic transducer, whose signal is picked up in receiver by microphone feeding triac frequency-sensitive relays energizing appropriate control solenoids or drive motors. Manual includes complete receiver circuit and describes operation in detail.—"704058 Series Remote Control Chassis," Magnavox Co., Fort Wayne, IN, Service Manual No. 7325, 1970.

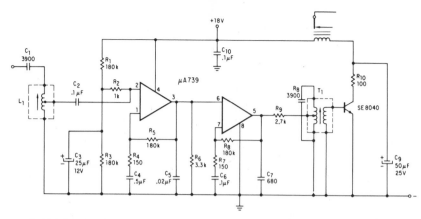

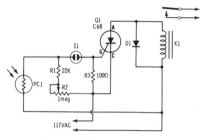

TV ULTRASONIC CONTROL—Receiver circuit shown, using μA739 pair connected in cascade, requires capacitor mike input for picking up 20–50 kHz ultrasonic signal from hand-held remote-control transmitter. After amplification, signal is applied to base of relay drive transistor through frequency-selective output transformer T1 (Magnavox 360944-1), so resulting rectified signal activates relay. L1 is 10 mH, with tap at 15% (Magnavox 360964-1). With tuned circuits shown, gain at 25 kHz is 105 dB. 3-dB bandwidth is 1 kHz, permitting use of four or more identical circuits tuned to different frequencies.—D. Campbell, W. Hoeft, and W. Votipka, "Applications of the μA739 and μA749 Dual Preamplifier Integrated Circuits in Home Entertainment Equipment," Fairchild Semiconductor, Mountain View, CA, No. 171, 1971, p 6–7.

COMMERICAL KILLER—Flick of flashlight beam on Clairex CL705 or equivalent photocell PC1 triggers scr Q1 through NE-83 neon I1, energizing latching relay K1 (Guardian IR-610L-A115), whose contacts stay closed after relay coil current is removed. Contacts of relay can be connected to kill sound of annoying radio or tv commercial. Next flash of light on PC1 energizes relay again to restore sound. Can also be used for on-off control of other devices.—H. Friedman, "99 Electronic Projects," Howard W. Sams, Indianapolis, IN, 1971, p 90–91.

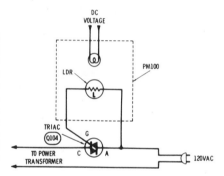

LAMP CONTROLS TRIAC—Used in RCA CTC54 color tv to turn entire receiver on or off with two-wire remote control containing only battery and switch for supplying power to lamp in light-tight housing in set, to lower resistance of light-dependent resistor LDR or cadmium sulfide cell. This triggers on triac to apply a-c to receiver. Triac power rating should be greater than power drain of set.—SCR's and Triacs—Testing and Theory of Operation, *Electronic Servicing*, Dec. 1971, p 42–46.

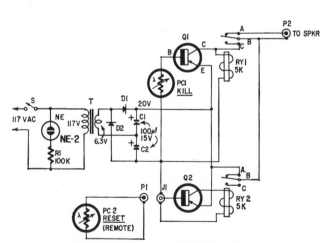

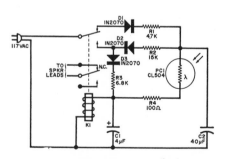

FLASHLIGHT SILENCES TV—Beam from flashlight on PC1 turns on Q1 to energize RY1 and open connection to speaker. Latching action of relay contacts keeps speaker open for duration of commercial. Flick of flashlight on PC2, mounted at another corner of tv set, makes Q2 conduct and restore speaker connection. 3-cell flashlight will give range up to 40 ft in room normally lit for tv viewing at night. Will not work in brightly lighted room or during daytime. Q1 and Q2 are any pnp audio transistors. Photocells are Lafayette 99R6306. D1 and D2 are general-purpose silicon or germanium diodes. Use 1 or 2 mA relays.—F. Blechman, Build a Flashlight Operated TV Silencer, *Radio-Electronics*, May 1969, p 49–50.

REMOTE CONTROL—Permits using flashlight to silence obnoxious radio or tv commercials, or turn on or off any other small appliance without getting up from chair. Will turn on garage lights when car enters driveway. One flash of light on Clairex CL504 photoresistor pulls in relay to open speaker leads or turn on device, and next flash releases relay. Make nonpolarized capacitor for C2 by connecting two 80-μF 150-V electrolytics in series positive to positive.—H. R. Mallory, Build A Photosensitive Switch, *Popular Electronics*, Dec. 1969, p 55–56.

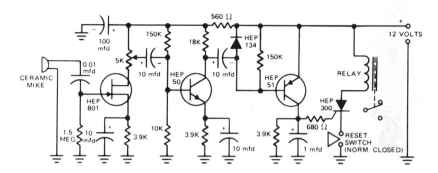

SOUND-ACTIVATED RELAY—Clap of hands or other sharp sound within range of mike energizes 12-V d-c relay for control of any desired device, such as turning off tv during commercials. Sensitivity is adjusted with 5K pot. Once energized, relay stays energized until normally closed reset switch is opened manually.—"Tips on Using FET's," Motorola Semiconductors, Phoenix, AZ, HMA-33, 1971.

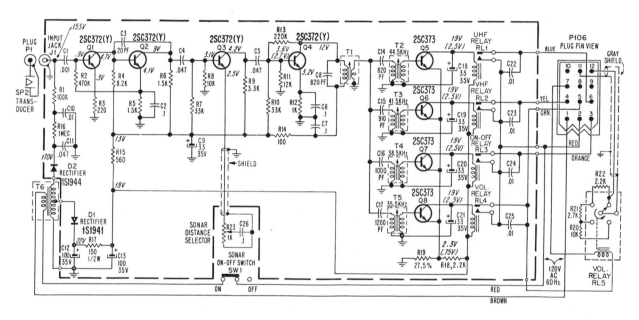

FOUR-FUNCTION REMOTE-CONTROL RECEIVER—Circuit of R300R Admiral Sonar receiver uses crystal transducer SP2 to pick up four different ultrasonic frequencies produced by hand-held remote-control oscillator, for energizing control relays through four-stage amplifier and frequency-selective relay drivers. R23 adjusts emitter degeneration in third preamp, to suppress triggering by undesired ultrasonic sources such as jiggling keys or coins.—"Remote Control," Admiral Corp., Bloomington, IL, Service Manual S1291, Jan. 1972.

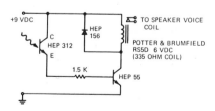

COMMERCIAL KILLER—Flashlight beam on HEP312 photodiode turns on transistor and energizes relay whose contacts are connected to terminals of radio or tv speaker, to short out speaker harmlessly for duration of obnoxious commercial. Circuit resets automatically when flashlight is turned off, to restore sound.—"Home Handyman's Construction Projects," Motorola Semiconductors, Phoenix, AZ, HMA-37, 1972.

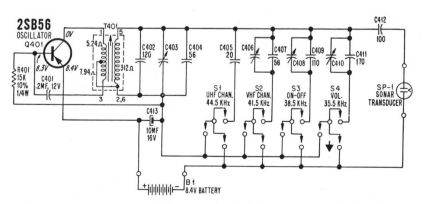

TV REMOTE-CONTROL TRANSMITTER—Pushbuttons S1–S4 insert appropriate capacitors in single-transistor oscillator circuit for generating four different ultrasonic frequencies in hand-held tv remote control. Signal is radiated to receiver in set by piezoelectric transducer SP1.—"Remote Control," Admiral Corp., Bloomington, IL, Service Manual S1291, Jan. 1972.

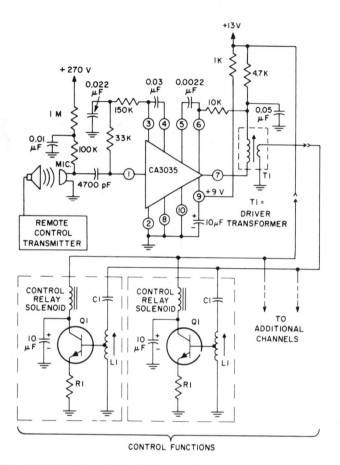

ULTRASONIC RECEIVER—Microphone feeds input signal of about 40 kHz to pin 1 of RCA CA3035 wideband amplifier which provides gain of 120 dB for capacitor mike used. Typical input voltage of 100 μV is required to pull in relay driver providing control function. In application covered, secondary winding of output transformer feeds eight tuned circuits scattered through range of 35 to 45 kHz, to provide total of eight different control functions for tv set.—"Linear Integrated Circuits," RCA, Somerville, NJ, IC-42, 1970, p 335–339.

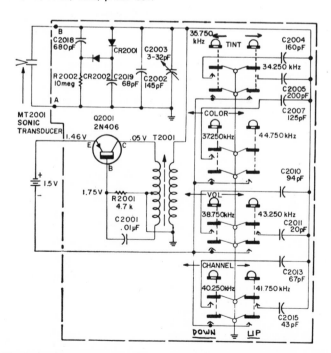

8-TONE ULTRASONIC TRANSMITTER—Transistor oscillator in hand-held 8-button remote-control unit for color tv set feeds crystal transducer at frequency between 34.250 and 44.750 kHz, depending on button pressed, for radiating ultrasonic note to mike in receiver. —L. Allen, Color TV by Remote Control, Radio-Electronics, Jan. 1971, p 45–49.

CHAPTER 119
Temperature Control Circuits

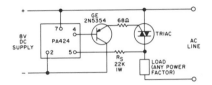

ANY POWER FACTOR—Uses GE PA424 IC zero-voltage switch operating from d-c supply. Thermistor for temperature control is connected to input section of IC as shown elsewhere in report. Choose triac to match load, which may be inductive or resistive.—R. W. Fox and R. E. Locher, "Solid State Electric Heating Controls," General Electric, Semiconductor Products Dept., Auburn, NY, No. 200.58, 1971, p 8–9.

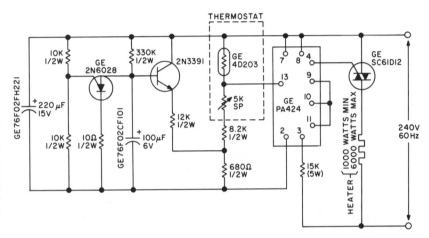

PROPORTIONAL CONTROL—Modulation generator is relaxation oscillator using 2N6028 put which produces sawtooth with period of 30 s. Modulation wave is coupled into sensor and reference circuit through 2N3392 buffer to get modulation amplitude equivalent to 2 F change in sensor temperature. Will control normal room temperature to within 0.5 F. For remote thermostat, three wires are needed. IC is zero-voltage switch.—R. W. Fox and R. E. Locher, "Solid State Electric Heating Controls," General Electric, Semiconductor Products Dept., Auburn, NY, No. 200.58, 1971, p 11–12.

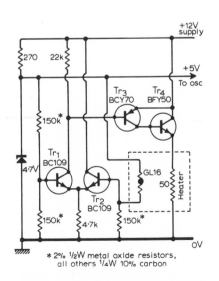

CRYSTAL OVEN CONTROL—Uses thermistor in bridge circuit supplied by 4.7-V zener to minimize effects of supply voltage variations. Resistance of GL16 is 1 meg at 20 C and drops to 150K at 60 C operating temperature of oven. Other three legs of bridge are accordingly 150K.—L. Nelson-Jones, Crystal Oven and Frequency Standard, *Wireless World*, June 1970, p 269–273.

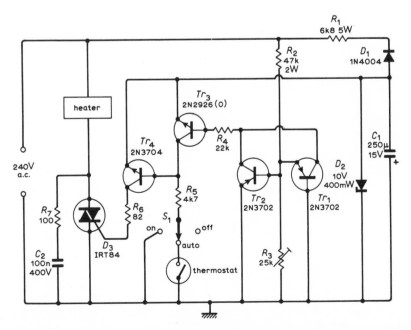

2.4-KW HEATER CONTROL—Uses zero-voltage switching to eliminate rfi. TR1 and TR2 are connected as zero-voltage detector driven from a-c line through voltage divider R2-R3. R3 controls reference voltage and width of triac gate pulse. Minimum load required is 300 W.—R. M. Marston, Electric Heater Control, *Wireless World*, June 1972, p 287–289.

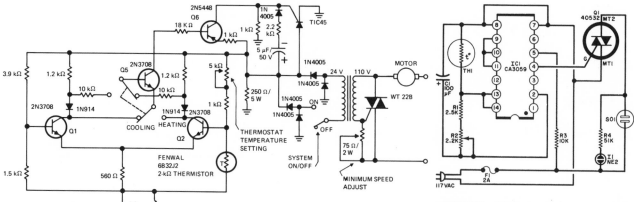

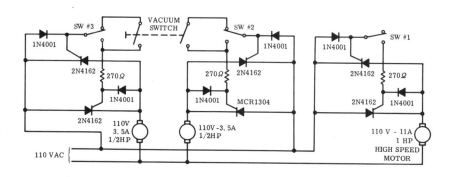

AIR-CONDITIONER SPEED CONTROL—Thermistor and 5K pot control speed of shaded-pole induction and permanent-capacitor a-c motors. Control circuit, limited to 24 V by power transformer, is isolated from a-c line.

Can be switched from cooling to heating. When initially activated, simple R-C timer automatically supplies full power to motor for about 30 s to give adequate starting torque. Speed range is generally limited from full to half-value.—J. M. Garrett, "Fan Motor Thermostatic Speed Control," Texas Instruments, Dallas, TX, Bulletin CA-121, 1969, p 4.

AQUARIUM CONTROL—With triac shown, will handle aquarium heaters up to 200 W. Sensor TH1 is Fenwall JA33J1 thermistor.—A. E. Donkin, Aquarium Heater Control For Fish Fanciers, *Popular Electronics*, Sept. 1972, p 70—72.

COOLING-WATER PUMP CONTROL—Used in measuring thermal properties of semiconductors. Scr circuit controls motors which drive cooling water pumps in hydraulic system. Gate signal is applied to scr's through microswitches located on or near levers attached to valves of hydraulic system, so positions of valves control motors and sequence of operation. Report covers complete system, including 1.2-kW water heater control.—R. Ivins, "Measurement of Thermal Properties of Semiconductors," Motorola Semiconductors, Phoenix, AZ, AN-226, 1972.

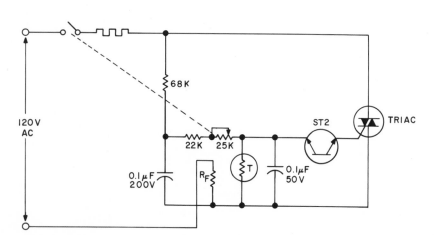

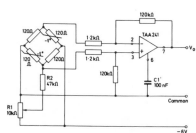

INDIRECTLY HEATED SENSOR—Used when it is not feasible to mount sensor in environment being controlled, as with electric blankets, vibrating systems, or corrosive baths. Load current through RF heats 50K NTC thermistor to achieve temperature regulation. Selection of RF is critical, since its temperature must be related to that of controlled space. Choice of triac depends on load. ST2 is 2N4992 diac used as back-to-back zener.—R. W. Fox and R. E. Locher, "Solid State Electric Heating Controls," General Electric, Semiconductor Products Dept., Auburn, NY, No. 200.58, 1971, p 5—6.

THERMISTOR-BRIDGE AMPLIFIER—Differential input for IC subtracting amplifier is derived from output of balanced thermistor bridge. Amplifier gain is 100. R1 and R2 balance bridge for zero output and cancel voltage offset.—"TAA241, TAA242 and TAA243 Differential Operational Amplifiers," Mullard, London, 1969, TP1086, p 14.

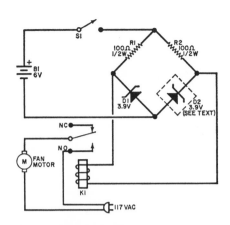

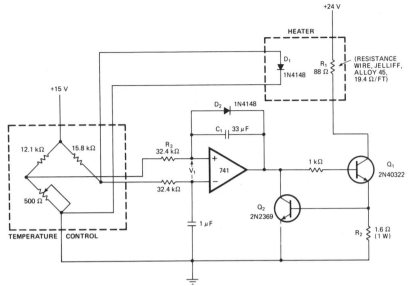

HI-FI COOLING CONTROL—Zener D2, which can be 1N748 or 1N4730, is mounted remotely in critical area near electrolytics or other components that are damaged or changed by excessive heat in high-power audio amplifier, tv set, or other electronic equipment. When heat rises sufficiently to unbalance bridge, 6-V relay K1 pulls in and starts cooling fan motor.—B. J. LaVaia, Reader's Circuit, *Popular Electronics*, Jan. 1969, p 87 and 99.

0.05 C ACCURACY—Tight temperature coefficient of silicon diodes permits precision control of temperature with low-cost opamp. When temperature is too low, Q1 passes heater current. When too high, D2 prevents opamp output from going negative, so Q1 turns off.—R. Koss, Diode Plus Low-Cost Op Amp Makes Accurate Thermostat, *Electronics*, June 19, 1972, p 90.

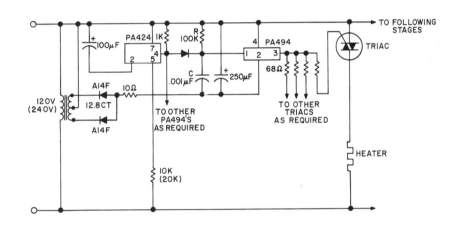

MULTIPLE-LOAD DRIVE—Uses PA494 IC threshold detector as buffer amplifier between PA424 IC zero-voltage switch and triac gates. With 50-mA IGT triacs, five can be driven by each PA494. Inductive loads are switched for full cycles, eliminating possibility of saturation.—R. W. Fox and R. E. Locher, "Solid State Electric Heating Controls," General Electric, Semiconductor Products Dept., Auburn, NY, No. 200.58, 1971, p 10.

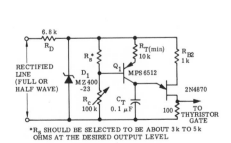

FEEDBACK CONTROL—Feedback is provided by sensing resistor Rs, which may respond to heat, light, moisture, pressure, or other stimuli. Rc establishes desired operating point. Will provide either half-wave or full-wave phase control of scr or triac, depending on rectified line.—D. A. Zinder, "Unijunction Trigger Circuits for Gated Thyristors," Motorola Semiconductors, Phoenix, AZ, AN-413, 1972.

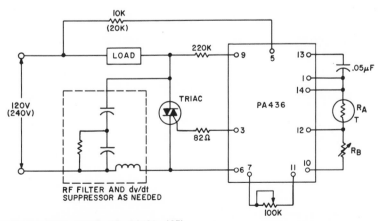

IC TRIGGER FOR TRIAC—GE PA436 (GEL 301F) IC phase control trigger converts analog input signal from 5K thermistor T to phase-control pulse for triggering triac chosen to handle required heater load. Signal is compared with reference, and phase angle of triggering is obtained by ramp and pedestal technique. RB varies temperature set point. Load voltage control range is 0 to 100%.—R. W. Fox and R. E. Locher, "Solid State Electric Heating Controls," General Electric Semiconductor Products Dept., Auburn, NY, No. 200.58, 1971, p 8.

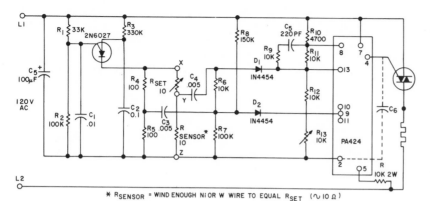

LOW-RESISTANCE SENSOR—Uses GE PA424 IC zero-voltage switch with 2N6027 put relaxation oscillator operating at 20 pps. Designed for high-temperature applications using low-impedance tungsten, platinum, or Nichrome sensors. Choose triac to match load. Report covers operation and adjustment of circuit.—R. W. Fox and R. E. Locher, "Solid State Electric Heating Controls," General Electric, Semiconductor Products Dept., Auburn, NY, No. 200.58, p 13.

* R_{SENSOR} = WIND ENOUGH NI OR W WIRE TO EQUAL R_{SET} ($\sim$10 Ω)

40-KHZ CARRIER POWER AMPLIFIER—Used in feedback system to control surface temperature of metallic body in boiling liquid. Input is 5-V signal from sine-wave oscillator, and output at O2 is 1 A for heater in liquid. T3 is pair of ITC model 7067 units connected for 3:1 voltage ratio. EC3 is 100 V from well-regulated ripple-free source. Q1 and Q2 are 2N3599; Q3 and Q5 are 2N3904; Q4 is 2N2904; diodes are 1N482.—W. C. Peterson, A. Thacker, and W. L. Avery, A Feedback System for Control of an Unstable Process, *IEEE Trans. on Industrial Electronics and Control Instruments*, Sept. 1969, p 165–171.

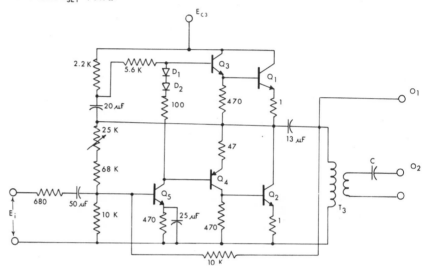

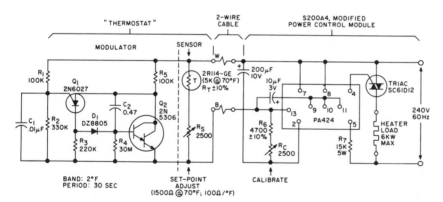

TWO-WIRE REMOTE THERMOSTAT—Modulating current for proportional control is introduced directly across reference arm of bridge in circuit using GE PA424 IC zero-voltage switch. Thermostat requires only two wires, from terminals W and B, for remote location.—R. W. Fox and R. E. Locher, "Solid State Electric Heating Controls," General Electric, Semiconductor Products Dept., Auburn, NY, No. 200.58, 1971, p 12–13.

LOW-RESISTANCE SENSOR—Designed for use with Nichrome, tungsten, and platinum sensors having low resistance and normally positive temperature coefficient for high-temperature applications. Uses IC zero-voltage switch to control triac capable of handling heater load. Circuit includes 2N6027 put relaxation oscillator operating at 20 pps to pulse resistor bridge.—"SCR Manual," General Electric, Syracuse, NY, 1972, 5th Ed., p 339–340.

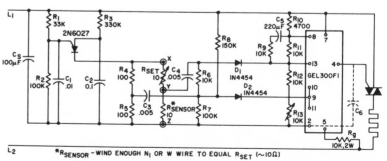

* R_{SENSOR} – WIND ENOUGH N_1 OR W WIRE TO EQUAL R_{SET} ($\sim$10 Ω)

NOTES:
1. ADJUST R_{13} TO MID POINT BETWEEN ON AND OFF WITH C_2 SHORTED, OSCILLATOR DISABLED.
2. PROPORTIONAL CONTROL BAND (GAIN) DETERMINED BY R_g.
3. WITH VALUES SHOWN, PROPORTIONAL BAND IS 1% R_{SENSOR} AND STROBE RATE IS 21/SECOND.
4. A PULSE TRANSFORMER CONNECTED BETWEEN (Y) AND (Z) GIVES A SENSOR ISOLATED FROM THE LINE.

AUTUMN-SPRING HOME CONTROL—Prevents temperature overshoot of hot-air or hot-water home-heating system during moderate weather when rooms would otherwise overheat if thermostat is set normally. Outdoor temperature sensor TH senses temperature differential and applies power to small heating element placed in room thermostat. Heater is energized only when thermostat is calling for heat, to make thermostat think room is a degree or two warmer than it actually is. Circuit is designed to make heater power vary linearly with outdoor temperature from 0.1 W at 20 deg to 1 W at 60 F.—L. B. Stein, Jr., Electronic Control Keeps Home Heater on the Mark, *Popular Science*, Sept. 1969, p 184–186 and 216.

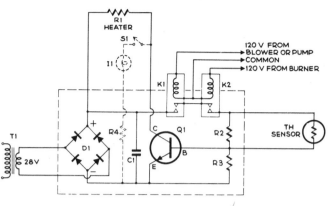

D1—Full-wave bridge rectifier, Mallory FW200
I1—Pilot lamp (night option), 6v, .040 amp #1483
K1, K2—Relay, 120v AC, SPST min., Sigma 41ROZ-5000CG-BSL or 11F-9000-ACS-SIL
Q1—Transistor, 2N2270
R1—Thermostat heater, carbon resistor, 560 ohm, 2w or 2 1100-ohm, 1w in parallel.
R2—Carb. res. 39K, ½ w
R3—Carb. resistor, 390 ohm, ½ w
R4—Carb. resistor (night option), 51 ohm, ½ w
C1—Pap. capacitor, .02 mfd. 400v
S1—Switch (night option), 0-12 hrs, M.H. Rhodes 90015 or 91054 (closed when timing)
T1—Transformer, 120/-28v, 0.1 amp (see text)
Th—Thermistor, 4 Fenwall RA41L3 in series parallel or 2 RA43L1 in parallel

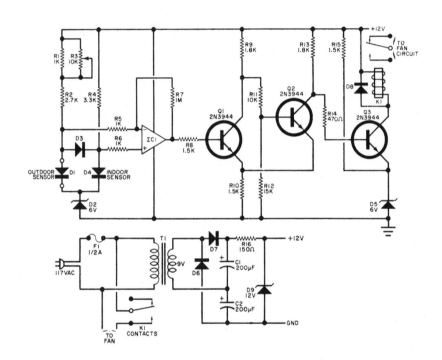

BASEMENT FAN CONTROL—Opamp (741 or similar) connected as differential amplifier senses difference between matched HEP134 germanium diodes used as indoor and outdoor temperature sensors, and feeds Schmitt trigger Q1-Q2 which converts slowly changing input signals to switch action. Used to pull in relay that turns on fan that lowers humidity of basement by drawing out cool damp air at floor level and replacing it with warmer air from outside. D8 is HEP134, D2 and D5 are HEPZ0214, D9 is HEP105, and other diodes are HEP154.—J. Ashe, Differential-Temperature Basement Ventilator, *Popular Electronics*, July 1972, p 31–33.

SEQUENTIAL ENERGIZING—Used to control multiple heater loads sequentially. RC time constant is set to be greater than period of 60-Hz line voltage, so if PA424 IC zero-voltage switch calls for load power in response to thermistor T, adjacent PA494 IC threshold detector will be on for at least next full cycle. When on, it discharges 1-μF capacitor quickly through diode around 10-meg resistor, to restart timing delay.—R. W. Fox and R. E. Locher, "Solid State Electric Heating Controls," General Electric, Semiconductor Products Dept., Auburn, NY, No. 200.58, 1971, p 11.

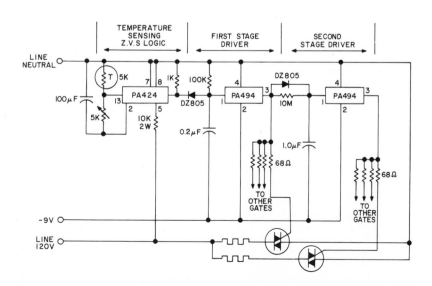

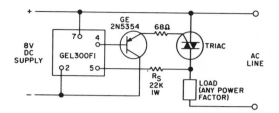

INDUCTIVE LOADS—Provides zero-voltage switching, controlled by IC, to keep emi at minimum for full-wave control of inductive load. Rs provides line synchronization with first trigger pulse, followed by d-c gate current, to eliminate objectionable random triggering point for initial turn-on. Book gives temperature sensor connections to IC. Control point repeatability is within 0.5% of sensor resistance. Thermistors used can range from 2.5K to 50K at operating temperature.—"SCR Manual," General Electric, Syracuse, NY, 1972, 5th Ed., p 336–337.

R_A = THERMISTOR FOR TEMPERATURE CONTROL APPLICATIONS

IC WITH TRIAC—Uses IC zero-voltage switch for triggering triac chosen to handle resistance heater load. Gating pulse produced by IC is centered on zero crossing and ends at specified time in each cycle.—"SCR Manual," General Electric, Syracuse, NY, 1972, 5th Ed., p 335.

600-W HIGH-PRECISION—Ramp-and-pedestal full-wave control requires only 2 C change in temperature of 5K GE ETRS-4942 thermistor for full range of power control. Includes built-in protection against transient voltages. CR1–CR4 and CR6 are GE504A, CR5 is series pair of GE-X11 zeners, and T1 is GE ETRS-4898 pulse transformer.—"Electronics Experimenters Circuit Manual," General Electric, Owensboro, KY, 1971, 3rd Ed., p 217–220.

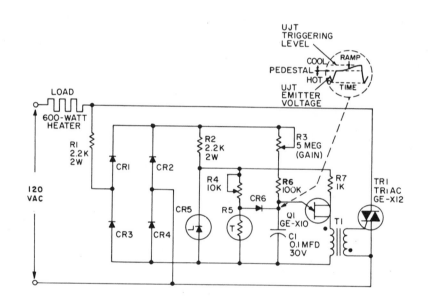

PUSH-PULL A-C—Used in feedback system to control surface temperature of metallic body in boiling liquid. Process is inherently unstable in one region of operation, but automatic feedback control system described in article maintains surface temperature accurately. Amplifier is used between comparator bridge and demodulator to provide voltage gain and optimum coupling. Collector supply EC2 should be −20 V from well-regulated ripple-free source. Q1 and Q2 are 2N2833; Q5 and Q6 are 2N3906; others are 2N327A.—W. C. Peterson, A. Thacker, and W. L. Avery, A Feedback System for Control of an Unstable Process, IEEE Trans. on Industrial Electronics and Control Instruments, Sept. 1969, p 165–171.

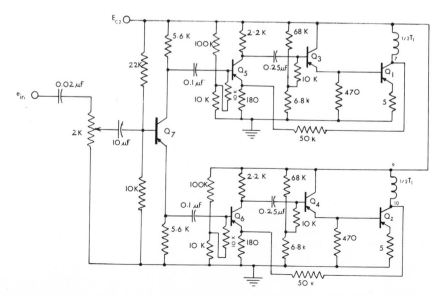

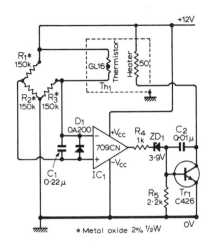

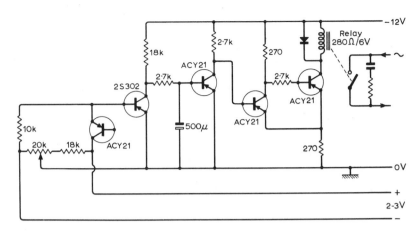

CRYSTAL-OVEN OPAMP CONTROL—Increased voltage gain of opamp and use of 12 V on thermistor bridge together increase loop gain, to improve performance of oven for 1-MHz square-wave frequency standard. Thermistor resistance is about 150K at 60 C control temperature. Article gives oven construction details.—L. Nelson-Jones, Crystal Oven and Frequency Standard, *Wireless World*, June 1970, p 269–273.

HEAT PUMP CONTROL—Sensing element is reverse-biased germanium transistor in bridge circuit. Covers range of 12–25 C. Sensitivity is so high that 500-μF capacitor is required to smooth out short-term fluctuations. Control is better than 0.1 C if application has negligible thermal lag.—A. Sewell, Sensitive Thermostat, *Wireless World*, July 1970, p 316.

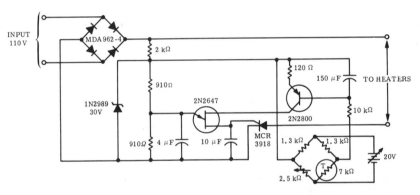

1.2-KW WATER HEATER—Handles three 400-W heating elements connected in parallel, generally in combination with two additional 400-W heaters operating directly from a-c line. Used in measuring thermal properties of semiconductors. Report covers complete system, including control for cooling-water pump.—R. Ivins, "Measurement of Thermal Properties of Semiconductors," Motorola Semiconductors, Phoenix, AZ, AN-226, 1972.

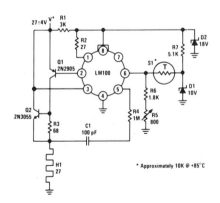

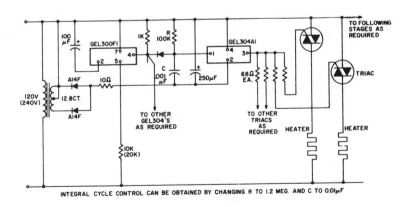

INTEGRAL CYCLE CONTROL CAN BE OBTAINED BY CHANGING R TO 1.2 MEG. AND C TO 0.01μF

OVEN CONTROL—Control accuracy is better than 1 C for wide range of ambient conditions, if thermistor sensor T and its reference diode D1 are placed in oven. LM100 IC voltage regulator need not be in oven. Thermistor senses temperature changes in oven and sends signal to LM100 to control power to heater by switching series-pass transistor Q2 on and off. Positive feedback around regulator gives variable-duty-cycle switching action. Thermistor should have temperature coefficient higher than 1% per deg C.—R. J. Widlar, "New Uses for the LM100 Regulator," National Semiconductor, Santa Clara, CA, 1968, AN-8.

MULTIPLE TRIAC TRIGGERING—Provides triggering of as many triacs as are required for separate heater loads, with full-cycle switching even for inductive loads so as to prevent saturation. Uses GEL304 IC threshold detector as buffer amplifier between GEL300 IC zero-voltage switch and triac gates. Provides pulse widths of 200 μs, to insure that all triacs are latched on; this eliminates need for selecting triacs for latching and pulse gate trigger current.—"SCR Manual," General Electric, Syracuse, NY, 1972, 5th Ed., p 340–341.

SOLDERING-IRON CONTROL—With Potter & Brumfield GP11 6-V d-c relay, having 5-A contact, provides on-off control of soldering irons up to 500 W when 1K thermistor R1 is mounted on soldering iron. Circuit is capable of controlling temperature to within 1 deg over range of 20 to 150 F determined by setting of R4. Can also be used to control temperature of water bath or other heated volume. To control cooling fan or air conditioner, use normally open contacts on relay or reverse connections to secondary winding W2 of T2.—"Electronics Experimenters Circuit Manual," General Electric, Owensboro, KY, 1971, 3rd Ed., p 213–216.

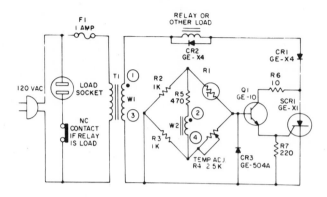

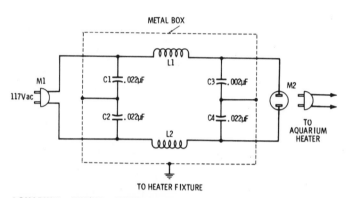

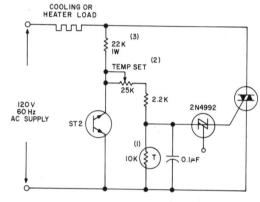

AQUARIUM HEATER FILTER—Prevents arc-generated interference of thermostat in aquarium heater from travelling over power lines and producing interference in neighborhood radio and tv sets. L1 and L2 are each single layer of No. 16 enamel wound on 1-inch form 5 inches long.—R. M. Brown and T. Kneitel, "49 Easy Entertainment & Science Projects," Howard W. Sams, Indianapolis, IN, 1971, p 39–40.

6-KW FULL-WAVE PHASE CONTROL—With 400-V 25-A SC60D triac, will control up to 6 kW of heating or cooling power. For cooling, interchange GE type 1D101 thermistor T and 25K reference pot. ST2 diac is used as back-to-back zener, for line voltage stabilization.—R. W. Fox and R. E. Locher, "Solid State Electric Heating Controls," General Electric, Semiconductor Products Dept., Auburn, NY, No. 200.58, 1971, p 5–6.

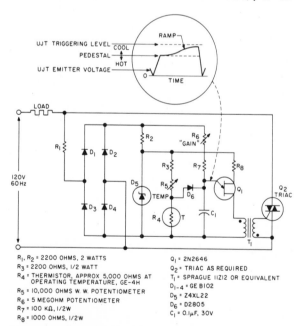

R₁, R₂ = 2200 OHMS, 2 WATTS
R₃ = 2200 OHMS, 1/2 WATT
R₄ = THERMISTOR, APPROX 5,000 OHMS AT
    OPERATING TEMPERATURE, GE-4H
R₅ = 10,000 OHMS W. W. POTENTIOMETER
R₆ = 5 MEGOHM POTENTIOMETER
R₇ = 100 KΩ, 1/2W
R₈ = 1000 OHMS, 1/2W

Q₁ = 2N2646
Q₂ = TRIAC AS REQUIRED
T₁ = SPRAGUE 11Z12 OR EQUIVALENT
D₁₋₄ = GE B102
D₅ = Z4XL22
D₆ = D2805
C₁ = 0.1μF, 30V

RAMP AND PEDESTAL—Provides precise proportional temperature control using low-resistance thermistor T (about 5K). System gain is varied with R6 to change amplitude of ramp. With 20-V zener and 1-V ramp, 22% change in thermistor resistance gives linear full-range change in output. Choose triac to match load.—R. W. Fox and R. E. Locher, "Solid State Electric Heating Controls," General Electric, Semiconductor Products Dept., Auburn, NY, No. 200.58, 1971, p 6.

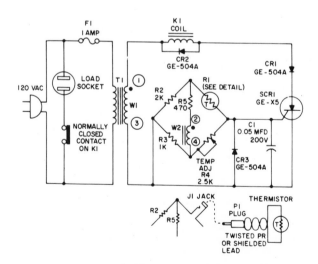

HIGH-CURRENT LOAD—Uses scr rated at 4 A to energize relay, with stage of transistor amplification to meet triggering requirement of scr. Thermistor R1 is 1K, such as GE type 1D303, located in volume controlled by heater.—"Electronics Experimenters Circuit Manual," General Electric, Owensboro, KY, 1971, 3rd Ed., p 213–216.

AQUARIUM HEATER—Uses 24 300-ohm ½-W resistors in parallel as heating element that can be hidden in sand of aquarium (R7), with GE 1819 28-V lamp across heater to indicate when it is on. Waterproof resistors and connections with epoxy. T1 is 2-A 25.2-V filament transformer. Temperature-dependent resistor TDR1, which senses temperature changes as small as 0.1 F, is Fenwall LP32J2. Article covers construction and calibration for 15-gallon and larger aquariums requiring up to 50 W.—S. Jarvin, Electronic Aquarium Heater, *Popular Electronics*, Jan. 1970, p 60–63.

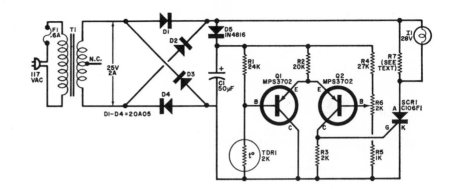

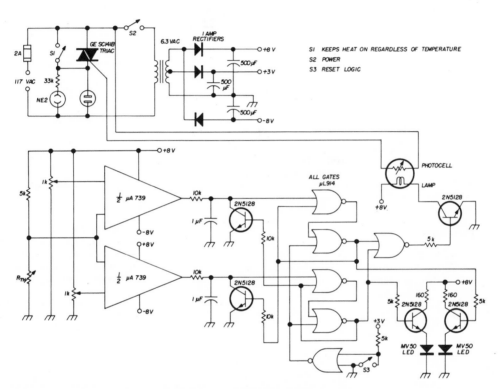

S1 KEEPS HEAT ON REGARDLESS OF TEMPERATURE
S2 POWER
S3 RESET LOGIC

DISTILLATION CONTROL—Automatically stops distillation when temperature rises above boiling point of pure material in pot (indicating presence of higher-boiling impurity) or falls below boiling point (when pot is nearly empty and not enough hot vapor flows past thermistor Rth in distillation head to keep it hot). Heating element is plugged into outlet in series with triac. Light-emitting diodes indicate whether shutdown was due to rise or fall in temperature.—L. Hutchinson, Practical Photofabrication of Printed Circuit Boards, *Ham Radio*, Sept. 1971, p 6–18.

CHAPTER 120
Test Circuits—General

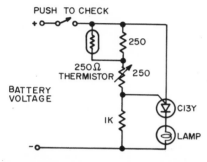

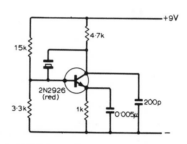

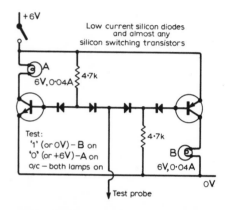

3-V BATTERY TESTER—Temperature-compensated circuit using complementary scr monitors voltage of 3-V battery. Lamp lights if voltage is above about 2.5 V, with about 0.1-V region of uncertainty below that. Operating temperature range is 40–110 F.—W. H. Sahm, III, and F. M. Matteson, "The Complementary SCR," General Electric, Semiconductor Products Dept., Auburn, NY, No. 90.94, 1970, p 14–15.

CRYSTAL TESTER—Simple Pierce oscillator can be coupled to digital frequency meter to provide quick frequency check of miscellaneous and surplus quartz crystals having frequencies between about 1 and 10 MHz.—P. Short, Ageing Crystals, *Wireless World*, Oct. 1969, p 473.

LOGIC TESTER—Indicates presence of logic level 1 or 0 when probe is held on terminal of logic system, and also indicates open-circuit (absence of either level). Developed for 6-V system using negative logic, but can be adapted to other systems. Will also indicate pulse waveforms if longer than 100 ms.—R. Williamson, Simple Logic Tester, *Wireless World*, Aug. 1969, p 379.

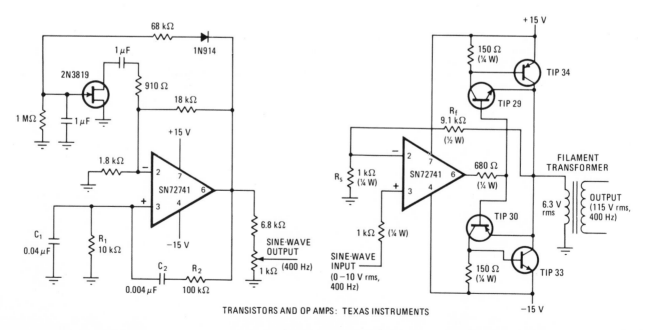

TRANSISTORS AND OP AMPS: TEXAS INSTRUMENTS

400-HZ 115-V SERVO TESTER—Uses 400-Hz audio oscillator (circuit at left) to feed amplifier using opamp for voltage gain and transistors for current gain and regulation. Use of filament transformer to step up output voltage cuts cost of supply. Output is 250 mA, with adequate regulation for testing servos and other aircraft equipment. Arrangement permits use of low-voltage transistors and opamps.—G. Coers, Filament Transformer Output Drops Cost of 400-Hz Supply, *Electronics*, April 10, 1972, p 100.

IC COUNTER TESTER—Test circuit simulates operation of decade counter by producing single pulse each time spdt switch is closed. Nixie tube is observed for proper operation. Change value of 22K resistor in series with Nixie as required to make all parts of each digit light up when corresponding cathode is grounded. Circuit also serves for testing Nixie drivers such as SN7441N.—R. Factor, 3 Versatile IC Testers, 73, Sept. 1970, p 38–40 and 42–44.

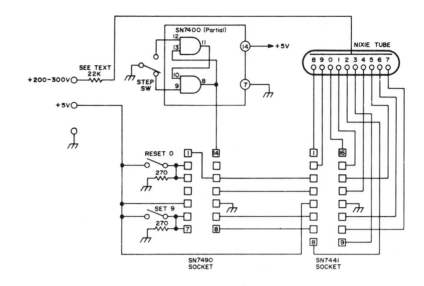

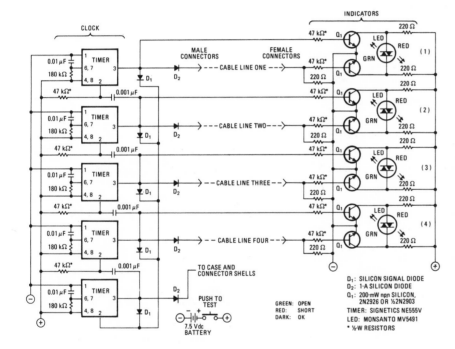

CABLE TESTER—Combination of type 555 IC timers and two-color led's will test cables having from 2 to 10 wires. Operates as ring timer which applies positive pulse in sequence to each of lines under test. Differential transistor pair for each line drives led. If same pulse is at both ends of same line, pair remains balanced and led will not glow. If clock pulse appears only at clock end of line, differential pair unbalances and forces current through green led section, to indicate open line. Pulse only at indicator end of line makes red led come on to indicate shorted line.—L. W. Herring, Timer ICs and LEDs Form Cable Tester, *Electronics*, May 10, 1973, p 115–116.

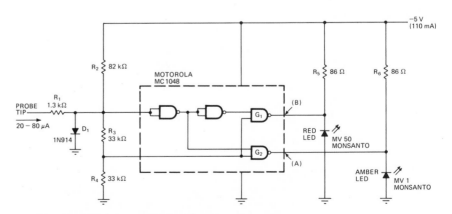

ECL TEST PROBE—Used for testing emitter-coupled logic circuits. Indicates logic high (−0.75 V) by lighting red led, logic low (−1.5 V) by lighting amber led, and open-circuit by leaving both dark. Entire circuit can be assembled inside discarded felt-tip pen.—W. Wilke, Logic Probe With LED Display Checks ECL Circuits, *Electronics*, July 31, 1972, p 76.

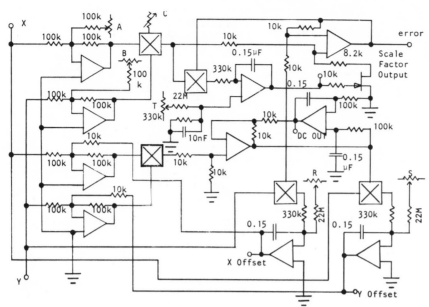

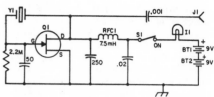

CRYSTAL CHECKER—Simple oscillator using 2N5486 or similar fet is used with GE 1869 10-V 14-mA lamp for checking crystals from 0.1 to 10 MHz. Lamp lights when crystal is bad or not in circuit. With good crystal, circuit oscillates and lamp current drops below 10 mA, so lamp dims or goes out. Can also be used as calibrator for generating markers 500 kHz apart if 500-kHz crystal is plugged in.—M. S. Robbins, A Simple Crystal Tester-Calibrator, QST, Feb. 1970, p 20.

MULTIPLIER TESTER—Used to test analog multipliers by comparing with standard multiplier having known low error performance. Uses ten opamps, which can be two OEI 9432 quintuple arrays if upper test frequency is not over 5 kHz. For tests up to 1 MHz, use individual OEI 9694 opamps. Multipliers can be 5485 up to 30 kHz and 5805 up to 1 MHz. Report covers adjustment and use. X and Y offset outputs are 10% of actual offset, while output level is equal to output offset. Fet is 2N5462.—"Automatic Multiplier Tester," Optical Electronics, Tucson, AZ, Application Tip 10229, 1971.

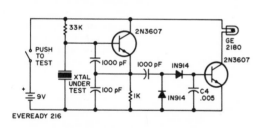

CRYSTAL CHECKER—If pilot lamp glows when button is pushed, crystal is good.—Circuits, 73, Nov. 1972, p 300.

SIGNAL TRACER—Designed for feeding into phono jack of working receiver, for troubleshooting an inoperable receiver. Set being repaired is tuned to strong local station, grounded input lead of signal tracer is clipped to ground of set, and other tracer lead is clipped to input and output of each stage in turn starting at antenna and working toward speaker. First stage at which signal becomes distorted, weak, or drops out is faulty stage.—R. M. Brown and T. Kneitel, "101 Easy Test Instrument Projects," Howard W. Sams, Indianapolis, IN, 1968, p 65–66.

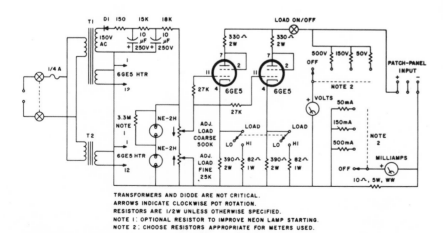

TRANSFORMERS AND DIODE ARE NOT CRITICAL.
ARROWS INDICATE CLOCKWISE POT ROTATION.
RESISTORS ARE 1/2W UNLESS OTHERWISE SPECIFIED.
NOTE I: OPTIONAL RESISTOR TO IMPROVE NEON LAMP STARTING.
NOTE 2: CHOOSE RESISTORS APPROPRIATE FOR METERS USED.

POWER-SUPPLY TESTER—Tubes serve as variable resistors for dissipating output power of supply while making load tests. Minimum resistance is 1,000 ohms. Designed for testing supplies having output voltages between 30 and 500 V.—J. Ashe, Vacuum Tube Load Box, 73, April 1970, p 96–98.

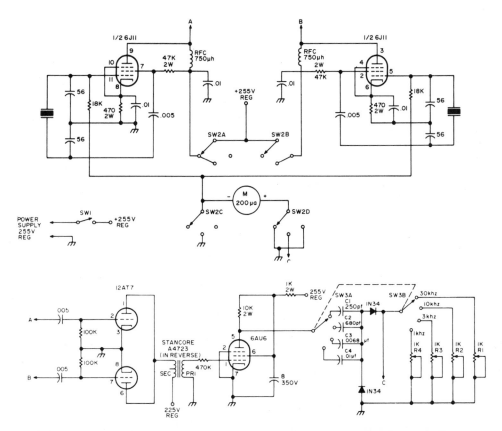

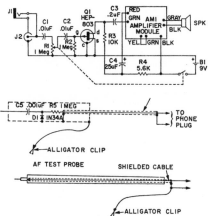

CRYSTAL CHECKER—Checks activity and measures difference in frequency of two crystals up to 30 kHz, directly on meter. Terminals A and B of mixer-limiter go to corresponding terminals of two identical Pierce oscillators having range of 3–9 MHz. Difference-frequency ranges of SW3 correspond to scales available for meter. Ideal for matching crystals for ssb filters.—N. Stinnette, Crystal Frequency and Activity Checker, 73, Dec. 1972, p 68–70.

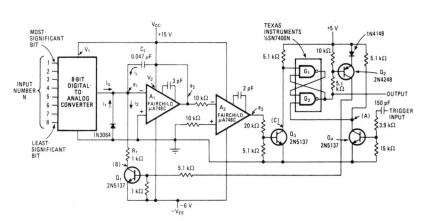

FET SIGNAL TRACER—Uses 0.1-W Lafayette 99F90425 or equivalent IC amplifier module, with fet ahead of it to increase input impedance and thereby decrease loading on circuit under test. Either jack may be used for either probe, because both r-f and a-f signals pass through C1 and R1. Article gives construction details and instructions for use in troubleshooting transistor radio.—G. McClellan, Mini-Trace Signal Sniffer, Elementary Electronics, Nov.-Dec. 1971, p 73–76.

COMPUTER - CONTROLLED TESTER — Digitally programmable mono uses 8-bit d-a converter, integrator, comparator, flip-flop, and set-reset circuit to provide controlled pulse width or time delay in which output pulse period drifts less than 0.005% for 1% change in either supply voltage. Gates G1 and G2 form set-reset flip-flop.—M. J. Shah, Programmable Monostable Is Immune to Supply Drift, Electronics, Feb. 1, 1973, p 98–99.

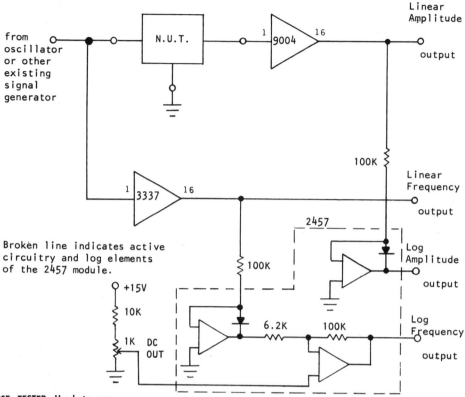

Broken line indicates active
circuitry and log elements
of the 2457 module.

NETWORK RESPONSE TESTER—Used to convert output frequency of amplifier, network, filter, or other frequency-sensitive device into linear d-c voltage for driving frequency-response plotter. Requires external signal generator for sweeping input frequency over range of interest. Uses OEI 3337 frequency-voltage transducer for linear frequency output, OEI 9004 absolute-value module for linear amplitude output, and OEI 2457 universal log module for plotting either output as log function.—"Frequency Response Measurements," Optical Electronics, Tucson, AZ, Application Tip 10202, 1971.

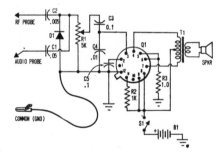

PISTOL-GRIP SIGNAL TRACER—RCA CA3020 wide-band amplifier operating from two penlight cells in series is mounted in pistol-shaped housing along with 2-inch speaker and other components. Probes made of sharpened coathanger wire project from muzzle end. Used in finding dead stage in radio that is getting signal from station or signal generator. Increase setting of sensitivity control R1 as signal weakens when moving toward input of set. Output transformer T1 has 125-ohm CT primary and 8-ohm secondary. —H. Davidson, Pistol Grip Signal Tracing Gun, *Elementary Electronics*, Jan.-Feb. 1970, p 55–58 and 107.

CHAPTER 121
Test Circuits—Solid State

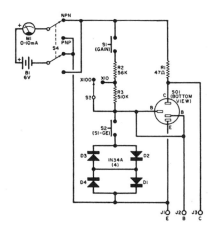

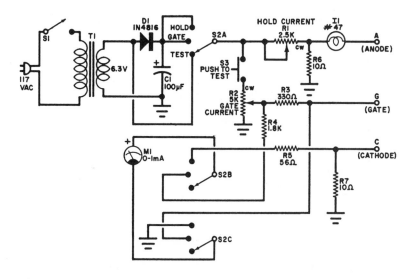

TRANSISTOR TESTER—Will check practically any npn or pnp silicon or germanium transistor or diode for shorts, opens, and leakage, and check transistors for gain. With S3 at X100 and S4 at NPN, insert transistor in socket. If meter deflects, transistor is shorted. Push S2; if meter deflects, transistor is pnp; if not, it is npn. If no reading for either, transistor is open. Set S4 to type of transistor; reading should be zero for silicon and less than 1 mA for germanium. Push S1 and adjust S3, to read d-c current gain (scale times multiplier).—D. Lancaster, Build the NGW Transistor Tester, *Popular Electronics*, Dec. 1967, p 57—59 and 98.

SCR TESTER—Gives quick good-bad test for scr's, which can only fail catastrophically. Article also tells how to determine gate current needed to fire scr and anode current needed to hold it in conduction. To use, place S2 in TEST position, rotate R2 full counterclockwise, set R1 full clockwise, connect scr, close S1 and advance R2 until lamp I1 comes on. Lamp should go off. If it stays lit, or does not light, scr is bad.—J. W. Cuccia, Build the SCR Tester, *Popular Electronics*, May 1969, p 47—49.

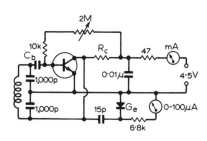

TRANSISTOR - TESTING OSCILLATOR — Permits measuring two parameters of transistor in addition to checking frequency capability. Frequency in MHz is approximately 160,000/CbRc, where Rc is in ohms (100 ohms) and Cb is in pF, switched in 12 steps from 200 pF to 3.3 pF to get frequency range of 10 to 500 MHz.—J. P. Holland, Oscillator Circuit For Measuring Beta and Ft, *Wireless World*, Dec. 1970, p 580.

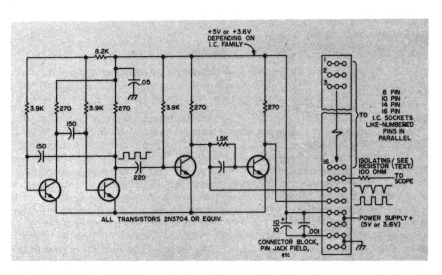

DIGITAL IC TESTER—Can be used for identifying and testing sections of unidentified TTL, DTL, or RTL IC units. Use with cro. Article covers method of use, starting with identifica-tion of gates by observing output current of power supply.—R. Factor, 3 Versatile IC Testers, *73*, Sept. 1970, p 38—40 and 42—44.

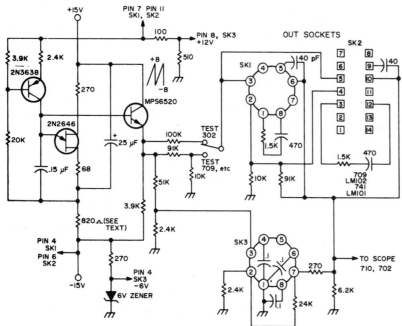

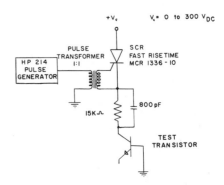

R-F POWER TRANSISTOR PULSER—Used in determining maximum r-f voltage that power transistor collector can withstand without failure. Test circuit shown is usually nondestructive. Provides pulses up to 300 V having nanosecond rise times. Article gives curve traces obtained.—B. Reich, E. B. Hakim, and G. J. Malinowski, Maximum RF Power Transistor Collector Voltage, *Proc. IEEE*, Oct 1969, p 1789–1791.

LINEAR IC TESTER—Designed primarily for testing 709 and 710 comparators. Will also test other opamps, such as LM101, 702, μA741, and LM102 voltage follower. Provides only go-no-go indications. Adjust 820-ohm re- sistor so voltage swing is symmetrical about ground at emitter of MPS6520 buffer.—R. Factor, 3 Versatile IC Testers, *73*, Sept. 1970, p 38–40 and 42–44.

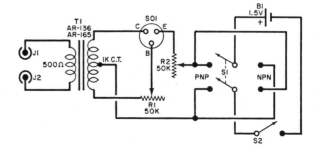

QUICK TEST FOR TRANSISTORS—Suitable for rapid tests of bargain batches of both pnp and npn transistors. Transistor under test is placed in modified Hartley audio oscillator. Use high-impedance phones. Adjust R1 and R2 for audio tone; if none is heard, transistor is bad. Will reject shorted, open, excessively leaky, and very low gain units.—T. Vanderelli, Reader's Circuit, *Popular Electronics*, Sept. 1968, p 86–87.

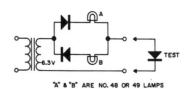

"A" & "B" ARE NO. 48 OR 49 LAMPS

ONE-SHOT TESTER—Developed for checking timing accuracy of industry-standard retriggerable mono mvbr's such as the 9601. Circuit reveals whether pulse width variations exceed permissible 10% for outputs above and below nominal 3.4-μs value. Circuit re- vealed wide variations in performance from vendor to vendor.—D. E. Green, One-Shot Timing Performance: Don't Take It for Granted, *Electronics*, March 27, 1972, p 101–103.

DIODE TESTER—If lamp A lights, diode under test is good. If lamp B lights, diode is good but is connected backward. Cathode of diode under test is thus identified. If both lamps light, diode is shorted. If neither lights, diode is open. Diode types for test circuit are not critical.—P. Franson, "Diode Circuits Handbook," 73 Inc., Peterborough, NH, 1967, p 23A.

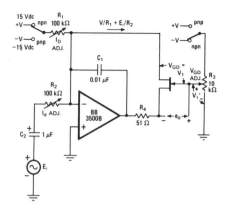

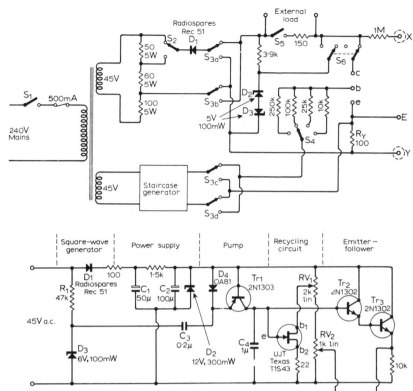

FET FORWARD TRANSCONDUCTANCE—Measurement is made with fixed drain bias and signal currents established by feedback around opamp. Gate source voltage developed by drain current is measured at output, and used to compute both static and dynamic values (gFS and gfs). Accuracy depends on temperature coefficient of fet under test. Test bandwidth is d-c to 10 kHz.—J. Graeme, Accurate Transistor Tests Can Be Made Inexpensively, *Electronics*, Feb. 28, 1972, p 84–89.

CURVE TRACER—Displays load lines and characteristic curves of zener diodes, junction transistors, and field-effect transistors on cro. Four-transistor staircase generator provides 1-V steps. Most component values are not critical. S5 places external load resistor in series with test resistor for load-line display.—A. J. Sargent, Electronically Stepped Curve Tracer, *Wireless World*, Dec. 1969, p 576–577.

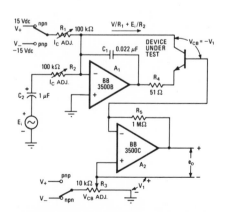

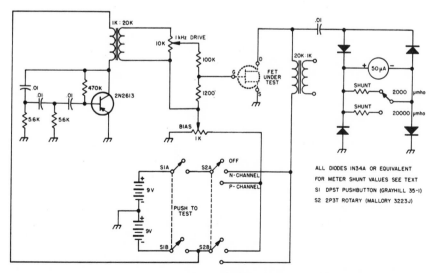

TRANSISTOR CURRENT GAIN—Collector bias and signal currents are established by feedback around opamp A1, which drives transistor under test so as to maintain nearly zero voltage and current at opamp inputs. A2 sets collector-base bias. Test accuracy depends on accuracies of R5, turn-counting pots R1 and R2, V, and Ei.—J. Graeme, Accurate Transistor Tests Can Be Made Inexpensively, *Electronics*, Feb. 28, 1972, p 84–89.

FET TESTER—Battery-operated circuit using 1,000-Hz 2N2613 oscillator provides simple transconductance test of field-effect transistors. Value of shunt resistor depends on meter used. With 100-μA meter, only one shunt is needed, for 20,000 μmhos full-scale.—J. Fisk, Field Effect Transistor Transconductance Tester, 73, Jan. 1967, p 16–19.

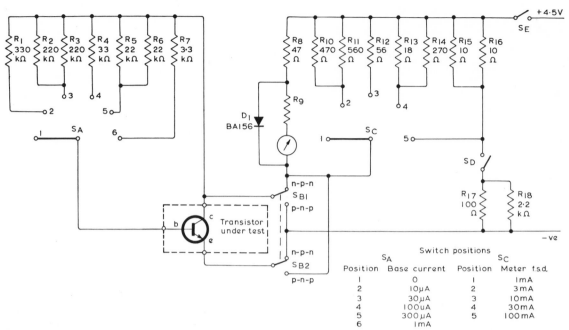

Switch positions			
S_A		S_C	
Position	Base current	Position	Meter f.s.d.
1	0	1	1mA
2	10μA	2	3mA
3	30μA	3	10mA
4	100uA	4	30mA
5	300μA	5	100mA
6	1mA		

TRANSISTOR TESTER—Uses 1-mA meter in series with R9 to give total resistance of 450 ohms, in combination with suitable shunts for measuring collector current of transistor under test. Base current of transistor may be switched to zero for measurement of collector-emitter reverse leakage current, or to choice of five current ranges for measuring static forward current transfer ratio. Switches SD and SE should be spring-loaded off to prevent battery drain when no measurement is being made.—"Transistor Audio and Radio Circuits," Mullard, London, 1972, TP1319, 2nd Ed., p 255—257.

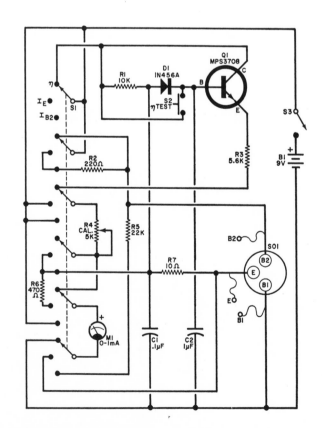

UJT TESTER—In S1 position shown, intrinsic standoff ratio is measured with peak voltage detector consisting of D1, Q1, and meter circuit. In middle position of S1, meter is in emitter circuit of ujt under test; R4 is adjusted for meter reading of 50 mA. S1 is placed in lowest position, and IB2 is measured on meter with scale having full-scale value of 100 mA.—J. W. Cuccia, Build the UJT Tester, *Popular Electronics*, June 1969, p 33—35 and 92.

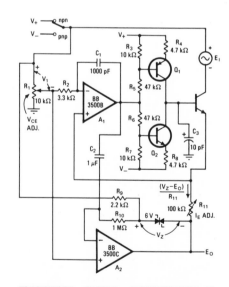

TRANSISTOR OUTPUT CONDUCTANCE—Circuit measures small slope of common-emitter characteristic curve accurately and quickly by providing output voltage proportional to change in emitter current induced in transistor under test (extreme right) by change in collector-emitter voltage. Differential current source Q1-Q2, controlled by opamp A1, biases base of test transistor. Article gives test procedure.—J. Graeme, Accurate Transistor Tests Can Be Made Inexpensively, *Electronics*, Feb. 28, 1972, p 84—89.

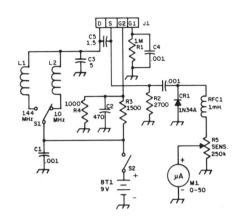

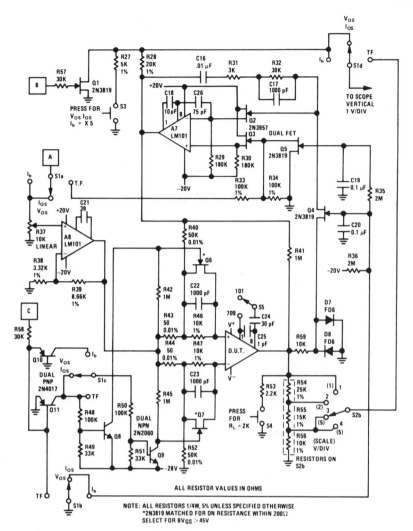

JFET AND MOSFET TESTER—Consists of common-gate r-f oscillator with provision for selecting coils for 10 or 144 MHz. If transistor under test oscillates, as indicated by meter reading, it is good. It may still be good, however, even though it fails to oscillate. To prevent damage by static electricity, wind thin bare wire around leads of mosfet before removing its shorting collar. Remove wire only after connecting to tester, and replace wire before removing from tester.—A Simple JFET and MOSFET Tester, *QST*, June 1970, p 42–43.

NOTE: ALL RESISTORS 1/4W, 5% UNLESS SPECIFIED OTHERWISE
*2N3819 MATCHED FOR ON RESISTANCE WITHIN 200Ω
SELECT FOR BV$_{GS}$ > 45V

ALL RESISTOR VALUES IN OHMS

OPAMP TESTER—Provides semiautomatic test of important opamp parameters over full power-supply and common-mode ranges. Report includes circuit of function generator which feeds ±19 V square wave to terminal C, −19 V to +19 V pulse with 1% duty cycle to terminal B, and ±5 V triangular wave to terminal A. D.U.T. is opamp under test. Designed specifically for testing LM709 and LM101 opamps, but report covers simple changes that permit testing any generally available opamp.—"A Simplified Test Set for Operational Amplifier Characterization," National Semiconductor, Santa Clara, CA, 1969, AN-24, p 6–9.

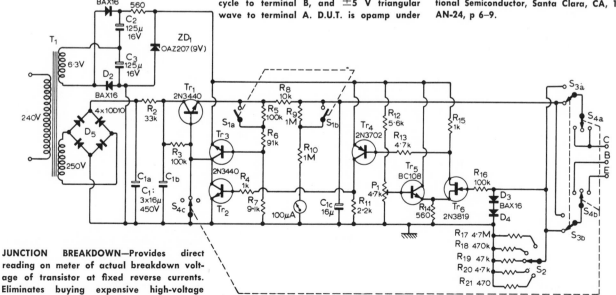

JUNCTION BREAKDOWN—Provides direct reading on meter of actual breakdown voltage of transistor at fixed reverse currents. Eliminates buying expensive high-voltage transistor when standard type will serve just as well. Transistor under test is not damaged, because circuit includes current limiter covering range of 0.1 μA to 1 mA in decade steps. Tr4, Tr5, and Tr6 form differential input amplifier operating as voltage comparator, with zener-stabilized low voltage serving as reference.—J. Langvad, Transistor Breakdown-Voltage Meter, *Wireless World*, Sept. 1970, p 443–444.

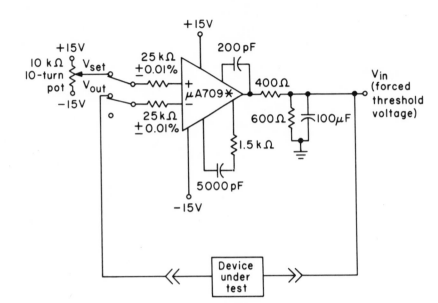

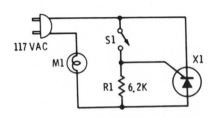

SCR CHECKER—Simple circuit provides good-bad and shorted-open tests for most general-purpose scr's. Open S1 after connecting scr as for X1; 25-W 117-V lamp M1 should be about half normal brilliance if scr is good. When S1 is closed, lamp should go out completely. If lamp has full brilliance with S1 open, scr is shorted. If lamp does not brighten regardless of switch position, scr is open.—R. M. Brown and T. Kneitel, "101 Easy Test Instrument Projects," Howard W. Sams, Indianapolis, IN, 1968, p 76.

IC THRESHOLD TESTER—Device under test is placed in feedback loop of Fairchild or equivalent opamp. Circuit then forces output voltage to equal preset voltage, in such a way that threshold voltage of IC can be read directly from its input terminal. Input switch changes from inverting to noninverting device. Eliminates tedious manual adjustments to get specific output voltage before measuring input voltage.—R. K. Repass, Threshold Testing Too Tedious? Automate, *The Electronic Engineer*, May 1968, p 71.

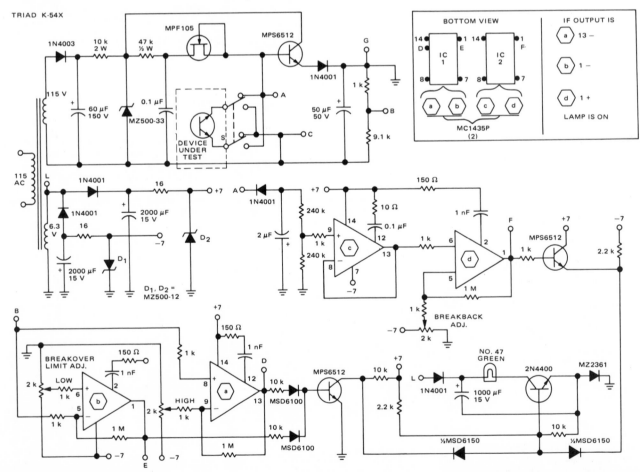

BILATERAL TRIGGER DIODE TESTER—Checks two limits on breakover voltage as well as breakback voltage limit on go-no-go basis. Dpdt switch permits testing trigger in both directions without reversing diode. Uses two opamp switching circuits, one acting at high breakover limit and other in opposite direction at low limit. If device under test is good, both amplifier outputs are low, and indicator lamp comes on. Report covers operation in detail.—D. A. Zinder, "Testers for Thyristors and Trigger Diodes," Motorola Semiconductors, Phoenix, AZ, AN-422, 1972.

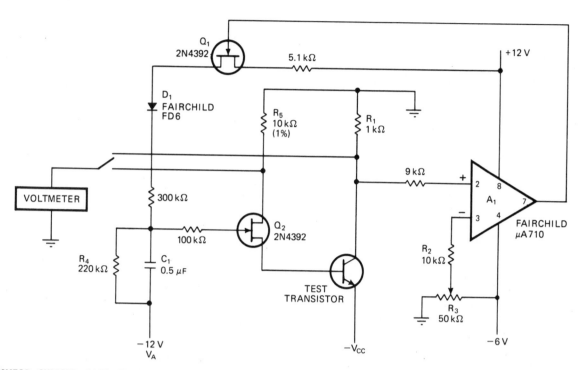

TRANSISTOR CURRENT GAIN—Circuit serves as variable constant-current source for measuring direct-current gain of bipolar transistor. Accuracy is better than 1%. Article tells how readings of voltmeter are used to compute current gain. Circuit is for npn transistor; for pnp, reverse supply polarities and use p-channel fet's. For go-no-go, use μA711 dual-limit detector in place of voltmeter.—J. B. Marshino, FET Supply Tests Bipolar Current Gain, *Electronics*, Sept. 27, 1971, p 61.

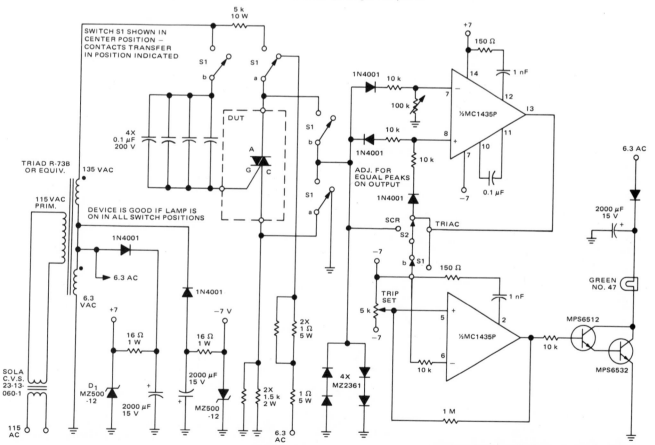

THYRISTOR TESTER—Checks important parameters on go-no-go basis, as required for incoming inspection at equipment production facility. Designed for testing thyristors rated at 200 V, such as 2N4154 and 2N4442 scr's and MAC1-4 triacs. May be adapted for other thyristors by changing voltage levels and a few components. Uses IC as adjustable-threshold switching circuit or detector. Report covers operation in detail. Checks forward and reverse leakage, forward voltage drop, and gate trigger current. Will also detect shorted or open units.—D. A. Zinder, "Testers for Thyristors and Trigger Diodes," Motorola Semiconductors, Phoenix, AZ, AN-422, 1972.

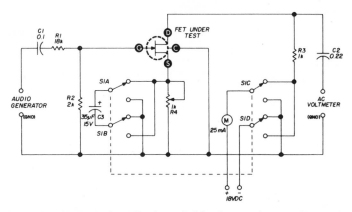

FET TESTER—Requires 1-V 1,000-Hz audio signal generator and either a-c voltmeter or cro. Forward transconductance in micromhos is equal to 10,000 times output voltage reading in volts. Adjust R4 to get drain current specified by fet manufacturer. S1 reverses power supply connections when changing from n-channel to p-channel fet.—R. S. Stein, Simple Transconductance Tester for Field-Effect Transistors, *Ham Radio*, Sept. 1971, p 44–47.

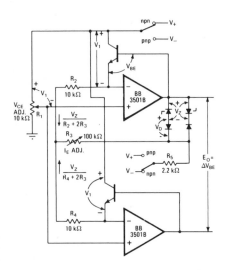

MATCHING TRANSISTORS—Opamp matching circuit makes current balance independent of emitter-base voltage difference. Bias currents and voltages are independently controlled. Transistor pairs can be matched either for VBE or VGS. Same circuit provides VGS match-testing for fet's. Article gives test procedure and analyzes measurement errors.—J. Graeme, Accurate Transistor Tests Can Be Made Inexpensively, *Electronics*, Feb. 28, 1972, p 84–89.

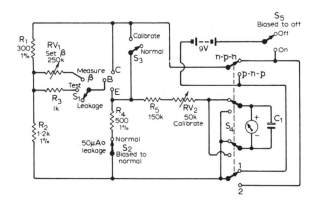

BETA TESTER—Checks good transistors for leakage and for beta from 20 to 1,000. Rectifiers and low-voltage zener diodes can also be checked for quality. Also indicates how faulty transistors have failed. Uses 50-μA meter with two linear scales, one with 50-μA meter with two linear scales, one with 50-μA full-scale at right and other with 9-V full-scale at left for zener voltage. Article gives construction details and instructions for making each type of test.—D. E. O'N. Waddington, Transistor Tester, *Wireless World*, June 1970, p 261–262.

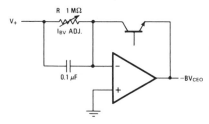

NPN TRANSISTOR BREAKDOWN—Opamp increases voltage on device under test until it breaks down and allows feedback current to flow to opamp input. Amplifier input voltage and current are returned to zero, and breakdown current is then equal to V+/R. Opamp is Burr-Brown 3501B.—J. Graeme, Accurate Transistor Tests Can Be Made Inexpensively, *Electronics*, Feb. 28, 1972, p 84–89.

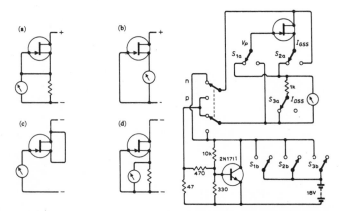

FET TESTER—Use with high-impedance voltmeter (20,000 ohms per V or higher). Switches are set to give connection (a) for measuring zero-bias drain current, (b) for gate cutoff voltage, and (c) for gate leakage current. Pressing VP and IDSS buttons simultaneously gives (d) for measuring drain current in mA and gate-source voltage in volts.—J. Skjelstad, Simple Tester for F.E.TS, *Wireless World*, Sept. 1972, p 443.

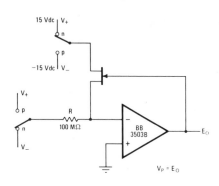

FET PINCH-OFF VOLTAGE—With source voltage of fet driven to zero at opamp input, output voltage Eo is VGS, which is approximately equal to pinch-off voltage Vp for small source current supplied through 100-meg resistor.—J. Graeme, Accurate Transistor Tests Can Be Made Inexpensively, *Electronics*, Feb. 28, 1972, p 84–89.

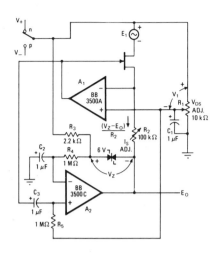

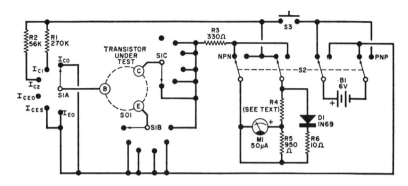

FET DRAIN-SOURCE RESISTANCE—Circuit checks rds as measure of ability of fet to drive high-resistance loads. Measurement bandwidth is about 20 Hz to 20 kHz. Article gives test procedure.—J. Graeme, Accurate Transistor Tests Can Be Made Inexpensively, *Electronics*, Feb. 28, 1972, p 84–89.

DIODE AND TRANSISTOR TESTER—Will measure leakage down to 10 μA, collector current to 10 mA, and other parameters for either npn or pnp transistors. Diodes are connected between collector and emitter pins of test socket for similar checks. Can also be used for matching transistors. Article covers construction and operation.—J. L. Keith, Nondestructive Transistor Tester, *Popular Electronics*, March 1971, p 47–49.

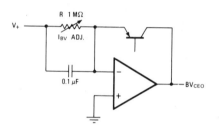

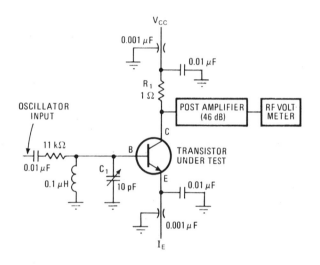

PNP TRANSISTOR BREAKDOWN—Opamp increases voltage on device under test until it breaks down and allows feedback current to flow to opamp input. Amplifier input voltage and current are returned to zero, and breakdown current is then equal to V+/R. Opamp is Burr-Brown 3501B.—J. Graeme, Accurate Transistor Tests Can Be Made Inexpensively, *Electronics*, Feb. 28, 1972, p 84–89.

TRANSISTOR H-F GAIN—Value of hfe for transistor under test is determined by measuring r-f voltage developed across small sampling resistor R1. Chief drawback is need to bypass R1 at high frequencies, at which capacitor impedance of several ohms causes large error. Noise voltage in required amplifier must be minimized because input can be as low as 10 μV.—G. Coers, A Simple Way of Measuring High-Frequency Transistor Gain, *Electronics*, April 24, 1972, p 117–118.

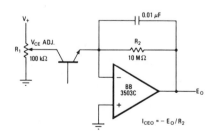

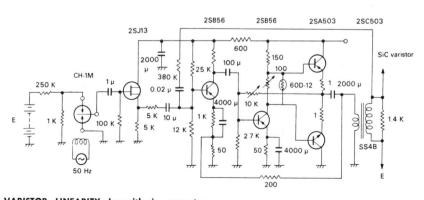

TRANSISTOR LEAKAGE CURRENT—Use of opamp in test circuit permits measuring leakages of 0.1 nA in silicon bipolar transistors. Leakage current flows through amplifier feedback path to develop output voltage directly proportional to leakage current. Bias voltage is established by pot. Use same circuit for collector-emitter leakage, but change R2 to 100K for collector-base leakage.—J. Graeme, Accurate Transistor Tests Can Be Made Inexpensively, *Electronics*, Feb. 28, 1972, p 84–89.

VARISTOR LINEARITY—Logarithmic converter changes d-c battery supply voltage to 50-Hz square wave with mechanical chopper and amplifies resulting a-c signal for application through output transformer to silicon-carbide varistor for measuring its nonlinearity index. Article gives other circuits used with converter for reading index value directly on digital voltmeter display, and gives theory of operation as based on voltage-current characteristic of varistor.—M. Uno, Nonlinearity-Index Meter for Silicon-Carbide Varistors, *Electronic Engineering*, Jan. 1971, p 41–43.

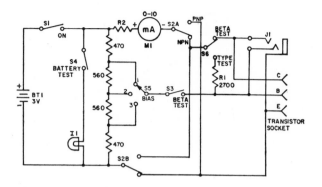

TRANSISTOR TESTER—Tells whether transistor is npn or pnp, and has beta test for d-c current gain. Supply uses two D cells; if they light No. 14 flashlight bulb in battery test circuit, batteries are good for transistor tester which draws only about 10 mA. For type test, position which gives meter reading identifies type. If both positions give same readings, transistor is bad. Make R2 large enough to prevent meter from being pinned if transistor is shorted.—H. J. Hanson, A Junk Box Transistor Checker, QST, Oct. 1969, p 18–19 and 122.

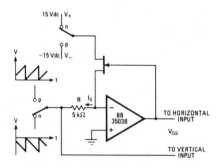

TRANSISTOR PARAMETER DISPLAY—Addition of ramp generator and scope to basic opamp transistor tester gives display of desired parameter dependency as function of desired parameter. With example shown, for displaying IS of fet vs VGS, temperature can be varied to locate zero temperature coefficient point. Opamp output voltage is VGS, displayed with respect to ramp voltage that sets source current.—J. Graeme, Accurate Transistor Tests Can Be Made Inexpensively, Electronics, Feb. 28, 1972, p 84–89.

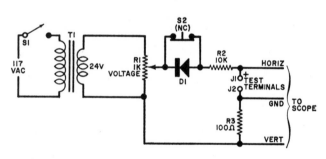

DIODE TESTER—Simple adapter for any cro permits sorting diodes quickly to tell which end is which and identify whether they are signal diodes, rectifiers, or zeners, and whether silicon or germanium. Will also test capacitors and led's for certain characteristics. Article gives cro traces for each type. T1 is filament transformer such as Lafayette 99T6266.—C. L. Andrew, Build Dynamic Diode Tester, Popular Electronics, July 1970, p 53 and 58–59.

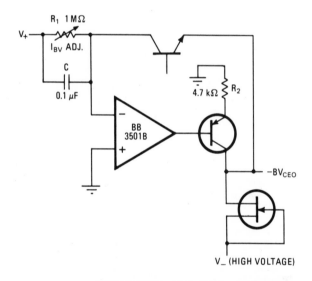

HIGH-VOLTAGE TRANSISTOR BREAKDOWN—When even a high-voltage opamp cannot supply sufficient voltage to break down transistor under test, voltage range can be extended by using common-emitter transistor to drive fet current-source load and thereby add voltage gain after opamp. Transistor types are not critical.—J. Graeme, Accurate Transistor Tests Can Be Made Inexpensively, Electronics, Feb. 28, 1972, p 84–89.

SCR AND TRIAC TESTER—Also checks power transistors and identifies device under test. Set S1 to SCR-NPN, S2 off and R1 to minimum V, connect device, and close S2. If bulb lights, device is shorted. Turn up R1 slowly; gradual brightening and gradual decrease as R1 is turned down again means device is npn transistor. Sudden lighting of bulb and no change in brilliance when R1 is turned down indicates scr or triac. Change S1; gradual brightening and darkening indicates pnp power transistor. No light indicates scr. Sudden brilliance, not changing when R1 is turned down, indicates triac.—SCR's and Triacs—Testing and Theory of Operation, Electronic Servicing, Dec. 1971, p 42–46.

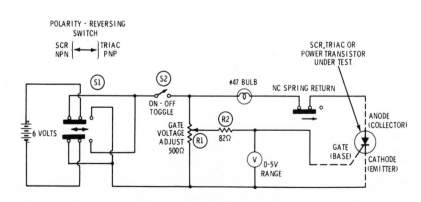

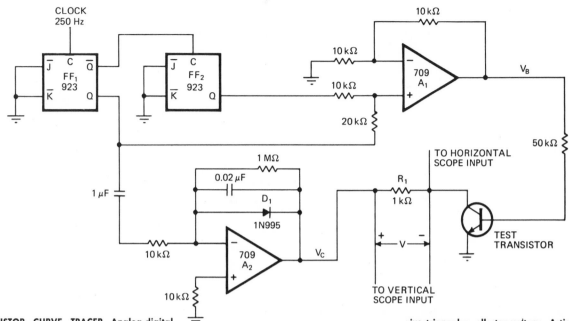

TRANSISTOR CURVE TRACER—Analog-digital converter circuit shown permits display of common-emitter characteristics of npn transistor on any standard scope having vertical differential amplifier input. Flip-flops FF form binary counter whose output is summed by opamp A1 to provide four-step staircase function for driving base of transistor. Output square wave of FF1 is integrated by A2 to give triangular collector voltage. Article gives modifications needed for pnp transistors.—R. D. Guyton, Analog-Digital Circuit Turns Scope Into Curve Tracer, Electronics, Oct. 25, 1971, p 80.

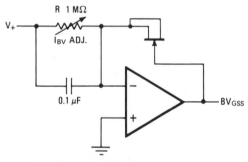

FET TRANSISTOR BREAKDOWN—Opamp increases voltage on device under test until it breaks down and allows feedback current to flow to opamp input. Amplifier input voltage and current are returned to zero, and breakdown current is then equal to $V+/R$. Opamp is Burr-Brown 3501B.—J. Graeme, Accurate Transistor Tests Can Be Made Inexpensively, Electronics, Feb. 28, 1972, p 84–89.

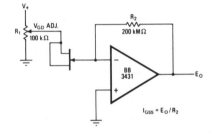

$$I_{GSS} = E_O/R_2$$

FET LEAKAGE CURRENT—Use of varactor-input opamp with very high feedback resistance permits measuring leakages of 0.1 pA. Leakage current flows through amplifier feedback path to develop output voltage directly proportional to leakage current. Bias voltage is established by pot.—J. Graeme, Accurate Transistor Tests Can Be Made Inexpensively, Electronics, Feb. 28, 1972, p 84–89.

R_1, R_4	1.5 K, ½ W
R_2, R_3	100 K, ½ W
R_5, R_6	33 K, ½ W
R_7	470 ohm, ½ W
R_8	100 ohm, 2 W
R_9	10 ohm, 10 W
R_{10}	1 ohm, 200 W
R_{11}	25 K, 4 W
R_{12}	2 K, 4 W
R_{13}	100 ohm, ½ W
R_{14}	150 ohm, 1 W
R_{15}, R_{16}	.05 ohm, 1%, 20 W
R_{17}	{ .2 ohm at 50 AIc { 1 ohm at 15 AIc
R_{18}	20 ohm, 5 W
C_1	.05 uF 50V
C_2	10 uF 15V
Q_1, Q_2	2N109
Q_3	2N1172
Q_4	2N553
Q_5, Q_6	2N278
CR_1	1N3491

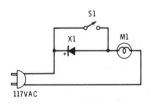

SILICON RECTIFIER CHECKER—Provides quick check of 120-V silicon rectifier diodes connected or plugged into circuit at X1. With S1 open, 25-W lamp M1 should be about half normal brilliance if diode is good. If lamp is full brilliance with S1 open, diode is shorted. Diode is open if lamp is dark. With S1 closed, lamp should have full brilliance regardless of diode.—R. M. Brown and T. Kneitel, "101 Easy Test Instrument Projects," Howard W. Sams, Indianapolis, IN, 1968, p 34.

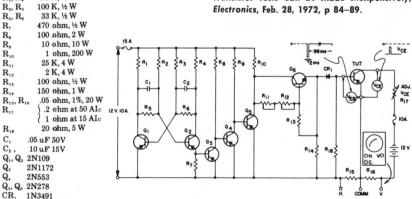

TRANSISTOR PULSE GAIN—Measures current gain at low duty cycle to prevent excessive heating of collector junction. Pulse generator provides 2-ms pulse for base circuit of transistor under test, at 2% duty cycle. Cro measures base and collector currents by observing voltage pulses across R15 and R16; currents are then calculated by $I = E/R$. R11 and R12 adjust base current of test transistor. Report gives complete measuring procedure.—"Germanium Alloy Power Transistor Electrical Test Procedures," Delco, Kokomo, IN, April 1972, Application Note 6-A.

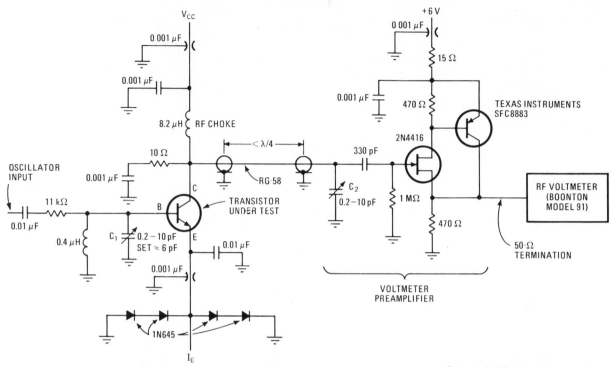

TRANSISTOR GAIN WITH TUNED LINE—Quarter-wave length of coax (slightly shortened) is tuned by C2 so collector of transistor under test looks like short-circuit and collector-emitter voltage remains constant over measuring frequency range. Method eliminates need for critical collector bypassing and reduces problems of ground loops and stray capacitances. Article gives test procedure and example of tuned line for 100 MHz.—G. Coers, A Simple Way of Measuring High-Frequency Transistor Gain, *Electronics*, April 24, 1972, p 117–118.

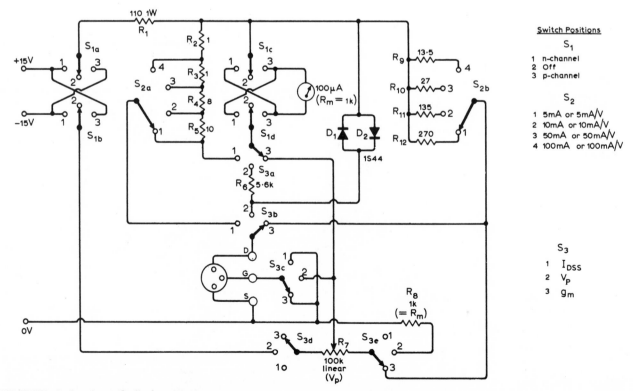

FET TESTER—Designed specifically for measuring mutual conductance, zero-bias drain current, and gate cutoff voltage of conventional depletion-mode junction field-effect transistors. Article covers construction and use.—D. E. O. Waddington, F.E.T Tester, *Wireless World*, Dec. 1971, p 579–582.

CHAPTER 122
Three-Phase Control Circuits

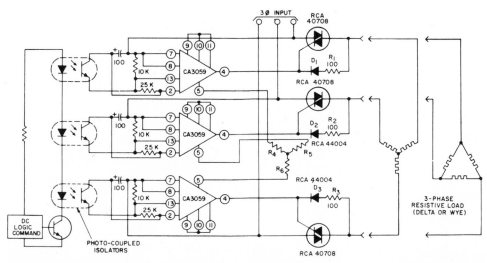

HEATER CONTROL—Provides zero-voltage synchronous switching in steady-state operating condition, with random starting. Logic command to turn system on or off is given through optoelectronic coupling in response to temperature sensor. Logic can be chosen to provide time-proportioning heat control if required. Uses CA3059 IC zero-voltage switches to provide positive pulse about 100 μs wide at each zero-voltage crossing for each phase. —"Thyristors, Rectifiers, and Diacs," RCA, Somerville, NJ, 1973 Edition, p 469–470.

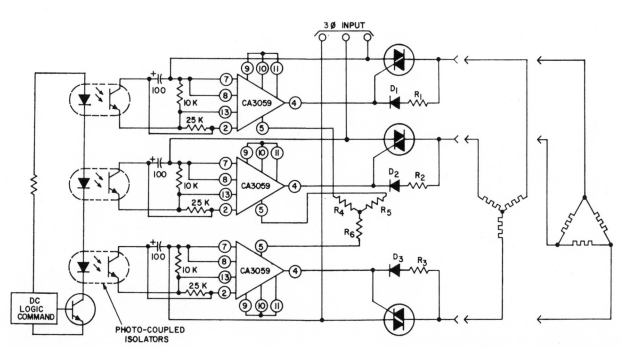

HEATER CONTROL—Provides zero-voltage synchronous switching for steady-state operating conditions, with random starting. Choice of triacs depends on load requirements; if 2.5-A capability of RCA 40530 triac is not sufficient, this can be used as trigger triac for turning on power triac in each phase. Book covers operation of circuit in detail.—"Solid State Power Circuits," RCA, Somerville, NJ, SP-52, 1971, p 364–366.

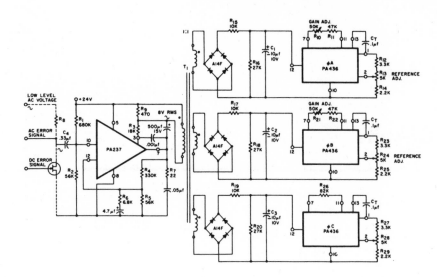

FLOATING-SUPPLY CONTROL FOR IC'S—Uses PA237 IC high-gain a-c amplifier to drive three bridge-rectifier supplies, each serving one PA436 IC phase-control trigger for three-phase d-c power supply using scr-rectifier combinations. Gain and reference adjustments are provided on two phases to insure proper tracking of control trigger angles between phases.—"SCR Manual," General Electric, Syracuse, NY, 1972, 5th Ed., p 283–284.

IC ZERO-VOLTAGE SWITCH—Three separate thermistors T provide control for three GEL300 IC zero-voltage switches, each triggering triac for one phase only when instantaneous phase voltage is under 5 V to meet NEMA WD-2 emi limit during both positive and negative half-cycles. Identical circuits are required for other two phases, connected as shown for delta load.—"SCR Manual," General Electric, Syracuse, NY, 1972, 5th Ed., p 321.

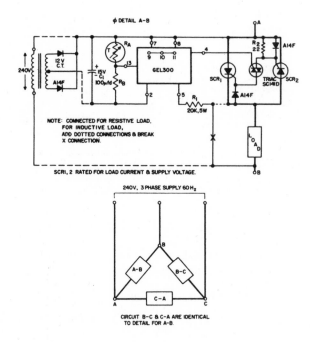

FREQUENCY CONVERTER LOGIC—Oscillator and six-stage ring counter are shown for logic of three-phase frequency converter delivering 750 W of three-phase power in frequency range from 380 to 1,250 Hz at 120/208-V rms. Uses three-phase bridge inverter supply from rectified single-phase or three-phase line at any frequency from 47 to 1,250 Hz. Bridge inverter uses pairs of 2N5805 switching transistors that are transformer-driven by diode matrix and driver connected to six outputs of ring counter. Report gives design data for all transformers. Efficiency is 75% at rated load current of 2.1 A.—"Power Transistors and Power Hybrid Circuits," RCA, Somerville, NJ, 1973 Ed., p 652–656.

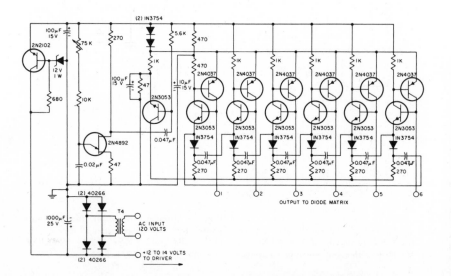

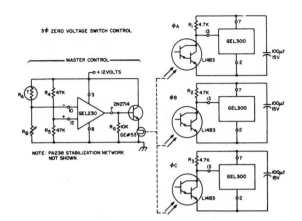

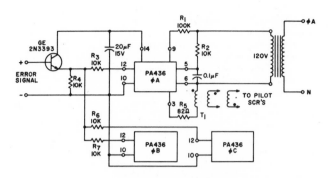

MASTER CONTROL FOR ZERO-VOLTAGE SWITCH—Photodarlingtons respond to single light source at output of control, to feed inputs of three isolated GEL300 IC zero-voltage switches in three-phase system. Each GEL300 controls scr pair handling one phase of delta or wye-connected load.—"SCR Manual," General Electric, Syracuse, NY, 1972, 5th Ed., p 322.

COMMON CONTROL INPUT—Used for phase control of three-phase resistive loads when it is desirable to connect together all control inputs of PA436 IC's and float only a-c timing line voltages. Provides rapid response to control input variations, but requires three pulse transformers and three control transformers to provide required a-c timing line voltages for all three phases. Circuits for control of inductive loads are also given.—"SCR Manual," General Electric, Syracuse, NY, 1972, 5th Ed., p 285–286.

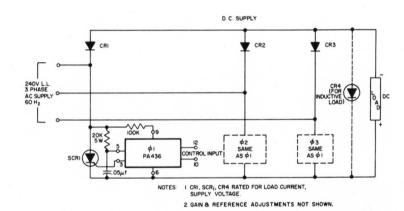

IC PHASE CONTROL—Three GE PA436 phase-control triggers provide smooth control of delta or wye-connected d-c loads with scr-rectifier combinations rated to handle load currents. Control range is 3 to 97% of full power.—"SCR Manual," General Electric, Syracuse, NY, 1972, 5th Ed., p 282–284.

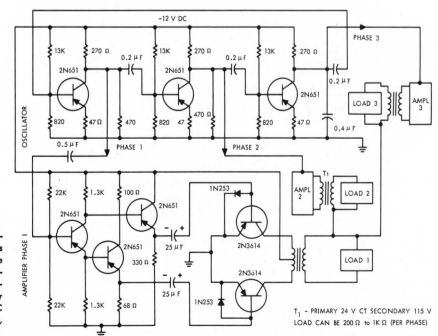

20-W INVERTER—Converts 12-V d-c to three-phase 115-V 400-Hz a-c. Transistor oscillators are R-C coupled so 120-deg phase difference exists at collectors of oscillators. Emitter-follower amplifier drives output power transistors operated in saturated switching mode for each phase.—R. J. Haver, "The ABC's of DC to AC Inverters," Motorola Semiconductors, Phoenix, AZ, AN-222, 1972.

FLOATING-SUPPLY CONTROL—Uses PA436 in single-phase mode with triac to provide smooth continuous control of a-c voltage supplied to three floating d-c supplies each consisting of bridge rectifier, filter, and PA436 for each phase serving either delta or wye-connected load.—"SCR Manual," General Electric, Syracuse, NY, 1972, 5th Ed., p 284.

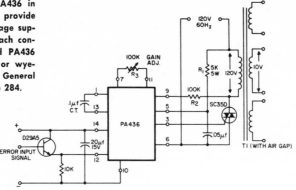

PHASE SEQUENCE DETECTOR—AA107 silicon controlled switch is normally off. If voltage of phase B lags that of phase A by 120 deg, currents I1 and I2 will be 180 deg out of phase and cancel, making I3 zero so scs does not trigger on. With opposite phase sequence, I2 will lag I1 by 60 deg and currents will add vectorially to make I3 2 mA, which triggers scs on and energizes lamp to indicate phase sequence reversal.—"Phase Sequence Detector for Three-Phase Systems," Unitrode Corp., Watertown, MA, New Design Idea No. 13.

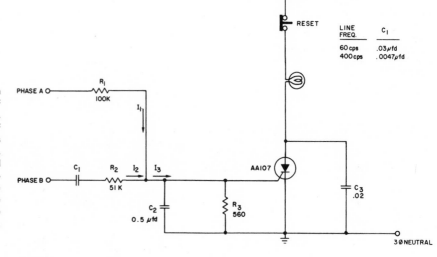

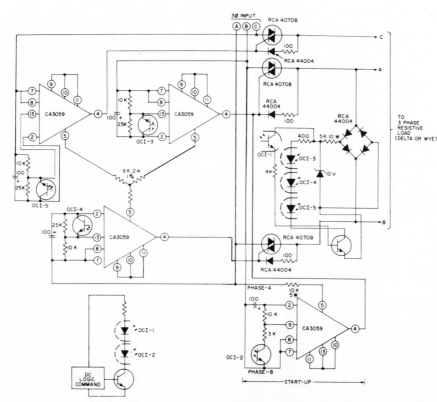

ZERO-VOLTAGE SWITCHING—Used in critical applications requiring suppression of all generated rfi. Includes zero-voltage starting circuit as well as synchronous switching for steady-state operation. Logic command is applied to three IC zero-voltage switches through optoelectronic coupling. When logic command is turned off, all control is ended and triacs automatically turn off when sinewave current decreases to zero. After first phase turns off, other two turn off simultaneously 90-deg later.—"Thyristors, Rectifiers, and Diacs," RCA, Somerville, NJ, 1973 Ed., p 470–471.

INDUCTIVE MOTOR—Triacs vary speed of three-phase motor, with triggering provided by CA3059 zero-voltage switch which in turn is controlled by d-c logic through optoelectronic coupling. IC is used in d-c mode to provide continuous d-c output instead of pulses at points of zero-voltage crossing, without utilizing zero-voltage turn-on because inductive current cannot increase instantaneously. Current-handling capacity of 40692 triac (2.5 A) is not sufficient for most three-phase motors, so this is used as trigger triac for turning on required larger power triac.—"Thyristors, Rectifiers, and Diacs," RCA, Somerville, NJ, 1973 Edition, p 471–472.

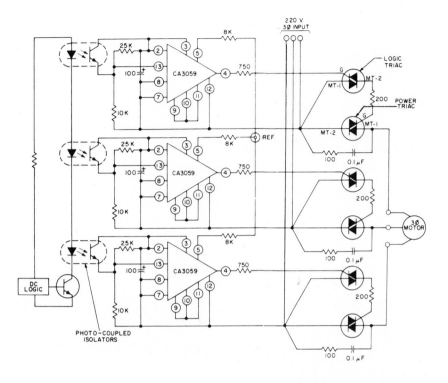

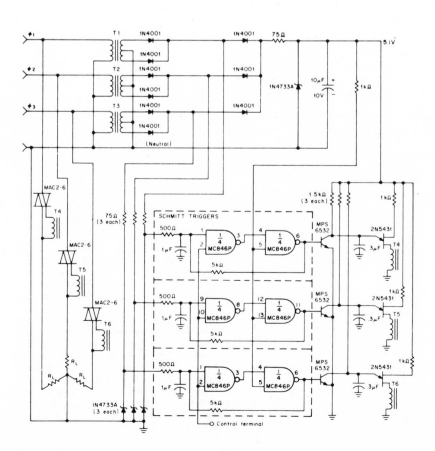

ELIMINATING SWITCHING TRANSIENTS—Three-phase high-power system suppresses emi by using triacs to give zero-point switching. System can be controlled with any logic that is compatible with DTL or TTL gates. Control terminal is kept at logic 1 level (5 V) to keep circuit off. Dropping level to 0 gives circuit turn-on. Raising it back to 5 V will then turn each triac off after the immediate half-cycle finishes, disconnecting its load RL.—R. A. Phillips, Zero-Point Switching Eliminates EMI, *The Electronic Engineer*, March 1969, p 92.

CHAPTER 123
Timer Circuits

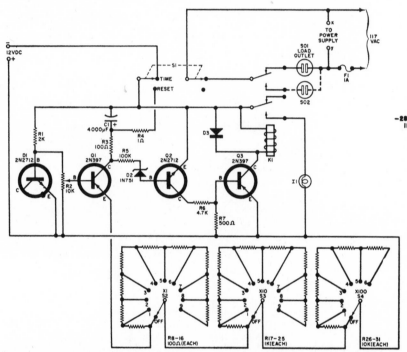

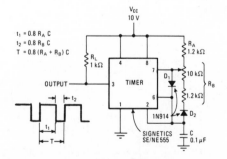

0.5 S TO 3 MIN—Amount of delay in energizing relay is determined by setting of 250K pot, in which each 10K of resistance gives about 1 s delay. Developed for use with autopatch circuit for making telephone calls through f-m repeater.—K. Sessions, Jr., For Emergencies: The Super Autopatch, *73*, July 1970, p 26–32 and 34–35.

OFF OR ON IN 1 S TO 10 MIN—Uses constant-current transistor Q1 to charge timing capacitor C1 essentially linearly regardless of line voltage. Decade switches provide calibrated time intervals in steps of 1 s (left), 10 s (center), and 100 s (right) up to 600 s, giving maximum timing interval of 11.65 min if all switches are maximum. Unmarked diodes are 750-mA 200-V rectifier types. Relay is 12 V at 80 mA and I1 is 14-V dial lamp. Load plugged into SO2 shuts off at end of timing interval, while load in SO1 is turned on at end of interval.—W. T. Lemen, Build a Stopclock, *Popular Electronics*, Dec. 1967, p 45–48.

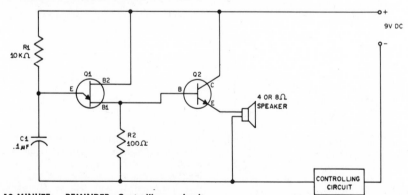

10-MINUTE REMINDER—Controlling circuit (not shown) uses scr to turn on 1,000-Hz audio oscillator Q1 and Q2 feeding small speaker, 10 minutes after timer switch is closed, to warn operator of ham transmitter that station identification must be made. Timing depends on charging of low-leakage tantalum capacitor at controlled rate with circuit using three ujt's, scr, diode, and one bipolar transistor. Q1 is relaxation oscillator, with R1 and C1 determining frequency, and Q2 is current amplifier. Combination generates train of current pulses through speaker to give crisp tone rich in harmonic content.—Unique Identiminder, *QST*, Nov. 1970, p 47–48.

99% DUTY-CYCLE RANGE—Achieved by providing independent charging and discharging paths for timing capacitor. Charging path is through RA and D1 to supply, while discharging path is through RB and D2 to ground. For output pulse frequency of 1 kHz, output period of 1 ms varies only 1% over adjustment range.—M. S. Robbins, IC Timer's Duty Cycle Can Stretch over 99%, *Electronics*, June 21, 1973, p 129.

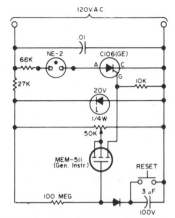

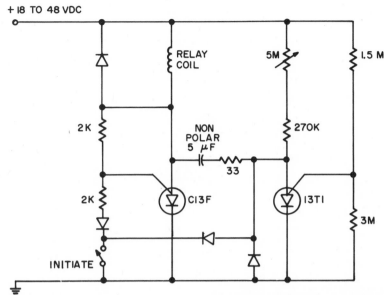

10-MINUTE TIMER—Designed for use with transmitters or transceivers, to provide visual reminder of need to identify station at 10-minute intervals, as required by FCC. Rectified line-voltage pulses charge 3-μF capacitor through 100-meg resistor. When voltage across capacitor reaches 6 V, MEM-511 mosfet is turned off, turning off scr and neon to indicate end of timing interval. Sensitive relay may be used in place of neon, to energize buzzer or other type of indicator. Pressing reset switch discharges capacitor, recycling circuit for 10 more minutes.—W. Pinner, A Solid-State 10-Minute Timer, 73, Oct. 1970, p 70–71.

DELAYED-DROPOUT RELAY—Provides controlled time delay regardless of when initiate switch is closed or opened. Switch is normally closed. Delay is adjustable from 1 to 30 s with 5-meg pot. All diodes are GE 55322.—W. H. Sahm, III, and F. M. Matteson, "The Complementary SCR," General Electric, Semiconductor Products Dept., Auburn, NY, No. 90.94, 1970, p 22–23.

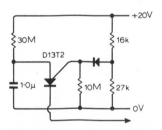

30-S TIMER—Chief advantage is ability to select circuit parameters in advance, because stand-off ratio of programmable ujt is determined by two external resistors. Diode type is not critical.—O. Greiter, PUT—The Programmable Unijunction, *Wireless World*, Sept. 1970, p 430–434.

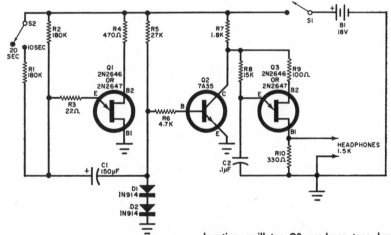

TIMER FOR RAPID-FIRE—Produces beep at precise intervals of 10 or 20 s, to pace rapid-fire sharpshooting while keeping eye on target instead of watching stopwatch. Ujt relaxation oscillator Q3 produces tone bursts that are gated by Q2 at intervals determined by setting of S2 in timing circuit.—C. F. Hadlock, Build a Firing-Range Timer, *Popular Electronics*, June 1967, p 37–39.

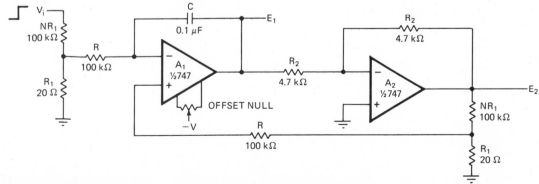

MINUTES WITH OPAMPS—Inexpensive general-purpose opamps permit use of low values of R and C for stretching one-shot output pulses up to several minutes. Network acts as multiplier up to 10,000 for R-C time constant. Values shown give 50-s constant. Can also serve as low-pass insertion filter for monitoring geoscientific phenomena where low-frequency noise is undesirable.—Q. Bristow, Op Amps Multiply RC Time Constants, *Electronics*, April 24, 1972, p 102–103.

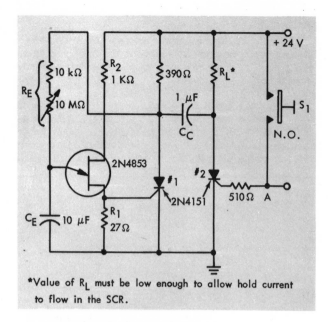

*Value of R_L must be low enough to allow hold current to flow in the SCR.

1 S–2.5 MIN SCR DELAY SWITCH—Uses additional scr in place of relay, which usually reduces circuit cost. After one cycle of operation, scr No. 1 will be on, lowering ujt emitter voltage and interrupting timing function. Next timing cycle is initiated by pushing S1 or applying positive pulse at point A. After delay determined by setting of 10-meg pot, ujt relaxation oscillator fires to turn scr No. 1 on and scr No. 2 off, interrupting load current.—"Unijunction Transistor Timers and Oscillators," Motorola Semiconductors, Phoenix, AZ, AN-294, 1972.

1 S–2.5 MIN DELAY—Ujt relaxation oscillator fires after delay determined by setting of 10-meg pot and turns on scr to energize relay. Relay stays energized until normally closed pushbutton S1 is pushed again to start new timing cycle. Design equation is given.—"Unijunction Transistor Timers and Oscillators," Motorola Semiconductors, Phoenix, AZ, AN-294, 1972.

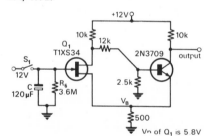

V_p of Q_1 is 5.8V

13-MIN DELAY—Schmitt trigger is used as long time-delay element. Use of fet with smaller value than 5.8 V for Vp of Q1 will give longer delay.—Some Applications of Field-Effect Transistors, *Electronic Engineering*, Sept. 1969, p 18–23.

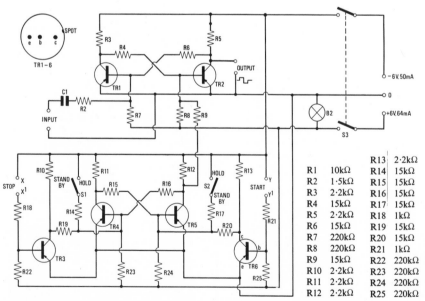

R1	10kΩ	R13	2·2kΩ
R2	1·5kΩ	R14	15kΩ
R3	2·2kΩ	R15	15kΩ
R4	15kΩ	R16	15kΩ
R5	2·2kΩ	R17	15kΩ
R6	15kΩ	R18	1kΩ
R7	220kΩ	R19	15kΩ
R8	220kΩ	R20	15kΩ
R9	15kΩ	R21	1kΩ
R10	2·2kΩ	R22	220kΩ
R11	2·2kΩ	R23	220kΩ
R12	2·2kΩ	R24	220kΩ
		R25	220kΩ

TIMER FOR DECADE SCALER—Input requires constant-frequency signal source with any waveform in range of 50 to 10,000 Hz at 1 to 7 V rms. TR1 and TR2 form bistable squarer providing square-wave output for scaler. TR4 and TR5 are second bistable or hold circuit, required to insure that start sequence is initiated when pulse arrives at base of TR5. Stop signal pulse is applied to base of TR4. Photodiodes may be connected to start and stop terminals to give control action when either light beam is interrupted. C1 is 0.22 μF. Diodes are Mullard OA81. Transistors are Mullard OC71. Connect either Mullard ORP60 photodiodes or press-to-break microswitches to XX' and YY'. B2 is 6-V 50mA lamp.—"An Electronic Timer," Educational Electronic Experiments, No. 17, Mullard, London.

30-S RELAY DELAY—Rheostat varies heater voltage of 6AL5 tube, to adjust time for cathode to reach emitting temperature and make tube conduct sufficiently to energize coil of relay. Delay ranges from about 10 s for zero resistance to maximum practical value of about 2 minutes, with highest accuracy around 30 s.—O. Wrench, Adjustable Time Delay Relay Circuit, 73, Oct. 1972, p 71–72.

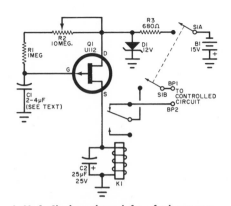

6-60 S—Single p-channel fet of almost any type in circuit shown gives inexpensive but accurate and stable interval timer for photographic darkroom and other uses. Timing rate depends on C1 and resistor values. Use any 8,000-ohm relay that pulls in at 1 mA.—B. W. Blachford, Build the FET Interval Timer, *Popular Electronics*, Sept. 1968, p 61.

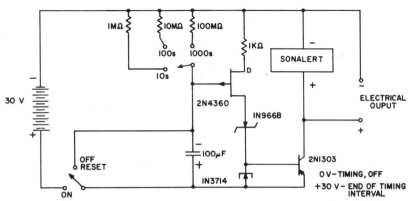

10-100-1,000 S TIMER—Draws minimum of power until audible alarm is energized at end of timing interval determined by switch position. Reset switch turns off alarm.—Simple Lab Timer, *Electro-Technology*, Oct. 1969, p 17.

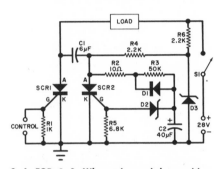

3 A FOR 1 S—When triggered by positive-going control pulse, supplies up to 3 A to load for 1 s, then switches off until triggered by another pulse. SCR1 is SSPI 3B3060; SCR2 is SSPI AA100; D1 is 1N483; D2 is 1N710A 6.8-V zener; D3 is 1N720A 18-V zener. Control pulse must be at least 1 V for at least 5 μs.—Manufacturer's Circuits, *Popular Electronics*, July 1971, p 81 and 87.

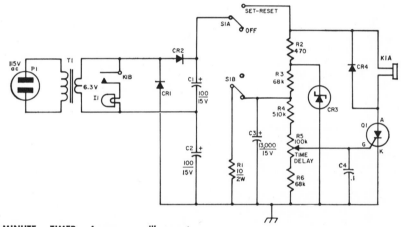

10-MINUTE TIMER — Accuracy will meet FCC requirements for station identification at 10-minute intervals. Diodes CR1 and CR2 are 50-piv 100-mA silicon, CR3 is 10-V 250-mW zener, and CR4 is 100-piv 100-mA silicon. K1 is 24-V 600-ohm d-c relay. Q1 is 2-A 30-V scr such as GE C106Y1.—R. B. Koehler, An Inexpensive Ten-Minute Timer, *QST*, Sept. 1969, p 40–41 and 75.

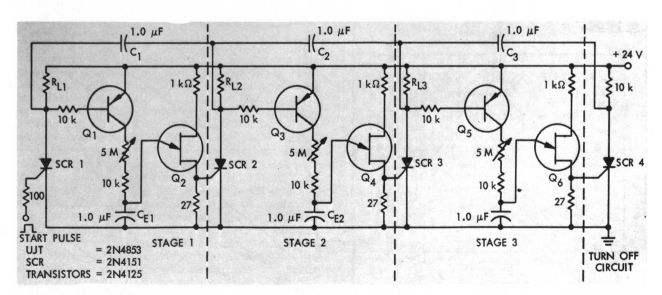

THREE-STAGE SEQUENTIAL—Each ujt relaxa-tion oscillator has independent control of time its scr is on. At end of timing interval for first stage, ujt relaxation oscillator Q2 fires and triggers scr to turn on second load RL2 for its required time interval. Useful in dishwasher and washing-machine controls, where different cycles of operation require different time durations. Additional stages can be added as required.—"Unijunction Transistor Timers and Oscillators," Motorola Semiconductors, Phoenix, AZ, AN-294, 1972.

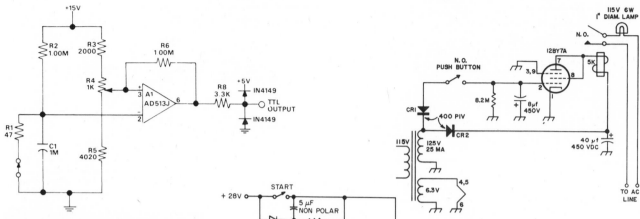

TIMER WITH OUTPUT CLAMP—Gives time delays ranging from 100 μs to over 10 s, with output clamped in range of −0.5 to +5.5 V for compatibility with TTL logic. Regenerative feedback through R6 provides hysteresis, speeding rise time and preventing retriggering on power supply noise. Input bridge circuit makes delay independent of low-frequency drift in supply voltage. Values shown for R1 and C1 give 1-s delay.—"AD513, AD516 I.C. Fet Input Operational Amplifiers," Analog Devices, Inc., Norwood, MA, Technical Bulletin, Aug. 1971.

1-MIN DELAY—Uses complementary scr to turn on load at interval after closing of switch required for capacitors to charge to 25 V and then discharge through 10-meg resistor to trigger scr on.—W. H. Sahm, III, and F. M. Matteson, "The Complementary SCR," General Electric, Semiconductor Products Dept., Auburn, NY, No. 90.94, 1970, p 12–13.

10-MINUTE TIMER—Tube is normally conducting and lamp is on. Pressing button momentarily biases tube beyond cutoff, extinguishes lamp, and charges 8-μF timing capacitor to about −175 V. Capacitor discharges slowly through 8.2-meg resistor for about 9.5 minutes, at which time relay pulls in and lamp comes on as reminder to identify station during ham radio ssb contact.—F. Davis, A Junkbox Ten Minute Timer, 73, Jan. 1967, p 106.

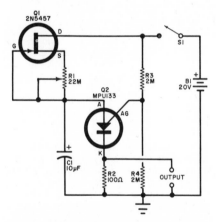

20-MIN DELAY—Ideal for darkroom timer, delayed alarm, process control, and similar applications. C1 is charged slowly by B1 through junction fet Q1 and delay control R1. When charge reaches firing voltage of programmable ujt Q2, it fires and discharges C1 through R2 to develop output pulse.—Useful Circuits, *Popular Electronics*, Dec. 1971, p 94.

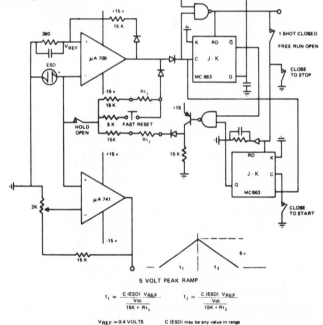

5 VOLT PEAK RAMP

$$t_1 \simeq \frac{C\,(ESD)\,V_{REF}}{\dfrac{V_{CC}}{15K + Rt_1}} \qquad t_2 \simeq \frac{C\,(ESD)\,V_{REF}}{\dfrac{V_{CC}}{15K + Rt_2}}$$

$V_{REF} \simeq 0.4$ VOLTS   C (ESD) may be any value in range

SECONDS TO HOURS WITH ESD—Based on use of electrochemical capacitor ESD (Gould Ionics) to store large amounts of charge at low voltage; thus type 1050C-1 can store 25 coulombs at 0.5 V, with leakage current under 1 pA. Circuit develops repeatable sawtooth waveforms which can have at least 100:1 adjustment of time period by changing one or two resistors (Rt1 and Rt2) or by using pots for them. Hold switch can interrupt ramp at any point and hold level. Fast reset serves for setup or abort situations. For continuous operation, open one-shot switch. Report includes circuit for programming unit.—"Programmable Timer," Gould Ionics Inc., Canoga Park, CA, Bulletin 71305.

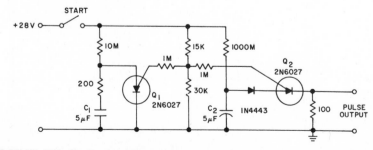

LONG-DELAY PUT—Circuit uses one put as timing element and other as sampling oscillator. C2 must be low-leakage film capacitor because of low current supplied to it. Provides precise time delay without requiring calibration pot.—"SCR Manual," General Electric, Syracuse, NY, 1972, 5th Ed., p 219.

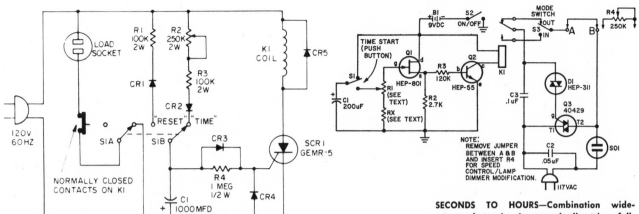

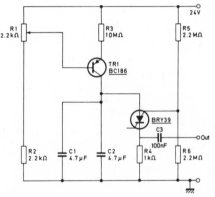

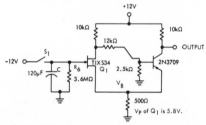

DELAYED-OFF LAMP—Will keep garage or driveway light on long enough after S1 is turned to TIME position for person to unlock door and turn on houselights. Delay is about 1 min. Diodes are GE-504A and K1 is Potter & Brumfield KA11AY 115-V a-c relay. Can also be used to control blender, mixer, or photoenlarger. R2 changes delay.—"Electronics Experimenters Circuit Manual," General Electric, Owensboro, KY, 1971, 3rd Ed., p 195–198.

SECONDS TO HOURS—Combination wide-range electronic timer and diac-triac full-wave power control will turn any device up to 600 W on or off after adjustable delay ranging from seconds to hours. Time-set pot R1 can be 5 or 10 meg. Rx sets lower limit; with 100K, minimum delay is 30 s. With C1 as shown and 10 meg for R1, maximum delay is over 1 hour. K1 can be Potter & Brumfield LM5 spdt relay.—S. Daniels, Hour Master . . . A Super Timer, *Elementary Electronics*, March-April 1972, p 57–58 and 93–94.

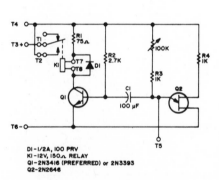

D1-1/2A, 100 PRV
K1-12V, 150Ω RELAY
Q1-2N3416 (PREFERRED) or 2N3393
Q2-2N2646

0–20 S—Uses 12-V 150-ohm relay. Time delay for pull-in of relay is controlled by 100K pot.—*Circuits*, 73, Aug. 1972, p 130.

30 MIN WITH SCS—Timing capacitors C1 are charged with constant current by BC186 transistor for interval determined by R1 up to 30 min, after which BRY39 scs is triggered on to give output pulse.—"Applications of the BRY39 Transistor," Mullard, London, 1973, TP1338, p 6.

13-MIN DELAY—High input impedance of fet gives long time delay in Schmitt trigger circuit. Capacitor C is initially charged to −12 V by closing S1. When S1 is open, C discharges through R6 until voltage on Q1 is small enough to make Schmitt trigger change state. Report gives design equations.—L. Delhom, "The Field-Effect Transistor in a Schmitt Trigger Circuit," Texas Instruments, Dallas, TX, Bulletin CA-88, 1969, p 6–7.

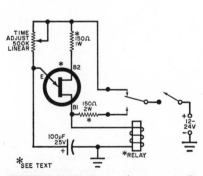

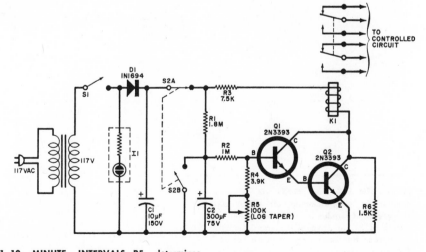

REPETITIVE TIMER—Turns relay on and off at desired intervals, adjustable up to about 5 s. Use 150-ohm relay. Ujt can be 2N1671B or 2N2160. Resistors are for 12-V supply; for 24 V, increase 150-ohm units to 330 ohms.—C. J. Schauers, Time Delay Circuit, *Popular Electronics*, July 1968, p 69.

1–10 MINUTE INTERVALS—R5 determines timing interval between momentary actuation of S2 and energizing of relay. I1 is 117-V neon pilot lamp with built-in resistor. K1 is 5,000-ohm relay such as Price Electric type 5509-24HS.—T. Wallace, Reader's Circuit, *Popular Electronics*, June 1969, p 79–80.

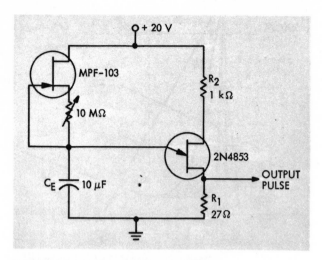

10-MIN UJT-JFET—Conventional delay-determining resistor of 2N4853 ujt is replaced by MPF-103 jfet, with 10-meg pot determining amount of off bias applied to it. Output pulse can be used to trigger appropriate scr for desired load. Long delays are made possible by constant-current charging of CE at rate of less than 1 μA.—"Unijunction Transistor Timers and Oscillators," Motorola Semiconductors, Phoenix, AZ, AN-294, 1972.

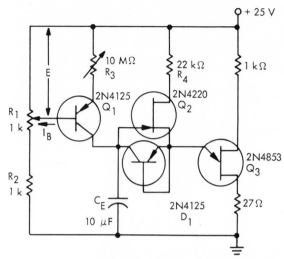

10-HR DELAY—Q1 and R1-R2-R3 form constant-current source whose charge current can be adjusted by R3 to as low as several nA. Since this current is not sufficient to fire ujt relaxation oscillator Q3, required peak current is supplied by fet Q2 acting as source follower. Delay varies linearly with R3.—"Unijunction Transistor Timers and Oscillators," Motorola Semiconductors, Phoenix, AZ, AN-294, 1972.

1-S TURN-OFF DELAY—Diodes CR1 and CR2 together supply about —20 V between MT1 and gate terminal of triac, but gate current can flow to trigger Q1 only when SW is closed to forward-bias Q2 into conduction. C2 takes about 1 s to discharge after SW is open, after which triac will commutate at next zero crossing and interrupt load current. Will handle up to 10 A.—"SCR Manual," General Electric, Syracuse, NY, 1972, 5th Ed., p 220–221.

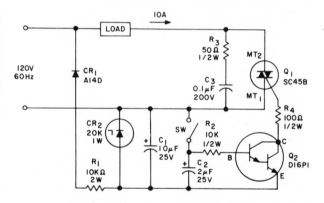

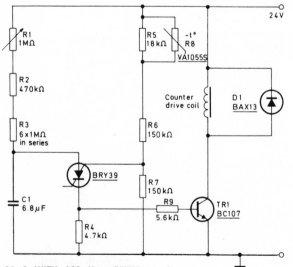

30 S WITH SCS—Uses BRY39 scs in same manner as programmable ujt which discharges C1 at regular intervals determined by setting of R1, to drive BC107 transistor into saturation for operating counting device. Thermistor minimizes changes in timing period with temperature.—"Applications of the BRY39 Transistor," Mullard, London, 1973, TP1338, p 4–6.

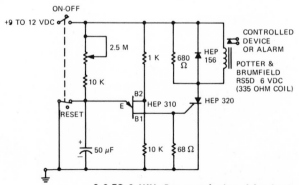

2 S TO 2 MIN—Pot controls time delay interval for energizing of relay after switch is closed. Relay contacts can be connected to turn controlled device on or off at end of interval. Suitable for kitchen or darkroom timer. Will operate from 9-V transistor radio battery.—"Home Handyman's Construction Projects," Motorola Semiconductors, Phoenix, AZ, HMA-37, 1972.

CHAPTER 124
Transceiver Circuits

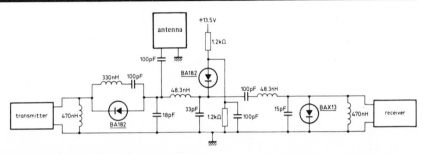

BIASED-DIODE SWITCH—Radar-type silicon-diode circuit provides 26 dB isolation for receiver whenever 160-MHz ssb transmitter is on. Bias is required only when carrier suppression is better than 30 dB, hence would not be needed with constant-carrier transmitters. Will handle up to 23-W transmitter power.—"Antenna Switching in the 160 MHz Communication Band Using Amperex BA182 Diodes," Amperex Electronic Corp., Slatersville, RI, Application Report S-156, April 1972.

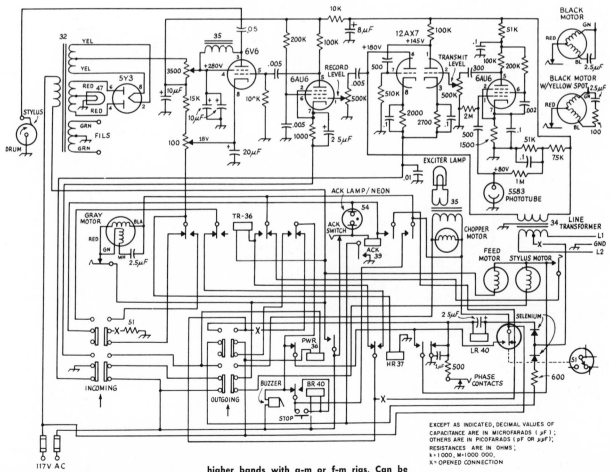

FAX ON AMATEUR BANDS—Modification shown for Western Union Telefax facsimile machine permits operation on 6 and 2 meters with a-m transmitter and on 220-MHz and higher bands with a-m or f-m rigs. Can be used either for transmitting or receiving line drawings and printed messages. Article covers modification procedure, including procedure for sending positive pictures and for sending or receiving sync information before picture scan begins.—H. King, Conversion of Telefax Transceivers to Amateur Service, QST, May 1972, p 23–26.

959

133-MHZ DUPLEXER—Uses two 1N5155 step-recovery varactors as switching devices in circuit using lumped-constant components. Chokes provide d-c conduction paths during transmission. Isolation between transmitter and receiver is about 25 dB, and insertion loss between transmitter and antenna 0.22 to 0.5 dB depending on frequency between 115 and 152 MHz. In receive mode, insertion loss between antenna and receiver varies from 0.15 to 0.25 dB.—J. Cochran, "Duplexing with Step Recovery Varactors," Motorola Semiconductors, Phoenix, AZ, AN-412, 1968.

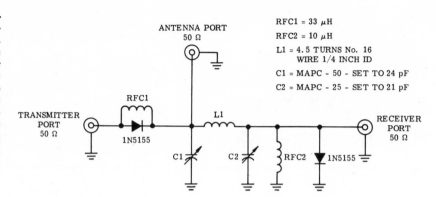

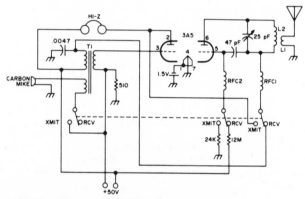

144-MHZ SINGLE-TUBE—Right-hand triode section functions as 144-MHz oscillator on transmit and as 2-meter superregenerative detector on receive. Left section is audio amplifier on receive and mike amplifier-modulator on transmit. For portable operation, use battery supply up to 90 V, although higher voltages may give excessive radiation on receive. Range is up to 100 miles. Article gives all coil data.—R. M. Brown, Experimenter's One-Tube $10 2-Meter Transceiver, 73, Feb. 1971, p 78–79.

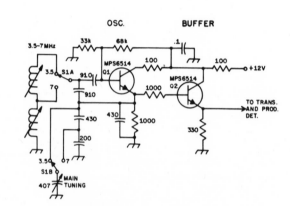

VFO FOR 3.5 AND 7 MHZ—Used in commercial transceiver in place of crystal for transmitting and for use as vfo for product detector when receiving. Gives chirp-free c-w note even when operating at output frequency of transmitter.—Ten Tec PM-2, QST, June 1970, p 51–54.

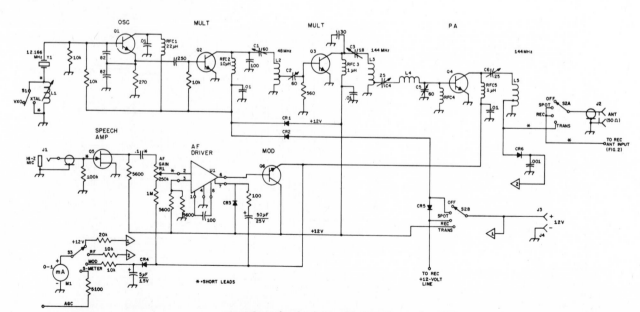

144-MHZ TRANSMITTER—Range is up to 200 miles in mountainous areas yet weight and size of transmitter and companion receiver covered in article permit transport in small mountain rucksack along with batteries and mike. Output is up to 1 W. Uses 12-MHz crystal oscillator, quadrupler, tripler, and collector-modulated final power amplifier. Q1 is 2N3563. Q2 and Q3 are 2N3866. Q4 is PT-3524. Q5 is 2N5459. Q6 is 2N4921. U1 is LM301A or equivalent. Agc voltage is obtained from receiver.—R. Preiss, The "2-Meter QRP Mountain Topper," QST, May 1970, p 11–15.

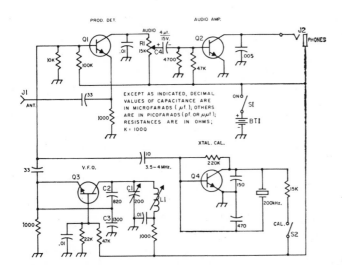

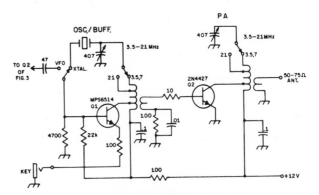

3-BAND TRANSMITTER—Used in commercial transceiver for operation on 3.5-, 7-, and 21-MHz bands. Q2 operates essentially class C. Article includes circuit of vfo that can be used in place of crystal.—Ten Tec PM-2, *QST*, June 1970, p 51–54.

MONITOR—Permits listening to quality of radiotelephone or c-w transmission from mobile transceiver by picking up radiated signal with short antenna; 6-inch antenna gives about 1.5 V of audio on all bands if transmitter has 100-W output. Use 2,000-ohm phones. All transistors are 2N706 or similar. Operates from 9-V battery. Includes calibrator providing 200-kHz markers when S2 is closed. L1 is about 2.6 μH.—W. L. North, A Transceiver Monitor Using Transistors, *QST*, Aug. 1968, p 40–41.

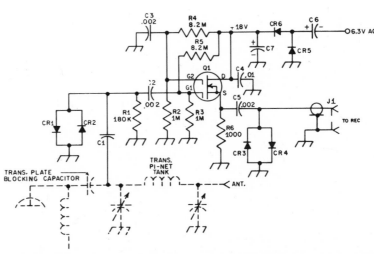

T-R SWITCH—Completely automatic switching circuit performs all required switching functions whenever transmitter is keyed. Only connection to transmitter is through C1, which is 5 to 10 pF. CR1–CR4 are 1N914 or equivalent, while CR5 and CR6 are 100-prv 100-mA silicon. Q1 is RCA 40673 or Motorola MPF121. J1 goes to receiver antenna terminal.—L. McCoy, Simplified Antenna Switching, *QST*, April 1971, p 30–33.

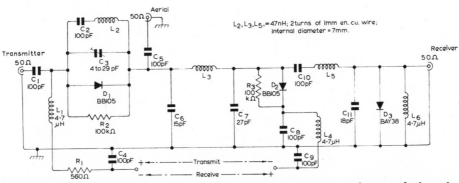

T-R SWITCH—Replaces conventional mechanical relay in 165-MHz mobile transceiver having transmitter power of 12 W. When transmitter is operating, D1 and D2 are forward-biased and have low resistance, to connect transmitter to antenna with power insertion loss of only 3.3%, while shorting receiver input. Any transmitting power leaking past D2 is limited by D3 which has zero bias. During reception, D1 and D2 are reverse-biased and antenna feeds receiver with power insertion loss of only 3.5%. D1 and D2 are variable-capacitance diodes.—J. M. Siemensma, Electronic Aerial Switch for Mobile Transceivers, *Mullard Technical Communications*, Jan. 1968, p 30–31.

25-W 2-METER AMPLIFIER—Designed for use with low-power f-m transceiver, either as intermediate amplifier stage or final power amplifier. Input and output impedances are 50 ohms. Article tells how to construct transmission-line output transformer from pair of 4-inch lengths of No. 20 enamel twisted to 16 twists per inch. RFC1 is ferrite choke and RFC3 is ferrite bead. Operates off auto battery.—R. C. Hejhall, Some 2-Meter Solid-State Rf Power-Amplifier Circuits, QST, May 1972, p 40–45 and 68.

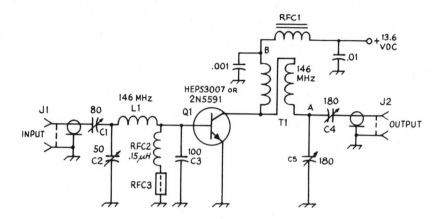

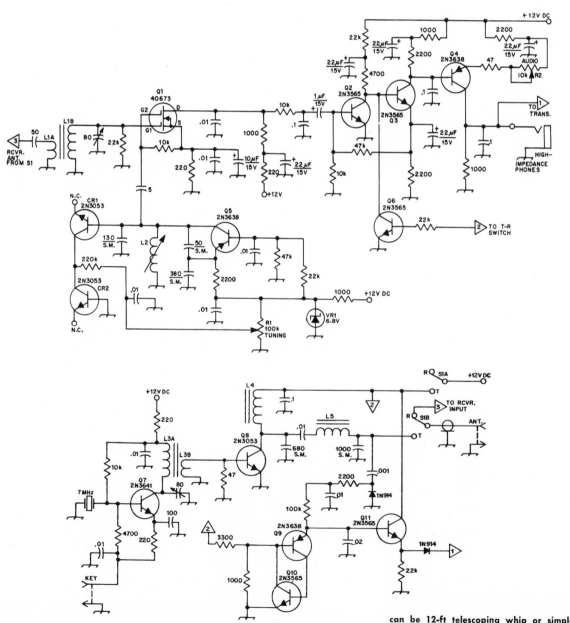

1-W 40-METER MOUNTAINTOP C-W—Operates from 12-V battery, with receiver drawing 14 mA and transmitter 100 mA with key down. Designed for minimum weight and single-control transmit-receive switching, for ease in carrying to remote locations. Antenna can be 12-ft telescoping whip or simple dipole. Article gives coil-winding data.—W. Hayward and T. White, The Mountaineer—An Ultraportable Cw Station, QST, Aug. 1972, p 23–26.

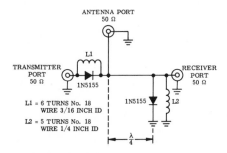

450-MHZ DUPLEXER—Circuit is 50-ohm microstrip line made from 1/16-inch Teflon-fiberglass board, with two 1N5155 step-recovery varactors serving as switching devices. Isolation between transmitter and receiver is 25 dB. Will handle 40-W c-w input power in transmit mode, with insertion loss of 0.3 dB. In receive mode, insertion loss is 0.25 dB.—J. Cochran, "Duplexing with Step Recovery Varactors," Motorola Semiconductors, Phoenix, AZ, AN-412, 1968.

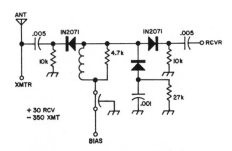

ANTENNA SWITCH—High-frequency transmit-receive switch uses only diodes.—*Circuits, 73*, Nov. 1972, p 303.

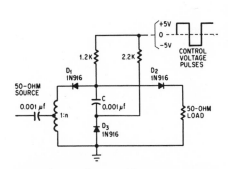

TRANSMIT-RECEIVE SWITCH—Three silicon diodes in simple wideband gate permit switching antenna from transmitter to receiver at rates up to 500 kHz. Isolation is 70 dB over 12.5-MHz band above and below 30 MHz.—J. Gilbert, High-Speed Wideband Gate Provides 70-dB Isolation, *Electronics*, Oct. 21, 1966, p 69–70.

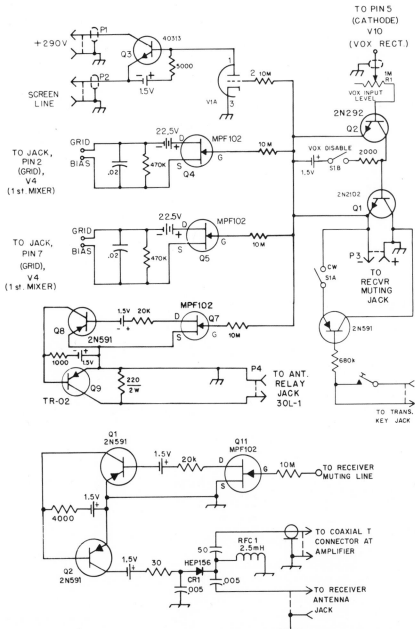

VOX CONTROL—Modification of circuit in Oct. 1968 *QST* uses voice-generated d-c pulse to forward-bias transistor controlling circuits that enable and disable transmitter and receiver alternately and switch antenna appropriately without use of relays. Electronic switching is fast enough so operator can hear between syllables or between dots or dashes. Adjust R1 for proper activation of Q1. Article discusses circuit operation in detail.—H. R. Hildreth, More on Instant Voice Interruption, *QST*, June 1972, p 19–21 and 37.

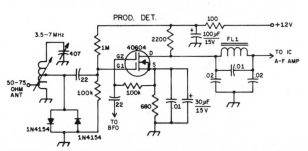

PRODUCT DETECTOR—Used in commercial transceiver covering amateur bands. RCA dual-gate mosfet provides good cross-modulation and overload immunity, along with good conversion gain and low noise figure. Detector is followed by audio filter having 2-kHz bandwidth for ssb reception.—Ten Tec PM-2, *QST*, June 1970, p 51–54.

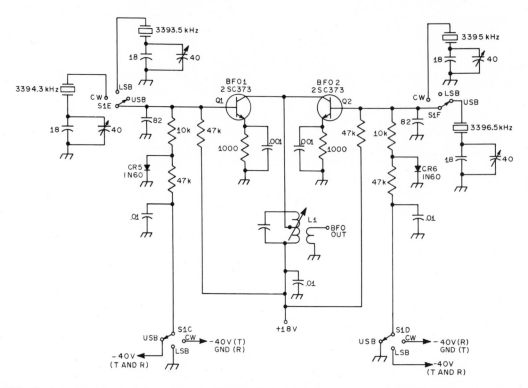

BFO INJECTION—Circuit uses teeter-totter arrangement of two separate oscillators which are activated individually by d-c switching, with S1 placing desired crystal in circuit for c-w, upper-sideband, and lower-sideband operation of 150-W commercial transceiver. American equivalent of transistors is HEP55.—Allied A-2517 Transceiver, *QST*, Nov. 1970, p 43–47.

BATTERY SAVER—Used in portable transceiver to disconnect battery in absence of received audio signal, for battery economy. Receiver supply is interrupted about 15 s after signal stops, and is restored within a few ms of receiving carrier or audio signal. Uses junction fet as voltage-controlled resistance.—R. M. Lea, The Junction F.E.T. as a Voltage-controlled Resistance, *Wireless World*, Aug. 1972, p 394–396.

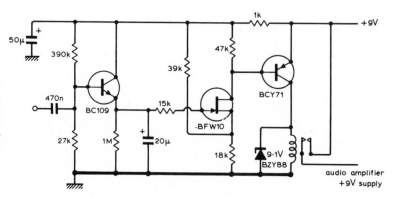

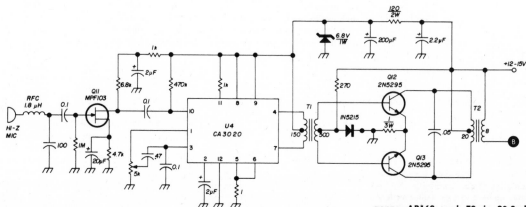

A-M FOR 2 METERS—Amplitude modulator was developed for mobile transceiver operating from 12-V storage battery. Use heat sinks for Q12 and Q13. T1 is 150:500-ohm Argonne AR163 and T2 is 20:8-ohm Stancor TA-12.—R. A. Thompson, Amplitude-Modulated Two-Meter Transmitter-Receiver, *Ham Radio*, Dec. 1971, p 55–63.

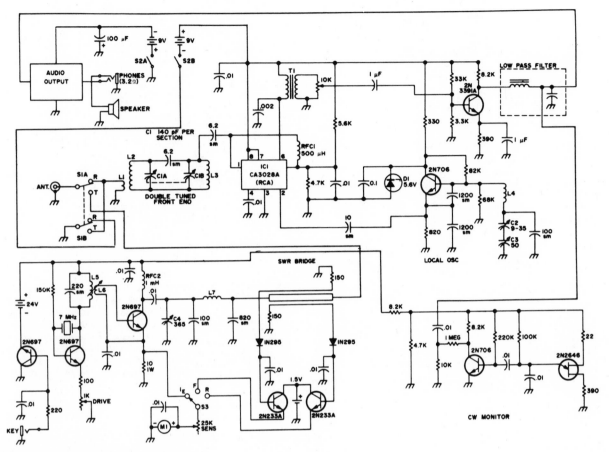

7-MHZ 2-W C-W—Operates from three mercury batteries, for portable use. Article covers construction and tuneup.—C. Sondgeroth, The SST-1 Solid State Transceiver For 40 Meters, 73, Nov. 1970, p 64–71.

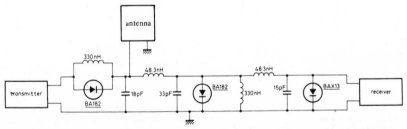

UNBIASED-DIODE SWITCH—Silicon diodes serve as switching elements utilizing radar principles, to provide 26 dB isolation for receiver whenever 160-MHz transmitter is on. Insertion loss between transmitter and antenna is then less than 0.6 dB. Will handle up to 23 W transmitter power. Circuit uses two quarter-wave sections between receiver and antenna.—"Antenna Switching in the 160 MHz Communication Band Using Amperex BA182 Diodes," Amperex Electronic Corp., Slatersville, RI, Application Report S-156, April 1972.

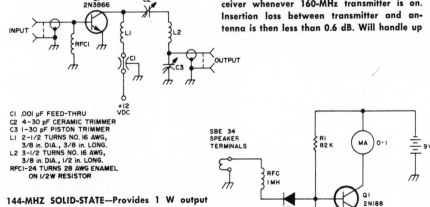

C1 .001 μF FEED-THRU
C2 4–30 pF CERAMIC TRIMMER
C3 1–30 pF PISTON TRIMMER
L1 2-1/2 TURNS NO. 16 AWG, 3/8 in. DIA., 3/8 in. LONG.
L2 3-1/2 TURNS NO. 16 AWG, 3/8 in. DIA., 1/2 in. LONG.
RFC1-24 TURNS 28 AWG ENAMEL ON 1/2W RESISTOR

144-MHZ SOLID-STATE—Provides 1 W output at 2 meters, sufficient for small transceiver used for local contacts. Article gives construction procedure for obtaining maximum efficiency.—C. Sondgeroth, Experimental Solid State VHF Amplifier, 73, July 1972, p 15–17.

ADDING S METER—Simple circuit can be connected directly to speaker of transceiver. R-f choke is required only when connecting to remote speaker whose leads may pick up r-f during transmit mode. Although meter reading depends somewhat on setting of receiver gain control, readings on strong signals will compare favorably with those of any other S meter when using normal room volume.—R. E. Barrington, SBE Improvements Made Easy, 73, Nov. 1969, p 122–124.

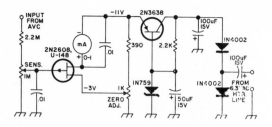

FET S METER—Includes regulated power supply operating from 6.3-V a-c heater line of tube-type receiver or transceiver. Input is negative agc voltage of receiver, driving p-channel fet. Circuit gives full-scale reading of 1 mA for input of —7.5 V d-c.—H. Olson, An FET S-Meter, 73, Jan. 1967, p 42–44.

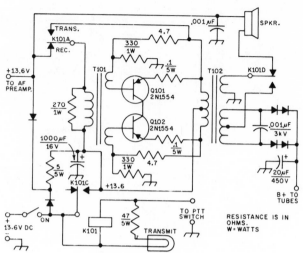

SWITCHING RELAY—Portion of 2-meter transceiver shows how relay K101 is used to transfer antenna and some of operating voltages between transmitter and receiver. Circuit is combination audio amplifier and d-c to d-c converter, with converter serving as audio amplifier during receive.—The Gladding 25 FM Transceiver, QST, Dec. 1971, p 42–44.

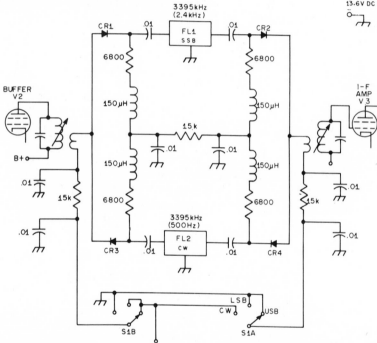

DIODE SWITCHING OF FILTERS—Filter FL1 provides 2.4-kHz bandwidth for ssb operation of 150-W transceiver, and FL2 provides 500-Hz bandwidth for c-w operation. Four diodes perform switching of filters without introducing capacitive and inductive coupling between filter input and output terminals. S1 simply applies +150 V with correct polarity for diodes that are to conduct. Diodes can be IN914.—Allied A-2517 Transceiver, QST, Nov. 1970, p 43–47.

LIGHTNING PROTECTION—Automatically grounds antenna when transceiver is turned off, for protection against lightning damage. Omit R2 initially; if K1 ungrounds antenna when turned-off transceiver is first plugged into SO1, use pot for R1 and adjust so K1 trips and ungrounds antenna only when transceiver is turned on. Power failure grounds antenna automatically.—H. Phillips, Automatic Lightning Protection, *Popular Electronics*, July 1970, p 61–64.

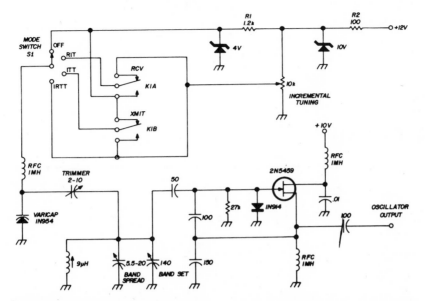

INCREMENTAL TUNING—Designed for adding to 5–5.5 MHz vfo of any transceiver, to permit instant switching for receiving only (RIT), transmitting only (ITT), or both (IRTT) without changing setting of main tuning dial. Uses varactor diode (VARICAP) connected as shown to trimmer and bandspread capacitors in set. Relay K1 is shown in receiver position, and must be switched with antenna-change-over relay. Adjust trimmer and bandset capacitors alternately to obtain amount of incremental tuning desired.—M. J. Goldstein, Adding Incremental Tuning to Your Transceiver, *Ham Radio*, Feb. 1971, p 66–68.

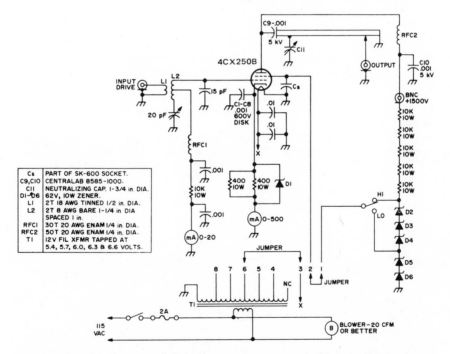

200-W AMPLIFIER FOR F-M TRANSCEIVER—Requires only 5-W drive, is not damaged if drive is lost, and requires no bias or screen supply. Zeners stabilize screen voltage and limit it to 310 V. Switch reduces screen voltage to 124 V for tuneup or local work. Tapped 12-V 5-A filament transformer is normally set at 6 V, but can be adjusted to compensate for line voltage and tube aging. Article covers construction and adjustment.—J. Townsend, Power Amplifiers for Two Meter FM, *73*, April 1972, p 17–23.

CHAPTER 125
Transmitter Circuits

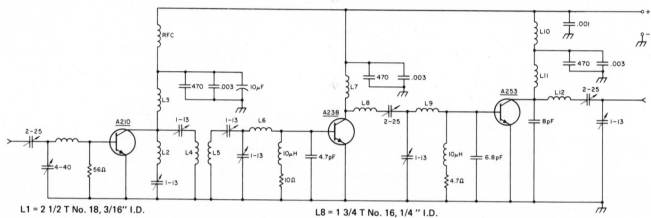

L1 = 2 1/2 T No. 18, 3/16" I.D.

L2 = 2 1/2 T No. 20, 3/16" I.D.

L3, 4, 5 = 1 1/2 T No. 16, 1/4" I.D.

L6 = STRIP LINE, 1 1/4 L X 1/2" W, ETCHED ONTO PC BOARD

L7 = 1 1/2 T No. 14, 1/4" I.D.

L8 = 1 3/4 T No. 16, 1/4" I.D.

L9 = STRIP LINE, 1" L X 7/16" W, ETCHED ONTO PC BOARD

L10 = 4 1/2 T, No. 24, 3/16" I.D.

L11 = 17 No. 14, 1/4" I.D.

L12 = 1 1/4 L, No. 14

RFC = FERROXCUBE No. VK20010/3B

450-MHZ 6-W AMPLIFIER—With first stage serving as frequency doubler, input of 100 mW at 225 MHz drives amplifier to maximum power rating of 6 W at 450 MHz for supply of 12.5 V. Increasing supply to 13.5 V boosts output to 7 W. Each stage operates as common-emitter class C amplifier.—"A Six Watt 450 MHz Amplifier," Amperex Electronic Corp., Slatersville, RI, Application Report S-148, Dec. 1969.

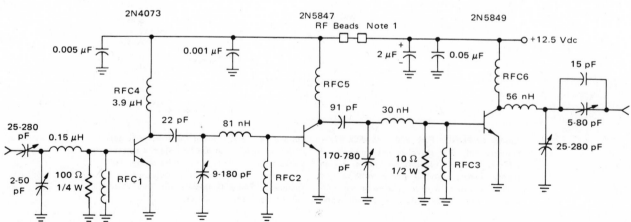

RFC1,2,3 are Ferroxcube Type VK 200 19/4B, ferrite chokes. RFC5,6 = 15T #16 AWG enameled wire, 3/16" ID, close wound in two layers. RF Beads are Ferroxcube Type 56-590-65/3B.

50-MHZ 40-W 12.5-V MOBILE—Designed for c-w or f-m operation, but not suitable for a-m because rating of output transistor would be exceeded. Operation is class C, with 50-ohm output impedance. Requires power input of 20 mW. Total current drain from storage battery is 5.4 A. Report covers design and construction.—C. Martens, "A 40-W 50-MHz Transmitter for 12.5-Volt Operation," Motorola Semiconductors, Phoenix, AZ, AN-502A, 1972.

968

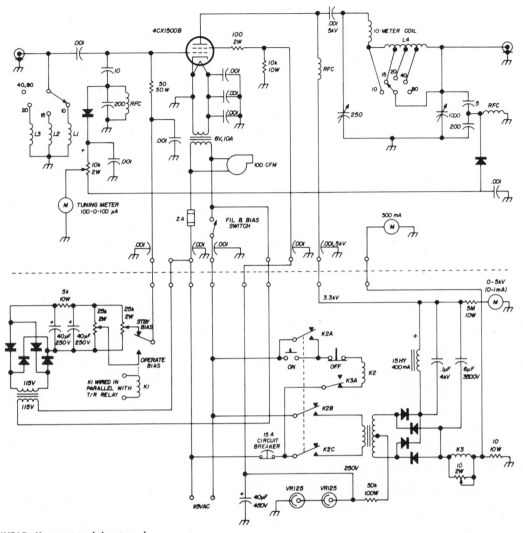

2-KW PEP LINEAR—Uses untuned input and pi-network output, requires no neutralization, and has low intermodulation distortion with Eimac ceramic triode. Covers five bands from 10 to 80 meters. Article gives coil data.—I. R. Wolfe, High-Power Linear for 80-10 Meters, *Ham Radio*, April 1971, p 56–60.

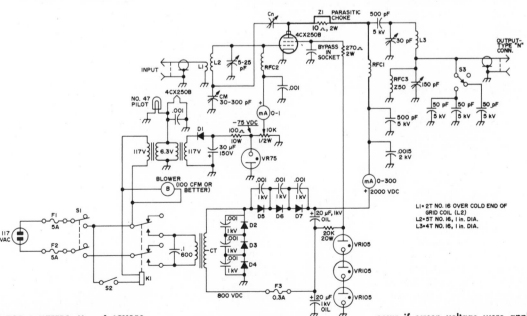

1-KW LINEAR FOR 6 METERS—Use of 4CX250 pentode makes compact construction possible. Transformer should deliver 900 V across full secondary, which feeds full-wave voltage doubler providing about 2,000 V for plate and 300 V regulated for screen. Power supply design prevents tube damage that might occur if screen voltage were applied before plate voltage. Article covers construction.—P. J. Bertini, A Compact Kilowatt for Six Meters, *73*, Nov. 1971, p 57–62.

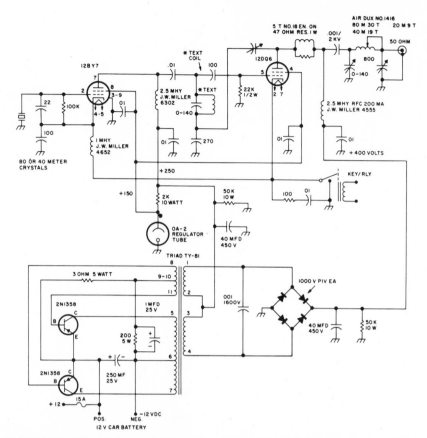

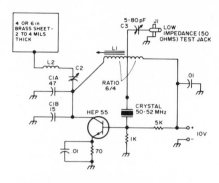

¼ W ON 6 METERS—Although only ¾ inch square and ⅛ inch thick, range is up to 15 miles with 6 inch metal-sheet antenna. Crystal is McCoy type MM glass-enclosed. L2 depends on size of antenna. C2 is 100 pF. Plastic-encased transistor is filed down to desired thickness. Article covers construction and tuneup.—B. Hoisington, Postage Stamp Transmitter for Six, 73, May 1970, p 80–87.

50-W MOBILE C-W—Can be installed in trunk of car, for operation from 12-V car battery. Can be operated on 80, 40, or 20-meter bands by changing plug-in coils. Keying is done by opening cathodes of tubes. Neutral-izing is not normally necessary, even though provisions are made for it. Inverter power supply provides 400 V for plate of final.—E. Marriner, A Mobile CW Transmitter, 73, May 1970, p 108–110.

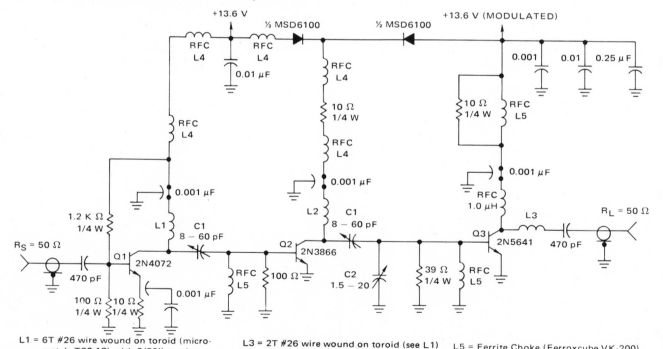

L1 = 6T #26 wire wound on toroid (micro-metals T30-13) with 3/32'' spacing

L2 = 2T #26 wire wound on toroid (see L1) with 1/8'' spacing

L3 = 2T #26 wire wound on toroid (see L1) with 5/16'' spacing

L4 = RF bead (one hole), 1/8''

L5 = Ferrite Choke (Ferroxcube VK-200)

C1 = 8-60 pF (Arco 404)

C2 = 1.5-20 pF (Arco 402)

2.5-W A-M AIRCRAFT—Will modulate up to 95% in frequency range of 118 to 136 MHz, with power input of 2 mW. Drain on 13.6-V power supply is about 360 mA. Does not require modulation transformer. Report includes circuits with additional amplifier stages for 7-W and 13-W power.—D. Brubaker, "A 13-W Broadband AM Aircraft Transmitter," Motorola Semiconductors, Phoenix, AZ, AN-507, 1970.

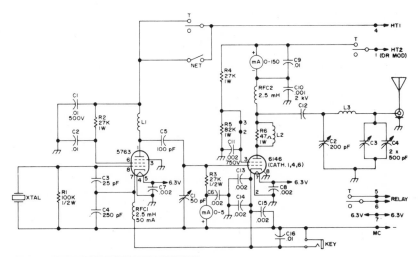

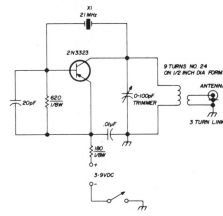

TABLE I—DATA				
BAND	DIAMETER	NO. OF TURNS	GAUGE	LENGTH
7–L1	1/2 in.	40	24	—
L3	1-1/4 in.	22	18	2 in.
14–L1	1/2 in.	20	24	—
L3	1-3/4 in.	10	16	1-1/8 in.
21–L1	1/2 in.	9	22	—
L3	1 in.	11	16	1-1/4 in.
28–L1	1/2 in.	9	22	—
L3	1-1/4 in.	5	14	1 in.
NOTE – L1 ALWAYS HAS TURNS SIDE BY SIDE				

60 W ON FOUR BANDS—Operates with excellent efficiency at 20 to 60 W on 7, 14, 21, and 28 MHz. Uses 7-MHz crystal in circuit that can be operated on multiples of fundamental. Article covers construction and tuneup. Oscillator draws 30 mA at 300 V. Supply for r-f amplifier can be from 300 to 600 V, with higher voltage giving higher output power. L2 is 5 turns No. 18 wound on R6.—F. G. Rayer, 20–60 W 1-4 Band TX, 73, Feb. 1972, p 59–62.

15-METER QRP—Experiment with feedback component values to get oscillation with good keying consistent with transmitter loading, for micropower amateur radio operation.—A. Wilson, Low-Power Transmitter and Indicating Wavemeter, *Ham Radio*, Dec. 1970, p 26–28.

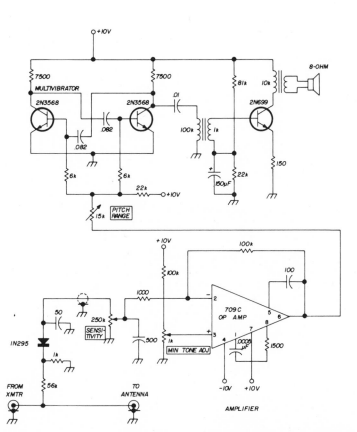

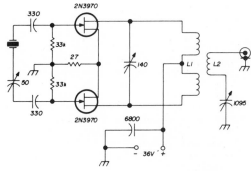

160
L1 65 turns no. 26 on 1-1/4" coil form, centertapped and divided

L2 20 turns no. 26 between halves of L1

80	**40**
L1 40 turns no. 24 on 1-1/4" coil form, centertapped and divided	21 turns no. 22 on 1-1/4" coil form, centertapped and divided
L2 13 turns no. 24 between halves of L1	7 turns no. 22 between halves of L1

AURAL TUNING METER—Rectified r-f current taken from output of transmitter controls frequency of mvbr audio oscillator driving speaker. Allows blind ham to tune transmitter for maximum output by adjusting for highest pitch of sound. Article includes simple power supply circuit.—D. C. Miller, Transmitter-Tuning Unit for the Blind, *Ham Radio*, June 1971, p 60–62.

1-W PUSH-PULL FET—Requires only two Siliconix transistors operating from three 12-V lantern batteries in series, for outputs above 1 W on 40, 80, and 160 meters.—E. Noll, Integrated Circuits, *Ham Radio*, June 1971, p 40–47.

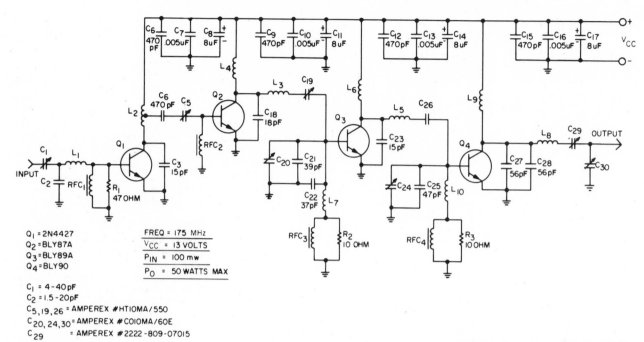

Q₁ = 2N4427
Q₂ = BLY87A
Q₃ = BLY89A
Q₄ = BLY90

FREQ = 175 MHz
V_{CC} = 13 VOLTS
P_{IN} = 100 mw
P_O = 50 WATTS MAX

C₁ = 4-40 pF
C₂ = 1.5-20 pF
C₅,₁₉,₂₆ = AMPEREX #HT10MA/550
C₂₀,₂₄,₃₀ = AMPEREX #C010MA/60E
C₂₉ = AMPEREX #2222-809-07015

50-W VHF AMPLIFIER—Designed for marine applications having FCC limit of 25 W, with extra power to compensate for losses in filter, antenna switches, and coax transmission line. Report gives winding data for all coils. Chokes are Ferroxcube VK 200-10/3B.—"Solid-State RF Power Amplifiers 146–175 MHz Land Fixed and Mobile and Marine Services," Amperex Electronic Corp., Slatersville, RI, Application Report S-152, Dec. 1971.

3.5–30 MHZ 15-W LINEAR—With proper harmonic filter at output, no tuning is necessary for complete coverage of each band when used as final amplifier for ssb or c-w transmitter. Requires 375-mW drive for full output. Input impedance is 50 ohms and gain is 16 dB minimum. Article also covers design of harmonic filter.—B. Lowe, A 15-Watt-Output Solid-State Linear Amplifier for 3.5 to 30 MHz, QST, Dec. 1971, p 11–14.

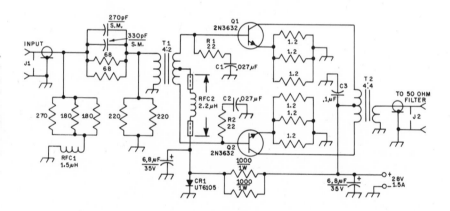

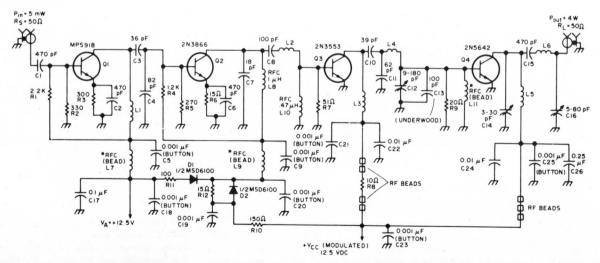

118–136 MHZ 4-W A-M—Broadband solid-state circuit for commercial applications requires no tuning for changing frequency. Article covers construction and testing.—D. Brubaker, A VHF AM Transmitter Using Low-Cost Transistors, 73, Aug. 1970, p 54–59.

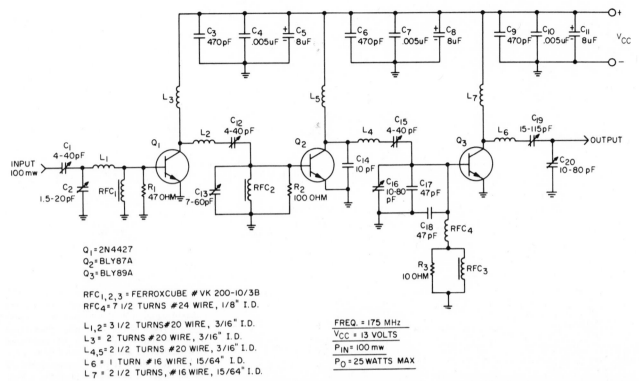

Q₁ = 2N4427
Q₂ = BLY87A
Q₃ = BLY89A

RFC₁,₂,₃ = FERROXCUBE # VK 200-10/3B
RFC₄ = 7 1/2 TURNS #24 WIRE, 1/8" I.D.

L₁,₂ = 3 1/2 TURNS #20 WIRE, 3/16" I.D.
L₃ = 2 TURNS #20 WIRE, 3/16" I.D.
L₄,₅ = 2 1/2 TURNS #20 WIRE, 3/16" I.D.
L₆ = 1 TURN #16 WIRE, 15/64" I.D.
L₇ = 2 1/2 TURNS, #16 WIRE, 15/64" I.D.

FREQ. = 175 MHz
V_{CC} = 13 VOLTS
P_{IN} = 100 mw
P_O = 25 WATTS MAX

25-W VHF AMPLIFIER—Class-C stages use series capacitors in L-type interstage coupling networks for simplified tuning. Circuit design stresses stablity and reliability, as required for all types of commercial service.—"Solid-State RF Power Amplifiers 146-175 MHz Land Fixed and Mobile and Marine Services," Amperex Electronic Corp., Slatersville, RI, Application Report S-152, Dec. 1971.

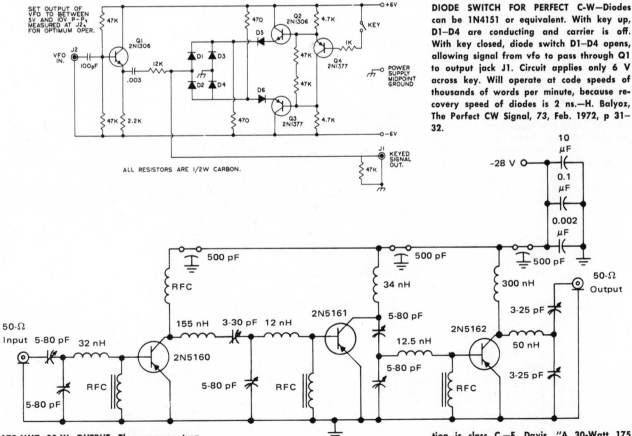

DIODE SWITCH FOR PERFECT C-W—Diodes can be 1N4151 or equivalent. With key up, D1–D4 are conducting and carrier is off. With key closed, diode switch D1–D4 opens, allowing signal from vfo to pass through Q1 to output jack J1. Circuit applies only 6 V across key. Will operate at code speeds of thousands of words per minute, because recovery speed of diodes is 2 ns.—H. Balyoz, The Perfect CW Signal, 73, Feb. 1972, p 31–32.

175-MHZ 30-W OUTPUT—Three power transistors operating on 28-V supply deliver 30-W r-f power to 50-ohm load with overall gain of 29 dB at efficiency of 50%. All spurious outputs are at least 40 dB below level for 175 MHz. Draws 2.1 A from supply. Operation is class C.—F. Davis, "A 30-Watt 175 MHz Power Amplifier Using PNP Transistors," Motorola Semiconductors, Phoenix, AZ, AN-477, 1971.

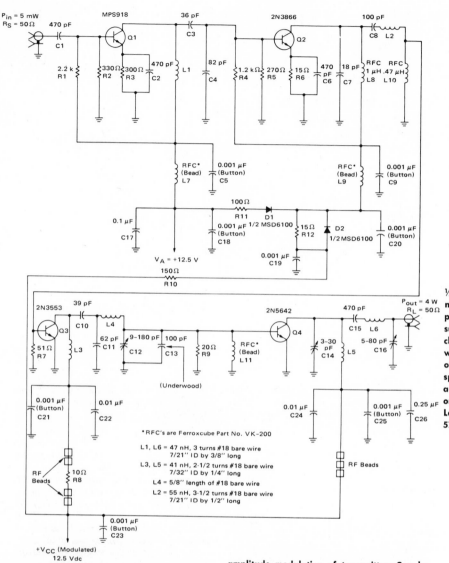

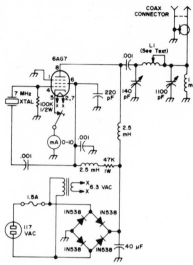

½-W C-W—Single pentode serves as combination of Pierce crystal oscillator and r-f amplifier drawing 150 V at 8 mA from power supply shown. Oscillator keying gives fairly clean c-w signal which can be monitored with receiver to make sure there is no chirp or frequency drift. L1 has 11 turns of No. 12 spaced about 3 turns per inch on 3½-inch-diameter form. Transmitter gave excellent DX on 40 meters with dipole 10 feet up.—D. J. Lazar, CQ DX on ½ Watt, 73, Feb. 1972, p 57–58.

4-W A-M LIGHT AIRCRAFT—Requires no tuning when changing frequency over design range of 118–136 MHz. Report gives circuit of four-transistor series modulator design for amplitude modulation of transmitter. Can be modulated in excess of 85%. Performance graphs are given.—D. Brubaker, "A Broadband 4-Watt Aircraft Transmitter," Motorola Semiconductors, Phoenix, AZ, AN-481, 1972.

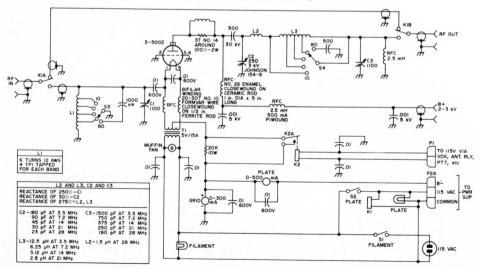

750-W LINEAR—Single triode with 3-kV d-c plate voltage can be switched in and out with plate voltage switch S2. Chief advantages are high efficiency and very low intermodulation products. Operates with zero bias for grounded-grid class B service. Article covers construction and includes high-voltage power supply circuit.—N. Ralph, Your Second Linear, 73, Dec. 1970, p 41–52 and 54–55.

10-METER DOUBLE-SIDEBAND—Uses crystal control and puts out about 1 W for driving antenna directly or driving r-f amplifier. Diode-type balanced modulator provides 40 dB of carrier suppression. Can be received on modern receivers with no difficulty.—R. L. Guard, Jr., A Transistorized 10 Meter DSB Transmitter, 73, May 1971, p 40–41.

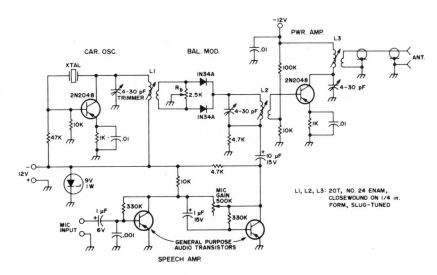

1-KHZ MODULATOR FOR 10-GHZ TRANSMITTER—Free-running mvbr TR4-TR5 drives base of TR7 through emitter-follower TR6. Resulting negative pulses on base of TR7 cut it off periodically, to interrupt 7-V supply to 10-GHz Gunn device and thereby provide square-wave modulation. Permits making qualitative measurements of signal with receiver having a-f amplifier.—"A Gunn Device Transmitter (10 GHz)," Educational Projects in Electronics, Mullard, London.

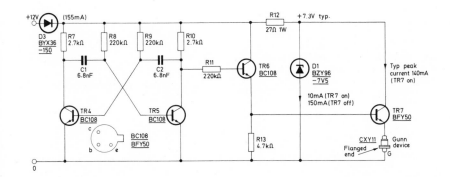

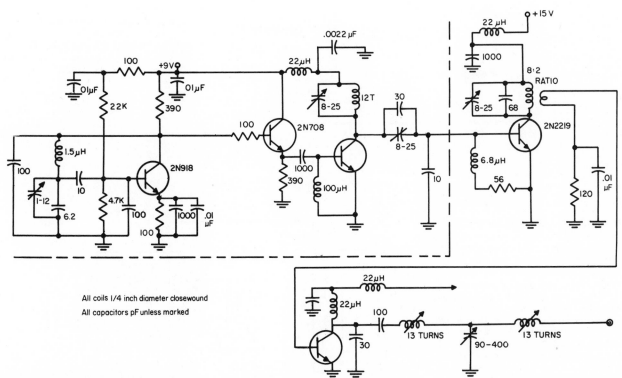

10 W AT 30 MHZ WITH VACKAR VFO—Provides tuning of 2.5:1 with inherent stability and essentially constant output amplitude. Article analyzes design and performance of circuit originally developed in Czechoslovakia by J. Vackar in 1949. Circuit is highly recommended over usual Colpitts, Hartley, and Clapp versions.—G. B. Jordan, The Vackar VFO, A Design to Try, The Electronic Engineer, Feb. 1968, p 56–59.

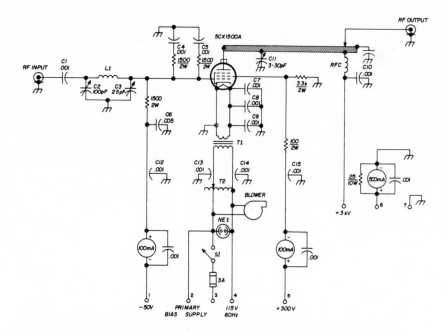

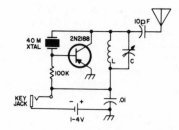

40-METER C-W—Entire transmitter, including watch-type mercury cell, can be fitted into crystal housing or other small case. For tuning, hold key down and adjust C for maximum signal at receiver being used. Operating range for c-w is up to 50 miles.—L. Zipin, Micro-Mitter ... Teensy-Weensy QRP on 40, 73, Feb. 1970, p 131.

1-KW LINEAR FOR 2 METERS—Uses Eimac pentode in grounded-cathode linear amplifier requiring no drive power. Tank circuit is strip line. Article gives coil data and power supply circuit. Requires no neutralization.—I. R. Wolfe, Low-Drive Kilowatt Linear for Two Meters, *Ham Radio*, July 1970, p 26–29.

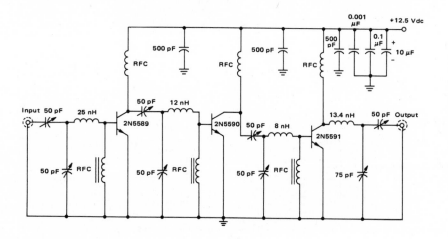

175 MHZ AT 25 W—Designed for operation from 12-V auto or boat storage battery. Overall gain is 21.2 dB at efficiency of 46.5%. All harmonics and spurious outputs are at least 40 dB below 175-MHz output signal. All base chokes are Ferroxcube ferrite type VK-200 19/4B. Input and output impedances are 50 ohms. Battery drain is 4.3 A. Power input requirement is 190 mW.—R. Hejhall, "A 25-Watt, 175 MHz Transmitter for 12.5-Volt Operation," Motorola Semiconductors, Phoenix, AZ, AN-495, 1972.

FIELD STRENGTH AND MODULATION MONITOR—To check modulation, antenna is short length of insulated wire projecting inside transmitter enclosure. Start with shortest possible length, because excessive r-f signal will burn out diodes and meter. Longer antenna used away from transmitter will provide check of field strength. Q1 and speaker permit use as code monitor when S2 is in CW position. L1 is 18-turn length of B & W 3015 Miniductor tapped 3 turns from end. T1 is output transformer having 10K C-T primary and 8-ohm secondary for 8-ohm speaker. No battery is needed.—R. M. Brown and T. Kneitel, "101 Easy Test Instrument Projects," Howard W. Sams, Indianapolis, IN, 1968, p 105–106.

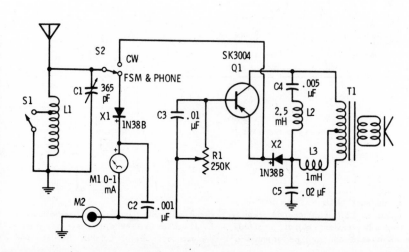

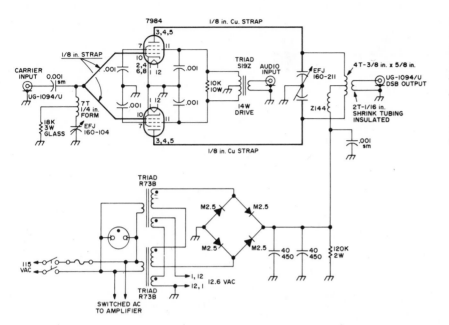

2-METER DOUBLE-SIDEBAND—Circuit provides carrier suppression, to approach performance of ssb without having ssb generating problems. Carrier suppression is —36 dB. Two power transformers in series give 400 to 450 V for bridge-rectifier supply. Article covers construction and gives alignment procedure. If receiving operator hears backward speech, ssb receiver sideband should be switched.—R. W. Campbell, VHF Double Sideband, *73*, Dec. 1971, p 85–91.

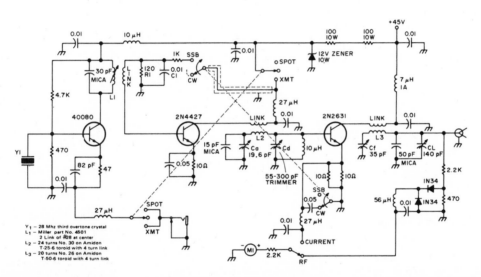

12 W FOR 10 METERS—Input stage operates either as third-overtone crystal oscillator for c-w or as linear amplifier for ssb on 10 meters. Input power to final is about 12 W on c-w, and about 8 W pep when driven as linear for low-distortion ssb. Oscillator and driver require 12 V, obtained from 45-V regulated supply with dropping resistor and 12-V zener. Article covers construction and gives coil data. Use 28-MHz third-overtone crystal and 1-mA meter.—R. E. Gould, The Transistor 12 Watter For 10, *73*, April 1971, p 131–134.

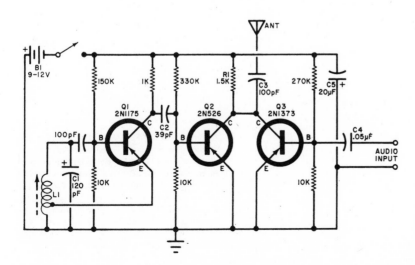

A-M WIRELESS PHONO—Modulated buffer amplifier Q2 is used between Hartley oscillator Q1 and modulator Q3 to improve stability of transmitter used to feed phonograph output to ordinary a-m broadcast receiver. L1 is standard tapped broadcast-band oscillator coil.—G. McClellan, Reader's Circuit, *Popular Electronics*, Feb. 1969, p 82 and 94.

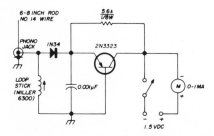

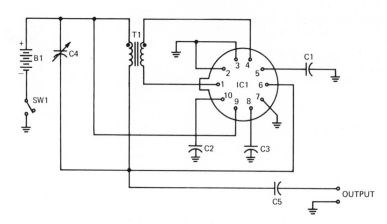

QRP TUNEUP—Single-transistor current amplifier increases sensitivity of milliammeter by factor of .10, for optimum tuning of micropower transmitter.—A. Wilson, Low-Power Transmitter and Indicating Wavemeter, *Ham Radio*, Dec. 1970, p 26–28.

PARTS LIST:

IC1	HEP 590	T1	21 TURNS · 7 TURNS
C1, 2, 3,5	0.1 µF		#36 WIRE ON MICROMETALS T-12-2 CORE
C4	170-780 pF		Lp = 1.3 µH, Ls = 0.1 µH
B1	6-12 VDC	SW1	SPST

40-METER VFO—Sine-wave output can be tuned with C4 over full IC oscillator range of 5 to 10 MHz.—"Radio Amateur's IC Projects," Motorola Semiconductors, Phoenix, AZ, HMA-36, 1971.

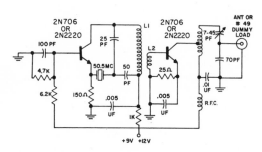

L₁—¼" dia. 7T #24
L₂—2T at cold end of L₁
L₃—6T Airdux 608 (B&W 3010) CT
RFC—20T #26 on 1W resistor

100 MW ON 6 METERS—Simple transmitter can be modulated with TA300 IC for a-m or with a couple of diodes for f-m.—*Circuits*, 73, Nov. 1972, p 300.

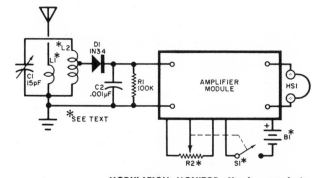

MODULATION MONITOR—Simple crystal detector, IC audio amplifier, and phones permit monitoring a-m phone transmissions for distortion, hum, overmodulation, and undermodulation from about 40 to 60 MHz. L2 is 14 turns of No. 18 on ⅝-inch coil form, spaced 6 turns per inch, tapped 4 turns from bottom. L1 is 2 turns wound around grounded end of L2. R2 is volume control with switch. Antenna is short wire or nothing at all. Increase C1 to 35 pF for citizens band.—R. L. Winklepleck, Build a Modulation Monitor, *Popular Electronics*, March 1968, p 35 and 99.

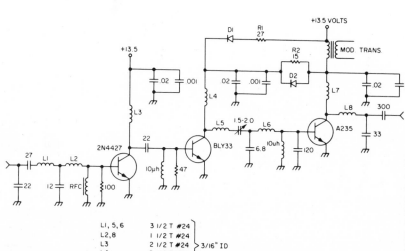

L1,5,6   3 1/2 T #24
L2,8   1 1/2 T #24
L3   2 1/2 T #24 } 3/16" ID
L4   2 1/2 T #20
L7   2 1/2 T #20
RFC   FERROCUBE BEAD #56 590 65/3B, WITH 5T #30

7-W A-M AIRCRAFT AMPLIFIER—Provides full output into 50 ohms from input of 20 mW across entire 118–136 MHz aircraft a-m route service band. Includes full protection against failure due to output mismatch. Diodes are 1N126. Alignment procedure is given.—"A Broad-Band Seven-Watt RF Solid-State Power Amplifier for Aircraft AM Transmitters," Amperex Electronic Corp., Slatersville, RI, Application Report S-153, Dec. 1971.

VFO FOR QRP—Four-band variable-frequency oscillator is chirp-free and stable, for use with low-power transmitter (under 5 W). Article gives tank circuit values for each band.—A. Weiss, A Multiband FET VFO QRPP Transmitter, *Ham Radio*, July 1972, p 39–45.

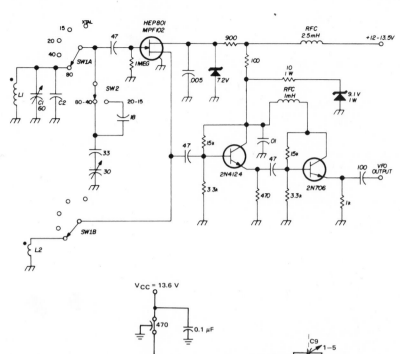

150–450 MHZ TRIPLER—Uses sharp filter having insertion loss of 0.6 dB at 450 MHz. Idler circuit L7-C12 provides necessary path for flow of unwanted current to ground at 300 MHz.—B. Becciolini, "Using Balanced Emitter Transistors in RF Applications." Motorola Semiconductors, Phoenix, AZ, AN-521, 1970.

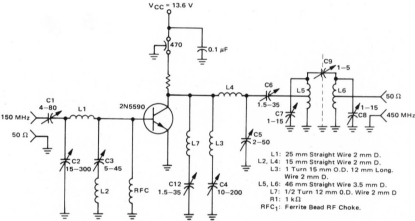

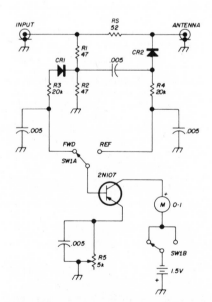

ANTENNA MATCHER—Permits adjusting T-match, gamma, or other antenna matching sections without creating interference by using station transmitter. Consists of resistance bridge and simple transistor amplifier. Signal source for input need be only 100 mW to give full-scale of meter on all bands from 10 to 80 meters. Rs should be close to output impedance of transmitter, usually 52 ohms. Article gives calibration procedure and method of use. Will make accurate swr measurements.—A. G. Shafer, One-Man Antenna Matcher, *Ham Radio*, June 1971, p 24–26.

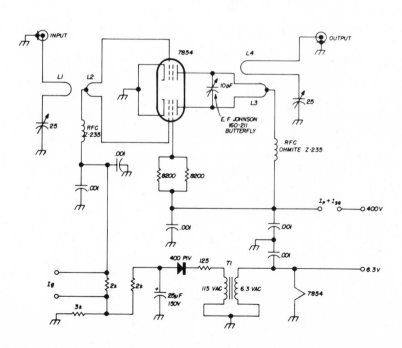

220-MHZ POWER AMPLIFIER—Requires no neutralization and minimal drive. Article covers construction. Can be used for ssb by de-creasing bias and increasing screen voltage.—M. O. Beck, An RF Power Amplifier for 220 MHz, *Ham Radio*, Jan. 1971, p 44–47.

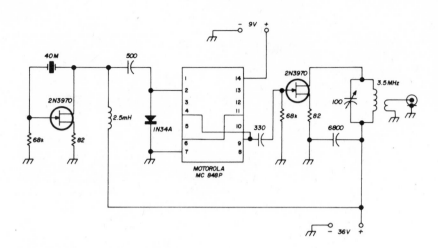

80-METER QRP—IC connected as flip-flop divides output of 40-meter Pierce crystal oscillator by 2 for low-power operation on 80-meter amateur band. Following fet boosts sine-wave r-f output to 200 mW.—E. Noll, Integrated Circuits, *Ham Radio*, July 1971, p 58–62.

175-MHZ 25-W—Transistors used have excellent resistance to burnout under conditions of mismatching and detuning in high-frequency power amplifiers. Operates from auto storage battery (12.5V).—B. Becciolini, "Using Balanced Emitter Transistors in RF Applications," Motorola Semiconductors, Phoenix, AZ, AN-521, 1970.

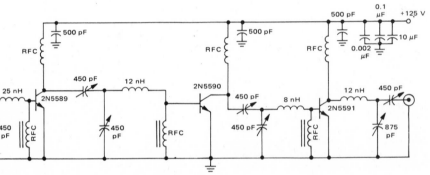

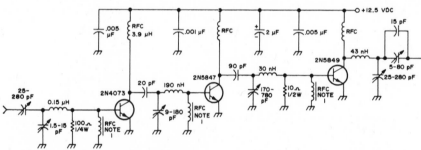

50-MHZ 40-W C-W MOBILE—Will operate continuously at full power on either c-w or f-m with transistors shown. Keyer or modulator, not shown, can easily be added to meet requirements for specific application. Total current is 5.4 A from 12.5-V auto storage battery. Article includes design and construction data for transmitter.—C. Martens, A 40W 6-Meter FM/CW Mobile Transmitter, *73*, May 1972, p 21–24.

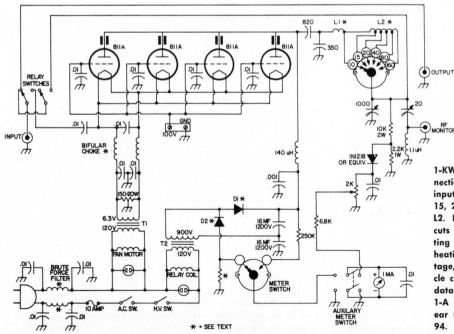

1-KW 6-BAND LINEAR—Grounded-grid connection of four triodes in parallel runs at input of about 1-kW pep, switchable to 10, 15, 20, 40, 80, and 160 meters with taps on L2. Relay activated by high-voltage switch cuts linear out of circuit to permit transmitting with low power while filaments are heating up. Switched meter reads plate voltage, plate current, and relative output. Article covers construction, including coil-winding data. D1 and D2 each consist of five 1,000-V 1-A diodes in series.—J. Barcz, Six Band Linear (At 5¢ Per Watt), *73*, Jan. 1973, p 89–94.

80-METER VFO FOR C-W—Low cost Vackar oscillator is easy to build, gives high-quality chirpless note when keyed, and gives freedom from fixed crystal frequencies. Article gives coil data, along with changes needed for operation in 40-meter band. Includes two buffer stages. If using 12-V lantern battery or eight D cells, connect 100-μF electrolytic across battery.—C. E. Galbreath, A VFO for Solid-State Transmitters, *Ham Radio*, Aug. 1970, p 36–40.

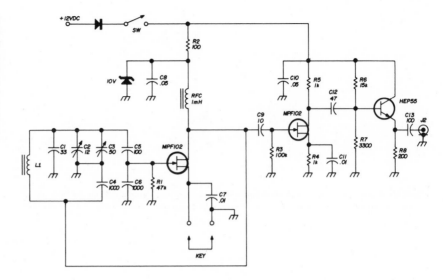

100-MW QRP—Will work several hundred miles on 40 meters, using only two Motorola IC's. First section of first IC operates as crystal oscillator and second as phase inverter. Provides equal-amplitude opposite-polarity r-f signals for driving push-pull output IC amplifier. Draws about 25 mA from 6-V lantern battery.—E. Noll, Integrated Circuits, *Ham Radio*, June 1971, p 40–47.

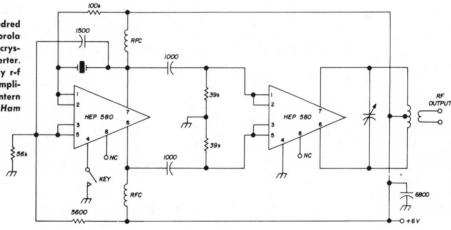

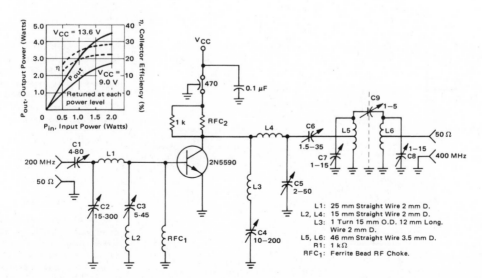

200–400 MHZ DOUBLER—Designed for use with 12.5-V or 28-V battery-powered transmitters. Uses sharp filter having insertion loss of 0.5 dB at 400 MHz and 0.6 dB at 450 MHz. Will operate satisfactorily on supplies down to 9 V.—B. Becciolini, "Using Balanced Emitter Transistors in RF Applications," Motorola Semiconductors, Phoenix, AZ, AN-521, 1970.

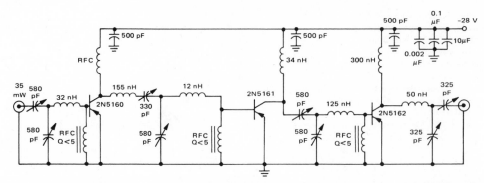

175-MHZ 30-W—Uses pnp balanced-emitter transistors that resist burnout during mis-matching and detuning. Requires 28-V supply.—B. Becciolini, "Using Balanced Emitter Transistors in RF Applications," Motorola Semiconductors, Phoenix, AZ, AN-521, 1970.

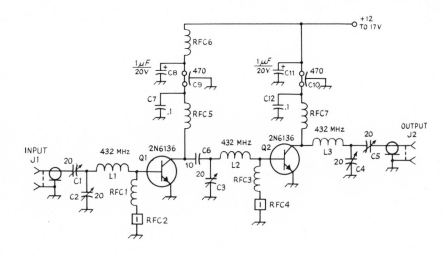

40 W AT 432 MHZ—Two Motorola 2N6136 transistors in strip-line configuration provide 12 dB overall gain while delivering 40 W. Requires regulated power supply rated at 5 A.—J. Buscemi, A 75-Watt Solid-State UHF Amplifier, QST, Oct. 1972, p 31–34.

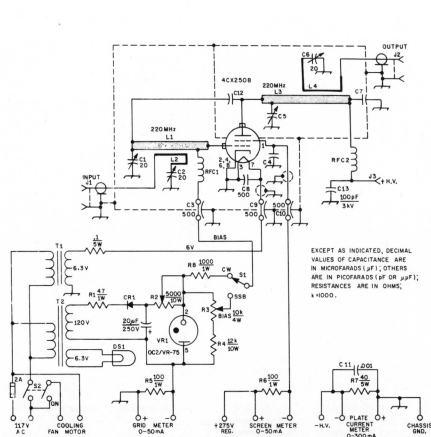

EXCEPT AS INDICATED, DECIMAL VALUES OF CAPACITANCE ARE IN MICROFARADS (μF); OTHERS ARE IN PICOFARADS (pF OR μμF); RESISTANCES ARE IN OHMS; k =1000.

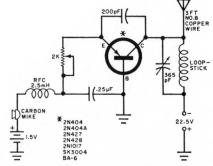

PARKING SCHOOL BUS—Developed to talk to driver of school bus via his a-m auto radio over range of 50 feet, to guide him in parking buses close together in too-small parking lot. Will not exceed FCC radiation limits.—C. J. Schauers, Wireless AM Mike, Popular Electronics, Feb. 1968, p 77.

230 W AT 220 MHZ—Uses single tube operating with 1,200 V on plate and 275 V on screen. Article covers construction of all coils. CR1 is 280-V 1-A silicon diode. Efficiency is over 70% for c-w. Requires 100-cfm cooling blower. Grid circuit is half-wave line, and plate circuit is quarter-wave coaxial tank.—D. V. Watters, A Coaxial-Line Amplifier for 220 MHz, QST, May 1972, p 27–29 and 47.

AUTOMATIC LINE TUNER—Eliminates need for correcting transmission-line match each time transmitter frequency is changed. Coil picks up coax current for summing with fraction of coax voltage. Resultant of these signals, detected by CR1 and CR2, produces d-c voltage that is positive, negative, or zero depending on current-voltage phase. Article gives instructions for winding coil on inner conductor of coax. Output of phase-sensitive demodulator is boosted by transistors sufficiently to energize relays K1 and K2 which control reversible d-c motor that drives tuning capacitor of matching network.—W. F. Stortz, The Automatic Transmission Line Tuner, 73, Feb. 1972, p 107–112.

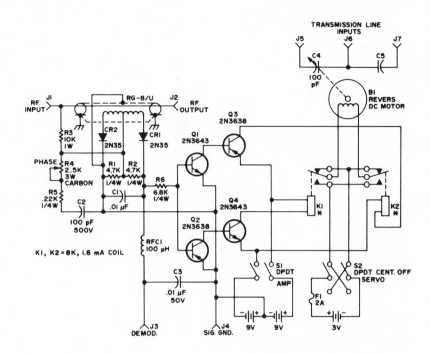

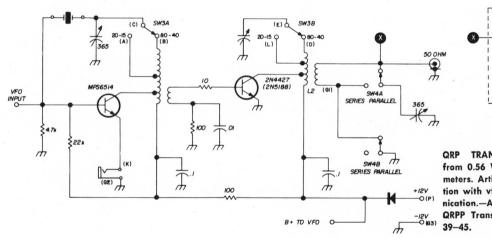

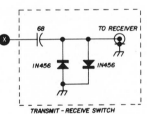

QRP TRANSMITTER—Power output ranges from 0.56 W on 15 meters to 1.87 W on 80 meters. Article covers construction and operation with vfo for low-power amateur communication.—A. Weiss, A Multiband FET VFO QRPP Transmitter, Ham Radio, July 1972, p 39–45.

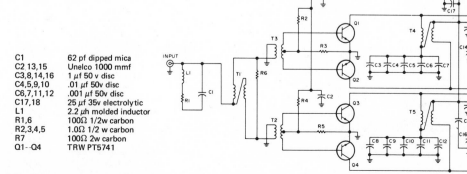

C1	62 pf dipped mica
C2,13,15	Unelco 1000 mmf
C3,8,14,16	1 μf 50 v disc
C4,5,9,10	.01 μf 50v disc
C6,7,11,12	.001 μf 50v disc
C17,18	25 μf 35v electrolytic
L1	2.2 μh molded inductor
R1,6	100Ω 1/2w carbon
R2,3,4,5	1.0Ω 1/2 w carbon
R7	100Ω 2w carbon
Q1--Q4	TRW PT5741

100-W SOLID-STATE LINEAR—Consists essentially of two 50-W push-pull amplifiers fed in phase with each other through T1. Can be driven by 25-W class A amplifier. Will operate over range of 1.5 to 30 MHz. Article gives transformer data.—100 Watts Solid State Linear!, 73, Nov. 1972, p 8–10.

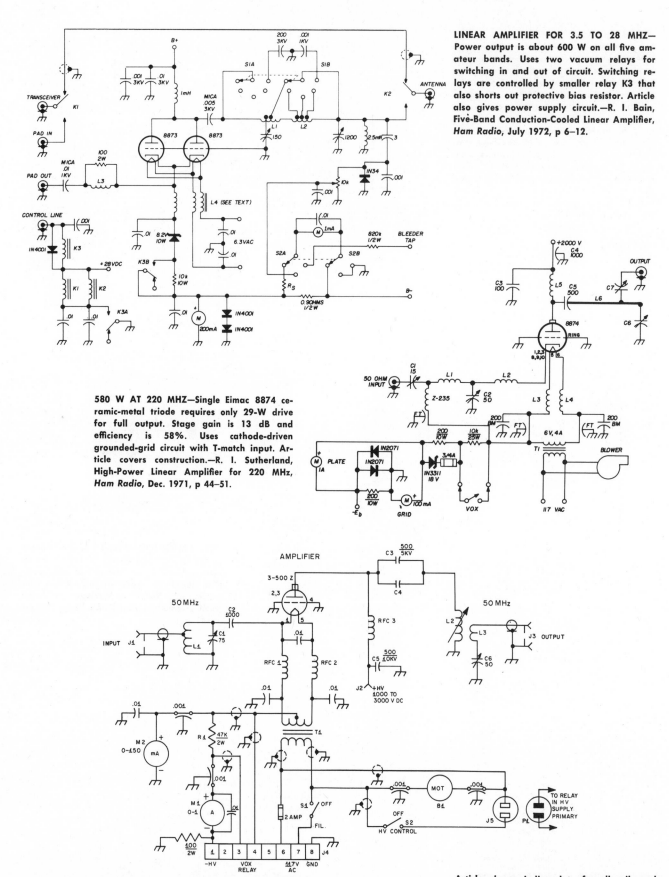

LINEAR AMPLIFIER FOR 3.5 TO 28 MHZ—Power output is about 600 W on all five amateur bands. Uses two vacuum relays for switching in and out of circuit. Switching relays are controlled by smaller relay K3 that also shorts out protective bias resistor. Article also gives power supply circuit.—R. I. Bain, Five-Band Conduction-Cooled Linear Amplifier, Ham Radio, July 1972, p 6–12.

580 W AT 220 MHZ—Single Eimac 8874 ceramic-metal triode requires only 29-W drive for full output. Stage gain is 13 dB and efficiency is 58%. Uses cathode-driven grounded-grid circuit with T-match input. Article covers construction.—R. I. Sutherland, High-Power Linear Amplifier for 220 MHz, Ham Radio, Dec. 1971, p 44–51.

600 W AT 50 MHZ—Simple circuit uses blower-cooled zero-bias triode capable of delivering 600 W on c-w with 30-W driving power. Maximum pep output is 750 W as class-B linear for single tone. If voice-controlled relay is used in exciter, it shunts out R1, allowing grid current to flow and make amplifier operative. Blower B1 should be at least 15 cfm.

Article gives winding data for all coils and chokes. T1 is 5-V 15-A filament transformer. Designed for 50-ohm load at J3.—T. F. McMullen, Jr. and E. P. Tilton, A 3-500Z Grounded-Grid Amplifier for 50 MHz, QST, Nov. 1970, p 24–27 and 58.

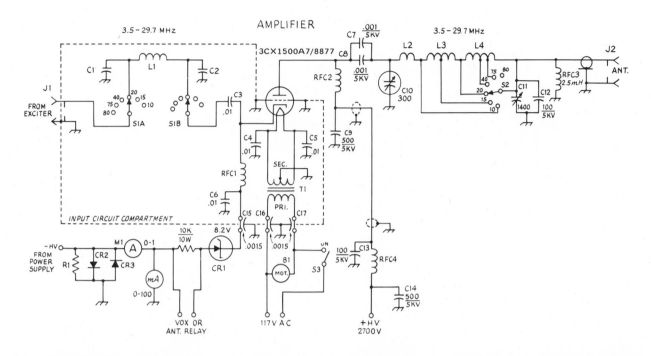

AMPLIFIER

2-KW GROUNDED-GRID AMPLIFIER—Article gives values of C1, C2, and L for operation at six different frequencies between 3.5 and 28 MHz. C10 is vacuum variable 5–300 pF. C11 is 4-section bc tuning with all sections in parallel. Article gives all coil data. Requires cooling blower.—M. B. Parten, Custom Design and Construction Techniques for Linear Amplifiers, QST, Sept. 1971, p 24–31.

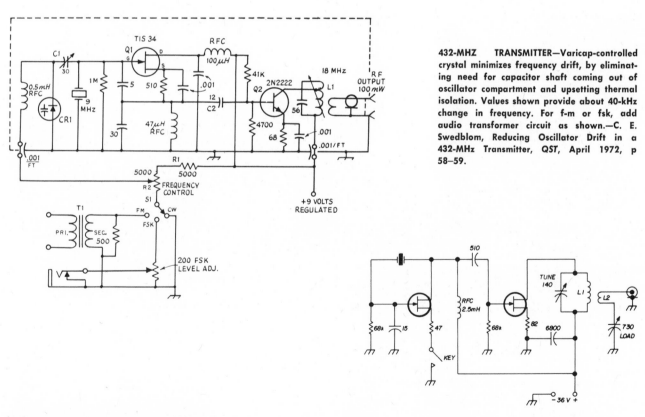

432-MHZ TRANSMITTER—Varicap-controlled crystal minimizes frequency drift, by eliminating need for capacitor shaft coming out of oscillator compartment and upsetting thermal isolation. Values shown provide about 40-kHz change in frequency. For f-m or fsk, add audio transformer circuit as shown.—C. E. Swedblom, Reducing Oscillator Drift in a 432-MHz Transmitter, QST, April 1972, p 58–59.

3-BAND QRP—Uses only two Siliconix 2N3970 switching fet's to give outputs up to ½ W on three bands shown and somewhat less on 20 meters. Current drain is about 75 mA from three 12-V lantern batteries in series.—E. Noll, Integrated Circuits, Ham Radio, June 1971, p 40–47.

	160	80	40
L1	65 turns no. 26 on 1-1/4" coil form	40 turns no. 24 on 1-1/4" coil form	21 turns no. 22 on 1-1/4" coil form
L2	20 turns no. 26, bifilar wound on cold end of L1	13 turns no. 24, bifilar wound on cold end of L1	7 turns no. 22, bifilar wound on cold end of L1

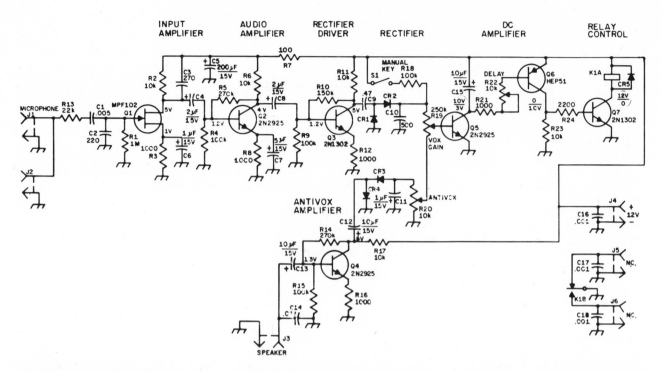

VOX—Voice-operated relay provides automatic transmit-receive switching. Delay circuit holds relay K1 closed for short time after audio signal ceases, to prevent relay from opening during short pauses between words or syllables. C15 and R22 set length of delay. Antivox amplifier prevents station receiver from operating vox. Use of 12-V reed relay (Magnecraft W104MX-2) for K1 minimizes turn-on delay, but antenna changeover relay of transmitter can still cause loss of first syllable or first c-w dot. CR1–CR4 are 1N67A or similar germanium. CR5 is 50-prv silicon. —D. A. Blakeslee, A Solid-State VOX, QST, Sept. 1970, p 11–14 and 49.

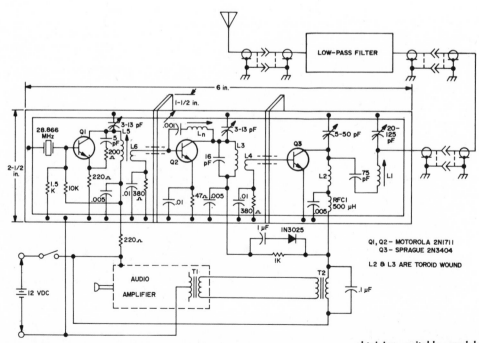

28–50 MHZ MOBILE—Will work on either 28-MHz or 50-MHz band. Final uses pair of tank circuits in parallel. Designed for operation from cigar lighter of car having 12.4-V battery with negative ground. Since most IC 1-W 12.4-V audio amplifiers have positive ground, shield connections for volume control and microphone are important. Difficulty in obtaining suitable modulation transformer was solved by wiring 1-W output transformer back to back with transformer of regular amplifier. Antenna is mobile loaded whip.—L. C. Maurer, Jr., 10M Solid State, 73, Nov. 1972, p 207–209.

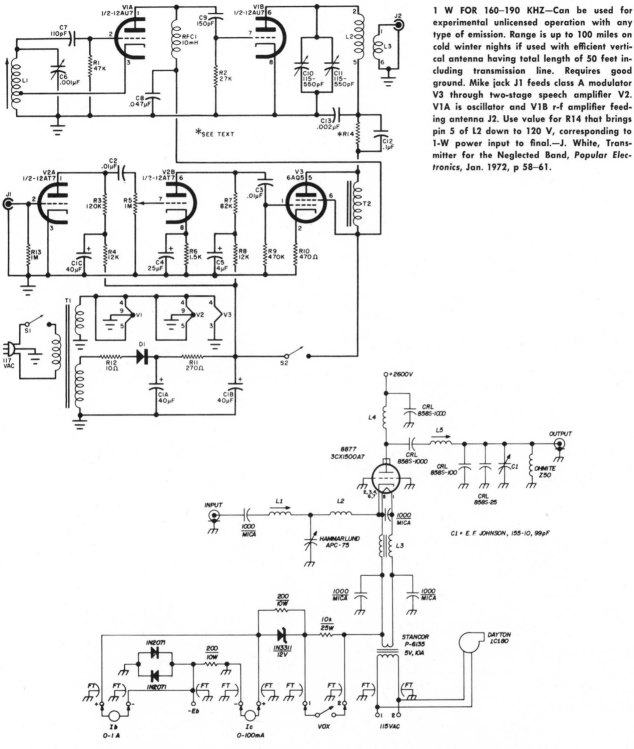

1 W FOR 160–190 KHZ—Can be used for experimental unlicensed operation with any type of emission. Range is up to 100 miles on cold winter nights if used with efficient vertical antenna having total length of 50 feet including transmission line. Requires good ground. Mike jack J1 feeds class A modulator V3 through two-stage speech amplifier V2. V1A is oscillator and V1B r-f amplifier feeding antenna J2. Use value for R14 that brings pin 5 of L2 down to 120 V, corresponding to 1-W power input to final.—J. White, Transmitter for the Neglected Band, *Popular Electronics*, Jan. 1972, p 58–61.

L1 6 turns no. 18 on a CTC 1538-4-3 form; coil length 7/8"

L2 6 turns no. 18, ½" diameter, 5/8" long, self-supporting

L3 Bifilar wound choke, ½" diameter core, 3" long, each coil 12 turns no. 10 Formvar; core is Indiana General CF-503

L4 54 turns no. 20 enameled on 1/2" diameter Teflon rod; winding length 1-13/16"

L5 3 turns 3/8" diameter copper tubing; inside diameter 1-7/8"; coil length 2-3/8"; shorted turn 2-¼" diameter 3/8" copper tubing ¼" from main coil

FT Erie 327 1000-pF feedthrough capacitors

6-METER 2-KW PEP LINEAR—Single Eimac tube in high-performance 50-MHz circuit provides excellent stability, good reliability, and minimum harmonic output even on 24-hour continuous key-down operation. Intended for maximum legal power input of 1,000-W d-c for amateur transmitters, corresponding to 2,000-W pep for ssb operation. Power gain is 14.8 dB.—R. I. Sutherland, Two-Kilowatt Linear Amplifier for Six Meters, *Ham Radio*, Feb. 1971, p 16–20.

CHAPTER 126
Trigger Circuits

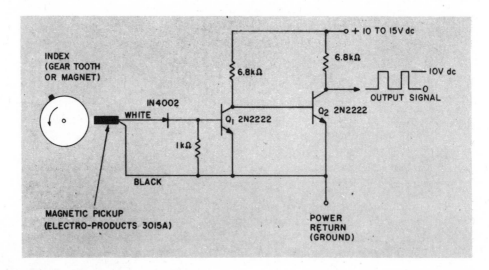

SHAPING MAGNETIC-PICKUP TRIGGER—Circuit is triggered by signal generated as gear tooth or other indexing device rotates past magnetic pickup. Output is fixed-amplitude square wave of about 10 V, with pulse width determined by rotational velocity and tooth size of indexing device.—E. N. Kaufman, Trigger Amplifies/Shapes Magnetic Pickup Output, *Electro-Technology*, Jan. 1970, p 30.

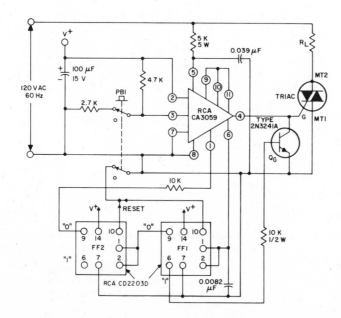

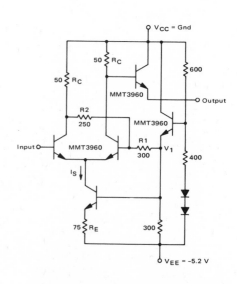

POWER ONE-SHOT—Provides zero-voltage switching logic required for solenoid drive of impulse hammer, stapling gun, and other applications where load current flow is required for only one complete half-cycle. Can also be adapted for applications requiring contact bounce suppression. Triggering of triac is initiated near zero voltage even though switch PB1 is pressed randomly during a-c cycle. Uses RCA IC comparator and two flip-flops. —"Linear Integrated Circuits," RCA, Somerville, NJ, IC-42, 1970, p 305–307.

CURRENT-MODE SCHMITT—Differential amplifier with feedback gives trigger having 500-mV input hysteresis for input thresholds of —1.5 V and —2 V. Used in driving high-speed flip-flops.—B. Broeker, "Micro-T Packaged Transistors for High Speed Logic Systems," Motorola Semiconductors, Phoenix, AZ, AN-536, 1970.

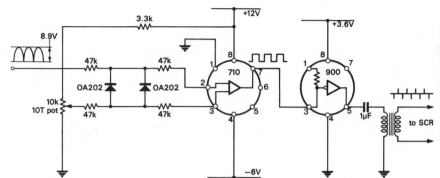

ADJUSTABLE-PHASE TRIGGER—Used to trigger scr synchronously with supply frequency, but with phase variable between 90 and 180 deg, in voltage sweep generator developed for transistor, diode, and thyristor curve-tracing. IC comparator detects time at which line voltage crosses selected level, and produces sharp pulse which is amplified and used to trigger thyristor through pulse transformer.—K. E. Forward, A Voltage Sweep Generator, Electronic Engineering, March 1970, p 44—46.

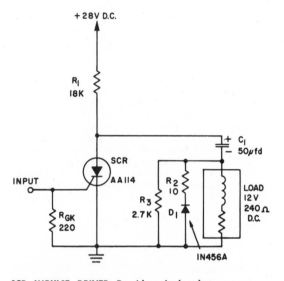

SCR IMPULSE DRIVER—Provides single electromechanical actuation as required by counter, print hammer, stapler, or other device requiring output current pulse of up to 10 A for 10 ms, as obtained by capacitor discharge. C1 is normally charged to supply voltage through R1. When scr is triggered on, C1 discharges through load. Scr turns off after input signal is terminated. Charging time constant of R1-C1 is about 0.9 s, which may be drawback in applications requiring high repetition rates.—"SCR Impulse Driver for Solenoid Actuation," Unitrode Corp., Watertown, MA, New Design Idea No. 15.

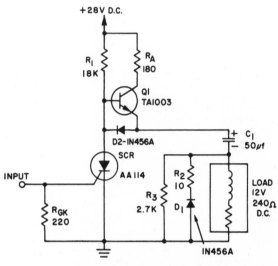

SCR IMPULSE DRIVER—Addition of transistor Q1 to scr circuit for providing single capacitor discharge through solenoid or other one-shot load speeds up recharging time of C1 to 10 ms, permitting repetition rates even faster than time required for counter solenoid to return to de-energized position. C1 is normally charged to supply voltage through R1, and discharges through load when scr is triggered on. Values shown give output current pulse up to 10 A for 10 ms.—"SCR Impulse Driver for Solenoid Actuation," Unitrode Corp., Watertown, MA, New Design Idea No. 15.

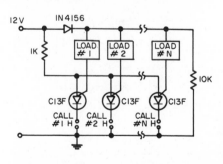

ANTICOINCIDENCE—Circuit provides ground only to load selected first. Load that is driven deprives all others of triggering voltage. Will handle any number of loads, at supply voltages up to 50 V and load currents up to 50 mA.—W. H. Sahm, III and F. M. Matteson, "The Complementary SCR," General Electric, Semiconductor Products Dept., Auburn, NY, No. 90.94, 1970, p 14.

SCHMITT—Can be used as regenerative squaring amplifier or d-c level-sensing circuit. When input voltage exceeds certain negative value, output switches to —12 V, and switches back to —6 V when input is reduced below threshold value. Hysteresis (difference between switch-on and switch-off) can be reduced by increasing values of R6 and R2, but switching time is then increased.—"P-N-P Practical Planar Transistors—BCY70 Family and BFX29 Family," Mullard, London, 1967, TP887, p 7.

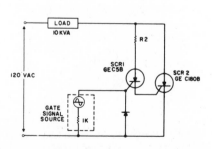

GATE TRIGGER AMPLIFIER—Uses scr with highly sensitive gate as amplifier for triggering high-power scr capable of handling 10-kVA load. Choose value of R2 to limit SCR1 to rated current.—"SCR Manual," General Electric, Syracuse, NY, 1972, 5th Ed., p 119—120.

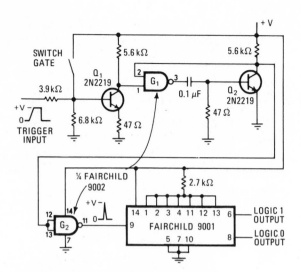

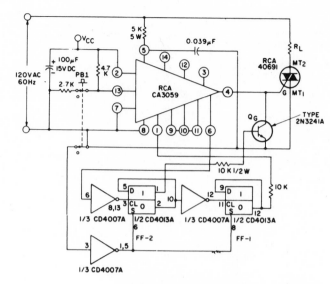

POSITIVE-GOING SPIKE—Hybrid one-shot uses NAND gates G1 and G2 to convert slowly rising input pulse to sharp, transient-free trigger having rise time of 10 to 100 ns and duration of 10 μs. Output from G2 may be used to trigger high-speed TTL logic such as 9001 J-K flip-flop shown.—R. J. Manco, One-Shot Makes Fast Trigger out of Slow Input Pulse, *Electronics*, Sept. 27, 1971, p 60.

**POWER ONE-SHOT—Pushing PB1 opens circuits of IC and flip-flop to produce triggering of triac near zero voltage on next alternation of a-c line for one complete half-cycle, which can be either positive or negative alternation. Used for one-shot solenoid drive of such devices as stapling guns and impulse hammers. Operation of circuit is described in detail in book.—"Linear Integrated Circuits and MOS Devices," RCA, Somerville, NJ, SSD-202A, 1973 Ed., p 232–233.

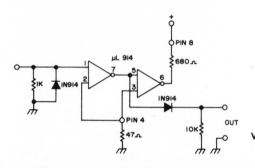

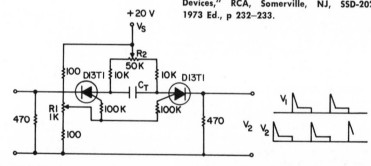

SCHMITT GIVES A-F SQUARE WAVES—Input can be 3-V or higher a-f source. Uses dual two-input gate, with power from pair of C cells. Supply should have on-off switch.—A. S. Joffe. IC Logic Doesn't Have to Be Obscure, *73*, Nov. 1972, p 217–219.

FLIP-FLOP TRIGGER—Two put relaxation oscillators coupled as shown can be used in inverter to deliver trigger pulses to two scr's alternately. R1 adjusts frequency and R2 trims waveforms for symmetry. One D13T1 put is always off when other is on, due to action of CT. Report gives tabulation of char-

acteristics of seven types of triggers. 2N6027 can be used for D13T1.—J. M. Reschovsky, "Design of Triggering Circuits for Power SCR'S," General Electric, Semiconductor Products Dept., Auburn, NY, No. 200.54, 1970, p 3.

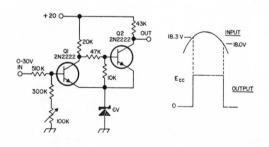

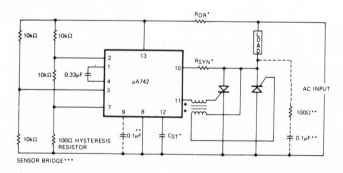

LOW-HYSTERESIS SCHMITT—Replacement of usual feedback resistor with 6-V zener reduces difference between turn-on and turn-off thresholds to 0.3 V or about 1.6%, as compared to 4% with resistor. Improvement is achieved without sacrifice of output-wave rectangularity.—D. Jensen, Some Ideas on Noise-Free CW Reception, *73*, Feb. 1966, p 84 and 86–88.

INVERSE-PARALLEL SCR FIRING—Single IC provides seven functions required for on-off control of inductive or resistive a-c load at zero-current point of a-c cycle, with minimum rfi generation. Choose scr types to handle load. Use dashed circuit only with inductive load. Triggering is provided by IC through pulse transformer. Arrangement handles much

larger loads than triacs. For 115-V a-c line, RDR and RSYN are 10K and CST is 0.47 μF. Report gives IC circuit and describes operation in detail.—A. Adamian and L. Blaser, "A Monolithic Zero-Crossing AC Trigger (Trigac) for Thyristor Power Controls," Fairchild Semiconductor, Mountain View, CA, No. 208, 1972, p 4–5.

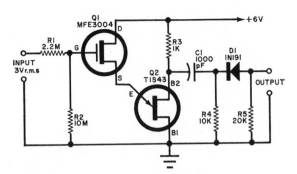

FAST SCHMITT—Will trigger on about 3-V rms, with driving power under 0.75 μW. Loading effect on input is negligible. Serves to convert a-c input into constant-level output pulse train of same frequency.—F. H. Tooker, Micro-Sensitive Schmitt Trigger, *Popular Electronics*, Jan. 1970, p 64–65.

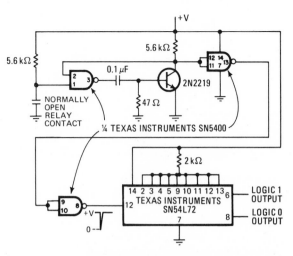

NEGATIVE-GOING SPIKE—Relay contact provides input trigger pulse for sharpening and inverting by three NAND gates to give 10-μs negative spike for triggering SN54L72 J-K master-slave flip-flop.—R. J. Manco, One-Shot Makes Fast Trigger out of Slow Input Pulse, *Electronics*, Sept. 27, 1971, p 60.

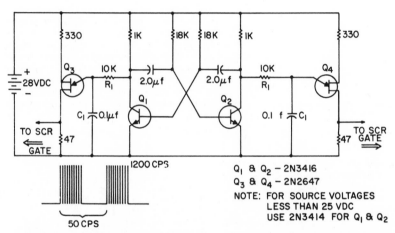

Q_1 & Q_2 – 2N3416
Q_3 & Q_4 – 2N2647
NOTE: FOR SOURCE VOLTAGES
LESS THAN 25 VDC
USE 2N3414 FOR Q_1 & Q_2

PULSE-TRAIN TRIGGER—Maintains trigger drive during conducting period of scr by means of pulse train, to reduce average gate dissipation. Mvbr Q1-Q2 provides alternate driving voltages to ujt oscillators which generate 50 trains per second of 1,200-Hz pulses. —J. M. Reschovsky, "Design of Triggering Circuits for Power SCR'S," General Electric, Semiconductor Products Dept., Auburn, NY, No. 200.54, 1970, p 7–8.

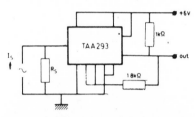

CURRENT-OPERATED SCHMITT IC—Uses general-purpose Amperex three-transistor IC amplifier. When input current Is is zero, circuit rests in high state at output of about 4 V. At input of about 26 μA, circuit switches to low state, and reverts to high state at about 9 μA. Output loading is 4 mA.—"The Amperex TAA 293 As a Schmitt Trigger and Multivibrator," Amperex Electronic Corp., Slatersville, RI, Application Report S-147.

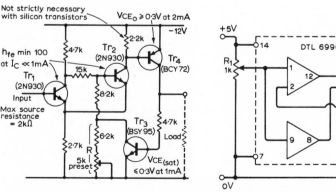

SCHMITT WITH BACKLASH CONTROL—Addition of level shifter Tr3 and electronic switch Tr4 to basic Schmitt trigger Tr1-Tr2 permits adjusting R to reduce backlash to zero. When input is above upper trip point, Tr1 is on, other three transistors are off, and R is in circuit. When input is below lower trip, Tr1 is off, others are on, and R is shorted out. —A. E. Crump, Schmitt Trigger with "Zero" Backlash, *Wireless World*, June 1969, p 269.

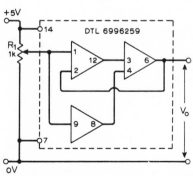

NAND-GATE TRIGGER—Output logic level Vo depends on setting of R1, and is at logic 1 for input of 5 V. At critical lower input voltage, gates switch over and output is at logic 0, staying there until input is raised above critical value of 1.35 V for circuit shown. Used in place of Schmitt trigger in voltage-level detector.—W. E. Price, Single I.C. Trip Unit, *Wireless World*, April 1971, p 204.

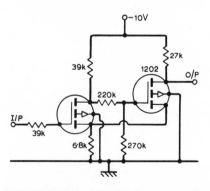

HIGH-IMPEDANCE SCHMITT—Uses two silicon-gate field-effect transistors (G.E.C. Semiconductors, England, or similar) in conventional Schmitt circuit to achieve high input impedance. Resistor in series with input allows input to have positive polarity. For negative-going input, input current is less than 100 pA. With values shown, upper trip point is 4 V and lower is 3.1 V, with rise and fall times under 1 μs.—J. A. Roberts and J. Driscoll, High Input-Impedance Schmitt Trigger, *Wireless World*, Sept. 1971, p 430.

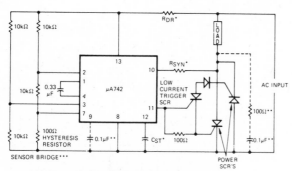

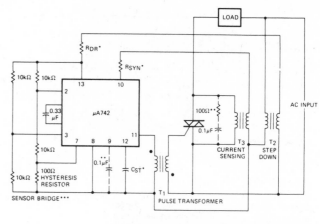

SCR TRIGGERS SCR'S—Single IC acts through low-current scr's for on-off control of power scr pair at zero-current point of a-c cycle, to minimize rfi. Dashed circuit is required with inductive load. Choose scr type to handle load. For 115-V a-c line, RDR and RSYN are 10K and CST is 0.47 μF. Report gives IC circuit and describes operation in detail.—A. Adamian and L. Blaser, "A Monolithic Zero-Crossing AC Trigger (Trigac) for Thyristor Power Controls," Fairchild Semiconductor, Mountain View, CA, No. 208, 1972, p 4–5.

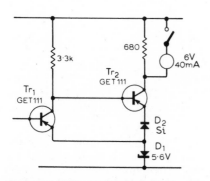

FEEDBACK WITH ZENER—With SX56 zener for D1 and surplus silicon diode for D2, trigger points are 5.35 V and 5.65 V with lamp, and 5.35 V and 5.5 V without lamp. Zener serves in place of common-emitter resistor in trigger for variable loads.—P. Gascoyne, Schmitt Triggers, *Wireless World*, July 1970, p 316.

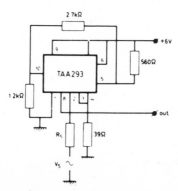

VOLTAGE-DRIVEN SCHMITT IC—Uses general-purpose Amperex three-transistor IC amplifier. With zero input voltage, circuit rests in high state with output of 6 V. If source impedance is 10K, circuit switches to low state when input Vs goes up to about 1.25 V, and reverts to high state when input drops back to about 0.8 V. Rise time is about 200 ns and fall time about 40 ns.—"The Amperex TAA 293 As a Schmitt Trigger and Multivibrator," Amperex Electronic Corp., Slatersville, RI, Application Report S-147.

ISOLATED INDUCTIVE LOAD—Single IC acts through pulse transformer for on-off control of triac at zero-current point of a-c cycle, to minimize rfi. Use of current-sensing transformer for supplying sync input to IC permits use with inductive or resistive loads. Inductive load requires dashed circuit. Choose triac to handle load. For 115-V a-c line, RDR and RSYN are 10K and CST is 0.47 μF. Report gives IC circuit and describes operation in detail.—A. Adamian and L. Blaser, "A Monolithic Zero-Crossing AC Trigger (Trigac) for Thyristor Power Controls," Fairchild Semiconductor, Mountain View, CA, No. 208, 1972, p 5.

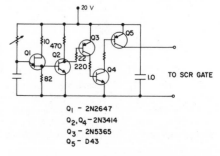

Q_1 — 2N2647
Q_2, Q_4 — 2N3414
Q_3 — 2N5365
Q_5 — D43

FAST-RISING CURRENT LOADS—Generates rectangular pulses 10 μs wide at rates up to 20 kHz without using inductive element. Provides 20 V for triggering most scr's. Q1 operates as conventional ujt relaxation oscillator whose frequency is controlled by pot. Output of ujt drives four-transistor amplifier which improves rise time of pulses while extending their width.—J. M. Rechovsky, "Design of Triggering Circuits for Power SCR'S," General Electric, Semiconductor Products Dept., Auburn, NY, No. 200.54, 1970, p 9.

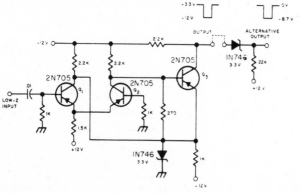

SYMMETRICAL SWITCH—Advantages over Schmitt trigger include excellent symmetry without hysteresis and higher sensitivity. Consists of current-mode inverter Q1-Q2 driving saturating inverter Q3. Rise time of output is only 50 ns.—D. Jensen, Some Ideas on Noise-Free CW Reception, 73, Feb. 1966, p 84 and 86–88.

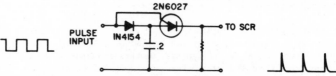

PUT PULSE SHARPENER—Converts slow-rising square-wave input pulse to trigger pulse having required fast rise time. When capacitor charges to peak voltage of input pulse, put switches on and delivers to scr gate a pulse having rise time of 50 to 100 ns.—"SCR Manual," General Electric, Syracuse, NY, 1972, 5th Ed., p 121.

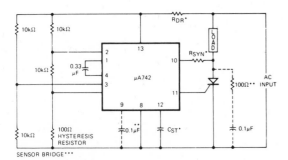

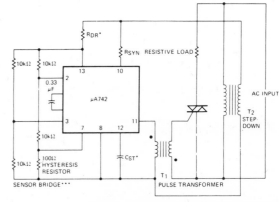

HALF-WAVE FIRING OF SCR—Single IC provides seven functions required for on-off control of inductive or resistive a-c load at zero-current point of a-c cycle, with minimum rfi generation. Choose scr to handle load. Use dashed circuit only with inductive load. For 115-V a-c line, RDR and RSYN are 10K and CST is 0.47 μF. Report gives IC circuit and describes operation in detail.—A. Adamian and L. Blaser, "A Monolithic Zero-Crossing AC Trigger (Trigac) for Thyristor Power Controls," Fairchild Semiconductor, Mountain View, CA, No. 208, 1972, p 4.

ISOLATED RESISTIVE LOAD—Single IC acts through pulse transformer for on-off control of triac at zero-current point of a-c cycle, to minimize rfi. Choose triac to handle load. For 115-V a-c line, RDR and RSYN are 10K and CST is 0.47 μF. Report gives IC circuit and describes operation in detail.—A. Adamian and L. Blaser, "A Monolithic Zero-Crossing AC Trigger (Trigac) for Thyristor Power Controls," Fairchild Semiconductor, Mountain View, CA, No. 208, 1972, p 5.

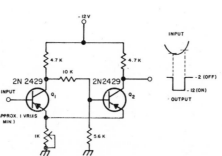

BASIC SCHMITT—Switching time is about 12 μs for 300-Hz input. Output has excellent rectangularity, but hysteresis causes about 10% difference between turn-on and turn-off thresholds.—D. Jensen, Some Ideas on Noise-Free CW Reception, 73, Feb. 1966, p 84 and 86–88.

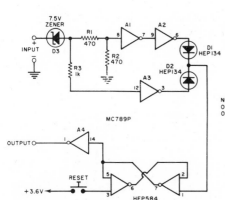

NOTE: DIODES D1, D2 OPERATE AS A 2-INPUT OR GATE

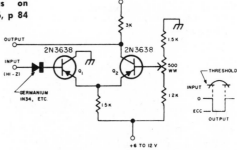

PROCESS - CONTROLLED TRIGGER—Designed to monitor d-c level of 9 V obtained from pressure, temperature, fluid, or other type of sensor. Circuit triggers on level increases or decreases of as little as 0.5 V, to make output terminal go from zero level to slightly more than 3 V positive. Once triggered, circuit remains latched until input voltage is returned to normal and reset button is pressed. Both IC's operate from same 3.6-V supply as reset button. Only four of inverters in MC789P hex inverter are used; inputs of others should be grounded. Dual 2-input gate is connected as R-S flip-flop used as latch.—F. H. Tooker, Up-Down Level-Sensitive Trigger, Electronics World, Oct. 1970, p 69.

AMPLITUDE-DIFFERENCE AMPLIFIER—Characteristics are similar to Schmitt trigger but with no hysteresis. Rise time of rectangular output wave is 800 μs at 300 Hz. Diode prevents shorting of power supply on negative half-cycles of input. Use 12-V supply for inputs of several volts, and 6 V for smaller inputs.—D. Jensen, Some Ideas on Noise-Free CW Reception, 73, Feb. 1966, p 84 and 86–88.

CHAPTER 127
Ultrasonic Circuits

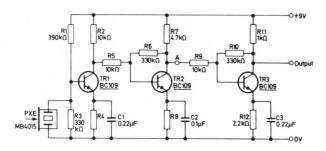

30–40 KHZ RECEIVER—Utilizes frequency-selective characteristic of Mullard PXE air-beam piezoelectric ceramic transducer to give bandwidth of about 3 kHz for center frequency in design frequency range. Gain at center frequency, near antiresonance frequency of transducer, is about 20 dB. R4 is 10K and R8 6.8K for three-stage receiver shown; for two stages (output at A), R4 is 12K and R8 is 15K.—"Designing Ultrasonic Systems Using PXE Air Beam Transducers," Mullard, London, 1972, TP1343, p 8–9.

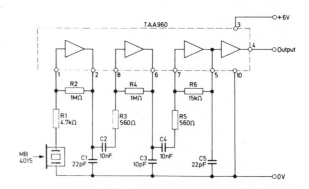

30–40 KHZ IC RECEIVER—Provides gain of about 20 dB, using Mullard IC in place of conventional three-transistor amplifier. Input is Mullard PXE air-beam piezoelectric ceramic transducer.—"Designing Ultrasonic Systems Using PXE Air Beam Transducers," Mullard, London, 1972, TP1343, p 8–9.

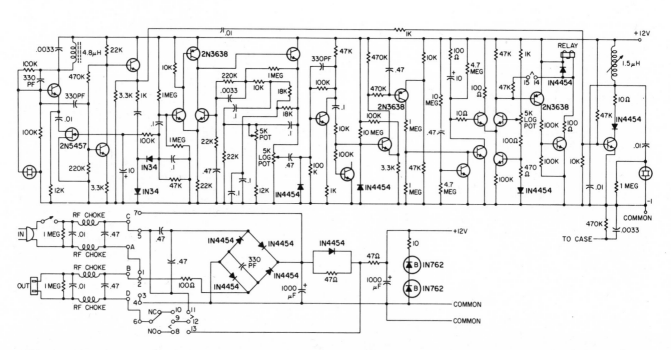

AREA INTRUDER ALARM—Combination ultrasonic transmitter-receiver having crystal transducers at opposite ends detects any movement in protected area. Used in Delta Products model 10,000 alarm system. Transistors not marked are 2N3414. Operates in range of 18 to 45 kHz. Movement of person shifts frequency about 40 Hz, enough to trigger alarm connected to contacts of relay in phase detector circuit. System is susceptible to false alarms caused by moving drapes, vibration of heavy street traffic, cats, or even mice.—J. Squires, Stop Burglars with Electronics, *Radio-Electronics*, Nov. 1971, p 23–27 and 94–95.

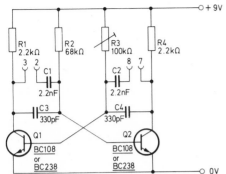

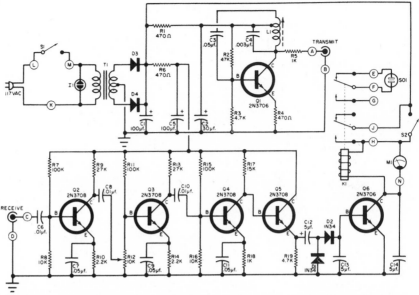

40-KHZ TRANSMITTER—Transducer used is 40-kHz piezoelectric ceramic bimorph plate connected to terminals 7 and 8, such as Mullard MB4015 39-kHz encapsulated element, which is less fragile than Mullard MB7010 bimorph plate. Can be used to demonstrate wave properties such as diffraction and interference, using gratings cut out of cardboard since wavelength at 40 kHz is just under 1 cm. With 6-kHz Mullard MB4013 piezoelectric ceramic transducer connected to terminals 2 and 3, circuit also produces 6-kHz sound waves for measuring effectiveness of soundproofing materials and in noise-pollution demonstration projects. Uses mvbr as square-wave generator for driving transducers at antiresonant frequencies.—"Sound and Ultrasound," Educational Projects in Electronics, Mullard, London, 1972.

ULTRASONIC-BEAM INTRUSION ALARM—Oscillator Q1 generates ultrasonic signal at 25 kHz that is beamed through area being guarded to receiver Q2–Q6 that releases alarm relay K1 when beam is interrupted. Alarm is fail-safe against cut wires or shorts to alarm leads, as well as for power-line carrier. For indoor use only, because wind can disturb beam and trip alarm.—D. Meyer, Build the Ultrasonic Omni-Alarm, *Popular Electronics*, April 1966, p 41–45 and 82.

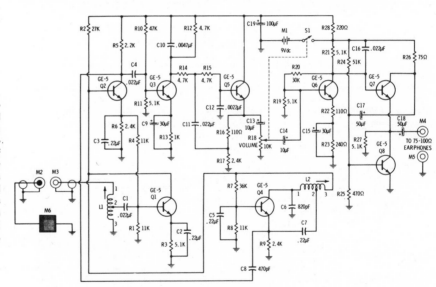

RECEIVER—Will pick up signals of keyed ultrasonic transmitter up to 1 mile away, and convert to normal audio range for monitoring with earphones. Can also be used for listening to sounds at frequencies above range of human audibility, but which can be heard by birds, dogs, cats, and other animals. Requires special ultrasonic mike M6, such as Admiral 78B147-1-G. L1 and L2 are ultrasonic coils such as Admiral 69C251-1-A.—R. M. Brown and T. Kneitel, "49 Easy Entertainment and Science Projects," Vol. 2, Howard W. Sams, Indianapolis, IN, 1969, p 18–21.

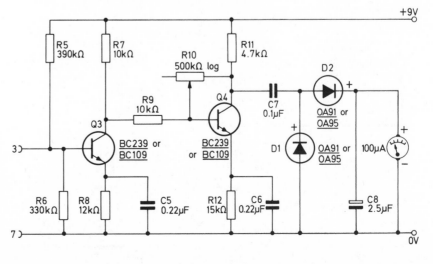

40-KHZ RECEIVER—Provides required high gain for low output from Mullard MB7010 bimorph plate connected to terminals 3 and 7. Designed for use in demonstration projects. Report covers construction and gives British sources for all parts. Can also be used with 6-kHz piezoelectric transducer for soundproofing and noise-pollution demonstrations.—"Sound and Ultrasound," Educational Projects in Electronics, Mullard, London, 1972.

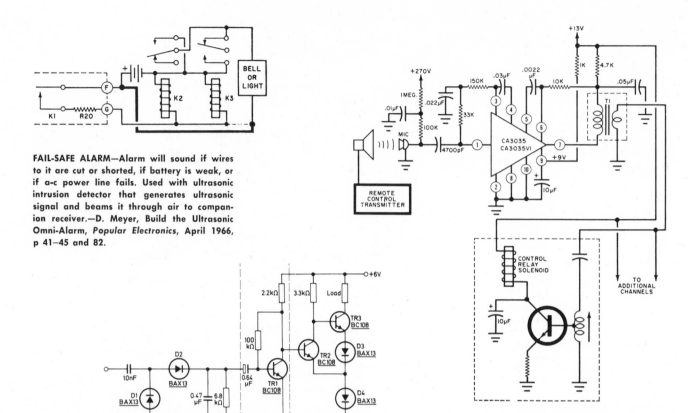

FAIL-SAFE ALARM—Alarm will sound if wires to it are cut or shorted, if battery is weak, or if a-c power line fails. Used with ultrasonic intrusion detector that generates ultrasonic signal and beams it through air to companion receiver.—D. Meyer, Build the Ultrasonic Omni-Alarm, *Popular Electronics*, April 1966, p 41–45 and 82.

LEVEL-CHANGE DETECTOR—Used in intrusion alarms and other ultrasonic receiver systems in which sudden change in received ultrasonic signal level, as by movement in protected area, will trigger alarm. Diode pump detector feeds differentiator giving output proportional to rate of change of signal level, for driving Schmitt trigger having alarm device as load. Unless load is self-latching, alarm stops when input level stops changing; add resettable flip-flop to keep alarm on until manually reset.—"Designing Ultrasonic Systems Using PXE Air Beam Transducers," Mullard, London, 1972, TP1343, p 12–13.

40-KHZ REMOTE-CONTROL RECEIVER—Pulls in control relay if 40-kHz audio transmitter or equivalent ultrasonic gong is activated within operational range. Dashed rectangle includes 40-kHz tuned circuit; similar circuits and relays for other ultrasonic frequencies may be added. Can be used as burglar alarm or for ultrasonic communication as well as for remote control.—D. Lancaster, Six Linear IC Applications, *Popular Electronics*, Dec. 1967, p 51.

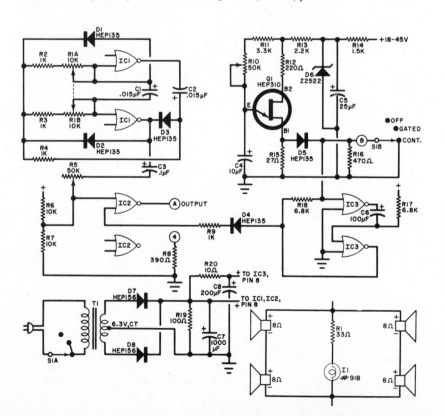

RAT EVICTOR—Generates 19.5-kHz 15-W ultrasonic signal that makes all rats within effective area of 225 square yards head for the hills. Also works on many types of insects, birds, and other small animals. Requires four 8-ohm tweeters connected in series-parallel to provide 8-ohm load for external power amplifier that should be capable of putting out at least 15 W of usable ultrasonic power. Single speaker will not carry full output without burning up in a few minutes; four speakers give three times as much acoustical output as one for same drive to amplifier. Author recommends University MS Supertweeter or equivalent, costing upwards of $20 each. All IC's are HEP584.—L. Greenlee, Electronic Pest Control, *Popular Electronics*, July 1972, p 47–50.

CHAPTER 128
Voltage-Controlled Oscillator Circuits

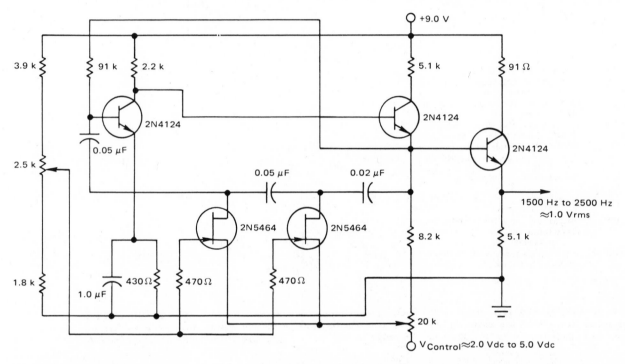

1.5–2.5 kHz PHASE-SHIFT FET—Produces good sine wave that varies linearly in frequency with control voltage.—B. Botos, "Low Frequency Applications of Field-Effect Transistors," Motorola Semiconductors, Phoenix, AZ, AN-511, 1971.

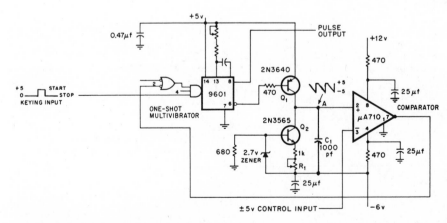

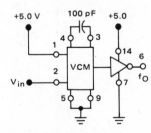

40:1 RANGE—Control input of −5 to +5 V changes pulse output frequency from 50 kHz to 4 MHz. Control level determines when IC comparator retriggers IC one-shot to start next cycle. For lower frequencies, down to 1 Hz, increase C1. Dropping keying input to zero stops oscillation. Can also be used as analog-digital converter or digital voltmeter. —E. Breeze, Comparator and Multivibrator Add up to a Linear VCO, *Electronics*, Aug. 17, 1970, p 90.

1–5 MHZ MVBR—Uses Motorola MC4324/4024 IC consisting of two independent voltage-controlled mvbr's with output buffers. Input d-c control voltage of 2.5 to 5.5 V covers entire frequency range. With smaller capacitor, maximum output frequency is 25 MHz.—"MC4324/MC4024 Dual Voltage-Controlled Multivibrators," Motorola Semiconductors, Phoenix, AZ, Data Sheet 9108, 1972.

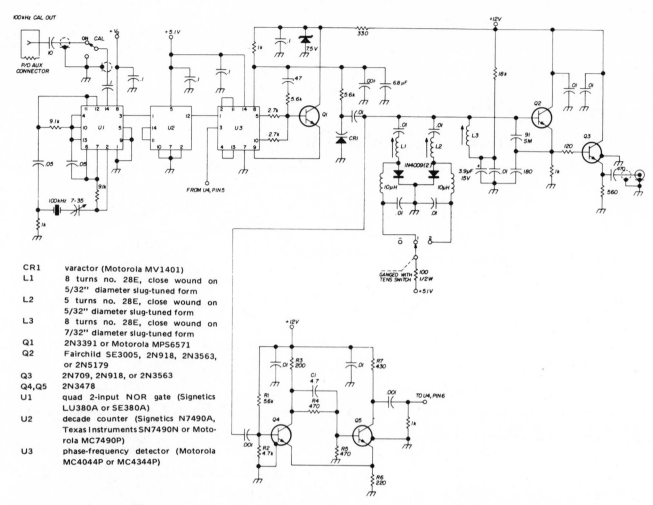

CR1   varactor (Motorola MV1401)
L1   8 turns no. 28E, close wound on 5/32" diameter slug-tuned form
L2   5 turns no. 28E, close wound on 5/32" diameter slug-tuned form
L3   8 turns no. 28E, close wound on 7/32" diameter slug-tuned form
Q1   2N3391 or Motorola MPS6571
Q2   Fairchild SE3005, 2N918, 2N3563, or 2N5179
Q3   2N709, 2N918, or 2N3563
Q4,Q5   2N3478
U1   quad 2-input NOR gate (Signetics LU380A or SE380A)
U2   decade counter (Signetics N7490A, Texas Instruments SN7490N or Motorola MC7490P)
U3   phase-frequency detector (Motorola MC4044P or MC4344P)

100-KHZ CRYSTAL AND VCO—Output of U1 is divided by ten in U2, and 20-kHz signal is applied to phase comparator U3. Q2 is vco using Colpitts circuit with varactor across tank. Used with frequency divider also shown in article, to cover entire tuning range from 1.5 to 30 MHz with only one crystal.—R. S. Stein, Frequency Synthesizer for the Drake R-4 Receiver, *Ham Radio*, Aug. 1972, p 6–19.

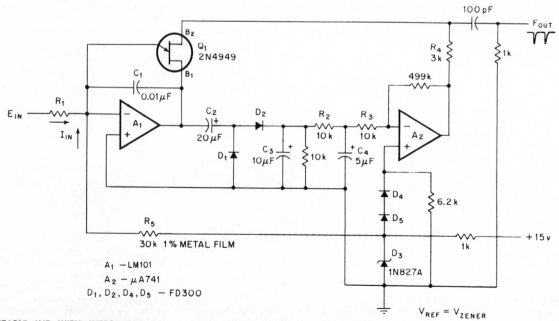

A₁ —LM101
A₂ —µA741
D₁,D₂,D₄,D₅ —FD300

3-KHZ STABLE UJT WITH INTEGRATOR—Although requiring no critical components, circuit gives better temperature stability than ujt vco and better linearity than ordinary integrating oscillator. Ujt Q1 is voltage-sensitive reset switch in integrator A1; comparator A2 stabilizes output, and keys period and level to reference voltage.—L. G. Smeins, Stable Unijunction VCO Needs No Critical Components, *Electronics*, Oct. 12, 1970, p 100.

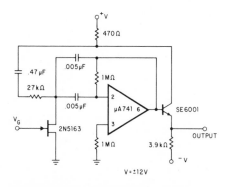

ACTIVE-FILTER VCO—Inverting output of opamp serving as active filter and feeding portion back to input gives a-f oscillator whose frequency can be varied between 200 and 3,200 Hz by applying negative d-c voltage of 0 to 5 V to terminal VG of fet. If control voltage is ramp voltage, sweep can be generated. Variations in output amplitude over tuning range can be minimized by inserting compressor amplifier in feedback loop. —N. Doyle, "Some Useful Signal Processing Circuits Using FET's and Operational Amplifiers," Fairchild Semiconductor, Mountain View, CA, No. 243, 1971, p 5.

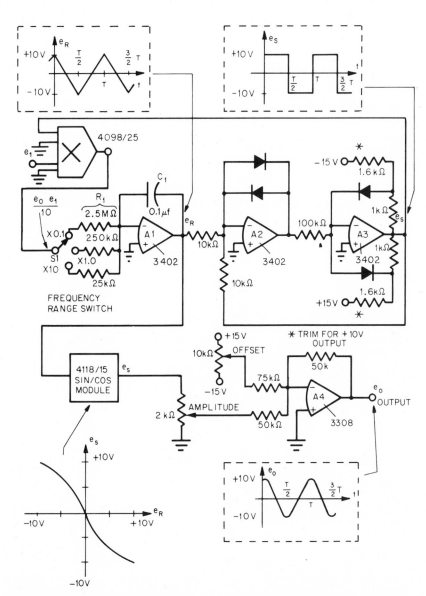

100:1 DYNAMIC RANGE—Burr-Brown opamp A1 generates triangle wave that can be controlled over 100:1 frequency range by A2 and A3 with 1% linearity, for any switch position. Frequency in Hz is equal to control voltage E1 divided by 40R1C1. Triangle wave is then put through shaping network that has sinusoidal gain. Operates from d-c up into high audio range.—J. G. Graeme, G. E. Tobey, and L. P. Huelsman, "Operational Amplifiers," McGraw-Hill Book Co., New York, 1971, p 403–404.

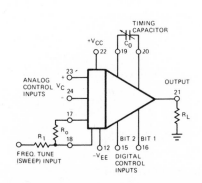

0.1 HZ TO 40 MHZ—Uses vco section of Exar XR-S200 IC, with frequency selected and controlled by three methods: (1) tuning to center frequency with C0, with free-running frequency inversely proportional to C0; (2) use of two digital control inputs to select four discrete frequencies at any center frequency; (3) applying sweep voltage through limiting 4K resistor R1 for frequency sweeping, on-off keying, and synchronization of vco to sync pulse. Voltage-to-frequency conversion of vco is highly linear. Conversion gain can be controlled through analog input, and is inversely proportional to Ro. Interfaces with ECL or TTL logic. Crystal may be used in place of C0.—"XR-S200 Multi-Function Integrated Circuit," Exar Integrated Systems Inc., Sunnyvale, CA, June 1972, p 5.

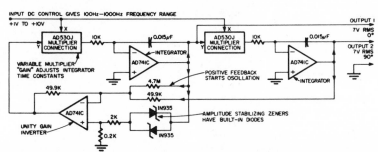

100–1,000 HZ WITH 1–10 V CONTROL—Cascaded integrators form two-phase oscillator, with unity-gain inverter providing additional 180-deg phase shift required for oscillation. Frequency is determined by time constants of IC multipliers, and 10:1 frequency range for d-c control can be anywhere in operating band from d-c to 1 MHz.—"AD530 Complete Monolithic MDSSR," Analog Devices, Inc., Norwood, MA, Technical Bulletin, July 1971.

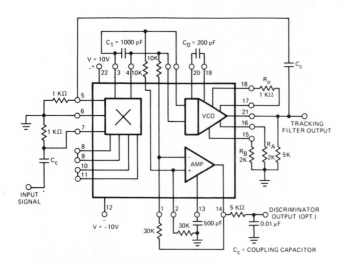

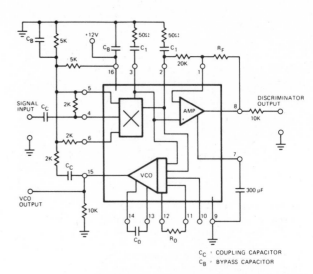

1-MHZ TRACKING FILTER—Uses Exar XR-S200 IC connected as phase-locked loop, to function as frequency filter when circuit locks on input signal. Produces filtered version of input frequency at vco output. Can track input over broad range of frequencies around free-running frequency of vco, for 3:1 input frequency range.—"XR-S200 Multi-Function Integrated Circuit," Exar Integrated Systems Inc., Sunnyvale, CA, June 1972, p 7.

TRACKING FILTER—Exar XR-215 phase-locked loop IC will track input signals over 3:1 frequency range centered about free-running frequency of vco. Ro is between 1K and 4K, and C1 is between 30 and 300 times value of timing capacitor C0 which depends on center frequency.—XR-215 Monolithic Phase-Locked Loop," Exar Integrated Systems Inc., Sunnyvale, CA, July 1972, p 8.

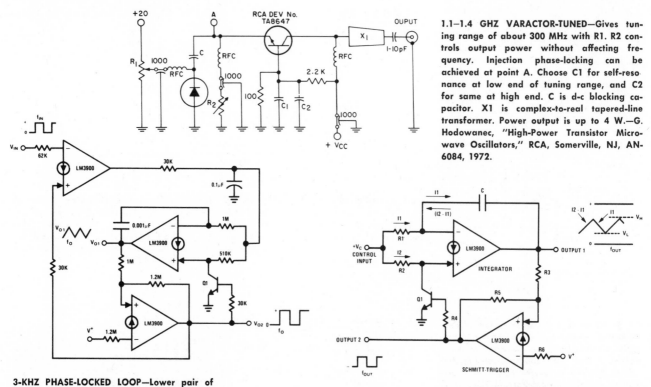

1.1—1.4 GHZ VARACTOR-TUNED—Gives tuning range of about 300 MHz with R1. R2 controls output power without affecting frequency. Injection phase-locking can be achieved at point A. Choose C1 for self-resonance at low end of tuning range, and C2 for same at high end. C is d-c blocking capacitor. X1 is complex-to-real tapered-line transformer. Power output is up to 4 W.—G. Hodowanec, "High-Power Transistor Microwave Oscillators," RCA, Somerville, NJ, AN-6084, 1972.

3-KHZ PHASE-LOCKED LOOP—Lower pair of Norton current-differencing opamps serve as vco and third opamp completes phase-locked loop having center frequency of about 3 kHz. Report covers design procedure for vco. Phase-locked loop can be used in tracking filters, frequency to d-c converters, f-m modulators and demodulators, and many control applications.—T. M. Frederiksen, W. M. Howard, and R. S. Sleeth, "The LM3900—A New Current-Differencing Quad of ± Input Amplifiers," National Semiconductor, Santa Clara, CA, 1972, AN-72, p 26.

VCO UP TO 10 KHZ—Uses Norton current-differencing opamps as integrator and Schmitt trigger providing square-wave output at frequency depending on d-c control voltage, trip voltages VH and VL of Schmitt circuit, and values of R1 and C. R2 is half of R1. Report gives design equations and describes operation in detail.—T. M. Frederiksen, W. M. Howard, and R. S. Sleeth, "The LM3900—A New Current-Differencing Quad of ± Input Amplifiers," National Semiconductor, Santa Clara, CA, 1972, AN-72, p 24—25.

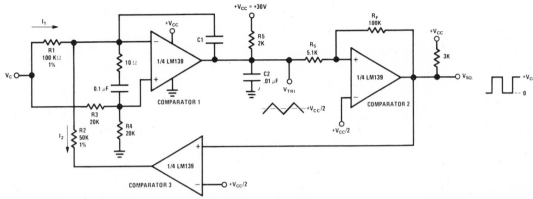

TRIANGLE AND SQUARE VCO—Covers frequency range of 670 Hz to 115 kHz with d-c control voltage from 250 mV to 50 V, with 1.5-V p-p triangle wave and 30-V p-p square wave when using 30-V supply for comparators. Report analyzes operation of circuit and tells how to increase or lower frequency limits.—R. T. Smathers, T. M. Frederiksen, and W. M. Howard, "LM139/LM239/LM339 A Quad of Independently Functioning Comparators," National Semiconductor, Santa Clara, CA, AN-74, 1973, p 8–9.

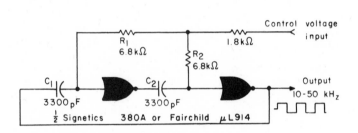

31.5-KHZ TV SYNC GENERATOR—Varying control voltage from 2.4 to 4.8 V changes output frequency from 10 to 50 kHz for component values shown. Choice of gates is not critical.—L. Toth, Two Gates = Voltage Controlled Oscillator, *The Electronic Engineer*, Jan. 1969, p 95.

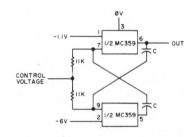

IC VCO—Frequency varies over 10:1 range as control voltage varies between 1 and 7 V. Both capacitors are same value and control center frequency. Minimum value is 110 pF, giving 2 MHz; maximum is 100 μF, giving 3 Hz. For 455-kHz i-f strip, use 470 pF.—J. Kyle, The Logical Approach to Surplus Buying, *73*, March 1970, p 80–92.

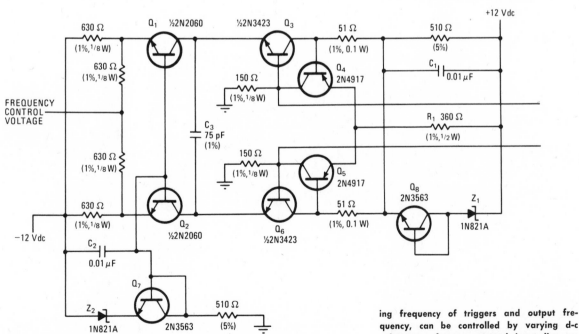

VOLTAGE-CONTROLLED TO 50 MHZ—Output frequency repeatability can be held, without adjustment, to within 5% range. Circuit is affected very little by temperature changes and power supply variations up to 15%. Consists of two Schmitt triggers, Q3-Q4 and Q5-Q6, which share two current sources Q1 and Q2. Q4 and Q5 form differential switch that allows only one trigger to be on at a time. Charge rate of C3, which determines switching frequency of triggers and output frequency, can be controlled by varying d-c voltage on frequency-control input line, as well as by changing value of C3. Q7, Q8, and zeners provide temperature compensation. Output lines at right provide complementary 3-V outputs.—S. F. Aldridge, Square-Wave Generator Stresses Frequency Stability, *Electronics*, May 10, 1973, p 98.

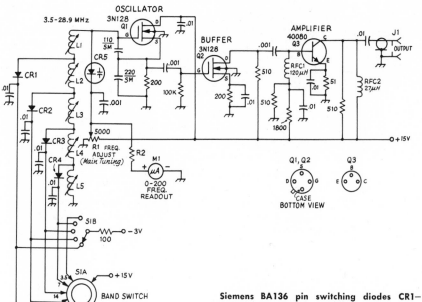

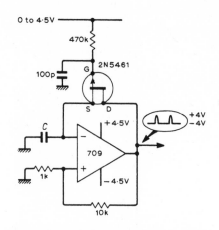

VARACTOR-TUNED OSCILLATOR—Provides stable fundamental-frequency output on five amateur bands from 3.5 to 30 MHz. Band-switching is accomplished electronically with

Siemens BA136 pin switching diodes CR1–CR4. CR5 is Siemens BB109 varactor. Article gives coil-winding data. Choose R2 to give full-scale reading on M1 when R1 is set for maximum voltage. Arrangement permits remote tuning.—D. M. Lee, A VTO for 80 Through 10 Meters, QST, Nov. 1970, p 21–23.

VCO FOR HIGH-PASS FILTER—Opamp and jfet form voltage-controlled oscillator giving logarithmic frequency-voltage relationship over five octaves, with pulsed output. Value of C is chosen to give desired frequency range.—D. G. Malham, Voltage-Controlled Filter and Oscillator, *Wireless World*, Sept. 1972, p 443.

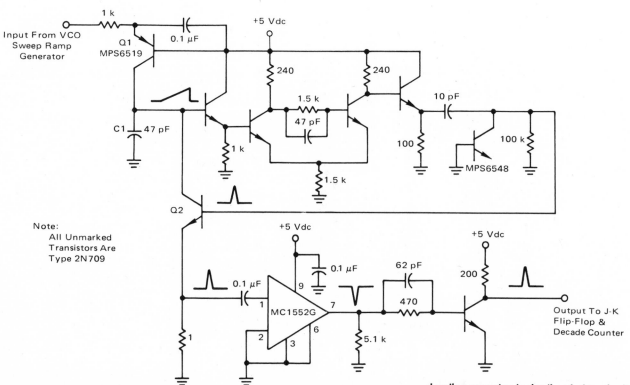

2–20 MHZ SQUARE-WAVE VCO—Provides square-wave input for divider under control of vco sweep ramp generator, for wide-range lab sweep oscillator covering audio and video ranges. Circuit is similar to relaxation oscillator design. Output frequency varies linearly within 1.5% with d-c control voltage of 6 to 12 V over 10 Hz–10 MHz range. Report describes operation in detail and gives circuit modification having even better linearity.—R. A. Botos and L. J. Newmire, "A Sychronously Gated N-Decade Sweep Oscillator," Motorola Semiconductors, Phoenix, AZ, AN-540, 1971.

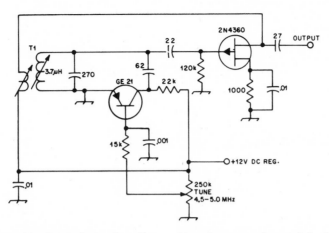

4.5–5 MHZ VOLTAGE-TUNED—Local oscillator circuit for receiver has transistor connected as synthesized Varicap tuning diode. Fet reduces drift and loading, thereby increasing Q of circuit. T1 is variable inductor with a few turns of enameled wire wound over it for feedback winding.—E. S. Cromartie, Synthesis of a Varicap, QST, Sept. 1972, p 44–45.

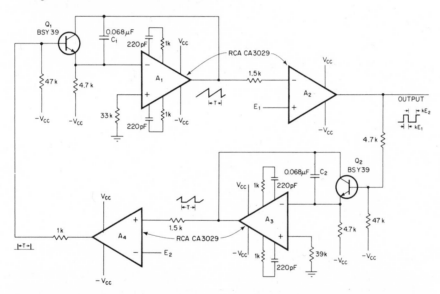

DUTY-CYCLE CONTROL—When sawtooth generated by A1 coincides with E1, comparator A2 changes state and gives pulse width proportional to E1. A3 and A4 take over to reset A1 when sawtooth of A3 coincides with E2. This makes time between pulses proportional to E2. Frequency range is 2–20 kHz. Supply voltages are 9 V and —6 V. 2N3261 may be used in place of Miniwatt BSY39 transistors. —A. Cavit and S. Bracho, Variable Oscillator Controls Pulse Width and Spacing, *Electronics*, July 5, 1971, p 59.

CHAPTER 129
Voltage-Level Detector Circuits

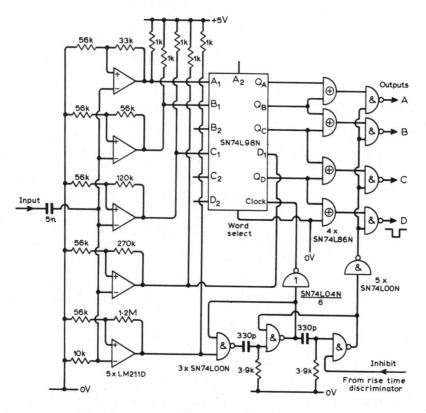

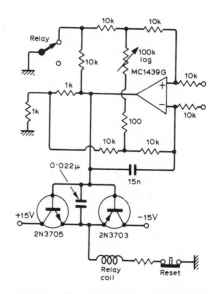

VOLTAGE TRIP—Setting of pot, controlling gain of opamp, determines input voltage (either positive or negative) at which saturated output energizes relay. Once tripped, amplifier state is held by feedback circuit, until reset, provided input overvoltage has been reduced or removed. Trip range is 50 mV to 5 V with 10K input resistor values shown.—N. Nicola, Sensitive Voltage Trip, *Wireless World*, Feb. 1971, p 66.

4-CHANNEL PULSE-HEIGHT ANALYZER—Used in rocket for solar x-ray research. Four-bit register stores levels because they are not acquired simultaneously. Timing circuit is initiated by lowest-level discriminator and strobes register output into four exclusive-OR gates, to give unique output for each pulse-height band. National LM211D comparators were selected for low power consumption and high input impedance.—R. W. Penny, Differential Discriminator Circuits, *Wireless World*, Jan. 1972, p 12–13.

LED NULL INDICATOR—Provides visual indication to within 0.2 mV of zero voltage at input, along with indication of polarity; with negative input (E1 less than 0 V), only led D1 lights, and with positive input only D2 lights. As input approaches null at 0 V, current through conducting led decreases and both led's appear dark. Sensitivity switch controls both brightness and width of null band; 50 ohms gives 0.2-mV repeatability and 500 ohms gives 2-mV repeatability in ambient light. Shielding with hoods improves sensitivity.—M. H. Loughnane, Light-Emitting Diode Pair Forms Null Indicator, *Electronics*, Aug. 2, 1971, p 58.

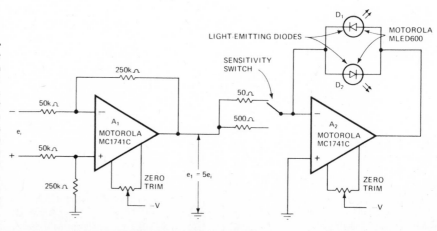

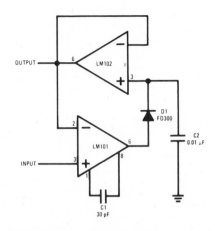

POSITIVE-PEAK DETECTOR—Diode conducts whenever input is greater than output, so output is always equal to peak value of input voltage. LM102 opamp is buffer for storage capacitor C2, giving low drift along with low output resistance. Difference between input and output voltages can change while circuit is holding.—R. J. Widlar, "Operational Amplifiers," National Semiconductor, Santa Clara, CA, 1968, AN-4.

LOW-VOLTAGE LAMP—Used when d-c supply voltage for transistor equipment is critical. Lamp I1 stays on for as long as voltage being monitored is below level determined by value of R7 and setting of R6. Choose value for R7 so small rotation of R6 from its center position turns lamp off. Vaules shown are for 12-V supply; for other supplies, change number of 22-meg resistors appropriately.—J. P. Hammes, Voltage Monitor, *Popular Electronics*, Nov. 1971, p 63–64.

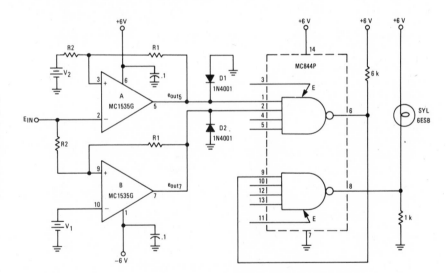

SIGNAL ENVELOPE DETECTOR—Uses two MC1535 dual opamps in conjunction with MC844 dual power gate. Pilot lamp is energized whenever input signal is out of range. Report includes alternative connection for holding lamp on until reset. R1 is 47K, R2 1K, V1 2.5 V, and V2 3.5 V. Design equations are given.—K. Wolf, "The MC1535 Monolithic Dual Operational Amplifier," Motorola Semiconductors, Phoenix, AZ, AN-411, 1972.

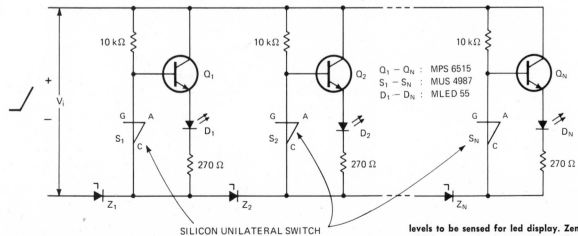

LED ANALOG-VOLTAGE SCANNER—Silicon unilateral switches S1–SN require 8 V to trigger but only 1 V to stay on, to provide sharp led transitions from dark to light for voltage-level sensing applications. Number of circuit sections depends on number of voltage levels to be sensed for led display. Zeners establish voltage levels sensed by each section. —T. Mazur, Analog Voltage Sensor Controls LED Threshold, *Electronics*, Aug. 14, 1972, p 112.

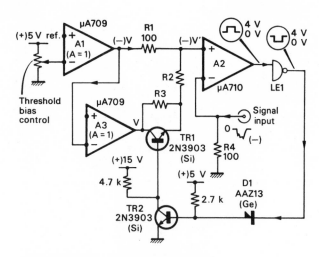

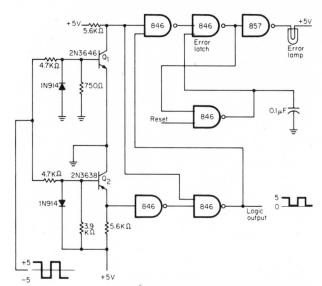

PULSE AMPLITUDE DISCRIMINATOR—When amplitude of input to Schmitt-type discriminator exceeds threshold bias, output changes abruptly to new level and remains there. To restore original no-signal level, input must be reduced to value below that causing first change. Difference between these *on* and *off* input levels, called backlash, is utilized in this circuit to eliminate noise. With 1.2K for R2, backlash is 8%; 820 ohms gives 12% and 560 ohms gives 18% backlash. Operates up to 1 MHz prr. Choose R3 to set emitter of TR1 to 0 V when collector current is zero.—H. A. Cole, Pulse Discriminator Takes Advantage of Backlash, *Electronic Engineering*, Sept. 1970, p 81–83.

LOGIC-LEVEL OUTPUT—Circuit ensures that each level of square-wave input exceeds minimum preset voltage, with predetermined transition-time limits and desired 5-V output logic level. Capacitor value determines allowable transition time, and transistor-base voltage dividers determine minimum allowable levels.—S. R. Martin, Circuit Detects Minimum Levels and Slow Transition Times, *The Electronic Engineer*, April 1971, p 67.

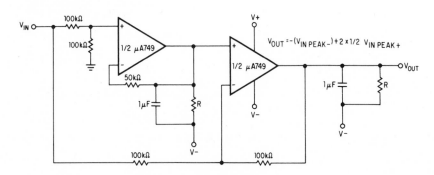

$$V_{OUT} = -(V_{IN\ PEAK-}) + 2 \times 1/2\ V_{IN\ PEAK+}$$

PEAK-TO-PEAK DETECTOR—Uses dual opamp, operating on supply voltages from 4 to 15 V. Left-hand opamp is connected as negative peak detector, obtaining its reference voltage from output of other opamp connected as positive peak detector. During capacitor charge, first opamp had gain of 2 from noninverting input; input to positive peak detector must therefore be divided by 2.—D. K. Long, "Applications of the μA749 Dual Operational Amplifier," Fairchild Semiconductor, Mountain View, CA, No. 268, 1971.

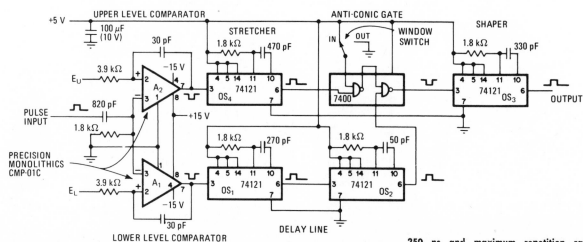

PULSE-HEIGHT ANALYZER—Combination of TTL and IC comparators speeds analysis of pulse heights when counting blood cells, studying white noise, and analyzing nuclear radiation, at fraction of usual cost. Accepts positive pulses having maximum risetime of 250 ns and maximum repetition rate of 500,000 pps. Comparators A1 and A2 set upper and lower voltage limits.—J. Laughter, TTL Gates Speed Up Pulse-Height Analysis, *Electronics*, Aug. 14, 1972, p 111.

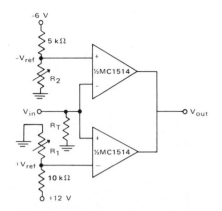

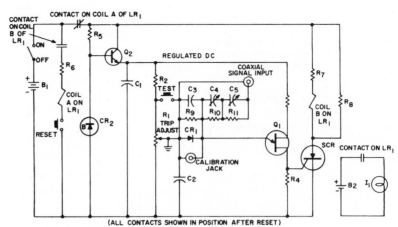

ZERO-CROSSING PULSE GENERATOR—Generates digital output pulse at zero crossover points of sinusoidal analog input signal, at frequencies up to 25.5 kHz for values shown. Uses two comparators, one set to detect minus voltage and the other to detect positive voltage. With reference voltages set at 10 mV above and below zero, comparator output is low only for interval during which input signal is between two reference voltages, to give pulse width of about 125 ns. Comparator output is high at all other times. —R. Brunner, "A High Speed Dual Differential Comparator—The MC1514," Motorola Semiconductors, Phoenix, AZ, AN-547, 1971.

SCR—GE C220F CONTROLLED RECTIFIER
Q₁— GE 2N490 UNIJUNCTION TRANSISTOR
Q₂— GE 2N3416 TRANSISTOR
CR₁—GE DT230H RECTIFIER
CR₂—Z4X18B ZENER DIODE, 17-21 V
B₁— 22 1/2 VOLT BATTERY, BURGESS 4156
B₂—1 1/2 VOLT BATTERY, BURGESS 2FBP
R₁—50,000 Ω HELIPOT, SERIES C, 3 TURN
R₂—4700 Ω 1/2 WATT
R₃—470 Ω 1/2 WATT
R₄—100 Ω 1/2 WATT

R₅—4700 Ω 1/2 WATT
R₆,R₇—47 Ω 1 WATT
R₈—200 Ω 1 WATT
R₉—2500 Ω 1 WATT, 1 % TOLERANCE
R₁₀,R₁₁—499,000 Ω 1 WATT, 1% TOLERANCE
LR₁—POTTER-BRUMFIELD LATCHING RELAY TYPE KE 17D-12V DC
I₁—GE TYPE 49 LAMP BULB, .06A, 2 VOLTS
C₁—100 MFD, 50V DC ELECTROLYTIC CAPACITOR GE 76FO2LNI0I
C₂—2.0 MFD, 200V DC PAPER CAPACITOR GE BAI7B205B
C₃—2000 MMFD MICA CAPACITOR
C₄, C₅—1 TO 7.5 MMF CERAMIC TRIMMER CAPACITORS

TRANSIENT PEAK INDICATOR—Circuit energizes indicating lamp through relay when voltage threshold set by precision pot R1 is exceeded by random transient on coax signal input line. Lamp stays on until reset button is pushed. Ujt Q1 compares input signal with reference. Voltage divider steps signal down to 1/400th of input value. Maximum error is under 2% of full scale for pulses from 1 μs duration up to pure d-c, using three-turn pot. —"SCR Manual," General Electric, Syracuse, NY, 1972, 5th Ed., p 474–475.

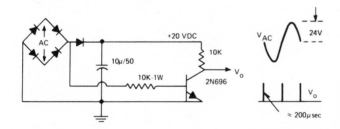

ZERO-CROSSING DETECTOR—Used in dealing with a-c power when synchronizing mvbr or other digital circuit to give repeatable firing. Reset diode is used to remove residual charge from capacitor at zero voltage.—"Silicon Controlled Rectifier Designers Handbook," Westinghouse, Youngwood, PA, 1970, 2nd Ed., p 6–18–6–19.

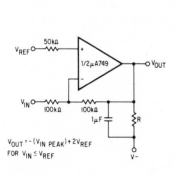

$$V_{OUT} = -(V_{IN\ PEAK}) + 2V_{REF}$$
FOR $V_{IN} \le V_{REF}$

NEGATIVE-PEAK DETECTOR—Circuit acts as voltage follower for inputs more negative than positive voltage of capacitor, so output is positive voltage equal to negative of peak input voltage shifted positive by twice reference voltage value.—D. K. Long, "Applications of the μA749 Dual Operational Amplifier," Fairchild Semiconductor, Mountain View, CA, No. 268, 1971.

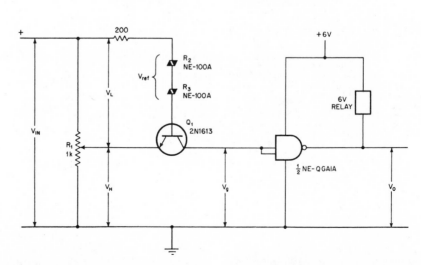

VOLTAGE DETECTOR—Nonlinear NAND gate generates low output when input voltage is too high or too low, and high output when input is within tolerance range determined by setting of R1 and values of components. Example uses tolerance of 1 V above and below 4 V, for range of 3 to 5 V.—R. N. Basu and A. Dvorak, Nonlinear Logic Detects Voltage Tolerance Levels, Electronics, March 15, 1971, p 78.

SQUARE-LAW FUNCTION—Each of four resistor-diode networks conducts at different predetermined level of input signal and places its resistor in parallel with the others. Resistor values are chosen so total network resistance decreases in approximately a square-law relation to input signal voltage, and opamp gain increases similarly. For values shown, 0–3 V pulse input gives 0–10 V pulse output.—R. P. Hennick, For a Square-Law Transfer Function, Try This Op Amp Connection, *The Electronic Engineer*, Aug. 1968, p 68.

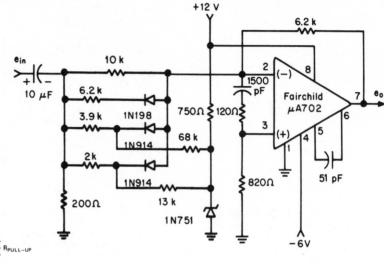

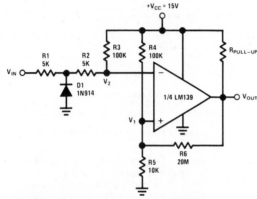

ZERO - CROSSING DETECTOR — Uses LM139 comparator to square up symmetrically a sine wave centered around 0 V. Small amount of positive feedback improves switching times and centers input threshold at ground.—R. T. Smathers, T. M. Frederiksen, and W. M. Howard, "LM139/LM239/LM339 A Quad of Independently Functioning Comparators," National Semiconductor, Santa Clara, CA, AN-74, 1973, p 5.

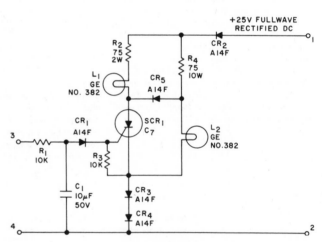

OPAMP STATUS INDICATOR—When d-c voltage applied between terminals 3 and 4 (3 positive) exceeds threshold of about 2.8 V, scr turns on and energizes lamp L1. This automatically turns off L2 (previously on) because of CR5. Threshold voltage can be increased by adding more diodes to CR1, CR3, and CR4 or by replacing all with appropriate zener.—"SCR Manual," General Electric, Syracuse, NY, 1972, 5th Ed., p 224–225.

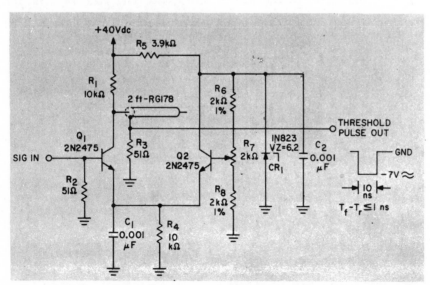

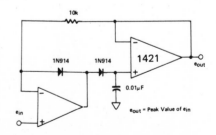

POSITIVE-PEAK DETECTOR—Uses two Philbrick 1421 fet opamps. First provides high impedance for input signal, and second acts as high-impedance buffer between storage capacitor and output signal which is peak value of input signal. Operates up to 1 MHz.—"General Purpose FET Operational Amplifier 1421/01/02," Teledyne Philbrick, Dedham, MA, 1972.

ADJUSTABLE-LEVEL DETECTOR—Basic reference element for threshold level is zener CR1, with R7 adjusting level by varying base voltage of Q2. When threshold level is reached at base of Q1, length of coax discharges its energy across R3 to turn Q1 off at same speed of less than 1 ns at which it was turned on. Width of output pulse is equal to twice time delay of coax, and can be adjusted by changing length of coax. Input signal consists of 3-ns spikes containing a-m components.—Detector Pulse Risetime Cut to 1 ns, *Electro-Technology*, Jan. 1970, p 10.

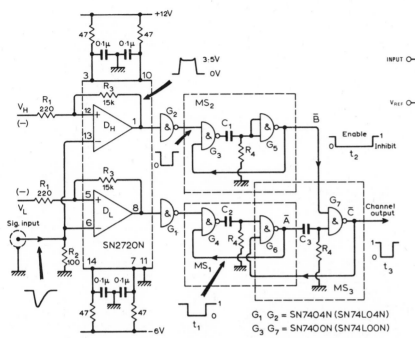

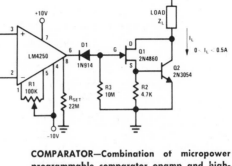

PULSE HEIGHT DISCRIMINATOR—Dual differential comparator feeds NAND-gate TTL elements. Used for pulse height analysis in nuclear research.—H. A. Cole, Differential Discriminator Circuits, *Wireless World*, Dec. 1971, p 603–604.

COMPARATOR—Combination of micropower programmable comparator opamp and high-current switch serves for monitoring input voltage while dissipating only 100 μW and switching 500-mA load when input exceeds reference voltage which can be anywhere between —8.5 V and +8.5 V. With minimum gain of 100,000, comparator can resolve input voltage differences as small as 0.2 mV. Load can be resistor, relay coil, or lamp.—M. K. Vander Kooi and G. Cleveland, "Micropower Circuits Using the LM4250 Programmable Op Amp," National Semiconductor, Santa Clara, CA, 1972, AN-71, p 4–5.

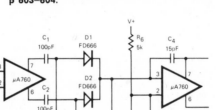

ZERO-CROSSING DETECTOR—Uses two IC comparators, with outputs of first one changing state each time input passes through zero. Network between comparators then generates short positive pulse for one input of second comparator. This gives positive pulse at corresponding output of second comparator and negative pulses at other output. Value of R6 determines output pulse width. Output drops from 7-V rms at 300 Hz to about 30 mV at 4 MHz.—P. Holtham, "The μA760—A High Speed Monolithic Voltage Comparator," Fairchild Semiconductor, Mounain View, CA, No. 311, 1972, p 5.

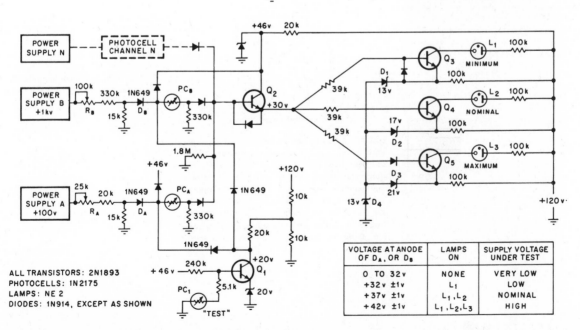

VOLTAGE AT ANODE OF D_A OR D_B	LAMPS ON	SUPPLY VOLTAGE UNDER TEST
0 TO 32v	NONE	VERY LOW
+32v ±1v	L_1	LOW
+37v ±1v	L_1, L_2	NOMINAL
+42v ±1v	L_1, L_2, L_3	HIGH

LAMPS MONITOR HIGH VOLTAGE—Set of three indicator lamps serves to indicate whether any one of large number of power supplies is within 15% of nominal value, when supply outputs are inaccessible or so high in voltage that they are dangerous to probe. Desired supply is checked by aiming flashlight beam at its photocell. Each voltage divider is adjusted (with RA, RB, etc) to 37 V, sufficient to turn on Q2 and light L1 and L2 if supply is nominal value. Table shows meanings of other lamp-on combinations.—R. S. Granchelli, Flashlight Helps Monitor Voltage Levels, *Electronics*, Aug. 17, 1970, p 88.

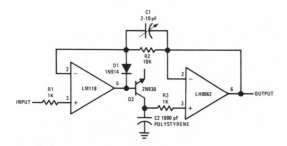

HIGH-SPEED—Will acquire +10 V peak signal in under 4 μs, with droop rates under 20 mV/s. Reversing polarity of diodes D1 and D2 allows detection of negative peaks. Any ultra-low-leakage diode may be substituted for 2N930 collector-base junction.—B. Siegel, "Applications for a High Speed FET Op Amp," National Semiconductor, Santa Clara, CA, 1972, AN-75.

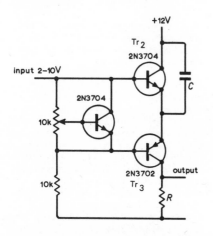

AMPLITUDE-INDEPENDENT PEAK DETECTOR—Operates independently of load. Set 10K pot initially for zero d-c output with no input. Values of R and C depend on frequency but are not critical, and for a-f can be 15K and 4.7 μF.—P. E. Pinnock, Peak Detector, *Wireless World*, July 1972, p 348–349.

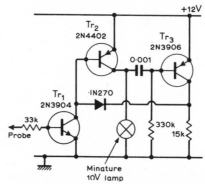

0.7-V INDICATOR—Lamp lights when pulse or level above 0.7 V exists at terminal touched with probe during troubleshooting of digital systems. High input impedance makes loading of circuit under test insignificant. Pulse triggers 1-ms mono mvbr TR2-TR3, lighting lamp momentarily. Steady input above 0.7 V holds lamp on.—J. M. Firth, Pulse and Voltage Level Indicator, *Wireless World*, June 1971, p 269.

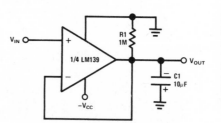

NEGATIVE PEAK—LM139 comparator is operated closed loop as unity-gain follower. Output transistor of comparator serves as required low-impedance current sink for detecting negative peaks. Value of R1 determines decay time.—R. T. Smathers, T. M. Frederiksen, and W. M. Howard, "LM139/LM239/LM339 A Quad of Independently Functioning Comparators," National Semiconductor, Santa Clara, CA, AN-74, 1973, p 16.

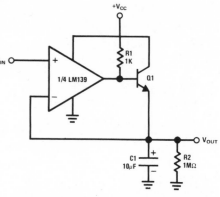

POSITIVE PEAK—Uses LM139 comparator operated closed loop as unity-gain follower, with large holding capacitor from output to ground. Transistor is added to output to give required low-impedance current source required for detecting positive peaks. When output of comparator goes high, C1 charges through Q1. Use output through high-impedance follower to avoid loading peak detector. Supply can be 5 V. Transistor type is not critical.—R. T. Smathers, T. M. Frederiksen, and W. M. Howard, "LM139/LM239/LM339 A Quad of Independently Functioning Comparators," National Semiconductor, Santa Clara, CA, AN-74, 1973, p 16.

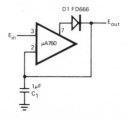

FAST PEAK DETECTOR—Operates with input pulse widths down to 50 ns. Discharge current for C1 is bias current for IC comparator, typically 14 μA. Decay time of voltage across C1 is about 50 ns/V.—P. Holtham, "The μA760—A High Speed Monolithic Voltage Comparator," Fairchild Semiconductor, Mountain View, CA, No. 311, 1972, p 7.

CHAPTER 130
Voltage Reference Circuits

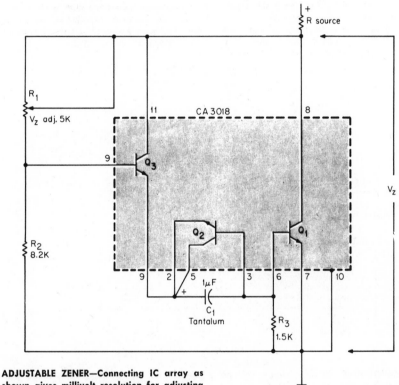

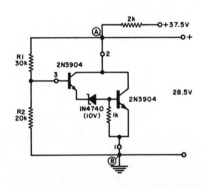

DISCRETE PROGRAMMABLE ZENER—Requires only 1 mA reverse current to start zener operation, as compared to 2 mA for General Electric D13V programmable zener device. Output reference voltage is determined by values of divider R1-R2, either of which may be variable.—R. L. Starliper, Programmable Zener Applications, *Electronics World*, June 1971, p 60–61.

ADJUSTABLE ZENER—Connecting IC array as shown gives millivolt resolution for adjusting reference output voltage Vz with R1. With 680-ohm source resistance and R1 set for 10 V, equivalent zener impedance was 10 ohms for current levels of 0 to 40 mA, and noise below 1-mV p-p.—W. Jung, IC Array As an Adjustable Zener, *The Electronic Engineer*, Sept. 1971, p 94.

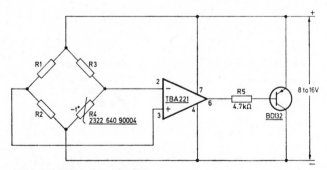

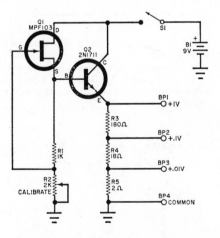

REFERENCE DIODE TEMPERATURE CONTROL—Used to minimize effects of temperature variations on reference diode by controlling temperature of small metal block in which diode is mounted. Will operate from low-power 8–16 V supply. R1–R3 and two-point NTC thermistor R4 form bridge across supply, with thermistor clamped to copper or aluminum block. Out-of-balance bridge voltage, varying with thermistor and block temperature, is applied to opamp operating open-loop as voltage comparator with very low hysteresis. Opamp output switches transistor on or off to change temperature of block on which transistor (serving as heater) is mounted with thermistor. For 70 C, R1–R3 are 2.2K.—"1N821 and BZX90 Series of High Stability Reference Diodes," Mullard, London, 1973, TP1339, p 3–4.

VOLTMETER CALIBRATOR—Provides 0.01, 0.1, and 1 V with accuracy better than 2% even when its 9-V battery has dropped to 5 V under load. Q1 operates as constant-current device, and Q2 as current amplifier to reduce effect of meter loading by 100:1. Can be used to check accuracy of any voltmeter from 1,000-ohms-per-volt vom to vtvm. Requires only one calibration, at 1 V.—F. H. Tooker, Accurate Low-Voltage Calibrator, *Popular Electronics*, Sept. 1968, p 66.

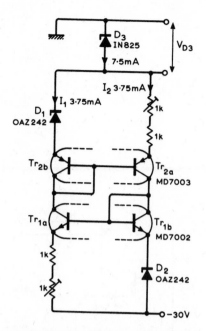

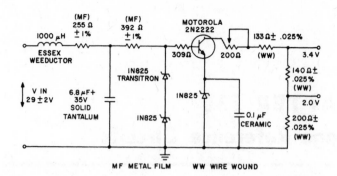

RADIATION-STABLE 2 AND 3.4 V—Uses zener reference rather than transistor regulator. Stability factor is 0.5% for combined conditions of —20 to +80 C, gamma dose of 1,000,000 rads, and above 10 keV. Developed for instrument calibration near nuclear reactors or weapons.—A. J. Sofia, Reference Voltage Stable to Radiation, *Electro-Technology*, May 1969, p 23.

0.08% STABILITY—Maintains stable reference at temperatures between 0 and 50 C and with supply voltages between 24 and 40 V. Uses zeners with negligible temperature coefficient, with ordinary transistor connected as forward-based diode in series with each. Use of dual transistors in ring-of-two voltage reference provides required temperature compensation. BD3 does not vary more than 1 mV from 6.338 V over supply range of 24 to 40 V.—H. A. Cole, Voltage Reference Source, *Wireless World*, Sept. 1971, p 446–447.

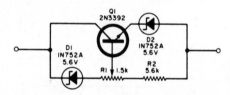

11–13 V VARIABLE REFERENCE—Used in place of zener when required voltage rating is not in stock or when reference must be adjusted at last minute.—D. H. Rogers, Variable Replacement for Zener Diode, *Electronics World*, Feb. 1971, p 72.

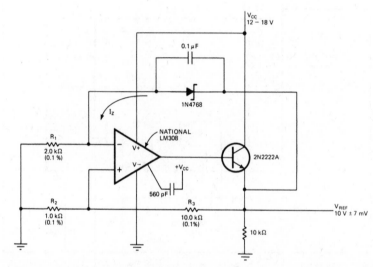

SELF-CONTROLLED ZENER—Voltage comparison circuit makes zener control its own current, eliminating need for separately regulated temperature-compensating current source. Provides precision 10-V reference source from single unregulated supply, stable within 7 mV for 12—18 V supply range. Output is 30 mA maximum. Temperature range is 0–75 C.—W. Goldfarb, Single-Supply Reference Source Uses Self-Regulated Zener, *Electronics*, June 7, 1973, p 107.

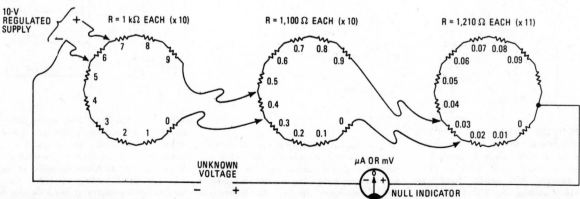

CIRCULAR VOLTAGE DIVIDER—Provides precision d-c voltages from 0 to 10 V in steps of 0.01 V by moving voltage-source leads rather than output leads, using only 31 precision resistors. Setup shown gives output of 6.43 V. Chief limitation is allowable power dissipation of resistor in first ring across which full supply voltage is applied.—D. Hileman, Circuit Voltage Divider Needs Fewer Resistors, *Electronics*, May 10, 1973, p 96–97.

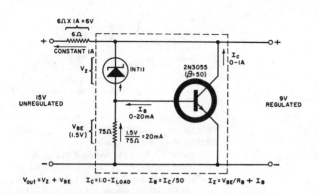

9 V WITH AMPLIFIED ZENER—Combination of transistor and zener in dashed box simulates higher-power (10 or 50 W) zener at much lower cost. Article gives design procedure for other voltages.—C. J. Ulrick, The Amplified Zener, *Electronics World*, Sept. 1970, p 42 and 63.

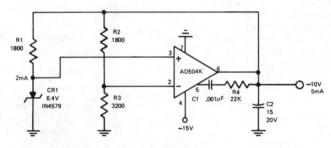

—10 V REFERENCE—Stability can be as low as 10 ppm per deg C, with up to 5-mA output. For positive reference output, reverse polarity of CR1 and C2, and apply power to amplifier from ground on pin 4 to +15 V on pin 7.—"Applying the AD504 Precision IC Operational Amplifier," Analog Devices, Inc., Norwood, MA, April 1972.

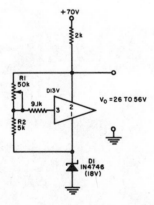

26 TO 56 V—Use of conventional zener D1 in series with General Electric D13V programmable zener boosts limits of reference voltage, varied with R1, by rating of zener.—R. L. Starliper, Programmable Zener Applications, *Electronics World*, June 1971, p 60–61.

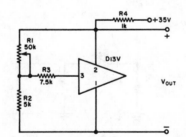

8.5 TO 30 V—R1 varies reference output voltage of General Electric D13V programmable zener. Temperature stability is 0.1% per deg C.—R. L. Starliper, Programmable Zener Applications, *Electronics World*, June 1971, p 60–61.

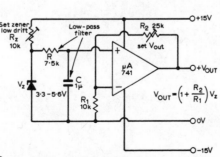

ADJUSTABLE ZENER—Opamp permits adjusting precise voltage from zener upward to desired higher precise voltage.—T. D. Towers, Elements of Linear Microcircuits, *Wireless World*, Feb. 1971, p 76–80.

$$V_{OUT} = \left(1 + \frac{R_2}{R_1}\right) V_Z$$

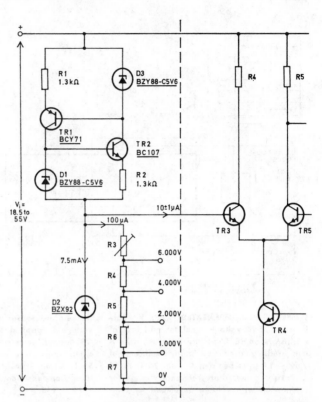

STABILIZER AS REFERENCE—BZX92 voltage reference diode rated at 6.5 V serves as reference voltage source for 1, 2, 4, and 6 V and as stabilized power supply through action of 7.5-mA constant-current source TR1-TR2-D1-D3. Input voltage is not critical. Temperature stability is equal to that of D2. Load resistor for D2 is highly stable pot tapped for required voltages.—"1N821 and BZX90 Series of High Stability Reference Diodes," Mullard, London, 1973, TP1339, p 10–11.

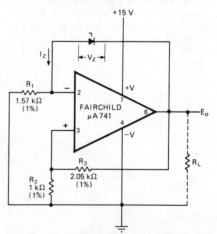

10 V WITH SINGLE SUPPLY—Grounding of one supply line of differential opamp permits operation with only the one supply that gives desired output polarity. Circuit shown gives positive output as long as noninverting input is positive. For negative output, ground positive supply line of IC and connect negative line to —15 V, then reverse zener polarity. With 6.4-V zener, output is 9.547 V with stability of 9.5 ppm/V for positive supply and —9.560 V at 2.6 ppm/V for negative supply. —M. J. Shah, Stable Voltage Reference Uses Single Power Supply, *Electronics*, March 13, 1972, p 74.

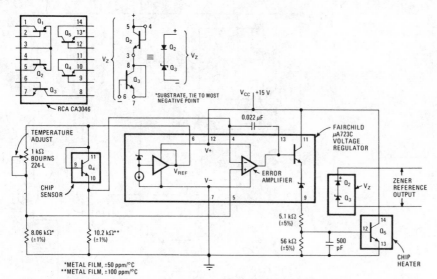

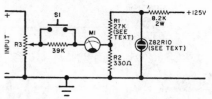

PRECISION VOLTMETER—Uses Signalite Z82R10 neon as voltage reference providing 1 V across R2 accurate within 0.012 V. Accuracy holds after R1 is initially adjusted to provide calibration. Unknown voltage is applied across R3, a 10-turn pot with calibrated dial adjusted for meter null. As null is approached, S1 is closed for maximum sensitivity. M1 can be 50-0-50 μA zero-center f-m tuning indicator.—J. Kyle, The Ubiquitous Neon Lamp, *Popular Electronics*, Nov. 1968, p 55–57.

TEMPERATURE-COMPENSATED 7.83 V—Combination of IC voltage regulator and four of transistors on IC five-transistor array (Q1 is not used) gives temperature coefficient of only —3.1 ppm per deg C from 33 C to 50 C. Pot sets temperature reference voltage for comparison with voltage developed by chip-temperature sensor transistor Q4. Difference voltage is applied to chip heater transistor Q5 through error amplifier to maintain constant temperature of transistor array.—M. J. Shah, Using Transistor Arrays for Temperature Compensation, *Electronics*, April 12, 1973, p 103.

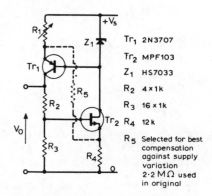

ADJUSTABLE REFERENCE—Circuit is equivalent to constant-current diode, with R1 chosen to provide 1 mA through R2 and R3. Use about 2.5K pot for R1. Provides maximum reference voltage of 20 V, constant within 0.1% for supply voltages between 24 and 34 V. May be used in voltage regulators.—P. Williams, Variable Voltage Reference Source for D. C. Regulators, *Wireless World*, Feb. 1970, p 83–84.

1-V DIODE REGULATOR—At voltages under 6 V, four-layer diode provides five times better regulation than zener. Although cost of diode version is higher, voltage regulation from no-load to full-load is only 6% at 1-V output, as compared to 30% for zener because of zener deviation from true avalanche effect below 6.2 V.—R. D. Clement and R. L. Starliper, Four-Layer Diode Circuit Out-Regulates Zener by 5:1, *Electronics*, June 7, 1971, p 81–82.

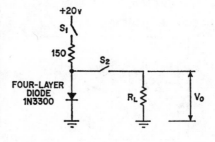

CHAPTER 131
Zero-Voltage Switching Circuits

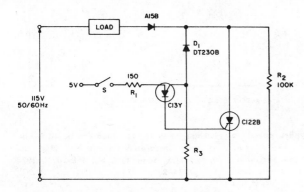

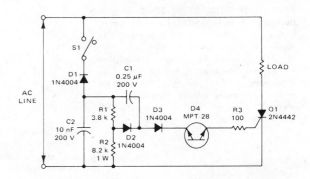

IDEAL CSCR SWITCH—Uses C13Y complementary or n-gate scr connected to turn on only when instantaneous line voltage is under 5 V, for zero-voltage switching to keep electromagnetic interference at minimum. Circuit may be used with any load power factor. Choose value of R3 so leakage current through D1 does not damage gate of C13Y. Circuit is easily scaled up for 220-V line.—"SCR Manual," General Electric, Syracuse, NY, 1972, 5th Ed., p 312–313.

ZERO-POINT SWITCH—Half-wave control for sensitive-gate scr uses repetitive exchange of charges from C1 to C2 to provide series of gate current pulses as line voltage crosses zero and starts to go positive. This turns on scr at extremely low current, to minimize emi.—R. J. Haver and L. T. Rees, "Zero Point Switching Techniques," Motorola Semiconductors, Phoenix, AZ, AN-453, 1972.

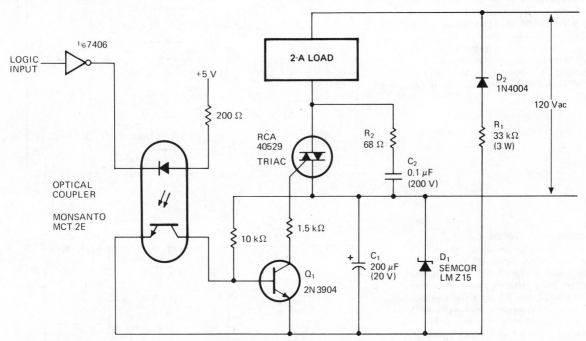

2-A A-C ZERO-VOLTAGE SWITCH—Optical coupler makes safe switching of a-c load possible with IC logic signals. Low logic input turns off coupler, making Q1 saturate and turn on triac to energize load. Triac stays on for complete half-cycle of line voltage once logic input goes low, and turns off at first zero crossing after logic input goes high. R2 and C2 suppress rfi and protect triac when driving inductive loads.—L. S. Bell, Switching Large AC Loads with Logic-Level Signals, *Electronics*, March 1, 1973, p 86.

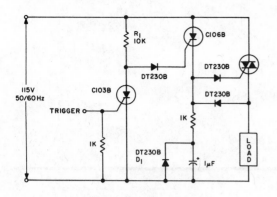

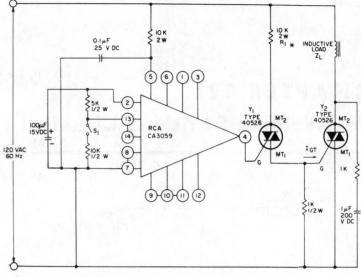

PILOT TRIGGERING OF TRIAC—Uses sensitive-gate scr to trigger triac reliably before instantaneous line voltage reaches NEMA WD-2 emi limits during positive half-cycles. Negative half-cycle triggering occurs satisfactorily with charged 1-μF capacitor. Maximum voltage for triggering C106B in circuit is 4.4 V. Requires selected gate triac.—"SCR Manual," General Electric, Syracuse, NY, 1972, 5th Ed., p 314.

* IF Y$_2$, FOR EXAMPLE, IS A 40-AMPERE TRIAC, R$_1$ MUST BE DECREASED TO SUPPLY SUFFICIENT I$_{GT}$ FOR Y$_2$.

TRANSIENT-FREE SWITCH—Supplies power to load at first zero crossing of a-c line after S1 is closed, to minimize rfi and emi for inductive or resistive loads. Maximum rms load current depends on rating of triac Y2; if this is changed to 2N5444, load can be 40 A rms.—"Linear Integrated Circuits and MOS Devices," RCA, Somerville, NJ, SSD-202A, 1973 Ed., p 234–235.

THREE-PHASE TRIGGER—IC at left, serving as zero-crossing trigger for scr or triac having lamp load, operates as master control driving three main power triacs through identical zero-crossing a-c trigger IC's having input photocells responding to master lamp. Load may have either wye or delta connection. Dashed circuits are required for inductive loads. RSYN is 10K for 115-V line and 22K for 230 V, while CST is 0.47 μF. Rfi generation is minimized.—A. Adamian and L. Blaser, "A Monolithic Zero-Crossing AC Trigger (Trigac) for Thyristor Power Controls," Fairchild Semiconductor, Mountain View, CA, No. 208, 1972, p 5.

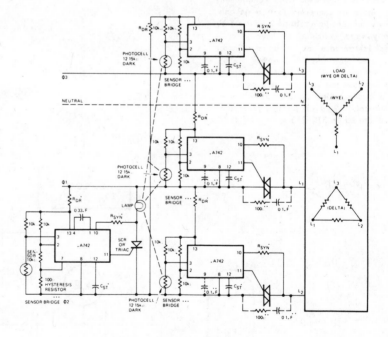

TWO-TRANSISTOR SWITCH—Uses Q1 along with S to provide gating signal for scr serving as load switch, while Q2 detects zero-voltage crossing for half-wave switching circuit that keeps electromagnetic interference at minimum. D1 and D2 protect low-voltage components during negative half-cycle. Additional scr, shown with dashed connections, is used for integral-cycle control to prevent saturation effects in critical load elements such as transformers having marginally designed magnetic cores. SCR1 is rated to handle loads up to 8 A.—"SCR Manual," General Electric, Syracuse, NY, 1972, 5th Ed., p 311–312.

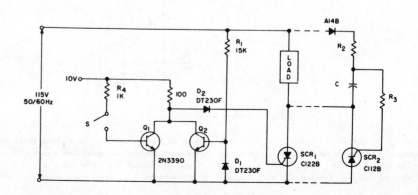

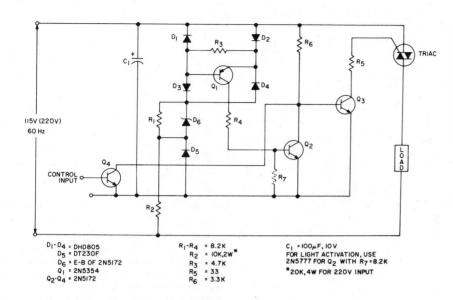

D1-D4 = DHD805
D5 = DT230F
D6 = E-B OF 2N5172
Q1 = 2N5354
Q2-Q4 = 2N5172

R1-R4 = 8.2K
R2 = 10K,2W*
R3 = 4.7K
R5 = 33
R6 = 3.3K

C1 =100μF, 10 V
FOR LIGHT ACTIVATION, USE
2N5777 FOR Q2 WITH R7 =8.2K
*20K, 4W FOR 220V INPUT

FOUR-TRANSISTOR TRIAC — Transistor circuit provides zero-voltage triggering with minimum emi for triac matching load. Includes regulated power supply for transistors. Q3 supplies negative gate drive for triggering triac. Q1 is turned on before line voltage reaches positive 5 V, to saturate Q2 and inhibit further gate drive. D1 and D4 conduct during negative half-cycle to provide zero-current switching.—"SCR Manual," General Electric, Syracuse, NY, 1972, 5th Ed., p 315–316.

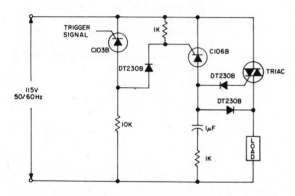

STANDARD FULL-WAVE TRIAC—Use of negative gate triggering provides reliable switching action with standard triac before instantaneous line voltage reaches NEMA WD-2 emi limits during positive and negative half-cycles.—"SCR Manual," General Electric, Syracuse, NY, 1972, 5th Ed., p 314–315.

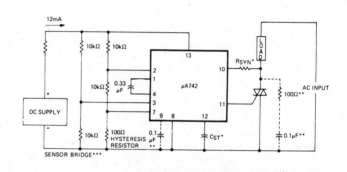

TRIGGER WITH D-C SUPPLY—Single IC provides seven functions required for on-off control of inductive or resistive a-c load at zero-current point of a-c cycle, with minimum rfi generation. Choose triac to handle load. Separate 24-V 12-mA supply keeps differential amplifier of IC in operation regardless of instantaneous polarity of a-c line; triac then conducts initially at start of positive half-cycle, and starts blocking only with negative half-cycle. Choose triac to match load. For 115-V a-c input, RSYN is 10K and CST 0.47 μF. Use dashed circuit only with inductive load. Report gives IC circuit and describes operation in detail.—A. Adamian and L. Blaser, "A Monolithic Zero-Crossing AC Trigger (Trigac) for Thyristor Power Controls," Fairchild Semiconductor, Mountain View, CA, No. 208, 1972, p 4.

*IF Y2, FOR EXAMPLE, IS A 40-AMPERE TRIAC, THEN R1 MUST BE DECREASED TO SUPPLY SUFFICIENT IGT FOR Y2

TRANSIENT-FREE SWITCH—Supplies power to load at first zero crossing of a-c line after opening of switch connected between pins 7 and 14 of RCA CA3059 zero-voltage switch. Will handle either resistive or inductive loads with absolute minimum of rfi and emi. Maximum rms load current depends on rating of triac Y2; if this is changed to 2N5444, load can be 40 A rms.—"Linear Integrated Circuits and MOS Devices," RCA, Somerville, NJ, SSD-202A, 1973 Ed., p 234–235.

TRIAC SWITCH—Triac with rating to match load requirement is gated on at start of positive half-cycle by current flow through 3-μF capacitor as long as scr is off. Load voltage then charges 1-μF capacitor so triac will be energized on next half-cycle. Provides zero-voltage switching with minimum electromagnetic interference. Chief problem is triggering triac during positive half-cycle; line voltage may read up to 15 V before it fires. Requires selected gate triac.—"SCR Manual," General Electric, Syracuse, NY, 1972, 5th Ed., p 313–314.

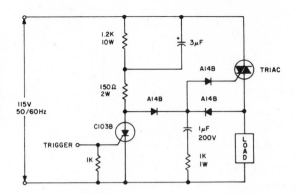

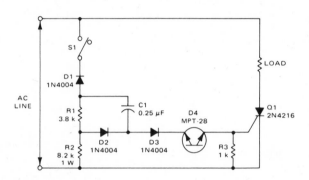

SENSITIVE-GATE SWITCH—Insures that control scr is turned on at start of each positive alternation, to minimize emi. Pulse is generated before zero crossing to provide small amount of gate current for scr when line voltage starts to go positive. Developed for sensitive-gate scr's. Provides half-wave control without excessive heating of scr while in reverse-blocking state.—R. J. Haver and L. T. Rees, "Zero Point Switching Techniques," Motorola Semiconductors, Phoenix, AZ, AN-453, 1972.

LIGHT-ACTIVATED—Photodarlington Q1 (L14A or L14B) is combined with GEL300 IC zero-voltage switch to give normally open solid-state relay using triac Q2 rated to meet load requirements. When light falls on Q1, triac is triggered on at next zero crossing of line voltage. Book also gives inverted logic connection for normally closed relay. Can be used as room temperature control that works only when light in room is on.—"SCR Manual," General Electric, Syracuse, NY, 1972, 5th Ed., p 435.

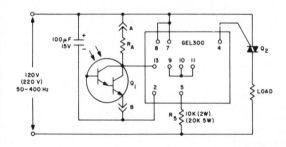

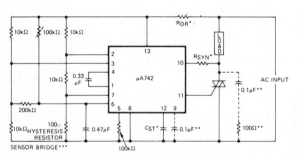

PROPORTIONAL CONTROL—Single IC provides seven functions required for on-off control of inductive or resistive a-c load at zero-current point of a-c cycle, with minimum rfi generation. Choose triac to handle load. For 115-V a-c line, RDR and RSYN are 10K and CST is 0.7 μF. Report gives IC circuit and describes operation in detail. Arrangement provides precise control needed in some temperature controls, as well as proportional control function needed for on-off a-c flashers. Use dashed circuit only with inductive load.—A. Adamian and L. Blaser, "A Monolithic Zero-Crossing AC Trigger (Trigac) For Thyristor Power Controls," Fairchild Semiconductor, Mountain View, CA, No. 208, 1972, p 3.

IDEAL HALF-WAVE SWITCH—Uses scr as switch that closes at instant when voltage across it is zero, and opens at instant when current through it is zero, to keep electromagnetic interference at minimum. Waveforms illustrate operation of circuit. Scr conducts only for complete half-cycles. Handles loads up to 4 A.—"SCR Manual," General Electric, Syracuse, NY, 1972, 5th Ed., p 310–311.

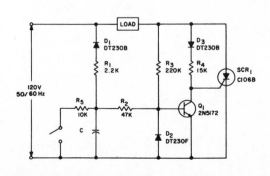

Name Index

Abelson, I. G., 27
Accardi, L., 527
Acuna, M. H., 253
Adamian, A., 990, 992, 993, 1016–1018
Adams, F. E., 790
Adams, G. P., 611
Adams, R. F., 608
Adey, W. R., 702
Agnew, J., 142
Aisenberg, S., 458
Albers, J. J., 70
Albisser, A. M., 470, 474
Alcindor, R. C., 878
Aldrich, R. A., 201, 202
Aldridge, S. F., 1001
Allen, D., 90
Allen, L., 926
Allen, P. E., 37, 40
Ambler, R., 624
Amick, J. A., 452
Anciaux, L. N., 567, 812
Anderson, L. T., 32, 35, 36
Anderson, R. A., 333
Anderton, C., 42, 250
Andes, C. B., 256
Andrews, C. L., 283, 944
Andriessen, G. J., 482
Antanaitis, B., Jr., 890
Apperson, R. C., Jr., 786
Archer, J. A., 629
Armer, H. L., 595
Aronhime, P., 575
Arp, R. C., Jr., 390
Arthur, P., 49
Artigo, R., 15, 862
Artusy, M., 254
Ashdown, I. E., 247
Ashe, J., 268, 795, 830, 925, 932
Assaly, R. N., 686
Assour, J., 507
Aumala, V., 602
Avery, W. L., 924, 926

Babcoke, C., 902
Babudro, E., 436
Bach, R. C., 97, 360, 361
Bailey, A. R., 523, 718
Bain, R., 48, 808, 984
Baker, L. E., 153, 542
Balakrishnan, B., 548, 550
Baldwin, C. A., 439
Ballerini, F., 89
Balph, T., 185, 538
Balyoz, H., 38, 455, 973

Banthorpe, C. H., 55, 716
Baranello, R. J., 9
Barcz, J., 980
Barnes, R. N., 371
Barrington, R. E., 965
Barton, D. M., 218
Basak, A., 596
Baskin, R. C., 545
Basu, R. N., 1007
Battes, R. J., 465, 477
Bauer, F. J., Jr., 90
Baxandall, P. J., 27
Beach, E., 543, 553
Beastall, H. R., 568
Becciolini, B., 979–982
Beck, G. J., 786
Beck, M. O., 979
Becker, D. F., 728
Becker, R. B. H., 659, 715
Beene, G., 460
Beiswenger, J., 77
Beitler, S., 487, 488
Belcher, D. K., 11, 70
Bell, J., 343
Bell, L. S., 1015
Benrey, R. M., 2, 65, 99, 100, 235, 278, 377
    452, 557, 562, 654
Benskin, G. D., 431
Bercher, D. K., 13
Beresford, A. C., 430, 441
Berry, J. B., Jr., 879, 880, 882
Bertini, P. J., 969
Beurrier, H. R., 764
Biancomano, V., 51
Bice, P. K., 151
Bik, R. J., 388
Bjornholt, J. E., 71
Blachford, B. W., 955
Black, M. F., 513
Blackburn, J., 219
Blakeslee, D. A., 133, 199, 201, 288, 291,
    306–308, 323, 399, 434, 438, 678, 722,
    742, 743, 802, 821, 879, 986
Blaser, L., 990, 992, 993, 1016-1018
Blechman, F., 918
Bliss, J., 92, 96, 185, 337, 343, 590, 630, 633,
    634, 678
Blomley, P., 27
Blood, B., 126, 315, 610, 612
Bly, D. W. J., 597
Boekhorst, A., 896
Bogart, R., 202
Boite, R., 262
Bollen, D., 108, 442
Bombi, F., 163, 164

Bongiorno, J., 475, 680, 713, 799, 817, 841
Borbely, E., 682
Borlase, W., 408, 410, 411
Botos, B., 60, 183, 286, 288, 289, 293, 328,
    370, 372, 374, 732–734, 855, 857, 860,
    997, 1002
Bottaro, D., 85
Bouchard, R. J., 183
Bowman, D. R., 287, 294, 297, 341, 716, 719
Bowman, G., 393
Boyd, W. T., 339
Bracho, S., 1003
Brackney, R. C., 549
Brandon, C. W., 482
Brandt, R., 200, 201, 371, 574, 777
Bratt, B., 491, 492, 494
Breeze, E. G., 226, 229, 528, 997
Breikss, I. P., 145, 560
Bremner, A., Jr., 138
Brier, H. S., 883
Briles, B., 906
Bristow, Q., 953
Broch-Tonolio, F., 552
Brockman, D. M., 21
Brockman, H. P., 490
Broeker, B., 128, 332, 988
Bronzite, M., 257
Brookstone, A., 135
Brown, K., 434
Brown, R. G., 181, 184
Brown, R. M., 960
Brown, T., 2
Brown, W. G. S., 89
Brubaker, D., 507, 513, 569, 571, 970, 972,
    974
Bruggemann, H., 536
Brunner, R., 492, 550, 1007
Bruton, L. T., 577
Bryan, D. J., 636
Bryson, C. E., 349
Buchanan, J., 202
Buckley, R., 209
Budak, A., 575
Budek, J., 629, 859, 861, 862
Bugg, R. E. F., 739–741, 802, 858
Burch, W. M., 489
Burgess, H., 444, 456
Burlingame, R. G., 185
Burns, A., 854
Burwen, R. S., 207, 368, 474, 537
Buscemi, J., 982
Busse, J. C., 632, 692
Bussell, D. M., 359
Butler, D., 134
Butler, F., 167

Buttschardt, C., 892
Buyton, R. D., 457
Byerly, J. E., 36, 375, 622, 624, 625, 627,
    846, 872, 873
Byers, C., 126, 610, 612
Byrne, J., 544

Cachia, S. L., 110, 111
Cadwallader, M., 198
Caggiano, A. C., 82, 84
Cairns, J. B. F., 854
Callahan, M. J., Jr., 196, 198
Camp, R. W., 331
Campbell, D., 20, 625, 835, 918
Campbell, R. W., 586–588, 977
Cancro, C. A., 668
Cantarano, S., 447
Caringella, C., 42, 45
Carlquist, B., 245
Carpenter, B., 127
Carr, J. J., 52, 65, 67, 71, 72, 74, 839, 868,
    873
Carroll, R. L., 76, 459
Cartwright, R., 252
Carvey, P., 266
Caso, L. F., 566
Catford, M. B., 42
Cathles, C. R., 17
Cavit, A., 1003
Cawlfield, B. F., 78
Centore, M., 545
Chace, A., 195
Chan, M., 70
Chang, K. W., 458
Chang, S., 549
Chapman, M. A., 601, 602, 711
Chapple, W. E., 713
Chari, S. L. V., 64
Chase, D. E., 772, 774
Chau, T., 193
Cherubini, F., 442
Chin, D., 334
Cho, S., 6, 14
Choice, L., 574
Christensen, R., 465
Chu, J. W., 40, 746, 898
Chua, L. O., 163, 165, 166, 168, 194, 861
Chuang, S. S., 612
Cicchiello, F., 673
Ciscato, D., 163
Ciuciura, A., 339, 904
Claasen, C. H., 735
Clack, J., 445
Claire, S., 74
Clark, B., 797
Clark, V. R., 208
Clarke, J. M., 793
Clayton, G. B., 17, 428, 474, 479, 575
Clement, R. D., 605, 666, 860, 1014
Clements, A., 696
Cleveland, G., 366, 424, 427, 474, 583, 673,
    756, 1009
Cochran, J., 309, 312–315, 960, 963
Cocke, W., 706
Cocking, W. T., 109, 908
Coers, G., 534, 553, 591, 672, 756, 822, 930,
    943, 946
Cohen, H., 237, 567, 722
Cole, H. A., 665, 1006, 1009, 1012
Collier, W., 727, 768, 770, 772, 773
Colt, J., 66, 161, 380
Conner, D. C., 761
Connolly, A. P., 282
Cook, G., 404
Cook, K. A., 338, 342
Cook, M. E., 508, 512, 808
Cooke, W. A., 112

Cooper, R. E., 254, 255
Cope, P. C., 137
Corney, A. C., 458
Couch, T., 277
Counts, L., 475, 818–820
Cox, J. R., 573
Cox, N. W., 164
Craig, B., 319
Cram, T. R., 449
Cramer, T. V., 403
Crank, B., 126, 203, 227, 233
Crapuchettes, J., 124
Crawford, J. W., 85, 172, 784
Creason, S. C., 430
Cromartie, E. S., 1003
Crouse, P. E., 916
Crowe, J. W., 571
Crowley, M. B., 566
Crump, A. E., 369, 991
Cuccia, J., 637, 694, 935, 938
Curtis, J. G., 138
Cushing, P., 258

Dahlem, R. K., 664
Dahlen, P., 453
Damora, V. A., 62
Daniels, S., 138, 173, 246, 247, 251, 305,
    404, 473, 629, 635, 640, 957
Dann, G., 439
Danz, L. W., 821, 823
Das, S., 124
David, E., 408, 410, 411
Davidson, H., 934
Davies, H. C., 368
Davis, D. D., 317, 321
Davis, W. F., 597, 667, 956, 973
De Beer, J. Du P., 48
Debronsky, M. J., 592
DeKold, D. F., 12, 15, 125, 259, 311, 314,
    506, 661, 825, 860, 878
Delagrange, A. D., 264
DeLaune, J. M., 148
Delhom, L., 534, 552, 859, 957
Deliyannis, T., 576
Demaw, D., 169, 170, 172, 174, 266, 301,
    302, 304, 342, 350, 355, 449, 454, 456,
    605, 609, 728
Deming, C., 777
Dempster, P., 524
Denker, J. B., 759
Dennison, R. C., 448, 523
Devine, R. E., 785
Diamond, J. M., 614–616, 676
Diamond, L., 573
Dickson, A. D., 890
Dielsi, F. J., 83
Dietrich, J., 20
Dighe, K. D., 414, 417
Dilger, T., 817
Director, S. W., 12
Dishington, R. H., 391, 392
Distefano, M. I., 449
Divilbiss, J. L., 113
Dobkin, B., 367, 369, 405–411, 478
Doeller, C. H., III, 309
Doering, M. D., 523
Donkin, A. E., 922
Donn, E. S., 333
Donovan, L. E., 158–161, 282, 383
Downs, R. F., 364
Doyle, N., 43, 46, 261, 896–900, 902–905,
    999
Drake, H., Jr., 894
Driscoll, J., 991
Dromgoole, M. V., 477
Drumeller, C. C., 703
Dudley, M., 785

Duncan, J. H., 550
Duncan, T., 846
Dunlop, J., 115
Dutton, R. H., 720, 721
Dvorak, A., 1007

Ebenhoech, H., 142
Edwards, J. A. H., 390, 853
Edwards, L., 281
Eggleton, R. C., 481, 484
Ehni, G., 80, 160
Ellis, J. N., 52
Ellison, J. H., 267
Embree, M. L., 742
Emerson, F. E., 114
Emerson, P., 460
English, M. J., 780, 781
Ennis, J., 734
Epp, V., 436, 438, 727
Erickson, B. K., 7, 543
Erst, S. J., 67
Esakov, D. A., 545
Eschmann, J., 137
Esformes, M., 683
Estaugh, T. E., 348
Evans, A. G., 724, 725
Evans, D., 266
Evans, F. C., 296
Ewer, P. S., 741
Ewins, A. J., 363, 364, 469, 593, 596, 676, 717

Factor, R., 46, 287, 325, 352, 931, 935–937
Fahnstock, D. E., 237
Faiman, M., 553
Fajardo, R. J., 249, 376
Fallon, J. M., 218, 223, 288, 530
Farell, C. L., 56
Farkas, Z. D., 596
Farmer, J. O., 654
Fay, G. V., 382, 515, 517, 518, 789
Fazio, J., 566
Feldman, L., 50, 829
Feldman, S., 873
Fell, R. B., 874, 875
Felstead, C., 104, 105
Fergus, R. W., 473
Ferguson, E. E., 875
Fine, S. B., 482, 487
Firth, J. M., 562, 1010
Fischer, H. P., 42
Fishback, W. H., 178
Fisher, M. J., 544
Fisher, P. D., 742
Fisk, J., 269, 283, 364, 418, 426, 427, 440,
    678, 856, 857
Fitzpatrick, V., 134
Fleming, L., 518–520, 526
Fleshren, D. F., 422
Fontaine, G., 152
Foot, N., 131
Foote, J. L., 317
Forward, K. E., 196, 989
Foster, J. A., 137
Fox, F. J., 488
Fox, R. W., 187, 190, 193, 276, 279, 280, 344,
    348, 386–390, 403, 558, 616, 631, 765,
    787, 887, 921–925, 928, 960
Franco, S., 113
Frank, R. W., 140
Franklin, C. A., 877
Franson, P., 346, 647, 814, 815, 936
Frater, R. H., 614
Fred, J. A., 736
Fredericks, W. D., 132
Frederiksen, T. M., 19, 114, 127, 142, 143,
    257, 261, 262, 264, 298, 299, 330, 333,

Frederiksen, T. M. (Cont.):
  335, 512, 552, 582, 597, 599, 667, 670,
  671, 760, 782, 818, 826, 851, 856, 1000,
  1001, 1008, 1010
Freeman, A. R., 110
Freesek, G. F., 74
French, A. S., 482
Frenzel, L. E., Jr., 413, 416
Freyling, N., 517, 518, 789, 790, 792
Friedman, H., 119, 250, 345, 379, 454, 639,
  645, 838, 880
Fullagar, D., 643
Fury, A. E., 851

Gabbey, R. C., 74
Galbreath, C. E., 12, 981
Gallegos, J., 772, 775
Galloway, J. H., 54, 558
Gant, R. P., 520, 523, 526
Garland, H., 544, 545, 785, 851
Garner, L., 243, 555, 620, 646, 680, 821
Garrett, J. M., 922
Garvin, G. L., 629
Garza, P. P., Jr., 579
Gascoyne, P., 992
Geckle, W. A., Jr., 549
Georgiou, V. J., 260
Gerard, V. B., 517
Gerdes, R. C., 197
Gervenack, R., 906
Getz, F., Jr., 126
Ghest, C., 818
Giannelli, J., 446
Gibbs, L. V., 603, 891
Gienger, M., 752
Gilbert, J., 963
Gilbert, T., 799
Giles, T. G., 688, 690
Gill, R. T., 532
Girling, F. E. J., 255, 256, 258
Giroux, J. R., 542
Glauber, J. J., 106
Gleason, M. R., 130
Gledhill, B., 480
Gohl, W. S., 785
Goldberg, B. C., 78, 81
Golden, D. P., 485
Golden, V. X., 280
Goldfarb, W., 1012
Goldsby, E., 798
Goldstein, M. J., 967
Goldstein, S., 569
Golio, G. P., 846
Good, E. F., 255, 256, 258
Goodman, B., 701
Goodrich, H. C., 687
Goodson, C., 746
Gordon, C., 193
Gordon, M. J., 145, 199
Gorgenyi, I., 111, 386, 854
Gottlieb, I. M., 422, 460, 498, 606
Gotwals, J. K., 727
Gould, R. E., 977
Goulden, C. H., 100
Gowthorpe, A., 194, 710
Graat, W., 902
Grabowski, R., 266
Graeme, J., 825, 937, 938, 942–945
Graf, C. R., 223
Graf, R. F., 2, 63, 64, 68, 88, 96, 99, 559,
  564, 800, 847
Grafham, D. R., 224, 277, 403, 515, 517, 670
Graham, M. H., 482, 487
Granchelli, R. S., 1009
Green, C., 122, 173, 528, 635, 691, 699, 852
Green, D. E., 936
Green, D. J., 693

Green, R. G., 875
Greenbank, J., 54
Greenlee, L. E., 171, 277, 347, 637, 654, 996
Gregg, G., 467
Greif, K., 248
Greiter, O., 593, 595, 953
Grenell, R. L., 132, 572
Griesinger, D., 450
Griffiths, D., 575
Griffiths, H. N., 17, 275, 763, 765
Grillet, A. C., 859
Grossi, P. C., 718, 870
Gruentzler, R. E., 570, 572
Grutzmann, S., 268
Guard, R. L., Jr., 85, 975
Gusdorf, D. M., 639
Guyton, R. D., 253, 945
Guzeman, D., 150, 413, 415, 416

Haas, E. L., 706
Hadlock, C. F., 303, 953
Hafler, D., 685
Hagen, J., 292
Haigh, L., 447
Hakim, E. B., 936
Hall, C., 41
Hall, J., 129, 134, 455, 524, 888, 889, 892,
  893
Hamaoui, M., 792
Hambly, S. C., 64
Hamilton, B. H., 742
Hammes, J. P., 1005
Hanchett, G. D., 684
Hansen, J. P., 391
Hanson, H. J., 944
Hanson, P., 44
Harbourt, C. O., 254, 255
Hardcastle, I., 30, 655, 717
Harms, P., 664
Harper, T., 287, 289
Harrick, W. F., 601
Harrington, W. D., 671, 864
Harris, J. R., 647
Harrop, G. B. C., 484
Hart, R. T., 570
Hartke, J. L., 58
Harvey, M. L., 83, 661
Hatch, V., 672
Hattaway, J. R., 11
Haver, R. J., 86, 298, 381, 389, 616, 618, 649,
  857, 949, 1015, 1018
Hawkins, W. J., 105, 209
Hayward, W., 134, 252, 439, 962
Haywood, A. J., 719, 834
Heinlein, W. E., 268
Heinzl, J., 547
Hejhall, R. C., 15, 291, 302, 306, 311, 313,
  314, 506, 508, 510, 511, 702, 751,
  805–807, 810, 962, 976
Hekimian, N. C., 600
Helfrick, A. D., 318
Heller, L. F., 604
Heller, W. G., 719
Hellwarth, G. A., 581
Hemingway, T. K., 735
Hennick, R. P., 1008
Herbert, J. W., 59
Herbst, C. A., 7, 126, 451, 566
Herrick, K. C., 493, 494
Herring, L. W., 931
Hertzler, R. A., 556
Hibberd, F., 688
Hilberg, R. P., 391, 392
Hilberman, D., 38
Hildreth, H. R., 963
Hileman, D., 420, 1012
Hillard, T., 816

Hirschfeld, R. A., 46, 49, 52, 56, 57, 60, 600,
  699, 703, 761, 765, 796, 810, 823
Hjorth, S. R., 116
Ho, C. F., 548
Hoberman, S., 107
Hodgson, J. B., 890
Hodowanec, G., 564, 611, 1000
Hoeft, W., 625, 835, 918
Hoff, D., 695
Hoffman, B. F., 486
Hoffman, G. B., 495
Hoffman, P., 771
Hoisington, B., 16, 40, 173, 267, 351, 378,
  598, 607-611, 658, 659, 700, 704, 799,
  970
Holcombe, W. H., 858
Holder, F. W., 76, 79, 81
Holford, D. J., 72
Holland, J. P., 935
Hollingberry, P. L., 345
Holmes, W. H., 268
Holtham, P., 23, 512, 1009, 1010
Holzman, S. E., 336
Hook, W. R., 391, 392
Hopkins, J., 913
Horn, R. E., 9
Horna, O. A., 14
Houck, T. E., 449
Houser, J. A., 454
Howard, W. M., 19, 114, 127, 142, 143, 257,
  261, 262, 264, 298, 299, 330, 333, 335,
  512, 552, 582, 599, 670, 671, 760, 782,
  818, 826, 851, 856, 1000, 1001, 1008,
  1010
Howell, E. K., 280
Hoyt, E. M., 87
Huang, V. K. L., 89
Huber, C. J., 23, 335
Huehne, K., 406, 407
Huffman, J. R., 139
Hunt, R., 584, 585
Hunter, C. A., 563
Hutchinson, L., 929
Hutchinson, P. B., 50
Hutton, L. I., 910, 915
Hyland, M. L., 730, 731, 747

Isengard, W., 234
Ishii, T. K., 393
Ivanov, A., 397, 665
Ivins, R., 922, 927
Iwakami, A., 700

Jaeger, R. C., 581
Jahn, M., 135
James, K. D., 454
Jameson, C., 787
Jansen, J., 200
Janssen, D. J. G., 153, 182, 186, 226, 233,
  334–336, 674
Jarrold, E. J., 911
Jarvin, S., 929
Jenkins, A., 50
Jenkins, J., 260
Jensen, D., 131, 990, 992, 993
Jessop, G. R., 13
Jeutter, D., 651
Joffe, A. S., 154, 293, 726, 990
Johnson, D., 751, 759
Johnson, F., 726
Johnson, H. D., 762, 764, 766, 859
Johnson, J. H., 15, 862
Johnson, N., 390, 429, 430, 432
Johnson, R. W., 421
Johnson, T., 19, 293
Johnston, G. I., 518, 520, 521

Johnstone, J., 457, 459, 477
Jolly, N. A., 28
Jones, B. E., 427
Jones, B. L., 911
Jones, D. V., 32, 35, 36, 148, 151, 152, 203, 204, 621, 709, 730, 731, 747, 832, 844, 897, 911, 914
Jones, F. C., 311, 316, 455, 456
Jones, G. W., 354, 890
Jones, H. E., 521
Jones, J. R., 604
Jonkuhn, G., 705
Jordan, G. B., 601, 975
Jorgensen, L., 657
Joy, R. C., 489
Jung, W. G., 51, 544, 684, 709, 737, 795, 797, 1011

Kadish, D., 889
Kado, R. T., 702
Kamnitsis, C., 18
Kampel, I. J., 462, 463, 465
Kaplan, L., 679
Karwoski, R. J., 398
Kashubosky, R., 646
Kasser, J. E., 508
Kaufman, B. M., 502, 504
Kaufman, E. N., 988
Kaufmann, J., 169, 355
Keedy, P., 49
Keenan, P., 300
Keeney, C. G., 741
Keith, J. L., 652, 943
Kelk, W. H. H., 603
Kell, V., 106, 377, 418
Kelland, R. E., 698
Kelley, B., 296
Kelly, S., 702
Kennedy, J. F., 556
Keohane, M., 690
Kerley, B. E., 462
Kertesz, B., 771
Kertzer, R., 874, 875
Kesner, D., 500, 504, 742, 748, 752
Kierstead, D. L., 480
Kilvington, J., 777
Kincaid, W. B., 773
Kindlmann, P. J., 186, 191
King, H., 959
King, T., 440
Kitsopoulos, S. C., 334
Klig, V., 487, 488
Klinert, C., 20, 154, 301, 769, 775
Klinikowski, J. J., 231, 233
Knapp, R., 580
Knausenberger, W. H., 358
Knibb, R. R., 262
Knight, R. E., 636
Knowles, C. H., 396
Knowlton, D. J., 373, 585
Koehler, R. B., 955
Kolody, O. A., 896
Kolodziej, J., 131
Korn, S. R., 448, 588, 590
Korth, B., 350–353
Koss, R., 929
Koval, B., 784
Kowols, G. J., 776
Krabbe, H., 207, 474, 537
Krause, V. R., 595
Krusberg, R. J., 341
Kubo, A. M., 674, 876
Kvamme, E. F., 200
Kyle, J., 71, 294, 297, 476, 527, 549, 640–644, 738, 862, 886, 1011, 1014

Lacey, P., 73, 714, 716, 739
Lafko, E., 662
Lamb, W., Jr., 664
Lambing, W. P., 767
Lampard, D. G., 16
Lamprecht, E. R., 891
Lampton, M., 329
Lancaster, D., 152, 181, 554, 714, 800, 814, 935, 996
Lane, B., 30
Langvad, J., 939
Larsen, E., 211
Latham, G. R., IV, 655
Laubach, R. T., 589
Laughlin, J. R., 832
Laughter, J., 1006
Laurino, A. J., 154
LaVaia, B. J., 923
Lavallee, J., 738
Lazar, D. J., 974
Lazarus, M. J., 698
Lea, R. M., 58, 822, 964
Leavey, M. I., 888
Lee, D. M., 1002
Lee, D. V., 910
Lee, H. B., 266
Lee, J. Y., 222, 223
Lehner, L., 384
Lembke, C. H., 705
Lemen, W. T., 347, 952
Lemons, W., 378
Leo, A., 242, 250
Leuenberger, K., 613
Levy, E., 177
Lewallen, R. W., 150
Lewart, C. R., 459, 788
Lewis, F. D., 437
Lieux, J. D., 737
Lin, W. C., 493
Lingane, P., 83
Linsley Hood, J. L., 7, 8, 10, 33, 38, 443, 478, 623, 718, 842
Linvill, J. G., 489
Lisle, L., 178
Locher, R. E., 632, 634, 921–925, 928
Lockridge, J. E., 4
Long, D. K., 141, 1006, 1007
Long, E. L., 624, 627, 872, 873
Lord, E., 222, 223
Loughnane, M. H., 1004
Lovallo, J., 892
Lovelock, P., 399, 440
Lowe, A. J., 639
Lowe, B., 972
Lubarsky, D., 736
Lucas, R. R., 133
Lukoff, H., 82, 422
Lumb, P., 436
Lynch, D. J., 437
Lynes, D. J., 146

McCaffrey, T., 371, 574, 777
McCartney, M. L., 628
McClellan, G., 933, 977
MacCluer, C. R., 375
McCormick, E. M., 54
McCoy, L. G., 132, 265, 269, 435, 452, 483, 961
MacDonald, H., 327, 425, 429
McDonald, M., 208, 210, 211, 290, 293
Macdonald, P., 561, 563, 565
McDowell, J., 495
McGee, A. E., Jr., 569
McGowan, E. J., Jr., 91
MacHattie, L. E., 458
McKinley, R. J., 413
Maclean, M. A., 877

MacLeish, K., 291, 723
McMullen, T., 177, 301
McMullen, T. F., Jr., 984
McProud, C. G., 836
McTaggart, T., 378
McWhorter, M., 741
Magee, W. A., 608
Malham, D. G., 252, 1002
Malinowski, G. J., 936
Mallory, H. R., 764–766, 918
Mammano, R. A., 758, 760
Manco, R. J., 990, 991
Mandl, M., 237–239
Mangieri, A. A., 89, 433, 637, 638, 695
Manly, E. P., 431
Mann, M., 260
Marcus, C., 718, 876
Marjanovic, S., 863
Markette, F., 795
Marovich, 460
Marriner, E., 701, 970
Marshino, J. B., 941
Marston, R. M., 858, 921
Martens, C., 968, 980
Martin, L., 295
Martin, P., 604
Martin, S. R., 567, 1006
Masson, C. R., 546
Matheson, R. J., 691
Matteis, R. M., 269
Matteson, F. M., 276, 278, 669, 930, 953, 956, 989
Mauch, H., 720
Mauldin, D. G., 485
Maurer, L. C., Jr., 986
Mawson, K. W., 87, 539
May, J. T., 11
Maynard, E. C., 762, 763
Maynard, F., 248, 404
Mazur, T., 1005
Means, J. A., 37, 40
Megirian, R., 353, 354, 723, 804, 855, 857
Melen, R., 259, 580, 785, 851
Mengel, B., 822
Mentler, S., 74, 82
Merritt, F., 887, 892, 893
Metz, A. J., 333
Meuron, S., 122, 123
Meyer, D., 96, 272, 295, 555, 647, 681, 833, 837, 995, 996
Meyerle, G., 78, 84, 273
Michael, M., 493
Michaels, R., 73, 242, 346, 389, 636, 643
Mickle, D. D., 403
Miler, G. G., 143, 327, 328, 791
Miller, D. C., 971
Mills, T. B., 907, 909, 914
Mims, F., 356, 357, 392, 394–396
Minton, R., 18
Mitchell, G., 471
Miwa, Y., 15
Montan'e, R. T., 555
Montevaldo, R., 186, 189
Moody, N. F., 470, 474
Moore, A. W., 793
Moore, R. D., 23
Mooring, E. E., 885
Morris, C., 86
Morris, E., 123, 235, 445
Morris, J., 86, 136
Morris, R. A., 181, 184
Morrison, R. D., 176
Mortensen, H. H., 155, 369
Moser, C., 149
Moser, D., 171
Moss, W., 495
Mouritsen, R. W., 678
Mrazek, D., 128, 146, 147, 202

Murray, E. J., 149, 187, 190
Myers, R. M., 134, 342, 721

Nagarajan, S., 330
Namekawa, T., 15, 900, 913
Neal, G., 823
Nelson, D. W., 174, 175, 724, 725
Nelson-Jones, L., 433, 446, 449, 741, 837,
    839, 844, 845, 921, 927
Nesbitt, D. R., 602, 603
Newell, A. F., 509, 537, 538, 801
Newmire, L. J., 153, 183, 374, 387, 650, 732,
    733, 855, 857, 860, 1002
Newton, D., 295
Nichols, B. B., 824
Nichols, F., 685
Nichols, J., 300, 321, 705
Nickel, L., 275
Nicola, N. S., 324, 1004
Nicosia, N. J., 136, 592, 712
Nightingale, D. E., 606
Niu, G., 144
Noaks, D. R., 863
Noll, E., 40, 129, 297, 298, 306, 439, 508,
    509, 602, 692, 696, 804, 887, 971, 980,
    981, 985
North, W. L., 961
Novak, S., 698
Nunley, J. A., 103, 236
Nurkka, G. A., 45
Nurse, H. L., 128
Nusbaum, A., 711

Oakes, M., 798
Obrda, P., 586
O'Brien, J. J., 822
Ockleshaw, R., 622
Oehlenschlager, J. G., 770
Oehmen, S., 809
O'Haver, T. C., 168
Okuno, K., 15
Olberg, S., 692, 893
Olson, H., 93, 95, 290, 435, 591–595, 604,
    608, 725–727, 803, 895, 966
Orr, W. J., 662
Osborn, R. R., 125
Osborne, W. E., 94
O'Toole, T., 331
Otsuka, B., 584, 585
Otsuka, W., 387
O'Veering, G. I., 679
Owen, T. R. E., 739

Pacak, M., 425, 427
Padalino, J. J., 260
Pafenberg, R. E., 784
Pallottino, G. V., 447
Papadopoulos, G. D., 310, 311
Pappot, G., 446
Pardoe, A. M., 687
Paris, D. A., 663
Parker, L., 472
Parten, M. B., 985
Pastoriza, J. J., 23
Pate, J. G., 202
Patterson, D. R., 711, 912
Patterson, R. P., 574
Pattison, H. O., 291
Pearson, E. E., 46, 124, 144, 697
Pecen, M., 308
Pederson, D. O., 608
Penny, R. W., 1004
Peri, R., 206
Perkins, D., 648
Perkins, P. J., 884

Persing, R., 347
Peterson, W. C., 924, 926
Peterson, W. E., 675
Phillips, D. H., 886
Phillips, H., 966
Phillips, M., 108–110, 112, 114
Phillips, R. A., 951
Phillips, V. J., 115
Pichulo, R., 771
Pierce, J. F., 11, 192
Pierce, R. D., 574
Pike, J. M., 259, 411, 425, 724
Pina, J., 882
Pinner, W., 135, 449, 848, 953
Pinnock, P. E., 1010
Piper, J. R. I., 653
Pippenger, D., 202
Pitzalis, O., 9
Poggi, G., 546
Polcyn, T. E., 686
Pollock, J. W., 549
Pongrance, D., 175, 648, 693
Poole, D. A., 801
Portus, R. T., 719, 834
Potter, K. E., 603
Pouliot, F., 818–820
Powell, G. A., 291
Powell, L., 378, 883
Preiss, R., 960
Prensky, S. D., 471, 476
Presley, H., 611
Price, D. J., 372
Price, W. E., 991
Prinn, A. E., 621
Prioste, J. E., 490
Proefrock, J., 132
Purland, D., 710
Pye, C. A., 592
Pye, J. D., 640

Quilter, P. M., 54

Rakes, C. D., 1, 92, 240, 245, 443, 497, 499,
    786, 882
Ralph, N., 974
Ramsey, J., 813, 823
Rankin, C. A., 138
Rao, M. R. K., 64
Rawlings, G. M., 272
Rayer, F. G., 807, 971
Reece, J. S., 738
Reedy, E. K., 192
Rees, L. T., 155, 635, 644, 1015, 1018
Reich, B., 936
Reich, S., 830
Reichel, A. H., Jr., 712
Reid, R., 575
Reinert, J., 500, 501
Renschler, E., 140, 500, 501, 539–541, 576,
    578, 582, 750
Repass, R. K., 940
Reschovsky, J. M., 598, 990–992
Reynold, J. F., 507
Reynolds, T., 534
Richards, B., 162, 284
Richardson, E., 4, 422
Richardson, T., 110
Richmond, G., 384
Richwell, G., 853
Ricks, B., 39, 40, 142, 737, 744, 746–748
Riff, J., 646
Rigby, G. A., 16
Riley, G. A., 452, 641, 642
Riley, T. P., 488
Ring, C. L., 694

Ritchie, J., 43
Robbins, K. W., 698
Robbins, M. S., 800, 932, 952
Roberts, A., 834
Roberts, J. A., 28, 316, 368, 991
Rogen, B. R., 374
Rogers, E. H., 414
Rogers, R. C., 286
Rohen, J. E., 45
Rohrer, R. A., 12
Rolf, J. A., 451
Ronald, M., 4, 859
Rosen, A., 507
Roshon, A. H., 545
Rothwell, G., 595
Roulston, J. F., 778
Routh, W. S., 118, 373, 475, 541, 599, 755
Rowe, C., 768
Rowe, W. M., Jr., 612
Rufer, R. P., 22
Rugg, H. H., 136
Ryan, O. K., 246

Saddler, J., 71
Sahm, W. H., III, 195, 245, 270, 276, 278,
    669, 729, 747, 930, 953, 956, 989
St. Laurent, H., 404, 883
Sakellakis, J., 887
Salomon, P. M., 543
Salvati, M. J., 421, 731, 795
Sargent, A. J., 937
Sastry, J. V., 294
Saw, J. E., 688, 690
Scarpulla, R. T., 544
Schaffner, G., 312, 315
Scharf, K., 418
Schauers, C. J., 3, 102, 139, 348, 701, 957,
    982
Schecner, A., 460
Scher, G. P., 227, 310
Schiers, R. A., Jr., 172
Schmitt, W. A., 391
Schoenbach, H., 420, 421
Schopp, W. W., 638, 639, 836
Schreyer, G., 722
Schrock, D., 750
Schuermann, J., 352
Schultz, J. J., 37, 268, 269, 271, 274, 353,
    536, 693, 805, 806, 809, 810, 812
Schwaneke, A. E., 483, 487
Scott, H., 90
Scott, M., 125, 128, 493–495
Scott, R. F., 273, 450
Sebol, R., 340
Seling, T. V., 601
Sessions, K. W., Jr., 18, 130, 379, 770, 773,
    774, 952
Sewell, A., 927
Shafer, A. G., 979
Shah, M. J., 463, 542, 671, 736, 933, 1013,
    1014
Sharp, D., 339
Shaw, I. M., 38
Shaw, W. S., 329
Sheatz, G. C., 620
Sheffer, E., 179
Sheingold, D. H., 206, 406, 408, 411, 510,
    779, 819
Shelton, G. B., 562
Shepler, J. E., 1
Sherwin, J. S., 197, 198
Shiner, B. C., 86, 298, 618, 647, 649, 857
Short, P., 930
Shorter, G., 681, 682, 684, 685
Shreve, J. S., 68
Shreve, R. B., 767, 881
Shubert, G. K., 266, 614

Shultz, J. J., 459
Siebert, R., 673
Siefert, A. V., 573
Siegel, B., 125, 128, 370, 473, 493–495, 782, 1010
Siemensma, J. M., 961
Silberstein, R., 43
Simciak, W., 153, 670
Simonton, J. S., Jr., 63, 95, 234, 237, 239, 241–244, 251, 277, 345, 761
Singh, D., 447
Sir, S., 60, 907
Skjelstad, J., 942
Skutt, H. R., 874, 875
Sleeth, R. S., 19, 114, 142, 143, 257, 261, 262, 264, 298, 299, 330, 552, 582, 599, 670, 671, 760, 782, 818, 826, 851, 856, 1000, 1001
Sloat, J. G., 271
Small, J., 387
Smathers, R. T., 127, 143, 333, 335, 512, 1008, 1010
Smeins, L. G., 998
Smith, B. R., 115–118
Smith, C., 342
Smith, D., 898–900, 903–904
Smith, D. A., 710
Smith, D. T., 249
Smith, F. E., 726
Smith, H. B., 697
Smith, J. H., 175, 704
Smith, K. L., 357
Smith, R., 705, 829
Smither, M. A., 22
Snaper, J., 298
Snow, B. C., 346
Snyder, R. V., 15
Sofia, A. J., 1012
Solomon, L., 675
Sonde, B. S., 510, 511
Sondgeroth, C., 607, 797, 856, 888, 965
Sowden, T. R., 697, 721
Spadaro, J. J., 44
Spencer, E., 471
Spencer, R. D., 714
Sperduti, A., 885
Sperling, L., 853
Splichal, W. F., Jr., 214, 275, 800
Spofford, W. R., Jr., 827
Spurgeon, C., 439
Stahler, A. F., 139
Stanaway, J. J., Jr., 404
Standke, W., 661
Stanton, W., 171
Stapp, I. P., Jr., 262
Stark, P. A., 3, 289, 292, 299, 321, 322, 434, 724, 731, 743, 774, 880
Starliper, R. L., 297, 547, 605, 666, 676, 718, 860, 1011, 1013, 1014
Stayton, J., 65
Steed, W. L., 438
Stein, L. B., Jr., 925
Stein, R. B., 482
Stein, R. S., 320, 712, 942, 998
Steinbach, D. L., 191, 297
Steine, M. L., 133
Steiner, N. A., 852
Steinman, A. J., 663
Stephenson, E., 487, 488
Sterner, J. F., 365
Stinehelfer, J., 300, 321, 561, 563–565, 705
Stinnette, N., 377, 713, 933
Stogel, J., 69
Stone, K., 131, 546, 805
Stopka, B., 650
Stortz, W. F., 383, 385, 983
Stover, H. H., 86
Strange, M., 825

Strangio, C., 279
Stratman, H., 656
Strauss, R. W., 334
Stremler, F. G., 262, 263
Stuart, J. R., 740, 864–867, 869, 870, 872
Su, K. L., 164
Subbarao, W. V., 310
Suding, R. T., 912
Sugata, E., 900, 913
Sula, S., 419
Sullivan, R. G., 663
Sum, K. K., 313
Summer, S. E., 66, 70
Susskind, C., 482, 487
Sutherland, B., 170, 984, 987
Suzuki, H. S., 907, 909, 914
Swedblom, C. E., 985
Swinnen, M. E., 485
Symons, L. R. P., 585
Szentirmai, G., 268

Tackett, R. P., 67
Taggart, R. E., 909, 910
Takesuye, J., 385, 749, 752
Tashetta, J. J., 815
Taylor, D., 798
Taylor, J., 716
Taylor, M., 368
Taylor, R. S., 812
Telewski, F., 806, 812
Tellefsen, R. N., 536
Temcor, W., 90
Terry, F. D., 662
Tesic, S., 166
Thacker, A., 924, 926
Thai, N. D., 472
Thickpenny, J., 444
Thiele, A. G., 6, 14
Thielking, R. C., 896, 903
Thiran, J. P. V., 262
Thomas, F. D., 133
Thomas, J. P., 80
Thomas, L. D., 593
Thomas, S., 280, 652
Thompson, C. N., Jr., 146
Thompson, R. A., 738, 964
Thompson, S. A., 22
Thorpe, D., 244
Tijou, J. A., 31
Tillotson, G., 437, 438, 441
Tilton, E. P., 984
Tirrell, J. C., 879
Tomsic, R. C., 687
Tong, D. A., 350, 546
Tooker, F. H., 3, 101, 184, 271, 273, 324, 347, 412, 413, 415–417, 432, 436, 480, 487, 547, 593, 608, 637, 638, 652, 664–666, 675, 677, 714, 800, 814, 991, 993, 1001
Torok, L. K., 414
Toth, L., 737, 1001
Towers, T. D., 6, 410, 429, 478, 622, 627, 656, 739, 867, 873, 1013
Townsend, J., 967
Travis, W. J., 459
Trietley, H. L., Jr., 367
Trinogga, L. A., 455
Troemel, C. P., 467
Trout, B., 60, 354, 708
Trusson, C. I. B., 130
Tucker, M. J., 563
Turino, J. L., 461
Turner, R. J., 5, 165, 332, 399, 400, 687, 689
Tuszynski, A., 260
Tweit, N., 674

Uhler, P. T., 651, 863

Ulrick, C., 853, 1013
Underwood, R. K., 141, 196, 366, 476, 583, 781, 782, 877
Uno, M., 943
Utz, C., 398

Vaceluke, R., 472
Vale, L. H., 132
Vanderelli, T., 936
Vander Kooi, M. K., 24, 36, 366, 375, 424, 427, 474, 583, 622, 625, 627, 673, 756, 757, 1009
Vanderkooy, J., 39, 511
Van Dick, R., 296
Van Duijn, J., 189
Van Geen, J., 895
Van Zant, F. N., 355
Van Zee, J., 584
Vaughan, D. E., 113
Venkateswarlu, V., 510, 511
Vicens, L., 786
Vicente, J., 525
Victor, A., 13
Vilardi, D., 7, 10, 19, 171
Vincent, W., 674
Vogt, J., 692
Volk, A. M., 253
Votipka, W., 625, 835, 918
Voulgaris, N. C., 59, 513

Waddington, D. E. O., 715, 828, 942, 946
Wade, J. M. A., 13
Waggoner, L., 771
Wald, S., 67
Walker, H. P., 51, 53, 621, 622, 717, 832, 833, 840–844
Walker, L., 210, 214, 216, 217
Walker, R. M., 201, 202
Wallace, R. J., 793
Wallace, T., 957
Walsmith, F., 613, 792, 794
Walton, R. A., 638, 719, 802
Ward, B., 75
Ward, C. C., 546
Ward, R. J., 798
Warren, F. A., Jr., 650
Watters, D. V., 811, 982
Webber, G., 172
Weber, H. F., 525, 749, 752
Weber, S. F., 856
Weinberger, R. G., 424, 426, 428
Weinstein, M. B., 880
Weir, J. P., Jr., 723
Weiss, A., 979, 983
Wendland, P. H., 563
West, R. P., Jr., 461
Westlake, R., 20
Wetherhold, E. E., 269, 886, 887
Whalen, G. J., 2, 63, 64, 68, 72, 88, 96, 99, 174, 559, 564, 800, 847
Wherry, D. M., 872
White, H. S., 796
White, J., 173, 176, 456, 987
White, T., 962
Wicklund, J. B., 238, 239
Widlar, R. J., 10, 21, 51–55, 64, 66, 74, 141, 143, 195–198, 207, 256, 261, 337, 338, 365–367, 372, 405, 406, 538, 541, 552, 731, 735, 736, 750–757, 760, 782, 783, 791, 804, 817, 927, 1005
Wiker, R. L., 412
Wilbur, R. L., 876
Wildenhin, B., 269, 445
Wilke, W., 653, 931
Wilkins, R. G., 78, 81
Wilkinson, J. H., 16, 19

Williams, D. A., 339, 865
Williams, F. D., 254, 255, 257–259, 262–264, 353, 355
Williams, M. W., 332
Williams, P., 596, 739, 1014
Williams, R. E., 901, 904
Williamson, K. H., 382
Williamson, R., 930
Wilnai, A., 207
Wilson, A., 177, 342, 971, 978
Wilson, G., 301
Wilt, R., 50
Winklepleck, R. L., 978
Wisseman, L. L., 479

Wittlinger, H. A., 419, 420
Wittmer, C. M., 578
Wolf, E., 516
Wolf, K., 1005
Wolfe, I. R., 969, 976
Wolff, R. A., 639
Wolfram, R., 274
Wollesen, D., 465
Wolthuis, R. A., 485
Wood, G. M., 109
Wood, V., 231
Woodward, R. P., 164
Woody, R., 246
Wrench, O., 654, 954

Yang, E. S., 59, 513
Young, J. B., 77
Young, R. G., 48, 415, 568, 858

Zadig, E., 85
Zicko, P., 115, 117, 118, 464, 477
Zimmer, S. R., 515, 517
Zinder, D. A., 87, 185, 387, 389, 390, 521, 616, 619, 751, 923, 940, 941
Zipin, L., 976
Zobel, D. W., 597, 667
Zwart, P., 263

Subject Index

Acceleration, measuring with pendulum-driven bridge, 72
Acoustic beacon, 1-pps pinger mvbr, 563
Acoustics, sound-level meter, 450
Active filter (see Filter, active)
Adapter:
  455-kHz narrow-band f-m, 305, 399
  slow-scan tv uses cro as monitor, 906
  vtvm for multimeter, 476
  (See also Converter, radio)
Adder:
  400-kHz and 25.767-kHz pulses, 184
  full using gates, 335
  half using gates, 335
  opamp inverting with resistive feedback, 575
  squaring 4-bit binary number, 818
Adding machine, binary demonstration, 147
A-f (see Audio)
Agc (see Automatic gain control)
Air conditioner, thermostatic speed control, 922
Air pollution, infrared auto exhaust gas analyzer, 359
Aircraft:
  75-MHz beacon receiver, 565
  75-MHz marker-beacon tone filter, 561, 565
  117-150 MHz superregenerative receiver, 698
  118-136 MHz 7-W a-m amplifier, 978
  audible rate-of-climb indicator, 562
  marker-beacon tone discriminator, 563
  receiver for pilot talk, 692
  vor signal conditioner, 560
Alarm:
  25-kHz ultrasonic intrusion detector, 995
  all-purpose with siren output, 2
  auto brake fluid level, 69
  auto idiot-light, 63, 72
  auto low-fuel, 61, 64, 66
  auto rear-lamp failure, 63
  auto speed, 63
  auto warning-lamp blinker, 70
  automatic siren, 815
  battery discharge, 651
  battery voltage monitor, 89
  beeper, 3
  blinking-light for message, 347
  boat theft on trailer or in water, 85
  burglar-scaring horn, 1
  carrier-operated, 3, 4
  clicker using relaxation oscillator, 4
  darkroom excess-light, 636
  fail-safe for cut wires and shorts, 996
  flame-failure resistance-sensing, 274

freezer failure, 2
frost for plant protection, 3
Hartley oscillator with speaker, 98
headlight-on, 65, 67, 71, 72, 74
headlight-on and idiot-light, 67
howler for obscene phone calls, 882
humidity, 450
icy-road, 66, 70
light-sensitive clicker, 93, 271
line-failure emergency light, 390
low a-c line voltage, 654
low radiator-water, 67
low-voltage monitor lamp for 12-V supply, 1005
low windshield-water, 73
mobile equipment theft, 82, 85
moisture, 1, 4
overvoltage with crowbar circuit, 650
pool splash, 403, 404
power-failure, 2-4
pushbutton siren, 814
rain, 1, 3, 4, 445
remote, for ringing telephone, 883
remote-control using power-line carrier, 761
storage-battery undervoltage warning, 86
storm warning, 1
sun-triggered clock, 632
temperature for high-power amplifier, 271, 274
three-tone door-identifying oscillator, 349
two-tone switched at 5Hz, 814
voice-actuated, for the deaf, 484
wailing electronic siren, 816
water or flood, 403
(See also Auto theft alarm; Automotive; Burglar alarm; Protection; Siren)
Alignment (see Signal generator; Tester)
Alternator:
  400-Hz 4-kVA regulator, 738
  voltage regulator, 73
Amateur radio (see Citizens band; Code; Repeater; Telephone; Teleprinter; entries for all types of circuits used in amateur radio receivers, transceivers, and transmitters)
Ammeter:
  1-μA using d-c opamp, 429
  1-μA using IC current amplifier, 425
  5-A a-c using thermocouple, 429
  10-range using two opamps, 479
  4-transistor microammeter in voltmeter, 473
  d-c micropower using programmable opamp, 427
  Hall-effect d-c transformer, 427

power-supply current monitor, 424
(See also Multimeter)
Amplifier:
  0.01-4,000 Hz with 89-dB step attenuator, 325
  0.1 Hz-1 MHz instrument output, 368
  0.16 Hz-800 kHz for capacitive sources, 10
  1.5 Hz-150 kHz IC driver, 13
  6 Hz-500 kHz 550-mW, 19
  20 Hz-7 MHz with photocoupler, 585
  30 Hz-3.5 MHz direct-coupled, 17
  50 Hz-2 MHz high-level instrument, 368
  800 Hz-32 MHz for frequency-counter input, 19
  10-10,000 kHz using three IC's, 6
  40-kHz carrier power, 924
  50-kHz at 200 W, 12
  100-kHz directly coupled, 16
  150-kHz opamp with 40 dB gain, 17
  470-kHz high-gain using Q multiplier, 539
  0-350 MHz crt cathode driver, 112
  0-500 MHz differential mosfet, 111
  0.65-MHz frequency-selective IC, 16
  1-MHz fet-input video, 15
  1-40 MHz broadband r-f, 16
  1.5-30 MHz 25-W ssb linear, 808
  1.5-30 MHz 100-W linear, 983
  2-30 MHz 60-W untuned linear broadband, 9
  2-30 MHz 80-W ssb linear, 810
  2-50 MHz 25-W ssb linear, 805
  3.5-28 MHz 200-W linear, 13
  3.5-28 MHz 600-W linear, 984
  3.5-28 MHz 2-kW linear, 985
  3.5-30 MHz 15-W linear, 972
  5-120 MHz wideband for frequency counter, 287
  10-MHz using active bandpass filter, 255
  14-MHz using active filter, 257
  20-MHz line driver with 17 dB gain, 5
  20-160 MHz untuned IC video, 18
  20-230 MHz CATV distributed, 910
  30-MHz 80-W linear ssb, 810
  30-MHz voltage-variable attenuator, 562
  40-MHz bandpass, 15
  45-MHz wideband, 57, 58
  50-MHz 600-W grounded-grid, 984
  87-MHz computer-designed, 12
  100-MHz cascode fet, 6
  100-MHz common-source fet, 16
  118-136 MHz 7-W a-m aircraft, 978
  127.333-MHz solid-state 1-W, 812
  144-MHz 1-W single-transistor, 965

Amplifier (Cont.):
146-MHz 40-W, 15
146-175 MHz 25-W, 973
146-175 MHz 50-W, 972
150-MHz low-drain wideband, 6, 14
150-325 MHz using impedance-matching filters, 15
175-MHz 30-W for 28-V supply, 973
200-MHz common-gate fet, 15
220-MHz 230-W coaxial-line, 982
220-MHz 580-W single-tube linear, 984
220-MHz fet, 700
220-MHz power, 979
225-400 MHz 17-W IC, 18
300-3,000 MHz temperature-compensated, 7
432-MHz 40-W, 982
432-MHz linear, 806
448-MHz 5-W for repeater, 770
450-MHz 6-W, 968
470-MHz common-gate fet, 17
2-meter 1-kW linear, 976
2-meter 3-W f-m driver, 306
2-meter 10-W, 18
2-meter 25-W f-m, 962
2-meter 50-W f-m, 302
2-meter 200-W transceiver, 967
2-meter fet preamp (146 MHz), 20
6-meter 1-kW linear, 969
10-meter 15-W ssb linear, 804
10-15-20-40-80-160 meter 1-kW linear, 980
3-W differential-output IC, 18
750-W grounded-grid linear, 974
4-meg input impedance, 17
automatic tuning of transmitter, 792, 794
bandpass using linearized TTL inverter, 7
bootstrapped, 16, 17, 19
class-B power with current limiting, 10
controllable-gain pnpn tetrode, 59
Darlington, 7, 8
diode-stabilized fet, 12, 15
directly coupled with differential input, 12
feedback opamp speeds phototransistor response, 628
floating chopper with common-mode voltage rejection, 115
frequency-response plotter drive, 934
gyrator without bias-level latch-up, 11
high-gain low-offset a-c opamp, 789
high-impedance buffer for vfo, 12
high-voltage opamp, 19, 114
hysteresis with high input impedance, 791
impedance converter with 1.5-meg input and 40-ohm output, 166
impedance converter with 1-teraohm input and 50-ohm output, 166
low-drift fet follower as opamp buffer, 18
low-frequency for pulsed photodiode, 360
low-noise fet for spacecraft magnetometer, 447
low-power complementary-transistor, 20
multiplier using long-tailed pairs, 538
overheating alarm, 271, 274
parametric variable-capacitance diode chopper, 116
protection against open-circuit damage, 650
push-pull a-c for feedback control, 926
remote gain control, 765
square-law transfer function, 1008
thermistor bridge, 922
track and hold, 781
vhf front-end using dual-gate mosfet, 60
voltage-controlled gain for analog multiplier, 59
voltage-follower with only 1-ns delay, 14
(See also Audio amplifier; Chopper; D-c amplifier; Differential amplifier;

Instrumentation amplifier; Intermediate-frequency amplifier; Operational amplifier; Preamplifier; Servo; Video; specialized circuits using amplifier stages)
Amplitude modulation (see Demodulator; Modulator; Receiver; Transmitter)
Analog:
bidirectional fet switch, 862
bipolar multiplexer, 535
converter to pulse width, 876
four-quadrant multiplier, 536
height-to-width converter, 165
integrator with jfet opamp, 371
integrator with reset mode, 371
logarithmic amplifier with five-decade range, 574
multiplexer 6-channel switch, 531
multiplexer 10-channel switch, 531
multiplier with adjustable scale factor, 538
multiplier using reciprocal circuit, 537
multiplier using single IC, 540
multiplier agc, 59
sample-and-hold with low capacitance leakage, 777
switch-multiplexer, 528
three-mode integrator, 373
(See also Digital-analog converter; all other Analog entries)
Analog commutator, expandable four-channel, 21
Analog comparator, square-wave generator, 661
Analog-digital converter:
8-bit tracking, 24
12-bit using 1-out-of-16 decoders, 25
commutator with buffered output, 21
digital voltmeter, 22
go-no-go low-cost tester, 23
Gray code, 22
height-to-width Gray code, 165
high-precision for digital power-factor meter, 163
high-speed sample and hold, 779
high-speed using seven IC comparators, 23
illegal code suppressor, 24
ladder-input voltage comparator, 143
ladder-network drive, 25
linear vco, 997
minimizing battery drain, 23
npn transistor curve tracer, 945
synchronous ramp generator, 21
trigger using zero-crossing detector, 22
voltage to pulse width, 163, 164
Analyzer:
infrared for auto exhaust gas, 359
sstv and rtty signals, 912
Anemometer, 0-30 mph wind gauge, 445
Antenna:
fast stopping of a-c rotator motor, 525
high-speed three-diode switching gate, 963
noise bridge, 570
protection from lightning, 966
t-r switch, 961, 963
Antenna matcher, swr bridge, 979
Antenna switch, 160-MHz, 959, 965
Antilog, voltage, 410
Antilog generator:
exponential output voltage, 408
opamps with transistors, 411
Applause meter, tube amplifier with a-c supply, 448
Apple grader, amplitude classifier, 631
Aquarium:
low-voltage heater, 929
temperature control, 922

Aquarium heater, arc suppression with power-line filter, 928
Arctangent, approximating with log module and opamp, 406
Astable multivibrator (see Multivibrator, astable)
Athlete, four-channel telemeter, 874, 875
Attenuator:
30-MHz voltage-variable with constant phase, 562
89-dB step for a-f function generator, 325
constant-impedance for audio oscillator, 596
electronic for remote volume control, 761
electronic remote-control for IC speech amplifier, 766
hi-fi headphones, 830
infinite for measuring vhf receiver sensitivity, 608
ladder-type receiver-input, 701
receiver input with 3-dB steps to 33 dB, 701
Audio:
300-2,000 Hz bandpass filter for hi-fi phones, 268
12-70 W universal quasi-complementary-symmetry amplifier, 36
amplified volume-unit meter, 443
applause meter, 448
Bauer cross-feed network for stereo headphones, 836
bipolar squelch, 822
bootstrapped tone control gives high s/n ratio, 17
cooling-fan control for hi-fi amplifier, 923
fet transient-free squelch, 822
f-m wireless mike buffer, 305
hard limiter multiplies 625 Hz by 40, 314
linear distortionless limiter, 398
rtty transform for cro display, 895
sine-square-triangle sweep generator, 329
switched two-cutoff filter for phone and c-w, 269
tunable notch filter, 260
vibratone 11-note keyboard, 244
vu meter in mike preamp, 655
(See also Amplifier; Electronic music; Intercom; Limiter; Preamplifier; Quadraphonic; Receiver; Recorder; Servo; Stereo; Tape recorder; Television; Transceiver; Transmitter)
Audio amplifier:
1.5 Hz-150 kHz IC driver, 13
10 Hz-100 kHz at 75 W, 837
10-20,000 Hz 100-W, 655
10-30,000 Hz low-drain, 36
30-25,000 Hz single-IC, 40
30-100,000 Hz 30-W class-B, 27
200-3,200 Hz voltage-tuned active filter with 80-Hz bandwidth, 261
500-Hz peak attenuation, 38
1.3-kHz signal injector for testing, 802
40-20,000 kHz 10-W using Zobel network, 29
10-V output with 20-dB gain, 37
200-μW IC hearing aid, 481
0.9-mW IC hearing aid, 486
1.5-mW IC hearing aid, 486
10-mW IC hearing aid with sliding bias, 485
300-mW using single IC, 28
1-W complementary-output, 30
1-W Darlington, 32
1-W four-transistor, 30
1-W IC, 31, 33, 35
2-W for color tv, 897
2-W with crossover for electrostatic and ribbon speakers, 27
2-W for crystal or ceramic input, 35
2-W four-transistor, 36

Audio amplifier (Cont.):
2-W IC for crystal pickup, 622
2-W IC with six parts, 38
2-W tv, 911
2.5-W push-pull class A, 31
3-W directly coupled, 29
3-W low-distortion, 39
5-W class A car-radio, 32
5-W Darlington with regulator, 40
5-W megaphone, 657
5-W portable stereo phono, 626
6-W class B car-radio, 28
10-W Dinsdale, 33
10-W hi-fi, 37
15-W with low distortion at low levels, 38
15-W class A or 20-W class AB, 26
15-W low-cost, 30
15-W silicon-transistor, 34
25-W, 31, 34, 39
35-W silicon-transistor, 26
50-W IC for quadraphonic system, 684
50-W silicon-transistor, 28
50-W stable linear hi-fi, 35
60-W push-pull class B, 656
100-W protected class B, 659
100-W using single module, 40
1,000-W solid-state, 679
agc with peak detection, 56, 60
bootstrapped low-noise, 28
bridge doubles power output, 36
combined with d-c to d-c converter, 966
eavesdropping with mike in parabolic
    reflector, 846
gain-controlled noise-reducing, 866
headphone adapter, 51
hearing aid, 483
linearity tester, 801
megaphone, 657, 660
microphone with adjustable gain, 660
mixing different microphone impedances, 49
nonmedical stethoscope, 847
opamp drives speaker directly, 39
phase inverter, 38
phone-jack with tunable a-f selectivity, 37
phono for crystal or ceramic input, 627
protecting class B output transistors, 653
push-pull class B with low crossover dis-
    tortion, 27
remote gain control, 761, 766
speech compressor and clipper, 41
telephone pickup coil, 847, 850
testing with digital white-noise generator,
    568
ultra-low-noise for instrumentation, 846
(See also Amplifier; Intercom; Instrumenta-
    tion amplifier; Music-controlled light;
    Phono; Preamplifier; Quadraphonic; Re-
    ceiver; Recorder; Servo; Stereo; Surveil-
    lance; Tape recorder; Transceiver; Trans-
    mitter)
Audio clipper, speech for ssb transmitter, 41
Audio compressor:
45-dB for recorder or transmitter, 45
agc, 41
agc for public-address preamp, 657
amplitude leveler, 46
dynamic mike, 43
electronic organ, 45
gain-controlled opamp, 46
microphone amplifier, 44
opamp with gain and feedback loop, 43
optoelectronic with expander, 42
optoelectronic coupling, 46
speech for a-f amplifier, 45
speech combined with clipping for ssb, 399
speech for recorder or transmitter, 42

speech for ssb transmitter, 41
speech in transceiver preamp, 42
transmitter or tape recorder, 43–45
voice-operated microphone, 46
Audio control:
6-dB presence, 55
1.8-kHz tone decoder, 880
2-kHz signal-controlled d-c switch, 48
9-MHz ssb generator, 48
27-MHz drive for Ionovac speaker, 47
200-W psychedelic lamp, 559
active four-input mixer with single output, 52
active tone control, 54
agc for voice amplifier, 58
agc-operated squelch, 823
amateur station tuneup and operation for
    the blind, 483
auto radio commercial killer, 54
automatic level control, 55
automatic slide changer, 53
background noise compensation, 658, 659
bass boost below 120 Hz, 48
battery saver for transceiver, 964
carrier-operated relay for f-m receiver, 3
Christmas-tree lights, 559
compressor and clipper for ssb transmitter,
    399
crossover network for bass and treble horns,
    54
current-summing mixer, 51
digital for tape recorder gain, 870
echo-chamber effect for tape recorder, 872
flasher for intercom, 378
four-channel mixer preamp, 53
guitar seven-circuit R-C tone control, 247
high-frequency active filter, 51
high-level tone, 52
lamp with triac, 558
low-distortion tone control, 54
low-pass and high-pass filter, 53
mixer with two inputs, 55
multicoupler for tape recorders, 872
muting for stereo, 50
noise and rumble filter, 47
noise-operated squelch, 821
presence-effect filter for soloist, 55
psychedelic strobe flash, 637
reducing volume control noise, 55
remote gain by d-c voltage, 765
remote volume, 761
r-f speech processor, 807
solid-state vox, 986
sound-actuated relay, 762, 919
sound-tripped photoflash, 641
speaker-phasing a-f transmitter and
    receiver, 656
speech clipper, 400
speech recognition with IC parameter
    extractor, 51
squelch for voice amplifier, 823
stereo mixer, 53
tape recorder, 49
telephone-bell amplifier with relay, 881
tone balance for old records, 624
tone-switch trigger for triac, 54
tone treble and bass cut and boost, 50
transceiver tuneup for the blind, 488
tv and radio commercial killer using
    flashlight, 919
variable-gain volume, 50
vibrato for hi-fi or recorder, 249
voice-operated preamp, 52
voice-operated relay, 787
voice-operated transmitter switching, 963
voice-operated turn-on, 848
volume expander, 50

woofer protection from transients, 49
(See also Electronic music; Lamp control;
    Limiter; Music-controlled light; Squelch;
    Voice)
Audio expander:
30-dB increase in dynamic range, 50
optoelectronic with compressor, 42
phono records, 42
Audio oscillator:
aural tuning aid for ssb transmitter, 488
dripping-sound generator, 348
fish caller, 349
frequency meter and square-wave shaper,
    676
glide tone with random six-neon blinker, 345
motorboat sound effect, 349
novelty squealing box, 346
three-tone door-identifying, 349
(See also Oscillator)
Audio preamplifier:
20-20,000 Hz mike, 656
20-20,000 Hz modular low-distortion, 38
carbon-mike, 32
ceramic or dynamic input, 620
ceramic-cartridge RIAA, 621
ceramic high-output pickup, 622, 623
crystal-mike, 32
crystal-pickup, 627
dynamic-mike using speaker, 660
hi-fi IC RIAA for phono, 620
high-gain using opamp, 33
high-impedance mike, 33
low-impedance dynamic mike, 32
magnetic-pickup, 622
magnetic-pickup filter, 623
magnetic-pickup low-noise, 621
micropower for capacitive transducer, 364
microphone with built-in vu meter, 655
microphone with clipper, 398
phone-mike with switched equalization, 621
resistance-coupled speech, 29
speech for low-impedance load, 33
squelch, 823
tape-playback, 873
transformer-coupled phase inverter, 33
variable-gain volume control, 50
(See also Amplifier; Audio amplifier; In-
    strumentation amplifier; Phono; Preamp-
    lifier; Quadraphonic; Receiver; Recorder;
    Servo; Stereo; Tape recorder; Transceiver;
    Transmitter)
Auditory stimulator:
modulator for burst generator, 487
trapezoidal generator, 488
Auto radio:
5-W class A audio amplifier, 32
6-W class B audio amplifier, 28
varactor-tuned a-m, 705
Auto theft alarm:
2-meter f-m paging transmitter, 85
battery-drain sensing, 82–84
boat on trailer, 85
chemical gas release, 85
delayed activation, 82
door-switch controlled, 83
hood and trunk latch, 84
mobile equipment, 82, 85
timed self-resetting, 83
timer-controlled, 83, 84
Autodyne tuner, uhf tv, 907
Automatic control, receiver i-f bandwidth, 353
Automatic frequency control:
adding to rtty receiver, 891
tv thin-film hybrid video i-f, 913
Automatic frequency-shift keyer:
single-transistor for Teletype, 888

Automatic frequency-shift keyer (Cont.):
two-pair tone generator, 892
Automatic gain control:
30 Hz-400 kHz with 30-dB dynamic range, 57
1-kHz twin-T oscillator, 600
1-22 MHz untuned video amplifier, 57
10-MHz video amplifier, 57, 60
45-MHz wideband, 57, 58
6-meter preamp, 56
analog multiplier, 59
audio, 41
cascode constant-bandwidth of 1 MHz, 56
cathode-follower improves performance, 59
color tv keyer-amplifier, 901
diode for IC video amplifier, 59
driving a-f squelch tube, 823
external diode for IC video amplifier, 59
fet pair with opamp, 60
hang circuit for ssb and c-w, 58
internal full-wave detection in opamp, 56
keyer for tv i-f, 909
lab sweep generator, 372, 374
negative-peak detector with opamp, 57
peak detection for IC audio amplifier, 56, 60
pnpn tetrode, 59
public-address preamp, 657
quadrature oscillator, 598
radar threshold detector, 689
S-meter drive in communication receiver, 59
series jfet for voice operation, 58
shunt jfet for voice operation, 58
tape recorder, 867
tv tuner agc delay, 914
tv video i-f amplifier, 907, 909
vhf receiver front-end, 60
video amplifier with 50-MHz bandwidth, 60
Wien network in oscillator, 801
Automotive:
0-8,000 rpm tachometer, 68
0-12,000 rpm noncontacting tachometer, 64
12-V combination electronic lock, 237
acceleration and braking measurement, 72
alternator voltage regulator, 73
a-m wireless mike as aid to parking buses, 982
audible idiot-light alarm, 72
automatic headlight turn-off, 73
barricade flasher, 280
battery monitor, 69
battery undervoltage warning lamp, 86
brake-fluid monitor, 69
charging two batteries with one generator, 647
combination lock for ignition, 77
dwell-angle booster using scr and delay, 65
dwellmeter, 67, 71
emergency flasher, 69, 277, 280
goo-hah horn, 815
headlight-off delay, 62, 65
headlight-on alarm, 65, 67, 71, 72, 74
headlight-operated garage-door switch, 766
headlights turn on porch light, 629
horn-actuated garage-door opener, 762
icy-road alarm, 66, 70
idiot-light blinker, 70
idiot-light tone alarm, 63, 67
ignition noise blanker for mobile converter, 570
infrared carbon monoxide analyzer for exhaust, 359
low-fuel indicating flasher, 61, 64, 66
miles-per-gallon meter, 64
neon-lamp tachometer calibrator, 68
oil-level checker, 65, 68

parked-car power-drain reminder, 62
photoelectric parking-light turn-on, 73
radar speed-detector preamp, 690
radar speedmeter nonblocking limiter, 688
radiator water-level alarm, 67
radio-antenna identifying blinker, 72
rear-lamp failure alarm, 63
sequential turn-signal flasher, 68, 70
solid-state vibrator replacement, 646
speed alarm, 63
strobe timing light, 637
tachometer, 67, 73
tachometer and dwellmeter, 66, 71
travel-time computer, 562
turn-signal audio-tone alarm, 67
turn-signal buzzer, 61
twilight turn-on road barrier lamp, 628
windshield-water alarm, 73
wiper slow-sweep for light drizzle, 70, 71, 74
(See also Auto radio; Auto theft alarm; Automotive ignition; Battery charger; Flasher; Tachometer)
Automotive ignition, capacitor-discharge, 76-81

Backlash, predictable constant percentage in pulse discriminator, 1012
Balance meter, stereo, 838
Bandwidth control, receiver i-f for QRM, 353
Bass boost, below 120 Hz, 48
Battery:
11-V regulator, 739
extending life in transistor radio with shunt electrolytic, 422
monitoring condition during starting, 69
nickel-cadmium protection from over-discharge, 653
protecting against complete discharge, 651
replacement with regulated power supply, 714
voltage monitoring with lamp, 472
Battery charger:
45-500 mA for dry cells, 89
500-mA for standby battery of digital clock, 208
550-mA for nickel-cadmium cells, 89
700-mA with 10% reverse current, 90
0-35V 750-mA for nickel-cadmium batteries, 91
1.5-V for battery radio, 90
1.5-90 V dry batteries, 87
3.6-V for nickel-cadmium D cells, 87
6-V standby for line-failure light, 390
12-V at 20 A with ujt control, 89
12-V with automatic start-stop, 88
12-V put-scr, 86
18-V for dry cells, 87
automatic standby using IC timer and zener, 91
charging two batteries in car with one generator, 647
constant-current adjustable to 3.1 A, 90
five-cell independently controlled, 86
flashlight-cell, 88, 90
high-speed with burp control, 88
nickel-cadmium with third-electrode sensing of charge, 87
pulsed standby for protecting ram data, 493
scr with voltage and rate control, 91
trickle for emergency light, 2
undervoltage warning lamp, 86
voltage monitor, 89
Battery power supply, minimizing drain for a-d converter, 23

Battery saver, audio-controlled transceiver switch, 964
Battery tester, 3-V scr with lamp, 930
Beacon:
75-MHz airborne marker receiver, 565
miniature CB, 850
Beeper:
2-pps Sonalert, 3
CB transmitter, 848
signal injector for radio troubleshooting, 795
Bicycle siren, 813
Binary:
d-a converter staircase generator display, 203
ripple multiplier, 538
Binary-coded decimal:
illegal code suppressor, 24
miltiplexer for ten inputs, 528
subtractor using adders and inverters, 151
Binary comparator:
serially transferred numbers, 149
wired OR gates, 416
Binary-hex decoder, mixing positive and negative gates, 413
Binary number, squaring 4-bit, 818
Binary-octal decoder, mixing positive and negative gates, 416
Bistable multivibrator (see Flip-flop)
Black-and-white television (see Television)
Blanker:
impulse noise in receiver output, 569
Nixie display, 233
Blender, speed control with timer, 515
Blind, aids for the:
amateur station control, 483
audible thermometer, 487
audio-tone tuneup for transceiver, 488
aural tuning meter for amateur transmitter, 483, 488, 971
liquid-level indicator for pouring coffee, 403
reading aid using tactile image, 489
receiver frequency spotter, 487
tactile tape recorder level indicator, 482
Blinker (see Flasher)
Blood pressure simulator, opamps with Schmitt triggers, 485
Boat:
combination lock for ignition, 77, 237
theft alarm, 85
Bolometer, infrared intruder alarm, 94
Bootstrapping (see Amplifier; Oscillator)
Bounce suppression, gate for logic pulses, 152
Breast tumor, Geiger counter for radiophosphorus, 489
Bridge:
10-Hz Wien sine-wave oscillator, 599
10-10,000 Hz Wien oscillator, 600
1,000-Hz Wien in passive filter, 269
23-kHz inverter, 161
2-30 MHz impedance using noise diode, 446
3.5-33 MHz R-X, 445
30-MHz variable attenuator with constant phase, 562
0-10 V in 0.01-V steps with circular voltage divider, 1012
500-W temperature control, 928
active circuit trimmer, 142
a-f amplifier power doubler, 36
ambient-immune overtemperature alarm, 271
amplifier for high-resistance thermistors, 365
amplifier for variable-resistance devices, 364
antenna noise, 570
audio amplifier leveler, 46
balanced transmission-line as f-m discriminator, 306

Bridge (Cont.):
balancing with double phase-sensitive detector, 614–616, 676
crt indicator, 446
diode series-shunt gate, 334
feedback-linearized resistance, 457
fet differential amplifier, 366
germanium-transistor temperature-sensing, 927
indoor-outdoor thermometer, 467
meter amplifier, 427
mos transistors as voltage-controlled resistors, 472
noninverting chopper-stabilized amplifier, 117
opamp power booster, 577
pendulum-driven acceleration-measuring, 72
portable impedance, 449
resistive for phone patch, 879
resonant inverter, 282
r-f admittance, 442
square-law diodes give linear-scale power meter, 454
swr for antenna matcher, 979
temperature control, 928
temperature control for reference diode, 1011
temperature-sensing for power amplifier, 923
thermistor for 60 C crystal oven control, 921, 927
thermistor with subtracting amplifier, 922
transducer instrumentation amplifier, 369
universal R-L-C using crt indicator, 449
voltage-variable capacitor as diode chopper, 116
Wheatstone, in thermocline-measuring fish finder, 349
Wheatstone seven-range, 461
Wien in 10 Hz-100 kHz audio oscillator, 593
Wien in oscillator for high temperatures, 596
Wien single-opamp, 598
Brightness control (see Lamp control; Lamp dimmer)
Buffer, unity-gain noninverting zero-offset opamp, 574
Bullet, velocity chronograph, 451
Bullhorn (see Megaphone)
Burglar alarm:
18-45 kHz ultrasonic, 994
40-kHz ultrasonic intrusion, 272
27-MHz transmitter, 95
50-W night-light control, 101
500-W night-light control, 93
1-kW night-light and alarm control, 98
5-50s adjustable-delay, 273
30-s entrance delay, 97
bistable amplifier for low light levels, 631
breakable-wire, 2
capacitance-type, 103
clicker driving speaker, 4
conductive-foil fail-safe, 96
conductive-loop lightning-immune, 100
doorknob capacitance, 106
Doppler using 10.69-GHz Gunn oscillator, 98
electronic siren, 815
flashlight-triggered, 766
Hartley oscillator with speaker, 98
high-sensitivity for sudden illumination, 271, 273
home security system, 99
IC sensor circuits with latch, 100
infrared, 93, 95, 271
infrared with 1,000-foot range, 94
infrared chopped-beam, 92, 360

infrared pulsed-beam antijamming counter, 361
intruder-tripped photoflash, 98
knock-over Sonalert, 96
light-beam photoelectric, 101
light-interruption, 96
light-sensitive clicker, 93, 271
light-triggered audio howler, 632
mvbr-driven horn, 1
passive infrared intruder, 97
photoelectric for doors and windows, 95
photoelectric intrusion without light beam, 92
photoelectric shadow-detecting, 99
protective home lighting control, 100
proximity for safe, 106
proximity detector, 104, 105
pulsed infrared, 97–99, 101
reduced phototransistor dark current for low light levels, 630
rising-wailing siren, 815
screamer oscillator with speaker, 815
sensing-wire, 103
siren, 813
sound-actuated, 271, 484, 787
store-home intercom with noise suppressor, 378
switch-opening, 96
thin-wire loop, 272
touch-contact using scs, 101
unattended transmitter or repeater, 379
vibration-triggered, 271
Waa-Waa solid-state sounder, 813
(See also Alarm; Auto theft alarm; Capacitance control; Photoelectric; Protection; Ultrasonic)
Butterworth filter, Fortran design program, 262

Cable tester, timers driving two-color led's, 931
Calculator:
binary adding machine for instruction, 147
driver for Panaplex II display, 222
Camera:
Plumbicon sstv circuits, 915
tv high-gain preamp, 907
tv preamp using cascoded fet's, 906
(See also Photography)
Camera tube, 25-MHz preamp and line driver, 907
Capacitance, measurement with R-L-C bridge, 449
Capacitance control:
0.16 Hz-800 kHz charge amplifier, 10
60-W with optoelectronic latch, 106
180-W lamp, 105
450-W touch-plate switch, 103
c-w break-in by touch, 137
doorknob alarm, 106
intruder proximity detector, 104
proximity, 103–105
proximity burglar alarm for safe, 106
relay sensing-wire antenna, 103
touch-plate switch, 102, 105, 107
touch-to-talk transmitter switch, 104, 105
wireless proximity detector, 102, 104
Capacitance measurement, frequency shift of 4-MHz oscillator, 422
Capacitance meter:
0-0.1 μF nonelectrolytic, 418
1,000-μF adapter, 420, 421
2 pF-0.1 μF, 419
15 pF-10 μF, 420
counter with relaxation oscillator, 421
five-decade, 421
instrument amplifier, 419

seven-frequency oscillator, 419
seven-range using IC transistor arrays, 420
Capacitance multiplier:
60-Hz active filter, 477
opamp giving 100,000 μF, 538
Capacitor leakage:
neon checker, 418
testing with microammeter, 422
Capacitor tester:
audio-tone, 418
neon indicator, 422
(See also Electrolytic capacitor)
Car (see Automotive)
Carrier, 500-kHz power-line remote control system, 761
Carrier control, pulsed twelve-load, 762
Carrier-operated relay:
f-m receiver squelch, 821, 822
f-m repeater, 774
receiver, 3, 4, 770, 859
split-site repeater, 770
Cassette, dynamic noise limiter for recorder tapes, 865
Catheter, cardiac with pulsed transducers, 481, 484
Cathode ray:
1 Hz-150 kHz linear sawtooth time base, 855
10 Hz-1 MHz preamp, 114
0-300 MHz vertical amplifier, 111
adapter changes cro to monitor, 906
automatic erasure for storage cro, 110
cathode driver, 112
converter provides four-channel display, 109
converting 9 MHz to 455 kHz for scope display, 172
d-c amplifier, 114
d-c/d-c converter for small scope, 339
focus-coil current regulator, 196
gamma correction, 110
high-voltage inverting opamp, 114
indicator for universal R-L-C bridge, 446, 449
level and polarity changer, 576
linear time base for scope, 854
pincushion and blurring correction, 113
push-pull electrostatic deflection amplifier, 113
ramp generator for color tv cro, 112
rtty monitor, 885
rtty phase-shift tuning indicator, 885
sawtooth horizontal sweep using scr with neons, 852
scope sensitivity booster, 475
square-wave calibrator for color tv cro, 108
sstv and rtty signal analyzer, 912
sweep generator for 100 kHz to 60 MHz, 856
switching waveform generator for dual-trace cro, 109
time-base trigger for color tv cro, 114
trace quadrupler for d-c scope, 108
transistor curve tracer, 937
triggered sweep for cro, 851, 853, 854
vertical amplifier for color tv cro, 109
vertical differential amplifier, 111
vertical preamp for color tv cro, 110
vertical preamp using IC, 111
voltage calibrator for scope, 114
weather satellite picture demodulator, 877
wideband deflection amplifier, 113
wideband gamma correction, 111
(See also Display; Oscilloscope)
CB (see Citizens band)
Ceramic cartridge, RIAA-equalized preamp, 621
Ceramic filter:
455-kHz modulated i-f signal generator, 796

Ceramic filter *(Cont.)*:
    stereo f-m i-f amplifier, 837
Character generator:
    64-character ASCII alphanumeric, 221, 222
    interface for composite video signal, 229
    MAN-3 led display drive, 535
    multiplexer for HP5082 series display, 529
    trapezoid generator for numeric art display, 230
    video numeric, 226
    (See also Display)
Character recognition, rtty printer turn-on, 888
Charge amplifier, 0.16 Hz-800 kHz for capacitive sources, 10
Charger (see Battery charger)
Chart recorder, time multiplexing with gated mosfets, 534
Chopper:
    1-kHz at 1 kW using gate-controlled switch, 117
    12-kHz in regulated d-c to d-c converter, 161
    balanced three-diode, 118
    bootstrapped-transistor for multiplexers, 116
    complementary-transistor series-shunt, 116
    differential thermocouple measurement, 464
    diode-bridge modulator for 300 to 3,500 Hz, 117
    fet with bipolar transistor, 117
    four-diode, 116
    infrared passive intruder alarm, 94
    integrated in digital-analog ladder, 117
    inverting as buffer for precision pot, 118
    isolated for d-c transfer, 116
    noninverting in 0.01% microvoltmeter, 477
    noninverting for bridge amplifier, 117
    noninverting as buffer for precision pot, 118
    noninverting floating amplifier, 115
    precision ramp generator, 853
    reset-stabilized amplifier, 118
    series-shunt most, 115
    stabilized d-c amplifier, 118
    summing-junction switch, 512
    two-fet for d-c input signals, 115
    variable-capacitance diode bridge, 116
Christmas tree, pseudo-random switching of lights, 275
Chroma processor, color tv three-IC, 898
Chronograph, bullet velocity, 451
Circuit breaker, automatic-reset solid-state, 652
Citizens band:
    beeper transmitter, 848
    bfo for ssb signals on a-m transceiver, 121, 123
    blanker for impulse noise in receiver audio, 569
    booster with 15 dB r-f gain, 122
    channel-9 override monitor, 119
    converter for 12-V car radio, 120
    converter for a-m auto radio, 119
    converter for broadcast radio, 122
    crystal activity tester and channel spotter, 123
    field-strength meter, 120
    headphone adapter, 51
    miniature transmitter, 850
    paging transmitter, 122
    phone jack for receiver, 122, 123
    phone patch for transceiver, 880
    portable tunable receiver, 120
    preamp, 122
    r-f wattmeter, 454
    search receiver for hidden transmitter, 849
    speech clipper, 400
    squelch for selective calling, 823
    transmitter tuner and power monitor, 121

wireless converter for a-m radio, 121
(See also Transceiver)
Clamp, 10-MHz fet, 686
Clipper:
    20-100,000 Hz square-wave generator, 677
    adjustable-level using programmable zener, 676
    combined with compressor for ssb transmitter, 399
    with compressor for ssb and a-m transmitters, 44
    frequency counter input amplifier, 291
    mike preamp, 398
    opamp with zeners, 579
    precision using fet-input opamp, 577
    r-f speech processor for ssb transmitter, 807
    speech for transmitter, 400
Clock:
    500-Hz narrow-band digital filter, 253
    1.4-kHz astable mvbr for Karnaugh map display, 126
    5-kHz optoelectronic coupler, 589
    600 kHz-10 MHz within 0.1% with c-mos inverters, 124
    1-MHz master using crystal with IC, 126
    15-ms strobe generator, 667
    80:1 Schmitt trigger with gate, 127
    automatic call-letter generator for f-m repeater, 769
    direct-coupled driver for ram, 493
    driver for two-phase mos, 127
    four-phase generator, 127
    frequency tripler for Schmitt pulses, 688
    gated in cascaded decade counter, 124
    high-speed driver for logic above 300-MHz, 128
    minimizing crosstalk, 125
    nine-overtone square-wave generator, 125
    overshoot limiting with diode clamp, 128
    solid-state keyer with switching-noise immunity, 126
    synchronizer for minicomputer interface, 124
    synchronizing with phase-locked loop, 126
    train overlap prevention with Schmitt trigger, 125
    two-frequency using quad gate, 128
    two-phase for digital storage, 128
    (See also Digital clock)
Clock driver:
    1-20 MHz crystal, 612
    50-100 MHz crystal, 610
    200-MHz crystal using doubler, 126
Code:
    100-Hz bandwidth active filter for c-w, 137
    750-Hz filter with 75-Hz bandwidth, 132
    750-Hz practice oscillator, 138
    800-Hz sidetone oscillator for keyer, 138
    1-kHz active filter, 262
    1-kHz tunable active c-w filter, 136
    1.1-kHz active filter for receiver, 256
    active IC filter, 139
    afsk-mcw-practice oscillator, 133
    automatic call-letter generator, 154, 769
    automatic keyer, 135
    automatic letter and word spacing for keyer, 131
    battery-portable IC keyer, 134
    beat-frequency oscillator, 809
    bfo injection for transceiver, 964
    break-in keyer without relays, 133
    call-letter generator for f-m repeater, 774
    CQ message generator using NAND logic, 136
    c-w monitor for transmitter, 132
    dial control of f-m repeater, 769

diode-controlled transceiver filter switching, 966
diode switch gives perfect keyed c-w, 973
dot-dash generator, 130
filter rejects hum and squeal, 134
high-current switch for keyer, 133
keyer, 131, 132, 135
keyer clock with noise immunity, 126
keyer tests c-w transmitter, 133
keying monitor for transmitter, 139
low-power a-m bc transmitter for keying practice, 138
message generator for c-w or rtty, 129
modulated-infrared detector, 362
monitor for transmitter, 976
practice oscillator, 98, 130, 131, 137, 139
practice oscillator and transmitter monitor, 134, 138
practice-oscillator conversion to metronome, 246
protection for illegal bcd, 24
radio-controlled Morse sounder, 132
regulated supply for automatic keyboard-driven generator, 726
relay driver for solid-state keyer, 138
Seiler keyed vfo, 129
slow-decay detector for Schmitt trigger, 131
solid-state switching for paddle, 132
tone-controlled switch, 139
tone keyer for transmitter, 131
tone-modulated practice oscillator, 136
touch-controlled c-w transmitter, 137
transmitter for practice, 134
ujt keyer with sidetone for ssb, 137
wireless practice oscillator for a-m radio, 135
Code converter, pcm 10-MHz seven-code, 200
Coder, 15-channel digital remote control, 763
Coincidence detector, signal-powered, 417
Color organ (see Electronic music; Music-controlled light)
Color television (see Television, color)
Combination lock, boat ignition, 77
Commercials:
    silencing in auto radio, 54
    silencing in flashlight-operated tv, 924
    silencing with touch-plate switch, 105
Community antenna television, vhf distributed amplifier, 910
Commutator:
    32-channel with 5-bit control, 527
    analog four-channel expandable, 21
Comparator:
    0.1 Hz-30 MHz phase using multiplier IC, 141
    60-kHz WWVB transmission frequency, 431
    3.58-MHz for calibrating signal sources, 317
    4-channel pulse-height analyzer, 1006
    500-meg input impedance, 143
    active resistance-bridge trimmer, 142
    a-d converter tester, 23
    adjustable-duty-cycle pulse generator, 671
    AND-gate with large fan-in, 335
    binary-weighted for a-d converter, 207
    converting synchronous motor to stepper, 523
    dual-level using dual opamp, 141
    dual opamp for ramp limit control, 852
    fast peak detector for 50-ns pulses, 1010
    fet with 1-min time constant, 144
    frequency-band using mono mvbrs and flip-flops, 144
    frequency-band limits, 144
    frequency doubler, 311
    high-speed for function generator, 328
    high-speed a-d converter, 23
    hysteresis amplifier, 791

Comparator (*Cont.*):
  illegal bcd code suppressor, 24
  input voltage of mono mvbr, 142
  ladder-input for a-d converter, 143
  lamp driver, 143
  light sources, 632
  linear vco, 997
  low-noise amplitude, 140
  magnitudes of opposite-polarity voltages, 143
  matched fet for thermocouple, 465
  micropower programmable opamp, 1009
  minimum-frequency detector, 142
  multibit binary numbers, 416
  negative-peak detector, 1010
  opamp with transistor drives 40-mA lamp, 143
  positive-peak detector, 1010
  power one-shot trigger for stapling gun, 988, 990
  precision current, 142
  precision LSI tester, 144
  precision voltage, 141
  probability density analyzer, 444
  pulse-width modulator, 512
  sample-hold, 782
  servo overshoot control, 794
  three-input AND gate, 333
  two binary numbers, 149
  two-phase mos clock driver, 127
  vco pulse width and spacing control, 1003
  voltage, 140, 141, 143
  voltage limit with lamp indicator, 143
  zero-crossing detector, 1009
  zero-crossing pulse generator, 1007
Compressor:
  audio using linear limiter, 398
  logarithmic for 100 μA to 10 V, 410
  tape recorder, 847
  (See also Audio compressor)
Computer:
  20-kHz pulsed 1-kW power supply, 381
  8-bit parity checker, 145
  1,991-bit shift register, 146
  ASCII encoder, 151
  asynchronous transfer register, 149
  auto travel-schedule, 562
  bcd subtractor using adders and inverters, 151
  binary adding demonstrator, 147
  binary-bcd converter, 147
  binary comparator, 149, 150
  clock synchronizer, 124
  core memory sense amplifier, 492
  double sequence generator, 150
  excess-three encoder with holding register, 150
  full adder using gates, 335
  full subtractor using gates, 335
  half-adder using gates, 335
  high-speed digit detector, 146
  idiot-proof logic switches, 145
  keyboard-to-binary converter, 150
  neon diagnostic display, 146
  odd-length shift register, 147
  relay driver for troubleshooting display, 146
  universal encoder, 148
  variable-delay shift register, 148
  wideband cro deflection amplifier for terminal display, 113
  (See also Memory)
Computer-aided design:
  87-MHz amplifier, 12
  Chebyshev filter, 266
  voltage-follower opamp, 575

Computer program, Fortran for Butterworth filter design, 262
Conductivity:
  plasma measurement, 458
  water measurement, 460
Constant-current drive, 60-mA teleprinter selector magnet, 887, 892, 893
Constant-current generator, dual-transistor, 198
Constant-current source:
  0.5-mA matched, 196
  7.5-mA within 0.08%, 1012
  differential amplifier, 198
Contact bounce suppression:
  auto ignition, 76
  call-letter generator, 154
  cross-coupled gates, 153
  dual two-input gate, 154
  flip-flop pair, 153
  keyboard encoder, 152
  latch for mvbr, 153, 670
  logic gate, 152
  pulse-counting generator, 672
  ring counter reset, 153
  set-reset flip-flop, 154
  single-pulse generator, 153
  switch-contact latch, 153
  tach for electronic ignition, 77
Contact resistance, measuring with low-range ohmmeter, 457
Contact time, prolonging with thyristor, 559
Continuous wave (see Code; Receiver; Transmitter)
Contourograph, real time ecg, 485
Control:
  600-W half-wave average-voltage feedback, 390
  600-W resistive half-wave scr, 390
  900-W resistive full-wave, 387
  line-voltage compensation for trigger of ujt-thyristor, 751
  liquid-level, 403
  scs-triggered triac full-wave a-c power, 617
  sequential using precision staircase generator, 825
  voltage-controlled negative resistance, 552
  voltage feedback with tachometer, 521
  (See also Audio control; Humidity control; Lamp control; Level control; Liquid level control; Motor control; Phase control; Photoelectric; Remote control; Switch; Television, remote control; Temperature control)
Controller, reducing a-c line voltage, 647
Converter:
  100-kHz mV signals to d-c, 478
  0-1 MHz sine-square, 677
  1.65 MHz to 135 kHz, 351, 704
  28-30 MHz i-f amplifier, 350
  136 MHz to 30 MHz, 875
  220-MHz from 28 MHz for ssb and c-w transmitter, 811
  1-V sine to 10-V square, 678
  1.5-V d-c to kV for cro or electric fence, 339
  2.4-V d-c to 500-V d-c for photoflash, 155, 644
  3-V d-c to 300-V d-c for laser, 160
  4-V d-c to 450-V d-c for photoflash, 644
  4-V d-c to 2,500-V d-c for photoflash, 640
  4.5-V d-c to 450-V d-c for photoflash, 644
  +5-V d-c to −15-V d-c, 155
  5-V to 180-V ringing-choke for numerical indicator, 383
  6-V d-c to 12-V d-c at 40 W, 162
  6-V d-c to 12-V d-c without rectifier, 157
  6-V d-c to 250-V d-c at 50 W, 159
  12-V d-c to 250-V d-c at 125 mA, 156

  12-V d-c to 300- and 600-V d-c, 162
  12-V d-c to 500-V d-c at 100 W, 158, 161
  12-V d-c to 1,000-V d-c at 500 W, 157
  14-V d-c to 300-V d-c for auto ignition, 80
  15-V d-c to 425-V d-c, 161
  20-32 V d-c to 28-V d-c at 10 A, 161
  28-V d-c to 68-V d-c at 23 W, 158
  28- or 180-V d-c to 69- or 172-V d-c, 383
  28-V d-c to 70-V d-c at 50 W, 159
  28-V d-c to 250-V d-c at 100 W, 158
  28-V d-c to 300-V d-c at 250 W, 159
  30-V d-c to 110-V d-c at 400 W, 160
  117-V a-c to 100-V d-c at 20 kHz and 1kW, 157
  150-V d-c to new value at 180 W, 159
  180-V d-c to 78-V d-c at 30 W, 160
  180-V d-c to 167-V d-c at 77 W, 161
  180-V d-c to lower voltage, 282
  275-V to 50-V pwm at 1 kW, 156
  2-kV IC voltage regulator, 338
  750-W three-phase frequency, 948
  4-channel for cro, 109
  15-key keyboard to binary, 150
  analog to pulse width, 876
  analog voltage to rms value, 820
  bcd to decimal with numerical readout, 289
  binary to bcd, 147, 150
  broadcast-band rejection filter for auto radio, 269
  capacitance to inductance with gyrator, 167
  C-R mutator, 165
  current to logarithmic voltage, 407, 411
  current to voltage with 600-kHz bandwidth, 426
  current to voltage with adjustable bandwidth, 429
  current to voltage with offset adjustments, 479
  current to voltage with opamp for microammeter, 428
  current to voltage in supply current monitor, 424
  d-c to d-c combined with a-f amplifier, 966
  digital ring counter to dice display, 348
  frequency-voltage for 1-30 MHz meter, 435
  frequency-voltage for a-c motor control, 382
  fsk or afsk to rtty using phase-locked loop, 893
  Gray-code analog-digital, 22
  ignition noise blanker, 570
  impedance with 1 teraohm input and 50 ohms output, 166
  impedance with 1.5 meg input and 40 ohms output, 166
  impedance for d-c to 10 kHz, 164
  logarithmic, 405, 409
  L-R mutator, 163
  micropower for capacitive-transducer preamp, 364
  radio: 5.5-1,100 kHz to 80 meters, 178
    500-kHz for communication receiver, 175
    1.65 MHz to 135 kHz, 267, 609
    3.5-4 MHz for transistor radio, 174
    9 MHz to 455 kHz, 172
    14-31 MHz for broadcast radio, 176
    27-MHz CB to broadcast band, 122
    28-54 or 108-176 MHz for auto radio, 172
    50-52 MHz to 28 MHz, 169
    135-175 MHz police-call for f-m receiver, 173
    144-MHz mosfet, 176
    144-146 MHz to 28 MHz, 170
    220-MHz f-m, 177
    1.296-GHz oscillator-doubler stages, 174
    2.304-GHz for amateur band, 171
    2-meter, 170, 172, 175, 178

Converter, radio (Cont.):
    6-meter to broadcast band, 177
    6-meter using high-efficiency tubes, 177
    6-meter mosfet, 175
    10-meter IC preamp, 16
    10-meter solid-state tuner, 173
    15-meter with 20-meter output, 173
    15-40-80 meter solid-state plug-in, 169
    CB to a-m for 12-V auto radio, 119, 120
    CB wireless for a-m transistor radio, 121
    short-wave to broadcast band, 173, 178
    vhf police calls for auto radio, 171
    vhf superregenerative mobile, 170
    WWV and CHU for a-m radio, 174
    WWV frequencies to ham bands, 175
    (See also Adapter; Mixer)
    rotator, 166, 168
    seven-code pcm, 200
    sine to spike, 665
    sine to square, 674
    synchronous motor to stepper, 523
    triangle-sine, 678
    triangle-sine fet, 675
    true rms to d-c for any waveform, 475
    variable-frequency drive for tape recorder, 864
    voltage to frequency, 163, 164, 167, 168
    voltage to frequency with pulse train output, 167
    voltage to frequency with temperature stabilization, 166
    voltage to frequency wide-range linear, 166
    voltage to Gray code, 165
    voltage to pulse width, 163, 164, 507, 509
    voltage-current for process-control instrumentation, 367
    X-Y to polar coordinates, 165
    (See also Analog-digital converter; Digital-analog converter; Inverter; Mixer)
Correlator, pseudo-random binary sequence, 568
Cosine, approximating with log module and opamp, 408
Counter:
    0.1-1-10 Hz time-base divider using 60-Hz reference, 297
    1.3-MHz bistable, 181
    1.5-MHz complementary-pair ring, 187
    5-MHz binary, 184
    5-MHz four-element binary up, 187
    6-MHz pnp-npn reversible ring, 190
    10-MHz frequency synthesizer, 317
    30-MHz bistable, 182
    15-channel digital remote-control, 763, 765
    12-hour digital clock, 208
    1-999 preset pulse generator, 672
    30-s timer drive, 958
    4:1 divider for 40 MHz, 296
    10:1 divider, 295–297, 299
    adding 400-kHz and 25.767-kHz pulses, 184
    bcd decoder drives seven-segment display, 191
    bcd decoder for numeric indicator, 231
    bcd decoder with scs store for indicator, 225
    bidirectional ripple, 185
    bidirectional with noise immunity, 189
    binary in four-phase clock generator, 127
    binary adding machine, 147
    binary ripple using NAND gates, 193
    breast-tumor radiophosphorus, 489
    capacitance meter, 421
    cascaded decade using gated clock, 124
    clocked R-S bistable with NAND gates, 192
    complementary driver for scaler, 192
    decade with Numitron display, 231
    decade tester, 937

decimal Gray-code, 191
digital display with zero suppression, 223
discrete NAND decoding gate, 336
divide by 196, 608 in crt display for ram, 492
divide-by-five, 190
divide-by-N programmable, 322
divide-by-three, 187
drive for nine-lamp decade display, 184
drive for numeric printout, 186
electronic watthour meter, 455
input pulse shaper, 674
L-C commutated ring, 187
level and polarity changer, 576
lockout for pulsed infrared burglar alarm, 361
lockout for pulsed infrared system, 98
minimum-logic for numeric display, 227
multiplexing to minimize decoder/drivers, 530
neon store for bistable, 188
numeric display with memory, 233
odd-even bcd decoder for numeric indicator, 232
odd-order division, 193
oscillator-driven ring for electronic dice, 344, 348
prevention of coincident pulses, 186
pulse-gobble with maximum divide modulus of 255, 185
pulse-shaping drive, 182
pulse subtractor, 181
reset pulse generator, 153
resettable decimal using lamps, 181
reversible Johnson decade, 192
rhythm generator, 241
ring with gate-driven load, 190
ring for switching loads sequentially, 185
ring 50-kHz scs, 179
ring 1-2-5 switchable scr, 183
ring complementary-pair, 181
ring scr, 193
ring scs for numeric display, 224, 226
ring scs without numeric indicator, 182
ripple-carry binary feedback divide-by-20, 180
scr impulse driver, 989
sstv sync generator, 912
staticiser bistable store, 189
synchronous AND with delayed AND, 336
synchronous AND gate, 335
synchronous AND gate with delayed OR, 334
testing with pulse burst generator, 179
transistor bistable store, 188
triggered bistable with speed-up capacitors, 180
TTL-compatible drive using scr, 332
tv lines for integral-field photography, 636
twisted-ring using clocked R-S bistable, 180
twisted-ring flip-flop as motor drive, 186, 191
up-down photoelectric, 189
up-down synchronous, 183
voltage-frequency converter, 167
(See also Frequency counter; Frequency divider)
Coupler, ham phone patch, 879
Crime, 6-V portable antimugging shocker, 343
Crossover distortion, eliminating with current booster for opamp, 573
Crossover network, bass and treble horns, 54
Crowbar:
    overvoltage for switching regulator, 756
    overvoltage protection of equipment, 650
Cryogenic, 144-MHz cooled preamp, 20

Cryptography, −15-V power-supply regulator, 759
Crystal, activity tester, 123, 449, 936
Crystal checker:
    frequency, 930
    frequency and activity, 933
    oscillator driving lamp, 932
Crystal detector, modulation monitor, 978
Crystal oscillator (see Oscillator)
Cubic-function generator:
    log converter, 410
    opamps with transistors, 409
Current booster, linear for opamp, 573
Current control:
    1-nA precision sink using fet opamp, 196
    2 μA-1 mA cascade fet source, 198
    5 μA-1 mA adjustable fet source, 197
    400-μA source for fet amplifier, 195
    0.05-mA matched sources, 196
    1-500 mA at 0-50 V in nine steps, 195
    5-mA constant current for thermistor, 452
    7.6-mA constant-current generator for voltage reference diode, 195, 198
    30-mA at 900 V for magnetron, 339
    50-200 mA regulator, 194
    60-mA quasi-constant load, 196
    200-mA for voltage regulator, 198
    1-A source using IC voltage regulator, 196
    20-A a-c overcurrent protector, 654
    2,000-A laser driver, 392
    10-Hz constant-current source for bipolar probe, 482
    adjustable for diode and transistor tests, 197
    adjustable for regulated supply, 724
    bilateral source using opamp, 197
    bipolar source for grounded load, 197
    differential amplifier, 198
    dual-transistor, 198
    focus coil, 196
    grounded load, 197
    microammeter amplifier with gain of 200, 425
    power supply protection, 194
    scalor, 194
    switchback current limiting, 754
    switching regulator, 195
    (See also Regulator)
Current measurement (see Ammeter; Galvanometer; Microammeter; Multimeter; Nanoammeter)
Current monitor, power supply using fet and opamp, 424
Current regulator:
    50-200 mA grounded-base, 194
    line resistance compensator, 760
    (See also Current control; Protection; Regulated power supply; Regulator)
Current scalor, two-port network element, 194
Current source, 455-kHz high-impedance, 691
Current switch, ripple-carry binary feedback counter, 180
Curve tracer:
    60-mA quasi-constant current load, 196
    npn transistors, 945
    transistor and zener for cro, 937
    trigger for voltage sweep generator, 989
C-w (see Code; Receiver; Transmitter)

Darlington, series-pass in 3-V regulated supply, 735
Data transmission:
    2.125-kHz active bandpass filter, 200, 201
    100-kHz fsk generator, 502
    4-MHz delay-line storage, 202
    0-MHz fsk generator, 505

Data transmission (Cont.):
  150-ohm line driver and receiver, 201
  400-bit 4-MHz shift register, 200
  adjustable delay for modem, 502
  fsk decoder, 503, 504
  fsk demodulator, 199
  fsk modem, 500, 503
  gate drives unterminated line, 202
  line receiver for zero input, 202
  linear time-shared multiplexer, 533
  multiplexer-decoder, 533
  optoelectronic coupling in high-speed line
    receiver, 199
  pam 4-channel multiplexer, 532
  parallel-gate line driver, 201
  pcm converter, 200
  serial at 20-MHz rate, 201
  serial-parallel converter, 199
  single-opamp data receiver, 202
  timer as low-cost line receiver, 202
  two-gate coax line driver, 202
  (See also Frequency-shift keying)
D-c amplifier:
  10-mA output for instrumentation, 369
  3-A power using Norton opamp, 582
  0-100 kHz differential with single-ended
    output, 20
  0-V output for 0-V input, 582
  6,500 voltage gain, 9
  70,000 voltage gain, 9
  100,000 voltage gain, 10
  150,000 voltage gain, 14
  bias-current compensation, 577–579
  chopper-stabilized, 118
  Darlington inverting bootstrapped, 10
  differential with differential output, 14
  differential-input with bias-current
    compensation, 581
  drift compensation, 367
  driving DTL or TTL with led, 584
  fet for cro or vtvm, 114
  instrumentation with decade-step gain
    switch, 368
  level-shifting with isolation, 581
  photodiode with switching action, 630
  voltage follower with bias-current
    compensation, 581
  (See also Amplifier; Operational amplifier;
    Preamplifier)
D-c restorer, 10-MHz video fet with clamp, 686
D-c transformer, Hall-effect current-measuring,
    427
Deaf, voice alarm for the, 484
Deceleration, measuring with pendulum-driven
    bridge, 72
Decision element, 8-gate asynchronous, 414
Decoder:
  107-Hz phase-locked loop for repeater, 775
  1-kHz for f-m repeater, 771
  1.8-kHz tone driving relay, 880
  2.3-kHz tone for f-m repeater, 774
  1-out-of-8 for 8-bit parity checker, 145
  1-out-of-16 for 12-bit a-d converter, 25
  1-out-of-64 using IC multiplex switches, 534
  15-channel digital remote control, 765
  afsk autostart for rtty, 889
  bcd to decimal for numeric indicator, 191,
    230–232
  bcd with diode gates, 416
  bcd combined with scs store for indicator,
    225
  binary to hex, 413
  binary to octal, 416
  Columbia SQ quadraphonic, 684
  f-m multiplex stereo, 831, 834, 836, 838
  multiplexer data-transmission, 533

  odd-even bcd for numeric indicator, 232
  pwm temperature-stable, 764
  quick-pulsing for f-m repeater phone patch,
    773
  single-tone for f-m repeater, 776
  SQ stereo-quadraphonic, 683
  stereo phase-locked loop, 832, 834
  tone using active filter, 254
  tone-burst for f-m repeater, 768
  universal quadraphonic, 685
Delay:
  100-ps gate using transistor, 333
  1-ns in high-speed voltage follower, 14
  7-27 ns digital using high-speed shift
    register, 148
  3-s for operating electronic lock, 237
  5-50 s in burglar alarm, 273
  20-s for penalty switches of electronic lock,
    234
  30-s for relay using heater warmup, 654,
    954
  30-s burglar-alarm entrance, 97
  adjustable for digital modem, 502
  auto theft alarm, 84
  carrier-operated relay for f-m repeater, 775
  commercial-killing touch switch, 105
  complementary driver for high-speed scaler,
    192
  headlight-off, 62, 65
  logic reset at power turn-on, 412
  power supply turn-on, 652
  scr trigger for auto dwell booster, 65
  sound-actuated control, 762
  temporary data storage in mos shift
    registers, 128, 202
  woofer protection from transients, 49
  (See also Timer)
Delay-line coupling, square-wave generator,
    547
Demodulation distortion, measuring for a-m
    receiver, 447
Demodulator:
  1-10 MHz f-m, 303, 305
  balanced ring in telephone scrambler, 882
  f-m with time multiplexing, 530
  f-m crystal-controlled phase-locked loop,
    305
  f-m low distortion, 300
  f-m narrow-band with carrier detection, 304
  f-m phase-locked loop, 302, 306, 308
  f-m stereo multiplex receiver, 839
  f-m telemetry using IC phase-locked loop,
    876
  fsk for 2.025 and 2.225 kHz, 501, 505
  fsk modem for Bell data sets, 503
  fsk single-supply IC, 502
  fsk split-supply IC, 501
  fsk using active filters, 199
  fsk using phase-locked loop, 503–505
  IRIG telemetry channel 13, 874
  pulse sampling, 779
  rtty using phase-locked loop as
    demodulator, 887
  synchronous a-m, 707
  weather satellite picture, 877
  (See also Detector; Discriminator; Modem,
    Ratio detector)
Detector:
  455-kHz with 120-dB dynamic range, 691
  28-MHz superregenerative, 694
  50-MHz photodiode for led communication,
    587
  144-MHz superregenerative, 694
  adjustable threshold level, 1008
  broad-area for led communication, 588
  changing-magnetic-field, 498

  frequency-band, 144
  frequency error for 5 to 20 MHz, 286
  liquid level with audio indication for the
    blind, 403
  minimum-frequency using IC comparator,
    142
  negative voltage peaks, 1007
  null using logamp, 410
  optoelectronic for slow-moving objects, 586
  peak amplitude-independent, 1010
  peak-to-peak voltage, 1006
  phase-sensitive using beam-switching tube,
    357
  position using photodiodes with
    perturbation, 404
  quadrature for led communication, 587
  reciprocating for dssc, 692
  slow-decay for c-w Schmitt trigger, 131
  synchronous for 144-MHz moon-bounce
    signals, 702
  synchronous a-m, 692, 705, 707
  vhf superregenerative, 170
  (See also Auto theft alarm; Burglar alarm;
    Demodulator; Discriminator; Fire alarm;
    Humidity alarm; Metal detector; Moisture
    detector; Phase control; Protection, Ratio
    detector; Voltage-level detector; Zero-
    voltage detector)
Deviation meter, f-m transmitter, 304
Diaper alarm, moisture detector, 4
Dice, oscillator-driven ring counter, 344, 348
Differential amplifier:
  0-100 kHz single-ended output, 20
  10-V single-supply voltage reference, 1013
  canceling offset with series opamps, 574
  fet input for opamp, 573
  instrumentation bridge, 366
  temperature-compensated current source,
    195
  (See also Comparator; D-c amplifier;
    Operational amplifier)
Differential opamp, low temperature drift for
    instrumentation, 368
Differentiation, reference input for
    frequency-error detector, 286
Differentiator:
  opamp converting triangle to square, 677
  swings above and below bias, 582
  zero-crossing detector as a-d converter
    trigger, 22
Digit detector, high-speed memory, 146
Digital-analog converter:
  0-10 V d-c output, 207
  5-bit word length with hex inverter, 207
  9-bit using cos-mos logic, 206
  10-bit using quint current switches, 204
  12-bit high-speed, 205
  12-bit using quad current switches, 204, 207
  analog switch-multiplexer, 528
  binary-weighted voltage comparator, 207
  computer-controlled programmable-mono
    tester, 933
  counter-type, 204
  low-cost ladder-network switch, 206
  multiplying, 203
  settling-time expander for cro, 205
  staircase generator for cro display, 203
  trimming for 12-bit accuracy, 206
  ultra-low power, 205
Digital-analog ladder, integrated chopper,
    117
Digital clock:
  12-hour, 209
  12-hour with minutes and seconds, 218
  12-hour with 19 IC's, 214
  12-hour gate-minimizing, 208

Digital clock (*Cont.*):
  12-hr or 24-hr with alarm and timer, 210, 211, 214, 216, 217
  12-hr or 24-hr six-digit, 215
  24-hour, 210, 211
  12:59 to 1:00 transition logic, 219
  a-c with standby battery, 208
  battery-powered led display, 212
  Digivac display, 213
  dividers for display, 217
  Morse-code time identifier for repeater, 215
  multiplexing led display, 218
  three-mode display, 209
  (See *also* Clock)
Digital computer (see Computer)
Digital differentiator, fsk receiver, 504
Digital filter:
  500-Hz active narrow-band, 253
  high-pass, 413
Digital frequency meter (see Frequency counter; Frequency divider; Frequency measurement)
Digital tape recorder, noise spike rejection, 566
Digital voltmeter:
  60-Hz active ripple filter, 477
  4½-digit using a-d converter, 22
  calibration with standard cell, 480
  conversion to ohmmeter, 461
  driver for numeric display, 472
  linear staircase generator, 824
  linear vco, 997
  time-averaging reduces noise uncertainty, 471
  trigger for measuring pulse amplitude, 476
Dimmer (see Lamp dimmer)
Diode tester:
  go-no-go, 936, 944
  simple nondestructive, 943
Dip meter:
  1-250 MHz single-transistor, 434
  1.5-30 MHz fet, 436
  1.5-50 MHz fet, 437
  1.7-300 MHz solid-state, 437
  1.8-150 MHz fet portable, 435
  2.9-155 MHz vfo/crystal, 438
  3-200 MHz jfet, 439
  a-f modulator, 506
  fet, 433, 440
  solid-state using lamp as indicator, 436
Direct-current (see *under* D-c)
Direction finder, hidden CB transmitter, 849
Discriminator:
  60-Hz using analog multiplier, 876
  1-MHz f-m transformerless, 307
  balanced f-m transmission-line bridge, 306
  color tv phase detector, 905
  f-m receiver, 301, 307
  pulse with constant relative backlash, 1006
  pulse height for nuclear research, 1009
  tracking filter for f-m, 1000
  (See *also* Demodulator; Detector; Frequency modulation; Ratio detector)
Dishwasher, sequential ujt-scr timer, 955
Display:
  12-V chaser for ad signs, 279
  20-W lamp chaser, 279
  4-channel on cro, 109
  64-character ASCII alphanumeric, 221, 222
  3-digit with overflow, 228
  5-digit strobed, 220
  6-digit with zero blanking, 228
  8-digit with parallel multiplexer, 529
  9-lamp decade counter, 184
  7-segment numeric, 191, 227
  airborne marker beacon, 561, 563, 565

bcd-decimal decoder, 220, 225, 230–233
binary 2-digit demonstration adding machine, 147
brightness regulator for incandescent readouts, 232
cathode-ray gamma correction, 110, 111
complementary-lamp switch, 861
counter drive for alphanumeric crt from ram, 492
crt pincuscion and blur correction, 113
data register driving alphanumeric crt, 494
digital with zero suppression, 223
digital thermometer, 464
drive for eight-decade Pandicon numeric indicator, 225
driver for dvom numeric, 472
driver for Panaplex II in calculator, 222
driver for Panaplex II gas-discharge, 223
fail-safe filament monitor, 221
floating-decimal, 229
inverting driver for neon, 146
Karnaugh map on cro screen, 227
Karnaugh map phase-shift oscillator, 233
led in analog voltage sensor, 1005
led in ECL test probe, 931
logic level indicator, 415
MAN-3 led drive, 535
mos interface, 224
multiplexer drive for HP5082 series characters, 529
multiplexing 16 seven-segment alphanumeric, 532
Nixie for 12-hour digital clock, 208
Nixie blanker, 233
Nixie dimmer, 231
Nixie driver using scr's, 224
numeric for frequency/period counter, 288
numeric with memory for counter, 233
numeric for ring counter, 226
numeric with scs ring counter drive, 224
numeric indicator drive with delay, 182
Numitron numeric with decade counter, 231
opamp status indicator, 1008
phase-shift monitor scope for rtty tuning, 885
rtty on cro using audio transform, 895
sample-and-hold using indicator tubes, 783
staircase generator for d-a converter driving cro, 203
switch-off indicating flasher, 280
three-mode digital clock, 209
timing and character generation from ram for crt, 491
trapezoid generator for numeric seven-segment crt, 230
tuning eye for color tv, 902
video numeric, 226
video 16-character, 229
zero-beat tuning with led's, 223
  (See *also* Calthode ray; Character generator; Music-controlled light)
Distillation, temperature control, 929
Distortion, guitar fuzz generator, 246
Distortion factor meter, audio amplifiers, 447
Diversity reception, audio combiner using photocell, 693
Divider:
  20:1 ripple-carry binary feedback counter, 180
  analog single-IC, 537, 540
  analog using log converters, 406
  analog using reciprocal circuit and multiplier, 537
  frequency-locking for servo, 793
  one-quadrant logarithmic, 407
  (See *also* Frequency divider)

Door lock:
  2-kHz modulated-light key, 237, 239
  2-kHz modulated-light receiver, 234
  8-bit binary code, 238, 239
  60-combination electronic, 234
  4-digit with 10,000 combinations, 236
  6-digit IC, 235
  coded scr-operated, 236
  combination four-digit, 235
  combination seven-pushbutton, 239
  crystal-oscillator key, 237
  key using diode in phone plug, 237, 238
  time-controlled, 235
Doorbell:
  three-tone door-identifying, 349
  touch switch with delay for long ring, 105
Doppler, 30-MHz variable attenuator with constant phase, 562
Doppler radar, self-detecting IC preamp, 690
Doubler (see Frequency multiplier)
Drum:
  electronic for guitar, 251
  rhythm generator, 241
Dry cell (see Battery)
Dual-trace scope, switching waveform generator, 109
Duplexer:
  133-MHz broadband, 960
  450-MHz using stripline, 963
  high-speed antenna switching gate, 963
Dwell-time extender, coil-and-breaker-point ignition, 78
Dwellmeter:
  auto, 65–67, 71
  auto and boat engine, 71

Ear temperature, 30-40 C infrared radiation thermometer, 466
Earphone (see Headphones)
Echo-chamber effect, tape recorder, 872
Electric blanket, temperature control using indirectly heated sensor, 922
Electric fence:
  6-V charger, 343, 348
  d-c/d-c converter, 339
Electric field, measurement with paper-lead probe, 444
Electric hammer, impulse driver, 988–990
Electrocardiograph:
  pen recorder driver amplifier, 484
  real-time contourograph display, 485
Electroencephalophone, stereo, 482, 487
Electrolytic capacitor:
  leakage tester, 418, 422
  measurement to 1,000 µF, 420, 421
  tester for high-capacitance low-voltage units, 422
  tester using neon indicator, 422
Electromagnetic interference (see Interference)
Electrometer:
  1-pA and 300-pA fast-response, 426
  5 pA to 500 µA, 424
  compensated mosfet, 425, 427
  high sensitivity, 428
  ion-chamber, 444
  simple mosfet, 427
Electromyograph, muscle-controlled audio tone generator, 785
Electronic ignition, automotive, capacitor-discharge, 76–81
Electronic music:
  2.1-kHz tone generator, 597
  13-V and 30-V supply for organ, 747
  broadcast-band metronome, 242
  digital 16-note melody generator, 245

Electronic music (Cont.):
  drum rhythm generator, 241
  drum using phase-shift oscillator, 249
  guitar 20-dB treble boost, 250
  guitar bass boost, 251
  guitar fuzz amplifier, 240, 242, 246, 250
  guitar rhythm generator, 251
  guitar 7-circuit tone control, 247
  guitar variable attack, 242
  guitar Waa-Waa sound, 244
  metronome, 130, 244, 246, 248–251
  mystery tone generator, 250
  organ with tremolo, 247
  organ using c-mos logic with diode gates, 246
  organ audio compressor, 45
  organ digital tuner, 248
  organ Leslie-effect and vibrato simulator, 243
  organ master tone generator, 245
  organ triggered-fuzz adapter, 250
  pitch-reference battery charger, 87
  Theremin, 243, 247
  touch-a-tone four-octave music 240
  toy organ, 349
  tuning-fork amplifier, 242
  vibrato tone generator, 245, 249
  vibratone toy, 244
  (See also Metronome; Music-controlled light)
Electronic organ (see Electronic music)
Electrooptical (see Optoelectronic;
    Photoelectric)
Electrostatic precipitator, 10-kV air-cleaning, 339
Encoder:
  87-112 Hz for f-m repeater access, 772
  107-Hz for repeater access, 775
  1-kHz for f-m transmitter, 771
  afsk rtty, 885
  ASCII teletypewriter, 151
  binary optoelectronic shaft-position, 590
  dice display for ring counter, 348
  excess-three with holding register, 150
  excess-three using diode matrix, 415
  universal keyboard, 148
Enlarger (see Photography)
Envelope signal level, detector with lamp
    indicator, 1005
Environment, three-level noise monitor, 567
Equalizer, delay for data modem, 502
Exciter, 52.5-MHz wide-band f-m, 303
Expander:
  music volume, 50
  phono records, 42
Exponential-function generator, 0.25-4
    adjustable, 411
Exposure meter:
  electronic flash, 635, 638
  low light levels, 643

Facsimile:
  Telefax modification for amateur service, 959
  tuning indicator, 889
Fan control, basement humidity, 925
Field-effect transistor:
  matching pairs, 942
  measuring drain-source resistance, 943
  measuring forward transconductance, 937
  measuring leakage current, 945
  measuring pinch-off voltage, 942
  tester, 942
Field-strength meter:
  10-GHz, 453
  general-purpose, 976

portable impedance bridge, 449
remote-reading, 456
Filter:
  455-kHz ceramic in i-f signal generator, 796
  active: 20-20,000 Hz tunable low-pass or
      high-pass, 258
    25-Hz switchable rumble for stereo, 833
    50-2,500 Hz voltage-tunable, 260
    60-Hz adjustable-Q notch, 259
    60-Hz hum on tape, 867
    60-Hz notch, 253, 261
    60-Hz ripple-suppressing, 477
    62-Hz a-f bandpass, 806
    100-Hz bandwidth for c-w receiver, 137
    100-Hz high-pass, 257, 261
    100-Hz low-pass, 560
    100-Hz low-Q bandpass, 263
    100-10,000 Hz voltage-controlled
        high-pass, 252
    150-Hz low-pass, 688
    150-1,500 Hz tunable bandpass with
        constant Q of 30, 259
    159-Hz 1/3rd-octave 2nd-order bandpass, 258
    160-Hz high-Q bandpass, 263
    200-Hz bandpass with Q of 5, 262
    200-3,200 Hz vco, 999
    200-3,200 Hz voltage-tuned with 80-Hz
        bandwidth, 261
    300-3,000 Hz voice communication, 263
    400-1,300-3,000 Hz marker beacon tones,
        561, 565
    1-kHz with 100-Hz bandwidth for c-w, 262
    1-kHz bandpass, 257, 261–263
    1-kHz bandstop, 264
    1-kHz high-pass, 257
    1-kHz low-pass, 261, 264
    1-kHz tunable for c-w, 136
    1-10 kHz adjustable bandpass, 260
    1.1-kHz for c-w, 256
    1.2-kHz bandstop, 263
    1.2-kHz twin-T bridge, 259
    1.37-kHz, 254, 255
    1.5-kHz high-pass, 258
    1.5-kHz low-pass, 258
    1.75-kHz low-pass and high-pass, 867
    2-kHz variable with 260-Hz constant
        bandwidth, 262
    2.025-kHz and 2.225-kHz in fsk
        demodulator, 199
    2.125-kHz bandpass for phone data line,
        200, 201
    2.125 and 2.975 kHz bandpass for rtty,
        893
    2.5-kHz bandpass, 257
    2.5-kHz high-pass, 257
    2.5-kHz low-pass, 259
    2.6-kHz bridged-T in phone patch, 882
    5-kHz low-pass synthesized lossless, 262
    9-kHz high-pass, 560
    10-kHz low-pass, 256
    10-200 kHz low-pass, 252
    455-kHz i-f amplifier, 353
    455-kHz twin-T notch, 263
    1-MHz tracking, 1000
    10-MHz bandpass in amplifier, 255
    14-MHz amplifier with 20 dB gain, 257
    60:1 power-supply ripple reduction, 262
    250:1 power-supply ripple reduction, 254
    adjustable R-C network, 253
    adjustable twin-T notch, 263
    audio noise and rumble, 47
    audio tone control, 54
    averaging for a-c voltmeter, 475
    cascaded twin-T c-w, 252
    coherent for YIG radiometer, 465

c-w IC, 139
digital for teleprinter, 891
digital bandpass, 253
digital high-low bandpass, 413
eighth-order linear-phase pulse, 255
elliptic ladder network, 577
fixed-tuned WWV receiver, 693
Fortran program for Butterworth design,
    262
high-frequency for stereo, 51
i-f notch for ssb receiver, 804, 805
low-pass with voltage-controlled cutoff,
    261
low-pass in voltage-frequency converter,
    167
low-pass using opamp time-constant
    multiplier, 953
low-pass adjustable for 1-10 kHz cutoff,
    255
low-pass Chebyshev in constant-phase
    limiter, 400
narrow-pass in 2:1 frequency divider, 296
R-C synthesized, 260
second-order Butterworth in fsk modem,
    500
signal tone, 7
sound-level meter, 450
speech clipper, 400
state-variable bandpass in modem tone
    oscillator, 504
third-order bandpass, 255, 256
tone signaling and decoding, 254
tracking for f-m discriminator, 1000
tunable with adjustable Q, 254
tunable 60-dB notch, 259
tunable a-f selectivity, 37
tunable notch and 500-2,000 Hz
    bandpass, 260
twin-T in distortion factor meter, 447
two-pole low-pass using linear IC, 264
variable-Q in 10.7-MHz tunable i-f
    amplifier, 355
vco with opamp, 1002
voltage-tuned low-pass, 256
voltage-tuned narrow-band, 264
frequency-response plotter drive, 934
integral notch in power amplifier protection
    circuit, 650
passive: 1-Hz Chebyshev low-pass computer-
    designed, 266
  80-Hz narrow-pass for c-w, 812
  100-2, 100 Hz speech, 267
  120-Hz ripple for transistor supply, 269
  300-2,000 Hz bandpass for hi-fi phones,
      268
  500-Hz and 5,000-Hz bandwidth in i-f
      amplifier, 352
  600-Hz center-frequency c-w, 134
  600-1,900 Hz speech, 267
  750-Hz c-w with 75-Hz bandwidth, 132
  800-Hz for c-w receiver, 265
  0-3.4 kHz low-pass seventh-order Cauer-
      parameter, 268
  1-kHz presence control, 55
  1-kHz Wien-bridge, 269
  1.275-1.445 kHz bandpass for rtty, 888,
      892
  1.275-2.125 kHz bandpass for rtty, 886,
      892, 893
  2.1-kHz bandpass mechanical for 455-kHz
      i-f, 267
  2.1-kHz low-pass elliptic-function speech,
      269
  2.125-2.975 kHz bandpass rtty, 887
  6.375-kHz notch for afsk rtty, 890
  8-kHz cutoff for 78-rpm records, 626

Filter, passive *(Cont.):*
    50-kHz T-notch, 268
    100-kHz synthesized narrow-band crystal, 268
    135-kHz bandpass, 267
    455-kHz L-C for f-m receiver, 266
    455-kHz tunable slot for i-f, 267
    500-1,600 kHz broadcast-band rejection, 266, 269
    3.8-MHz trap in luminance amplifier, 901
    22-MHz and 30-MHz low-pass, 268
    40-MHz high-pass for tvi, 268, 269
    400-MHz bandpass, 313
    448-MHz bandpass, 768
    adjustable audio low-pass and high-pass, 53
    adjustable audio notch for hi-fi, 265
    a-f low-pass m-derived c-w, 355
    Butterworth uhf impedance-matching, 15
    cascaded half-lattice crystal, 265
    Chebyshev computer-designed all-pass, 266
    Chebyshev uhf impedance-matching, 15
    fluorescent-lamp rfi, 283
    four-crystal 9-MHz ssb, 808
    i-f notch, 809, 810
    low-pass and high-pass for magnetic pickup, 623
    power-line for thermostat arc interference, 928
    presence-effect for soloist, 55
    scratch and rumble, 626
    scratch for stereo phono, 621
    ssb crystal lattice, 809
    switched two-cutoff audio for phone and c-w, 269
    varactor i-f notch, 269
    tracking using zero-forcing of gate output, 332
Fire alarm:
    battery-operated, 273
    high-sensitivity for a-c power, 273
    high-sensitivity for battery operation, 271
    infrared, 270, 271
    light-sensitive clicker, 93, 271
    silicon-diode sensor, 273
    thin-wire loop, 272
Fire calls:
    135-175 MHz converter for f-m receiver, 173
    auto radio converter, 171, 172
Fish caller:
    audio tone generator, 349
    click-generating oscillator, 349
Fish finder, thermocline temperature sensor, 349
Fish lure, buzzer and flasher, 344
Flame failure:
    infrared-sensing for gas furnace, 272
    sensing flame resistance, 274
Flash:
    audio-controlled strobe, 637
    duration sensor, 634
    measurement with fet voltmeter, 638
    sound-fired for stop-motion photography, 639
    (See also Photoflash; Photography; Strobe)
Flash circuit, pulsing led with scr, 359
Flash exposure meter, insulated-gate fet, 635
Flasher:
    1-Hz using IC square-wave source, 275
    2-kHz as key for photoelectric door lock, 237, 239
    4.8-V d-c, 279
    6-V photoelectric turn-on, 278
    6-12 V with adjustable on and off times, 276
    12-V auto emergency, 69, 277, 280

12-V chaser, 279
12-V single-lamp, 278
20-W sequential using ring counter, 279
150-W IC trigger for triac, 276
200-W two-lamp, 277
1-kW two-lamp driven by flip-flop, 279
500-cp Christmas-tree lamp, 277
astable mvbr driving led pair, 544
audio-controlled psychedelic fluorescent, 281
auto idiot light, 70
auto low-fuel indicator, 61, 66
auto rear-lamp failure, 63
barricade photocell-controlled, 280
dual-rate emergency, 277
fish lure with buzzer, 344
five-neon for twinkling star, 275
flip-flop mvbr, 280
fluorescent lamp for 6 V, 282
L-C commutated ring counter, 187
led for intercom message alert, 378
low-drain switch-off indicator, 280
monostable mvbr driving led pair, 545
parked-car identifier, 72
random three-neon, 280
ring counter with gate-driven load, 190
sequential auto turn-signal, 68, 70
sequential turn-on delay, 276
solid-state switch, 279
switching receiver channels, 771
turn-signal left-on alarm, 67
xenon high-intensity, 278
zero-crossing IC proportional control, 1018
Flashover protection, testing with triggered spark gap, 339
Flip-flop:
    500-Hz narrow-band digital active filter, 253
    60-mA using quad gates, 331
    450-W touch-plate switch, 103
    bounceless switch, 154
    clocked R-S, 547
    coin flipper, 784
    complementary-transistor low-dissipation, 552
    converter for 28 V d-c to 68 V d-c, 158
    counter range extender to 100 MHz, 297
    current-mode high-speed Schmitt trigger, 988
    decimal Gray-code counter, 191
    digital 16-note melody generator, 245
    dot-dash keyer, 135
    flasher drive, 280
    Gray-code generator without glitches, 149
    initializing after line transients or failure, 542
    inverting driver for neon, 146
    J-K pair as pulse separator, 542
    logic-controlled 5:1 or 6:1 divider, 294
    noise-immune bidirectional counter, 189
    opamp with led status indicators, 544
    predictable R-S, 545
    put trigger for scr's in inverter, 990
    suppressing spurious triggering in one-shot, 546
    triac drive for 1-kW flasher, 279
    trigger for scr pair in inverter, 550, 551
    trigger frequency-dividing, 298, 299
    triggered lab sweep, 853
    twisted-ring counter, 186, 191
    voltage-controlled negative resistance, 552
    (See also Logic; Multivibrator)
Flood alarm, conductive-salt, 403
Fluorescent lamp:
    6-V flasher, 282
    12-V inverter supply, 284

22-W from 12-V battery, 284
100-W mercury-arc ballast, 281, 283
175-W mercury-arc ballast, 282
400-W mercury-arc ballast, 284, 285
500-W resonant-bridge inverter, 282
d-c converter, 282
music-controlled brightness, 283
psychedelic flasher, 281
rfi filter, 283
Flutter meter, phono turntable, 622
Flux monitor, ferrite phased-array radar, 687
Flying-spot scanner:
    crt for sstv, 909
    sync pulse generator, 908
    test generator for sstv, 910
F-m (see Frequency modulation)
Focus blur, correction in precision crt, 113
Focus coil, current-control regulator, 196
Foghorn, automatic control, 564
Foldback protection, regulated power supply, 650
Follow and hold, voltage measurement, 778
Free-running multivibrator (see Multivibrator, astable)
Frequency calibrator:
    10-100 kHz square-wave, 434
    10-100-1,000 kHz crystal, 437
    25-50-100-200 kHz crystal, 435
    15-V regulated supply, 727
    crystal for f-m up to 450 MHz, 438
Frequency comparator (see Comparator)
Frequency control, 60-Hz discriminator, 876
Frequency counter:
    4 Hz-42 MHz prescaler and preamp, 286
    10 Hz-220 MHz with 0.0001% accuracy, 290, 293
    20 Hz-30 MHz input conditioner, 290
    60-Hz demonstrator, 293
    100-kHz time base using 1.4-MHz crystal, 293
    200-kHz crystal standard, 290
    0-20 MHz input shaper, 292
    1-MHz crystal-clock reference, 289, 293
    5-120 MHz, 287
    10-MHz input amplifier with clipper, 291
    32-MHz input amplifier, 19
    200-MHz divide-by-10 prescaler, 292
    5/9/180-V regulated supply, 723
    adapter for receiver, 291
    bcd-decimal decoder for numerical readout, 289
    digital display, 291
    digital frequency readout for receiver, 292
    error detection, 286
    five-Nixie for transmitter vfo, 287
    mono mvbr for holding reset pulse, 293
    prescaler for digital frequency display, 288
    pulse shaper for J-K flip-flops, 288
    quad latch for display, 288
    shaper for input signal, 287
    time base for digital clock, 210
    vhf with 99,999 display, 289
    (See also Counter; Frequency divider)
Frequency detector, minimum using IC comparator, 142
Frequency divider:
    25-kHz ham-band calibrator, 430
    100-kHz standard with 1-kHz markers, 430
    100-kHz steps from 100-MHz marker, 438
    100 kHz to 25 kHz, 432
    200-kHz crystal calibrator, 435
    1-MHz crystal calibrator, 437
    1.4 MHz to 100 kHz, 293
    1.5-30 MHz synthesizer for amateur receiver, 320

Frequency divider (Cont.):
200-MHz:20-MHz prescaler for frequency counter, 292
2:1 to 11:1 for square-wave a-f, 298
2:1 to 30:1 adjustable even-order digital, 295
2:1 to 30:1 for 1.1 MHz, 298
2:1 for 80-meter qrp transmitter, 980
2:1 with pulse shaper, 297
2:1 sine-wave with variable tuning, 296
3:1-5:1-7:1-9:1 using phase-locked loop, 294
4:1 for 40 MHz, 296
5:1 or 6:1 logic-controlled, 294
6:1 and 10:1 with 60-Hz reference, 297
10:1 for 100-MHz input, 296
10:1 for 175 MHz, 295
10:1 for 300 MHz, 299
10:1 for counting to 100 MHz, 297
10:1 with bcd-decimal decoder, 289
50:1 for digital frequency meter, 294
100:1 for 1-MHz crystal mvbr, 297, 298
counter decade using staircase, 825
digital 90-deg ssb phase shifter, 614
digital phase-locked loop, 295
mono IC mvbr, 297
odd-order symmetrical counter, 193
regulated supply, 716
square-wave for 200-kHz crystal, 434
ssb exciter using 2-MHz crystal, 805
sstv sync pulse generator, 908
trigger for triac flasher, 276
trigger flip-flop, 298, 299
tunnel-diode staircase generator, 827
variable-ratio for 2-meter frequency synthesizer, 321
(See also Frequency counter)
Frequency doubler (see Frequency multiplier)
Frequency locking, multiplier-divider for servo, 792
Frequency marker:
25-kHz ham-band calibrator, 430
100-kHz crystal with 10-kHz mvbr, 432
100-kHz crystal with harmonics, 436
100-kHz crystal with harmonics to 150 MHz, 434
100-kHz crystal-controlled square-wave with harmonics, 437
100-kHz with 1-kHz markers, 430
100-kHz with 100-kHz markers to 100 MHz, 437
100 kHz-1,296 MHz, 439
200-kHz intervals for transceiver, 961
500-kHz crystal with 8-Hz warble indentification, 439
1-MHz band-edge with harmonics up to 30 MHz and 1-kHz modulation, 440
1-MHz crystal with harmonics to 30 MHz, 437, 441
1-MHz square-wave, 439
3-MHz with 30-kHz markers to 150 MHz, 436
2-meter f-m receiver, 439
short-wave receiver, 433
Frequency measurement:
60-Hz generator, 455
100-kHz standard with 10-kHz markers, 441
200-kHz markers from crystal calibrator, 961
10:1 divider for counter, 299
band comparator using one-shots and flip-flops, 144
frequency-error counter, 286
Frequency meter:
10 Hz-100 kHz for audio oscillator, 676
60-Hz for emergency-power a-c generators, 434
1-30 MHz using frequency-voltage converter, 435

1-250 MHz dip oscillator, 434
1.5-30 MHz fet dip oscillator, 436, 437
1.7-300 MHz solid-state dip oscillator, 437
1.8-150 MHz fet dip oscillator, 435
2.9-155 MHz vfo/crystal dip oscillator, 438
3-200 MHz gate-dip oscillator, 439
dip oscillator, 433, 440
dipper using lamp as indicator, 436
Frequency modulation:
60-Hz discriminator using analog multiplier, 876
700-Hz carrier for stereo electroencephalophone, 487
1.8-450 MHz signal generator, 796
30-MHz signal generator, 796
144-148 MHz receiver aligner, 798
1-kHz decoder for repeater, 771
1-kHz encoder for repeater use, 771
1.8-kHz and 2-kHz tone-burst keyer for repeater, 767
455-kHz i-f filter, 266
455-kHz narrow-band adapter, 305, 399
1-10 MHz demodulator, 303, 305
45-MHz oscillator and tripler for 2-meter receiver, 607
50-MHz 40-W 12.5-V transmitter, 968
50-MHz 40-W c-w mobile transmitter, 980
52-MHz modulated oscillator, 308
52.5-MHz wide-band exciter, 303
135-175 MHz police-call converter, 173
146-MHz preamp, 398
146.34-MHz transmitter, 301
220-MHz converter, 177
220-MHz varactor quintupler, 301
432-MHz varicap-controlled crystal oscillator, 985
2-meter 3-W driver, 306
2-meter 25-W amplifier, 962
2-meter 50-W amplifier, 302
2-meter converter for broadcast radio, 178
2-meter f-m frequency marker, 439
2-meter frequency synthesizer, 321, 322
2-meter receiver, 302
2-meter solid-state transmitter, 304
2-meter synthesizer gives 30-kHz steps, 319
2-meter tunable vfo for receiver, 301
1-min timer for Morse-code identifier, 768
6-V and 12-V low-ripple regulated supply, 714
16-V regulator for transmitter supply, 738
200-W transceiver power amplifier, 967
a-m/f-m i-f with detectors, 351
automatic call-letter generator for repeater, 154, 769
automatic monitoring of two repeater channels, 771
bridge discriminator, 306
carrier-operated relay, 3, 770, 775
carrier-operated squelch relay, 826, 827
carrier switching for split-site repeater, 770
ceramic-filter stereo i-f amplifier, 837
crystal discriminator, 301
demodulator using phase-locked loop, 302, 305, 306, 308
dial control of repeater, 769
digital frequency display, 288, 291
discriminator, 307
fet tuner with 2-gang capacitor, 306
frequency synthesizer for tuning, 321
fsk demodulator, 501, 502
i-f with Foster-Seeley discriminator, 353
i-f and quadrature detector, 352
i-f and ratio detector, 350
i-f amplifier updating, 352
i-f single-IC, 354
i-f sweep generator, 851

limiter using IC, 401
low-distortion demodulator, 300
mobile repeater control with Touch-tone pad, 767
moisture alarm using wireless mike, 4
mosfet tuner, 303
narrow-band demodulator with carrier detect, 304
noise-operated squelch, 821, 822
peak deviation meter, 304
pentode limiter, 398
phase modulator for a-m transmitter, 308
ratio detector, 307
ratio detector for telemetry system, 876
reactance modulator, 303, 305
repeater tone-burst decoder, 768
repeater turn-off timer, 770
SCA background-music adapter using phase-locked loop, 527
SCA multiplex adapter, 528
squelch with high sensitivity, 821
squelch for SCA adapter, 821
squelch add-on for receiver, 823
stereo adapter for standard receiver, 840
stereo buzzer or lamp indicator, 843
stereo indicator lamp drive, 839
stereo multiplex adapter, 830
stereo multiplex decoder, 831
stereo mulitplex demodulator, 839
stereo tuner, 845
stereo tuning meter, 829
telemeter for athletes, 874, 875
telemetry demodulator using phase-locked loop, 876
time-multiplexing between two f-m channels, 530
tracking filter for discriminator, 1000
transformerless discriminator, 307
transmitter-receiver aligner, 798
tuner with 3-gang capacitor, 307
tuner using varactors, 300, 308
tv sound i-f and demodulator, 912
tv sound locked-oscillator detector and limiter, 911
waveform generator as modulator, 326
wireless mike, 305
(See also Receiver; Repeater, Squelch; Transceiver; Transmitter)
Frequency multiplier:
7.074 MHz to 382 MHz, 809
112-448 MHz for repeater, 768
150 MHz to 450 MHz for transmitter, 979
200 MHz to 400 MHz for transmitter, 981
0.5 GHz to 1 GHz with varactor, 314
2 to 10 times using phase-locked loop, 294
7 times for 5 MHz in K-band radiometer, 311
8 times for 50 MHz with push-push varactor, 313
8 times for 500 MHz using varactor, 312
24 times for f-m receiver aligner, 798
40 times with hard limiter for 625-Hz input, 314
doubler: for 10 Hz-500 kHz, 313
for 200 Hz-1 MHz, 310
for 1 MHz down to audio, 314
for 7 MHz to 112 MHz in four stages, 772
for 14 MHz, 311, 316
for 35 MHz in K-band radiometer, 310
for 50 MHz at 200 W, 312
for 100-MHz crystal, 315
for 150 MHz using balanced modulator, 311
for 500 MHz using varactor, 315
for any wave shape, 311
for square wave, 310
for squaring with IC multiplier, 818

Frequency multiplier, doubler *(Cont.)*:
  using cascaded mos transistors, 316
  using IC balanced modulator, 313, 314
  using IC multiplier, 312
  using only digital logic, 309
  linearized TTL inverter, 7
  quintupler for 220-MHz f-m repeater, 301
  sextupler for 232-MHz local oscillator, 607
  tripler: for 150 MHz using varactor, 315
    for 200 MHz using varactor, 312
    for 333 MHz, 309, 312
    for 0.4 GHz using varactor, 309
    using gated clock generator, 688
  tripler driving counter, 690
Frequency response, drive for linear or log
  plotter, 934
Frequency-shift keying:
  2.025-2.225 kHz demodulator using active
    filters, 199
  2.125 and 2.975 kHz with slope and voltage
    detection, 500
  30-MHz signal generator, 796
  decoder with vco, 503
  demodulator: for 2.025 and 2.225 kHz, 501,
    505
    for 300 or 1,800 baud, 505
    as modem, 503
    using single supply, 502
    using split supply, 501
  generator: for carrier up to 100 kHz, 502
    for carrier up to 10 MHz, 505
    with two output levels from pll, 503
  low-speed modem, 500
  modem receiver, 504
  modem tone oscillator, 504
  phase-locked loop as demodulator, 887
  phase-locked loop converter for rtty, 886,
    893
  self-generating 2.125 and 2.975 kHz, 501
  waveform generator, 326
Frequency standard:
  440-Hz tuning-fork amplifier, 242
  50-kHz from 500-kHz crystal with mvbr, 432
  100-kHz crystal, 440
  100-500-1,000 kHz crystal with harmonics,
    435
  1-MHz with frequency dividers, 434
  1-MHz crystal oven control, 921, 927
  1-MHz modified Pierce crystal oscillator with
    harmonics, 440
  1-MHz square-wave, 433
  3-MHz with dividers, 438
  10-MHz crystal, 434
  monostable crystal-controlled mvbr, 549
Frequency synthesizer:
  300-10,000 kHz in 1-kHz steps, 318
  1.5-30 MHz for amateur reciever, 320, 998
  3.2-MHz reference crystal oscillator, 321
  3.58-MHz from 5-MHz reference, 321
  5-MHz using 3.58-MHz tv subcarrier as
    reference, 317
  10-MHz crystal-reference counter, 317
  105-120 MHz navigation receiver, 561, 564
  200-MHz crystal oscillator using doubler, 126
  2-meter f-m, 319, 321, 322
  12-V regulated power supply, 712
  divide-by-N programmable counter, 322
  fsk generator with divider, 318
  multiplier and vco with divider, 321
  phase-locked loop with divider, 320
  varactor-tuned a-m front end, 705
  varactor-tuned f-m front end, 300
Frost alarm, plants, 3
Fuel cell, high-current inverter, 385
Fuel consumption, auto miles-per-gallon meter,
  64

Function generator:
  0-725 Hz sine-square-triangle, 328
  0-1,250 Hz sawtooth and pulse, 330
  0.002-0.25 Hz square-triangle, 324
  0.01 Hz-2 MHz, 329
  0.01-4,000 Hz sine-square-triangle, 325
  0.1 Hz-30 kHz programmable, 327
  1 Hz-5 MHz sine-square-triangle-pulse, 323
  1 Hz-100 kHz square-triangle, 327
  2.5 Hz-250 kHz square-triangle, 328
  3.5 Hz-1.2 MHz sine-square-triangle, 323
  20-5,000 Hz sine-square-triangle, 329
  30-25,000 Hz square-triangle, 326
  50-Hz triangle at 3 V from square wave, 330
  80-800 Hz triangle, 324
  100-Hz square-triangle, 324
  120-3,000 Hz square-triangle, 548
  670 Hz-115 kHz sine-square voltage-
    controlled, 1001
  1.8-ms square-triangle, 330
  89-dB step attenuator, 325
  converting triangle to sine, 675
  high-speed comparator with level
    transmission, 328
  pulse-square-triangle, 324
  sine-square-triangle for f-m and fsk, 326
  sine-square-triangle ulf, 325
  square-triangle with single opamp, 326
  square-triangle-spike, 328
  triangle with inverting integrator, 370
  triangle from square wave, 371, 373, 677
  triangle using direct-coupled astable mvbr,
    549
  triangle 450-kHz with hysteresis switch, 371
  triangle constant-amplitude, 372
  triangle variable-amplitude, 327
  triangle variable-frequency constant-
    amplitude, 373
  triangle voltage-controlled a-f with sine
    output, 999
  (See also Oscillator; Signal generator;
    Square-wave generator; Sweep)
Furnace, flame-failure protection, 272, 274
Fuzz box, guitar, 240, 242, 246, 250

Gain controller, 2-quadrant multiplier, 539
Galvanometer, amplifier using long-tailed
  pair, 424, 425
Garage-door opener:
  26.995-MHz radio control, 763
  headlight-operated switch, 766
  horn-actuated, 762
Gas:
  measuring thermal conductivity with
    thermistors, 452
  miles-per-gallon meter, 64
Gas alarm, low-fuel flasher, 61, 64, 66
Gas analyzer, infrared for auto exhaust, 359
Gate:
  60-mA quad flip-flop, 331
  500-Hz narrow-band digital active filter,
    253
  10-50 kHz voltage-controlled oscillator, 1001
  500-kHz antenna-switching, 963
  8.5-12 MHz remote-control oscillator, 601
  12-V bias supply for n-mos, 332
  adjustable even-order frequency divider,
    295
  AND with large fan-in, 335
  asynchronous transfer register, 149
  bidirectional fet switch, 862
  bidirectional one-shot, 331
  clock in cascaded decade counter, 124
  c-mos squaring, 678
  contact bounce suppression, 152, 153

control for adjustable frequency divider, 298
discrete NAND for counter output, 336
dot-dash keyer, 135
driving coax data line, 201, 202
driving unterminated line, 202
dual analog spst make-before-break, 333
fast-switching reset for TTL-compatible
  counter drive, 332
frequency doubler for 100-MHz crystal, 315
frequency tripler passing three clock pulses,
  688
full adder, 335
full subtractor, 335
half-adder, 335
high-noise-immunity with diode switching
  matrix, 413
increasing 8-input multiplexer to 10-input,
  528
jfet for high-speed digital data, 333
logic-controlled 5:1 or 6:1 divider, 294
logic pulser for testing, 415
mixing positive and negative gates in
  decoder, 413, 416
monostable or astable mvbr, 543
monostable mvbr with bipolar output, 544
mosfet time-multiplexing for recorder, 534
NAND using Schottky diodes, 335
offset error elimination in sense amplifier,
  334
protection against setting wrong switches,
  145
pulse width sorter, 333
quad for two-frequency clock, 128
sample-hold for opamp zero correction to
  $3\mu V$, 581
sampling with zener reset, 331
series-shunt diode bridge, 334
signal-powered exclusive-OR, 417
signal-powered 3-input, 414
single-transistor with only 100-ps delay, 333
strobe pulse generator with noise immunity,
  662
suppressing transients by forcing output to
  zero, 332
switch contact latch, 153
synchronous AND with delayed AND for
  counter, 336
synchronous AND with delayed OR for
  counter, 334
synchronous AND for ring counter, 335
three-input AND using comparator, 333
transmission using opamp, 334
trigger-sharpening NAND, 990, 991
TTL turns off IC audio amplifier, 761
video transient-suppressing, 336
voltage level detector, 1007
wired OR binary comparator, 416
(See also Logic)
Gate dip meter *(see* Dip meter)
Gating, preventing saturation in ferrite-core
  memory sense amplifier, 490
Geiger counter:
  breast-tumor radiophosphorus, 489
  d-c/d-c converter, 339
  two-transistor portable, 423
  uranium prospecting, 428
Generator *(see* Oscillator; Pulse generator;
  Signal generator)
Glider, audible rate-of-climb indicator, 562
Glitches, eliminating in Gray-code generator,
  149
Gobble counter, divide modulus of 255
  maximum, 185
Gold detector *(see* Metal detector)
Golf cart, pwm motor control for 200 A at 36 V,
  525

Graphic display (see Cathode ray; Display)
Gray code:
  analog-digital converter, 22
  analog-input converter to pulse width, 165
  decimal counter, 191
Gray-code generator, glitch-free flip-flops, 149
Grid-dip meter (see Dip meter)
Guarding, full differential amplifier input, 583
Guided missile, 960-Hz tuning-fork time-base generator, 597
Guitar (see Electronic music)
Gunn oscillator:
  10.69-GHz Doppler intruder detector, 98
  150-Hz low-pass modulation filter, 688
  (See also Oscillator)
Gyrator:
  converting capacitance to inductance, 167
  eliminating bias-level latch-up, 11

Hall effect:
  d-c transformer for measuring current, 427
  magnetic-field-change detector, 499
Ham radio (see Citizens band; Code; Repeater, Telephone, Teleprinter, entries for all types of circuits used in amateur radio receivers, transceivers, and transmitters)
Hammer, impulse driver, 994-996
Hardening, radiation-triggered signal-shunting switch, 571
Harmonic distortion meter, audio amplifiers, 447
Harmonic generator, nine-overtone clock-driven, 125
Headlight:
  automatic off switch, 73
  off delay, 62, 65
  on alarm, 65, 67, 71, 72, 74
Headphone adapter, receiver, 51
Headphone jack, CB receiver, 122, 123
Headphones:
  300-2,000 Hz bandpass filter, 268
  attenuator for power amplifier, 830
  Bauer cross-feed network for stereo, 836
Hearing aid:
  200-μW IC, 481
  0.9-mW simplified using IC, 486
  1.5-mW IC, 486
  10-mW IC using sliding bias, 485
  three-transistor, 483
Heart beat rate, display on ecg contourograph, 485
Heat alarm, light-sensitive clicker, 93, 271
Heat control (see Temperature control)
Heater, 28-V aquarium, 935
Helium leak detector, ten-range electrometer, 424
High voltage:
  2-kV regulator for converter, 338
  2.5-kV at 2 kW, 340
  3.5-kV photomultiplier supply, 341
  10-kV electrostatic precipitator, 339
  12-kV supply for diode image converter, 338
  16-kV supply for diode image converter, 342
  20-kV laser image-converter supply, 393
  50-kV voltmeter, 470, 474
  750-V supply at 125 mA, 341
  800-V and 800-V universal ham transmitter supply, 342
  900-V at constant 30 mA for magnetron, 339
  900-V sextupler power supply, 343
  1,000-V d-c supply at 100 W, 338
  2,500-V photoflash supply using 4-V battery, 640

3,000-V supply at 700 mA, 342
5,000-V supply for laser, 392
converter from 1.5 V d-c to kV, 339
electric-fence 6-V charger, 343
flashlight-controlled voltage level detector, 1009
light-triggered 6-kV series scr switch, 343
optoelectronic regulator, 337
photomultiplier tube supply, 337
triggered spark gap tests crt flashover protection, 339
High-voltage pulser, 300-V with low standby drain, 662
Hobby circuit:
  700-Hz bell-tone generator, 347
  2-kHz flasher as key for photoelectric door lock, 237, 239
  2-kHz modulated-light receiver for door lock, 234
  19.5-kHz 15-W ultrasonic pest chaser, 996
  6-V electric fence shocker, 348
  10-kV electrostatic precipitator, 339
  200-W sound-controlled lamp, 556, 559
  500-W lamp modulator for color organ, 555
  60-combination electronic lock, 234
  120-combination electronic lock, 237
  audible auto idiot-light alarm, 72
  audible light meter, 787
  audio-controlled lamp, 556, 558
  audio dancing light, 558
  auto travel-schedule computer, 562
  automatic bedside night light, 629
  baseball velocity timer, 346
  bicycle siren, 813
  buzzing and flashing fish lure, 344
  chirping siren, 815
  Christmas light twinkler, 275
  coin flipper, 784
  color organ with 450 W per channel, 556
  combination four-digit lock for power-tool outlet, 235
  combination lock for boat ignition, 77
  combination 6-digit IC lock, 235
  commercial silencer for auto radio, 54
  dancing lights, 788
  DO-NOT-TOUCH squealing box, 346
  dripping-sound generator, 348
  electronic coin-flipper, 786
  fish caller, 349
  fish finder, 349
  five-neon flasher for twinkling star, 275
  flasher for intercom, 378
  flashlight motor control, 784
  flickering-flame lamp control, 787
  goo-hah horn, 815
  headlights turn on porch light, 629
  infrared light-beam communicator, 356, 357
  knock-over alarm in can, 96
  lamp puzzle, 786
  lie detector, 785, 786, 788
  light-controlled tone generator, 554
  light-controlled toy, 785
  message alarm, 347
  model train speed control, 345, 347, 348
  mosquito-repelling audio oscillator, 347
  motor for science fair, 786
  motorboat sound effect, 349
  muscle-controlled whistle, 785
  music-controlled light display, 557
  mystery music box, 250
  night-light using nine-lamp counter display, 184
  nonmedical stethoscope, 847
  photoelectric rifle range, 346
  police siren, 813

programmed vacant-home lighting control, 100
psychedelic fluorescent flasher, 281
pushbutton siren, 814
random six-neon blinker with glide tones, 345
ring counter for electronic dice, 344, 348
science-fair beeper with meter, 788
scr-operated coded lock, 236
screamer or alarm siren, 816
siren with clicker, 815
siren using mvbr, 813
siren drives earphone for child, 814
slot-car controller, 346
slot-car photoelectric win detector, 347
sound-activated relay, 919
sound-controlled Christmas lights, 559
sound-controlled lamp using scr, 556
steam-whistle generator, 345
sunlight-controlled neon relaxation oscillator, 788
three-tone doorbell, 349
three-transmitter power-line carrier intercom, 376, 377
tool magnetizer, 344
touch switch with time delay, 105
twilight turn-on lamp, 628
voice-operated switch, 46, 787
Waa-Waa siren, 813
wailing siren, 815, 816
(See also Alarm; Audio control; Auto theft alarm; Burglar alarm; Capacitance control; Citizens band; Digital clock; Door lock; Electronic music; Flasher; Lamp control; Lamp dimmer; Metal detector; Music-controlled light; Photography; Science fair; Siren; Transceiver)
Horn, mvbr-driven burglar-scaring, 1
Horn loudspeaker, crossover network, 54
Hum filter, 60-Hz active for tape, 867
Humidity alarm, clicking-loudspeaker, 271
Humidity control:
  400-Hz hygrometer bridge, 446
  alarm or controller, 450
  differential temperature sensor for basement, 925
  thermistor hygrometer, 442
Humidity detector, alarm or control, 1
Hygrometer, thermistor with opamp, 442
Hysteresis amplifier, high input impedance with opamp, 791

Ice warning, auto air temperature alarm, 66, 70
I-f amplifier (see Intermediate-frequency amplifier)
Ignition:
  combination lock, 77
  (See also Automotive ignition)
Illumination control (see Fluorescent lamp; Lamp control; Lamp dimmer; Photoelectric; Photography)
Image converter, 20-kV supply, 393
Impedance bridge:
  2-30 MHz using noise diode, 446
  3.5-33 MHz using r-f source, 445
  portable, 446
Impedance converter:
  d-c to 10 kHz, 164
  micropower for preamp, 364
Impedance meter, 1 ohm to 1 meg, 443
Impulse response, pseudo-random binary sequence correlator, 568

Indicator:
  three-lamp for multiple supply voltages,
    1009
  (See also Character generator; Display;
    Measurement; Voltmeter)
Inductance measurement:
  0.3 μH to 7 mH, 451
  0.25-10 mH direct-reading, 445
  R-L-C bridge, 449
Induction heater:
  300-kHz 240-kW ceramic-triode, 606
  400-kHz 60-kW ceramic-triode, 606
Induction receiver, museum commentary,
  848
Inductor, simulation with opamp, 37, 40
Inertial guidance, H switch controls direction of
  motor drive, 560
Infrared:
  700-Hz chopped-led receiver, 95
  700-Hz drive for led, 93
  2.7-kHz coded-beam detector, 362
  12-kV supply for diode image converter, 338
  16-kV supply for diode image converter, 342
  30-40 C radiation thermometer for auditory
    canal, 466
  amplifier for uncooled lead sulfide detector,
    358, 359, 362
  bias for cooled detector, 357, 359-362
  bias for RPY77 uncooled detector, 360
  fire and burglar alarm, 271
  fire alarm with interference discrimination,
    270
  fire alarm using bistable switch, 271
  frequency-sensitive chopped-beam photo-
    alarm, 92
  gallium arsenide diode modulator, 361
  gas analyzer for air pollution, 359
  gas-furnace flame-failure protection, 272
  lockout counter for jam-proof pulsed-beam
    burglar alarm, 361
  low-frequency amplifier for photodiode, 360
  low-level image detector, 358
  low-noise amplifier for ORP10 uncooled
    detector, 360
  modulated-radiation detector, 361
  passive intruder alarm, 94, 97
  phase detector for spectograph, 357
  preamp for radiometer, 463
  preamp for thermal-scanning tv, 468
  pulsed receiver, 101
  pulsed receiver lockout counter, 98
  pulsed transmitter, 97, 99
  pulsing led with scr, 359
  radiation thermometer for 100-500 C, 464
  receiver for voice communication, 358
  receiver using silicon solar cells, 357
  transmitter for voice communication, 356
  transmitter using led, 356
Infrared microscope, radiation thermometer for
  transistors, 467
Instrument amplifier:
  1-μA micrommeter, 429
  100-nA nanoammeter, 424
  bridge-type for microammeter, 427
  capacitance meter, 419
  d-c with decade-step gain switch, 368
  d-c ammeter, 427
  galvanometer long-tailed pair, 424, 425
  microammeter, 425, 426
Instrumentation:
  100-kHz to d-c converter, 478
  transferring d-c with isolated chopper,
    116
  ultra-low-noise a-f amplifier, 846
  voltage-current converter for chart recorder,
    367

Instrumentation amplifier:
  0.1 Hz-1 MHz instrument output, 368
  0-15 MHz unity-gain high-accuracy, 365
  20-dB gain, 365
  1,000 voltage gain, 366
  adjustable-gain d-c input amplifier, 365
  bridge for variable-resistance devices, 364
  d-c to 100 kHz, 363, 364
  d-c with 10-mA output, 369
  differential-input, 366, 367, 369
  drift-compensated d-c opamp, 367
  fet differential, 366
  high-level for solid-state lab instruments, 368
  high-resistance thermistor bridge, 365
  low temperature drift, 363, 368
  micropower for capacitive transducer, 364
  micropower with gain of 100, 366
  null for zero-center meter, 364
  piezoelectric-transducer, 364
  small differential signals, 369
  transducer bridge, 369
Insulation resistance, measuring with follower
  voltmeter, 458
Integrated circuit:
  automatic threshold tester, 940
  temperature measurement with infrared
    microscope, 467
  tester for unmarked units, 935
  (See also entries for type of circuit desired)
Integrator:
  100-3,000 Hz sawtooth generator, 372
  3-kHz vco, 998
  10-100 kHz for lab sweep generator, 374
  450-kHz triangle generator, 371
  0.1-10 MHz for lab sweep generator, 372
  analog with reset mode, 371
  analog three-mode, 373
  continuous sampling, 373
  delay for counter drive, 182
  double using single opamp, 373
  fast low-error using slow opamps, 372
  fsk modem, 500
  inverting for function generator, 370
  jfet opamp, 371
  opamp with low offset voltage drift, 374
  opamp converting square wave to triangle,
    371
  programmable, 370
  square-triangle converter, 373, 374
  triangle generator, 373
  TTL-compatible pulse output, 332
  wideband opamp, 373
Intercom:
  6-V battery-powered, 375
  9-V regulated power supply, 713
  0.5-W IC two-station, 377
  0.5-W portable amplifier, 376
  1-W IC with one transistor, 376
  a-c powered, 377
  master-remote using single IC, 375
  master-remote with single IC, 378
  message-signaling led flasher, 378
  radio over a-c lines, 378
  repeater break-in alarm, 379
  single-IC, 379
  store-home burglar-detecting, 378
  three-transmitter power-line carrier, 376,
    377
  two-way 250-mW IC, 379
  two-way 5-transistor, 375
  two-way megaphone, 658
  (See also Audio amplifier)
Interference:
  40-MHz high-pass tv filter, 268, 269
  blocking with 30-kHz window in f-m
    repeater, 776

  broadcast-band stop filter, 266
  fluorescent-lamp power-line filter, 283
  power-line filter for arcing thermostat, 928
  suppressing three-phase with zero-point
    switching, 951
  vacuum-cleaner suppressed with diode
    limiter, 566
Intermediate frequency:
  2.1-kHz mechanical bandpass filter, 267
  50-100 kHz T-notch filter, 268
  455-kHz tunable slot filter, 267
  agc for tv video i-f, 907
  broadband phase-transparent limiter, 687
  cascaded half-lattice crystal filter, 265
  noise silencer, 569
  notch filter for ssb receiver, 805, 806, 809
  passive notch filter, 809, 810
  varactor notch filter, 269
Intermediate-frequency amplifier:
  135-kHz for 1.65-MHz converter, 351
  455-kHz or 10.7-MHz with fet, 351
  455-kHz a-m signal generator, 796
  455-kHz Q multiplier, 536
  455-kHz using active filter, 353
  470-kHz high-gain using Q multiplier, 539
  9-MHz IC in 20-meter receiver, 697
  9-MHz panoramic, 352
  10.7-MHz tunable using variable-Q active
    filter, 355
  28-MHz alignment signal generator, 799
  28-30 MHz for 2-meter converter, 350
  60-MHz IC with 1.5-MHz bandwidth, 354
  a-m/f-m with detectors, 351
  a-m single-IC, 353
  automatic bandwidth control, 353
  f-m with Foster-Seeley discriminator, 353
  f-m quadrature detector with IC i-f for auto
    radio, 352
  f-m ratio detector with IC i-f, 350
  f-m single-IC, 354
  f-m stereo ceramic filter, 837
  matching to noncompatible i-f strip, 696
  modernizing in communication receiver,
    352
  noise blanker, 355
  ssb and c-w single-IC, 804
  tunable for 80-40-15 meter plug-in con-
    verters, 355
  wobbulator for tv alignment, 908
  (See also Receiver; Transceiver)
Intrusion alarm:
  25-kHz ultrasonic, 995
  passive using infrared detector, 97
  (See also Alarm; Auto theft alarm; Burglar
    alarm; Infrared; Photoelectric; Protection)
Inverter:
  20-kHz at 1 kW, 157
  20-kHz quiet switching-transistor, 384
  25-kHz 200-W, 380
  2-V at 50 A from solar cell to high voltage,
    385
  6 V d-c to 115 V a-c, 380
  12 V d-c to 110 V a-c: at 90 W, 383
    at 200 W, 385
  12 V d-c to 115 V at 400 Hz for auto, 381
  12 V d-c to 115 V a-c, 381
    at 115 W, 384
  12 V d-c to 117 V a-c, 380
    for 22-W fluorescent, 284
  12 V d-c to three-phase 400-Hz at 20 W,
    949
  12-V fluorescent lamp, 284
  18 V d-c to 60 V a-c for neon, 384
  24 V d-c to 115 V a-c at 400 W, 385
  32 V d-c to 150 V p-p at 2.5 kHz, 383
  75 V d-c to 110 V at 5 kHz, 282

Inverter (Cont.):
  300 V d-c to 20-kHz pulsed 1 kW for computer, 381
  20-kV supply for laser image converter, 393
  doubling 6-V d-c without rectifier, 157
  flip-flop driving scr pair, 551
  flip-flop put trigger for scr's, 990
  frequency-voltage converter, 382
  hex IC in variable pulse generator, 661
  induction motor speed control, 382
  logic-controlled for 750-W three-phase frequency converter, 948
  opamp with resistive feedback, 575
  protecting 15-A from 20-A overload, 653
  transistor flip-flop driving scr pair, 550
  (See also Converter)
Inverting adder, opamp with resistive feedback, 575
Ion chamber, measuring polarity of air ionization, 444
Ion collector, high-sensitivity electrometer, 428
Ionization, measurement with ion chamber, 444
Ionosphere topside sounder:
  ionogram telemetry converter, 875
  sweep calibrator, 877
Ionovac speaker, 27-MHz drive with modulator, 47
Isolation amplifier, 50-dB isolation up to 6 MHz, 5
Isolator (see Optoelectronic)

Jitter distortion, reducing in fsk modem tone oscillators, 504

Karnaugh map display:
  1.4-kHz astable-mvbr clock generator, 126
  22-kHz phase-shift oscillator, 233
  generating on cro screen, 227
Keyboard:
  ASCII encoder, 151
  binary encoder, 150
  contact bounce suppression, 152
  excess-three encoder, 150, 415
  universal encoder, 148
Keyer:
  1-kHz audio oscillator driving speaker, 592
  1.8-kHz and 2-kHz tone-burst for f-m repeater, 767
  1-A solid-state switch, 133
  10-50 wpm using IC flip-flops, 131
  afsk narrow-shift and wide-shift, 890
  automatic code, 135
  automatic letter and word spacing, 131
  battery-portable IC, 134
  break-in without relays, 133
  clock with noise immunity, 126
  dot-dash IC, 130, 132, 135, 138
  relay driver for solid-state, 138
  tone-actuated for transmitter, 131
  touch-controlled c-w break-in, 137
  ujt with sidetone for ssb, 137

Lamp, operating neon from 6-V battery, 380
Lamp control:
  2-A logic controlled, 1005
  2-kHz flasher as key for photoelectric door lock, 237, 239
  1-min delay for turnoff, 957
  50-W automatic night light, 101
  100-W projection-lamp regulator, 630
  200-W psychedelic sound, 559
  500-W gradual-on photoelectric, 632
  500-W photoelectric night light, 93

  500-W remote, 764, 766
  600-W full-range scr, 616
  600-W full-wave a-c triac, 617
  900-W full-wave, 619
  1-kW darkness-on switch, 98, 629
  adjustable turn-off delay, 389
  auto parking-light turn-on, 73
  comparator with transistor switch, 143
  complementary-load for display, 861
  constant-brightness led source, 387
  dancing effect with put, 788
  diode and switched capacitors, 390
  emergency light with power-failure relay, 2, 390
  engineering puzzle, 786
  flickering flame effect, 787
  fluorescent two-level from 12-V battery, 284
  high-voltage regulated power supply, 337
  hysteresis-free phase control, 616
  light intensity regulator, 590
  photoelectric for bedside, 629
  photoelectric servo, 794
  photoelectric trigger for scr's, 631
  programmed for lights in vacant home, 100
  regulator monitoring lamp power, 390
  science-fair exhibit, 784
  scs-triggered triac full-wave a-c, 617
  single-channel rhythm light, 555
  sound-dimmed lover's lamp, 787
  touch switch, 105
  turn-off delay, 387
  twilight turn-on, 98, 628, 629
  (See also Flash; Flasher; Fluorescent lamp; Lamp dimmer; Mercury-arc lamp; Music-controlled light; Photoelectric)
Lamp dimmer:
  40-A phase control, 618
  6-V low-loss 0-100%, 388
  150-W half-wave scr, 387
  150-W high-intensity study lamp, 387
  360-W triac, 390
  400-W triac, 386
  500-W combined with brightener, 386
  500-W half-wave, 514
  500-W photoflood, 642
  500-W triac, 386
  600-W full-wave a-c triac, 617
  600-W half-wave average-voltage feedback, 390
  600-W half-wave ujt-scr, 390
  600-W time-dependent, 388
  900-W full-wave ujt-triac, 387
  900-W limited-range scr, 389
  1-kW soft-start triac, 389
  1-kW full-wave sbs-triac, 387
  1.2-kW ujt-triac, 388
  3-kW scs-triggered triac full-wave, 618
  combined with dimming-rate timer, 388
  half-wave using transistor as zener, 390
  high-gain IC with light feedback, 389
  line voltage compensation, 386
  minimum-hysteresis sbs-triac, 387
  Nixie display, 231
  remote low-voltage control, 765, 766
  (See also Lamp control; Music-controlled light; Photoelectric)
Lamp flasher (see Flasher)
Large-scale integration, tester, 144
Laser:
  30-A 4-kHz pulser for 20-diode array, 396
  50-A GaAs diode drive, 391
  75-A general-purpose driver, 395
  75-A krytron pulser, 395
  2,000-A krytron driver, 392
  3-V to 300-V d-c converter for pulser, 160
  300-V pulse generator, 662

  20-kV image-converter supply, 393
  diac-scr trigger, 396
  diode modulator, 396
  diode pulse driver, 392
  measuring pulse power with photocell, 393
  one-way communication, 396
  pfm voice transmitter, 391
  pulsed diode, 394
  pulser using four-layer diode, 395
  pulser using parallel avalanche transistors, 394
  Q-switching: with krytron and Pockels cell, 391
    with scr and transformer, 392
  receiver, 394
    using fotofet input for 200-ns pulses, 395
    using IC, 395
  scr pulser with 1-10,000 Hz prr, 393
  voice receiver using pfm, 392
Latch:
  frequency error detector, 286
  holding register for excess-three encoder, 150
  light-controlled, 924
  multichannel pulse-width converter, 543
  noise-immune IC, 567
  power-on reset, 654
  reducing wire-ORed turn-off time, 495
Lawn moisture meter, transistor with probes, 449
Leakage tester, transistors and diodes, 948
Led (see Light-emitting diode)
Leslie effect, simulator for electronic organ, 243
Level control:
  liquid two-probe, 403
  photoscr bin or reservoir feed, 402
  shifting with isolating fet between opamps, 581
  (See also Liquid level; Liquid level control)
Level detector, auto engine oil, 68
Level indicator, stereo tape recorder, 865
Level monitor, 9-V up-down trigger, 993
Lie detector:
  measuring skin resistance, 785
  modulated-tone, 786
  opamp driving microammeter, 788
Light (see Lamp control; Lamp dimmer; Photoelectric)
Light-beam communication:
  infrared, 356–358, 361
  laser, 391, 392, 394–396
  (See also Infrared; Laser; Optoelectronic)
Light control (see Lamp control; Lamp dimmer; Music-controlled light)
Light-emitting diode:
  astable mvbr status indicator, 544
  d-c amplifier for driving DTL or TTL, 584
  flip-flop status indicators, 544
  isolating relay from DTL, 585
  modulator, 586
  mono mvbr status indicator, 545
  phototransistor drive for TTL or DTL, 585
  pulsed 1.25-W power supply, 586
  pulsed 900-mW power supply, 587
  pulsing with scr, 359
  voice modulator, 586
  (See also Optoelectronic)
Lightning detector, 9-V battery-operated fire alarm, 271
Lightning protection, transceiver antenna, 966
Limiter:
  10-Hz IC pulse squarer, 677
  455-kHz narrow-band f-m, 399
  500-kHz two-IC, 401

Limiter (Cont.):
  0-30 MHz broadband phase-transparent
    radar, 687
  28-MHz two-stage IC, 397
  146-MHz f-m preamp, 398
  50-dB dynamic range for 2-60 kHz, 400
  200-μV for protecting opamp, 397
  audio linear distortionless, 398
  audio noise using crystal diode, 566
  CB speech clipper, 400
  counter range extender to 100 MHz, 297
  f-m high-gain IC, 401
  f-m pentode, 398
  multiplying 625 Hz by 40, 314
  nonblocking for radar speed meter, 688
  nonlinear with 16-Hz quadrature oscillator,
    371
  opamp with zeners, 579
  overload current, 652
  series low-cost for ten video channels, 399
  shunt for individual video channel, 399
  speech-compressing for ssb transmitter, 399
  tv sound with locked-oscillator detector, 911
Line driver, 50-dB isolation up to 6 MHz, 5
Line voltage regulator, 105 V a-c for Motorola
  tv, 645
Liquid level:
  alarm with siren output, 2
  auto brake-fluid monitor, 69
  auto oil checker, 65
  low radiator-water alarm, 67
  low windshield-water alarm, 73
  pool splash alarm, 403, 404
Liquid level control:
  photodiode with IC, 402
  pump-motor or drain-valve load, 402, 404
  two-probe, 403
Liquid level detector, photodiodes with perturbation, 404
Liquid-level indicator, audio for the blind, 403
Load line, transistor curve tracer for cro, 937
Lock:
  120-combination electronic, 237
  combination for boat ignition, 77
  combination four-digit for power-tool outlet,
    235
  (See also Door lock)
Locked-oscillator detector, intercarrier tv
  sound, 911
Locomotive, three-transistor steam whistle, 345
Logarithm:
  antilog of voltage, 410
  current or voltage ratios, 408
  current ratio, 408, 410
Logarithmic:
  adjustable-power function generator, 411
  analog multiplier-divider, 406
  antilog generator, 408
  approximating arctangent, 406
  approximating cosine of angle, 408
  approximating sine of angle, 408
  bipolar signal compression of $\sinh^{-1}$
    function, 411
  cube generator, 410
  filterless mosfet squarer, 310
  frequency-response plotter drive, 934
  light meter for low levels, 643
  multiplier/divider, 406
  one-quadrant divider, 407
  root sum of squares, 819
  square-root computation, 817, 818
  true rms computation, 819
  vector difference, 820
  vector length computer, 818
  vector summation for root sum of squares,
    820

Logarithmic amplifier:
  five-decade range, 411, 574
  low-cost opamp, 407
  null detector, 410
  opamp without bias current compensation,
    406
Logarithmic compression, 100-μA to 10-V
  range, 410
Logarithmic converter, one-quadrant temperature-compensated, 405
Logarithmic detector, tape recorder, 864
Logarithmic generator:
  80-dB dynamic range, 409, 411
  100-dB dynamic range, 407, 409
Logic:
  1-10 pps pulser for testing gates, 415
  1-out-of-64 decoder, 534
  8-bit parity checker, 145
  asynchronous 8-gate decision element, 414
  bcd decoder-counter with diode gates, 416
  binary-hex decoder, 413
  binary-octal decoder, 416
  bounce-free pushbutton, 152
  clock driver above 300 MHz, 128
  comparison of binary numbers, 149, 416
  current-mode high-speed Schmitt trigger,
    988
  d-c amplifier with led drives DTL or TTL, 584
  detecting allowable levels and transition
    times, 1006
  digit detector for high-speed memory, 146
  digital filter, 413
  diode switching matrix with noise-immune
    gate, 413
  disable gate, 417
  ECL gates double 100-MHz crystal, 315
  emergency buffer with fan-out of 60, 412,
    416
  emergency 4-input gate from 2-input gates,
    415
  emergency 8-input gate from 2-input gates,
    413
  enable gate, 415
  excess-three encoder, 150, 415
  exclusive-or behaving like half-adder, 413
  high-speed jfet switch, 333
  inverting driver between flip-flop and relay,
    146
  led-phototransistor drive for TTL and DTL,
    585
  level-indicating probe, 414, 415
  mos-indicator interface, 224
  mos output interface, 415
  multiplexing bipolar analog signals, 535
  NAND gate controls triac, 861
  noise-immune parity checker, 150
  optical driver, 633
  power-on reset for latch, 654
  power-supply monitor, 584
  power-transformer tap-changing with triacs,
    862
  protection against setting wrong switches,
    145
  pseudo-random binary sequence correlator,
    568
  pulse-width converter, 543
  relay isolation from DTL with led, 585
  resetting correctly at power turn-on, 412
  seven-code pcm converter, 200
  shunt switch for digitally set pot, 868
  signal-powered coincidence detector, 417
  signal-powered exclusive-OR, 417
  signal-powered 3-input gate, 414
  single-pulse generator using latch and
    one-shot mvbr, 153, 670
  testing unmarked digital IC unit, 941

TTL-mos interface, 412–414, 416
  voltage-level changer, 415
  voltage-tolerance detector, 1007
  (See also Computer; Flip-flop; Gate; Trigger)
Logic level, indicator probe, 930
Long-tailed pair, 1.6-V regulator, 739
Loran, trf receiver, 703
Loudhailer (see Megaphone)
Loudspeaker:
  phasing with a-f transmitter and receiver for
    p-a, 656
  rear added to 2-channel stereo, 685
  sound pressure-level meter, 443
Lover's lamp, sound-dimmed, 787

Machine-tool control, binary optoelectronic
  shaft encoder, 590
Magnetic amplifier, 1-kW regulated power
  converter, 156
Magnetic field, measurement with paper-lead
  probe, 444
Magnetic field detector, moving ferrous ob-
  jects, 498
Magnetic field stabilizer:
  pulse adder, 184
  pulse subtractor, 181
Magnetic pickup:
  combination low-pass and high-pass filter,
    623
  low-noise preamp, 621
Magnetic tape (see Tape recorder)
Magnetizer, tools and steel rods, 344
Magnetometer, low-noise fet spacecraft amplifier, 447
Magnetoresistance, current-measuring d-c
  transformer, 427
Magnetron, 900-V supply at constant 30 mA,
  339
Marine (see Navigation)
Marker, 10-kHz in frequency standard, 441
Marker beacon:
  75-MHz airborne receiver, 565
  active tone filters for display, 561, 565
  digital tone discriminator, 563
Marker generator:
  500-kHz crystal oscillator, 932
  1-MHz with 10-kHz markers, 431
  band-edge providing 25-kHz to 200-kHz in-
    tervals, 801
  (See also Frequency marker)
Mass spectrometer, ten-range electrometer, 424
Matching, transistor pairs, 942
Matrix:
  diode for excess-three encoder, 150, 415
  diode switching with high-noise-immunity
    gate, 413
Measurement:
  0-5 W power for 10-80 meters, 456
  0-10 W r-f wattmeter, 454
  0.25-10 W r-f wattmeter, 456
  5-300 W r-f wattmeter, 455
  1,000-W in-line r-f wattmeter, 454
  a-c noise in power supplies, 566
  a-c power monitor for appliances, 453
  acceleration of auto, 72
  antenna resonant frequency and radiation
    resistance, 570
  applause meter, 448
  apple size, 631
  auto battery voltage during starting, 69
  auto engine speed with neon tachometer
    calibrator, 68
  auto miles per gallon, 64
  baseball velocity timer, 346
  braking deceleration of auto, 72

Measurement (Cont):
bullet velocity, 451
carbon monoxide with infrared gas
analyzer, 359
CB transceiver output power, 454
contact resistance with 500-milliohm ohmmeter, 457
crystal frequency, 930
current: with Hall-effect d-c transformer, 427
(See also Ammeter; Current; Galvanometer; Multimeter)
demodulation distortion in a-m detector
transistor, 447
direct-current gain of transistor, 941
dwell angle of auto and boat engines, 71
electrometer: for 5 pA to 500μA, 424
with fast response, 426
with high sensitivity, 428
using mosfet, 425, 427
fet drain-source resistance, 943
fet forward transconductance, 937
fet leakage current, 945
fet pinch-off voltage, 942
fet transistor breakdown current, 945
field strength at 10 GHz, 453
field strength for CB, 120
flame resistance alarm, 274
flash exposure, 635
frequency: of a-c power generator, 455
(See also Frequency entries)
frequency deviation of f-m transmitter, 304
frost alarm for plants, 3
harmonic distortion meter, 447
humidity with oscillator-driven bridge, 446
humidity with thermistor hygrometer, 442
humidity alarm and control, 450
impedance and field strength, 449
impedance with noise bridge, 446
impedance with r-f bridge, 445
impedance meter, 443
inductance from 0.25 to 10 mH, 445
inductance from 0.3 μH to 7mH, 451
instantaneous wattmeter, 455
ionization polarity of air, 444
junction fet characteristics, 946
laser pulse power, 393
light at low levels, 585
light for triggering lascr, 448
light comparator, 632
light flash with fet voltmeter, 638
light flash duration, 634
light intensity for photography, 643
low-level wideband video amplifier, 706
magnetic and electric fields, 444
magnetic fields in interplanetary space, 447
meter amplifier for a-c or d-c, 363, 364
meter resistance, 460
moisture in lawn, 449
moisture in wood with ohmmeter, 452
moisture alarm, 1, 4
noise figure, 572
noise figure up to 432 MHz, 567
noise tester for 10-10,000 Hz opamps, 571
nonlinearity index for silicon-carbide varistors, 943
npn transistor breakdown current, 942
null amplifier for zero-center meter, 364
optoelectronic r-f wattmeter, 456
pH of solutions, 448
phase difference between input signals, 451
plasma conductivity, 458
pnp transistor breakdown current, 943
power with level-shifting multiplier, 541
power factor for digital display, 163
power meter with square-law diodes, 454
(See also Power)

probability density analyzer, 444
pulse width variations of mono mvbr, 942
radiation with Geiger counter, 423, 428
radiation with high-gain preamp, 423
radiation with solid-state detector, 423
rain alarm, 1, 4
rate-of-climb with audible indicator, 562
recording level for stereo tape, 869
resistance (see Megohmmeter; Multimeter;
Ohmmeter; Resistance; Teraohmmeter)
r-f admittance bridge, 442
rms value of analog voltage, 817
rotational speed with noncontacting
tachometer, 64
semiconductor thermal properties, 922, 927
sensitivity of vhf receiver with oscillator in
waveguide, 608
skin resistance with lie detector, 785
S meter for receiver, 966
S meter connects to speaker leads, 965
sound level for pure tones and third-octave
noise, 450
sound pollution, 567
sound pressure-level meter, 443
starlight magnitude with photometer, 449
swr for low-power transmitter, 452
temperature (see Temperature control;
Temperature measurement; Thermometer)
thermal conductivity of gas, 452
transient peaks, 1007
transistor breakdown voltage, 944
transistor current gain, 935, 937, 945
transistor high-frequency gain: with sampling resistor, 943
with tuned-line setup, 946
transistor leakage current, 943
transistor output conductance, 938
transistor temperature with infrared microscope, 467
transistor thermal resistance, 468
transmitter power with remote-reading field
meter, 456
true rms value of voltage, 819
true rms value of any waveform, 475
uhf transmitter power, 454
universal R-L-C bridge, 449
voltage: with follow and hold, 778
with precision neon-lamp reference, 1014
(See also Digital voltmeter; Voltage measurement; Voltmeter)
volume-unit meter with amplifier, 443
vswr for 2-W c-w transmitter, 449
vu meter for stereo, 832
water conductivity, 460
water temperature in fish-attracting thermocline, 349
wattage with multiple neons, 454
watthour meter, 455
wattmeter for uhf transmitter power, 455
white-noise generator, 568
wind direction, 450
wind velocity with anemometer, 445
wow and flutter of turntable, 622
(See also Capacitance measurement;
Counter; Digital voltmeter; Electrometer;
Frequency measurement; Meter; Multimeter; Ohmmeter; Photography; Pulse height
analyzer; Resistance; Telemetry; Temperature measurement; Temperature control;
Tester; Thermometer; Timer; Voltage measurement; Voltmeter)
Medical:
10-Hz constant-current source for bipolar
probe, 482
200-μW IC hearing aid, 481
0.9-mW simplified IC hearing aid, 486

1.5-mW IC hearing aid, 486
10-mW hearing aid with sliding bias, 485
1-s blinking darkroom timer for the deaf,
638, 639
air ionization polarity measurement, 444
amateur station control for the blind, 483
audible and flashing timer for the blind or
deaf, 638
audio-tone tuning aid for the blind, 488
auditory stimulator, 487, 488
aural tuning aid for the blind, 488, 971
beeping timer for the deaf, 635
blood pressure simulator, 485
breast-tumor radiophosphorus Geiger
counter, 489
cardiac catheter, 481, 484
dangerous-appliance ground tester, 654
ecg real-time contourograph, 485
ecg recorder drive amplifier, 484
four-channel telemeter for athletes, 874, 875
galvanic skin resistance meter, 785
hearing aid, 483
helping the blind pour coffee, 403
infrared radiation thermometer for auditory
canal, 466
lie detector, 785, 788
modulated-tone lie detector, 786
neural analog, 482
pulse generator for d-c torque motor of
cardiac-bypass roller pump, 486
reading aid for the blind, 489
receiver frequency spotter for the blind, 487
sleep-inducing metronome, 251
stereo electroencephalophone, 482, 487
stethoscope, 850
tactile tape recorder level indicator for the
blind, 482
transmitter tuning aid for the blind, 483
voice alarm for the deaf, 484
Megaphone:
5-W using IC, 657
one-transistor, 657, 660
two-transistor 12-V, 657, 658
two-way 12-V portable, 658
Megohmmeter, 1,000-megohm, 459
Memory:
15-ms strobe generator driven by clock, 667
4,096-bit static read-only, 493
bistable switch using sus, 860
core sense amplifier, 492
counter and retrace control for crt display
drive by ram, 492
data register driving alphanumeric crt, 494
delay-line using mos shift registers, 128, 202
direct-coupled clock driver for ram, 493
direct-coupled ram driver, 494, 495
divide-by-N programmable counter, 322
five-digit strobed display, 220
high-speed digit detector, 146
manual read-only programmer, 490
negative-level converter for ram drive, 493,
495
nonvolatile low-capacity, 493
numeric display for counter, 233
offset error elimination in sense amplifier,
334
protecting ram data from power failure,
493, 494
read-write for ram, 495
reducing wire-ORed turn-off time, 495
sense amplifier for laminated ferrite-core,
490
timing and character generation for ram crt
display, 491
Mercury-arc lamp:
100-W ballast, 281, 283

Mercury-arc lamp (Cont):
  175-W switching-regulator ballast, 282
  400-W ballast, 284, 285
Message alarm, blinking-light and Sonalert, 347
Metal detector:
  beat-frequency, 497, 498
  changing-magnetic-field, 498
  concealed pipes, 498
  crystal-filter for buried coins, 497
  Hall generator for magnetic field changes, 499
  pipe and stud locator, 499
  proximity switch, 499
  single-transistor using portable a-m radio, 499
  three-stage beat-frequency, 496
  three-transistor, 496
  tuned-loop oscillator with crystal filter, 497, 499
  two-oscillator using transistor a-m radio, 498
Meteorology:
  frost alarm, 3
  rain alarm, 3
  storm-warning alarm, 1
Meter, measuring resistance accurately, 460
Meter amplifier, d-c to 100 kHz, 363, 364
Metronome:
  30-240 beats per min, 249
  40-208 beats per min, 248
  adjustable-accent, 244
  adjustable-rate, 130, 251
  audible flashing darkroom timer, 638
  code oscillator conversion, 246
  dripping-sound, 348
  transistor-radio, 242
  two-transistor driving speaker, 250
  visible and audible, 249
Microammeter:
  1-μA full-scale using opamp, 429
  2-100 μA adjustable-sensitivity amplifier, 426
  bridge-type meter amplifier, 427
  current-voltage converter, 426, 428, 429, 479
  d-c using opamp, 425, 426
  (See also Ammeter; Galvanometer; Multimeter)
Microphone:
  a-m wireless, 849
  beeper for testing p-a system, 655
  f-m wireless, 846, 849
  mixing impedances in summing amplifier, 49
  preamp with clipper, 398
  voice-operated control, 46
Microphone amplifier (see Audio preamplifier)
Micropower:
  100-nA nanoammeter, 424
  1-55 Hz 600-mW opamp, 576
  360-6,000 Hz pulse generator, 673
  5-kHz astable mvbr, 548
  5-V regulator, 756
  500-nW inverting programmable opamp, 583
  d-c ammeter, 427
  instrumentation amplifier with gain of 100, 366
  inverting 20-dB amplifier, 582
  monostable mvbr with zero quiescent power, 548
  noninverting 20-dB amplifier, 582
  preamp for capacitive transducer input, 364
  voltage comparator with high-current switch, 1009
Milliammeter (see Ammeter; Multimeter)
Milliohmmeter, checking for bad contacts, 459

Mixer:
  0.1-100 Hz beat-note oscillator, 802
  1.65 MHz to 135 kHz converter, 704
  30-MHz i-f output using fet, 696
  100-MHz IC for 30-MHz i-f, 708
  220-MHz fet, 700
  432-MHz ssb, 806
  6-meter heterodyne vfo, 704
  1-min adjustable timer, 957
  active four-input audio with single output, 52
  audio for repeater, 771
  combining swept and fixed r-f to give swept audio, 802
  current-summing audio, 51
  doubly balanced for 9-MHz output, 510, 702
  dual-conversion for 18-30 MHz, 699
  external marker for sweep generator, 855
  four-channel audio preamp, 53
  loss-free tape-recording, 873
  millimeter-wave Gunn, 698
  speed control with timer, 515
  stereo for synthetic quadraphonic sound, 682
  stereo for virtual ground, 53
  summing opamp for high- and low-impedance mikes, 49
  two-input unity-gain, 55
  (See also Converter, radio)
Mobile (see Automotive)
Model train, speed control for varying load, 345
Modem:
  adjustable delay, 502
  fsk data receiver decoder, 503, 504
  fsk demodulator for Bell data sets, 503
  fsk generator, 502
  fsk receiver, 504
  fsk single-supply, 502
  fsk two-tone generator, 500
  low-jitter tone oscillator, 504
  mark-space generator, 505
  split-supply, 501
Modulation checker, neon tone generator, 137
Modulation monitor:
  crystal detector and IC audio amplifier, 978
  field-strength meter, 976
Modulator:
  220-Hz sine-wave in 470-kHz oscillator, 603
  1-kHz for 10-GHz Gunn device, 981
  1-kHz for oscillators up to 1,300 MHz, 598
  10-kHz micropower a-m, 511
  10-kHz sine-wave, 511
  100-kHz a-m pnpn tetrode, 513
  500-kHz balanced, 511
  500-kHz micropower a-m, 510
  52-MHz f-m, 308
  2-meter hybrid, 508
  30-dB linear for 60-150 MHz, 513
  a-m for 2-meter transceiver, 970
  a-m/f-m/c-w waveform generator, 323
  amplitude for 6 meters, 512
  amplitude up to 100%, 509
  amplitude using balanced IC, 506
  amplitude using fet, 508
  amplitude using IC multiplier, 510, 513
  auditory stimulator, 487
  balanced amplitude using fet, 512
  balanced IC, 508
    for single 12-V d-c supply, 508
    for ssb, 804
    for ssb or a-m, 510
  binary r-f phase for radar, 689
  bootstrapped-transistor chopper, 116
  chopper-type high-linearity, 506
  chopper using summing junction switch, 512
  diode bridge for 300 to 3,500 Hz, 117

Dome phase shifter for ssb, 808
double-sideband amplitude, 507
doubly balanced mixer for 9 MHz, 510
dual-fet balanced, 509
f-m for a-m transmitter, 308
fsk for carrier up to 10 MHz, 505
fsk for carrier up to 100 kHz, 502
grid dipper, 506
hi-fi a-m tuner tester, 795
infrared, 356, 361
laser, 396
mosfet reactance for f-m transmitter, 303
multiplier with index control, 511
product detector with high sensitivity, 810
pulse sample demodulator, 779
pulse-width, 877
pulse width down to 1 Hz, 512
pulse-width using comparator, 512
pulse-width using R-C phase-shift network, 509
pulse-width using switched integrator, 507
pulser for avalanche diodes, 507
radar magnetron, 690
series for transmitters up to 35 W, 513
series amplitude for 4-W aircraft transmitter, 507
single-IC, 40
suppressed-carrier a-m, 513
switching regulator for ssb transmitter, 804
three-state uhf radar, 686
varactor reactance for f-m transmitter, 305
variable-capacitance diode chopper, 116
voltage to pulse width, 509
YIG for radiometer, 463
(See also Audio amplifier; Frequency modulation; Infrared; Modem; Pulse-width modulator; Single sideband; Transceiver; Transmitter)
Moisture control, animal-tripped photoflash, 641
Moisture detector:
  alarm or control, 1
  conductive-grid rain alarm, 445
  conductive-salt alarm, 403
Moisture meter, lawn watering guide, 449
Monitor:
  carrier alarm for communication receiver, 4
  c-w with code-practice oscillator, 134
  ssb transmitter, 807
  transmitter keying, 138, 139
Monostable multivibrator (see Multivibrator)
Moon-bounce:
  144-MHz preamp cooled in liquid nitrogen, 20
  synchronous detection for 144-MHz, 702
Morse code:
  1-min timer for f-m repeater indentifier, 768
  dot-dash generator, 130
  message generator, 129
  supply for automatic keyboard generator, 726
  time-identifying clock, 215
  (See also Code)
Morse telegraph sounder, radio-controlled, 132
Mosquito, electronic repeller, 347
Motor, transistor d-c for science fair, 786
Motorboat, sound effect with oscillator, 349
Motor control:
  1.5-A for universal blender motor, 519
  2-A with timer for large food mixer, 515, 517
  2-A logic-controlled, 1015
  5-A half-wave universal, 521, 523
  5-A series universal, 519
  8-A a-c drill or saw speed, 519
  40-A universal a-c/d-c, 618

Motor control (Cont):

40-A zero-voltage switch, 1016, 1017
200-A at 36-V pwm for golf cart, 525
1/15-hp series half-wave scr, 526
1/15-hp series or universal speed and direction, 518
1/15-hp shunt-wound speed and direction, 515
1/6-hp induction using inverter, 382
1/6-hp p-m or shunt half-wave scr, 518
1/3-hp d-c lathe motor on a-c line, 520
1/3-hp d-c shunt, 519
1/2-hp series a-c, 520
400-Hz supply using long-tailed pair in damping-controlled oscillator, 521
10-V 0.2-A X-Y servo drive, 793
12-V d-c speed by pwm, 517, 518
24-V d-c parallel-wound, 690
29-W servo at 80 to 400 Hz, 790
150-W a-c shaded-pole fan, 387
360-W triac, 390
400-W servo, 791
500-W triac, 386
500-W universal, 514
600-W full-range scr, 616
600-W full-wave a-c, 617
750-W triac for induction motors, 519
2-kW squirrel-cage capacitor start-run fan, 515
5:1 speed-range half-wave scr, 520
200:1 speed range for 2.5-V d-c motor, 517
armature-voltage feedback for d-c motor to 10 hp, 522
automatic transmission-line tuner, 983
automatic tuning of r-f amplifier, 792, 794
battery tape recorder speed, 864
constant-speed for variable-load p-m, 345
crystal-controlled detent tuner, 704
d-c motor in electric drill, 523
fast stopping with capacitor bank, 525
flashlight turn-on and turn-off, 784
frequency-voltage converter, 382
full-wave d-c universal series, 525
full-wave diac-triac for power tools, 524
high torque at slow speed for drill, 526
H switch for inertial guidance motors, 560
liquid-level-sensing, 404
magnetic stirrer, 523
model-train, 347, 348
phase detector for servo automatic tuning, 613
phase-locked loop regulates speed to 0.002%, 793
photoelectric for toy sports car, 785
p-m d-c sampling regulator, 524
put-triggered universal series, 516
quadrature signals for synchronous induction, 518, 520, 521
scr for low speeds, 523, 526
scr for shunt or p-m motor, 516
slot-car speed and braking, 346
stall protection, 270
stepper for two-phase synchronous, 523
stepper torque booster, 516
sus-triggered universal series, 524
tachometer system for d-c motor to 10 hp, 522
tactile tape recorder level indicator for the blind, 482
tape player, 868, 873
thermostatic speed for air conditioner, 922
thyristor system for d-c shunt motor, 514

translator for stepping motor, 515
(See also Control; Servo)
Motor drive, twisted-ring counter, 186, 191
Movie, optical sound track readout, 631
Multicoupler, tape recorders, 872
Multimeter:
current amplifier, 429
d-c electronic, 477
electronic a-d/d-c, 478
high input impedance using two opamps, 479
voltage amplifier, 478
vtvm adapter, 476
Multiplex:
bidirectional fet switch, 862
f-m stereo decoder, 830, 834, 836, 838
Multiplexer:
1-MHz wideband differential, 528
32-channel with 5-bit switching, 527
analog switch, 528, 531
bcd 10-input from 8-input IC, 528
bipolar analog signal, 535
bootstrapped-transistor chopper, 116
character generator, 529, 535
decoder output lines, 534
digital display, 530, 532
diode choppers with common supply, 118
f-m phase-locked loop, 530
keyboard-to-binary converter, 150
led display for digital clock, 218
pam 4-channel data transmission, 532
parallel 8-digit, 529
SCA adapter for f-m receiver, 527, 528
single-fet analog switch, 534
switching gates for recorder, 534
three-channel using opamps, 533
time-shared data transmission, 533
two-channel data transmission with decoder, 533
two-channel using opamps, 530
Multiplier:
0.1 Hz-30 MHz phase comparator, 141
1.6-MHz preamp, 540
adjustable scale factor, 538
analog with voltage-controlled gain, 59
analog four-quadrant, 536, 540
analog optoelectronic using opamps, 541
analog quotient, 537
analog single-IC, 540
analog using log converters, 406
analog using long-tailed pairs, 538
automatic tester, 932
binary 4-bit by 4-bit in 23 ns, 538
capacitance to 100,000 μF, 538
divider using single-IC multiplier, 537
frequency-locking for servo, 792
level-shifting, 539–541
preamp with gain of 50, 537
pulse width with fet one-shot, 543
Q for 455-kHz i-f, 536
Q for 470-kHz amplifier, 539
resistance for inverting opamp, 541
squarer and frequency doubler, 818
time-constant with opamps, 953
two-quadrant as gain controller, 539
voltage-variable low-pass filter, 256
(see Frequency multiplier; Q multiplier)
Multiplier/divider:
30-dB dynamic range for agc, 57
log converter, 406
Multivibrator:
1 Hz-1 MHz, 544
1 Hz-5 MHz using IC one-shots, 663
1-12,000 Hz voltage-controlled, 544
38-114 Hz square-wave temperature-frequency converter, 878

50-Hz pulse train generator at 1,200 Hz, 991
57-Hz dpdt switch, 860
100-Hz free-running opamp, 552
100-10,000 Hz in a-f sweep generator, 798
500-1,000 Hz switched at 5 Hz, 814
1-kHz in a-f and r-f signal injector, 795
1-kHz series in pulsed standby battery for ram, 493
1.4-kHz astable, 126
2.5 kHz-5 MHz crystal-controlled, 545
3.3-kHz using 4-gate IC, 547, 548
5-kHz astable square-wave in keyer, 133
6-kHz piezoelectric sound source, 995
40-kHz ultrasonic transmitter, 995
50-kHz in marker generator, 430
50-kHz frequency standard using 500-kHz crystal, 432
93-kHz tape erase-bias, 870
100-kHz crystal-controlled marker generator, 437
107-kHz tape erase-bias, 872
200-kHz crystal in marker generator, 801
500-kHz in power-line carrier remote control, 761
1-MHz crystal-controlled, 297, 298, 439
1-5 MHz voltage-controlled, 997
2-min square-wave with mosfets, 551
1-pps underwater acoustic pinger, 563
1-10 pps for testing logic gates, 415
2-pps beeper for Sonalert, 3
12-V proportional voltage regulator, 745
12-V switching voltage regulator, 745
15-channel digital remote-control coder, 763
1-99% duty cycle range, 542
1,000:1 duty cycle, 545
a-f and r-f signal injector, 795
astable 1-Hz single-capacitor, 553
astable 1-30 Hz square-wave, 546
astable 1 Hz-1 MHz low-cost, 553
astable 120-3,000 Hz, 548
astable 1-kHz IC, 551
astable 7-kHz square-wave, 550
astable 120-kHz voltage-controlled, 549
astable 1-28 MHz crystal, 549
astable 5-MHz IC, 550
astable 10-Mhz IC, 544
astable 12-V as neon supply, 550
astable adjustable long-recovery, 546
astable with fast rise time, 546
astable with led status indicators, 544
astable with low quiescent power, 548
astable or monostable, 543
astable in square-wave generator, 662
astable with 3:1 frequency control pot, 546
astable direct-coupled triangular-output, 549
astable IC, 550
audio in c-w monitor, 132
aural tuning meter for transmitter, 971
bounce-free mechanical switch, 152
burglar-scaring horn, 1
cascaded monos generate pulse train, 665
controlling phase of relay switch-on, 858
delay-line coupling, 547
digital bandpass filter, 253
digital 16-note melody generator, 245
dual retriggerable mono for contact bounce suppression, 152
electronic watthour meter, 455
flasher drive, 280
flip-flop 500-Hz, 551
flip-flop 1-kHz, 550
flip-flop with led status indicators, 544
flip-flop clocked R-S, 547
flip-flop complementary-transistor, 552

Multivibrator (Cont):
  flip-flop suppresses spurious triggering, 546
  four-function programmable, 3fsinitializing
    after transient or power turn-on, 542
  linear vconou
  model-train speed con
    n er8
  mono with adjustable divider for 1.1 MHz,
    298
  mono or astable, 543
  mono with duty cycles up to 92%, 545
  mono for frequency-counter reset, 293
  mono as frequency divider, 297
  mono in frequency tripler, 690
  mono with input comparator, 142
  mono with latch for single pulse, 153, 670
  mono with led status indicators, 545
  mono as light-flash-width sensor, 634
  mono with negative output, 552
  mono with positive output, 552
  mono with pulse width multiplier, 543
  mono with zero quiescent power, 548
  mono 10-ms pulse-duration, 547
  mono 10-100 μs using three opamps, 548
  mono accuracy tester, 936
  mono adjustable-delay, 550
  mono adjustable-ujt, 549
  mono bias-insensitive ujt, 553
  mono bidirectional using exclusive-OR gate,
    331
  mono crystal-controlled, 549
  mono fast-retriggering, 552
  mono hybrid with bipolar output, 544
  mono IC, 551
  mono lascr-triggered self-clearing, 633
  mono noise-rejecting for data channel, 566
  mono pulse generator, 664
  mono trigger-sharpening, 990, 991
  optically driven pulse stretcher, 678
  overload protection for series regulator, 749
  parked-car identifying blinker, 72
  predictable R-S reset flip-flop, 545
  programmable mono, 933
  programmable-zener, 547
  pulse separator with J-K flip-flops, 542
  pulse-stretching latch, 543
  rtty signal generator, 887
  siren with speaker, 815
  tactile tape recorder level indicator for the
    blind, 482
  temperature-stabliized voltage-frequency
    converter, 166
  voltage-controlled negative resistance, 552
  wailing-siren, 816
  (See also Flip-flop; Pulse generator)
Muscle-controlled whistle, audio tone
  generator, 785
Museum, induction receiver for commentary,
  848
Music (see Electronic music; Music-controlled
  light)
Music-controlled light:
  150-W per color, 558
  200-W per color, 554, 556, 559
  300-W for rhythm, 555
  450-W per color, 556
  500-W per color, 555, 557
  Christmas-tree string, 559
  fluorescent flasher, 281
  fluorescent lamps, 283
  stereo, 555
  three-color display, 557
  triac, 558
Mutator:
  C-R using opamps, 165
  L-R using opamps, 163

Muting, power turn-on transients for stereo, 50

Nanoammeter, 100-nA full-scale, 424
Navigation:
  75-MHz airborne receiver, 565
  105-120 MHz receiver, 561
  1.68-GHz radiosonde oscillator, 564
  1-pps underwater acoustic beacon, 563
  automatic foghorn control, 564
  digital frequency synthesizer for receiver,
    564
  inertial-guidance H-switch, 560
  marker-beacon tone discriminator for air-
    craft, 563
  marker-beacon tone filter for aircraft, 561,
    565
  optical-tracking 30-MHz voltage-variable at-
    tenuator, 562
  rate-of-climb indicator, 562
  star tracker, 563
  travel-time computer for autos, 562
  vhf omnirange signal conditioner, 560
Navy, 13-28 kHz c-w receiver, 697
Negative resistance, voltage-controlled, 552
Neon:
  1-s blinking darkroom timer, 638
  12-V mvbr supply, 550
  audio tone generator, 137
  capacitor leakage checker, 418, 422
  electronic chirper, 815
  inverter supply for 18 V d-c, 384
  inverting driver, 146
  operating from 6-V battery, 380
  random blinker with random glide tones, 345
  sawtooth horizontal sweep using scr with
    neons, 852
  tachometer calibrator, 68
  trigger for phase-controlled scr, 616, 617
Network, frequency-response plotter drive, 934
Neural analog, single nerve cell, 482
Night light, nine-lamp decade counter display,
  184
Nixie:
  blanker reduces cathode voltage, 233
  dimmer using on-off control 231
  tester for driver, 931
Noise:
  10-10,000 Hz p-p measurement, 571
  200-MHz test circuit, 569
  450-MHz test circuit, 571
  active adjustable low-pass filter, 255
  agc-controlled squelch, 823
  antenna bridge, 570
  audio active filter, 47
  blanker for audio output of short-wave re-
    ceiver, 569
  blanker for transceiver or receiver, 572
  blanking in store-home intercom burglar
    alarm, 378
  bounceless switch using set-reset flip-flop,
    154
  burglar-scaring mvbr-driven horn, 1
  cancelling in tv sync separator, 910
  contact bounce suppression, 154
  crystal-diode limiter for a-f, 566
  flip-flop suppresses spurious triggering in
    one-shot, 546
  gain-controlled amplifier minimizes in tape
    recorder, 866
  i-f silencer, 569
  ignition blanker for mobile receiver, 570
  ignoring in parity checker, 150
  initializing flip-flop after line transient, 542
  measurement up to 432 MHz, 567
  measurement in d-c power supplies, 566

muting stereo turn-on transients, 50
  optimizing in audio amplifier, 28
  power-line filter for arcing aquarium heater,
    928
  radiation-triggered switch, 571
  reducing with dual-fet input for opamp,
    578
  reducing in volume control, 55
  reduction in tape casette players, 865
  rejection in bidirectional counter, 189
  rejection in latch circuit, 567
  solid-state i-f blanker, 355
  spike rejection in data channel, 566
  squelch for a-f voice amplifier, 823
  squelch for f-m receiver, 823
  squelch for SCA adapter, 821
  suppressing fluorescent-lamp interference,
    283
  suppression with dssc reciprocating detector,
    692
  third-octave measurement with sound-level
    meter, 450
  three-level sound monitor, 567
  time-averaging circuit for dvm, 471
  white approximation, 568
  (See also Squelch)
Noise bridge, 2-30 MHz impedance-measur-
  ing, 446
Noise-figure meter, 432-MHz using diode as
  noise source, 567
Noise generator:
  digital white, 568
  diode, 569, 570
  lamp, 572
Noise immunity, IC gate with diode switching
  matrix, 413, 415
Nonlinear network, rotator, 166, 168
Notch filter:
  20-18,000 Hz adjustable, 265
  60-Hz active, 253
  60-Hz adjustable-Q, 259
  60-Hz high-Q, 261
  455-kHz tunable passive, 267
  passive varactor i-f, 269
  tunable, 260
Nucleonics:
  measuring thermal conductivity of gas, 452
  pulse height discriminator, 1009
  (See also Radiation)
Null detector, logarithmic amplifier, 410
Null indicator, led pair with opamps, 1004
Numeric indicator:
  180-V inverter supply from 5 V, 383
  bcd-decimal decoder, 230, 232
  drive for eight-decade Pandicon, 225
  interface with mos array, 224

Offset, eliminating with series-connected
  opamps, 574
Ohmmeter:
  0-1 and 0-10 ohms full-scale, 460
  1.5-V regulated supply, 719
  500-milliohm full-scale, 457
  conversion from digital voltmeter, 461
  four-range using two opamps, 479
  linear-scale high-precision, 459
  linearizing scale for vtvm, 460
  low-resistance for bad contacts, 459
  low-voltage for solid-state circuits, 460
  Wheatstone-bridge, 461
  wood moisture measurement, 452
Oil, auto level checker, 65, 68
Omnirange, signal conditioner for light planes,
  560
One-shot (see Multivibrator, mono)

Operational amplifier:
3-A power using Norton opamp, 582
1-55 Hz at 600 mW, 576
2 Hz-20 kHz square-wave generator, 801
100-Hz astable mvbr, 552
0-30 kHz with 60-V output swing at 22 W peak, 579
850-kHz noninverting with resistive feedback, 575
1-MHz feed-forward unity-gain, 578
10-MHz summing using feedforward compensation, 577
200-μV protective limiter, 397
0-V output for 0-V input, 582
110-V output swing, 574
500-W transistor-bridge booster, 577
adjustable-gain, 18
antilog generator, 411
balanced differential, 580
bias-current compensation, 577–579, 581
blood pressure simulator, 485
chopper-stabilized 0.01% microvoltmeter, 477
clipper using zeners, 579
computer-aided design, 575
cubic-function generator, 409
current-voltage converter, 428, 479
delay function with active R-C networks, 576
differential-fet, 8, 580
differential input and output, 578
differentiator, 582
dual-fet instrumentation for low temperature drift, 368
educational using discrete components, 575
elliptic low-pass ladder filter transfer function, 577
fet input for high source resistance, 576
fet input buffer, 18
flip-flop with led status indicators, 544
full-wave small-signal rectifier, 580
guarded full differential, 583
inductor simulator, 37, 40
integrators in 16-Hz quadrature oscillator, 371
inverter with resistive feedback, 575
inverting 20-dB micropower, 582
inverting 500-nW micropower, 583
inverting with resistance multiplier, 541
inverting adder, 575
keyer, 132
level and polarity changer, 576
level-shifting with isolation, 581
linear current booster, 573
logarithmic with five-decade range, 574
logarithmic-response five-decade, 411
logarithmic-voltage generator, 409
low input current and offset drift with dual-fet input, 579
low-noise with 5-kHz bandwidth, 578
low-noise with dual-fet input, 578
low-offset a-c with 1,000 gain, 790
measuring T-attenuator resistors, 459
monostable 10-100 μs mvbr, 548
multiplexer for 2 channels, 530
multiplexer for 3 channels, 533
multiplier with level shifting, 539–541
noise tester, 571
noninverting 20-dB micropower, 582
nonlinear transfer function with sharp breakpoint, 581
optimizing differential fet input, 573
optoelectronic analog multiplier, 541
parameter tester, 324, 945
power booster, 580
power-supply splitter, 574
precision clipper using fet input, 577

reciprocal-function generator, 405
rezeroing by command, 877
second-order all-pass transfer function, 575
series pair cancels offset voltage, 574
series pair as zero-offset buffer, 574
speeding phototransistor response, 628
square-law signal-controlled gain, 1008
square-root, 817, 818
squarer for vector length, 817, 818
status indicator, 1008
subtracter with resistive feedback, 575
summing closed-loop, 582
summing-scaling IC, 580
tester, 936
voltage follower, 576
zero correction to 3 μV, 581
(See also Amplifier; Integrator; Logarithmic; Servo; Squarer; entries for desired applications)
Optical character recognition, photodiode-opamp replace photomultiplier, 563
Optoelectronic:
2-A logic-controlled a-c switch, 1015
10-A normally closed relay, 590
10-A normally open relay, 588
20 Hz-7 MHz isolation amplifier, 585
10-MHz isolator, 588
50-MHz photodiode for led communication, 587
900-mW pulsed supply for led, 587
1.25-W pulsed supply for led, 586
100-W projection-lamp brightness regulator, 630
analog multiplier using opamps, 541
audio compressor, 46
audio compressor-expander, 42
broad-area detector for led, 588
color tv remote control, 924
constant-current 900-V supply for magnetron, 339
coupler for 5-kHz clock pulse, 589
coupler for three-phase heater control, 947
d-c amplifier for driving TTL or DTL with led, 584
distillation temperature control, 929
diversity audio combiner, 693
fiber-optic speed feedback for pwm speed control of d-c motor, 518
guitar Waa-Waa sound effect, 244
isolator for high-speed line receiver, 199
isolator with unity coupling, 589
latch for 60-W scr capacitance control, 106
led display for analog voltage sensor, 1005
led isolates DTL from relay, 585
light-intensity regulator, 590
light-triggered 6-kV series scr switch, 343
low-level isolator, 588
master control for three-phase zero-voltage switch, 949
measuring low light levels, 585
message alarm, 347
modulator, 586
motion detector, 589
phototransistor drive for TTL or DTL, 585
power supply monitor, 584
pulse-stretching mvbr with isolation, 678
pulsing led with scr, 359
quadrature detector for led communication, 587
reading aid for blind, 489
regulator for high-voltage supply, 337
r-f low-power wattmeter, 456
shaft encoder, 590
slow-motion detector, 586
three-phase heater control, 947
voice modulator using led, 586

volume expander for phono, 42
Organ (see Electronic music)
Oscillator:
0.01-4,000 Hz sine-square-triangle, 325
0.1-100 Hz beat-note, 802
0.1 Hz-30 kHz current-controlled, 327
0.5-Hz bipolar-pulse ujt relaxation, 664
1-Hz sine-wave two-opamp, 592
5-60,000 Hz sine-wave signal generator, 799
8-Hz for identifying crystal marker signals, 439
10-Hz variable phase-shift, 591
10-Hz Wien-bridge sine-wave, 599
10-4,990 Hz square-wave, 798
10-10,000 Hz precision Wien bridge, 600
10 Hz-100 kHz frequency meter and square-wave shaper, 676
10 Hz-100 kHz ujt relaxation, 664
10 Hz-100 kHz Wien-bridge, 593
15 Hz-1.5 MHz fet, 595
15 Hz-200 kHz Wien-bridge, 800
16-Hz quadrature with limiting, 371
20-5,000 Hz sine-square-triangle sweep, 329
20 Hz-70 kHz for high temperatures, 596
23 Hz-460 kHz for capacitance meter, 419
30-20,000 Hz Wien-bridge, 592
35-Hz Wien-bridge sine-wave, 597
55-7,040 Hz Hartley in electronic organ, 246
60-Hz stabilized ujt relaxation, 592
87-112 Hz continuous-tone encoder for f-m repeater access, 772
100-1,000 Hz voltage-tunable, 598
100-1,000-10,000 Hz square-wave for amplifier linearity test, 798
130-2,000 Hz tone generator with vibrato, 245
150-Hz sine-wave relaxation, 596
160-1,600 Hz Wien in tunable variable-Q active filter, 254
300-10,000 Hz audible rate-of-climb indicator, 562
400-Hz for servo-testing supply, 930
400-Hz damping-controlled motor supply, 521
400-Hz twin-T, 446, 779
440-Hz tone generator, 809
440-Hz tuning-fork, 242
500-Hz flip-flop trigger for inverter scr's, 551
550-Hz side-tone for IC keyer, 135
650-Hz f-m repeater tone control, 776
700-Hz modulated bell-tone, 347
750-Hz phase-shift code practice, 138
800-Hz and 2 kHz two-tone test generator, 802, 803
800-Hz 3-gate for electronic keyer, 138
920-Hz phase-shift sine-cosine, 599
960-Hz tuning-fork, 597
0-50 kHz variable sine-wave, 593
1-kHz in 10-min timer, 952
1-kHz driving speaker, 592
1-kHz encoder for f-m transmitter, 771
1-kHz flip-flop trigger for inverter scr's, 550
1-kHz jfet Hartley, 600
1-kHz modulator, 598
1-kHz operating on 3-V supply, 593
1-kHz signal generator, 799
1-kHz sine-wave, 591, 599
1-kHz tone pulse, 595
1-kHz twin-T with agc, 600
1-kHz Wien-bridge, 593, 597
1-500 kHz with subharmonic sync, 661
1.005-kHz Hartley with 360-deg phase shifter, 615
1.1-kHz sine-wave tone, 597
1.2-kHz and 2.2-kHz for modem, 504

Oscillator (Cont):

1.24-kHz f-m in stereo electro-encephalophone, 487
1.3-kHz a-f signal injector, 802
1.5-2.5 kHz voltage-controlled, 997
1.75-kHz repeater turn-on, 775
1.8-kHz and 2-kHz in tone-burst keyer for f-m repeater, 767
2-kHz driving flashlight-lamp key for electronic lock, 237, 239
2-kHz mosquito repeller, 347
2.1-kHz sine-wave tone, 597
2.125-kHz and 2.975-kHz afsk, 888
2.125-kHz and 2.975-kHz code practice and afsk-mcw, 133
2.295-kHz and 2.975-kHz afsk for rtty, 890
5-kHz constant-amplitude, 596
5-kHz phase-shift for drums, 249
6-kHz with 60-Hz divide-and-sample loop, 777
10-kHz sine-wave two-opamp, 600
10-kHz and 100-kHz frequency standard, 432
10-100 kHz ultraminiature quartz crystal, 612
17.00-22.25 kHz ultrasonic remote control, 917
20-kHz at 13 W, 595
20-kHz class-C Hartley, 116
20-kHz ujt pulse generator, 992
22-kHz phase-shift, 233
25-kHz in d-c converter, 158
35.5-kHz ultrasonic remote control, 919
40-kHz all-diode relaxation, 605
40-kHz ultrasonic intrusion alarm, 272
50-kHz power-supply for 50-kV voltmeter, 474
50-kHz ujt square-wave generator, 663
59.5-kHz IC for digital afsk, 884
93-kHz tape erase-bias, 870
99.925-kHz ujt relaxation using crystal, 605
100-kHz with 1-kHz markers, 430
100-kHz with 10-kHz markers, 441
100-kHz with divide-by-ten for vco, 1004
100-kHz crystal in 50-kHz marker generator, 430
100-kHz crystal with 100-kHz markers to 100 MHz, 437
100-kHz crystal calibrator, 436, 440
100-500-1,000 kHz crystal calibrator with harmonics, 435
100-kHz crystal-controlled square-wave, 437
100-kHz crystal with divider for 25-kHz calibrator, 432
100-kHz crystal with harmonics to 1,296 Hz, 439
100-kHz crystal with harmonics to 150 MHz, 434
100-kHz crystal scs, 608
100-kHz crystal standard, 440
100-kHz marker for communications set, 601
100-kHz using opamp in twin-T, 604
100-kHz time base for frequency counter, 293
107-kHz tape erase-bias, 872
130-kHz Wien IC with Miller compensation, 608
160-190 kHz 1-W experimental transmitter, 987
200-kHz crystal in 25-kHz ham-band calibrator, 430
200-kHz crystal with frequency dividers, 435
200-kHz crystal driving square-wave frequency dividers, 434
200-kHz crystal mvbr in marker generator, 801

200-kHz crystal standard for frequency counter, 290
300-kHz 240-kW induction heater, 606
307-kHz crystal, 608
400-kHz 60-kW induction heater, 606
455-kHz or 10.7-MHz with fet, 351
455-kHz modulated constant-output, 796
500-kHz crystal in 50-kHz standard, 432
500-kHz relaxation using 4-layer diode, 666
600 kHz-10 MHz within 0.1% with c-mos inverters, 124
600 kHz-30 MHz inductance checker, 451
999.663-kHz ujt relaxation using crystal, 605
1-MHz with 10-kHz markers, 431
1-MHz with frequency dividers, 434
1-MHz crystal, 126, 293
1-MHz crystal with dividers in calibrator, 437
1-MHz crystal in marker generator, 437, 441
1-MHz crystal standard with harmonics, 440
1-MHz modulated crystal band-edge marker, 440
1-MHz square-wave crystal mvbr, 439
1-MHz square-wave with divide-by-100, 298
1-MHz square-wave reference in crystal oven, 433
1 MHz to vhf using fet pair, 604
1-MHz weak-signal source for alignment, 797
1-10 MHz Pierce in crystal checker, 930
1-20 MHz crystal, 612
1-28 MHz crystal using TTL logic, 549
1-250 MHz dip meter, 436
1.1-MHz crystal in pulse divider, 298
1.5-30 MHz dip meter, 436
1.5-50 MHz dip meter, 437
1.7-300 MHz solid-state dip meter, 437
1.785 MHz for converter, 609
1.8-150 MHz dip meter, 435
2-30 MHz crystal, 608
2.2-2.5 MHz beat-frequency for amateur receiver, 355
2.9-155 MHz vfo/crystal dip meter, 438
3-MHz with 30-kHz markers to 150 MHz, 436
3-MHz frequency standard with dividers, 438
3-MHz square-wave with divide-by-100, 297, 298
3-9 MHz Pierce in crystal checker, 933
3-30 MHz crystal frequency spotter for blind operator, 487
3-200 MHz dip meter, 439
3.045-3.545 MHz for communication receiver, 602
3.2-MHz crystal reference for frequency synthesizer, 321
3.5-MHz transistorized vfo, 607
3.5-4 MHz vfo for transceiver monitor, 961
3.58-MHz injection-locked color tv, 902
4-MHz for measuring capacitance by frequency shift, 422
4.5-5 MHz voltage-tuned, 1003
5-5.5 MHz Colpitts in 20-meter receiver, 697
5-6 MHz vfo, 609
5-10 MHz sine-wave IC, 978
5.88-6.38 MHz solid-state Vackar, 604
6-MHz variable crystal, 612
6-MHz voltage-controlled for 2-meter synthesizer, 322
6.1-MHz variable-frequency crystal, 605
7-MHz crystal for 448-MHz repeater, 773
7-7.3 MHz Seiler, 602
7-7.3 MHz Vackar, 603
8-MHz tunable crystal, 611
8-43 MHz phase-locked crystal, 698
9.0015-MHz for ssb product detector, 890
10-MHz crystal reference for counter, 434

10-MHz decade-counter frequency synthesizer, 317
10-MHz relaxation, 666
11.215-MHz crystal, 354
21-MHz lumped-line, 611
24-MHz using low-activity crystal, 603
26.670-MHz crystal in electronic door lock, 237
27-MHz drive for Ionovac speaker, 47
28-MHz for i-f alignment, 799
30-MHz Vackar in 10-W transmitter, 975
30-MHz Vackar transmitter vfo, 601
30-MHz voltage-controlled with crystal stability, 549
40-75 MHz for 6-meter superhet, 609
45-MHz crystal, 607, 608
48-MHz crystal, 609
50-MHz with infinite attenuator, 608
50-100 MHz crystal, 610
52.5-MHz crystal in wide-band f-m exciter, 303
127.333-MHz crystal with transistor amplifier, 812
140-300 MHz strap-line, 611
144-MHz crystal in 2-meter amplifier tester, 797
146-MHz modulated crystal, 170
200-MHz crystal IC with doubler, 126
200-MHz using ECL gates as crystal doubler, 315
232-MHz crystal sextupler for 220-MHz receiver, 607
382-MHz multiplying from 7.074 MHz, 809
423-MHz varicap-controlled crystal, 985
850-1,100 MHz half-wave strap-line, 610
1.1-1.4 GHz 4-W varactor-tuned, 1000
1.2-1.3 GHz, 610
1.68-GHz radiosonde, 564
1.7-GHz at 6 W, 611
8-GHz fast-warmup Gunn-diode, 601
10.69-GHz Gunn in Doppler intruder detector, 98
10.69-GHz Gunn in doppler radar, 688
6- and 80-meter vfo, 610
40-meter and 80-meter vfo, 609, 611
40-meter crystal IC, 981
40-meter driving 2:1 IC frequency divider, 980
40-meter ½-W single-tube c-w transmitter, 974
80-meter in code-practice transmitter, 134
80-meter vfo with buffer, 981
160-meter push-pull fet crystal, 602
a-f: Baxandall Wien-bridge phase-shift, 594
   bridged-T opamp, 593
   constant-impedance attenuator, 596
   IC with mosfet feedback control, 591
   interrupted frequency-modulated tone for alarm sounder, 817
   phase-shift with four R-C sections, 595
   phase-shift opamp, 594
   quadrature 5-opamp with amplitude control, 598
   relaxation using put, 598
   signal injector for troubleshooting, 795, 800
   twin-T opamp, 592
   voltage-sensitive tuning aid for blind, 483
   Wein-bridge with diode control, 594
   Wein-bridge with fet feedback control, 594
   Wien-bridge IC, 595
   Wien-bridge opamp, 595
   Wien-bridge single-opamp, 598

Oscillator, a-f (Cont):

(See also Code; Converter; Electronic music; Inverter; Modulator; Signal generator; Siren; Transceiver)
audible light meter, 787
audible thermometer for blind, 487
audio fsk for radioteletype, 886
audio mike-testing beeper, 655
audio for modulating grid dipper, 506
automatic level control, 55
bfo for ssb signals on a-m CB transceiver, 121, 123
bias-erase for stereo tape recorder, 873
blocking neon in photoflash converter, 155
capacitor tester using audio tone, 418
clicker driving speaker for intrusion alarm, 4
clock two-frequency crystal using quad gate, 128
code practice, 98, 130, 137, 139
code practice converted to metronome, 246
Colpitts crystal in CB channel spotter, 123
crystal with neon regulator for screen grid, 738
crystal ringing for color tv subcarrier regenerator, 897, 903
diode switching of crystals, 762, 764, 766, 865
dipper using lamp as indicator, 436
dot generator for c-w transmitter test, 133
electrolytic capacitor tester using neon, 422
emg voltage-controlled audio, 785
fet dip meter, 433, 440
four-band for qrp transmitter, 979
Hartley a-f, 98
Hartley in auto idiot-light alarm, 63
Hartley in f-m wireless mike, 846, 849
injection-locked subcarrier for color tv, 902
laser-triggered ujt relaxation, 633
light-sensitive clicker as alarm, 93, 271
light-sensitive in photoresist exposure control, 641
light-triggered audio alarm, 632
locked bias-erase for stereo tape recorder, 864
metronome for transistor radio, 242
modulated-tone lie detector, 786
neon in flickering-flame lamp control, 787
organ R-C master tone, 245
ovonic-switch relaxation, 860
put sampling with long-delay timer, 956
r-f: 470-kHz with 220-Hz modulation, 603
1.7-9 MHz vfo, 612
5-5.5 MHz stable 2-band vfo, 602
8.5-12 MHz gated remote-control, 601
beat-frequency for ssb and c-w on a-m, 809
phase-shift with distributed R-C line, 604
scr relaxation for damped sine waves, 605
signal injector for toubleshooting, 800
temperature compensation with thermistor, 608
vhf f-m, 603
(See also Capacitance control; Converter; Dip meter; Frequency multiplier; Function generator; Inverter; Metal detector; Pulse generator; Radar; Receiver; Remote control; Signal generator; Square- wave generator; Staircase generator; Sweep; Telemetry; Television; Transceiver; Transmitter; Ultrasonic; Video; Voltage-controlled oscillator)
rtty tone generator, 894
science-fair beeper with meter, 788
slow-kick windshield wiper, 70

speaker-driving overtemperature alarm, 274
sunlight-controlled neon relaxation, 788
telephone obscene-call squelcher, 888
Theremin, 243, 247
tone for automatic call-letter generator, 769
vibratone 11-note, 244
Wien network with automatic amplitude control, 801
wireless code practice for a-m radio, 135
(See also Capacitance control; Code; Converter; Dip meter; Electronic music; Frequency multiplier; Function generator; Inverter; Metal detector; Pulse generator; Radar; Receiver; Remote control; Signal generator; Square-wave generator; Staircase generator; Sweep; Telemetry; Television; Transceiver; Transmitter; Ultrasonic; Voltage-controlled oscillator)

Oscilloscope:
10 Hz-1 MHz preamp, 114
0-300 MHz vertical amplifier, 111
automatic erasure, 110
color-tv vertical preamp, 110
d-c amplifier, 114
dual-trace switching waveform generator, 109
four-channel display, 109
npn transistor curve tracer, 945
time-base trigger, 114
trace quadrupler, 108
transistor parameter display, 944
vertical amplifier, 109
vertical preamp, 111
voltage calibrator, 114
(See also Cathode-ray)
Oven control (see Temperature control)
Overload (see Protection)
Overvoltage protection, single scr for two supplies, 651
Ovonic-threshold switch, relaxation oscillator, 860

Paddle, solid-state switching of keyer, 132
Pager:
CB transmitter, 122
noise-actuated squelch, 822
Panic button:
pushbutton siren, 814
self-cycling siren, 815
Panoramic receiver, 9-MHz i-f amplifier for 2 meters, 352
Paper tape, photoelectric reader, 633
Paper-tape punch, tone-controlled switch, 139
Parity checker:
eight-bit using four decoders, 145
noise-immune for binary words, 150
Passive filter (see Filter, passive)
Patch, push-to-talk telephone, 880
Peak detector:
agc for IC audio amplifier, 56, 60
amplitude-independent, 1010
comparator for 50-ns pulses, 1010
high-speed voltage, 1010
negative and positive combined, 1006
negative peaks with comparator, 1010
negative voltages, 1007
positive peaks to 1 MHz, 1008
positive peaks with comparator, 1010
positive-voltage, 1005
(See also Voltage level detector)
Peak rectifier, precision half-wave, 480
Perturbation, optical in photoelectric position detector, 404
pH meter, low-cost, 448
Phase comparator (see Comparator)

Phase control:
40-A power, 618
0-360 deg shifter with jfets, 619
360-deg shifter for 1,005-Hz oscillator, 615
1-10 kHz constant-angle shifter, 615
1.005-kHz bridge-balancing, 616
1-kW full-wave sbs-triac, 387
1-kW lamp dimmer, 616
1.2-kW ujt-triac manual, 388
3-kW full-wave scs, 618
6-kW full-wave cooling or heating, 928
0-10 MHz phase detector using logamps with chopper, 619
30-MHz phase splitter, 614
150-W half-wave scr for lamp, 387
600-W full-wave, 616, 617
600-W half-wave average-voltage feedback, 390
900-W full-wave, 619
automatic tuning of r-f amplifier, 613
digital 90-deg phase shifter, 614
digital with counter and comparators, 613
direct-conversion ssb receiver, 810
full-wave light-activated, 632
full-wave neon for scr, 617
full-wave scs, 617
full-wave single-scr with put, 618
half-wave light-activated, 634
half-wave neon for scr, 616
half-wave photoelectric for 900-W lamp, 389
hysteresis-free for lamp load, 616
IC trigger for triac, 929
inductive or resistive load, 921
line-voltage compensation, 386
phase-sensitive detector, 614
put trigger for universal series motor, 516
radar ferrite phase shifter, 687
relay switch-on using mono mvbr, 858
splitter for double phase-sensitive detector, 615
sus trigger for universal series motor, 524
(See also Lamp control; Motor control; Power control; Three-phase control)
Phase detector:
automatic tuning of r-f amplifier, 791, 792
discriminator-type for color tv, 905
product-type for color tv, 902
Phase-locked loop:
3-kHz using vco, 1000
broadcast and amateur receiver, 696
digital divider, 295
f-m demodulator, 302, 306
f-m multiplexer, 530
frequency multiplier or divider, 294
fsk converter for rtty, 892, 899
fsk decoder, 503, 504
fsk generator with two levels, 503
IRIG channel 13 demodulator, 874
rtty demodulator, 887
satellite picture demodulator, 877
SCA background-music adapter, 527
stereo decoder, 832, 834
synchronizing clock signals, 126
synchronous a-m detector, 692
tone decoder for f-m repeater, 776
(See also Telemetry)
Phase-locked oscillator, 8-43 MHz crystal, 698
Phase meter, 360-deg digital differential, 451
Phase-sensitive demodulator, automatic transmission-line tuner, 983
Phase-sensitive detector:
beam-switching tube, 357
double circuit for low-frequency bridge-balancing, 614
Phase-shift oscillator, 920-Hz sine-cosine, 599

Phase shifter:
90-deg Dome for ssb modulator, 808
twisted-ring counter, 186, 191
Phasing, p-a speakers, 656
Phone (see Headphones; Telephone)
Phone patch, amateur radio station, 881
Phono:
8-kHz-cutoff scratch filter for 78-rpm records, 626
9-V regulated supply with negative output resistance, 739
2-W a-f amplifier for crystal pickup, 622
3-W amplifier for crystal pickup, 623
3-W amplifier with tone controls, 627
5-W stereo amplifier, 626
a-m wireless transmitter, 977
ceramic-cartridge preamp, 620, 621, 623, 832
combination low-pass and high-pass filter for magnetic pickup, 623
common-mode volume and tone controls, 625
common-mode volume control, 622
crystal-pickup IC amplifier, 625
crystal pickup preamp, 627
dual-supply single-IC stereo preamp, 625
hi-fi preamp, 620
low-noise RIAA preamp, 841
magnetic-pickup preamp, 621, 622, 624, 627
preamp for ceramic or crystal pickup, 627
preamp for high-output ceramic pickup, 622
preamp with switched equalization, 621
RIAA-equalized IC amplifier, 627
scratch and rumble filter, 626
scratch filter with continuously variable control, 621
simple four-transistor amplifier, 695
single-supply single-IC stereo preamp, 835
stereo with fet preamp, 625
tone-balance control for 78-rpm records, 624
volume expander, 42
wow and flutter meter, 622
(See also Audio amplifier; Audio preamplifier; Quadraphonic; Stereo)
Phonograph (see Phono)
Photocoupler (see Optoelectronic)
Photoelectric:
16-A power control for 230 V, 628
0.1 Hz-30 kHz light-controlled oscillator, 327
2-kHz flashing key for door lock, 237, 239
2-kHz modulated-light receiver for door lock, 234
1-kW darkness-on switch, 98, 629
50-MHz photodiode for led communication, 587
50-W automatic night light, 101
500-W gradual-on outdoor light control, 632
500-W night light, 93
600-W darkness-turn-on triac, 617
900-W half-wave scr lamp control, 389
alarm with siren output, 2
animal-tripped photoflash, 641
aperture-control servo, 794
apple grader to size, 631
audible light meter, 787
auto parking-light turn-on, 73
automatic night light, 629
automatic welcome light for porch, 629
barricade flasher with ambient-light control, 280
bistable amplifier for low light levels, 631
broad-area detector for led, 588
burglar alarm, 92, 95, 96, 99, 101
chopped-beam infrared alarm, 92
commercial-killer with flashlight, 924

commercial killer for tv or radio, 925
dark-off trigger for scr, 628
dark-on trigger for scr, 630
darkness-turn-on 6-V flasher, 278
darkroom excess-light control, 636
flash measurement, 638
flash-width sensor, 634
flashlight-controlled motor in toy, 785
flashlight-monitored power-supply voltages, 1009
flashlight-operated bistable switch, 761
flashlight-operated tv silencer, 918
full-wave phase control, 632
half-wave phase control, 634
headlight-operated garage-door switch, 766
infrared modulated-beam detector, 361
laser pulse power measurement, 393
level control for bin or reservoir, 402
level detector, 402
light-activated zero-voltage switch, 1018
light comparator, 632
light-controlled tone generator, 554
light-deenergized 5-mA relay, 634
light-energized 5-mA relay, 634
light-feedback phase control for lamp, 389
light meter for low levels, 643
light trigger for triac, 629
measuring light for triggering lascr, 448
motion detector, 589
motor-controlled demonstrator for science fair, 784
opamp speeds phototransistor response, 628
optical logic driver, 633
optical sound track readout, 631
photodiode and opamp replace photomultiplier, 563
photodiode switching amplifier, 630
photoresist exposure control, 641, 642
position detector using perturbation, 404
punched paper-tape reader, 633
quadrature detector for led communication, 587
reading aid for blind, 489
reduced dark current for low light levels, 630
regulator for 100-W light source, 630
rifle range, 346
saturating preamp for solar cell, 791
scr trigger control, 631
self-clearing lascr monostable mvbr, 633
slot-car win detector, 347
starlight photometer, 449
start-stop control for scaler, 954
sunlight-operated alarm clock, 632
sun-powered receiver, 706
tv and radio commercial killer, 918
twilight turn-on road barrier lamp, 628
up-down object counter, 189
volume expander for phono, 42
(See also Lamp control; Optoelectronic; Photography)
Photoflash:
2.4 V to 500 V converter, 155
450-V supply using 3 D cells, 644
450-V supply using 4-V battery, 644
500-V supply using 2.4-V battery, 644
500-V trigger, 635
2,500-V supply using 4-V battery, 640
animal-tripped photographic, 641
burglar alarm, 98
exposure meter with holding circuit, 638
fast-recharge, 639
fill-in for camera close-ups, 637
scr trigger, 641
slave, 637, 640, 643
thyratron trigger, 641
trigger for 450-V flashtube, 642

trigger for 900-V flashtube, 644
trigger for 2,500-V flashtube, 643
(See also Flash; Photography)
Photography:
3-300 click-per-minute timer, 249
0.1-99 s darkroom timer, 638
1-s blinking darkroom timer, 638, 639
2 s-2 min darkroom timer, 958
5-60 s darkroom timer, 637
6-60 s timer, 955
450-V flash supply, 644
450-V photoflash supply using 4-V battery, 644
500-V flash supply, 644
2,500-V photoflash supply using 4-V battery, 640
500-W photoflood dimmer, 642
animal-tripped photoflash, 641
audible and flashing darkroom timer, 638
audible light meter, 787
automatic room-light dimmer for movies, 386
beeping timer for darkroom, 635
burglar-tripped photoflash, 98
darkroom excess-light alarm, 636
enlarger lamp voltage regulator, 639
enlarger phototimer, 644
fast-recharge flashgun, 639
fill-in flash for close-ups, 637
flash exposure meter, 635
flash-reading meter with holding circuit, 638
integral number of tv fields, 636
light meter for low levels, 643
light-sensitive oscillator in photoresist exposure control, 641
light-triggered photoflash slave, 643
photoflash trigger, 635
photoresist exposure control, 642
remote shutter release, 637
scr trigger for flashtube, 641
slave flash, 637, 640
stop-motion using sound-fired strobe, 639
thyratron trigger for flashtube, 641
timer for enlarger, 957
trigger for 450-V flashtube, 642
trigger for 900-V flashtube, 644
trigger for 2,500-V flashtube, 643
two-flashtube strobe, 640
universal enlarge-develop timer, 643
(See also Flash)
Photometer, starlight magnitude, 449
Photomultiplier:
3.5-kV regulated supply, 341
power supply using IC voltage regulator, 337
Photoresist, exposure control, 641, 642
Piezoelectric transducer, amplifier, 364
Pincushion correction, cathode-ray display, 113
Pinger, 1-pps underwater acoustic beacon, 563
Pipe finder, metal-detecting, 498
Pipe locator, beat-frequency metal detector, 499
Plasma, measuring conductivity with probe, 458
Pockels cell, Q-switched laser, 391, 392
Polar coordinates, converter from X-Y, 165
Police, lie detector, 785
Police calls:
135-175 MHz converter for f-m receiver, 173
auto radio converter, 171, 172
Pollution, water conductivity tester, 460
Pool, splash alarm, 403, 404
Position detector, photodiode using perturbation, 404
Potentiometer, digitally set with shunt switch, 863
Power:
0-5 W meter for 10-80 meters, 456

Power (Cont):
   0-10 W r-f wattmeter, 454
   0.25-10 W wattmeter for up to 450 MHz, 456
   5-300 W wattmeter for up to 450 MHz, 455
   1,000-W wattmeter for 3.5-30 MHz, 454
   CB transceiver output, 454
   instantaneous wattmeter, 455
   measurement with level-shifting multiplier, 541
   monitoring for a-c appliance, 453
   neon-bulb wattmeter, 454
   r-f optoelectronic wattmeter, 456
   transmitter field-strength meter, 456
   transmitter output meter, 454
   uhf wattmeter, 455
Power control:
   600-W diac-triac with timer, 957
   NAND gate controls triac, 861
   transformer tap-changing with triacs, 862
Power-factor meter, a-d converter, 163
Power failure:
   batteryless alarm, 3
   emergency light and alarm relay, 2
Power-line carrier, remote control, 761
Power meter, square-law diodes give linear scale, 454
Power monitor, CB transmitter, 121
Power supply:
   20-A a-c overcurrent and overvoltage protection, 654
   60:1 active ripple reduction filter, 262
   250:1 active ripple reduction filter, 254
   400-Hz 115-V servo-testing, 930
   2.5-kV photoflash using 4-V battery, 640
   3-kV at 700 mA, 342
   10-kV electrostatic precipitator, 339
   12-kV for image converter, 338
   16-kV for image converter, 342
   0-120 V a-c electronic Variac, 646
   1.5-V d-c battery with charger, 90
   6-V for battery radio, 648
   12-V dropped to 6 V in auto, 647
   12-V gate bias from 5 V, 332
   14-V bipolar, 645
   25-V and 36-V bipolar for 30-W and 60-W stereo, 646
   80 V at 3 A, 647
   90-V a-c regulator, 649
   105-V a-c regulator for tv receiver, 645
   120-V a-c regulator, 646
   450-V photoflash using 4-V battery, 644
   500-V photoflash using 2.4-V battery, 644
   700-V at 200 mA for transceiver, 648
   700-V and 250-V for ssb transceiver, 649
   750-V for transceiver, 341
   800-V and 300-V universal ham transmitter, 342
   900-V at 30 mA, 339
   900-V sextupler transformerless, 343
   a-c line voltage reducer, 647
   charging two batteries with one generator, 647
   current monitor, 424
   diode and switched capacitors for lamp control, 390
   flashlight-controlled voltage monitor, 1009
   full-wave 20-A 3.2-V, 649
   half-wave 20-A 3.2-V rectifier, 647
   measuring ripple and spikes, 566
   neon using 12-V mvbr, 550
   pulsed for infrared led, 97
   solid-state vibrator replacement, 646
   (See also Battery charger; Converter; High voltage; Inverter; Regulated power supply; Regulator)

Power supply protection (see Protection; Regulated power supply)
Preamplifier:
   1.296-GHz, 7, 19
   2.304-GHz with 8-dB gain, 10
   4 Hz-42 MHz for frequency counter, 286
   9 Hz-250 kHz low-noise, 11
   10 Hz-1 MHz preamp, 114
   30 Hz-100 kHz with 20-dB gain for instruments, 365
   250-kHz a-c using resistive feedback, 11
   2-meter cascoded jfet, 692
   2-meter single-fet, 18
   6-meter with 30-dB gain, 56
   10-meter IC, 16
   1-MHz with gain of 50 for IC multiplier, 537
   1-100 MHz broadband, 13
   1.6-MHz opamp for fast multiplier, 540
   144-MHz cooled in liquid nitrogen, 20
   144-432 MHz low-noise, 11
   146-MHz for f-m receiver, 398
   150-MHz for 2-meter receiver, 20
   275-MHz bootstrapped vhf, 5
   CB receiver, 122
   CB r-f booster, 122
   fet with current control, 8
   radiation detector, 423
   scope vertical input, 111
   (See also Amplifier; Audio amplifier; Audio preamplifier; Phono; Quadraphonic; Receiver; Servo; Stereo; Video amplifier)
Precipitator, 10-kV electrostatic air-cleaning, 339
Prescaler (see Frequency counter; Frequency divider)
Preselector, self-contained for amateur receiver, 703
Presence control, 6-dB accentuation at 1,000 Hz, 55
Pressure-level meter, loudspeaker enclosure design, 443
Printer, control driven by ring counter, 186
Probability density analyzer, opamps with window comparator, 444
Probe, magnetic and electric field, 444
Process control:
   automatic signal level control, 55
   level-sensitive up-down trigger, 993
   voltage-current converter, 367
Product detector:
   ssb high-sensitivity, 810
   ssb using IC balanced modulator, 807
Programmable frequency divider, 2-meter f-m frequency synthesizer, 321
Programmable log module, rms computation, 823, 824
Programmable zener (see Voltage reference)
Programmer, manual 512-bit read-only memory, 490
Proportional voltage regulator, 12-V at 2 A using pwm, 745
Protection:
   20-A a-c overcurrent and overvoltage, 654
   100-mA current limit for 15-V regulator, 194
   200-μV limiter for opamp, 397
   448-MHz 5-W power amplifier against open or shorted antenna, 773
   5-V overvoltage crowbar for switching regulator, 756
   6-V and 12-V low-ripple regulated supply for f-m, 714
   12-V regulated supply, 652
   12-V voltage regulator, 745
   50-V regulated power supply, 715
   100-W class-B a-f amplifier, 659
   amplifier overheating alarm, 274

area with 700-Hz chopped infrared, 93, 95
auto storage battery with lights-on alarm, 65, 67, 71, 72, 74
automatic-reset solid-state circuit breaker, 652
automatic shutdown of regulator, 750
battery life in transceiver with audio-controlled switch, 964
brownout alarm for low line voltage, 654
class B a-f output transistors, 653
class C power amplifier from loss of load, 650
darkroom excess-light alarm, 636
d-c motor undervoltage, 270
electronic combination lock, 77, 235, 237
equipment against overvoltage with crowbar, 650
fail-safe digital display, 221
flame failure in furnace, 272, 274
flashover at color tv picture tube, 904
freezer failure alarm, 2
grounded-appliance tester, 654
headlight turn-off control, 73
illegal bcd code in a-d converter, 24
initializing flip-flop after power failure, 542
inverter from d-c overload, 653
line-failure emergency light, 390
Ni-Cd battery discharge, 653
optoelectronic power supply monitor, 584
overload for regulated supply, 709, 730
overload shutoff for 3-A regulator, 753
overvoltage in 5-V regulated supply, 735
overvoltage in two supplies with single scr, 651
parked-car power-drain reminder, 62
portable antimugging shocker, 343
power-failure with self-switched standby battery, 95
power-failure alarm and emergency light, 2
power-off flasher, 280
power-on reset for latch, 654
power-supply with turn-on delay, 652
power-supply turn-on delay, 652
programmed home lighting control, 100
pulsed standby battery for ram data, 493
radiation-induced noise blocker, 571
ram chip-select line during power failure, 494
regulator overload, 749
resetting logic after power interruption, 412
semiconductor overload, 652
setting wrong switches, 145
shadow-detecting alarm, 99
short-circuit in 15-V regulated supply, 715
short-circuit for 30-V IC regulator, 737
short-circuit for IC regulator, 746
short-circuit of regulator by switchback current limiting, 754
short-proof regulated power supply, 714
speaker against fast voltage buildup at switch-on, 716
storage-battery undervoltage warning lamp, 86
swimming pool splash alarm, 403, 404
transceiver antenna from lightning, 966
voltage-regulator against battery discharge, 651
voltage-regulator transistor with foldback, 650
woofers in hi-fi system, 49
(See also Alarm; Auto theft alarm; Burglar alarm; Photoelectric)
Proton resonance magnet stabilizer, pulse adder, 184
Proximity detector, capacitance-type, 103-107

Proximity switch, threshold switch module detects moving metal strip, 499
Pseudo-random binary sequence, fet switch and level changer, 858
Psychedelic:
  200-W sound-controlled lamp, 559
  audio-controlled fluorescent flasher, 281
  audio-controlled strobe flash, 637
  touch-contact vibrator tone generator, 245
  (See also Music-controlled light)
Public address:
  6-V megaphone, 660
  12-V megaphone, 657
  agc for mike preamp, 657
  background noise compensation, 658, 659
  megaphone using one transistor, 657
  mike beeper for tests, 655
  phasing speakers with a-f transmitter and receiver, 656
  portable 5-W megaphone, 657
  remote volume control, 761
  two-way megaphone, 658
  vu meter, 655
  (See also Audio amplifier; Audio preamplifier; Megaphone; Squelch)
Pulse adder, preventing cancellation or overlap, 184
Pulse amplifier:
  trigger for high-power scr, 989
  (See also D-c amplifier)
Pulse amplitude modulation, 4-channel multiplexer, 532
Pulse code modulation, seven-code converter, 200
Pulse converter, analog amplitude to Gray-code width, 165
Pulse discriminator, predictable constant backlash, 1006
Pulse divider, 2:1 to 30:1 for 1.1 MHz, 298
Pulse duration modulation, combined with f-m in 16-V regulator, 738
Pulse generator:
  50-A GaAs laser diode drive, 391
  1-hr spacing, 672
  0.5-Hz bipolar ujt, 664
  1 Hz-5 MHz with IC one-shots, 663
  1-Hz mvbr, 666
  1-10,000 Hz at 30 A for laser, 393
  10 Hz-100 kHz, 664
  50-Hz train of 1,200-Hz pulses, 991
  100-Hz 75-A krytron for laser, 395
  360-6,000 Hz micropower mvbr, 673
  1-kHz with 100-μs pulse width, 671
  1-75 kHz voltage-controlled, 671
  4-kHz 30-A for laser array, 396
  10-10,000 kHz with voltage-controlled duty cycle, 671
  20-kHz with 10-μs pulse width, 998
  500-kHz relaxation oscillator, 666
  1-MHz maximum with high prr stability, 673
  10-MHz relaxation oscillator, 666
  10-MHz variable using four inverters, 661
  20-MHz strobe, 663
  2.5-min spacing, 672
  15-ms strobe driven by clock, 667
  50-ns tunnel-diode low-drain, 668
  1-per-min for f-m repeater identifier, 768
  2-25,000 pps positive-spike, 666
  100-10,000 pps spike from sine wave, 665
  1-999 pulse-counting, 672
  300-V for laser research, 662
  adjustable duty cycle, 671
  a-f sine-spike converter, 665
  beeping timer for darkroom, 635
  cascaded monos generate pulse train, 665
  clock using Schmitt trigger with gate, 127

complementary square-wave outputs, 665
drive for d-c torque motor of roller pump, 486
electronic drum, 251
monostable or astable mvbr, 543
monostable mvbr with high duty cycle, 545
monostable mvbr with put timer, 670
predetermined number in burst, 179
programming for rom, 669
reset for logic at power turn-on, 412
ring counter reset, 153
single 3-ns test pulse, 153
single-pulse using latch and mono mvbr, 153, 670
space-mark with voltage-controlled duty cycle, 668
strobe with noise immunity, 662
transistor current gain measurement, 945
trapezoidal for auditory stimulator, 488
voltage-frequency converter, 166–168
zero-crossing using comparator, 1007
(See also Audio oscillator; Chopper; Clock; Code; Computer; Electronic music; Flash; Function generator; Inverter; Latch; Multivibrator; Noise; Oscillator; Radar; Remote control; Signal generator; Square-wave generator; Staircase generator; Sweep; Telemetry)
Pulse-height analyzer:
  500,000-pps with comparators and TTL, 1012
  four-channel, 1004
Pulse measurement, 0.7-V with lamp, 1010
Pulse repetition rate, doubling with IC, 310
Pulse separator:
  J-K flip-flops, 542
  noise-free pushbutton switch, 153
Pulse shaper:
  2:1 frequency divider, 297
  10-Hz squarer, 677
  1-MHz long-tailed pair, 677
  5-MHz sine-square converter, 674
  clipper using programmable zener, 676
  optically driven stretcher, 678
  Schmitt trigger for frequency counter, 288
  scs driver for scs ring counter, 674
  sine to square, 674–678
  triangle to sine, 678
  trigger sharpener for scr gate, 992
Pulse stretcher:
  50-s using opamp time-constant multiplier, 959
  multichannel data transfer, 543
Pulse subtractor, preventing cancellation or overlap, 181
Pulse width:
  measuring variations in mono mvbr, 936
  multiplying with fet one-shot, 543
Pulse-width converter:
  analog voltage input, 163
  multichannel using latch and one-shot, 543
Pulse-width modulator:
  1-Hz up using comparator, 512
  1-kHz adjustable, 509
  12-V d-c motor speed control, 517, 518
  12-V proportional voltage regulator, 745
  12-V switching voltage regulator, 745
  150-V 2-A switching regulator, 743
  500-V 1-kW switching regulator, 744
  45-W servo amplifier, 789
  analog-input, 877
  comparator with ramp generator, 512
  converter for 275 V to 50 V at 1 kW, 156
  decoder for radio remote control, 764
  switched integrator, 507
Pulse-width sorter, gates and one-shot, 333

Pulser, 500-V at 10 A for avalanche-diode modulator, 507
Punched card, optoelectronic motion detector, 586
Punched paper tape, photoelectric reader, 633
Puzzle, lamp control with diodes, 786

Q multiplier:
  455-kHz i-f, 536
  i-f image-rejecting, 350
Quadraphonic:
  50-W with one IC per channel, 684
  60-W amplifier, 680
  adding rear speaker to 2-channel system, 685
  automatic blending for SQ decoder, 684
  Columbia SQ decoder IC, 684
  decoder with 10-40 blending, 681, 682
  equalized IC preamp, 679
  four-channel stereo decoder, 681
  four-channel synthesizer for stereo, 680
  matrix adapter for two-channel stereo, 683
  SQ decoder, 683
  surround-sound from stereo records, 685
  synthetic from stereo with IC mixers, 682
  universal decoder, 685
  (See also Audio amplifier; Stereo)
Quadrature detector, light-beam communication, 587

Radar:
  10.69-GHz Gunn oscillator, 98
  150-Hz low-pass modulation filter, 688
  10-MHz clamp, 686
  30-MHz logamp for ppi display, 687
  binary r-f phase modulator, 689
  broadband phase-transparent limiter, 687
  doppler-radar tester, 690
  fast-switching three-state uhf modulator, 686
  ferrite phase shifter control for phased array, 687
  magnetron driver using pulse-modulator scr, 690
  nonblocking limiter for speed meter, 688
  self-detecting preamp for doppler, 690
  threshold detector for receiver agc loop, 689
  trigger drive for doppler, 688
  tripler for doppler, 688, 690
Radiation:
  Geiger counter, 423, 428
  high-gain preamp, 423
  preamp for solid-state detector, 423
  pulse height discriminator, 1009
  signal-shunting switch triggered by radiation, 571
  stable voltage reference, 1012
Radio:
  commercial-killing touch switch, 105
  light-controlled commercial killer, 918
Radio-frequency oscillator (see Oscillator, r-f)
Radio remote control, pwm decoder, 764
Radio teleprinter, automatic turn-on by call letters, 888
Radiometer:
  5-35 MHz frequency multiplier, 311
  35-70 MHz frequency multiplier, 310
  coherent filter, 465
  infrared preamp, 463
  surface temperature controller, 462
  YIG modulator, 463
Radiosonde, 1.68-GHz 1.9-W oscillator, 564
Radioteletype:
  afc for receiver, 891

Radioteletype (Cont):
  solid-state r-f and i-f circuits, 354
  (See also Teleprinter)
Railroad, steam whistle using three transistors, 345
Rain alarm, 1, 3, 4, 445
Raindrop simulator, sleep-inducing metronome, 251
Ramp and hold, low-drift, 851
Ramp generator:
  1/8th-Hz for sstv, 910
  15-Hz for sstv, 911
  pwm with comparator, 512
  solid-state cro for color tv, 112
  synchronous for a-d converter, 21
  (See also Sweep)
Ramp limit control, 0.33-1,000 Hz sawtooth generator, 852
Random-access memory, high-speed read-write, 495
Rat chaser, 19.5-kHz 15-W ultrasonic, 996
Rate-of-climb indicator, audible for glider pilots, 562
Ratio detector:
  f-m receiver, 307
  f-m telemetry, 876
Reactance modulator, f-m transmitter, 303, 305
Read-only memory, programming pulse generator, 669
Reading aid, tactile image for the blind, 489
Receiver:
  10-GHz d-c amplifier for field-strength meter, 453
  80-Hz narrow-pass filter for c-w, 812
  800-Hz filter improves c-w, 265
  1-kHz active filter for c-w, 262
  1.1-kHz active filter for c-w, 256
  2-kHz for modulated-light door lock, 234
  13-28 kHz Navy-station, 697
  20-kHz vlf for WWVL, 697
  45-55-65 kHz power-line carrier intercom, 377
  50-kHz marker generator with harmonics, 430
  500-1,600 kHz rejection filter, 266
  1.5-30 MHz frequency synthesizer, 320, 998
  7-MHz regenerative two-transistor, 704
  14-31 MHz converter for broadcast radio, 176
  26.995-MHz for remote control, 763
  28-MHz superregenerative detector, 694
  50-MHz and 144-MHz superregenerative, 700
  75-MHz marker beacon, 565
  80-150 MHz vhf for airport listening, 692
  105-120 MHz navigation using digital frequency synthesizer, 561, 564
  117-150 MHz aircraft-band, 698
  144-MHz superregenerative detector, 694
  150-MHz preamp, 20
  220-MHz fet mixer, 700
  220-MHz fet r-f stage, 700
  232-MHz local oscillator chain, 607
  2-meter cascoded jfet preamp, 692
  2-meter f-m, 302
  2-meter f-m oscillator and tripler, 607
  6-meter IC superhet, 700
  6-meter preamp, 56
  15-40-80 meter solid state plug-in converter, 169
  20-meter and 80-meter, 697
  40-meter and 80-meter mobile c-w, 701
  adapter for digital frequency counter, 291
  adding S meter, 965, 966
  afc for rtty, 891
  agc for c-w and ssb, 59

agc and S-meter, 59
a-m a-c transistor, 703
a-m five-transistor line-operated, 706
a-m/f-m i-f amplifier with detectors, 351
a-m hi-fi tuner, 694
a-m three-transistor, 705
a-m varactor-tuned using frequency synthesizer, 705
automatic bandwidth control, 353
automatic switching between channels, 771
background noise compensation, 658, 659
batteryless single-transistor a-m, 701
blanker for impulse noise, 569
broadcast-band stop filter, 266
carrier-operated relay, 859, 864
CB preamp, 122
CB tunable, 120
crystal-controlled detent tuner, 704
crystal diode, 707, 708
data line for zero input, 202
differential digital data, 202
digital frequency readout, 292
direct-conversion vhf for car radio, 170
direction-finding for hidden transmitter, 849
diversity audio combiner, 693
dual-conversion mixer for 18-30 MHz, 699
extending battery life with shunt electrolytic, 422
fixed-tuned WWV, 693
four-transistor, 695
frequency synthesizer for tuning, 321
fsk data decoder, 503, 504
fsk modem, 504
full-wave detector with 12-dB dynamic range, 691
hang agc for ssb and c-w, 58
headphone adapter, 51
headphone jack for CB, 122, 123
i-f and a-f for 80-40-15 meter plug-in converters, 355
i-f noise silencer, 569
i-f phase-transparent limiter, 687
image-rejecting i-f Q multiplier, 350
induction for museum commentary, 848
input attenuator with 3-dB steps to 33 dB, 701
instant vox control for transmitter, 963
Japanese a-m a-c, 707
Japanese a-m portable, 708
ladder-type input attenuator, 701
laser, 396
led zero-beat tuning indicator, 223
matching tuner to noncompatible i-f strip, 696
measuring a-m detector demodulation distortion, 447
measuring vhf sensitivity, 608
mixer for 30-MHz i-f output, 696
noise blanker, 572
noise generator for testing, 570, 572
oscillator using digital phase-locked loop, 295
panoramic 9-MHz i-f amplifier, 352
phase-locked local oscillator, 698
phase-locked loop, 696
phase-locked loop as rtty demodulator, 887
preselector improves selectivity, 703
product detector for ssb and rtty, 890
push-to-talk control with Styleline phone, 879
Q multiplier for 455-kHz i-f, 536
reciprocating detector for dssc, 692
reflex broadcast-band two-tube, 699
regenerative, 691, 693, 695, 702
relay control for a-m radio, 102
rtty transform for cro display, 895

self-oscillating Gunn mixer, 698
short-wave marker-pip generator, 433
signal tracer, 932-934
silencing auto radio commercials, 54
small-signal full-wave single-opamp, 580
solid-state r-f and i-f for ssb, c-w, and rtty, 354
speech filter, 269
storm-warning alarm, 1
sun-powered, 706
synchronous a-m, 705, 707
synchronous a-m detector using phase-locked loop, 692
synchronous detection for 144-MHz moonbounce, 702
trf broadcast using single IC, 703
trf using single IC, 699
tuning aid for the blind, 487
ultrasonic animal-range sounds, 995
updating communication i-f amplifier, 352
varactor-tuned a-m auto, 705
varicap-tuned a-m regenerative for headphones, 695
warble identification for crystal marker signals, 439
(See also Audio; Auto radio; Citizens band; Converter, radio; Frequency modulation; Intermediate-frequency amplifier; Limiter; Marker generator; Oscillator, r-f; Preamplifier; Remote control; Single sideband; Transceiver; Tuner)
Reciprocal circuit, analog divider using multiplier, 537
Reciprocal-function generator, one-quadrant multiplier-divider, 405
Record player (see Audio amplifier; Audio preamplifier; Phono; Quadraphonic; Stereo)
Recorder:
  ecg drive amplifier, 484
  frequency-response linear or log plotter, 934
  level indicator for stereo, 867
  speech compressor, 42
  S-Y servo drive, 793
  telephone pickup coil, 847, 850
  voice-operated turn-on, 848
  voltage-current converter, 367
  (See also Tape recorder)
Rectangular-wave generator (see Flip-flop; Multivibrator; Oscillator; Pulse generator; Square-wave generator)
Rectifier:
  100-kHz mV signals to d-c, 478
  3.2-V 20-A synchronous, 647, 649
  precision full-wave for a-c voltmeter, 475
  precision half-wave peak, 480
Reed switch, 2-kV pulse-actuated, 858
Regenerative receiver (see Receiver)
Regulated power supply:
  50-kHz for 50-kV voltmeter, 474
  1-kV 100-W linear d-c, 338
  2.5-kV at 2 kW, 340
  3.5-kV photomultiplier, 341
  −10-V at 5 mA, 1013
  −43-V at 100 mA, 713
  0-10 V at 0-500 mA short-proof, 714
  0-20 V at 1 A, 728
  0-50 V at 1-500 mA, 195
  0-250 V 0.1-A lab, 720
  0.5-31 V adjustable, 720
  0.7-30 V at 1 A, 728
  0.8-18 V adjustable, 724
  1-15 V at 500 mA, 744
  1.4-32 V at 420 mA, 725
  1.5-V replacing ohmmeter battery, 719
  2-36 V at 0-10 A, 722

Regulated power supply (Cont):
3/6/9/12-V at 100/130/240-mA current limits, 726
3.4-28 V d-c at 2 A for lab, 711
3.5-V at 1.5 A and 5-V at 0.5 A, 710
3.5-21 V at 1 A, 722
3.5-21 V at 1.5-A, 729
3.6-V at 500 mA and 6-V at 100 mA, 733
3.6-V at 700 mA, 726
3.6-V at 1.5 A, 714
3.6-V for sstv sync generator, 711
4-V and −7-V, 711
4.1-V at 35 mA, 712
5-V at 500 mA, 733
5-V at 1.5 A, 724, 731
5-V at 4 A using opamp, 716
5-V for digital afsk, 884
5-V for digital clock, 208
5-V for IC's, 725
5-V shunt-type, 741
5/9/180-V for frequency counter, 723
5/9-18 V bipolar for breadboards, 731
5/12-V for tone burst generator, 709
5/15-V at 40 mA, 723
6-V bipolar, 719
6-V dual-polarity servo, 721
6/12-V low-ripple for f-m, 714
6/15-V bipolar, 734
6-18 V current-limited, 724
6-24 V variable, 720
7-18 V at 500 mA, 724
9-V at 30 mA for portable radio, 725
9-V at 250 mA, 723
9-V from car battery, 716
9-V for f-m test set, 727
9-V for IC intercom, 713
9-V for solid-state projects, 732
9-V for transistor radio, 710
9-V using amplified zener, 1013
10-V at 300-mA, 722
10-12 V at 0.5 A, 729
10-25 V with programmable zener, 718
10-28 V at 300 mA, 730
10-34 V at 1 A, 730
11-V series-stabilized, 719
12-V, 718
12-V at 65 mA, 712
12-V for 80-20 meter receiver, 721
12-V from 6 V in converter, 162
12-V with automatic standby battery, 95
12-V alternator voltage regulator, 73
12-V bipolar, 711, 725
12-V dual, 712
12/24-V at 500 mA, 733
12/28-V bipolar, 721
12/37-V, 727
13-V for a-c operation of mobile transceiver, 723
13.5-V at 850 mA, 728
15-V at 100 mA, 727
15-V with battery discharge protection, 651
15-V for crystal standard, 726
15-V bipolar at 100 mA, 727
15-V bipolar at 250 mA, 731
15-V bipolar at 400 mA, 732
15-V bipolar with short-circuit protection, 715
15-V dual at 100 mA, 710
15/30-V with single-scr protection, 651
18-V at 100 mA, 717
18/30-V for stereo, 717
20-V at 1.5 A series-pass, 721
20-V at 4 A for $13, 734
20-V with 35-V unregulated, 729
22-V programmable-zener IC, 832
28-V at 250 mA, 728

30-V with 50 V unregulated, 732
34-V at 1 A protected, 709
40-45 V at 1 A, 731
47-V with short-circuit protection, 718
50-V at 3.5 A, 717
50-V fully protected, 715
60-V at 1.6 A, 718
60-V at 1.2 kW, 730
75-150 V at 20 mA using neon reference, 738
115-V 400-Hz using audio oscillator, 930
120-V d-c for enlarger lamp, 639
600-V scr series regulator, 713
constant-current reference diode, 1013
foldback protection against shorts, 650
optoelectronic control for high voltage, 337
slowing voltage buildup, 716
three-phase floating for phase-control triggers, 948
three voltages for logic, 726
tuned-choke ripple filter, 269
universal multi-voltage, 719
voltage splitter, 574, 725
(See also High voltage; Power supply; Regulator)
Regulator:
50-200 mA current for 50-V input, 194
1-A current source, 196
2-A IC, 198
2.1-A IC with switchback current limiting, 754
3-A negative shunt using IC, 751
3-A switching with overload cutoff, 753
0.1-100 Hz beat-note oscillator, 802
400-Hz 4-kVA alternator, 738
−50-V with floating bias for IC, 757
−15 V at 200 mA from 5 V, 155
−15-V series-shunt, 759
−10-V at 5 mA, 1013
−10-V high-stability at 1 A, 758
−5-V driven switching, 757
−5-V switching with current limiting, 755
0-150 V, 748
0.5-7 V at 15 W, 746
1-15 V at 500 mA, 744
1.6-V at 400 mA, 739
2-V using IC, 740
2-37 V high-stability IC, 751
2.5-5 V readout brightness control, 232
3-25 V tracking, 747
3.2-V at 2 A, 735
3.6-V for 6-V lantern battery, 736
5-V and −15-V tracking, 760
5-V at 5 A with current-sharing, 757
5-V combination switching and linear, 756
5-V driven switching, 752
5-V high-current switching, 753
5-V micropower, 756
5-V with overvoltage protection, 735, 756
5-V positive-negative IC switching, 754
5-V remote-shutdown, 737
5-V shunt-type, 741
5-V switching with current-limiting, 750
5/12-V, 743
5/15-V tracking, 758
6/12-V with bidirectional current flow, 736
6.8-V at 100 mA from 12 V, 741
7-30 V at 15 W, 746
9-V, 737–739
9-25 V at 100 mA using opamp, 746
10-V, 742, 752, 759
10-V using self-controlled zener, 1012
11-V for battery supply, 739
12-V at 2 A, 745
12-V and −6-V, 760
12-V d-c IC worst-case design, 735, 736

12-V with load on-off switching, 742
12.6-V, 742, 743
13/30-V for electronic organ, 747
15-V, 748, 754, 755, 759
15-V dual 100-mA with current limiter, 194
15-V dual-polarity IC, 748, 751, 758
16-V from 28 V using pdm f-m, 738
19.5-V d-c worst-case design, 735
20-V at 500 mA, 740
20-V adjustable reference, 1020
22-24 V for color tv, 747
28-V from 60 V, 744
28-V 100-W Darlington switching, 742
30-V with remote shutdown, 737
40-V switching, 749
48-V at 3 A, 741
50-250 V using 24-V zener, 756
75-V neon, 738
90-V rms for 500-W load, 649
100-V, 737, 746
105-V a-c output for varying line voltages, 645
120-V 600-W a-c, 648
150-V at 400 mA, 740
150-V pwm switching at 2 A, 743
185-V at 800 mA, 739, 741
250-V shunt using 90-V transistor, 736
290-V at 600 mA, 745
300-V 1.5-kW Darlington switching, 740
500-V 1-kW switching, 744
2-kV for converter, 338
2.5-kV scr for 2-kW supply, 340
100-W projection-lamp brightness, 630
automatic shutdown by short-circuit, 750
average load voltage by full-wave feedback control, 751
basic series-pass, 752
battery supply, 739
current with line resistance compensator, 760
current-limiting for class-B power amplifier, 10
current source with floating load, 195
current source with grounded load, 197
diodes permit parallel VR strings, 746
floating high-precision, 747
focus coil current control, 196
high output current at 15 W, 746
input-voltage protection, 760
lamp voltage, 390
light intensity using IC and phototransistor, 590
negative output voltage, 746
neon photomultiplier in starlight photometer, 449
oven temperature with switching current control, 927
photomultiplier supply, 337
reducing a-c line voltage, 647
series with overload protection, 749, 753, 755
switching for high-voltage input, 754
tracking positive-negative, 755
voltage for screen grid of crystal oscillator, 738
(See also Current control; Regulated power supply; Temperature control)
Relaxation oscillator (see Oscillator)
Relay:
10-A solid-state optoelectronic control, 588
1,800-Hz tone decoder, 880
5-s adjustable repetitive timer, 957
30-s delay using heater warmup, 654, 954
bilateral voltage-sensing, 861
carrier-operated, 3, 770, 774, 775, 859
delayed-dropout scr timer, 953
inverting driver for flip-flop, 146

Relay (Cont):
  phase-controlled, 858
  power-failure alarm, 2
  remote-control latching, 764–766
  r-f powered for transmitter, 862
  slow-kick wiper for VW, 74
  voice-controlled, 787
  voice-operated transmit-receive, 49
  voltage-level detector drive, 1004
  (See also Alarm; Auto theft alarm; Burglar
    alarm; Capacitance control; Door lock;
    Fire alarm; Lamp control; Latch; Motor
    control; Photoelectric; Remote control;
    Temperature control; Zero-voltage switch)
Remote control:
  100-1,000 Hz oscillator, 598
  650-Hz tone for f-m repeater, 776
  40-kHz ultrasonic receiver, 996
  8.5-12 MHz gated oscillator, 601
  26.995-MHz receiver, 763
  26.995-MHz transmitter, 763
  40-MHz ultrasonic tv, 920
  9-V remote battery for 500-W triac, 766
  500-W using bell wire to switch, 764
  15-channel digital, 763, 765
  6-digit IC lock, 235
  a-c push on and push off, 764–766
  audio amplifier gain with electronic
    attenuator, 766
  audio amplifier using IC, 761
  audio volume, 761
  automatic slide changer with audio tone, 53
  battery-operated push on and push off, 765
  camera shutter release, 637
  color tv optoelectronic with triac, 918
  commercial-killer with flashlight, 918
  dimmer for lamp load, 765, 766
  diode switching of crystals, 762, 764, 766,
    859
  flashlight-operated commercial killer, 918
  gain of audio amplifier, 765
  headlight-operated garage door, 766
  headlights turn on porch light, 629
  level of twin-T 1-kHz oscillator, 600
  line resistance compensator for current
    regulator, 760
  photoelectric using flashlight, 761
  power-line carrier system, 761
  pulsed radio or carrier, 762
  pwm radio decoder, 764
  slave photoflash trigger, 637
  sound-actuated, 762
  teleprinter turn-on by call letters, 888
  two-way d-c on single wire, 878
  ultrasonic color tv transmitter, 917
  ultrasonic receiver for tv, 918
  (See also Control; Motor control;
    Photoelectric; Receiver; Telemetry;
    Television, remote control; Transmitter)
Remote shutdown, 30-V regulator, 737
Remote tuning, 3.5-30 MHz varactor-tuned
  oscillator, 1002
Repeater:
  87-112 Hz continuous-tone encoder, 772
  107-Hz encoder for transmitter, 775
  107-Hz phase-locked loop IC decoder, 775
  1-kHz decoder, 771
  1-kHz encoder for f-m transmitter, 771
  1.75-kHz tone oscillator for mobile
    transmitter, 775
  2.3-kHz tone decoder, 774
  30-kHz window blocks over-deviation and
    off-frequency inputs, 776
  146.34-MHz portable transmitter, 301
  220-MHz f-m using varactor quintupler,
    301

448-MHz with four doubler stages for 7 MHz
    to 112 MHz, 772
  448-MHz using 7-MHz crystal oscillator, 773
  448-MHz 5-W power amplifier, 770
  448-MHz protection circuit for open or
    shorted antenna, 773
  448-MHz quadrupler stage, 768
  448-MHz regulated power supply, 727
  1-min timer for Morse-code identifier, 768
  3-10 min control timer, 776
  audio mixer, 771
  automatic call-letter generator, 154, 769
  automatic monitoring of two channels, 771
  break-in alarm, 379
  call-letter generator, 774
  carrier-operated relay, 770, 774, 775
  decoder for pulsing phone line, 772
  fast audio-decoder, 773
  f-m crystal calibrator up to 450 MHz, 438
  f-m tone-burst decoder, 768
  Morse-code clock, 215
  phone-line autopatch, 774
  remote control by telephone ringing, 880
  solid-state controller, 773
  split-site carrier-switching system, 770
  telephone-dial control for mobile f-m, 769
  timer for automatic turn-off, 770
  tone-burst keyer for f-m transceiver, 767
  tone decoder, 776
  Touch-tone control, 767
  transmitter dropout delay, 775
  turn-on 650-Hz tone generator, 776
Resistance:
  500-milliohm ohmmeter for soldered joints,
    457
  high-precision linear-scale ohmmeter, 459
  insulation measurement with follower
    voltmeter, 458
  linear bridge, 457
  low-value ohmmeter, 459
  low-voltage ohmmeter, 460
  measuring d-c meter movement, 460
  measuring with opamps in encapsulated T
    attenuator, 459
  measuring under 10 ohms, 460
  measuring with R-L-C bridge, 449
  megohmmeter for 1,000 megohms, 459
  seven-range Wheatstone bridge, 461
  teraohmmeter with medium precision, 458
  (See also Megohmmeter; Milliohmmeter;
    Multimeter; Ohmmeter; Teraohmmeter;
    Voltohmmeter)
Resistance box, four-resistor decade, 459
Resistance multiplier, inverting opamp, 541
Resonator, active in 2.125-kHz bandpass filter,
  200, 201
R-f amplifier (see Amplifier)
R-f oscillator (see Oscillator)
Rhythm generator, counter with tone
  generators, 241
Rhythm light, 300-W single-channel with mike,
  555
Ring counter:
  12-V chaser for ad signs, 279
  20-W sequential flasher lamp load, 279
  (See also Counter; Scalor)
Ring-of-two, 6.3-V voltage reference within
  0.08%, 1012
Ripple counter, bidirectional IC, 185
Ripple filter, 60-Hz active, 477
Ripple tester, d-c power supplies, 566
Root extractor, log converter and antilog
  generator, 817
Rotator, two-terminal network element, 166,
  168
Rtty (see Radioteletype)

Rumble filter:
  25-Hz switchable for stereo, 833
  45-Hz active, 47

Sample and hold:
  100-μs hold time, 780
  40,000 hold-sample ratio with series stages,
    778
  continuous integrator, 373
  counter with digital display, 783
  eliminating fet switch leakage, 783
  high-speed using fet opamps, 782
  jfet opamp reduces capacitance leakage,
    777
  low-cost using programmable opamp, 780
  low-drift opamp-fet, 782
  low-drift with jfet's, 783
  matched jfet's, 783
  offset and transient compensation, 781
  offset compensation, 780
  opamp zero correction to 3 μV, 581
  precision using fet opamp buffer, 781
  transient compensation, 781
  voltage subtractor for automatic tester, 782
  wide input voltage range, 777
Sampling:
  60-Hz reference to control 6-kHz vco, 777
  amplitude demodulator, 779
  automatic gate reset with zener, 331
  follow and hold, 778
  high-speed with current-amplifying switch,
    779
  high-speed with hold for a-d conversion, 779
  hold with continuous comparison, 782
  stereo decoder, 828
  track and hold amplifier, 779, 781
Satellite:
  ionosphere topside sounder, 877
  weather picture demodulator, 877
Sawtooth generator (see Sweep)
SCA, multiplex adapter for f-m receiver, 527,
  528
SCA adapter, squelch, 821
Scaling, summing IC amplifier, 580
Scalor:
  complementary driver with 0.25-ns rise time,
    192
  current, 194
  voltage with current gives power, 194
Schmitt trigger (see Trigger, Schmitt)
Science fair:
  audible light meter, 787
  beeper with meter, 788
  binary adding demonstrator, 147
  coin flipper, 784, 786
  dancing-light effect, 788
  flashlight motor control, 784
  lamp puzzle, 786
  lie detector, 785, 786, 788
  light-controlled toy, 785
  muscle-controlled whistle, 785
  mystery lamp-switching circuit, 784
  neon in flickering-flame effect with
    three-way lamp, 787
  sound-dimmed lamp, 787
  sunlight-controlled neon relaxation
    oscillator, 788
  transistor d-c motor, 786
  voice-controlled relay, 787
  (See also Hobby circuit)
Scope (see Cathode ray; Oscilloscope)
Scrambler:
  single-inversion for phone, 882
  telephone conversation, 882
Scratch filter, continuously variable, 621

Secrecy, speech scrambler, 882
Security:
    communication with binary r-f phase
        modulator, 689
    (See also Alarm; Auto theft alarm; Burglar
        alarm; Photoelectric; Protection)
Selective calling, solid-state squelch for CB
    receiver, 823
Sense amplifier, laminated ferrite-core
    memory, 490
Separator, J-K flip-flop for pulses, 542
Servo:
    80-400 Hz 29-W control amplifier, 790
    400-Hz supply using audio oscillator, 930
    6-V dual-polarity regulated supply, 721
    10-V 0.2-A X-Y recorder, 793
    7.5-W transformer-coupled amplifier, 792
    10-W a-c amplifier, 793
    10-W complementary-transistor amplifier,
        790
    10-W transformer-coupled amplifier, 789
    45-W pwm power amplifier, 789
    400-W speed control, 791
    closed-loop velocity system with tach
        feedback, 791
    frequency-locking with analog
        multiplier-divider, 792
    hysteresis opamp, 791
    loop phase comparator, 790
    overshoot control, 794
    phase-locked loop regulates speed to
        0.002%, 793
    photoelectric aperture control, 794
    r-f amplifier tank capacitor for automatic
        tuning, 792, 794
    saturating opamp with rate feedback, 791
    (See also Amplifier; Audio amplifier; Motor
        control)
Settling-time expander, d-a converter, 205
Shadow detector, photoelectric alarm, 99
Shaft encoder, binary optoelectronic, 590
Sharpshooting, 20-s beeping timer for rapid
    fire, 953
Shift register:
    4-MHz 400-bit dynamic, 200
    4-MHz mos delay line for temporary data
        storage, 202
    16-MHz mos delay line for temporary data
        storage, 128
    7-27 ns digital pulse delay, 148
    asynchronous transfer, 149
    divide-by-seven counter, 193
    double sequence generator, 147, 150
    five-digit strobe display, 220
    four-channel analog commutator, 21
    odd-length double-clocked, 146
    odd-length mask-programmed, 147
    pseudo-random switching of Christmas tree
        lights, 275
    serial-parallel converter for data, 199
    testing with predetermined-number pulse
        burst generator, 179
    (See also Digital clock; Logic)
Shocker, 6-V portable antimugging, 343
Signal generator:
    0.01-4,000 Hz sine-square-triangle, 325
    0.1-100 Hz beat-note, 802
    0.1 Hz-30 MHz a-m/f-m/sweep, 796
    1 Hz-5 MHz a-m/f-m/c-w/fsk/psk, 323
    1.2 Hz-1,000 kHz sweep, 851
    2 Hz-20 kHz five-range opamp, 801
    3.5 Hz-1.2 MHz sine-square-triangle, 323
    5-60,000 Hz sine-wave, 799
    6 Hz-60 kHz precision square-wave, 802
    10-4,990 Hz square-wave, 798
    15 Hz-200 kHz Wien-bridge, 800

100-1,000-10,000 Hz amplifier linearity
    checker, 801
100-10,000 Hz linear sweep, 798
400-Hz twin-T feedback, 799
750-Hz audio, 138
800-Hz and 2-kHz for two-tone test, 802,
    803
1-kHz audio testing, 799
1-kHz Wien-bridge opamp, 591
1.3-kHz a-f amplifier signal injector, 802
30.5-42.5 kHz wobbulator for tv i-f
    alignment, 908
400 kHz-30 MHz five-band sweep, 855, 857
455-kHz modulated constant-output, 796
455-kHz modulated i-f, 796
1-MHz weak-signal source, 797
1.8-450 MHz single-transistor crystal, 796
10-MHz frequency synthesizer, 317
28-MHz i-f alignment, 799
144-MHz for testing 2-meter amplifiers, 797
144-148 MHz aligner for f-m receiver, 798
a-f and r-f for a-m radio troubleshooting,
    795, 800
a-f and r-f injector, 800
a-f and r-f using mvbr, 800
a-f adjustable ujt, 180
audio-tone neon, 137
band-edge marker, 801
f-m/a-m transmitter-receiver aligner, 798
linear ramp, 856
r-f modulated, 800
rtty for transmitter modulation check, 887
swept audio from swept and fixed r-f, 795
tone-burst, 795, 797
Wien network with automatic amplitude
    control, 801
(See also Function generator; Marker
    generator; Noise generator; Pulse
    generator; Square-wave generator;
    Staircase generator; Sweep; Tester)
Signal tracer, a-m radio troubleshooting, 800,
    932-934
Silicon-carbide varistor, nonlinearity index
    meter, 943
Silicon controlled rectifier, tester, 935, 940
Simulation, neural analog, 482
Simulator:
    blood pressure, 485
    inductor with opamp circuit, 37, 40
Sine, approximating with log module and
    opamp, 408
Sine-square converter, frequency-doubling,
    311
Single sideband:
    440-Hz tone generator for netting, 809
    800-Hz and 2-kHz two-tone test generator,
        803
    2-kHz audio filter with product detector, 963
    1.5-30 MHz 25-W amplifier, 808
    2-30 MHz 80-W linear amplifier, 806
    2-50 MHz 25-W linear amplifier, 805
    9-MHz filter-type generator, 810
    30-MHz 80-W linear amplifier, 810
    127.333-MHz oscillator and solid-state
        amplifier, 812
    220-MHz converter from 28 MHz, 811
    220-MHz power amplifier, 979
    382-MHz oscillator chain, 809
    432-MHz linear amplifier, 806
    432-MHz mixer, 806
    10-meter 12-W transmitter, 977
    10-meter 15-W single-transistor amplifier,
        804
    6-W pep class-A broadband driver, 812
    50-W pep class AB broadband amplifier,
        811

afsk tone generator for rtty, 892
approximation with double-sideband
    suppressed-carrier, 975, 977
audio fsk oscillator for radioteletype, 886
audio phasing network, 810
aural tuning aid for the blind, 488
automatic keyer with sidetone for ssb, 137
balanced IC modulator, 510, 804
beat-frequency oscillator, 809
bfo for a-m CB transceiver, 121, 123
bfo injection for transceiver, 964
biased-diode antenna switch, 959
cathode follower improves agc action, 59
compressor and clipper for transmitter, 399
crystal lattice filter, 809
digital 90-deg phase shifter, 614
diode-controlled transceiver filter switching,
    966
Dome 90-deg phase shift circuit, 808
frequency divider for 2-MHz crystal, 805
hang agc circuit, 58
i-f amplifier updating, 352
i-f single-IC, 804
product detector with high sensitivity, 810
product detector using IC balanced
    modulator, 807
product detector used also for rtty and c-w,
    890
r-f speech processor, 807
solid-state r-f and i-f circuits, 354
speech compressor, 45
    and clipper, 41, 44
speech processor, 43
switching regulator as modulator, 804
transceiver monitor, 807
transmitter automatic load control, 810
Sinh^{-1} function, synthesizing with antilog
    transconductors, 411
Sink, 1-nA precision current using fet opamp,
    196
Siren:
    all-purpose alarm, 2
    automatic rising and wailing, 815
    automatic self-cycling, 815
    bicycle two-transistor, 813
    burglar alarm or bicycle, 813
    combined with variable-rate clicker, 815
    earphone for child's toys, 814
    electronic chirper, 815
    portable with speaker, 816
    pushbutton-operated two-transistor, 814
    twee-dell twee-dell two-tone, 814
    two-transistor, 813
    Waa-Waa alarm, 813
    wailing, 813, 816
    (See also Alarm; Auto theft alarm)
Slide changer, automatic control with audio
    tone and triac, 53
Slot car:
    photoelectric win detector, 347
    speed and braking control, 346
Slow-scan television:
    ⅛th-Hz ramp generator, 910
    15-Hz ramp generator, 911
    adapter for cro as monitor, 906
    flying-spot scanner module, 909
    light-bar tuning indicator, 910
    Plumbicon camera circuits, 915
    signal analyzer, 912
    sync pulse generator, 908, 910
    test generator for flying-spot scanner, 910
S-meter, communication receiver, 59
Snoring, wakeup alarm, 484
Solar cell:
    high-current inverter, 385
    infrared receiver, 357

Soldered joint, measuring with low-range ohmmeter, 457
Soldering iron, temperature control for heat drift test, 463
Solenoid:
  control with thyristor, 859
  liquid level control valve, 404
  scr impulse driver, 989
Sonar:
  remote-control receiver for tv, 917, 919
  (See also Television, remote control; Ultrasonic)
Sorter:
  apples to five sizes, 631
  pulse width, 333
Sound control:
  adjustable release time, 762
  telephone-bell amplifier, 881
Sound effects (see Alarm; Electronic music; Circuit breaker; Hobby; Oscillator; Siren)
Sound-level meter, 20-20,000 Hz with third-octave-noise filter, 450
Sound pollution:
  6-kHz demonstration, 995
  three-level monitor, 567
Sound pressure meter, loudspeaker enclosure design, 443
Sound track, optical photoelectric readout, 631
Spacecraft, magnetometer amplifier, 447
Space-mark generator, voltage-controlled duty cycle, 668
Speaker (see Loudspeaker)
Spectrum analyzer, rtty transform for cro display, 895
Speech amplifier (see Audio amplifier)
Speech compressor (see Audio compressor)
Speech control (see Audio compressor; Audio control; Limiter; Voice control)
Speech processor, ssb transmitter, 43, 807
Speech recognition, parameter extractor, 51
Speed alarm, auto, 63
Speed control (see Motor control)
Spike, measurement in power supplies, 566
Splitter, voltage for regulated power supply, 725
Square-law:
  true rms operation, 820
  vector length computer, 818
Square-law function, signal-controlled opamp, 1008
Square root:
  computing with opamps, 820
  square-law module with opamp, 820
  sum of squares, 820
  of sum of squares, analog module with opamps, 819
  vector difference between squares, 820
Square-root extractor, log converter and antilog generator, 817
Square rooter, analog using single IC, 819
Square wave:
  converting from triangle with integrator, 677
  doubling prr with IC, 310
Square-wave generator:
  0.002-0.25 Hz, 324
  0.01-4,000 Hz, 325
  0.3 Hz-10 MHz, 673
  1-Hz IC for flasher, 275
  1-Hz up pwm, 512
  1-30 Hz astable mvbr, 546
  1 Hz-100 kHz with triangle, 327
  2 Hz-2 MHz, 661
  2 Hz-20 kHz five-range opamp, 801
  2.5 Hz-250 kHz with triangle, 328
  5 Hz-50 kHz, 664
  6 Hz-60 kHz with 1% accuracy, 802

10 Hz-4.99-kHz, 798
10-10,000 Hz opamp, 667
20-5,000 Hz sweep, 329
20-100-000 Hz sine-wave clipper, 677
50-15,000 Hz for sine-wave input, 675
60-Hz counter demonstrator, 293
100-Hz adjustable, 324
100-Hz free-running mvbr using opamp, 552
100-1,000-10,000 Hz amplifier linearity checker, 801
120-3,000 Hz, 548
1-kHz astable mvbr, 551
1-kHz opamp, 670
1-500 kHz with subharmonic sync, 661
1.005-kHz with sine-wave input, 676
7-kHz astable mvbr, 550
10-kHz in solid-state cro, 108
10-kHz vco, 1000
10-50 kHz voltage-controlled, 1001
10-100 kHz divided from 200-kHz crystal, 434
20-kHz with 10-μs width, 992
30-kHz crystal-controlled, 297
50-kHz and 100-kHz crystal in marker generator, 437
50-kHz high-stability, 663
100-kHz crystal calibrator, 440
100-kHz crystal scs, 608
0-1 MHz sine-square converter, 677
1-MHz crystal mvbr, 439
1-28 MHz crystal using TTL logic, 549
2-MHz using IC, 670
4-72 MHz, 668
5-MHz astable mvbr, 550
5-MHz sine-wave input, 674
50-MHz high-stability, 1001
10-100 μs mono mvbr, 548
2-min using mosfet mvbr, 551
add-on for audio sine-wave generator, 817
adjustable-width using programmable zener, 667
a-f and r-f signal injector, 800
a-f Schmitt trigger, 990
current-differencing opamps, 330
delay-line coupling stabilizes frequency, 547
frequency-doubling for sine input, 311
linear opamp, 664
monostable crystal, 549
nine-over tone clock-driven, 125
sine-square converter, 674
single-supply opamp, 667
squaring sine waves with scs, 675
symmetrical low-hysteresis trigger, 992
triggered fuzz for guitar, 250
variable-frequency, 328
variable rate and duty cycle, 669
variable rate and width, 664
variable width and duty cycles, 662
(See also Flip-flop; Function generator; Multivibrator; Oscillator; Pulse generator)
Square-wave shaper, 10 Hz-100 kHz audio oscillator, 676
Squarer:
  10-Hz IC, 677
  1.005-kHz for sine waves, 676
  0-1 MHz sine-square converter, 677
  5-MHz sine waves, 674
  1-V sine to 10-V square, 678
  4-bit binary number, 818
  analog using single IC multiplier, 817
  current-differencing opamp with hysteresis, 818
  filterless dual-gate mosfet, 310
  frequency-doubling with IC multiplier, 818
  inverter and clamp for signal generator, 678

rms converter for analog input, 820
root extractor using log converter, 817
root-mean for rms voltmeter, 474
sine-wave audio generator, 678
single-quadrant, 817
two-quadrant, 819
Squelch:
  add-on circuit for solid-state vhf f-m receiver, 823
  adjustable-width using programmable zener, 667
  a-f voice amplifier, 822
  agc-controlled for receiver a-f, 823
  audio preamp with hysteresis, 823
  audio two-fet transient-free, 822
  audio two-transistor bipolar, 822
  carrier-operated relay for f-m communication set, 822
  f-m high-sensitivity, 821
  narrow-band f-m demodulator, 304
  noise-actuated for f-m pager, 822
  noise-controlled for f-m receiver, 823
  SCA adapter for f-m, 821
  signal-operated opamp relay, 822
  solid-state replacing relay for selective-calling CB receiver, 823
Stabilization:
  bootstrapping in 60-Hz ujt oscillator, 592
  diodes in fet amplifier, 12, 15
Staircase generator:
  5-mV steps for digital voltmeter, 824
  5-step using tunnel-diode frequency divider, 827
  8-step, 826
  counter decade divider, 825
  d-a converter display, 203
  decade counter with decoder/driver, 824
  free-running four-opamp, 826
  low-drift analog, 826
  precision for sequential control, 825
  up-down, 826
Stall, protection for d-c motor, 270
Standard cell, calibrating digital voltmeter up to 1,000 V, 480
Standing-wave ratio:
  indicator for 2-W c-w transmitter, 449
  meter for low-power transmitter, 452
Stapling gun, power one-shot trigger, 988, 990
Star tracker, photodiode with opamp, 563
Static, storm-warning alarm, 1
Staticiser, bistable store for counter, 189
Steam whistle, three-transistor, 345
Stepper:
  conversion from synchronous motor, 523
  torque booster for four-phase 28-V, 516
Stepping motor, translator for clock input, 515
Stereo:
  20-18,000 Hz adjustable notch filter, 265
  18-V and 30-V regulated supply, 717
  3.7-W a-f amplifier without heat sinks, 836
  5-W IC amplifier, 844
  5-W portable phonograph, 626
  10-W solid-state amplifier, 839
  15-W low-cost amplifier, 838
  20-W class AB amplifier, 842
  30-W high-quality power amplifier, 833
  30-W or 60-W for 2 or 4 channels, 680
  50-W power amplifier, 829
  75-W rms audio amplifier per channel, 837
  80-W hi-fi transistor amplifier, 829
  adding rear speaker to 2-channel system, 685
  aural null indicator, 842
  auxiliary preamp for wide input level range, 844

Stereo (Cont):
  balance control, 835
    and mono-stereo switch, 843
  balance meter, 832, 840
  Bauer cross-feed network for headphones, 836
  buzzer or lamp indicator, 843
  ceramic-cartridge Darlington preamp, 832
  ceramic-cartridge preamp, 621
  ceramic-filter f-m i-f amplifier, 837
  decoder for f-m multiplex, 831
  decoder using IC phase-locked loop, 832
  decoder using sample-and-hold, 828
  dual-supply phono preamp, 625
  electroencephalophone modulator, 482, 487
  fet phono preamp, 625
  f-m multiplex adapter, 830
  f-m multiplex decoder, 834, 836, 838
  f-m multiplex demodulator, 839
  f-m tuner using dual-gate fet's, 845
  f-m tuning meter for IC, 829
  four-channel decoder, 681
  four-channel matrix enhancer adapter, 683
  four-channel synthesizer, 680
  headphone preamp, 838
  high-frequency active filter, 51
  IC adapter for standard f-m receiver, 840
  image width control, 834
  indicator lamp drive, 839
  low-distortion tone control, 54
  low-noise preamp, 841, 844
  magnetic-pickup preamp, 624
  microphone preamp with built-in vu meter, 655
  mixer for synthetic four-channel, 682
  monitor amplifier for phones or speaker, 842
  muting power turn-on transients, 50
  network improves spatial effect with phones, 829, 840
  phase-locked decoder, 834
  phono preamp with switched equalization, 621
  preamp: for high-output ceramic cartridge, 622
    for low-impedance mike, 840, 843
    for medium-impedance mike, 831
    for radio and ceramic cartridge, 828
    with tone controls, 830
  push-pull amplifier for 600-ohm line, 841
  rumble filter and tone control, 833
  scratch filter for phone, 621
  single-ended amplifier for 600-ohm line, 841
  single-supply phono preamp, 835
  sound source with control, 835
  tape playback amplifier, 869
  tape recorder bias-erase oscillator, 864, 873
  tape recorder bias-erase power from common oscillator, 866
  tape recorder level indicator, 867, 869
    stereo tape recorder, 865
  tone-balance control, 624
  tone controls, 832
  virtual-ground mixer, 53
  volume expander, 50
  vu meter, 832
  (See also Audio; Audio amplifier; Audio preamplifier; Music-controlled light; Phono; Quadraphonic; Tape recorder)
Stethoscope:
  contact mike with a-f amplifier module, 850
  nonmedical for vibration analysis, 847
Stirrer, magnetic variable-speed triac-controlled, 523
Storage (see Memory)
Storm, warning alarm, 1

Strobe:
  20-MHz with 10-ns pulse widths, 663
  adjustable and audio-controlled, 637
  music-controlled fluorescent, 283
  noise-immune four-gate, 662
  sound-fired for flash photography, 639
  two-flashtube, 640
  (See also Flash)
Strain gage, noninverting chopper-stabilized opamp, 117
Subharmonic sync, square-wave generator, 661
Subtracter:
  bcd using adders and inverters, 151
  full using gates, 335
  opamp with resistive feedback, 575
  pulse in magnetic field stabilizer, 181
  voltage in sample-and-hold automatic tester, 782
Subtracting amplifier, thermistor bridge, 922
Summer:
  bias-current compensation, 577–579
  closed-loop IC opamp, 582
  differential-input opamp with bias-current compensation, 581
Summing:
  binary adding machine, 147
  IC video amplifier with scaling, 580
Summing amplifier:
  100-kHz directly coupled, 16
  10-MHz feedforward-compensated opamp, 577
  analog multiplier-divider, 406
Superregenerative, 50-MHz and 144-MHz receiver, 700
Superregenerative detector:
  28-MHz, 694
  144-MHz, 694
Suppressed carrier, amplitude modulator, 513
Surface temperature, measurement, 462–465
Surveillance:
  a-m wireless mike, 849
  compressor for tape recorder, 847
  eavesdropping with crystal mike in parabolic reflector, 846
  f-m wireless mike, 846, 849
  induction receiver, 848
  stethoscope for walls, 847, 850
  telephone monitor, 847, 850
  ultra-low-noise a-f amplifier, 846
  voice-controlled recorder turn-on, 848
Sweep:
  0.33-1,000 Hz with ramp limit control, 852
  1 Hz-150 kHz linear cro time base, 855
  1.2 Hz-1,000 kHz IC, 851
  10 Hz-10 MHz integrator-shaper-agc, 372, 374
  10 Hz-10 MHz linear ramp generator, 855
  10 Hz-10 MHz ring counter, 183
  10 Hz-10 MHz vco for ramp, 857
  20-5,000 Hz sine-square-triangle, 329
  25 Hz-30 kHz linear, 856
  100-3,000 Hz constant-amplitude, 372
  1-3 kHz high-linearity, 857
  100 kHz to 60 MHz, 856
  400 kHz-30 MHz, 855, 857
  455-kHz and 1.65-kHz for i-f response on cro, 852
  2-20 MHz vco, 1002
  7-MHz and 14-MHz adjustable, 856
  10 μs-10 s triggered by 100 ns pulse, 853
  5-s constant-current ramp generator, 854
  100 s/V with 0.7 s/V reset, 856
  capacitance meter, 421
  chopper-discharged, 853
  linear ramp, 854, 856
  low-drift with hold, 851

  pulse width and spacing control in vco, 1003
  silicon planar transistor as zener for ramp, 853
  triggered for any cro, 851, 854
  triggered for lab scope, 853
  variable-phase trigger, 989
  voltage-controlled ramp, 857
  voltage-frequency converter, 168
  (See also Integrator; Pulse generator)
Sweep calibrator, blocking oscillator and gate, 877
Swimming pool, splash alarm, 403, 404
Switch:
  1-A solid-state switch, 133
  2-kHz audio-controlled, 48
  1-s turn-off delay, 958
  2-kV reed actuated by pulse, 858
  180-W touch-plate, 105
  450-W touch-plate, 103
  32-channel multiplexer with 5-bit control, 527
  a-c latching using tone-triggered triac, 54
  analog multiplexer, 528, 531
  audio-controlled transceiver battery-saver, 964
  balanced-modulator for tone-burst generator, 795
  bidirectional fet, 862
  bilateral voltage-sensing, 861
  bistable using sus pair, 860
  bistable flashlight-operated, 761
  carrier-operated relay, 859
  complementary-load, 861
  contact bounce suppression, 152, 154
  current-amplifiying for fast sample-hold, 779
  d-c voltage-controlled diodes in transceiver, 966
  differential multiplexer, 528
  digital-analog ladder with integrated chopper, 117
  diode for remote changing of crystals, 762, 764, 766, 859
  dpdt solid-state, 860
  drive for d-a converter ladder network, 206
  gate turn-off scr for Nixie driver, 224
  H for inertial-guidance motor, 560
  high-frequency jfet, 861
  high-speed replacement for triac, 861
  jfet in 5-band sweep generator, 855
  knock-over with scr and Sonalert, 96
  light-triggered 6-kV series scr, 343
  multiplex analog using single fet per channel, 534
  multiplier with voltage level changer, 858
  noise-free pushbutton for pulse separator, 153
  phase-controlled scr, 858
  power-transformer tap-changing with triacs, 862
  prolonging contact time, 859
  Q for laser using krytron and Pockels cell, 391
  quad current for 12-bit dac, 207
  quint current for 10-bit dac, 204
  radiation-triggered signal-shunting, 571
  r-f powered for transmitter, 862
  saturating cascode, 863
  series-shunt fet analog, 859
  shunt for digitally set pot, 863
  solenoid control with thyristor, 859
  stepper for pulsed remote control, 762
  summing junction as chopper for modulator, 512
  symmetrical low-hysteresis, 992
  three-way with neon indicator, 862
  tone-controlled for c-w and rtty, 139
  touch-plate, 102, 105, 107

Switch (Cont):
    touch-to-talk transmitter, 104, 105
    triac operated by light, 629
    triac operated by NAND gate, 861
    uhf radar modulator, 686
    video gate with transient suppression, 336
    voice-controlled for ssb generator, 48
    voice-operated, 46
    (See also Alarm; Auto theft alarm; Burglar
        alarm; Capacitance control; Door lock;
        Fire alarm; Integrator; Lamp control;
        Latch; Motor control; Photoelectric; Remote
        control; Temperature control; Zero-voltage
        switch)
Switched current amplifier, voltage to
    pulse-width converter, 164
Switching regulator:
    5-V, 750–754, 756
    10-V at 3 A, 752
    15-V at 500 mA, 754
    28-V 100-W Darlington, 742
    150-V 2-A pwm, 743
    300-V 1.5-kW Darlington, 740
    500-V 1-kW, 744
    100-W mercury-arc ballast, 281
    400-W mercury-arc ballast, 285
    converting 5 V to −15 V, 155
    high-voltage input, 754
    short-circuit protection above 3 A, 753
    (See also Regulator)
Switching waveform generator, dual-trace cro,
    109
Sync separator, noise canceller, 910
Synchronizer, tone-burst generator, 797
Synchronous detection, 144-MHz moon-bounce
    signals, 702
Synchronous detector, a-m using phase-locked
    loop, 692
Synchronous motor, conversion to stepper, 523
Synchronous rectifier:
    3.2-V 20-A full-wave, 649
    3.2-V 20-A half-wave, 647
Synthesizer:
    four-channel stereo, 680
    (See also Frequency synthesizer)

T attenuator, measuring resistor with opamps,
    459
Tachometer:
    60-Hz frequency meter for emergency a-c
        generator, 434
    0-8,000 rpm auto, 68
    0-12,000 rpm noncontacting, 64
    auto, 67, 71, 73
        with dwellmeter, 66
    d-c motor speed control, 522
    neon-lamp calibrator, 68
    voltage feedback for thyristor trigger, 521
Tape deck, theft alarm, 82, 85
Tape player:
    Japanese portable, 869
    motor speed control, 868, 873
    (See also Tape recorder)
Tape recorder:
    60-Hz active hum filter, 867
    9-V regulated supply with negative output
        resistance, 739
    4-W four-speed, 871
    45-dB audio compressor, 45
    audio compressor, 44, 45
    audio control, 49
    automatic gain control, 867
    automatic level control, 55
    automatic phone message repeater, 883
    bias-erase oscillator, 864, 870, 872, 873

carrier-operated relay, 859
cassette Japanese portable, 868
cassette for 6 V or a-c, 871
common oscillator for bias and erase power,
    866
compressor for conversation groups, 847
digital gain control, 870
dynamic noise limiter for cassettes, 865
echo-chamber effect, 872
gain-controlled amplifier minimizes noise,
    866
Japanese 3-inch reel-type, 866
level indicator for stereo, 865
loss-free mixer, 873
multicoupler, 872
noise-reducing active filters, 867
noise-reducing logarithmic detector, 864
playback preamp, 867, 872, 873
recording amplifier, 873
speed control, 865
speed control by changing power frequency,
    864
stereo level indicator, 867
stereo playback amplifier, 869
stereo recording level indicator, 869
tactile level indicator for the blind, 482
tone-controlled switch, 139
tone keyer for transmitter, 131
voice-operated control, 46, 52, 848
Telefax, modification for amateur service, 959
Telegraph sounder, radio-controlled, 132
Telemetry:
    60-Hz discriminator, 876
    audio amplitude leveler, 46
    converter for satellite ionogram signals,
        875
    f-m demodulator using IC phase-locked
        loop, 876
    four-channel f-m for athletes, 874, 875
    ionosphere topside sounder, 877
    IRIG channel 13 demodulator, 874
    pam 4-channel multiplexer, 532
    pwm decoder, 764
    ratio detector, 876
    rezeroing opamp, 877
    sine-square converter, 674
    temperature-frequency converter, 878
    two-way d-c on single wire, 878
    weather satellite picture demodulator, 877
Telephone:
    700-Hz bell-tone oscillator, 347
    1.8-kHz Touch-tone decoder, 880
    2.125-kHz active bandpass filter for data
        line, 200, 201
    automatic level control, 55
    automatic message repeater, 883
    autopatch for f-m repeater, 774
    bell amplifier, 881
    CB transceiver patch, 880
    conversation scrambler, 882
    data line receiver for zero input, 202
    d-c isolator for radio patch, 880
    decoder for pulsing phone line with f-m
        repeater, 772
    dial control of f-m repeater, 769
    fast audio decoder for f-m repeater
        autopatch, 773
    fsk modem, 500
    ham phone patch, 879, 881, 883
    mobile f-m repeater control with Touch-tone
        pad, 767
    obscene-call squelcher, 882
    patch using bridged-T filter, 882
    patch using resistive bridge, 879
    pickup-coil amplifier, 847, 850
    push-to-talk patch, 880

push-to-talk transmitter control in Styleline,
    879
remote control of repeater by ringing signal,
    880
ringer light and remote bell, 883
ringing pickup for power-line carrier
    intercom, 376
scrambler providing single inversion, 882
serial data transmission at 20 MHz on wire
    line, 201
tone oscillator for fsk modem, 504
Teleprinter:
    1.275-1.445 kHz bandpass filter, 888, 892
    1.275-2.125 kHz bandpass filter, 886, 892,
        893
    2.025-kHz and 2.225-kHz fsk demodulator,
        199
    2.125-2.975 kHz bandpass filter, 887, 893
    6.375-kHz afsk notch filter, 890
    afsk encoder, 885
    afsk keyer, 890
    afsk tone generator, 888
    ASCII encoder, 151
    automatic message generator, 129
    autostart decoder, 889
    autostart for unattended rtty, 892
    call-letter monitor, 888
    constant-current selector-magnet drive, 887,
        892, 893
    crystal-controlled fsk tone generator, 894
    demodulator using phase-locked loop, 887
    digital active filter, 891
    digital afsk signal generator, 884
    fsk converter using phase-locked loop, 886,
        893
    fsk demodulator for 2.025 kHz and 2.225
        kHz, 501, 505
    keyboard-operated transmission, 889, 893
    monitor scope, 885
    phase-shift monitor scope, 885
    product detector used also for ssb and c-w,
        890
    rtty signal generator, 887
    signal analyzer, 912
    space-mark generator with voltage-con-
        trolled duty cycle, 668
    synchronous-phase audio fsk oscillator, 886
    tone-controlled switch, 139
    transform for cro display, 895
    tuning indicator, 889
    two-pair tone generator for afsk, 892
    universal keyboard encoder, 148
    (See also Frequency-shift keying; Modem)
Television:
    30-MHz gamma-matching logamp, 687
    40-MHz high-pass tvi filter, 268, 269
    2,000-VA horizontal output, 914
    2-W audio amplifier, 911
    a-c line voltage regulator, 645
    agc keyer for i-f, 909
    autodyne uhf tuner, 907
    camera preamp, 906, 907
    CATV distributed vhf amplifier, 910
    closed-circuit infrared scanner, 468
    color: 3.58-MHz subcarrier as frequency
        synthesizer reference, 317, 321
        3.58-MHz vco, 902
        22-24 V regulated supply, 747
        1-W sound system, 900
        2-W audio amplifier, 897
        agc keyer-amplifier, 901
        auxiliary pulse generator, 896
        chroma processor, 898, 899
        Colpitts subcarrier oscillator, 898, 899
        Colpitts-Pierce subcarrier oscillator, 903
        compensated luminance drive, 904

Televison, color (Cont):
  crystal ringing subcarrier regenerator, 897, 903
  difference amplifier, 896, 897
  difference amplifier linearity measurement, 905
  discriminator-type phase detector, 905
  injection-locked oscillator, 902
  luminance amplifier, 904
  luninance drive with 3.8-MHz trap, 901
  matrixing with demodulation transformer, 896
  optoelectronic remote control with triac, 918
  picture-tube flashover protection, 904
  product-type phase detector, 902
  reference oscillator using IC's, 900
  Rennick subcarrier oscillator, 896
  RGB amplifier, 903
  solid-state cro, 108, 109, 112
  solid-state time-base trigger, 114
  sound i-f using opamp, 901
  sync signal processor, 900
  tuning-eye drive, 902
  ultrasonic remote-control transmitter, 917
  vertical preamp for cro, 110
  video i-f amplifier, 898, 899
commercial-killing touch switch, 105
final video i-f and detector, 914
f-m sound i-f and demodulator, 912
focus coil current regulator, 196
gain-controlled video i-f, 909
horizontal drive using single IC, 914
locked-oscillator detector and limiter for sound, 911
noise canceller for sync separator, 910
photographing integral number of tv fields, 636
Plumbicon slow-scan circuits, 915
remote control:
  flashlight-operated, 918
  light-controlled commercial killer, 918, 919
  photoelectric commercial killer, 919
  sound-activated relay kills commercials, 919
  two-wire for power, 918
  ultrasonic, 917–920, 923
series limiter, 399
shunt limiter, 399
slow-scan sync generator, 912
sstv ⅛th-Hz ramp generator, 910
sstv 15-Hz ramp generator, 911
sstv monitor cro, 906
sstv scanning crt, 909
sstv signal analyzer, 912
sstv sync pulse generator, 908
sstv tuning indicator, 910
test generator for sstv, 910
thin-film hybrid video i-f, 913
triggered spark gap tests flashover protection, 339
tuner agc delay, 914
varactor vhf tuner, 913
vertical deflection, 914, 916
vhf mixer, 911
video amplifier for camera, 911
video i-f with low-cost agc, 907
wobbulator for i-f alignment, 908
Temperature compensation:
  400-μA current source, 195
  300-3,000 MHz power transistor, 7
  7.83-V voltage reference, 1014
  diodes in fet amplifier, 12, 15
Temperature control:
  0.05 C accuracy with opamp, 923

0-100 F using diode probe, 467
28-V aquarium heater, 929
360-W triac, 390
500-W soldering iron, 928
600-W high-precision, 928
1.2-kW water heater, 927
2.4-kW with zero-voltage switching, 921
6-kW full-wave cooling or heating, 928
6-kW proportional, 921, 924
air conditioner motor speed, 922
amplifier for high-resistance thermistor bridge, 365
amplifier overheating alarm, 271, 274
aquarium, 922
auto oil level checker with thermistor, 65
comparator using matched fet's, 465
cooling fan for electronic equipment, 923
cooling-water pump motors, 922
differential for basement ventilator, 925
digital thermometer, 464
distillation, 929
driver for YIG modulator of radiometer, 463
fall-spring overshoot-preventing for home, 925
feedback with sensor for thyristor trigger, 923
freezer failure alarm, 2
frost alarm for plants, 3
high-current, 928
house heating opamp, 927
IC for triac, 923, 926
indirectly heated sensor, 922
indoor-outdoor thermometer, 467
inductive or resistive load, 921
light-activated zero-voltage switch, 1018
low-resistance sensor, 924
metallic surface in boiling liquid, 924, 926
multiple-load drive, 922
multiple-triac trigger, 927
oven using switching current regulator, 927
passive filter for aquarium thermostat, 928
preamp for infrared detector of radiometer, 463
proportional ramp and pedestal, 928
reference diode using transistor as heater, 1011
soldering iron for checking heat drift, 463
staged sequentially energized, 925
surface-temperature thermometer, 462
thermistor bridge, 922
thermistor bridge for 60 C crystal oven, 921, 927
thermocouple amplifier, 465, 468
thermocouple voltmeter, 464
three-phase with zero-voltage switching, 947
three-phase heater, 947
YIG surface radiometer, 462
zero-voltage switching for inductive load, 926
Temperature measurement:
  37-115F temperature-frequency converter, 878
  audible, for the blind, 487
  icy road alarm, 66, 70
  thermocline with fish finder, 349
  transistor thermal resistance, 468
Teraohmmeter, medium-precision low-cost, 458
Tester:
  20-5,000 Hz sine-square-triangle sweep generator, 329
  2.1-kHz tone generator, 597
  200-MHz noise figure, 569
  450-MHz noise figure, 571
  0.7-V lamp indicator for digital troubleshooting, 1010
  3-V battery monitor, 930

5-V logic-level probe, 414
115-V 400-Hz servo supply, 930
500-milliohm ohmmeter for contacts and joints, 457
a-c power monitor for appliances, 453
a-d converter go-no-go, 23
add-on squarer for audio testing, 817
a-f amplifier linearity, 801
a-f transmitter and receiver for phasing p-a speakers, 656
a-m hi-fi tuner, 795
analog multiplier by comparison, 932
appliance grounds, 654
automatic IC threshold, 940
balanced-modulator switch for tone-burst generator, 795
bilateral trigger diode, 940
cable fault, 931
capacitor using audio tone, 418
capacitor using neon bulb, 422
capacitor leakage, 418, 422
computer-controlled digitally programmable pulse-width generator, 933
constant-current power supply, 197
crystal activity, 123, 932
crystal frequency, 930
crystal frequency and activity, 932, 933
decade counter and Nixie driver, 931
diode go-no-go, 936
diode sorting, 944
dot generator for c-w transmitter, 133
dry cells during charge, 86
ECL test probe, 931
electrolytic capacitor, 422
electrolytic capacitor leakage, 418
fet breakdown, 945
fet drain-source resistance, 943
fet leakage current, 945
fet pinch-off voltage, 942
fet transconductance, 937, 942
flashover protection with triggered spark gap, 339
f-m repeaters and mobile f-m sets, 438
frequency response, 934
harmonic distortion factor meter, 447
impulse response with pseudo-random binary signal, 568
jfet and mosfet, 939
junction fet, 942, 946
lascr, 448
linear IC opamps, 936
logic-level indicator probe, 930
logic pulser for gates, 415
LSI using precision comparator, 144
matching transistor pairs, 942
mike beeper for audio system, 655
multiple-level precision staircase generator, 825
noise generator for receivers, 570, 572
npn and pnp transistor, 936
npn transistor curve tracer, 945
opamp using function generator, 324
opamp parameters, 939
opamp p-p noise, 571
phono turntable wow and flutter, 622
pnp transistor breakdown, 943
portable impedance bridge, 449
power-supply ripple, 566
pulse burst generator for counters, 179
r-f power transistor breakdown voltage, 936
scr, 935, 940
  and triac, 944
signal tracer for radios, 932–934
signal tracer and signal injector, 800
silicon-carbide varistor nonlinearity index meter, 943

Tester *(Cont)*:
  silicon rectifier, 945
  single-pulse generator for digital circuitry,
    153
  sstv and rtty signal analyzer, 912
  temperature stability of component with
    controlled soldering iron, 463
  thyristor go-no-go, 941
  timing accuracy of mono mvbr, 936
  transistor, 948
    and diode, 942, 943
    using oscillator, 935
    with scope display, 944
  transistor breakdown, 939, 942, 944
  transistor current gain, 937, 941, 945
  transistor curve tracer for cro, 937
  transistor gain, 945, 946
  transistor high-frequency gain, 943
  transistor leakage current, 943
  transistor output conductance, 938
  transistor type and beta, 944
  transmitter field strength, 976
  tv color difference amplifier linearity, 905
  unijunction transistor, 938
  unmarked digital IC units, 935
  variable load for power supply, 932
  voltage-subtracting sample and hold, 782
  water conductivity, 460
  wobbulator for tv i-f alignment, 908
  (See *also* Measurement; Signal generator)
Theremin:
  7-transistor battery-operated, 243
  three-transistor, 247
Thermal resistance, measurement for
  transistors, 468
Thermal scanner, infrared closed-circuit tv,
  468
Thermistor, 5-mA constant-current supply, 452
Thermocline, measuring with fish finder, 349
Thermocouple:
  electronic watthour meter, 455
  remote measuring circuit, 464
Thermocouple amplifier, matched fet, 465
Thermometer:
  30-40 C radiation for auditory canal, 466
  0-100 F using diode probe, 467
  audible, for the blind, 487
  digital display, 464
  electronic with digital voltmeter readout, 465
  indoor-outdoor resistance-bridge, 467
  infrared microscope for transistors, 467
  radiation for 100-500 C surfaces, 464
  surface-temperature, 462
Three-phase control:
  1-kW inverter, 381
  a-c voltage serving three floating d-c
    supplies, 950
  common control inputs, 949
  floating supply for phase-control triggers,
    948
  frequency converter using logic-controlled
    inverter, 948
  heater with zero-voltage synchronous
    switching, 947
  induction motor speed with triacs, 951
  inverter for 12 V d-c to 400 Hz, 949
  optoelectronic photodarlington, 949
  phase-sequence reversal indicator, 950
  scr-rectifier combination, 949
  zero-voltage starting and synchronous
    switching, 950
  zero-voltage switching of triacs, 948
  zero-voltage trigger, 1016
Threshold detector (see Voltage level detector)
Thunderstorm, static-triggered alarm, 1
Thyristor, tester, 941, 944

Time base:
  0.1-1-10 Hz using 60-Hz reference, 297
  100-kHz using 1.4-MHz crystal, 293
Time-base generator, 960-Hz tuning-fork for
  missile, 597
Time-constant multiplier, opamp for pulse
  stretcher, 953
Time delay (see Delay; Timer)
Timer:
  100-μs-10 s with clamped output, 956
  0-20 s adjustable, 957
  0.1-99 s darkroom, 638
  0.5 s-3 min, 952
  1-s for 3-A load, 955
  1-s blinking neon for darkroom, 638, 639
  1-s turn-off delay, 958
  1 s-2.5 min ujt-scr, 954
  1 s-10 min with three-decade control, 952
  1-30 s relay dropout delay, 953
  2-s repeater dropout delay, 775
  2 s-2 min, 958
  2 s-4.5 min for f-m repeater, 770
  5-s adjustable repetitive, 957
  5-60 s darkroom, 637
  6-60 s fet, 955
  10-s or 20-s beeper for rapid-fire sharp-
    shooting, 953
  10-100-1,000 s with alarm, 955
  20-s and 2.5 min headlight turn-off, 73
  20-s delay for auto theft alarm, 82
  30-s for mixer motor speed control, 517
  30-s using programmable ujt, 953
  30-s relay delay using heater warmup, 654,
    954
  30-s scs driving counter coil, 958
  30 s-15 min lamp turn-off delay, 389
  50-s using time-constant multiplier, 953
  3-300 clicks per minute for darkroom, 249
  1-min for f-m repeater identifier, 768
  1-min for mixer motor speed control, 515
  1-min with motor speed control, 521
  1-min lamp-turnoff delay, 957
  1-min load turn-on delay, 956
  1-10 min, 957
  3-10 min f-m repeater control, 776
  10-min, 955
  10-min repeater transmitter turnoff, 755
  10-min station-identification reminder, 952,
    953, 956
  10-min ujt-jfet, 958
  13-min Schmitt trigger, 954, 957
  20-min adjustable with fet and put, 956
  30-min scs, 957
  10-hr using constant-current nA source, 958
  300-hour using ESD electrochemical
    capacitor, 956
  99% duty-cycle range, 952
  arming-disarming delay for auto theft
    alarm, 83
  audible flashing darkroom, 638
  auto wiper, 71
  automatic standby battery charger, 91
  baseball velocity, 346
  beeping for darkroom, 635
  bullet velocity, 451
  commercial-killing touch switch, 105
  dimming-rate for lamp, 388
  enlarging and developing, 643
  headlight-off, 62, 65
  lamp turn-off delay, 387
  long-delay put, 956
  low-cost line receiver, 202
  photographic enlarger, 644
  put for one-shot pulse generator, 670
  resetting for auto theft alarm, 83
  ring-connected cable fault tester, 931

seconds to hours with power control, 957
sequential delay for multiple lamps, 276
sequential ujt-scr, 955
start-stop control for scaler, 954
  (See *also* Counter; Delay)
Tone-burst generator:
  balanced-modulator switch, 795
  regulated power supply, 709
  synchronizer, 797
Tone control:
  active in stereo preamp, 844
  balance for old records, 624
  bootstrapping gives high s/n ratio, 17
  separate bass and treble stereo, 832
  (See *also* Audio amplifier; Audio control;
    Electronic music)
Tone generator, electronic drums, 241
Tool magnetizer, capacitor-discharge, 344
Topside sounder:
  ionogram telemetry converter, 875
  sweep calibrator, 877
Touch control (see Capacitance control)
Touch-tone:
  mobile f-m repeater control, 767
  push-to-talk connection for transmitter, 879
Track and hold, 4-kHz sampling amplifier, 779
Track and hold amplifier, 0-4 kHz sampling
  amplifier, 781
Tracking filter:
  1-MHz using phase-locked loop, 1000
  phase-locked loop for f-m discriminator,
    1000
Transceiver:
  440-Hz tone generator for network
    operations, 809
  25-kHz calibrator, 432
  3.5-4 MHz monitor with 200-kHz markers,
    961
  3.5-7-21 MHz two-transistor transmitter, 961
  7-MHz 2-W c-w, 965
  133-MHz duplexer, 960
  144-MHz 1-W portable transmitter, 960
  144-MHz single-tube, 960
  450-MHz duplexer, 963
  2-meter 1-W, 965
  2-meter 1-W portable c-w, 962
  2-meter 200-W power amplifier, 967
  2-meter amplitude modulator, 964
  2-meter low-power f-m, 962
  2-meter switching circuit, 966
  13-V supply for fixed-base operation, 723
  700-V and 250-V power supply, 649
  750-V power supply, 341, 648
  900-V sextupler power supply, 343
  adding S meter, 965, 966
  audio-controlled battery saver, 964
  audio-tone tuneup for the blind, 488
  bfo for ssb signals on a-m CB, 121, 123
  bfo injection for c-w or ssb, 964
  channel-9 CB override monitor, 119
  converter for 12-V tube-type mobile, 162
  diode-controlled filter switching, 966
  instant vox control for transmitter, 963
  lightning protection for antenna, 966
  noise blanker, 572
  phone-jack amplifier with tunable a-f
    selectivity, 37
  product detector and 2-kHz ssb audio filter,
    963
  serial data transmission at 20 MHz on wire
    line, 201
  speech compression in preamp, 42
  ssb monitor, 807
  Telefax modification for amateur service,
    959
  theft alarm for car, 82

Transceiver *(Cont):*
  three-diode t-r switch, 961
  tone-burst keyer for f-m repeater, 767
  t-r switch, 961, 963, 965
  varactor-controlled incremental tuning, 967
  vfo and bfo for 3.5 and 7 MHz, 960
  voice-operated control, 46, 49, 52
  (See *also* Amplifier; Audio amplifier;
    Citizens band; Intermediate-frequency
    amplifier; Oscillator; Receiver; Repeater;
    Transmitter; Tuner)
Transconductance tester, fet, 937, 942
Transfer function, nonlinear opamp with sharp
  breakpoint, 581
Transformer:
  360-W solid-state variable, 390
  tap-changing with triacs, 862
Transients:
  muting power turn-on for stereo, 50
  peak voltage indicator, 1007
Transistor:
  breakdown-voltage meter, 939
  current gain measurement, 945
  curve tracer for cro, 937, 945
  extending range of breakdown tester, 944
  matching pairs, 942
  measuring breakdown current, 942, 943, 945
  measuring current gain, 937
  measuring direct-current gain, 941
  measuring high-frequency gain, 943, 946
  measuring leakage current, 943
  measuring output conductance, 938
  parameter display on scope, 944
  temperature measurement with infrared
    microscope, 467
  tester, 935–939, 942–944, 946
Transistor multimeter:
  a-c/d-c, 478
  d-c to 300 V and 300 mA, 477
Transistor voltmeter:
  100-mV fet opamp, 476
  250 mV-500 V fet with 8 ranges, 470
  combined with microammeter, 473
  eight-range using fet, 473
  six-range d-c, 480
Transition time, allowable-limit circuit for logic
  pulses, 1006
Transmission gate, opamp with switching fet,
  334
Transmission line:
  automatic tuner, 983
  driving with parallel gates for 5-MHz data
    rate, 202
  single-gate data drive, 202
Transmit-receive switch:
  160-MHz, 959, 965
  165-MHz transceiver, 961
  high-frequency using diodes, 963
  high-speed three-diode gate, 963
  solid-state automatic for c-w, 961
  voice-operated, 49
Transmitter:
  800-Hz and 2 kHz two-tone test generator,
    802
  45-dB audio compressor, 45
  50-kHz marker generator with harmonics,
    430
  65-kHz power-line carrier intercom, 376
  160-190 kHz 1-W experimental unlicensed,
    987
  550-1,500 kHz c-w low-power, 138
  1.5-30 MHz 100-W amplifier, 983
  220-MHz converter from 28 MHz for ssb and
    c-w, 811
  3.5-28 MHz 600-W linear amplifier, 984
  3.5-28 MHz 2-kW linear amplifier, 985

3.5-30 MHz 15-W linear amplifier, 972
5-10 MHz vfo, 978
7-14-21-28 MHz at 60 W, 971
26.995-MHz for remote control, 763
27-MHz burglar-alarm and security, 95
28-50 MHz mobile solid-state, 986
30-MHz 10-W using Vackar vfo, 975
40-60 MHz modulation monitor, 978
50-MHz 40-W 12.5-V c-w or f-m, 968
50-MHz 40-W f-m c-w mobile, 980
52.5-MHz wide-band f-m exciter, 303
118-136 MHz 2.5-W a-m aircraft, 970
118-136 MHz 4-W a-m, 972
118-136 MHz 4-W series modulator, 507
118-136 MHz 4-W wideband a-m aircraft,
  974
118-136 MHz 7-W a-m aircraft amplifier,
  978
146-MHz 40-W power amplifier, 15
146-175 MHz 25-W amplifier, 973
146-175 MHz 50-W amplifier, 972
146.34-MHz f-m portable for repeater, 301
160-MHz biased-diode ssb antenna switch,
  959
160-MHz unbiased-diode antenna switch,
  965
175-MHz 25-W mobile for 12.5 V, 980
175-MHz 30-W mobile for 28 V, 982
175-MHz 30-W three-transistor power
  amplifier, 973
175-MHz power amplifier, 976
220-MHz 230-W coaxial-line amplifier, 982
220-MHz 580-W single-tube linear
  amplifier, 984
220-MHz power amplifier, 979
225-400 MHz 17-W IC power amplifier, 18
382-MHz ssb oscillator chain, 809
432-MHz 40-W two-transistor amplifier, 982
432-MHz varicap-controlled crystal
  oscillator, 985
448-MHz 5-W power amplifier, 770
450-MHz 6-W amplifier, 968
10-GHz Gunn device with 1 KHz modulator,
  975
300-V and 800-V universal power supply,
  342
0.5-W three-band qrp, 985
25-W 2-meter f-m amplifier, 962
750-W grounded-grid linear amplifier, 974
2-meter 3-W f-m driver, 306
2-meter 50-W f-m power amplifier, 302
2-meter 200-W f-m power amplifier, 967
2-meter 1-kW linear amplifier, 976
2-meter double-sideband suppressed-
  carrier, 977
2-meter solid-state f-m, 304
6-meter 100-mW two-transistor, 978
6-meter 250-mW postage-stamp size, 970
6-meter 1-kW linear amplifier, 969
6-meter 2-kW pep linear amplifier, 987
10-meter 1-W double-sideband, 975
10-meter 12-W c-w or ssb, 977
10-15-20-40-80-160 meter 1-kW linear, 980
10-80 meter 2-kW pep linear amplifier, 969
15-meter, 971
15-80 meter low-power four-band, 983
20-40-80 meter 50-W mobile c-w, 970
40-meter 100-mW two-IC qrp, 981
40-meter 500-mW c-w, 974
40-meter c-w micropower, 976
40-80-160 meter 1-W push-pull fet, 971
80-meter qrp using 40-meter crystal, 980
80-meter vfo with buffer, 981
10-min station-identification timer, 952, 953
aligning with receiver, 798
a-m wireless mike, 849, 982

a-m wireless phonograph, 977
antenna matcher using swr bridge, 979
audio agc for preventing overmodulation, 41
audio-tone tuneup for the blind, 488
aural tuning meter, 971
auto theft alarm paging, 85
automatic transmission-line tuner, 983
automatic tuning of final amplifier, 792, 794
break-in alarm, 379
CB beacon, 850
CB beeper, 848
CB pager, 122
CB tuning meter and power monitor, 121
charging separate mobile battery with
  regular generator, 647
code practice, 134
compressor and clipper improve speech
  clarity, 399
c-w monitor, 132
digital frequency counter, 287
diode remote-control switching of crystals,
  762, 764, 766, 859
diode switch gives perfect c-w, 973
doubler for 200 MHz to 400 MHz, 981
f-m wireless mike, 846, 849
instant voice interruption, 963
keying monitor, 138, 139
low-power for wireless code practice, 135
modulation and field-strength meter, 976
phase detector for automatic r-f tuning, 613
power-supply turn-on delay, 654
protection against open or shorted antenna,
  773
pulsed infrared, 97
push-to-talk control with Styleline phone,
  879
remote-reading field-strength meter, 456
r-f powered switch, 862
solid-state vibrator for vhf mobile, 646
speech clipper, 400
speech clipper-amplifier, 398
speech clipper-compressor, 44
speech compressor, 42
speech processor for ssb, 43
swr measurement at low power, 452
testing with dot generator, 133
tone keyer, 131
touch-controlled c-w break-in, 137
touch-to-talk switch, 104, 105
t-r switch for c-w, 961
t-r switch using diodes, 963
tripler for 150 MHz to 450 MHz, 979
tuning aid for the blind, 483
tuning meter for qrp operation, 978
uhf wattmeter, 455
varactor-controlled incremental tuning, 967
vfo for multiband low-power operation, 979
voice-operated control, 46, 52, 986
vswr indicator for 2-W c-w, 449
(See *also* Amplifier; Audio amplifier;
  Frequency modulation; Modulator;
  Oscillator; Radar; Remote control;
  Repeater; Single sideband; Telemetry;
  Television; Transceiver; Ultrasonic)
Transponder, pulser for avalanche-diode
  modulator, 507
Trapezoid generator:
  auditory stimulator, 488
  numeric crt display, 230
Treasure finder (see Metal detector)
Tremolo, simulator for electronic organ, 243
Triangle, converting from square wave with
  integrator, 371
Triangle generator:
  300-hr programmable timer, 956
  pulse-width modulator, 509

Triangle generator (Cont):
  (See also Function generator; Sweep)
Trickle charger (see Battery charger)
Trigger:
  50-Hz train of 1,200-Hz pulses, 991
  500-Hz flip-flop for inverter scr pair, 551
  1-kHz flip-flop for inverter scr pair, 550
  amplitude-difference amplifier, 993
  anticoincidence, 989
  fast-rising current loads, 992
  firing scr with zener-connected transistor, 390
  firing scr pair with third scr, 992
  frequency-dividing flip-flop, 298, 299
  full-wave neon phase control for scr, 617
  half-wave neon phase control for scr, 616
  hysteresis amplifier, 791
  IC-triac: for isolated inductive load, 992
    for isolated resistive load, 993
  level-sensitive up-down process control, 993
  magnetic-pickup with square-wave output, 988
  multiple-triac for temperature control, 927
  NAND-gate for voltage-level detector, 991
  photocell-controlled for scr's, 631
  photoflash firing, 635
  power one-shot for stapling gun, 988, 990
  preventing clocked train overlap, 125
  pulse amplifier, 989
  pulse-sharpening for scr gate, 992
  put flip-flop for inverter scr's, 990
  Schmitt: in 0-20 MHz input for frequency
      counter, 292
    with adjustable backlash, 991
    in battery discharge alarm, 651
    as clock pulse generator, 127
    for d-c level-sensing, 989
    in frequency-counter pulse shaper, 288
    with high input impedance, 991
    in optoelectronic motion detector, 586
    for squaring sine waves, 678
    using zener, 992
  Schmitt 150-Hz feeding counter, 688
  Schmitt 300-Hz basic, 993
  Schmitt 13-min time delay, 954
  Schmitt a-f square-wave generator, 990
  Schmitt blood-pressure simulator, 485
  Schmitt current-mode high-speed, 988
  Schmitt current-operated IC, 991
  Schmitt low-hysteresis, 990
  Schmitt micropower, 991
  Schmitt voltage-operated IC, 992
  scr impulse driver for solenoid, 989
  scs for triac, 617
  sharpening with NAND-gate mono mvbr, 990,
      991
  slow-decay detector for c-w reception, 131
  symmetrical low-hysteresis, 992
  three-phase IC for triacs, 1016
  three-phase zero-point switching, 951
  variable-phase for scr of voltage sweep
      generator, 989
  voltmeter for measuring pulse amplitude, 476
  zero-crossing a-c for triac, 1017
  zero-crossing a-c half-wave scr, 993
  zero-crossing detector for a-d converter, 22
  zero-crossing IC for inverse-parallel scr's, 990
  (See also Lamp control; Lamp dimmer; Motor
      control; Temperature control)
Trimmer, active using resistance bridge and
    comparator, 142
Tripler (see Frequency multiplier)
Troubleshooting:
  0.7-V indicator lamp for digital IC, 1010
  signal injector, 795
  signal tracer for radios, 932–934
  signal tracer and signal injector, 800

Tuned line, transistor gain measurement, 946
Tuner:
  10-meter for vhf and uhf converters, 173
  a-m for frequency synthesizer, 705
  a-m hi-fi, 694
  a-m hi-fi tester, 795
  autodyne uhf tv, 907
  automatic transmission-line, 983
  crystal-controlled detent for communication
      receiver, 704
  digital for electronic organs, 248
  electronic a-m auto radio using varactors,
      705
  f-m using varactors, 300, 308
  f-m fet 2-gang capacitor, 306
  f-m 3-gang capacitor, 307
  f-m stereo, 845
  matching to noncompatible i-f strip, 696
  tv with agc delay, 914
  vhf tv varactor, 913
  (See also Receiver; Television; Transceiver)
Tuning, automatic with d-c servo, 792, 794
Tuning eye, color tv, 902
Tuning fork, 960-Hz time-base generator, 597
Tuning indicator:
  facsimile and rtty, 889
  led zero-beat, 223
  light-bar for sstv, 910
Tuning meter:
  aural for blind operator, 971
  transmitter for qrp operation, 978
Turn signal, audio-tone alarm, 67
Turntable, wow and flutter meter, 622

Ultrasonic:
  5-60,000 Hz sine-wave signal generator, 799
  19.5-kHz 15-W pest control, 996
  25-kHz intrusion alarm, 995
  30-40 kHz air-beam receiver, 994
  40-kHz intrusion alarm, 272
  40-kHz receiver for demonstrations, 995
  40-kHz remote-control receiver, 996
  40-kHz transmitter for demonstrations, 995
  burglar alarm, 994
  cardiac catheter, 481, 484
  change-of-level detector for alarm receiver,
      996
  fail-safe alarm, 996
  induction receiver for museum commentary,
      854
  receiver with aural converter, 995
  remote-control receiver for tv, 919
  tv remote control, 917–920
Underwater acoustic beacon, 1-pps mvbr, 563
Uranium prospecting, Geiger counter, 428

Vackar oscillator, 30-MHz transmitter vfo, 975
Vacuum gage, ten-range electrometer, 424
Vacuum-tube voltmeter:
  1.5-V regulated supply for ohmmeter, 719
  adapter for capacitance meter, 421
  d-c amplifier, 114
  linearizing ohmmeter scale, 460
  low-cost neon-indicating for 9 and 12 V d-c,
      476
  sensitivity booster, 475
  (See also Transistor voltmeter; Voltmeter)
Varactor:
  52-MHz f-m oscillator, 308
  a-m car radio tuning, 705
  a-m regenerative tuner for phones, 695
  diode-bridge chopper, 116
  f-m tuning, 300, 308
  incremental tuning, 967

passive ssb i-f notch filter, 269
  vhf tv tuner, 913
Variac, 0-120 V a-c electronic, 646
Varicap (see Varactor)
Varistor, nonlinearity-index meter, 949
Vector, log circuit for root sum of squares, 820
Vector difference, analog multiplier-divider,
    820
Vibration, stethoscope, 847
Vibrato:
  IC oscillator in vibratone, 244
  simulator for electronic organ, 243
  touch-contact musical tone generator, 245
Vibrator, hi-fi tape recorder, 249
Vibrator replacement, solid-state for mobile vhf
    transmitter, 646
Video:
  10-MHz clamp and d-c level restorer, 686
  30-MHz logarithmic amplifier, 687
  camera amplifier, 911
  gate with switching transient suppression,
      336
  interface for character generator, 229
  low-level wideband amplifier, 706
  phase-transparent broadband limiter, 687
  preamp and line driver for camera tube, 907
  RGB amplifier for color tv, 903
  series limiter for ten channels, 399
  shunt limiter for single channel, 399
  weather satellite picture demodulator, 877
  (See also Amplifier; Radar; Receiver; Televi-
      sion)
Video amplifier:
  1-22 MHz untuned using two IC's, 57
  10-MHz IC, 57, 60
  50-MHz bandwidth IC, 60
  agc control with diode, 59
  tv thin-film hybrid i-f, 913
Voice control:
  model train or lamp, 787
  speech parameter extractor, 51
  ssb generator, 48
  tape recorder, 848
    or transmitter, 46, 52
  (See also Audio control)
Voice-operated relay, transmit-receive, 49, 986
Voice-operated transmitter, 9-MHz ssb
    generator, 810
Voltage calibrator, 5-V square-wave for cro,
    114
Voltage comparator (see Comparator)
Voltage control:
  1-12,000 Hz astable mvbr, 544
  transistor pair for a-c line voltage, 647
  (See also Regulated power supply)
Voltage-controlled amplifier, analog multi-
    plier, 59
Voltage-controlled oscillator:
  0.1 Hz-30 kHz two-opamp, 327
  0.1 Hz-40 MHz, 999
  1 Hz-2 MHz with comparator and mvbr, 997
  200-3,200 Hz using active filter, 999
  670 Hz-115 kHz with 250 mV-50 V d-c con-
      trol, 1001
  1.5-2.5 kHz phase-shift fet, 997
  1-75 kHz with adjustable pulse width, 671
  3-kHz phase-locked loop, 1000
  3-kHz ujt with integrator, 998
  6-kHz with 60-Hz divide-and-sample loop,
      777
  10-kHz square-wave, 1000
  10-50 kHz using two gates, 1001
  0-1 MHz two-phase, 999
  1-5 MHz mvbr, 997
  1.5-30 MHz frequency synthesizer for re-
      ceiver, 998

Voltage-controlled oscillator (Cont):
  2-20 MHz square-wave for lab sweep
    generator, 1002
  3.5-30 MHz varactor-tuned, 1002
  3.58-MHz color tv, 98
  30-MHz astable mvbr, 549
  1.1-1.4 GHz varactor-tuned, 1000
  2-meter frequency synthesizer, 322
  a-f sine from triangular, 999
  five-octave active filter, 1002
  f-m tuner, 308
  four-layer diode as comparator, 168
  frequency synthesizer, 318, 320, 321
  IC with 10:1 range, 1001
  loop phase comparator, 790
  phase-locked loop receiver, 696
  pulse width and spacing control, 1003
  sweep generator, 857
  synchronizing clock signals, 126
  tracking filter for 3:1 frequency range, 1000
  transmitter tuning aid for the blind, 483
  (See also Oscillator)
Voltage divider:
  0-10 V circular resistor bridge, 1012
  splitter for regulated power supply, 725
Voltage follower:
  analysis by computer, 575
  ladder-network drive for a-d converter, 25
Voltage-level changer, unequal to equal logic
  levels, 415
Voltage-level detector:
  500,000-pps pulse-height analyzer, 1006
  0.7-V pulse and level indicator lamp,
    1010
  2.8-V opamp status indicator, 1008
  9-V up-down level-sensitive trigger, 993
  10-V peaks, 1010
  12-V driving pilot lamp, 1005
  adjustable-level pulse generator, 1008
  amplitude-independent peak, 1010
  analog with led display, 1005
  analog-digital converter trigger, 22
  apple size sorter, 631
  auto theft alarm, 82, 84
  comparator for −8.5 to 8.5 V, 1009
  differential discriminator, 1009
  fast peak for 50-ns pulses, 1010
  flashlight-controlled three-lamp high-
    voltage, 1009
  four-channel pulse, 1004
  led null indicator, 1004
  logic level tolerance, 1007
  logic pulses, 1006
  low a-c line-voltage alarm, 654
  monitoring auto battery condition during
    starting, 69
  NAND-gate, 991
  negative peaks, 1007, 1010
  peak-to-peak using dual opamp, 1006
  positive peaks, 1005, 1008, 1010
  pulse discriminator with constant relative
    backlash, 1006
  saturable output driving relay, 1004
  Schmitt trigger, 989
  signal envelope, 1005
  square-law signal-controlled gain, 1008
  transient peaks, 1007
  zero crossing of sine wave, 1007–1009
  (See also Comparator; Peak detector; Pulse
    height analyzer; Voltage measurement;
    Zero-voltage switch)
Voltage measurement:
  a-c line overvoltage detector, 654
  battery-monitoring lamp, 472
  low-level across remote current shunt, 115
  precision half-wave peak rectifier, 480

(See also Digital voltmeter; Vacuum-tube
  voltmeter; Voltage-level detector; Voltmeter; Zero-voltage switch)
Voltage monitor, battery level, 89
Voltage null detector, opamp-driven led pair,
  1004
Voltage reference:
  7.6-mA constant-current source, 195, 198
  −10-V at 5 mA, 1013
  1-2-4-6 V using reference diode and
    constant-current source, 1013
  2-V and 3.4-V radiation-stable, 1012
  6.3-V within 0.08%, 1012
  7.83-V temperature-compensated, 1014
  8.5-30 V programmable zener, 1013
  9-V zener, 1013
  10-V with opamp single supply, 1013
  10-V self-controlled zener, 1012
  11-13 V variable replacement for zener,
    1012
  26-56 V programmable zener, 1013
  28.5-V discrete programmable zener, 1011
  adjustable up to 20 V, 1014
  adjustable zener, 1011, 1013
  neon lamp in precision voltmeter, 1014
  temperature stabilizer for reference diode,
    1011
  voltmeter calibrator, 1011
Voltage regulator (see Regulated power supply; Regulator)
Voltage regulator tube, diodes permit parallel
  operation, 746
Voltage scalor, cascading with current for
  power, 194
Voltage sensor, storage-battery undervoltage
  warning lamp, 86
Voltage subtractor, sample-and-hold automatic
  tester, 782
Voltage sweep generator:
  60-mA quasi-constant current load, 196
  variable-phase trigger for scr, 989
Voltage trip, adjustable 50-mV to 5-V driving
  relay, 1004
Voltage-tunable filter, active narrow-band fet,
  264
Voltage-variable capacitor (see Varactor)
Voltammeter, four-transistor eight-range, 473
Voltmeter:
  10 Hz-200 kHz a-c at 200 mV rms full-scale,
    474
  60-Hz digital active filter, 477
  100 Hz-500 kHz six-range a-c, 473
  1.005-kHz bridge-balancing White follower,
    616
  10-mV to 100-V full-scale d-c micropower,
    474
  10-mV a-c using opamp, 474
  50-mV a-c using opamp, 477
  100-mV fet opamp, 476
  250 mV-500 V fet with 8 ranges, 470
  9-V and 12-V meterless using neon indi-
    cators, 476
  50-kV using 50-kHz power-oscillator supply,
    474
  50-kV a-c and d-c, 470
  0.01% microvoltmeter, 477
  average-reading rms-calibrated a-c, 475
  calibrator for 0.01-1 V, 1011
  d-c 3-range with 0.5% accuracy, 479
  d-c 5-range using mos IC logic, 472
  d-c 8-range 0.5-1,000 V full-scale, 469
  d-c using high-impedance opamps, 479
  digital conversion to ohmmeter, 461
  driver for numeric display, 472
  fet in photoflash meter, 638
  follower with teraohm input impedance, 458

  four-transistor high-impedance, 473
  high-impedance adapter for multimeter, 476
  linear-scale a-c to 300 mV, 469
  opamp amplifier, 478
  precision using neon-lamp reference, 1014
  root-mean-square, 474
  true rms value for a-c or d-c, 820
  wideband rms to 200 kHz, 480
  (See also Digital voltmeter; Vacuum-tube
    voltmeter; Voltage measurement)
Voltohmmeter:
  adapter for capacitance meter, 420
  boosting sensitivity with fet adapter, 471,
    476
Volume control:
  reducing contact noise, 55
  variable-gain for a-f preamp, 50
Volume expander (see Audio expander)
Volume-unit meter, amplified high-sensitivity,
  443
Vox, instant transmitter-receiver switching, 963
Vswr detector, no-load protection for power
  amplifier, 650
Vu meter:
  built into microphone preamp, 655
  stereo system, 832

Waa-Waa sound effect, optoelectronic for
  guitar, 244
Washing machine, sequential ujt-scr timer, 955
Water conductivity, measurement, 460
Water level (see Liquid level)
Watthour meter, thermocouple multiplier drives
  electromagnetic counter, 455
Wattmeter:
  2-500 MHz at up to 1 W, 455
  3.5-30 MHz 1,000-W in-line, 454
  0-10 W r-f, 454
  5-300 W to 450 MHz, 455
  in-line r-f CB transceiver, 454
  instantaneous using analog IC multiplier,
    455
  neon-bulb, 454
  optoelectronic up to 148 MHz, 456
Weather satellite, picture demodulator, 877
Wheatstone bridge, seven-range, 461
Whistle, steam for model railroad, 345
White noise:
  digital generator for a-f amplifier testing,
    568
  substitution with pseudo-random binary se-
    quence correlator, 568
Wien bridge:
  10 Hz-100 kHz audio oscillator, 593
  1-kHz oscillator, 591, 597
  5-kHz constant-amplitude oscillator, 596
  synchronous motor speed control, 520
  (See also Bridge)
Wind indicator, reed switches on weather vane,
  450
Wind velocity, 0-30 mph anemometer, 445
Window comparator, probability density
  analyzer, 444
Windshield, low wiper-water alarm, 73
Wiper, slow-sweep auto, 70, 71, 74
Wire-ORed memory, reducing turn-off time,
  495
Wireless converter, CB for a-m radio, 121
Wireless microphone:
  a-m to help park school buses, 982
  a-m bc single-transistor, 849
  fet buffer stops f-m detuning, 305
  f-m three-transistor transmitter, 846, 849
Wireless phonograph, a-m transmitter, 977
Wobbulator, television i-f alignment, 914

Wood, moisture measurement with ohmmeter, 452
Woofer, protection from transients, 49
Wow meter, phono turntable, 622
WWV:
   converter for ham bands, 175
   fixed-tuned receiver, 693
WWVB, 60-kHz frequency-comparator receiver, 431

X-Y coordinates, converting to polar, 165

Zener:
   adjustable with mV resolution, 1011
   amplified in 9-V regulated supply, 1013
   automatic reset of sampling gate, 331
   tester, 942
   (See also Regulated power supply; Voltage reference)
Zero correction, opamp to 3 μV, 581
Zero-crossing detector:
   300 Hz-4 MHz using comparators, 1009
   comparator for squaring sine wave, 1008
   sync for digital circuit, 1007
Zero suppression, digital display for counter, 223
Zero-voltage detector:
   firing scr pair with third scr, 992
   half-wave trigger for scr, 993
   pulse-transformer IC trigger for inverse-parallel scr's, 990
   triac trigger for isolated inductive load, 992
   triac trigger for isolated resistive load, 993
   trigger for a-d converter, 22
Zero-voltage switch:
   2-A logic-controlled, 1015
   4-A half-wave, 1018
   8-A, 1015, 1016
   40-A, 1016, 1017
   2.4-kW heater control, 921
   6-kW proportional temperature control, 924
   a-c trigger with separate d-c supply, 1017
   divider-triggered triac flasher, 276
   four-transistor full-wave triac, 1017
   half-wave sensitive-gate scr, 1015, 1018
   IC with triac for heater, 926
   inductive or resistive load, 921
   light-activated, 1018
   low-resistance temperature sensor for triac control, 924
   multiple-load drive, 923
   multiple triac triggering, 927
   negative-gate triggering for standard triac, 1017
   power one-shot trigger for stapling gun, 988, 990
   proportional temperature control, 1018
   sequentially energized temperature control, 925
   temperature control with inductive load, 926
   temperature control using low-resistance sensor, 924
   three-phase for delta load, 948
   three-phase IC triggers for triacs, 1016
   three-phase photodarlington control, 949
   triac full-wave with scr control, 1016, 1018
   triacs suppress emi, 951
   (See also Voltage-level detector)